Springer Handbook of Enzymes · Index A

Springer
Berlin
Heidelberg
New York
Barcelona
Hong Kong
London
Milan
Paris
Tokyo

Dietmar Schomburg
Ida Schomburg (Eds.)

Springer Handbook of Enzymes

Synonym Index

Second Edition

Springer

Professor Dietmar Schomburg
e-mail: d.schomburg@uni-koeln.de

Dr. Ida Schomburg
e-mail: i.schomburg@uni-koeln.de

University to Cologne
Institute for Biochemistry
Zülpicher Strasse 47
50674 Cologne
Germany

ISBN 3-540-41830-X 2nd Edition Springer-Verlag Berlin Heidelberg New York

Library of Congress Cataloging-in-Publication Data applied for

Springer handbook of enzymes / Dietmar Schomburg and Ida Schomburg (ed.). - Berlin ; Heidelberg ; New York ; Barcelona ; Hong Kong ; London ; Milan ; Paris ; Tokyo : Springer

Index A. Synonym index. - 2. ed. - 2001
ISBN 3-540-41830-X

Springer-Verlag Berlin Heidelberg New York
a member of BertelsmannSpringer Science+Business Media GmbH

http://www.springer.de

Cover design: Erich Kirchner, Heidelberg
Typesetting: medio Technologies AG, Berlin

SPIN: 10833497 2/3020xv-5 4 3 2 1 0 – printed on acid-free paper

Attention all Users of the "Springer Handbook of Enzymes"

Information on this handbook can be found on the internet at
http://www.springer.de/enzymes/

A complete list of all enzyme entries either as an alphabetical Name Index or as the EC-Number Index is available at the above mentioned URL. You can download and print them free of charge.

Preface and Notes for the User

Today, as the entire information contained in the genome is becoming available for a rapidly increasing number of organisms and transcriptome and proteome analyses are beginning to provide us with a much wider image of protein regulation and function, it is obvious that there are limitations to our ability to access functional data for the gene products, viz., the proteins and, in particular, the enzymes. These data are inherently very difficult to collect, interpret and standardize, as they are widely distributed among journals from different fields and are often subject to experimental conditions. Nevertheless, a systematic collection is essential for our interpretation of genome information and, more so, for the applications of this knowledge in the fields of medicine, agriculture, etc. Progress regarding enzyme immobilization, enzyme production, enzyme inhibition, coenzyme regeneration and enzyme engineering has opened up fascinating new fields for the potential application of enzymes in a wide range of different areas.

The development of the enzyme data information system BRENDA was started in 1987 at the German National Research Centre for Biotechnology in Braunschweig (GBF) and now continues at the **Institute of Biochemistry, University at Cologne**. The present book "Springer Handbook of Enzymes" represents the printed version of this data bank. The information system has been developed into a full metabolic data base.

The enzymes in this Handbook are arranged according to the Enzyme Commission list of enzymes. Some **3,700** "different" enzymes are covered, for which there exist about **25,000 synonyms.** This **Synonym Index** helps in eliciting the recommended name for a given enzyme. It is also correlated with the EC number and the corresponding volume and page number in the Springer Handbook of Enzymes, of which Volumes 1–5 have so far been published. As this Synonym Index features the current status, also very newly classified enzymes and their synonyms are included which are not yet covered in Vol. 1–5. These will appear in the first Supplement of the Springer Handbook of Enzymes. For some enzymes you might only know the old EC number which you can look up at the very end of this index (listed as the synonym). There you will find the Recommended Name as well as the new EC number (in the first column).

When all volumes (approximately 25) appear in the second edition, an updated Synonym Index will be published with some 100–200 additional synonyms for newly classified enzymes and the volume/page number supplemented for the whole Handbook.

Cologne
Spring 2001

Dietmar Schomburg, Ida Schomburg

List of Abbreviations

A	adenine
Ac	acetyl
ADP	adenosine 5'-diphosphate
Ala	alanine
All	allose
Alt	altrose
AMP	adenosine 5'-monophosphate
Ara	arabinose
Arg	arginine
Asn	asparagine
Asp	aspartic acid
ATP	adenosine 5'-triphosphate
Bicine	N,N'-bis(2-hydroxyethyl)glycine
C	cytosine
cal	calorie
CDP	cytidine 5'-diphosphate
CDTA	trans-1,2-diaminocyclohexane-N,N,N,N-tetraacetic acid
CMP	cytidine 5'-monophosphate
CoA	coenzyme A
CTP	cytidine 5'-triphosphate
Cys	cysteine
d	deoxy-
D-	(and L-) prefixes indicating configuration
DFP	diisopropyl fluorophosphate
DNA	deoxyribonucleic acid
DPN	diphosphopyridinium nucleotide (now NAD)
DTNB	5,5'-dithiobis(2-nitrobenzoate)
DTT	dithiothreitol (i.e. Cleland's reagent)
EC	number of enzyme in Enzyme Commission's system
E. coli	Escherichia coli
EDTA	ethylene diaminetetraacetate
EGTA	ethylene glycol bis(-aminoethyl ether) tetraacetate
ER	endoplasmic reticulum
Et	ethyl
EXAFS	extended X-ray absorption fine structure
FAD	flavin-adenine dinucleotide
FMN	flavin mononucleotide (riboflavin 5'-monophosphate)
Fru	fructose
Fuc	fucose
G	guanine
Gal	galactose
GDP	guanosine 5'-diphosphate
Glc	glucose

GlcN	glucosamine
GlcNAc	N-acetylglucosamine
Gln	glutamine
Glu	glutamic acid
Gly	glycine
GMP	guanosine 5'-monophosphate
GSH	glutathione
GSSG	oxidized glutathione
GTP	guanosine 5'-triphosphate
Gul	gulose
h	hour
H4	tetrahydro
HEPES	4-(2-hydroxyethyl)-1-piperazineethane sulfonic acid
His	histidine
HPLC	high performance liquid chromatography
Hyl	hydroxylysine
Hyp	hydroxyproline
IAA	iodoacetamide
Ig	immunoglobulin
Ile	isoleucine
Ido	idose
IDP	inosine 5'-diphosphate
IMP	inosine 5'-monophosphate
ITP	inosine 5'-triphosphate
K_m	Michaelis constant
L-	(and D-) prefixes indicating configuration
Leu	leucine
Lys	lysine
Lyx	lyxose
M	mol/l
mM	millimol/l
m-	*meta-*
Man	mannose
MES	2-(N-morpholino)ethane sulfonate
Met	methionine
min	minute
MOPS	3-(N-morpholino)propane sulfonate
Mur	muramic acid
MW	molecular weight
NAD^+	nicotinamide-adenine dinucleotide
NADH	reduced NAD
$NADP^+$	NAD phosphate
NADPH	reduced NADP
NAD(P)H	indicates either NADH or NADPH
NDP	nucleoside 5'-diphosphate
NEM	N-ethylmaleimide
Neu	Neuraminic acid
NMN	nicotinamide mononucleotide
NMP	nucleoside 5'-monophosphate

NTP	nucleoside 5'-triphosphate
o-	*ortho-*
Orn	ornithine
p-	*para-*
PCMB	*p*-chloromercuribenzoate
PEP	phosphoenolpyruvate
pH	$-\log10[H^+]$
Ph	phenyl
Phe	phenylalanine
PIXE	proton-induced X-ray emission
PMSF	phenylmethane-sulfonylfluoride
Pro	proline
Q_{10}	factor for the change in reaction rate for a 10 °C temperature increase
Rha	rhamnose
Rib	ribose
RNA	ribonucleic acid
mRNA	messenger RNA
rRNA	ribosomal RNA
tRNA	transfer RNA
Sar	N-methylglycine (sarcosine)
SDS-PAGE	sodium dodecyl sulfate polyacrylamide gel electrophoresis
Ser	serine
T	thymine
t_H	time for half-completion of reaction
Tal	talose
TDP	thymidine 5'-diphosphate
TEA	triethanolamine
Thr	threonine
TMP	thymidine 5'-monophosphate
Tos-	tosyl-(*p*-toluenesulfonyl-)
TPN	triphosphopyridinium nucleotide (now NADP)
Tris	tris(hydroxymethyl)-aminomethane
Trp	tryptophan
TTP	thymidine 5'-triphosphate
Tyr	tyrosine
U	uridine
U/mg	µmol/(mg*min)
UDP	uridine 5'-diphosphate
UMP	uridine 5'-monophosphate
UTP	uridine 5'-triphosphate
Val	valine
Xaa	symbol for an amino acid of unknown constitution in peptide formula
XAS	X-ray absorption spectroscopy
Xyl	xylose

Index of Synonyms

EC-No.	Synonym Recommended Name	Vol., Page
2.3.1.26	ACAT Sterol O-acyltransferase	
6.4.1.2	ACCase Acetyl-CoA carboxylase	Vol. 2, Page 721
4.1.99.4	ACC deaminase 1-aminocyclopropane-1-carboxylate deaminase	
3.1.1.7	AcCholE acetylcholinesterase	
4.4.1.14	ACC synthase 1-aminocyclopropane-1-carboxylate synthase	
3.4.15.1	ACE peptidyl-dipeptidase A	
3.1.4.45	2-Acet-amido-2-deoxy-α-D-glucose 1-phosphodiester acetamidodeoxyglucohydrolase N-Acetylglucosamine-1-phosphodiester N-acetylglucosaminidase	
1.2.1.10	Acetaldehyde:NAD^+ oxidoreductase (CoA-acetylating) Acetaldehyde dehydrogenase (acetylating)	
1.2.1.10	Acetaldehyde dehydrogenase (acetylating) Acetaldehyde dehydrogenase (acetylating)	
5.4.2.3	2-Acetamido-2-deoxy-D-glucose-1,6-bisphosphate:2-acetamido-2-deoxy-D-glucose 1-phosphate phosphotransferase Phosphoacetylglucosamine mutase	Vol. 1, Page 515
3.5.1.25	2-Acetamido-2-deoxy-D-glucose-6-phosphate amidohydrolase N-Acetylglucosamine-6-phosphate deacetylase	
3.1.6.14	2-Acetamido-2-deoxy-D-glucose 6-sulfate sulfatase N-Acetylglucosamine-6-sulfatase	
3.2.1.50	α-D-2-acetamido-2-deoxyglucosidase α-N-acetylglucosaminidase	
2.6.1.65	6-Acetamido-3-aminohexanoate:2-oxoglutarate aminotransferase N^6-Acetyl-β-lysine transaminase	
3.2.1.92	β-2-acetamido-3-O-(D-1-carboxyethyl)-2-deoxy-D-glucoside acetamidodeoxyglucohydrolase Peptidoglycan β-N-acetylmuramidase	
3.5.1.63	4-Acetamidobutanoate amidohydrolase 4-Acetamidobutyrate deacetylase	
3.5.1.51	4-Acetamidobutanoyl-CoA amidohydrolase 4-Acetamidobutyryl-CoA deacetylase	

EC-No.	Synonym Recommended Name	Vol., Page
3.5.1.63	4-Acetamidobutyrate deacetylase 4-Acetamidobutyrate deacetylase	
3.5.1.51	4-Acetamidobutyryl-CoA deacetylase 4-Acetamidobutyryl-CoA deacetylase	
3.5.1.29	2-(Acetamidomethylene)succinate amidohydrolase (deaminating, decarboxylating) 2-(Acetamidomethylene)succinate hydrolase	
3.5.1.29	2-(Acetamidomethylene)succinate hydrolase 2-(Acetamidomethylene)succinate hydrolase	
6.2.1.1	Acetate-CoA ligase Acetate-CoA ligase	Vol. 2, Page 186
6.2.1.13	Acetate-CoA ligase (ADP-forming) Acetate-CoA ligase (ADP-forming)	Vol. 2, Page 267
6.2.1.22	Acetate:citrate-(pro-3S)-lyase(thiol form) ligase (AMP-forming) Citrate (pro-3S)-lyase ligase	Vol. 2, Page 304
6.2.1.13	Acetate:CoA ligase (ADP-forming) Acetate-CoA ligase (ADP-forming)	Vol. 2, Page 267
6.2.1.1	Acetate:CoA ligase (AMP-forming) Acetate-CoA ligase	Vol. 2, Page 186
6.2.1.22	Acetate:HS-citrate lyase ligase Citrate (pro-3S)-lyase ligase	Vol. 2, Page 304
2.8.3.8	Acetate CoA-transferase Acetate CoA-transferase	
2.8.3.8	Acetate coenzyme A-transferase Acetate CoA-transferase	
2.7.2.1	Acetate kinase Acetate kinase	
2.7.2.12	Acetate kinase (PPi) Acetate kinase (pyrophosphate)	
2.7.2.12	Acetate kinase (pyrophosphate) Acetate kinase (pyrophosphate)	
6.2.1.1	Acetate thiokinase Acetate-CoA ligase	Vol. 2, Page 186
3.1.1.6	acetic-ester acetylhydrolase acetylesterase	

EC-No.	Synonym Recommended Name	Vol., Page
3.1.1.72	Acetic ester hydrolase Acetylxylan esterase	
3.1.1.6	acetic ester hydrolase acetylesterase	
2.7.2.1	Acetic kinase Acetate kinase	
3.6.1.7	Acetic phosphatase Acylphosphatase	
6.2.1.1	Acetic thiokinase Acetate-CoA ligase	Vol. 2, Page 186
6.2.1.16	Acetoacetate-CoA ligase Acetoacetate-CoA ligase	Vol. 2, Page 282
6.2.1.16	Acetoacetate:CoA ligase (AMP-forming) Acetoacetate-CoA ligase	Vol. 2, Page 282
4.1.1.4	Acetoacetate carboxy-lyase Acetoacetate decarboxylase	
4.1.1.4	Acetoacetate decarboxylase Acetoacetate decarboxylase	
2.8.3.5	Acetoacetate succinyl-CoA transferase 3-Oxoacid CoA-transferase	
4.1.1.4	Acetoacetic decarboxylase Acetoacetate decarboxylase	
2.3.1.16	Acetoacetyl-CoA β-ketothiolase Acetyl-CoA C-acyltransferase	
3.1.2.11	acetoacetyl-CoA hydrolase acetoacetyl-CoA hydrolase	
3.1.2.11	acetoacetyl-CoA hydrolase acetoacetyl-CoA hydrolase	
6.2.1.16	Acetoacetyl-CoA ligase Acetoacetate-CoA ligase	Vol. 2, Page 282
1.1.1.36	Acetoacetyl-CoA reductase Acetoacetyl-CoA reductase	
6.2.1.16	Acetoacetyl-CoA synthase Acetoacetate-CoA ligase	Vol. 2, Page 282
6.2.1.16	Acetoacetyl-CoA synthetase Acetoacetate-CoA ligase	Vol. 2, Page 282

EC-No.	Synonym Recommended Name	Vol., Page
2.3.1.9	Acetoacetyl-CoA thiolase Acetyl-CoA C-acetyltransferase	
6.2.1.16	Acetoacetyl-coenzyme A synthetase Acetoacetate-CoA ligase	Vol. 2, Page 282
3.1.2.11	acetoacetyl CoA deacylase acetoacetyl-CoA hydrolase	
6.2.1.16	Acetoacetyl CoA synthetase Acetoacetate-CoA ligase	Vol. 2, Page 282
2.3.1.9	Acetoacetyl CoA thiolase Acetyl-CoA C-acetyltransferase	
2.8.3.5	Acetoacetyl coenzyme A-succinic thiophorase 3-Oxoacid CoA-transferase	
3.1.2.11	acetoacetyl coenzyme A deacylase acetoacetyl-CoA hydrolase	
3.1.2.11	acetoacetyl coenzyme A hydrolase acetoacetyl-CoA hydrolase	
1.1.1.36	Acetoacetyl coenzyme A reductase Acetoacetyl-CoA reductase	
2.3.1.9	β-Acetoacetyl coenzyme A thiolase Acetyl-CoA C-acetyltransferase	
4.1.3.5	Acetoacetyl coenzyme A transacetase Hydroxymethylglutaryl-CoA synthase	
3.1.2.6	Acetoacetylglutathione hydrolase Hydroxyacylglutathione hydrolase	
3.1.2.9	S-acetoacetylhydrolipoate hydrolase S-acetoacetylhydrolipoate hydrolase	
3.1.2.8	S-acetoacylglutathione hydrolase S-acetoacylglutathione hydrolase	
4.2.1.9	acetohydroxyacid dehydratase dihydroxy-acid dehydratase	
5.4.99.3	Acetohydroxy acid isomerase 2-Acetolactate mutase	Vol. 1, Page 597
1.1.1.86	Acetohydroxy acid isomeroreductase Ketol-acid reductoisomerase	
1.1.1.86	Acetohydroxy acid reductoisomerase Ketol-acid reductoisomerase	

EC-No.	Synonym Recommended Name	Vol., Page
4.1.3.18	acetohydroxy acid synthase acetolactate synthase	
4.1.3.18	acetohydroxyacid synthase acetolactate synthase	
4.1.3.18	α-acetohydroxyacid synthase acetolactate synthase	
4.1.3.18	acetohydroxy acid synthetase acetolactate synthase	
4.1.3.18	α-acetohydroxy acid synthetase acetolactate synthase	
2.2.1.4	Acetoin-ribose-5-phosphate transaldolase Acetoin-ribose-5-phosphate transaldolase	
1.1.1.5	Acetoin:NAD$^+$ oxidoreductase Acetoin dehydrogenase	
1.1.1.5	Acetoin dehydrogenase Acetoin dehydrogenase	
5.1.2.4	Acetoin racemase Acetoin racemase	Vol. 1, Page 85
5.1.2.4	Acetoin racemase Acetoin racemase	Vol. 1, Page 85
2.7.2.1	Acetokinase Acetate kinase	
4.1.1.5	Acetolactate decarboxylase Acetolactate decarboxylase	
4.1.1.5	α-Acetolactate decarboxylase Acetolactate decarboxylase	
5.4.99.3	2-Acetolactate mutase 2-Acetolactate mutase	Vol. 1, Page 597
5.4.99.3	2-Acetolactate mutase 2-Acetolactate mutase	Vol. 1, Page 597
5.4.99.3	Acetolactate mutase 2-Acetolactate mutase	Vol. 1, Page 597
4.1.3.18	acetolactate pyruvate-lyase (carboxylating) acetolactate synthase	
1.1.1.86	Acetolactate reductoisomerase Ketol-acid reductoisomerase	

EC-No.	Synonym Recommended Name	Vol., Page
4.1.3.18	acetolactate synthase acetolactate synthase	
4.1.3.18	α-acetolactate synthase acetolactate synthase	
4.1.3.18	acetolactate synthetase acetolactate synthase	
4.1.3.18	α-acetolactate synthetase acetolactate synthase	
4.1.3.18	acetolactic synthetase acetolactate synthase	
2.7.1.29	Acetol kinase Glycerone kinase	
4.1.2.37	Acetone-cyanhydrin acetone-lyase Acetone-cyanhydrin lyase	
4.1.2.37	Acetone-cyanhydrin lyase Acetone-cyanhydrin lyase	
4.1.2.37	Acetone-cyanohydrin lyase Acetone-cyanhydrin lyase	
4.1.2.39	Acetone-cyanohydrin lyase Hydroxynitrilase	
4.1.2.37	Acetone cyanohydrin lyase Acetone-cyanhydrin lyase	
3.5.5.1	Acetonitrilase Nitrilase	
3.1.1.54	5-(4-Acetoxy-1-butynyl)-2,2'-bithiophene:acetate esterase Acetoxybutynylbithiophene deacetylase	
3.1.1.54	5-(4-Acetoxybut-1-ynyl)-2,2'-bithiophene O-acetylhydrolase Acetoxybutynylbithiophene deacetylase	
3.1.1.54	Acetoxybutynylbithiophene deacetylase Acetoxybutynylbithiophene deacetylase	
3.5.1.21	N-Acetyl-β-alanine amidohydrolase N-Acetyl-β-alanine deacetylase	
3.5.1.21	N-Acetyl-β-alanine deacetylase N-Acetyl-β-alanine deacetylase	
3.2.1.49	N-Acetyl-α-D-galactosaminidase α-N-acetylgalactosaminidase	

EC-No.	Synonym Recommended Name	Vol., Page
3.2.1.53	N-Acetyl-β-D-galactosaminidase β-N-acetylgalactosaminidase	
3.2.1.50	N-Acetyl-α-D-glucosaminidase α-N-acetylglucosaminidase	
3.2.1.52	N-Acetyl-β-D-hexosaminidase β-N-acetylhexosaminidase	
3.2.1.49	N-Acetyl-α-galactosaminidase α-N-acetylgalactosaminidase	
3.2.1.53	N-Acetyl-β-galactosaminidase β-N-acetylgalactosaminidase	
3.2.1.50	N-Acetyl-α-glucosaminidase α-N-acetylglucosaminidase	
1.2.1.38	N-Acetyl-γ-glutamyl-phosphate reductase N-Acetyl-γ-glutamyl-phosphate reductase	
3.2.1.52	N-Acetyl-β-hexosaminidase β-N-acetylhexosaminidase	
3.1.1.71	2-Acetyl-1-alkyl-sn-glycerol acetylhydrolase Acetylalkylglycerol acetylhydrolase	
1.14.18.1	N-Acetyl-6-hydroxytryptophan oxidase Monophenol monooxygenase	
3.5.1.15	Acetyl-aspartic deaminase Aspartoacylase	
2.3.1.107	Acetyl-CoA-17-O-deacetylvindoline 17-O-acetyltransferase 17-O-Deacetylvindoline O-acetyltransferase	
2.3.1.7	Acetyl-CoA-carnitine O-acetyltransferase Carnitine O-acetyltransferase	
2.3.1.34	Acetyl-CoA-D-tryptophan-α-N-acetyltransferase D-Tryptophan N-acetyltransferase	
2.3.1.53	Acetyl-CoA-L-phenylalanine α-N-acetyltransferase Phenylalanine N-acetyltransferase	
2.8.3.12	Acetyl-CoA:(E)-glutaconate CoA-transferase Glutaconate CoA-transferase	
2.3.1.18	Acetyl-CoA:β-D-galactoside 6-acetyltransferase Galactoside O-acetyltransferase	
2.3.1.78	Acetyl-CoA:α-glucosaminide N-acetyltransferase Heparan-α-glucosaminide N-acetyltransferase	

EC-No.	Synonym Recommended Name	Vol., Page
2.3.1.108	Acetyl-CoA:α-tubulin-L-lysine N^{ε}-acetyltransferase Tubulin N-acetyltransferase	
2.3.1.67	Acetyl-CoA:1-alkyl-2-lyso-sn-glycero-3-phosphocholine 2-O-acetyltransferase 1-Alkylglycerophosphocholine O-acetyltransferase	
2.3.1.105	Acetyl-CoA:1-alkyl-sn-glycero-3-phosphate 2-O-acetyltransferase Alkylglycerophosphate 2-O-acetyltransferase	
2.3.1.67	Acetyl-CoA:1-alkyl-sn-glycero-3-phosphocholine 2-O-acetyltransferase 1-Alkylglycerophosphocholine O-acetyltransferase	
2.3.1.107	Acetyl-CoA:17-O-deacetylvindoline 17-O-acetyltransferase 17-O-Deacetylvindoline O-acetyltransferase	
2.3.1.81	Acetyl-CoA:2-deoxystreptamine-antibiotic $N^{3'}$-acetyltransferase Aminoglycoside $N^{3'}$-acetyltransferase	
2.8.3.14	Acetyl-CoA:5-hydroxypentanoate CoA-transferase 5-Hydroxypentanoate CoA-transferase	
2.3.1.38	Acetyl-CoA:[acyl-carrier-protein] S-acetyltransferase [Acyl-carrier-protein] S-acetyltransferase	
2.3.1.9	Acetyl-CoA:acetyl-CoA C-acetyltransferase Acetyl-CoA C-acetyltransferase	
2.3.1.84	Acetyl-CoA:alcohol O-acetyltransferase Alcohol O-acetyltransferase	
2.3.1.57	Acetyl-CoA:alkane-α,ω-diamine N-acetyltransferase Diamine N-acetyltransferase	
2.3.1.87	Acetyl-CoA:aralkylamine N-acetyltransferase Aralkylamine N-acetyltransferase	
2.3.1.5	Acetyl-CoA:arylamine N-acetyltransferase Arylamine N-acetyltransferase	
6.4.1.2	Acetyl-CoA:carbon-dioxide ligase (ADP-forming) Acetyl-CoA carboxylase	Vol. 2, Page 721
3.1.3.44	[Acetyl-CoA:carbon-dioxide ligase (ADP-forming)]-phosphate phosphohydrolase [Acetyl-CoA carboxylase]-phosphatase	
2.3.1.7	Acetyl-CoA:carnitine O-acetyltransferase Carnitine O-acetyltransferase	
2.3.1.28	Acetyl-CoA:chloramphenicol 3-O-acetyltransferase Chloramphenicol O-acetyltransferase	

EC-No.	Synonym Recommended Name	Vol., Page
2.3.1.6	Acetyl-CoA:choline-O-acetyltransferase Choline O-acetyltransferase	
2.3.1.6	Acetyl-CoA:choline O-acetyltransferase Choline O-acetyltransferase	
2.8.3.11	Acetyl-CoA:citramalate CoA-transferase Citramalate CoA-transferase	
2.8.3.10	Acetyl-CoA:citrate CoA-transferase Citrate-CoA transferase	
2.3.1.27	Acetyl-CoA:cortisol O-acetyltransferase Cortisol O-acetyltransferase	
2.3.1.36	Acetyl-CoA:D-amino-acid N-acetyltransferase D-Amino-acid N-acetyltransferase	
2.3.1.4	Acetyl-CoA:D-glucosamine-6-phosphate N-acetyltransferase Glucosamine-phosphate N-acetyltransferase	
2.3.1.3	Acetyl-CoA:D-glucosamine N-acetyltransferase Glucosamine N-acetyltransferase	
2.3.1.34	Acetyl-CoA:D-tryptophan N-acetyltransferase D-Tryptophan N-acetyltransferase	
2.3.1.12	Acetyl-CoA:dihydrolipoamide S-acetyltransferase Dihydrolipoamide S-acetyltransferase	
2.3.1.54	Acetyl-CoA:formate C-acetyltransferase Formate C-acetyltransferase	
2.3.1.59	Acetyl-CoA:gentamicin-C1a $N^{2'}$-acetyltransferase Gentamicin 2'-N-acetyltransferase	
2.3.1.60	Acetyl-CoA:gentamicin-C $N^{3'}$-acetyltransferase Gentamicin 3'-N-acetyltransferase	
2.3.1.29	Acetyl-CoA:glycine C-acetyltransferase Glycine C-acetyltransferase	
2.3.1.78	Acetyl-CoA:heparan-α-D-glucosaminide N-acetyltransferase Heparan-α-glucosaminide N-acetyltransferase	
2.3.1.48	Acetyl-CoA:histone acetyltransferase Histone acetyltransferase	
2.3.1.10	Acetyl-CoA:hydrogen-sulfide S-acetyltransferase Hydrogen-sulfide S-acetyltransferase	
2.3.1.2	Acetyl-CoA:imidazole N-acetyltransferase Imidazole N-acetyltransferase	

EC-No.	Synonym Recommended Name	Vol., Page
2.3.1.82	Acetyl-CoA:kanamycin-B N6'-acetyltransferase Aminoglycoside N6'-acetyltransferase	
2.3.1.89	Acetyl-CoA:L-2,3,4,5-tetrahydrodipicolinate N^2-acetyltransferase Tetrahydrodipicolinate N-acetyltransferase	
2.3.1.17	Acetyl-CoA:L-aspartate N-acetyltransferase Aspartate N-acetyltransferase	
2.3.1.1	Acetyl-CoA:L-glutamate N-acetyltransferase Amino-acid N-acetyltransferase	
2.3.1.33	Acetyl-CoA:L-histidine N-acetyltransferase Histidine N-acetyltransferase	
2.3.1.31	Acetyl-CoA:L-homoserine O-acetyltransferase Homoserine O-acetyltransferase	
2.3.1.66	Acetyl-CoA:L-leucine N-acetyltransferase Leucine N-acetyltransferase	
2.3.1.53	Acetyl-CoA:L-phenylalanine N-acetyltransferase Phenylalanine N-acetyltransferase	
2.3.1.30	Acetyl-CoA:L-serine O-acetyltransferase Serine O-acetyltransferase	
2.3.1.67	Acetyl-CoA:Lyso-PAF acetyltransferase 1-Alkylglycerophosphocholine O-acetyltransferase	
2.8.3.3	Acetyl-CoA:malonate CoA-transferase Malonate CoA-transferase	
2.3.1.79	Acetyl-CoA:maltose O-acetyltransferase Maltose O-acetyltransferase	
2.3.1.69	Acetyl-CoA:monoterpenol O-acetyltransferase Monoterpenol O-acetyltransferase	
2.3.1.45	Acetyl-CoA:N-acetylneuraminate-7- or/and 8-O-acetyltransferase N-Acetylneuraminate 7-O(or 9-O-)-acetyltransferase	
2.3.1.45	Acetyl-CoA:N-acetylneuraminate-7- or 8-O-acetyltransferase N-Acetylneuraminate 7-O(or 9-O-)-acetyltransferase	
2.3.1.45	Acetyl-CoA:N-acetylneuraminate-9(7)-O-acetyltransferase N-Acetylneuraminate 7-O(or 9-O-)-acetyltransferase	
2.3.1.45	Acetyl-CoA:N-acetylneuraminate-9(or7)-O-acetyltransferase N-Acetylneuraminate 7-O(or 9-O-)-acetyltransferase	
2.3.1.44	Acetyl-CoA:N-acetylneuraminate 4-O-acetyltransferase N-Acetylneuraminate 4-O-acetyltransferase	

EC-No.	Synonym Recommended Name	Vol., Page
2.3.1.45	Acetyl-CoA:N-acetylneuraminate 7-O(or 9-O)-acetyltransferase N-Acetylneuraminate 7-O(or 9-O-)-acetyltransferase	
2.3.1.9	Acetyl-CoA:N-acetyltransferase Acetyl-CoA C-acetyltransferase	
2.3.1.118	Acetyl-CoA:N-hydroxyarylamine O-acetyltransferase N-Hydroxyarylamine O-acetyltransferase	
2.3.1.102	Acetyl-CoA:N^6-hydroxy-L-lysine 6-acetyltransferase N^6-Hydroxylysine O-acetyltransferase	
2.3.1.8	Acetyl-CoA:orthophosphate acetyltransferase Phosphate acetyltransferase	
2.3.1.88	Acetyl-CoA:peptide N^α-acetyltransferase Peptide α-N-acetyltransferase	
2.3.1.136	Acetyl-CoA:polysialic-acid O-acetyltransferase Polysialic-acid O-acetyltransferase	
2.8.3.1	Acetyl-CoA:propanoate CoA-transferase Propionate CoA-transferase	
2.3.1.128	Acetyl-CoA:ribosomal-protein-L-alanine N-acetyltransferase Ribosomal-protein-alanine N-acetyltransferase	
2.3.1.80	Acetyl-CoA:S-substituted L-cysteine N-acetyltransferase Cysteine-S-conjugate N-acetyltransferase	
2.3.1.150	Acetyl-CoA:salutaridinol-7-O-acetyltransferase Salutaridinol 7-O-acetyltransferase	
2.3.1.150	Acetyl-CoA:salutaridinol 7-O-acetyltransferase Salutaridinol 7-O-acetyltransferase	
2.3.1.11	Acetyl-CoA:thioethanolamine S-acetyltransferase Thioethanolamine S-acetyltransferase	
2.3.1.9	Acetyl-CoA acetyltransferase Acetyl-CoA C-acetyltransferase	
3.1.2.1	acetyl-CoA acylase acetyl-CoA hydrolase	
2.3.1.16	Acetyl-CoA acyltransferase Acetyl-CoA C-acyltransferase	
2.3.1.9	Acetyl-CoA C-acetyltransferase Acetyl-CoA C-acetyltransferase	
2.3.1.16	Acetyl-CoA C-acyltransferase Acetyl-CoA C-acyltransferase	

EC-No.	Synonym Recommended Name	Vol., Page
2.3.1.155	acetyl-CoA C-myristoyltransferase acetyl-CoA C-myristoyltransferase	
6.4.1.2	Acetyl-CoA carboxylase Acetyl-CoA carboxylase	Vol. 2, Page 721
3.1.3.44	[Acetyl-CoA carboxylase]-phosphatase [Acetyl-CoA carboxylase]-phosphatase	
2.7.1.111	[Acetyl-CoA carboxylase] kinase [Acetyl-CoA carboxylase] kinase	
2.7.1.128	[Acetyl-CoA carboxylase] kinase [Acetyl-CoA carboxylase] kinase	
2.7.1.128	Acetyl-CoA carboxylase kinase [Acetyl-CoA carboxylase] kinase	
2.7.1.128	Acetyl-CoA carboxylase kinase-2 [Acetyl-CoA carboxylase] kinase	
3.1.2.1	acetyl-CoA deacylase acetyl-CoA hydrolase	
3.1.2.1	acetyl-CoA decarbonylase synthase enzyme complex acetyl-CoA hydrolase	
3.1.2.1	acetyl-CoA hydrolase acetyl-CoA hydrolase	
3.1.2.1	acetyl-CoA hydrolase acetyl-CoA hydrolase	
6.2.1.1	Acetyl-CoA synthetase Acetate-CoA ligase	Vol. 2, Page 186
6.2.1.13	Acetyl-CoA synthetase (ADP-forming) Acetate-CoA ligase (ADP-forming)	Vol. 2, Page 267
3.1.2.1	acetyl-CoA thiol esterase acetyl-CoA hydrolase	
2.3.1.78	Acetyl-coenzyme A-α-glucosaminide N-acetyltransferase Heparan-α-glucosaminide N-acetyltransferase	
2.3.1.57	Acetyl-coenzyme A-1,4-diaminobutane N-acetyltransferase Diamine N-acetyltransferase	
2.3.1.136	Acetyl-coenzyme A-polysialic acid O-acetyltransferase Polysialic-acid O-acetyltransferase	
4.1.3.21	Acetyl-coenzyme A:2-ketoglutarate C-acetyl transferase Homocitrate synthase	

EC-No.	Synonym Recommended Name	Vol., Page
6.4.1.2	Acetyl-coenzyme A carboxylase Acetyl-CoA carboxylase	Vol. 2, Page 721
2.7.1.128	Acetyl-coenzyme A carboxylase kinase [Acetyl-CoA carboxylase] kinase	
6.2.1.1	Acetyl-coenzyme A synthase Acetate-CoA ligase	Vol. 2, Page 186
3.1.6.12	N-Acetyl-D-galactosamine-4-sulfate 4-sulfohydrolase N-Acetylgalactosamine-4-sulfatase	
3.5.1.80	N-Acetyl-D-galactosamine-6-phosphate deacetylase N-Acetyl-D-galactosamine-6-phosphate deacetylase	
3.1.6.4	N-Acetyl-D-galactosamine-6-sulfate 6-sulfohydrolase N-Acetylgalactosamine-6-sulfatase	
3.5.1.80	N-Acetyl-D-galactosamine 6-phosphate aminohydrolase N-Acetyl-D-galactosamine-6-phosphate deacetylase	
3.2.1.49	α-N-acetyl-D-galactosaminide N-acetylgalactosaminohydrolase α-N-acetylgalactosaminidase	
3.2.1.53	β-N-acetyl-D-galactosaminide N-acetylgalactosaminohydrolase β-N-acetylgalactosaminidase	
3.1.1.58	N-Acetyl-D-galactosaminoglycan acetylhydrolase N-Acetylgalactosaminoglycan deacetylase	
3.1.4.45	α-N-acetyl-D-glucosamine-1-phosphodiester N-acetylglucosaminidase N-Acetylglucosamine-1-phosphodiester N-acetylglucosaminidase	
3.5.1.25	N-Acetyl-D-glucosamine-6-phosphate amidohydrolase N-Acetylglucosamine-6-phosphate deacetylase	
3.1.6.14	N-Acetyl-D-glucosamine-6-sulfate 6-sulfohydrolase N-Acetylglucosamine-6-sulfatase	
5.4.2.3	N-Acetyl-D-glucosamine 1,6-phosphomutase Phosphoacetylglucosamine mutase	Vol. 1, Page 515
3.5.1.33	N-Acetyl-D-glucosamine amidohydrolase N-Acetylglucosamine deacetylase	
3.2.1.50	α-N-acetyl-D-glucosaminidase α-N-acetylglucosaminidase	
2.4.1.38	β-N-Acetyl-D-glucosaminide β-1,4-galactosyltransferase β-N-Acetylglucosaminylglycopeptide β-1,4-galactosyltransferase	
3.2.1.50	α-N-acetyl-D-glucosaminide N-acetylglucosaminohydrolase α-N-acetylglucosaminidase	

EC-No.	Synonym Recommended Name	Vol., Page
3.5.1.33	N-Acetyl-D-glucosaminyl N-deacetylase N-Acetylglucosamine deacetylase	
3.1.1.69	N-Acetyl-D-glucosaminylphosphatidylinositol acetylhydrolase N-Acetylglucosaminylphosphatidylinositol deacetylase	
2.4.1.198	Acetyl-D-glucosaminyltransferase, uridine diphosphoacetylglucosamine α1,6- Phosphatidylinositol N-acetylglucosaminyltransferase	
3.1.1.33	6-acetyl-D-glucose acetylhydrolase 6-acetylglucose deacetylase	
1.1.1.240	N-Acetyl-D-hexosamine:NAD$^+$ 1-oxidoreductase N-Acetylhexosamine 1-dehydrogenase	
1.1.1.240	N-Acetyl-D-hexosamine dehydrogenase N-Acetylhexosamine 1-dehydrogenase	
3.2.1.52	β-N-acetyl-D-hexosaminidase β-N-acetylhexosaminidase	
3.2.1.52	β-N-acetyl-D-hexosaminide N-acetylhexosaminohydrolase β-N-acetylhexosaminidase	
1.1.1.233	N-Acetyl-D-mannosamine dehydrogenase N-Acylmannosamine 1-dehydrogenase	
3.5.1.83	N-Acetyl-L-aspartate amidohydrolase N-Acyl-D-aspartate deacylase	
3.5.1.47	N-Acetyl-L-diaminopimelic acid deacylase Acetyldiaminopimelate deacetylase	
1.2.1.38	N-Acetyl-L-glutamate-5-semialdehyde:NADP$^+$ 5-oxidoreductase (phosphorylating) N-Acetyl-γ-glutamyl-phosphate reductase	
1.2.1.38	N-Acetyl-L-glutamate γ-semialdehyde:NADP oxidoreductase (phosphorylating) N-Acetyl-γ-glutamyl-phosphate reductase	
2.3.1.1	N-Acetyl-L-glutamate synthetase Amino-acid N-acetyltransferase	
2.3.1.35	N-Acetyl-L-glutamate synthetase Glutamate N-acetyltransferase	
4.2.99.10	O-acetyl-L-homoserine acetate-lyase (adding methanethiol) O-acetylhomoserine (thiol)-lyase	
4.2.99.10	O-acetyl-L-homoserine sulfhydrylase O-acetylhomoserine (thiol)-lyase	

EC-No.	Synonym Recommended Name	Vol., Page
2.3.1.35	α-N-Acetyl-L-ornithine:L-glutamate N-acetyltransferase Glutamate N-acetyltransferase	
4.2.99.8	O-acetyl-L-serine acetate-lyase (adding hydrogen sulfide) Cysteine synthase	
4.2.99.13	O-acetyl-L-serine acetate-lyase (adding N^6-substituted adenine) β-(9-cytokinin)-alanine synthase	
4.2.99.8	O-acetyl-L-serine sulfhydrylase Cysteine synthase	
4.2.99.8	O-acetyl-L-serine sulfohydrolase Cysteine synthase	
3.5.1.47	N-Acetyl-LL-2,6-diaminoheptanedioate amidohydrolase Acetyldiaminopimelate deacetylase	
3.5.1.47	N-Acetyl-LL-diaminopimelate deacylase Acetyldiaminopimelate deacetylase	
3.1.1.7	acetyl β-methylcholinesterase acetylcholinesterase	
2.3.1.32	Acetyl-phosphate:L-lysine N^6-acetyltransferase Lysine N-acetyltransferase	
6.2.1.1	Acetyl activating enzyme Acetate-CoA ligase	Vol. 2, Page 186
3.1.1.71	Acetylalkylglycerol acetylhydrolase Acetylalkylglycerol acetylhydrolase	
2.7.1.60	Acetylamidodeoxymannokinase N-Acylmannosamine kinase	
2.7.1.59	2-Acetylamino-2-deoxy-D-glucose kinase N-Acetylglucosamine kinase	
2.7.1.59	Acetylaminodeoxyglucokinase N-Acetylglucosamine kinase	
3.5.1.33	Acetylaminodeoxyglucose acetylhydrolase N-Acetylglucosamine deacetylase	
2.7.1.9	acetylaminodeoxyglucose kinase acetylaminodeoxyglucose kinase	
3.5.1.80	Acetylaminodeoxyglucosephosphate acetylhydrolase N-Acetyl-D-galactosamine-6-phosphate deacetylase	
3.5.1.25	Acetylaminodeoxyglucosephosphate acetylhydrolase N-Acetylglucosamine-6-phosphate deacetylase	

EC-No.	Synonym Recommended Name	Vol., Page
5.4.2.3	Acetylaminodeoxyglucose phosphomutase Phosphoacetylglucosamine mutase	Vol. 1, Page 515
3.2.1.52	β-acetylaminodeoxyhexosidase β-N-acetylhexosaminidase	
3.5.1.4	N-Acetylaminohydrolase Amidase	
3.5.1.29	α-(N-acetylaminomethylene)succinic acid hydrolase 2-(Acetamidomethylene)succinate hydrolase	
3.5.1.76	N-Acetylarylalkylamine amidohydrolase Arylalkyl acylamidase	
3.5.1.15	N-Acetylaspartate amidohydrolase Aspartoacylase	
3.5.1.83	N-Acetylaspartate amidohydrolase N-Acyl-D-aspartate deacylase	
3.5.1.83	Acetylaspartic deaminase N-Acyl-D-aspartate deacylase	
3.4.17.21	Acetylaspartylglutamate dipeptidase Glutamate carboxypeptidase II	
3.4.17.21	N-Acetylated-α-linked-acidic dipeptidase Glutamate carboxypeptidase II	
3.4.17.21	N-Acetylated-α-linked-amino dipeptidase Glutamate carboxypeptidase II	
3.4.17.21	N-Acetylated-α-linked acidic dipeptidase Glutamate carboxypeptidase II	
3.4.17.21	N-Acetylated α-linked acidic dipeptidase Glutamate carboxypeptidase II	
2.3.1.7	Acetylcarnitine transferase Carnitine O-acetyltransferase	
3.1.1.7	acetylcholine acetylhydrolase acetylcholinesterase	
3.1.1.7	acetylcholine esterase acetylcholinesterase	
3.1.1.7	acetylcholine hydrolase acetylcholinesterase	
3.1.1.7	acetylcholinesterase acetylcholinesterase	

EC-No.	Synonym Recommended Name	Vol., Page
2.3.1.5	Acetyl CoA-arylamine N-acetyltransferase Arylamine N-acetyltransferase	
2.3.1.6	Acetyl CoA:choline-O-acetyltransferase Choline O-acetyltransferase	
6.4.1.2	Acetyl CoA carboxylase Acetyl-CoA carboxylase	Vol. 2, Page 721
6.3.4.15	Acetyl CoA holocarboxylase synthetase Biotin-[acetyl-CoA-carboxylase] ligase	Vol. 2, Page 638
6.2.1.1	Acetyl CoA ligase Acetate-CoA ligase	Vol. 2, Page 186
6.2.1.1	Acetyl CoA synthase Acetate-CoA ligase	Vol. 2, Page 186
2.3.1.38	Acetyl coenzyme A-acyl-carrier-protein transacylase [Acyl-carrier-protein] S-acetyltransferase	
2.3.1.107	Acetylcoenzyme A-deacetylvindoline 4-O-acetyltransferase 17-O-Deacetylvindoline O-acetyltransferase	
2.3.1.107	Acetylcoenzyme A:deacetylvindoline 4-O-acetyltransferase 17-O-Deacetylvindoline O-acetyltransferase	
2.3.1.107	Acetylcoenzyme A:deacetylvindoline O-acetyltransferase 17-O-Deacetylvindoline O-acetyltransferase	
3.1.2.1	acetyl coenzyme A acylase acetyl-CoA hydrolase	
6.4.1.2	Acetyl coenzyme A carboxylase Acetyl-CoA carboxylase	Vol. 2, Page 721
3.1.2.1	acetyl coenzyme A deacylase acetyl-CoA hydrolase	
6.3.4.15	Acetyl coenzyme A holocarboxylase synthetase Biotin-[acetyl-CoA-carboxylase] ligase	Vol. 2, Page 638
3.1.2.1	acetyl coenzyme A hydrolase acetyl-CoA hydrolase	
6.2.1.1	Acetyl coenzyme A synthetase Acetate-CoA ligase	Vol. 2, Page 186
2.3.1.9	Acetyl coenzyme A thiolase Acetyl-CoA C-acetyltransferase	
3.5.1.47	Acetyldiaminopimelate deacetylase Acetyldiaminopimelate deacetylase	

EC-No.	Synonym Recommended Name	Vol., Page
3.1.1.62	N-acetyldiaminopimelate deacylase N-acetyldiaminopimelate deacylase	
5.3.3.8	Acetylene-allene isomerase Dodecenoyl-CoA δ-isomerase	Vol. 1, Page 413
4.2.1.71	Acetylenecarboxylate hydratase Acetylenecarboxylate hydratase	
4.1.1.78	acetylenedicarboxylate carboxy-lyase acetylenedicarboxylate decarboxylase	
4.1.1.78	acetylenedicarboxylate decarboxylase acetylenedicarboxylate decarboxylase	
4.2.1.72	Acetylenedicarboxylate hydrase Acetylenedicarboxylate hydratase	
4.2.1.72	Acetylenedicarboxylate hydratase Acetylenedicarboxylate hydratase	
4.2.1.71	Acetylenemonocarboxylate hydrase Acetylenecarboxylate hydratase	
4.2.1.71	Acetylenemonocarboxylate hydratase Acetylenecarboxylate hydratase	
4.2.1.71	Acetylenemonocarboxylic acid hydrase Acetylenecarboxylate hydratase	
4.2.1.27	acetylenemonocarboxylic acid hydrase malonate-semialdehyde dehydratase	
3.1.1.53	9(4)-O-acetylesterase sialate O-acetylesterase	
3.1.1.72	Acetyl esterase Acetylxylan esterase	
3.1.1.6	acetylesterase acetylesterase	
3.1.6.12	N-Acetylgalactosamine-4-sulfatase N-Acetylgalactosamine-4-sulfatase	
3.1.6.4	N-Acetylgalactosamine-6-sulfatase N-Acetylgalactosamine-6-sulfatase	
3.1.6.4	N-Acetylgalactosamine-6-sulfate sulfatase N-Acetylgalactosamine-6-sulfatase	
3.1.6.12	Acetylgalactosamine 4-sulfatase N-Acetylgalactosamine-4-sulfatase	

EC-No.	Synonym Recommended Name	Vol., Page
3.1.6.12	N-Acetylgalactosamine 4-sulfate sulfohydrolase N-Acetylgalactosamine-4-sulfatase	
3.1.6.4	Acetylgalactosamine 6-sulfatase N-Acetylgalactosamine-6-sulfatase	
3.1.6.4	N-Acetylgalactosamine 6-sulfatase N-Acetylgalactosamine-6-sulfatase	
3.1.6.4	N-Acetylgalactosamine 6-sulphate sulphatase N-Acetylgalactosamine-6-sulfatase	
3.2.1.53	β-N-acetylgalactosaminidase β-N-acetylgalactosaminidase	
3.2.1.53	β-acetylgalactosaminidase β-N-acetylgalactosaminidase	
3.2.1.53	N-Acetylgalactosaminidase β-N-acetylgalactosaminidase	
3.2.1.49	α-N-acetylgalactosaminidase α-N-acetylgalactosaminidase	
3.2.1.49	α-acetylgalactosaminidase α-N-acetylgalactosaminidase	
3.2.1.53	β-N-acetylgalactosaminidase β-N-acetylgalactosaminidase	
2.4.99.3	α-N-Acetylgalactosaminide α-2,6-sialyltransferase α-N-Acetylgalactosaminide α-2,6-sialyltransferase	
3.1.1.58	N-Acetyl galactosaminoglycan deacetylase N-Acetylgalactosaminoglycan deacetylase	
3.1.1.58	N-Acetylgalactosaminoglycan deacetylase N-Acetylgalactosaminoglycan deacetylase	
2.4.1.147	Acetylgalactosaminyl-O-glycosyl-glycoprotein β-1,3-N-acetylglucosaminyltransferase Acetylgalactosaminyl-O-glycosyl-glycoprotein β-1,3-N-acetylglucosaminyltransferase	
2.4.1.148	Acetylgalactosaminyl-O-glycosyl-glycoprotein β-1,6-N-acetylglucosaminyltransferase Acetylgalactosaminyl-O-glycosyl-glycoprotein β-1,6-N-acetylglucosaminyltransferase	
2.4.99.3	α-N-Acetylgalactosaminylprotein α2→6 sialyltransferase α-N-Acetylgalactosaminide α-2,6-sialyltransferase	

EC-No.	Synonym Recommended Name	Vol., Page
2.4.1.40	α-3-N-Acetylgalactosaminyltransferase Fucosylgalactose α-N-acetylgalactosaminyltransferase	
2.4.1.40	Acetylgalactosaminyltransferase, [blood-group substance] α- Fucosylgalactose α-N-acetylgalactosaminyltransferase	
2.4.1.40	Acetylgalactosaminyltransferase, fucosylgalactose Fucosylgalactose α-N-acetylgalactosaminyltransferase	
2.4.1.165	Acetylgalactosaminyltransferase, uridine diphosphoacetylgalactosamine-acetylneuraminyl(α2→3)galactosyl(β1→4)glucosyl β1→4- N-Acetylneuraminylgalactosylglucosylceramide β-1,4-N-acetylgalactosaminyltransferase	
2.4.1.174	Acetylgalactosaminyltransferase, uridine diphosphoacetylgalactosamine-chondroitin, I Glucuronylgalactosylproteoglycan β-1,4-N-acetylgalactosaminyltransferase	
2.4.1.175	Acetylgalactosaminyltransferase, uridine diphosphoacetylgalactosamine-chondroitin, II Glucuronyl-N-acetylgalactosaminylproteoglycan β-1,4-N-acetylgalactosaminyltransferase	
2.4.1.79	Acetylgalactosaminyltransferase, uridine diphosphoacetylgalactosamine-galactosylgalactosylglucosylceramide Galactosylgalactosylglucosylceramide β-D-acetylgalactosaminyltransferase	
2.4.1.92	Acetylgalactosaminyltransferase, uridine diphosphoacetylgalactosamine-ganglioside GM3 (N-Acetylneuraminyl)-galactosylglucosylceramide N-acetylgalactosaminyltransferase	
2.4.1.88	Acetylgalactosaminyltransferase, uridine diphosphoacetylgalactosamine-globoside α- Globoside α-N-acetylgalactosaminyltransferase	
2.4.1.41	Acetylgalactosaminyltransferase, uridine diphosphoacetylgalactosamine-glycoprotein Polypeptide N-acetylgalactosaminyltransferase	
2.4.1.154	Acetylgalactosaminyltransferase, uridine diphosphoacetylgalactosamine-glycosphingolipid Globotriosylceramide β-1,6-N-acetylgalactosaminyltransferase	
2.4.1.174	N-Acetylgalactosaminyltransferase I Glucuronylgalactosylproteoglycan β-1,4-N-acetylgalactosaminyltransferase	
2.4.1.175	N-Acetylgalactosaminyltransferase II Glucuronyl-N-acetylgalactosaminylproteoglycan β-1,4-N-acetylgalactosaminyltransferase	

EC-No.	Synonym Recommended Name	Vol., Page
3.1.1.72	Acetylglucomannan esterase Acetylxylan esterase	
3.1.4.45	N-Acetylglucosamine-1-phosphodiester N-acetylglucosaminidase N-Acetylglucosamine-1-phosphodiester N-acetylglucosaminidase	
2.7.8.15	Acetylglucosamine-1-phosphotransferase, uridine diphosphoacetylglucosamine-dolichyl phosphate UDP-N-acetylglucosamine-dolichyl-phosphate N-acetylglucosaminephosphotransferase	
2.7.8.17	Acetylglucosamine-1-phosphotransferase, uridine diphosphoacetylglucosamine-glycoprotein UDP-N-acetylglucosamine-lysosomal-enzyme N-acetylglucosaminephosphotransferase	
2.7.8.17	Acetylglucosamine-1-phosphotransferase, uridine diphosphoacetylglucosamine-lysosomal enzyme precursor UDP-N-acetylglucosamine-lysosomal-enzyme N-acetylglucosaminephosphotransferase	
3.5.1.80	N-Acetylglucosamine-6-phosphate deacetylase N-Acetyl-D-galactosamine-6-phosphate deacetylase	
3.5.1.25	N-Acetylglucosamine-6-phosphate deacetylase N-Acetylglucosamine-6-phosphate deacetylase	
2.3.1.4	N-Acetylglucosamine-6-phosphate synthase Glucosamine-phosphate N-acetyltransferase	
3.1.6.14	N-Acetylglucosamine-6-sulfatase N-Acetylglucosamine-6-sulfatase	
2.4.1.197	Acetylglucosamine-oligosaccharide acetylglucosaminyltransferase High-mannose-oligosaccharide β-1,4-N-acetyl-glucosaminyltransferase	
2.7.7.23	Acetylglucosamine 1-phosphate uridylyltransferase UDP-N-acetylglucosamine pyrophosphorylase	
5.1.3.8	N-Acetylglucosamine 2-epimerase N-Acylglucosamine 2-epimerase	Vol. 1, Page 140
3.1.6.14	Acetylglucosamine 6-sulfatase N-Acetylglucosamine-6-sulfatase	
3.1.6.14	N-Acetylglucosamine 6-sulfate sulfatase N-Acetylglucosamine-6-sulfatase	
3.5.1.33	N-Acetylglucosamine deacetylase N-Acetylglucosamine deacetylase	
2.7.1.59	N-Acetylglucosamine kinase N-Acetylglucosamine kinase	

EC-No.	Synonym Recommended Name	Vol., Page
3.5.1.80	Acetylglucosamine phosphate deacetylase N-Acetyl-D-galactosamine-6-phosphate deacetylase	
3.5.1.25	Acetylglucosamine phosphate deacetylase N-Acetylglucosamine-6-phosphate deacetylase	
5.3.1.11	acetylglucosaminephosphate isomerase acetylglucosaminephosphate isomerase	Vol. 1, Page 321
5.4.2.3	Acetylglucosamine phosphomutase Phosphoacetylglucosamine mutase	Vol. 1, Page 515
2.7.5.2	acetylglucosamine phosphomutase acetylglucosamine phosphomutase	
3.2.1.50	α-acetylglucosaminidase α-N-acetylglucosaminidase	
3.2.1.52	β-N-acetylglucosaminidase β-N-acetylhexosaminidase	
3.2.1.50	α-N-acetylglucosaminidase α-N-acetylglucosaminidase	
3.2.1.30	β-D-acetylglucosaminidase β-D-acetylglucosaminidase	
3.2.2.11	Acetylglucosaminidase,.β.-aspartyl β-aspartyl-N-acetylglucosaminidase	
3.2.1.96	Acetylglucosaminidase, endo-.β. Mannosyl-glycoprotein endo-β-N-acetyl glucosaminidase	
2.4.1.90	β-N-Acetylglucosaminide β1−4 − galactosyltransferase N-Acetyllactosamine synthase	
2.4.1.141	N-Acetylglucosaminyldiphosphodolichol N-acetylglucosaminyltransferase N-Acetylglucosaminyldiphosphodolichol N-acetylglucosaminyltransferase	
2.4.1.188	N-Acetylglucosaminyldiphosphoundecaprenol glucosyltransferase N-Acetylglucosaminyldiphosphoundecaprenol glucosyltransferase	
2.4.1.187	N-Acetylglucosaminyldiphosphoundecaprenol N-acetyl-β-D-mannosaminyltransferase N-Acetylglucosaminyldiphosphoundecaprenol N-acetyl-β-D-mannosaminyltransferase	
2.4.1.38	β-N-Acetylglucosaminylglycopeptide β-1,4-galactosyltransferase β-N-Acetylglucosaminylglycopeptide β-1,4-galactosyltransferase	
3.1.1.69	N-Acetylglucosaminylphosphatidylinositol de-N-acetylase N-Acetylglucosaminylphosphatidylinositol deacetylase	

EC-No.	Synonym Recommended Name Vol., Page
3.1.1.69	Acetylglucosaminylphosphatidylinositol deacetylase N-Acetylglucosaminylphosphatidylinositol deacetylase
3.1.1.69	N-Acetylglucosaminylphosphatidylinositol deacetylase N-Acetylglucosaminylphosphatidylinositol deacetylase
3.1.4.45	α-N-acetylglucosaminyl phosphodiesterase N-Acetylglucosamine-1-phosphodiester N-acetylglucosaminidase
2.7.8.17	N-Acetylglucosaminyl phosphotransferase UDP-N-acetylglucosamine-lysosomal-enzyme N-acetylglucosaminepho-sphotransferase
2.7.8.17	N-Acetylglucosaminylphosphotransferase UDP-N-acetylglucosamine-lysosomal-enzyme N-acetylglucosaminepho-sphotransferase
2.4.1.65	β-Acetylglucosaminylsaccharide fucosyltransferase Galactoside 3(4)-L-fucosyltransferase
2.4.1.150	N-Acetylglucosaminyltransferase N-Acetyllactosaminide β-1,6-N-acetylglucosaminyltransferase
2.4.1.138	α-N-Acetylglucosaminyltransferase Mannotetraose 2-α-N-acetylglucosaminyltransferase
2.4.1.206	β1→3-N-Acetylglucosaminyltransferase Lactosylceramide 1,3-N-acetyl-β-D-glucosaminyltransferase
2.4.1.102	β6−N − Acetylglucosaminyltransferase β-1,3-Galactosyl-O-glycosyl-glycoprotein β-1,6-N-acetylglucosaminyl-transferase
2.4.1.16	Acetylglucosaminyltransferase, chitin-uridine diphosphate Chitin synthase
2.4.1.39	Acetylglucosaminyltransferase, hydroxy steroid Steroid N-acetylglucosaminyltransferase
2.4.1.101	Acetylglucosaminyltransferase, uridine diphosphoacetylglucosamine-α-1,3-mannosylglycoprotein β-1,2-N- α-1,3-Mannosyl-glycoprotein β-1,2-N-acetylglucosaminyltransferase
2.4.1.143	Acetylglucosaminyltransferase, uridine diphosphoacetylglucosamine-α-1,6-mannosylglycoprotein β-1-2-N- α-1,6-Mannosyl-glycoprotein β-1,2-N-acetylglucosaminyltransferase
2.4.1.143	Acetylglucosaminyltransferase, uridine diphosphoacetylglucosamine-α-D-mannoside β1-2- α-1,6-Mannosyl-glycoprotein β-1,2-N-acetylglucosaminyltransferase

EC-No.	Synonym Recommended Name	Vol., Page
2.4.1.155	Acetylglucosaminyltransferase, uridine diphosphoacetylglucosamine-α-mannoside β1→6- α-1,3(6)-Mannosylglycoprotein β-1,6-N-acetylglucosaminyltransferase	
2.4.1.164	Acetylglucosaminyltransferase, uridine diphosphoacetylglucosamine-acetyllactosaminide β1→6- Galactosyl-N-acetylglucosaminylgalactosylglucosylceramide β-1,6-N-acetylglucosaminyltransferase	
2.4.1.149	Acetylglucosaminyltransferase, uridine diphosphoacetylglucosamine-acetyllactosaminide β1→3- N-Acetyllactosaminide β-1,3-N-acetylglucosaminyltransferase	
2.4.1.163	Acetylglucosaminyltransferase, uridine diphosphoacetylglucosamine-acetyllactosaminide β1→3- β-Galactosyl-N-acetylglucosaminylgalactosylglucosylceramide β-1,3-acetylglucosaminyltransferase	
2.4.1.150	Acetylglucosaminyltransferase, uridine diphosphoacetylglucosamine-acetyllactosaminide β1→6- N-Acetyllactosaminide β-1,6-N-acetylglucosaminyltransferase	
2.4.1.150	Acetylglucosaminyltransferase, uridine diphosphoacetylglucosamine-acetyltransaminide β1→6- N-Acetyllactosaminide β-1,6-N-acetylglucosaminyltransferase	
2.4.1.153	Acetylglucosaminyltransferase, uridine diphosphoacetylglucosamine-dolichol phosphate Dolichyl-phosphate α-N-acetylglucosaminyltransferase	
2.4.1.141	Acetylglucosaminyltransferase, uridine diphosphoacetylglucosamine-dolichylacetylglucosamine pyrophosphate N-Acetylglucosaminyldiphosphodolichol N-acetylglucosaminyltransferase	
2.4.1.201	Acetylglucosaminyltransferase, uridine diphosphoacetylglucosamine-glycopeptide β-1→4-, VI Mannosyl-glycoprotein β-1,4-N-acetylglucosaminyltransferase	
2.4.1.201	Acetylglucosaminyltransferase, uridine diphosphoacetylglucosamine-glycopeptide β-1.4-, VI Mannosyl-glycoprotein β-1,4-N-acetylglucosaminyltransferase	
2.4.1.144	Acetylglucosaminyltransferase, uridine diphosphoacetylglucosamine-glycopeptide β4-, III β-1,4-Mannosyl-glycoprotein β-1,4-N-acetylglucosaminyltransferase	
2.4.1.145	Acetylglucosaminyltransferase, uridine diphosphoacetylglucosamine-glycopeptide β4-, IV α-1,3-Mannosylglycoprotein β-1,4-N-acetylglucosaminyltransferase	

EC-No.	Synonym Recommended Name	Vol., Page
2.4.1.206	Acetylglucosaminyltransferase, uridine diphosphoacetylglucosamine-lactosylceramide β- Lactosylceramide 1,3-N-acetyl-β-D-glucosaminyltransferase	
2.4.1.56	Acetylglucosaminyltransferase, uridine diphosphoacetylglucosamine-lipopolysaccharide Lipopolysaccharide N-acetylglucosaminyltransferase	
2.4.1.143	Acetylglucosaminyltransferase, uridine diphosphoacetylglucosamine-mannoside α1→6- α-1,6-Mannosyl-glycoprotein β-1,2-N-acetylglucosaminyltransferase	
2.4.1.147	Acetylglucosaminyltransferase, uridine diphosphoacetylglucosamine-mucin β(1→3)- Acetylgalactosaminyl-O-glycosyl-glycoprotein β-1,3-N-acetylglucosaminyltransferase	
2.4.1.146	Acetylglucosaminyltransferase, uridine diphosphoacetylglucosamine-mucin β(1→3)- (elongating) β-1,3-Galactosyl-O-glycosyl-glycoprotein β-1,3-N-acetylglucosaminyltransferase	
2.4.1.148	Acetylglucosaminyltransferase, uridine diphosphoacetylglucosamine-mucin β(1→6)-, B Acetylgalactosaminyl-O-glycosyl-glycoprotein β-1,6-N-acetylglucosaminyltransferase	
2.4.1.102	Acetylglucosaminyltransferase, uridine diphosphoacetylglucosamine-mucin β-(1→6)- β-1,3-Galactosyl-O-glycosyl-glycoprotein β-1,6-N-acetylglucosaminyltransferase	
2.4.1.102	Acetylglucosaminyltransferase, uridine diphosphoacetylglucosamine-mucin β-(1→6)-, A β-1,3-Galactosyl-O-glycosyl-glycoprotein β-1,6-N-acetylglucosaminyltransferase	
2.4.1.197	Acetylglucosaminyltransferase, uridine diphosphoacetylglucosamine-oligosaccharide High-mannose-oligosaccharide β-1,4-N-acetyl-glucosaminyltransferase	
2.4.1.70	Acetylglucosaminyltransferase, uridine diphosphoacetylglucosamine-poly(-ribitol phosphate) Poly(ribitol-phosphate) N-acetylglucosaminyltransferase	
2.4.1.94	Acetylglucosaminyltransferase, uridine diphosphoacetylglucosamine-protein Protein N-acetylglucosaminyltransferase	
2.4.1.138	Acetylglucosaminyltransferase, uridine diphosphoacetylglucosamine mannoside α1→2- Mannotetraose 2-α-N-acetylglucosaminyltransferase	

EC-No.	Synonym Recommended Name	Vol., Page
2.4.1.101	N-Acetylglucosaminyltransferase I α-1,3-Mannosyl-glycoprotein β-1,2-N-acetylglucosaminyltransferase	
2.4.1.143	Acetylglucosaminyltransferase II α-1,6-Mannosyl-glycoprotein β-1,2-N-acetylglucosaminyltransferase	
2.4.1.143	N-Acetylglucosaminyltransferase II α-1,6-Mannosyl-glycoprotein β-1,2-N-acetylglucosaminyltransferase	
2.4.1.144	N-Acetylglucosaminyltransferase III β-1,4-Mannosyl-glycoprotein β-1,4-N-acetylglucosaminyltransferase	
2.4.1.145	N-Acetylglucosaminyltransferase IV α-1,3-Mannosylglycoprotein β-1,4-N-acetylglucosaminyltransferase	
2.4.1.145	β-Acetylglucosaminyltransferase IV α-1,3-Mannosylglycoprotein β-1,4-N-acetylglucosaminyltransferase	
3.1.1.33	6-O-acetylglucose deacetylase 6-acetylglucose deacetylase	
3.1.1.33	6-acetylglucose deacetylase 6-acetylglucose deacetylase	
2.7.2.8	N-Acetylglutamate-5-phosphotransferase Acetylglutamate kinase	
2.3.1.35	Acetylglutamate-acetylornithine transacetylase Glutamate N-acetyltransferase	
2.7.2.8	N-Acetylglutamate 5-phosphotransferase Acetylglutamate kinase	
1.2.1.38	N-Acetylglutamate 5-semialdehyde dehydrogenase N-Acetyl-γ-glutamyl-phosphate reductase	
2.7.2.8	Acetylglutamate kinase Acetylglutamate kinase	
2.7.2.8	N-Acetylglutamate kinase Acetylglutamate kinase	
2.7.2.8	Acetylglutamate phosphokinase Acetylglutamate kinase	
2.7.2.8	N-Acetylglutamate phosphokinase Acetylglutamate kinase	
2.3.1.1	Acetylglutamate synthase Amino-acid N-acetyltransferase	
2.3.1.1	N-Acetylglutamate synthase Amino-acid N-acetyltransferase	

EC-No.	Synonym Recommended Name	Vol., Page
2.3.1.35	N-Acetylglutamate synthase Glutamate N-acetyltransferase	
2.3.1.1	Acetylglutamate synthetase Amino-acid N-acetyltransferase	
2.3.1.35	Acetylglutamate synthetase Glutamate N-acetyltransferase	
2.3.1.1	N-Acetylglutamate synthetase Amino-acid N-acetyltransferase	
2.3.1.35	N-Acetylglutamate synthetase Glutamate N-acetyltransferase	
2.3.1.35	Acetylglutamic-acetylornithine transacetylase Glutamate N-acetyltransferase	
1.2.1.38	N-Acetylglutamic γ-semialdehyde dehydrogenase N-Acetyl-γ-glutamyl-phosphate reductase	
2.7.2.8	N-Acetylglutamic 5-phosphotransferase Acetylglutamate kinase	
2.3.1.1	Acetylglutamic synthetase Amino-acid N-acetyltransferase	
2.3.1.35	Acetylglutamic synthetase Glutamate N-acetyltransferase	
1.1.1.240	N-Acetylhexosamine 1-dehydrogenase N-Acetylhexosamine 1-dehydrogenase	
3.2.1.52	β-acetylhexosaminidase β-N-acetylhexosaminidase	
3.2.1.52	N-Acetylhexosaminidase β-N-acetylhexosaminidase	
3.2.1.52	β-N-acetylhexosaminidase β-N-acetylhexosaminidase	
3.2.1.52	β-acetylhexosaminidinase β-N-acetylhexosaminidase	
3.4.13.5	N^{α}-acetylhistidine deacetylase X-methyl-His dipeptidase	
3.4.13.5	acetylhistidine deacetylase X-methyl-His dipeptidase	
3.5.1.34	acetylhistidine deacetylase acetylhistidine deacetylase	

EC-No.	Synonym Recommended Name	Vol., Page
2.3.1.33	Acetylhistidine synthetase Histidine N-acetyltransferase	
4.2.99.10	O-acetylhomoserine (thiol)-lyase O-acetylhomoserine (thiol)-lyase	
4.2.99.10	O-acetylhomoserine sulfhydrylase O-acetylhomoserine (thiol)-lyase	
3.1.1.47	acetylhydrolase 1-alkyl-2-acetylglycerophosphocholine esterase	
1.7.3.2	N-Acetylindoxyl:oxygen oxidoreductase Acetylindoxyl oxidase	
1.7.3.2	Acetylindoxyl oxidase Acetylindoxyl oxidase	
2.4.1.149	N-Acetyllactosamine β(1-3)N-acetylglucosaminyltransferase N-Acetyllactosaminide β-1,3-N-acetylglucosaminyltransferase	
2.4.1.124	N-Acetyllactosamine 3-α-galactosyltransferase N-Acetyllactosamine 3-α-galactosyltransferase	
2.4.1.22	N-Acetyllactosamine synthase Lactose synthase	
2.4.1.90	N-Acetyllactosamine synthase N-Acetyllactosamine synthase	
2.4.1.90	Acetyllactosamine synthetase N-Acetyllactosamine synthase	
2.4.1.90	N-Acetyllactosamine synthetase N-Acetyllactosamine synthase	
2.4.1.151	N-Acetyllactosaminide α-1,3-galactosyltransferase N-Acetyllactosaminide α-1,3-galactosyltransferase	
2.4.1.149	N-Acetyllactosaminide β-1,3-N-acetylglucosaminyltransferase N-Acetyllactosaminide β-1,3-N-acetylglucosaminyltransferase	
2.4.1.150	N-Acetyllactosaminide β-1,6-N-acetylglucosaminyltransferase N-Acetyllactosaminide β-1,6-N-acetylglucosaminyltransferase	
2.4.99.6	N-Acetyllactosaminide α-2,3-sialyltransferase N-Acetyllactosaminide α-2,3-sialyltransferase	
2.7.1.60	Acetylmannosamine kinase N-Acylmannosamine kinase	
2.7.1.60	N-Acetylmannosamine kinase N-Acylmannosamine kinase	

EC-No.	Synonym Recommended Name	Vol., Page
2.4.1.180	Acetylmannosaminuronosyltransferase, uridine diphosphoacetylmannosaminuronate-acetylglucosaminylpyrophosphorylundecaprenol Lipopolysaccharide N-acetylmannosaminouronosyltransferase	
2.4.1.187	N-Acetylmannosaminyltransferase N-Acetylglucosaminyldiphosphoundecaprenol N-acetyl-β-D-mannosaminyltransferase	
2.4.1.187	Acetylmannosaminyltransferase, uridine diphosphoacetyl-mannosamineacetylglucosaminylpyrophosphorylundecaprenol N-Acetylglucosaminyldiphosphoundecaprenol N-acetyl-β-D-mannosaminyltransferase	
1.8.4.5	Acetylmethionine sulfoxide reductase Methionine-S-oxide reductase	
3.4.19.4	N-acetylmethionylpeptide peptidase N-acetylmethionylpeptide peptidase	
5.1.2.4	Acetylmethylcarbinol racemase Acetoin racemase	Vol. 1, Page 85
3.5.1.28	N-Acetylmuramic acid L-alanine amidase N-Acetylmuramoyl-L-alanine amidase	
3.2.1.17	1,4-N-Acetylmuramidase Lysozyme	
3.2.1.92	Acetylmuramidase, exo-.β.- Peptidoglycan β-N-acetylmuramidase	
3.5.1.28	Acetylmuramoyl-alanine amidase N-Acetylmuramoyl-L-alanine amidase	
3.5.1.28	N-Acetylmuramoyl-L-alanine amidase N-Acetylmuramoyl-L-alanine amidase	
3.5.1.28	N-Acetylmuramoyl-L-alanine amidase Type I N-Acetylmuramoyl-L-alanine amidase	
3.5.1.28	N-Acetylmuramoyl-L-alanine amidase Type II N-Acetylmuramoyl-L-alanine amidase	
3.5.1.28	Acetylmuramyl-alanine amidase N-Acetylmuramoyl-L-alanine amidase	
3.5.1.28	Acetylmuramyl-L-alanine amidase N-Acetylmuramoyl-L-alanine amidase	
3.5.1.28	N-Acetylmuramyl-L-alanine amidase N-Acetylmuramoyl-L-alanine amidase	

EC-No.	Synonym Recommended Name	Vol., Page
3.5.1.28	N-Acetylmuramylalanine amidase N-Acetylmuramoyl-L-alanine amidase	
3.1.1.72	Acetylnaphthylesterase Acetylxylan esterase	
1.14.99.18	N-Acetylneuraminate,ascorbate(or NADPH):O_2 – oxidoreductase(N – acetyl – hydroxylating) N-Acetylneuraminate monooxygenase	
1.14.99.18	N-Acetylneuraminate,hydrogen-donor:oxygen oxidoreductase (N-acetyl-hydroxylating) N-Acetylneuraminate monooxygenase	
2.4.99.8	α-N-Acetylneuraminate α-2,8-sialyltransferase α-N-Acetylneuraminate α-2,8-sialyltransferase	
3.1.3.29	N-acetylneuraminate-9-phosphate phosphohydrolase N-acylneuraminate-9-phosphatase	
4.1.3.20	N-Acetylneuraminate-9-phosphate pyruvate-lyase (pyruvate-phosphorylating) N-Acetylneuraminate-9-phosphate synthase	
4.1.3.20	N-Acetylneuraminate-9-phosphate synthase N-Acetylneuraminate-9-phosphate synthase	
2.3.1.44	N-Acetylneuraminate 4-O-acetyltransferase N-Acetylneuraminate 4-O-acetyltransferase	
2.3.1.45	N-Acetylneuraminate 7(8)-O-acetyltransferase N-Acetylneuraminate 7-O(or 9-O-)-acetyltransferase	
2.3.1.45	N-Acetylneuraminate 7,8-O-acetyltransferase N-Acetylneuraminate 7-O(or 9-O-)-acetyltransferase	
2.3.1.45	N-Acetylneuraminate 7-O(or 9-O-)-acetyltransferase N-Acetylneuraminate 7-O(or 9-O-)-acetyltransferase	
4.1.3.3	N-Acetylneuraminate aldolase N-acetylneuraminate lyase	
4.1.3.3	N-Acetylneuraminate lyase N-acetylneuraminate lyase	
4.1.3.3	N-acetylneuraminate lyase N-acetylneuraminate lyase	
1.14.99.18	N-Acetylneuraminate monooxygenase N-Acetylneuraminate monooxygenase	
2.3.1.45	N-Acetylneuraminate O7-(or O9-)acetyltransferase N-Acetylneuraminate 7-O(or 9-O-)-acetyltransferase	

EC-No.	Synonym Recommended Name Vol., Page
4.1.3.3	Acetylneuraminate pyruvate-lyase N-acetylneuraminate lyase
4.1.3.3	N-Acetylneuraminate pyruvate-lyase N-acetylneuraminate lyase
4.1.3.3	N-acetylneuraminate pyruvate-lyase N-acetylneuraminate lyase
4.1.3.19	N-acetylneuraminate pyruvate-lyase (pyruvate-phosphorylating) N-acetylneuraminate synthase
4.1.3.19	N-acetylneuraminate synthase N-acetylneuraminate synthase
4.1.3.3	N-Acetylneuraminic acid aldolase N-acetylneuraminate lyase
4.1.3.19	N-acetylneuraminic acid condensing enzyme N-acetylneuraminate synthase
4.1.3.3	N-Acetylneuraminic acid lyase N-acetylneuraminate lyase
4.1.3.3	N-Acetylneuraminic aldolase N-acetylneuraminate lyase
4.1.3.3	N-Acetylneuraminic lyase N-acetylneuraminate lyase
1.14.99.18	N-Acetylneuraminic monooxygenase N-Acetylneuraminate monooxygenase
3.2.1.18	Acetylneuraminidase Sialidase
2.4.1.92	(N-Acetylneuraminyl)-galactosylglucosylceramide N-acetylgalactosaminyl-transferase (N-Acetylneuraminyl)-galactosylglucosylceramide N-acetylgalactosaminyl-transferase
2.4.99.7	(α-N-Acetylneuraminyl-2,3-β-galactosyl-1,3)-N-acetylgalactosaminide α-2,6-sialyltransferase (α-N-Acetylneuraminyl-2,3-β-galactosyl-1,3)-N-acetylgalactosaminide α-2,6-sialyltransferase
2.4.1.165	N-Acetylneuraminylgalactosylglucosylceramide β-1,4-N-acetylgalactosaminyltransferase N-Acetylneuraminylgalactosylglucosylceramide β-1,4-N-acetylgalactosaminyltransferase
2.3.1.35	Acetylornithinase Glutamate N-acetyltransferase

EC-No.	Synonym Recommended Name	Vol., Page
3.5.1.16	N-Acetylornithinase Acetylornithine deacetylase	
2.6.1.11	N-Acetylornithine-δ-transaminase Acetylornithine transaminase	
2.6.1.11	Acetylornithine δ-transaminase Acetylornithine transaminase	
2.6.1.11	Acetylornithine 5-aminotransferase Acetylornithine transaminase	
2.6.1.11	Acetylornithine aminotransferase Acetylornithine transaminase	
2.6.1.11	N-Acetylornithine aminotransferase Acetylornithine transaminase	
3.5.1.16	Acetylornithine deacetylase Acetylornithine deacetylase	
2.3.1.35	Acetylornithine glutamate acetyltransferase Glutamate N-acetyltransferase	
2.6.1.11	Acetylornithine transaminase Acetylornithine transaminase	
3.6.1.7	Acetylphosphatase Acylphosphatase	
3.1.1.72	N-Acetylphosphinothricin deacetylase Acetylxylan esterase	
2.3.1.49	S-Acetylphosphopantetheine:deacetyl-[citrate-oxaloacetate-lyase((pro-3S)-CH_2COO–acetate)] S-acetyltransferase Deacetyl-[citrate-(pro-3S)-lyase] S-acetyltransferase	
2.3.1.49	S-Acetyl phosphopantetheine:deacetyl citrate lyase S-acetyltransferase Deacetyl-[citrate-(pro-3S)-lyase] S-acetyltransferase	
3.5.1.62	N-Acetylputrescine acetylhydrolase Acetylputrescine deacetylase	
3.5.1.62	Acetylputrescine deacetylase Acetylputrescine deacetylase	
3.7.1.6	Acetylpyruvate hydrolase Acetylpyruvate hydrolase	
3.1.1.55	Acetylsalicylate deacetylase Acetylsalicylate deacetylase	

EC-No.	Synonym Recommended Name	Vol., Page
3.1.1.55	Acetylsalicylate O-acetylhydrolase Acetylsalicylate deacetylase	
3.1.1.55	Acetylsalicylic acid esterase Acetylsalicylate deacetylase	
4.2.99.8	O-acetylserine (thiol)-lyase Cysteine synthase	
4.2.99.8	O-acetylserine (thiol)-lyase A Cysteine synthase	
4.2.99.8	O-acetylserine-O-acetylhomoserine sulfhydro-lyase Cysteine synthase	
4.2.99.8	O-acetylserine sulfhydrylase Cysteine synthase	
4.2.99.8	acetylserine sulfhydrylase Cysteine synthase	
4.2.99.8	O-acetylserine sulfhydrylase A Cysteine synthase	
2.1.1.4	Acetylserotonin methyltransferase Acetylserotonin N-methyltransferase	
2.1.1.4	Acetylserotonin N-methyltransferase Acetylserotonin N-methyltransferase	
2.1.1.4	N-Acetylserotonin O-methyltransferase Acetylserotonin N-methyltransferase	
3.1.1.53	9-O-acetylsialic acid esterase sialate O-acetylesterase	
3.1.1.53	O-acetylsialic acid esterase sialate O-acetylesterase	
3.5.1.48	Acetylspermidine deacetylase Acetylspermidine deacetylase	
3.1.1.7	acetylthiocholinesterase acetylcholinesterase	
2.3.1.87	N-Acetyltransferase Aralkylamine N-acetyltransferase	
2.3.1.5	N-Acetyltransferase Arylamine N-acetyltransferase	
2.3.1.88	N^{α}-Acetyltransferase Peptide α-N-acetyltransferase	

EC-No.	Synonym Recommended Name	Vol., Page
2.3.1.78	Acetyltransferase, α-glucosaminide Heparan-α-glucosaminide N-acetyltransferase	
2.3.1.108	Acetyltransferase, α-tubulin Tubulin N-acetyltransferase	
2.3.1.67	Acetyltransferase, 1-alkyl-2-lysolecithin 1-Alkylglycerophosphocholine O-acetyltransferase	
2.3.1.67	Acetyltransferase, 1-alkylglycerophosphocholine 1-Alkylglycerophosphocholine O-acetyltransferase	
2.3.1.5	Acetyltransferase, 2-naphthylamine N- Arylamine N-acetyltransferase	
2.3.1.5	Acetyltransferase, 4-aminobiphenyl Arylamine N-acetyltransferase	
2.3.1.38	Acetyltransferase, [acyl-carrier-protein] [Acyl-carrier-protein] S-acetyltransferase	
2.3.1.9	Acetyltransferase, acetyl coenzyme A Acetyl-CoA C-acetyltransferase	
2.3.1.84	Acetyltransferase, alcohol Alcohol O-acetyltransferase	
2.3.1.1	Acetyltransferase, amino acid Amino-acid N-acetyltransferase	
2.3.1.81	Acetyltransferase, aminoglycoside 3-N- Aminoglycoside $N^{3'}$-acetyltransferase	
2.3.1.82	Acetyltransferase, aminoglycoside 6'-N- Aminoglycoside $N^{6'}$-acetyltransferase	
2.3.1.56	Acetyltransferase, aromatic hydroxylamine Aromatic-hydroxylamine O-acetyltransferase	
2.3.1.87	Acetyltransferase, arylalkylamine N- Aralkylamine N-acetyltransferase	
2.3.1.5	Acetyltransferase, arylamine Arylamine N-acetyltransferase	
2.3.1.17	Acetyltransferase, aspartate Aspartate N-acetyltransferase	
2.3.1.7	Acetyltransferase, carnitine Carnitine O-acetyltransferase	
2.3.1.28	Acetyltransferase, chloramphenicol Chloramphenicol O-acetyltransferase	

EC-No.	Synonym Recommended Name	Vol., Page
2.3.1.6	Acetyltransferase, choline Choline O-acetyltransferase	
2.3.1.49	Acetyltransferase, citrate lyase Deacetyl-[citrate-(pro-3S)-lyase] S-acetyltransferase	
2.3.1.27	Acetyltransferase, corticosteroid Cortisol O-acetyltransferase	
2.3.1.27	Acetyltransferase, cortisol Cortisol O-acetyltransferase	
2.3.1.80	Acetyltransferase, cysteine S-conjugate N- Cysteine-S-conjugate N-acetyltransferase	
2.3.1.36	Acetyltransferase, D-amino acid D-Amino-acid N-acetyltransferase	
2.3.1.34	Acetyltransferase, D-tryptophan D-Tryptophan N-acetyltransferase	
2.3.1.107	Acetyltransferase, deacetylvindoline 17-O-Deacetylvindoline O-acetyltransferase	
2.3.1.54	Acetyltransferase, formate Formate C-acetyltransferase	
2.3.1.18	Acetyltransferase, galactoside Galactoside O-acetyltransferase	
2.3.1.59	Acetyltransferase, gentamicin 2'- Gentamicin 2'-N-acetyltransferase	
2.3.1.60	Acetyltransferase, gentamicin 3- Gentamicin 3'-N-acetyltransferase	
2.3.1.3	Acetyltransferase, glucosamine Glucosamine N-acetyltransferase	
2.3.1.4	Acetyltransferase, glucosamine phosphate Glucosamine-phosphate N-acetyltransferase	
2.3.1.35	Acetyltransferase, glutamate Glutamate N-acetyltransferase	
2.3.1.29	Acetyltransferase, glycine Glycine C-acetyltransferase	
2.3.1.33	Acetyltransferase, histidine Histidine N-acetyltransferase	
2.3.1.48	Acetyltransferase, histone Histone acetyltransferase	

EC-No.	Synonym Recommended Name	Vol., Page
2.3.1.31	Acetyltransferase, homoserine Homoserine O-acetyltransferase	
2.3.1.10	Acetyltransferase, hydrogen sulfide Hydrogen-sulfide S-acetyltransferase	
2.3.1.2	Acetyltransferase, imidazole Imidazole N-acetyltransferase	
2.3.1.82	Acetyltransferase, kanamycin Aminoglycoside $N^{6'}$-acetyltransferase	
2.3.1.66	Acetyltransferase, leucine Leucine N-acetyltransferase	
2.3.1.12	Acetyltransferase, lipoate Dihydrolipoamide S-acetyltransferase	
2.3.1.32	Acetyltransferase, lysine Lysine N-acetyltransferase	
2.3.1.79	Acetyltransferase, maltose Maltose O-acetyltransferase	
2.3.1.69	Acetyltransferase, monoterpenol Monoterpenol O-acetyltransferase	
2.3.1.44	Acetyltransferase, N-acetylneuraminate 4-O- N-Acetylneuraminate 4-O-acetyltransferase	
2.3.1.45	Acetyltransferase, N-acetylneuraminate 9(7)-O- N-Acetylneuraminate 7-O(or 9-O-)-acetyltransferase	
2.3.1.118	Acetyltransferase, N-hydroxyarylamine O- N-Hydroxyarylamine O-acetyltransferase	
2.3.1.102	Acetyltransferase, N^{ε}-hydroxylysine N^{6}-Hydroxylysine O-acetyltransferase	
2.3.1.5	Acetyltransferase, *p*-aminosalicylate N- Arylamine N-acetyltransferase	
2.3.1.88	Acetyltransferase, peptide N-terminal Peptide α-N-acetyltransferase	
2.3.1.53	Acetyltransferase, phenylalanine Phenylalanine N-acetyltransferase	
2.3.1.8	Acetyltransferase, phosphate Phosphate acetyltransferase	
2.3.1.136	Acetyltransferase, poly(N-acetylneuraminate) Polysialic-acid O-acetyltransferase	

EC-No.	Synonym Recommended Name	Vol., Page
2.3.1.5	Acetyltransferase, procainamide N- Arylamine N-acetyltransferase	
2.3.1.57	Acetyltransferase, putrescine Diamine N-acetyltransferase	
2.3.1.128	Acetyltransferase, ribosomal protein S 18 Ribosomal-protein-alanine N-acetyltransferase	
2.3.1.150	Acetyltransferase, salutaridinol 7-O- Salutaridinol 7-O-acetyltransferase	
2.3.1.30	Acetyltransferase, serine Serine O-acetyltransferase	
2.3.1.5	Acetyltransferase, serotonin N- Arylamine N-acetyltransferase	
2.3.1.89	Acetyltransferase, tetrahydrodipicolinate Tetrahydrodipicolinate N-acetyltransferase	
2.3.1.11	Acetyltransferase, thioethanolamine Thioethanolamine S-acetyltransferase	
3.1.1.72	Acetylxylan esterase Acetylxylan esterase	
3.1.1.72	Acetylxylan esterase Acetylxylan esterase	
3.1.1.72	Acetyte esterase Acetylxylan esterase	
3.1.1.7	AChE acetylcholinesterase	
3.4.21.50	Achrombacter protease I lysyl endopeptidase	
3.4.24.8	Achromobacter iophagus collagenase Achromobacter iophagus collagenase	
3.4.24.3	Achromobacter iophagus collagenase microbial collagenase	
3.4.21.50	Achromobacter proteinase I lysyl endopeptidase	
3.4.21.50	achromopeptidase lysyl endopeptidase	
3.4.24.32	Achromopeptidase component β-Lytic metalloendopeptidase	

EC-No.	Synonym Recommended Name	Vol., Page
1.7.3.5	3-aci-Nitropropanoate:oxygen oxidoreductase 3-aci-Nitropropanoate oxidase	
1.7.3.5	3-aci-Nitropropanoate oxidase 3-aci-Nitropropanoate oxidase	
3.1.3.19	(acid) β-glycerophosphatase glycerol-2-phosphatase	
6.2.1.10	Acid-CoA ligase (GDP) Acid-CoA ligase (GDP-forming)	Vol. 2, Page 249
6.2.1.10	Acid-CoA ligase (GDP-forming) Acid-CoA ligase (GDP-forming)	Vol. 2, Page 249
3.2.1.45	Acid β-glucosidase Glucosylceramidase	
6.2.1.3	Acid:CoA ligase (AMP-forming) Long-chain-fatty-acid-CoA ligase	Vol. 2, Page 206
6.2.1.10	Acid:CoA ligase (GDP) Acid-CoA ligase (GDP-forming)	Vol. 2, Page 249
6.2.1.10	Acid:CoA ligase (GDP-forming) Acid-CoA ligase (GDP-forming)	Vol. 2, Page 249
3.4.18.1	acid carboxypeptidase cysteine-type carboxypeptidase	
3.1.22.1	Acid deoxyribonuclease Deoxyribonuclease II	
3.1.22.1	Acid DNase Deoxyribonuclease II	
3.2.1.26	Acid invertase β-fructofuranosidase	
3.2.1.3	Acid maltase Glucan 1,4-α-glucosidase	
3.4.24.39	Acid metalloproteinase Deuterolysin	
3.1.3.2	Acid monophosphatase Acid phosphatase	
3.1.3.2	Acid nucleoside diphosphate phosphatase Acid phosphatase	
3.1.3.31	acid nucleotidase nucleotidase	

EC-No.	Synonym Recommended Name	Vol., Page
3.1.3.2	Acid phosphatase Acid phosphatase	
3.1.3.10	acid phosphatase glucose-1-phosphatase	
3.1.3.4	Acid phosphatidyl phosphatase Phosphatidate phosphatase	
3.6.1.11	Acid phosphoanhydride phosphohydrolase Exopolyphosphatase	
3.1.3.2	Acid phosphohydrolase Acid phosphatase	
3.1.3.2	Acid phosphomoesterase Acid phosphatase	
3.1.3.2	Acid phosphomonoester hydrolase Acid phosphatase	
3.4.23.40	Acid protease Phytepsin	
3.4.23.20	Acid protease A Penicillopepsin	
3.1.27.1	Acid ribonuclease Ribonuclease T2	
3.1.27.1	Acid RNase Ribonuclease T2	
4.1.3.8	ACL ATP citrate (pro-3S)-lyase	
5.1.1.15	ACL racemase 2-Aminohexano-6-lactam racemase	Vol. 1, Page 61
4.1.1.45	ACMSD Aminocarboxymuconate-semialdehyde decarboxylase	
2.6.1.11	ACOAT Acetylornithine transaminase	
4.2.1.3	aconitase aconitate hydratase	
5.3.3.7	Aconitate Δ-isomerase Aconitate Δ-isomerase	Vol. 1, Page 409
5.3.3.7	Aconitate Δ^2-Δ^3-isomerase Aconitate Δ-isomerase	Vol. 1, Page 409

EC-No.	Synonym Recommended Name	Vol., Page
4.1.1.6	Aconitate decarboxylase Aconitate decarboxylase	
4.1.1.6	Aconitate decarboxylase Aconitate decarboxylase	
4.2.1.3	aconitate hydratase aconitate hydratase	
5.3.3.7	Aconitate isomerase Aconitate Δ-isomerase	Vol. 1, Page 409
2.3.1.15	ACP:sn-glycerol-3-phosphate acyltransferase Glycerol-3-phosphate O-acyltransferase	
2.3.1.38	[ACP]acetyltransferase [Acyl-carrier-protein] S-acetyltransferase	
4.1.99.4	ACPC deaminase 1-aminocyclopropane-1-carboxylate deaminase	
3.1.2.21	12:0-ACP TE Dodecanoyl-[acyl-carrier protein] hydrolase	
3.1.2.21	12-ACP thioesterase Dodecanoyl-[acyl-carrier protein] hydrolase	
3.5.4.11	Acrasinase Pteridin deaminase	
3.4.23.28	Acrocylindricum proteinase Acrocylindropepsin	
3.4.99.1	acrocylindricum proteinase acrocylindricum proteinase	
3.4.23.28	Acrocylindrium acid proteinase Acrocylindropepsin	
3.4.23.28	Acrocylindrium proteinase Acrocylindropepsin	
3.4.23.28	Acrocylindropepsin Acrocylindropepsin	
3.4.21.10	acrosin acrosin	
3.4.21.10	α-acrosin acrosin	
3.4.21.10	β-acrosin acrosin	

EC-No.	Synonym Recommended Name	Vol., Page
3.4.21.10	acrosin amidase acrosin	
3.4.21.10	acrosomal protease acrosin	
3.4.21.10	acrosomal proteinase acrosin	
3.4.21.10	acrozonase acrosin	
3.5.1.40	Acrylagmatine amidohydrolase Acylagmatine amidase	
4.2.1.84	acrylonitrile hydratase nitrile hydratase	
4.2.1.54	acrylyl coenzyme A hydratase lactoyl-CoA dehydratase	
6.2.1.1	ACS Acetate-CoA ligase	Vol. 2, Page 186
6.2.1.13	ACS Acetate-CoA ligase (ADP-forming)	Vol. 2, Page 267
6.2.1.3	ACS3 Long-chain-fatty-acid-CoA ligase	Vol. 2, Page 206
3.4.21.7	actase plasmin	
2.1.1.85	Actin-specific histidine methyltransferase Protein-histidine N-methyltransferase	
3.4.11.22	Actinase AS Aminopeptidase I	
3.4.24.31	Actinase E Mycolysin	
3.4.22.14	actinidain actinidain	
3.4.22.14	actinidia anionic protease actinidain	
3.4.22.14	actinidin actinidain	
3.1.1.39	actinomycin lactonase actinomycin lactonase	

EC-No.	Synonym Recommended Name	Vol., Page
3.1.1.39	actinomycin lactonohydrolase actinomycin lactonase	
1.97.1.4	Activase, pyruvate formate-lyase Formate acetyltransferase activating enzyme	
3.4.21.6	activated blood-coagulation factor X coagulation factor Xa	
3.4.21.27	activated blood-coagulation factor XI coagulation factor XIa	
3.4.21.21	activated blood coagulation factor VII coagulation factor VIIa	
3.4.21.69	Activated blood coagulation factor XIV Protein C (activated)	
3.4.21.22	activated Christmas factor coagulation factor IXa	
3.4.21.41	activated complement C1r complement subcomponent C1r	
3.4.21.42	activated complement C1s complement subcomponent C1sbar	
3.4.21.22	activated factor IX coagulation factor IXa	
3.4.21.6	activated factor X coagulation factor Xa	
3.4.21.27	activated plasma thromboplastin antecedent coagulation factor XIa	
3.4.21.69	Activated protein C Protein C (activated)	
3.4.21.6	activated Stuart-Prower factor coagulation factor Xa	
3.6.1.32	Actomyosin Myosin ATPase	
3.5.1.70	Aculeacin-A amidohydrolase Aculeacin-A deacylase	
3.5.1.70	Aculeacin-A deacylase Aculeacin-A deacylase	
3.5.1.70	Aculeacin A acylase Aculeacin-A deacylase	

EC-No.	Synonym Recommended Name	Vol., Page
3.4.24.11	acute lymphoblastic leukemia antigen neprilysin	
1.1.3.29	N-Acyl-β-D-hexosamine:oxygen 1-oxidoreductase N-Acylhexosamine oxidase	
1.14.13.26	1-Acyl-2-oleoyl-sn-glycero-3-phosphocholine,NADH:oxygen oxidoreductase (12-hydroxylating) Phosphatidylcholine 12-monooxygenase	
1.3.1.35	1-Acyl-2-oleoyl-sn-glycero-3-phosphocholine:NAD^+ Δ^{12}-oxidoreductase Phosphatidylcholine desaturase	
1.14.99.6	Acyl-[acyl-carrier-protein],hydrogen-donor:oxygen oxidoreductase Acyl-[acyl-carrier-protein] desaturase	
1.14.19.2	acyl-[acyl-carrier-protein], hydrogen-donor:oxygen oxidoreductase acyl-[acyl-carrier-protein] desaturase	
2.3.1.40	Acyl-[acyl-carrier-protein]-phospholipid O-acyltransferase Acyl-[acyl-carrier-protein]-phospholipid O-acyltransferase	
2.3.1.129	Acyl-[acyl-carrier-protein]-UDP-N-acetylglucosamine O-acyltransferase Acyl-[acyl-carrier-protein]-UDP-N-acetylglucosamine O-acyltransferase	
2.3.1.41	Acyl-[acyl-carrier-protein]:malonyl-[acyl-carrier-protein] C-acyltransferase (decarboxylating) 3-Oxoacyl-[acyl-carrier-protein] synthase	
1.3.1.9	Acyl-[acyl-carrier-protein]: NAD^+ oxidoreductase Enoyl-[acyl-carrier-protein] reductase (NADH)	
1.3.1.39	Acyl-[acyl-carrier-protein]: $NADP^+$ oxidoreductase (A-specific) Enoyl-[acyl-carrier-protein] reductase (NADPH, A-specific)	
1.3.1.10	Acyl-[acyl-carrier-protein]: $NADP^+$ oxidoreductase (B-specific) Enoyl-[acyl-carrier-protein] reductase (NADPH, B-specific)	
2.3.1.40	Acyl-[acyl-carrier-protein]:O-(2-acyl-sn-glycero-3-phospho)-ethanolamine O-acyltransferase Acyl-[acyl-carrier-protein]-phospholipid O-acyltransferase	
1.14.99.6	Acyl-[acyl-carrier-protein] desaturase Acyl-[acyl-carrier-protein] desaturase	
1.14.19.2	acyl-[acyl-carrier-protein] desaturase acyl-[acyl-carrier-protein] desaturase	
3.1.2.14	acyl-[acyl-carrier-protein] hydrolase oleoyl-[acyl-carrier-protein] hydrolase	
6.2.1.20	Acyl-[acyl-carrier-protein]synthetase Long-chain-fatty-acid-[acyl-carrier-protein] ligase	Vol. 2, Page 296

EC-No.	Synonym Recommended Name	Vol., Page
2.3.1.141	Acyl-[acyl-carrier protein]:D-galactosylacylglycerol O-acyltransferase Galactosylacylglycerol O-acyltransferase	
6.2.1.20	Acyl-[acyl carrier protein] synthetase Long-chain-fatty-acid-[acyl-carrier-protein] ligase	Vol. 2, Page 296
3.1.2.14	acyl-ACP-hydrolase oleoyl-[acyl-carrier-protein] hydrolase	
2.3.1.141	Acyl-ACP:lyso-MGDG acyltransferase Galactosylacylglycerol O-acyltransferase	
6.2.1.20	Acyl-ACP synthetase Long-chain-fatty-acid-[acyl-carrier-protein] ligase	Vol. 2, Page 296
6.2.1.1	Acyl-activating enzyme Acetate-CoA ligase	Vol. 2, Page 186
6.2.1.2	Acyl-activating enzyme Butyrate-CoA ligase	Vol. 2, Page 199
6.2.1.3	Acyl-activating enzyme Long-chain-fatty-acid-CoA ligase	Vol. 2, Page 206
2.3.1.141	Acyl-acyl-carrier protein:lysomonogalactosyldiacylglycerol acyltransferase Galactosylacylglycerol O-acyltransferase	
3.1.2.14	acyl-acyl carrier protein hydrolase oleoyl-[acyl-carrier-protein] hydrolase	
6.2.1.20	Acyl-acyl carrier protein synthetase Long-chain-fatty-acid-[acyl-carrier-protein] ligase	Vol. 2, Page 296
3.1.4.14	[Acyl-carrier-protein] 4'-pantetheine-phosphohydrolase [Acyl-carrier-protein] phosphodiesterase	
2.3.1.38	[Acyl-carrier-protein]acetyltransferase [Acyl-carrier-protein] S-acetyltransferase	
3.1.4.14	[Acyl-carrier-protein] phosphodiesterase [Acyl-carrier-protein] phosphodiesterase	
2.3.1.38	[Acyl-carrier-protein] S-acetyltransferase [Acyl-carrier-protein] S-acetyltransferase	
2.3.1.39	[Acyl-carrier-protein] S-malonyltransferase [Acyl-carrier-protein] S-malonyltransferase	
1.3.99.3	Acyl-CoA:(acceptor) 2,3-oxidoreductase Acyl-CoA dehydrogenase	
2.3.1.20	Acyl-CoA:1,2-diacylglycerol O-acyltransferase Diacylglycerol O-acyltransferase	

EC-No.	Synonym Recommended Name	Vol., Page
2.3.1.23	Acyl-CoA:1-acyl-glycero-3-phosphocholine transacylase 1-Acylglycerophosphocholine O-acyltransferase	
2.3.1.23	Acyl-CoA:1-acyl-sn-glycero-3-phosphocholine O-acyltransferase 1-Acylglycerophosphocholine O-acyltransferase	
2.3.1.51	Acyl-CoA:1-acyl-sn-glycerol-3-phosphate 2-O-acyltransferase 1-Acylglycerol-3-phosphate O-acyltransferase	
2.3.1.104	Acyl-CoA:1-alkenylglycerophosphocholine O-acyltransferase 1-Alkenylglycerophosphocholine O-acyltransferase	
2.3.1.121	Acyl-CoA:1-alkenylglycerophosphoethanolamine O-acyltransferase 1-Alkenylglycerophosphoethanolamine O-acyltransferase	
2.3.1.67	Acyl-CoA:1-alkyl-sn-glycero-3-phosphocholine acyltransferase 1-Alkylglycerophosphocholine O-acetyltransferase	
2.3.1.63	Acyl-CoA:1-alkyl-sn-glycero-3-phosphocholine O-acyltransferase 1-Alkylglycerophosphocholine O-acyltransferase	
2.3.1.25	Acyl-CoA:1-O-alk-1-enyl-glycero-3-phosphocholine 2-O-acyltransferase Plasmalogen synthase	
2.3.1.125	Acyl-CoA:1-O-alkyl-2-acetyl-sn-glycerol O-acyltransferase 1-Alkyl-2-acetylglycerol O-acyltransferase	
2.3.1.62	Acyl-CoA:2-acyl-sn-glycero-3-phosphocholine O-acyltransferase 2-Acylglycerophosphocholine O-acyltransferase	
2.3.1.52	Acyl-CoA:2-acyl-sn-glycerol 3-phosphate O-acyltransferase 2-Acylglycerol-3-phosphate O-acyltransferase	
2.3.1.22	Acyl-CoA:2-acylglycerol O-acyltransferase 2-Acylglycerol O-acyltransferase	
2.8.3.8	Acyl-CoA:acetate CoA-transferase Acetate CoA-transferase	
2.3.1.16	Acyl-CoA:acetyl-CoA C-acyltransferase Acetyl-CoA C-acyltransferase	
2.3.1.26	Acyl-CoA:cholesterol acyltransferase Sterol O-acyltransferase	
2.3.1.26	Acyl-CoA:cholesterol O-acyltransferase Sterol O-acyltransferase	
2.3.1.123	Acyl-CoA:dolichol acyltransferase Dolichol O-acyltransferase	
2.3.1.139	Acyl-CoA:ecdysone acyltransferase Ecdysone O-acyltransferase	

EC-No.	Synonym Recommended Name	Vol., Page
2.3.1.42	Acyl-CoA:glycerone-phosphate O-acyltransferase Glycerone-phosphate O-acyltransferase	
2.3.1.13	Acyl-CoA:glycine N-acyltransferase Glycine N-acyltransferase	
2.3.1.68	Acyl-CoA:L-glutamine N-acyltransferase Glutamine N-acyltransferase	
2.3.1.75	Acyl-CoA:long-chain-alcohol O-acyltransferase Long-chain-alcohol O-fatty-acyltransferse	
4.1.1.9	Acyl-CoA: malonate CoA transferase/malonyl-CoA decarboxylase Malonyl-CoA decarboxylase	
2.3.1.86	Acyl-CoA:malonyl-CoA C-acyltransferase (decarboxylating, oxoacyl- and enoyl-reducing) Fatty-acyl-CoA synthase	
2.3.1.85	Acyl-CoA:malonyl-CoA C-acyltransferase (decarboxylating, oxoacyl- and enoyl-reducing and thioester-hydrolysing) Fatty-acid synthase	
2.3.1.111	Acyl-CoA:methylmalonyl-CoA C-acyltransferase (decarboxylating, oxoacyl- and enoyl-reducing) Mycocerosate synthase	
1.3.1.44	Acyl-CoA:NAD^+ trans-2-oxidoreductase trans-2-Enoyl-CoA reductase (NAD^+)	
1.3.1.8	Acyl-CoA: $NADP^+$ 2-oxidoreductase Acyl-CoA dehydrogenase ($NADP^+$)	
1.3.1.37	Acyl-CoA: $NADP^+$ cis-2-oxidoreductase cis-2-Enoyl-CoA reductase (NADPH)	
1.3.1.38	Acyl-CoA: $NADP^+$ trans-2-oxidoreductase trans-2-Enoyl-CoA reductase (NADPH)	
1.3.3.6	Acyl-CoA:oxygen 2-oxidoreductase Acyl-CoA oxidase	
2.3.1.76	Acyl-CoA:retinol O-acyltransferase Retinol O-fatty-acyltransferase	
2.3.1.50	Acyl-CoA:serine C-2 acyltransferase decarboxylating Serine C-palmitoyltransferase	
2.3.1.15	Acyl-CoA:sn-glycerol-3-phosphate 1-O-acyltransferase Glycerol-3-phosphate O-acyltransferase	
2.3.1.24	Acyl-CoA:sphingosine N-acyltransferase Sphingosine N-acyltransferase	

EC-No.	Synonym Recommended Name	Vol., Page
1.3.99.3	Acyl-CoA dehydrogenase Acyl-CoA dehydrogenase	
1.3.2.2	acyl-CoA dehydrogenase acyl-CoA dehydrogenase	
1.3.1.8	Acyl-CoA dehydrogenase ($NADP^+$) Acyl-CoA dehydrogenase ($NADP^+$)	
1.14.99.5	Acyl-CoA desaturase Stearoyl-CoA desaturase	
1.14.19.1	acyl-CoA desaturase stearoyl-CoA 9-desaturase	
2.3.1.119	Acyl-CoA elongase Icosanoyl-CoA synthase	
3.1.2.20	Acyl-CoA hydrolase Acyl-CoA hydrolase	
3.1.2.20	Acyl-CoA hydrolase Acyl-CoA hydrolase	
6.2.1.3	Acyl-CoA ligase Long-chain-fatty-acid-CoA ligase	Vol. 2, Page 206
1.3.3.6	Acyl-CoA oxidase Acyl-CoA oxidase	
1.2.1.50	Acyl-CoA reductase Long-chain-fatty-acyl-CoA reductase	
6.2.1.3	Acyl-CoA synthetase Long-chain-fatty-acid-CoA ligase	Vol. 2, Page 206
6.2.1.10	Acyl-CoA synthetase (GDP-forming) Acid-CoA ligase (GDP-forming)	Vol. 2, Page 249
6.2.1.3	Acyl-CoA synthetase 3 Long-chain-fatty-acid-CoA ligase	Vol. 2, Page 206
3.1.2.20	Acyl-CoA thioesterase Acyl-CoA hydrolase	
6.2.1.3	Acyl-coenzyme A ligase Long-chain-fatty-acid-CoA ligase	Vol. 2, Page 206
3.5.1.81	N-Acyl-D-amino-acid deacylase N-Acyl-D-amino-acid deacylase	
3.5.1.81	N-Acyl-D-amino acid amidohydrolase N-Acyl-D-amino-acid deacylase	

EC-No.	Synonym Recommended Name	Vol., Page
3.5.1.81	N-Acyl-D-amino acid amidohydrolase N-Acyl-D-amino-acid deacylase	
3.5.1.83	N-Acyl-D-aspartate amidohydrolase N-Acyl-D-aspartate deacylase	
3.5.1.83	N-Acyl-D-aspartate amidohydrolase N-Acyl-D-aspartate deacylase	
3.5.1.83	N-Acyl-D-aspartate deacylase N-Acyl-D-aspartate deacylase	
5.1.3.9	N-Acyl-D-glucosamine-6-phosphate 2-epimerase N-Acylglucosamine-6-phosphate 2-epimerase	Vol. 1, Page 144
5.1.3.8	N-Acyl-D-glucosamine 2-epimerase N-Acylglucosamine 2-epimerase	Vol. 1, Page 140
3.5.1.82	N-Acyl-D-glutamate amidohydrolase N-Acyl-D-glutamate deacylase	
3.5.1.82	N-Acyl-D-glutamate deacylase N-Acyl-D-glutamate deacylase	
1.1.3.29	N-Acyl-D-hexosamine:oxygen 1-oxidoreductase N-Acylhexosamine oxidase	
1.1.3.29	N-Acyl-D-hexosamine oxidase N-Acylhexosamine oxidase	
1.1.1.233	N-Acyl-D-mannosamine:NAD^+ 1-oxidoreductase N-Acylmannosamine 1-dehydrogenase	
1.1.1.233	N-Acyl-D-mannosamine dehydrogenase N-Acylmannosamine 1-dehydrogenase	
2.7.1.60	N-Acyl-D-mannosamine kinase N-Acylmannosamine kinase	
2.5.1.26	1-Acyl-glycerone-3-phosphate:long-chain-alcohol O-3-phospho-2-oxopropanyltransferase Alkylglycerone-phosphate synthase	
3.5.1.14	N-Acyl-L-amino-acid amidohydrolase Aminoacylase	
3.5.1.15	N-Acyl-L-aspartate amidohydrolase Aspartoacylase	
3.5.1.17	Acyl-lysine deacylase Acyl-lysine deacylase	

EC-No.	Synonym Recommended Name	Vol., Page
2.3.1.41	Acyl-malonyl(acyl-carrier-protein)-condensing enzyme 3-Oxoacyl-[acyl-carrier-protein] synthase	
3.1.1.53	N-acyl-O-acetylneuraminate O-acetylhydrolase sialate O-acetylesterase	
2.7.1.61	Acyl-phosphate-hexose phosphotransferase Acyl-phosphate-hexose phosphotransferase	
2.7.1.61	Acyl-phosphate:D-hexose phosphotransferase Acyl-phosphate-hexose phosphotransferase	
6.2.1.19	Acyl-protein synthetase Long-chain-fatty-acid-luciferin-component ligase	Vol. 2, Page 293
2.3.1.100	Acyl-protein synthetase Myelin-proteolipid O-palmitoyltransferase	
2.3.1.51	1-Acyl-sn-glycero-3-phosphate acyltransferase 1-Acylglycerol-3-phosphate O-acyltransferase	
2.3.1.23	1-Acyl-sn-glycero-3-phosphocholine acyltransferase 1-Acylglycerophosphocholine O-acyltransferase	
2.3.1.51	1-Acyl-sn-glycerol 3-phosphate acyltransferase 1-Acylglycerol-3-phosphate O-acyltransferase	
3.5.1.40	Acylagmatine amidase Acylagmatine amidase	
3.5.1.40	Acylagmatine amidohydrolase Acylagmatine amidase	
3.5.1.40	Acylagmatine deacylase Acylagmatine amidase	
3.5.1.4	Acylamidase Amidase	
3.5.1.4	Acylamide amidohydrolase Amidase	
3.5.1.11	α-acylamino-β-lactam acylhydrolase Penicillin amidase	
3.4.14.3	acylamino-acid-releasing enzyme acylamino-acid-releasing enzyme	
3.4.19.1	acylamino-acid-releasing enzyme acylaminoacyl-peptidase	
3.5.1.14	α-N-acylaminoacid hydrolase Aminoacylase	

EC-No.	Synonym Recommended Name	Vol., Page
3.4.19.1	acylaminoacyl-peptidase acylaminoacyl-peptidase	
3.4.19.1	N-acylaminoacyl-peptide hydrolase acylaminoacyl-peptidase	
2.7.1.60	Acylaminodeoxymannokinase N-Acylmannosamine kinase	
3.5.1.4	Acylase Amidase	
3.5.1.14	Acylase Aminoacylase	
3.5.1.11	Acylase, penicillin Penicillin amidase	
3.5.1.14	Acylase I Aminoacylase	
3.5.1.15	Acylase II Aspartoacylase	
3.5.1.83	Acylase II N-Acyl-D-aspartate deacylase	
3.1.1.28	O-acylcarnitine acylhydrolase acylcarnitine hydrolase	
3.1.1.28	acylcarnitine hydrolase acylcarnitine hydrolase	
2.3.1.21	Acylcarnitine transferase Carnitine O-palmitoyltransferase	
2.3.1.39	[Acyl carrier protein]malonyltransferase [Acyl-carrier-protein] S-malonyltransferase	
2.7.8.7	Acyl carrier protein holoprotein (holo-ACP) synthetase Holo-[acyl-carrier-protein] synthase	
2.7.8.7	Acyl carrier protein synthetase Holo-[acyl-carrier-protein] synthase	
3.1.1.13	Acylcholesterol lipase Cholesterol esterase	
3.1.1.8	acylcholine acyl-hydrolse cholinesterase	
3.1.1.8	acylcholine acylhydrolase cholinesterase	

EC-No.	Synonym Recommended Name	Vol., Page
1.3.99.3	Acyl CoA dehydrogenase Acyl-CoA dehydrogenase	
1.14.99.25	Δ^6-Acyl CoA desaturase Linoleoyl-CoA desaturase	
1.14.19.3	Δ^6-acyl CoA desaturase linoleoyl-CoA desaturase	
1.14.99.5	Δ^9-Acyl CoA desaturase Stearoyl-CoA desaturase	
3.1.2.20	Acyl CoA hydrolase Acyl-CoA hydrolase	
2.3.1.26	Acyl coenzyme A-cholesterol-O-acyltransferase Sterol O-acyltransferase	
2.3.1.23	Acyl coenzyme A-monoacylphosphatidylcholine acyltransferase 1-Acylglycerophosphocholine O-acyltransferase	
2.3.1.22	Acyl coenzyme A-monoglyceride acyltransferase 2-Acylglycerol O-acyltransferase	
2.3.1.26	Acylcoenzyme A:cholesterol O-acyltransferase Sterol O-acyltransferase	
2.3.1.76	Acyl coenzyme A:retinol O-acyltransferase Retinol O-fatty-acyltransferase	
2.3.1.148	Acylcoenzyme A acyltransferase Glycerophospholipid acyltransferase (CoA-dependent)	
1.3.99.3	Acyl coenzyme A dehydrogenase Acyl-CoA dehydrogenase	
1.3.1.8	Acyl coenzyme A dehydrogenase (nicotinamide adenine dinucleotide phosphate) Acyl-CoA dehydrogenase ($NADP^+$)	
1.14.99.5	Acyl coenzyme A desaturase Stearoyl-CoA desaturase	
4.2.1.17	acyl coenzyme A hydrase enoyl-CoA hydratase	
3.1.2.20	Acyl coenzyme A hydrolase Acyl-CoA hydrolase	
1.3.3.6	Acyl coenzyme A oxidase Acyl-CoA oxidase	

EC-No.	Synonym Recommended Name	Vol., Page
6.2.1.3	Acyl coenzyme A synthetase Long-chain-fatty-acid-CoA ligase	Vol. 2, Page 206
6.2.1.10	Acyl coenzyme A synthetase (GDP-forming) Acid-CoA ligase (GDP-forming)	Vol. 2, Page 249
3.1.2.20	Acyl coenzyme A thioesterase Acyl-CoA hydrolase	
1.3.99.3	Acyl dehydrogenase Acyl-CoA dehydrogenase	
1.1.1.101	Acyldihydroxyacetone phosphate reductase Acylglycerone-phosphate reductase	
3.5.1.60	N-Acylethanolamine amidohydrolase N-(Long-chain-acyl)ethanolamine deacylase	
5.1.3.9	Acylglucosamine-6-phosphate 2-epimerase N-Acylglucosamine-6-phosphate 2-epimerase	Vol. 1, Page 144
5.1.3.9	N-Acylglucosamine-6-phosphate 2-epimerase N-Acylglucosamine-6-phosphate 2-epimerase	Vol. 1, Page 144
5.1.3.8	Acylglucosamine 2-epimerase N-Acylglucosamine 2-epimerase	Vol. 1, Page 140
5.1.3.8	N-Acylglucosamine 2-epimerase N-Acylglucosamine 2-epimerase	Vol. 1, Page 140
5.1.3.9	Acylglucosamine phosphate 2-epimerase N-Acylglucosamine-6-phosphate 2-epimerase	Vol. 1, Page 144
3.1.2.7	S-acylglutathione hydrolase glutathione thiolesterase	
2.3.1.51	1-Acylglycero-3-phosphate acyltransferase 1-Acylglycerol-3-phosphate O-acyltransferase	
2.3.1.51	1-Acylglycerol-3-phosphate O-acyltransferase 1-Acylglycerol-3-phosphate O-acyltransferase	
2.3.1.52	2-Acylglycerol-3-phosphate O-acyltransferase 2-Acylglycerol-3-phosphate O-acyltransferase	
2.3.1.62	2-Acylglycerol-3-phosphorylcholine acyltransferase 2-Acylglycerophosphocholine O-acyltransferase	
2.7.1.94	Acylglycerol kinase Acylglycerol kinase	
3.1.1.23	acylglycerol lipase acylglycerol lipase	

EC-No.	Synonym Recommended Name	Vol., Page
2.3.1.22	2-Acylglycerol O-acyltransferase 2-Acylglycerol O-acyltransferase	
2.3.1.22	Acylglycerol palmitoyltransferase 2-Acylglycerol O-acyltransferase	
2.3.1.51	1-Acylglycerolphosphate acyltransferase 1-Acylglycerol-3-phosphate O-acyltransferase	
1.1.1.101	Acylglycerone-phosphate reductase Acylglycerone-phosphate reductase	
2.3.1.51	1-Acylglycerophosphate acyltransferase 1-Acylglycerol-3-phosphate O-acyltransferase	
2.3.1.52	2-Acylglycerophosphate acyltransferase 2-Acylglycerol-3-phosphate O-acyltransferase	
2.3.1.62	2-Acylglycerophosphocholine acyltransferase 2-Acylglycerophosphocholine O-acyltransferase	
2.3.1.23	1-Acylglycerophosphocholine O-acyltransferase 1-Acylglycerophosphocholine O-acyltransferase	
2.3.1.62	2-Acylglycerophosphocholine O-acyltransferase 2-Acylglycerophosphocholine O-acyltransferase	
1.1.3.29	N-Acylhexosamine oxidase N-Acylhexosamine oxidase	
1.1.1.233	N-Acylmannosamine 1-dehydrogenase N-Acylmannosamine 1-dehydrogenase	
1.1.1.233	N-Acylmannosamine dehydrogenase N-Acylmannosamine 1-dehydrogenase	
2.7.1.60	Acylmannosamine kinase N-Acylmannosamine kinase	
2.7.1.60	N-Acylmannosamine kinase N-Acylmannosamine kinase	
3.4.17.7	acylmuramoyl-alanine carboxypeptidase acylmuramoyl-alanine carboxypeptidase	
3.4.19.10	acylmuramoyl-Ala peptidase acylmuramoyl-Ala peptidase	
3.5.1.28	N-Acylmuramyl-L-alanine amidase N-Acetylmuramoyl-L-alanine amidase	
3.1.3.29	N-acylneuraminate-9-phosphatase N-acylneuraminate-9-phosphatase	

EC-No.	Synonym Recommended Name	Vol., Page
3.1.3.29	N-acylneuraminate-9-phosphate phosphohydrolase N-acylneuraminate-9-phosphatase	
3.1.3.29	acylneuraminate 9-phosphatase N-acylneuraminate-9-phosphatase	
4.1.3.20	N-Acylneuraminate 9-phosphate lyase N-Acetylneuraminate-9-phosphate synthase	
4.1.3.20	N-Acylneuraminate 9-phosphate synthase N-Acetylneuraminate-9-phosphate synthase	
4.1.3.20	N-Acylneuraminate 9-phosphate synthetase N-Acetylneuraminate-9-phosphate synthase	
2.7.7.43	Acylneuraminate cytidyltransferase N-Acylneuraminate cytidylyltransferase	
2.7.7.43	N-Acylneuraminate cytidylyltransferase N-Acylneuraminate cytidylyltransferase	
3.1.3.29	N-acylneuraminic (sialic) acid 9-phosphatase N-acylneuraminate-9-phosphatase	
3.1.3.29	N-acylneuraminic acid 9-phosphate phosphatase N-acylneuraminate-9-phosphatase	
4.1.3.20	N-Acylneuraminic acid 9-phosphate synthetase N-Acetylneuraminate-9-phosphate synthase	
3.4.19.1	N-acylpeptide hydrolase acylaminoacyl-peptidase	
3.4.19.1	α-N-acylpeptide hydrolase acylaminoacyl-peptidase	
3.6.1.7	Acylphosphatase Acylphosphatase	
3.6.1.7	Acylphosphatase phosphohydrolase Acylphosphatase	
2.7.1.61	Acyl phosphate-hexose phosphotransferase Acyl-phosphate-hexose phosphotransferase	
2.7.1.61	Acyl phosphate:hexose phosphotransferase Acyl-phosphate-hexose phosphotransferase	
3.6.1.20	5'-Acylphosphoadenosine acylhydrolase 5'-Acylphosphoadenosine hydrolase	
3.6.1.20	5'-Acylphosphoadenosine hydrolase 5'-Acylphosphoadenosine hydrolase	

EC-No.	Synonym Recommended Name	Vol., Page
6.2.1.19	Acyl protein synthetase Long-chain-fatty-acid-luciferin-component ligase	Vol. 2, Page 293
3.1.2.22	Acyl protein thioester hydrolase Palmitoyl-(protein) hydrolase	
3.7.1.5	3-Acylpyruvate acylhydrolase Acylpyruvate hydrolase	
3.7.1.5	Acylpyruvate hydrolase Acylpyruvate hydrolase	
3.5.1.23	N-Acylsphingosine amidohydrolase Acylsphingosine deacylase	
3.5.1.23	Acylsphingosine deacylase Acylsphingosine deacylase	
2.4.1.47	N-Acylsphingosine galactosyltransferase N-Acylsphingosine galactosyltransferase	
2.7.1.138	Acylsphingosine kinase Ceramide kinase	
2.3.1.51	Acyltransferase, 1-acylglycerol phosphate 1-Acylglycerol-3-phosphate O-acyltransferase	
2.3.1.147	Acyltransferase, 1-alkylglycerophosphocholine Glycerophospholipid arachidonoyl-transferase (CoA-independent)	
2.3.1.125	Acyltransferase, 1-hexadecyl-2-acetylglycerol 1-Alkyl-2-acetylglycerol O-acyltransferase	
2.3.1.52	Acyltransferase, 2-acylglycerol phosphate 2-Acylglycerol-3-phosphate O-acyltransferase	
2.3.1.62	Acyltransferase, 2-acylglycerophosphocholine 2-Acylglycerophosphocholine O-acyltransferase	
2.3.1.16	Acyltransferase, acetyl coenzyme A Acetyl-CoA C-acyltransferase	
2.3.1.40	Acyltransferase, acyl-[acyl carrier protein]-phospholipid Acyl-[acyl-carrier-protein]-phospholipid O-acyltransferase	
2.3.1.104	Acyltransferase, alkenylglycerophosphocholine 1-Alkenylglycerophosphocholine O-acyltransferase	
2.3.1.121	Acyltransferase, alkenylglycerophosphoethanolamine 1-Alkenylglycerophosphoethanolamine O-acyltransferase	
2.3.1.153	Acyltransferase, anthocyanin Anthocyanin 5-aromatic acyltransferase	

EC-No.	Synonym Recommended Name Vol., Page
2.3.1.26	Acyltransferase, cholesterol Sterol O-acyltransferase
2.3.1.20	Acyltransferase, diacylglycerol Diacylglycerol O-acyltransferase
2.3.1.73	Acyltransferase, diacylglycerol-sterol Diacylglycerol-sterol O-acyltransferase
2.3.1.42	Acyltransferase, dihydroxyacetone phosphate Glycerone-phosphate O-acyltransferase
2.3.1.123	Acyltransferase, dolichol Dolichol O-acyltransferase
2.3.1.139	Acyltransferase, ecdysone Ecdysone O-acyltransferase
2.3.1.134	Acyltransferase, galactolipid-galactolipid Galactolipid O-acyltransferase
2.3.1.15	Acyltransferase, glycerol phosphate Glycerol-3-phosphate O-acyltransferase
2.3.1.13	Acyltransferase, glycine Glycine N-acyltransferase
2.3.1.65	Acyltransferase, glycine-taurine N- Glycine N-choloyltransferase
2.3.1.43	Acyltransferase, lecithin-cholesterol Phosphatidylcholine-sterol O-acyltransferase
2.3.1.135	Acyltransferase, lecithin-retinol Phosphatidylcholine-retinol O-acyltransferase
2.3.1.75	Acyltransferase, long-chain alcohol Long-chain-alcohol O-fatty-acyltransferse
2.3.1.23	Acyltransferase, lysolecithin 1-Acylglycerophosphocholine O-acyltransferase
2.3.1.141	Acyltransferase, lysomonogalactosyldiacylglycerol Galactosylacylglycerol O-acyltransferase
2.3.1.96	Acyltransferase, mucus glycoprotein Glycoprotein N-palmitoyltransferase
2.3.1.142	Acyltransferase, protein Glycoprotein O-fatty-acyltransferase
2.3.1.76	Acyltransferase, retinol Retinol O-fatty-acyltransferase

EC-No.	Synonym Recommended Name	Vol., Page
2.3.1.24	Acyltransferase, sphingosine Sphingosine N-acyltransferase	
2.3.1.77	Acyltransferase, triacylglycerol-sterol Triacylglycerol-sterol O-acyltransferase	
2.3.1.129	Acyltransferase, uridine diphosphoacetylglucosamine Acyl-[acyl-carrier-protein]-UDP-N-acetylglucosamine O-acyltransferase	
3.2.1.18	Acyneuraminyl hydrolase Sialidase	
3.4.24.46	Adamalysin Adamalysin	
3.4.24.46	Adamalysin II Adamalysin	
3.5.4.2	ADase Adenine deaminase	
4.1.1.19	ADC Arginine decarboxylase	
2.7.7.31	Addase DNA nucleotidylexotransferase	
3.4.22.39	adenain adenain	
3.5.4.2	Adenase Adenine deaminase	
2.1.1.36	Adenine-1-methylase tRNA (adenine-N^1-)-methyltransferase	
3.5.4.2	Adenine aminase Adenine deaminase	
3.5.4.2	Adenine aminohydrolase Adenine deaminase	
3.5.4.2	Adenine deaminase Adenine deaminase	
3.5.4.17	Adenine nucleotide deaminase Adenosine-phosphate deaminase	
2.4.2.7	Adenine phosphoribosylpyrophosphate transferase Adenine phosphoribosyltransferase	
2.4.2.7	Adenine phosphoribosyltransferase Adenine phosphoribosyltransferase	

EC-No.	Synonym Recommended Name	Vol., Page
3.2.2.7	Adenosinase Adenosine nucleosidase	
3.5.4.17	Adenosine (phosphate) deaminase Adenosine-phosphate deaminase	
3.1.3.7	adenosine-3'(2'),5'-bisphosphate 3'(2')-phosphohydrolase 3'(2'),5'-bisphosphate nucleotidase	
3.1.3.7	adenosine-3',5'-bisphosphate 3'-phosphohydrolase 3'(2'),5'-bisphosphate nucleotidase	
2.7.1.25	Adenosine-5'-phosphosulfate-3'-phosphokinase Adenylylsulfate kinase	
2.7.7.4	Adenosine-5'-triphosphate sulfurylase Sulfate adenylyltransferase	
3.5.4.17	Adenosine-phosphate aminohydrolase Adenosine-phosphate deaminase	
3.5.4.17	Adenosine-phosphate deaminase Adenosine-phosphate deaminase	
3.6.1.14	Adenosine-tetraphosphatase Adenosine-tetraphosphatase	
3.6.1.14	Adenosine-tetraphosphate phosphohydrolase Adenosine-tetraphosphatase	
1.8.4.8	adenosine 3',5'-bisphosphate,sulfite:oxidized-thioredoxin oxidoreductase (3'-phosphoadenosine-5'-phosphosulfate -forming) phosphoadenylyl-sulfate reductase (thioredoxin)	
1.8.99.4	Adenosine 3'-phosphate 5'-phosphosulfate:thioredoxin oxidoreductase (sulfite-forming) Phosphoadenosine-phosphosulfate reductase	
1.8.99.4	Adenosine 3'-phosphate 5'-phosphosulfate reductase Phosphoadenosine-phosphosulfate reductase	
2.8.2.16	Adenosine 3'-phosphate 5'-sulphatophosphate sulfotransferase Thiol sulfotransferase	
3.1.3.5	adenosine 5'-monophosphatase 5'-nucleotidase	
3.1.3.5	adenosine 5'-phosphatase 5'-nucleotidase	
2.7.1.25	Adenosine 5'-phosphosulfate kinase Adenylylsulfate kinase	

EC-No.	Synonym Recommended Name	Vol., Page
1.8.99.2	Adenosine 5'-phosphosulfate reductase Adenylylsulfate reductase	
3.6.1.3	Adenosine 5'-triphosphatase Adenosinetriphosphatase	
2.7.3.3	Adenosine 5'-triphosphate-arginine phosphotransferase Arginine kinase	
2.7.3.3	Adenosine 5'-triphosphate: L-arginine phosphotransferase Arginine kinase	
3.5.4.6	Adenosine 5-monophosphate deaminase AMP deaminase	
3.5.4.6	Adenosine 5-phosphate aminohydrolase AMP deaminase	
3.5.4.6	Adenosine 5-phosphate deaminase AMP deaminase	
3.6.2.1	Adenosine 5-phosphosulfate sulfohydrolase Adenylylsulfatase	
3.5.4.4	Adenosine aminohydrolase Adenosine deaminase	
3.5.4.4	Adenosine deaminase Adenosine deaminase	
3.6.1.5	Adenosine diphosphatase Apyrase	
3.6.1.6	Adenosine diphosphatase Nucleoside-diphosphatase	
3.5.4.7	Adenosine diphosphate deaminase ADP deaminase	
2.4.1.21	Adenosine diphosphate glucose-starch glucosyltransferase Starch synthase	
2.7.7.36	Adenosine diphosphate glucose:orthophosphate adenylyltransferase Aldose-1-phosphate adenylyltransferase	
2.7.7.27	Adenosine diphosphate glucose pyrophosphorylase Glucose-1-phosphate adenylyltransferase	
3.1.3.28	adenosine diphosphate phosphoglycerate phosphatase ADPphosphoglycerate phosphatase	
2.4.2.30	Adenosine diphosphate ribosyltransferase NAD^+ ADP-ribosyltransferase	

EC-No.	Synonym Recommended Name	Vol., Page
2.7.7.5	Adenosine diphosphate sulfurylase Sulfate adenylyltransferase (ADP)	
2.4.1.21	Adenosine diphosphoglucose-starch glucosyltransferase Starch synthase	
2.7.7.27	Adenosine diphosphoglucose pyrophosphorylase Glucose-1-phosphate adenylyltransferase	
5.1.3.20	Adenosine diphosphoglyceromannoheptose 6-epimerase ADPglyceromanno-heptose 6-epimerase	Vol. 1, Page 175
5.1.3.20	Adenosine diphosphoglyceromannoheptose 6-epimerase ADPglyceromanno-heptose 6-epimerase	Vol. 1, Page 175
2.4.2.31	(Adenosine diphosphoribose)transferase, nicotinamide adenine dinucleotide-arginine $NAD(P)^+$-arginine ADP-ribosyltransferase	
2.4.2.37	(Adenosine diphosphoribose)transferase, nicotinamide adenine dinucleotide-azoferredoxin NAD^+-dinitrogen-reductase ADP-D-ribosyltransferase	
2.4.2.36	(Adenosine diphosphoribose)transferase, nicotinamide adenine dinucleotide-elongation factor 2 NAD^+-diphthamide ADP-ribosyltransferase	
2.4.2.30	(Adenosine diphosphoribose)transferase, nicotinamide adenine dinucleotide-protein NAD^+ ADP-ribosyltransferase	
2.7.7.35	Adenosine diphosphoribose phosphorylase Ribose-5-phosphate adenylyltransferase	
3.6.1.13	Adenosine diphosphoribose pyrophosphatase ADPribose diphosphatase	
2.7.7.36	Adenosine diphosphosugar phosphorylase Aldose-1-phosphate adenylyltransferase	
3.6.1.21	Adenosine diphosphosugar pyrophosphatase ADPsugar diphosphatase	
3.2.2.7	Adenosine hydrolase Adenosine nucleosidase	
2.7.1.20	Adenosine kinase Adenosine kinase	
2.7.1.20	Adenosine kinase (phosphorylating) Adenosine kinase	

EC-No.	Synonym Recommended Name	Vol., Page
3.1.3.5	adenosine monophosphatase 5'-nucleotidase	
3.5.4.6	Adenosine monophosphate deaminase AMP deaminase	
3.2.2.4	Adenosine monophosphate nucleosidase AMP nucleosidase	
3.2.2.7	Adenosine nucleosidase Adenosine nucleosidase	
2.4.2.7	Adenosine phosphoribosyltransferase Adenine phosphoribosyltransferase	
2.7.1.25	Adenosine phosphosulfate kinase Adenylylsulfate kinase	
1.8.99.2	Adenosine phosphosulfate reductase Adenylylsulfate reductase	
2.7.1.25	Adenosine phosphosulfokinase Adenylylsulfate kinase	
3.6.1.6	Adenosinepyrophosphatase Nucleoside-diphosphatase	
3.5.4.7	Adenosinepyrophosphate deaminase ADP deaminase	
3.2.2.7	Adenosine ribohydrolase Adenosine nucleosidase	
3.6.1.14	Adenosinetetraphosphatase Adenosine-tetraphosphatase	
3.6.1.41	Adenosine tetraphosphate phosphodiesterase Bis(5'-nucleosyl)-tetraphosphatase (symmetrical)	
3.6.1.3	Adenosine triphosphatase Adenosinetriphosphatase	
3.6.1.3	Adenosinetriphosphatase Adenosinetriphosphatase	
3.6.1.4	adenosinetriphosphatase (Mg-activated) adenosinetriphosphatase (Mg-activated)	
2.7.3.2	Adenosine triphosphate-creatine transphosphorylase Creatine kinase	
3.1.21.3	Adenosine triphosphate-dependent deoxyribonuclease Type I site-specific deoxyribonuclease	

EC-No.	Synonym Recommended Name	Vol., Page
2.7.7.1	Adenosine triphosphate-nicotinamide mononucleotide transadenylase Nicotinamide-nucleotide adenylyltransferase	
2.7.7.2	Adenosine triphosphate-riboflavine mononucleotide transadenylase FMN adenylyltransferase	
2.7.7.2	Adenosine triphosphate-riboflavin mononucleotide transadenylase FMN adenylyltransferase	
2.7.7.42	Adenosine triphosphate:glutamine synthetase adenylyltransferase [Glutamate-ammonia-ligase] adenylyltransferase	
2.7.7.19	Adenosine triphosphate:ribonucleic acid adenylyltransferase Polynucleotide adenylyltransferase	
4.1.3.8	adenosine triphosphate citrate lyase ATP citrate (pro-3S)-lyase	
3.5.4.18	Adenosine triphosphate deaminase ATP deaminase	
2.4.2.17	Adenosine triphosphate phosphoribosyltransferase ATP phosphoribosyltransferase	
3.6.1.8	Adenosine triphosphate pyrophosphatase ATP diphosphatase	
2.7.7.4	Adenosinetriphosphate sulfurylase Sulfate adenylyltransferase	
3.5.4.28	S-Adenosyl-L-homocysteine aminohydrolase S-Adenosylhomocysteine deaminase	
3.2.2.9	S-Adenosyl-L-homocysteine homocysteinylribohydrolase Adenosylhomocysteine nucleosidase	
3.3.1.1	S-Adenosyl-L-homocysteine hydrolase Adenosylhomocysteinase	
3.3.1.1	S-Adenosyl-L-homocysteine hydrolase Adenosylhomocysteinase	
2.1.1.133	S-Adenosyl-L-methionine-dependent cobalt-precorrin-4 transmethylase Precorrin-4 C_{11}-methyltransferase	
2.1.1.101	S-Adenosyl-L-methionine-macrocin O-methyltransferase Macrocin O-methyltransferase	
2.1.1.125	S-Adenosyl-L-methionine:(histone)-arginine N^{ω}-methyltransferase (Histone)-arginine N-methyltransferase	

EC-No.	Synonym Recommended Name	Vol., Page
2.1.1.115	S-Adenosyl-L-methionine:(RS)-1-benzyl-1,2,3,4-tetrahydroisoquinoline N-methyltransferase (RS)-1-Benzyl-1,2,3,4-tetrahydroisoquinoline N-methyltransferase	
2.1.1.128	S-Adenosyl-L-methionine:(RS)-norcoclaurine 6-O-methyltransferase (RS)-Norcoclaurine 6-O-methyltransferase	
2.1.1.122	S-Adenosyl-L-methionine: (S)-7,8,13,14-tetrahydroberberine-cis-N-methyltransferase (S)-Tetrahydroprotoberberine N-methyltransferase	
2.1.1.122	S-Adenosyl-L-methionine:(S)-7,8,13,14-tetrahydroprotoberberine cis-N-methyltransferase (S)-Tetrahydroprotoberberine N-methyltransferase	
2.1.1.128	S-Adenosyl-L-methionine:(S)-coclaurine-N-methyltransferase (RS)-Norcoclaurine 6-O-methyltransferase	
2.1.1.117	S-Adenosyl-L-methionine:(S)-scoularine 9-O-methyltransferase (S)-Scoulerine 9-O-methyltransferase	
2.1.1.95	S-Adenosyl-L-methionine:γ-tocopherol 5-O-methyltransferase Tocopherol O-methyltransferase	
2.1.1.18	S-Adenosyl-L-methionine:1,4-α-D-glucan 6-O-methyltransferase Polysaccharide O-methyltransferase	
2.1.1.132	S-Adenosyl-L-methionine:1-precorrin-6Y $C^{5,15}$-methyltransferase (C12-decarboxylating) Precorrin-6Y C5,15-methyltransferase (decarboxylating)	
2.1.1.119	S-Adenosyl-L-methionine:10-hydroxydihydrosanguinarine 10-O-methyltransferase 10-Hydroxydihydrosanguinarine 10-O-methyltransferase	
2.1.1.94	S-Adenosyl-L-methionine:11-O-demethyl-17-O-deacetylvindoline 11-O-methyltransferase 11-O-Demethyl-17-O-deacetylvindoline O-methyltransferase	
2.1.1.120	S-Adenosyl-L-methionine:12-hydroxydihydrochelirubine 12-O-methyltransferase 12-Hydroxydihydrochelirubine 12-O-methyltransferase	
2.1.1.99	S-Adenosyl-L-methionine:16-methoxy-2,3-dihydro-3-hydroxytabersonine N-methyltransferase 16-Methoxy-2,3-dihydro-3-hydroxytabersonine N-methyltransferase	
2.1.1.98	S-Adenosyl-L-methionine:2-(3-carboxy-3-aminopropyl)-L-histidine methyltransferase Diphthine synthase	
2.1.1.26	S-Adenosyl-L-methionine:2-iodophenol O-methyltransferase Iodophenol O-methyltransferase	

EC-No.	Synonym Recommended Name Vol., Page
2.1.1.91	S-Adenosyl-L-methionine:2-methylpropanal-oxime O-methyltrans-ferase Isobutyraldoxime O-methyltransferase
2.1.1.64	S-Adenosyl-L-methionine:2-nonaprenyl-3-methyl-5-hydroxy-6-methoxy-1,4-benzoquinone 3-O-methyltransferase 3-Demethylubiquinone-9 3-O-methyltransferase
2.1.1.64	S-Adenosyl-L-methionine:2-octaprenyl-3-methyl-5-hydroxy-6-methoxy-1,4-benzoquinone-O-methyltransferase 3-Demethylubiquinone-9 3-O-methyltransferase
2.1.1.84	S-Adenosyl-L-methionine:3',4',5,6-tetrahydroxy-3,7-dimethoxyflavone 6-O-methyltransferase Methylquercetagetin 6-O-methyltransferase
2.1.1.82	S-Adenosyl-L-methionine:3',4',5,7-tetrahydroxy-3-methoxyflavone 7-O-methyltransferase 3-Methylquercitin 7-O-methyltransferase
2.1.1.83	S-Adenosyl-L-methionine:3',4',5-trihydroxy-3,7-dimethoxyflavone 4'-O-methyltransferase 3,7-Dimethylquercitin 4'-O-methyltransferase
2.1.1.139	S-adenosyl-L-methionine:3'-demethylstaurosporine O-methyltransferase 3'-demethylstaurosporine O-methyltransferase
2.1.1.116	S-Adenosyl-L-methionine: 3'-hydroxy-N-methyl-(S)-coclaurine-4'-O-methyl transferase 3'-Hydroxy-N-methyl-(S)-coclaurine 4'-O-methyltransferase
2.1.1.116	S-Adenosyl-L-methionine:3'-hydroxy-N-methyl-(S)-coclaurine 4'-O-methyltransferase 3'-Hydroxy-N-methyl-(S)-coclaurine 4'-O-methyltransferase
2.1.1.88	S-Adenosyl-L-methionine:3,3',4',5,7,8-hexahydroxyflavone 8-O-methyl-transferase 8-Hydroxyquercitin 8-O-methyltransferase
2.1.1.68	S-Adenosyl-L-methionine:3,4-dihydroxy-trans-cinnamate 3-O-methyltrans-ferase Caffeate O-methyltransferase
2.1.1.76	S-Adenosyl-L-methionine:3,5,7,3',4'-pentahydroxyflavone 3-O-methyltrans-ferase Quercetin 3-O-methyltransferase
2.1.1.114	S-Adenosyl-L-methionine:3-hexaprenyl-4,5-dihydroxylate O-methyltrans-ferase Hexaprenyldihydroxybenzoate methyltransferase
2.1.1.97	S-Adenosyl-L-methionine:3-hydroxyanthranilate 4-C-methyltransferase 3-Hydroxyanthranilate 4-C-methyltransferase

EC-No.	Synonym Recommended Name Vol., Page
2.1.1.42	S-Adenosyl-L-methionine:5,7,3',4'-tetrahydroxyflavone 3'-O-methyltransferase Luteolin O-methyltransferase
2.1.1.75	S-Adenosyl-L-methionine:5,7,4'-trihydroxyflavone 4'-O-methyltransferase Apigenin 4'-O-methyltransferase
2.1.1.89	S-Adenosyl-L-methionine:5,8,13,13a-tetrahydrocolumbamine 2-O-methyltransferase Tetrahydrocolumbamine 2-O-methyltransferase
2.1.1.69	S-Adenosyl-L-methionine:5-hydroxyfuranocoumarin 5-O-methyltransferase 5-Hydroxyfuranocoumarin 5-O-methyltransferase
2.1.1.69	S-Adenosyl-L-methionine:5-hydroxyfuranocoumarin O-methyltransferase 5-Hydroxyfuranocoumarin 5-O-methyltransferase
2.1.1.109	S-Adenosyl-L-methionine:6-demethylsterigmatocystin 6-O-methyltransferase Demethylsterigmatocystin 6-O-methyltransferase
2.1.1.108	S-Adenosyl-L-methionine:6-hydroxymellein 6-O-methyltransferase 6-Hydroxymellein O-methyltransferase
2.1.1.108	S-Adenosyl-L-methionine:6-hydroxymellein O-methyltransferase 6-Hydroxymellein O-methyltransferase
2.1.1.121	S-Adenosyl-L-methionine:6-O-methylnorlaudanosoline 5'-O-methyltransferase 6-O-Methylnorlaudanosoline 5'-O-methyltransferase
2.6.1.62	S-Adenosyl-L-methionine:8-amino-7-oxononanoate aminotransferase Adenosylmethionine-8-amino-7-oxononanoate transaminase
2.1.1.70	S-Adenosyl-L-methionine:8-hydroxyfuranocoumarin 8-O-methyltransferase 8-Hydroxyfuranocoumarin 8-O-methyltransferase
2.1.1.70	S-Adenosyl-L-methionine:8-hydroxyfuranocoumarin O-methyltransferase 8-Hydroxyfuranocoumarin 8-O-methyltransferase
2.1.1.127	S-Adenosyl-L-methionine:[3-phospho-D-glycerate-carboxy-lyase(dimerizing)]-lysine N^6-methyltransferase [Ribulose-bisphosphate-carboxylase]-lysine N-methyltransferase
2.1.1.124	S-Adenosyl-L-methionine:[cytochrome-c]-arginine N^{ω}-methyltransferase [Cytochrome-c]-arginine N-methyltransferase
2.1.1.123	S-Adenosyl-L-methionine:[cytochrome c]-methionine S-methyltransferase [Cytochrome-c]-methionine S-methyltransferase

EC-No.	Synonym Recommended Name	Vol., Page
2.1.1.126	S-Adenosyl-L-methionine:[myelin-basic-protein]-arginine N-ω-methyltransferase [Myelin basic protein]-arginine N-methyltransferase	
2.1.1.49	S-Adenosyl-L-methionine:amine N-methyltransferase Amine N-methyltransferase	
2.1.1.111	S-Adenosyl-L-methionine:anthranilate N-methyltransferase Anthranilate N-methyltransferase	
2.1.1.137	S-adenosyl-L-methionine:arsenite As-methyltransferase arsenite methyltransferase	
2.1.1.92	S-Adenosyl-L-methionine:bergaptol O-methyltransferase Bergaptol O-methyltransferase	
2.1.1.104	S-Adenosyl-L-methionine:caffeoyl-CoA 3-O-methyltransferase Caffeoyl-CoA O-methyltransferase	
2.1.1.60	S-Adenosyl-L-methionine:calmodulin-L-lysine N^6-methyltransferase Calmodulin-lysine N-methyltransferase	
2.1.1.22	S-Adenosyl-L-methionine:carnosine N-methyltransferase Carnosine N-methyltransferase	
2.1.1.6	S-Adenosyl-L-methionine:catechol O-methyltransferase Catechol O-methyltransferase	
2.1.1.118	S-Adenosyl-L-methionine:columbamine O-methyltransferase Columbamine O-methyltransferase	
2.1.1.59	S-Adenosyl-L-methionine:cytochrome-c-L-lysine N^6-methyltransferase Cytochrome-c-lysine N-methyltransferase	
2.1.1.102	S-Adenosyl-L-methionine:demethylmacrocin 2'''-O-methyltransferase Demethylmacrocin O-methyltransferase	
2.1.1.96	S-Adenosyl-L-methionine:dimethyl-sulfide S-methyltransferase Thioether S-methyltransferase	
2.1.1.37	S-Adenosyl-L-methionine:DNA (cytosine-5-)-methyltransferase DNA (cytosine-5-)-methyltransferase	
2.1.1.98	S-Adenosyl-L-methionine:elongation factor 2 methyltransferase Diphthine synthase	
2.1.1.103	S-Adenosyl-L-methionine:ethanolamine-phosphate N-methyltransferase Phosphoethanolamine N-methyltransferase	
2.1.1.15	S-Adenosyl-L-methionine:fatty-acid O-methyltransferase Fatty-acid O-methyltransferase	

EC-No.	Synonym Recommended Name Vol., Page
2.1.1.112	S-Adenosyl-L-methionine:glucuronoxylan-D-glucuronate 4-O-methyltransferase Glucuronoxylan 4-O-methyltransferase
2.1.1.20	S-Adenosyl-L-methionine:glycine methyltransferase Glycine N-methyltransferase
2.1.1.20	S-Adenosyl-L-methionine:glycine N-methyltransferase Glycine N-methyltransferase
2.1.1.8	S-Adenosyl-L-methionine:histamine N-tele-methyltransferase Histamine N-methyltransferase
2.1.1.125	S-Adenosyl-L-methionine:histone-arginine N-methyltransferase (Histone)-arginine N-methyltransferase
2.1.1.43	S-Adenosyl-L-methionine:histone-L-lysine N^6-methyltransferase Histone-lysine N-methyltransferase
2.1.1.47	S-Adenosyl-L-methionine:indolepyruvate C-methyltransferase Indolepyruvate C-methyltransferase
2.1.1.46	S-Adenosyl-L-methionine:isoflavone 4'-O-methyltransferase Isoflavone 4'-O-methyltransferase
2.1.1.78	S-Adenosyl-L-methionine:isoorientin 3'-O-methyltransferase Isoorientin 3'-O-methyltransferase
2.1.1.10	S-Adenosyl-L-methionine: L-homocysteine methyltransferase Homocysteine S-methyltransferase
2.1.1.10	S-Adenosyl-L-methionine:L-homocysteine S-methyltransferase Homocysteine S-methyltransferase
2.1.1.12	S-Adenosyl-L-methionine:L-methionine S-methyltransferase Methionine S-methyltransferase
2.1.1.106	S-Adenosyl-L-methionine:L-tryptophan 2-C-methyltransferase Tryptophan 2-C-methyltransferase
2.1.1.65	S-Adenosyl-L-methionine:licodione 2'-O-methyltransferase Licodione 2'-O-methyltransferase
2.1.1.50	S-Adenosyl-L-methionine:loganate 11-O-methyltransferase Loganate O-methyltransferase
2.1.1.50	S-Adenosyl-L-methionine:loganic acid methyltransferase Loganate O-methyltransferase
2.1.1.101	S-Adenosyl-L-methionine:macrocin 3'''-O-methyltransferase Macrocin O-methyltransferase

EC-No.	Synonym Recommended Name	Vol., Page
2.1.1.11	(-)-S-Adenosyl-L-methionine:magnesium-protoporphyrin IX methyltransferase Magnesium protoporphyrin O-methyltransferase	
2.1.1.11	S-Adenosyl-L-methionine:magnesium protoporphyrin O-methyltransferase Magnesium protoporphyrin O-methyltransferase	
2.1.1.138	S-adenosyl-L-methionine:methylarsonite As-methyltransferase methylarsonite methyltransferase	
2.1.1.11	S-Adenosyl-L-methionine:Mg protoporphyrin methyltransferase Magnesium protoporphyrin O-methyltransferase	
2.1.1.62	S-Adenosyl-L-methionine:mRNA (2'-O-methyladenosine-N^6-)-methyltransferase mRNA (2'-O-methyladenosine-N^6-)-methyltransferase	
2.1.1.56	S-Adenosyl-L-methionine:mRNA (guanine-N^7-)-methyltransferase mRNA (guanine-N^7-)-methyltransferase	
2.1.1.57	S-Adenosyl-L-methionine:mRNA (nucleoside-2'-O-)-methyltransferase mRNA (nucleoside-2'-O-)-methyltransferase	
2.1.1.39	S-Adenosyl-L-methionine:myo-inositol 1-O-methyltransferase myo-Inositol 1-O-methyltransferase	
2.1.1.40	S-Adenosyl-L-methionine:myo-inositol 3-O-methyltransferase myo-Inositol 3-O-methyltransferase	
2.1.1.129	S-Adenosyl-L-methionine:myo-inositol 4-O-methyltransferase myo-Inositol 4-O-methyltransferase	
2.1.1.134	S-Adenosyl-L-methionine:myo-inositol 6-O-methyltransferase myo-Inositol 6-O-methyltransferase	
2.1.1.44	S-Adenosyl-L-methionine:N$^\alpha$;N$^\alpha$-dimethyl-L-histidine N$^\alpha$-methyltransferase Dimethylhistidine N-methyltransferase	
2.1.1.4	S-Adenosyl-L-methionine:N-acetylserotonin O-methyltransferase Acetylserotonin N-methyltransferase	
2.1.1.105	S-Adenosyl-L-methionine:N-benzoyl-4-O-hydroxyanthranilate 4-O-methyltransferase N-Benzoyl-4-hydroxyanthranilate 4-O-methyltransferase	
2.1.1.2	S-Adenosyl-L-methionine:N-guanidinoacetate methyltransferase Guanidinoacetate N-methyltransferase	
2.5.1.24	S-Adenosyl-L-methionine:N^6-(Δ^2-isopentenyl)-adenine 3-(3-amino-3-carboxypropyl)-transferase Discadenine synthase	

EC-No.	Synonym Recommended Name	Vol., Page
2.1.1.1	S-Adenosyl-L-methionine:nicotinamide N-methyltransferase Nicotinamide N-methyltransferase	
2.1.1.7	S-Adenosyl-L-methionine:nicotinate N-methyltransferase Nicotinate N-methyltransferase	
2.5.1.38	S-Adenosyl-L-methionine:nocardicin-E 3-amino-3-carboxypropyltransferase Isonocardicin synthase	
2.1.1.128	S-Adenosyl-L-methionine:norcoclaurine 6-O-methyltransferase (RS)-Norcoclaurine 6-O-methyltransferase	
2.1.1.38	S-Adenosyl-L-methionine:O-demethylpuromycin O-methyltransferase O-Demethylpuromycin O-methyltransferase	
2.1.1.25	S-Adenosyl-L-methionine:phenol O-methyltransferase Phenol O-methyltransferase	
2.1.1.28	S-Adenosyl-L-methionine:phenylethanolamine N-methyltransferase Phenylethanolamine N-methyltransferase	
2.1.1.71	S-Adenosyl-L-methionine:phosphatidyl-N-methylethanolamine N-methyltransferase Phosphatidyl-N-methylethanolamine N-methyltransferase	
2.1.1.17	S-Adenosyl-L-methionine:phosphatidylethanolamine N-methyltransferase Phosphatidylethanolamine N-methyltransferase	
2.1.1.130	S-Adenosyl-L-methionine:precorrin-2 methyltransferase Precorrin-2 C^{20}-methyltransferase	
2.1.1.131	S-Adenosyl-L-methionine:precorrin-3B C^{17}-methyltransferase Precorrin-3B C^{17}-methyltransferase	
2.1.1.133	S-Adenosyl-L-methionine:precorrin-4 C^{11}-methyltransferase Precorrin-4 C_{11}-methyltransferase	
2.1.1.130	S-Adenosyl-L-methionine:precorrin-4 C^{20}-methyltransferase Precorrin-2 C^{20}-methyltransferase	
2.1.1.100	S-Adenosyl-L-methionine:protein-C-terminal-S-farnesyl-L-cysteine O-methyltransferase Protein-S-isoprenylcysteine O-methyltransferase	
2.1.1.80	S-Adenosyl-L-methionine:protein-L-glutamate O-methyltransferase Protein-glutamate O-methyltransferase	
2.1.1.85	S-Adenosyl-L-methionine:protein-L-histidine N-tele-methyltransferase Protein-histidine N-methyltransferase	
2.1.1.77	S-Adenosyl-L-methionine:protein-L-isoaspartate O-methyltransferase Protein-L-isoaspartate(D-aspartate) O-methyltransferase	

EC-No.	Synonym Recommended Name	Vol., Page
2.1.1.125	S-Adenosyl-L-methionine:protein arginine N-methyltransferase (Histone)-arginine N-methyltransferase	
2.1.1.124	S-Adenosyl-L-methionine:protein arginine N-methyltransferase [Cytochrome-c]-arginine N-methyltransferase	
2.1.1.126	S-Adenosyl-L-methionine:protein arginine N-methyltransferase [Myelin basic protein]-arginine N-methyltransferase	
2.1.1.53	S-Adenosyl-L-methionine:putrescine N-methyltransferase Putrescine N-methyltransferase	
2.1.1.87	S-Adenosyl-L-methionine:pyridine N-methyltransferase Pyridine N-methyltransferase	
2.1.1.48	S-Adenosyl-L-methionine:rRNA (adenine-N^6-)-methyltransferase rRNA (adenine-N^6-)-methyltransferase	
2.1.1.66	S-Adenosyl-L-methionine:rRNA (adenosine-2'-O-)-methyltransferase rRNA (adenosine-2'-O-)-methyltransferase	
2.1.1.51	S-Adenosyl-L-methionine:rRNA (guanine-N^1-)-methyltransferase rRNA (guanine-N^1-)-methyltransferase	
2.1.1.52	S-Adenosyl-L-methionine:rRNA (guanine-N^2-)-methyltransferase rRNA (guanine-N^2-)-methyltransferase	
2.5.1.43	S-Adenosyl-L-methionine:S-adenosyl-L-methionine:S-adenosyl-methionine alanyltransferase Nicotianamine synthase	
2.1.1.110	S-Adenosyl-L-methionine:sterigmatocystin 7-O-methyltransferase Sterigmatocystin 7-O-methyltransferase	
2.1.1.122	S-Adenosyl-L-methionine:tetrahydroberberine-cis-N-methyltransferase (S)-Tetrahydroprotoberberine N-methyltransferase	
2.1.1.96	S-Adenosyl-L-methionine:thioether S-methyltransferase Thioether S-methyltransferase	
2.1.1.9	S-Adenosyl-L-methionine:thiol S-methyltransferase Thiol S-methyltransferase	
2.1.1.67	S-Adenosyl-L-methionine:thiopurine S-methyltransferase Thiopurine S-methyltransferase	
2.1.1.136	S-adenosyl-L-methionine:trichlorophenol O-methyltransferase chlorophenol O-methyltransferase	
2.1.1.61	S-Adenosyl-L-methionine:tRNA (5-methylaminomethyl-2-thiouridylate)-methyltransferase tRNA(5-methylaminomethyl-2-thiouridylate)-methyltransferase	

EC-No.	Synonym Recommended Name Vol., Page
2.1.1.36	S-Adenosyl-L-methionine:tRNA (adenine-N^1-)-methyltransferase tRNA (adenine-N^1-)-methyltransferase
2.1.1.55	S-Adenosyl-L-methionine:tRNA (adenine-N^6)-methyltransferase tRNA (adenine-N^6-)-methyltransferase
2.1.1.29	S-Adenosyl-L-methionine:tRNA (cytosine-5-)-methyltransferase tRNA (cytosine-5-)-methyltransferase
2.1.1.31	S-Adenosyl-L-methionine:tRNA (guanine-N^1-)-methyltransferase tRNA (guanine-N^1-)-methyltransferase
2.1.1.32	S-Adenosyl-L-methionine:tRNA (guanine-N^2-)-methyltransferase tRNA (guanine-N^2-)-methyltransferase
2.1.1.33	S-Adenosyl-L-methionine:tRNA (guanine-N^7-)-methyltransferase tRNA (guanine-N^7-)-methyltransferase
2.1.1.34	S-Adenosyl-L-methionine:tRNA (guanosine-2'-O-)-methyltransferase tRNA (guanosine-2'-O-)-methyltransferase
2.1.1.35	S-Adenosyl-L-methionine:tRNA (uracil-5-)-methyltransferase tRNA (uracil-5-)-methyltransferase
2.5.1.25	S-Adenosyl-L-methionine:tRNA-uridine 3-(3-amino-3-carboxypropyl)-transferase tRNA-uridine aminocarboxypropyltransferase
2.1.1.27	S-Adenosyl-L-methionine:tyramine N-methyltransferase Tyramine N-methyltransferase
2.1.1.79	S-Adenosyl-L-methionine:unsaturated-phospholipid methyltransferase (cyclizing) Cyclopropane-fatty-acyl-phospholipid synthase
2.1.1.16	S-Adenosyl-L-methionine:unsaturated-phospholipid methyltransferase (methenylating) Methylene-fatty-acyl-phospholipid synthase
2.1.1.107	S-Adenosyl-L-methionine:uroporphyrin-III C-methyltransferase Uroporphyrin-III C-methyltransferase
2.1.1.93	S-Adenosyl-L-methionine:xanthotoxol O-methyltransferase Xanthotoxol O-methyltransferase
2.1.1.41	S-Adenosyl-L-methionine:zymosterol C-methyltransferase 24-Sterol C-methyltransferase
2.5.1.4	S-Adenosyl-L-methionine alkyltransferase (cyclizing) Adenosylmethionine cyclotransferase
4.1.1.50	S-Adenosyl-L-methionine carboxy-lyase Adenosylmethionine decarboxylase

EC-No.	Synonym Recommended Name	Vol., Page
4.1.1.50	S-Adenosyl-L-methionine decarboxylase Adenosylmethionine decarboxylase	
2.1.1.107	S-Adenosyl-L-methionine dependent uroporphyrinogen III methylase Uroporphyrin-III C-methyltransferase	
3.3.1.2	S-Adenosyl-L-methionine hydrolase Adenosylmethionine hydrolase	
4.4.1.14	S-adenosyl-L-methionine methylthioadenosine-lyase 1-aminocyclopropane-1-carboxylate synthase	
2.5.1.6	S-Adenosyl-L-methionine synthetase Methionine adenosyltransferase	
2.1.1.27	S-Adenosyl-methionine:tyramine N-methyltransferase Tyramine N-methyltransferase	
4.2.1.28	adenosylcobalamin-dependent diol dehydratase propanediol dehydratase	
3.3.1.1	Adenosylhomocysteinase Adenosylhomocysteinase	
3.3.1.1	S-Adenosylhomocysteinase Adenosylhomocysteinase	
3.2.2.9	S-Adenosylhomocysteine/5'-methylthioadenosine nucleosidase Adenosylhomocysteine nucleosidase	
3.5.4.28	S-Adenosylhomocysteine deaminase S-Adenosylhomocysteine deaminase	
3.3.1.1	Adenosylhomocysteine hydrolase Adenosylhomocysteinase	
3.3.1.1	S-Adenosylhomocysteine hydrolase Adenosylhomocysteinase	
3.2.2.9	S-Adenosylhomocysteine hydrolase Adenosylhomocysteine nucleosidase	
3.2.2.9	Adenosylhomocysteine nucleosidase Adenosylhomocysteine nucleosidase	
3.2.2.9	S-Adenosylhomocysteine nucleosidase Adenosylhomocysteine nucleosidase	
3.3.1.1	S-Adenosylhomocysteine synthase Adenosylhomocysteinase	

EC-No.	Synonym Recommended Name	Vol., Page
2.5.1.23	S-Adenosylmethioninamine:propane-1,3-diamine 3-aminopropyltransferase Sym-Norspermidine synthase	
2.5.1.16	S-Adenosylmethioninamine:putrescine 3-aminopropyltransferase Spermidine synthase	
2.5.1.22	S-Adenosylmethioninamine:spermidine 3-aminopropyltransferase Spermine synthase	
2.5.1.4	Adenosylmethioninase Adenosylmethionine cyclotransferase	
2.6.1.62	Adenosylmethionine-8-amino-7-oxononanoate transaminase Adenosylmethionine-8-amino-7-oxononanoate transaminase	
2.1.1.80	S-Adenosylmethionine-glutamyl methyltransferase Protein-glutamate O-methyltransferase	
2.1.1.8	S-Adenosylmethionine-histamine N-methyltransferase Histamine N-methyltransferase	
2.1.1.10	S-Adenosylmethionine-homocysteine transmethylase Homocysteine S-methyltransferase	
2.1.1.12	S-Adenosylmethionine-methionine methyltransferase Methionine S-methyltransferase	
2.1.1.130	S-Adenosylmethionine-precorrin-2-methyltransferase Precorrin-2 C_{20}-methyltransferase	
2.1.1.107	Adenosylmethionine-uroporphyrinogen III methyltransferase Uroporphyrin-III C-methyltransferase	
2.1.1.128	S-Adenosylmethionine:(R),(S)-norlaudanosoline-6-O-methyltransferase (RS)-Norcoclaurine 6-O-methyltransferase	
2.1.1.135	S-Adenosylmethionine:[methionine synthase]-cobalamin methyltransferase (CoII reducing) [Methionine synthase]-cobalamin methyltransferase (cob(II)alamin reducing)	
2.1.1.91	S-Adenosylmethionine:aldoxime O-methyltransferase Isobutyraldoxime O-methyltransferase	
2.1.1.60	S-Adenosylmethionine:calmodulin N-methyltransferase (lysine) Calmodulin-lysine N-methyltransferase	
2.1.1.42	S-Adenosylmethionine:flavone/flavonol 3'-O-methyltransferase Luteolin O-methyltransferase	
2.1.1.10	Adenosylmethionine:homocysteine methyltransferase Homocysteine S-methyltransferase	

EC-No.	Synonym Recommended Name	Vol., Page
2.1.1.10	S-Adenosylmethionine:homocysteine methyltransferase Homocysteine S-methyltransferase	
2.1.1.12	S-Adenosyl methionine:methionine methyl transferase Methionine S-methyltransferase	
2.1.1.39	S-Adenosylmethionine:myo-inositol 1-methyltransferase myo-Inositol 1-O-methyltransferase	
2.1.1.40	S-Adenosylmethionine:myo-inositol 3-methyltransferase myo-Inositol 3-O-methyltransferase	
2.1.1.38	S-Adenosylmethionine:O-demethylpuromycin O-methyltransferase O-Demethylpuromycin O-methyltransferase	
2.1.1.80	S-Adenosylmethionine:protein-carboxyl O-methyltransferase Protein-glutamate O-methyltransferase	
2.1.1.85	S-Adenosyl methionine:protein-histidine N-methyltransferase Protein-histidine N-methyltransferase	
2.1.1.122	S-Adenosylmethionine:tetrahydroberberine-N-methyltransferase (S)-Tetrahydroprotoberberine N-methyltransferase	
2.1.1.106	S-Adenosylmethionine:tryptophan 2-methyltransferase Tryptophan 2-C-methyltransferase	
3.3.1.2	S-Adenosylmethionine cleaving enzyme Adenosylmethionine hydrolase	
2.5.1.4	Adenosylmethionine cyclotransferase Adenosylmethionine cyclotransferase	
4.1.1.50	Adenosylmethionine decarboxylase Adenosylmethionine decarboxylase	
4.1.1.50	S-Adenosylmethionine decarboxylase Adenosylmethionine decarboxylase	
2.1.1.10	S-Adenosylmethionine homocysteine transmethylase Homocysteine S-methyltransferase	
3.3.1.2	Adenosylmethionine hydrolase Adenosylmethionine hydrolase	
3.3.1.2	Adenosylmethionine lyase Adenosylmethionine hydrolase	
2.5.1.6	S-Adenosylmethionine synthase Methionine adenosyltransferase	
2.5.1.6	Adenosylmethionine synthetase Methionine adenosyltransferase	

EC-No.	Synonym Recommended Name	Vol., Page
2.5.1.6	S-Adenosylmethionine synthetase Methionine adenosyltransferase	
2.1.1.10	Adenosylmethionine transmethylase Homocysteine S-methyltransferase	
2.1.1.12	S-Adenosylmethionine transmethylase Methionine S-methyltransferase	
2.5.1.6	Adenosyltransferase, methionine Methionine adenosyltransferase	
2.5.1.17	Adenosyltransferase, vitamin B_{12s} Cob(I)alamin adenosyltransferase	
3.4.22.39	adenovirus endopeptidase adenain	
3.4.22.39	adenovirus type 2 protease adenain	
2.7.1.114	Adenylate-nucleoside phosphotransferase AMP-thymidine kinase	
3.5.4.6	Adenylate aminohydrolase AMP deaminase	
4.6.1.1	adenylate cyclase adenylate cyclase	
3.5.4.6	5-Adenylate deaminase AMP deaminase	
3.5.4.6	Adenylate deaminase AMP deaminase	
3.5.4.17	Adenylate deaminase Adenosine-phosphate deaminase	
3.5.4.6	Adenylate desaminase AMP deaminase	
2.5.1.27	Adenylate isopentenyltransferase Adenylate isopentenyltransferase	
2.7.4.3	Adenylate kinase Adenylate kinase	
3.2.2.4	Adenylate nucleosidase AMP nucleosidase	
2.4.2.7	Adenylate pyrophosphorylase Adenine phosphoribosyltransferase	

EC-No.	Synonym Recommended Name	Vol., Page
4.6.1.1	adenyl cyclase adenylate cyclase	
3.5.4.6	Adenyl deaminase AMP deaminase	
2.7.1.114	Adenylic acid:deoxythymidine 5'-phosphotransferase AMP-thymidine kinase	
3.5.4.6	5-Adenylic acid deaminase AMP deaminase	
3.5.4.6	Adenylic acid deaminase AMP deaminase	
3.5.4.6	Adenylic deaminase AMP deaminase	
2.7.4.3	Adenylic kinase Adenylate kinase	
3.1.3.5	5'-adenylic phosphatase 5'-nucleotidase	
2.4.2.7	Adenylic pyrophosphorylase Adenine phosphoribosyltransferase	
2.7.4.3	Adenylokinase Adenylate kinase	
4.3.2.2	adenylosuccinase adenylosuccinate lyase	
4.3.2.2	adenylosuccinate AMP-lyase adenylosuccinate lyase	
4.3.2.2	adenylosuccinate lyase adenylosuccinate lyase	
6.3.4.4	Adenylosuccinate synthase Adenylosuccinate synthase	Vol. 2, Page 579
6.3.4.4	Adenylosuccinate synthase Adenylosuccinate synthase	Vol. 2, Page 579
6.3.4.4	Adenylosuccinate synthetase Adenylosuccinate synthase	Vol. 2, Page 579
3.6.1.3	Adenylpyrophosphatase Adenosinetriphosphatase	
3.1.4.15	adenylyl(glutamine synthetase) hydrolase adenylyl-[glutamate-ammonia ligase] hydrolase	

EC-No.	Synonym Recommended Name	Vol., Page
3.1.4.15	adenylyl-[glutamate-ammonia ligase] hydrolase adenylyl-[glutamate-ammonia ligase] hydrolase	
3.1.4.15	adenylyl-[glutamine-synthetase] hydrolase adenylyl-[glutamate-ammonia ligase] hydrolase	
3.1.4.15	adenylyl-[L-glutamate:ammonia ligase (ADP-forming)] adenylylhydrolase adenylyl-[glutamate-ammonia ligase] hydrolase	
1.8.4.9	adenylyl-sulfate reductase (glutathione) adenylyl-sulfate reductase (glutathione)	
4.6.1.1	adenylyl cyclase adenylate cyclase	
4.6.1.1	adenylylcyclase adenylate cyclase	
1.8.99.4	Adenylyl phosphosulfate reductase Phosphoadenosine-phosphosulfate reductase	
3.6.2.1	Adenylylsulfatase Adenylylsulfatase	
2.7.7.51	Adenylylsulfate-ammonia adenylyltransferase Adenylylsulfate-ammonia adenylyltransferase	
2.7.7.51	Adenylyl sulfate:ammonia adenylyl transferase Adenylylsulfate-ammonia adenylyltransferase	
2.7.7.51	Adenylylsulfate:ammonia adenylyltransferase Adenylylsulfate-ammonia adenylyltransferase	
2.7.1.25	Adenylylsulfate kinase Adenylylsulfate kinase	
2.7.7.4	Adenylylsulfate pyrophosphorylase Sulfate adenylyltransferase	
1.8.4.9	5'-adenylylsulfate reductase adenylyl-sulfate reductase (glutathione)	
1.8.99.2	Adenylylsulfate reductase Adenylylsulfate reductase	
3.6.2.1	Adenylylsulfate sulfohydrolase Adenylylsulfatase	
2.7.7.53	Adenylyltransferase, adenine triphosphate ATP adenylyltransferase	
2.7.7.51	Adenylyltransferase, adenylylsulfate-ammonia Adenylylsulfate-ammonia adenylyltransferase	

EC-No.	Synonym Recommended Name	Vol., Page
2.7.7.55	Adenylyltransferase, anthranilate Anthranilate adenylyltransferase	
2.7.7.46	Adenylyltransferase, gentamicin 2- Gentamicin 2-nucleotidyltransferase	
2.7.7.27	Adenylyltransferase, glucose 1-phosphate Glucose-1-phosphate adenylyltransferase	
2.7.7.42	Adenylyltransferase, glutamine synthetase [Glutamate-ammonia-ligase] adenylyltransferase	
2.7.7.1	Adenylyltransferase, nicotinamide mononucleotide Nicotinamide-nucleotide adenylyltransferase	
2.7.7.18	Adenylyltransferase, nicotinate mononucleotide Nicotinate-nucleotide adenylyltransferase	
2.7.7.3	Adenylyltransferase, pantetheine phosphate Pantetheine-phosphate adenylyltransferase	
2.7.7.54	Adenylyltransferase, phenylalanine Phenylalanine adenylyltransferase	
2.7.7.2	Adenylyltransferase, riboflavin mononucleotide FMN adenylyltransferase	
2.7.7.35	Adenylyltransferase, ribose 5-phosphate Ribose-5-phosphate adenylyltransferase	
2.7.7.47	Adenylyltransferase, streptomycin 3- Streptomycin 3-adenylyltransferase	
2.7.7.36	Adenylyltransferase, sugar 1-phosphate Aldose-1-phosphate adenylyltransferase	
2.7.7.4	Adenylyltransferase, sulfate Sulfate adenylyltransferase	
2.7.7.5	Adenylyltransferase, sulfate (adenosine diphosphate) Sulfate adenylyltransferase (ADP)	
2.7.7.25	Adenylyltransferase, transfer ribonucleate tRNA adenylyltransferase	
2.7.7.21	Adenylyltransferase, transfer ribonucleate tRNA cytidylyltransferase	
1.5.1.17	ADH Alanopine dehydrogenase	
1.1.1.1	ADH Alcohol dehydrogenase	

EC-No.	Synonym Recommended Name	Vol., Page
3.2.2.9	AdoHcy/MTA nucleosidase Adenosylhomocysteine nucleosidase	
3.3.1.1	AdoHcyase Adenosylhomocysteinase	
3.4.22.2	Adolph's Meat Tenderizer papain	
2.1.1.125	AdoMet:histone-arginine N-methyltransferase (Histone)-arginine N-methyltransferase	
2.1.1.126	AdoMet:myelin basic protein-arginine N-methyltransferase [Myelin basic protein]-arginine N-methyltransferase	
4.1.1.50	AdoMetDC Adenosylmethionine decarboxylase	
2.5.1.6	AdoMet synthetase Methionine adenosyltransferase	
1.1.1.56	Adonitol dehydrogenase Ribitol 2-dehydrogenase	
3.1.3.28	3-(ADP)-2-phosphoglycerate phosphohydrolase ADPphosphoglycerate phosphatase	
3.5.4.7	ADP-aminohydrolase ADP deaminase	
3.2.2.24	ADP-D-ribosyl-[dinitrogen reductase] ADPribosylhydrolase ADP-ribosyl-[dinitrogen reductase] hydrolase	
3.1.2.19	ADP-dependent-medium-chain-acyl-CoA hydrolase ADP-dependent medium-chain-acyl-CoA hydrolase	
3.1.2.18	ADP-dependent-short-chain-acyl-CoA hydrolase ADP-dependent short-chain-acyl-CoA hydrolase	
3.1.2.19	ADP-dependent medium-chain-acyl-CoA hydrolase ADP-dependent medium-chain-acyl-CoA hydrolase	
3.1.2.18	ADP-dependent short-chain-acyl-CoA hydrolase ADP-dependent short-chain-acyl-CoA hydrolase	
2.4.1.113	ADP-glucose-protein glucosyltransferase α-1,4-Glucan-protein synthase (ADP-forming)	
2.4.1.21	ADP-glucose-starch glucosyltransferase Starch synthase	
2.7.7.27	ADP-glucose pyrophosphorylase Glucose-1-phosphate adenylyltransferase	

EC-No.	Synonym Recommended Name	Vol., Page
2.4.1.21	ADP-glucose starch synthase Starch synthase	
2.7.7.27	ADP-glucose synthase Glucose-1-phosphate adenylyltransferase	
2.4.1.21	ADP-glucose synthase Starch synthase	
2.7.7.27	ADP-glucose synthetase Glucose-1-phosphate adenylyltransferase	
2.4.1.21	ADP-glucose transglucosylase Starch synthase	
5.1.3.20	ADP-L-glycero-D-manno-heptose-6-epimerase ADPglyceromanno-heptose 6-epimerase	Vol. 1, Page 175
5.1.3.20	ADP-L-glycero-D-manno-heptose-epimerase, Aquifex aelicis gene rfaD ADPglyceromanno-heptose 6-epimerase	Vol. 1, Page 175
5.1.3.20	ADP-L-glycero-D-manno-heptose 6-epimerase ADPglyceromanno-heptose 6-epimerase	Vol. 1, Page 175
5.1.3.20	ADP-L-glycero-D-manno-heptose 6-epimerase ADPglyceromanno-heptose 6-epimerase	Vol. 1, Page 175
5.1.3.20	ADP-L-glycero-D-mannoheptose-6-epimerase ADPglyceromanno-heptose 6-epimerase	Vol. 1, Page 175
5.1.3.20	ADP-L-glycero-D-mannoheptose-6-epimerase (Escherichia coli strain K-12 substrain MG1655 clone EC19-98 gene rfaD) ADPglyceromanno-heptose 6-epimerase	Vol. 1, Page 175
5.1.3.20	ADP-L-glycero-D-mannoheptose 6-epimerase ADPglyceromanno-heptose 6-epimerase	Vol. 1, Page 175
5.1.3.20	ADP-L-glycero-D-mannoheptose 6-epimerase ADPglyceromanno-heptose 6-epimerase	Vol. 1, Page 175
5.1.3.20	ADP-L-glycero-D-mannoheptose 6-epimerase (Neisseria gonorrhoeae gene gme) ADPglyceromanno-heptose 6-epimerase	Vol. 1, Page 175
3.1.3.28	ADP-phosphoglycerate phosphatase ADPphosphoglycerate phosphatase	
3.2.2.19	ADP-ribose-L-arginine cleavage enzyme ADPribosylarginine hydrolase	
3.6.1.13	ADP-ribose phosphohydrolase ADPribose diphosphatase	

EC-No.	Synonym Recommended Name	Vol., Page
3.2.2.24	ADP-ribosyl-[dinitrogen reductase] hydrolase ADP-ribosyl-[dinitrogen reductase] hydrolase	
3.2.2.19	ADP-ribosylarginine hydrolase ADPribosylarginine hydrolase	
3.2.2.24	ADP-ribosyl glycohydrolase ADP-ribosyl-[dinitrogen reductase] hydrolase	
2.4.2.31	ADP-ribosyltransferase NAD(P)$^+$-arginine ADP-ribosyltransferase	
2.4.2.30	ADP-ribosyltransferase NAD$^+$ ADP-ribosyltransferase	
2.4.2.36	ADP-ribosyltransferase NAD$^+$-diphthamide ADP-ribosyltransferase	
2.4.2.30	ADP-ribosyltransferase (polymerizing) NAD$^+$ ADP-ribosyltransferase	
2.7.7.5	ADP-sulfurylase Sulfate adenylyltransferase (ADP)	
2.7.1.118	ADP-thymidine kinase ADP-thymidine kinase	
2.7.7.36	ADP:aldose-1-phosphate adenylytransferase Aldose-1-phosphate adenylyltransferase	
2.7.7.53	ADP:ATP adenylyltransferase ATP adenylyltransferase	
2.7.7.35	ADP:D-ribose-5-phosphate adenylyltransferase Ribose-5-phosphate adenylyltransferase	
2.7.1.118	ADP:dThd phosphotransferase ADP-thymidine kinase	
2.7.7.5	ADP:sulfate adenylyltransferase Sulfate adenylyltransferase (ADP)	
2.7.1.118	ADP:thymidine 5'-phosphotransferase ADP-thymidine kinase	
2.7.7.36	ADPaldose phosphorylase Aldose-1-phosphate adenylyltransferase	
3.5.4.7	ADP aminohydrolase ADP deaminase	
3.6.1.5	ADPase Apyrase	

EC-No.	Synonym Recommended Name	Vol., Page
3.6.1.6	ADPase Nucleoside-diphosphatase	
3.5.4.7	ADP deaminase ADP deaminase	
2.4.1.21	ADPG-starch glucosyltransferase Starch synthase	
2.4.1.21	ADPglucose-starch glucosyltransferase Starch synthase	
2.4.1.21	ADPglucose:1,4-α-D-glucan 4-α-D-glucosyltransferase Starch synthase	
2.4.1.113	ADPglucose:protein 4-α-D-glucosyltransferase α-1,4-Glucan-protein synthase (ADP-forming)	
2.4.1.113	ADPglucose:protein glucosyltransferase α-1,4-Glucan-protein synthase (ADP-forming)	
2.7.7.27	ADPglucose pyrophosphorylase Glucose-1-phosphate adenylyltransferase	
5.1.3.20	ADPglyceromanno-heptose 6-epimerase ADPglyceromanno-heptose 6-epimerase	Vol. 1, Page 175
2.7.7.27	ADPG pyrophosphorylase Glucose-1-phosphate adenylyltransferase	
2.4.1.21	ADPG starch synthetase Starch synthase	
3.1.3.28	ADPphosphoglycerate phosphatase ADPphosphoglycerate phosphatase	
3.6.1.13	ADPR-PPase ADPribose diphosphatase	
1.17.4.1	ADP reductase Ribonucleoside-diphosphate reductase	
3.6.1.13	ADPribose diphosphatase ADPribose diphosphatase	
2.7.7.35	ADPribose phosphorylase Ribose-5-phosphate adenylyltransferase	
3.6.1.13	ADPribose pyrophosphatase ADPribose diphosphatase	
3.6.1.13	ADPribose ribophosphohydrolase ADPribose diphosphatase	

EC-No.	Synonym Recommended Name	Vol., Page
3.2.2.19	ADPribosylarginine hydrolase ADPribosylarginine hydrolase	
3.6.1.21	ADPsugar diphosphatase ADPsugar diphosphatase	
2.7.7.36	ADP sugar phosphorylase Aldose-1-phosphate adenylyltransferase	
3.6.1.21	ADPsugar pyrophosphatase ADPsugar diphosphatase	
3.6.1.21	ADPsugar sugarphosphohydrolase ADPsugar diphosphatase	
2.7.7.5	ADP sulfurylase Sulfate adenylyltransferase (ADP)	
4.1.2.30	Adrenal 17,20-lyase 17α-Hydroxyprogesterone aldolase	
1.4.3.4	Adrenaline oxidase Amine oxidase (flavin-containing)	
2.7.1.126	β-Adrenergic-receptor kinase β-Adrenergic-receptor kinase	
1.18.1.2	Adrenodoxin reductase Ferredoxin-NADP$^+$ reductase	
1.13.12.5	Aequorin Renilla-luciferin 2-monooxygenase	
1.2.3.4	Aero-oxalo dehydrogenase Oxalate oxidase	
3.2.1.87	Aerobacter-capsular-polysaccharide galactohydrolase Capsular-polysaccharide endo-1,3-α-galactosidase	
3.4.24.25	Aeromonas neutral protease Aeromonolysin	
3.4.11.10	Aeromonas proteolytica aminopeptidase bacterial leucyl aminopeptidase	
3.4.24.25	Aeromonas proteolytica neutral proteinase Aeromonolysin	
3.4.24.25	Aeromonolysin Aeromonolysin	
3.4.21.55	Afaacytin Venombin AB	

EC-No.	Synonym Recommended Name Vol., Page
1.6.5.4	AFR Monodehydroascorbate reductase (NADH)
1.6.5.4	AFR-reductase Monodehydroascorbate reductase (NADH)
3.2.1.81	Agarase Agarase
3.2.1.81	β-agarase Agarase
2.3.2.9	Agaritine γ-glutamyltransferase Agaritine γ-glutamyltransferase
3.2.1.81	Agarose 3-glycanohydrolase Agarase
2.3.1.1	AGAS Amino-acid N-acetyltransferase
3.5.1.82	D-AGase N-Acyl-D-glutamate deacylase
3.4.99.2	agavain agavain
3.4.24.72	Agkistrodon contortrix contortrix metalloproteinase Fibrolase
3.4.21.28	Agkistrodon serine proteinase Agkistrodon serine proteinase
3.5.3.11	Agmatinase Agmatinase
3.5.3.11	Agmatine amidinohydrolase Agmatinase
3.5.3.12	Agmatine amidinohydrolase Agmatine deiminase
2.3.1.64	Agmatine coumaroyltransferase Agmatine N^4-coumaroyltransferase
3.5.3.12	Agmatine deiminase Agmatine deiminase
3.5.3.12	Agmatine iminohydrolase Agmatine deiminase
2.7.3.10	Agmatine kinase Agmatine kinase

EC-No.	Synonym Recommended Name	Vol., Page
2.3.1.64	Agmatine N^4-coumaroyltransferase Agmatine N^4-coumaroyltransferase	
3.5.3.11	Agmatine ureohydrolase Agmatinase	
2.6.1.44	AGT Alanine-glyoxylate transaminase	
3.1.2.14	AH oleoyl-[acyl-carrier-protein] hydrolase	
4.1.3.18	AHAS acetolactate synthase	
4.2.99.10	AHS O-acetylhomoserine (thiol)-lyase	
5.3.3.7	AI Aconitate Δ-isomerase	Vol. 1, Page 409
2.1.2.3	AICAR formyltransferase Phosphoribosylaminoimidazolecarboxamide formyltransferase	
2.1.2.3	AICAR transformylase Phosphoribosylaminoimidazolecarboxamide formyltransferase	
4.1.1.21	AIR carboxylase Phosphoribosylaminoimidazole carboxylase	
6.3.3.1	AIR synthetase Phosphoribosylformylglycinamidine cyclo-ligase	Vol. 2, Page 530
2.7.2.1	AK Acetate kinase	
2.7.2.4	AK Aspartate kinase	
1.2.4.2	AKGDH Oxoglutarate dehydrogenase (lipoamide)	
4.2.1.24	Al-D porphobilinogen synthase	
4.2.1.24	ALA-D porphobilinogen synthase	
6.3.2.10	D-Ala-D-Ala-adding enzyme UDP-N-acetylmuramoylalanyl-D-glutamyl-lysine-D-alanyl-D-alanine ligase	Vol. 2, Page 458

EC-No.	Synonym Recommended Name	Vol., Page
6.3.2.4	D-Ala-D-Ala synthetase D-Alanine-D-alanine ligase	Vol. 2, Page 423
3.4.13.20	β-Ala-His dipeptidase β-Ala-His dipeptidase	
6.1.1.7	Ala-tRNA synthetase Alanine-tRNA ligase	Vol. 2, Page 51
3.4.17.6	D-Ala carboxypeptidase IB-C alanine carboxypeptidase	
4.2.1.24	ALAD porphobilinogen synthase	
1.4.1.1	AlaDH Alanine dehydrogenase	
6.3.2.8	L-Ala ligase UDP-N-acetylmuramate-alanine ligase	Vol. 2, Page 442
2.6.1.43	Alanine-γ,δ-dioxovalerate aminotransferase Aminolevulinate transaminase	
2.6.1.18	β-Alanine-α-alanine transaminase β-Alanine-pyruvate transaminase	
2.6.1.12	L-Alanine-α-keto acid aminotransferase Alanine-oxo-acid transaminase	
4.1.1.64	L-Alanine-α-ketobutyrate aminotransferase 2,2-Dialkylglycine decarboxylase (pyruvate)	
2.6.1.2	Alanine-α-ketoglutarate aminotransferase Alanine transaminase	
2.6.1.2	L-Alanine-α-ketoglutarate aminotransferase Alanine transaminase	
2.6.1.43	L-Alanine-4,5-dioxovalerate aminotransferase Aminolevulinate transaminase	
6.1.1.13	D-Alanine-activating enzyme D-Alanine-poly(phosphoribitol) ligase	Vol. 2, Page 97
6.3.2.8	Alanine-adding enzyme UDP-N-acetylmuramate-alanine ligase	Vol. 2, Page 442
6.3.2.16	D-Alanine-alanyl-poly(glycerolphosphate) ligase D-Alanine-alanyl-poly(glycerolphosphate) ligase	Vol. 2, Page 485
3.4.17.8	D-alanine-D-alanine-carboxypeptidase muramoylpentapeptide carboxypeptidase	

EC-No.	Synonym Recommended Name	Vol., Page
6.3.2.4	D-Alanine-D-alanine ligase D-Alanine-D-alanine ligase	Vol. 2, Page 423
6.1.1.13	D-Alanine-D-alanyl carrier protein ligase D-Alanine-poly(phosphoribitol) ligase	Vol. 2, Page 97
2.6.1.43	Alanine-dioxovalerate aminotransferase Aminolevulinate transaminase	
2.3.2.14	D-Alanine γ-glutamyltransferase D-Alanine γ-glutamyltransferase	
2.6.1.44	L-Alanine-glycine transaminase Alanine-glyoxylate transaminase	
2.6.1.44	Alanine-glyoxylate aminotransferase Alanine-glyoxylate transaminase	
2.6.1.44	L-Alanine-glyoxylate aminotransferase Alanine-glyoxylate transaminase	
2.6.1.44	Alanine-glyoxylate transaminase Alanine-glyoxylate transaminase	
2.6.1.44	Alanine-glyoxylic aminotransferase Alanine-glyoxylate transaminase	
2.6.1.12	Alanine-keto acid aminotransferase Alanine-oxo-acid transaminase	
2.6.1.47	Alanine-ketomalonate (mesoxalate) transaminase Alanine-oxomalonate transaminase	
2.6.1.47	L-Alanine-ketomalonate transaminase Alanine-oxomalonate transaminase	
6.3.2.16	D-Alanine-membrane acceptor-ligase D-Alanine-alanyl-poly(glycerolphosphate) ligase	Vol. 2, Page 485
6.1.1.13	D-Alanine-membrane acceptor ligase D-Alanine-poly(phosphoribitol) ligase	Vol. 2, Page 97
2.6.1.56	L-Alanine-N-amidino-3-(or 5-)keto-scyllo-inosamine transaminase 1D-1-Guanidino-3-amino-1,3-dideoxy-scyllo-inositol transaminase	
2.6.1.12	Alanine-oxo-acid transaminase Alanine-oxo-acid transaminase	
2.6.1.12	Alanine-oxo acid aminotransferase Alanine-oxo-acid transaminase	
2.6.1.19	β-Alanine-oxoglutarate aminotransferase 4-Aminobutyrate transaminase	

EC-No.	Synonym Recommended Name	Vol., Page
2.6.1.19	β-Alanine-oxoglutarate transaminase 4-Aminobutyrate transaminase	
2.6.1.47	Alanine-oxomalonate aminotransferase Alanine-oxomalonate transaminase	
2.6.1.47	Alanine-oxomalonate transaminase Alanine-oxomalonate transaminase	
3.4.11.4	alanine-phenylalanine-proline arylamidase tripeptide aminopeptidase	
6.1.1.13	D-Alanine-poly(phosphoribitol) ligase D-Alanine-poly(phosphoribitol) ligase	Vol. 2, Page 97
2.6.1.2	Alanine-pyruvate aminotransferase Alanine transaminase	
2.6.1.18	β-Alanine-pyruvate aminotransferase β-Alanine-pyruvate transaminase	
2.6.1.18	β-Alanine-pyruvate transaminase β-Alanine-pyruvate transaminase	
3.4.11.2	alanine-specific aminopeptidase membrane alanyl aminopeptidase	
6.1.1.8	D-alanine-sRNA sythetase D-alanine-sRNA sythetase	Vol. 2, Page 58
6.1.1.7	Alanine-transfer RNA ligase Alanine-tRNA ligase	Vol. 2, Page 51
6.1.1.7	Alanine-tRNA ligase Alanine-tRNA ligase	Vol. 2, Page 51
2.6.1.32	Alanine-valine transaminase Valine-3-methyl-2-oxovalerate transaminase	
2.1.2.7	D-Alanine 2-hydroxymethyltransferase D-Alanine 2-hydroxymethyltransferase	
2.6.1.12	L-Alanine:2-oxo-acid aminotransferase Alanine-oxo-acid transaminase	
2.6.1.21	D-Alanine:2-oxoglutarate aminotransferase D-Alanine transaminase	
2.6.1.2	L-Alanine:2-oxoglutarate aminotransferase Alanine transaminase	
2.6.1.18	L-Alanine:3-oxopropanoate aminotransferase β-Alanine-pyruvate transaminase	

EC-No.	Synonym Recommended Name	Vol., Page
2.6.1.43	Alanine:4,5-dioxovalerate aminotransferase Aminolevulinate transaminase	
2.6.1.43	L-Alanine:4,5-dioxovaleric acid transaminase Aminolevulinate transaminase	
6.3.2.4	Alanine:alanine ligase (ADP-forming) D-Alanine-D-alanine ligase	Vol. 2, Page 423
6.3.2.16	D-Alanine:alanyl-poly(glycerolphosphate) ligase (ADP-forming) D-Alanine-alanyl-poly(glycerolphosphate) ligase	Vol. 2, Page 485
2.6.1.74	L-Alanine:cephalosporin-C aminotransferase Cephalosporin-C transaminase	
6.3.2.4	D-Alanine:D-alanine ligase D-Alanine-D-alanine ligase	Vol. 2, Page 423
5.1.1.1	L-Alanine:D-alanine racemase Alanine racemase	Vol. 1, Page 1
2.6.1.43	L-Alanine:dioxovalerate transaminase Aminolevulinate transaminase	
2.6.1.44	L-Alanine:glyoxylate aminotransferase Alanine-glyoxylate transaminase	
6.3.2.16	D-Alanine:membrane-acceptor ligase D-Alanine-alanyl-poly(glycerolphosphate) ligase	Vol. 2, Page 485
6.1.1.13	D-Alanine: membrane acceptor ligase D-Alanine-poly(phosphoribitol) ligase	Vol. 2, Page 97
6.3.2.16	D-Alanine:membrane acceptor ligase D-Alanine-alanyl-poly(glycerolphosphate) ligase	Vol. 2, Page 485
1.4.1.1	L-Alanine:NAD^+ oxidoreductase (deaminating) Alanine dehydrogenase	
2.6.1.47	L-Alanine:oxomalonate aminotransferase Alanine-oxomalonate transaminase	
6.1.1.13	D-Alanine:poly(phosphoribitol) ligase (AMP-forming) D-Alanine-poly(phosphoribitol) ligase	Vol. 2, Page 97
6.1.1.7	L-Alanine:$tRNA^{Ala}$ ligase (AMP-forming) Alanine-tRNA ligase	Vol. 2, Page 51
6.3.2.8	L-Alanine adding enzyme UDP-N-acetylmuramate-alanine ligase	Vol. 2, Page 442
3.4.11.2	L-alanine aminopeptidase membrane alanyl aminopeptidase	

EC-No.	Synonym Recommended Name	Vol., Page
3.4.11.2	alanine aminopeptidase membrane alanyl aminopeptidase	
2.6.1.2	Alanine aminotransferase Alanine transaminase	
2.6.1.21	D-Alanine aminotransferase D-Alanine transaminase	
2.6.1.2	L-Alanine aminotransferase Alanine transaminase	
2.6.1.19	β-Alanine aminotransferase 4-Aminobutyrate transaminase	
2.6.1.2	β-Alanine aminotransferase Alanine transaminase	
3.4.17.8	D-alanine carboxypeptidase muramoylpentapeptide carboxypeptidase	
3.4.16.4	D-alanine carboxypeptidase serine-type D-Ala-D-Ala carboxypeptidase	
3.4.17.6	alanine carboxypeptidase alanine carboxypeptidase	
3.4.17.8	D-alanine carboxypeptidase I muramoylpentapeptide carboxypeptidase	
3.4.17.8	D-alanine carboxypeptidase I muramoylpentapeptide carboxypeptidase	
3.4.16.4	D-alanine carboxypeptidase I serine-type D-Ala-D-Ala carboxypeptidase	
1.4.1.1	Alanine dehydrogenase Alanine dehydrogenase	
1.4.1.1	L-Alanine dehydrogenase Alanine dehydrogenase	
1.4.1.1	α-Alanine dehydrogenase Alanine dehydrogenase	
1.4.1.1	Alanine oxidoreductase Alanine dehydrogenase	
5.1.1.1	Alanine racemase Alanine racemase	Vol. 1, Page 1
5.1.1.1	Alanine racemase Alanine racemase	Vol. 1, Page 1

EC-No.	Synonym Recommended Name	Vol., Page
5.1.1.1	L-Alanine racemase Alanine racemase	Vol. 1, Page 1
6.3.2.23	β-Alanine specific hGSH synthetase Homoglutathione synthase	Vol. 2, Page 518
2.6.1.2	Alanine transaminase Alanine transaminase	
2.6.1.21	D-Alanine transaminase D-Alanine transaminase	
2.6.1.2	L-Alanine transaminase Alanine transaminase	
6.1.1.7	Alanine transfer RNA synthetase Alanine-tRNA ligase	Vol. 2, Page 51
6.1.1.7	Alanine translase Alanine-tRNA ligase	Vol. 2, Page 51
6.1.1.7	Alanine tRNA synthetase Alanine-tRNA ligase	Vol. 2, Page 51
1.5.1.17	Alanopine dehydrogenase Alanopine dehydrogenase	
1.5.1.26	β-Alanopine dehydrogenase β-Alanopine dehydrogenase	
6.3.2.16	D-Alanyl-alanyl-polyglycerolphosphate synthetase D-Alanine-alanyl-poly(glycerolphosphate) ligase	Vol. 2, Page 485
4.3.1.6	β-alanyl-CoA ammonia-lyase β-alanyl-CoA ammonia-lyase	
4.3.1.6	β-alanyl-CoA ammonia-lyase β-alanyl-CoA ammonia-lyase	
6.3.2.15	D-Alanyl-D-alanine-adding enzyme UDP-N-acetylmuramoylalanyl-D-glutamyl-2,6-diaminopimelate-D-alanyl-D-alanine ligase	Vol. 2, Page 480
3.4.16.4	D-alanyl-D-alanine-carboxypeptidase serine-type D-Ala-D-Ala carboxypeptidase	
3.4.16.4	D-alanyl-D-alanine-cleaving-peptidase serine-type D-Ala-D-Ala carboxypeptidase	
3.4.17.14	D-Alanyl-D-alanine-cleaving carboxypeptidase Zinc D-Ala-D-Ala carboxypeptidase	

EC-No.	Synonym Recommended Name	Vol., Page
3.4.16.4	D-alanyl-D-alanine-cleaving peptidase serine-type D-Ala-D-Ala carboxypeptidase	
6.3.2.15	D-Alanyl-D-alanine adding enzyme UDP-N-acetylmuramoylalanyl-D-glutamyl-2,6-diaminopimelate-D-alanyl-D-alanine ligase	Vol. 2, Page 480
3.4.17.8	D-alanyl-D-alanine carboxypeptidase muramoylpentapeptide carboxypeptidase	
3.4.17.14	D-Alanyl-D-alanine hydrolase Zinc D-Ala-D-Ala carboxypeptidase	
3.4.16.4	D-alanyl-D-alanine hydrolase serine-type D-Ala-D-Ala carboxypeptidase	
3.4.17.8	D-alanyl-D-alanine peptidase muramoylpentapeptide carboxypeptidase	
3.4.16.4	D-alanyl-D-alanine peptidase serine-type D-Ala-D-Ala carboxypeptidase	
6.3.2.4	D-Alanyl-D-alanine synthetase D-Alanine-D-alanine ligase	Vol. 2, Page 423
6.3.2.16	D-Alanyl-poly(phosphoglycerol) synthetase D-Alanine-alanyl-poly(glycerolphosphate) ligase	Vol. 2, Page 485
6.1.1.13	D-Alanyl-poly-(phosphoribitol) synthetase D-Alanine-poly(phosphoribitol) ligase	Vol. 2, Page 97
2.3.2.10	Alanyl-transfer ribonucleate-uridine diphosphoacetylmuramoylpentapeptide transferase UDP-N-acetylmuramoylpentapeptide-lysine N^6-alanyltransferase	
6.1.1.7	Alanyl-transfer ribonucleate synthetase Alanine-tRNA ligase	Vol. 2, Page 51
6.1.1.7	Alanyl-transfer ribonucleic acid synthetase Alanine-tRNA ligase	Vol. 2, Page 51
6.1.1.7	Alanyl-transfer RNA synthetase Alanine-tRNA ligase	Vol. 2, Page 51
2.3.2.11	L-Alanyl-tRNA:phosphatidylglycerol alanyltransferase Alanylphosphatidylglycerol synthase	
2.3.2.10	L-Alanyl-tRNA:UDP-N-acetylmuramoyl-L-alanyl-D-glutamyl-L-lysyl-D-alanyl-D-alanine N^6-alanyltransferase UDP-N-acetylmuramoylpentapeptide-lysine N^6-alanyltransferase	
6.1.1.7	Alanyl-tRNA synthetase Alanine-tRNA ligase	Vol. 2, Page 51

EC-No.	Synonym Recommended Name	Vol., Page
6.3.2.4	D-Alanylalanine synthetase D-Alanine-D-alanine ligase	Vol. 2, Page 423
3.4.11.2	alanyl aminopeptidase membrane alanyl aminopeptidase	
3.4.16.4	D-alanyl carboxypeptidase serine-type D-Ala-D-Ala carboxypeptidase	
4.3.1.6	β-alanyl coenzyme A ammonia-lyase β-alanyl-CoA ammonia-lyase	
2.3.2.11	Alanylphosphatidylglycerol synthase Alanylphosphatidylglycerol synthase	
2.3.2.11	Alanyl phosphatidylglycerol synthetase Alanylphosphatidylglycerol synthase	
2.3.2.10	Alanyltransferase, uridine diphosphoacetylmuramoylpentapeptide lysine N^6- UDP-N-acetylmuramoylpentapeptide-lysine N^6-alanyltransferase	
6.1.1.7	AlaRS Alanine-tRNA ligase	Vol. 2, Page 51
2.3.1.37	ALAS 5-Aminolevulinate synthase	
2.3.1.37	ALA synthase 5-Aminolevulinate synthase	
2.3.1.37	ALA synthetase 5-Aminolevulinate synthase	
4.2.1.24	ALA synthetase porphobilinogen synthase	
1.14.13.32	Albendazole,NADPH:oxygen oxidoreductase (sulfoxide-forming) Albendazole monooxygenase	
1.14.13.32	Albendazole monooxygenase Albendazole monooxygenase	
1.14.13.32	Albendazole sulfoxidase Albendazole monooxygenase	
3.4.21.62	Alcalase Subtilisin	
3.4.21.62	Alcalase 0.6L Subtilisin	

EC-No.	Synonym Recommended Name	Vol., Page
3.4.21.62	Alcalase 2.5L Subtilisin	
2.8.2.2	Alcohol/hydroxysteroid sulfotransferase Alcohol sulfotransferase	
1.1.99.8	Alcohol:(acceptor) oxidoreductase Alcohol dehydrogenase (acceptor)	
1.1.1.71	Alcohol:NAD(P)$^+$ oxidoreductase Alcohol dehydrogenase (NAD(P)$^+$)	
1.1.1.1	Alcohol:NAD$^+$ oxidoreductase Alcohol dehydrogenase	
1.1.1.2	Alcohol:NADP$^+$ oxidoreductase Alcohol dehydrogenase (NADP$^+$)	
1.1.3.13	Alcohol:oxygen oxidoreductase Alcohol oxidase	
2.3.1.84	Alcohol acetyltransferase Alcohol O-acetyltransferase	
1.1.1.1	Alcohol dehydrogenase Alcohol dehydrogenase	
1.1.99.8	Alcohol dehydrogenase (acceptor) Alcohol dehydrogenase (acceptor)	
1.1.1.71	Alcohol dehydrogenase (NAD(P)$^+$) Alcohol dehydrogenase (NAD(P)$^+$)	
1.1.1.1	Alcohol dehydrogenase (NAD) Alcohol dehydrogenase	
1.1.1.2	Alcohol dehydrogenase (NADP$^+$) Alcohol dehydrogenase (NADP$^+$)	
2.3.1.84	Alcohol O-acetyltransferase Alcohol O-acetyltransferase	
2.3.1.152	Alcohol O-cinnamoyltransferase Alcohol O-cinnamoyltransferase	
1.1.3.13	Alcohol oxidase Alcohol oxidase	
2.8.2.2	Alcohol sulfotransferase Alcohol sulfotransferase	
4.1.1.5	ALDC Acetolactate decarboxylase	

EC-No.	Synonym Recommended Name	Vol., Page
1.2.99.6	Aldehyde:(acceptor) oxidoreductase Carboxylate reductase	
1.2.99.3	Aldehyde:(pyrroloquinoline-quinone) oxidoreductase Aldehyde dehydrogenase (pyrroloquinoline-quinone)	
1.2.1.5	Aldehyde:NAD(P)$^+$ oxidoreductase Aldehyde dehydrogenase (NAD(P)$^+$)	
1.2.1.3	Aldehyde:NAD$^+$ oxidoreductase Aldehyde dehydrogenase (NAD$^+$)	
1.2.1.4	Aldehyde:NADP$^+$ oxidoreductase Aldehyde dehydrogenase (NADP$^+$)	
1.2.3.1	Aldehyde:oxygen oxidoreductase Aldehyde oxidase	
4.1.99.5	Aldehyde decarbonylase Octadecanal decarbonylase	
1.2.99.3	Aldehyde dehydrogenase (acceptor) Aldehyde dehydrogenase (pyrroloquinoline-quinone)	
1.2.1.10	Aldehyde dehydrogenase (acylating) Acetaldehyde dehydrogenase (acetylating)	
1.2.1.5	Aldehyde dehydrogenase (NAD(P)$^+$) Aldehyde dehydrogenase (NAD(P)$^+$)	
1.2.1.3	Aldehyde dehydrogenase (NAD$^+$) Aldehyde dehydrogenase (NAD$^+$)	
1.2.1.4	Aldehyde dehydrogenase (NADP$^+$) Aldehyde dehydrogenase (NADP$^+$)	
1.2.99.3	Aldehyde dehydrogenase (pyrroloquinoline-quinone) Aldehyde dehydrogenase (pyrroloquinoline-quinone)	
1.2.99.4	Aldehyde dismutase Formaldehyde dismutase	
1.2.99.6	Aldehyde ferredoxin oxidoreductase Carboxylate reductase	
1.14.14.3	Aldehyde monooxygenase Alkanal monooxygenase (FMN-linked)	
1.2.3.1	Aldehyde oxidase Aldehyde oxidase	
1.2.99.6	Aldehyde oxidoreductase Carboxylate reductase	

EC-No.	Synonym Recommended Name	Vol., Page
1.1.1.1	Aldehyde reductase Alcohol dehydrogenase	
1.1.1.21	Aldehyde reductase Aldehyde reductase	
1.2.99.6	Aldehyde reductase Carboxylate reductase	
1.1.1.19	Aldehyde reductase Glucuronate reductase	
1.1.1.2	Aldehyde reductase (NADPH) Alcohol dehydrogenase ($NADP^+$)	
1.6.2.4	Aldehyde reductase (NADPH-dependent) NADPH-ferrihemoprotein reductase	
1.1.1.71	Aldehyde reductase (NADPH/NADH) Alcohol dehydrogenase ($NAD(P)^+$)	
1.1.1.184	Aldehyde reductase 1 Carbonyl reductase (NADPH)	
1.1.1.184	Aldehyde reductase I Carbonyl reductase (NADPH)	
1.1.1.19	Aldehyde reductase II Glucuronate reductase	
1.2.1.5	ALDH Aldehyde dehydrogenase ($NAD(P)^+$)	
1.1.1.200	Alditol 6-phosphate:NADP 1-oxidoreductase Aldose-6-phosphate reductase (NADPH)	
1.1.1.21	Alditol:$NAD(P)^+$ 1-oxidoreductase Aldehyde reductase	
1.1.1.21	Alditol:$NADP^+$ 1-oxidoreductase Aldehyde reductase	
1.1.1.21	Alditol:NADP oxidoreductase Aldehyde reductase	
1.1.99.13	D-Aldohexopyranoside dehydrogenase Glucoside 3-dehydrogenase	
1.1.1.118	D-Aldohexose dehydrogenase Glucose 1-dehydrogenase (NAD^+)	
1.1.99.13	D-Aldohexoside:(acceptor) 3-oxidoreductase Glucoside 3-dehydrogenase	

EC-No.	Synonym Recommended Name	Vol., Page
1.1.99.13	D-Aldohexoside:cytochrome c oxidoreductase Glucoside 3-dehydrogenase	
4.4.1.5	aldoketomutase lactoylglutathione lyase	
4.1.2.13	Aldolase Fructose-bisphosphate aldolase	
4.1.2.30	Aldolase, 17α-hydroxyprogesterone 17α-Hydroxyprogesterone aldolase	
4.1.2.34	Aldolase, 2'-carboxybenzalpyruvate 4-(2-Carboxyphenyl)-2-oxobut-3-enoate aldolase	
4.1.2.20	Aldolase, 2-keto-3-deoxy-D-glucarate 2-Dehydro-3-deoxyglucarate aldolase	
4.1.2.28	Aldolase, 2-keto-3-deoxy-D-pentonate 2-Dehydro-3-deoxy-D-pentonate aldolase	
4.1.2.18	Aldolase, 2-keto-3-deoxy-L-arabonate 2-Dehydro-3-deoxy-L-pentonate aldolase	
4.1.2.18	Aldolase, 2-keto-3-deoxy-L-pentonate 2-Dehydro-3-deoxy-L-pentonate aldolase	
4.1.2.23	aldolase, 2-keto-3-deoxyoctonate 3-deoxy-D-manno-octulosonate aldolase	
4.1.3.14	Aldolase, 3-hydroxyaspartate 3-Hydroxyaspartate aldolase	
4.1.3.16	Aldolase, 4-hydroxy-2-oxoglutarate 4-Hydroxy-2-oxoglutarate aldolase	
4.1.3.17	aldolase, 4-hydroxy-4-methyl-2-oxoglutarate 4-hydroxy-4-methyl-2-oxoglutarate aldolase	
4.1.2.4	Aldolase, deoxyribo Deoxyribose-phosphate aldolase	
4.1.2.25	Aldolase, dihydroneopterin Dihydroneopterin aldolase	
4.1.2.27	Aldolase, dihydrosphingosine 1-phosphate Sphinganine-1-phosphate aldolase	
4.1.2.24	Aldolase, dimethylaniline N-oxide Dimethylaniline-N-oxide aldolase	
4.1.2.13	Aldolase, fructose diphosphate Fructose-bisphosphate aldolase	

EC-No.	Synonym Recommended Name	Vol., Page
4.1.2.2	Aldolase, ketotetrose phosphate Ketotetrose-phosphate aldolase	
4.1.2.17	Aldolase, L-fuculose phosphate L-Fuculose-phosphate aldolase	
4.1.2.26	Aldolase, phenylserine Phenylserine aldolase	
4.1.2.21	Aldolase, phospho-2-keto-3-deoxygalactonate 2-Dehydro-3-deoxyphosphogalactonate aldolase	
4.1.2.14	Aldolase, phospho-2-keto-3-deoxygluconate 2-Dehydro-3-deoxyphosphogluconate aldolase	
4.1.2.15	Aldolase, phospho-2-keto-3-deoxyheptanoate 2-Dehydro-3-deoxyphosphoheptanoate aldolase	
4.1.2.16	Aldolase, phospho-2-keto-3-deoxyoctonate 2-Dehydro-3-deoxyphosphooctonate aldolase	
4.1.2.29	Aldolase, phospho-5-keto-2-deoxygluconate Phospho-5-dehydro-2-deoxygluconate aldolase	
4.1.2.19	Aldolase, rhamnulose phosphate Rhamnulose-1-phosphate aldolase	
4.1.2.40	Aldolase, tagatose 1,6-diphosphate Tagatose-bisphosphate aldolase	
4.1.2.5	Aldolase, threonine Threonine aldolase	
1.1.1.45	L-3-Aldonate dehydrogenase L-Gulonate 3-dehydrogenase	
1.1.1.45	L-3-Aldonic dehydrogenase L-Gulonate 3-dehydrogenase	
3.1.1.18	aldonolactonase aldonolactonase	
3.1.1.17	aldonolactonase gluconolactonase	
1.2.1.33	D-Aldopantoate dehydrogenase (R)-Dehydropantoate dehydrogenase	
2.7.7.36	Aldose-1-phosphate adenylyltransferase Aldose-1-phosphate adenylyltransferase	
2.7.7.37	Aldose-1-phosphate nucleotidyltransferase Aldose-1-phosphate nucleotidyltransferase	

EC-No.	Synonym Recommended Name	Vol., Page
1.1.1.200	D-Aldose-6-phosphate:NADP$^+$ 1-oxidoreductase Aldose-6-phosphate reductase (NADPH)	
1.1.1.200	Aldose-6-phosphate reductase Aldose-6-phosphate reductase (NADPH)	
1.1.1.200	Aldose-6-phosphate reductase (NADPH) Aldose-6-phosphate reductase (NADPH)	
1.1.1.200	Aldose-6-P reductase Aldose-6-phosphate reductase (NADPH)	
2.4.1.162	Aldose β-D-fructosyltransferase Aldose β-D-fructosyltransferase	
1.1.1.121	Aldose 1-dehydrogenase Aldose 1-dehydrogenase	
5.1.3.3	Aldose 1-epimerase Aldose 1-epimerase	Vol. 1, Page 113
5.1.3.3	Aldose 1-epimerase Aldose 1-epimerase	Vol. 1, Page 113
1.1.1.121	D-Aldose:NAD$^+$ 1-oxidoreductase Aldose 1-dehydrogenase	
5.1.3.3	Aldose mutarotase Aldose 1-epimerase	Vol. 1, Page 113
1.1.1.21	Aldose reductase Aldehyde reductase	
2.4.1.162	α-D-Aldosyl-β-D-fructoside:aldose 1-β-D-fructosyltransferase Aldose β-D-fructosyltransferase	
2.1.1.91	Aldoxime methyltransferase Isobutyraldoxime O-methyltransferase	
3.4.22.15	Aldrichina grahami cysteine proteinase cathepsin L	
3.4.24.75	ALE-1 Lysostaphin	
3.4.22.16	Aleurain Cathepsin H	
3.2.1.16	alginase alginase	
4.2.2.11	Alginase II Poly(α-L-guluronate) lyase	

EC-No.	Synonym Recommended Name	Vol., Page
4.2.99.4	alginate lyase alginate lyase	
4.2.2.3	alginate lyase I poly(β-D-mannuronate) lyase	
4.2.2.3	alginate lyase VI poly(β-D-mannuronate) lyase	
2.4.1.33	Alginate synthase Alginate synthase	
4.2.2.3	AlgL poly(β-D-mannuronate) lyase	
3.1.1.1	Ali-esterase Carboxylesterase	
3.1.1.1	Aliesterase Carboxylesterase	
1.1.1.1	Aliphatic alcohol dehydrogenase Alcohol dehydrogenase	
3.5.5.7	Aliphatic nitrilase Aliphatic nitrilase	
3.5.5.7	Aliphatic nitrile aminohydrolase Aliphatic nitrilase	
4.2.1.84	aliphatic nitrile hydratase nitrile hydratase	
2.4.1.103	Alizarin 2-β-glucosyltransferase Alizarin 2-β-glucosyltransferase	
3.4.21.62	ALK-enzyme Subtilisin	
3.2.2.21	AlkA DNA-3-methyladenine glycosidase II	
3.2.1.4	Alkali cellulase Cellulase	
3.1.21.1	Alkaline deoxyribonuclease Deoxyribonuclease I	
3.1.21.1	Alkaline DNase Deoxyribonuclease I	
3.2.1.26	Alkaline invertase β-fructofuranosidase	

EC-No.	Synonym Recommended Name	Vol., Page
3.1.3.1	alkaline phenyl phosphatase alkaline phosphatase	
3.1.3.1	alkaline phosphatase alkaline phosphatase	
3.1.3.1	alkaline phosphohydrolase alkaline phosphatase	
3.1.3.1	alkaline phosphomonoesterase alkaline phosphatase	
3.1.27.5	Alkaline ribonuclease Pancreatic ribonuclease	
1.1.99.20	Alkan-1-ol:(acceptor) oxidoreductase Alkan-1-ol dehydrogenase (acceptor)	
1.1.99.20	Alkan-1-ol dehydrogenase (acceptor) Alkan-1-ol dehydrogenase (acceptor)	
1.14.14.3	Alkanal,reduced-FMN: oxygen oxidoreductase (1-hydroxylating, luminescing) Alkanal monooxygenase (FMN-linked)	
1.14.14.3	Alkanal monooxygenase (FMN-linked) Alkanal monooxygenase (FMN-linked)	
1.14.15.3	Alkane,reduced-rubredoxin:oxygen 1-oxidoreductase Alkane 1-monooxygenase	
1.14.15.3	Alkane 1-hydroxylase Alkane 1-monooxygenase	
1.14.15.3	Alkane 1-monooxygenase Alkane 1-monooxygenase	
1.14.15.3	Alkane hydroxylase Alkane 1-monooxygenase	
1.14.15.3	Alkane monooxygenase Alkane 1-monooxygenase	
1.3.1.27	2-Alkenal reductase 2-Hexadecenal reductase	
3.3.2.2	1-(1-Alkenyl)-sn-glycero-3-phosphocholine aldehydohydrolase Alkenylglycerophosphocholine hydrolase	
3.3.2.5	1-(1-Alkenyl)-sn-glycero-3-phosphoethanolamine aldehydohydrolase Alkenylglycerophosphoethanolamine hydrolase	

EC-No.	Synonym Recommended Name	Vol., Page
2.7.8.22	1-Alkenyl-2-acylglycerol cholinephosphotransferase 1-Alkenyl-2-acylglycerol cholinephosphotransferase	
2.3.1.25	1-Alkenyl-glycero-3-phosphorylcholine:acyl-CoA acyltransferase Plasmalogen synthase	
2.3.1.25	O-1-Alkenylglycero-3-phosphorylcholine acyltransferase Plasmalogen synthase	
3.3.2.2	Alkenylglycerophosphocholine hydrolase Alkenylglycerophosphocholine hydrolase	
2.3.1.104	1-Alkenylglycerophosphocholine O-acyltransferase 1-Alkenylglycerophosphocholine O-acyltransferase	
3.3.2.5	Alkenylglycerophosphoethanolamine hydrolase Alkenylglycerophosphoethanolamine hydrolase	
2.3.1.121	1-Alkenylglycerophosphoethanolamine O-acyltransferase 1-Alkenylglycerophosphoethanolamine O-acyltransferase	
4.4.1.4	S-alkyl(en)yl-L-cysteine lyase alliin lyase	
2.7.8.2	1-Alkyl-2-acetyl-*m*-glycerol:CDPcholine choline phosphotransferase Diacylglycerol cholinephosphotransferase	
3.1.3.59	1-Alkyl-2-acetyl-sn-glycero-3-phosphate phosphohydrolase Alkylacetylglycerophosphatase	
3.1.1.47	1-alkyl-2-acetyl-sn-glycero-3-phosphocholine: acetylhydrolase 1-alkyl-2-acetylglycerophosphocholine esterase	
3.1.1.47	1-alkyl-2-acetyl-sn-glycero-3-phosphocholine acetohydrolase 1-alkyl-2-acetylglycerophosphocholine esterase	
2.7.8.2	1-Alkyl-2-acetyl-sn-glycerol cholinephosphotransferase Diacylglycerol cholinephosphotransferase	
2.7.8.2	1-Alkyl-2-acetylglycerol cholinephosphotransferase Diacylglycerol cholinephosphotransferase	
2.7.8.16	1-alkyl-2-acetylglycerol choline phosphotransferase 1-alkyl-2-acetylglycerol choline phosphotransferase	
2.3.1.125	1-Alkyl-2-acetylglycerol O-acyltransferase 1-Alkyl-2-acetylglycerol O-acyltransferase	
3.1.1.47	1-alkyl-2-acetylglycerophosphocholine esterase 1-alkyl-2-acetylglycerophosphocholine esterase	
3.1.1.47	1-alkyl-2-acetyllecithin deacetylase 1-alkyl-2-acetylglycerophosphocholine esterase	

EC-No.	Synonym Recommended Name	Vol., Page
2.3.1.149	1-Alkyl-2-acyl-sn-glycero-3-phosphocholine:1-organyl-2-lyso-sn-glycero-3-phospholipid acetyltransferase Platelet-activating factor acetyltransferase	
1.14.99.19	O-1-Alkyl-2-acyl-sn-glycero-3-phosphoethanolamine,hydrogen–donor:oxygen oxidoreductase Plasmanylethanolamine desaturase	
1.14.99.19	1-O-Alkyl-2-acyl-sn-glycero-3-phosphorylethanolamine desaturase Plasmanylethanolamine desaturase	
2.3.1.67	1-Alkyl-2-lyso-sn-glycero-3-phosphocholine acetyltransferase 1-Alkylglycerophosphocholine O-acetyltransferase	
2.5.1.26	Alkyl-DHAP Alkylglycerone-phosphate synthase	
3.8.1.1	Alkyl-halide halidohydrolase Alkylhalidase	
4.4.1.6	S-alkyl-L-cysteinase S-alkylcysteine lyase	
4.4.1.6	S-alkyl-L-cysteine alkylthiol-lyase (deaminating) S-alkylcysteine lyase	
4.4.1.6	S-alkyl-L-cysteine lyase S-alkylcysteine lyase	
4.4.1.6	S-alkyl-L-cysteine sulfoxide lyase S-alkylcysteine lyase	
1.5.3.4	ε-Alkyl-L-lysine:oxygen oxidoreductase N^6-Methyl-lysine oxidase	
3.1.4.39	1-Alkyl-sn-glycero-3-phosphoethanolamine ethanolaminehydrolase Alkylglycerophosphoethanolamine phosphodiesterase	
1.14.16.5	1-Alkyl-sn-glycerol,tetrahydrobiopterin:oxygen oxidoreductase Glyceryl-ether monooxygenase	
1.14.99.19	1-O-Alkyl 2-acyl-sn-glycero-3-phosphorylethanolamine desaturase Plasmanylethanolamine desaturase	
3.1.1.47	alkylacetyl-GPC:acetylhydrolase 1-alkyl-2-acetylglycerophosphocholine esterase	
3.1.1.71	Alkylacetylglycerol acetylhydrolase Acetylalkylglycerol acetylhydrolase	
3.1.3.59	Alkylacetylglycerophosphatase Alkylacetylglycerophosphatase	

EC-No.	Synonym Recommended Name	Vol., Page
1.14.99.19	Alkylacylglycero-phosphorylethanolamine dehydrogenase Plasmanylethanolamine desaturase	
2.7.8.2	Alkylacylglycerol choline phosphotransferase Diacylglycerol cholinephosphotransferase	
2.7.8.2	Alkylacylglycerol cholinephosphotransferase Diacylglycerol cholinephosphotransferase	
1.14.99.19	Alkylacylglycerophosphoethanolamine desaturase Plasmanylethanolamine desaturase	
3.5.1.39	Alkylamidase Alkylamidase	
3.2.2.20	Alkylated-DNA glycohydrolase (releasing methyladenine and methylguanine) DNA-3-methyladenine glycosidase I	
3.2.2.21	Alkylated-DNA glycohydrolase (releasing methyladenine and methylguanine) DNA-3-methyladenine glycosidase II	
1.13.11.25	3-Alkylcatechol 2,3-dioxygenase 3,4-Dihydroxy-9,10-secoandrosta-1,3,5(10)-triene-9,17-dione 4,5-dioxygenase	
4.4.1.6	S-alkylcysteinase S-alkylcysteine lyase	
4.4.1.6	S-alkylcysteine α,β-lyase S-alkylcysteine lyase	
4.4.1.6	S-alkylcysteine lyase S-alkylcysteine lyase	
4.4.1.6	alkyl cysteine lyase S-alkylcysteine lyase	
4.4.1.6	alkylcysteine lyase S-alkylcysteine lyase	
4.4.1.4	S-alkylcysteine sulfoxide lyase alliin lyase	
4.4.1.4	alkylcysteine sulfoxide lyase alliin lyase	
2.5.1.26	Alkyl DHAP synthetase Alkylglycerone-phosphate synthase	
2.7.1.84	Alkyldihydroxyacetone kinase Alkylglycerone kinase	

EC-No.	Synonym Recommended Name	Vol., Page
2.5.1.26	Alkyl dihydroxyacetone phosphate synthase Alkylglycerone-phosphate synthase	
2.5.1.26	Alkyldihydroxyacetone phosphate synthase Alkylglycerone-phosphate synthase	
2.5.1.26	Alkyldihydroxyacetonephosphate synthase Alkylglycerone-phosphate synthase	
2.5.1.26	Alkyl dihydroxyacetone phosphate synthetase Alkylglycerone-phosphate synthase	
2.7.1.93	Alkylglycerol kinase Alkylglycerol kinase	
1.14.16.5	Alkylglycerol monooxygenase Glyceryl-ether monooxygenase	
1.14.16.5	O-Alkylglycerol monooxygenase Glyceryl-ether monooxygenase	
2.7.1.93	Alkylglycerol phosphotransferase Alkylglycerol kinase	
2.5.1.26	Alkylglycerone-phosphate synthase Alkylglycerone-phosphate synthase	
2.7.1.84	Alkylglycerone kinase Alkylglycerone kinase	
2.5.1.26	Alkylglycerone phosphate synthase Alkylglycerone-phosphate synthase	
2.3.1.105	Alkylglycerophosphate 2-O-acetyltransferase Alkylglycerophosphate 2-O-acetyltransferase	
2.3.1.67	1-Alkylglycerophosphocholine O-acetyltransferase 1-Alkylglycerophosphocholine O-acetyltransferase	
2.3.1.63	1-Alkylglycerophosphocholine O-acyltransferase 1-Alkylglycerophosphocholine O-acyltransferase	
3.1.4.39	Alkylglycerophosphoethanolamine phosphodiesterase Alkylglycerophosphoethanolamine phosphodiesterase	
3.8.1.1	Alkylhalidase Alkylhalidase	
1.5.3.4	ε-Alkyllysinase N^6-Methyl-lysine oxidase	
2.3.1.105	Alkyllyso-GP:acetyl-CoA acetyltransferase Alkylglycerophosphate 2-O-acetyltransferase	

EC-No.	Synonym Recommended Name	Vol., Page
4.99.1.2	alkylmercury lyase alkylmercury lyase	
4.99.1.2	alkylmercury mercuric-lyase alkylmercury lyase	
1.1.1.165	2-Alkyn-1-ol dehydrogenase 2-Alkyn-1-ol dehydrogenase	
5.2.1.7	all-trans→11cis-Retinoid isomerase Retinol isomerase	Vol. 1, Page 215
2.5.1.30	All-trans-Heptaprenyl-diphosphate synthase trans-Hexaprenyltranstransferase	
2.5.1.30	All-trans-Hexaprenyl-diphosphate:isopentenyl-diphosphate hexaprenyl-transtransferase trans-Hexaprenyltranstransferase	
2.5.1.33	All-trans-Hexaprenyl-diphosphate synthase trans-Pentaprenyltranstransferase	
2.5.1.11	All-trans-Nonaprenyl-diphosphate synthase trans-Octaprenyltranstransferase	
2.5.1.33	All-trans-Pentaprenyl-diphosphate:isopentenyl-diphosphate pentaprenyl-transtransferase trans-Pentaprenyltranstransferase	
5.2.1.3	all-trans-Retinal 11-cis-trans-isomerase Retinal isomerase	Vol. 1, Page 202
5.2.1.7	all-trans-Retinol 11-cis-trans-isomerase Retinol isomerase	Vol. 1, Page 215
5.2.1.3	all-trans-Retinol isomerase Retinal isomerase	Vol. 1, Page 202
5.2.1.7	all-trans-Retinol isomerase Retinol isomerase	Vol. 1, Page 215
3.1.1.64	all-trans-Retinyl-palmitate acylhydrolase all-trans-Retinyl-palmitate hydrolase	
3.1.1.64	all-trans-Retinyl-palmitate hydrolase all-trans-Retinyl-palmitate hydrolase	
1.1.1.105	All-trans retinol dehydrogenase Retinol dehydrogenase	
3.5.3.4	Allantoate amidinohydrolase Allantoicase	

EC-No.	Synonym Recommended Name	Vol., Page
3.5.3.9	Allantoate amidinohydrolase (decarboxylating) Allantoate deiminase	
3.5.3.9	Allantoate amidohydrolase Allantoate deiminase	
3.5.3.9	Allantoate deiminase Allantoate deiminase	
3.5.3.4	Allantoicase Allantoicase	
3.5.2.5	Allantoin amidohydrolase Allantoinase	
3.5.2.5	Allantoinase Allantoinase	
5.1.99.3	Allantoin racemase Allantoin racemase	Vol. 1, Page 185
5.1.99.3	Allantoin racemase Allantoin racemase	Vol. 1, Page 185
5.3.99.6	Allene-oxide cyclase Allene-oxide cyclase	Vol. 1, Page 483
5.3.99.6	Allene oxide cyclase Allene-oxide cyclase	Vol. 1, Page 483
4.4.1.4	alliin alkyl-sulfenate-lyase alliin lyase	
4.4.1.4	alliinase alliin lyase	
4.4.1.4	alliin lyase alliin lyase	
2.7.1.55	Allokinase Allose kinase	
2.7.1.55	D-Allokinase Allose kinase	
3.5.1.54	Allophanate hydrolase Allophanate hydrolase	
3.5.1.54	Allophanate lyase Allophanate hydrolase	
2.7.1.55	D-Allose-6-kinase Allose kinase	

EC-No.	Synonym Recommended Name	Vol., Page
2.7.1.55	Allose kinase Allose kinase	
2.1.2.1	Allothreonine aldolase Glycine hydroxymethyltransferase	
4.1.2.6	allothreonine aldolase allothreonine aldolase	
1.1.1.54	Allyl-alcohol:NADP$^+$ oxidoreductase Allyl-alcohol dehydrogenase	
1.1.1.54	Allyl-alcohol dehydrogenase Allyl-alcohol dehydrogenase	
1.1.1.54	Allyl alcohol dehydrogenase Allyl-alcohol dehydrogenase	
2.5.1.20	Allyltransferase, rubber Rubber cis-polyprenylcistransferase	
3.2.1.111	Almond emulsin fucosidase I 1,3-α-L-Fucosidase	
3.1.3.1	ALP alkaline phosphatase	
1.5.1.17	ALPDH Alanopine dehydrogenase	
1.1.1.2	ALR 1 Alcohol dehydrogenase (NADP$^+$)	
1.1.1.21	ALR2 Aldehyde reductase	
1.1.1.184	ALR3 Carbonyl reductase (NADPH)	
4.1.3.18	ALS acetolactate synthase	
4.1.3.18	α-ALS acetolactate synthase	
2.6.1.2	ALT Alanine transaminase	
2.4.1.140	Alternansucrase Alternansucrase	
3.4.21.16	Alternaria serine proteinase Alternaria serine proteinase	

EC-No.	Synonym Recommended Name	Vol., Page
3.4.21.47	alternative-complement-pathway C3/C5convertase alternative-complement-pathway C3/C5 convertase	
3.4.21.47	alternative complement pathway C3(C5) convertase alternative-complement-pathway C3/C5 convertase	
1.1.1.58	D-Altronate:NAD$^+$ 3-oxidoreductase Tagaturonate reductase	
4.2.1.7	altronate dehydratase altronate dehydratase	
1.1.1.58	Altronate dehydrogenase Tagaturonate reductase	
4.2.1.7	D-altronate hydro-lyase altronate dehydratase	
4.2.1.8	Altronate hydrolase Mannonate dehydratase	
1.1.1.58	Altronate oxidoreductase Tagaturonate reductase	
4.2.1.8	Altronic hydro-lyase Mannonate dehydratase	
1.1.1.58	Altronic oxidoreductase Tagaturonate reductase	
3.1.1.3	amano AP triacylglycerol lipase	
3.1.1.3	amano B triacylglycerol lipase	
3.1.1.3	amano CE triacylglycerol lipase	
3.1.1.3	amano CES triacylglycerol lipase	
3.1.1.3	amano P triacylglycerol lipase	
4.1.1.76	AMDASE Arylmalonate decarboxylase	
4.1.1.76	AMDase Arylmalonate decarboxylase	
4.1.1.50	AMDC Adenosylmethionine decarboxylase	

EC-No.	Synonym Recommended Name	Vol., Page
3.4.22.35	Amebapain Histolysain	
3.5.1.4	Amidase Amidase	
3.5.1.3	ω-amidase ω-amidase	
3.5.1.68	Amidase, β-citrylglutamate N-Formylglutamate deformylase	
3.5.1.3	Amidase,.ω. ω-amidase	
3.5.1.67	Amidase, 4-methyleneglutamine 4-Methyleneglutaminase	
3.5.1.60	Amidase, acylethanolamine N-(Long-chain-acyl)ethanolamine deacylase	
3.5.1.72	Amidase, benzoyl-D-arginine aryl- D-Benzoylarginine-4-nitroanilide amidase	
3.5.1.59	Amidase, carbamoylsarcosine N-Carbamoylsarcosine amidase	
3.10.1.2	Amidase, cyclohexylsulfamate sulf- Cyclamate sulfohydrolase	
3.5.1.56	Amidase, dimethylform- N,N-Dimethylformamidase	
3.5.1.78	Amidase, glutathionylspermidine Glutathionylspermidine amidase	
3.5.1.73	Amidase, L-carnitin- Carnitinamidase	
3.5.1.55	Amidase, long-chain acylglutamate Long-chain-fatty-acyl-glutamate deacylase	
3.5.1.82	Amidase, long-chain acylglutamate N-Acyl-D-glutamate deacylase	
3.5.2.14	Amidase, methylhydantoin N-Methylhydantoinase (ATP-hydrolyzing)	
3.5.1.77	Amidase, N-carbamoyl-D-amino acid N-Carbamoyl-D-amino acid hydrolase	
3.5.1.53	Amidase, N-carbamoylputrescine N-Carbamoylputrescine amidase	

EC-No.	Synonym Recommended Name	Vol., Page
3.5.1.79	Amidase, phthalyl Phthalyl amidase	
3.5.3.19	Amidase, ureidoglycolate Ureidoglycolate hydrolase	
3.5.1.50	Amidase, valer- Pentanamidase	
3.5.3.14	N-Amidino-L-aspartate amidinohydrolase Amidinoaspartase	
3.5.3.14	Amidinoaspartase Amidinoaspartase	
3.5.3.14	Amidinoaspartic amidinohydrolase Amidinoaspartase	
2.1.4.1	Amidinotransferase, glycine Glycine amidinotransferase	
2.1.4.2	Amidinotransferase, inosamine phosphate Inosamine-phosphate amidinotransferase	
3.5.1.14	Amido acid deacylase Aminoacylase	
3.5.1.3	ω-amidodicarboxylate amidohydrolase ω-amidase	
3.5.1.4	Amidohydrolase Amidase	
3.5.1.12	Amidohydrolase biotinidase Biotinidase	
2.4.2.14	Amidophosphoribosyltransferase Amidophosphoribosyltransferase	
2.4.2.14	Amidotransferase, phosphoribosyl pyrophosphate Amidophosphoribosyltransferase	
2.6.1.29	Amine-ketoacid transaminase Diamine transaminase	
1.4.99.3	Amine: oxidoreductase (acceptor deaminating) Amine dehydrogenase	
1.4.3.6	Amine:oxygen oxidoreductase (deaminating, copper-containing) Amine oxidase (copper-containing)	
1.4.3.4	Amine:oxygen oxidoreductase (deaminating, flavin-containing) Amine oxidase (flavin-containing)	

EC-No.	Synonym Recommended Name	Vol., Page
1.4.99.3	Amine dehydrogenase Amine dehydrogenase	
2.1.1.49	Amine N-methyltransferase Amine N-methyltransferase	
2.8.2.3	Amine N-sulfotransferase Amine sulfotransferase	
1.4.3.6	Amine oxidase (copper-containing) Amine oxidase (copper-containing)	
1.4.3.4	Amine oxidase (flavin-containing) Amine oxidase (flavin-containing)	
1.4.3.6	Amine oxidase (pyridoxal containing) Amine oxidase (copper-containing)	
1.4.3.6	Amine oxygen oxidoreductase Amine oxidase (copper-containing)	
2.8.2.3	Amine sulfotransferase Amine sulfotransferase	
2.6.1.29	Amine transaminase Diamine transaminase	
5.1.1.15	α-Amino-ε-caprolactam racemase 2-Aminohexano-6-lactam racemase	Vol. 1, Page 61
5.1.1.15	α-Amino-ε-caprolactam racemase 2-Aminohexano-6-lactam racemase	Vol. 1, Page 61
4.1.1.45	α-Amino-β-carboxymuconate-ε-semialdehade decarboxylase Aminocarboxymuconate-semialdehyde decarboxylase	
4.1.1.45	α-Amino-β-carboxymuconate-ε-semialdehyde β-decarboxylase Aminocarboxymuconate-semialdehyde decarboxylase	
5.1.1.15	α-Amino-δ-valerolactam racemase 2-Aminohexano-6-lactam racemase	Vol. 1, Page 61
2.1.2.3	5-Amino-1-ribosyl-4-imidazolecarboxamide 5'-phosphate transformylase Phosphoribosylaminoimidazolecarboxamide formyltransferase	
6.3.2.6	N-(5-Amino-1-ribosyl-4-imidazolylcarbonyl)-L-aspartic acid 5′-phosphate synthetase Phosphoribosylaminoimidazolesuccinocarboxamide synthase	Vol. 2, Page 434
4.1.1.21	5-Amino-1-ribosylimidazole 5-phosphate carboxylase Phosphoribosylaminoimidazole carboxylase	

EC-No.	Synonym Recommended Name	Vol., Page
4.2.1.26	2-amino-2-deoxy-D-gluconate hydro-lyase (deaminating) aminodeoxygluconate dehydratase	
3.5.99.6	2-amino-2-deoxy-D-glucose-6-phosphate aminohydrolase (ketol isomerizing) glucosamine-6-phosphate deaminase	
2.6.1.16	2-Amino-2-deoxy-D-glucose-6-phosphate ketol-isomerase Glutamine-fructose-6-phosphate transaminase (isomerizing)	
5.3.1.10	2-Amino-2-deoxy-D-glucose-6-phosphate ketol isomerase (deaminating) Glucosamine-6-phosphate isomerase	Vol. 1, Page 312
3.5.4.20	1-(4-Amino-2-methylpyrimid-5-yl-methyl)-3-(β-hydroxy-ethyl)-2-methylpyrimidinium-bromide aminohydrolase Pyrithiamin deaminase	
1.1.1.4	D-1-Amino-2-propanol:NAD^+ oxidoreductase (R,R)-Butanediol dehydrogenase	
1.1.1.4	1-Amino-2-propanol dehydrogenase (R,R)-Butanediol dehydrogenase	
1.1.1.4	D-1-Amino-2-propanol dehydrogenase (R,R)-Butanediol dehydrogenase	
1.1.1.4	1-Amino-2-propanol oxidoreductase (R,R)-Butanediol dehydrogenase	
2.3.1.29	2-Amino-3-ketobutyrate-CoA ligase Glycine C-acetyltransferase	
2.3.1.29	2-Amino-3-ketobutyrate CoA ligase Glycine C-acetyltransferase	
2.3.1.29	2-Amino-3-ketobutyrate coenzyme A ligase Glycine C-acetyltransferase	
1.13.11.36	5-Amino-4-chloro-2-(2,3-dihydroxyphenyl)-3(2H)-pyridazinone 1,2-oxidoreductase (decyclizing) Chloridazon-catechol dioxygenase	
4.5.1.4	L-2-Amino-4-chloropent-4-enoate chloride-lyase (deaminating) L-2-Amino-4-chloropent-4-enoate dehydrochlorinase	
4.5.1.4	L-2-Amino-4-chloropent-4-enoate dehydrochlorinase L-2-Amino-4-chloropent-4-enoate dehydrochlorinase	
5.4.99.5	4-Amino-4-deoxychorismate mutase Chorismate mutase	Vol. 1, Page 604

EC-No.	Synonym Recommended Name	Vol., Page
4.1.2.25	2-Amino-4-hydroxy-6-(D-erythro-1,2,3-trihydroxypropyl)-7,8-dihydropteridine glycolaldehyde-lyase Dihydroneopterin aldolase	
2.5.1.15	2-Amino-4-hydroxy-6-hydroxymethyl-7,8-dihydropteridine-diphosphate:4-aminobenzoate 2-amino-4-hydroxydihydropteridine-6-methenyltransferase Dihydropteroate synthase	
2.7.6.3	2-Amino-4-hydroxy-6-hydroxymethyldihydropteridine pyrophosphokinase 2-Amino-4-hydroxy-6-hydroxymethyldihydropteridine pyrophosphokinase	
1.17.3.1	2-Amino-4-hydroxypteridine:oxygen oxidoreductase (7-hydroxylating) Pteridine oxidase	
3.5.4.11	2-Amino-4-hydroxypteridine aminohydrolase Pteridin deaminase	
2.1.2.3	5-Amino-4-imidazolecarboxamide ribonucleotide transformylase Phosphoribosylaminoimidazolecarboxamide formyltransferase	
2.1.2.3	5-Amino-4-imidazolecarboxamide ribotide transformylase Phosphoribosylaminoimidazolecarboxamide formyltransferase	
4.2.3.12	2-amino-4-oxo-6-(erythro-1',2',3'-trihydroxypropyl)-7,8-dihydroxypterdine triphosphate lyase 6-pyruvoyltetrahydropterin synthase	
4.6.1.10	2-Amino-4-oxo-6-(erythro-1',2',3'-trihydroxypropyl)-7,8-dihydroxypteridine triphosphate triphosphate lyase 6-Pyruvoyltetrahydtropterin synthase	
1.1.1.193	5-Amino-6-(5-phosphoribitylamino)uracil:NADP$^+$ 1'-oxidoreductase 5-Amino-6-(5-phosphoribosylamino)uracil reductase	
1.1.1.193	5-Amino-6-(5-phosphoribosylamino)uracil reductase 5-Amino-6-(5-phosphoribosylamino)uracil reductase	
2.3.1.47	8-Amino-7-oxononanoate synthase 8-Amino-7-oxononanoate synthase	
2.3.1.47	8-Amino-7-oxopelargonate synthase 8-Amino-7-oxononanoate synthase	
3.1.1.43	α-amino-acid-ester aminoacylhydrolase α-amino-acid esterase	
1.4.99.1	D-Amino-acid:(acceptor) oxidoreductase (deaminating) D-Amino acid dehydrogenase	
1.4.1.5	L-Amino-acid:NAD$^+$ oxidoreductase (deaminating) L-Amino-acid dehydrogenase	

EC-No.	Synonym Recommended Name	Vol., Page
1.4.3.3	D-Amino-acid:oxygen oxidoreductase (deaminating) D-Amino-acid oxidase	
1.4.3.2	L-Amino-acid:oxygen oxidoreductase (deaminating) L-Amino-acid oxidase	
3.5.1.14	L-Amino-acid acylase Aminoacylase	
1.4.1.5	L-Amino-acid dehydrogenase L-Amino-acid dehydrogenase	
3.1.1.43	α-amino-acid esterase α-amino-acid esterase	
2.3.1.1	Amino-acid N-acetyltransferase Amino-acid N-acetyltransferase	
2.3.1.36	D-Amino-acid N-acetyltransferase D-Amino-acid N-acetyltransferase	
1.4.3.3	D-Amino-acid oxidase D-Amino-acid oxidase	
1.4.3.2	L-Amino-acid oxidase L-Amino-acid oxidase	
5.1.1.10	Amino-acid racemase Amino-acid racemase	Vol. 1, Page 41
5.1.1.10	Amino-acid racemase Amino-acid racemase	Vol. 1, Page 41
3.4.13.18	L-Amino-acyl-L-amino-acid hydrolase Cytosol nonspecific dipeptidase	
4.1.1.45	Amino-carboxymuconate-semialdehyde decarboxylase Aminocarboxymuconate-semialdehyde decarboxylase	
6.3.4.13	2-Amino-N-ribosylacetamide 5'-phosphate kinosynthase Phosphoribosylamine-glycine ligase	Vol. 2, Page 626
2.1.2.2	2-Amino-N-ribosylacetamide 5'-phosphate transformylase Phosphoribosylglycinamide formyltransferase	
3.4.11.2	amino-oligopeptidase membrane alanyl aminopeptidase	
2.3.1.88	Amino-terminal amino acid-acetylating enzyme Peptide α-N-acetyltransferase	
2.3.1.36	D-Amino acid-α-N-acetyltransferase D-Amino-acid N-acetyltransferase	

EC-No.	Synonym Recommended Name	Vol., Page
1.4.3.2	L-Amino acid:O_2 oxidoreductase L-Amino-acid oxidase	
2.3.1.1	Amino acid acetyltransferase Amino-acid N-acetyltransferase	
2.3.1.36	D-Amino acid acetyltransferase D-Amino-acid N-acetyltransferase	
1.4.99.1	D-Amino acid dehydrogenase D-Amino acid dehydrogenase	
3.1.1.43	α-amino acid ester hydrolase α-amino-acid esterase	
3.1.1.43	α-amino acid ester hydrolase α-amino-acid esterase	
2.3.1.65	Amino acid N-choloyltransferase Glycine N-choloyltransferase	
1.4.3.3	D-Aminoacid oxidase D-Amino-acid oxidase	
5.1.1.10	L-Amino acid racemase Amino-acid racemase	Vol. 1, Page 41
3.4.11.4	α-aminoacyl-dipeptide hydrolase tripeptide aminopeptidase	
3.4.3.3	aminoacyl-histidine dipeptidase, carnosinase aminoacyl-histidine dipeptidase, carnosinase	
3.4.13.3	Aminoacyl-L-histidine hydrolase Aminoacylhistidine dipeptidase	
3.4.13.4	aminoacyl-L-lysine(-L-arginine) hydrolase X-Arg dipeptidase	
3.4.13.4	aminoacyl-lysine dipeptidase X-Arg dipeptidase	
3.4.13.5	aminoacyl-methylhistidine dipeptidase X-methyl-His dipeptidase	
3.4.13.5	aminoacyl-methylhistidine dipeptidase X-methyl-His dipeptidase	
3.4.3.4	aminoacyl-methylhistidine dipeptidase, anserinase aminoacyl-methylhistidine dipeptidase, anserinase	
3.4.13.5	aminoacyl-Npi-methyl-L-histidine hydrolase X-methyl-His dipeptidase	

EC-No.	Synonym Recommended Name	Vol., Page
3.4.11.16	α-aminoacyl-peptide hydrolase Aminopeptidase X-Trp	
3.4.11.22	α-Aminoacyl-peptide hydrolase Aminopeptidase I	
3.4.11.14	α-aminoacyl-peptide hydrolase aminopeptidase, human liver	
3.4.11.3	α-aminoacyl-peptide hydrolase cystinyl aminopeptidase	
3.4.11.1	α-aminoacyl-peptide hydrolase (cytosol) leucyl aminopeptidase	
3.4.14.5	amino acyl-prolyl dipeptidyl aminopeptidase dipeptidyl-peptidase IV	
3.1.1.29	aminoacyl-tRNA aminoacylhydrolase aminoacyl-tRNA hydrolase	
3.1.1.29	aminoacyl-tRNA hydrolase aminoacyl-tRNA hydrolase	
3.5.1.14	Aminoacylase Aminoacylase	
3.5.1.81	D-Amino acylase N-Acyl-D-amino-acid deacylase	
3.5.1.81	D-Aminoacylase N-Acyl-D-amino-acid deacylase	
3.5.1.14	L-Aminoacylase Aminoacylase	
3.5.1.14	Aminoacylase I Aminoacylase	
3.5.1.15	Aminoacylase II Aspartoacylase	
3.5.1.83	Aminoacylase II N-Acyl-D-aspartate deacylase	
3.4.13.3	Aminoacylhistidine dipeptidase Aminoacylhistidine dipeptidase	
3.4.13.4	aminoacyllysine dipeptidase X-Arg dipeptidase	
3.4.11.9	aminoacylproline aminopeptidase X-Pro aminopeptidase	

EC-No.	Synonym Recommended Name	Vol., Page
3.4.16.2	aminoacylproline carboxypeptidase lysosomal Pro-X carboxypeptidase	
3.4.11.9	aminoacylprolyl-peptide hydrolase X-Pro aminopeptidase	
1.2.1.31	L-2-Aminoadipate-6-semialdehyde:NAD(P)$^+$ 6-oxidoreductase L-Aminoadipate-semialdehyde dehydrogenase	
1.2.1.31	L-α-Aminoadipate δ-semialdehyde:NAD oxidoreductase L-Aminoadipate-semialdehyde dehydrogenase	
1.2.1.31	L-α-Aminoadipate δ-semialdehyde:nicotinamide adenine dinucleotide oxidoreductase L-Aminoadipate-semialdehyde dehydrogenase	
1.2.1.31	L-Aminoadipate-semialdehyde dehydrogenase L-Aminoadipate-semialdehyde dehydrogenase	
1.2.1.31	α-Aminoadipate-semialdehyde dehydrogenase L-Aminoadipate-semialdehyde dehydrogenase	
1.2.1.31	L-α-Aminoadipate δ-semialdehyde oxidoreductase L-Aminoadipate-semialdehyde dehydrogenase	
2.6.1.39	L-2-Aminoadipate:2-oxoglutarate aminotransferase 2-Aminoadipate transaminase	
2.6.1.39	2-Aminoadipate aminotransferase 2-Aminoadipate transaminase	
2.6.1.39	α-Aminoadipate aminotransferase 2-Aminoadipate transaminase	
1.2.1.31	α-Aminoadipate reductase L-Aminoadipate-semialdehyde dehydrogenase	
1.5.1.10	Aminoadipate semialdehyde-glutamate reductase Saccharopine dehydrogenase (NADP$^+$, L-glutamate forming)	
1.2.1.31	2-Aminoadipate semialdehyde dehydrogenase L-Aminoadipate-semialdehyde dehydrogenase	
2.6.1.39	2-Aminoadipate transaminase 2-Aminoadipate transaminase	
2.6.1.39	2-Aminoadipic aminotransferase 2-Aminoadipate transaminase	
1.5.1.10	Aminoadipic semialdehyde-glutamate reductase Saccharopine dehydrogenase (NADP$^+$, L-glutamate forming)	

EC-No.	Synonym Recommended Name	Vol., Page
1.5.1.10	Aminoadipic semialdehyde-glutamic reductase Saccharopine dehydrogenase (NADP$^+$, L-glutamate forming)	
1.2.1.31	2-Aminoadipic semialdehyde dehydrogenase L-Aminoadipate-semialdehyde dehydrogenase	
1.5.1.9	Aminoadipic semialdehyde synthase Saccharopine dehydrogenase (NAD$^+$, L-glutamate-forming)	
4.2.3.2	amino alcohol O-phosphate phospholyase ethanolamine-phosphate phospho-lyase	
4.2.99.7	amino alcohol O-phosphate phospholyase ethanolamine-phosphate phospho-lyase	
1.6.6.7	*p*-Aminoazobenzene reductase Azobenzene reductase	
1.14.12.14	2-Aminobenzenesulfonate, NADH:oxygen oxidoreductase (2,3-hydroxylating, ammonia-forming) 2-Aminobenzenesulfonate 2,3-dioxygenase	
1.14.12.14	2-Aminobenzenesulfonate 2,3-dioxygenase 2-Aminobenzenesulfonate 2,3-dioxygenase	
1.14.12.14	2-Aminobenzenesulfonate dioxygenase 2-Aminobenzenesulfonate 2,3-dioxygenase	
1.14.13.27	4-Aminobenzoate,NAD(P)H:oxygen oxidoreductase (1-hydroxylating, decarboxylating) 4-Aminobenzoate 1-monooxygenase	
6.2.1.32	2-Aminobenzoate-CoA ligase Anthranilate-CoA ligase	Vol. 2, Page 336
6.2.1.32	2-Aminobenzoate-coenzyme A ligase Anthranilate-CoA ligase	Vol. 2, Page 336
1.14.13.27	4-Aminobenzoate 1-monooxygenase 4-Aminobenzoate 1-monooxygenase	
4.1.1.24	Aminobenzoate carboxy-lyase Decarboxylase, aminobenzoate	
6.2.1.32	2-Aminobenzoate coenzyme A ligase Anthranilate-CoA ligase	Vol. 2, Page 336
1.14.13.27	4-Aminobenzoate hydroxylase 4-Aminobenzoate 1-monooxygenase	
1.14.13.40	2-Aminobenzoyl-CoA,NAD(P)H:oxygen oxidoreductase (dearomatizing) Anthraniloyl-CoA monooxygenase	

EC-No.	Synonym Recommended Name	Vol., Page
1.14.13.40	2-Aminobenzoyl-CoA monooxygenase/reductase Anthraniloyl-CoA monooxygenase	
2.3.1.5	4-Aminobiphenyl N-acetyltransferase Arylamine N-acetyltransferase	
1.2.1.19	4-Aminobutanal:NAD$^+$ 1-oxidoreductase Aminobutyraldehyde dehydrogenase	
1.2.1.19	4-Aminobutanal dehydrogenase Aminobutyraldehyde dehydrogenase	
2.6.1.19	4-Aminobutanoate:2-oxoglutarate aminotransferase 4-Aminobutyrate transaminase	
1.2.1.19	γ-Aminobutyraldehyde dehydroganase Aminobutyraldehyde dehydrogenase	
1.2.1.19	4-Aminobutyraldehyde dehydrogenase Aminobutyraldehyde dehydrogenase	
1.2.1.19	Aminobutyraldehyde dehydrogenase Aminobutyraldehyde dehydrogenase	
1.2.1.19	γ-aminobutyraldehyde dehydrogenase Aminobutyraldehyde dehydrogenase	
2.6.1.19	γ-Aminobutyrate-α-ketoglutarate aminotransferase 4-Aminobutyrate transaminase	
2.6.1.19	γ-Aminobutyrate-α-ketoglutarate transaminase 4-Aminobutyrate transaminase	
2.6.1.19	4-Aminobutyrate-2-ketoglutarate aminotransferase 4-Aminobutyrate transaminase	
2.6.1.19	4-Aminobutyrate-2-oxoglutarate aminotransferase 4-Aminobutyrate transaminase	
2.6.1.19	4-Aminobutyrate-2-oxoglutarate transaminase 4-Aminobutyrate transaminase	
2.6.1.19	γ-Aminobutyrate:α-oxoglutarate aminotransferase 4-Aminobutyrate transaminase	
2.6.1.19	γ-Aminobutyrate aminotransaminase 4-Aminobutyrate transaminase	
2.6.1.19	4-Aminobutyrate aminotransferase 4-Aminobutyrate transaminase	
2.6.1.19	Aminobutyrate aminotransferase 4-Aminobutyrate transaminase	

EC-No.	Synonym Recommended Name	Vol., Page
2.6.1.19	4-Aminobutyrate transaminase 4-Aminobutyrate transaminase	
2.6.1.19	Aminobutyrate transaminase 4-Aminobutyrate transaminase	
2.6.1.19	γ-Aminobutyrate transaminase 4-Aminobutyrate transaminase	
2.6.1.19	γ-Aminobutyric acid-α-ketoglutarate transaminase 4-Aminobutyrate transaminase	
2.6.1.19	γ-Aminobutyric acid-α-ketoglutaric acid aminotransferase 4-Aminobutyrate transaminase	
2.6.1.19	γ-Aminobutyric acid-2-oxoglutarate transaminase 4-Aminobutyrate transaminase	
2.6.1.19	4-Aminobutyric acid 2-ketoglutaric acid aminotransferase 4-Aminobutyrate transaminase	
2.6.1.19	4-Aminobutyric acid aminotransferase 4-Aminobutyrate transaminase	
2.6.1.19	γ-Aminobutyric acid aminotransferase 4-Aminobutyrate transaminase	
2.6.1.19	γ-Aminobutyric acid pyruvate transaminase 4-Aminobutyrate transaminase	
2.6.1.19	γ-Aminobutyric acid transaminase 4-Aminobutyrate transaminase	
2.6.1.22	β-Aminobutyric transaminase (S)-3-Amino-2-methylpropionate transaminase	
2.6.1.19	γ-Aminobutyric transaminase 4-Aminobutyrate transaminase	
4.3.1.14	3-Aminobutyryl-CoA ammonia-lyase 3-Aminobutyryl-CoA ammonia-lyase	
4.3.1.14	L-3-Aminobutyryl-CoA ammonia-lyase 3-Aminobutyryl-CoA ammonia-lyase	
4.3.1.14	L-3-Aminobutyryl-CoA deaminase 3-Aminobutyryl-CoA ammonia-lyase	
3.5.1.51	Aminobutyryl-CoA thiolesterase 4-Acetamidobutyryl-CoA deacetylase	
4.3.1.14	L-3-Aminobutyryl coenzyme A deaminase 3-Aminobutyryl-CoA ammonia-lyase	

EC-No.	Synonym Recommended Name	Vol., Page
3.5.2.11	L-α-aminocaprolactam hydrolase L-Lysine-lactamase	
5.1.1.15	Aminocaprolactam racemase 2-Aminohexano-6-lactam racemase	Vol. 1, Page 61
4.1.1.45	Aminocarboxymuconate-semialdehyde decarboxylase Aminocarboxymuconate-semialdehyde decarboxylase	
2.5.1.38	Aminocarboxypropyltransferase, nocardicin Isonocardicin synthase	
4.1.99.4	1-aminocyclopropane-1-carboxylate deaminase 1-aminocyclopropane-1-carboxylate deaminase	
4.1.99.4	1-aminocyclopropane-1-carboxylate endolyase (deaminating) 1-aminocyclopropane-1-carboxylate deaminase	
4.4.1.14	1-aminocyclopropane-1-carboxylate synthase 1-aminocyclopropane-1-carboxylate synthase	
4.4.1.14	aminocyclopropane-1-carboxylate synthase 1-aminocyclopropane-1-carboxylate synthase	
4.4.1.14	1-aminocyclopropane-1-carboxylate synthetase 1-aminocyclopropane-1-carboxylate synthase	
4.1.99.4	1-aminocyclopropane-1-carboxylic acid deaminase 1-aminocyclopropane-1-carboxylate deaminase	
4.4.1.14	1-aminocyclopropane-1-carboxylic acid synthase 1-aminocyclopropane-1-carboxylate synthase	
4.4.1.14	aminocyclopropanecarboxylate synthase 1-aminocyclopropane-1-carboxylate synthase	
4.4.1.14	aminocyclopropanecarboxylic acid synthase 1-aminocyclopropane-1-carboxylate synthase	
4.2.1.26	aminodeoxygluconate dehydratase aminodeoxygluconate dehydratase	
2.7.1.8	Aminodeoxyglucose kinase Glucosamine kinase	
2.3.1.4	Aminodeoxyglucosephosphate acetyltransferase Glucosamine-phosphate N-acetyltransferase	
5.3.1.10	Aminodeoxyglucosephosphate isomerase Glucosamine-6-phosphate isomerase	Vol. 1, Page 312
3.5.99.6	aminodeoxyglucosephosphate isomerase glucosamine-6-phosphate deaminase	

EC-No.	Synonym Recommended Name	Vol., Page
1.1.1.193	Aminodioxyphosphoribosylaminopyrimidine reductase 5-Amino-6-(5-phosphoribosylamino)uracil reductase	
3.4.13.19	Aminodipeptidase Membrane dipeptidase	
1.14.11.17	2-aminoethanesulfonate dioxygenase taurine dioxygenase	
2.6.1.37	(2-Aminoethyl)phosphonate-pyruvate transaminase (2-Aminoethyl)phosphonate-pyruvate transaminase	
2.6.1.37	(2-Aminoethyl)phosphonate:pyruvate aminotransferase (2-Aminoethyl)phosphonate-pyruvate transaminase	
2.6.1.37	(2-Aminoethyl)phosphonate aminotransferase (2-Aminoethyl)phosphonate-pyruvate transaminase	
2.6.1.37	(2-Aminoethyl)phosphonate transaminase (2-Aminoethyl)phosphonate-pyruvate transaminase	
2.6.1.37	(2-Aminoethyl)phosphonic acid aminotransferase (2-Aminoethyl)phosphonate-pyruvate transaminase	
1.14.17.1	4-(2-Aminoethyl)pyrocatechol β-oxidase Dopamine β-monooxygenase	
2.6.1.37	2-Aminoethylphosphonate-pyruvate aminotransferase (2-Aminoethyl)phosphonate-pyruvate transaminase	
2.6.1.37	2-Aminoethylphosphonate aminotransferase (2-Aminoethyl)phosphonate-pyruvate transaminase	
3.4.11.4	aminoexotripeptidase tripeptide aminopeptidase	
2.3.1.82	Aminoglycoside-6'-acetyltransferase Aminoglycoside $N^{6'}$-acetyltransferase	
2.3.1.82	Aminoglycoside-6-N-acetyltransferase Aminoglycoside $N^{6'}$-acetyltransferase	
2.7.7.47	Aminoglycoside 3-adenylyltransferase Streptomycin 3-adenylyltransferase	
2.3.1.82	Aminoglycoside 6'-N-acetyltransferase Aminoglycoside $N^{6'}$-acetyltransferase	
2.3.1.81	3'-Aminoglycoside acetyltransferase Aminoglycoside $N^{3'}$-acetyltransferase	
2.3.1.81	3-N-Aminoglycoside acetyltransferase Aminoglycoside $N^{3'}$-acetyltransferase	

EC-No.	Synonym Recommended Name	Vol., Page
2.3.1.60	Aminoglycoside acetyltransferase AAC(3)-1 Gentamicin 3'-N-acetyltransferase	
2.7.7.46	Aminoglycoside adenylyltransferase Gentamicin 2-nucleotidyltransferase	
2.3.1.81	Aminoglycoside $N^{3'}$-acetyltransferase Aminoglycoside $N^{3'}$-acetyltransferase	
2.3.1.82	Aminoglycoside $N^{6'}$-acetyltransferase Aminoglycoside $N^{6'}$-acetyltransferase	
5.1.1.15	2-Aminohexano-6-lactam racemase 2-Aminohexano-6-lactam racemase	Vol. 1, Page 61
5.1.1.15	2-Aminohexano-6-lactamracemase 2-Aminohexano-6-lactam racemase	Vol. 1, Page 61
3.5.2.12	6-Aminohexanoate-cyclic-dimer hydrolase 6-Aminohexanoate-cyclic-dimer hydrolase	
3.5.1.46	6-Aminohexanoate-dimer hydrolase 6-Aminohexanoate-dimer hydrolase	
2.6.1.67	L-2-Aminohexanoate:2-oxoglutarate aminotransferase 2-Aminohexanoate transaminase	
2.6.1.67	2-Aminohexanoate transaminase 2-Aminohexanoate transaminase	
3.5.1.46	6-Aminohexanoic acid oligomer hydrolase 6-Aminohexanoate-dimer hydrolase	
3.5.1.46	N-(6-Aminohexanoyl)-6-aminohexanoate amidohydrolase 6-Aminohexanoate-dimer hydrolase	
3.5.4.8	Aminoimidazolase Aminoimidazolase	
6.3.2.6	5-Aminoimidazole-4-N-succinocarboxamide ribonucleotide synthetase Phosphoribosylaminoimidazolesuccinocarboxamide synthase	Vol. 2, Page 434
3.5.4.8	4-Aminoimidazole aminohydrolase Aminoimidazolase	
2.1.2.3	Aminoimidazolecarboxamide ribonucleotide transformylase Phosphoribosylaminoimidazolecarboxamide formyltransferase	
3.5.4.8	4-Aminoimidazole hydrolase Aminoimidazolase	

EC-No.	Synonym Recommended Name	Vol., Page
6.3.3.1	5'-Aminoimidazole ribonucleotide synthetase Phosphoribosylformylglycinamidine cyclo-ligase	Vol. 2, Page 530
2.6.1.40	D-3-Aminoisobutyrate-pyruvate aminotransferase (R)-3-Amino-2-methylpropionate-pyruvate transaminase	
2.6.1.40	β-Aminoisobutyrate-pyruvate aminotransferase (R)-3-Amino-2-methylpropionate-pyruvate transaminase	
2.6.1.40	D-3-Aminoisobutyrate-pyruvate transaminase (R)-3-Amino-2-methylpropionate-pyruvate transaminase	
2.6.1.22	L-3-Aminoisobutyrate transaminase (S)-3-Amino-2-methylpropionate transaminase	
2.6.1.22	L-3-Aminoisobutyric aminotransferase (S)-3-Amino-2-methylpropionate transaminase	
2.6.1.43	5-Aminolevulinate:pyruvate aminotransferase Aminolevulinate transaminase	
4.2.1.24	5-aminolevulinate dehydrase porphobilinogen synthase	
4.2.1.24	aminolevulinate dehydrase porphobilinogen synthase	
4.2.1.24	δ-aminolevulinate dehydrase porphobilinogen synthase	
4.2.1.24	5-aminolevulinate dehydratase porphobilinogen synthase	
4.2.1.24	aminolevulinate dehydratase porphobilinogen synthase	
4.2.1.24	δ-aminolevulinate dehydratase porphobilinogen synthase	
4.2.1.24	5-aminolevulinate hydro-lyase (adding 5-aminolevulinate and cyclizing) porphobilinogen synthase	
4.2.1.24	5-aminolevulinate hydro-lyase (adding 5-aminolevulinate and cyclizing) porphobilinogen synthase	
2.3.1.37	5-Aminolevulinate synthase 5-Aminolevulinate synthase	
2.3.1.37	Aminolevulinate synthase 5-Aminolevulinate synthase	
2.3.1.37	δ-Aminolevulinate synthase 5-Aminolevulinate synthase	

EC-No.	Synonym Recommended Name	Vol., Page
2.3.1.37	5-Aminolevulinate synthetase 5-Aminolevulinate synthase	
2.3.1.37	Aminolevulinate synthetase 5-Aminolevulinate synthase	
2.3.1.37	δ-Aminolevulinate synthetase 5-Aminolevulinate synthase	
2.6.1.43	Aminolevulinate transaminase Aminolevulinate transaminase	
4.2.1.24	5-aminolevulinic acid dehydrase porphobilinogen synthase	
4.2.1.24	δ-aminolevulinic acid dehydrase porphobilinogen synthase	
4.2.1.24	5-aminolevulinic acid dehydratase porphobilinogen synthase	
4.2.1.24	δ-aminolevulinic acid dehydratase porphobilinogen synthase	
4.2.1.24	γ-aminolevulinic acid dehydratase porphobilinogen synthase	
2.3.1.37	Aminolevulinic acid synthase 5-Aminolevulinate synthase	
2.3.1.37	δ-Aminolevulinic acid synthase 5-Aminolevulinate synthase	
2.3.1.37	α-Aminolevulinic acid synthase 5-Aminolevulinate synthase	
2.3.1.37	5-Aminolevulinic acid synthetase 5-Aminolevulinate synthase	
2.3.1.37	Aminolevulinic acid synthetase 5-Aminolevulinate synthase	
2.3.1.37	δ-Aminolevulinic acid synthetase 5-Aminolevulinate synthase	
2.6.1.43	5-Aminolevulinic acid transaminase Aminolevulinate transaminase	
2.6.1.43	Aminolevulinic acid transaminase Aminolevulinate transaminase	
4.2.1.24	aminolevulinic dehydratase porphobilinogen synthase	

EC-No.	Synonym Recommended Name	Vol., Page
4.2.1.24	δ-aminolevulinic dehydratase porphobilinogen synthase	
2.3.1.37	Aminolevulinic synthetase 5-Aminolevulinate synthase	
2.3.1.37	δ-Aminolevulinic synthetase 5-Aminolevulinate synthase	
4.1.1.10	aminomalonate decarboxylase aminomalonate decarboxylase	
4.1.1.12	Aminomalonic decarboxylase Aspartate 4-decarboxylase	
2.1.2.10	S-Aminomethyldihydrolipoylprotein:(6S)-tetrahydrofolate aminomethyl-transferase (ammonia-forming) Aminomethyltransferase	
2.1.2.10	Aminomethyltransferase Aminomethyltransferase	
1.2.1.32	2-Aminomuconate-6-semialdehyde:NAD^+ oxidoreductase Aminomuconate-semialdehyde dehydrogenase	
1.2.1.32	Aminomuconate-semialdehyde dehydrogenase Aminomuconate-semialdehyde dehydrogenase	
3.5.99.5	2-aminomuconate aminohydrolase 2-aminomuconate deaminase	
3.5.99.5	2-aminomuconate deaminase 2-aminomuconate deaminase	
1.2.1.32	Aminomuconate semialdehyde dehydrogenase Aminomuconate-semialdehyde dehydrogenase	
1.2.1.32	α-Aminomuconic ε-semialdehyde dehydrogenase Aminomuconate-semialdehyde dehydrogenase	
5.4.3.3	Aminomutase, β-lysine 5,6- β-Lysine 5,6-aminomutase	Vol. 1, Page 558
5.4.3.4	Aminomutase, D-lysine 5,6- D-Lysine 5,6-aminomutase	Vol. 1, Page 562
5.4.3.5	Aminomutase, D-ornithine 4,5- D-Ornithine 4,5-aminomutase	Vol. 1, Page 565
5.4.3.7	Aminomutase, leucine 2,3- Leucine 2,3-aminomutase	Vol. 1, Page 571

EC-No.	Synonym Recommended Name	Vol., Page
5.4.3.2	Aminomutase, lysine 2,3- Lysine 2,3-aminomutase	Vol. 1, Page 553
5.4.3.6	Aminomutase, tyrosine 2,3- Tyrosine 2,3-aminomutase	Vol. 1, Page 568
3.5.1.30	5-Aminonorvaleramidase 5-Aminopentanamidase	
3.4.11.22	Aminooligopeptidase Aminopeptidase I	
3.5.1.30	5-Aminopentanamidase 5-Aminopentanamidase	
3.5.1.30	5-Aminopentanamide amidohydrolase 5-Aminopentanamidase	
2.6.1.48	5-Aminopentanoate:2-oxoglutarate aminotransferase 5-Aminovalerate transaminase	
1.4.4.1	5-Aminopentanoate:lipoate oxidoreductase (cyclizing) D-Proline reductase (dithiol)	
3.4.11.22	Aminopeptidase Aminopeptidase I	
3.4.11.19	D-Aminopeptidase D-Stereospecific aminopeptidase	
3.4.11.22	L-aminopeptidase Aminopeptidase I	
3.4.1.2	aminopeptidase aminopeptidase	
3.4.11.11	aminopeptidase aminopeptidase	
3.4.11.15	aminopeptidase (cobalt-activated) Aminopeptidase Y	
3.4.11.9	aminopeptidase, aminoacylproline X-Pro aminopeptidase	
3.4.11.6	aminopeptidase, arginine aminopeptidase B	
3.4.11.7	aminopeptidase, aspartate glutamyl aminopeptidase	
3.4.11.3	aminopeptidase, cystyl cystinyl aminopeptidase	

EC-No.	Synonym Recommended Name	Vol., Page
3.4.11.14	aminopeptidase, cytosol alanyl aminopeptidase, human liver	
3.4.19.7	Aminopeptidase, formylmethionine N-Formylmethionyl-peptidase	
3.4.14.5	aminopeptidase, glycylproline dipeptidyl-peptidase IV	
3.4.11.14	aminopeptidase, human liver aminopeptidase, human liver	
3.4.11.18	Aminopeptidase, methionine Methionyl aminopeptidase	
3.4.11.2	aminopeptidase, microsomal membrane alanyl aminopeptidase	
3.4.11.5	aminopeptidase, proline prolyl aminopeptidase	
3.4.19.3	aminopeptidase, pyroglutamate pyroglutamyl-peptidase I	
3.4.19.6	Aminopeptidase, thyrotropin-releasing factor pyroglutamate Pyroglutamyl-peptidase II	
3.4.14.9	Aminopeptidase, tripeptidyl, I Tripeptidyl-peptidase I	
3.4.14.10	Aminopeptidase, tripeptidyl, II Tripeptidyl-peptidase II	
3.5.1.57	Aminopeptidase, tryptophan Tryptophanamidase	
3.4.11.7	aminopeptidase A glutamyl aminopeptidase	
3.4.11.6	aminopeptidase B aminopeptidase B	
3.4.22.40	aminopeptidase C bleomycin hydrolase	
3.4.11.15	aminopeptidase Co Aminopeptidase Y	
3.4.11.20	Aminopeptidase Ey Aminopeptidase Ey	
3.4.22.40	aminopeptidase H bleomycin hydrolase	

EC-No.	Synonym Recommended Name	Vol., Page
3.4.22.40	aminopeptidase H bleomycin hydrolase	
3.4.11.22	Aminopeptidase I Aminopeptidase I	
3.4.11.1	aminopeptidase I leucyl aminopeptidase	
3.4.11.1	aminopeptidase II leucyl aminopeptidase	
3.4.11.22	Aminopeptidase III Aminopeptidase I	
3.4.11.1	aminopeptidase III leucyl aminopeptidase	
3.4.11.2	aminopeptidase M membrane alanyl aminopeptidase	
3.4.11.2	aminopeptidase M II membrane alanyl aminopeptidase	
3.4.11.2	aminopeptidase N membrane alanyl aminopeptidase	
3.4.11.9	aminopeptidase P X-Pro aminopeptidase	
3.4.11.16	Aminopeptidase W Aminopeptidase X-Trp	
3.4.11.16	Aminopeptidase X-Trp Aminopeptidase X-Trp	
3.4.11.15	Aminopeptidase Y Aminopeptidase Y	
3.4.11.15	aminopeptidase Y Aminopeptidase Y	
3.4.11.22	Aminopeptidase yscI Aminopeptidase I	
3.4.11.22	Aminopeptidase yscII Aminopeptidase I	
3.4.11.22	Aminopeptidase yscXVI Aminopeptidase I	
1.10.3.4	2-Aminophenol:oxygen oxidoreductase *o*-Aminophenol oxidase	

EC-No.	Synonym Recommended Name	Vol., Page
1.10.3.4	*o*-Aminophenol oxidase *o*-Aminophenol oxidase	
3.4.11.22	Aminopolypeptidase Aminopeptidase I	
3.4.11.14	aminopolypeptidase aminopeptidase, human liver	
3.4.24.14	aminoprocollagen peptidase procollagen N-endopeptidase	
1.1.1.75	1-Aminopropan-2-ol-dehydrogenase (R)-Aminopropanol dehydrogenase	
1.1.1.75	1-Aminopropan-2-ol-NAD^+ dehydrogenase (R)-Aminopropanol dehydrogenase	
1.1.1.4	D-Aminopropanol dehydrogenase (R,R)-Butanediol dehydrogenase	
1.1.1.74	D-aminopropanol dehydrogenase D-aminopropanol dehydrogenase	
1.1.1.75	L-Aminopropanol dehydrogenase (R)-Aminopropanol dehydrogenase	
1.1.1.4	Aminopropanol oxidoreductase (R,R)-Butanediol dehydrogenase	
2.5.1.16	Aminopropyltransferase Spermidine synthase	
2.5.1.16	Aminopropyltransferase, putrescine Spermidine synthase	
2.5.1.22	Aminopropyltransferase, spermidine Spermine synthase	
2.3.1.5	*p*-Aminosalicylate N-acetyltransferase Arylamine N-acetyltransferase	
1.14.12.14	2-Aminosulfobenzene 2,3-dioxygenase 2-Aminobenzenesulfonate 2,3-dioxygenase	
3.4.24.14	aminoterminal procollagen peptidase procollagen N-endopeptidase	
2.6.1.37	Aminotransferase, (2-aminoethyl)phosphonate (2-Aminoethyl)phosphonate-pyruvate transaminase	
2.6.1.65	Aminotransferase, ε-acetyl-β-lysine N^6-Acetyl-β-lysine transaminase	

EC-No.	Synonym Recommended Name	Vol., Page
2.6.1.18	Aminotransferase, β-alanine-pyruvate β-Alanine-pyruvate transaminase	
2.6.1.39	Aminotransferase, 2-aminoadipate 2-Aminoadipate transaminase	
2.6.1.23	Aminotransferase, 4-hydroxyglutamate 4-Hydroxyglutamate transaminase	
2.6.1.48	Aminotransferase, 5-aminovalerate 5-Aminovalerate transaminase	
2.6.1.11	Aminotransferase, acetylornithine Acetylornithine transaminase	
2.6.1.2	Aminotransferase, alanine Alanine transaminase	
2.6.1.44	Aminotransferase, alanine-glyoxylate Alanine-glyoxylate transaminase	
2.6.1.12	Aminotransferase, alanine-keto acid Alanine-oxo-acid transaminase	
2.6.1.66	Aminotransferase, alanine-oxoisovalerate Valine-pyruvate transaminase	
2.6.1.47	Aminotransferase, alanine-oxomalonate Alanine-oxomalonate transaminase	
2.6.1.19	Aminotransferase, aminobutyrate 4-Aminobutyrate transaminase	
2.6.1.43	Aminotransferase, aminolevulinate Aminolevulinate transaminase	
2.6.1.57	Aminotransferase, aromatic amino acid Aromatic-amino-acid transaminase	
2.6.1.60	Aminotransferase, aromatic amino acid-glyoxylate Aromatic-amino-acid-glyoxylate transaminase	
2.6.1.14	Aminotransferase, asparagine-keto acid Asparagine-oxo-acid transaminase	
2.6.1.1	Aminotransferase, aspartate Aspartate transaminase	
2.6.1.70	Aminotransferase, aspartate-phenylpyruvate Aspartate-phenylpyruvate transaminase	
2.6.1.42	Aminotransferase, branched-chain amino acid Branched-chain-amino-acid transaminase	

EC-No.	Synonym Recommended Name	Vol., Page
2.6.1.74	Aminotransferase, cephalosporin C Cephalosporin-C transaminase	
2.6.1.3	Aminotransferase, cysteine Cysteine transaminase	
2.6.1.75	Aminotransferase, cysteine conjugate Cysteine-conjugate transaminase	
2.6.1.40	Aminotransferase, D-3-aminoisobutyrate-pyruvate (R)-3-Amino-2-methylpropionate-pyruvate transaminase	
2.6.1.21	Aminotransferase, D-alanine D-Alanine transaminase	
2.6.1.72	Aminotransferase, D-hydroxyphenylglycine D-4-Hydroxyphenylglycine transaminase	
2.6.1.41	Aminotransferase, D-methionine D-Methionine-pyruvate transaminase	
2.6.1.29	Aminotransferase, diamine Diamine transaminase	
2.6.1.8	Aminotransferase, diamino acid 2,5-Diaminovalerate transaminase	
2.6.1.46	Aminotransferase, diaminobutyrate-pyruvate Diaminobutyrate-pyruvate transaminase	
2.6.1.49	Aminotransferase, dihydroxyphenylalanine Dihydroxyphenylalanine transaminase	
2.6.1.24	Aminotransferase, diiodotyrosine Diiodotyrosine transaminase	
5.4.3.8	Aminotransferase, glutamate semialdehyde Glutamate-1-semialdehyde 2,1-aminomutase	Vol. 1, Page 575
2.6.1.15	Aminotransferase, glutamine-keto acid Glutamine-pyruvate transaminase	
2.6.1.64	Aminotransferase, glutamine-phenylpyruvate Glutamine-phenylpyruvate transaminase	
2.6.1.50	Aminotransferase, glutamine-scyllo-inosose Glutamine-scyllo-inosose transaminase	
2.6.1.4	Aminotransferase, glycine Glycine transaminase	
2.6.1.35	Aminotransferase, glycine-oxalacetate Glycine-oxaloacetate transaminase	

EC-No.	Synonym Recommended Name	Vol., Page
2.6.1.56	Aminotransferase, guanidinoaminodideoxy-scyllo-inositol-pyruvate 1D-1-Guanidino-3-amino-1,3-dideoxy-scyllo-inositol transaminase	
2.6.1.38	Aminotransferase, histidine Histidine transaminase	
2.6.1.9	Aminotransferase, histidinol phosphate Histidinol-phosphate transaminase	
2.6.1.7	Aminotransferase, kynurenine Kynurenine-oxoglutarate transaminase	
2.6.1.63	Aminotransferase, kynurenine-glyoxylate Kynurenine-glyoxylate transaminase	
2.6.1.22	Aminotransferase, L-3-aminoisobutyrate (S)-3-Amino-2-methylpropionate transaminase	
2.6.1.6	Aminotransferase, leucine Leucine transaminase	
2.6.1.71	Aminotransferase, lysine-pyruvate Lysine-pyruvate 6-transaminase	
2.6.1.36	Aminotransferase, lysine 6- L-Lysine 6-transaminase	
2.6.1.73	Aminotransferase, methionine-glyoxylate Methionine-glyoxylate transaminase	
2.6.1.67	Aminotransferase, norleucine (leucine) 2-Aminohexanoate transaminase	
2.6.1.68	Aminotransferase, ornithine (lysine) Ornithine(lysine) transaminase	
2.6.1.13	Aminotransferase, ornithine-keto acid Ornithine-oxo-acid transaminase	
2.6.1.58	Aminotransferase, phenylalanine (histidine) Phenylalanine(histidine) transaminase	
2.6.1.52	Aminotransferase, phosphoserine Phosphoserine transaminase	
2.6.1.31	Aminotransferase, pyridoxamine-oxalacetate Pyridoxamine-oxaloacetate transaminase	
2.6.1.30	Aminotransferase, pyridoxamine-pyruvate Pyridoxamine-pyruvate transaminase	
2.6.1.54	Aminotransferase, pyridoxamine phosphate Pyridoxamine-phosphate transaminase	

EC-No.	Synonym Recommended Name	Vol., Page
2.6.1.45	Aminotransferase, serine-glyoxylate Serine-glyoxylate transaminase	
2.6.1.17	Aminotransferase, succinyldiaminopimelate Succinyldiaminopimelate transaminase	
2.6.1.55	Aminotransferase, taurine Taurine transaminase	
2.6.1.33	Aminotransferase, thymidine diphospho-4-amino-4,6-dideoxyglucose dTDP-4-amino-4,6-dideoxy-D-glucose transaminase	
2.6.1.59	Aminotransferase, thymidine diphosphoaminodideoxygalactose dTDP-4-amino-4,6-dideoxygalactose transaminase	
2.6.1.26	Aminotransferase, thyroid hormone Thyroid-hormone transaminase	
2.6.1.27	Aminotransferase, tryptophan Tryptophan transaminase	
2.6.1.5	Aminotransferase, tyrosine Tyrosine transaminase	
2.6.1.34	Aminotransferase, uridine diphospho-4-amino-2-acetamido-2,4,6-trideoxyglucose UDP-2-acetamido-4-amino-2,4,6-trideoxyglucose transaminase	
2.6.1.32	Aminotransferase, valine-3-methyl-2-oxovalerate Valine-3-methyl-2-oxovalerate transaminase	
2.6.1.66	Aminotransferase, valine-pyruvate Valine-pyruvate transaminase	
3.4.1.3	aminotripeptidase aminotripeptidase	
3.4.11.4	aminotripeptidase tripeptide aminopeptidase	
3.5.1.30	5-Aminovaleramidase 5-Aminopentanamidase	
2.6.1.48	δ-Aminovalerate aminotransferase 5-Aminovalerate transaminase	
2.6.1.48	5-Aminovalerate transaminase 5-Aminovalerate transaminase	
2.6.1.48	δ-Aminovalerate transaminase 5-Aminovalerate transaminase	

EC-No.	Synonym Recommended Name	Vol., Page
4.3.1.6	ammonia-lyase, β-alanyl coenzyme A β-alanyl-CoA ammonia-lyase	
4.3.1.1	ammonia-lyase, aspartate aspartate ammonia-lyase	
4.3.1.13	ammonia-lyase, carbamoylserine carbamoyl-serine ammonia-lyase	
4.3.1.15	Ammonia-lyase, diaminopropionate Diaminopropionate ammonia-lyase	
4.3.1.15	Ammonia-lyase, diaminopropionate Diaminopropionate ammonia-lyase	
4.3.1.11	Ammonia-lyase, dihydroxyphenylalanine Dihydroxyphenylalanine ammonia-lyase	
4.3.1.7	ammonia-lyase, ethanolamine ethanolamine ammonia-lyase	
4.3.1.9	Ammonia-lyase, glucosaminate Glucosaminate ammonia-lyase	
4.3.1.3	ammonia-lyase, histidine histidine ammonia-lyase	
4.3.1.2	ammonia-lyase, methylaspartate methylaspartate ammonia-lyase	
4.3.1.5	ammonia-lyase, phenylalanine phenylalanine ammonia-lyase	
4.3.1.10	ammonia-lyase, serine sulfate serine-sulfate ammonia-lyase	
6.3.4.7	Ammonia-ribose 5-phosphate aminotransferase Ribose-5-phosphate-ammonia ligase	Vol. 2, Page 608
1.7.99.1	Ammonia:(acceptor) oxidoreductase Hydroxylamine reductase	
1.7.7.1	Ammonia:ferredoxin oxidoreductase Ferredoxin-nitrite reductase	
2.7.3.8	Ammonia kinase Ammonia kinase	
4.3.1.14	Ammonia lyase, L-3-aminobutyryl coenzyme A 3-Aminobutyryl-CoA ammonia-lyase	
1.6.6.11	Ammonium dehydrogenase Hydroxylamine reductase (NADH)	

EC-No.	Synonym Recommended Name	Vol., Page
3.1.1.3	amno N-AP triacylglycerol lipase	
3.4.11.9	AmP X-Pro aminopeptidase	
1.8.99.2	AMP,sulfite:(acceptor) oxidoreductase Adenylylsulfate reductase	
1.8.4.9	AMP,sulfite:oxidized-glutathione oxidoreductase (adenosine-5'-phospho-sulfate-forming) adenylyl-sulfate reductase (glutathione)	
2.7.1.109	AMP-activated protein kinase [Hydroxymethylglutaryl-CoA reductase (NADPH)] kinase	
2.7.1.110	AMP-activated protein kinase kinase Dephospho-[reductase kinase] kinase	
2.7.4.3	5'-AMP-kinase Adenylate kinase	
2.4.2.7	AMP-pyrophosphate phosphoribosyltransferase Adenine phosphoribosyltransferase	
2.7.1.114	AMP-thymidine kinase AMP-thymidine kinase	
2.7.1.114	AMP:deoxythymidine 5'-phosphotransferase AMP-thymidine kinase	
2.7.1.114	AMP:deoxythymidine kinase AMP-thymidine kinase	
2.7.1.114	AMP:dThd kinase AMP-thymidine kinase	
2.4.2.7	AMP:pyrophosphate phospho-D-ribosyltransferase Adenine phosphoribosyltransferase	
2.7.1.114	AMP:thymidine 5'-phosphotransferase AMP-thymidine kinase	
3.5.4.6	AMP aminase AMP deaminase	
3.5.4.6	5-AMP aminohydrolase AMP deaminase	
3.5.4.6	AMP aminohydrolase AMP deaminase	

EC-No.	Synonym Recommended Name	Vol., Page
3.5.4.6	AMP aminohydrolase AMP deaminase	
3.1.3.5	5'-AMPase 5'-nucleotidase	
3.1.3.5	AMPase 5'-nucleotidase	
3.5.4.6	5-AMP deaminase AMP deaminase	
3.5.4.6	AMP deaminase AMP deaminase	
3.5.1.11	Ampicillin acylase Penicillin amidase	
3.5.2.6	Ampicillinase β-lactamase	
3.2.2.4	AMP nucleosidase AMP nucleosidase	
3.1.3.5	5'-AMP nucleotidase 5'-nucleotidase	
3.1.3.5	AMP phosphatase 5'-nucleotidase	
3.1.3.5	AMP phosphohydrolase 5'-nucleotidase	
3.2.2.4	AMP phosphoribohydrolase AMP nucleosidase	
2.7.7.19	AMP polynucleotidylexotransferase Polynucleotide adenylyltransferase	
2.4.2.7	AMP pyrophosphorylase Adenine phosphoribosyltransferase	
4.3.2.2	AMPS lyase adenylosuccinate lyase	
3.2.1.117	Amygdalase Amygdalin β-glucosidase	
3.2.1.21	Amygdalase β-glucosidase	
3.2.1.117	Amygdalin β-D-glucohydrolase Amygdalin β-glucosidase	

EC-No.	Synonym Recommended Name	Vol., Page
3.2.1.117	Amygdalin β-glucosidase Amygdalin β-glucosidase	
3.2.1.117	Amygdalinase Amygdalin β-glucosidase	
3.2.1.21	Amygdalinase β-glucosidase	
3.2.1.117	Amygdalin glucosidase Amygdalin β-glucosidase	
3.2.1.117	Amygdalin hydrolase Amygdalin β-glucosidase	
3.2.1.1	α-amylase α-amylase	
3.2.1.2	β amylase β-amylase	
3.2.1.2	β-amylase β-amylase	
3.2.1.3	γ-amylase Glucan 1,4-α-glucosidase	
3.2.1.1	Amylase,.α.- α-amylase	
3.2.1.2	Amylase,.β.- β-amylase	
2.4.1.161	Amylase III Oligosaccharide 4-α-D-glucosyltransferase	
2.4.1.18	Amylo-(1,4-1,6)-transglycosylase 1,4-α-Glucan branching enzyme	
3.2.1.33	Amylo-1,6-glucosidase Amylo-1,6-glucosidase	
3.2.1.33	Amylo-1,6-glucosidase Amylo-1,6-glucosidase	
3.2.1.3	Amyloglucosidase Glucan 1,4-α-glucosidase	
2.4.1.25	Amylomaltase 4-α-Glucanotransferase	
2.4.1.3	amylomaltase amylomaltase	

EC-No.	Synonym Recommended Name	Vol., Page
3.2.1.9	amylopectin-1,6-glucosidase amylopectin-1,6-glucosidase	
3.2.1.142	amylopectin-1,6-glucosidase limit dextrinase	
3.2.1.33	Amylopectin 1,6-glucosidase Amylo-1,6-glucosidase	
3.2.1.41	Amylopectin 6-glucanohydrolase α-dextrin endo-1,6-α-glucosidase	
3.2.1.69	amylopectin 6-glucanohydrolase amylopectin 6-glucanohydrolase	
2.4.1.1	Amylopectin phosphorylase Phosphorylase	
2.4.1.1	Amylophosphorylase Phosphorylase	
3.4.11.22	Amylorhizin aminopeptidase Aminopeptidase I	
3.4.11.22	Amylorizin Aminopeptidase I	
2.4.1.18	Amylose isomerase 1,4-α-Glucan branching enzyme	
2.4.1.4	Amylosucrase Amylosucrase	
3.4.22.31	Ananain Ananain	
3.4.22.33	Ananase Fruit bromelain	
3.4.17.3	anaphylatoxin inactivator lysine carboxypeptidase	
3.2.2.7	ANase Adenosine nucleosidase	
3.4.21.74	Ancrod Venombin A	
2.7.1.37	β-Andrenergic receptor kinase Protein kinase	
3.4.21.61	Andrenorphin-Gly-generating enzyme Kexin	

EC-No.	Synonym Recommended Name	Vol., Page
1.14.13.54	Androgen hydroxylase Ketosteroid monooxygenase	
1.14.99.12	Androst-4-ene-3,17-dione,hydrogen-donor:oxygen oxidoreductase (13-hydroxylating, lactonizing) 4-Androstene-3,17-dione monooxygenase	
1.14.99.12	Androst-4-ene-3,17-dione 17-oxidoreductase 4-Androstene-3,17-dione monooxygenase	
1.14.99.12	Androst-4-ene-3,17-dione hydroxylase 4-Androstene-3,17-dione monooxygenase	
1.14.99.12	Androstene-3,17-dione hydroxylase 4-Androstene-3,17-dione monooxygenase	
1.14.99.12	4-Androstene-3,17-dione monooxygenase 4-Androstene-3,17-dione monooxygenase	
1.14.13.54	Androstenedione, NADPH2:oxygen oxidoreductase (17-hydroxylating, lactonizing) Ketosteroid monooxygenase	
1.3.1.23	Androstenedione 5β-reductase Cholestenone 5β-reductase	
1.14.99.12	Androstenedione monooxygenase 4-Androstene-3,17-dione monooxygenase	
2.8.2.2	5α-Androstenol sulfotransferase Alcohol sulfotransferase	
3.4.15.1	angiotensin-converting enzyme peptidyl-dipeptidase A	
3.4.23.15	angiotensin-forming enzyme renin	
3.4.15.1	angiotensin 1 converting enzyme peptidyl-dipeptidase A	
3.4.99.3	angiotensinase angiotensinase	
3.4.11.7	angiotensinase A glutamyl aminopeptidase	
3.4.16.2	angiotensinase C lysosomal Pro-X carboxypeptidase	
3.4.15.1	angiotensin converting enzyme peptidyl-dipeptidase A	

EC-No.	Synonym Recommended Name Vol., Page
3.4.15.1	angiotensin I¯converting enzyme peptidyl-dipeptidase A
3.4.23.15	angiotensinogenase renin
4.2.1.1	anhydrase carbonate dehydratase
1.1.1.263	1,5-anhydro-D-fructose reductase 1,5-anhydro-D-fructose reductase
1.1.1.263	1,5-anhydro-D-glucitol:NADP oxidoreductase 1,5-anhydro-D-fructose reductase
3.2.1.138	Anhydroneuraminidase Anhydrosialidase
3.2.1.138	Anhydrosialidase Anhydrosialidase
1.14.13.38	Anhydrotetracycline,NADPH:oxygen oxidoreductase (6-hydroxylating) Anhydrotetracycline monooxygenase
1.14.13.38	Anhydrotetracycline monooxygenase Anhydrotetracycline monooxygenase
1.14.13.38	Anhydrotetracycline oxygenase Anhydrotetracycline monooxygenase
3.4.24.28	Anilozyme P 10 Bacillolysin
2.4.2.4	Animal growth regulators, blood platelet-derived endothelial cell growth factors Thymidine phosphorylase
1.14.99.15	*p*-Anisic O-demethylase 4-Methoxybenzoate monooxygenase (O-demethylating)
3.4.13.5	anserinase X-methyl-His dipeptidase
2.3.1.153	Anthocyanidin 3,5-diglucoside 5-O-glucoside-6-O-hydroxycinnamoyltransferase Anthocyanin 5-aromatic acyltransferase
2.4.1.115	Anthocyanidin 3-O-glucosyltransferase Anthocyanidin 3-O-glucosyltransferase
2.3.1.153	Anthocyanin 5-aromatic acyltransferase Anthocyanin 5-aromatic acyltransferase

EC-No.	Synonym Recommended Name	Vol., Page
2.3.1.153	Anthocyanin acyltransferase Anthocyanin 5-aromatic acyltransferase	
2.3.1.153	Anthocyanin hydroxycinnamoyltransferase Anthocyanin 5-aromatic acyltransferase	
1.14.12.1	Anthranilate,NAD(P)H:oxygen oxidoreductase (1,2-hydroxylating, deaminating, decarboxylating) Anthranilate 1,2-dioxygenase (deaminating, decarboxylating)	
1.14.13.35	Anthranilate,NADPH:oxygen oxidoreductase (3-hydroxylating, deaminating) Anthranilate 3-monooxygenase (deaminating)	
1.14.16.3	Anthranilate,tetrahydrobiopterine:oxygen oxidoreductase (3-hydroxylating) Anthranilate 3-monooxygenase	
2.4.2.18	Anthranilate-5-phosphoribosylphosphate phosphoribosyltransferase Anthranilate phosphoribosyltransferase	
6.2.1.32	Anthranilate-CoA ligase Anthranilate-CoA ligase	Vol. 2, Page 336
2.4.2.18	Anthranilate-PP-ribose-P phosphoribosyltransferase Anthranilate phosphoribosyltransferase	
1.14.12.1	Anthranilate 1,2-dioxygenase (deaminating, decarboxylating) Anthranilate 1,2-dioxygenase (deaminating, decarboxylating)	
1.14.12.2	anthranilate 2,3-dioxygenase (deaminating) anthranilate 2,3-dioxygenase (deaminating)	
1.14.13.35	Anthranilate 2,3-hydroxylase (deaminating) Anthranilate 3-monooxygenase (deaminating)	
1.14.16.3	Anthranilate 3-hydroxylase Anthranilate 3-monooxygenase	
1.14.16.3	Anthranilate 3-monooxygenase Anthranilate 3-monooxygenase	
1.14.13.35	Anthranilate 3-monooxygenase (deaminating) Anthranilate 3-monooxygenase (deaminating)	
2.4.2.18	Anthranilate 5-phosphoribosylpyrophosphate phosphoribosyltransferase Anthranilate phosphoribosyltransferase	
6.2.1.32	Anthranilate:CoA ligase (AMP-forming) Anthranilate-CoA ligase	Vol. 2, Page 336
2.7.7.55	Anthranilate adenylyltransferase Anthranilate adenylyltransferase	

EC-No.	Synonym Recommended Name Vol., Page
1.14.12.1	Anthranilate hydroxylase Anthranilate 1,2-dioxygenase (deaminating, decarboxylating)
1.14.16.3	Anthranilate hydroxylase Anthranilate 3-monooxygenase
1.14.13.35	Anthranilate hydroxylase (deaminating) Anthranilate 3-monooxygenase (deaminating)
2.3.1.144	Anthranilate N-benzoyltransferase Anthranilate N-benzoyltransferase
2.3.1.113	Anthranilate N-malonyltransferase Anthranilate N-malonyltransferase
2.1.1.111	Anthranilate N-methyltransferase Anthranilate N-methyltransferase
2.4.2.18	Anthranilate phosphoribosylpyrophosphate phosphoribosyltransferase Anthranilate phosphoribosyltransferase
2.4.2.18	Anthranilate phosphoribosyltransferase Anthranilate phosphoribosyltransferase
4.1.3.27	anthranilate synthase anthranilate synthase
4.1.3.27	anthranilate synthase anthranilate synthase
4.1.3.27	anthranilate synthetase anthranilate synthase
2.7.7.55	Anthranilic acid adenylyltransferase Anthranilate adenylyltransferase
1.14.12.1	Anthranilic acid hydroxylase Anthranilate 1,2-dioxygenase (deaminating, decarboxylating)
1.14.16.3	Anthranilic acid hydroxylase Anthranilate 3-monooxygenase
2.1.1.111	Anthranilic acid N-methyltransferase Anthranilate N-methyltransferase
1.14.12.1	Anthranilic hydroxylase Anthranilate 1,2-dioxygenase (deaminating, decarboxylating)
1.14.16.3	Anthranilic hydroxylase Anthranilate 3-monooxygenase
1.14.13.35	Anthranilic hydroxylase Anthranilate 3-monooxygenase (deaminating)

EC-No.	Synonym Recommended Name	Vol., Page
1.14.13.40	Anthraniloyl-CoA monooxygenase Anthraniloyl-CoA monooxygenase	
2.4.1.181	Anthraquinone-specific glucosyltransferase Hydroxyanthraquinone glucosyltransferase	
3.1.1.8	anticholinesterase cholinesterase	
3.4.21.77	P-30 antigen Semenogelase	
3.4.21.77	Antigen (human clone lambdaHPSA-1 prostate-specific protein moiety reduced) Semenogelase	
3.4.24.11	antigen, CALLA (common acute lymphoblastic leukemia-associated) neprilysin	
3.4.11.7	antigen BP-1/6C3 of mouse B lymphocytes glutamyl aminopeptidase	
2.7.1.37	T-antigen kinase Protein kinase	
3.4.21.77	Antigen PSA (human clone 5P1 protein moiety reduced) Semenogelase	
3.4.21.77	Antigen PSA (human prostate-specific) Semenogelase	
2.4.99.1	Antigens, CD75 β-Galactoside α-2,6-sialyltransferase	
1.2.99.6	AOR Carboxylate reductase	
1.2.99.6	W-AOR Carboxylate reductase	
5.3.99.4	Aortic cytochrom P450 Prostacyclin synthase	Vol. 1, Page 465
4.2.99.18	AP-endonuclease DNA-(apurinic or apyrimidinic site) lyase	
3.4.11.9	AP-P X-Pro aminopeptidase	
3.6.1.41	Ap4A hydrolase Bis(5'-nucleosyl)-tetraphosphatase (symmetrical)	

EC-No.	Synonym Recommended Name	Vol., Page
2.7.1.37	AP50 kinase Protein kinase	
3.4.21.69	APC Protein C (activated)	
4.2.99.18	AP endonuclease Class I DNA-(apurinic or apyrimidinic site) lyase	
6.3.2.19	APF-1-protein amide synthetase Ubiquitin-protein ligase	Vol. 2, Page 506
3.4.19.1	APH acylaminoacyl-peptidase	
3.4.11.22	API Aminopeptidase I	
3.4.21.50	API lysyl endopeptidase	
3.4.21.63	API 21 Oryzin	
2.1.1.75	Apigenin 4'-O-methyltransferase Apigenin 4'-O-methyltransferase	
1.1.1.114	D-Apiitol:NAD$^+$ 1-oxidoreductase Apiose 1-reductase	
1.1.1.114	D-Apiitol reductase Apiose 1-reductase	
1.1.1.114	Apiose 1-reductase Apiose 1-reductase	
1.1.1.114	D-Apiose reductase Apiose 1-reductase	
2.4.2.25	Apiosyltransferase, uridine diphosphoapiose-flavone Flavone apiosyltransferase	
4.2.99.18	AP lyase DNA-(apurinic or apyrimidinic site) lyase	
1.13.12.12	apo-β-carotenoid-14',13'-dioxygenase apo-β-carotenoid-14',13'-dioxygenase	
1.13.12.12	8'-apo-β-carotenol:O_2 oxidoreductase apo-β-carotenoid-14',13'-dioxygenase	
1.11.1.5	Apocytochrome c peroxidase Cytochrome-c peroxidase	

EC-No.	Synonym Recommended Name	Vol., Page
3.4.22.29	2A Protease Picornain 2A	
3.4.22.29	2A Proteinase Picornain 2A	
2.4.2.7	APRT Adenine phosphoribosyltransferase	
1.8.99.2	APS-reductase Adenylylsulfate reductase	
2.7.7.51	APSAT Adenylylsulfate-ammonia adenylyltransferase	
4.2.99.18	AP site-DNA 5'-phosphomonoester-lyase DNA-(apurinic or apyrimidinic site) lyase	
2.7.1.25	APS kinase Adenylylsulfate kinase	
1.8.99.2	APS reductase Adenylylsulfate reductase	
4.2.99.18	Apurinic-apyrimidinic DNA endonuclease DNA-(apurinic or apyrimidinic site) lyase	
4.2.99.18	Apurinic-apyrimidinic endodeoxyribonuclease DNA-(apurinic or apyrimidinic site) lyase	
4.2.99.18	Apurinic-apyrimidinic endonuclease DNA-(apurinic or apyrimidinic site) lyase	
4.2.99.18	Apurinic/apyrimidinic lyase DNA-(apurinic or apyrimidinic site) lyase	
4.2.99.18	Apurinic/apyrimidinic specific endonuclease DNA-(apurinic or apyrimidinic site) lyase	
4.2.99.18	Apurinic DNA endonuclease DNA-(apurinic or apyrimidinic site) lyase	
4.2.99.18	Apurinic endodeoxyribonuclease DNA-(apurinic or apyrimidinic site) lyase	
4.2.99.18	Apurinic endonuclease DNA-(apurinic or apyrimidinic site) lyase	
3.6.1.5	Apyrase Apyrase	
4.2.99.18	Apyrimidinic endonuclease DNA-(apurinic or apyrimidinic site) lyase	

EC-No.	Synonym Recommended Name	Vol., Page
2.5.1.17	Aquacob(I)alamin adenosyltransferase Cob(I)alamin adenosyltransferase	
1.6.99.8	Aquacobalamin reductase Aquacobalamin reductase	
1.6.99.11	Aquacobalamin reductase (NADPH) Aquacobalamin reductase (NADPH)	
2.5.1.17	Aquocob(I)alamin adenosyltransferase Cob(I)alamin adenosyltransferase	
1.6.99.8	Aquocobalamin reductase Aquacobalamin reductase	
1.14.13.28	(6aR,11aR)-3,9-Dihydroxypterocarpan,NADPH:oxygen oxidoreductase (6a-hydroxylating) 3,9-Dihydroxypterocarpan 6a-monooxygenase	
2.7.1.74	Ara-C kinase Deoxycytidine kinase	
3.2.1.99	1,5-α-L-arabinan 1,5-α-L-arabinanohydrolase Arabinan endo-1,5-α-L-arabinosidase	
3.2.1.55	α-L-arabinanase α-L-arabinofuranosidase	
3.2.1.99	Arabinan endo-1,5-α-L-arabinosidase Arabinan endo-1,5-α-L-arabinosidase	
3.2.1.99	Arabinase, endo-1,5-.α.-L- Arabinan endo-1,5-α-L-arabinosidase	
1.1.1.250	D-Arabinitol 2-dehydrogenase D-Arabinitol 2-dehydrogenase	
1.1.1.13	L-Arabinitol 2-dehydrogenase L-Arabinitol 2-dehydrogenase	
1.1.1.250	D-Arabinitol 2-dehydrogenase (ribulose-forming) D-Arabinitol 2-dehydrogenase	
1.1.1.11	D-Arabinitol 4-dehydrogenase D-Arabinitol 4-dehydrogenase	
1.1.1.12	L-Arabinitol 4-dehydrogenase L-Arabinitol 4-dehydrogenase	
1.1.1.13	L-Arabinitol:NAD$^+$ 2-oxidoreductase (L-ribulose-forming) L-Arabinitol 2-dehydrogenase	

EC-No.	Synonym Recommended Name	Vol., Page
1.1.1.11	D-Arabinitol:NAD$^+$ 4-oxidoreductase D-Arabinitol 4-dehydrogenase	
1.1.1.12	L-Arabinitol:NAD$^+$ 4-oxidoreductase (L-xylulose-forming) L-Arabinitol 4-dehydrogenase	
1.1.1.250	D-Arabinitol:NAD 2-oxidoredictase (D-ribulose-forming) D-Arabinitol 2-dehydrogenase	
1.1.1.250	D-Arabinitol dehydrogenase D-Arabinitol 2-dehydrogenase	
1.1.1.13	L-Arabinitol dehydrogenase (ribulose-forming) L-Arabinitol 2-dehydrogenase	
3.2.1.55	α-arabinofuranosidase α-L-arabinofuranosidase	
3.2.1.55	α-L-arabinofuranosidase α-L-arabinofuranosidase	
3.2.1.55	α-L-arabinofuranoside arabinofuranohydrolase α-L-arabinofuranosidase	
3.2.1.55	α-L-arabinofuranoside hydrolase α-L-arabinofuranosidase	
3.2.1.79	α-L-arabinofuranoside hydrolase α-L-arabinofuranoside hydrolase	
2.7.1.74	Arabinofuranosylcytosine kinase Deoxycytidine kinase	
3.2.1.90	Arabinogalactan 3-β-D-galactanohydrolase Arabinogalactan endo-1,3-β-galactosidase	
3.2.1.89	Arabinogalactan 4-β-D-galactanohydrolase Arabinogalactan endo-1,4-β-galactosidase	
3.2.1.90	Arabinogalactanase Arabinogalactan endo-1,3-β-galactosidase	
3.2.1.89	Arabinogalactanase Arabinogalactan endo-1,4-β-galactosidase	
3.2.1.90	Arabinogalactan endo-1,3-β-galactosidase Arabinogalactan endo-1,3-β-galactosidase	
3.2.1.89	Arabinogalactan endo-1,4-β-galactosidase Arabinogalactan endo-1,4-β-galactosidase	
2.7.1.54	D-Arabinokinase D-Arabinokinase	

EC-No.	Synonym Recommended Name	Vol., Page
2.7.1.46	L-Arabinokinase L-Arabinokinase	
3.1.1.15	L-arabino lactonase L-arabinonolactonase	
4.2.1.5	Arabinonate dehydratase Arabinonate dehydratase	
4.2.1.25	L-Arabinonate dehydratase L-Arabinonate dehydratase	
4.2.1.5	D-Arabinonate hydro-lyase Arabinonate dehydratase	
4.2.1.25	L-Arabinonate hydro-lyase L-Arabinonate dehydratase	
1.1.3.37	D-Arabinono-1,4-lactone:oxygen oxidoreductase D-Arabinono-1,4-lactone oxidase	
3.1.1.30	D-arabinono-1,4-lactone lactonohydrolase D-arabinonolactonase	
3.1.1.15	L-arabinono-1,4-lactone lactonohydrolase L-arabinonolactonase	
1.1.3.37	D-Arabinono-1,4-lactone oxidase D-Arabinono-1,4-lactone oxidase	
3.1.1.30	D-arabinonolactonase D-arabinonolactonase	
3.1.1.15	L-arabinonolactonase L-arabinonolactonase	
1.1.1.116	Arabinose(fucose)dehydrogenase D-Arabinose 1-dehydrogenase	
5.3.1.25	D-Arabinose (L-Fucose) isomerase L-Fucose isomerase	Vol. 1, Page 359
5.3.1.3	D-Arabinose(L-fucose) isomerase Arabinose isomerase	Vol. 1, Page 249
5.3.1.13	Arabinose-5-phosphate isomerase Arabinose-5-phosphate isomerase	Vol. 1, Page 325
5.3.1.13	D-Arabinose-5-phosphate ketol-isomerase Arabinose-5-phosphate isomerase	Vol. 1, Page 325
1.1.1.116	D-Arabinose 1-dehydrogenase D-Arabinose 1-dehydrogenase	

EC-No.	Synonym Recommended Name	Vol., Page
1.1.1.46	L-Arabinose 1-dehydrogenase L-Arabinose 1-dehydrogenase	
1.1.1.117	D-Arabinose 1-dehydrogenase $(NAD(P)^+)$ D-Arabinose 1-dehydrogenase $(NAD(P)^+)$	
5.3.1.13	D-Arabinose 5-phosphate isomerase Arabinose-5-phosphate isomerase	Vol. 1, Page 325
1.1.1.117	D-Arabinose:$NAD(P)^+$ 1-oxidoreductase D-Arabinose 1-dehydrogenase $(NAD(P)^+)$	
1.1.1.116	D-Arabinose:NAD^+ 1-oxidoreductase D-Arabinose 1-dehydrogenase	
1.1.1.46	L-Arabinose:NAD^+ 1-oxidoreductase L-Arabinose 1-dehydrogenase	
5.3.1.3	Arabinose isomerase Arabinose isomerase	Vol. 1, Page 249
5.3.1.3	D-Arabinose isomerase Arabinose isomerase	Vol. 1, Page 249
5.3.1.25	D-Arabinose isomerase L-Fucose isomerase	Vol. 1, Page 359
5.3.1.4	L-Arabinose isomerase L-Arabinose isomerase	Vol. 1, Page 254
5.3.1.3	D-Arabinose ketol-isomerase Arabinose isomerase	Vol. 1, Page 249
5.3.1.4	L-Arabinose ketol-isomerase L-Arabinose isomerase	Vol. 1, Page 254
5.3.1.13	Arabinose phosphate isomerase Arabinose-5-phosphate isomerase	Vol. 1, Page 325
5.3.1.13	Arabinose phosphate isomerase, arabinose phosphate Arabinose-5-phosphate isomerase	Vol. 1, Page 325
3.2.1.88	.β.-L-Arabinosidase β-L-arabinosidase	
3.2.1.55	α-L-arabinosidase α-L-arabinofuranosidase	
3.2.1.55	α-arabinosidase α-L-arabinofuranosidase	
3.2.1.55	Arabinosidase α-L-arabinofuranosidase	

EC-No.	Synonym Recommended Name	Vol., Page
3.2.1.55	L-Arabinosidase α-L-arabinofuranosidase	
3.2.1.88	β-L-arabinosidase β-L-arabinosidase	
3.2.1.88	Arabinosidase,.β.-L- β-L-arabinosidase	
3.2.1.88	β-L-arabinoside arabinohydrolase β-L-arabinosidase	
2.4.2.34	Arabinosylindolylacetylinositol synthase Indolylacetylinositol arabinosyltransferase	
2.4.2.34	Arabinosyltransferase, uridine diphosphoarabinose-indolylacetylinositol Indolylacetylinositol arabinosyltransferase	
1.1.1.11	Arabitol dehydrogenase D-Arabinitol 4-dehydrogenase	
1.1.1.11	D-Arabitol dehydrogenase D-Arabinitol 4-dehydrogenase	
1.1.1.12	L-Arabitol dehydrogenase L-Arabinitol 4-dehydrogenase	
4.2.1.25	L-Arabonate dehydrase L-Arabinonate dehydratase	
4.2.1.25	L-Arabonate dehydratase L-Arabinonate dehydratase	
6.2.1.15	Arachidonate-CoA ligase Arachidonate-CoA ligase	Vol. 2, Page 277
1.13.11.31	Arachidonate 12-lipoxygenase Arachidonate 12-lipoxygenase	
1.13.11.33	Arachidonate 15-lipoxygenase Arachidonate 15-lipoxygenase	
1.13.11.34	Arachidonate 5-lipoxygenase Arachidonate 5-lipoxygenase	
1.13.11.40	Arachidonate 8-lipoxygenase Arachidonate 8-lipoxygenase	
6.2.1.15	Arachidonate:CoA ligase (AMP-forming) Arachidonate-CoA ligase	Vol. 2, Page 277
1.13.11.31	Arachidonate:oxygen 12-oxidoreductase Arachidonate 12-lipoxygenase	

EC-No.	Synonym Recommended Name	Vol., Page
1.13.11.33	Arachidonate:oxygen 15-oxidoreductase Arachidonate 15-lipoxygenase	
1.13.11.34	Arachidonate:oxygen 5-oxidoreductase Arachidonate 5-lipoxygenase	
1.13.11.40	Arachidonate:oxygen 8-oxidoreductase Arachidonate 8-lipoxygenase	
1.13.11.34	Arachidonic 5-lipoxygenase Arachidonate 5-lipoxygenase	
1.13.11.34	Arachidonic acid 5-lipoxygenase Arachidonate 5-lipoxygenase	
6.2.1.15	Arachidonoyl-CoA synthetase Arachidonate-CoA ligase	Vol. 2, Page 277
2.7.1.107	Arachidonoyl-specific diacylglycerol kinase Diacylglycerol kinase	
2.3.1.70	Arachidonyl-CoA:CDPacylglycerol O-arachidonyltransferase CDPacylglycerol O-arachidonyltransferase	
2.3.1.70	Arachidonyltransferase, cytidine diphosphomonoacylglycerol CDPacylglycerol O-arachidonyltransferase	
1.4.99.4	Aralkylamine:(acceptor) oxidoreductase (deaminating) Aralkylamine dehydrogenase	
1.4.99.4	Aralkylamine dehydrogenase Aralkylamine dehydrogenase	
2.3.1.87	Aralkylamine N-acetyltransferase Aralkylamine N-acetyltransferase	
2.6.1.57	ArAT Aromatic-amino-acid transaminase	
3.2.1.21	Arbutinase β-glucosidase	
3.4.22.2	arbuz papain	
1.1.1.250	ArDH D-Arabinitol 2-dehydrogenase	
3.3.2.3	Arene-oxide hydratase Epoxide hydrolase	
4.2.1.64	arene-oxide hydratase arene-oxide hydratase	

EC-No.	Synonym Recommended Name	Vol., Page
1.14.99.8	arene monooxygenase (epoxidizing) arene monooxygenase (epoxidizing)	
3.3.2.3	Arene oxide hydratase Epoxide hydrolase	
3.4.22.37	Arg-gingipain Gingipain R	
3.4.22.37	Arg-gingivain-55 proteinase Gingipain R	
3.4.22.37	Arg-gingivain-70 proteinase Gingipain R	
3.4.22.37	Arg-gingivain-75 proteinase Gingipain R	
3.4.13.4	X-Arg dipeptidase X-Arg dipeptidase	
3.5.3.1	Arginase Arginase	
3.5.3.10	D-Arginase D-Arginase	
3.5.3.1	L-Arginase Arginase	
3.5.3.10	D-Arginase amidinohydrolase D-Arginase	
3.4.22.37	Argingipain Gingipain R	
1.14.13.39	L-Arginine,NADPH:oxygen oxidoreductase (nitric-oxide-forming) Nitric-oxide synthase	
2.1.4.1	Arginine-glycine amidinotransferase Glycine amidinotransferase	
2.1.4.1	Arginine-glycine transamidinase Glycine amidinotransferase	
3.4.22.37	Arginine-specific cysteine protease Gingipain R	
3.4.22.37	Arginine-specific gingipain Gingipain R	
3.4.22.37	Arginine-specific gingivain Gingipain R	

EC-No.	Synonym Recommended Name	Vol., Page
4.3.2.1	arginine-succinate lyase argininosuccinate lyase	
6.1.1.19	Arginine-tRNA ligase Arginine-tRNA ligase	Vol. 2, Page 146
6.1.1.19	Arginine-tRNA synthetase Arginine-tRNA ligase	Vol. 2, Page 146
1.13.12.1	Arginine 2-monooxygenase Arginine 2-monooxygenase	
2.1.4.2	L-Arginine:1-amino-1-deoxy-scyllo-inositol-4-phosphate amidinotransfer-ase Inosamine-phosphate amidinotransferase	
2.1.4.1	L-Arginine:glycine amidinotransferase Glycine amidinotransferase	
2.1.4.2	L-Arginine:inosamine-P-amidinotransferase Inosamine-phosphate amidinotransferase	
2.1.4.2	L-Arginine:inosamine phosphate amidinotransferase Inosamine-phosphate amidinotransferase	
1.13.12.1	L-Arginine:oxygen 2-oxidoreductase (decarboxylating) Arginine 2-monooxygenase	
6.1.1.19	L-Arginine:tRNA^{Arg} ligase (AMP-forming) Arginine-tRNA ligase	Vol. 2, Page 146
3.5.3.1	Arginine amidinase Arginase	
3.5.3.1	L-Arginine amidinohydrolase Arginase	
3.4.11.6	L-arginine aminopeptidase aminopeptidase B	
3.4.11.6	arginine aminopeptidase aminopeptidase B	
4.1.1.19	L-Arginine carboxy-lyase Arginine decarboxylase	
3.4.17.20	Arginine carboxypeptidase Carboxypeptidase U	
3.4.17.3	arginine carboxypeptidase lysine carboxypeptidase	

EC-No.	Synonym Recommended Name	Vol., Page
1.13.12.1	Arginine decarboxy-oxidase Arginine 2-monooxygenase	
1.13.12.1	Arginine decarboxylase Arginine 2-monooxygenase	
4.1.1.19	Arginine decarboxylase Arginine decarboxylase	
4.1.1.19	L-Arginine decarboxylase Arginine decarboxylase	
3.5.3.6	Arginine deiminase Arginine deiminase	
3.5.3.6	L-Arginine deiminase Arginine deiminase	
3.4.24.61	N-Arginine dibasic convertase Nardilysin	
3.5.3.6	Arginine dihydrolase Arginine deiminase	
3.5.3.6	L-Arginine iminohydrolase Arginine deiminase	
2.7.3.3	Arginine kinase Arginine kinase	
2.1.1.125	Arginine methylase (Histone)-arginine N-methyltransferase	
2.1.1.124	Arginine methylase [Cytochrome-c]-arginine N-methyltransferase	
2.1.1.126	Arginine methylase [Myelin basic protein]-arginine N-methyltransferase	
1.13.12.1	Arginine monooxygenase Arginine 2-monooxygenase	
2.3.1.109	Arginine N-succinyltransferase Arginine N-succinyltransferase	
1.13.12.1	Arginine oxygenase (decarboxylating) Arginine 2-monooxygenase	
2.7.3.3	Arginine phosphokinase Arginine kinase	
5.1.1.9	Arginine racemase Arginine racemase	Vol. 1, Page 37

EC-No.	Synonym Recommended Name	Vol., Page
5.1.1.9	Arginine racemase Arginine racemase	Vol. 1, Page 37
2.4.2.31	Arginine specific ADP-ribosyltransferase NAD(P)$^+$-arginine ADP-ribosyltransferase	
2.4.2.31	Arginine specific mono-ADP-ribosyltransferase NAD(P)$^+$-arginine ADP-ribosyltransferase	
6.3.4.5	Arginine succinate synthetase Argininosuccinate synthase	Vol. 2, Page 595
2.3.1.109	Arginine succinyl transferase Arginine N-succinyltransferase	
3.5.3.1	Arginine transamidinase Arginase	
2.3.2.8	Arginine transferase Arginyltransferase	
6.1.1.19	Arginine translase Arginine-tRNA ligase	Vol. 2, Page 146
4.3.2.1	argininosuccinase argininosuccinate lyase	
4.3.2.1	L-argininosuccinate arginine-lyase argininosuccinate lyase	
4.3.2.1	argininosuccinate lyase argininosuccinate lyase	
6.3.4.5	Argininosuccinate synthase Argininosuccinate synthase	Vol. 2, Page 595
6.3.4.5	Argininosuccinate synthetase Argininosuccinate synthase	Vol. 2, Page 595
4.3.2.1	argininosuccinic acid lyase argininosuccinate lyase	
6.3.4.5	Argininosuccinic acid synthetase Argininosuccinate synthase	Vol. 2, Page 595
3.4.11.6	L-arginyl(L-lysyl)-peptide hydrolase aminopeptidase B	
2.3.2.8	Arginyl-transfer ribonucleate-protein aminoacyltransferase Arginyltransferase	
2.3.2.8	Arginyl-transfer ribonucleate-protein transferase Arginyltransferase	

EC-No.	Synonym Recommended Name	Vol., Page
6.1.1.19	Arginyl-transfer RNA synthetase Arginine-tRNA ligase	Vol. 2, Page 146
2.3.2.8	L-Arginyl-tRNA:protein arginyltransferase Arginyltransferase	
2.3.2.8	Arginyl-tRNA protein transferase Arginyltransferase	
6.1.1.19	Arginyl-tRNA synthetase Arginine-tRNA ligase	Vol. 2, Page 146
3.4.11.6	arginyl aminopeptidase aminopeptidase B	
2.3.2.8	Arginyltransferase Arginyltransferase	
6.1.1.19	Arginyl transfer ribonucleic acid synthetase Arginine-tRNA ligase	Vol. 2, Page 146
2.5.1.40	Aristolochene synthase Aristolochene synthase	
4.2.3.9	aristolochene synthase aristolochene synthase	
2.7.1.126	β-ARK 1 β-Adrenergic-receptor kinase	
2.7.1.126	β-ARK 2 β-Adrenergic-receptor kinase	
3.4.99.32	Armillaria mellea neutral proteinase Armillaria mellea neutral proteinase	
3.4.24.20	Armillaria mellea neutral proteinase Peptidyl-Lys metalloendopeptidase	
4.2.1.91	Arogenate dehydratase Carboxycyclohexadienyl dehydratase	
1.3.1.43	Arogenate dehydrogenase Pretyrosine dehydrogenase	
4.2.1.91	L-Arogenate hydro-lyase Carboxycyclohexadienyl dehydratase	
1.3.1.43	Arogenic dehydrogenase Pretyrosine dehydrogenase	
2.6.1.57	Arom.-amino-acid transaminase Aromatic-amino-acid transaminase	

EC-No.	Synonym Recommended Name	Vol., Page
1.1.3.7	Arom. alcohol oxidase Aryl-alcohol oxidase	
2.6.1.60	Arom. amino acid-glyoxylate aminotransferase Aromatic-amino-acid-glyoxylate transaminase	
2.6.1.57	Arom. amino acid transferase Aromatic-amino-acid transaminase	
2.6.1.60	Aromatic-amino-acid-glyoxylate transaminase Aromatic-amino-acid-glyoxylate transaminase	
2.6.1.57	Aromatic-amino-acid:2-oxoglutarate aminotransferase Aromatic-amino-acid transaminase	
2.6.1.60	Aromatic-amino-acid:glyoxylate aminotransferase Aromatic-amino-acid-glyoxylate transaminase	
2.6.1.57	Aromatic-amino-acid transaminase Aromatic-amino-acid transaminase	
2.3.1.56	Aromatic-hydroxylamine O-acetyltransferase Aromatic-hydroxylamine O-acetyltransferase	
1.1.1.96	Aromatic α-keto acid reductase Diiodophenylpyruvate reductase	
4.1.1.28	Aromatic-L-amino-acid carboxy-lyase Aromatic-L-amino-acid decarboxylase	
4.1.1.28	Aromatic-L-amino-acid decarboxylase Aromatic-L-amino-acid decarboxylase	
1.2.1.30	Aromatic acid reductase Aryl-aldehyde dehydrogenase ($NADP^+$)	
1.4.99.4	Aromatic amine dehydrogenase Aralkylamine dehydrogenase	
2.6.1.60	Aromatic amino acid-glyoxylate aminotransferase Aromatic-amino-acid-glyoxylate transaminase	
2.6.1.57	Aromatic amino acid aminotransferase Aromatic-amino-acid transaminase	
4.1.1.28	Aromatic amino acid decarboxylase Aromatic-L-amino-acid decarboxylase	
4.1.1.28	L-Aromatic amino acid decarboxylase Aromatic-L-amino-acid decarboxylase	
2.6.1.57	Aromatic amino acid transaminase Aromatic-amino-acid transaminase	

EC-No.	Synonym Recommended Name	Vol., Page
2.6.1.57	Aromatic aminotransferase Aromatic-amino-acid transaminase	
3.1.1.49	aromatic choline esterase sinapine esterase	
3.1.1.2	aromatic esterase arylesterase	
4.1.1.28	Aromatic L-amino acid decarboxylase Aromatic-L-amino-acid decarboxylase	
4.1.1.53	Aromatic L-amino acid decarboxylase Phenylalanine decarboxylase	
1.97.1.6	arsenate:(acceptor) oxidoreductase arsenate reductase (donor)	
1.97.1.6	arsenate reductase (donor) arsenate reductase (donor)	
1.97.1.5	arsenate reductase (glutaredoxin) arsenate reductase (glutaredoxin)	
2.1.1.137	arsenite methyltransferase arsenite methyltransferase	
3.4.21.17	Arthrobacter serine proteinase Arthrobacter serine proteinase	
3.1.1.2	ary-ester hydrolase arylesterase	
3.2.1.21	Aryl-β-glucosidase β-glucosidase	
4.1.1.76	2-Aryl-2-methylmalonate carboxy-lyase Arylmalonate decarboxylase	
1.14.14.1	Aryl-4-monooxygenase Unspecific monooxygenase	
3.5.1.13	Aryl-acylamide amidohydrolase Arylacylamidase	
1.1.1.90	Aryl-alcohol:NAD^+ oxidoreductase Aryl-alcohol dehydrogenase	
1.1.1.91	Aryl-alcohol:$NADP^+$ oxidoreductase Aryl-alcohol dehydrogenase ($NADP^+$)	
1.1.3.7	Aryl-alcohol:oxygen oxidoreductase Aryl-alcohol oxidase	

EC-No.	Synonym Recommended Name	Vol., Page
1.1.1.90	Aryl-alcohol dehydrogenase Aryl-alcohol dehydrogenase	
1.1.1.91	Aryl-alcohol dehydrogenase ($NADP^+$) Aryl-alcohol dehydrogenase ($NADP^+$)	
1.1.3.7	Aryl-alcohol oxidase Aryl-alcohol oxidase	
1.2.1.29	Aryl-aldehyde:NAD^+ oxidoreductase Aryl-aldehyde dehydrogenase	
1.2.1.30	Aryl-aldehyde:$NADP^+$ oxidoreductase (ATP-forming) Aryl-aldehyde dehydrogenase ($NADP^+$)	
1.2.3.9	Aryl-aldehyde:oxygen oxidoreductase Aryl-aldehyde oxidase	
1.2.1.29	Aryl-aldehyde dehydrogenase Aryl-aldehyde dehydrogenase	
1.2.1.30	Aryl-aldehyde dehydrogenase ($NADP^+$) Aryl-aldehyde dehydrogenase ($NADP^+$)	
1.2.3.9	Aryl-aldehyde oxidase Aryl-aldehyde oxidase	
3.5.1.9	Aryl-formylamine amidohydrolase Arylformamidase	
3.2.1.139	Aryl α-glucuronidase α-Glucosiduronase	
3.1.6.1	Aryl-sulfate sulfohydrolase Arylsulfatase	
3.5.5.5	Arylacetonitrilase Arylacetonitrilase	
3.5.5.5	Arylacetonitrile aminohydrolase Arylacetonitrilase	
3.5.1.13	Arylacylamidase Arylacylamidase	
1.1.1.91	Aryl alcohol dehydrogenase (nicotinamide adenine dinucleotide phosphate) Aryl-alcohol dehydrogenase ($NADP^+$)	
1.1.3.7	Aryl alcohol oxidase Aryl-alcohol oxidase	

EC-No.	Synonym Recommended Name	Vol., Page
3.5.1.76	Arylalkyl acylamidase Arylalkyl acylamidase	
3.5.1.76	Arylalkyl acylamidase Arylalkyl acylamidase	
2.3.1.87	Arylalkylamine N-acetyltransferase Aralkylamine N-acetyltransferase	
3.4.11.6	arylamidase II aminopeptidase B	
2.3.1.5	Arylamine acetylase Arylamine N-acetyltransferase	
2.3.1.5	Arylamine acetyltransferase Arylamine N-acetyltransferase	
2.4.1.71	Arylamine glucosyltransferase Arylamine glucosyltransferase	
2.3.1.5	Arylamine N-acetyltransferase Arylamine N-acetyltransferase	
2.3.1.118	Arylamine N-acetyltransferase N-Hydroxyarylamine O-acetyltransferase	
2.1.1.49	Arylamine N-methyltransferase Amine N-methyltransferase	
2.8.2.3	Arylamine sulfotransferase Amine sulfotransferase	
3.1.8.1	Aryldialkylphosphatase Aryldialkylphosphatase	
3.3.2.3	Aryl epoxide hydrase Epoxide hydrolase	
3.1.1.2	arylesterase arylesterase	
3.5.1.9	Arylformamidase Arylformamidase	
1.14.14.1	Aryl hydrocarbon hydroxylase Unspecific monooxygenase	
2.3.1.56	Arylhydroxamate acyltransferase Aromatic-hydroxylamine O-acetyltransferase	
2.3.1.56	Arylhydroxamate N,O-acetyltransferase Aromatic-hydroxylamine O-acetyltransferase	

EC-No.	Synonym Recommended Name	Vol., Page
2.3.1.118	Arylhydroxamate N,O-acetyltransferase N-Hydroxyarylamine O-acetyltransferase	
2.3.1.56	Arylhydroxamic acid acyltransferase Aromatic-hydroxylamine O-acetyltransferase	
2.3.1.56	Arylhydroxamic acid N,O-acetyltransferase Aromatic-hydroxylamine O-acetyltransferase	
2.3.1.56	Arylhydroxamic acyltransferase Aromatic-hydroxylamine O-acetyltransferase	
4.1.1.76	Arylmalonate decarboxylase Arylmalonate decarboxylase	
4.1.1.76	Arylmalonate decarboxylase Arylmalonate decarboxylase	
4.1.1.76	Arylmalonate decarboxylase (Alcaligenes bronchisepticus strain KU 1201) Arylmalonate decarboxylase	
3.1.6.1	Arylsulfatase Arylsulfatase	
3.1.6.1	Arylsulfatase Arylsulfatase	
3.1.6.1	Arylsulfatase A Arylsulfatase	
3.1.6.8	Arylsulfatase A Cerebroside-sulfatase	
3.1.6.1	Arylsulfatase B Arylsulfatase	
3.1.6.12	Arylsulfatase B N-Acetylgalactosamine-4-sulfatase	
3.1.6.1	Arylsulfatase C Arylsulfatase	
3.1.6.2	Arylsulfatase C Steryl-sulfatase	
2.8.2.22	Arylsulfate-phenol sulfotransferase Arylsulfate sulfotransferase	
2.8.2.22	Arylsulfate:phenol sulfotransferase Arylsulfate sulfotransferase	
2.8.2.22	Arylsulfate sulfotransferase Arylsulfate sulfotransferase	

EC-No.	Synonym Recommended Name	Vol., Page
3.1.6.1	Arylsulfohydrolase Arylsulfatase	
2.8.2.1	Aryl sulfotransferase Aryl sulfotransferase	
2.8.2.1	Arylsulfotransferase Aryl sulfotransferase	
2.8.2.22	Arylsulfotransferase Arylsulfate sulfotransferase	
2.8.2.9	Aryl sulfotransferase IV Tyrosine-ester sulfotransferase	
3.1.8.1	Aryltriphosphatase Aryldialkylphosphatase	
3.1.8.1	Aryltriphosphate dialkylphosphohydrolase Aryldialkylphosphatase	
6.3.5.4	AS-B Asparagine synthase (glutamine-hydrolysing)	Vol. 2, Page 672
1.2.1.11	ASA dehydrogenase Aspartate-semialdehyde dehydrogenase	
4.3.2.1	ASAL argininosuccinate lyase	
3.4.22.7	asclepain asclepain	
1.10.3.3	Ascorbase L-Ascorbate oxidase	
1.10.2.1	L-Ascorbate-cytochrome-b_5 reductase L-Ascorbate-cytochrome-b_5 reductase	
1.10.2.1	Ascorbate-cytochrome b_5 reductase L-Ascorbate-cytochrome-b_5 reductase	
1.13.11.13	Ascorbate 2,3-dioxygenase Ascorbate 2,3-dioxygenase	
1.10.2.1	L-Ascorbate:ferricytochrome-b_5 oxidoreductase L-Ascorbate-cytochrome-b_5 reductase	
1.11.1.11	L-Ascorbate:hydrogen-peroxide oxidoreductase L-Ascorbate peroxidase	
1.10.3.3	L-Ascorbate:O_2 oxidoreductase L-Ascorbate oxidase	

EC-No.	Synonym Recommended Name	Vol., Page
1.13.11.13	Ascorbate:oxygen 2,3-oxidoreductase (bond-cleaving) Ascorbate 2,3-dioxygenase	
1.10.3.3	L-Ascorbate:oxygen oxidoreductase L-Ascorbate oxidase	
1.10.3.3	Ascorbate dehydrogenase L-Ascorbate oxidase	
1.6.5.4	Ascorbate free-radical reductase Monodehydroascorbate reductase (NADH)	
1.6.5.4	Ascorbate free radical reductase Monodehydroascorbate reductase (NADH)	
1.10.3.3	Ascorbate oxidase L-Ascorbate oxidase	
1.10.3.3	L-Ascorbate oxidase L-Ascorbate oxidase	
1.11.1.11	Ascorbate peroxidase L-Ascorbate peroxidase	
1.11.1.11	L-Ascorbate peroxidase L-Ascorbate peroxidase	
1.11.1.11	L-Ascorbic acid-specific peroxidase L-Ascorbate peroxidase	
1.10.3.3	Ascorbic acid oxidase L-Ascorbate oxidase	
1.10.3.3	L-Ascorbic acid oxidase L-Ascorbate oxidase	
1.11.1.11	Ascorbic acid peroxidase L-Ascorbate peroxidase	
1.11.1.11	L-Ascorbic acid peroxidase L-Ascorbate peroxidase	
1.6.5.4	Ascorbic free radical reductase Monodehydroascorbate reductase (NADH)	
1.10.3.3	Ascorbic oxidase L-Ascorbate oxidase	
1.14.12.14	2AS dioxygenase 2-Aminobenzenesulfonate 2,3-dioxygenase	
4.3.2.2	ASL adenylosuccinate lyase	

EC-No.	Synonym Recommended Name	Vol., Page
4.3.2.1	ASL argininosuccinate lyase	
2.1.1.4	ASMT Acetylserotonin N-methyltransferase	
6.3.5.4	AsnB Asparagine synthase (glutamine-hydrolysing)	Vol. 2, Page 672
3.5.1.1	Asparaginase Asparaginase	
3.5.1.1	L-Asparaginase Asparaginase	
3.5.1.1	α-asparaginase Asparaginase	
3.5.1.1	Asparaginase II Asparaginase	
2.6.1.14	Asparagine-oxo-acid transaminase Asparagine-oxo-acid transaminase	
6.1.1.22	Asparagine-tRNA ligase Asparagine-tRNA ligase	Vol. 2, Page 178
2.6.1.14	L-Asparagine:2-oxo-acid aminotransferase Asparagine-oxo-acid transaminase	
2.3.2.7	L-Asparagine:hydroxylamine γ-aspartyltransferase Aspartyltransferase	
6.1.1.22	L-Asparagine:tRNA^{Asn} ligase (AMP-forming) Asparagine-tRNA ligase	Vol. 2, Page 178
3.5.1.1	L-Asparagine amidohydrolase Asparaginase	
4.2.1.65	L-Asparagine hydro-lyase 3-Cyanoalanine hydratase	
2.4.1.119	Asparagine N-glycosyltransferase Dolichyl-diphosphooligosaccharide-protein glycotransferase	
6.3.5.4	Asparagine synthase (glutamine-hydrolysing) Asparagine synthase (glutamine-hydrolysing)	Vol. 2, Page 672
6.3.1.1	Asparagine synthetase Aspartate-ammonia ligase	Vol. 2, Page 344
6.3.5.4	L-Asparagine synthetase Asparagine synthase (glutamine-hydrolysing)	Vol. 2, Page 672

EC-No.	Synonym Recommended Name	Vol., Page
6.3.1.1	L-Asparagine synthetase Aspartate-ammonia ligase	Vol. 2, Page 344
6.3.1.4	Asparagine synthetase (adenosine diphosphate-forming) Aspartate-ammonia ligase	Vol. 2, Page 375
6.3.1.4	Asparagine synthetase (ADP-forming) Aspartate-ammonia ligase	Vol. 2, Page 375
6.3.5.4	Asparagine synthetase (glutamine) Asparagine synthase (glutamine-hydrolysing)	Vol. 2, Page 672
6.3.5.4	Asparagine synthetase (glutamine-hydrolysing) Asparagine synthase (glutamine-hydrolysing)	Vol. 2, Page 672
6.3.5.4	Asparagine synthetase (glutamine hydrolyzing) Asparagine synthase (glutamine-hydrolysing)	Vol. 2, Page 672
6.3.1.1	Asparagine synthetase A Aspartate-ammonia ligase	Vol. 2, Page 344
6.3.5.4	Asparagine synthetase B Asparagine synthase (glutamine-hydrolysing)	Vol. 2, Page 672
6.1.1.22	Asparagine translase Asparagine-tRNA ligase	Vol. 2, Page 178
6.1.1.22	Asparaginyl-transfer ribonucleate synthetase Asparagine-tRNA ligase	Vol. 2, Page 178
6.1.1.22	Asparaginyl-tRNA synthetase Asparagine-tRNA ligase	Vol. 2, Page 178
3.4.22.34	Asparaginyl endopeptidase Legumain	
6.1.1.22	Asparaginyl transfer ribonucleic acid synthetase Asparagine-tRNA ligase	Vol. 2, Page 178
6.1.1.22	Asparaginyl transfer RNA synthetase Asparagine-tRNA ligase	Vol. 2, Page 178
1.6.4.7	Asparagusate dehydrogenase Asparagusate reductase (NADH)	
1.6.4.7	Asparagusate reductase (NADH) Asparagusate reductase (NADH)	
1.6.4.7	Asparagusic dehydrogenase Asparagusate reductase (NADH)	
6.1.1.22	Asparagyl-transfer RNA synthetase Asparagine-tRNA ligase	Vol. 2, Page 178

EC-No.	Synonym Recommended Name	Vol., Page
4.1.1.12	L-Asparate 4-carboxy-lyase Aspartate 4-decarboxylase	
4.3.1.1	L-aspartase aspartate ammonia-lyase	
4.3.1.1	aspartase aspartate ammonia-lyase	
4.1.1.11	L-Aspartate-α-decarboxylase Aspartate 1-decarboxylase	
4.1.1.15	L-Aspartate-α-decarboxylase Glutamate decarboxylase	
2.6.1.1	L-Aspartate-α-ketoglutarate transaminase Aspartate transaminase	
1.2.1.11	L-Aspartate-β-semialdehyde:NADP oxidoreductase (phosporylating) Aspartate-semialdehyde dehydrogenase	
2.6.1.1	L-Aspartate-2-ketoglutarate aminotransferase Aspartate transaminase	
2.6.1.1	L-Aspartate-2-oxoglutarate-transaminase Aspartate transaminase	
2.6.1.1	L-Aspartate-2-oxoglutarate aminotransferase Aspartate transaminase	
2.6.1.1	Aspartate-2-oxoglutarate transaminase Aspartate transaminase	
1.2.1.11	L-Aspartate-4-semialdehyde:NADP$^+$ oxidoreductase (phosphorylating) Aspartate-semialdehyde dehydrogenase	
4.2.1.52	L-aspartate-4-semialdehyde hydro-lyase (adding pyruvate and cyclizing) dihydrodipicolinate synthase	
6.3.1.1	Aspartate-ammonia ligase Aspartate-ammonia ligase	Vol. 2, Page 344
6.3.1.4	Aspartate-ammonia ligase Aspartate-ammonia ligase	Vol. 2, Page 375
4.1.1.12	Aspartate β-decarboxylase Aspartate 4-decarboxylase	
4.1.1.12	Aspartate ω-decarboxylase Aspartate 4-decarboxylase	
4.1.1.11	L-Aspartate α-decarboxylase Aspartate 1-decarboxylase	

EC-No.	Synonym Recommended Name	Vol., Page
4.1.1.12	L-Aspartate β-decarboxylase Aspartate 4-decarboxylase	
2.6.1.49	Aspartate-DOPP transaminase (ADT) Dihydroxyphenylalanine transaminase	
1.14.11.16	Aspartate β-hydroxylase Peptide-aspartate β-dioxygenase	
2.6.1.1	Aspartate α-ketoglutarate transaminase Aspartate transaminase	
2.6.1.70	Aspartate-phenylpyruvate transaminase Aspartate-phenylpyruvate transaminase	
1.2.1.11	Aspartate-semialdehyde dehydrogenase Aspartate-semialdehyde dehydrogenase	
6.1.1.12	Aspartate-tRNA ligase Aspartate-tRNA ligase	Vol. 2, Page 86
4.1.1.11	L-Aspartate 1-carboxy-lyase Aspartate 1-decarboxylase	
4.1.1.11	Aspartate 1-decarboxylase Aspartate 1-decarboxylase	
4.1.1.15	Aspartate 1-decarboxylase Glutamate decarboxylase	
4.1.1.12	L-Aspartate 4-carboxy-lyase Aspartate 4-decarboxylase	
4.1.1.12	Aspartate 4-decarboxylase Aspartate 4-decarboxylase	
2.6.1.1	Aspartate:2-oxoglutarate aminotransferase Aspartate transaminase	
2.6.1.1	L-Aspartate:2-oxoglutarate aminotransferase Aspartate transaminase	
6.3.1.4	L-Aspartate:ammonia ligase (ADP-forming) Aspartate-ammonia ligase	Vol. 2, Page 375
6.3.1.1	L-Aspartate:ammonia ligase (AMP-forming) Aspartate-ammonia ligase	Vol. 2, Page 344
6.3.5.4	L-Aspartate:L-glutamine amido-ligase (AMP-forming) Asparagine synthase (glutamine-hydrolysing)	Vol. 2, Page 672
1.4.3.1	D-Aspartate:oxygen oxidoreductase (deaminating) D-Aspartate oxidase	

EC-No.	Synonym Recommended Name	Vol., Page
1.4.3.16	L-Aspartate:oxygen oxidoreductase (deaminating) L-Aspartate oxidase	
2.6.1.70	L-Aspartate:phenylpyruvate aminotransferase Aspartate-phenylpyruvate transaminase	
6.1.1.12	L-Aspartate:tRNA^{Asp} ligase (AMP-forming) Aspartate-tRNA ligase	Vol. 2, Page 86
2.3.1.17	Aspartate acetyltransferase Aspartate N-acetyltransferase	
3.4.11.7	L-aspartate aminopeptidase glutamyl aminopeptidase	
3.4.11.7	aspartate aminopeptidase glutamyl aminopeptidase	
2.6.1.1	Aspartate aminotransferase Aspartate transaminase	
4.3.1.1	L-aspartate ammonia-lyase aspartate ammonia-lyase	
4.3.1.1	aspartate ammonia-lyase aspartate ammonia-lyase	
2.1.3.2	Aspartate carbamoyltransferase Aspartate carbamoyltransferase	
2.1.3.2	Aspartate carbamyltransferase Aspartate carbamoyltransferase	
3.4.17.5	aspartate carboxypeptidase aspartate carboxypeptidase	
2.7.2.4	Aspartate kinase Aspartate kinase	
2.3.1.17	Aspartate N-acetyltransferase Aspartate N-acetyltransferase	
2.3.1.17	L-Aspartate N-acetyltransferase Aspartate N-acetyltransferase	
1.4.3.1	D-Aspartate oxidase D-Aspartate oxidase	
1.4.3.16	L-Aspartate oxidase L-Aspartate oxidase	
5.1.1.13	Aspartate racemase Aspartate racemase	Vol. 1, Page 55

EC-No.	Synonym Recommended Name	Vol., Page
5.1.1.13	Aspartate racemase Aspartate racemase	Vol. 1, Page 55
5.1.1.13	Aspartate racemase Aspartate racemase	Vol. 1, Page 55
5.1.1.13	D-Aspartate racemase Aspartate racemase	Vol. 1, Page 55
1.2.1.11	Aspartate semialdehyde dehydrogenase Aspartate-semialdehyde dehydrogenase	
2.6.1.1	Aspartate transaminase Aspartate transaminase	
2.6.1.21	D-Aspartate transaminase D-Alanine transaminase	
2.6.1.10	D-aspartate transaminase D-aspartate transaminase	
2.6.1.1	L-Aspartate transaminase Aspartate transaminase	
2.1.3.2	Aspartate transcarbamoylase Aspartate carbamoyltransferase	
2.1.3.2	L-Aspartate transcarbamoylase Aspartate carbamoyltransferase	
2.1.3.2	Aspartate transcarbamylase Aspartate carbamoyltransferase	
2.1.3.2	L-Aspartate transcarbamylase Aspartate carbamoyltransferase	
4.1.1.11	Aspartic α-decarboxylase Aspartate 1-decarboxylase	
4.1.1.15	Aspartic α-decarboxylase Glutamate decarboxylase	
4.1.1.12	Aspartic β-decarboxylase Aspartate 4-decarboxylase	
4.1.1.12	Aspartic ω-decarboxylase Aspartate 4-decarboxylase	
1.2.1.11	Aspartic β-semialdehyde dehydrogenase Aspartate-semialdehyde dehydrogenase	
2.6.1.1	Aspartic acid aminotransferase Aspartate transaminase	

EC-No.	Synonym Recommended Name	Vol., Page
2.1.3.2	Aspartic acid transcarbamoylase Aspartate carbamoyltransferase	
6.1.1.12	Aspartic acid translase Aspartate-tRNA ligase	Vol. 2, Page 86
2.6.1.1	Aspartic aminotransferase Aspartate transaminase	
2.6.1.21	D-Aspartic aminotransferase D-Alanine transaminase	
2.6.1.1	L-Aspartic aminotransferase Aspartate transaminase	
2.1.3.2	Aspartic carbamyltransferase Aspartate carbamoyltransferase	
3.4.23.38	Aspartic hemoglobinase I Plasmepsin I	
3.4.23.39	Aspartic hemoglobinase II Plasmepsin II	
2.7.2.4	Aspartic kinase Aspartate kinase	
1.4.3.1	Aspartic oxidase D-Aspartate oxidase	
1.4.3.1	D-Aspartic oxidase D-Aspartate oxidase	
3.4.23.41	aspartic protease yapsin I yapsin I	
3.4.23.25	Aspartic proteinase yscA Saccharopepsin	
1.2.1.11	Aspartic semialdehyde dehydrogenase Aspartate-semialdehyde dehydrogenase	
2.1.3.2	Aspartic transcarbamylase Aspartate carbamoyltransferase	
3.5.1.15	Aspartoacylase Aspartoacylase	
3.5.1.83	Aspartoacylase N-Acyl-D-aspartate deacylase	
2.7.2.4	Aspartokinase Aspartate kinase	

EC-No.	Synonym Recommended Name	Vol., Page
2.7.2.4	β-Aspartokinase Aspartate kinase	
2.3.2.7	Aspartotransferase Aspartyltransferase	
3.4.11.7	L-α-aspartyl(L-α-glutamyl)-peptide hydrolase glutamyl aminopeptidase	
3.5.1.26	N-Aspartyl-β-glucosaminidase N^4-(β-N-acetylglucosaminyl)-L-asparaginase	
3.2.2.11	1-β-aspartyl-N-acetyl-D-glucosaminylamine L-asparaginohydrolase β-aspartyl-N-acetylglucosaminidase	
3.2.2.11	β-aspartyl-N-acetylglucosaminidase β-aspartyl-N-acetylglucosaminidase	
3.4.19.5	β-Aspartyl-peptidase β-Aspartyl-peptidase	
6.1.1.12	Aspartyl-transfer ribonucleic acid synthetase Aspartate-tRNA ligase	Vol. 2, Page 86
6.1.1.12	Aspartyl-transfer RNA synthetase Aspartate-tRNA ligase	Vol. 2, Page 86
6.1.1.12	Aspartyl-tRNA synthetase Aspartate-tRNA ligase	Vol. 2, Page 86
2.1.1.77	D-Aspartyl/L-isoaspartyl methyltransferase Protein-L-isoaspartate(D-aspartate) O-methyltransferase	
2.1.1.77	L-Aspartyl/L-isoaspartyl protein methyltransferase Protein-L-isoaspartate(D-aspartate) O-methyltransferase	
3.2.2.11	.β.-Aspartylacetylglucosaminidase β-aspartyl-N-acetylglucosaminidase	
2.6.1.1	Aspartyl aminotransferase Aspartate transaminase	
3.4.19.5	β-Aspartyl dipeptidase β-Aspartyl-peptidase	
3.4.13.10	β-aspartyldipeptidase β-aspartyldipeptidase	
3.4.99.4	aspartylendopeptidase aspartylendopeptidase	
3.4.23.40	Aspartyl endoproteinase Phytepsin	

EC-No.	Synonym Recommended Name	Vol., Page
3.5.1.26	Aspartylglucosaminidase N^4-(β-N-acetylglucosaminyl)-L-asparaginase	
3.5.1.26	Aspartylglucosylaminase N^4-(β-N-acetylglucosaminyl)-L-asparaginase	
3.5.1.26	β-aspartylglucosylamine amidohydrolase N^4-(β-N-acetylglucosaminyl)-L-asparaginase	
3.5.1.26	Aspartylglucosylamine deaspartylase N^4-(β-N-acetylglucosaminyl)-L-asparaginase	
3.5.1.37	4-L-aspartylglycosylamine amidohydrolase 4-L-aspartylglycosylamine amidohydrolase	
3.5.1.26	Aspartylglycosylamine amidohydrolase N^4-(β-N-acetylglucosaminyl)-L-asparaginase	
3.4.19.5	β-Aspartyl peptidase β-Aspartyl-peptidase	
3.4.13.16	aspartylphenylalanine dipeptidase aspartylphenylalanine dipeptidase	
6.1.1.12	Aspartyl ribonucleate synthetase Aspartate-tRNA ligase	Vol. 2, Page 86
6.1.1.12	Aspartyl ribonucleic synthetase Aspartate-tRNA ligase	Vol. 2, Page 86
2.3.2.7	Aspartyltransferase Aspartyltransferase	
2.3.2.7	β-Aspartyl transferase Aspartyltransferase	
3.4.23.18	Aspergillopepsin A Aspergillopepsin I	
3.4.21.63	Aspergillopepsin B Oryzin	
3.4.21.63	Aspergillopepsin F Oryzin	
3.4.23.18	Aspergillopepsin I Aspergillopepsin I	
3.4.23.19	Aspergillopepsin II Aspergillopepsin II	
3.4.23.18	Aspergillopeptidase A Aspergillopepsin I	

EC-No.	Synonym Recommended Name Vol., Page
3.4.4.17	aspergillopeptidase A aspergillopeptidase A
3.4.21.63	Aspergillopeptidase B Oryzin
3.4.21.32	aspergillopeptidase C brachyurin
3.4.24.3	aspergillopeptidase C microbial collagenase
3.4.23.18	Aspergillus acid protease Aspergillopepsin I
3.4.23.18	Aspergillus acid proteinase Aspergillopepsin I
3.4.21.15	Aspergillus alkaline proteinase Aspergillus alkaline proteinase
3.4.21.63	Aspergillus alkaline proteinase Oryzin
3.4.23.18	Aspergillus aspartic proteinase Aspergillopepsin I
3.4.23.18	Aspergillus awamori acid proteinase Aspergillopepsin I
3.1.3.16	Aspergillus awamori acid protein phosphatase Phosphoprotein phosphatase
3.4.21.63	Aspergillus candidus alkaline proteinase Oryzin
3.4.23.18	Aspergillus carboxyl proteinase Aspergillopepsin I
3.1.22.2	Aspergillus deoxyribonuclease K1 Aspergillus deoxyribonuclease K1
3.4.21.63	Aspergillus flavus alkaline proteinase Oryzin
3.4.21.63	Aspergillus melleus semi-alkaline proteinase Oryzin
3.4.23.18	Aspergillus niger acid proteinase Aspergillopepsin I
3.4.23.19	Aspergillus niger var. macrosporus aspartic proteinase Aspergillopepsin II

EC-No.	Synonym Recommended Name	Vol., Page
3.1.30.1	Aspergillus nuclease S1 Aspergillus nuclease S1	
3.4.21.63	Aspergillus oryzae alkaline proteinase Oryzin	
3.1.4.8	Aspergillus oryzae ribonuclease Aspergillus oryzae ribonuclease	
3.1.27.3	Aspergillus oryzae ribonuclease Ribonuclease T1	
3.4.21.63	Aspergillus parasiticus alkaline proteinase Oryzin	
3.4.23.18	Aspergillus saitoi acid proteinase Aspergillopepsin I	
3.4.21.63	Aspergillus serine proteinase Oryzin	
3.1.1.55	Aspirin esterase Acetylsalicylate deacetylase	
3.1.1.55	Aspirin hydrolase Acetylsalicylate deacetylase	
6.1.1.12	AspRS Aspartate-tRNA ligase	Vol. 2, Page 86
2.6.1.1	AspT Aspartate transaminase	
2.5.1.35	Aspulvinone dimethylallyltransferase Aspulvinone dimethylallyltransferase	
6.3.4.5	ASS Argininosuccinate synthase	Vol. 2, Page 595
3.4.21.97	assemblin assemblin	
3.4.21.97	assembly protein precursor-processing proteinase assemblin	
1.8.99.1	Assimilatory-type sulfite reductase Sulfite reductase	
1.7.7.2	Assimilatory ferredoxin-nitrate reductase Ferredoxin-nitrate reductase	
1.6.6.2	Assimilatory NAD(P)H-nitrate reductase Nitrate reductase (NAD(P)H)	

EC-No.	Synonym Recommended Name	Vol., Page
1.6.6.1	Assimilatory NADH: nitrate reductase Nitrate reductase (NADH)	
1.6.6.3	Assimilatory NADPH-nitrate reductase Nitrate reductase (NADPH)	
1.7.7.2	Assimilatory nitrate reductase Ferredoxin-nitrate reductase	
1.6.6.2	Assimilatory nitrate reductase Nitrate reductase (NAD(P)H)	
1.6.6.1	Assimilatory nitrate reductase Nitrate reductase (NADH)	
1.6.6.3	Assimilatory nitrate reductase Nitrate reductase (NADPH)	
1.6.6.4	Assimilatory nitrite reductase Nitrite reductase (NAD(P)H)	
1.6.6.3	Assimilatory reduced nicotinamide adenine dinucleotide phosphate-nitrate reductase Nitrate reductase (NADPH)	
1.8.99.1	Assimilatory sulfite reductase Sulfite reductase	
2.8.2.22	ASST Arylsulfate sulfotransferase	
2.6.1.1	AST Aspartate transaminase	
3.4.24.21	Astacin Astacin	
3.4.24.21	Astacus proteinase Astacin	
2.1.3.2	ATCase Aspartate carbamoyltransferase	
2.1.3.2	ATC domain of CAD Aspartate carbamoyltransferase	
1.14.13.38	ATC oxygenase Anhydrotetracycline monooxygenase	
2.7.7.25	ATP(CTP)-tRNA nucleotidyltransferase tRNA adenylyltransferase	

EC-No.	Synonym Recommended Name	Vol., Page
2.7.7.21	ATP(CTP)-tRNA nucleotidyltransferase tRNA cytidylyltransferase	
2.7.7.25	ATP(CTP):tRNA nucleotidyltransferase tRNA adenylyltransferase	
2.7.7.21	ATP(CTP):tRNA nucleotidyltransferase tRNA cytidylyltransferase	
2.7.2.3	ATP-3-phospho-D-glycerate-1-phosphotransferase Phosphoglycerate kinase	
2.7.1.93	ATP-alkylglycerol phosphotransferase Alkylglycerol kinase	
4.1.3.8	ATP-citric lyase ATP citrate (pro-3S)-lyase	
2.7.1.1	ATP-D-hexose 6-phosphotransferase Hexokinase	
3.4.21.92	ATP-dependent Clp protease Endopeptidase Clp	
3.1.21.3	ATP-dependent deoxyribonuclease Type I site-specific deoxyribonuclease	
3.1.21.3	ATP-dependent DNase Type I site-specific deoxyribonuclease	
4.2.1.93	ATP-dependent $H_4NAD(P)OH$ dehydratase ATP-dependent $H_4NAD(P)OH$ dehydratase	
2.7.1.1	ATP-dependent hexokinase Hexokinase	
3.4.21.53	ATP-dependent Lon proteinase Endopeptidase La	
2.7.1.11	ATP-dependent phosphofructokinase 6-Phosphofructokinase	
3.4.21.53	ATP-dependent protease La Endopeptidase La	
3.4.21.53	ATP-dependent serine proteinase Endopeptidase La	
3.6.1.5	ATP-diphosphatase Apyrase	
2.7.6.5	ATP-GTP 3'-diphosphotransferase GTP pyrophosphokinase	

EC-No.	Synonym Recommended Name	Vol., Page
2.7.2.11	ATP-L-glutamate 5-phosphotransferase Glutamate 5-kinase	
2.5.1.6	ATP-methionine adenosyltransferase Methionine adenosyltransferase	
2.7.7.19	ATP-polynucleotide adenylyltransferase Polynucleotide adenylyltransferase	
2.7.1.37	ATP-protein transphosphorylase Protein kinase	
2.7.7.4	ATP-sulfurylase Sulfate adenylyltransferase	
2.7.1.102	ATP/hamamelose 2'-phosphotransferase Hamamelose kinase	
2.7.4.11	ATP:(d)AMP phosphotransferase (Deoxy)adenylate kinase	
2.7.4.8	ATP:(d)GMP phosphotransferase Guanylate kinase	
2.7.4.12	ATP:(d)NMP phosphotransferase T2-induced deoxynucleotide kinase	
4.1.1.33	ATP:(R)-5-Diphosphomevalonate carboxy-lyase (dehydrating) Diphosphomevalonate decarboxylase	
2.7.4.2	ATP:(R)-5-phosphomevalonate phosphotransferase Phosphomevalonate kinase	
2.7.1.31	ATP:(R)-glycerate 3-phosphotransferase Glycerate kinase	
2.7.1.36	ATP:(R)-mevalonate 5-phosphotransferase Mevalonate kinase	
2.7.1.126	ATP:β-adrenergic-receptor phosphotransferase β-Adrenergic-receptor kinase	
2.7.7.27	ATP:α-D-glucose-1-phosphate adenylyltransferase Glucose-1-phosphate adenylyltransferase	
2.7.2.11	ATP:γ-L-glutamate phosphotransferase Glutamate 5-kinase	
2.7.1.107	ATP:1,2-diacylglycerol 3-phosphotransferase Diacylglycerol kinase	
2.7.1.93	ATP: 1-alkyl-sn-glycerol phosphotransferase Alkylglycerol kinase	

EC-No.	Synonym Recommended Name	Vol., Page
2.7.1.65	ATP:1-amino-1-deoxy-scyllo-inositol 4-phosphotransferase scyllo-Inosamine 4-kinase	
2.7.1.93	ATP:1-O-alkyl-sn-glycerol 3-phosphotransferase Alkylglycerol kinase	
2.7.1.68	ATP:1-phosphatidyl-1D-myo-inositol-4-phosphate 5-phosphotransferase 1-Phosphatidylinositol-4-phosphate 5-kinase	
2.7.1.137	ATP:1-phosphatidyl-1D-myo-inositol 3-phosphotransferase 1-Phosphatidylinositol 3-kinase	
2.7.1.67	ATP:1-phosphatidyl-1D-myo-inositol 4-phosphotransferase 1-Phosphatidylinositol 4-kinase	
2.7.1.140	ATP:1D-myo-inositol-1,3,4,6-tetrakisphosphate 5-phosphotransferase 1D-myo-Inositol-tetrakisphosphate 5-kinase	
2.7.1.139	ATP:1D-myo-inositol-1,3,4-trisphosphate 5-phosphotransferase 1D-myo-Inositol-trisphosphate 5-kinase	
2.7.1.133	ATP:1D-myo-inositol-1,3,4-trisphosphate 6-phosphotransferase 1D-myo-Inositol-trisphosphate 6-kinase	
2.7.1.127	ATP:1D-myo-inositol-1,4,5-trisphosphate 3-phosphotransferase 1D-myo-Inositol-trisphosphate 3-kinase	
2.7.1.134	ATP:1D-myo-inositol-3,4,5,6-tetrakisphosphate 1-phosphotransferase 1D-myo-Inositol-tetrakisphosphate 1-kinase	
2.7.1.130	ATP:2,3,2',3'-tetrakis(3-hydroxytetradecanoyl)-D-glucosaminyl-β-D-1,6-glucosaminyl-β-phosphate 4'-O-phosphotransferase Tetraacyldisaccharide 4'-kinase	
2.7.7.58	ATP:2,3-dihydroxybenzoate adenylyltransferase (2,3-Dihydroxybenzoyl)adenylate synthase	
2.7.1.59	ATP:2-acetylamino-2-deoxy-D-glucose 6-phosphotransferase N-Acetylglucosamine kinase	
2.7.1.8	ATP:2-amino-2-deoxy-D-glucose-6-phosphotransferase Glucosamine kinase	
2.7.6.3	ATP:2-amino-4-hydroxy-6-hydroxymethyl-7,8-dihydropteridine 6'-pyrophosphotransferase 2-Amino-4-hydroxy-6-hydroxymethyldihydropteridine pyrophosphokinase	
2.7.1.58	ATP:2-dehydro-3-deoxy-D-galactonate 6-phosphotransferase 2-Dehydro-3-deoxygalactonokinase	
2.7.1.45	ATP:2-dehydro-3-deoxy-D-gluconate 6-phosphotransferase 2-Dehydro-3-deoxygluconokinase	

EC-No.	Synonym Recommended Name	Vol., Page
2.7.1.13	ATP:2-dehydro-D-gluconate 6-phosphotransferase Dehydrogluconokinase	
2.7.2.3	ATP:3-phospho-D-glycerate 1-phosphotransferase Phosphoglycerate kinase	
2.7.4.7	ATP:4-amino-2-methyl-5-phosphomethylpyrimidine phosphotransferase Phosphomethylpyrimidine kinase	
2.7.1.49	ATP:4-amino-5-hydroxymethyl-2-methylpyrimidine 5-phosphotransferase Hydroxymethylpyrimidine kinase	
2.7.1.50	ATP:4-methyl-5-(2-hydroxyethyl)-thiazole 2-phosphotransferase Hydroxyethylthiazole kinase	
2.7.1.78	ATP:5'-dephosphopolynucleotide 5'-phosphatase Polynucleotide 5'-hydroxy-kinase	
2.7.1.78	ATP:5'-dephosphopolynucleotide 5'-phosphotransferase Polynucleotide 5'-hydroxy-kinase	
2.7.1.92	ATP:5-dehydro-2-deoxy-D-gluconate 6-phosphotransferase 5-Dehydro-2-deoxygluconokinase	
2.7.4.19	ATP:5-methyldeoxycytidine-5'-phosphate phosphotransferase 5-Methyldeoxycytidine-5'-phosphate kinase	
2.7.4.2	ATP:5-phosphomevalonate phosphotransferase Phosphomevalonate kinase	
2.7.1.52	ATP:6-deoxy-L-galactose 1-phosphotransferase Fucokinase	
2.7.1.115	ATP:[3-methyl-2-oxobutanoate dehydrogenase (lipoamide)] phosphotrans-ferase [3-Methyl-2-oxobutanoate dehydrogenase (lipoamide)] kinase	
2.7.1.128	ATP:[acetyl-CoA carboxylase] phosphotransferase [Acetyl-CoA carboxylase] kinase	
2.7.1.141	ATP:[DNA-directed RNA polymerase] phosphotransferase [RNA-polymerase]-subunit kinase	
2.7.1.109	ATP:[hydroxymethylglutaryl-CoA reductase (NADPH)] phosphotransferase [Hydroxymethylglutaryl-CoA reductase (NADPH)] kinase	
2.7.1.116	ATP:[isocitrate dehydrogenase ($NADP^+$)] phosphotransferase [Isocitrate dehydrogenase ($NADP^+$)] kinase	
2.7.7.42	ATP:[L-glutamate:ammonia ligase (ADP-forming)] adenylyltransferase [Glutamate-ammonia-ligase] adenylyltransferase	

EC-No.	Synonym Recommended Name	Vol., Page
2.7.1.124	ATP:[L-tyrosine,tetrahydropteridine:oxygen oxidoreductase (3-hydroxylating)] O-phosphotransferase [Tyrosine 3-monooxygenase] kinase	
2.7.1.99	ATP:[pyruvate dehydrogenase (lipoamide)] phosphotransferase [Pyruvate dehydrogenase (lipoamide)] kinase	
2.7.2.1	ATP:acetate phosphotransferase Acetate kinase	
2.7.1.94	ATP:acylglycerol 3-phosphotransferase Acylglycerol kinase	
2.7.1.20	ATP:adenosine 5'-phosphotransferase Adenosine kinase	
2.7.1.25	ATP:adenylylsulfate 3'-phosphotransferase Adenylylsulfate kinase	
2.7.3.10	ATP:agmatine N^4-phosphotransferase Agmatine kinase	
2.7.3.8	ATP:ammonia phosphotransferase Ammonia kinase	
2.7.4.3	ATP:AMP phosphotransferase Adenylate kinase	
2.7.7.55	ATP:anthranilate N-adenylyltransferase Anthranilate adenylyltransferase	
2.7.2.14	ATP:branched-chain-fatty-acid 1-phosphotransferase Branched-chain-fatty-acid kinase	
2.7.2.7	ATP:butanoate 1-phosphotransferase Butyrate kinase	
2.7.1.120	ATP:caldesmon O-phosphotransferase Caldesmon kinase	
2.7.2.2	ATP:carbamate phosphotransferase Carbamate kinase	
2.7.1.85	ATP:cellobiose 6-phosphotransferase *β*-Glucoside kinase	
2.7.1.138	ATP:ceramide 1-phosphotransferase Ceramide kinase	
2.7.1.32	ATP:choline phosphotransferase Choline kinase	

EC-No.	Synonym Recommended Name Vol., Page
4.1.3.8	ATP:citrate lyase ATP citrate (pro-3S)-lyase
4.1.3.8	ATP:citrate oxaloacetate-lyase (pro-3S-$CH_2COO \rightarrow$acetyl-CoA, ATP dephosphorylating) ATP citrate (pro-3S)-lyase
4.1.3.8	ATP:citrate oxaloacetate-lyase CoA-acetylating and ATP-dephosphorylating ATP citrate (pro-3S)-lyase
4.1.3.8	ATP:citrate oxaloacetate lyase ((pro-3S)-CH_2COO-$\rightarrow$ acetyl-CoA) (ATP-dephosphorylating) ATP citrate (pro-3S)-lyase
2.7.4.14	ATP:CMP phosphotransferase Cytidylate kinase
2.5.1.17	ATP:cob(I)alamin Coβ-adenosyltransferase Cob(I)alamin adenosyltransferase
2.7.3.2	ATP:creatine N-phosphotransferase Creatine kinase
2.7.3.2	ATP:creatine phosphotransferase Creatine kinase
2.7.2.3	ATP:D-3-phosphoglycerate 1-phosphotransferase Phosphoglycerate kinase
2.7.1.55	ATP:D-allose 6-phosphotransferase Allose kinase
2.7.1.54	ATP:D-arabinose 5-phosphotransferase D-Arabinokinase
2.7.1.56	ATP:D-fructose-1-phosphate 6-phosphotransferase 1-Phosphofructokinase
2.7.1.11	ATP:D-fructose-6-phosphate 1-phosphotransferase 6-Phosphofructokinase
2.7.1.105	ATP:D-fructose-6-phosphate 2-phosphotransferase 6-Phosphofructo-2-kinase
2.7.1.3	ATP:D-fructose 1-phosphotransferase Ketohexokinase
2.7.1.4	ATP:D-fructose 6-phosphotransferase Fructokinase
2.7.1.6	ATP:D-galactose-1-phosphotransferase Galactokinase

EC-No.	Synonym Recommended Name	Vol., Page
2.7.1.6	ATP:D-galactose 1-phosphotransferase Galactokinase	
2.7.1.44	ATP:D-galacturonate 1-phosphotransferase Galacturonokinase	
2.7.1.12	ATP:D-gluconate 6-phosphotransferase Gluconokinase	
2.7.1.8	ATP:D-glucosamine phosphotransferase Glucosamine kinase	
2.7.1.10	ATP:D-glucose-1-phosphate 6-phosphotransferase Phosphoglucokinase	
2.7.1.2	ATP:D-glucose 6-phosphotransferase Glucokinase	
2.7.1.43	ATP:D-glucuronate 1-phosphotransferase Glucuronokinase	
2.7.1.28	ATP:D-glyceraldehyde 3-phosphotransferase Triokinase	
2.7.1.31	ATP:D-glycerate 2-phosphotransferase Glycerate kinase	
2.7.1.102	ATP:D-hamamelose 2'-phosphotransferase Hamamelose kinase	
2.7.1.1	ATP:D-hexose 6-phosphotransferase Hexokinase	
2.7.1.7	ATP:D-mannose 6-phosphotransferase Mannokinase	
2.7.1.18	ATP:D-ribose-5-phosphate 1-phosphotransferase Phosphoribokinase	
2.7.6.1	ATP:D-ribose-5-phosphate pyrophosphotransferase Ribose-phosphate pyrophosphokinase	
2.7.1.15	ATP:D-ribose 5-phosphotransferase Ribokinase	
2.7.1.19	ATP:D-ribulose-5-phosphate 1-phosphotransferase Phosphoribulokinase	
2.7.1.47	ATP:D-ribulose 5-phosphotransferase D-Ribulokinase	
2.7.1.144	ATP:D-tagatose-6-phosphate 1-phosphotransferase Tagatose-6-phosphate kinase	

EC-No.	Synonym Recommended Name	Vol., Page
2.7.1.101	ATP:D-tagatose 6-phosphotransferase Tagatose kinase	
2.7.1.17	ATP:D-xylulose 5-phosphotransferase Xylulokinase	
2.7.1.76	ATP:deoxyadenosine 5'-phosphotransferase Deoxyadenosine kinase	
2.7.1.113	ATP:deoxyguanosine 5'-phosphotransferase Deoxyguanosine kinase	
2.7.4.13	ATP:deoxynucleoside-phosphate phosphotransferase (Deoxy)nucleoside-phosphate kinase	
2.7.1.24	ATP:dephospho-CoA 3'-phosphotransferase Dephospho-CoA kinase	
2.7.1.110	ATP:dephospho[[hydroxymethylglutaryl-CoA reductase (NADPH)] kinase] phosphotransferase Dephospho-[reductase kinase] kinase	
2.7.1.107	ATP:diacylglycerol phosphotransferase Diacylglycerol kinase	
2.7.1.88	ATP:dihydrostreptomycin-6-P 3'α-phosphotransferase Dihydrostreptomycin-6-phosphate 3'α-kinase	
2.7.1.88	ATP:dihydrostreptomycin-6-phosphate 3'α-phosphotransferase Dihydrostreptomycin-6-phosphate 3'α-kinase	
2.7.4.9	ATP:dTMP phosphotransferase dTMP kinase	
2.7.1.27	ATP:erythritol 4-phosphotransferase Erythritol kinase	
2.7.1.82	ATP:ethanolamine O-phosphotransferase Ethanolamine kinase	
2.7.4.18	ATP:farnesyl-diphosphate phosphotransferase Farnesyl-diphosphate kinase	
2.7.7.2	ATP:FMN adenylyltransferase FMN adenylyltransferase	
2.7.2.6	ATP:formate phosphotransferase Formate kinase	
2.7.7.42	ATP:glutamine synthetase adenylyltransferase [Glutamate-ammonia-ligase] adenylyltransferase	

EC-No.	Synonym Recommended Name	Vol., Page
2.7.1.30	ATP:glycerol-3-phosphotransferase Glycerol kinase	
2.7.1.30	ATP:glycerol 3-phosphotransferase Glycerol kinase	
2.7.1.29	ATP:glycerone phosphotransferase Glycerone kinase	
2.7.4.8	ATP:GMP phosphotransferase Guanylate kinase	
2.7.6.5	ATP:GTP 3'-pyrophosphotransferase GTP pyrophosphokinase	
2.7.3.1	ATP:guanidinoacetate N-phosphotransferase Guanidinoacetate kinase	
2.7.3.7	ATP:guanidinoethyl-methyl-phosphate phosphotransferase Opheline kinase	
2.7.3.7	ATP:guanidinoethylmethylphosphate phosphotransferase Opheline kinase	
2.7.1.119	ATP:hygromycin-B 7-O-phosphotransferase Hygromycin-B kinase	
2.7.3.6	ATP:hypotaurocyamine N-phosphotransferase Hypotaurocyamine kinase	
2.7.1.65	ATP:inosamine phosphotransferase scyllo-Inosamine 4-kinase	
2.7.1.73	ATP:inosine 5'-phosphotransferase Inosine kinase	
2.7.1.95	ATP:kanamycin 3'-O-phosphotransferase Kanamycin kinase	
2.7.1.16	ATP:L(or D)-ribulose 5-phosphotransferase Ribulokinase	
2.7.1.46	ATP:L-arabinose 1-phosphotransferase L-Arabinokinase	
2.7.3.3	ATP:L-arginine N-phosphotransferase Arginine kinase	
2.7.2.4	ATP:L-aspartate 4-phosphotransferase Aspartate kinase	
2.7.1.51	ATP:L-fuculose 1-phosphotransferase L-Fuculokinase	

EC-No.	Synonym Recommended Name	Vol., Page
2.7.2.13	ATP:L-glutamate 1-phosphotransferase Glutamate 1-kinase	
2.7.2.11	ATP:L-glutamate 5-phosphotransferase Glutamate 5-kinase	
2.7.1.39	ATP:L-homoserine O-phosphotransferase Homoserine kinase	
2.5.1.6	ATP:L-methionine-S-adenosyltransferase Methionine adenosyltransferase	
2.5.1.6	ATP:L-methionine S-adenosyltransferase Methionine adenosyltransferase	
2.7.7.54	ATP:L-phenylalanine adenylyltransferase Phenylalanine adenylyltransferase	
2.7.1.5	ATP:L-rhamnulose 1-phosphotransferase Rhamnulokinase	
2.7.1.53	ATP:L-xylulose 5-phosphotransferase L-Xylulokinase	
2.7.3.5	ATP:lombricine N-phosphotransferase Lombricine kinase	
2.7.1.131	ATP:low-density-lipoprotein-L-serine O-phosphotransferase Low-density-lipoprotein kinase	
2.7.1.136	ATP:macrolide 2'-O-phosphotransferase Macrolide 2'-kinase	
2.7.1.36	ATP:mevalonate 5-phosphotransferase Mevalonate kinase	
2.7.1.64	ATP:myo-inositol 1-phosphotransferase myo-Inositol 1-kinase	
2.7.1.129	ATP:myosin-heavy-chain O-phosphotransferase Myosin-heavy-chain kinase	
2.7.1.117	ATP:myosin-light-chain O-phosphotransferase Myosin-light-chain kinase	
2.7.1.59	ATP:N-acetyl-D-glucosamine 6-phosphotransferase N-Acetylglucosamine kinase	
2.7.2.8	ATP:N-acetyl-L-glutamate 5-phosphotransferase Acetylglutamate kinase	
2.7.1.60	ATP:N-acetylmannosamine 6-phosphotransferase N-Acylmannosamine kinase	

EC-No.	Synonym Recommended Name	Vol., Page
2.7.1.60	ATP:N-acyl-D-mannosamine 6-phosphotransferase N-Acylmannosamine kinase	
2.7.1.22	ATP:N-ribosylnicotinamide 5'-phosphotransferase Ribosylnicotinamide kinase	
2.7.1.23	ATP:NAD$^+$ 2'-phosphotransferase NAD$^+$ kinase	
2.7.1.86	ATP:NADH 2'-phosphotransferase NADH kinase	
2.7.7.1	ATP:nicotinamide-nucleotide adenylyltransferase Nicotinamide-nucleotide adenylyltransferase	
2.7.7.18	ATP:nicotinate-ribonucleotide adenylyltransferase Nicotinate-nucleotide adenylyltransferase	
2.7.7.1	ATP:NMN adenylyltransferase Nicotinamide-nucleotide adenylyltransferase	
2.7.6.4	ATP:nucleoside-5'-phosphate pyrophosphotransferase Nucleotide pyrophosphokinase	
2.7.4.6	ATP:nucleoside-diphosphate phosphotransferase Nucleoside-diphosphate kinase	
2.7.4.4	ATP:nucleoside-phosphate phosphotransferase Nucleoside-phosphate kinase	
2.7.6.4	ATP:nucleotide pyrophosphotransferase Nucleotide pyrophosphokinase	
2.7.1.84	ATP:O-alkylglycerone phosphotransferase Alkylglycerone kinase	
4.1.1.49	ATP:oxaloacetate carboxy-lyase (transphosphorylating) Phosphoenolpyruvate carboxykinase (ATP)	
2.7.7.3	ATP:pantetheine-4'-phosphate adenylyltransferase Pantetheine-phosphate adenylyltransferase	
2.7.1.34	ATP:pantetheine 4'-phosphotransferase Pantetheine kinase	
2.7.1.33	ATP:pantothenate 4'-phosphotransferase Pantothenate kinase	
2.7.1.38	ATP:phosphorylase-b phosphotransferase Phosphorylase kinase	
2.7.7.19	ATP:polynucleotide adenylyltransferase Polynucleotide adenylyltransferase	

EC-No.	Synonym Recommended Name	Vol., Page
2.7.7.19	ATP:polynucleotidylexotransferase Polynucleotide adenylyltransferase	
2.7.4.1	ATP:polyphosphate phosphotransferase Polyphosphate kinase	
2.7.1.70	ATP:protamine O-phosphotransferase Protamine kinase	
2.7.3.11	ATP:protein-L-histidine N-pros-phosphotransferase Protein-histidine pros-kinase	
2.7.3.12	ATP:protein-L-histidine N-tele-phosphotransferase Protein-histidine tele-kinase	
2.7.1.112	ATP:protein-tyrosine O-phosphotransferase Protein-tyrosine kinase	
2.7.1.123	ATP:protein O-phosphotransferase (calmodulin-dependent) Ca^{2+}/calmodulin-dependent protein kinase	
2.7.1.37	ATP:protein phosphotransferase Protein kinase	
2.7.1.83	ATP:pseudouridine 5'-phosphotransferase Pseudouridine kinase	
2.7.1.35	ATP:pyridoxal 5'-phosphotransferase Pyridoxal kinase	
2.7.9.1	ATP:pyruvate,orthophosphate phosphotransferase phosphate dikinase pyruvate	
2.7.9.2	ATP:pyruvate,water phosphotransferase water dikinase pyruvate	
2.7.1.40	ATP:pyruvate 2-O-phosphotransferase Pyruvate kinase	
3.1.3.49	[ATP:pyruvate O_2-phosphotransferase]-phosphate phosphohydrolase [Pyruvate kinase]-phosphatase	
2.7.1.125	ATP:rhodopsin phosphotransferase Rhodopsin kinase	
2.7.1.26	ATP:riboflavin 5'-phosphotransferase Riboflavin kinase	
2.7.1.100	ATP:S5-methyl-5-thio-D-ribose 1-phosphotransferase 5-Methylthioribose kinase	
2.7.1.14	ATP:sedoheptulose 7-phosphotransferase Sedoheptulokinase	

EC-No.	Synonym Recommended Name	Vol., Page
2.7.9.3	ATP:selenide, water phosphotransferase selenide, water dikinase	
2.7.1.71	ATP:shikimate 3-phosphotransferase Shikimate kinase	
2.7.1.91	ATP:sphinganine 1-phosphotransferase Sphinganine kinase	
2.7.7.47	ATP:streptomycin 3-adenylyltransferase Streptomycin 3-adenylyltransferase	
2.7.1.87	ATP:streptomycin 3-phosphotransferase Streptomycin 3-kinase	
2.7.1.72	ATP:streptomycin 6-phosphotransferase Streptomycin 6-kinase	
2.7.7.4	ATP:sulfate adenylyltransferase Sulfate adenylyltransferase	
2.7.1.135	ATP:tau-protein O-phosphotransferase τ-protein kinase	
2.7.3.4	ATP:taurocyamine N-phosphotransferase Taurocyamine kinase	
2.7.3.4	ATP:taurocyamine phosphotransferase Taurocyamine kinase	
2.7.4.15	ATP:thiamin-diphosphate phosphotransferase Thiamine-diphosphate kinase	
2.7.4.16	ATP:thiamin-phosphate phosphotransferase Thiamine-phosphate kinase	
2.7.4.15	ATP:thiamine-diphosphate phosphotransferase Thiamine-diphosphate kinase	
2.7.4.16	ATP:thiamine-phosphate phosphotransferase Thiamine-phosphate kinase	
2.7.1.89	ATP:thiamine phosphotransferase Thiamine kinase	
2.7.6.2	ATP:thiamine pyrophosphotransferase Thiamine pyrophosphokinase	
2.7.1.89	ATP:thiamin phosphotransferase Thiamine kinase	
2.7.6.2	ATP:thiamin pyrophosphotransferase Thiamine pyrophosphokinase	

EC-No.	Synonym Recommended Name	Vol., Page
2.7.1.21	ATP:thymidine 5'-phosphotransferase Thymidine kinase	
2.7.7.25	ATP:tRNA adenylyltransferase tRNA adenylyltransferase	
2.7.7.21	ATP:tRNA adenylyltransferase tRNA cytidylyltransferase	
2.7.7.25	ATP:tRNA nucleotidyltransferase (CTP) tRNA adenylyltransferase	
2.7.7.21	ATP:tRNA nucleotidyltransferase (CTP) tRNA cytidylyltransferase	
2.7.1.132	ATP:tropomyosin O-phosphotransferase Tropomyosin kinase	
2.7.4.14	ATP:UMP-CMP phosphotransferase Cytidylate kinase	
2.7.1.66	ATP:undecaprenol phosphotransferase Undecaprenol kinase	
6.3.4.6	ATP:urea amidolyase Urea carboxylase	Vol. 2, Page 603
2.7.1.48	ATP:uridine 5'-phosphotransferase Uridine kinase	
2.7.1.103	ATP:viomycin O-phosphotransferase Viomycin kinase	
2.7.1.122	ATP:xylitol 5-phosphotransferase Xylitol kinase	
2.7.7.53	ATP adenylyltransferase ATP adenylyltransferase	
3.5.4.18	ATP aminohydrolase ATP deaminase	
3.6.1.8	ATPase ATP diphosphatase	
3.6.1.3	ATPase Adenosinetriphosphatase	
4.1.3.8	ATP citrate (pro-3S)-lyase ATP citrate (pro-3S)-lyase	
3.5.4.18	ATP deaminase ATP deaminase	

EC-No.	Synonym Recommended Name	Vol., Page
3.6.1.8	ATP diphosphatase ATP diphosphatase	
3.6.1.5	ATP diphosphohydrolase Apyrase	
3.6.1.3	ATP hydrolase Adenosinetriphosphatase	
3.6.1.3	ATP monophosphatase Adenosinetriphosphatase	
2.7.6.4	ATP nucleotide 3'-pyrophosphokinase Nucleotide pyrophosphokinase	
3.6.1.3	ATP phosphohydrolase Adenosinetriphosphatase	
3.6.1.35	ATP phosphohydrolase H^+-transporting ATPase	
3.6.1.38	ATP phosphohydrolase (Ca^{2+}-transporting) Ca^{2+}-transporting ATPase	
3.6.1.34	ATP phosphohydrolase (H^+-transporting) H^+-transporting ATP synthase	
3.6.1.35	ATP phosphohydrolase (H^+-transporting, acyl-phosphate-linked) H^+-transporting ATPase	
3.6.1.36	ATP phosphohydrolase (H^+/K^+-transporting) H^+/K^+-transporting ATPase	
3.6.1.37	ATP phosphohydrolase (Na^+/K^+-transporting) Na^+/K^+-transporting ATPase	
2.4.2.17	ATP phosphoribosyltransferase ATP phosphoribosyltransferase	
3.6.1.8	ATP pyrophosphatase ATP diphosphatase	
4.6.1.1	ATP pyrophosphate-lyase (cyclizing) adenylate cyclase	
3.6.1.8	ATP pyrophosphohydrolase ATP diphosphatase	
2.7.7.4	ATP sulfurylase Sulfate adenylyltransferase	
3.8.1.8	atrazine chlorohydrolase atrazine chlorohydrolase	

EC-No.	Synonym Recommended Name	Vol., Page
3.8.1.8	atrazine chlorohydrolase atrazine chlorohydrolase	
3.4.15.4	Atrial di-(tri-)peptidyl carboxyhydrolase Peptidyl-dipeptidase B	
3.4.15.4	Atrial dipeptidyl carboxyhydrolase Peptidyl-dipeptidase B	
3.4.15.4	Atrial peptide convertase Peptidyl-dipeptidase B	
3.4.15.4	Atriopeptin convertase Peptidyl-dipeptidase B	
3.4.24.1	atrolysin A atrolysin A	
3.4.24.41	Atrolysin B Atrolysin B	
3.4.24.42	Atrolysin C Atrolysin C	
3.4.24.44	Atrolysin E Atrolysin E	
3.4.24.45	Atrolysin F Atrolysin F	
3.1.1.10	Atropinase Tropinesterase	
3.1.1.10	Atropine acylhydrolase Tropinesterase	
3.1.1.10	Atropine esterase Tropinesterase	
3.1.1.10	Atropinesterase Tropinesterase	
3.1.1.10	Atropinesterase ((-)hyoscyamine acylhydrolase, E.C. 3.1.1.10) Tropinesterase	
3.4.24.43	Atroxase Atroxase	
3.4.24.43	Atroxase (Crotalus atrox) Atroxase	
3.8.1.8	AtzA atrazine chlorohydrolase	

EC-No.	Synonym Recommended Name	Vol., Page
3.5.99.3	AtzB hydroxydechloroatrazine ethylaminohydrolase	
3.5.99.4	AtzC N-isopropylammelide isopropylaminohydrolase	
3.4.24.29	Aureolysin Aureolysin	
3.4.24.38	Autolysin Autolysin	
3.4.21.6	autoprothrombin C coagulation factor Xa	
3.4.21.69	Autoprothrombin II-A Protein C (activated)	
3.4.23.18	Avamorin Aspergillopepsin I	
3.4.21.1	Avazyme Chymotrypsin	
3.2.1.4	Avicelase Cellulase	
3.2.1.91	Avicelase Cellulose 1,4-β-cellobiosidase	
3.4.23.18	Awamorin Aspergillopepsin I	
1.6.6.7	Azo-dye reductase Azobenzene reductase	
1.6.6.7	Azobenzene reductase Azobenzene reductase	
3.4.21.32	azocollase brachyurin	
3.4.24.7	azocollase interstitial collagenase	
3.4.24.3	azocollase microbial collagenase	
1.6.6.7	Azo dye reductase Azobenzene reductase	
3.2.2.24	Azoferredoxin-activating enzymes ADP-ribosyl-[dinitrogen reductase] hydrolase	

EC-No.	Synonym Recommended Name	Vol., Page
2.4.2.37	Azoferredoxin ADP-ribosyltransferase NAD^+-dinitrogen-reductase ADP-D-ribosyltransferase	
1.6.6.7	Azo reductase Azobenzene reductase	
1.6.6.7	Azoreductase Azobenzene reductase	
1.6.99.8	B_{12a} reductase Aquacobalamin reductase	
2.1.1.13	B_{12} N^5-Methyltetrahydrofolate homocysteine methyltransferase 5-Methyltetrahydrofolate-homocysteine S-methyltransferase	
1.6.99.9	B_{12r} reductase Cob(II)alamin reductase	
3.4.24.74	Baceroides fragilis enterotoxin Fragilysin	
3.4.24.28	Bacillolysin Bacillolysin	
3.4.21.62	Bacillopeptidase A Subtilisin	
3.4.21.62	Bacillopeptidase B Subtilisin	
2.4.1.19	Bacillus macerans amylase Cyclomaltodextrin glucanotransferase	
3.4.24.28	Bacillus metalloendopeptidase Bacillolysin	
3.4.24.28	Bacillus metalloproteinase Bacillolysin	
3.4.24.28	Bacillus neutral proteinase Bacillolysin	
3.4.21.62	Bacillus subtilis alkaline proteinase Bioprase Subtilisin	
3.4.24.28	Bacillus subtilis neutral proteinase Bacillolysin	
3.1.27.2	Bacillus subtilis ribonuclease Bacillus subtilis ribonuclease	
3.4.24.27	Bacillus thermoproteolyticus neutral proteinase Thermolysin	

EC-No.	Synonym Recommended Name	Vol., Page
3.6.1.34	Bacterial Ca^{2+}/Mg^{2+} ATPase H^+-transporting ATP synthase	
3.4.21.89	Bacterial leader peptidase 1 Signal peptidase I	
3.4.11.10	bacterial leucyl aminopeptidase bacterial leucyl aminopeptidase	
1.14.14.3	Bacterial luciferase Alkanal monooxygenase (FMN-linked)	
3.1.11.4	bacteriophage SP3 deoxyribonuclease exodeoxyribonuclease (phage SP3-induced)	
3.1.25.1	Bacteriophage T4 endodeoxyribonuclease V Deoxyribonuclease (pyrimidine dimer)	
3.4.24.74	Bacteroides fragilis toxin Fragilysin	
2.5.1.31	Bactoprenyl-diphosphate synthase Di-trans,poly-cis-Decaprenylcistransferase	
2.4.1.129	Bactoprenyldiphospho-N-acetylmuramoyl-(N-acetyl-D-glucosaminyl)-pentapeptide:peptidoglycan N-acetylmuramoyl-N-acetyl-D-glucosaminyl-transferase Peptidoglycan glycosyltransferase	
1.2.1.57	BAD Butanal dehydrogenase	
4.1.1.19	bADC Arginine decarboxylase	
1.2.1.8	BADH betaine-aldehyde dehydrogenase	
3.4.21.48	baker's yeast proteinase cerevisin	
3.4.21.48	baker's yeast proteinase B cerevisin	
3.4.22.40	BANA-hydrolase bleomycin hydrolase	
3.5.1.72	D-BAPA-ase D-Benzoylarginine-4-nitroanilide amidase	
3.5.2.1	Barbiturase Barbiturase	

EC-No.	Synonym Recommended Name	Vol., Page
3.5.2.1	Barbiturate amidohydrolase Barbiturase	
3.4.23.40	Barley grain aspartic proteinase Phytepsin	
3.1.30.2	Barley nuclease Serratia marcescens nuclease	
3.4.23.35	Bar proteinase Barrierpepsin	
3.4.23.35	Barrier Barrierpepsin	
3.4.23.35	Barrierpepsin Barrierpepsin	
3.4.23.35	Barrier proteinase Barrierpepsin	
3.5.3.7	G-Base Guanidinobutyrase	
3.1.27.1	Base-non-specific ribonuclease Ribonuclease T_2	
3.4.21.74	Batroxobin Venombin A	
2.7.3.2	BB-CK Creatine kinase	
1.5.3.9	BBE Reticuline oxidase	
5.3.1.16	BBM II ketolisomerase N-(5'-Phospho-D-ribosylformimino)-5-amino-1-(5-phosphoribosyl)-4-imidazole carboxamide isomerase	Vol. 1, Page 335
6.3.4.14	BC Biotin carboxylase	Vol. 2, Page 632
3.1.1.8	BChE cholinesterase	
4.1.1.72	BCKA Branched-chain-2-oxoacid decarboxylase	
1.2.4.4	BCKDH 3-Methyl-2-oxobutanoate dehydrogenase (lipoamide)	

EC-No.	Synonym Recommended Name	Vol., Page
3.1.3.52	BCKDH [3-Methyl-2-oxobutanoate dehydrogenase (lipoamide)]-phosphatase	
3.1.3.16	BCKDH phosphatase Phosphoprotein phosphatase	
1.2.4.4	BCOAD 3-Methyl-2-oxobutanoate dehydrogenase (lipoamide)	
1.1.1.76	L-BDH (S,S)-Butanediol dehydrogenase	
3.4.22.34	Bean endopeptidase Legumain	
3.4.21.9	BEK enteropeptidase	
1.2.1.28	Benzaldehyde:NAD$^+$ oxidoreductase Benzaldehyde dehydrogenase (NAD$^+$)	
1.2.1.7	Benzaldehyde:NADP$^+$ oxidoreductase Benzaldehyde dehydrogenase (NADP$^+$)	
1.2.1.6	benzaldehyde dehydrogenase benzaldehyde dehydrogenase	
1.2.1.28	Benzaldehyde dehydrogenase (NAD) Benzaldehyde dehydrogenase (NAD$^+$)	
1.2.1.28	Benzaldehyde dehydrogenase (NAD$^+$) Benzaldehyde dehydrogenase (NAD$^+$)	
1.2.1.7	Benzaldehyde dehydrogenase (NADP$^+$) Benzaldehyde dehydrogenase (NADP$^+$)	
4.1.2.38	Benzaldehyde lyase Benzoin aldolase	
3.5.1.14	Benzamidase Aminoacylase	
1.14.12.3	Benzene,NADH:oxygen oxidoreductase (1,2-hydroxylating) Benzene 1,2-dioxygenase	
1.13.11.37	Benzene-1,2,4-triol:oxygen 1,2-oxidoreductase (decyclizing) Hydroxyquinol 1,2-dioxygenase	
1.14.12.15	Benzene-1,4-dicarboxylate, NADH:oxygen oxidoreductase (1,2-hydroxylating) Terephthalate 1,2-dioxygenase	

EC-No.	Synonym Recommended Name	Vol., Page
1.14.12.15	Benzene-1,4-dicarboxylate 1,2-dioxygenase Terephthalate 1,2-dioxygenase	
1.14.12.3	Benzene 1,2-dioxygenase Benzene 1,2-dioxygenase	
1.10.3.1	1,2-Benzenediol:oxygen oxidoreductase Catechol oxidase	
1.10.3.2	Benzenediol:oxygen oxidoreductase Laccase	
1.14.12.3	Benzene dioxygenase Benzene 1,2-dioxygenase	
1.14.12.3	Benzene hydroxylase Benzene 1,2-dioxygenase	
3.3.2.3	Benzo(a)pyrene-4,5-epoxide hydratase Epoxide hydrolase	
3.3.2.3	Benzo(a)pyrene-4,5-oxide hydratase Epoxide hydrolase	
3.3.2.3	Benzo[a]pyrene-4,5-oxide hydratase Epoxide hydrolase	
1.14.12.10	Benzoate,NADH:oxygen oxidoreductase (1,2-hydroxylating, decarboxylating) Benzoate 1,2-dioxygenase	
1.14.13.12	Benzoate,NADPH:oxygen oxidoreductase (4-hydroxylating) Benzoate 4-monooxygenase	
1.14.13.12	Benzoate-4-hydroxylase Benzoate 4-monooxygenase	
6.2.1.25	Benzoate-CoA ligase Benzoate-CoA ligase	Vol. 2, Page 314
6.2.1.25	Benzoate-CoA ligase (AMP-forming) Benzoate-CoA ligase	Vol. 2, Page 314
1.14.13.12	Benzoate-*p*-hydroxylase Benzoate 4-monooxygenase	
1.14.12.10	Benzoate 1,2-dioxygenase Benzoate 1,2-dioxygenase	
1.13.99.2	benzoate 1,2-dioxygenase benzoate 1,2-dioxygenase	

EC-No.	Synonym Recommended Name	Vol., Page
1.14.13.12	Benzoate 4-hydroxylase Benzoate 4-monooxygenase	
1.14.13.12	Benzoate 4-monooxygenase Benzoate 4-monooxygenase	
6.2.1.25	Benzoate:CoA ligase (AMP-forming) Benzoate-CoA ligase	Vol. 2, Page 314
6.2.1.25	Benzoate coenzyme A ligase Benzoate-CoA ligase	Vol. 2, Page 314
1.14.12.10	Benzoate dioxygenase Benzoate 1,2-dioxygenase	
1.14.12.10	Benzoate hydroxylase Benzoate 1,2-dioxygenase	
1.14.13.12	Benzoic 4-hydroxylase Benzoate 4-monooxygenase	
1.14.13.12	Benzoic acid 4-hydroxylase Benzoate 4-monooxygenase	
1.14.12.10	Benzoic hydroxylase Benzoate 1,2-dioxygenase	
4.1.2.38	Benzoin aldolase Benzoin aldolase	
3.5.5.1	Benzonitrilase Nitrilase	
2.3.1.151	Benzophenone synthase Benzophenone synthase	
1.14.14.2	benzopyrene 3-monooxygenase benzopyrene 3-monooxygenase	
1.6.99.2	*p*-Benzoquinone reductase NAD(P)H dehydrogenase (quinone)	
1.6.5.5	*p*-Benzoquinone reductase NADPH:quinone reductase	
1.6.5.6	*p*-benzoquinone reductase (NADPH) *p*-benzoquinone reductase (NADPH)	
2.1.1.105	N-Benzoyl-4-hydroxyanthranilate 4-O-methyltransferase N-Benzoyl-4-hydroxyanthranilate 4-O-methyltransferase	
1.14.13.58	Benzoyl-CoA, NADPH2:oxygen oxidoreductase (3-hydroxylating) Benzoyl-CoA 3-monooxygenase	

EC-No.	Synonym Recommended Name	Vol., Page
1.14.13.58	Benzoyl-CoA 3-hydroxylase Benzoyl-CoA 3-monooxygenase	
1.14.13.58	Benzoyl-CoA 3-monooxygenase Benzoyl-CoA 3-monooxygenase	
2.3.1.144	Benzoyl-CoA:anthranilate N-benzoyltransferase Anthranilate N-benzoyltransferase	
2.3.1.71	Benzoyl-CoA:glycine N-acyltransferase Glycine N-benzoyltransferase	
2.3.1.71	Benzoyl-CoA:glycine N-benzoyltransferase Glycine N-benzoyltransferase	
2.3.1.127	Benzoyl-CoA:L-ornithine N-benzoyltransferase Ornithine N-benzoyltransferase	
6.2.1.25	Benzoyl-CoA ligase Benzoate-CoA ligase	Vol. 2, Page 314
1.3.99.15	Benzoyl-CoA reductase Benzoyl-CoA reductase	
1.3.99.15	Benzoyl-CoA reductase (dearomatizing) Benzoyl-CoA reductase	
6.2.1.25	Benzoyl-CoA synthetase Benzoate-CoA ligase	Vol. 2, Page 314
3.5.1.72	N-Benzoyl-D-arginine-4-nitroanilide amidohydrolase D-Benzoylarginine-4-nitroanilide amidase	
3.4.17.6	N-benzoyl-L-alanine-aminohydrolase alanine carboxypeptidase	
3.4.24.18	N-Benzoyl-L-tyrosyl-*p*-aminobenzoic acid hydrolase Meprin A	
3.5.1.40	Benzoylagmatine amidohydrolase Acylagmatine amidase	
3.5.1.32	N-Benzoylamino-acid amidohydrolase Hippurate hydrolase	
3.4.22.16	N-Benzoylarginine-β-naphthylamide hydrolase Cathepsin H	
3.5.1.72	D-Benzoylarginine-4-nitroanilide amidase D-Benzoylarginine-4-nitroanilide amidase	
3.1.1.9	benzoylcholinesterase benzoylcholinesterase	

EC-No.	Synonym Recommended Name	Vol., Page
3.1.1.8	benzoylcholinesterase cholinesterase	
2.3.1.71	Benzoyl CoA-amino acid N-acyltransferase Glycine N-benzoyltransferase	
6.2.1.25	Benzoyl coenzyme A synthetase Benzoate-CoA ligase	Vol. 2, Page 314
4.1.1.7	Benzoylformate carboxy-lyase Benzoylformate decarboxylase	
4.1.1.7	Benzoylformate decarboxylase Benzoylformate decarboxylase	
2.3.1.144	Benzoyltransferase, anthranilate N- Anthranilate N-benzoyltransferase	
2.3.1.71	Benzoyltransferase, glycine Glycine N-benzoyltransferase	
2.3.1.127	Benzoyltransferase, ornithine N- Ornithine N-benzoyltransferase	
1.1.1.217	Benzyl-(2R,3S)-2-methyl-3-hydroxybutanoate:$NADP^+$ 3-oxidoreductase Benzyl-2-methyl-hydroxybutyrate dehydrogenase	
1.1.1.217	Benzyl-2-methyl-hydroxybutyrate dehydrogenase Benzyl-2-methyl-hydroxybutyrate dehydrogenase	
5.99.1.1	Benzyl-thiocyanate isomerase Thiocyanate isomerase	Vol. 1, Page 719
1.1.1.217	Benzyl 2-methyl-3-hydroxybutyrate dehydrogenase Benzyl-2-methyl-hydroxybutyrate dehydrogenase	
1.1.1.90	Benzyl alcohol dehydrogenase Aryl-alcohol dehydrogenase	
1.4.3.6	Benzylamine oxidase Amine oxidase (copper-containing)	
3.5.1.64	N^{α}-Benzyloxycarbonyl-L-leucine urethanehydrolase N^{α}-Benzyloxycarbonylleucine hydrolase	
3.5.1.58	N^{α}-Benzyloxycarbonyl amino acid urethane hydrolase N-Benzyloxycarbonylglycine hydrolase	
3.5.1.58	N^{α}-Benzyloxycarbonyl amino acid urethane hydrolase I N-Benzyloxycarbonylglycine hydrolase	
3.5.1.64	N^{α}-Benzyloxycarbonyl amino acid urethane hydrolase IV N^{α}-Benzyloxycarbonylleucine hydrolase	

EC-No.	Synonym Recommended Name	Vol., Page
3.5.1.58	N-Benzyloxycarbonylglycine hydrolase N-Benzyloxycarbonylglycine hydrolase	
3.5.1.58	N-Benzyloxycarbonylglycine urethanehydrolase N-Benzyloxycarbonylglycine hydrolase	
3.5.1.64	N^{α}-Benzyloxycarbonylleucine hydrolase N^{α}-Benzyloxycarbonylleucine hydrolase	
3.5.1.11	Benzylpenicillin acylase Penicillin amidase	
4.1.99.11	benzylsuccinate fumarate-lyase benzylsuccinate synthase	
4.1.99.11	benzylsuccinate synthase benzylsuccinate synthase	
1.1.3.34	Benzyltetrahydroisoquinoline oxidase Berbamunine synthase	
1.9.6.1	Benzyl viologen-nitrate reductase Nitrate reductase (cytochrome)	
1.1.3.34	Berbamunine synthase Berbamunine synthase	
1.5.3.9	Berberine-bridge-forming enzyme Reticuline oxidase	
1.5.3.9	Berberine bridge enzyme Reticuline oxidase	
1.1.3.26	Berberine synthase Columbamine oxidase	
2.1.1.92	Bergaptol O-methyltransferase Bergaptol O-methyltransferase	
4.1.1.7	BFD Benzoylformate decarboxylase	
1.1.1.157	BHBD 3-Hydroxybutyryl-CoA dehydrogenase	
1.1.1.198	Bicyclic monoterpenol dehydrogenase (+)-Borneol dehydrogenase	
1.18.99.1	Bidirectional hydrogenase Hydrogenase	

EC-No.	Synonym Recommended Name	Vol., Page
1.1.1.83	Bifunctional L(+)-tartrate dehydrogenase-D(+)-malate dehydrogenase (decarboxylating, EC 1.1.1.93 and EC 1.1.1.83) D-Malate dehydrogenase (decarboxylating)	
2.8.2.14	Bile-salt sulfotransferase Bile-salt sulfotransferase	
6.2.1.7	Bile acid-coenzyme A ligase Cholate-CoA ligase	Vol. 2, Page 236
6.2.1.7	Bile acid:CoA ligase Cholate-CoA ligase	Vol. 2, Page 236
6.2.1.7	Bile acid coenzyme A synthetase Cholate-CoA ligase	Vol. 2, Page 236
3.5.1.24	Bile salt hydrolase Cholylglycine hydrolase	
2.8.2.14	Bile salt sulfotransferase Bile-salt sulfotransferase	
2.4.1.95	Bilirubin-glucuronoside:bilirubin-glucuronoside D-glucuronosyltransferase Bilirubin-glucuronoside glucuronosyltransferase	
2.4.1.95	Bilirubin-glucuronoside glucuronosyltransferase Bilirubin-glucuronoside glucuronosyltransferase	
1.3.1.24	Bilirubin:NAD(P)$^+$ oxidoreductase Biliverdin reductase	
1.3.3.5	Bilirubin:oxygen oxidoreductase Bilirubin oxidase	
2.4.1.95	Bilirubin glucuronoside glucuronosyltransferase Bilirubin-glucuronoside glucuronosyltransferase	
2.4.1.17	Bilirubin glucuronyltransferase Glucuronosyltransferase	
2.4.1.17	Bilirubin monoglucuronide glucuronyltransferase Glucuronosyltransferase	
2.4.1.95	Bilirubin monoglucuronide transglucuronidase Bilirubin-glucuronoside glucuronosyltransferase	
1.3.3.5	Bilirubin oxidase Bilirubin oxidase	
1.3.3.5	Bilirubin oxidase (EC 1.3.3.5) Bilirubin oxidase	

EC-No.	Synonym Recommended Name	Vol., Page
1.3.3.5	Bilirubin oxidase M-1 Bilirubin oxidase	
2.4.1.17	Bilirubin UDP-glucuronosyltransferase Glucuronosyltransferase	
2.4.1.17	Bilirubin UDPglucuronosyltransferase Glucuronosyltransferase	
2.4.1.17	Bilirubin UDPGT Glucuronosyltransferase	
2.4.1.17	Bilirubin uridine diphosphoglucuronyltransferase Glucuronosyltransferase	
1.3.1.24	Biliverdin reductase Biliverdin reductase	
3.1.27.3	Binase Ribonuclease T1	
2.8.1.6	BIO2 protein (Arabidopsis thaliana clone pMP101 gene BIO2 reduced) Biotin synthase	
1.3.1.46	Biochanin-A reductase Biochanin-A reductase	
3.4.21.62	Bioprase AL 15 Subtilisin	
3.4.21.62	Bioprase APL 30 Subtilisin	
4.1.1.19	Biosynthetic arginine decarboxylase Arginine decarboxylase	
6.3.4.11	Biotin-β-methylcrotonyl coenzyme A carboxylase synthetase Biotin-[methylcrotonyl-CoA-carboxylase] ligase	Vol. 2, Page 622
6.3.4.15	Biotin-[acetyl-CoA-carboxylase] ligase Biotin-[acetyl-CoA-carboxylase] ligase	Vol. 2, Page 638
6.3.4.15	Biotin-[acetyl-CoA carboxylase] synthetase Biotin-[acetyl-CoA-carboxylase] ligase	Vol. 2, Page 638
6.3.4.15	Biotin-[acetyl coenzyme A carboxylase] synthetase Biotin-[acetyl-CoA-carboxylase] ligase	Vol. 2, Page 638
6.3.4.11	Biotin-[methylcrotonyl-CoA-carboxylase] ligase Biotin-[methylcrotonyl-CoA-carboxylase] ligase	Vol. 2, Page 622
6.3.4.11	Biotin-[methylcrotonyl-CoA-carboxylase] synthetase Biotin-[methylcrotonyl-CoA-carboxylase] ligase	Vol. 2, Page 622

EC-No.	Synonym Recommended Name	Vol., Page
6.3.4.9	Biotin-[methylmalonyl-CoA-carboxyltransferase] ligase Biotin-[methylmalonyl-CoA-carboxyltransferase] ligase	Vol. 2, Page 613
6.3.4.9	Biotin-[methylmalonyl-CoA-carboxyltransferase] synthetase Biotin-[methylmalonyl-CoA-carboxyltransferase] ligase	Vol. 2, Page 613
6.3.4.10	Biotin-[propionyl-CoA-carboxylase (ATP-hydrolysing)] ligase Biotin-[propionyl-CoA-carboxylase (ATP-hydrolysing)] ligase	Vol. 2, Page 617
6.3.4.10	Biotin-[propionyl-CoA-carboxylase (ATP-hydrolysing)] synthetase Biotin-[propionyl-CoA-carboxylase (ATP-hydrolysing)] ligase	Vol. 2, Page 617
3.5.1.12	Biotin-amide amidohydrolase Biotinidase	
6.3.4.9	Biotin-apotranscarboxylase synthetase Biotin-[methylmalonyl-CoA-carboxyltransferase] ligase	Vol. 2, Page 613
6.3.4.14	Biotin-carboxyl-carrier-protein:carbon-dioxide ligase (ADP-forming) Biotin carboxylase	Vol. 2, Page 632
6.2.1.11	Biotin-CoA ligase Biotin-CoA ligase	Vol. 2, Page 252
6.3.4.9	Biotin-methylmalonyl coenzyme A carboxyltransferase synthetase Biotin-[methylmalonyl-CoA-carboxyltransferase] ligase	Vol. 2, Page 613
6.3.4.10	Biotin-propionyl coenzyme A carboxylase synthetase Biotin-[propionyl-CoA-carboxylase (ATP-hydrolysing)] ligase	Vol. 2, Page 617
6.3.4.9	Biotin-transcarboxylase synthetase Biotin-[methylmalonyl-CoA-carboxyltransferase] ligase	Vol. 2, Page 613
6.3.4.11	Biotin:apo-[3-methylcrotonyl-CoA:carbon-dioxide ligase (ADP-forming)] ligase(AMP-forming) Biotin-[methylcrotonyl-CoA-carboxylase] ligase	Vol. 2, Page 622
6.3.4.15	Biotin:apo-[acetyl-CoA:carbon-dioxide ligase (ADP-forming)] ligase (AMP-forming) Biotin-[acetyl-CoA-carboxylase] ligase	Vol. 2, Page 638
6.3.4.10	Biotin:apo-[propanoyl-CoA:carbon-dioxide ligase (ADP-forming)] ligase (AMP-forming) Biotin-[propionyl-CoA-carboxylase (ATP-hydrolysing)] ligase	Vol. 2, Page 617
6.3.4.9	Biotin:apo[methylmalonyl-CoA:pyruvate carboxyltransferase] ligase (AMP-forming) Biotin-[methylmalonyl-CoA-carboxyltransferase] ligase	Vol. 2, Page 613

EC-No.	Synonym Recommended Name	Vol., Page
6.3.4.15	Biotin:apocarboxylase ligase Biotin-[acetyl-CoA-carboxylase] ligase	Vol. 2, Page 638
6.2.1.11	Biotin:CoA ligase (AMP-forming) Biotin-CoA ligase	Vol. 2, Page 252
6.3.4.9	(+)-Biotin:methylmalonyl-CoA-pyruvate apocarboxyltransferase ligase (AMP) Biotin-[methylmalonyl-CoA-carboxyltransferase] ligase	Vol. 2, Page 613
6.3.4.14	Biotin carboxylase Biotin carboxylase	Vol. 2, Page 632
6.3.4.15	Biotin holoenzyme synthetase Biotin-[acetyl-CoA-carboxylase] ligase	Vol. 2, Page 638
3.5.1.12	Biotinidase Biotinidase	
2.8.1.6	Biotin synthase Biotin synthase	
2.8.1.6	Biotin synthase (Saccharomyces cerevisiae strain 20B-12 clone pUCH2.4 gene BIO2) Biotin synthase	
2.8.1.6	Biotin synthase (Treponema pallidum gene TP0228) Biotin synthase	
2.8.1.6	Biotin synthetase Biotin synthase	
2.8.1.6	Biotin synthetase (Aquifex aeolicus gene bioB) Biotin synthase	
2.8.1.6	Biotin synthetase (Arabidopsis thaliana clone pMB101 gene BIO2 reduced) Biotin synthase	
2.8.1.6	Biotin synthetase (Bacillus subtilis gene bioB) Biotin synthase	
2.8.1.6	Biotin synthetase (Chlamydia trachomatis gene birA) Biotin synthase	
6.2.1.11	Biotinyl-CoA synthetase Biotin-CoA ligase	Vol. 2, Page 252
6.2.1.11	Biotinyl CoA synthetase Biotin-CoA ligase	Vol. 2, Page 252
6.2.1.11	Biotinyl coenzyme A synthetase Biotin-CoA ligase	Vol. 2, Page 252

EC-No.	Synonym Recommended Name	Vol., Page
1.13.11.39	Biphenyl-2,3-diol 1,2-dioxygenase Biphenyl-2,3-diol 1,2-dioxygenase	
1.13.11.39	Biphenyl-2,3-diol:oxygen 1,2-oxidoreductase (decyclizing) Biphenyl-2,3-diol 1,2-dioxygenase	
1.13.11.39	Biphenyl-2,3-diol dioxygenase Biphenyl-2,3-diol 1,2-dioxygenase	
5.4.2.4	Biphosphoglycerate synthase Bisphosphoglycerate mutase	Vol. 1, Page 520
3.1.1.60	Bis(2-ethylhexyl)phthalate acylhydrolase Bis(2-ethylhexyl)phthalate esterase	
3.1.1.60	Bis(2-ethylhexyl)phthalate esterase Bis(2-ethylhexyl)phthalate esterase	
1.13.11.43	1,2-Bis(4-hydroxy-3-methoxyphenyl)ethylene:oxygen oxidoreductase ($\alpha\beta$-bond-cleaving) Lignostilbene $\alpha\beta$-dioxygenase	
3.6.1.17	Bis(5'-adenosyl)-tetraphosphatase Bis(5'-nucleosyl)-tetraphosphatase (asymmetrical)	
3.6.1.29	Bis(5'-adenosyl)-triphosphatase Bis(5'-adenosyl)-triphosphatase	
3.6.1.41	Bis(5'-adenosyl) tetraphosphatase Bis(5'-nucleosyl)-tetraphosphatase (symmetrical)	
3.6.1.17	Bis(5'-guanosyl)-tetraphosphatase Bis(5'-nucleosyl)-tetraphosphatase (asymmetrical)	
3.6.1.17	Bis(5'-nucleosyl)-tetraphosphatase (asymmetrical) Bis(5'-nucleosyl)-tetraphosphatase (asymmetrical)	
3.6.1.41	Bis(5'-nucleosyl)-tetraphosphatase (symmetrical) Bis(5'-nucleosyl)-tetraphosphatase (symmetrical)	
2.7.7.53	Bis(5'-nucleosyl)-tetraphosphate phosphorylase (NDP-forming) ATP adenylyltransferase	
1.6.4.9	Bis-γ-glutamylcystine reductase (NADPH) Bis-γ-glutamylcystine reductase (NADPH)	
3.2.1.134	Bis-D-fructose-2',1:2,1'-dianhydride fructohydrolase Difructose-anhydride synthase	
3.1.3.7	3'(2'),5'-bisphosphate nucleotidase 3'(2'),5'-bisphosphate nucleotidase	

EC-No.	Synonym Recommended Name	Vol., Page
3.1.3.13	2,3-bisphospho-D-glycerate 2-phosphohydrolase bisphosphoglycerate phosphatase	
5.4.2.4	2,3-Bisphosphoglycerate mutase Bisphosphoglycerate mutase	Vol. 1, Page 520
5.4.2.4	Bisphosphoglycerate mutase Bisphosphoglycerate mutase	Vol. 1, Page 520
3.1.3.13	2,3-bisphosphoglycerate phosphatase bisphosphoglycerate phosphatase	
3.1.3.13	bisphosphoglycerate phosphatase bisphosphoglycerate phosphatase	
5.4.2.4	Bisphosphoglyceromutase Bisphosphoglycerate mutase	Vol. 1, Page 520
2.7.5.4	bisphosphoglyceromutase bisphosphoglyceromutase	
3.1.3.7	3'(2'),5'-bisphosphonucleoside 3'(2')-phosphohydrolase 3'(2'),5'-bisphosphate nucleotidase	
1.8.99.3	Bisulfite reductase Hydrogensulfite reductase	
3.5.1.84	biuret amidohydrolase biuret amidohydrolase	
3.5.1.84	biuret amidohydrolase biuret amidohydrolase	
3.4.21.19	BLase glutamyl endopeptidase	
3.5.4.23	Blasticidin-S aminohydrolase Blasticidin-S deaminase	
3.5.4.23	Blasticidin-S deaminase Blasticidin-S deaminase	
3.4.22.40	bleomycin hydrolase bleomycin hydrolase	
3.4.22.40	Blh1p bleomycin hydrolase	
3.4.21.5	blood-coagulation factor II, activated thrombin	
3.4.21.5	blood-coagulation factor IIa thrombin	

EC-No.	Synonym Recommended Name	Vol., Page
3.4.21.21	blood-coagulation factor VII, activated coagulation factor VIIa	
3.4.21.21	blood-coagulation factor VIIa coagulation factor VIIa	
3.4.21.6	blood-coagulation factor X, activated coagulation factor Xa	
3.4.24.58	Blood-coagulation factor X activating enzyme Russellysin	
3.4.21.27	blood-coagulation factor XI, activated coagulation factor XIa	
3.4.21.27	blood-coagulation factor XIa coagulation factor XIa	
3.4.21.38	blood-coagulation factor XII, activated β coagulation factor XIIa	
3.4.21.38	blood-coagulation factor XIIaβ coagulation factor XIIa	
3.4.21.38	blood-coagulation factor XIIf coagulation factor XIIa	
3.4.21.69	Blood-coagulation factor XIV, activated Protein C (activated)	
3.4.21.69	Blood-coagulation factor XIVa Protein C (activated)	
2.4.1.40	Blood-group substance α-acetyltransferase Fucosylgalactose α-N-acetylgalactosaminyltransferase	
2.4.1.40	Blood-group substance A-dependent acetylgalactosaminyltransferase Fucosylgalactose α-N-acetylgalactosaminyltransferase	
2.4.1.37	Blood-group substance B-dependent galactosyltransferase Fucosylglycoprotein 3-α-galactosyltransferase	
2.4.1.69	Blood-group substance H-dependent fucosyltransferase Galactoside 2-α-L-fucosyltransferase	
2.4.1.65	Blood-group substance Lea-dependent fucosyltransferase Galactoside 3(4)-L-fucosyltransferase	
3.4.21.22	blood coagulation factor IXa coagulation factor IXa	
3.4.21.27	blood coagulation factor XIa coagulation factor XIa	

EC-No.	Synonym Recommended Name	Vol., Page
3.2.1.102	Blood group-substance 1,4-β-D-galactanohydrolase Blood group-substance endo-1,4-β-galactosidase	
3.2.1.102	Blood group-substance endo-1,4-β-galactosidase Blood group-substance endo-1,4-β-galactosidase	
2.4.1.69	Blood group H α-2-fucosyltransferase Galactoside 2-α-L-fucosyltransferase	
2.4.1.65	Blood group Lewis α-4-fucosyltransferase Galactoside 3(4)-L-fucosyltransferase	
2.4.1.37	[Blood group substance] α-galactosyltransferase Fucosylglycoprotein 3-α-galactosyltransferase	
3.1.1.47	blood platelet-activating factor-acetyl hydrolase 1-alkyl-2-acetylglycerophosphocholine esterase	
2.3.1.67	Blood platelet-activating factor acetyltransferase 1-Alkylglycerophosphocholine O-acetyltransferase	
2.4.2.4	Blood platelet-derived endothelial cell growth factor Thymidine phosphorylase	
2.4.1.19	BMA Cyclomaltodextrin glucanotransferase	
3.1.13.1	BN ribonuclease Exoribonuclease II	
4.1.1.17	bODC Ornithine decarboxylase	
3.4.24.69	BoNT Bontoxilysin	
3.4.24.69	Bontoxilysin Bontoxilysin	
1.14.15.1	2-Bornanone 5-exo-hydroxylase Camphor 5-monooxygenase	
1.14.15.1	Bornanone 5-exo-hydroxylase Camphor 5-monooxygenase	
1.1.1.198	(+)-Borneol:NAD$^+$ oxidoreductase (+)-Borneol dehydrogenase	
1.1.1.227	(-)-Borneol:NAD$^+$ oxidoreductase (-)-Borneol dehydrogenase	
1.1.1.198	(+)-Borneol dehydrogenase (+)-Borneol dehydrogenase	

EC-No.	Synonym Recommended Name	Vol., Page
1.1.1.227	(-)-Borneol dehydrogenase (-)-Borneol dehydrogenase	
5.5.1.8	(+)-Bornyl-diphosphate lyase (decyclizing) Geranyl-diphosphate cyclase	Vol. 1, Page 705
5.5.1.8	(+)-Bornylpyrophosphate cyclase Geranyl-diphosphate cyclase	Vol. 1, Page 705
3.1.7.3	bornyl pyrophosphate hydrolase monoterpenyl-diphosphatase	
5.5.1.8	Bornyl pyrophosphate synthase Geranyl-diphosphate cyclase	Vol. 1, Page 705
5.5.1.8	Bornyl pyrophosphate synthetase Geranyl-diphosphate cyclase	Vol. 1, Page 705
3.4.24.50	Bothrolysin Bothrolysin	
3.4.24.49	Bothropasin Bothropasin	
3.4.24.1	bothropasin atrolysin A	
3.4.21.29	Bothrops atrox serine proteinase Bothrops atrox serine proteinase	
3.4.24.50	Bothrops metalloendopeptidase J Bothrolysin	
3.4.24.69	Botulinum neurotoxin Bontoxilysin	
3.1.26.1	Bovine adrenal cortex RNase Physarum polycephalum ribonuclease	
3.4.24.62	Bovine neurosecretory granule protease cleaving pro-oxytocin/neurophysin Magnolysin	
4.2.99.14	BPA-synthase β-Pyrazolylalanine synthase (acetylserine)	
5.4.2.4	BPGM Bisphosphoglycerate mutase	Vol. 1, Page 520
5.5.1.8	(+)-BPP cyclase Geranyl-diphosphate cyclase	Vol. 1, Page 705
5.5.1.8	(-)-BPP cyclase Geranyl-diphosphate cyclase	Vol. 1, Page 705

EC-No.	Synonym Recommended Name	Vol., Page
3.1.7.3	BPP hydrolase monoterpenyl-diphosphatase	
3.4.21.32	brachyurin brachyurin	
3.4.17.3	bradykinase lysine carboxypeptidase	
3.4.17.3	bradykinin-decomposing enzyme lysine carboxypeptidase	
3.4.17.3	bradykininase lysine carboxypeptidase	
3.4.21.34	bradykininogenase plasma kallikrein	
3.4.21.35	bradykininogenase tissue kallikrein	
3.5.1.13	Brain acetylcholinesterase Arylacylamidase	
3.4.17.17	Brain I carboxypeptidase Tubulinyl-Tyr carboxypeptidase	
1.2.4.4	Branched-chain (-2-oxoacid BCD) dehydrogenase 3-Methyl-2-oxobutanoate dehydrogenase (lipoamide)	
4.1.1.72	Branched-chain-2-oxoacid decarboxylase Branched-chain-2-oxoacid decarboxylase	
2.6.1.42	Branched-chain-amino-acid:2-oxoglutarate aminotransferase Branched-chain-amino-acid transaminase	
2.6.1.42	Branched-chain-amino-acid transaminase Branched-chain-amino-acid transaminase	
2.7.2.14	Branched-chain-fatty-acid kinase Branched-chain-fatty-acid kinase	
4.1.1.72	Branched-chain α-keto acid decarboxylase Branched-chain-2-oxoacid decarboxylase	
1.2.4.4	Branched-chain α-keto acid dehydrogenase 3-Methyl-2-oxobutanoate dehydrogenase (lipoamide)	
2.7.1.115	Branched-chain α-keto acid dehydrogenase kinase [3-Methyl-2-oxobutanoate dehydrogenase (lipoamide)] kinase	
2.7.1.115	Branched-chain α-ketoacid dehydrogenase kinase [3-Methyl-2-oxobutanoate dehydrogenase (lipoamide)] kinase	

EC-No.	Synonym Recommended Name	Vol., Page
3.1.3.16	Branched-chain α-keto acid dehydrogenase phosphatase Phosphoprotein phosphatase	
3.1.3.52	Branched-chain α-keto acid dehydrogenase phosphatase [3-Methyl-2-oxobutanoate dehydrogenase (lipoamide)]-phosphatase	
1.2.4.4	Branched-chain α-oxo acid dehydrogenase 3-Methyl-2-oxobutanoate dehydrogenase (lipoamide)	
1.2.4.4	Branched-chain 2-keto acid dehydrogenase 3-Methyl-2-oxobutanoate dehydrogenase (lipoamide)	
4.1.1.72	Branched-chain 2-oxoacid decarboxylase Branched-chain-2-oxoacid decarboxylase	
1.2.4.4	Branched-chain 2-oxo acid dehydrogenase 3-Methyl-2-oxobutanoate dehydrogenase (lipoamide)	
2.7.1.115	Branched-chain 2-oxo acid dehydrogenase kinase [3-Methyl-2-oxobutanoate dehydrogenase (lipoamide)] kinase	
1.3.99.12	Branched-chain acyl-CoA dehydrogenase 2-Methylacyl-CoA dehydrogenase	
2.6.1.42	Branched-chain amino acid-glutamate transaminase Branched-chain-amino-acid transaminase	
2.6.1.42	Branched-chain amino acid aminotransferase Branched-chain-amino-acid transaminase	
2.6.1.42	Branched-chain aminotransferase Branched-chain-amino-acid transaminase	
2.7.2.14	Branched-chain fatty acid kinase Branched-chain-fatty-acid kinase	
4.1.1.72	Branched-chain keto acid decarboxylase Branched-chain-2-oxoacid decarboxylase	
1.2.4.4	Branched-chain keto acid dehydrogenase 3-Methyl-2-oxobutanoate dehydrogenase (lipoamide)	
1.2.4.4	Branched-chain ketoacid dehydrogenase 3-Methyl-2-oxobutanoate dehydrogenase (lipoamide)	
2.7.1.115	Branched-chain keto acid dehydrogenase kinase [3-Methyl-2-oxobutanoate dehydrogenase (lipoamide)] kinase	
3.1.3.52	Branched-chain oxo-acid dehydrogenase phosphatase [3-Methyl-2-oxobutanoate dehydrogenase (lipoamide)]-phosphatase	
4.1.1.72	Branched-chain oxo acid decarboxylase Branched-chain-2-oxoacid decarboxylase	

EC-No.	Synonym Recommended Name	Vol., Page
3.2.1.115	Branched-dextran exo-1,2-α-glucosidase Branched-dextran exo-1,2-α-glucosidase	
2.6.1.42	L-Branched chain amino acid aminotransferase Branched-chain-amino-acid transaminase	
1.2.4.4	Branched chain keto acid dehydrogenase 3-Methyl-2-oxobutanoate dehydrogenase (lipoamide)	
2.4.1.18	Branching enzyme 1,4-α-Glucan branching enzyme	
2.4.1.18	Branching factor, enzymatic 1,4-α-Glucan branching enzyme	
2.4.1.18	Branching glycosyltransferase 1,4-α-Glucan branching enzyme	
3.4.21.48	brewer's yeast proteinase cerevisin	
3.4.22.32	Bromelain Stem bromelain	
3.4.22.5	bromelain (juice) bromelain (juice)	
3.4.22.4	bromelain (stem) bromelain (stem)	
3.4.22.33	Bromelain, juice Fruit bromelain	
3.4.22.32	Bromelain, stem Stem bromelain	
3.4.22.33	Bromelase Fruit bromelain	
3.4.22.33	Bromelin Fruit bromelain	
2.6.1.75	S-(4-Bromophenyl)-L-cysteine:2-oxoglutarate aminotransferase Cysteine-conjugate transaminase	
3.5.5.6	Bromoxynil nitrilase Bromoxynil nitrilase	
5.3.3.12	Brown (b) locus protein Dopachrome Δ-isomerase	Vol. 1, Page 432
3.1.1.8	BtChoEase cholinesterase	

EC-No.	Synonym Recommended Name	Vol., Page
3.1.2.14	BTE oleoyl-[acyl-carrier-protein] hydrolase	
3.1.1.8	BuChE cholinesterase	
1.1.1.160	Bunolol reductase Dihydrobunolol dehydrogenase	
1.2.1.57	Butanal:NAD(P)$^+$ oxidoreductase (CoA-acylating) Butanal dehydrogenase	
1.2.1.57	Butanal dehydrogenase Butanal dehydrogenase	
1.1.1.4	2,3-Butanediol dehydrogenase (R,R)-Butanediol dehydrogenase	
1.1.1.4	D-(-)-Butanediol dehydrogenase (R,R)-Butanediol dehydrogenase	
6.2.1.2	Butanoate:CoA ligase (AMP-forming) Butyrate-CoA ligase	Vol. 2, Page 199
1.3.1.31	Butanoate: NAD$^+$ Δ^2-oxidoreductase 2-Enoate reductase	
1.3.99.2	Butanoyl-CoA: (acceptor) 2,3-oxidoreductase Butyryl-CoA dehydrogenase	
2.8.3.9	Butanoyl-CoA:acetoacetate CoA-transferase Butyrate-acetoacetate CoA-transferase	
2.3.1.19	Butanoyl-CoA:orthophosphate butanoyltransferase Phosphate butyryltransferase	
1.1.1.4	Butylene glycol dehydrogenase (R,R)-Butanediol dehydrogenase	
1.1.1.4	Butyleneglycol dehydrogenase (R,R)-Butanediol dehydrogenase	
1.1.1.165	2-Butyne-1,4-diol:NAD$^+$ 1-oxidoreductase 2-Alkyn-1-ol dehydrogenase	
1.2.1.57	Butyraldehyde dehydrogenase Butanal dehydrogenase	
2.8.3.9	Butyrate-acetoacetate CoA-transferase Butyrate-acetoacetate CoA-transferase	
6.2.1.2	Butyrate-CoA ligase Butyrate-CoA ligase	Vol. 2, Page 199

EC-No.	Synonym Recommended Name	Vol., Page
2.8.3.4	butyrate CoA-transferase butyrate CoA-transferase	
3.1.1.1	Butyrate esterase Carboxylesterase	
2.7.2.7	Butyrate kinase Butyrate kinase	
3.1.1.3	butyrinase triacylglycerol lipase	
1.14.11.1	γ-Butyrobetaine dioxygenase γ-Butyrobetaine dioxygenase	
1.14.11.1	Butyrobetaine hydroxylase γ-Butyrobetaine dioxygenase	
1.14.11.1	α-Butyrobetaine hydroxylase γ-Butyrobetaine dioxygenase	
1.14.11.1	γ-Butyrobetaine hydroxylase γ-Butyrobetaine dioxygenase	
2.8.3.9	Butyryl-CoA-acetoacetate CoA-transferase Butyrate-acetoacetate CoA-transferase	
5.4.99.13	Butyryl-CoA:isobutyryl-CoA mutase Isobutyryl-CoA mutase	Vol. 1, Page 646
1.3.99.2	Butyryl-CoA dehydrogenase Butyryl-CoA dehydrogenase	
1.3.2.1	butyryl-CoA dehydrogenase butyryl-CoA dehydrogenase	
6.2.1.2	Butyryl-CoA synthetase Butyrate-CoA ligase	Vol. 2, Page 199
6.2.1.2	Butyryl-coenzyme A synthetase Butyrate-CoA ligase	Vol. 2, Page 199
3.1.1.8	butyrylcholine-hydrolyzing enzyme cholinesterase	
3.1.1.8	butyrylcholine esterase cholinesterase	
3.1.1.8	butyrylcholinesterase cholinesterase	
2.8.3.8	Butyryl CoA:acetate CoA transferase Acetate CoA-transferase	

EC-No.	Synonym Recommended Name	Vol., Page
1.3.99.2	Butyryl coenzyme A dehydrogenase Butyryl-CoA dehydrogenase	
2.8.3.8	Butyryl coenzyme A transferase Acetate CoA-transferase	
1.3.99.2	Butyryl dehydrogenase Butyryl-CoA dehydrogenase	
3.1.1.1	Butyryl esterase Carboxylesterase	
2.3.1.19	Butyryltransferase, phosphate Phosphate butyryltransferase	
1.2.99.6	BV-AIDH Carboxylate reductase	
4.1.2.30	C-17/C-20 lyase 17α-Hydroxyprogesterone aldolase	
2.7.7.25	-C-C-A pyrophosphorylase tRNA adenylyltransferase	
2.7.7.21	-C-C-A pyrophosphorylase tRNA cytidylyltransferase	
3.4.21.42	C1-esterase complement subcomponent C1sbar	
3.1.2.21	C12: 0 – ACPthioesterase Dodecanoyl-[acyl-carrier protein] hydrolase	
3.1.2.14	C12:0 ACP TE oleoyl-[acyl-carrier-protein] hydrolase	
4.1.2.30	C17(20) lyase 17α-Hydroxyprogesterone aldolase	
2.3.1.119	C_{18}-CoA elongase Icosanoyl-CoA synthase	
3.1.2.14	C18:1-ACP thioesterase oleoyl-[acyl-carrier-protein] hydrolase	
3.2.1.91	C1 cellulase Cellulose 1,4-β-cellobiosidase	
3.4.21.41	C1rbar-esterase complement subcomponent C1r	
1.14.15.6	C27-Side chain cleavage enzyme Cholesterol monooxygenase (side-chain-cleaving)	

EC-No.	Synonym Recommended Name	Vol., Page
3.4.21.47	(C3b)n,Bb alternative-complement-pathway C3/C5 convertase	
3.4.21.47	C3b,Bb alternative-complement-pathway C3/C5 convertase	
3.4.21.45	C3b/C4b inactivator complement factor I	
3.4.21.45	C3bINA complement factor I	
3.4.21.45	C3bINA complement factor I	
3.4.21.45	C3b inactivator complement factor I	
3.4.21.47	C3 convertase alternative-complement-pathway C3/C5 convertase	
3.4.21.43	C3 convertase classical-complement-pathway C3/C5 convertase	
2.4.2.30	C3 exoenzyme NAD^+ ADP-ribosyltransferase	
3.4.21.47	C3 proactivator alternative-complement-pathway C3/C5 convertase	
3.4.21.46	C3 proactivator convertase complement factor D	
3.4.21.43	C4 classical-complement-pathway C3/C5 convertase	
3.4.21.43	C423 classical-complement-pathway C3/C5 convertase	
3.4.21.43	C4b,2a classical-complement-pathway C3/C5 convertase	
3.4.21.43	C4b,2a,3b classical-complement-pathway C3/C5 convertase	
2.7.1.66	C55-Isoprenoid alcohol kinase Undecaprenol kinase	
2.7.1.66	C55-Isoprenoid alcohol phosphokinase Undecaprenol kinase	
2.7.1.66	C55-Isoprenyl alcohol phosphokinase Undecaprenol kinase	

EC-No.	Synonym Recommended Name	Vol., Page
3.6.1.27	C_{55}-Isoprenyl pyrophosphatase Undecaprenyl-diphosphatase	
3.4.21.47	C5 convertase alternative-complement-pathway C3/C5 convertase	
3.4.21.43	C5 convertase classical-complement-pathway C3/C5 convertase	
5.1.3.17	C5 uronosyl epimerase Heparosan-N-sulfate-glucuronate 5-epimerase	Vol. 1, Page 167
4.2.1.1	CA carbonate dehydratase	
3.6.1.3	$(Ca^{2+} + Mg^{2+})$-ATPase Adenosinetriphosphatase	
3.4.11.7	Ca^{2+}-activated glutamate aminopeptidase glutamyl aminopeptidase	
3.4.22.17	Ca^{2+}-dependent protease calpain	
3.6.1.38	Ca^{2+}-pumping ATPase Ca^{2+}-transporting ATPase	
3.6.1.38	Ca^{2+}-transporting ATPase Ca^{2+}-transporting ATPase	
2.7.1.123	Ca^{2+}/calmodulin-dependent protein kinase Ca^{2+}/calmodulin-dependent protein kinase	
1.14.13.11	CA4H trans-Cinnamate 4-monooxygenase	
3.1.1.3	cacordase triacylglycerol lipase	
1.1.1.195	CAD Cinnamyl-alcohol dehydrogenase	
1.1.1.194	CAD Coniferyl-alcohol dehydrogenase	
4.1.1.29	CAD Sulfinoalanine decarboxylase	
4.1.1.29	CADCase Sulfinoalanine decarboxylase	
4.6.1.11	δ-Cadinene cyclase (+)-δ-Cadinene synthase	

EC-No.	Synonym Recommended Name	Vol., Page
4.2.3.13	(+)-d-cadinene synthase (+)-d-cadinene synthase	
4.6.1.11	(+)-δ-Cadinene synthase (+)-δ-Cadinene synthase	
4.6.1.11	δ-Cadinene synthase (+)-δ-Cadinene synthase	
1.16.3.1	Caeruloplasmin Ferroxidase	
1.13.11.22	Caffeate 3,4-dioxygenase Caffeate 3,4-dioxygenase	
2.1.1.68	Caffeate 3-O-methyltransferase Caffeate O-methyltransferase	
2.1.1.68	Caffeate methyltransferase Caffeate O-methyltransferase	
2.1.1.68	Caffeate O-methyltransferase Caffeate O-methyltransferase	
2.1.1.68	Caffeic acid-O-methyltransferase Caffeate O-methyltransferase	
6.2.1.12	Caffeolyl coenzyme A synthetase 4-Coumarate-CoA ligase	Vol. 2, Page 256
2.1.1.104	Caffeoyl-CoA 3-O-methyltransferase Caffeoyl-CoA O-methyltransferase	
2.3.1.140	Caffeoyl-CoA:3-(3,4-dihydroxyphenyl)lactate 2'-O-caffeoyltransferase Rosmarinate synthase	
2.3.1.126	Caffeoyl-CoA:isocitrate 3-O-(3,4-dihydroxycinnamoyl)transferase Isocitrate O-dihydroxycinnamoyltransferase	
2.3.1.138	Caffeoyl-CoA:putrescine N-(3,4-dihydroxycinnamoyl)transferase Putrescine N-hydroxycinnamoyltransferase	
2.1.1.104	Caffeoyl-CoA O-methyltransferase Caffeoyl-CoA O-methyltransferase	
2.3.1.138	Caffeoyl-CoA putrescine N-caffeoyl transferase Putrescine N-hydroxycinnamoyltransferase	
2.3.1.140	Caffeoyl-Coenzyme A:3,4-dihydroxyphenyllactic acid caffeoyltransferase Rosmarinate synthase	
2.3.1.98	Caffeoyltransferase, chlorogenate-glucarate Chlorogenate-glucarate O-hydroxycinnamoyltransferase	

EC-No.	Synonym Recommended Name	Vol., Page
3.4.21.39	CAGE chymase	
1.14.13.13	Calcidiol,NADPH:oxygen oxidoreductase (1-hydroxylating) Calcidiol 1-monooxygenase	
1.14.13.13	Calcidiol 1-monooxygenase Calcidiol 1-monooxygenase	
3.1.3.16	Calcineurin Phosphoprotein phosphatase	
3.4.22.17	calcium-activated factor calpain	
3.4.22.17	calcium-activated neutral protease calpain	
3.4.22.17	calcium-activated neutral proteinase calpain	
2.7.1.37	Calcium-dependent protein kinase C Protein kinase	
2.7.1.117	Calcium/calmodulin-dependent myosin light chain kinase Myosin-light-chain kinase	
2.7.1.37	Calcium/phospholipid-dependent protein kinase Protein kinase	
3.6.1.38	Calcium pump Ca^{2+}-transporting ATPase	
2.7.1.120	Caldesmon Caldesmon kinase	
3.1.3.55	Caldesmon-phosphatase Caldesmon-phosphatase	
3.1.3.55	Caldesmon-phosphate phosphohydrolase Caldesmon-phosphatase	
2.7.1.120	Caldesmon kinase Caldesmon kinase	
3.1.3.55	Caldesmon phosphatase Caldesmon-phosphatase	
3.1.26.4	Calf thymus ribonuclease H Calf thymus ribonuclease H	
3.4.24.11	CALLA neprilysin	

EC-No.	Synonym Recommended Name	Vol., Page
3.4.24.11	CALLA (common acute lymphoblastic leukemia-associated) antigens neprilysin	
3.4.24.11	CALLA antigen neprilysin	
3.4.24.11	CALLA glycoproteins neprilysin	
3.2.1.39	Callase Glucan endo-1,3-β-glucosidase	
3.4.21.34	callicrein plasma kallikrein	
3.4.21.35	callicrein tissue kallikrein	
2.4.1.34	Callose synthase 1,3-β-Glucan synthase	
2.4.1.34	Callose synthetase 1,3-β-Glucan synthase	
2.1.1.60	Calmodulin-lysine N-methyltransferase Calmodulin-lysine N-methyltransferase	
6.3.2.21	Calmodulin:ubiquitin ligase (AMP-forming) Ubiquitin-calmodulin ligase	Vol. 2, Page 513
2.1.1.60	Calmodulin lysine N-methyltransferase Calmodulin-lysine N-methyltransferase	
3.4.22.17	calpain calpain	
3.4.22.17	*m*-calpain calpain	
2.7.1.37	cAMP-dependent protein kinase Protein kinase	
2.7.1.37	cAMP-dependent protein kinase A Protein kinase	
1.14.15.2	(+)-Camphor,reduced-rubredoxin:oxygen oxidoreductase (1,2-lactonizing) Camphor 1,2-monooxygenase	
1.14.15.1	(+)-Camphor,reduced putidaredoxin:oxygen oxidoreductase (5-hydroxylating) Camphor 5-monooxygenase	

EC-No.	Synonym Recommended Name	Vol., Page
1.14.15.1	D-Camphor-exo-hydroxylase Camphor 5-monooxygenase	
1.14.15.2	Camphor 1,2-monooxygenase Camphor 1,2-monooxygenase	
1.14.15.1	Camphor 5-exo-hydroxylase Camphor 5-monooxygenase	
1.14.15.1	Camphor 5-exo-methylene hydroxylase Camphor 5-monooxygenase	
1.14.15.1	Camphor 5-exohydroxylase Camphor 5-monooxygenase	
1.14.15.1	Camphor 5-monooxygenase Camphor 5-monooxygenase	
1.14.15.1	Camphor hydroxylase Camphor 5-monooxygenase	
1.14.15.1	Camphor methylene hydroxylase Camphor 5-monooxygenase	
1.14.15.1	d-Camphor monooxygenase Camphor 5-monooxygenase	
1.1.3.36	Canadine synthase (S)-Canadine synthase	
3.5.3.1	Canavanase Arginase	
3.4.22.26	Cancer procoagulant Cancer procoagulant	
3.4.23.24	Candida albicans aspartic proteinase Candidapepsin	
3.4.23.24	Candida albicans carboxyl proteinase Candidapepsin	
3.4.23.24	Candida albicans secretory acid proteinase Candidapepsin	
3.4.23.24	Candida olea acid proteinase Candidapepsin	
3.4.23.24	Candidapepsin Candidapepsin	
1.2.99.4	Cannizzanase Formaldehyde dismutase	

EC-No.	Synonym Recommended Name	Vol., Page
3.4.22.17	CANP1 calpain	
3.4.22.17	CANP2 calpain	
3.4.22.17	CANPI calpain	
3.4.22.17	CANPII calpain	
3.4.11.3	CAP-I cystinyl aminopeptidase	
3.4.11.3	CAP-II cystinyl aminopeptidase	
3.1.1.3	capalase L triacylglycerol lipase	
1.11.1.6	Caperase Catalase	
3.4.22.17	CAPN 3 calpain	
2.7.1.103	Capreomycin phosphotransferase Viomycin kinase	
3.2.1.124	Capsular-polysaccharide 3-deoxy-D-manno-2-octulosonohydrolase 3-Deoxy-2-octulosonidase	
3.2.1.87	Capsular-polysaccharide endo-1,3-α-galactosidase Capsular-polysaccharide endo-1,3-α-galactosidase	
2.7.2.2	Carbamate kinase Carbamate kinase	
3.5.1.6	N-Carbamoyl-β-alanine amidohydrolase β-ureidopropionase	
3.5.1.77	N-Carbamoyl-D-amino acid amidase N-Carbamoyl-D-amino acid hydrolase	
3.5.1.77	N-Carbamoyl-D-amino acid amidohydrolase N-Carbamoyl-D-amino acid hydrolase	
3.5.1.77	N-Carbamoyl-D-amino acid hydrolase N-Carbamoyl-D-amino acid hydrolase	
3.5.1.7	N-Carbamoyl-L-aspartate amidohydrolase Ureidosuccinase	

EC-No.	Synonym Recommended Name	Vol., Page
4.3.1.13	O-carbamoyl-L-serine ammonia-lyase (pyruvate forming) carbamoyl-serine ammonia-lyase	
4.3.1.13	O-carbamoyl-L-serine deaminase carbamoyl-serine ammonia-lyase	
4.3.1.13	O-carbamoyl-L-serine deaminase (OCS) carbamoyl-serine ammonia-lyase	
2.1.3.7	Carbamoyl-phosphate:3-hydroxymethylceph-3-em-4-carboxylate carbamoyltransferase 3-Hydroxymethylcephem carbamoyltransferase	
2.1.3.2	Carbamoyl-phosphate:L-aspartate carbamoyltransferase Aspartate carbamoyltransferase	
2.1.3.8	Carbamoyl-phosphate:L-lysine carbamoyltransferase Lysine carbamoyltransferase	
2.1.3.3	Carbamoyl-phosphate:L-ornithine carbamoyltransferase Ornithine carbamoyltransferase	
2.1.3.5	Carbamoyl-phosphate:oxamate carbamoyltransferase Oxamate carbamoyltransferase	
2.1.3.6	Carbamoyl-phosphate:putrescine carbamoyltransferase Putrescine carbamoyltransferase	
6.3.4.16	Carbamoyl-phosphate synthase (ammonia) Carbamoyl-phosphate synthase (ammonia)	Vol. 2, Page 641
2.7.2.5	carbamoyl-phosphate synthase (ammonia) carbamoyl-phosphate synthase (ammonia)	
2.7.2.9	carbamoyl-phosphate synthase (glutamine) carbamoyl-phosphate synthase (glutamine)	
6.3.5.5	Carbamoyl-phosphate synthase (glutamine-hydrolysing) Carbamoyl-phosphate synthase (glutamine-hydrolysing)	Vol. 2, Page 689
6.3.5.5	Carbamoyl-phosphate synthetase (glutamine-hydrolysing) Carbamoyl-phosphate synthase (glutamine-hydrolysing)	Vol. 2, Page 689
6.3.4.16	Carbamoyl-phosphate synthetase I Carbamoyl-phosphate synthase (ammonia)	Vol. 2, Page 641
4.3.1.13	carbamoyl-serine ammonia-lyase carbamoyl-serine ammonia-lyase	
4.1.1.13	carbamoylaspartate decarboxylase carbamoylaspartate decarboxylase	

EC-No.	Synonym Recommended Name	Vol., Page
3.5.2.3	Carbamoylaspartic dehydrase Dihydroorotase	
2.1.3.2	Carbamoylaspartotranskinase Aspartate carbamoyltransferase	
6.3.4.16	Carbamoylphosphate synthase Carbamoyl-phosphate synthase (ammonia)	Vol. 2, Page 641
6.3.5.5	Carbamoylphosphate synthase Carbamoyl-phosphate synthase (glutamine-hydrolysing)	Vol. 2, Page 689
6.3.4.16	Carbamoylphosphate synthase (ammonia) Carbamoyl-phosphate synthase (ammonia)	Vol. 2, Page 641
6.3.5.5	Carbamoyl phosphate synthase (glutamine) Carbamoyl-phosphate synthase (glutamine-hydrolysing)	Vol. 2, Page 689
6.3.5.5	Carbamoylphosphate synthetase Carbamoyl-phosphate synthase (glutamine-hydrolysing)	Vol. 2, Page 689
6.3.4.16	Carbamoylphosphate synthetase (ammonia) Carbamoyl-phosphate synthase (ammonia)	Vol. 2, Page 641
6.3.5.5	Carbamoyl phosphate synthetase (glutamine-hydrolyzing) Carbamoyl-phosphate synthase (glutamine-hydrolysing)	Vol. 2, Page 689
6.3.4.16	Carbamoylphosphate synthetase I Carbamoyl-phosphate synthase (ammonia)	Vol. 2, Page 641
6.3.5.5	Carbamoylphosphate synthetase II Carbamoyl-phosphate synthase (glutamine-hydrolysing)	Vol. 2, Page 689
2.7.2.2	Carbamoyl phosphokinase Carbamate kinase	
3.5.1.53	N-Carbamoylputrescine amidase N-Carbamoylputrescine amidase	
3.5.1.53	N-Carbamoylputrescine amidohydrolase N-Carbamoylputrescine amidase	
3.5.1.59	N-Carbamoylsarcosine amidase N-Carbamoylsarcosine amidase	
3.5.1.59	N-Carbamoylsarcosine amidohydrolase N-Carbamoylsarcosine amidase	
4.3.1.13	carbamoylserine deaminase carbamoyl-serine ammonia-lyase	
2.1.3.7	Carbamoyltransferase, 3-hyroxymethylcephem 3-Hydroxymethylcephem carbamoyltransferase	

EC-No.	Synonym Recommended Name	Vol., Page
2.1.3.2	Carbamoyltransferase, aspartate Aspartate carbamoyltransferase	
2.1.3.8	Carbamoyltransferase, lysine Lysine carbamoyltransferase	
2.1.3.3	Carbamoyltransferase, ornithine Ornithine carbamoyltransferase	
2.1.3.5	Carbamoyltransferase, oxamate Oxamate carbamoyltransferase	
2.1.3.6	Carbamoyltransferase, putrescine Putrescine carbamoyltransferase	
3.5.1.77	N-Carbamyl-D-amino acid amidohydrolase N-Carbamoyl-D-amino acid hydrolase	
3.5.1.77	N-Carbamyl amino acid amidohydrolase N-Carbamoyl-D-amino acid hydrolase	
2.1.3.2	Carbamylaspartotranskinase Aspartate carbamoyltransferase	
2.1.3.3	Carbamylphosphate-ornithine transcarbamylase Ornithine carbamoyltransferase	
6.3.4.16	Carbamyl phosphate synthase I Carbamoyl-phosphate synthase (ammonia)	Vol. 2, Page 641
6.3.4.16	Carbamylphosphate synthetase Carbamoyl-phosphate synthase (ammonia)	Vol. 2, Page 641
6.3.5.5	Carbamyl phosphate synthetase (glutamine) Carbamoyl-phosphate synthase (glutamine-hydrolysing)	Vol. 2, Page 689
6.3.4.16	Carbamylphosphate synthetase I Carbamoyl-phosphate synthase (ammonia)	Vol. 2, Page 641
6.3.5.5	Carbamyl phosphate sythetase II Carbamoyl-phosphate synthase (glutamine-hydrolysing)	Vol. 2, Page 689
2.7.2.2	Carbamyl phosphokinase Carbamate kinase	
4.2.1.96	4a-Carbinolamine dehydratase 4a-Hydroxytetrahydrobiopterin dehydratase	
6.3.4.16	Carbmoylphosphate synthetase Carbamoyl-phosphate synthase (ammonia)	Vol. 2, Page 641
3.5.1.58	N-α-Carbobenzoxyamino acid amidohydrolase N-Benzyloxycarbonylglycine hydrolase	

EC-No.	Synonym Recommended Name	Vol., Page
4.1.3.15	carboligase 2-hydroxy-3-oxoadipate synthase	
6.3.4.16	Carbon-dioxide-ammonia ligase Carbamoyl-phosphate synthase (ammonia)	Vol. 2, Page 641
6.3.4.16	Carbon-dioxide:ammonia ligase (ADP-forming, carbamate-phosphorylating) Carbamoyl-phosphate synthase (ammonia)	Vol. 2, Page 641
6.3.5.5	Carbon-dioxide:L-glutamine amido-ligase (ADP forming, carbamate-phosphorylating) Carbamoyl-phosphate synthase (glutamine-hydrolysing)	Vol. 2, Page 689
1.2.99.2	Carbon-monoxide:(acceptor) oxidoreductase Carbon-monoxide dehydrogenase	
1.2.3.10	Carbon-monoxide:oxygen oxidoreductase Carbon-monoxide oxidase	
1.2.99.2	Carbon-monoxide dehydrogenase Carbon-monoxide dehydrogenase	
1.2.3.10	Carbon-monoxide oxidase Carbon-monoxide oxidase	
1.2.2.4	Carbon-monoxide oxygenase (cytochrome b-561) Carbon-monoxide oxygenase (cytochrome b-561)	
4.2.99.12	carbon-oxygen lyase carboxymethyloxysuccinate lyase	
4.2.1.1	carbonate anhydrase carbonate dehydratase	
4.2.1.1	carbonate dehydratase carbonate dehydratase	
4.2.1.1	carbonate hydro-lyase carbonate dehydratase	
6.3.4.16	Carbonate kinase (phosphorylating) Carbamoyl-phosphate synthase (ammonia)	Vol. 2, Page 641
1.2.99.2	Carbon dioxide/carbon monoxide oxidoreductase Carbon-monoxide dehydrogenase	
1.2.2.4	Carbon dioxide/carbon monoxide oxidoreductase Carbon-monoxide oxygenase (cytochrome b-561)	
4.2.1.1	carbonic acid anhydrase carbonate dehydratase	

EC-No.	Synonym Recommended Name	Vol., Page
4.2.1.1	carbonic anhydrase carbonate dehydratase	
4.2.1.1	carbonic dehydratase carbonate dehydratase	
1.2.3.10	Carbon monoxide:methylene blue oxidoreductase Carbon-monoxide oxidase	
1.2.99.2	Carbon monoxide dehydrogenase Carbon-monoxide dehydrogenase	
1.2.2.4	Carbon monoxide dehydrogenase Carbon-monoxide oxygenase (cytochrome b-561)	
1.2.99.2	Carbon monoxide dehydrogenase-corrinoid enzyme complex Carbon-monoxide dehydrogenase	
1.2.3.10	Carbon monoxide oxidase Carbon-monoxide oxidase	
1.2.99.2	Carbon monoxide oxidoreductase Carbon-monoxide dehydrogenase	
1.2.2.4	Carbon monoxide oxidoreductase Carbon-monoxide oxygenase (cytochrome b-561)	
1.2.2.4	Carbon monoxide water:cytochrome $b_5$61 oxidoreductase Carbon-monoxide oxygenase (cytochrome b-561)	
1.1.1.184	Carbonyl reductase Carbonyl reductase (NADPH)	
1.1.1.184	Carbonyl reductase (NADPH) Carbonyl reductase (NADPH)	
3.4.19.2	carboxamidopeptidase peptidyl-glycinamidase	
5.5.1.2	2-Carboxy-2,5-dihydro-5-oxofuran-2-acetate lyase (decyclizing) 3-Carboxy-cis,cis-muconate cycloisomerase	Vol. 1, Page 668
1.1.1.85	3-Carboxy-2-hydroxy-4-methylpentanoate:NAD$^+$ oxidoreductase 3-Isopropylmalate dehydrogenase	
1.2.1.45	4-Carboxy-2-hydroxy-cis,cis-muconate-6-semialdehyde:NADP$^+$ oxidoreductase 2-Hydroxy-4-carboxymuconate-6-semialdehyde dehydrogenase	
1.1.1.87	3-Carboxy-2-hydroxyadipate:NAD$^+$ oxidoreductase (decarboxylating) 3-Carboxy-2-hydroxyadipate dehydrogenase	

EC-No.	Synonym Recommended Name	Vol., Page
1.1.1.87	3-Carboxy-2-hydroxyadipate dehydrogenase 3-Carboxy-2-hydroxyadipate dehydrogenase	
1.2.1.45	4-Carboxy-2-hydroxymuconate-6-semialdehyde dehydrogenase 2-Hydroxy-4-carboxymuconate-6-semialdehyde dehydrogenase	
4.2.1.83	4-Carboxy-2-oxobutane-1,2,4-tricarboxylate 2,3-hydro-lyase 4-Oxalmesaconate hydratase	
4.2.1.83	4-Carboxy-2-oxohexenedioate hydratase 4-Oxalmesaconate hydratase	
1.3.1.43	3-(1-Carboxy-4-hydroxycyclohexa-2,5-dien-1-yl)-L-alanine: NAD^+ oxidoreductase (decarboxylating) Pretyrosine dehydrogenase	
5.5.1.5	3-Carboxy-5-oxo-2,5-dihydrofuran-2-acetate (decyclizing) Carboxy-cis,cis-muconate cyclase	Vol. 1, Page 686
5.5.1.5	Carboxy-cis,cis-muconate cyclase Carboxy-cis,cis-muconate cyclase	Vol. 1, Page 686
5.5.1.2	3-Carboxy-cis,cis-muconate cycloisomerase 3-Carboxy-cis,cis-muconate cycloisomerase	Vol. 1, Page 668
5.5.1.5	3-Carboxy-cis,cis-muconate lactonizing enzyme Carboxy-cis,cis-muconate cyclase	Vol. 1, Page 686
5.5.1.2	*β*-Carboxy-cis,cis-muconate lactonizing enzyme 3-Carboxy-cis,cis-muconate cycloisomerase	Vol. 1, Page 668
5.5.1.5	3-Carboxy-cis-cis-muconate cyclase Carboxy-cis,cis-muconate cyclase	Vol. 1, Page 686
5.5.1.2	3-Carboxy-cis-cis-muconate cycloisomerase 3-Carboxy-cis,cis-muconate cycloisomerase	Vol. 1, Page 668
3.1.3.63	2-Carboxy-D-arabinitol-1-phosphatase 2-Carboxy-D-arabinitol-1-phosphatase	
3.1.3.63	2-Carboxy-D-arabinitol-1-phosphate 1-phosphohydrolase 2-Carboxy-D-arabinitol-1-phosphatase	
3.1.3.63	2-Carboxy-D-arabinitol 1-phosphate phosphohydrolase 2-Carboxy-D-arabinitol-1-phosphatase	
3.4.24.19	Carboxy-terminal endopeptidase Procollagen C-endopeptidase	
3.4.19.2	carboxyamidase peptidyl-glycinamidase	

EC-No.	Synonym Recommended Name	Vol., Page
4.2.1.1	carboxyanhydrase carbonate dehydratase	
3.1.3.63	Carboxyarabinitol 1-phosphatase 2-Carboxy-D-arabinitol-1-phosphatase	
4.1.2.34	2'-Carboxybenzalpyruvate aldolase 4-(2-Carboxyphenyl)-2-oxobut-3-enoate aldolase	
3.4.15.1	carboxycathepsin peptidyl-dipeptidase A	
4.2.1.91	Carboxycyclohexadienyl dehydratase Carboxycyclohexadienyl dehydratase	
4.1.1.39	Carboxydismutase Ribulose-bisphosphate carboxylase	
3.1.1.1	Carboxyesterase Carboxylesterase	
1.13.11.16	3-Carboxyethylcatechol 2,3-dioxygenase 3-Carboxyethylcatechol 2,3-dioxygenase	
6.2.1.14	6-Carboxyhexanoate-CoA ligase 6-Carboxyhexanoate-CoA ligase	Vol. 2, Page 273
6.2.1.14	6-Carboxyhexanoate-CoA synthetase 6-Carboxyhexanoate-CoA ligase	Vol. 2, Page 273
6.2.1.14	6-Carboxyhexanoate:CoA ligase (AMP-forming) 6-Carboxyhexanoate-CoA ligase	Vol. 2, Page 273
2.3.1.47	6-Carboxyhexanoyl-CoA:L-alanine C-carboxyhexanoyltransferase (decarboxylating) 8-Amino-7-oxononanoate synthase	
6.2.1.14	6-Carboxyhexanoyl-CoA synthetase 6-Carboxyhexanoate-CoA ligase	Vol. 2, Page 273
4.1.1.49	Carboxykinase, phosphopyruvate (adenosine triphosphate) Phosphoenolpyruvate carboxykinase (ATP)	
4.1.1.32	Carboxykinase, phosphopyruvate (guanosine triphosphate) Phosphoenolpyruvate carboxykinase (GTP)	
4.1.1.38	carboxykinase, phosphopyruvate (pyrophosphate) phosphoenolpyruvate carboxykinase (pyrophosphate)	
4.1.1.1	α-Carboxylase Pyruvate decarboxylase	

EC-No.	Synonym Recommended Name	Vol., Page
6.4.1.2	Carboxylase, acetyl coenzyme A Acetyl-CoA carboxylase	Vol. 2, Page 721
6.3.4.14	Carboxylase, biotin Biotin carboxylase	Vol. 2, Page 632
6.4.1.5	Carboxylase, geranoyl coenzyme A Geranoyl-CoA carboxylase	Vol. 2, Page 752
6.4.1.4	Carboxylase, methylcrotonyl coenzyme A Methylcrotonoyl-CoA carboxylase	Vol. 2, Page 744
4.1.1.31	Carboxylase, phosphopyruvate (phosphate) Phosphoenolpyruvate carboxylase	
6.4.1.3	Carboxylase, propional coenzyme A (adenosine triphosphate-hydrolyzing) Propionyl-CoA carboxylase	Vol. 2, Page 738
6.4.1.1	Carboxylase, pyruvate Pyruvate carboxylase	Vol. 2, Page 708
3.1.1.1	Carboxylate esterase Carboxylesterase	
1.2.99.6	Carboxylate reductase Carboxylate reductase	
3.1.1.1	Carboxylesterase Carboxylesterase	
3.1.1.1	α-Carboxylesterase Carboxylesterase	
3.1.1.1	Carboxyl ester hydrolase Carboxylesterase	
3.1.1.1	Carboxylic-ester hydrolase Carboxylesterase	
3.1.1.1	Carboxylic acid esterase Carboxylesterase	
1.2.99.6	Carboxylic acid reductase Carboxylate reductase	
1.2.99.6	Carboxylic acid reductase/aldehyde reductase Carboxylate reductase	
3.1.1.1	Carboxylic esterase Carboxylesterase	
3.4.24.19	Carboxyl procollagen peptidase Procollagen C-endopeptidase	

EC-No.	Synonym Recommended Name	Vol., Page
3.4.23.18	Carboxyl proteinase Aspergillopepsin I	
3.4.23.40	Carboxyl proteinase Phytepsin	
2.1.3.1	Carboxyltransferase, methylmalonyl coenzyme A Methylmalonyl-CoA carboxyltransferase	
4.1.1.21	Carboxylyase, phosphoribosylaminoimidazole Phosphoribosylaminoimidazole carboxylase	
6.2.1.23	Carboxylyl-CoA synthetase Dicarboxylate-CoA ligase	Vol. 2, Page 308
5.3.3.10	5-(Carboxymethyl)-2-hydroxymuconate isomerase 5-Carboxymethyl-2-hydroxymuconate Δ-isomerase	Vol. 1, Page 427
5.3.3.10	5-Carboxymethyl-2-hydroxymuconate Δ-isomerase 5-Carboxymethyl-2-hydroxymuconate Δ-isomerase	Vol. 1, Page 427
5.3.3.10	5-Carboxymethyl-2-hydroxymuconate Δ^2,Δ^4-2-oxo,Δ^3-isomerase 5-Carboxymethyl-2-hydroxymuconate Δ-isomerase	Vol. 1, Page 427
5.3.3.10	5-Carboxymethyl-2-hydroxymuconate isomerase 5-Carboxymethyl-2-hydroxymuconate Δ-isomerase	Vol. 1, Page 427
1.2.1.60	5-carboxymethyl-2-hydroxymuconic-semialdehyde:NAD oxidoreductase 5-carboxymethyl-2-hydroxymuconic-semialdehyde dehydrogenase	
1.2.1.60	5-carboxymethyl-2-hydroxymuconic-semialdehyde dehydrogenase 5-carboxymethyl-2-hydroxymuconic-semialdehyde dehydrogenase	
3.1.1.16	4-carboxymethyl-4-hydroxyisocrotonolactonase 4-carboxymethyl-4-hydroxyisocrotonolactonase	
5.4.99.14	4-Carboxymethyl-4-methylbut-2-en-1,4-olide methylmutase 4-Carboxymethyl-4-methylbutenolide mutase	Vol. 1, Page 648
5.4.99.14	4-Carboxymethyl-4-methylbutenolide mutase 4-Carboxymethyl-4-methylbutenolide mutase	Vol. 1, Page 648
4.5.1.5	S-Carboxymethyl-L-cysteine synthase S-Carboxymethylcysteine synthase	
3.1.1.24	4-carboxymethylbut-3-en-4-olide enol-lactonohydrolase 3-oxoadipate enol-lactose	
3.1.1.24	carboxymethylbutenolide lactonase 3-oxoadipate enol-lactose	
4.5.1.5	S-Carboxymethylcysteine synthase S-Carboxymethylcysteine synthase	

EC-No.	Synonym Recommended Name	Vol., Page
3.1.1.45	4-carboxymethylenebut-2-en-4-olide hydrolase carboxymethylenebutenolidase	
3.1.1.45	4-carboxymethylenebut-2-en-4-olide lactonohydrolase carboxymethylenebutenolidase	
3.1.1.45	carboxymethylene butenolidase carboxymethylenebutenolidase	
3.1.1.45	carboxymethylenebutenolidase carboxymethylenebutenolidase	
3.1.1.45	carboxymethylene butenolide hydrolase carboxymethylenebutenolidase	
3.5.2.4	L-5-Carboxymethylhydantoin amidohydrolase Carboxymethylhydantoinase	
3.5.2.4	Carboxymethylhydantoinase Carboxymethylhydantoinase	
1.2.1.60	carboxymethylhydroxymuconic semialdehyde dehydrogenase 5-carboxymethyl-2-hydroxymuconic-semialdehyde dehydrogenase	
4.2.99.12	carboxymethyloxysuccinate glycolate-lyase carboxymethyloxysuccinate lyase	
4.2.99.12	carboxymethyloxysuccinate lyase carboxymethyloxysuccinate lyase	
5.5.1.5	3-Carboxymuconate cyclase Carboxy-cis,cis-muconate cyclase	Vol. 1, Page 686
5.5.1.2	β-Carboxymuconate lactonizing enzyme 3-Carboxy-cis,cis-muconate cycloisomerase	Vol. 1, Page 668
4.1.1.44	4-carboxymuconolactone carboxy-lyase 4-carboxymuconolactone decarboxylase	
4.1.1.44	4-carboxymuconolactone decarboxylase 4-carboxymuconolactone decarboxylase	
4.1.1.44	γ-carboxymuconolactone decarboxylase 4-carboxymuconolactone decarboxylase	
5.5.1.2	3-Carboxymuconolactone hydrolase 3-Carboxy-cis,cis-muconate cycloisomerase	Vol. 1, Page 668
3.4.16.2	carboxypeptidase, aminoacylproline lysosomal Pro-X carboxypeptidase	
3.4.17.3	carboxypeptidase, arginine lysine carboxypeptidase	

EC-No.	Synonym Recommended Name	Vol., Page
3.4.17.3	carboxypeptidase, arginine lysine carboxypeptidase	
3.4.15.1	carboxypeptidase, dipeptidyl peptidyl-dipeptidase A	
3.4.17.10	carboxypeptidase, enkephalin precursor carboxypeptidase H	
3.4.17.13	Carboxypeptidase, lysyl-D-alanine Muramoyltetrapeptide carboxypeptidase	
3.4.17.8	carboxypeptidase, muramoylpentapeptide muramoylpentapeptide carboxypeptidase	
3.4.17.13	Carboxypeptidase, muramoyltetrapeptide Muramoyltetrapeptide carboxypeptidase	
3.4.16.2	carboxypeptidase, peptidylprolylamino acid lysosomal Pro-X carboxypeptidase	
3.4.17.17	Carboxypeptidase, tubulin-tyrosine Tubulinyl-Tyr carboxypeptidase	
3.4.17.17	Carboxypeptidase-tubulin Tubulinyl-Tyr carboxypeptidase	
3.4.17.20	Carboxypeptidase-U Carboxypeptidase U	
3.4.17.1	carboxypeptidase A carboxypeptidase A	
3.4.17.1	carboxypeptidase A carboxypeptidase A	
3.4.2.1	carboxypeptidase A carboxypeptidase A	
3.4.17.4	carboxypeptidase a Gly-X carboxypeptidase	
3.4.17.15	Carboxypeptidase A2 Carboxypeptidase A2	
3.4.17.2	carboxypeptidase B carboxypeptidase B	
3.4.2.2	carboxypeptidase B carboxypeptidase B	
3.4.17.20	Carboxypeptidase B, pro- Carboxypeptidase U	

EC-No.	Synonym Recommended Name	Vol., Page
3.4.17.2	carboxypeptidase BII carboxypeptidase B	
3.4.16.5	Carboxypeptidase C Carboxypeptidase C	
3.4.16.6	Carboxypeptidase D Carboxypeptidase D	
3.4.17.22	Carboxypeptidase D Metallocarboxypeptidase D	
3.4.17.8	carboxypeptidase D-alanyl-D-alanine muramoylpentapeptide carboxypeptidase	
3.4.17.10	carboxypeptidase E carboxypeptidase H	
3.4.17.11	Carboxypeptidase G Glutamate carboxypeptidase	
3.4.19.9	Carboxypeptidase G γ-Glu-X carboxypeptidase	
3.4.17.11	Carboxypeptidase G1 Glutamate carboxypeptidase	
3.4.17.11	Carboxypeptidase G2 Glutamate carboxypeptidase	
3.4.17.10	carboxypeptidase H carboxypeptidase H	
3.4.17.8	carboxypeptidase I muramoylpentapeptide carboxypeptidase	
3.4.16.4	carboxypeptidase I serine-type D-Ala-D-Ala carboxypeptidase	
3.4.17.13	Carboxypeptidase II Muramoyltetrapeptide carboxypeptidase	
3.4.17.13	Carboxypeptidase IIW Muramoyltetrapeptide carboxypeptidase	
3.4.16.6	Carboxypeptidase Kex1 Carboxypeptidase D	
3.4.17.12	Carboxypeptidase M Carboxypeptidase M	
3.4.17.3	carboxypeptidase N lysine carboxypeptidase	

EC-No.	Synonym Recommended Name	Vol., Page
3.4.17.16	Carboxypeptidase P Membrane Pro-X carboxypeptidase	
3.4.16.2	carboxypeptidase P lysosomal Pro-X carboxypeptidase	
3.4.17.20	Carboxypeptidase R Carboxypeptidase U	
3.4.17.4	carboxypeptidase S Gly-X carboxypeptidase	
3.4.17.9	carboxypeptidase S carboxypeptidase S	
3.4.17.18	Carboxypeptidase T Carboxypeptidase T	
3.4.17.19	Carboxypeptidase Taq Carboxypeptidase Taq	
3.4.17.20	Carboxypeptidase U Carboxypeptidase U	
3.4.16.5	Carboxypeptidase Y Carboxypeptidase C	
4.1.2.34	4-(2-Carboxyphenyl)-2-oxobut-3-enoate aldolase 4-(2-Carboxyphenyl)-2-oxobut-3-enoate aldolase	
4.1.1.48	1-(2-Carboxyphenylamino)-1-deoxy-D-ribulose-5-phosphate carboxy-lyase (cyclizing) Indole-3-glycerol phosphate synthase	
2.7.8.23	Carboxyphosphoenolpyruvate phosphonomutase Carboxyvinyl-carboxyphosphonate phosphorylmutase	
5.4.2.9	Carboxyphosphoenolpyruvate phosphonomutase Phosphoenolpyruvate mutase	Vol. 1, Page 546
3.4.17.1	carboxypolypeptidase carboxypeptidase A	
3.4.24.19	Carboxyprocollagen peptidase Procollagen C-endopeptidase	
1.3.99.19	2-Carboxyquinoline 4-monooxygenase Quinoline-4-carboxylate 2-oxidoreductase	
3.4.14.2	carboxytripeptidase dipeptidyl-peptidase II	

EC-No.	Synonym Recommended Name	Vol., Page
4.2.3.5	5-O-(1-carboxyvinyl)-3-phosphoshikimate phosphate-lyase chorismate synthase	
2.7.8.23	Carboxyvinyl-carboxyphosphonate phosphorylmutase Carboxyvinyl-carboxyphosphonate phosphorylmutase	
2.7.8.23	1-Carboxyvinyl carboxyphosphonate phosphorylmutase (decarboxylating) Carboxyvinyl-carboxyphosphonate phosphorylmutase	
3.4.23.40	Cardosin Phytepsin	
3.4.22.30	Caricain Caricain	
3.2.1.59	Cariogenanase Glucan endo-1,3-α-glucosidase	
3.2.1.59	Cariogenase Glucan endo-1,3-α-glucosidase	
3.5.1.73	Carnitinamidase Carnitinamidase	
3.5.1.73	L-Carnitinamide amidohydrolase Carnitinamidase	
2.3.1.7	Carnitine-acetyl-CoA transferase Carnitine O-acetyltransferase	
1.1.1.108	Carnitine 3-dehydrogenase Carnitine 3-dehydrogenase	
1.1.1.108	Carnitine:NAD$^+$ 3-oxidoreductase Carnitine 3-dehydrogenase	
2.3.1.7	Carnitine acetylase Carnitine O-acetyltransferase	
2.3.1.7	Carnitine acetyl coenzyme A transferase Carnitine O-acetyltransferase	
2.3.1.7	Carnitine acetyltransferase Carnitine O-acetyltransferase	
3.5.1.73	Carnitine amidase Carnitinamidase	
3.5.1.73	L-Carnitine amidase Carnitinamidase	
4.1.1.42	Carnitine carboxy-lyase Carnitine decarboxylase	

EC-No.	Synonym Recommended Name	Vol., Page
4.1.1.42	Carnitine decarboxylase Carnitine decarboxylase	
4.2.1.89	carnitine dehydratase carnitine dehydratase	
1.1.1.254	D-Carnitine dehydrogenase (S)-Carnitine 3-dehydrogenase	
3.1.1.28	carnitine ester hydrolase acylcarnitine hydrolase	
4.2.1.89	L-carnitine hydro-lyase carnitine dehydratase	
2.3.1.137	Carnitine medium-chain acyltransferase Carnitine O-octanoyltransferase	
2.3.1.7	Carnitine O-acetyltransferase Carnitine O-acetyltransferase	
2.3.1.137	Carnitine O-octanoyltransferase Carnitine O-octanoyltransferase	
2.3.1.21	Carnitine O-palmitoyltransferase Carnitine O-palmitoyltransferase	
2.3.1.21	Carnitine palmitoyltransferase Carnitine O-palmitoyltransferase	
2.3.1.21	L-Carnitine palmitoyltransferase Carnitine O-palmitoyltransferase	
2.3.1.21	Carnitine palmitoyltransferase-A Carnitine O-palmitoyltransferase	
2.3.1.21	Carnitine palmitoyltransferase I Carnitine O-palmitoyltransferase	
2.3.1.21	Carnitine palmitoyltransferase II Carnitine O-palmitoyltransferase	
3.4.13.3	Carnosinase Aminoacylhistidine dipeptidase	
6.3.2.11	Carnosine-anserine synthetase Carnosine synthase	Vol. 2, Page 460
6.3.2.11	Carnosine-homocarnosine synthetase Carnosine synthase	Vol. 2, Page 460
2.1.1.22	Carnosine N-methyltransferase Carnosine N-methyltransferase	

EC-No.	Synonym Recommended Name	Vol., Page
6.3.2.11	Carnosine synthase Carnosine synthase	Vol. 2, Page 460
6.3.2.11	Carnosine synthetase Carnosine synthase	Vol. 2, Page 460
1.14.99.30	Carotene, hydrogen-donor:oxygen oxidoreductase Carotene 7,8-desaturase	
1.13.11.21	Carotene 15,15'-dioxygenase *β*-Carotene 15,15'-dioxygenase	
1.13.11.21	*β*-Carotene 15,15'-dioxygenase *β*-Carotene 15,15'-dioxygenase	
1.14.99.30	Carotene 7,8-desaturase Carotene 7,8-desaturase	
1.13.11.21	*β*-Carotene:oxygen 15,15'-oxidoreductase (bond-cleaving) *β*-Carotene 15,15'-dioxygenase	
1.13.11.21	Carotene dioxygenase *β*-Carotene 15,15'-dioxygenase	
1.13.11.12	Carotene oxidase Lipoxygenase	
3.2.1.83	Carrageenase,.*κ*.- Kappa-carrageenase	
1.1.1.243	Carveol dehydrogenase Carveol dehydrogenase	
6.3.2.11	CAS Carnosine synthase	Vol. 2, Page 460
4.6.1.7	Casbene synthase Casbene synthase	
4.2.3.8	casbene synthase casbene synthase	
4.6.1.7	Casbene synthetase Casbene synthase	
4.2.3.8	casbene synthetase casbene synthase	
2.7.1.37	Casein kinase Protein kinase	
2.7.1.37	Casein kinase-2 Protein kinase	

EC-No.	Synonym Recommended Name	Vol., Page
2.7.1.37	Casein kinase 2 Protein kinase	
2.7.1.37	Casein kinase I Protein kinase	
2.7.1.37	Casein kinase II Protein kinase	
3.4.21.92	Caseinolytic protease Endopeptidase Clp	
3.1.3.16	Casein phosphatase Phosphoprotein phosphatase	
1.11.1.6	CAT Catalase	
1.11.1.6	Catalase Catalase	
1.11.1.6	Catalase-Peroxidase Catalase	
2.3.1.7	CATC Carnitine O-acetyltransferase	
1.13.11.1	Catechase Catechol 1,2-dioxygenase	
1.13.11.1	Catechol-oxygen 1,2-oxidoreductase Catechol 1,2-dioxygenase	
1.13.11.1	Catechol 1,2-dioxygenase Catechol 1,2-dioxygenase	
1.13.11.1	Catechol 1,2-oxygenase Catechol 1,2-dioxygenase	
1.13.11.2	Catechol 2,3-dioxygenase Catechol 2,3-dioxygenase	
1.13.11.2	Catechol 2,3-oxygenase Catechol 2,3-dioxygenase	
1.13.11.1	Catechol:oxygen 1,2-oxidoreductase Catechol 1,2-dioxygenase	
1.13.11.2	Catechol:oxygen 2,3-oxidoreductase (decyclizing) Catechol 2,3-dioxygenase	
1.1.3.14	Catechol:oxygen oxidoreductase (dimerizing) Catechol oxidase (dimerizing)	

EC-No.	Synonym Recommended Name	Vol., Page
2.1.1.6	Catecholamine O-methyltransferase Catechol O-methyltransferase	
1.10.3.1	Catecholase Catechol oxidase	
1.14.18.1	Catecholase Monophenol monooxygenase	
1.13.11.1	Catechol dioxygenase Catechol 1,2-dioxygenase	
2.1.1.6	Catechol methyltransferase Catechol O-methyltransferase	
2.1.1.6	Catechol O-methyltransferase Catechol O-methyltransferase	
1.10.3.1	Catechol oxidase Catechol oxidase	
1.14.18.1	Catechol oxidase Monophenol monooxygenase	
1.1.3.14	Catechol oxidase (dimerizing) Catechol oxidase (dimerizing)	
1.13.11.2	Catechol oxygenase Catechol 2,3-dioxygenase	
3.4.16.5	Cathepsin A Carboxypeptidase C	
3.4.22.1	cathepsin B cathepsin B	
3.4.22.1	cathepsin B1 cathepsin B	
3.4.18.1	cathepsin B2 cysteine-type carboxypeptidase	
3.4.22.16	Cathepsin Ba, Cathepsin B3 Cathepsin H	
3.4.4.9	cathepsin C cathepsin C	
3.4.14.1	cathepsin C dipeptidyl-peptidase I	
3.4.23.5	cathepsin D cathepsin D	

EC-No.	Synonym Recommended Name	Vol., Page
3.4.23.34	Cathepsin D-like acid proteinase Cathepsin E	
3.4.23.34	Cathepsin D-type proteinase Cathepsin E	
3.4.23.34	Cathepsin E Cathepsin E	
3.4.23.34	Cathepsin E-like acid proteinase Cathepsin E	
3.4.22.41	cathepsin F cathepsin F	
3.4.21.20	cathepsin G cathepsin G	
3.4.22.16	Cathepsin H Cathepsin H	
3.4.22.1	cathepsin II cathepsin B	
3.4.11.1	cathepsin III leucyl aminopeptidase	
3.4.18.1	cathepsin IV cysteine-type carboxypeptidase	
3.4.14.1	cathepsin J dipeptidyl-peptidase I	
3.4.22.38	Cathepsin K Cathepsin K	
3.4.22.15	cathepsin L cathepsin L	
3.4.22.43	Cathepsin L2 cathepsin V	
3.4.22.42	cathepsin O cathepsin O	
3.4.22.38	Cathepsin O2 Cathepsin K	
3.4.21.52	cathepsin R cathepsin R	
3.4.99.33	cathepsin R cathepsin R	

EC-No.	Synonym Recommended Name	Vol., Page
3.4.22.27	Cathepsin S Cathepsin S	
3.4.22.24	Cathepsin T Cathepsin T	
3.4.22.43	cathepsin U cathepsin V	
3.4.22.43	cathepsin V cathepsin V	
2.3.1.28	CAT I Chloramphenicol O-acetyltransferase	
2.3.1.28	CAT II Chloramphenicol O-acetyltransferase	
2.3.1.28	CAT III Chloramphenicol O-acetyltransferase	
6.2.1.33	4-CBA:CoA ligase 4-Chlorobenzoate-CoA ligase	Vol. 2, Page 339
3.2.1.91	CBH 1 Cellulose 1,4-β-cellobiosidase	
4.4.1.8	CBL cystathionine β-lyase	
3.4.21.70	CBPP Pancreatic endopeptidase E	
4.2.1.22	CBS Cystathionine β-synthase	
3.8.1.7	4-CBS-CoA dehalogenase 4-Chlorobenzoyl-CoA dehalogenase	
3.1.21.6	CC-preferring endodeoxyribonuclease CC-preferring endodeoxyribonuclease	
3.4.21.79	CCPII Granzyme B	
1.7.99.3	Cd-cytochrome nitrite reductase Nitrite reductase	
3.4.24.11	CD10 neprilysin	
3.4.11.2	CD13 membrane alanyl aminopeptidase	

EC-No.	Synonym Recommended Name	Vol., Page
3.1.3.48	CD45 protein-tyrosine-phosphatase	
1.2.99.2	Cdh Carbon-monoxide dehydrogenase	
1.13.11.1	CD I Catechol 1,2-dioxygenase	
1.13.11.1	CD II Catechol 1,2-dioxygenase	
1.17.1.1	CDP-4-dehydro-3,6-dideoxy-D-glucose:NAD(P)$^+$ 3-oxidoreductase CDP-4-dehydro-6-deoxyglucose reductase	
1.17.1.1	CDP-4-dehydro-6-deoxyglucose reductase CDP-4-dehydro-6-deoxyglucose reductase	
1.17.1.1	CDP-4-keto-6-deoxy-D-glucose-3-dehydrogenase system CDP-4-dehydro-6-deoxyglucose reductase	
1.17.1.1	CDP-4-keto-deoxy-glucose reductase CDP-4-dehydro-6-deoxyglucose reductase	
2.7.8.22	CDP-choline-1-alkenyl-2-acyl-glycerol phosphocholinetransferase 1-Alkenyl-2-acylglycerol cholinephosphotransferase	
2.7.8.10	CDP-choline-sphingosine cholinephosphotransferase Sphingosine cholinephosphotransferase	
2.7.8.2	CDP-choline diglyceride phosphotransferase Diacylglycerol cholinephosphotransferase	
2.7.7.15	CDP-choline pyrophosphorylase Choline-phosphate cytidylyltransferase	
2.7.7.15	CDP-choline synthetase Choline-phosphate cytidylyltransferase	
5.1.3.10	CDP-D-abequose 2-epimerase CDPabequose 2-epimerase	Vol. 1, Page 146
4.2.1.45	CDP-D-glucose-4,6-dehydratase CDPglucose 4,6-dehydratase	
4.2.1.45	CDP-D-glucose 4,6-dehydratase CDPglucose 4,6-dehydratase	
2.7.7.41	CDP-DG Phosphatidate cytidylyltransferase	
2.7.8.11	CDP-DG:inositol transferase CDPdiacylglycerol-inositol 3-phosphatidyltransferase	

EC-No.	Synonym Recommended Name	Vol., Page
2.7.7.41	CDP-diacylglyceride synthetase Phosphatidate cytidylyltransferase	
2.7.8.11	CDP-diacylglycerol-inositol phosphatidyltransferase CDPdiacylglycerol-inositol 3-phosphatidyltransferase	
2.7.8.8	CDP-diacylglycerol:L-serine O-phosphatidyltransferase CDPdiacylglycerol-serine O-phosphatidyltransferase	
2.7.8.11	CDP-diacylglycerol:myo-inositol-3-phosphatidyltransferase CDPdiacylglycerol-inositol 3-phosphatidyltransferase	
2.7.7.41	CDP-diacylglycerol synthase Phosphatidate cytidylyltransferase	
2.7.8.11	CDP-diglyceride-inositol transferase CDPdiacylglycerol-inositol 3-phosphatidyltransferase	
2.7.8.8	CDP-diglyceride-L-serine phosphatidyltransferase CDPdiacylglycerol-serine O-phosphatidyltransferase	
2.7.8.11	CDP-diglyceride:inositol transferase CDPdiacylglycerol-inositol 3-phosphatidyltransferase	
2.7.8.8	CDP-diglyceride:L-serine phosphatidyltransferase CDPdiacylglycerol-serine O-phosphatidyltransferase	
2.7.8.8	CDP-diglyceride:serine phosphatidyltransferase CDPdiacylglycerol-serine O-phosphatidyltransferase	
2.7.7.33	CDP-glucose pyrophosphorylase Glucose-1-phosphate cytidylyltransferase	
2.7.7.39	CDP-glycerol pyrophosphorylase Glycerol-3-phosphate cytidylyltransferase	
2.7.7.40	CDP-ribitol pyrophosphorylase D-Ribitol-5-phosphate cytidylyltransferase	
5.1.3.10	CDPabequose 2-epimerase CDPabequose 2-epimerase	Vol. 1, Page 146
2.4.1.60	CDPabequose:D-mannosyl-L-rhamnosyl-D-galactose-1-diphospholipid D-abequosyltransferase Abequosyltransferase	
5.1.3.10	CDP abequose epimerase CDPabequose 2-epimerase	Vol. 1, Page 146
2.3.1.70	CDPacylglycerol O-arachidonyltransferase CDPacylglycerol O-arachidonyltransferase	

EC-No.	Synonym Recommended Name	Vol., Page
3.6.1.6	CDPase Nucleoside-diphosphatase	
2.7.8.2	CDPcholine:1,2-diacylglycerol cholinephosphotransferase Diacylglycerol cholinephosphotransferase	
2.7.8.22	CDPcholine:1-alkenyl-2-acylglycerol cholinephosphotransferase 1-Alkenyl-2-acylglycerol cholinephosphotransferase	
2.7.8.3	CDPcholine:N-acylsphingosine cholinephosphotransferase Ceramide cholinephosphotransferase	
2.7.8.10	CDPcholine:sphingosine cholinephosphotransferase Sphingosine cholinephosphotransferase	
2.7.8.5	CDPdiacylglycerol-glycerol-3-phosphate 3-phosphatidyltransferase CDPdiacylglycerol-glycerol-3-phosphate 3-phosphatidyltransferase	
2.7.8.11	CDPdiacylglycerol-inositol 3-phosphatidyltransferase CDPdiacylglycerol-inositol 3-phosphatidyltransferase	
2.7.8.8	CDPdiacylglycerol-L-serine O-phosphatidyltransferase CDPdiacylglycerol-serine O-phosphatidyltransferase	
2.7.8.8	CDPdiacylglycerol-serine O-phosphatidyltransferase CDPdiacylglycerol-serine O-phosphatidyltransferase	
2.7.8.5	CDPdiacylglycerol-sn-glycerol-3-phosphate 3-phosphatidyltransferase CDPdiacylglycerol-glycerol-3-phosphate 3-phosphatidyltransferase	
2.7.8.5	CDPdiacylglycerol:glycerol-3-phosphate phosphatidyltransferase CDPdiacylglycerol-glycerol-3-phosphate 3-phosphatidyltransferase	
2.7.8.8	CDPdiacylglycerol:L-serine 3-O-phosphatidyltransferase CDPdiacylglycerol-serine O-phosphatidyltransferase	
2.7.8.8	CDPdiacylglycerol:L-serine O-phosphatidyltransferase CDPdiacylglycerol-serine O-phosphatidyltransferase	
2.7.8.11	CDPdiacylglycerol:myo-inositol 3-phosphatidyltransferase CDPdiacylglycerol-inositol 3-phosphatidyltransferase	
2.7.8.5	CDPdiacylglycerol:sn-glycero-3-phosphate phosphatidyltransferase CDPdiacylglycerol-glycerol-3-phosphate 3-phosphatidyltransferase	
2.7.8.5	CDPdiacylglycerol:sn-glycerol-3-phosphate 3-phosphatidyltransferase CDPdiacylglycerol-glycerol-3-phosphate 3-phosphatidyltransferase	
3.6.1.26	CDPdiacylglycerol diphosphatase CDPdiacylglycerol diphosphatase	
3.6.1.26	CDP diacylglycerol hydrolase CDPdiacylglycerol diphosphatase	

EC-No.	Synonym Recommended Name	Vol., Page
3.6.1.26	CDPdiacylglycerol phosphatidylhydrolase CDPdiacylglycerol diphosphatase	
3.6.1.26	CDPdiacylglycerol pyrophosphatase CDPdiacylglycerol diphosphatase	
2.7.8.11	CDPdiglyceride-inositol phosphatidyltransferase CDPdiacylglycerol-inositol 3-phosphatidyltransferase	
2.7.8.8	CDPdiglyceride-serine O-phosphatidyltransferase CDPdiacylglycerol-serine O-phosphatidyltransferase	
2.7.7.41	CDPdiglyceride pyrophosphorylase Phosphatidate cytidylyltransferase	
2.7.8.1	CDPethanolamine:1,2-diacylglycerol ethanolaminephosphotransferase Ethanolaminephosphotransferase	
2.7.8.4	CDPethanolamine:L-serine ethanolaminephosphotransferase Serine-phosphoethanolamine synthase	
2.7.8.1	CDPethanolamine diglyceride phosphotransferase Ethanolaminephosphotransferase	
4.2.1.45	CDPglucose 4,6-dehydratase CDPglucose 4,6-dehydratase	
4.2.1.45	CDPglucose 4,6-hydro-lyase CDPglucose 4,6-dehydratase	
2.7.7.33	CDPglucose pyrophosphorylase Glucose-1-phosphate cytidylyltransferase	
2.7.8.12	CDPglycerol:poly(glycerophosphate) glycerophosphotransferase CDPglycerol glycerophosphotransferase	
3.6.1.16	CDP glycerol diphosphatase CDP glycerol diphosphatase	
2.7.8.12	CDPglycerol glycerophosphotransferase CDPglycerol glycerophosphotransferase	
3.6.1.16	CDP glycerol phosphoglycerohydrolase CDP glycerol diphosphatase	
3.6.1.16	CDP glycerol pyrophosphatase CDP glycerol diphosphatase	
2.7.7.39	CDPglycerol pyrophosphorylase Glycerol-3-phosphate cytidylyltransferase	
5.1.3.10	CDP paratose epimerase CDPabequose 2-epimerase	Vol. 1, Page 146

EC-No.	Synonym Recommended Name	Vol., Page
1.17.4.1	CDP reductase Ribonucleoside-diphosphate reductase	
2.7.8.14	CDPribitol:poly(ribitol phosphate) ribitolphosphotransferase CDPribitol ribitolphosphotransferase	
2.7.7.40	CDPribitol pyrophosphorylase D-Ribitol-5-phosphate cytidylyltransferase	
2.7.8.14	CDPribitol ribitolphosphotransferase CDPribitol ribitolphosphotransferase	
1.1.1.225	CDR Chlordecone reductase	
3.4.21.96	Cell-envelope-located serine proteinase Lactocepin	
3.4.21.96	Cell-envelope proteinase Lactocepin	
3.4.21.96	Cell-wall-bound proteinase Lactocepin	
3.4.21.78	T-cell associated protease 1 Granzyme A	
3.4.21.78	T-cell derived serine proteinase 1 Granzyme A	
3.2.1.21	Cellobiase β-glucosidase	
3.2.1.91	Cellobiohydrolase Cellulose 1,4-β-cellobiosidase	
3.2.1.91	Cellobiohydrolase, exo- Cellulose 1,4-β-cellobiosidase	
3.2.1.91	Cellobiohydrolase I Cellulose 1,4-β-cellobiosidase	
1.1.5.1	Cellobiose-quinone oxidoreductase Cellobiose dehydrogenase (quinone)	
5.1.3.11	Cellobiose 2-epimerase Cellobiose epimerase	Vol. 1, Page 148
1.1.99.18	Cellobiose:(acceptor) 1-oxidoreductase Cellobiose dehydrogenase (acceptor)	
2.4.1.20	Cellobiose:orthophosphate α-D-glucosyltransferase Cellobiose phosphorylase	

EC-No.	Synonym Recommended Name	Vol., Page
1.1.3.25	Cellobiose:oxygen 1-oxidoreductase Cellobiose oxidase	
1.1.5.1	Cellobiose:quinone 1-oxidoreductase Cellobiose dehydrogenase (quinone)	
1.1.5.1	Cellobiose:quinone oxidoreductase Cellobiose dehydrogenase (quinone)	
1.1.99.18	Cellobiose dehydrogenase Cellobiose dehydrogenase (acceptor)	
1.1.99.18	Cellobiose dehydrogenase (acceptor) Cellobiose dehydrogenase (acceptor)	
1.1.5.1	Cellobiose dehydrogenase (quinone) Cellobiose dehydrogenase (quinone)	
5.1.3.11	Cellobiose epimerase Cellobiose epimerase	Vol. 1, Page 148
1.1.3.25	Cellobiose oxidase Cellobiose oxidase	
2.4.1.20	Cellobiose phosphorylase Cellobiose phosphorylase	
3.2.1.91	Cellobiosidase Cellulose 1,4-β-cellobiosidase	
3.2.1.91	Cellobiosidase, 1,4-.β.-glucan Cellulose 1,4-β-cellobiosidase	
3.2.1.74	Cellodextrin glucohydrolase Glucan 1,4-β-glucosidase	
2.4.1.49	Cellodextrin phosphorylase Cellodextrin phosphorylase	
3.2.1.4	Celluase A Cellulase	
3.2.1.4	Celludextrinase Cellulase	
3.4.21.73	Cellular plasminogen activator u-Plasminogen activator	
3.2.1.4	9.5 Cellulase Cellulase	
3.2.1.4	Cellulase Cellulase	

EC-No.	Synonym Recommended Name	Vol., Page
5.4.99.15	Cellulase (Arthrobacter strain Q36 clone pBMT13 trehalose oligosaccharide-forming) (1->4)-α-Glucan 1-α-glucosylmutase	Vol. 1, Page 652
5.4.99.15	Cellulase (Rhizobium strain M11 clone pBMT7 trehalose oligosaccharide-forming) (1->4)-α-Glucan 1-α-glucosylmutase	Vol. 1, Page 652
3.2.1.91	Cellulase, C1 Cellulose 1,4-β-cellobiosidase	
3.2.1.4	Cellulase A 3 Cellulase	
3.1.6.7	cellulose-sulfate sulfohydrolase cellulose polysulfatase	
3.2.1.91	Cellulose 1,4-β-cellobiosidase Cellulose 1,4-β-cellobiosidase	
3.1.6.7	cellulose polysulfatase cellulose polysulfatase	
3.1.6.7	cellulose polysulphatase cellulose polysulfatase	
2.4.1.29	Cellulose synthase (GDP-forming) Cellulose synthase (GDP-forming)	
2.4.1.29	Cellulose synthase (guanosine diphosphate-forming) Cellulose synthase (GDP-forming)	
2.4.1.12	Cellulose synthase (UDP-forming) Cellulose synthase (UDP-forming)	
2.4.1.29	Cellulose synthetase Cellulose synthase (GDP-forming)	
3.2.1.4	Cellulosin AP Cellulase	
2.4.1.5	CEP Dextransucrase	
3.4.21.96	CEP Lactocepin	
3.5.2.6	Cephalosporin-β-lactamase β-lactamase	
3.1.1.41	Cephalosporin-C acetylhydrolase Cephalosporin-C deacetylase	

EC-No.	Synonym Recommended Name	Vol., Page
3.1.1.41	Cephalosporin-C deacetylase Cephalosporin-C deacetylase	
2.6.1.74	Cephalosporin-C transaminase Cephalosporin-C transaminase	
3.1.1.41	Cephalosporin acetylesterase Cephalosporin-C deacetylase	
3.5.2.6	Cephalosporinase β-lactamase	
3.5.2.8	cephalosporinase cephalosporinase	
3.1.1.41	Cephalosporin C acetyl-hydrolase Cephalosporin-C deacetylase	
3.1.1.41	Cephalosporin C acetylase Cephalosporin-C deacetylase	
3.1.1.41	Cephalosporin C acetylesterase Cephalosporin-C deacetylase	
3.1.1.41	Cephalosporin C acetylhydrolase Cephalosporin-C deacetylase	
3.5.1.23	Ceramidase Acylsphingosine deacylase	
3.2.1.123	Ceramidase, endoglycosyl- Endoglycosylceramidase	
3.2.1.47	Ceramidase, galactosylgalactosylglucosyl- Galactosylgalactosylglucosylceramidase	
3.2.1.46	Ceramidase, galcatosyl- Galactoceramidase	
3.2.1.45	Ceramidase, glucosyl- Glucosylceramidase	
3.2.1.62	Ceramidase, glycosyl- Glycosylceramidase	
2.4.1.80	Ceramide:UDP-glucose glucosyltransferase Ceramide glucosyltransferase	
2.4.1.80	Ceramide:UDPGlc glucosyltransferase Ceramide glucosyltransferase	
2.7.8.3	Ceramide cholinephosphotransferase Ceramide cholinephosphotransferase	

EC-No.	Synonym Recommended Name	Vol., Page
3.2.1.46	Ceramide galactosidase Galactoceramidase	
3.2.1.45	Ceramide glucosidase Glucosylceramidase	
2.4.1.80	Ceramide glucosyltransferase Ceramide glucosyltransferase	
2.7.1.138	Ceramide kinase Ceramide kinase	
2.3.1.24	Ceramide synthetase Sphingosine N-acyltransferase	
3.2.1.47	Ceramide trihexosidase Galactosylgalactosylglucosylceramidase	
3.2.1.47	Ceramidetrihexosidase Galactosylgalactosylglucosylceramidase	
3.2.1.47	Ceramidetrihexoside-.α.-galactosidase Galactosylgalactosylglucosylceramidase	
3.1.27.5	Ceratitis capitata alkaline ribonuclease Pancreatic ribonuclease	
3.4.16.6	Cereal serine carboxypeptidase II Carboxypeptidase D	
3.2.1.62	Cerebrosidase Glycosylceramidase	
3.1.6.8	Cerebroside-3-sulfate 3-sulfohydrolase Cerebroside-sulfatase	
3.1.6.8	Cerebroside-sulfatase Cerebroside-sulfatase	
3.2.1.46	Cerebroside. β.-galactosidase Galactoceramidase	
3.2.1.45	Cerebroside.β.-glucosidase Glucosylceramidase	
3.2.1.46	Cerebroside galactosidase Galactoceramidase	
3.1.6.8	Cerebroside sulfatase Cerebroside-sulfatase	
3.1.6.8	Cerebroside sulfate sulfatase Cerebroside-sulfatase	

EC-No.	Synonym Recommended Name	Vol., Page
3.4.21.48	cerevisin cerevisin	
1.16.3.1	Ceruloplasmin Ferroxidase	
3.1.1.70	Cetraxate-benzyl-ester benzylhydrolase Cetraxate benzylesterase	
3.1.1.70	Cetraxate benzylesterase Cetraxate benzylesterase	
4.2.1.24	CF-2 porphobilinogen synthase	
2.1.1.79	CFA synthase Cyclopropane-fatty-acyl-phospholipid synthase	
3.1.21.3	CfrAI Type I site-specific deoxyribonuclease	
2.7.1.37	cGMP-dependent protein kinase Protein kinase	
3.1.4.35	cGMP-PDE 3',5'-Cyclic-GMP phosphodiesterase	
3.1.4.35	cGMP phosphodiesterase 3',5'-Cyclic-GMP phosphodiesterase	
2.7.8.12	CGPTase CDPglycerol glycerophosphotransferase	
2.6.1.3	CGT Cysteine transaminase	
2.4.1.19	CGTase Cyclomaltodextrin glucanotransferase	
6.3.3.2	CH^+-THF synthetase 5-Formyltetrahydrofolate cyclo-ligase	Vol. 2, Page 535
1.5.1.20	5,10-CH_2-H_4folate reductase Methylenetetrahydrofolate reductase (NADPH)	
2.3.1.6	ChAcT Choline O-acetyltransferase	
5.5.1.6	Chalcone-flavanone isomerase Chalcone isomerase	Vol. 1, Page 691
5.5.1.6	Chalcone isomerase Chalcone isomerase	Vol. 1, Page 691

EC-No.	Synonym Recommended Name	Vol., Page
2.3.1.74	Chalcone synthase Naringenin-chalcone synthase	
2.3.1.74	Chalcone synthetase Naringenin-chalcone synthase	
4.2.2.5	A-Chase chondroitin AC lyase	
4.2.2.5	F-Chase chondroitin AC lyase	
3.1.1.8	CHE cholinesterase	
3.1.1.8	ChE cholinesterase	
3.1.1.61	CheB methylesterase Protein-glutamate methylesterase	
1.1.3.33	S-Cheilanthifoline synthase S-Cheilanthifoline synthase	
4.99.1.1	chelatase, ferro- ferrochelatase	
4.99.1.1	chelatase, ferro-protoporphyrin ferrochelatase	
3.1.1.61	Chemotaxis-specific methylesterase Protein-glutamate methylesterase	
3.5.1.74	Chenodeoxycholoyltaurine amidohydrolase Chenodeoxycholoyltaurine hydrolase	
3.5.1.74	Chenodeoxycholoyltaurine hydrolase Chenodeoxycholoyltaurine hydrolase	
5.5.1.6	CHI Chalcone isomerase	Vol. 1, Page 691
2.4.1.16	Chitin-UDP N-acetylglucosaminyltransferase Chitin synthase	
3.5.1.41	Chitin amidohydrolase Chitin deacetylase	
3.2.1.14	Chitinase Chitinase	
3.5.1.41	Chitin deacetylase Chitin deacetylase	

EC-No.	Synonym Recommended Name	Vol., Page
2.4.1.16	Chitin synthase Chitin synthase	
2.4.1.16	Chitin synthetase Chitin synthase	
3.2.1.29	chitobiase chitobiase	
2.4.1.142	Chitobiosyldiphosphodolichol α-mannosyltransferase Chitobiosyldiphosphodolichol α-mannosyltransferase	
2.7.8.15	Chitobiosylpyrophosphoryldolichol synthase UDP-N-acetylglucosamine-dolichyl-phosphate N-acetylglucosamine-phosphotransferase	
3.2.1.14	Chitodextrinase Chitinase	
3.2.1.132	Chitosanase Chitosanase	
3.2.1.132	Chitosan N-acetylglucosaminohydrolase Chitosanase	
3.4.24.38	Chlamydomonas cell wall degrading protease Autolysin	
3.4.24.38	Chlamydomonas reinhardtii metalloproteinase Autolysin	
3.1.1.14	Chlase chlorophyllase	
2.3.1.28	Chloramphenicol acetylase Chloramphenicol O-acetyltransferase	
2.3.1.28	Chloramphenicol acetyltransferase Chloramphenicol O-acetyltransferase	
2.3.1.28	Chloramphenicol O-acetyltransferase Chloramphenicol O-acetyltransferase	
2.3.1.28	Chloramphenicol transacetylase Chloramphenicol O-acetyltransferase	
1.97.1.1	Chlorate reductase Chlorate reductase	
1.97.1.1	Chlorate reductase C Chlorate reductase	

EC-No.	Synonym Recommended Name	Vol., Page
1.1.1.225	Chlordecone-alcohol:NADP$^+$ 2-oxidoreductase Chlordecone reductase	
1.1.1.225	Chlordecone reductase Chlordecone reductase	
1.13.11.36	Chloridazon-catechol dioxygenase Chloridazon-catechol dioxygenase	
1.11.1.10	Chloride:hydrogen-peroxide oxidoreductase Chloride peroxidase	
1.11.1.10	Chloride peroxidase Chloride peroxidase	
1.97.1.1	Chlorite:acceptor oxidoreductase Chlorate reductase	
5.2.1.10	2-Chloro-4-carboxymethylenebut-2-en-1,4-olide cis-trans-isomerase 2-Chloro-4-carboxymethylenebut-2-en-1,4-olide isomerase	Vol. 1, Page 231
5.2.1.10	2-Chloro-4-carboxymethylenebut-2-en-1,4-olide isomerase 2-Chloro-4-carboxymethylenebut-2-en-1,4-olide isomerase	Vol. 1, Page 231
5.5.1.7	2-Chloro-5-oxo-2,5-dihydrofuran-2-acetate lyase (decyclizing) Chloromuconate cycloisomerase	Vol. 1, Page 699
4.5.1.2	3-chloro-D-alanine chloride-lyase 3-chloro-D-alanine dehydrochlorinase	
4.5.1.2	3-chloro-D-alanine dehydrochlorinase 3-chloro-D-alanine dehydrochlorinase	
4.5.1.2	β-chloro-D-alanine dehydrochlorinase 3-chloro-D-alanine dehydrochlorinase	
4.5.1.2	3-chloro-D-alanine hydrogen-chloride-lyase (deaminating) 3-chloro-D-alanine dehydrochlorinase	
4.5.1.5	3-Chloro-L-alanine chloride-lyase (adding thioglycolate) S-Carboxymethylcysteine synthase	
3.1.1.72	Chloroacetate esterase Acetylxylan esterase	
1.14.12.13	2-Chlorobenzoate,NADH:oxygen oxidoreductase (1,2-hydroxylating, dechlorinating, decarboxylating) 2-Chlorobenzoate 1,2-dioxygenase	
1.14.12.13	2-Chlorobenzoate-1,2-dioxygenase 2-Chlorobenzoate 1,2-dioxygenase	

EC-No.	Synonym Recommended Name	Vol., Page
6.2.1.33	4-Chlorobenzoate-CoA ligase 4-Chlorobenzoate-CoA ligase	Vol. 2, Page 339
3.8.1.7	4-Chlorobenzoate-coenzyme A dehalogenase 4-Chlorobenzoyl-CoA dehalogenase	
1.14.12.13	2-Chlorobenzoate 1,2-dioxygenase 2-Chlorobenzoate 1,2-dioxygenase	
6.2.1.33	4-Chlorobenzoate:CoA ligase 4-Chlorobenzoate-CoA ligase	Vol. 2, Page 339
6.2.1.33	4-Chlorobenzoate:coenzyme A ligase 4-Chlorobenzoate-CoA ligase	Vol. 2, Page 339
3.8.1.6	4-Chlorobenzoate chlorohydrolase 4-Chlorobenzoate dehalogenase	
3.8.1.6	4-Chlorobenzoate dehalogenase 4-Chlorobenzoate dehalogenase	
1.3.1.63	4-chlorobenzoyl-CoA:NADP oxidoreductase (halogenating) 2,4-dichlorobenzoyl-CoA reductase	
3.8.1.7	4-Chlorobenzoyl-CoA dehalogenase 4-Chlorobenzoyl-CoA dehalogenase	
3.8.1.7	4-Chlorobenzoyl-CoA dehalogenase 4-Chlorobenzoyl-CoA dehalogenase	
6.2.1.33	4-Chlorobenzoyl-CoA ligase 4-Chlorobenzoate-CoA ligase	Vol. 2, Page 339
3.8.1.7	4-Chlorobenzoyl-coenzyme A dehalogenase 4-Chlorobenzoyl-CoA dehalogenase	
6.2.1.33	4-Chlorobenzoyl-coenzyme A ligase 4-Chlorobenzoate-CoA ligase	Vol. 2, Page 339
3.8.1.7	4-Chlorobenzoyl CoA chlorohydrolase 4-Chlorobenzoyl-CoA dehalogenase	
3.8.1.7	4-Chlorobenzoyl CoA dehalogenase 4-Chlorobenzoyl-CoA dehalogenase	
5.2.1.10	2-Chlorocarboxymethylenebutenolide isomerase 2-Chloro-4-carboxymethylenebut-2-en-1,4-olide isomerase	Vol. 1, Page 231
5.2.1.10	Chlorodienelactone isomerase 2-Chloro-4-carboxymethylenebut-2-en-1,4-olide isomerase	Vol. 1, Page 231

EC-No.	Synonym Recommended Name	Vol., Page
3.1.1.72	Chloroesterase Acetylxylan esterase	
3.1.1.6	chloroesterase acetylesterase	
3.1.1.42	chlorogenase chlorogenate hydrolase	
2.3.1.98	Chlorogenate-glucarate O-hydroxycinnamoyltransferase Chlorogenate-glucarate O-hydroxycinnamoyltransferase	
2.3.1.98	Chlorogenate:glucarate caffeoyltransferase Chlorogenate-glucarate O-hydroxycinnamoyltransferase	
2.3.1.98	Chlorogenate:glucarate O-(hydroxycinnamoyl)transferase Chlorogenate-glucarate O-hydroxycinnamoyltransferase	
3.1.1.42	chlorogenate hydrolase chlorogenate hydrolase	
3.1.1.42	chlorogenate hydrolase chlorogenate hydrolase	
2.3.1.98	Chlorogenic acid:glucaric acid O-caffeoyltransferase Chlorogenate-glucarate O-hydroxycinnamoyltransferase	
3.1.1.42	chlorogenic acid esterase chlorogenate hydrolase	
1.14.18.1	Chlorogenic acid oxidase Monophenol monooxygenase	
1.14.18.1	Chlorogenic oxidase Monophenol monooxygenase	
3.8.1.5	1-Chlorohexane halidohydrolase Haloalkane dehalogenase	
5.5.1.7	Chloromuconate cycloisomerase Chloromuconate cycloisomerase	Vol. 1, Page 699
1.11.1.10	Chloroperoxidase Chloride peroxidase	
2.1.1.136	chlorophenol O-methyltransferase chlorophenol O-methyltransferase	
1.14.12.9	4-Chlorophenylacetate,NADH:oxygen oxidoreductase (3,4-hydroxylating, dechlorinating) 4-Chlorophenylacetate 3,4-dioxygenase	

EC-No.	Synonym Recommended Name	Vol., Page
1.14.12.9	4-Chlorophenylacetate 3,4-dioxygenase 4-Chlorophenylacetate 3,4-dioxygenase	
1.13.99.4	4-chlorophenylacetate 3,4-dioxygenase 4-chlorophenylacetate 3,4-dioxygenase	
3.1.1.14	chlorophyllase chlorophyllase	
3.1.1.14	chlorophyll chlorophyllido-hydrolyase chlorophyllase	
3.1.1.14	chlorophyll chlorophyllidohydrolyase chlorophyllase	
1.3.1.33	Chlorophyllide-a:$NADP^+$ 7,8-oxidoreductase Protochlorophyllide reductase	
3.6.1.34	Chloroplast ATPase H^+-transporting ATP synthase	
5.3.3.10	CHMI 5-Carboxymethyl-2-hydroxymuconate Δ-isomerase	Vol. 1, Page 427
1.14.13.60	27CHO 7α-OHase 27-Hydroxycholesterol 7α-monooxygenase	
6.2.1.7	Cholate-CoA ligase Cholate-CoA ligase	Vol. 2, Page 236
6.2.1.7	Cholate:CoA ligase (AMP-forming) Cholate-CoA ligase	Vol. 2, Page 236
6.2.1.7	Cholate thiokinase Cholate-CoA ligase	Vol. 2, Page 236
3.1.1.8	CholE cholinesterase	
1.3.1.22	Cholest-4-en-3-one 5α-reductase Cholestenone 5α-reductase	
1.1.1.181	Cholest-5-ene-3β,7α-diol 3β-dehydrogenase Cholest-5-ene-3β,7α-diol 3β-dehydrogenase	
1.1.1.181	Cholest-5-ene-3β,7α-diol:NAD^+ 3-oxidoreductase Cholest-5-ene-3β,7α-diol 3β-dehydrogenase	
1.3.3.2	5α-Cholest-7-en-3β-ol:oxygen Δ^5-oxidoreductase Lathosterol oxidase	
1.1.1.161	5β-Cholestane-3α,7α,12α,26-tetraol:NAD^+ 26-oxidoreductase Cholestanetetraol 26-dehydrogenase	

EC-No.	Synonym Recommended Name	Vol., Page
1.1.1.161	5β-Cholestane-3 α,7α,12α,26-tetrol dehydrogenase Cholestanetetraol 26-dehydrogenase	
1.14.13.15	5β-Cholestane-3α,7α,12α-triol,NADPH:oxygen oxidoreductase (26-hydroxylating) Cholestanetriol 26-monooxygenase	
1.14.13.15	5β-Cholestane-3 α,7α,12α-triol 26-hydroxylase Cholestanetriol 26-monooxygenase	
1.14.13.15	5β-Cholestane-3α,7α,12α-triol hydroxylase Cholestanetriol 26-monooxygenase	
1.1.1.161	Cholestanetetraol 26-dehydrogenase Cholestanetetraol 26-dehydrogenase	
1.14.13.15	Cholestanetriol 26-hydroxylase Cholestanetriol 26-monooxygenase	
1.14.13.15	Cholestanetriol 26-monooxygenase Cholestanetriol 26-monooxygenase	
5.3.3.5	Δ^8-Cholestenol-Δ^7-cholestenol isomerase Cholestenol Δ-isomerase	Vol. 1, Page 404
5.3.3.5	Cholestenol Δ-isomerase Cholestenol Δ-isomerase	Vol. 1, Page 404
5.3.3.5	Δ^7-Cholestenol Δ^7-Δ^8-isomerase Cholestenol Δ-isomerase	Vol. 1, Page 404
1.3.1.22	Cholestenone 5α-reductase Cholestenone 5α-reductase	
1.3.1.23	Cholestenone 5β-reductase Cholestenone 5β-reductase	
3.1.1.13	Cholesterase Cholesterol esterase	
1.14.13.17	Cholesterol,NADPH:oxygen oxidoreductase (7α-hydroxylating) Cholesterol 7α-monooxygenase	
1.14.15.6	Cholesterol,reduced-adrenal-ferredoxin:oxygen oxidoreductase (side-chain-cleaving) Cholesterol monooxygenase (side-chain-cleaving)	
3.4.21.70	Cholesterol-binding pancreatic proteinase Pancreatic endopeptidase E	
3.4.21.70	Cholesterol-binding proteinase Pancreatic endopeptidase E	

EC-No.	Synonym Recommended Name	Vol., Page
3.4.21.70	Cholesterol-binding serine proteinase Pancreatic endopeptidase E	
1.1.3.6	Cholesterol-O_2 oxidoreductase Cholesterol oxidase	
1.14.15.6	Cholesterol 20-22-desmolase Cholesterol monooxygenase (side-chain-cleaving)	
1.14.13.17	Cholesterol 7α-hydroxylase Cholesterol 7α-monooxygenase	
1.14.13.17	Cholesterol 7α-monooxygenase Cholesterol 7α-monooxygenase	
1.3.1.21	Cholesterol: $NADP^+$ Δ^7-oxidoreductase 7-Dehydrocholesterol reductase	
1.1.3.6	Cholesterol:oxygen oxidoreductase Cholesterol oxidase	
2.3.1.26	Cholesterol acyltransferase Sterol O-acyltransferase	
1.14.15.6	Cholesterol C_{20}-$_{22}$ desmolase Cholesterol monooxygenase (side-chain-cleaving)	
1.14.15.6	Cholesterol desmolase Cholesterol monooxygenase (side-chain-cleaving)	
3.1.1.13	Cholesterol esterase Cholesterol esterase	
3.1.1.13	Cholesterol ester hydrolase Cholesterol esterase	
2.3.1.26	Cholesterol ester synthase Sterol O-acyltransferase	
2.3.1.26	Cholesterol ester synthetase Sterol O-acyltransferase	
1.14.15.6	Cholesterol monooxygenase (side-chain-cleaving) Cholesterol monooxygenase (side-chain-cleaving)	
1.1.3.6	Cholesterol oxidase Cholesterol oxidase	
1.14.15.6	Cholesterol side-chain-cleaving enzyme Cholesterol monooxygenase (side-chain-cleaving)	
1.14.15.6	Cholesterol side-chain cleavage enzyme Cholesterol monooxygenase (side-chain-cleaving)	

EC-No.	Synonym Recommended Name	Vol., Page
2.8.2.2	Cholesterol sulfotransferase Alcohol sulfotransferase	
3.2.1.104	Cholesteryl-β-D-glucoside glucohydrolase Steryl-β-glucosidase	
3.1.1.13	Cholesteryl esterase Cholesterol esterase	
3.1.1.13	Cholesteryl ester hydrolase Cholesterol esterase	
2.3.1.26	Cholesteryl ester synthetase Sterol O-acyltransferase	
6.2.1.7	Cholic acid:CoA ligase Cholate-CoA ligase	Vol. 2, Page 236
6.2.1.7	Cholic thiokinase Cholate-CoA ligase	Vol. 2, Page 236
1.14.14.4	choline, reduced ferredoxin: oxygen oxidoreductase choline monooxygenase	
1.1.99.1	Choline-cytochrome c reductase Choline dehydrogenase	
1.1.3.17	Choline-cytochrome c reductase Choline oxidase	
2.7.1.32	Choline-ethanolamine kinase Choline kinase	
2.7.7.15	Choline-phosphate cytidylyltransferase Choline-phosphate cytidylyltransferase	
3.1.6.6	Choline-sulfate sulfohydrolase Cholinesulfatase	
2.7.1.32	Choline/ethanolamine kinase Choline kinase	
1.1.99.1	Choline:(acceptor) 1-oxidoreductase Choline dehydrogenase	
1.1.99.1	Choline:(acceptor) oxidoreductase Choline dehydrogenase	
1.1.3.17	Choline:oxygen 1-oxidoreductase Choline oxidase	
2.3.1.6	Choline acetylase Choline O-acetyltransferase	

EC-No.	Synonym Recommended Name	Vol., Page
2.3.1.6	Choline acetyltransferase Choline O-acetyltransferase	
1.1.99.1	Choline dehydrogenase Choline dehydrogenase	
1.1.3.17	Choline dehydrogenase Choline oxidase	
3.1.1.8	choline esterase cholinesterase	
3.1.1.7	choline esterase I acetylcholinesterase	
3.1.1.8	choline esterase II (unspecific) cholinesterase	
2.7.1.32	Choline kinase Choline kinase	
1.14.14.4	choline monooxygenase choline monooxygenase	
2.3.1.6	Choline O-acetyltransferase Choline O-acetyltransferase	
1.1.99.1	Choline oxidase Choline dehydrogenase	
1.1.3.17	Choline oxidase Choline oxidase	
3.1.4.4	Choline phosphatase Phospholipase D	
2.7.7.15	Choline phosphate cytidylyltransferase Choline-phosphate cytidylyltransferase	
2.7.1.32	Choline phosphokinase Choline kinase	
2.7.8.2	Cholinephosphotransferase Diacylglycerol cholinephosphotransferase	
2.7.8.22	Cholinephosphotransferase, 1-alkenyl-2-acylglycerol 1-Alkenyl-2-acylglycerol cholinephosphotransferase	
2.7.8.2	Cholinephosphotransferase, 1-alkyl-2-acetylglycerol Diacylglycerol cholinephosphotransferase	
2.7.8.3	Cholinephosphotransferase, ceramide Ceramide cholinephosphotransferase	

EC-No.	Synonym Recommended Name	Vol., Page
2.7.8.2	Cholinephosphotransferase, diacylglycerol Diacylglycerol cholinephosphotransferase	
2.7.8.10	Cholinephosphotransferase, sphingosine Sphingosine cholinephosphotransferase	
3.1.1.7	cholinesterase acetylcholinesterase	
3.1.1.8	cholinesterase cholinesterase	
3.1.6.6	Choline sulfatase Cholinesulfatase	
3.1.6.6	Cholinesulfatase Cholinesulfatase	
2.8.2.6	Choline sulfotransferase Choline sulfotransferase	
3.1.6.6	Choline sulphatase Cholinesulfatase	
2.8.2.6	Choline sulphokinase Choline sulfotransferase	
2.3.1.65	Choloyl-CoA:glycine N-choloyltransferase Glycine N-choloyltransferase	
6.2.1.7	Choloyl-CoA synthetase Cholate-CoA ligase	Vol. 2, Page 236
6.2.1.7	Choloyl coenzyme A synthetase Cholate-CoA ligase	Vol. 2, Page 236
3.5.1.24	Cholylglycine hydrolase Cholylglycine hydrolase	
3.5.1.24	Cholylglycine hydrolase Cholylglycine hydrolase	
3.1.6.9	Chondro-4-sulfatase Chondro-4-sulfatase	
3.1.6.10	Chondro-6-sulfatase Chondro-6-sulfatase	
3.1.6.9	Chondroitin-4-sulfatase Chondro-4-sulfatase	
5.1.3.19	Chondroitin-D-glucuronate 5-epimerase Chondroitin-glucuronate 5-epimerase	Vol. 1, Page 172

EC-No.	Synonym Recommended Name	Vol., Page
5.1.3.19	Chondroitin-glucuronate 5-epimerase Chondroitin-glucuronate 5-epimerase	Vol. 1, Page 172
2.8.2.5	Chondroitin 4-sulfotransferase Chondroitin 4-sulfotransferase	
2.8.2.17	Chondroitin 6-O-sulfotransferase Chondroitin 6-sulfotransferase	
2.8.2.17	Chondroitin 6-sulfotransferase Chondroitin 6-sulfotransferase	
4.2.2.4	chondroitin ABC eliminase chondroitin ABC lyase	
4.2.2.4	chondroitin ABC lyase chondroitin ABC lyase	
4.2.2.4	chondroitin ABC lyase chondroitin ABC lyase	
4.2.2.5	chondroitin AC eliminase chondroitin AC lyase	
4.2.2.5	chondroitin AC lyase chondroitin AC lyase	
4.2.2.5	chondroitin AC lyase chondroitin AC lyase	
3.2.1.35	Chondroitinase Hyaluronoglucosaminidase	
3.1.6.12	Chondroitinase N-Acetylgalactosamine-4-sulfatase	
3.1.6.4	Chondroitinase N-Acetylgalactosamine-6-sulfatase	
4.2.2.4	chondroitinase chondroitin ABC lyase	
4.2.2.5	chondroitinase chondroitin AC lyase	
3.2.1.34	chondroitinase chondroitinase	
4.2.2.4	chondroitinase ABC chondroitin ABC lyase	
4.2.2.5	chondroitinase AC chondroitin AC lyase	

EC-No.	Synonym Recommended Name	Vol., Page
3.2.1.35	Chondroitinase I Hyaluronoglucosaminidase	
5.1.3.19	Chondroitin D-glucuronosyl 5-epimerase Chondroitin-glucuronate 5-epimerase	Vol. 1, Page 172
3.1.6.4	Chondroitin sulfatase N-Acetylgalactosamine-6-sulfatase	
3.1.6.13	Chondroitinsulfatase Iduronate 2-sulfatase	
3.1.6.12	Chondroitinsulfatase N-Acetylgalactosamine-4-sulfatase	
3.1.6.4	Chondroitinsulfatase N-Acetylgalactosamine-6-sulfatase	
3.1.6.14	Chondroitinsulfatase N-Acetylglucosamine-6-sulfatase	
3.1.6.15	chondroitinsulfatase N-sulfoglucosamine-3-sulfatase	
4.2.99.6	Chondroitin sulfate lyase Chondroitin sulfate lyase	
4.2.2.5	chondroitin sulfate lyase chondroitin AC lyase	
2.8.2.5	Chondroitin sulfotransferase Chondroitin 4-sulfotransferase	
3.4.24.67	Choriolysin H Choriolysin H	
3.4.24.66	Choriolysin L Choriolysin L	
3.4.24.12	chorion-digesting proteinase envelysin	
3.4.24.12	chorionase envelysin	
4.1.3.27	chorismate lyase anthranilate synthase	
5.4.99.5	Chorismate mutase Chorismate mutase	Vol. 1, Page 604
5.4.99.5	Chorismate mutase/prephenate dehydratase Chorismate mutase	Vol. 1, Page 604

EC-No.	Synonym Recommended Name	Vol., Page
4.1.3.27	chorismate pyruvate-lyase (amino-accepting) anthranilate synthase	
5.4.99.5	Chorismate pyruvatemutase Chorismate mutase	Vol. 1, Page 604
4.2.3.5	chorismate synthase chorismate synthase	
4.6.1.4	chorismate synthase chorismate synthase	
4.6.1.4	chorismate synthetase chorismate synthase	
3.1.3.32	chromatin 3'-phosphatase polynucleotide 3'-phosphatase	
2.3.1.74	CHS Naringenin-chalcone synthase	
4.2.2.5	ChSase-AC chondroitin AC lyase	
4.2.1.66	CHT cyanide hydratase	
3.4.21.1	α chymar Chymotrypsin	
3.4.21.1	Chymar Chymotrypsin	
3.4.21.1	α-chymar ophth Chymotrypsin	
3.4.21.39	chymase chymase	
3.4.23.4	chymase chymosin	
3.4.22.25	Chymopapain Glycyl endopeptidase	
3.4.22.6	chymopapain chymopapain	
3.4.4.11	chymopapain chymopapain	
3.4.22.6	chymopapain A chymopapain	

EC-No.	Synonym Recommended Name	Vol., Page
3.4.22.6	chymopapain B chymopapain	
3.4.22.25	Chymopapain M Glycyl endopeptidase	
3.4.22.30	Chymopapain S Caricain	
3.4.22.6	chymopapain S chymopapain	
3.4.23.4	chymosin chymosin	
3.4.24.12	chymostrypsin envelysin	
3.4.21.1	Chymotest Chymotrypsin	
3.4.21.1	.α.-Chymotrypsin Chymotrypsin	
3.4.21.1	α-chymotrypsin Chymotrypsin	
3.4.21.1	Chymotrypsin Chymotrypsin	
3.4.4.5	chymotrypsin chymotrypsin	
3.4.21.20	chymotrypsin-like proteinase cathepsin G	
3.4.21.1	Chymotrypsin A Chymotrypsin	
3.4.21.1	α-Chymotrypsin A Chymotrypsin	
3.4.21.1	Chymotrypsin B Chymotrypsin	
3.4.4.6	chymotrypsin B chymotrypsin B	
3.4.21.2	chymotrypsin C chymotrypsin C	
5.5.1.6	CI Chalcone isomerase	Vol. 1, Page 691

EC-No.	Synonym Recommended Name	Vol., Page
3.1.1.73	cinnAE feruloyl esterase	
1.2.1.44	Cinnamaldehyde:NADP$^+$ oxidoreductase (CoA-cinnamoylating) Cinnamoyl-CoA reductase	
2.4.1.177	Cinnamate β-D-glucosyltransferase Cinnamate β-D-glucosyltransferase	
1.14.13.11	Cinnamate 4-hydroxylase trans-Cinnamate 4-monooxygenase	
1.14.13.11	Cinnamate 4-monooxygenase trans-Cinnamate 4-monooxygenase	
1.14.13.11	Cinnamate hydroxylase trans-Cinnamate 4-monooxygenase	
1.14.13.14	Cinnamic 2-hydroxylase trans-Cinnamate 2-monooxygenase	
1.14.13.11	Cinnamic 4-hydroxylase trans-Cinnamate 4-monooxygenase	
1.14.13.14	Cinnamic acid 2-hydroxylase trans-Cinnamate 2-monooxygenase	
1.14.13.11	Cinnamic acid 4-hydroxylase trans-Cinnamate 4-monooxygenase	
1.14.13.11	Cinnamic acid 4-monooxygenase trans-Cinnamate 4-monooxygenase	
1.14.13.11	t-Cinnamic acid hydroxylase trans-Cinnamate 4-monooxygenase	
1.14.13.11	Cinnamic acid *p*-hydroxylase trans-Cinnamate 4-monooxygenase	
1.2.1.44	Cinnamoyl-CoA:NADPH reductase Cinnamoyl-CoA reductase	
1.2.1.44	Cinnamoyl-CoA reductase Cinnamoyl-CoA reductase	
1.2.1.44	Cinnamoyl-coenzyme A reductase Cinnamoyl-CoA reductase	
1.1.1.195	Cinnamyl-alcohol:NADP$^+$ oxidoreductase Cinnamyl-alcohol dehydrogenase	
1.1.1.195	Cinnamyl-alcohol dehydrogenase Cinnamyl-alcohol dehydrogenase	

EC-No.	Synonym Recommended Name	Vol., Page
1.1.1.195	Cinnamyl alcohol dehydrogenase Cinnamyl-alcohol dehydrogenase	
2.4.1.17	Ciramadol UDP-glucuronyltransferase Glucuronosyltransferase	
5.3.3.8	Δ^3-cis,Δ^2-trans-Enoyl-CoA isomerase Dodecenoyl-CoA δ-isomerase	Vol. 1, Page 413
5.5.1.1	cis,cis-Muconate-lactonizing enzyme Muconate cycloisomerase	Vol. 1, Page 660
5.5.1.1	cis,cis-Muconate cycloisomerase Muconate cycloisomerase	Vol. 1, Page 660
1.3.1.29	cis-1,2-Dihydro-1,2-dihydroxynaphthalene dehydrogenase cis-1,2-Dihydro-1,2-dihydroxynaphthalene dehydrogenase	
1.3.1.66	cis-1,2-dihydro-3-ethylcatechol:NAD oxidoreductase cis-dihydroethylcatechol dehydrogenase	
1.3.1.19	cis-1,2-Dihydrobenzene-1,2-diol:NAD^+ oxidoreductase cis-1,2-Dihydrobenzene-1,2-diol dehydrogenase	
1.3.1.19	cis-1,2-Dihydrobenzene-1,2-diol dehydrogenase cis-1,2-Dihydrobenzene-1,2-diol dehydrogenase	
1.3.1.19	cis-1,2-Dihydrocyclohexa-3,5-diene oxidoreductase (nicotinamide adenine dinucleotide) cis-1,2-Dihydrobenzene-1,2-diol dehydrogenase	
1.3.1.29	cis-1,2-Dihydronaphthalene-1,2-diol: NAD^+ 1,2-oxidoreductase cis-1,2-Dihydro-1,2-dihydroxynaphthalene dehydrogenase	
1.3.1.60	cis-1,2-dihydroxy-1,2-dihydrodibenzothiophene:NAD oxidoreductase dibenzothiophene dihydrodiol dehydrogenase	
1.3.1.67	cis-1,2-dihydroxy-4-methylcyclohexa-3,5-diene-1-carboxylate:NAD(P)oxidoreductase (decarboxylating) cis-1,2-dihydroxy-4-methylcyclohexa-3,5-diene-1-carboxylate dehydrogenase	
1.3.1.67	cis-1,2-dihydroxy-4-methylcyclohexa-3,5-diene-1-carboxylate dehydrogenase cis-1,2-dihydroxy-4-methylcyclohexa-3,5-diene-1-carboxylate dehydrogenase	
1.3.1.55	cis-1,2-Dihydroxycyclohexa-3,5-diene-1-carboxylate:NAD oxidoreductase cis-1,2-Dihydroxycyclohexa-3,5-diene-1-carboxylate dehydrogenase	
1.3.1.55	cis-1,2-Dihydroxycyclohexa-3,5-diene-1-carboxylate dehydrogenase cis-1,2-Dihydroxycyclohexa-3,5-diene-1-carboxylate dehydrogenase	

EC-No.	Synonym Recommended Name	Vol., Page
1.3.1.56	cis-2,3-dihydrobiphenyl-2,3-diol dehydrogenase cis-2,3-dihydrobiphenyl-2,3-diol dehydrogenase	
1.3.1.58	cis-2,3-dihydroxy-2,3-dihydro-*p*-cumate:NAD oxidoreductase 2,3-dihydroxy-2,3-dihydro-*p*-cumate dehydrogenase	
1.3.1.37	cis-2-Enoyl-CoA reductase (NADPH) cis-2-Enoyl-CoA reductase (NADPH)	
1.3.1.37	cis-2-Enoyl-coenzyme A reductase cis-2-Enoyl-CoA reductase (NADPH)	
5.3.3.8	3-cis-2-trans-Enoyl-CoA isomerase Dodecenoyl-CoA δ-isomerase	Vol. 1, Page 413
5.3.3.3	Δ^3-cis-Δ^2-trans-enoyl-CoA isomerase Vinylacetyl-CoA Δ-isomerase	Vol. 1, Page 395
5.3.3.8	δ^3-cis-δ^2-trans-enoyl-CoA isomerase Dodecenoyl-CoA δ-isomerase	Vol. 1, Page 413
1.3.1.49	cis-3,4-Dihydro-3,4-dihydroxyphenanthrene dehydrogenase cis-3,4-Dihydrophenanthrene-3,4-diol dehydrogenase	
1.3.1.49	(+)-cis-3,4-Dihydrophenanthrene-3,4-diol:NAD^+ 3,4-oxidoreductase cis-3,4-Dihydrophenanthrene-3,4-diol dehydrogenase	
1.3.1.49	cis-3,4-Dihydrophenanthrene-3,4-diol dehydrogenase cis-3,4-Dihydrophenanthrene-3,4-diol dehydrogenase	
1.1.1.219	cis-3,4-Leucopelargonidin:$NADP^+$ 4-oxidoreductase Dihydrokaempferol 4-reductase	
5.3.3.3	cis-3,trans-2-Enoyl-CoA-isomerase Vinylacetyl-CoA Δ-isomerase	Vol. 1, Page 395
1.3.1.56	cis-3-phenylcyclohexa-3,5-diene-1,2-diol:NAD oxidoreductase cis-2,3-dihydrobiphenyl-2,3-diol dehydrogenase	
1.3.1.64	cis-4,5-dihydroxycyclohexa-1(6),2-diene-1,2-dicarboxylate:NAD oxidoreductase phthalate 4,5-cis-dihydrodiol dehydrogenase	
1.3.1.61	cis-4,5-dihydroxycyclohexa-1(6),2-diene-1,4-dicarboxylate:NAD oxidoreductase (decarboxylating) terephthalate 1,2-cis-dihydrodiol dehydrogenase	
4.2.1.3	cis-aconitase aconitate hydratase	
4.1.1.6	cis-Aconitate carboxy-lyase Aconitate decarboxylase	

EC-No.	Synonym Recommended Name	Vol., Page
4.1.1.6	cis-Aconitic acid decarboxylase Aconitate decarboxylase	
1.3.1.19	cis-Benzene glycol dehydrogenase cis-1,2-Dihydrobenzene-1,2-diol dehydrogenase	
1.3.1.29	cis-Dihydrodiol naphthalene dehydrogenase cis-1,2-Dihydro-1,2-dihydroxynaphthalene dehydrogenase	
1.3.1.66	cis-dihydroethylcatechol dehydrogenase cis-dihydroethylcatechol dehydrogenase	
1.3.1.55	cis-Diol dehydrogenase (Acinetobacter calcoaceticus reduced) cis-1,2-Dihydroxycyclohexa-3,5-diene-1-carboxylate dehydrogenase	
1.3.1.55	cis-Diol dehydrogenase (Pseudomonas putida reduced) cis-1,2-Dihydroxycyclohexa-3,5-diene-1-carboxylate dehydrogenase	
3.3.2.3	Cis-epoxide hydrolase Epoxide hydrolase	
4.2.1.36	cis-Homoaconitase Homoaconitate hydratase	
1.3.1.29	(+)-cis-Naphthalene dihydrodiol dehydrogenase cis-1,2-Dihydro-1,2-dihydroxynaphthalene dehydrogenase	
1.3.1.29	cis-Naphthalene dihydrodiol dehydrogenase cis-1,2-Dihydro-1,2-dihydroxynaphthalene dehydrogenase	
2.4.1.209	cis-*p*-coumarate glucosyltransferase cis-*p*-coumarate glucosyltransferase	
1.3.1.49	cis-Phenanthrene 3,4-dihydrodiol dehydrogenase cis-3,4-Dihydrophenanthrene-3,4-diol dehydrogenase	
1.1.1.9	2,3-cis-Polyol(DPN) dehydrogenase (C3-5) D-Xylulose reductase	
2.5.1.20	cis-Prenyl transferase Rubber cis-polyprenylcistransferase	
3.1.1.63	11-cis-Retinyl-palmitate acylhydrolase 11-cis-Retinyl-palmitate hydrolase	
3.1.1.63	11-cis-Retinyl-palmitate hydrolase 11-cis-Retinyl-palmitate hydrolase	
1.1.1.228	(+)-cis-Sabinol:NAD$^+$ oxidoreductase (+)-Sabinol dehydrogenase	
4.2.1.35	Citraconase (R)-2-Methylmalate dehydratase	

EC-No.	Synonym Recommended Name	Vol., Page
4.2.1.35	Citraconate hydratase (R)-2-Methylmalate dehydratase	
2.8.3.11	Citramalate CoA-transferase Citramalate CoA-transferase	
2.8.3.7	Citramalate CoA-transferase Succinate-citramalate CoA-transferase	
4.2.1.35	Citramalate hydro-lyase (R)-2-Methylmalate dehydratase	
4.1.3.22	citramalate lyase citramalate lyase	
4.1.3.22	(+)-citramalate pyruvate-lyase citramalate lyase	
4.1.3.22	citramalate synthase citramalate lyase	
4.1.3.22	citramalate synthetase citramalate lyase	
4.1.3.22	citramalic-condensing enzyme citramalate lyase	
4.2.1.34	(+)-Citramalic hydro-lyase (S)-2-Methylmalate dehydratase	
4.1.3.22	citramalic synthase citramalate lyase	
4.2.1.56	Citramalyl-CoA hydro-lyase Itaconyl-CoA hydratase	
4.1.3.25	citramalyl-CoA lyase citramalyl-CoA lyase	
4.1.3.25	citramalyl-CoA lyase citramalyl-CoA lyase	
4.1.3.25	citramalyl-thio-acyl carrier protein lyase citramalyl-CoA lyase	
4.1.3.25	citramalyl coenzyme A lyase citramalyl-CoA lyase	
4.1.3.6	citrase citrate lyase	
4.1.3.6	citratase citrate lyase	

EC-No.	Synonym Recommended Name	Vol., Page
4.2.1.3	citrate (isocitrate)hydro-lyase aconitate hydratase	
4.1.3.6	citrate (pro-3S)-lyase citrate lyase	
6.2.1.22	Citrate (pro-3S)-lyase ligase Citrate (pro-3S)-lyase ligase	Vol. 2, Page 304
4.1.3.28	Citrate (re)-synthase Citrate (re)-synthase	
4.1.3.7	Citrate (si)-synthase Citrate (si)-synthase	
3.1.2.16	[Citrate-(pro-3S)-lyase](acetyl-form) hydrolase [Citrate-(pro-3S)-lyase] thiolesterase	
3.1.2.16	[Citrate-(pro-3S)-lyase] thiolesterase [Citrate-(pro-3S)-lyase] thiolesterase	
4.1.3.8	citrate-ATP lyase ATP citrate (pro-3S)-lyase	
6.2.1.18	Citrate-CoA ligase Citrate-CoA ligase	Vol. 2, Page 291
2.8.3.10	Citrate-CoA transferase Citrate-CoA transferase	
6.2.1.18	Citrate:CoA ligase (ADP-forming) Citrate-CoA ligase	Vol. 2, Page 291
4.1.3.6	citrate aldolase citrate lyase	
4.1.3.8	citrate cleavage enzyme ATP citrate (pro-3S)-lyase	
4.1.3.7	Citrate condensing enzyme Citrate (si)-synthase	
4.2.1.4	Citrate dehydratase Citrate dehydratase	
4.2.1.4	Citrate hydro-lyase Citrate dehydratase	
4.2.1.3	citrate hydro-lyase aconitate hydratase	
4.1.3.6	citrate lyase citrate lyase	

EC-No.	Synonym Recommended Name	Vol., Page
4.1.3.6	citrate lyase citrate lyase	
6.2.1.22	Citrate lyase acetylating enzyme Citrate (pro-3S)-lyase ligase	Vol. 2, Page 304
3.1.2.16	Citrate lyase deacetylase [Citrate-(pro-3S)-lyase] thiolesterase	
6.2.1.22	Citrate lyase ligase Citrate (pro-3S)-lyase ligase	Vol. 2, Page 304
6.2.1.22	Citrate lyase synthetase Citrate (pro-3S)-lyase ligase	Vol. 2, Page 304
4.1.3.28	Citrate oxaloacetate-lyase ((pro-3R)-$CH_2COO^- \rightarrow$acetyl-CoA) Citrate (re)-synthase	
4.1.3.7	Citrate oxaloacetate-lyase ((pro-3S)-$CH_2COO^- \rightarrow$ acetyl-CoA) Citrate (si)-synthase	
4.1.3.6	citrate oxaloacetate-lyase ((pro-3S)-$CH_2COO^- \rightarrow$ acetate) citrate lyase	
4.1.3.7	Citrate oxaloacetate-lyase (CoA-acetylating) Citrate (si)-synthase	
4.1.3.6	citrate oxaloacetate-lyase(pro-3S-$COO^- \rightarrow$ acetate) citrate lyase	
4.1.3.7	Citrate oxaloacetate-lyase, CoA-acetylating Citrate (si)-synthase	
4.1.3.7	Citrate synthase Citrate (si)-synthase	
4.1.3.7	Citrate synthetase Citrate (si)-synthase	
4.1.3.7	Citric-condensing enzyme Citrate (si)-synthase	
4.1.3.6	citric aldolase citrate lyase	
4.1.3.8	citric cleavage enzyme ATP citrate (pro-3S)-lyase	
4.1.3.7	Citric synthase Citrate (si)-synthase	
4.1.3.6	citridesmolase citrate lyase	

EC-No.	Synonym Recommended Name	Vol., Page
4.1.3.6	citritase citrate lyase	
4.1.3.7	Citrogenase Citrate (si)-synthase	
3.5.4.9	Citrovorum factor cyclodehydrase Methenyltetrahydrofolate cyclohydrolase	
3.5.1.20	Citrullinase Citrullinase	
6.3.4.5	Citrulline-aspartate ligase Argininosuccinate synthase	Vol. 2, Page 595
6.3.4.5	L-Citrulline:L-aspartate ligase (AMP-forming) Argininosuccinate synthase	Vol. 2, Page 595
3.5.1.20	Citrulline hydrolase Citrullinase	
3.5.3.6	Citrulline iminase Arginine deiminase	
3.5.1.20	L-Citrulline N^5-carbamoyldihydrolase Citrullinase	
2.1.3.3	Citrulline phosphorylase Ornithine carbamoyltransferase	
3.5.1.20	Citrulline ureidase Citrullinase	
3.1.1.72	Citrus acetylesterase Acetylxylan esterase	
3.1.1.6	citrus acetylesterase acetylesterase	
4.1.3.34	citryl-CoA lyase citryl-CoA lyase	
3.1.2.7	citryl-glutathione thioesterhydrolase glutathione thiolesterase	
3.5.1.68	β-Citryl-L-glutamate-hydrolyzing enzyme N-Formylglutamate deformylase	
3.5.1.68	β-Citryl-L-glutamate amidase N-Formylglutamate deformylase	
3.5.1.68	β-Citryl-L-glutamate amidohydrolase N-Formylglutamate deformylase	

EC-No.	Synonym Recommended Name	Vol., Page
3.5.1.68	β-Citryl-L-glutamate hydrolase N-Formylglutamate deformylase	
4.1.3.34	citryl-S-ACP lyase citryl-CoA lyase	
3.5.1.68	β-Citrylglutamate amidase N-Formylglutamate deformylase	
2.7.3.2	CK Creatine kinase	
2.7.1.37	CK-2 Protein kinase	
2.7.3.2	CK-BB Creatine kinase	
2.7.3.2	CK-MB Creatine kinase	
2.7.3.2	CK-MM Creatine kinase	
2.7.1.37	CK2 Protein kinase	
2.7.2.2	CKase Carbamate kinase	
2.7.1.37	CKI Protein kinase	
2.7.1.37	CKII Protein kinase	
2.7.3.2	$CK_{M}iMi$ Creatine kinase	
6.2.1.12	4CL 4-Coumarate-CoA ligase	Vol. 2, Page 256
3.4.11.6	Cl^{-}-activated arginine aminopeptidase aminopeptidase B	
3.4.23.26	Cladosporium acid protease Rhodotorulapepsin	
3.4.23.26	Cladosporium acid proteinase Rhodotorulapepsin	
3.4.21.43	classical-complement-pathway C3/C5 convertase classical-complement-pathway C3/C5 convertase	

EC-No.	Synonym Recommended Name	Vol., Page
3.1.1.34	clearing factor lipoprotein lipase	
3.4.11.13	Clostridial aminopeptidase Clostridial aminopeptidase	
3.4.22.8	clostridiopeptidase clostripain	
3.4.21.32	clostridiopeptidase A brachyurin	
3.4.4.19	clostridiopeptidase A clostridiopeptidase A	
3.4.24.3	clostridiopeptidase A microbial collagenase	
3.4.4.20	clostridiopeptidase B clostridiopeptidase B	
3.4.22.8	clostridiopeptidase B clostripain	
3.4.21.32	clostridiopeptidase I brachyurin	
3.4.24.3	clostridiopeptidase I microbial collagenase	
3.4.21.32	clostridiopeptidase II brachyurin	
3.4.24.3	clostridiopeptidase II microbial collagenase	
3.4.22.8	α-clostridipain clostripain	
3.4.11.13	Clostridium histolyticum aminopeptidase Clostridial aminopeptidase	
3.4.24.3	Clostridium histolyticum collagenase microbial collagenase	
3.4.99.5	Clostridium histolyticum collagenase 2 Clostridium histolyticum collagenase 2	
3.4.22.8	Clostridium histolyticum proteinase B clostripain	
3.1.4.3	Clostridium oedematiens β- and γ-toxins Phospholipase C	

EC-No.	Synonym Recommended Name	Vol., Page
3.1.4.3	Clostridium welchii α-toxin Phospholipase C	
3.4.22.8	clostripain clostripain	
3.4.21.86	Clotting enzyme Limulus clotting enzyme	
3.4.21.92	ClpP Endopeptidase Clp	
3.4.21.92	Clp protease Endopeptidase Clp	
1.14.13.22	Clyclohexanone oxygenase Cyclohexanone monooxygenase	
4.2.1.51	CM-PD prephenate dehydratase	
5.5.1.7	2CMA-cycloisomerase Chloromuconate cycloisomerase	Vol. 1, Page 699
5.5.1.7	CMC Chloromuconate cycloisomerase	Vol. 1, Page 699
5.3.3.10	CMH isomerase 5-Carboxymethyl-2-hydroxymuconate Δ-isomerase	Vol. 1, Page 427
5.5.1.5	CMLE Carboxy-cis,cis-muconate cyclase	Vol. 1, Page 686
2.7.7.38	CMP-3-deoxy-D-manno-octulosonate pyrophosphorylase 3-Deoxy-manno-octulosonate cytidylyltransferase	
2.7.7.38	CMP-3-deoxy-D-manno-octulosonate synthetase 3-Deoxy-manno-octulosonate cytidylyltransferase	
2.4.99.1	CMP-acetylneuraminate-galactosylglycoprotein sialyltransferase β-Galactoside α-2,6-sialyltransferase	
2.4.99.1	CMP-acetylneuraminate-glycoprotein sialyltransferase β-Galactoside α-2,6-sialyltransferase	
2.4.99.9	CMP-acetylneuraminate-lactosylceramide-sialyltransferase Lactosylceramide α-2,3-sialyltransferase	
2.4.99.11	CMP-acetylneuraminate-lactosylceramide-sialyltransferase Lactosylceramide α-2,6-N-sialyltransferase	
2.4.99.9	CMP-acetylneuraminic acid:lactosylceramide sialyltransferase Lactosylceramide α-2,3-sialyltransferase	

EC-No.	Synonym Recommended Name	Vol., Page
2.7.7.38	CMP-KDO synthetase 3-Deoxy-manno-octulosonate cytidylyltransferase	
1.14.99.18	CMP-N–acetylneuraminic acid:NADH oxidoreductase (N-acetyl-hydroxylating) N-Acetylneuraminate monooxygenase	
1.14.13.45	CMP-N-acetylneuraminate,NAD(P)H:oxygen oxidoreductase (hydroxylating) CMP-N-acetylneuraminate monooxygenase	
2.4.99.7	CMP-N-acetylneuraminate:(α-N-acetylneuraminyl-2,3-β-D-galactosyl-1,3)-N-acetyl-D-galactosaminide α-2,6-N-acetylneuraminyltransferase (α-N-Acetylneuraminyl-2,3-β-galactosyl-1,3)-N-acetylgalactosaminide α-2,6-sialyltransferase	
2.4.99.4	CMP-N-acetylneuraminate:β-D-galactoside α-2,3-N-acetylneuraminyltransferase β-Galactoside α-2,3-sialyltransferase	
2.4.99.1	CMP-N-acetylneuraminate:β-D-galactosyl-1,4-N-acetyl-β-D-glucosamine α-2,6-N-acetylneuraminyltransferase β-Galactoside α-2,6-sialyltransferase	
2.4.99.6	CMP-N-acetylneuraminate:β-D-galactosyl-1,4-N-acetyl-D-glucosaminyl-glycoprotein α-2,3-N-acetylneuraminyltransferase N-Acetyllactosaminide α-2,3-sialyltransferase	
2.4.99.8	CMP-N-acetylneuraminate:α-N-acetylneuraminyl-2,3-β-D-galactoside α-2,8-N-acetylneuraminyltransferase α-N-Acetylneuraminate α-2,8-sialyltransferase	
2.4.99.5	CMP-N-acetylneuraminate:1,2-diacyl-3-β-D-galactosyl-sn-glycerol N-acetylneuraminyltransferase Galactosyldiacylglycerol α-2,3-sialyltransferase	
2.4.99.2	CMP-N-acetylneuraminate:D-galactosyl-N-acetyl-D-galactosaminyl-(N-acetylneuraminyl)-D-galactosyl-D-glucosylceramide N-acetylneuraminyltransferase Monosialoganglioside sialyltransferase	
2.4.99.3	CMP-N-acetylneuraminate:glycano-1,3-(N-acetyl-α-D-galactosaminyl)-glycoprotein α-2,6-N-acetylneuraminyltransferase α-N-Acetylgalactosaminide α-2,6-sialyltransferase	
2.4.99.9	CMP-N-acetylneuraminate:lactosylceramide α-2,3-N-acetylneuraminyltransferase Lactosylceramide α-2,3-sialyltransferase	
2.4.99.11	CMP-N-acetylneuraminate:lactosylceramide α-2,6-N-acetylneuraminyltransferase Lactosylceramide α-2,6-N-sialyltransferase	

EC-No.	Synonym	Recommended Name	Vol., Page
2.4.99.10	CMP-N-acetylneuraminate:neolactotetraosylceramide α-2,3-sialyltransferase	Neolactotetraosylceramide α-2,3-sialyltransferase	
3.1.4.40	CMP-N-acetylneuraminate hydrolase	CMP-N-acylneuraminate phosphodiesterase	
1.14.13.45	CMP-N-acetylneuraminate monooxygenase	CMP-N-acetylneuraminate monooxygenase	
2.7.7.43	CMP-N-acetylneuraminate synthase	N-Acylneuraminate cytidylyltransferase	
2.7.7.43	CMP-N-acetylneuraminate synthetase	N-Acylneuraminate cytidylyltransferase	
2.4.99.1	CMP-N-acetylneuraminic acid-glycoprotein sialyltransferase	β-Galactoside α-2,6-sialyltransferase	
2.4.99.11	CMP-N-acetylneuraminic acid:lactosylceramide sialyltransferase	Lactosylceramide α-2,6-N-sialyltransferase	
1.14.13.45	CMP-N-acetylneuraminic acid hydroxylase	CMP-N-acetylneuraminate monooxygenase	
2.7.7.43	CMP-N-acetylneuraminic acid synthase	N-Acylneuraminate cytidylyltransferase	
3.1.4.40	CMP-N-acylneuraminate N-acylneuramidohydrolase	CMP-N-acylneuraminate phosphodiesterase	
3.1.4.40	CMP-N-acylneuraminate phosphodiesterase	CMP-N-acylneuraminate phosphodiesterase	
3.1.4.40	CMP-N-acylneuraminic acid hydrolase	CMP-N-acylneuraminate phosphodiesterase	
2.7.7.43	CMP-NANA synthetase	N-Acylneuraminate cytidylyltransferase	
1.14.13.45	CMP-Neu5Ac hydroxylase	CMP-N-acetylneuraminate monooxygenase	
1.14.99.18	CMP-Neu5Ac hydroxylase	N-Acetylneuraminate monooxygenase	
2.7.7.43	CMP-Neu5Ac synthetase	N-Acylneuraminate cytidylyltransferase	
2.4.99.2	CMP-NeuAc:GM1α2>-3-sialyltransferase	Monosialoganglioside sialyltransferase	

EC-No.	Synonym Recommended Name	Vol., Page
2.4.99.8	CMP-NeuAc:LM1(α2-8) sialyltranferase α-N-Acetylneuraminate α-2,8-sialyltransferase	
2.7.7.43	CMP-NeuAc synthetase N-Acylneuraminate cytidylyltransferase	
3.1.4.40	CMP-sialate hydrolase CMP-N-acylneuraminate phosphodiesterase	
2.7.7.43	CMP-sialate synthase N-Acylneuraminate cytidylyltransferase	
2.7.7.43	CMP-sialate synthetase N-Acylneuraminate cytidylyltransferase	
2.4.99.9	CMP-sialic acid:lactosylceramide-sialyltransferase Lactosylceramide α-2,3-sialyltransferase	
2.4.99.11	CMP-sialic acid:lactosylceramide sialyltransferase Lactosylceramide α-2,6-N-sialyltransferase	
3.1.4.40	CMP-sialic acid hydrolase CMP-N-acylneuraminate phosphodiesterase	
2.7.7.43	CMP-sialic synthetase N-Acylneuraminate cytidylyltransferase	
2.7.4.14	CMP kinase Cytidylate kinase	
2.7.7.43	CMP sialate pyrophosphorylase N-Acylneuraminate cytidylyltransferase	
2.7.7.43	CMPsialate pyrophosphorylase N-Acylneuraminate cytidylyltransferase	
2.7.7.43	CMPsialate synthase N-Acylneuraminate cytidylyltransferase	
3.1.3.5	cN-I 5'-nucleotidase	
4.2.1.65	β-CNAla hydrolase 3-Cyanoalanine hydratase	
4.2.1.65	β-CNA nitrilase 3-Cyanoalanine hydratase	
3.1.4.37	2':3'-CNMP-3'-ase 2',3'-Cyclic-nucleotide 3'-phosphodiesterase	
3.1.4.37	CNPase 2',3'-Cyclic-nucleotide 3'-phosphodiesterase	

EC-No.	Synonym Recommended Name	Vol., Page
1.2.99.2	Co-DG Carbon-monoxide dehydrogenase	
3.1.4.38	Co^{2+}-glycerophosphocholine cholinphosphodiesterase glycerophosphocholine cholinephosphodiesterase	
1.2.3.10	CO:MB oxidoreductase Carbon-monoxide oxidase	
1.6.4.10	CoA-disulfide reductase (NADH) CoA-disulfide reductase (NADH)	
1.6.4.6	CoA-glutathione reductase (NADPH) CoA-glutathione reductase (NADPH)	
1.2.1.3	CoA-independent aldehyde dehydrogenase Aldehyde dehydrogenase (NAD^+)	
2.3.1.147	CoA-independent transacylase Glycerophospholipid arachidonoyl-transferase (CoA-independent)	
2.7.8.7	CoA:apo-[acyl-carrier-protein] pantetheinephosphotransferase Holo-[acyl-carrier-protein] synthase	
3.4.21.22	coagulation factor IXa coagulation factor IXa	
3.4.21.21	coagulation factor VIIa coagulation factor VIIa	
3.4.21.6	coagulation factor Xa coagulation factor Xa	
3.4.21.6	coagulation factor Xa coagulation factor Xa	
3.4.21.27	coagulation factor XIa coagulation factor XIa	
3.4.21.38	coagulation factor XIIa coagulation factor XIIa	
2.5.1.17	Cob(I)alamin adenosyltransferase Cob(I)alamin adenosyltransferase	
1.6.99.9	Cob(II)alamin reductase Cob(II)alamin reductase	
2.1.1.133	Cobalt-precorrin-4 methyltransferase Precorrin-4 C_{11}-methyltransferase	
3.4.17.10	cobalt stimulated chromaffin granule carboxypeptidase carboxypeptidase H	

EC-No.	Synonym Recommended Name	Vol., Page
3.4.21.47	cobra venom factor-dependent C_3 convertase alternative-complement-pathway C3/C5 convertase	
3.1.1.1	Cocaine esterase Carboxylesterase	
3.4.24.30	Coccolysin Coccolysin	
3.4.21.4	cocoonase trypsin	
1.2.99.2	CO dehydrogenase Carbon-monoxide dehydrogenase	
1.2.2.4	CO dehydrogenase Carbon-monoxide oxygenase (cytochrome b-561)	
1.2.99.2	CO dehydrogenase/acetyl-CoA synthase Carbon-monoxide dehydrogenase	
1.2.99.2	CO dehydrogenase complex Carbon-monoxide dehydrogenase	
1.1.1.247	Codeine: NADP oxidoreductase Codeinone reductase (NADPH)	
1.1.1.247	Codeine:NADP oxidoreductase Codeinone reductase (NADPH)	
1.1.1.247	Codeine:NADP oxidoreductase Codeinone reductase (NADPH)	
1.1.1.247	Codeine oxidase Codeinone reductase (NADPH)	
1.1.1.247	Codeinone reductase Codeinone reductase (NADPH)	
1.1.1.247	Codeinone reductase (NADPH) Codeinone reductase (NADPH)	
1.2.99.2	CODH Carbon-monoxide dehydrogenase	
1.12.99.2	Coenzyme-M-7-mercaptoheptanoylthreonine-phosphate-heterodisulfide hydrogenase Coenzyme-M-7-mercaptoheptanoylthreonine-phosphate-heterodisulfide hydrogenase	
2.3.1.147	Coenzyme A-independent transacylase Glycerophospholipid arachidonoyl-transferase (CoA-independent)	

EC-No.	Synonym Recommended Name	Vol., Page
2.8.3.5	Coenzyme A-transferase, 3-oxoacid 3-Oxoacid CoA-transferase	
2.8.3.6	Coenzyme A-transferase, 3-oxoadipate 3-Oxoadipate CoA-transferase	
2.8.3.14	Coenzyme A-transferase, 5-hydroxyvalerate 5-Hydroxypentanoate CoA-transferase	
2.8.3.8	Coenzyme A-transferase, acetate Acetate CoA-transferase	
2.8.3.9	Coenzyme A-transferase, butyryl coenzyme A-acetoacetate Butyrate-acetoacetate CoA-transferase	
2.8.3.11	Coenzyme A-transferase, citramalate Citramalate CoA-transferase	
2.8.3.7	Coenzyme A-transferase, citramalate Succinate-citramalate CoA-transferase	
2.8.3.10	Coenzyme A-transferase, citrate Citrate-CoA transferase	
2.8.3.12	Coenzyme A-transferase, glutaconate Glutaconate CoA-transferase	
2.8.3.13	Coenzyme A-transferase, hydroxymethylglutarate Succinate-hydroxymethylglutarate CoA-transferase	
2.8.3.3	Coenzyme A-transferase, malonate Malonate CoA-transferase	
2.8.3.2	Coenzyme A-transferase, oxalate Oxalate CoA-transferase	
2.7.8.7	Coenzyme A:fatty acid synthetase apoenzyme 4'-phosphopantetheine transferase Holo-[acyl-carrier-protein] synthase	
1.8.4.3	Coenzyme A:oxidized-glutathione oxidoreductase Glutathione-CoA-glutathione transhydrogenase	
1.6.4.6	Coenzyme A disulfide-glutathione reductase CoA-glutathione reductase (NADPH)	
1.6.4.6	Coenzyme A glutathione disulfide reductase CoA-glutathione reductase (NADPH)	
2.8.3.1	Coenzyme A transferase, propionate Propionate CoA-transferase	

EC-No.	Synonym Recommended Name	Vol., Page
4.2.1.28	coenzyme B_{12}-dependent diol dehydrase propanediol dehydratase	
1.12.99.1	Coenzyme F420-dependent hydrogenase Coenzyme F420 hydrogenase	
1.5.99.11	coenzyme F420-dependent N5,N^{10}-methenyltetrahydromethanopterin reductase coenzyme F420-dependent N5,N^{10}-methenyltetrahydromethanopterin reductase	
1.12.99.1	Coenzyme F420 hydrogenase Coenzyme F420 hydrogenase	
1.10.2.2	Coenzyme Q-cytochrome c reductase Ubiquinol-cytochrome-c reductase	
1.10.2.2	Coenzyme QH_2−cytochrome c reductase Ubiquinol-cytochrome-c reductase	
1.6.5.3	Coenzyme Q reductase NADH dehydrogenase (ubiquinone)	
1.1.1.235	Coformycin:$NADP^+$ 8-oxidoreductase 8-Oxocoformycin reductase	
3.5.1.1	Colaspase Asparaginase	
3.4.21.62	Colistinase Subtilisin	
3.4.24.17	Collagen-activating protein Stromelysin 1	
3.4.21.32	collagenase brachyurin	
3.4.24.7	collagenase interstitial collagenase	
3.4.24.3	collagenase microbial collagenase	
3.4.21.32	collagenase A brachyurin	
3.4.24.7	collagenase A interstitial collagenase	
3.4.24.3	collagenase A microbial collagenase	

EC-No.	Synonym Recommended Name	Vol., Page
3.4.24.17	Collagenase activating protein Stromelysin 1	
3.4.24.3	collagenase I microbial collagenase	
3.4.24.24	Collagenase IV Gelatinase A	
3.4.24.35	Collagenase IV Gelatinase B	
3.4.21.32	collagenase MMP-1 brachyurin	
3.4.24.7	collagenase MMP-1 interstitial collagenase	
3.4.24.3	collagenase MMP-1 microbial collagenase	
3.4.24.24	Collagenase type IV Gelatinase A	
3.4.24.35	Collagenase type IV Gelatinase B	
2.4.1.50	Collagen galactosyltransferase Procollagen galactosyltransferase	
2.4.1.66	Collagen glucosyltransferase Procollagen glucosyltransferase	
2.4.1.50	Collagen hydroxylysyl galactosyltransferase Procollagen galactosyltransferase	
2.4.1.66	Collagen hydroxylysyl glucosyltransferase Procollagen glucosyltransferase	
1.14.11.4	Collagen lysine hydroxylase Procollagen-lysine 5-dioxygenase	
3.4.21.32	collagenolytic protease brachyurin	
3.4.21.32	collagen peptidase brachyurin	
3.4.24.7	collagen peptidase interstitial collagenase	
3.4.24.3	collagen peptidase microbial collagenase	

EC-No.	Synonym Recommended Name	Vol., Page
1.14.11.2	Collagen proline hydroxylase Procollagen-proline dioxygenase	
3.4.24.7	collagen protease interstitial collagenase	
3.4.24.3	collagen protease microbial collagenase	
1.1.3.26	Columbamine:oxygen oxidoreductase (cyclizing) Columbamine oxidase	
2.1.1.118	Columbamine O-methyltransferase Columbamine O-methyltransferase	
1.1.3.26	Columbamine oxidase Columbamine oxidase	
3.4.24.11	common acute lymphoblastic leukemia-associated antigens neprilysin	
3.4.24.11	common acute lymphoblastic leukemia antigen neprilysin	
3.4.21.42	complement C.hivin. 1s complement subcomponent C1sbar	
3.4.21.41	complement C1r, activated complement subcomponent C1r	
3.4.21.42	complement C1s, activated complement subcomponent C1sbar	
3.4.21.47	complement C 3(C 5) convertase (amplification) alternative-complement-pathway C3/C5convertase	
3.4.21.45	complement C3b/C4b inactivator complement factor I	
3.4.21.45	complement C3b inactivator complement factor I	
3.4.21.43	complement C3 convertase classical-complement-pathway C3/C_5 convertase	
3.4.21.45	complement C4bi complement factor I	
3.4.21.45	complement C4b inactivator complement factor I	
3.4.21.45	complement compoment C3b inactivator complement factor I	

EC-No.	Synonym Recommended Name	Vol., Page
3.4.21.47	complement component C3/C5 convertase (alternative) alternative-complement-pathway C3/C5 convertase	
3.4.21.44	complement component C5 convertase complement component C5 convertase	
3.4.21.46	complement factor D complement factor D	
3.4.21.45	complement factor I complement factor I	
3.4.21.41	complement subcomponent C1r complement subcomponent C1r	
3.4.21.42	complement subcomponent C1sbar complement subcomponent C1sbar	
1.6.5.3	Complex I (electron transport chain) NADH dehydrogenase (ubiquinone)	
1.6.5.3	Complex I (mitochondrial electron transport) NADH dehydrogenase (ubiquinone)	
1.6.5.3	Complex I (NADH:Q1 oxidoreductase) NADH dehydrogenase (ubiquinone)	
1.6.5.3	Complex I dehydrogenase NADH dehydrogenase (ubiquinone)	
1.3.5.1	Complex II Succinate dehydrogenase (ubiquinone)	
1.9.3.1	Complex IV (mitochondrial electron transport) Cytochrome-c oxidase	
3.6.1.3	Complex V (mitochondrial electron transport) Adenosinetriphosphatase	
3.5.1.66	Compound B hydrolase 2-(Hydroxymethyl)-3-(acetamidomethylene)succinate hydrolase	
2.1.1.6	S-COMT Catechol O-methyltransferase	
2.1.1.6	COMT I Catechol O-methyltransferase	
2.1.1.6	COMT II Catechol O-methyltransferase	
2.3.1.41	Condensing enzyme 3-Oxoacyl-[acyl-carrier-protein] synthase	

EC-No.	Synonym Recommended Name	Vol., Page
4.1.3.7	Condensing enzyme Citrate (si)-synthase	
3.4.21.45	conglutinogen-activating factor C complement factor I	
3.2.1.126	Coniferin β-D-glucosidase Coniferin β-glucosidase	
3.2.1.126	Coniferin β-glucosidase Coniferin β-glucosidase	
3.2.1.126	Coniferin-hydrolyzing β-glucosidase Coniferin β-glucosidase	
1.1.1.194	Coniferyl-alcohol:$NADP^+$ oxidoreductase Coniferyl-alcohol dehydrogenase	
1.1.1.194	Coniferyl-alcohol dehydrogenase Coniferyl-alcohol dehydrogenase	
2.4.1.111	Coniferyl-alcohol glucosyltransferase Coniferyl-alcohol glucosyltransferase	
1.2.1.68	coniferyl-aldehyde dehydrogenase coniferyl-aldehyde dehydrogenase	
1.2.1.68	coniferyl 2-hydroxy-1,4-benzoquinonealdehyde dehydrogenase coniferyl-aldehyde dehydrogenase	
1.1.1.90	Coniferyl alcohol dehydrogenase Aryl-alcohol dehydrogenase	
1.1.1.91	Coniferyl alcohol dehydrogenase Aryl-alcohol dehydrogenase ($NADP^+$)	
1.2.1.68	coniferyl aldehyde:NAD(P) oxidoreductase coniferyl-aldehyde dehydrogenase	
3.4.19.9	Conjugase γ-Glu-X carboxypeptidase	
3.4.21.46	convertase, C3 proactivator complement factor D	
3.4.21.47	convertase, complement C3(C5) (amplification) alternative-complement-pathway C3/C5 convertase	
1.15.1.1	Copper-zinc superoxide dismutase Superoxide dismutase	
1.3.3.3	Coproporphyrinogen:oxygen oxidoreductase (decarboxylating) Coproporphyrinogen oxidase	

EC-No.	Synonym Recommended Name	Vol., Page
1.3.3.3	Coproporphyrinogenase Coproporphyrinogen oxidase	
1.3.3.3	Coproporphyrinogen III oxidase Coproporphyrinogen oxidase	
1.3.3.3	Coproporphyrinogen oxidase Coproporphyrinogen oxidase	
1.10.2.2	$CoQH_2$–cytochrome c oxidoreductase Ubiquinol-cytochrome-c reductase	
2.4.1.102	Core 2 acetylglucosaminyltransferase β-1,3-Galactosyl-O-glycosyl-glycoprotein β-1,6-N-acetylglucosaminyltransferase	
2.4.1.147	Core 3β-GlcNAc-transferase Acetylgalactosaminyl-O-glycosyl-glycoprotein β-1,3-N-acetylglucosaminyltransferase	
2.4.1.148	Core 4 β6-GalNAc-transferase Acetylgalactosaminyl-O-glycosyl-glycoprotein β-1,6-N-acetylglucosaminyltransferase	
2.4.1.102	Core 6-β-GlcNAc-transferase A β-1,3-Galactosyl-O-glycosyl-glycoprotein β-1,6-N-acetylglucosaminyltransferase	
2.4.1.148	Core 6β-GalNAc-transferase B Acetylgalactosaminyl-O-glycosyl-glycoprotein β-1,6-N-acetylglucosaminyltransferase	
3.1.25.1	Correndonuclease II Deoxyribonuclease (pyrimidine dimer)	
2.3.1.27	Corticosteroid-21-O-acetyltransferase Cortisol O-acetyltransferase	
1.1.1.146	Corticosteroid 11-reductase 11β-Hydroxysteroid dehydrogenase	
5.3.1.21	Corticosteroid side-chain-isomerase Corticosteroid side-chain-isomerase	Vol. 1, Page 345
5.3.1.21	Corticosteroid side chain isomerase Corticosteroid side-chain-isomerase	Vol. 1, Page 345
1.14.15.5	Corticosterone,reduced-adrenal-ferredoxin:oxygen oxidoreductase (18-hydroxylating) Corticosterone 18-monooxygenase	
1.14.15.5	Corticosterone-18-hydroxylase Corticosterone 18-monooxygenase	

EC-No.	Synonym Recommended Name	Vol., Page
1.14.15.5	Corticosterone 18-hydroxylase Corticosterone 18-monooxygenase	
1.14.15.5	Corticosterone 18-monooxygenase Corticosterone 18-monooxygenase	
1.14.15.5	Corticosterone methyl oxidase Corticosterone 18-monooxygenase	
2.3.1.27	Cortisol O-acetyltransferase Cortisol O-acetyltransferase	
2.8.2.18	Cortisol sulfotransferase Cortisol sulfotransferase	
1.3.1.4	Cortisone α-reductase Cortisone α-reductase	
1.3.1.4	Cortisone α-reductase Cortisone α-reductase	
1.3.1.3	Cortisone β-reductase Cortisone β-reductase	
1.3.1.3	Cortisone β-reductase Cortisone β-reductase	
1.3.1.4	Cortisone Δ 4-5αreductase Cortisone α-reductase	
1.3.1.3	Cortisone 5β-reductase Cortisone β-reductase	
1.1.1.53	Cortisone reductase 3α(or 20β)-Hydroxysteroid dehydrogenase	
1.1.3.4	Corylophyline Glucose oxidase	
2.3.1.137	COT Carnitine O-octanoyltransferase	
6.2.1.12	4-Coumarate-CoA ligase 4-Coumarate-CoA ligase	Vol. 2, Page 256
1.14.17.2	4-coumarate 3-monooxygenase 4-coumarate 3-monooxygenase	
6.2.1.12	4-Coumarate:CoA ligase (AMP-forming) 4-Coumarate-CoA ligase	Vol. 2, Page 256
6.2.1.12	4-Coumarate:coenzyme A ligase 4-Coumarate-CoA ligase	Vol. 2, Page 256

EC-No.	Synonym Recommended Name	Vol., Page
2.4.1.114	2-Coumarate O-β-glucosyltransferase 2-Coumarate O-β-glucosyltransferase	
1.3.1.11	Coumarate reductase Coumarate reductase	
1.14.13.36	5-O-(4-Coumaroyl)-D-quinate/shikimate 3'-hydroxylase 5-O-(4-Coumaroyl)-D-quinate 3'-monooxygenase	
1.14.13.36	5-O-(4-Coumaroyl)-D-quinate 3'-monooxygenase 5-O-(4-Coumaroyl)-D-quinate 3'-monooxygenase	
2.3.1.64	*p*-Coumaroyl-CoA-agmatine N-*p*-coumaroyltransferase Agmatine N^4-coumaroyltransferase	
2.3.1.140	4-Coumaroyl-CoA:4-hydroxyphenyllactic acid 4-coumaroyl transferase Rosmarinate synthase	
2.3.1.64	4-Coumaroyl-CoA:agmatine N^4-coumaroyltransferase Agmatine N^4-coumaroyltransferase	
2.3.1.133	4-Coumaroyl-CoA:shikimate O-(hydroxycinnamoyl)transferase Shikimate O-hydroxycinnamoyltransferase	
6.2.1.12	*p*-Coumaroyl CoA ligase 4-Coumarate-CoA ligase	Vol. 2, Page 256
2.3.1.64	Coumaroyltransferase, agmatine Agmatine N^4-coumaroyltransferase	
6.2.1.12	*p*-Coumaryl-CoA ligase 4-Coumarate-CoA ligase	Vol. 2, Page 256
6.2.1.12	4-Coumaryl-CoA synthetase 4-Coumarate-CoA ligase	Vol. 2, Page 256
6.2.1.12	*p*-Coumaryl-CoA synthetase 4-Coumarate-CoA ligase	Vol. 2, Page 256
6.2.1.12	*p*-Coumaryl coenzyme A synthetase 4-Coumarate-CoA ligase	Vol. 2, Page 256
6.3.2.5	Coupling enzyme Phosphopantothenoate-cysteine ligase	Vol. 2, Page 431
3.6.1.34	Coupling factors (F_0, F_1 and CF_1) H^+-transporting ATP synthase	
3.4.22.28	Coxsackievirus 3C proteinase Picornain 3C	
3.4.17.15	CPA2 Carboxypeptidase A2	

EC-No.	Synonym Recommended Name	Vol., Page
3.4.17.3	CPase N lysine carboxypeptidase	
3.4.17.2	CPB carboxypeptidase B	
3.4.17.22	CPD Metallocarboxypeptidase D	
3.4.16.6	CPDW-II Carboxypeptidase D	
2.7.8.23	CPEP mutase Carboxyvinyl-carboxyphosphonate phosphorylmutase	
2.7.8.23	CPEP phosphonomutase Carboxyvinyl-carboxyphosphonate phosphorylmutase	
5.4.2.9	CPEP phosphonomutase Phosphoenolpyruvate mutase	Vol. 1, Page 546
5.2.1.8	CPH Peptidylprolyl isomerase	Vol. 1, Page 218
3.4.17.12	CPM Carboxypeptidase M	
3.4.17.3	CPN lysine carboxypeptidase	
3.4.21.98	Cpro-2 hepacivirin	
3.4.21.98	Cpro-2 proteinase hepacivirin	
3.4.22.28	3C Protease Picornain 3C	
3.4.22.28	3C Proteinase Picornain 3C	
6.3.5.5	CPS Carbamoyl-phosphate synthase (glutamine-hydrolysing)	Vol. 2, Page 689
6.3.5.5	CPSase Carbamoyl-phosphate synthase (glutamine-hydrolysing)	Vol. 2, Page 689
6.3.5.5	CPSase-A Carbamoyl-phosphate synthase (glutamine-hydrolysing)	Vol. 2, Page 689
6.3.5.5	CPSase-P Carbamoyl-phosphate synthase (glutamine-hydrolysing)	Vol. 2, Page 689

EC-No.	Synonym Recommended Name	Vol., Page
6.3.4.16	CPS I Carbamoyl-phosphate synthase (ammonia)	Vol. 2, Page 641
3.4.17.18	CPT Carboxypeptidase T	
2.3.1.21	CPT Carnitine O-palmitoyltransferase	
2.7.8.2	CPT Diacylglycerol cholinephosphotransferase	
2.3.1.21	CPT-A Carnitine O-palmitoyltransferase	
2.3.1.21	CPT-B Carnitine O-palmitoyltransferase	
2.3.1.21	CPT I Carnitine O-palmitoyltransferase	
2.3.1.21	CPTi Carnitine O-palmitoyltransferase	
2.3.1.21	CPTo Carnitine O-palmitoyltransferase	
3.4.17.20	CPU Carboxypeptidase U	
3.4.21.32	crab protease I brachyurin	
3.4.21.32	crab protease II brachyurin	
3.5.1.1	Crasnitin Asparaginase	
3.4.99.6	crayfish low-molecular-weight proteinase crayfish low-molecular-weight proteinase	
3.4.24.21	Crayfish small-molecule proteinase Astacin	
3.5.3.3	Creatinase Creatinase	
3.5.3.3	Creatine amidinohydrolase Creatinase	
2.7.3.2	Creatine kinase Creatine kinase	

EC-No.	Synonym Recommended Name	Vol., Page
3.4.17.3	creatine kinase conversion factor lysine carboxypeptidase	
3.9.1.1	Creatine phosphatase Phosphoamidase	
2.7.3.2	Creatine phosphokinase Creatine kinase	
2.7.3.2	Creatine phosphotransferase Creatine kinase	
3.5.2.10	Creatininase Creatininase	
3.5.2.10	Creatinine amidohydrolase Creatininase	
3.5.4.21	Creatinine deiminase Creatinine deiminase	
3.5.4.21	Creatinine desiminase Creatinine deiminase	
3.5.2.10	Creatinine hydrolase Creatininase	
3.5.4.21	Creatinine hydrolase Creatinine deiminase	
3.5.4.21	Creatinine iminohydrolase Creatinine deiminase	
3.4.17.3	creatinine kinase convertase lysine carboxypeptidase	
1.14.99.33	crepenynate synthase Δ^{12}-fatty acid dehydrogenase	
1.17.99.1	*p*-Cresol-(acceptor) oxidoreductase (hydroxylating) 4-Cresol dehydrogenase (hydroxylating)	
1.17.99.1	4-Cresol:(acceptor) oxidoreductase (methyl-hydroxylating) 4-Cresol dehydrogenase (hydroxylating)	
1.14.18.1	Cresolase Monophenol monooxygenase	
1.17.99.1	4-Cresol dehydrogenase (hydroxylating) 4-Cresol dehydrogenase (hydroxylating)	
1.17.99.1	*p*-Cresol methylhydroxylase 4-Cresol dehydrogenase (hydroxylating)	

EC-No.	Synonym Recommended Name	Vol., Page
2.7.7.6	C RNA formation factors DNA-directed RNA polymerase	
3.1.22.4	Crossover junction endodeoxyribonuclease Crossover junction endodeoxyribonuclease	
3.4.21.74	Crotalase Venombin A	
3.4.24.46	Crotalus adamanteus metalloendopeptidase Adamalysin	
3.4.21.30	Crotalus adamanteus serine proteinase Crotalus adamanteus serine proteinase	
3.4.24.46	Crotalus adamanteus venom proteinase II Adamalysin	
3.4.24.1	crotalus atrox α-proteinase atrolysin A	
3.4.24.1	crotalus atrox.α.-proteinase atrolysin A	
3.4.24.41	Crotalus atrox metalloendopeptidase b Atrolysin B	
3.4.24.42	Crotalus atrox metalloendopeptidase c Atrolysin C	
3.4.24.44	Crotalus atrox metalloendopeptidase e Atrolysin E	
3.4.24.45	Crotalus atrox metalloendopeptidase f Atrolysin F	
3.4.24.1	crotalus atrox metalloproteinase atrolysin A	
3.4.24.1	crotalus atrox proteinase atrolysin A	
3.4.24.47	Crotalus horridus metalloendopeptidase Horrilysin	
3.4.24.48	Crotalus ruber metalloendopeptidase II Ruberlysin	
4.2.1.17	crotonase enoyl-CoA hydratase	
4.2.1.58	Crotonoyl-[acyl-carrier-protein] hydratase Crotonoyl-[acyl-carrier-protein] hydratase	

EC-No.	Synonym Recommended Name	Vol., Page
1.3.1.8	Crotonyl-CoA reductase Acyl-CoA dehydrogenase ($NADP^+$)	
4.2.1.58	Crotonyl acyl carrier protein hydratase Crotonoyl-[acyl-carrier-protein] hydratase	
1.3.1.8	Crotonyl coenzyme A reductase Acyl-CoA dehydrogenase ($NADP^+$)	
4.2.1.17	crotonyl hydrase enoyl-CoA hydratase	
4.3.2.1	δ^2-crystallin argininosuccinate lyase	
4.2.99.9	CS O-succinylhomoserine (thiol)-lyase	
4.1.1.29	CSAD Sulfinoalanine decarboxylase	
4.1.1.29	CSAD/CAD Sulfinoalanine decarboxylase	
4.1.1.29	CSADCase Sulfinoalanine decarboxylase	
5.3.1.21	CSC isomerase Corticosteroid side-chain-isomerase	Vol. 1, Page 345
4.1.1.29	CSD Sulfinoalanine decarboxylase	
3.1.1.53	CSE sialate O-acetylesterase	
2.7.1.141	CTD kinase [RNA-polymerase]-subunit kinase	
4.4.1.1	γ-CTL cystathionine γ-lyase	
3.4.21.79	CTLA1 Granzyme B	
3.4.21.78	CTLA3 Granzyme A	
2.7.7.25	CTP(ATP):tRNA nucleotidyltransferase tRNA adenylyltransferase	
2.7.7.21	CTP(ATP):tRNA nucleotidyltransferase tRNA cytidylyltransferase	

EC-No.	Synonym Recommended Name	Vol., Page
2.7.7.41	CTP-diacylglycerol synthetase Phosphatidate cytidylyltransferase	
2.7.7.15	CTP-phosphocholine cytidylyltransferase Choline-phosphate cytidylyltransferase	
2.7.7.14	CTP-phosphoethanolamine cytidylyltransferase Ethanolamine-phosphate cytidylyltransferase	
2.7.7.41	CTP:1,2-diacylglycerophosphate-cytidyl transferase Phosphatidate cytidylyltransferase	
2.7.7.38	CTP:3-deoxy-D-manno-octulosonate cytidylyltransferase 3-Deoxy-manno-octulosonate cytidylyltransferase	
2.7.7.15	CTP:choline-phosphate cytidylyltransferase Choline-phosphate cytidylyltransferase	
2.7.7.15	CTP:cholinephosphate cytidylyltransferase Choline-phosphate cytidylyltransferase	
2.7.7.38	CTP:CMP-3-deoxy-D-manno-octulosonate cytidylyltransferase 3-Deoxy-manno-octulosonate cytidylyltransferase	
2.7.4.14	CTP:CMP phosphotransferase Cytidylate kinase	
2.7.7.33	CTP:D-glucose-1-phosphate cytidylyltransferase Glucose-1-phosphate cytidylyltransferase	
2.7.7.40	CTP:D-ribitol-5-phosphate cytidylyltransferase D-Ribitol-5-phosphate cytidylyltransferase	
2.7.1.108	CTP:dolichol O-phosphotransferase Dolichol kinase	
2.7.7.14	CTP:ethanolamine-phosphate cytidylyltransferase Ethanolamine-phosphate cytidylyltransferase	
2.7.7.39	CTP:glycerol 3-phosphate cytidylyltransferase Glycerol-3-phosphate cytidylyltransferase	
2.7.7.43	CTP:N-acylneuraminate cytidylyltransferase N-Acylneuraminate cytidylyltransferase	
2.7.7.57	CTP:N-methylethanolamine-phosphate cytidylyltransferase N-Methylphosphoethanolamine cytidylyltransferase	
2.7.7.57	CTP:P-MEA cytidylyltransferase N-Methylphosphoethanolamine cytidylyltransferase	
2.7.7.41	CTP:phosphatidate cytidylyltransferase Phosphatidate cytidylyltransferase	

EC-No.	Synonym Recommended Name	Vol., Page
2.7.7.15	CTP:phosphorylcholine cytidylyltransferase Choline-phosphate cytidylyltransferase	
2.7.7.39	CTP:sn-glycerol-3-phosphate cytidylyltransferase Glycerol-3-phosphate cytidylyltransferase	
2.7.7.25	CTP:tRNA cytidylyltransferase tRNA adenylyltransferase	
2.7.7.21	CTP:tRNA cytidylyltransferase tRNA cytidylyltransferase	
4.6.1.6	CTP pyrophosphate-lyase (cyclizing) Cytidylate cyclase	
6.3.4.2	CTP synthase CTP synthase	Vol. 2, Page 559
1.15.1.1	Cu,Zn-SOD Superoxide dismutase	
1.4.3.6	Cu-Amine oxidase Amine oxidase (copper-containing)	
1.15.1.1	Cu-Zn superoxide dismutase Superoxide dismutase	
3.4.21.25	cucumisin cucumisin	
1.3.1.5	Cucurbitacin.δ.23 reductase Cucurbitacin Δ^{23}-reductase	
1.3.1.5	Cucurbitacin Δ^{23}-reductase Cucurbitacin Δ^{23}-reductase	
1.3.1.5	Cucurbitacin B Δ^{23}-reductase Cucurbitacin Δ^{23}-reductase	
1.15.1.1	Cuprein Superoxide dismutase	
3.1.1.74	cutinase cutinase	
3.1.1.74	cutin hydrolase cutinase	
3.4.21.47	(CVF)-dependent glycine-rich-β-glucoprotein alternative-complement-pathway C3/C5 convertase	
3.4.21.47	CVF,Bb alternative-complement-pathway C3/C5 convertase	

EC-No.	Synonym Recommended Name	Vol., Page
4.2.1.69	cyanamide-hydratase cyanamide hydratase	
4.2.1.69	cyanamide hydratase cyanamide hydratase	
4.3.99.1	Cyanase Cyanate lyase	
4.3.99.1	Cyanate aminohydrolase Cyanate lyase	
4.3.99.1	Cyanate C-N-lyase Cyanate lyase	
4.3.99.1	Cyanate hydrolase Cyanate lyase	
3.5.5.3	cyanate hydrolase cyanate hydrolase	
4.3.99.1	Cyanate lyase Cyanate lyase	
4.2.1.66	cyanide hydratase cyanide hydratase	
2.4.1.116	Cyanidin-3-rhamnosylglucoside 5-O-glucosyltransferase Cyanidin-3-rhamnosylglucoside 5-O-glucosyltransferase	
3.5.5.4	3-Cyano-L-alanine aminohydrolase Cyanoalanine nitrilase	
4.4.1.9	β-cyano-L-alanine synthase L-3-cyanoalanine synthase	
4.2.1.65	3-Cyanoalanine hydratase 3-Cyanoalanine hydratase	
4.2.1.65	β-Cyanoalanine hydratase 3-Cyanoalanine hydratase	
4.2.1.65	β-Cyanoalanine hydrolase 3-Cyanoalanine hydratase	
3.5.5.4	Cyanoalanine nitrilase Cyanoalanine nitrilase	
4.4.1.9	L-3-cyanoalanine synthase L-3-cyanoalanine synthase	
4.4.1.9	β-cyanoalanine synthase L-3-cyanoalanine synthase	

EC-No.	Synonym Recommended Name	Vol., Page
1.6.99.12	Cyanocobalamin reductase (NADPH, cyanide-eliminating) Cyanocobalamin reductase (NADPH, cyanide-eliminating)	
2.4.1.85	Cyanohydrin β-glucosyltransferase Cyanohydrin β-glucosyltransferase	
2.4.1.178	Cyanohydrin glucosyltransferase Hydroxymandelonitrile glucosyltransferase	
4.2.1.84	3-Cyanopyridine hydratase nitrile hydratase	
2.4.1.171	Cycasin synthase Methyl-ONN-azoxymethanol β-D-glucosyltransferase	
3.10.1.2	Cyclamate sulfamatase Cyclamate sulfohydrolase	
3.10.1.2	Cyclamate sulfamidase Cyclamate sulfohydrolase	
3.10.1.2	Cyclamate sulfohydrolase Cyclamate sulfohydrolase	
4.6.1.8	Cyclase, (-)-endo-fenchol (-)-endo-Fenchol synthase	
4.6.1.11	Cyclase, δ-cadinene (+)-δ-Cadinene synthase	
5.4.99.8	Cyclase, 2,3-oxidosqualene-cycloartenol Cycloartenol synthase	Vol. 1, Page 631
5.4.99.7	Cyclase, 2,3-oxidosqualene-lanosterol Lanosterol synthase	Vol. 1, Page 624
4.6.1.1	cyclase, adenylate adenylate cyclase	
5.3.99.6	Cyclase, allene oxide Allene-oxide cyclase	Vol. 1, Page 483
5.5.1.5	Cyclase, carboxy-cis,cis-muconate- Carboxy-cis,cis-muconate cyclase	Vol. 1, Page 686
4.6.1.6	Cyclase, cytidylate Cytidylate cyclase	
4.2.1.48	cyclase, D-glutamate D-glutamate cyclase	
2.5.1.40	Cyclase, farnesyl pyrophosphate Aristolochene synthase	

EC-No.	Synonym Recommended Name	Vol., Page
4.3.1.12	cyclase, ornithine (deaminating) ornithine cyclodeaminase	
4.3.2.4	Cyclase, purine imidazole ring Purine imidazole-ring cyclase	
4.6.1.9	Cyclase, sabinene hydrate Sabinene-hydrate synthase	
3.1.4.16	Cyclic-2',3'-nucleotide 2'-phosphohydrolase 2',3'-Cyclic-nucleotide 2'-phosphodiesterase	
3.1.4.37	Cyclic-CMP phosphodiesterase 2',3'-Cyclic-nucleotide 3'-phosphodiesterase	
4.6.1.6	3',5'-Cyclic-CMP synthase Cytidylate cyclase	
3.1.4.35	3',5'-Cyclic-GMP 5'-nucleotidohydrolase 3',5'-Cyclic-GMP phosphodiesterase	
3.1.4.35	3',5'-Cyclic-GMP phosphodiesterase 3',5'-Cyclic-GMP phosphodiesterase	
3.1.4.36	1,2-Cyclic-inositol-phosphate phosphodiesterase 1,2-Cyclic-inositol-phosphate phosphodiesterase	
3.1.4.16	2',3'-Cyclic-nucleotide 2'-phosphodiesterase 2',3'-Cyclic-nucleotide 2'-phosphodiesterase	
3.1.4.37	2',3'-Cyclic-nucleotide 3'-phosphodiesterase 2',3'-Cyclic-nucleotide 3'-phosphodiesterase	
3.1.4.17	3',5'-Cyclic-nucleotide 5'-nucleotidohydrolase 3',5'-Cyclic-nucleotide phosphodiesterase	
3.1.4.17	3',5'-Cyclic-nucleotide phosphodiesterase 3',5'-Cyclic-nucleotide phosphodiesterase	
3.1.4.37	Cyclic 2',3'-nucleotide 3'-phosphodiesterase 2',3'-Cyclic-nucleotide 3'-phosphodiesterase	
3.1.4.16	Cyclic 2',3'-nucleotide phosphodiesterase 2',3'-Cyclic-nucleotide 2'-phosphodiesterase	
3.1.4.37	Cyclic 2',3'-nucleotide phosphodiesterase 2',3'-Cyclic-nucleotide 3'-phosphodiesterase	
3.1.4.35	Cyclic 3',5'-GMP phosphodiesterase 3',5'-Cyclic-GMP phosphodiesterase	
3.1.4.17	Cyclic 3',5'-mononucleotide phosphodiesterase 3',5'-Cyclic-nucleotide phosphodiesterase	

EC-No.	Synonym Recommended Name	Vol., Page
3.1.4.17	Cyclic 3',5'-nucleotide phosphodiesterase 3',5'-Cyclic-nucleotide phosphodiesterase	
3.1.4.17	Cyclic 3',5'-phosphodiesterase 3',5'-Cyclic-nucleotide phosphodiesterase	
3.1.4.17	Cyclic 3',5-nucleotide monophosphate phosphodiesterase 3',5'-Cyclic-nucleotide phosphodiesterase	
2.7.1.37	Cyclic AMP-dependent protein kinase Protein kinase	
2.7.1.37	Cyclic AMP-dependent protein kinase A Protein kinase	
3.1.4.16	2',3'-Cyclic AMP 2'-phosphohydrolase 2',3'-Cyclic-nucleotide 2'-phosphodiesterase	
3.1.4.16	2',3'-Cyclic AMP phosphodiesterase 2',3'-Cyclic-nucleotide 2'-phosphodiesterase	
3.1.4.37	2',3'-Cyclic AMP phosphodiesterase 2',3'-Cyclic-nucleotide 3'-phosphodiesterase	
4.6.1.1	3',5'-cyclic AMP synthetase adenylate cyclase	
3.1.4.35	Cyclic GMP phosphodiesterase 3',5'-Cyclic-GMP phosphodiesterase	
3.1.4.35	Cyclic guanosine 3',5'-monophosphate phosphodiesterase 3',5'-Cyclic-GMP phosphodiesterase	
3.1.4.35	Cyclic guanosine 3',5'-phosphate phosphodiesterase 3',5'-Cyclic-GMP phosphodiesterase	
2.7.1.37	Cyclic monophosphate-dependent protein kinase Protein kinase	
3.1.4.16	2',3'-Cyclic nucleoside monophosphate phosphodiesterase 2',3'-Cyclic-nucleotide 2'-phosphodiesterase	
3.1.4.37	2',3'-Cyclic nucleoside monophosphate phosphodiesterase 2',3'-Cyclic-nucleotide 3'-phosphodiesterase	
3.1.4.17	3', 5'-Cyclic nucleoside monophosphate phosphodiesterase 3',5'-Cyclic-nucleotide phosphodiesterase	
3.1.4.16	2',3'-Cyclic nucleotidase 2',3'-Cyclic-nucleotide 2'-phosphodiesterase	
2.7.1.37	Cyclic nucleotide-dependent protein kinase Protein kinase	

EC-No.	Synonym Recommended Name	Vol., Page
3.1.4.16	2',3'-Cyclic nucleotide 2'-phosphodiesterase 2',3'-Cyclic-nucleotide 2'-phosphodiesterase	
3.1.4.37	2':3'-Cyclic nucleotide 3'-phosphodiesterase 2',3'-Cyclic-nucleotide 3'-phosphodiesterase	
3.1.4.37	2',3'-Cyclic nucleotide 3'-phosphohydrolase 2',3'-Cyclic-nucleotide 3'-phosphodiesterase	
3.1.4.17	3':5'-Cyclic nucleotide 5'-nucleotidohydrolase 3',5'-Cyclic-nucleotide phosphodiesterase	
3.1.4.17	Cyclic nucleotide phosphodiesterase 3',5'-Cyclic-nucleotide phosphodiesterase	
3.1.4.16	2':3'-Cyclic nucleotide phosphodiesterase:3'-nucleotidase 2',3'-Cyclic-nucleotide 2'-phosphodiesterase	
3.1.4.16	2',3'-Cyclic nucleotide phosphohydrolase 2',3'-Cyclic-nucleotide 2'-phosphodiesterase	
3.1.4.37	2',3'-Cyclic nucleotide phosphohydrolase 2',3'-Cyclic-nucleotide 3'-phosphodiesterase	
3.1.4.16	2',3'-Cyclic phosphodiesterase 2',3'-Cyclic-nucleotide 2'-phosphodiesterase	
3.1.4.16	2':3'-Cyclic phosphodiesterase 2',3'-Cyclic-nucleotide 2'-phosphodiesterase	
3.1.4.16	Cyclic Phosphodiesterase:3'-nucleotidase 2',3'-Cyclic-nucleotide 2'-phosphodiesterase	
2.7.1.37	Cyclin-dependent kinase Protein kinase	
3.5.2.13	Cyclo(Gly-Gly) hydrolase 2,5-Dioxopiperazine hydrolase	
5.4.99.8	Cycloartenol synthase Cycloartenol synthase	Vol. 1, Page 631
3.2.1.54	Cyclodextrinase Cyclomaltodextrinase	
2.4.1.19	Cyclodextrin glucanotransferase Cyclomaltodextrin glucanotransferase	
2.4.1.19	α-Cyclodextrin glucanotransferase Cyclomaltodextrin glucanotransferase	
2.4.1.19	β-Cyclodextrin glucanotransferase Cyclomaltodextrin glucanotransferase	

EC-No.	Synonym Recommended Name	Vol., Page
2.4.1.19	Cyclodextrin glycosyltransferase Cyclomaltodextrin glucanotransferase	
2.4.1.19	α-Cyclodextrin glycosyltransferase Cyclomaltodextrin glucanotransferase	
2.4.1.19	β-Cyclodextrin glycosyltransferase Cyclomaltodextrin glucanotransferase	
2.4.1.19	γ-Cyclodextrin glycosyltransferase Cyclomaltodextrin glucanotransferase	
5.5.1.9	Cycloeucalenol-obtusifoliol isomerase Cycloeucalenol cycloisomerase	Vol. 1, Page 710
5.5.1.9	Cycloeucalenol-obtusifoliol isomerase Cycloeucalenol cycloisomerase	Vol. 1, Page 710
5.5.1.9	Cycloeucalenol cycloisomerase Cycloeucalenol cycloisomerase	Vol. 1, Page 710
5.5.1.9	Cycloeucalenol isomerase Cycloeucalenol cycloisomerase	Vol. 1, Page 710
5.5.1.9	Cycloeucalenol lyase (cyclopropane-decyclizing) Cycloeucalenol cycloisomerase	Vol. 1, Page 710
3.2.1.54	Cycloheptaglucanase Cyclomaltodextrinase	
3.2.1.12	cycloheptaglucanase cycloheptaglucanase	
1.3.99.15	Cyclohexa-1,5-diene-1-carbonyl-CoA:acceptor oxidoreductase (aromatizing, ATP-forming) Benzoyl-CoA reductase	
4.2.1.100	cyclohexa-1,5-diene-1-carbonyl-CoA hydratase cyclohexa-1,5-dienecarbonyl-CoA hydratase	
4.2.1.100	cyclohexa-1,5-dienecarbonyl-CoA hydratase cyclohexa-1,5-dienecarbonyl-CoA hydratase	
4.2.1.100	cyclohexa-1,5-dienecarbonyl-CoA hydro-lyase cyclohexa-1,5-dienecarbonyl-CoA hydratase	
1.3.1.25	3,5-Cyclohexadiene-1,2-diol-1-carboxylate dehydrogenase 1,6-Dihydroxycyclohexa-2,4-diene-1-carboxylate dehydrogenase	
1.3.1.25	3,5-Cyclohexadiene-1,2-diol-1-carboxylic acid dehydrogenase 1,6-Dihydroxycyclohexa-2,4-diene-1-carboxylate dehydrogenase	

EC-No.	Synonym Recommended Name	Vol., Page
4.2.1.91	Cyclohexadienyl dehydratase Carboxycyclohexadienyl dehydratase	
3.2.1.54	Cyclohexaglucanase Cyclomaltodextrinase	
3.2.1.13	cyclohexaglucanase cyclohexaglucanase	
1.1.1.174	Cyclohexane-1,2-diol dehydrogenase Cyclohexane-1,2-diol dehydrogenase	
3.7.1.10	Cyclohexane-1,3-dione acylhydrolase (decyclizing) Cyclohexane-1,3-dione hydrolase	
3.7.1.10	Cyclohexane-1,3-dione hydrolase Cyclohexane-1,3-dione hydrolase	
1.1.1.245	Cyclohexanol:NAD$^+$ oxidoreductase Cyclohexanol dehydrogenase	
1.1.1.245	Cyclohexanol dehydrogenase Cyclohexanol dehydrogenase	
1.14.13.22	Cyclohexanone,NADPH:oxygen oxidoreductase (6-hydroxylating, 1,2-lactonizing) Cyclohexanone monooxygenase	
1.3.99.14	Cyclohexanone:(acceptor) 2-oxidoreductase Cyclohexanone dehydrogenase	
1.3.99.14	Cyclohexanone dehydrogenase Cyclohexanone dehydrogenase	
1.14.13.22	Cyclohexanone monooxygenase Cyclohexanone monooxygenase	
4.2.1.51	cyclohexydienyl dehydratase prephenate dehydratase	
1.4.3.12	Cyclohexylamine:oxygen oxidoreductase (deaminating) Cyclohexylamine oxidase	
1.4.3.12	Cyclohexylamine oxidase Cyclohexylamine oxidase	
3.10.1.2	Cyclohexylsulfamate sulfohydrolase Cyclamate sulfohydrolase	
3.5.4.9	Cyclohydrolase Methenyltetrahydrofolate cyclohydrolase	

EC-No.	Synonym Recommended Name	Vol., Page
3.2.1.54	Cyclomaltodextrinase Cyclomaltodextrinase	
3.2.1.54	Cyclomaltodextrin dextrin-hydrolase (decyclizing) Cyclomaltodextrinase	
2.4.1.19	Cyclomaltodextrin glucanotransferase Cyclomaltodextrin glucanotransferase	
2.4.1.19	Cyclomaltodextrin glucotransferase Cyclomaltodextrin glucanotransferase	
2.4.1.19	Cyclomaltodextrin glycosyltransferase Cyclomaltodextrin glucanotransferase	
3.1.4.17	3',5'-Cyclonucleotide phosphodiesterase 3',5'-Cyclic-nucleotide phosphodiesterase	
1.1.1.163	Cyclopentanol:NAD^+ oxidoreductase Cyclopentanol dehydrogenase	
1.1.1.163	Cyclopentanol dehydrogenase Cyclopentanol dehydrogenase	
1.14.13.16	Cyclopentanone,NADPH:oxygen oxidoreductase (5-hydroxylating,lactonizing) Cyclopentanone monooxygenase	
1.14.13.16	Cyclopentanone monooxygenase Cyclopentanone monooxygenase	
1.14.13.16	Cyclopentanone oxygenase Cyclopentanone monooxygenase	
5.2.1.8	Cyclophilin Peptidylprolyl isomerase	Vol. 1, Page 218
5.2.1.8	Cyclophilin 18 Peptidylprolyl isomerase	Vol. 1, Page 218
5.2.1.8	Cyclophilin A Peptidylprolyl isomerase	Vol. 1, Page 218
5.2.1.8	Cyclophilin B Peptidylprolyl isomerase	Vol. 1, Page 218
5.2.1.8	Cyclophilin C Peptidylprolyl isomerase	Vol. 1, Page 218
1.3.99.9	β-Cyclopiazinate:(acceptor) oxidoreductase (cyclizing) β-Cyclopiazonate dehydrogenase	

EC-No.	Synonym Recommended Name	Vol., Page
1.3.99.9	β-Cyclopiazonate dehydrogenase β-Cyclopiazonate dehydrogenase	
1.3.99.9	β-Cyclopiazonate oxidocyclase β-Cyclopiazonate dehydrogenase	
1.3.99.9	β-Cyclopiazonic oxidocyclase β-Cyclopiazonate dehydrogenase	
2.1.1.79	Cyclopropane-fatty-acyl-phospholipid synthase Cyclopropane-fatty-acyl-phospholipid synthase	
2.1.1.79	Cyclopropane fatty acid synthase Cyclopropane-fatty-acyl-phospholipid synthase	
2.1.1.79	Cyclopropane fatty acid synthetase Cyclopropane-fatty-acyl-phospholipid synthase	
2.1.1.79	Cyclopropane synthase Cyclopropane-fatty-acyl-phospholipid synthase	
2.3.2.4	Cyclotransferase, γ-glutamyl γ-Glutamylcyclotransferase	
2.5.1.4	Cyclotransferase, adenosylmethionine Adenosylmethionine cyclotransferase	
2.3.2.5	Cyclotransferase, glutaminyl-transfer ribonucleate Glutaminyl-peptide cyclotransferase	
3.4.23.40	Cynarase Phytepsin	
5.2.1.8	Cyp3 PPIase Peptidylprolyl isomerase	Vol. 1, Page 218
1.1.3.34	CYP80 Berbamunine synthase	
5.2.1.8	CyPA Peptidylprolyl isomerase	Vol. 1, Page 218
5.2.1.8	CyPB Peptidylprolyl isomerase	Vol. 1, Page 218
1.13.12.6	Cypridina-luciferin 2-monooxygenase Cypridina-luciferin 2-monooxygenase	
1.13.12.6	Cypridina-luciferin:oxygen 2-oxidoreductase (decarboxylating) Cypridina-luciferin 2-monooxygenase	
1.13.12.6	Cypridina-type luciferase Cypridina-luciferin 2-monooxygenase	

EC-No.	Synonym Recommended Name	Vol., Page
1.13.12.6	Cypridina luciferase Cypridina-luciferin 2-monooxygenase	
1.13.12.6	Cypridina luciferin 2-monooxygenase Cypridina-luciferin 2-monooxygenase	
3.4.21.26	cyproase I prolyl oligopeptidase	
3.4.13.6	Cys-Gly dipeptidase Cys-Gly dipeptidase	
4.4.1.1	cystalysin cystathionine γ-lyase	
4.4.1.8	β-cystathionase cystathionine β-lyase	
4.4.1.8	β-cystathionase cystathionine β-lyase	
4.4.1.1	cystathionase cystathionine γ-lyase	
4.4.1.1	γ-cystathionase cystathionine γ-lyase	
4.4.1.1	cystathioninase cystathionine γ-lyase	
4.2.99.9	cystathionine-γ-synthase O-succinylhomoserine (thiol)-lyase	
4.4.1.8	cystathionine β-lyase cystathionine β-lyase	
4.4.1.1	cystathionine γ-lyase cystathionine γ-lyase	
4.4.1.1	cystathionine γ-lyase cystathionine γ-lyase	
4.2.1.22	Cystathionine β-synthase Cystathionine β-synthase	
4.2.1.21	cystathionine β-synthase cystathionine β-synthase	
4.4.1.1	L-cystathionine cysteine-lyase (deaminating) cystathionine γ-lyase	
4.4.1.8	cystathionine L-homocysteine-lyase (deaminating) cystathionine β-lyase	

EC-No.	Synonym Recommended Name	Vol., Page
4.2.99.9	cystathionine synthase O-succinylhomoserine (thiol)-lyase	
4.2.99.9	cystathionine synthetase O-succinylhomoserine (thiol)-lyase	
1.13.11.19	Cysteamine:oxygen oxidoreductase Cysteamine dioxygenase	
1.8.1.1	cysteamine dehydrogenase cysteamine dehydrogenase	
1.13.11.19	Cysteamine dioxygenase Cysteamine dioxygenase	
1.13.11.19	Cysteamine oxygenase Cysteamine dioxygenase	
4.1.1.15	Cysteic acid decarboxylase Glutamate decarboxylase	
4.1.1.29	Cysteic acid decarboxylase Sulfinoalanine decarboxylase	
4.1.1.29	Cysteic decarboxylase Sulfinoalanine decarboxylase	
2.6.1.3	L-Cysteine-2-oxoglutarate aminotransferase Cysteine transaminase	
2.6.1.75	Cysteine-conjugate α-ketoglutarate transaminase Cysteine-conjugate transaminase	
4.4.1.13	L-cysteine-conjugate thiol-lyase (deaminating) cysteine-S-conjugate β-lyase	
2.6.1.75	Cysteine-conjugate transaminase Cysteine-conjugate transaminase	
4.4.1.13	cysteine-S-conjugate β-lyase cysteine-S-conjugate β-lyase	
2.3.1.80	Cysteine-S-conjugate N-acetyltransferase Cysteine-S-conjugate N-acetyltransferase	
4.1.1.29	Cysteine-sulfinate decarboxylase Sulfinoalanine decarboxylase	
6.1.1.16	Cysteine-tRNA ligase Cysteine-tRNA ligase	Vol. 2, Page 121
3.4.18.1	cysteine-type carboxypeptidase cysteine-type carboxypeptidase	

EC-No.	Synonym Recommended Name	Vol., Page
2.6.1.3	L-Cysteine:2-oxoglutarate aminotransferase Cysteine transaminase	
1.13.11.20	L-Cysteine:oxygen oxidoreductase Cysteine dioxygenase	
6.1.1.16	L-Cysteine:tRNA^{Cys} ligase (AMP-forming) Cysteine-tRNA ligase	Vol. 2, Page 121
2.8.1.4	L-Cysteine:tRNA sulfurtransferase tRNA sulfurtransferase	
3.4.11.3	cysteine aminopeptidase cystinyl aminopeptidase	
2.6.1.3	Cysteine aminotransferase Cysteine transaminase	
2.6.1.3	L-Cysteine aminotransferase Cysteine transaminase	
4.4.1.13	cysteine conjugate β-lyase cysteine-S-conjugate β-lyase	
4.4.1.13	cysteine conjugate. β.-lyase cysteine-S-conjugate β-lyase	
4.4.1.15	D-cysteine desulfhydrase D-cysteine desulfhydrase	
4.4.1.1	cysteine desulfhydrase cystathionine γ-lyase	
1.13.11.20	Cysteine dioxygenase Cysteine dioxygenase	
3.4.22.39	cysteine endopeptidase adenain	
3.4.22.39	cysteine endoprotease adenain	
4.4.1.9	L-cysteine hydrogen-sulfide-lyase (adding HCN) L-3-cyanoalanine synthase	
4.4.1.10	L-cysteine hydrogen-sulfide-lyase (adding sulfite) cysteine lyase	
4.4.1.8	cysteine lyase cystathionine β-lyase	
4.4.1.1	cysteine lyase cystathionine γ-lyase	

EC-No.	Synonym Recommended Name	Vol., Page
4.4.1.10	cysteine lyase cysteine lyase	
1.13.11.20	Cysteine oxidase Cysteine dioxygenase	
3.4.22.39	cysteine peptidase adenain	
3.4.22.39	cysteine protease adenain	
3.4.22.39	L-cysteine proteinase adenain	
3.4.22.39	cysteine proteinase adenain	
3.4.22.28	Cysteine proteinase 3C Picornain 3C	
4.4.1.15	D-cysteine sulfide-lyase D-cysteine desulfhydrase	
4.1.1.12	L-Cysteine sulfinate acid desulfinase Aspartate 4-decarboxylase	
4.1.1.29	L-Cysteine sulfinate carboxy-lyase Sulfinoalanine decarboxylase	
4.1.1.29	Cysteine sulfinate decarboxylase Sulfinoalanine decarboxylase	
4.1.1.29	Cysteinesulfinate decarboxylase Sulfinoalanine decarboxylase	
4.1.1.29	Cysteine sulfinic acid decarboxylase Sulfinoalanine decarboxylase	
4.1.1.29	Cysteinesulfinic acid decarboxylase Sulfinoalanine decarboxylase	
4.1.1.29	L-Cysteinesulfinic acid decarboxylase Sulfinoalanine decarboxylase	
4.1.1.12	Cysteine sulfinic desulfinase Aspartate 4-decarboxylase	
4.4.1.4	L-cysteine sulfoxide lyase alliin lyase	
4.4.1.4	cysteine sulfoxide lyase alliin lyase	

EC-No.	Synonym Recommended Name	Vol., Page
4.2.1.22	Cysteine synthase Cystathionine β-synthase	
4.2.99.8	Cysteine synthase Cysteine synthase	
4.2.99.8	cysteine synthetase Cysteine synthase	
2.6.1.3	Cysteine transaminase Cysteine transaminase	
6.1.1.16	Cysteine translase Cysteine-tRNA ligase	Vol. 2, Page 121
4.2.1.50	Cystein synthase Pyrazolylalanine synthase	
3.4.3.5	cysteinyl-glycine dipeptidase cysteinyl-glycine dipeptidase	
6.1.1.16	Cysteinyl-transfer ribonucleate synthetase Cysteine-tRNA ligase	Vol. 2, Page 121
6.1.1.16	Cysteinyl-transfer RNA synthetase Cysteine-tRNA ligase	Vol. 2, Page 121
2.9.1.1	Cysteinyl-tRNASec-selenium transferase L-Seryl-tRNASec selenium transferase	
6.1.1.16	Cysteinyl-tRNA synthetase Cysteine-tRNA ligase	Vol. 2, Page 121
2.9.1.1	CysteinyltRNASel-selenium transferase L-Seryl-tRNASec selenium transferase	
3.4.11.3	L-cystine aminopeptidase cystinyl aminopeptidase	
3.4.11.3	cystine aminopeptidase cystinyl aminopeptidase	
4.4.1.1	cystine desulfhydrase cystathionine γ-lyase	
4.4.1.8	cystine lyase cystathionine β-lyase	
1.6.4.1	Cystine reductase Cystine reductase (NADH)	
1.6.4.1	Cystine reductase (NADH) Cystine reductase (NADH)	

EC-No.	Synonym Recommended Name	Vol., Page
3.4.11.3	cystinyl aminopeptidase cystinyl aminopeptidase	
3.4.11.3	cystyl aminopeptidase cystinyl aminopeptidase	
1.14.99.18	Cytidine-5'-monophosphate-N-acetylneuraminic acid hydroxylase N-Acetylneuraminate monooxygenase	
2.7.1.37	Cytidine 3',5'-cyclic monophosphate-responsive protein kinase Protein kinase	
3.1.4.17	Cytidine 3':5'-monophosphate phosphodiesterase (cyclic CMP) 3',5'-Cyclic-nucleotide phosphodiesterase	
2.7.8.8	Cytidine 5'-diphospho-1,2-diacyl-sn-glycerol:L-serine O-phosphatidyl-transferase CDPdiacylglycerol-serine O-phosphatidyltransferase	
2.7.8.8	Cytidine 5'-diphospho-1,2-diacyl-sn-glycerol:L-serine O-phosphatidyl-transferase (CDPdiglyceride) CDPdiacylglycerol-serine O-phosphatidyltransferase	
2.7.8.11	Cytidine 5'-diphospho-1,2-diacyl-sn-glycerol:myo-inositol 3-phosphatidyl-transferase CDPdiacylglycerol-inositol 3-phosphatidyltransferase	
2.7.8.5	Cytidine 5'-diphospho-1,2-diacyl-sn-glycerol:sn-glycerol-3-phosphate phosphatidyltransferase (CDPdiglyceride) CDPdiacylglycerol-glycerol-3-phosphate 3-phosphatidyltransferase	
2.1.2.8	d-Cytidine 5'-monophosphate hydroxymethylase Deoxycytidylate 5-hydroxymethyltransferase	
2.7.7.43	Cytidine 5'-monophospho-N-acetylneuraminic acid synthetase N-Acylneuraminate cytidylyltransferase	
2.7.7.43	Cytidine 5'-monophosphosialic acid synthetase N-Acylneuraminate cytidylyltransferase	
6.3.4.2	Cytidine 5'-triphosphate synthetase CTP synthase	Vol. 2, Page 559
2.7.7.43	Cytidine 5-monophosphate N-acetylneuraminic acid synthetase N-Acylneuraminate cytidylyltransferase	
3.5.4.5	Cytidine aminohydrolase Cytidine deaminase	
3.5.4.5	Cytidine deaminase Cytidine deaminase	

EC-No.	Synonym Recommended Name	Vol., Page
2.7.7.33	Cytidine diphosphate-D-glucose pyrophosphorylase Glucose-1-phosphate cytidylyltransferase	
1.17.1.1	Cytidine diphosphate 4-keto-6-deoxy-D-glucose-3-dehydrogenase CDP-4-dehydro-6-deoxyglucose reductase	
2.7.7.33	Cytidine diphosphate glucose pyrophosphorylase Glucose-1-phosphate cytidylyltransferase	
2.7.7.39	Cytidine diphosphate glycerol pyrophosphorylase Glycerol-3-phosphate cytidylyltransferase	
5.1.3.10	Cytidine diphosphate paratose-2-epimerase CDPabequose 2-epimerase	Vol. 1, Page 146
2.7.7.40	Cytidine diphosphate ribitol pyrophosphorylase D-Ribitol-5-phosphate cytidylyltransferase	
5.1.3.10	Cytidine diphosphoabequose epimerase CDPabequose 2-epimerase	Vol. 1, Page 146
2.7.8.10	Cytidine diphosphocholine-sphingosine cholinephosphotransferase Sphingosine cholinephosphotransferase	
2.7.8.2	Cytidine diphosphocholine glyceride transferase Diacylglycerol cholinephosphotransferase	
2.7.7.15	Cytidine diphosphocholine pyrophosphorylase Choline-phosphate cytidylyltransferase	
5.1.3.10	Cytidine diphosphodideoxyglucose epimerase CDPabequose 2-epimerase	Vol. 1, Page 146
4.2.1.45	cytidine diphosphoglucose oxidoreductase CDPglucose 4,6-dehydratase	
2.7.7.33	Cytidine diphosphoglucose pyrophosphorylase Glucose-1-phosphate cytidylyltransferase	
2.7.8.11	Cytidine diphosphoglyceride-inositol phosphatidyltransferase CDPdiacylglycerol-inositol 3-phosphatidyltransferase	
2.7.8.11	Cytidine diphosphoglyceride-inositol transferase CDPdiacylglycerol-inositol 3-phosphatidyltransferase	
2.7.7.41	Cytidine diphosphoglyceride pyrophosphorylase Phosphatidate cytidylyltransferase	
2.7.7.39	Cytidine diphosphoglycerol pyrophosphorylase Glycerol-3-phosphate cytidylyltransferase	
5.1.3.10	Cytidine diphosphoparatose epimerase CDPabequose 2-epimerase	Vol. 1, Page 146

EC-No.	Synonym Recommended Name	Vol., Page
2.7.7.40	Cytidine diphosphoribitol pyrophosphorylase D-Ribitol-5-phosphate cytidylyltransferase	
2.7.8.2	Cytidine diphosphorylcholine diglyceride transferase Diacylglycerol cholinephosphotransferase	
3.1.4.40	Cytidine monophosphate-N-acetylneuraminic acid hydrolase CMP-N-acylneuraminate phosphodiesterase	
2.7.7.43	Cytidine monophosphate-N-acetylneuraminic acid synthetase N-Acylneuraminate cytidylyltransferase	
2.7.7.38	Cytidine monophospho-3-deoxy-D-manno-octulosonate pyrophosphorylase 3-Deoxy-manno-octulosonate cytidylyltransferase	
2.7.7.43	Cytidine monophospho-sialic acid synthetase N-Acylneuraminate cytidylyltransferase	
2.4.99.1	Cytidine monophosphoacetylneuraminate-galactosylglycoprotein sialyltransferase β-Galactoside α-2,6-sialyltransferase	
2.4.99.9	Cytidine monophosphoacetylneuraminate-lactosylceramide sialyltransferase Lactosylceramide α-2,3-sialyltransferase	
2.4.99.11	Cytidine monophosphoacetylneuraminate-lactosylceramide sialyltransferase Lactosylceramide α-2,6-N-sialyltransferase	
2.7.7.43	Cytidine monophosphoacetylneuraminic synthetase N-Acylneuraminate cytidylyltransferase	
2.7.7.43	Cytidine monophosphosialate pyrophosphorylase N-Acylneuraminate cytidylyltransferase	
2.7.7.43	Cytidine monophosphosialate synthetase N-Acylneuraminate cytidylyltransferase	
3.1.4.40	Cytidine monophosphosialic hydrolase CMP-N-acylneuraminate phosphodiesterase	
6.3.4.2	Cytidine triphosphate synthetase CTP synthase	Vol. 2, Page 559
4.6.1.6	Cytidylate cyclase Cytidylate cyclase	
2.7.4.14	Cytidylate kinase Cytidylate kinase	

EC-No.	Synonym Recommended Name	Vol., Page
4.6.1.6	Cytidyl cyclase Cytidylate cyclase	
2.7.7.43	Cytidyltransferase, acylneuraminate N-Acylneuraminate cytidylyltransferase	
4.6.1.6	Cytidylyl cyclase Cytidylate cyclase	
2.7.7.38	Cytidylyltransferase, 2-keto-3-deoxyoctonate 3-Deoxy-manno-octulosonate cytidylyltransferase	
2.7.7.43	Cytidylyltransferase, acetylneuraminate N-Acylneuraminate cytidylyltransferase	
2.7.7.15	Cytidylyltransferase, choline phosphate Choline-phosphate cytidylyltransferase	
2.7.7.14	Cytidylyltransferase, ethanolamine phosphate Ethanolamine-phosphate cytidylyltransferase	
2.7.7.33	Cytidylyltransferase, glucose 1-phosphate Glucose-1-phosphate cytidylyltransferase	
2.7.7.39	Cytidylyltransferase, glycerol 3-phosphate Glycerol-3-phosphate cytidylyltransferase	
2.7.7.57	Cytidylyltransferase, monomethylethanolamine phosphate N-Methylphosphoethanolamine cytidylyltransferase	
2.7.7.41	Cytidylyltransferase, phosphatidate Phosphatidate cytidylyltransferase	
2.7.7.40	Cytidylyltransferase, ribitol 5-phosphate D-Ribitol-5-phosphate cytidylyltransferase	
2.7.7.25	Cytidylyltransferase, transfer ribonucleate tRNA adenylyltransferase	
2.7.7.21	Cytidylyltransferase, transfer ribonucleate tRNA cytidylyltransferase	
1.6.2.2	Cytochrome-b_5 reductase Cytochrome-b_5 reductase	
2.1.1.59	Cytochrome-c-lysine N-methyltransferase Cytochrome-c-lysine N-methyltransferase	
1.12.2.1	Cytochrome-c3 hydrogenase Cytochrome-c3 hydrogenase	
2.1.1.124	[Cytochrome-c]-arginine N-methyltransferase [Cytochrome-c]-arginine N-methyltransferase	

EC-No.	Synonym Recommended Name	Vol., Page
2.1.1.123	[Cytochrome-c]-methionine S-methyltransferase [Cytochrome-c]-methionine S-methyltransferase	
1.9.3.1	Cytochrome-c oxidase Cytochrome-c oxidase	
1.11.1.5	Cytochrome-c peroxidase Cytochrome-c peroxidase	
1.1.2.4	Cytochrome-dependent D-(-)-lactate dehydrogenase D-Lactate dehydrogenase (cytochrome)	
1.9.3.1	Cytochrome a_3 Cytochrome-c oxidase	
1.9.3.1	Cytochrome aa_3 Cytochrome-c oxidase	
1.1.2.3	Cytochrome b_2 L-Lactate dehydrogenase (cytochrome)	
1.6.2.2	Cytochrome b_5 reductase Cytochrome-b_5 reductase	
1.10.99.1	Cytochrome b_6-f complex Plastoquinol-plastocyanin reductase	
1.10.99.1	Cytochrome b_6/f complex Plastoquinol-plastocyanin reductase	
1.9.3.2	Cytochrome c-551:O_2, NO_2^- oxidoreductase Pseudomonas cytochrome oxidase	
1.11.1.5	Cytochrome c-551 peroxidase Cytochrome-c peroxidase	
1.11.1.5	Cytochrome c-H_2O oxidoreductase Cytochrome-c peroxidase	
2.1.1.59	Cytochrome c-specific protein-lysine methyltransferase Cytochrome-c-lysine N-methyltransferase	
2.1.1.59	Cytochrome c-specific protein methylase III Cytochrome-c-lysine N-methyltransferase	
1.6.2.5	Cytochrome c_2 reductase (reduced nicotinamide adinine dinucleotide phosphate, NADPH) NADPH-cytochrome c_2 reductase	
1.12.2.1	Cytochrome c_3 hydrogenase Cytochrome-c_3 hydrogenase	

EC-No.	Synonym Recommended Name	Vol., Page
1.12.2.1	Cytochrome c_3 reductase Cytochrome-c_3 hydrogenase	
1.9.3.2	Cytochrome c_d Pseudomonas cytochrome oxidase	
1.2.3.10	Cytochrome cd_1 Carbon-monoxide oxidase	
1.9.3.2	Cytochrome cd_1 Pseudomonas cytochrome oxidase	
4.4.1.17	Cytochrome c heme-lyase Holocytochrome-c synthase	
2.1.1.59	Cytochrome c methyltransferase Cytochrome-c-lysine N-methyltransferase	
1.9.3.1	Cytochrome c oxidase Cytochrome-c oxidase	
1.11.1.5	Cytochrome c peroxidase Cytochrome-c peroxidase	
1.6.99.3	Cytochrome c reductase NADH dehydrogenase	
1.6.2.4	Cytochrome c reductase (reduced nicotinamide adenine dinucleotide phosphate, NADPH, NADPH-dependent) NADPH-ferrihemoprotein reductase	
4.4.1.17	Cytochrome c synthase Holocytochrome-c synthase	
1.12.2.1	Cytochrome hydrogenase Cytochrome-c_3 hydrogenase	
1.9.3.1	Cytochrome oxidase Cytochrome-c oxidase	
1.14.99.9	Cytochrome P-450 (P-45017 α,lyase) Steroid 17α-monooxygenase	
1.14.15.6	Cytochrome P-450_{scc} Cholesterol monooxygenase (side-chain-cleaving)	
1.14.99.9	Cytochrome P45017α Steroid 17α-monooxygenase	
1.14.13.11	Cytochrome P450 cinnamate 4-hydroxylase trans-Cinnamate 4-monooxygenase	

EC-No.	Synonym Recommended Name	Vol., Page
1.14.13.68	cytochrome P450II-dependent monooxygenase 4-hydroxyphenylacetaldehyde oxime monooxygenase	
1.11.1.5	Cytochrome peroxidase Cytochrome-c peroxidase	
1.6.2.3	cytochrome reductase (NADPH) cytochrome reductase (NADPH)	
1.15.1.1	Cytocuprein Superoxide dismutase	
4.2.99.13	β-(9-cytokinin)-alanine synthase β-(9-cytokinin)-alanine synthase	
4.2.99.13	β-(9-cytokinin)alanine synthase β-(9-cytokinin)-alanine synthase	
2.4.1.118	Cytokinin 7-β-glucosyltransferase Cytokinin 7-β-glucosyltransferase	
2.4.1.118	Cytokinin 7-glucosyltransferase Cytokinin 7-β-glucosyltransferase	
1.4.3.18	cytokinin:oxygen oxidoreductase cytokinin oxidase	
1.4.3.18	cytokinin oxidase cytokinin oxidase	
2.5.1.27	Cytokinin synthase Adenylate isopentenyltransferase	
3.4.21.97	cytomeglovirus protease assemblin	
2.1.1.37	Cytosine-specific DNA methyltransferase DNA (cytosine-5-)-methyltransferase	
2.1.1.37	Cytosine 5-methyltransferase DNA (cytosine-5-)-methyltransferase	
3.5.4.1	Cytosine aminohydrolase Cytosine deaminase	
3.5.4.21	Cytosine deaminase Creatinine deiminase	
3.5.4.1	Cytosine deaminase Cytosine deaminase	
2.1.1.37	Cytosine DNA methylase DNA (cytosine-5-)-methyltransferase	

EC-No.	Synonym Recommended Name	Vol., Page
2.1.1.37	Cytosine DNA methyltransferase DNA (cytosine-5-)-methyltransferase	
3.5.4.5	Cytosine nucleoside deaminase Cytidine deaminase	
3.4.11.6	cytosol aminopeptidase aminopeptidase B	
3.4.11.1	cytosol aminopeptidase leucyl aminopeptidase	
3.4.11.5	cytosol aminopeptidase V prolyl aminopeptidase	
1.2.1.36	Cytosolic retinal dehydrogenase Retinal dehydrogenase	
3.1.1.53	cytosolic sialate:O-acetylesterase sialate O-acetylesterase	
3.1.1.53	cytosolic sialic acid 9-O-acetylesterase sialate O-acetylesterase	
3.4.13.18	Cytosol nonspecific dipeptidase Cytosol nonspecific dipeptidase	
3.4.21.79	Cytotoxic cell proteinase-1 Granzyme B	
3.4.21.78	Cytotoxic T lymphocyte serine protease Granzyme A	
3.4.11.14	cytyosol aminopeptidase III aminopeptidase, human liver	
1.1.1.254	D(+)-Carnitine dehydrogenase (S)-Carnitine 3-dehydrogenase	
1.5.1.16	D(+)-Lysopine dehydrogenase D-Lysopine dehydrogenase	
1.1.1.36	D(-)-β-Hydroxybutyryl CoA-NADP oxidoreductase Acetoacetyl-CoA reductase	
1.1.2.4	D(-)-Lactic cytochrome c reductase D-Lactate dehydrogenase (cytochrome)	
4.2.1.81	D(-)-Tartrate dehydratase D(-)-Tartrate dehydratase	
1.5.1.26	N-(D-1-Carboxyethyl)-β-alanine:NAD^+ oxidoreductase (β-alanine-forming) β-Alanopine dehydrogenase	

EC-No.	Synonym Recommended Name	Vol., Page
1.5.1.28	N-(1-D-Carboxyethyl)-L-norvaline dehydrogenase Opine dehydrogenase	
3.2.1.107	2-O-α-D-glucopyranosyl-5-O-β-D-galactopyranosylhydroxy-L-lysine glucohydrolase Protein-glucosylgalactosylhydroxylysine glucosidase	
3.2.2.1	N-D-Ribosyl-purine ribohydrolase Purine nucleosidase	
3.4.24.60	Dactylysin Dactylysin	
4.1.1.19	dADC Arginine decarboxylase	
2.7.7.41	DAG synthetase Phosphatidate cytidylyltransferase	
4.1.2.15	DAH7-P synthase (phe) 2-Dehydro-3-deoxyphosphoheptanoate aldolase	
4.1.2.15	DAHP(Phe) 2-Dehydro-3-deoxyphosphoheptanoate aldolase	
4.1.2.15	DAHP synthase 2-Dehydro-3-deoxyphosphoheptanoate aldolase	
4.1.2.15	DAHP synthase-phe 2-Dehydro-3-deoxyphosphoheptanoate aldolase	
4.1.2.15	DAHP synthase-trp 2-Dehydro-3-deoxyphosphoheptanoate aldolase	
4.1.2.15	DAHP synthase-tyr 2-Dehydro-3-deoxyphosphoheptanoate aldolase	
1.4.3.3	DAO D-Amino-acid oxidase	
4.1.1.20	DAP-decarboxylase Diaminopimelate decarboxylase	
5.1.1.7	DAP-epimerase Diaminopimelate epimerase	Vol. 1, Page 27
2.6.1.62	DAPA aminotransferase Adenosylmethionine-8-amino-7-oxononanoate transaminase	
4.3.1.15	DAP ammonia-lyase Diaminopropionate ammonia-lyase	

EC-No.	Synonym Recommended Name	Vol., Page
2.6.1.62	DAPA transaminase Adenosylmethionine-8-amino-7-oxononanoate transaminase	
5.1.1.7	DAP epimerase Diaminopimelate epimerase	Vol. 1, Page 27
3.4.14.1	DAP I dipeptidyl-peptidase I	
2.3.1.107	DAT 17-O-Deacetylvindoline O-acetyltransferase	
2.6.99.1	dATP(dGTP)-DNA purinetransferase dATP(dGTP)-DNA purinetransferase	
2.6.99.1	dATP(dGTP):depurinated-DNA purinetransferase dATP(dGTP)-DNA purinetransferase	
6.3.2.14	DBS synthetase 2,3-Dihydroxybenzoate-serine ligase	Vol. 2, Page 478
3.5.1.77	Dcase N-Carbamoyl-D-amino acid hydrolase	
5.3.3.12	DCF Dopachrome Δ-isomerase	Vol. 1, Page 432
6.1.1.13	Dcl D-Alanine-poly(phosphoribitol) ligase	Vol. 2, Page 97
3.5.4.12	dCMP aminohydrolase dCMP deaminase	
3.5.4.12	dCMP deaminase dCMP deaminase	
2.1.2.8	dCMP hydroxymethylase Deoxycytidylate 5-hydroxymethyltransferase	
2.7.4.14	dCMP kinase Cytidylate kinase	
2.1.1.54	dCMP methyltransferase Deoxycytidylate C-methyltransferase	
4.2.1.96	DCoH/PCD 4a-Hydroxytetrahydrobiopterin dehydratase	
3.4.15.1	DCP peptidyl-dipeptidase A	
5.3.3.12	DCT Dopachrome Δ-isomerase	Vol. 1, Page 432

EC-No.	Synonym Recommended Name	Vol., Page
3.5.4.13	dCTP aminohydrolase dCTP deaminase	
3.6.1.12	dCTPase dCTP diphosphatase	
3.5.4.13	dCTP deaminase dCTP deaminase	
3.6.1.12	dCTP diphosphatase dCTP diphosphatase	
3.6.1.12	dCTP nucleotidohydrolase dCTP diphosphatase	
3.6.1.12	dCTP pyrophosphatase dCTP diphosphatase	
3.4.17.14	DD-Carboxypeptidase Zinc D-Ala-D-Ala carboxypeptidase	
3.4.17.8	DD-carboxypeptidase muramoylpentapeptide carboxypeptidase	
3.4.16.4	DD-carboxypeptidase serine-type D-Ala-D-Ala carboxypeptidase	
3.4.17.14	DD-Carboxypeptidase-transpeptidase Zinc D-Ala-D-Ala carboxypeptidase	
3.4.17.8	DD-peptidase muramoylpentapeptide carboxypeptidase	
3.4.16.4	DD-peptidase serine-type D-Ala-D-Ala carboxypeptidase	
3.4.16.4	DD-transpeptidase serine-type D-Ala-D-Ala carboxypeptidase	
4.1.1.28	DDC Aromatic-L-amino-acid decarboxylase	
4.5.1.1	DDT-ase DDT-dehydrochlorinase	
4.5.1.1	DDT-dehydrochlorinase DDT-dehydrochlorinase	
2.3.1.49	Deacetyl-[citrate-(pro-3S)-lyase] acetyltransferase Deacetyl-[citrate-(pro-3S)-lyase] S-acetyltransferase	
2.3.1.49	Deacetyl-[citrate-(pro-3S)-lyase] S-acetyltransferase Deacetyl-[citrate-(pro-3S)-lyase] S-acetyltransferase	

EC-No.	Synonym Recommended Name	Vol., Page
3.1.1.47	deacetylase, 1-alkyl-2-acetyllecithin 1-alkyl-2-acetylglycerophosphocholine esterase	
3.5.1.63	Deacetylase, 4-acetamidobutyrate 4-Acetamidobutyrate deacetylase	
3.1.1.33	deacetylase, 6-O-acetylglucose 6-acetylglucose deacetylase	
3.5.1.80	Deacetylase, acetylglucosaminephosphate N-Acetyl-D-galactosamine-6-phosphate deacetylase	
3.5.1.25	Deacetylase, acetylglucosaminephosphate N-Acetylglucosamine-6-phosphate deacetylase	
3.1.1.69	Deacetylase, acetylglucosaminylphosphatidylinositol N-Acetylglucosaminylphosphatidylinositol deacetylase	
3.4.13.5	deacetylase, acetylhistidine X-methyl-His dipeptidase	
3.5.1.48	Deacetylase, acetylspermidine Acetylspermidine deacetylase	
3.1.1.41	Deacetylase, cephalosporin C Cephalosporin-C deacetylase	
3.1.2.16	Deacetylase, citrate lyase [Citrate-(pro-3S)-lyase] thiolesterase	
3.1.1.58	Deacetylase, polysaccharide N-Acetylgalactosaminoglycan deacetylase	
3.5.1.51	Deacetylase-thiolesterase 4-Acetamidobutyryl-CoA deacetylase	
4.3.3.4	deacetylipecoside dopamine-lyase deacetylipecoside synthase	
4.3.3.4	deacetylipecoside synthase deacetylipecoside synthase	
4.3.3.3	deacetylisoipecoside dopamine-lyase deacetylisoipecoside synthase	
4.3.3.3	deacetylisoipecoside synthase deacetylisoipecoside synthase	
2.3.1.107	17-O-Deacetylvindoline-17-O-acetyltransferase 17-O-Deacetylvindoline O-acetyltransferase	
2.3.1.107	17-O-Deacetylvindoline O-acetyltransferase 17-O-Deacetylvindoline O-acetyltransferase	

EC-No.	Synonym Recommended Name	Vol., Page
3.5.1.70	Deacylase, aculeacin A Aculeacin-A deacylase	
3.5.1.69	Deacylase, glycosphingolipid ceramide Glycosphingolipid deacylase	
3.5.1.55	Deacylase, long-chain-fatty-acyl-glutamate Long-chain-fatty-acyl-glutamate deacylase	
3.5.1.82	Deacylase, long-chain-fatty-acyl-glutamate N-Acyl-D-glutamate deacylase	
3.5.1.55	Deacylase, N-acyl-D-glutamate Long-chain-fatty-acyl-glutamate deacylase	
3.5.1.82	Deacylase, N-acyl-D-glutamate N-Acyl-D-glutamate deacylase	
3.4.16.5	Deamidase Carboxypeptidase C	
6.3.1.5	Deamido-NAD^+:ammonia ligase (AMP-forming) NAD^+ synthase	Vol. 2, Page 377
6.3.5.1	Deamido-NAD^+:L-glutamine amido-ligase (AMP-forming) NAD^+ synthase (glutamine-hydroylysing)	Vol. 2, Page 651
2.7.7.18	Deamido-NAD^+ pyrophosphorylase Nicotinate-nucleotide adenylyltransferase	
2.7.7.18	Deamidonicotinamide adenine dinucleotide pyrophosphorylase Nicotinate-nucleotide adenylyltransferase	
3.5.1.4	Deaminase Amidase	
4.1.99.4	deaminase, 1-aminocyclopropane-1-carboxylate 1-aminocyclopropane-1-carboxylate deaminase	
3.5.4.2	Deaminase, adenine Adenine deaminase	
3.5.4.4	Deaminase, adenosine Adenosine deaminase	
3.5.4.7	Deaminase, adenosine diphosphate ADP deaminase	
3.5.4.28	Deaminase, adenosylhomocysteine S-Adenosylhomocysteine deaminase	
3.5.4.6	Deaminase, adenylate AMP deaminase	

EC-No.	Synonym Recommended Name	Vol., Page
3.5.4.23	Deaminase, blasticidin S Blasticidin-S deaminase	
3.5.4.5	Deaminase, cytidine Cytidine deaminase	
3.5.4.1	Deaminase, cytosine Cytosine deaminase	
3.5.4.14	Deaminase, deoxycytidine Deoxycytidine deaminase	
3.5.4.12	Deaminase, deoxycytidylate dCMP deaminase	
4.3.1.4	deaminase, formiminotetrahydrofolate cyclo- formiminotetrahydrofolate cyclodeaminase	
3.5.4.3	Deaminase, guanine Guanine deaminase	
4.3.1.12	deaminase, ornithine cyclo- ornithine cyclodeaminase	
2.4.1.25	Debranching enzyme 4-α-Glucanotransferase	
3.2.1.68	Debranching enzyme Isoamylase	
3.2.1.41	Debranching enzyme α-dextrin endo-1,6-α-glucosidase	
3.4.22.3	debricin ficain	
3.6.1.30	Decapase $m^7G(5)pppN$ diphosphatase	
4.1.99.5	Decarbonylase Octadecanal decarbonylase	
4.1.1.75	Decarboxylase, 2-oxoarginine 5-Guanidino-2-oxopentanoate decarboxylase	
4.1.1.69	decarboxylase, 3,4-dihydroxyphthalate 3,4-dihydroxyphthalate decarboxylase	
4.1.1.51	Decarboxylase, 3-hydroxy-2-methylpyridine-4,5-dicarboxylate 4- 3-Hydroxy-2-methylpyridine-4,5-dicarboxylate 4-decarboxylase	
4.1.1.56	Decarboxylase, 3-oxolaurate 3-Oxolaurate decarboxylase	

EC-No.	Synonym Recommended Name	Vol., Page
4.1.1.55	Decarboxylase, 4,5-dihydroxyphthalate 4,5-Dihydroxyphthalate decarboxylase	
4.1.1.77	Decarboxylase, 4-oxalocrotonate 4-Oxalocrotonate decarboxylase	
4.1.1.68	Decarboxylase, 5-oxopenta-3-ene-1,2,5-tricarboxylate 5-Oxopent-3-ene-1,2,5-tricarboxylate decarboxylase	
4.1.1.52	Decarboxylase, 6-methylsalicylate 6-Methylsalicylate decarboxylase	
4.1.1.4	Decarboxylase, acetoacetate Acetoacetate decarboxylase	
4.1.1.5	Decarboxylase, acetolactate Acetolactate decarboxylase	
4.1.1.6	Decarboxylase, aconitate Aconitate decarboxylase	
4.1.1.24	Decarboxylase, aminobenzoate Decarboxylase, aminobenzoate	
4.1.1.45	Decarboxylase, aminocarboxymuconate semialdehyde Aminocarboxymuconate-semialdehyde decarboxylase	
4.1.1.19	Decarboxylase, arginine Arginine decarboxylase	
4.1.1.28	Decarboxylase, aromatic amino acid Aromatic-L-amino-acid decarboxylase	
4.1.1.76	Decarboxylase, arylmalonate Arylmalonate decarboxylase	
4.1.1.76	Decarboxylase, arylmalonate (Bordetella bronchiseptica clone pAMD100 reduced) Arylmalonate decarboxylase	
4.1.1.76	Decarboxylase, arylmalonate (Bordetella bronchiseptica strain KU1201 reduced) Arylmalonate decarboxylase	
4.1.1.12	Decarboxylase, aspartate 4- Aspartate 4-decarboxylase	
4.1.1.7	Decarboxylase, benzoylformate Benzoylformate decarboxylase	
4.1.1.72	Decarboxylase, branched-chain oxo acid Branched-chain-2-oxoacid decarboxylase	

EC-No.	Synonym Recommended Name	Vol., Page
4.1.1.29	Decarboxylase, cysteinesulfinate Sulfinoalanine decarboxylase	
4.1.1.64	Decarboxylase, dialkyl amino acid (pyruvate) 2,2-Dialkylglycine decarboxylase (pyruvate)	
4.1.1.20	Decarboxylase, diaminopimelate Diaminopimelate decarboxylase	
4.1.1.54	Decarboxylase, dihydroxyfumarate Dihydroxyfumarate decarboxylase	
4.1.1.59	Decarboxylase, gallate Gallate decarboxylase	
4.1.1.62	Decarboxylase, gentisate Gentisate decarboxylase	
4.1.1.70	decarboxylase, glutaconyl coenzyme A glutaconyl-CoA decarboxylase	
4.1.1.15	Decarboxylase, glutamate Glutamate decarboxylase	
1.4.4.2	Decarboxylase, glycine Glycine dehydrogenase (decarboxylating)	
4.1.1.22	Decarboxylase, histidine Histidine decarboxylase	
4.1.1.16	Decarboxylase, hydroxyglutamate Hydroxyglutamate decarboxylase	
4.1.1.40	Decarboxylase, hydroxypyruvate Hydroxypyruvate decarboxylase	
4.1.1.74	Decarboxylase, indolepyruvate Indolepyruvate decarboxylase	
4.1.1.74	Decarboxylase, indolepyruvate Indolepyruvate decarboxylase	
4.1.1.34	Decarboxylase, keto-L-gulonate Dehydro-L-gulonate decarboxylase	
4.1.1.14	Decarboxylase, leucine Valine decarboxylase	
4.1.1.9	Decarboxylase, malonyl coenzyme A Malonyl-CoA decarboxylase	
4.1.1.57	Decarboxylase, methionine Methionine decarboxylase	

EC-No.	Synonym Recommended Name	Vol., Page
4.1.1.41	Decarboxylase, methylmalonyl coenzyme A Methylmalonyl-CoA decarboxylase	
4.1.1.46	Decarboxylase, *o*-pyrocatechuate *o*-Pyrocatechuate decarboxylase	
4.1.1.17	Decarboxylase, ornithine Ornithine decarboxylase	
4.1.1.23	Decarboxylase, orotidine 5'-phosphate Orotidine-5'-phosphate decarboxylase	
4.1.1.58	Decarboxylase, orsellinate Orsellinate decarboxylase	
4.1.1.2	Decarboxylase, oxalate Oxalate decarboxylase	
4.1.1.8	Decarboxylase, oxalyl coenzyme A Oxalyl-CoA decarboxylase	
4.1.1.71	Decarboxylase, oxoglutarate 2-Oxoglutarate decarboxylase	
4.1.1.61	Decarboxylase, *p*-hydroxybenzoate 4-Hydroxybenzoate decarboxylase	
4.1.1.30	Decarboxylase, pantothenoylcysteine Pantothenoylcysteine decarboxylase	
4.1.1.53	Decarboxylase, phenylalanine Phenylalanine decarboxylase	
4.1.1.43	Decarboxylase, phenylpyruvate Phenylpyruvate decarboxylase	
4.1.1.65	Decarboxylase, phosphatidylserine Phosphatidylserine decarboxylase	
4.1.1.36	Decarboxylase, phosphopantothenoylcysteine Phosphopantothenoylcysteine decarboxylase	
4.1.1.63	Decarboxylase, protocatechuate Protocatechuate decarboxylase	
4.1.1.33	Decarboxylase, pyrophosphomevalonate Diphosphomevalonate decarboxylase	
4.1.1.1	Decarboxylase, pyruvate Pyruvate decarboxylase	
4.1.1.60	Decarboxylase, stipitatonate Stipitatonate decarboxylase	

EC-No.	Synonym Recommended Name	Vol., Page
4.1.1.73	Decarboxylase, tartrate Tartrate decarboxylase	
4.1.1.25	Decarboxylase, tyrosine Tyrosine decarboxylase	
4.1.1.66	Decarboxylase, uracil-5-carboxylate Uracil-5-carboxylate decarboxylase	
4.1.1.35	Decarboxylase, uridine diphosphoglucuronate UDPglucuronate decarboxylase	
4.1.1.37	Decarboxylase, uroporphyrinogen Uroporphyrinogen decarboxylase	
4.1.1.14	Decarboxylase, valine Valine decarboxylase	
1.14.13.50	Dechlorinase, pentachlorophenol Pentachlorophenol monooxygenase	
4.1.3.23	2-Decylcitrate synthase Decylcitrate synthase	
4.1.3.23	Decylcitrate synthase Decylcitrate synthase	
4.1.3.29	2-Decylhomocitrate synthase Decylhomocitrate synthase	
4.1.3.29	Decylhomocitrate synthase Decylhomocitrate synthase	
3.5.1.8	Deformylase, formylaspartate Formylaspartate deformylase	
3.5.1.68	Deformylase, formylglutamate N-Formylglutamate deformylase	
3.8.1.7	Dehalogenase, 4-chlorobenzoyl coenzyme A 4-Chlorobenzoyl-CoA dehalogenase	
3.8.1.7	Dehalogenase, 4-chlorobenzoyl coenzyme A (Pseudomonas strain CBS-3 clone pMMB22 reduced) 4-Chlorobenzoyl-CoA dehalogenase	
4.5.1.3	Dehalogenase, dichloromethane Dichloromethane dehalogenase	
3.8.1.5	Dehalogenase, haloalkane Haloalkane dehalogenase	

EC-No.	Synonym Recommended Name	Vol., Page
3.8.1.6	Dehalogenase, halobenzoate 4-Chlorobenzoate dehalogenase	
4.5.1.4	Dehalogenase, L-2-amino-4-chloro-4-pentenoate L-2-Amino-4-chloropent-4-enoate dehydrochlorinase	
3.1.1.60	DEHP esterase Bis(2-ethylhexyl)phthalate esterase	
4.2.2.13	Dehydratase, α-1,4-glucan exo-(1,4)-α-D-Glucan lyase	
4.2.1.59	Dehydratase, β-hydroxyoctanoyl thioester 3-Hydroxyoctanoyl-[acyl-carrier-protein] dehydratase	
4.2.1.61	Dehydratase, β-hydroxypalmitoyl thioester 3-Hydroxypalmitoyl-[acyl-carrier-protein] dehydratase	
4.2.1.33	dehydratase,β-isopropylmalate 3-isopropylmalate dehydratase	
4.2.1.43	dehydratase, 2-keto-3-deoxyL-arabonate 2-dehydro-3-deoxy-L-arabinonate dehydratase	
4.2.1.79	Dehydratase, 2-methylcitrate 2-Methylcitrate dehydratase	
4.2.1.99	Dehydratase, 2-methylisocitrate 2-Methylisocitrate dehydratase	
4.2.1.10	dehydratase, 3-dehydroquinate 3-dehydroquinate dehydratase	
4.2.1.38	Dehydratase, 3-hydroxyaspartate Erythro-3-hydroxyaspartate dehydratase	
4.2.1.96	Dehydratase, 4a-carbinolamine 4a-Hydroxytetrahydrobiopterin dehydratase	
4.2.1.62	Dehydratase, 5α-hydroxysterol 5-α-Hydroxysteroid dehydratase	
4.2.1.7	dehydratase, altronate altronate dehydratase	
4.2.1.26	dehydratase, aminodeoxygluconate aminodeoxygluconate dehydratase	
4.2.1.5	Dehydratase, arabinonate Arabinonate dehydratase	
4.2.1.91	Dehydratase, arogenate Carboxycyclohexadienyl dehydratase	

EC-No.	Synonym Recommended Name	Vol., Page
4.2.1.1	dehydratase, carbonate carbonate dehydratase	
4.2.1.89	dehydratase, carnitine carnitine dehydratase	
4.2.1.34	Dehydratase, citramalate (S)-2-Methylmalate dehydratase	
4.2.1.4	Dehydratase, citrate Citrate dehydratase	
4.2.1.45	dehydratase, cytidine diphosphoglucose CDPglucose 4,6-dehydratase	
4.2.1.55	Dehydratase, D-3-hydroxybutyryl coenzyme A 3-Hydroxybutyryl-CoA dehydratase	
4.2.1.60	Dehydratase, D-3-hydroxydecanoyl-[acyl-carrier protein] 3-Hydroxydecanoyl-[acyl-carrier-protein] dehydratase	
4.2.1.59	Dehydratase, D-3-hydroxyoctanoyl-[acyl carrier protein] 3-Hydroxyoctanoyl-[acyl-carrier-protein] dehydratase	
4.2.1.61	Dehydratase, D-3-hydroxypalmitoyl-[acyl-carrier-protein] 3-Hydroxypalmitoyl-[acyl-carrier-protein] dehydratase	
4.2.1.67	Dehydratase, D-fuconate D-Fuconate dehydratase	
4.2.1.14	Dehydratase, D-serine D-Serine dehydratase	
4.2.1.81	Dehydratase, D-tartrate D(-)-Tartrate dehydratase	
4.2.1.82	Dehydratase, D-xylonate Xylonate dehydratase	
4.2.1.41	Dehydratase, deoxyketoglucarate 5-Dehydro-4-deoxyglucarate dehydratase	
4.2.1.9	dehydratase, dihydroxy acid dihydroxy-acid dehydratase	
4.2.1.28	dehydratase, diol propanediol dehydratase	
4.2.1.42	Dehydratase, galactarate Galactarate dehydratase	
4.2.1.6	dehydratase, galactonate galactonate dehydratase	

EC-No.	Synonym Recommended Name	Vol., Page
4.2.1.40	dehydratase, glucarate glucarate dehydratase	
4.2.1.39	dehydratase, gluconate Gluconate dehydratase	
4.2.1.30	dehydratase, glycerol glycerol dehydratase	
4.4.1.1	dehydratase, homoserine cystathionine γ-lyase	
4.2.1.19	dehydratase, imidazoleglycerol phosphate imidazoleglycerol-phosphate dehydratase	
4.2.1.29	Dehydratase, indoleacetaldoxime Indoleacetaldoxime dehydratase	
4.2.1.44	Dehydratase, inosose 2,3- Myo-inosose-2 dehydratase	
4.2.1.25	Dehydratase, L-arabinonate L-Arabinonate dehydratase	
4.2.1.68	dehydratase, L-fuconate L-fuconate hydratase	
4.2.1.90	Dehydratase, L-rhamnonate L-Rhamnonate dehydratase	
4.2.1.13	Dehydratase, L-serine L-Serine dehydratase	
4.2.1.54	dehydratase, lactoyl-coenzyme A lactoyl-CoA dehydratase	
4.2.1.27	dehydratase, malonate semialdehyde malonate-semialdehyde dehydratase	
4.2.1.8	Dehydratase, mannonate Mannonate dehydratase	
4.2.1.87	Dehydratase, octopamine Octopamine dehydratase	
4.2.1.12	dehydratase, phosphogluconate phosphogluconate dehydratase	
4.2.1.51	dehydratase, prephenate prephenate dehydratase	
4.2.1.28	dehydratase, propanediol propanediol dehydratase	

EC-No.	Synonym Recommended Name	Vol., Page
4.2.1.73	Dehydratase, protoaphin aglucone (cyclizing) Protoaphin-aglucone dehydratase (cyclizing)	
4.2.1.93	Dehydratase, reduced nicotinamide adenine dinucleotide hydrate ATP-dependent $H_4$21NAD(P)OH dehydratase	
4.2.1.94	Dehydratase, scytalone Scytalone dehydratase	
4.2.1.32	Dehydratase, tartrate L(+)-Tartrate dehydrogenase	
2.7.1.92	5-Dehydro-2-deoxygluconokinase 5-Dehydro-2-deoxygluconokinase	
4.1.2.15	2-Dehydro-3-deoxy-D-arabino-heptonate-7-phosphate D-erythrose-4-phosphate-lyase (pyruvate-phosphorylating) 2-Dehydro-3-deoxyphosphoheptanoate aldolase	
4.1.2.21	2-Dehydro-3-deoxy-D-galactonate-6-phosphate D-glyceraldehyde-3-phosphate-lyase 2-Dehydro-3-deoxyphosphogalactonate aldolase	
4.1.2.20	2-Dehydro-3-deoxy-D-glucarate tartronate-semialdehyde-lyase 2-Dehydro-3-deoxyglucarate aldolase	
1.1.1.127	2-Dehydro-3-deoxy-D-gluconate 5-dehydrogenase 2-Dehydro-3-deoxy-D-gluconate 5-dehydrogenase	
1.1.1.126	2-Dehydro-3-deoxy-D-gluconate 6-dehydrogenase 2-Dehydro-3-deoxy-D-gluconate 6-dehydrogenase	
1.1.1.127	2-Dehydro-3-deoxy-D-gluconate:NAD^+ 5-oxidoreductase 2-Dehydro-3-deoxy-D-gluconate 5-dehydrogenase	
1.1.1.126	2-Dehydro-3-deoxy-D-gluconate: $NADP^+$ 6-oxidoreductase 2-Dehydro-3-deoxy-D-gluconate 6-dehydrogenase	
4.1.2.16	2-Dehydro-3-deoxy-D-octonate-8-phosphate D-arabinose-5-phosphate-lyase (pyruvate-phosphorylating) 2-Dehydro-3-deoxyphosphooctonate aldolase	
4.1.2.28	2-Dehydro-3-deoxy-D-pentonate aldolase 2-Dehydro-3-deoxy-D-pentonate aldolase	
4.1.2.28	2-Dehydro-3-deoxy-D-pentonate glycolaldehyde-lyase 2-Dehydro-3-deoxy-D-pentonate aldolase	
4.2.1.43	2-dehydro-3-deoxy-L-arabinonate dehydratase 2-dehydro-3-deoxy-L-arabinonate dehydratase	
4.2.1.43	2-dehydro-3-deoxy-L-arabinonate hydro-lyase 2-dehydro-3-deoxy-L-arabinonate dehydratase	

EC-No.	Synonym Recommended Name	Vol., Page
4.1.2.18	2-Dehydro-3-deoxy-L-pentonate aldolase 2-Dehydro-3-deoxy-L-pentonate aldolase	
4.1.2.18	2-Dehydro-3-deoxy-L-pentonate glycolaldehyde-lyase 2-Dehydro-3-deoxy-L-pentonate aldolase	
2.7.1.58	2-Dehydro-3-deoxygalactonokinase 2-Dehydro-3-deoxygalactonokinase	
4.1.2.20	2-Dehydro-3-deoxyglucarate aldolase 2-Dehydro-3-deoxyglucarate aldolase	
2.7.1.45	2-Dehydro-3-deoxygluconokinase 2-Dehydro-3-deoxygluconokinase	
4.1.2.21	2-Dehydro-3-deoxyphosphogalactonate aldolase 2-Dehydro-3-deoxyphosphogalactonate aldolase	
4.1.2.14	2-Dehydro-3-deoxyphosphogluconate aldolase 2-Dehydro-3-deoxyphosphogluconate aldolase	
4.1.2.15	2-Dehydro-3-deoxyphosphoheptanoate aldolase 2-Dehydro-3-deoxyphosphoheptanoate aldolase	
4.1.2.16	2-Dehydro-3-deoxyphosphooctonate aldolase 2-Dehydro-3-deoxyphosphooctonate aldolase	
4.2.1.41	5-Dehydro-4-deoxy-D-glucarate hydro-lyase (decarboxylating) 5-Dehydro-4-deoxyglucarate dehydratase	
4.2.1.41	5-Dehydro-4-deoxyglucarate dehydratase 5-Dehydro-4-deoxyglucarate dehydratase	
1.1.99.4	2-Dehydro-D-gluconate:acceptor 2-oxidoreductase Dehydrogluconate dehydrogenase	
1.1.1.130	3-Dehydro-L-gulonate 2-dehydrogenase 3-Dehydro-L-gulonate 2-dehydrogenase	
1.1.1.130	3-Dehydro-L-gulonate:NAD(P)$^+$ 2-oxidoreductase 3-Dehydro-L-gulonate 2-dehydrogenase	
4.1.1.34	3-Dehydro-L-gulonate carboxy-lyase Dehydro-L-gulonate decarboxylase	
4.1.1.34	Dehydro-L-gulonate decarboxylase Dehydro-L-gulonate decarboxylase	
1.8.5.1	Dehydroascorbate reductase Glutathione dehydrogenase (ascorbate)	
1.8.5.1	Dehydroascorbic acid reductase Glutathione dehydrogenase (ascorbate)	

EC-No.	Synonym Recommended Name	Vol., Page
1.8.5.1	Dehydroascorbic reductase Glutathione dehydrogenase (ascorbate)	
4.5.1.1	dehydrochlorinase DDT-dehydrochlorinase	
1.3.1.21	7-Dehydrocholesterol reductase 7-Dehydrocholesterol reductase	
3.1.6.2	Dehydroepiandrosterone sulfatase Steryl-sulfatase	
3.1.6.2	Dehydroepiandrosterone sulfate sulfatase Steryl-sulfatase	
2.8.2.2	Dehydroepiandrosterone sulfotransferase Alcohol sulfotransferase	
1.3.99.5	Δ^4-5α-dehydrogenase 3-Oxo-5α-steroid Δ^4-dehydrogenase	
1.3.99.4	δ^1-dehydrogenase 3-Oxosteroid Δ^1-dehydrogenase	
1.1.1.228	Dehydrogenase, (+)-cis-sabinol (+)-Sabinol dehydrogenase	
1.1.1.227	Dehydrogenase, (-)-borneol (-)-Borneol dehydrogenase	
1.1.1.243	Dehydrogenase, (-)-trans-carveol Carveol dehydrogenase	
1.1.3.35	Dehydrogenase, (R)-reticuline Salutaridine synthase	
1.5.1.26	Dehydrogenase, β-alanopine β-Alanopine dehydrogenase	
1.2.1.54	Dehydrogenase, γ-guanidinobutyraldehyde γ-Guanidinobutyraldehyde dehydrogenase	
1.1.1.66	Dehydrogenase, ω-hydroxydecanoate ω-Hydroxydecanoate dehydrogenase	
1.1.1.51	Dehydrogenase, β-hydroxy steroid 3(or 17)β-Hydroxysteroid dehydrogenase	
1.3.1.55	Dehydrogenase, 1,2-dihydro-1,2-dihydroxybenzoate (Acinetobacter calcoaceticus strain ADP1 gene benD reduced) cis-1,2-Dihydroxycyclohexa-3,5-diene-1-carboxylate dehydrogenase	

EC-No.	Synonym Recommended Name Vol., Page
1.3.1.55	Dehydrogenase, 1,2-dihydro-1,2-dihydroxybenzoate (plasmid pWWO clone pPL392 gene xylL reduced) cis-1,2-Dihydroxycyclohexa-3,5-diene-1-carboxylate dehydrogenase
1.3.1.29	Dehydrogenase, 1,2-dihydro-1,2-dihydroxynaphthalene cis-1,2-Dihydro-1,2-dihydroxynaphthalene dehydrogenase
1.3.1.53	Dehydrogenase, 1,2-dihydroxy-3,5-cyclohexadiene-1,4-dicarboxylate (3S,4R)3,4-Dihydroxycyclohexa-1,5-diene-1,4-dicarboxylate dehydrogenase
1.1.1.202	Dehydrogenase, 1,3-propanediol 1,3-Propanediol 1-dehydrogenase
1.5.1.12	Dehydrogenase, 1-pyrroline-5-carboxylate 1-Pyrroline-5-carboxylate dehydrogenase
1.1.1.146	Dehydrogenase, 11β-hydroxy steroid 11β-Hydroxysteroid dehydrogenase
1.1.1.176	Dehydrogenase, 12α-hydroxy steroid 12α-Hydroxysteroid dehydrogenase
1.1.1.238	Dehydrogenase, 12β-hydroxy steroid (nicotinamide adenine dinucleotide phosphate) 12β-Hydroxysteroid dehydrogenase
1.1.1.232	Dehydrogenase, 15-hydroxyeicosatetraenoate 15-Hydroxyeicosatetraenoate dehydrogenase
1.1.1.141	Dehydrogenase, 15-hydroxyprostaglandin 15-Hydroxyprostaglandin dehydrogenase (NAD^+)
1.1.1.197	Dehydrogenase, 15-hydroxyprostaglandin (nicotinamide adenine dinucleotide phosphate) 15-Hydroxyprostaglandin dehydrogenase ($NADP^+$)
1.1.1.196	Dehydrogenase, 15-hydroxyprostaglandin (nicotinamide adenine dinucleotide phosphate) 15-Hydroxyprostaglandin-D dehydrogenase ($NADP^+$)
1.1.1.147	Dehydrogenase, 16α-hydroxy steroid 16α-Hydroxysteroid dehydrogenase
1.1.1.51	Dehydrogenase, 17β-hydroxy steroid 3(or 17)β-Hydroxysteroid dehydrogenase
1.3.1.28	Dehydrogenase, 2,3-dihydro-2,3-dihydroxybenzoate 2,3-Dihydro-2,3-dihydroxybenzoate dehydrogenase
1.1.1.165	Dehydrogenase, 2-alkyn-1-ol 2-Alkyn-1-ol dehydrogenase

EC-No.	Synonym Recommended Name	Vol., Page
1.1.1.125	Dehydrogenase, 2-deoxygluconate 2-Deoxy-D-gluconate 3-dehydrogenase	
1.2.1.45	Dehydrogenase, 2-hydroxy-4-carboxymuconate 6-semialdehyde 2-Hydroxy-4-carboxymuconate-6-semialdehyde dehydrogenase	
1.1.99.2	Dehydrogenase, 2-hydroxyglutarate 2-Hydroxyglutarate dehydrogenase	
1.2.1.32	Dehydrogenase, 2-hydroxymuconate semialdehyde Aminomuconate-semialdehyde dehydrogenase	
1.1.1.126	Dehydrogenase, 2-keto-3-deoxygluconate 2-Dehydro-3-deoxy-D-gluconate 6-dehydrogenase	
1.1.1.127	Dehydrogenase, 2-keto-3-deoxygluconate (nicotinamide adenine dinucleotide (phosphate)) 2-Dehydro-3-deoxy-D-gluconate 5-dehydrogenase	
1.1.1.178	Dehydrogenase, 2-methyl-3-hydroxybutyryl coenzyme A 3-Hydroxy-2-methylbutyryl-CoA dehydrogenase	
1.2.4.4	Dehydrogenase, 2-oxoisocaproate 3-Methyl-2-oxobutanoate dehydrogenase (lipoamide)	
1.2.1.25	Dehydrogenase, 2-oxoisovalerate (acylating) 2-Oxoisovalerate dehydrogenase (acylating)	
1.2.4.4	Dehydrogenase, 2-oxoisovalerate (lipoate) 3-Methyl-2-oxobutanoate dehydrogenase (lipoamide)	
1.1.1.149	Dehydrogenase, 20α-hydroxy steroid 20α-Hydroxysteroid dehydrogenase	
1.1.1.53	Dehydrogenase, 20β-hydroxy steroid 3α(or 20β)-Hydroxysteroid dehydrogenase	
1.1.1.150	Dehydrogenase, 21-hydroxy steroid 21-Hydroxysteroid dehydrogenase (NAD^+)	
1.1.1.151	Dehydrogenase, 21-hydroxy steroid (nicotinamide adenine dinucleotide phosphate) 21-Hydroxysteroid dehydrogenase ($NADP^+$)	
1.1.1.209	Dehydrogenase, 3(17)α-hydroxy steroid 3(or 17)α-Hydroxysteroid dehydrogenase	
1.1.1.239	Dehydrogenase, 3α,17β-hydroxy steroid 3α(17β)-Hydroxysteroid dehydrogenase (NAD^+)	
1.1.1.210	Dehydrogenase, 3β,20α-hydroxy steroid 3β(or 20α)-Hydroxysteroid dehydrogenase	

EC-No.	Synonym Recommended Name	Vol., Page
1.3.1.25	Dehydrogenase, 3,5-cyclohexadiene-1,2-diol-1-carboxylate 1,6-Dihydroxycyclohexa-2,4-diene-1-carboxylate dehydrogenase	
1.1.1.170	Dehydrogenase, 3β-hydroxy-4β-methylcholestenoate 3β-Hydroxy-4β-methylcholestenecarboxylate 3-dehydrogenase (decarboxylating)	
1.1.1.145	Dehydrogenase, 3β-hydroxy-Δ^5-steroid 3β-Hydroxy-Δ^5-steroid dehydrogenase	
1.1.1.97	Dehydrogenase, 3-hydroxybenzyl alcohol 3-Hydroxybenzyl-alcohol dehydrogenase	
1.1.1.30	Dehydrogenase, 3-hydroxybutyrate 3-Hydroxybutyrate dehydrogenase	
1.1.1.52	Dehydrogenase, 3α-hydroxycholanate 3α-Hydroxycholanate dehydrogenase	
1.1.99.26	Dehydrogenase, 3-hydroxycyclohexanone 3-Hydroxycyclohexanone dehydrogenase	
1.1.1.230	Dehydrogenase, 3α-hydroxyglycyrrhetinate 3α-Hydroxyglycyrrhetinate dehydrogenase	
1.1.1.31	Dehydrogenase, 3-hydroxyisobutyrate 3-Hydroxyisobutyrate dehydrogenase	
1.1.1.50	Dehydrogenase, 3α-hydroxy steroid 3α-Hydroxysteroid dehydrogenase (B-specific)	
1.1.1.51	Dehydrogenase, 3β-hydroxy steroid 3(or 17)β-Hydroxysteroid dehydrogenase	
1.1.1.130	Dehydrogenase, 3-ketogulonate 3-Dehydro-L-gulonate 2-dehydrogenase	
1.3.99.5	Dehydrogenase, 3-oxo-5αsteroid δ^4- 3-Oxo-5α-steroid Δ^4-dehydrogenase	
1.3.99.4	Dehydrogenase, 3-oxo steroid δ 1- 3-Oxosteroid Δ1-dehydrogenase	
1.1.1.181	Dehydrogenase, 5-cholestene-3β,7α-diol-3β- Cholest-5-ene-3β,7α-diol 3β-dehydrogenase	
1.1.1.241	Dehydrogenase, 6-endo-hydroxycineole 6-endo-Hydroxycineole dehydrogenase	
1.3.1.18	Dehydrogenase, 7,8-dihydro-7,8-dihydroxykynurenate Kynurenate-7,8-dihydrodiol dehydrogenase	

EC-No.	Synonym Recommended Name	Vol., Page
1.1.1.159	Dehydrogenase, 7α-hydroxy steroid 7α-Hydroxysteroid dehydrogenase	
1.1.1.201	Dehydrogenase, 7β-hydroxy steroid (nicotinamide adenine dinucleotide phosphate) 7β-Hydroxysteroid dehydrogenase ($NADP^+$)	
1.3.99.3	Dehydrogenase, acyl coenzyme A Acyl-CoA dehydrogenase	
1.3.1.8	Dehydrogenase, acyl coenzyme A (nicotinamide adenine dinucleotide phosphate) Acyl-CoA dehydrogenase ($NADP^+$)	
1.4.1.1	Dehydrogenase, alanine Alanine dehydrogenase	
1.5.1.17	Dehydrogenase, alanopine Alanopine dehydrogenase	
1.1.1.1	Dehydrogenase, alcohol Alcohol dehydrogenase	
1.1.99.8	Dehydrogenase, alcohol (acceptor) Alcohol dehydrogenase (acceptor)	
1.1.1.71	Dehydrogenase, alcohol (nicotinamide adenine dinucleotide (phosphate)) Alcohol dehydrogenase ($NAD(P)^+$)	
1.1.1.2	Dehydrogenase, alcohol (nicotinamide adenine dinucleotide phosphate) Alcohol dehydrogenase ($NADP^+$)	
1.2.99.3	Dehydrogenase, aldehyde (acceptor) Aldehyde dehydrogenase (pyrroloquinoline-quinone)	
1.2.1.5	Dehydrogenase, aldehyde (nicotinamide adenine dinucleotide (phosphate)) Aldehyde dehydrogenase ($NAD(P)^+$)	
1.2.1.4	Dehydrogenase, aldehyde (nicotinamide adenine dinucleotide phosphate) Aldehyde dehydrogenase ($NADP^+$)	
1.1.1.121	Dehydrogenase, aldose Aldose 1-dehydrogenase	
1.14.99.19	Dehydrogenase, alkyl-acylglycerophosphorylethanolamine Plasmanylethanolamine desaturase	
1.1.1.54	Dehydrogenase, allyl alcohol Allyl-alcohol dehydrogenase	
1.4.99.3	Dehydrogenase, amine Amine dehydrogenase	

EC-No.	Synonym Recommended Name	Vol., Page
1.2.1.31	Dehydrogenase, aminoadipate semialdehyde L-Aminoadipate-semialdehyde dehydrogenase	
1.2.1.19	Dehydrogenase, aminobutyraldehyde Aminobutyraldehyde dehydrogenase	
1.2.1.32	Dehydrogenase, aminomuconate semialdehyde Aminomuconate-semialdehyde dehydrogenase	
1.1.1.90	Dehydrogenase, aryl alcohol Aryl-alcohol dehydrogenase	
1.1.1.91	Dehydrogenase, aryl alcohol (nicotinamide adenine dinucleotide phosphate) Aryl-alcohol dehydrogenase ($NADP^+$)	
1.4.99.4	Dehydrogenase, arylamine Aralkylamine dehydrogenase	
1.2.1.11	Dehydrogenase, aspartate semialdehyde Aspartate-semialdehyde dehydrogenase	
1.1.1.217	Dehydrogenase, benzyl 2-methyl-3-hydroxybutyrate Benzyl-2-methyl-hydroxybutyrate dehydrogenase	
1.2.4.4	Dehydrogenase, branched chain α-keto acid 3-Methyl-2-oxobutanoate dehydrogenase (lipoamide)	
1.3.99.2	Dehydrogenase, butyryl coenzyme A Butyryl-CoA dehydrogenase	
1.2.99.2	Dehydrogenase, carbon monoxide Carbon-monoxide dehydrogenase	
1.2.2.4	Dehydrogenase, carbon monoxide Carbon-monoxide oxygenase (cytochrome b-561)	
1.1.99.18	Dehydrogenase, cellobiose Cellobiose dehydrogenase (acceptor)	
1.1.5.1	Dehydrogenase, cellobiose Cellobiose dehydrogenase (quinone)	
1.1.1.161	Dehydrogenase, cholestanetetrol 26- Cholestanetetraol 26-dehydrogenase	
1.2.1.40	Dehydrogenase, cholestanetriol-26-al 26- $3\alpha,7\alpha,12\alpha$-Trihydroxycholestan-26-al 26-dehydrogenase	
1.1.99.1	Dehydrogenase, choline Choline dehydrogenase	

EC-No.	Synonym Recommended Name	Vol., Page
1.1.3.17	Dehydrogenase, choline Choline oxidase	
1.1.1.195	Dehydrogenase, cinnamyl alcohol Cinnamyl-alcohol dehydrogenase	
1.3.1.19	Dehydrogenase, cis-1,2-dihydrobenzene-1,2-diol cis-1,2-Dihydrobenzene-1,2-diol dehydrogenase	
1.3.1.49	Dehydrogenase, cis-3,4-dihydro-3,4-dihydroxyphenanthrene cis-3,4-Dihydrophenanthrene-3,4-diol dehydrogenase	
1.1.1.245	Dehydrogenase, cyclohexanol Cyclohexanol dehydrogenase	
1.3.99.14	Dehydrogenase, cyclohexanone Cyclohexanone dehydrogenase	
1.1.1.163	Dehydrogenase, cyclopentanol Cyclopentanol dehydrogenase	
1.1.99.6	Dehydrogenase, D-2-hydroxy acid D-2-Hydroxy-acid dehydrogenase	
1.1.1.98	Dehydrogenase, D-2-hydroxy fatty acid (R)-2-Hydroxy-fatty-acid dehydrogenase	
1.1.1.121	Dehydrogenase, D-aldohexose Aldose 1-dehydrogenase	
1.2.1.33	Dehydrogenase, D-aldopantoate (R)-Dehydropantoate dehydrogenase	
1.4.99.1	Dehydrogenase, D-amino acid D-Amino acid dehydrogenase	
1.1.1.4	Dehydrogenase, D-aminopropanol (R,R)-Butanediol dehydrogenase	
1.1.1.11	Dehydrogenase, D-arabinitol D-Arabinitol 4-dehydrogenase	
1.1.1.250	Dehydrogenase, D-arabinitol (Candida albicans clone pEMBLYe23 gene arDH reduced) D-Arabinitol 2-dehydrogenase	
1.1.1.116	Dehydrogenase, D-arabinose D-Arabinose 1-dehydrogenase	
1.1.1.117	Dehydrogenase, D-arabinose (nicotinamide adenine dinucleotide (phosphate)) D-Arabinose 1-dehydrogenase ($NAD(P)^+$)	

EC-No.	Synonym Recommended Name	Vol., Page
1.1.1.4	Dehydrogenase, D-butanediol (R,R)-Butanediol dehydrogenase	
1.1.1.254	Dehydrogenase, D-carnitine (S)-Carnitine 3-dehydrogenase	
1.1.99.13	Dehydrogenase, D-glucoside 3- Glucoside 3-dehydrogenase	
1.1.1.15	Dehydrogenase, D-iditol D-Iditol 2-dehydrogenase	
1.1.1.78	Dehydrogenase, D-lactaldehyde D-Lactaldehyde dehydrogenase	
1.1.1.28	Dehydrogenase, D-lactate D-Lactate dehydrogenase	
1.1.2.4	Dehydrogenase, D-lactate (cytochrome) D-Lactate dehydrogenase (cytochrome)	
1.5.1.16	Dehydrogenase, D-lysopine D-Lysopine dehydrogenase	
1.1.1.83	Dehydrogenase, D-malate (decarboxylating) D-Malate dehydrogenase (decarboxylating)	
1.1.1.142	Dehydrogenase, D-pinitol D-Pinitol dehydrogenase	
1.1.1.140	Dehydrogenase, D-sorbitol 6-phosphate Sorbitol-6-phosphate 2-dehydrogenase	
1.1.1.175	Dehydrogenase, D-xylose D-Xylose 1-dehydrogenase	
1.1.1.179	Dehydrogenase, D-xylose (nicotinamide adenine dinucleotide phosphate) D-Xylose 1-dehydrogenase ($NADP^+$)	
1.1.1.160	Dehydrogenase, dihydrobunolol Dihydrobunolol dehydrogenase	
1.3.1.1	Dehydrogenase, dihydrouracil Dihydrouracil dehydrogenase (NAD^+)	
1.3.1.2	Dehydrogenase, dihydrouracil (nicotinamide adenine dinucleotide phosphate) Dihydropyrimidine dehydrogenase ($NADP^+$)	
1.1.1.166	Dehydrogenase, dihydroxycyclohexanecarboxylate Hydroxycyclohexanecarboxylate dehydrogenase	

EC-No.	Synonym Recommended Name Vol., Page
1.1.1.86	Dehydrogenase, dihydroxyisovalerate (isomerizing) Ketol-acid reductoisomerase
1.5.99.10	Dehydrogenase, dimethylamine Dimethylamine dehydrogenase
1.5.99.10	Dehydrogenase, dimethylamine (Hyphomicrobium strain X clone pUHA65/ pUHB49 gene dmd) Dimethylamine dehydrogenase
1.5.99.10	Dehydrogenase, dimethylamine (Hyphomicrobium strain X clone pUHA65/ pUHB49 gene dmd precursor) Dimethylamine dehydrogenase
1.1.1.62	Dehydrogenase, estradiol 17β- Estradiol 17β-dehydrogenase
1.1.1.152	Dehydrogenase, etiocholanolone 3α- 3α-Hydroxy-5β-androstan-17-one 3α-dehydrogenase
1.1.1.216	Dehydrogenase, farnesol (nicotinamide adenine dinucleotide phosphate) Farnesol dehydrogenase
1.2.1.46	Dehydrogenase, formaldehyde Formaldehyde dehydrogenase
1.2.1.1	Dehydrogenase, formaldehyde Formaldehyde dehydrogenase (glutathione)
1.2.1.2	Dehydrogenase, formate Formate dehydrogenase
1.2.2.1	Dehydrogenase, formate (cytochrome) Formate dehydrogenase (cytochrome)
1.2.1.43	Dehydrogenase, formate (nicotinamide adenine dinucleotide phosphate) Formate dehydrogenase ($NADP^+$)
1.2.99.5	Dehydrogenase, formylmethanofuran Formylmethanofuran dehydrogenase
1.5.1.6	Dehydrogenase, formyltetrahydrofolate Formyltetrahydrofolate dehydrogenase
1.1.99.11	Dehydrogenase, fructose 5- (acceptor) Fructose 5-dehydrogenase
1.1.1.124	Dehydrogenase, fructose 5-(nicotinamide adenine dinucleotide phosphate) Fructose 5-dehydrogenase ($NADP^+$)
1.1.1.16	Dehydrogenase, galactitol Galactitol 2-dehydrogenase

EC-No.	Synonym Recommended Name	Vol., Page
1.3.2.3	Dehydrogenase, galactonolactone Galactonolactone dehydrogenase	
1.1.1.120	Dehydrogenase, galactose (nicotinamide adenine dinucleotide phosphate) Galactose 1-dehydrogenase ($NADP^+$)	
1.3.1.36	Dehydrogenase, geissoschizine Geissoschizine dehydrogenase	
1.1.99.3	Dehydrogenase, gluconate 2- Gluconate 2-dehydrogenase	
1.1.1.69	Dehydrogenase, gluconate 5- Gluconate 5-dehydrogenase	
1.1.1.47	Dehydrogenase, glucose Glucose 1-dehydrogenase	
1.1.99.10	Dehydrogenase, glucose (acceptor) Glucose dehydrogenase (acceptor)	
1.1.99.10	Dehydrogenase, glucose (Aspergillus) Glucose dehydrogenase (acceptor)	
1.1.1.118	Dehydrogenase, glucose (nicotinamide adenine dinucleotide) Glucose 1-dehydrogenase (NAD^+)	
1.1.1.119	Dehydrogenase, glucose (nicotinamide adenine dinucleotide phosphate) Glucose 1-dehydrogenase ($NADP^+$)	
1.1.99.17	Dehydrogenase, glucose (pyrroloquinoline-quinone) Glucose dehydrogenase (pyrroloquinoline-quinone)	
1.4.1.2	Dehydrogenase, glutamate Glutamate dehydrogenase	
1.4.3.11	Dehydrogenase, glutamate (acceptor) L-Glutamate oxidase	
1.4.1.3	Dehydrogenase, glutamate (nicotinamide adenine dinucleotide (phosphate)) Glutamate dehydrogenase ($NAD(P)^+$)	
1.4.1.4	Dehydrogenase, glutamate (nicotinamide adenine dinucleotide phosphate) Glutamate dehydrogenase ($NADP^+$)	
1.2.1.41	Dehydrogenase, glutamate semialdehyde Glutamate-5-semialdehyde dehydrogenase	
1.2.1.20	Dehydrogenase, glutarate semialdehyde Glutarate-semialdehyde dehydrogenase	

EC-No.	Synonym Recommended Name	Vol., Page
1.8.5.1	Dehydrogenase, glutathione (ascorbate) Glutathione dehydrogenase (ascorbate)	
1.2.1.12	Dehydrogenase, glyceraldehyde phosphate Glyceraldehyde-3-phosphate dehydrogenase	
1.2.1.59	Dehydrogenase, glyceraldehyde phosphate (nicotinamide adenine dinucleotide (phosphate)) Glyceraldehyde-3-phosphate dehydrogenase (NAD(P)) (phosphorylating)	
1.2.1.9	Dehydrogenase, glyceraldehyde phosphate (nicotinamide adenine dinucleotide phosphate) Glyceraldehyde-3-phosphate dehydrogenase ($NADP^+$)	
1.2.1.13	Dehydrogenase, glyceraldehyde phosphate (nicotinamide adenine dinucleotide phosphate phosphorylating) Glyceraldehyde-3-phosphate dehydrogenase ($NADP^+$) (phosphorylating)	
1.1.1.29	Dehydrogenase, glycerate Glycerate dehydrogenase	
1.1.1.8	Dehydrogenase, glycerol phosphate Glycerol-3-phosphate dehydrogenase (NAD^+)	
1.1.99.5	Dehydrogenase, glycerol phosphate (acceptor) Glycerol-3-phosphate dehydrogenase	
1.1.1.94	Dehydrogenase, glycerol phosphate (nicotinamide adenine dinucleotide (phosphate)) Glycerol-3-phosphate dehydrogenase ($NAD(P)^+$)	
1.1.1.177	Dehydrogenase, glycerol phosphate (nicotinamide adenine dinucleotide phosphate) Glycerol-3-phosphate 1-dehydrogenase ($NADP^+$)	
1.4.1.10	Dehydrogenase, glycine Glycine dehydrogenase	
1.4.2.1	Dehydrogenase, glycine (cytochrome) Glycine dehydrogenase (cytochrome)	
1.1.1.185	Dehydrogenase, glycol (nicotinamide adenine dinucleotide (phosphate)) L-Glycol dehydrogenase	
1.2.1.21	Dehydrogenase, glycol aldehyde Glycolaldehyde dehydrogenase	
1.1.99.14	Dehydrogenase, glycolate Glycolate dehydrogenase	
1.1.1.135	Dehydrogenase, guanosine diphospho-6-deoxy-D-talose GDP-6-deoxy-D-talose 4-dehydrogenase	

EC-No.	Synonym Recommended Name	Vol., Page
1.1.1.132	Dehydrogenase, guanosine diphosphomannose GDPmannose 6-dehydrogenase	
1.2.1.42	Dehydrogenase, hexadecanal (acylating) Hexadecanal dehydrogenase (acylating)	
1.1.1.164	Dehydrogenase, hexadecanol Hexadecanol dehydrogenase	
1.1.1.23	Dehydrogenase, histidinol Histidinol dehydrogenase	
1.1.1.155	Dehydrogenase, homoisocitrate Homoisocitrate dehydrogenase	
1.12.1.2	Dehydrogenase, hydrogen Hydrogen dehydrogenase	
1.1.1.167	Dehydrogenase, hydroxymalonate Hydroxymalonate dehydrogenase	
1.2.1.8	Dehydrogenase, betaine aldehyde *β*ine-aldehyde dehydrogenase	
1.1.1.205	Dehydrogenase, inosinate IMP dehydrogenase	
1.1.1.18	Dehydrogenase, inositol 2- myo-Inositol 2-dehydrogenase	
1.1.1.223	Dehydrogenase, isopiperitenol Isopiperitenol dehydrogenase	
1.1.1.80	Dehydrogenase, isopropanol (nicotinamide adenine dinucleotide phosphate) Isopropanol dehydrogenase ($NADP^+$)	
1.1.99.4	Dehydrogenase, ketogluconate Dehydrogluconate dehydrogenase	
1.1.99.27	Dehydrogenase, L-(+)-pantoyllactone (R)-Pantolactone dehydrogenase (flavin)	
1.1.1.99	Dehydrogenase, L-2-hydroxy fatty acid (S)-2-Hydroxy-fatty-acid dehydrogenase	
1.1.1.157	Dehydrogenase, L-3-hydroxybutyryl coenzyme A (nicotinamide adenine dinucleotide phosphate) 3-Hydroxybutyryl-CoA dehydrogenase	
1.4.1.5	Dehydrogenase, L-amino acid L-Amino-acid dehydrogenase	

EC-No.	Synonym Recommended Name	Vol., Page
1.1.1.75	Dehydrogenase, L-aminopropanol (R)-Aminopropanol dehydrogenase	
1.1.1.12	Dehydrogenase, L-arabinitol L-Arabinitol 4-dehydrogenase	
1.1.1.13	Dehydrogenase, L-arabinitol (ribulose-forming) L-Arabinitol 2-dehydrogenase	
1.1.1.46	Dehydrogenase, L-arabinose L-Arabinose 1-dehydrogenase	
1.1.1.76	Dehydrogenase, L-butanediol (S,S)-Butanediol dehydrogenase	
1.1.1.122	Dehydrogenase, L-fucose D-threo-Aldose 1-dehydrogenase	
1.1.1.45	Dehydrogenase, L-gulonate L-Gulonate 3-dehydrogenase	
1.1.1.14	Dehydrogenase, L-iditol L-Iditol 2-dehydrogenase	
1.1.1.128	Dehydrogenase, L-idonate L-Idonate 2-dehydrogenase	
1.1.1.113	Dehydrogenase, L-xylose L-Xylose 1-dehydrogenase	
1.2.1.22	Dehydrogenase, lactaldehyde Lactaldehyde dehydrogenase	
1.1.1.27	Dehydrogenase, lactate L-Lactate dehydrogenase	
1.1.2.3	Dehydrogenase, lactate (cytochrome) L-Lactate dehydrogenase (cytochrome)	
1.4.1.9	Dehydrogenase, leucine Leucine dehydrogenase	
1.8.1.4	Dehydrogenase, lipoamide Dihydrolipoamide dehydrogenase	
1.1.1.211	Dehydrogenase, long-chain 3-hydroxyacyl coenzyme A Long-chain-3-hydroxyacyl-CoA dehydrogenase	
1.1.1.192	Dehydrogenase, long-chain alcohol Long-chain-alcohol dehydrogenase	
1.2.1.48	Dehydrogenase, long-chain aliphatic aldehyde Long-chain-aldehyde dehydrogenase	

EC-No.	Synonym Recommended Name	Vol., Page
1.1.99.16	Dehydrogenase, malate (acceptor) Malate dehydrogenase (acceptor)	
1.1.1.82	Dehydrogenase, malate (nicotinamide adenine dinucleotide phosphate) Malate dehydrogenase ($NADP^+$)	
1.2.1.15	Dehydrogenase, malonate semialdehyde Malonate-semialdehyde dehydrogenase	
1.1.1.17	Dehydrogenase, mannitol 1-phosphate Mannitol-1-phosphate 5-dehydrogenase	
1.1.1.131	Dehydrogenase, mannonate (nicotinamide adenine dinucleotide (phosphate)) Mannuronate reductase	
1.3.1.7	Dehydrogenase, meso-tartrate meso-Tartrate dehydrogenase	
1.1.1.244	Dehydrogenase, methanol Methanol dehydrogenase	
1.5.1.5	Dehydrogenase, methylenetetrahydrofolate Methylenetetrahydrofolate dehydrogenase ($NADP^+$)	
1.5.1.15	Dehydrogenase, methylenetetrahydrofolate (nicotinamide adenine dinucleotide) Methylenetetrahydrofolate dehydrogenase (NAD^+)	
1.5.99.9	Dehydrogenase, methylenetetrahydromethanopterin Methylenetetrahydromethanopterin dehydrogenase	
1.12.99.4	Dehydrogenase, methylenetetrahydromethanopterin N^5,N^{10}-Methenyltetrahydromethanopterin hydrogenase	
1.1.1.218	Dehydrogenase, morphine 6- Morphine 6-dehydrogenase	
1.1.1.240	Dehydrogenase, N-acetylhexosamine N-Acetylhexosamine 1-dehydrogenase	
1.1.1.233	Dehydrogenase, N-acylmannosamine N-Acylmannosamine 1-dehydrogenase	
1.5.1.13	Dehydrogenase, nicotinate Nicotinate dehydrogenase	
1.5.1.19	Dehydrogenase, nopaline D-Nopaline dehydrogenase	
1.5.1.11	Dehydrogenase, octopine D-Octopine dehydrogenase	

EC-No.	Synonym Recommended Name	Vol., Page
1.5.1.28	Dehydrogenase, opine Opine dehydrogenase	
1.2.4.2	Dehydrogenase, oxoglutarate Oxoglutarate dehydrogenase (lipoamide)	
1.1.1.106	Dehydrogenase, pantoate Pantoate 4-dehydrogenase	
1.1.1.144	Dehydrogenase, perillyl alcohol Perillyl-alcohol dehydrogenase	
1.2.1.39	Dehydrogenase, phenylacetaldehyde Phenylacetaldehyde dehydrogenase	
1.4.1.20	Dehydrogenase, phenylalanine Phenylalanine dehydrogenase	
1.2.1.58	Dehydrogenase, phenylglyoxylate (CoA-benzoylating) Phenylglyoxylate dehydrogenase (acylating)	
1.1.1.43	Dehydrogenase, phosphogluconate Phosphogluconate 2-dehydrogenase	
1.1.1.44	Dehydrogenase, phosphogluconate (decarboxylating) Phosphogluconate dehydrogenase (decarboxylating)	
1.1.1.95	Dehydrogenase, phosphoglycerate Phosphoglycerate dehydrogenase	
1.1.99.20	Dehydrogenase, polyethylene glycol Alkan-1-ol dehydrogenase (acceptor)	
1.1.99.23	Dehydrogenase, polyvinyl alcohol Polyvinyl-alcohol dehydrogenase (acceptor)	
1.1.3.30	Dehydrogenase, polyvinyl alcohol Polyvinyl-alcohol oxidase	
1.3.1.12	Dehydrogenase, prephenate Prephenate dehydrogenase	
1.3.1.13	Dehydrogenase, prephenate (nicotinamide adenine dinucleotide phosphate) Prephenate dehydrogenase ($NADP^+$)	
1.3.1.43	Dehydrogenase, pretyrosine Pretyrosine dehydrogenase	
1.1.1.7	Dehydrogenase, propanediol phosphate Propanediol-phosphate dehydrogenase	
1.1.1.231	Dehydrogenase, prostacyclin 15-Hydroxyprostaglandin-I dehydrogenase ($NADP^+$)	

EC-No.	Synonym Recommended Name	Vol., Page
1.1.1.196	Dehydrogenase, prostaglandin D_2 15-Hydroxyprostaglandin-D dehydrogenase ($NADP^+$)	
1.1.1.65	Dehydrogenase, pyridoxol 4- Pyridoxine 4-dehydrogenase	
1.1.99.9	Dehydrogenase, pyridoxol 5- Pyridoxine 5-dehydrogenase	
1.2.4.1	Dehydrogenase, pyruvate Pyruvate dehydrogenase (lipoamide)	
1.1.99.25	Dehydrogenase, quinate (pyrroloquinoline-quinone) Quinate dehydrogenase (pyrroloquinoline-quinone)	
1.3.99.17	Dehydrogenase, quinoline 2- Quinoline 2-oxidoreductase	
1.6.99.3	Dehydrogenase, reduced nicotinamide adenine dinucleotide NADH dehydrogenase	
1.6.5.5	Dehydrogenase, reduced nicotinamide adenine dinucleotide (phosphate) (quinone) NADPH:quinone reductase	
1.6.99.2	Dehydrogenase, reduced nicotinamide adenine dinucleotide (phosphate, quinone) NAD(P)H dehydrogenase (quinone)	
1.6.99.5	Dehydrogenase, reduced nicotinamide adenine dinucleotide (quinone) NADH dehydrogenase (quinone)	
1.6.99.1	Dehydrogenase, reduced nicotinamide adenine dinucleotide phosphate NADPH dehydrogenase	
1.6.99.6	Dehydrogenase, reduced nicotinamide adenine dinucleotide phosphate (quinone) NADPH dehydrogenase (quinone)	
1.2.1.36	Dehydrogenase, retinal Retinal dehydrogenase	
1.1.1.105	Dehydrogenase, retinol Retinol dehydrogenase	
1.1.1.137	Dehydrogenase, ribitol 5-phosphate Ribitol-5-phosphate 2-dehydrogenase	
1.1.1.115	Dehydrogenase, ribose (nicotinamide adenine dinucleotide phosphate) Ribose 1-dehydrogenase ($NADP^+$)	

EC-No.	Synonym Recommended Name Vol., Page
1.5.1.9	Dehydrogenase, saccharopine (nicotinamide adenine dinucleotide, glutamate-forming) Saccharopine dehydrogenase (NAD^+, L-glutamate-forming)
1.5.1.7	Dehydrogenase, saccharopine (nicotinamide adenine dinucleotide, lysine forming) Saccharopine dehydrogenase (NAD^+, L-lysine forming)
1.5.1.10	Dehydrogenase, saccharopine (nicotinamide adenine dinucleotide phosphate, glutamate-forming) Saccharopine dehydrogenase ($NADP^+$, L-glutamate forming)
1.5.1.8	Dehydrogenase, saccharopine (nicotinamide adenine dinucleotide phosphate, lysine-forming) Saccharopine dehydrogenase ($NADP^+$, L-lysine forming)
1.1.1.143	Dehydrogenase, sequoyitol Sequoyitol dehydrogenase
1.4.1.7	Dehydrogenase, serine Serine dehydrogenase
1.1.99.12	Dehydrogenase, sorbose Sorbose dehydrogenase
1.1.1.123	Dehydrogenase, sorbose (nicotinamide adenine dinucleotide phosphate) Sorbose 5-dehydrogenase ($NADP^+$)
1.5.1.22	Dehydrogenase, strombine Strombine dehydrogenase
1.3.99.1	Dehydrogenase, succinate Succinate dehydrogenase
1.2.1.24	Dehydrogenase, succinate semialdehyde Succinate semialdehyde dehydrogenase
1.2.1.16	Dehydrogenase, succinate semialdehyde (nicotinamide adenine dinucleotide (phosphate)) Succinate-semialdehyde dehydrogenase ($NAD(P)^+$)
1.8.2.1	Dehydrogenase, sulfite Sulfite dehydrogenase
1.1.1.93	Dehydrogenase, tartrate Tartrate dehydrogenase
1.4.99.2	Dehydrogenase, taurine Taurine dehydrogenase
1.1.1.63	Dehydrogenase, testosterone 17β- Testosterone 17β-dehydrogenase

EC-No.	Synonym Recommended Name	Vol., Page
1.1.1.64	Dehydrogenase, testosterone 17β- (nicotinamide adenine dinucleotide phosphate) Testosterone 17β-dehydrogenase ($NADP^+$)	
1.5.1.3	Dehydrogenase, tetrahydrofolate Dihydrofolate reductase	
1.1.1.129	Dehydrogenase, threonate L-Threonate 3-dehydrogenase	
1.1.1.103	Dehydrogenase, threonine 3- L-Threonine 3-dehydrogenase	
1.1.1.134	Dehydrogenase, thymidine diphospho-6-deoxy-L-talose dTDP-6-deoxy-L-talose 4-dehydrogenase	
1.3.1.20	Dehydrogenase, trans-1,2-dihydrobenzene-1,2-diol trans-1,2-Dihydrobenzene-1,2-diol dehydrogenase	
1.1.1.226	Dehydrogenase, trans-4-hydroxycyclohexanecarboxylate 4-Hydroxycyclohexanecarboxylate dehydrogenase	
1.10.1.1	Dehydrogenase, trans-acenaphthylene-1,2-diol trans-Acenaphthene-1,2-diol dehydrogenase	
1.2.1.47	Dehydrogenase, trimethylaminobutyraldehyde 4-Trimethylammoniobutyraldehyde dehydrogenase	
1.4.1.19	Dehydrogenase, tryptophan Tryptophan dehydrogenase	
1.1.99.19	Dehydrogenase, uracil Uracil dehydrogenase	
1.1.1.154	Dehydrogenase, ureidoglycolate Ureidoglycolate dehydrogenase	
1.1.1.136	Dehydrogenase, uridine diphosphoacetylglucosamine UDP-N-acetylglucosamine 6-dehydrogenase	
1.1.1.22	Dehydrogenase, uridine diphosphoglucose UDPglucose 6-dehydrogenase	
1.1.1.203	Dehydrogenase, uronate Uronate dehydrogenase	
1.4.1.8	Dehydrogenase, valine (nicotinamide adenine dinucleotide phosphate) Valine dehydrogenase ($NADP^+$)	
1.1.1.204	Dehydrogenase, xanthine Xanthine dehydrogenase	

EC-No.	Synonym Recommended Name	Vol., Page
1.3.1.55	Dehydrogenase 1,2-dihydro-1,2-dihydroxybenzoate (Acinetobacter strain ADP1 gene benD) cis-1,2-Dihydroxycyclohexa-3,5-diene-1-carboxylate dehydrogenase	
1.1.99.4	Dehydrogluconate dehydrogenase Dehydrogluconate dehydrogenase	
2.7.1.13	Dehydrogluconokinase Dehydrogluconokinase	
1.8.1.4	Dehydrolipoate dehydrogenase Dihydrolipoamide dehydrogenase	
1.1.1.169	2-Dehydropantoate 2-reductase 2-Dehydropantoate 2-reductase	
4.1.2.12	2-Dehydropantoate formaldehyde-lyase Ketopantoaldolase	
2.1.2.11	Dehydropantoate hydroxymethyltransferase 3-Methyl-2-oxobutanoate hydroxymethyltransferase	
1.1.99.27	2-Dehydropantolactone reductase (flavin) (R)-Pantolactone dehydrogenase (flavin)	
1.1.1.168	2-Dehydropantoyl-lactone reductase 2-Dehydropantoyl-lactone reductase (A-specific)	
1.1.1.168	2-Dehydropantoyl-lactone reductase (A-specific) 2-Dehydropantoyl-lactone reductase (A-specific)	
1.1.1.214	2-Dehydropantoyl-lactone reductase (B-specific) 2-Dehydropantoyl-lactone reductase (B-specific)	
1.1.99.27	2-Dehydropantoyl-lactone reductase (flavin) (R)-Pantolactone dehydrogenase (flavin)	
3.4.13.19	Dehydropeptidase I (DPH I) Membrane dipeptidase	
3.5.1.14	Dehydropeptidase II Aminoacylase	
4.2.1.86	16-Dehydroprogesterone hydratase 16-Dehydroprogesterone hydratase	
4.2.1.86	16-Dehydroprogesterone reductase 16-Dehydroprogesterone hydratase	
4.2.1.10	3-dehydroquinase 3-dehydroquinate dehydratase	

EC-No.	Synonym Recommended Name	Vol., Page
4.2.1.10	5-dehydroquinase 3-dehydroquinate dehydratase	
4.2.1.10	dehydroquinase 3-dehydroquinate dehydratase	
4.2.1.10	3-dehydroquinate dehydratase 3-dehydroquinate dehydratase	
4.2.1.10	5-dehydroquinate dehydratase 3-dehydroquinate dehydratase	
4.2.1.10	dehydroquinate dehydratase 3-dehydroquinate dehydratase	
4.2.1.10	3-dehydroquinate hydro-lyase 3-dehydroquinate dehydratase	
4.2.1.10	5-dehydroquinate hydro-lyase 3-dehydroquinate dehydratase	
4.2.3.4	3-dehydroquinate synthase 3-dehydroquinate synthase	
4.6.1.3	3-dehydroquinate synthase 3-dehydroquinate synthase	
4.2.3.4	5-dehydroquinate synthase 3-dehydroquinate synthase	
4.6.1.3	5-dehydroquinate synthase 3-dehydroquinate synthase	
4.2.3.4	dehydroquinate synthase 3-dehydroquinate synthase	
4.6.1.3	dehydroquinate synthase 3-dehydroquinate synthase	
4.2.3.4	3-dehydroquinate synthetase 3-dehydroquinate synthase	
4.6.1.3	3-dehydroquinate synthetase 3-dehydroquinate synthase	
4.6.1.3	dehydroquinate synthetase 3-dehydroquinate synthase	
4.6.1.3	3-dehydroquinic acid synthetase 3-dehydroquinate synthase	
4.2.3.4	5-dehydroquinic acid synthetase 3-dehydroquinate synthase	

EC-No.	Synonym Recommended Name	Vol., Page
4.6.1.3	5-dehydroquinic acid synthetase 3-dehydroquinate synthase	
1.5.1.27	1,2-Dehydroreticuline reductase 1,2-Dehydroreticulinium reductase ($NADPH_2$)	
1.5.1.27	1,2-Dehydroreticulinium ion reductase 1,2-Dehydroreticulinium reductase ($NADPH_2$)	
1.5.1.27	1,2-Dehydroreticulinium reductase ($NADPH_2$) 1,2-Dehydroreticulinium reductase ($NADPH_2$)	
1.1.1.25	5-Dehydroshikimate reductase Shikimate 5-dehydrogenase	
1.1.1.25	Dehydroshikimic reductase Shikimate 5-dehydrogenase	
1.1.1.102	3-Dehydrosphinganine reductase 3-Dehydrosphinganine reductase	
1.1.1.102	D-3-Dehydrosphinganine reductase 3-Dehydrosphinganine reductase	
4.2.1.98	Dehydroxylase, hydroxy steroid 16α 16α-Hydroxyprogesterone dehydratase	
3.5.3.12	Deiminase, agmatine Agmatine deiminase	
3.5.3.9	Deiminase, allantoate Allantoate deiminase	
3.5.3.6	Deiminase, arginine Arginine deiminase	
3.5.3.5	Deiminase, formiminoaspartate Formiminoaspartate deiminase	
3.5.3.21	Deiminase, methylenediurea Methylenediurea deaminase	
2.1.1.139	3'-demethoxy-3'-hydroxystaurosporine O-methyltransferase 3'-demethylstaurosporine O-methyltransferase	
2.1.1.94	11-O-Demethyl-17-O-deacetylvindoline 11-O-methyltransferase 11-O-Demethyl-17-O-deacetylvindoline O-methyltransferase	
2.1.1.94	11-O-Demethyl-17-O-deacetylvindoline O-methyltransferase 11-O-Demethyl-17-O-deacetylvindoline O-methyltransferase	
2.1.1.102	Demethylmacrocin O-methyltransferase Demethylmacrocin O-methyltransferase	

EC-No.	Synonym Recommended Name	Vol., Page
2.1.1.38	O-Demethylpuromycin methyltransferase O-Demethylpuromycin O-methyltransferase	
2.1.1.38	O-Demethylpuromycin O-methyltransferase O-Demethylpuromycin O-methyltransferase	
2.1.1.139	3'-demethylstaurosporine O-methyltransferase 3'-demethylstaurosporine O-methyltransferase	
2.1.1.109	Demethylsterigmatocystin 6-O-methyltransferase Demethylsterigmatocystin 6-O-methyltransferase	
2.1.1.64	3-Demethylubiquinone-9 3-O-methyltransferase 3-Demethylubiquinone-9 3-O-methyltransferase	
1.5.3.2	Demeythylase N-Methyl-L-amino-acid oxidase	
3.4.21.98	Den2 protease hepacivirin	
3.4.23.18	Denapsin Aspergillopepsin I	
3.4.23.18	Denapsin XP 271 Aspergillopepsin I	
1.1.3.4	Deoxin-1 Glucose oxidase	
5.99.1.2	Deoxiribonucleate topoisomerase DNA topoisomerase	Vol. 1, Page 721
5.99.1.2	Deoxiribonucleic topoisomerase DNA topoisomerase	Vol. 1, Page 721
3.1.21.2	Deoxriboendonuclease Deoxyribonuclease IV (Phage T_4-induced)	
2.7.4.11	(Deoxy)adenylate kinase (Deoxy)adenylate kinase	
2.7.4.13	(Deoxy)nucleoside-phosphate kinase (Deoxy)nucleoside-phosphate kinase	
3.1.6.9	4-Deoxy-β-D-gluc-4-enuronosyl-(1,4)-N-acetyl-D-galactosamine-4-sulfate 4-sulfohydrolase Chondro-4-sulfatase	
3.1.6.10	4-Deoxy-β-D-gluc-4-enuronosyl-(1,4)-N-acetyl-D-galactosamine-6-sulfate 6-sulfohydrolase Chondro-6-sulfatase	

EC-No.	Synonym Recommended Name	Vol., Page
3.2.1.112	2-Deoxy-α-D-glucosidase 2-Deoxyglucosidase	
3.2.1.112	2-Deoxy-α-D-glucoside deoxyglucohydrolase 2-Deoxyglucosidase	
3.2.1.124	3-Deoxy-2-octulosonidase 3-Deoxy-2-octulosonidase	
2.4.2.28	5'-Deoxy-5'-methylthioadenosine phosphorylase 5'-Methylthioadenosine phosphorylase	
4.2.1.41	D-4-Deoxy-5-ketoglucarate hydro-lyase 5-Dehydro-4-deoxyglucarate dehydratase	
4.2.1.41	D-4-Deoxy-5-ketoglucarate hydro-lyase (decarboxylating) 5-Dehydro-4-deoxyglucarate dehydratase	
4.2.1.41	4-Deoxy-5-oxoglucarate hydro-lyase (decarboxylating) 5-Dehydro-4-deoxyglucarate dehydratase	
4.2.1.41	D-4-Deoxy-5-oxoglucarate hydro-lyase (decarboxylating) 5-Dehydro-4-deoxyglucarate dehydratase	
4.2.3.4	3-deoxy-arabino-heptulosonate-7-phosphate phosphate-lyase (cyclizing) 3-dehydroquinate synthase	
3.5.4.12	Deoxy-CMP-deaminase dCMP deaminase	
3.6.1.12	Deoxy-CTPase dCTP diphosphatase	
2.2.1.4	1-Deoxy-D-altro-heptulose-7-phosphate synthase Acetoin-ribose-5-phosphate transaldolase	
2.2.1.4	1-Deoxy-D-altro-heptulose-7-phosphate synthetase Acetoin-ribose-5-phosphate transaldolase	
4.1.2.15	3-Deoxy-D-arabino-2-heptulosonic acid 7-phosphate synthetase 2-Dehydro-3-deoxyphosphoheptanoate aldolase	
4.1.2.15	3-Deoxy-D-arabino-heptolosonate-7-phosphate synthetase 2-Dehydro-3-deoxyphosphoheptanoate aldolase	
4.1.2.15	3-Deoxy-D-arabino-heptulosonate-7-phosphate synthase 2-Dehydro-3-deoxyphosphoheptanoate aldolase	
4.1.2.15	3-Deoxy-D-arabino-heptulosonate 7-phosphate synthetase 2-Dehydro-3-deoxyphosphoheptanoate aldolase	
4.1.2.15	Deoxy-D-arabinoheptulosonate-7-phosphate synthetase 2-Dehydro-3-deoxyphosphoheptanoate aldolase	

EC-No.	Synonym Recommended Name	Vol., Page
1.1.1.125	2-Deoxy-D-gluconate 3-dehydrogenase 2-Deoxy-D-gluconate 3-dehydrogenase	
1.1.1.125	2-Deoxy-D-gluconate:NAD$^+$ 3-oxidoreductase 2-Deoxy-D-gluconate 3-dehydrogenase	
3.1.3.68	2-Deoxy-D-glucose-6-phosphate phosphohydrolase 2-Deoxyglucose-6-phosphatase	
3.1.3.45	3-deoxy-D-manno-octulosonate-8-phosphate 8-phosphohydrolase 3-deoxy-manno-octulosonate-8-phosphatase	
3.1.3.45	3-deoxy-D-manno-octulosonate-8-phosphate phosphatase 3-deoxy-manno-octulosonate-8-phosphatase	
3.1.3.45	3-deoxy-D-manno-octulosonate-8-phosphate phosphohydrolase 3-deoxy-manno-octulosonate-8-phosphatase	
4.1.2.16	3-Deoxy-D-manno-octulosonate-8-phosphate synthase 2-Dehydro-3-deoxyphosphooctonate aldolase	
4.1.2.16	3-Deoxy-D-manno-octulosonate 8-phosphate synthetase 2-Dehydro-3-deoxyphosphooctonate aldolase	
4.1.2.23	3-deoxy-D-manno-octulosonate aldolase 3-deoxy-D-manno-octulosonate aldolase	
2.7.7.38	3-Deoxy-D-manno-octulosonate cytidylyltransferase 3-Deoxy-manno-octulosonate cytidylyltransferase	
4.1.2.23	3-deoxy-D-manno-octulosonate D-arabinose-lyase 3-deoxy-D-manno-octulosonate aldolase	
4.1.2.23	3-deoxy-D-manno-octulosonic aldolase 3-deoxy-D-manno-octulosonate aldolase	
4.1.2.16	3-Deoxy-D-mannooctulosonate-8-phosphate synthetase 2-Dehydro-3-deoxyphosphooctonate aldolase	
4.1.2.28	3-Deoxy-D-pentulosonic acid aldolase 2-Dehydro-3-deoxy-D-pentonate aldolase	
4.1.2.18	3-Deoxy-D-penulosonic acid aldolase 2-Dehydro-3-deoxy-L-pentonate aldolase	
3.2.2.17	Deoxy-D-ribocyclobutadipyridine polynucleotidodeoxyribohydrolase Deoxyribodipyrimidine endonucleosidase	
4.1.2.4	2-Deoxy-D-ribose-5-phosphate acetaldehyde-lyase Deoxyribose-phosphate aldolase	
3.1.5.1	Deoxy-GTPase dGTPase	

EC-No.	Synonym Recommended Name	Vol., Page
5.3.1.17	4-Deoxy-L-threo-5-hexosulose-uronate ketol-isomerase 4-Deoxy-L-threo-5-hexosulose-uronate ketol-isomerase	Vol. 1, Page 338
5.3.1.17	4-Deoxy-L-threo-5-hexosulose-uronate ketol-isomerase 4-Deoxy-L-threo-5-hexosulose-uronate ketol-isomerase	Vol. 1, Page 338
5.3.1.17	4-Deoxy-L-threo-5-hexosulose uronic acid isomerase 4-Deoxy-L-threo-5-hexosulose-uronate ketol-isomerase	Vol. 1, Page 338
3.1.3.45	3-deoxy-manno-octulosonate-8-phosphatase 3-deoxy-manno-octulosonate-8-phosphatase	
2.7.7.38	3-Deoxy-manno-octulosonate cytidylyltransferase 3-Deoxy-manno-octulosonate cytidylyltransferase	
3.5.4.4	Deoxyadenosine deaminase Adenosine deaminase	
2.7.1.76	Deoxyadenosine kinase Deoxyadenosine kinase	
2.3.1.74	6'-Deoxychalcone synthase Naringenin-chalcone synthase	
2.3.1.120	6'-deoxychalcone synthase 6'-deoxychalcone synthase	
2.1.2.8	DeoxyCMP hydroxymethylase Deoxycytidylate 5-hydroxymethyltransferase	
5.3.1.21	11-Deoxycorticosterone ketol-isomerase Corticosteroid side-chain-isomerase	Vol. 1, Page 345
3.5.4.12	Deoxycytidine-5'-monophosphate aminohydrolase dCMP deaminase	
3.5.4.12	Deoxycytidine-5'-phosphate deaminase dCMP deaminase	
2.7.1.74	Deoxycytidine-cytidine kinase Deoxycytidine kinase	
3.5.4.14	Deoxycytidine aminohydrolase Deoxycytidine deaminase	
3.5.4.14	Deoxycytidine deaminase Deoxycytidine deaminase	
2.7.1.74	2'-Deoxycytidine kinase Deoxycytidine kinase	
2.7.1.74	Deoxycytidine kinase Deoxycytidine kinase	

EC-No.	Synonym Recommended Name	Vol., Page
3.5.4.12	Deoxycytidine monophosphate deaminase dCMP deaminase	
2.7.4.14	Deoxycytidine monophosphokinase Cytidylate kinase	
3.6.1.12	Deoxycytidine triphosphatase dCTP diphosphatase	
3.5.4.13	Deoxycytidine triphosphate deaminase dCTP deaminase	
2.1.2.8	Deoxycytidylate 5-hydroxymethyltransferase Deoxycytidylate 5-hydroxymethyltransferase	
3.5.4.12	Deoxycytidylate aminohydrolase dCMP deaminase	
2.1.1.54	Deoxycytidylate C-methyltransferase Deoxycytidylate C-methyltransferase	
3.5.4.12	Deoxycytidylate deaminase dCMP deaminase	
2.1.2.8	Deoxycytidylate hydroxymethylase Deoxycytidylate 5-hydroxymethyltransferase	
2.1.2.8	Deoxycytidylate hydroxymethyltransferase Deoxycytidylate 5-hydroxymethyltransferase	
2.7.4.14	Deoxycytidylate kinase Cytidylate kinase	
2.7.4.5	deoxycytidylate kinase deoxycytidylate kinase	
2.1.1.54	Deoxycytidylate methyltransferase Deoxycytidylate C-methyltransferase	
2.1.2.8	Deoxycytidylic hydroxymethylase Deoxycytidylate 5-hydroxymethyltransferase	
1.1.1.125	2-Deoxygluconate dehydrogenase 2-Deoxy-D-gluconate 3-dehydrogenase	
3.1.3.68	2-Deoxyglucose-6-phosphatase 2-Deoxyglucose-6-phosphatase	
3.1.3.68	2-Deoxyglucose-6-phosphate phosphatase 2-Deoxyglucose-6-phosphatase	
3.1.3.68	2-Deoxyglucose 6-phosphate phosphatase 2-Deoxyglucose-6-phosphatase	

EC-No.	Synonym Recommended Name	Vol., Page
3.2.1.112	2-Deoxyglucosidase 2-Deoxyglucosidase	
3.1.5.1	Deoxyguanosine 5-triphosphate triphosphohydrolase dGTPase	
2.7.1.113	2'-Deoxyguanosine kinase Deoxyguanosine kinase	
2.7.1.113	Deoxyguanosine kinase Deoxyguanosine kinase	
3.1.5.1	Deoxyguanosine triphosphatase dGTPase	
3.1.5.1	Deoxyguanosine triphosphate triphosphohydrolase dGTPase	
2.7.4.8	Deoxyguanylate kinase Guanylate kinase	
1.14.99.29	Deoxyhypusine,hydrogen-donor:oxygen oxidoreductase (2-hydroxylating) Deoxyhypusine monooxygenase	
1.14.99.29	Deoxyhypusine hydroxylase Deoxyhypusine monooxygenase	
1.14.99.29	Deoxyhypusine monooxygenase Deoxyhypusine monooxygenase	
3.1.3.31	deoxyinosine-activated nucleotidase nucleotidase	
4.2.1.41	Deoxyketoglucarate dehydratase 5-Dehydro-4-deoxyglucarate dehydratase	
3.1.1.46	Deoxylimonate A-ring-lactonase Deoxylimonate A-ring-lactonase	
3.1.1.46	Deoxylimonate A-ring-lactonohydrolase Deoxylimonate A-ring-lactonase	
3.1.1.46	Deoxylimonate A-ring lactone hydrolase Deoxylimonate A-ring-lactonase	
2.7.7.7	Deoxynucleate polymerase DNA-directed DNA polymerase	
2.7.4.13	Deoxynucleoside-5'-monophosphate kinase (Deoxy)nucleoside-phosphate kinase	

EC-No.	Synonym Recommended Name	Vol., Page
2.7.7.7	Deoxynucleoside-triphosphate:DNA deoxynucleotidyltransferase (DNA-directed) DNA-directed DNA polymerase	
2.7.7.49	Deoxynucleoside-triphosphate:DNA deoxynucleotidyltransferase (RNA-directed) RNA-directed DNA polymerase	
2.7.4.13	Deoxynucleoside monophosphate kinase (Deoxy)nucleoside-phosphate kinase	
3.1.3.34	3'-deoxynucleotidase deoxynucleotide 3'-phosphatase	
3.1.3.34	deoxynucleotide 3'-phosphatase deoxynucleotide 3'-phosphatase	
2.7.4.12	Deoxynucleotide kinase T_2-induced deoxynucleotide kinase	
2.7.7.31	Deoxynucleotidyl terminal transferase DNA nucleotidylexotransferase	
3.2.1.144	3-deoxyoctulosonase 3-deoxyoctulosonase	
4.1.2.16	3-Deoxyoctulosonic 8-phosphate synthetase 2-Dehydro-3-deoxyphosphooctonate aldolase	
4.1.2.23	3-deoxyoctulosonic aldolase 3-deoxy-D-manno-octulosonate aldolase	
3.2.1.144	3-deoxyoctulosonyl-lipopolysaccharide hydrolase 3-deoxyoctulosonase	
4.2.1.41	Deoxyoxoglucarate dehydratase 5-Dehydro-4-deoxyglucarate dehydratase	
4.1.2.4	Deoxyriboaldolase Deoxyribose-phosphate aldolase	
4.1.99.3	deoxyribocyclobutadipyrimidine pyrimidine-lyase deoxyribodipyrimidine photolyase	
3.2.2.17	Deoxyribodipyrimidine endonucleosidase Deoxyribodipyrimidine endonucleosidase	
4.1.99.3	deoxyribodipyrimidine photolyase deoxyribodipyrimidine photolyase	
4.1.99.3	deoxyribodipyrimidine photolyase deoxyribodipyrimidine photolyase	

EC-No.	Synonym Recommended Name	Vol., Page
3.1.21.6	Deoxyriboendonuclease CC-preferring endodeoxyribonuclease	
2.7.1.15	Deoxyribokinase Ribokinase	
5.4.2.7	Deoxyribomutase Phosphopentomutase	Vol. 1, Page 535
4.2.99.18	Deoxyribonuclease DNA-(apurinic or apyrimidinic site) lyase	
3.1.21.1	Deoxyribonuclease Deoxyribonuclease I	
3.1.4.5	deoxyribonuclease deoxyribonuclease	
3.1.21.3	Deoxyribonuclease (ATP- and S-adenosyl-L-methionine dependent) Type I site-specific deoxyribonuclease	
3.1.21.3	Deoxyribonuclease (ATP-dependent) Type I site-specific deoxyribonuclease	
3.1.21.1	Deoxyribonuclease (pancreatic) Deoxyribonuclease I	
3.1.25.1	Deoxyribonuclease (pyrimidine dimer) Deoxyribonuclease (pyrimidine dimer)	
3.1.21.1	Deoxyribonuclease A Deoxyribonuclease I	
3.1.21.1	Deoxyribonuclease I Deoxyribonuclease I	
3.1.22.1	Deoxyribonuclease II Deoxyribonuclease II	
3.1.4.6	deoxyribonuclease II deoxyribonuclease II	
3.1.21.2	Deoxyribonuclease IV (Phage T_4-induced) Deoxyribonuclease IV (Phage T_4-induced)	
3.1.30.1	Deoxyribonuclease S1 Aspergillus nuclease S1	
3.1.22.3	Deoxyribonuclease V Deoxyribonuclease V	

EC-No.	Synonym Recommended Name	Vol., Page
3.1.22.5	Deoxyribonuclease X Deoxyribonuclease X	
3.1.22.1	Deoxyribonucleate 3'-nucleotidohydrolase Deoxyribonuclease II	
3.1.3.32	deoxyribonucleate 3'-phosphatase polynucleotide 3'-phosphatase	
3.1.11.4	deoxyribonucleate 5-dinucleotidase exodeoxyribonuclease (phage SP3-induced)	
6.5.1.1	Deoxyribonucleate ligase DNA ligase (ATP)	Vol. 2, Page 755
6.5.1.2	Deoxyribonucleate ligase DNA ligase (NAD^+)	Vol. 2, Page 773
2.1.1.37	Deoxyribonucleate methylase DNA (cytosine-5-)-methyltransferase	
2.1.1.37	Deoxyribonucleate methyltransferase DNA (cytosine-5-)-methyltransferase	
2.7.7.7	Deoxyribonucleate nucleotidyltransferase DNA-directed DNA polymerase	
4.1.99.3	deoxyribonucleate pyrimidine dimer lyase (photosensitive) deoxyribodipyrimidine photolyase	
2.1.1.37	Deoxyribonucleic (cytosine-5-)-methyltransferase DNA (cytosine-5-)-methyltransferase	
6.5.1.1	Deoxyribonucleic-joining enzyme DNA ligase (ATP)	Vol. 2, Page 755
6.5.1.2	Deoxyribonucleic-joining enzyme DNA ligase (NAD^+)	Vol. 2, Page 773
3.1.11.4	deoxyribonucleic 5-dinucleotidohydrolase exodeoxyribonuclease (phage SP3-induced)	
2.1.1.37	Deoxyribonucleic acid (cytosine-5-)-methyltransferase DNA (cytosine-5-)-methyltransferase	
2.7.7.6	Deoxyribonucleic acid-dependent ribonucleic acid polymerase DNA-directed RNA polymerase	
6.5.1.1	Deoxyribonucleic acid-joining enzyme DNA ligase (ATP)	Vol. 2, Page 755
2.7.7.7	Deoxyribonucleic acid duplicase DNA-directed DNA polymerase	

EC-No.	Synonym Recommended Name	Vol., Page
3.2.2.15	Deoxyribonucleic acid glycosylase DNA-deoxyinosine glycosidase	
6.5.1.1	Deoxyribonucleic acid joinase DNA ligase (ATP)	Vol. 2, Page 755
6.5.1.2	Deoxyribonucleic acid joinase DNA ligase (NAD^+)	Vol. 2, Page 773
6.5.1.1	Deoxyribonucleic acid ligase DNA ligase (ATP)	Vol. 2, Page 755
6.5.1.2	Deoxyribonucleic acid ligase DNA ligase (NAD^+)	Vol. 2, Page 773
2.1.1.37	Deoxyribonucleic acid methylase DNA (cytosine-5-)-methyltransferase	
2.1.1.37	Deoxyribonucleic acid methyltransferase DNA (cytosine-5-)-methyltransferase	
2.1.1.37	Deoxyribonucleic acid modification methylase DNA (cytosine-5-)-methyltransferase	
2.7.7.31	Deoxyribonucleic acid nucleotidyltransferase DNA nucleotidylexotransferase	
2.7.7.7	Deoxyribonucleic acid polymerase DNA-directed DNA polymerase	
6.5.1.1	Deoxyribonucleic acid repair enzyme DNA ligase (ATP)	Vol. 2, Page 755
4.1.99.3	deoxyribonucleic cyclobutane dipyrimidine photolyase deoxyribodipyrimidine photolyase	
2.7.7.7	Deoxyribonucleic duplicase DNA-directed DNA polymerase	
6.5.1.1	Deoxyribonucleic joinase DNA ligase (ATP)	Vol. 2, Page 755
6.5.1.2	Deoxyribonucleic joinase DNA ligase (NAD^+)	Vol. 2, Page 773
6.5.1.1	Deoxyribonucleic ligase DNA ligase (ATP)	Vol. 2, Page 755
6.5.1.2	Deoxyribonucleic ligase DNA ligase (NAD^+)	Vol. 2, Page 773
2.1.1.37	Deoxyribonucleic methylase DNA (cytosine-5-)-methyltransferase	

EC-No.	Synonym Recommended Name	Vol., Page
2.7.7.31	Deoxyribonucleic nucleotidyltransferase DNA nucleotidylexotransferase	
3.1.21.1	Deoxyribonucleic phosphatase Deoxyribonuclease I	
4.1.99.3	deoxyribonucleic photolyase deoxyribodipyrimidine photolyase	
2.7.7.7	Deoxyribonucleic polymerase DNA-directed DNA polymerase	
2.7.7.7	Deoxyribonucleic polymerase I DNA-directed DNA polymerase	
6.5.1.1	Deoxyribonucleic repair enzyme DNA ligase (ATP)	Vol. 2, Page 755
6.5.1.2	Deoxyribonucleic repair enzyme DNA ligase (NAD^+)	Vol. 2, Page 773
3.1.3.31	deoxyribonucleoside-activated nucleotidase nucleotidase	
1.17.4.1	2'-Deoxyribonucleoside-diphosphate:oxidized-thioredoxin 2'-oxidoreductase Ribonucleoside-diphosphate reductase	
1.17.4.2	2'-Deoxyribonucleoside-triphosphate:oxidized-thioredoxin 2'-oxidoreductase Ribonucleoside-triphosphate reductase	
2.7.4.13	Deoxyribonucleoside monophosphokinase (Deoxy)nucleoside-phosphate kinase	
3.1.3.34	deoxyribonucleotide 3'-phosphohydrolase deoxynucleotide 3'-phosphatase	
4.1.2.4	2-Deoxyribose-5-phosphate aldolase Deoxyribose-phosphate aldolase	
4.1.2.4	Deoxyribose-5-phosphate aldolase Deoxyribose-phosphate aldolase	
4.1.2.4	Deoxyribose-phosphate aldolase Deoxyribose-phosphate aldolase	
5.4.2.7	Deoxyribose phosphomutase Phosphopentomutase	Vol. 1, Page 535
2.4.2.6	Deoxyribose transferase Nucleoside deoxyribosyltransferase	

EC-No.	Synonym Recommended Name Vol., Page
2.4.2.6	Deoxyribosyltransferase, nucleoside Nucleoside deoxyribosyltransferase
3.6.1.39	Deoxythymidine-5'-triphosphatase Thymidine triphosphatase
2.7.4.9	Deoxythymidine 5'-monophosphate kinase dTMP kinase
2.7.1.21	2'-Deoxythymidine kinase Thymidine kinase
2.4.2.4	Deoxythymidine phosphorylase Thymidine phosphorylase
3.1.3.35	deoxythymidylate 5'-nucleotidase thymidylate 5'-phosphatase
3.1.3.35	deoxythymidylate phosphohydrolase thymidylate 5'-phosphatase
3.1.3.35	deoxythymidylic 5'-nucleotidase thymidylate 5'-phosphatase
1.14.11.10	2-Deoxyuridine,2-oxoglutarate:oxygen oxidoreductase (1'-hydroxylating) Pyrimidine-deoxynucleoside 1'-dioxygenase
1.14.11.3	2-Deoxyuridine,2-oxoglutarate:oxygen oxidoreductase (2'-hydroxylating) Pyrimidine-deoxynucleoside 2'-dioxygenase
1.14.11.3	Deoxyuridine 2'-hydroxylase Pyrimidine-deoxynucleoside 2'-dioxygenase
2.4.2.23	Deoxyuridine:orthophosphate deoxy-D-ribosyltransferase Deoxyuridine phosphorylase
2.4.2.23	Deoxyuridine phosphorylase Deoxyuridine phosphorylase
2.7.1.110	Dephospho-[reductase kinase] kinase Dephospho-[reductase kinase] kinase
2.7.1.24	3'-Dephospho-CoA kinase Dephospho-CoA kinase
2.7.1.24	Dephospho-CoA kinase Dephospho-CoA kinase
2.7.7.3	3'-Dephospho-CoA pyrophosphorylase Pantetheine-phosphate adenylyltransferase
2.7.7.3	Dephospho-CoA pyrophosphorylase Pantetheine-phosphate adenylyltransferase

EC-No.	Synonym Recommended Name	Vol., Page
2.7.7.3	Dephospho-coenzyme A pyrophosphorylase Pantetheine-phosphate adenylyltransferase	
2.7.1.24	Dephosphocoenzyme A kinase Dephospho-CoA kinase	
2.7.1.38	Dephosphophosphorylase kinase Phosphorylase kinase	
3.2.1.121	Depolymerase, polymannuronate Polymannuronate hydrolase	
3.4.21.34	depot-Padutin plasma kallikrein	
3.4.21.35	depot-Padutin tissue kallikrein	
4.1.2.4	DERA Deoxyribose-phosphate aldolase	
5.1.3.19	Dermatan-sulfate 5-epimerase Chondroitin-glucuronate 5-epimerase	Vol. 1, Page 172
1.14.99.25	Δ-6-Desaturase Linoleoyl-CoA desaturase	
1.14.99.25	Δ^6-desaturase Linoleoyl-CoA desaturase	
1.14.19.3	Δ^6-desaturase linoleoyl-CoA desaturase	
1.14.99.5	Δ^9-desaturase Stearoyl-CoA desaturase	
1.14.19.1	Δ^9-desaturase stearoyl-CoA 9-desaturase	
1.14.99.6	Desaturase, acyl-[acyl carrier protein] Acyl-[acyl-carrier-protein] desaturase	
1.14.99.5	Desaturase, acyl coenzyme A Stearoyl-CoA desaturase	
1.14.99.19	Desaturase, alkylacylglycerophosphorylethanolamine Plasmanylethanolamine desaturase	
1.14.99.25	Desaturase, fatty acid Δ^6- Linoleoyl-CoA desaturase	
1.14.99.5	Desaturase, fatty acid Δ^9- Stearoyl-CoA desaturase	

EC-No.	Synonym Recommended Name	Vol., Page
1.14.99.25	Desaturase, linoleate Linoleoyl-CoA desaturase	
1.3.1.35	Desaturase, oleate Phosphatidylcholine desaturase	
1.14.99.30	Desaturase, zeta-carotene Carotene 7,8-desaturase	
1.14.99.30	Desaturase, zeta-carotene (Capsicum annuum clone pCapZDS precursor reduced) Carotene 7,8-desaturase	
1.14.99.30	Desaturase, zeta-carotene (Nostoc muscorum clone pZDS1A reduced) Carotene 7,8-desaturase	
1.14.99.30	Desaturase, zeta-carotene (Synechocystis strain PCC 6803 clone cs0223/cs0128/ps0014/cs0681/cs0294 gene ctrQ-2 reduced) Carotene 7,8-desaturase	
1.14.15.6	Desmolase, steroid 20-22 Cholesterol monooxygenase (side-chain-cleaving)	
3.10.1.1	2-Desoxy-D-glucoside-2-sulphamate sulphohydrolase N-Sulfoglucosamine sulfohydrolase	
3.1.21.1	Desoxyribonuclease Deoxyribonuclease I	
3.6.1.23	Desoxyuridine-triphosphatase dUTP diphosphatase	
3.6.1.23	Desoxyuridine 5'-triphosphatase dUTP diphosphatase	
3.6.1.23	Desoxyuridine 5'-triphosphate nucleotidohydrolase dUTP diphosphatase	
6.3.3.3	Desthiobiotin synthetase Dethiobiotin synthase	Vol. 2, Page 542
4.4.1.3	desulfhydrase dimethylpropiothetin dethiomethylase	
4.4.1.1	desulfhydrase, cysteine cystathionine γ-lyase	
4.4.1.2	desulfhydrase, homocysteine homocysteine desulfhydrase	
4.1.1.12	Desulfinase Aspartate 4-decarboxylase	

EC-No.	Synonym Recommended Name	Vol., Page
1.8.99.3	Desulfofuscidin Hydrogensulfite reductase	
2.8.2.24	Desulfoglucosinolate sulfotransferase Desulfoglucosinolate sulfotransferase	
2.8.2.8	Desulfoheparin sulfotransferase Desulfoheparin sulfotransferase	
1.8.99.3	Desulforubidin Hydrogensulfite reductase	
1.8.99.3	Desulfoviridin Hydrogensulfite reductase	
2.8.1.6	Dethiobiotin:sulfur sulfurtransferase Biotin synthase	
6.3.3.3	Dethiobiotin synthase Dethiobiotin synthase	Vol. 2, Page 542
6.3.3.3	Dethiobiotin synthetase Dethiobiotin synthase	Vol. 2, Page 542
4.4.1.3	dethiomethylase, dimethylpropiothetin dimethylpropiothetin dethiomethylase	
3.4.24.39	Deuterolysin Deuterolysin	
3.2.1.115	Dextran α-1,2 debranching enzyme Branched-dextran exo-1,2-α-glucosidase	
2.4.1.5	Dextran-sucrase Dextransucrase	
3.2.1.95	Dextran 1,6-α-isomaltotriosidase Dextran 1,6-α-isomaltotriosidase	
3.2.1.11	Dextranase Dextranase	
3.2.1.94	Dextranase, isomalto- Glucan 1,6-α-isomaltosidase	
3.2.1.61	Dextranase, myco- Mycodextranase	
3.2.1.11	Dextranase DL 2 Dextranase	
2.4.1.2	Dextran dextrinase Dextrin dextranase	

EC-No.	Synonym Recommended Name	Vol., Page
3.2.1.70	Dextran glucosidase Glucan 1,6-α-glucosidase	
3.2.1.11	Dextran hydrolase Dextranase	
2.4.1.5	Dextransucrase Dextransucrase	
3.2.1.142	dextrin α-1,6-glucanohydrolase limit dextrinase	
3.2.1.33	Dextrin-1,6-glucosidase Amylo-1,6-glucosidase	
3.2.1.33	Dextrin 6-α-D-glucanohydrolase Amylo-1,6-glucosidase	
3.2.1.10	Dextrin 6-α-D-glucanohydrolase Oligo-1,6-glucosidase	
3.2.1.33	Dextrin 6-α-D-glucosidase Amylo-1,6-glucosidase	
3.2.1.10	Dextrin 6-glucanohydrolase Oligo-1,6-glucosidase	
3.2.1.41	α-dextrin 6-glucanohydrolase α-dextrin endo-1,6-α-glucosidase	
2.4.1.2	Dextrin 6-glucosyltransferase Dextrin dextranase	
3.2.1.10	Dextrin 6.α.-glucanohydrolase Oligo-1,6-glucosidase	
3.2.1.10	Dextrinase, limit Oligo-1,6-glucosidase	
2.4.1.2	Dextrin dextranase Dextrin dextranase	
3.2.1.41	α-dextrin endo-1,6-α-glucosidase α-dextrin endo-1,6-α-glucosidase	
2.4.1.25	Dextrin glycosyltransferase 4-α-Glucanotransferase	
2.4.1.25	Dextrin transglycosylase 4-α-Glucanotransferase	
3.1.8.2	DFPase Diisopropyl-fluorophosphatase	

EC-No.	Synonym Recommended Name	Vol., Page
4.1.1.64	DGD 2,2-Dialkylglycine decarboxylase (pyruvate)	
2.7.1.107	DGK Diacylglycerol kinase	
2.7.1.107	DG kinase Diacylglycerol kinase	
2.4.1.208	DGlcDAG synthase Diglucosyl diacylglycerol synthase	
3.1.5.1	dGTPase dGTPase	
3.1.5.1	dGTP triphosphohydrolase dGTPase	
4.2.1.9	DHAD dihydroxy-acid dehydratase	
1.1.1.156	DHA oxidoreductase Glycerol 2-dehydrogenase ($NADP^+$)	
2.5.1.26	DHAP-AT Alkylglycerone-phosphate synthase	
1.8.5.1	DHA reductase Glutathione dehydrogenase (ascorbate)	
4.1.1.46	2,3-DHBA decarboxylase *o*-Pyrocatechuate decarboxylase	
4.1.1.46	DHBA decarboxylase *o*-Pyrocatechuate decarboxylase	
4.1.1.46	DHBD *o*-Pyrocatechuate decarboxylase	
1.3.1.28	2,3-DHB dehydrogenase 2,3-Dihydro-2,3-dihydroxybenzoate dehydrogenase	
1.3.1.25	DHBDH 1,6-Dihydroxycyclohexa-2,4-diene-1-carboxylate dehydrogenase	
1.5.3.12	DHBP Dihydrobenzophenanthridine oxidase	
6.3.2.14	DHBS synthase 2,3-Dihydroxybenzoate-serine ligase	Vol. 2, Page 478
6.2.1.28	DHCA-CoA ligase 3α,7α-Dihydroxy-5β-cholestanate-CoA ligase	Vol. 2, Page 326

EC-No.	Synonym Recommended Name	Vol., Page
4.2.1.52	DHDPS dihydrodipicolinate synthase	
4.1.1.54	DHF decarboxylase Dihydroxyfumarate decarboxylase	
1.5.1.3	DHFR Dihydrofolate reductase	
2.1.1.114	DHHB methyltransferase Hexaprenyldihydroxybenzoate methyltransferase	
4.1.2.25	DHNA Dihydroneopterin aldolase	
1.3.3.1	(DHO) dehydrogenase Dihydroorotate oxidase	
1.3.3.1	DHOdehase Dihydroorotate oxidase	
3.4.13.19	DHP-I Membrane dipeptidase	
4.1.1.55	DHP decarboxylase 4,5-Dihydroxyphthalate decarboxylase	
1.3.1.2	DHPDH Dihydropyrimidine dehydrogenase ($NADP^+$)	
1.6.99.7	DHPR Dihydropteridine reductase	
2.5.1.15	DHPS Dihydropteroate synthase	
4.2.1.10	DHQase 3-dehydroquinate dehydratase	
4.2.1.10	DHQ synthase 3-dehydroquinate dehydratase	
1.3.1.2	DHU dehydrogenase Dihydropyrimidine dehydrogenase ($NADP^+$)	
3.8.2.1	di-isopropyl-fluorophosphatase di-isopropyl-fluorophosphatase	
3.2.1.96	Di-N-acetylchitobiosyl β-N-acetylglucosaminidase Mannosyl-glycoprotein endo-β-N-acetyl glucosaminidase	

EC-No.	Synonym Recommended Name	Vol., Page
2.5.1.31	Di-trans,poly-cis-Decaprenyl-diphosphate:isopentenyl-diphosphate undecaprenylcistransferase Di-trans,poly-cis-Decaprenylcistransferase	
2.5.1.31	Di-trans,poly-cis-Decaprenylcistransferase Di-trans,poly-cis-Decaprenylcistransferase	
2.5.1.31	Di-trans,poly-cis-Undecaprenyl-diphosphate synthase Di-trans,poly-cis-Decaprenylcistransferase	
3.1.1.66	5-(3,4-Diacetoxybut-1-ynyl)-2,2'-bithiophene acetylhydrolase 5-(3,4-Diacetoxybut-1-ynyl)-2,2'-bithiophene deacetylase	
3.1.1.66	5-(3,4-Diacetoxybut-1-ynyl)-2,2'-bithiophene deacetylase 5-(3,4-Diacetoxybut-1-ynyl)-2,2'-bithiophene deacetylase	
3.1.1.66	3,4-Diacetoxybutinylbithiophene:4-acetate esterase 5-(3,4-Diacetoxybut-1-ynyl)-2,2'-bithiophene deacetylase	
1.1.1.5	Diacetyl reductase Acetoin dehydrogenase	
1.1.1.4	Diacetyl reductase (acetoin) (R,R)-Butanediol dehydrogenase	
3.1.1.26	1,2-diacyl-3-β-D-galactosyl-sn-glycerol acylhydrolase galactolipase	
2.3.1.73	1,2-Diacyl-sn-glycerol:sterol acyl transferase Diacylglycerol-sterol O-acyltransferase	
2.3.1.73	1,2-Diacyl-sn-glycerol:sterol O-acyltransferase Diacylglycerol-sterol O-acyltransferase	
2.3.1.73	Diacylglycerol-sterol O-acyltransferase Diacylglycerol-sterol O-acyltransferase	
2.4.1.46	1,2-Diacylglycerol 3-β-galactosyltransferase 1,2-Diacylglycerol 3-β-galactosyltransferase	
2.4.1.157	1,2-Diacylglycerol 3-glucosyltransferase 1,2-Diacylglycerol 3-glucosyltransferase	
2.7.1.107	Diacylglycerol:ATP kinase Diacylglycerol kinase	
2.3.1.20	1,2-Diacylglycerol acyltransferase Diacylglycerol O-acyltransferase	
2.3.1.20	Diacylglycerol acyltransferase Diacylglycerol O-acyltransferase	

EC-No.	Synonym Recommended Name Vol., Page
2.3.1.124	diacylglycerol acyltransferase diacylglycerol acyltransferase
2.7.8.2	Diacylglycerol choline phosphotransferase Diacylglycerol cholinephosphotransferase
2.7.8.2	Diacylglycerol cholinephosphotransferase Diacylglycerol cholinephosphotransferase
2.7.8.1	Diacylglycerol ethanolaminephosphotransferase Ethanolaminephosphotransferase
3.1.1.34	diacylglycerol hydrolase lipoprotein lipase
2.7.1.107	1,2-Diacylglycerol kinase Diacylglycerol kinase
2.7.1.107	Diacylglycerol kinase Diacylglycerol kinase
3.1.1.34	diacylglycerol lipase lipoprotein lipase
2.3.1.20	Diacylglycerol O-acyltransferase Diacylglycerol O-acyltransferase
3.1.1.51	12,13-diacylphorbate 12-acylhydrolase phorbol-diester hydrolase
3.1.1.51	diacylphorbate 12-hydrolase phorbol-diester hydrolase
3.6.1.41	Diadenosine 5',5'''-P1,P4-tetraphosphatase Bis(5'-nucleosyl)-tetraphosphatase (symmetrical)
2.7.7.53	Diadenosine 5',5'''-P1,P4-tetraphosphate α,β-phosphorylase (ADP-forming) ATP adenylyltransferase
3.6.1.29	Diadenosine 5,5-P1,P3triphosphatase Bis(5'-adenosyl)-triphosphatase
3.6.1.17	Diadenosine P1,P4-tetraphosphatase Bis(5'-nucleosyl)-tetraphosphatase (asymmetrical)
3.6.1.41	Diadenosine polyphosphate hydrolase Bis(5'-nucleosyl)-tetraphosphatase (symmetrical)
3.6.1.41	Diadenosinetetraphosphatase (symmetrical) Bis(5'-nucleosyl)-tetraphosphatase (symmetrical)

EC-No.	Synonym Recommended Name	Vol., Page
2.7.7.53	Diadenosinetetraphosphate $\alpha\beta$-phosphorylase ATP adenylyltransferase	
3.6.1.41	Diadenosine tetraphosphate hydrolase Bis(5'-nucleosyl)-tetraphosphatase (symmetrical)	
4.1.1.64	2,2-Dialkyl-2-amino acid-pyruvate aminotransferase 2,2-Dialkylglycine decarboxylase (pyruvate)	
4.1.1.64	Dialkylamino-acid decarboxylase (pyruvate) 2,2-Dialkylglycine decarboxylase (pyruvate)	
4.1.1.64	α-Dialkylamino acid aminotransferase 2,2-Dialkylglycine decarboxylase (pyruvate)	
4.1.1.64	α-Dialkyl amino acid transaminase 2,2-Dialkylglycine decarboxylase (pyruvate)	
4.1.1.64	α-Dialkylamino acid transaminase 2,2-Dialkylglycine decarboxylase (pyruvate)	
3.1.8.2	Dialkylfluorophosphatase Diisopropyl-fluorophosphatase	
4.1.1.64	2,2-Dialkylglycine carboxy-lyase (amino-transferring) 2,2-Dialkylglycine decarboxylase (pyruvate)	
4.1.1.64	2,2-Dialkylglycine decarboxylase (pyruvate) 2,2-Dialkylglycine decarboxylase (pyruvate)	
2.6.1.29	Diamine-ketoglutaric transaminase Diamine transaminase	
2.6.1.29	Diamine:2-oxoglutarate aminotransferase Diamine transaminase	
1.4.3.6	Diamine:O2 oxidoreductase (deaminating) Amine oxidase (copper-containing)	
2.3.1.57	Diamine acetyltransferase Diamine N-acetyltransferase	
2.6.1.29	Diamine aminotransferase Diamine transaminase	
2.3.1.57	Diamine N-acetyltransferase Diamine N-acetyltransferase	
2.6.1.29	Diamine transaminase Diamine transaminase	
3.2.2.23	2,6-Diamino-4-hydroxy-^{5}N-formamidopyrimidine-DNA glycosylase Formamidopyrimidine-DNA glycosidase	

EC-No.	Synonym Recommended Name	Vol., Page
3.2.2.23	2,6-Diamino-4-hydroxy-^{5}N-methyl-formamidopyrimidine-DNA glycosylase Formamidopyrimidine-DNA glycosidase	
3.5.4.26	2,5-Diamino-6-hydroxy-4-(5-phosphoribosylamino)-pyrimidine 2-amino-hydrolase Diaminohydroxyphosphoribosylaminopyrimidine deaminase	
2.6.1.8	Diamino-acid transaminase 2,5-Diaminovalerate transaminase	
2.6.1.8	Diamino acid aminotransferase 2,5-Diaminovalerate transaminase	
2.6.1.46	L-2,4-Diaminobutanoate:pyruvate aminotransferase Diaminobutyrate-pyruvate transaminase	
2.6.1.76	diaminobutyrate-2-oxoglutarate transaminase diaminobutyrate-2-oxoglutarate transaminase	
2.6.1.46	Diaminobutyrate-pyruvate aminotransferase Diaminobutyrate-pyruvate transaminase	
2.6.1.46	Diaminobutyrate-pyruvate transaminase Diaminobutyrate-pyruvate transaminase	
2.6.1.76	2,4-diaminobutyrate 4-aminotransferase diaminobutyrate-2-oxoglutarate transaminase	
2.6.1.76	L-2,4-diaminobutyrate:2-ketoglutarate 4-aminotransferase diaminobutyrate-2-oxoglutarate transaminase	
2.6.1.76	L-2,4-diaminobutyrate:2-oxoglutarate 4-aminotransferase diaminobutyrate-2-oxoglutarate transaminase	
2.6.1.46	L-Diaminobutyric acid transaminase Diaminobutyrate-pyruvate transaminase	
5.4.3.4	D-2,6-Diaminohexanoate 5,6-aminomutase D-Lysine 5,6-aminomutase	Vol. 1, Page 562
1.4.1.11	L-3,5-Diaminohexanoate dehydrogenase L-erythro-3,5-Diaminohexanoate dehydrogenase	
3.5.4.26	Diaminohydroxyphosphoribosylaminopyrimidine deaminase Diaminohydroxyphosphoribosylaminopyrimidine deaminase	
6.3.3.3	7,8-Diaminononanoate:carbon-dioxide cyclo-ligase (ADP-forming) Dethiobiotin synthase	Vol. 2, Page 542
2.6.1.62	7,8-Diaminononanoate transaminase Adenosylmethionine-8-amino-7-oxononanoate transaminase	

EC-No.	Synonym Recommended Name	Vol., Page
1.4.3.6	Diamino oxhydrase Amine oxidase (copper-containing)	
1.4.3.6	Diaminooxidase Amine oxidase (copper-containing)	
2.6.1.62	Diaminopelargonate synthase Adenosylmethionine-8-amino-7-oxononanoate transaminase	
2.6.1.62	7,8-Diaminopelargonic acid aminotransferase Adenosylmethionine-8-amino-7-oxononanoate transaminase	
2.6.1.8	2,5-Diaminopentanoate:2-oxoglutarate aminotransferase 2,5-Diaminovalerate transaminase	
1.4.1.12	2,4-Diaminopentanoate:NAD(P)$^+$ oxidoreductase (deaminating) 2,4-Diaminopentanoate dehydrogenase	
1.4.1.12	2,4-Diaminopentanoate dehydrogenase 2,4-Diaminopentanoate dehydrogenase	
1.4.1.12	2,4-Diaminopentanoic acid C_4 dehydrogenase 2,4-Diaminopentanoate dehydrogenase	
4.1.1.20	Diaminopimelate decarboxylase Diaminopimelate decarboxylase	
1.4.1.16	Diaminopimelate dehydrogenase Diaminopimelate dehydrogenase	
5.1.1.7	Diaminopimelate epimerase Diaminopimelate epimerase	Vol. 1, Page 27
6.3.2.13	Diaminopimelic-adding enzyme UDP-N-acetylmuramoylalanyl-D-glutamate-2,6-diaminopimelate ligase	Vol. 2, Page 473
4.1.1.20	Diaminopimelic acid decarboxylase Diaminopimelate decarboxylase	
5.1.1.7	Diaminopimelic acid epimerase Diaminopimelate epimerase	Vol. 1, Page 27
5.1.1.7	Diaminopimelic epimerase Diaminopimelate epimerase	Vol. 1, Page 27
4.3.1.15	Diaminopropionatase Diaminopropionate ammonia-lyase	
4.3.1.15	2,3-Diaminopropionate:ammonia-lyase Diaminopropionate ammonia-lyase	

EC-No.	Synonym Recommended Name	Vol., Page
4.3.1.15	2,3-Diaminopropionate ammonia-lyase Diaminopropionate ammonia-lyase	
4.3.1.15	Diaminopropionate ammonia-lyase Diaminopropionate ammonia-lyase	
4.3.1.15	L-Diaminopropionate ammonia-lyase Diaminopropionate ammonia-lyase	
4.3.1.15	α,β-Diaminopropionate ammonia-lyase Diaminopropionate ammonia-lyase	
2.3.1.58	2,3-Diaminopropionate N-oxalyltransferase 2,3-Diaminopropionate N-oxalyltransferase	
2.6.1.8	2,5-Diaminovalerate transaminase 2,5-Diaminovalerate transaminase	
3.1.3.31	DIAN nucleotidase	
1.6.99.5	D-Diaphorase NADH dehydrogenase (quinone)	
1.8.1.4	Diaphorase Dihydrolipoamide dehydrogenase	
1.6.99.2	Diaphorase NAD(P)H dehydrogenase (quinone)	
1.6.99.3	Diaphorase NADH dehydrogenase	
1.6.99.1	Diaphorase NADPH dehydrogenase	
1.11.1.14	Diarylpropane:oxygen,hydrogen-peroxide oxidoreductase (C-C-bond-cleaving) Diarylpropane peroxidase	
1.11.1.14	Diarylpropane oxygenase Diarylpropane peroxidase	
1.11.1.14	Diarylpropane peroxidase Diarylpropane peroxidase	
3.5.2.12	1,8-Diazacyclotetradecane-2,9-dione lactamhydrolase 6-Aminohexanoate-cyclic-dimer hydrolase	
3.4.21.75	Dibasic processing enzyme Furin	

EC-No.	Synonym Recommended Name	Vol., Page
1.3.1.60	dibenzothiophene dihydrodiol dehydrogenase dibenzothiophene dihydrodiol dehydrogenase	
2.1.1.26	DIB O-methyltransferase (3,5-diiodo-4-hydroxy-benzoic acid) Iodophenol O-methyltransferase	
3.5.5.6	3,5-Dibromo-4-hydroxybenzonitrile aminohydrolase Bromoxynil nitrilase	
1.6.6.7	Dibromopropylaminophenylazobenzoic azoreductase Azobenzene reductase	
1.14.12.15	1,4-Dicarboxybenzoate 1,2-dioxygenase Terephthalate 1,2-dioxygenase	
2.8.3.13	Dicarboxyl-CoA:dicarboxylic acid coenzyme A transferase Succinate-hydroxymethylglutarate CoA-transferase	
6.2.1.23	Dicarboxylate-CoA ligase Dicarboxylate-CoA ligase	Vol. 2, Page 308
6.2.1.23	ω-Dicarboxylate:CoA ligase (AMP-forming) Dicarboxylate-CoA ligase	Vol. 2, Page 308
5.5.1.11	2,4-Dichloro-2,5-dihydro-5-oxofuran-2-acetate lyase (decyclizing) Dichloromuconate cycloisomerase	Vol. 1, Page 716
2.3.1.114	3,4-Dichloroaniline N-malonyltransferase 3,4-Dichloroaniline N-malonyltransferase	
1.3.1.63	2,4-dichlorobenzoyl-CoA reductase 2,4-dichlorobenzoyl-CoA reductase	
4.5.1.3	Dichloromethane chloride-lyase Dichloromethane dehalogenase	
4.5.1.3	Dichloromethane dehalogenase Dichloromethane dehalogenase	
5.5.1.11	Dichloromuconate cycloisomerase Dichloromuconate cycloisomerase	Vol. 1, Page 716
5.5.1.11	Dichloromuconate cycloisomerase Dichloromuconate cycloisomerase	Vol. 1, Page 716
1.14.13.20	2,4-Dichlorophenol, NADPH:oxygen oxidoreductase (6-hydroxylating) 2,4-Dichlorophenol 6-monooxygenase	
1.14.13.20	2,4-Dichlorophenol 6-monooxygenase 2,4-Dichlorophenol 6-monooxygenase	
1.14.13.20	2,4-Dichlorophenol hydroxylase 2,4-Dichlorophenol 6-monooxygenase	

EC-No.	Synonym Recommended Name	Vol., Page
1.5.1.21	1,2-Didehydropipecolate reductase $\Delta^1$1-Piperideine-2-carboxylate reductase	
1.5.1.14	1,2-didehydropipecolate reductase 1,2-didehydropipecolate reductase	
1.5.1.21	1,2-Didehydropipecolic reductase $\Delta^1$1-Piperideine-2-carboxylate reductase	
3.1.1.45	dienelactone hydrolase carboxymethylenebutenolidase	
4.2.1.100	dienoyl-CoA hydratase cyclohexa-1,5-dienecarbonyl-CoA hydratase	
1.3.1.34	2,4-Dienoyl-CoA reductase (NADPH) 2,4-Dienoyl-CoA reductase (NADPH)	
1.1.1.229	Diethyl-(2R,3R)-2-methyl-3-hydroxysuccinate:$NADP^+$ 3-oxidoreductase Diethyl 2-methyl-3-oxosuccinate reductase	
1.1.1.229	Diethyl 2-methyl-3-oxosuccinate reductase Diethyl 2-methyl-3-oxosuccinate reductase	
1.16.1.2	Diferric-transferrin reductase Diferric-transferrin reductase	
1.16.1.2	Diferric transferrin reductase Diferric-transferrin reductase	
3.2.1.134	Difructose-anhydride synthase Difructose-anhydride synthase	
2.4.1.208	Diglucosyl diacylglycerol synthase Diglucosyl diacylglycerol synthase	
2.4.1.208	Diglucosyldiacylglycerol synthase Diglucosyl diacylglycerol synthase	
2.3.1.20	Diglyceride acyltransferase Diacylglycerol O-acyltransferase	
2.7.1.107	Diglyceride kinase Diacylglycerol kinase	
3.1.1.34	diglyceride lipase lipoprotein lipase	
2.3.1.20	Diglyceride O-acyltransferase Diacylglycerol O-acyltransferase	
3.4.13.18	Diglycinase Cytosol nonspecific dipeptidase	

EC-No.	Synonym Recommended Name	Vol., Page
3.5.3.20	Diguanidinobutanase Diguanidinobutanase	
3.5.3.20	1,4-Diguanidinobutane amidinohydrolase Diguanidinobutanase	
3.6.1.17	Diguanosine tetraphosphatase Bis(5'-nucleosyl)-tetraphosphatase (asymmetrical)	
3.6.1.17	Diguanosinetetraphosphatase (asymmetrical) Bis(5'-nucleosyl)-tetraphosphatase (asymmetrical)	
2.7.7.45	Diguanosine tetraphosphate synthetase Guanosine-triphosphate guanylyltransferase	
1.3.3.1	Dihydoorotic acid dehydrogenase Dihydroorotate oxidase	
2.1.1.42	*o*-Dihydric phenol *meta*-O-methyltransferase Luteolin O-methyltransferase	
2.1.1.42	*o*-Dihydric phenol methyltransferase Luteolin O-methyltransferase	
1.3.1.29	1,2-Dihydro-1,2-dihydroxynaphthalene dehydrogenase cis-1,2-Dihydro-1,2-dihydroxynaphthalene dehydrogenase	
1.3.1.28	2,3-Dihydro-2,3-dihydroxybenzoate: NAD^+ oxidoreductase 2,3-Dihydro-2,3-dihydroxybenzoate dehydrogenase	
1.3.1.28	2,3-Dihydro-2,3-dihydroxybenzoate dehydrogenase 2,3-Dihydro-2,3-dihydroxybenzoate dehydrogenase	
1.1.1.109	2,3-dihydro-2,3-dihydroxybenzoate dehydrogenase 2,3-dihydro-2,3-dihydroxybenzoate dehydrogenase	
3.3.2.1	2,3-Dihydro-2,3-dihydroxybenzoate synthase Isochorismatase	
1.3.1.56	2,3-dihydro-2,3-dihydroxybiphenyl dehydrogenase cis-2,3-dihydrobiphenyl-2,3-diol dehydrogenase	
4.2.1.49	4,5-Dihydro-4-oxo-5-imidazole-propanoate hydro-lyase Urocanate hydratase	
5.5.1.1	2,5-Dihydro-5-oxofuran-2-acetate lyase (decyclizing) Muconate cycloisomerase	Vol. 1, Page 660
2.7.6.3	7,8-Dihydro-6-hydroxymethylpterin pyrophosphokinase 2-Amino-4-hydroxy-6-hydroxymethyldihydropteridine pyrophosphokinase	

EC-No.	Synonym Recommended Name	Vol., Page
1.3.1.18	7,8-Dihydro-7,8-dihydroxykynurenate:NAD$^+$ oxidoreductase Kynurenate-7,8-dihydrodiol dehydrogenase	
1.3.1.14	L-5,6-Dihydro-orotate:NAD oxidoreductase Orotate reductase (NADH)	
1.3.1.14	Dihydro-orotic dehydrogenase Orotate reductase (NADH)	
4.1.1.63	3,4-Dihydrobenzoate decarboxylase Protocatechuate decarboxylase	
1.5.3.12	Dihydrobenzophenanthridine:oxygen oxidoreductase Dihydrobenzophenanthridine oxidase	
1.5.3.12	Dihydrobenzophenanthridine oxidase Dihydrobenzophenanthridine oxidase	
1.3.1.46	Dihydrobiochanin-A:NADP$^+$ Δ^2-oxidoreductase Biochanin-A reductase	
1.1.1.153	7,8-Dihydrobiopterin:NADP$^+$ oxidoreductase Sepiapterin reductase	
1.1.1.160	Dihydrobunolol dehydrogenase Dihydrobunolol dehydrogenase	
1.14.13.57	Dihydrochelirubine, NADPH2:oxygen oxidoreductase (12-hydroxylating) Dihydrochelirubine 12-monooxygenase	
1.14.13.57	Dihydrochelirubine 12-hydroxylase Dihydrochelirubine 12-monooxygenase	
1.14.13.57	Dihydrochelirubine 12-hydroxylase Dihydrochelirubine 12-monooxygenase	
1.14.13.57	Dihydrochelirubine 12-monooxygenase Dihydrochelirubine 12-monooxygenase	
1.6.99.3	Dihydrocodehydrogenase I dehydrogenase NADH dehydrogenase	
1.10.2.2	Dihydrocoenzyme Q-cytochrome c reductase Ubiquinol-cytochrome-c reductase	
1.3.1.4	4,5α-Dihydrocortisone:NADP$^+$ Δ^4-oxidoreductase Cortisone α-reductase	
1.3.1.3	4,5β-Dihydrocortisone:NADP$^+$ Δ^4-oxidoreductase Cortisone β-reductase	
3.1.1.35	3,4-dihydrocoumarin hydrolase dihydrocoumarin hydrolase	

EC-No.	Synonym Recommended Name	Vol., Page
3.1.1.35	dihydrocoumarin hydrolase dihydrocoumarin hydrolase	
3.1.1.35	dihydrocoumarin lactonohydrolase dihydrocoumarin hydrolase	
1.3.1.5	23,24-Dihydrocucurbitacin:NAD(P)$^+$ Δ^{23}-oxidoreductase Cucurbitacin $\Delta^2$3-reductase	
1.3.1.25	Dihydrodihydroxybenzoate dehydrogenase 1,6-Dihydroxycyclohexa-2,4-diene-1-carboxylate dehydrogenase	
1.3.1.20	Dihydrodiol dehydrogenase trans-1,2-Dihydrobenzene-1,2-diol dehydrogenase	
1.3.1.26	Dihydrodipicolinate reductase Dihydrodipicolinate reductase	
4.2.1.52	dihydrodipicolinate synthase dihydrodipicolinate synthase	
1.3.1.26	Dihydrodipicolinic acid reductase Dihydrodipicolinate reductase	
4.2.1.52	dihydrodipicolinic acid synthase dihydrodipicolinate synthase	
1.1.1.219	Dihydroflavanol 4-reductase Dihydrokaempferol 4-reductase	
1.5.1.4	dihydrofolate dehydrogenase dihydrofolate dehydrogenase	
6.3.4.17	Dihydrofolate formyltransferase Formate-dihydrofolate ligase	Vol. 2, Page 649
1.5.1.3	7,8-Dihydrofolate reductase Dihydrofolate reductase	
1.5.1.3	Dihydrofolate reductase Dihydrofolate reductase	
1.5.1.3	Dihydrofolate reductase:thymidylate synthase Dihydrofolate reductase	
6.3.2.12	Dihydrofolate synthase Dihydrofolate synthase	Vol. 2, Page 466
6.3.2.12	7,8-Dihydrofolate synthetase Dihydrofolate synthase	Vol. 2, Page 466
6.3.2.12	Dihydrofolate synthetase Dihydrofolate synthase	Vol. 2, Page 466

EC-No.	Synonym Recommended Name	Vol., Page
1.5.1.3	Dihydrofolic acid reductase Dihydrofolate reductase	
1.5.1.3	Dihydrofolic reductase Dihydrofolate reductase	
1.1.1.219	Dihydrokaempferol 4-reductase Dihydrokaempferol 4-reductase	
1.8.1.4	Dihydrolipoamide:NAD^+ oxidoreductase Dihydrolipoamide dehydrogenase	
1.8.1.4	Dihydrolipoamide dehydrogenase Dihydrolipoamide dehydrogenase	
1.6.4.3	dihydrolipoamide reductase (NAD^+) dihydrolipoamide reductase (NAD^+)	
2.3.1.12	Dihydrolipoamide S-acetyltransferase Dihydrolipoamide S-acetyltransferase	
2.3.1.61	Dihydrolipoamide S-succinyltransferase Dihydrolipoamide S-succinyltransferase	
2.3.1.61	Dihydrolipoamide succinyltransferase Dihydrolipoamide S-succinyltransferase	
2.3.1.12	Dihydrolipoate acetyltransferase Dihydrolipoamide S-acetyltransferase	
1.8.1.4	Dihydrolipoic dehydrogenase Dihydrolipoamide dehydrogenase	
2.3.1.12	Dihydrolipoic transacetylase Dihydrolipoamide S-acetyltransferase	
2.3.1.61	Dihydrolipoic transsuccinylase Dihydrolipoamide S-succinyltransferase	
2.3.1.61	Dihydrolipolyl transsuccinylase Dihydrolipoamide S-succinyltransferase	
2.3.1.12	Dihydrolipoyl acetyltransferase Dihydrolipoamide S-acetyltransferase	
1.8.1.4	Dihydrolipoyl dehydrogenase Dihydrolipoamide dehydrogenase	
2.3.1.61	Dihydrolipoyl transsuccinylase Dihydrolipoamide S-succinyltransferase	
4.1.2.25	Dihydroneopterin aldolase Dihydroneopterin aldolase	

EC-No.	Synonym Recommended Name	Vol., Page
3.5.4.16	Dihydroneopterin triphosphate synthase GTP cyclohydrolase I	
1.6.5.3	Dihydronicotinamide adenine dinucleotide-coenzyme Q reductase NADH dehydrogenase (ubiquinone)	
1.6.2.2	Dihydronicotinamide adenine dinucleotide-cytochrome b_5 reductase Cytochrome-b_5 reductase	
1.18.1.1	Dihydronicotinamide adenine dinucleotide-rubredoxin reductase Rubredoxin-NAD^+ reductase	
1.6.99.3	Dihydronicotinamide adenine dinucleotide dehydrogenase NADH dehydrogenase	
1.6.99.1	Dihydronicotinamide adenine dinucleotide phosphate dehydrogenase NADPH dehydrogenase	
3.5.2.3	Dihydroorotase Dihydroorotase	
1.3.3.1	4,5-L-Dihydroorotate: oxygen oxidoreductase Dihydroorotate oxidase	
1.3.99.11	Dihydroorotate:ubiquinone oxidoreductase Dihydroorotate dehydrogenase	
1.3.99.11	Dihydroorotate dehydrogenase Dihydroorotate dehydrogenase	
1.3.3.1	Dihydroorotate dehydrogenase Dihydroorotate oxidase	
1.3.1.14	Dihydroorotate dehydrogenase Orotate reductase (NADH)	
3.5.2.3	Dihydroorotate hydrolase Dihydroorotase	
1.3.3.1	Dihydroorotate oxidase Dihydroorotate oxidase	
1.3.1.57	dihydrophloroglucinol:NADP oxidoreductase phloroglucinol reductase	
4.2.1.52	dihydropicolinate synthetase dihydrodipicolinate synthase	
1.6.99.7	Dihydropteridine reductase Dihydropteridine reductase	
1.6.99.7	Dihydropteridine reductase (NADH) Dihydropteridine reductase	

EC-No.	Synonym Recommended Name	Vol., Page
1.6.99.10	dihydropteridine reductase (NADH) dihydropteridine reductase (NADH)	
6.3.2.12	7,8-Dihydropteroate:L-glutamate ligase (ADP) Dihydrofolate synthase	Vol. 2, Page 466
6.3.2.12	Dihydropteroate:L-glutamate ligase (ADP-forming) Dihydrofolate synthase	Vol. 2, Page 466
2.5.1.15	Dihydropteroate pyrophosphorylase Dihydropteroate synthase	
2.5.1.15	7,8-Dihydropteroate synthase Dihydropteroate synthase	
2.5.1.15	Dihydropteroate synthase Dihydropteroate synthase	
2.5.1.15	7,8-Dihydropteroate synthetase Dihydropteroate synthase	
2.5.1.15	Dihydropteroate synthetase Dihydropteroate synthase	
2.5.1.15	7,8-Dihydropteroic acid synthetase Dihydropteroate synthase	
2.5.1.15	Dihydropteroic synthetase Dihydropteroate synthase	
3.5.2.2	5,6-Dihydropyridine amidohydrolase Dihydropyrimidinase	
3.5.2.2	Dihydropyrimidinase Dihydropyrimidinase	
1.3.1.1	Dihydropyrimidine dehydrogenase Dihydrouracil dehydrogenase (NAD^+)	
1.3.1.2	Dihydropyrimidine dehydrogenase ($NADP^+$) Dihydropyrimidine dehydrogenase ($NADP^+$)	
1.14.13.56	Dihydrosanguinarine,NADPH:oxygen oxidoreductase (10-hydroxylating) Dihydrosanguinarine 10-monooxygenase	
1.14.13.56	Dihydrosanguinarine 10-hydroxylase Dihydrosanguinarine 10-monooxygenase	
1.14.13.56	Dihydrosanguinarine 10-monooxygenase Dihydrosanguinarine 10-monooxygenase	
1.3.1.47	1,2-Dihydrosantonin:NAD(P)$^+$ 1,2-oxidoreductase α-Santonin 1,2-reductase	

EC-No.	Synonym Recommended Name	Vol., Page
4.1.2.27	Dihydrosphingosine 1-phosphate aldolase Sphinganine-1-phosphate aldolase	
2.7.1.91	Dihydrosphingosine kinase Sphinganine kinase	
2.7.1.88	Dihydrostreptomycin-6-phosphate 3'α-kinase Dihydrostreptomycin-6-phosphate 3'α-kinase	
2.4.2.27	Dihydrostreptosyltransferase, thymidine diphosphodihydrostreptose-streptidine 6-phosphate dTDPdihydrostreptose-streptidine-6-phosphate dihydrostreptosyltransferase	
1.8.1.4	Dihydrothioctic dehydrogenase Dihydrolipoamide dehydrogenase	
1.3.1.2	4,5-Dihydrothymine: oxidoreductase Dihydropyrimidine dehydrogenase ($NADP^+$)	
1.3.1.2	Dihydrothymine dehydrogenase Dihydropyrimidine dehydrogenase ($NADP^+$)	
1.3.1.1	Dihydrothymine dehydrogenase Dihydrouracil dehydrogenase (NAD^+)	
1.3.1.1	5,6-Dihydrouracil:NAD^+ oxidoreductase Dihydrouracil dehydrogenase (NAD^+)	
1.3.1.2	5,6-Dihydrouracil:$NADP^+$ oxidoreductase Dihydropyrimidine dehydrogenase ($NADP^+$)	
1.3.3.7	5,6-Dihydrouracil:oxygen oxidoreductase Dihydrouracil oxidase	
1.3.1.1	Dihydrouracil dehydrogenase (NAD^+) Dihydrouracil dehydrogenase (NAD^+)	
1.3.1.2	Dihydrouracil dehydrogenase (NADP) Dihydropyrimidine dehydrogenase ($NADP^+$)	
1.3.1.2	Dihydrouracil dehydrogenase ($NADP^+$) Dihydropyrimidine dehydrogenase ($NADP^+$)	
1.3.3.7	Dihydrouracil oxidase Dihydrouracil oxidase	
1.3.1.41	5,12-Dihydroxanthommatin:NAD^+ oxidoreductase Xanthommatin reductase	
1.13.11.16	2,3-Dihydroxy-β-phenylpropionate oxygenase 3-Carboxyethylcatechol 2,3-dioxygenase	

EC-No.	Synonym Recommended Name	Vol., Page
1.13.11.16	2,3-Dihydroxy-β-phenylpropionic dioxygenase 3-Carboxyethylcatechol 2,3-dioxygenase	
1.3.1.58	2,3-dihydroxy-2,3-dihydro-*p*-cumate dehydrogenase 2,3-dihydroxy-2,3-dihydro-*p*-cumate dehydrogenase	
3.3.2.1	2,3-Dihydroxy-2,3-dihydrobenzoate synthase Isochorismatase	
3.3.2.1	2,3-Dihydroxy-2,3-dihydrobenzoic synthase Isochorismatase	
1.3.1.65	5,6-dihydroxy-3-methyl-2-oxo-1,2,5,6-tetrahydroquinoline:NAD oxidoreductase 5,6-dihydroxy-3-methyl-2-oxo-1,2,5,6-tetrahydroquinoline dehydrogenase	
1.3.1.65	5,6-dihydroxy-3-methyl-2-oxo-1,2,5,6-tetrahydroquinoline dehydrogenase 5,6-dihydroxy-3-methyl-2-oxo-1,2,5,6-tetrahydroquinoline dehydrogenase	
6.2.1.28	3α,7α-Dihydroxy-5β-cholestanate-CoA ligase 3α,7α-Dihydroxy-5β-cholestanate-CoA ligase	Vol. 2, Page 326
6.2.1.28	3α,7α-Dihydroxy-5β-cholestanate:CoA ligase (AMP-forming) 3α,7α-Dihydroxy-5β-cholestanate-CoA ligase	Vol. 2, Page 326
2.1.1.114	3,4-Dihydroxy-5-hexaprenylbenzoate methyltransferase Hexaprenyldihydroxybenzoate methyltransferase	
1.3.1.59	1,6-dihydroxy-5-methylcyclohexa-2,4-dienecarboxylate:NAD oxidoreductase (decarboxylating) 1,6-dihydroxy-5-methylcyclohexa-2,4-dienecarboxylate dehydrogenase	
1.3.1.59	1,6-dihydroxy-5-methylcyclohexa-2,4-dienecarboxylate dehydrogenase 1,6-dihydroxy-5-methylcyclohexa-2,4-dienecarboxylate dehydrogenase	
1.3.1.68	1,2-dihydroxy-6-methylcyclohexa-3,5-dienecarboxylate:NAD oxidoreductase (decarboxylating) 1,2-dihydroxy-6-methylcyclohexa-3,5-dienecarboxylate dehydrogenase	
1.3.1.68	1,2-dihydroxy-6-methylcyclohexa-3,5-dienecarboxylate dehydrogenase 1,2-dihydroxy-6-methylcyclohexa-3,5-dienecarboxylate dehydrogenase	
2.4.1.202	2,4-Dihydroxy-7-methoxy-2H-1,4-benzoxazin-3(4H)-one 2-D-glucosyltransferase 2,4-Dihydroxy-7-methoxy-2H-1,4-benzoxazin-3(4H)-one 2-D-glucosyltransferase	
1.13.11.25	3,4-Dihydroxy-9,10-secoandrosta-1,3,5(10)-triene-9,17-dione 4,5-dioxygenase 3,4-Dihydroxy-9,10-secoandrosta-1,3,5(10)-triene-9,17-dione 4,5-dioxygenase	

EC-No.	Synonym Recommended Name	Vol., Page
1.13.11.25	3,4-Dihydroxy-9,10-secoandrosta-1,3,5(10)-triene-9,17-dione:oxygen 4,5-oxidoreductase (decyclizing) 3,4-Dihydroxy-9,10-secoandrosta-1,3,5(10)-triene-9,17-dione 4,5-dioxygenase	
1.1.1.141	11α,15-Dihydroxy-9-oxoprost-13-enoate:NAD^+ 15-oxidoreductase 15-Hydroxyprostaglandin dehydrogenase (NAD^+)	
4.2.1.9	dihydroxy-acid dehydratase dihydroxy-acid dehydratase	
4.2.1.9	2,3-dihydroxy-acid hydro-lyase dihydroxy-acid dehydratase	
2.6.1.49	3,4-Dihydroxy-L-phenylalanine:2-oxoglutarate aminotransferase Dihydroxyphenylalanine transaminase	
1.13.11.30	3,4-Dihydroxy-L-phenylalanine:oxygen 2,3-oxidoreductase (recyclizing) Stizolobinate synthase	
1.13.11.29	3,4-Dihydroxy-L-phenylalanine:oxygen 4,5-oxidoreductase (recyclizing) Stizolobate synthase	
4.3.1.11	3,4-Dihydroxy-L-phenylalanine ammonia-lyase Dihydroxyphenylalanine ammonia-lyase	
6.3.2.14	2,3-Dihydroxy-N-benzoyl-L-serine synthetase 2,3-Dihydroxybenzoate-serine ligase	Vol. 2, Page 478
1.13.11.22	3,4-Dihydroxy-trans-cinnamate:oxygen 3,4-oxidoreductase (decyclizing) Caffeate 3,4-dioxygenase	
2.5.1.26	Dihydroxyacetone-phosphate acyltransferase Alkylglycerone-phosphate synthase	
2.7.1.29	Dihydroxyacetone kinase Glycerone kinase	
2.3.1.42	Dihydroxyacetone phosphate acyltransferase Glycerone-phosphate O-acyltransferase	
1.1.1.156	Dihydroxyacetone reductase Glycerol 2-dehydrogenase ($NADP^+$)	
1.1.1.156	Dihydroxyacetone reductase (NADPH) Glycerol 2-dehydrogenase ($NADP^+$)	
2.2.1.3	Dihydroxyacetone synthase Formaldehyde transketolase	
2.2.1.2	Dihydroxyacetonetransferase Transaldolase	

EC-No.	Synonym Recommended Name	Vol., Page
1.13.11.41	2,4'-Dihydroxyacetophenone:oxygen oxidoreductase (C-C-bond-cleaving) 2,4'-Dihydroxyacetophenone dioxygenase	
1.13.11.41	2,4'-Dihydroxyacetophenone dioxygenase 2,4'-Dihydroxyacetophenone dioxygenase	
4.2.1.9	dihydroxy acid dehydrase dihydroxy-acid dehydratase	
4.2.1.9	α,β-dihydroxy acid dehydratase dihydroxy-acid dehydratase	
4.2.1.9	dihydroxy acid dehydratase dihydroxy-acid dehydratase	
4.2.1.9	dihydroxyacid dehydratase dihydroxy-acid dehydratase	
2.7.7.58	2,3-Dihydroxybenzoate-AMP ligase (2,3-Dihydroxybenzoyl)adenylate synthase	
6.3.2.14	2,3-Dihydroxybenzoate-serine ligase 2,3-Dihydroxybenzoate-serine ligase	Vol. 2, Page 478
1.13.11.14	2,3-Dihydroxybenzoate 1,2-dioxygenase 2,3-Dihydroxybenzoate 3,4-dioxygenase	
1.13.11.28	2,3-Dihydroxybenzoate 2,3-dioxygenase 2,3-Dihydroxybenzoate 2,3-dioxygenase	
1.13.11.28	2,3-Dihydroxybenzoate 2,3-oxygenase 2,3-Dihydroxybenzoate 2,3-dioxygenase	
1.13.11.14	2,3-Dihydroxybenzoate 3,4-dioxygenase 2,3-Dihydroxybenzoate 3,4-dioxygenase	
6.3.2.14	2,3-Dihydroxybenzoate:L-serine ligase 2,3-Dihydroxybenzoate-serine ligase	Vol. 2, Page 478
1.13.11.28	2,3-Dihydroxybenzoate:oxygen 2,3-oxidoreductase (decyclizing) 2,3-Dihydroxybenzoate 2,3-dioxygenase	
1.13.11.14	2,3-Dihydroxybenzoate: oxygen 3,4-oxidoreductase (decyclizing) 2,3-Dihydroxybenzoate 3,4-dioxygenase	
4.1.1.46	2,3-Dihydroxybenzoate carboxy-lyase *o*-Pyrocatechuate decarboxylase	
4.1.1.62	2,5-Dihydroxybenzoate decarboxylase Gentisate decarboxylase	
1.13.11.4	2,5-Dihydroxybenzoate dioxygenase Gentisate 1,2-dioxygenase	

EC-No.	Synonym Recommended Name	Vol., Page
1.13.11.14	2,3-Dihydroxybenzoate oxygenase 2,3-Dihydroxybenzoate 3,4-dioxygenase	
4.1.1.46	2,3-Dihydroxybenzoic acid decarboxylase *o*-Pyrocatechuate decarboxylase	
1.13.11.14	2,3-Dihydroxybenzoic oxygenase 2,3-Dihydroxybenzoate 3,4-dioxygenase	
6.3.2.14	N-(2,3-Dihydroxybenzoyl)-serine synthetase 2,3-Dihydroxybenzoate-serine ligase	Vol. 2, Page 478
2.7.7.58	(2,3-Dihydroxybenzoyl)adenylate synthase (2,3-Dihydroxybenzoyl)adenylate synthase	
6.3.2.14	2,3-Dihydroxybenzoylserine synthetase 2,3-Dihydroxybenzoate-serine ligase	Vol. 2, Page 478
1.13.11.39	2,3-Dihydroxybiphenyl dioxygenase Biphenyl-2,3-diol 1,2-dioxygenase	
2.4.1.104	*o*-Dihydroxycoumarin 7-O-glucosyltransferase *o*-Dihydroxycoumarin 7-O-glucosyltransferase	
1.3.1.25	1,6-Dihydroxycyclohexa-2,4-diene-1-carboxylate: NAD^+ oxidoreductase (decarboxylating) 1,6-Dihydroxycyclohexa-2,4-diene-1-carboxylate dehydrogenase	
1.3.1.25	1,6-Dihydroxycyclohexa-2,4-diene-1-carboxylate dehydrogenase 1,6-Dihydroxycyclohexa-2,4-diene-1-carboxylate dehydrogenase	
1.3.1.55	1,2-Dihydroxycyclohexa-3,5-diene dehydrogenase cis-1,2-Dihydroxycyclohexa-3,5-diene-1-carboxylate dehydrogenase	
1.1.1.166	Dihydroxycylo-hexanecarboxylate dehydrogenase Hydroxycyclohexanecarboxylate dehydrogenase	
4.1.1.54	Dihydroxyfumarate carboxy-lyase Dihydroxyfumarate decarboxylase	
4.1.1.54	Dihydroxyfumarate decarboxylase Dihydroxyfumarate decarboxylase	
2.1.1.114	Dihydroxyhexaprenylbenzoate methyltransferase Hexaprenyldihydroxybenzoate methyltransferase	
1.13.11.23	2,3-Dihydroxyindole 2,3-dioxygenase 2,3-Dihydroxyindole 2,3-dioxygenase	
1.13.11.23	2,3-Dihydroxyindole:oxygen 2,3-oxidoreductase (decyclizing) 2,3-Dihydroxyindole 2,3-dioxygenase	

EC-No.	Synonym Recommended Name	Vol., Page
4.2.1.9	2,3-dihydroxyisovalerate dehydratase dihydroxy-acid dehydratase	
4.2.1.9	α,β-dihydroxyisovalerate dehydratase dihydroxy-acid dehydratase	
1.1.1.86	Dihydroxyisovalerate dehydrogenase (isomerizing) Ketol-acid reductoisomerase	
1.1.1.89	dihydroxyisovalerate dehydrogenase (isomerizing) dihydroxyisovalerate dehydrogenase (isomerizing)	
1.13.11.10	7,8-Dihydroxykynurenate 8,8α-dioxygenase 7,8-Dihydroxykynurenate 8,8a-dioxygenase	
1.13.11.10	7,8-Dihydroxykynurenate 8,8a-dioxygenase 7,8-Dihydroxykynurenate 8,8a-dioxygenase	
1.13.11.10	7,8-Dihydroxykynurenate:oxygen 8,8a-oxidoreductase (decyclizing) 7,8-Dihydroxykynurenate 8,8a-dioxygenase	
1.13.11.10	7,8-Dihydroxykynurenate oxygenase 7,8-Dihydroxykynurenate 8,8a-dioxygenase	
1.3.1.18	7,8-Dihydroxykynurenic acid 7,8-diol dehydrogenase Kynurenate-7,8-dihydrodiol dehydrogenase	
2.7.6.3	7,8-Dihydroxymethylpterin-pyrophosphokinase 2-Amino-4-hydroxy-6-hydroxymethyldihydropteridine pyrophosphokinase	
4.1.3.36	dihydroxynaphthoate synthase naphthoate synthase	
1.6.2.4	Dihydroxynicotinamide adenine dinucleotide phosphate-cytochrome c reductase NADPH-ferrihemoprotein reductase	
1.14.17.1	(3,4-Dihydroxyphenethylamine)β-mono-oxygenase Dopamine β-monooxygenase	
1.14.17.1	3,4-Dihydroxyphenethylamine,ascorbate:oxygen oxidoreductase (β-hydroxylating) Dopamine β-monooxygenase	
1.14.17.1	3,4-Dihydroxyphenethylamine β-oxidase Dopamine β-monooxygenase	
4.3.1.11	β-(3,4-dihydroxyphenyl)-L-alanine (DOPA)ammonia-lyase Dihydroxyphenylalanine ammonia-lyase	
1.13.11.16	3-(2,3-Dihydroxyphenyl)propanoate:oxygen 1,2-oxidoreductase 3-Carboxyethylcatechol 2,3-dioxygenase	

EC-No.	Synonym Recommended Name	Vol., Page
1.13.11.16	3-(2-Dihydroxyphenyl)propanoate:oxygen 1,2-oxidoreductase (decyclizing) 3-Carboxyethylcatechol 2,3-dioxygenase	
1.13.11.15	3,4-Dihydroxyphenylacetate 2,3-dioxygenase 3,4-Dihydroxyphenylacetate 2,3-dioxygenase	
1.13.11.7	3,4-dihydroxyphenylacetate 3,4-dioxygenase 3,4-dihydroxyphenylacetate 3,4-dioxygenase	
1.13.11.15	3,4-Dihydroxyphenylacetate:oxygen 2,3-oxidoreductase (decyclizing) 3,4-Dihydroxyphenylacetate 2,3-dioxygenase	
1.13.11.15	3,4-Dihydroxyphenylacetic acid 2,3-dioxygenase 3,4-Dihydroxyphenylacetate 2,3-dioxygenase	
4.1.1.28	Dihydroxyphenylalanine-5-hydroxytryptophan decarboxylase Aromatic-L-amino-acid decarboxylase	
2.6.1.49	Dihydroxyphenylalanine aminotransferase Dihydroxyphenylalanine transaminase	
4.3.1.11	Dihydroxyphenylalanine ammonia-lyase Dihydroxyphenylalanine ammonia-lyase	
4.1.1.28	3,4-Dihydroxyphenylalanine decarboxylase Aromatic-L-amino-acid decarboxylase	
4.1.1.28	L-3,4-Dihydroxyphenylalanine decarboxylase Aromatic-L-amino-acid decarboxylase	
2.6.1.49	Dihydroxyphenylalanine transaminase Dihydroxyphenylalanine transaminase	
4.1.1.69	3,4-dihydroxyphthalate 2-decarboxylase 3,4-dihydroxyphthalate decarboxylase	
4.1.1.69	3,4-dihydroxyphthalate carboxy-lyase 3,4-dihydroxyphthalate decarboxylase	
4.1.1.55	4,5-Dihydroxyphthalate carboxy-lyase 4,5-Dihydroxyphthalate decarboxylase	
4.1.1.69	3,4-dihydroxyphthalate decarboxylase 3,4-dihydroxyphthalate decarboxylase	
4.1.1.55	4,5-Dihydroxyphthalate decarboxylase 4,5-Dihydroxyphthalate decarboxylase	
2.7.1.113	(Dihydroxypropoxymethyl)guanine kinase Deoxyguanosine kinase	
1.14.13.28	3,9-Dihydroxypterocarpan 6a-hydroxylase 3,9-Dihydroxypterocarpan 6a-monooxygenase	

EC-No.	Synonym Recommended Name	Vol., Page
1.14.13.28	3,9-Dihydroxypterocarpan 6a-monooxygenase 3,9-Dihydroxypterocarpan 6a-monooxygenase	
1.14.13.10	2,6-Dihydroxypyridine,NADH:oxygen oxidoreductase (3-hydroxylating) 2,6-Dihydroxypyridine 3-monooxygenase	
1.14.13.10	2,6-Dihydroxypyridine 3-monooxygenase 2,6-Dihydroxypyridine 3-monooxygenase	
1.13.11.9	2,5-Dihydroxypyridine 5,6-dioxygenase 2,5-Dihydroxypyridine 5,6-dioxygenase	
1.13.11.9	2,5-Dihydroxypyridine:oxygen 5,6-oxidoreductase 2,5-Dihydroxypyridine 5,6-dioxygenase	
1.14.13.10	2,6-Dihydroxypyridine oxidase 2,6-Dihydroxypyridine 3-monooxygenase	
1.13.11.9	2,5-Dihydroxypyridine oxygenase 2,5-Dihydroxypyridine 5,6-dioxygenase	
1.13.99.5	3,4-Dihydroxyquinoline 2,4-dioxygenase 3,4-Dihydroxyquinoline 2,4-dioxygenase	
1.1.1.242	Dihydrozeatin:$NADP^+$ oxidoreductase Zeatin reductase	
1.1.1.96	3-(3,5-Diiodo-4-hydroxyphenyl)lactate:NAD^+ oxidoreductase Diiodophenylpyruvate reductase	
2.6.1.24	3,5-Diiodo-L-tyrosine:2-oxoglutarate aminotransferase Diiodotyrosine transaminase	
1.1.1.96	Diiodophenylpyruvate reductase Diiodophenylpyruvate reductase	
3.8.1.4	Diiodothyronine 5'-deiodinase Thyroxine deiodinase	
2.6.1.24	Diiodotyrosine aminotransferase Diiodotyrosine transaminase	
2.6.1.24	Diiodotyrosine transaminase Diiodotyrosine transaminase	
3.1.8.2	Diisopropyl-fluorophosphatase Diisopropyl-fluorophosphatase	
3.1.8.2	Diisopropyl-fluorophosphate fluorohydrolase Diisopropyl-fluorophosphatase	
3.1.8.2	Diisopropylfluorophosphatase Diisopropyl-fluorophosphatase	

EC-No.	Synonym Recommended Name	Vol., Page
3.1.8.2	Diisopropylfluorophosphonate dehalogenase Diisopropyl-fluorophosphatase	
3.1.8.2	Diisopropylphosphofluoridase Diisopropyl-fluorophosphatase	
3.1.8.2	Diisopropyl phosphorofluoridate hydrolase Diisopropyl-fluorophosphatase	
1.14.15.2	2,5-Diketocamphane lactonizing enzyme Camphor 1,2-monooxygenase	
3.7.1.2	β-diketonase Fumarylacetoacetase	
3.7.1.7	β-Diketone hydrolase β-Diketone hydrolase	
3.4.21.35	Dilminal D tissue kallikrein	
3.4.21.34	dilminal D plasma kallikrein	
4.2.1.96	Dimerization factor of HNF1/pterin-4α-carbinolamine dehydratase 4a-Hydroxytetrahydrobiopterin dehydratase	
1.14.99.16	4,4-Dimethyl-5α-cholest-7-en-3β-ol,hydrogen-donor: oxygen oxidoreductase Methylsterol monooxygenase	
2.5.1.9	6,7-Dimethyl-8-(1-D-ribityl)lumazine:6,7-dimethyl-8-(1-D-ribityl)lumazine 2,3-butanediyltransferase Riboflavin synthase	
2.5.1.34	4-(γ,γ-Dimethylallyl)tryptophan synthase Tryptophan dimethylallyltransferase	
2.5.1.36	Dimethylallyl-diphosphate:(6aS,11aS)-3,6a,9-trihydroxypterocarpan dimethylallyltransferase Trihydroxypterocarpan dimethylallyltransferase	
2.5.1.35	Dimethylallyl-diphosphate:aspulvinone-E dimethylallyltransferase Aspulvinone dimethylallyltransferase	
2.5.1.28	Dimethylallyl-diphosphate:isopentenyl-diphosphate dimethylallylcistransferase Dimethylallylcistransferase	
2.5.1.1	Dimethylallyl-diphosphate:isopentenyl-diphosphate dimethylallyltranstransferase Dimethylallyltranstransferase	

EC-No.	Synonym Recommended Name	Vol., Page
2.5.1.34	Dimethylallyl-diphosphate:L-tryptophan dimethylallyltransferase Tryptophan dimethylallyltransferase	
2.5.1.28	Dimethylallylcistransferase Dimethylallylcistransferase	
2.5.1.36	Dimethylallylpyrophosphate:3,6a,9-trihydroxypterocarpan dimethylallyl-transferase Trihydroxypterocarpan dimethylallyltransferase	
2.5.1.35	Dimethylallyl pyrophosphate:aspulvinone dimethylallyltransferase Aspulvinone dimethylallyltransferase	
2.5.1.34	Dimethylallylpyrophosphate:L-tryptophan dimethylallyltransferase Tryptophan dimethylallyltransferase	
2.5.1.36	Dimethylallylpyrophosphate:trihydroxypterocarpan dimethylallyl transferase Trihydroxypterocarpan dimethylallyltransferase	
2.5.1.34	Dimethylallylpyrophosphate:tryptophan dimethylallyl transferase Tryptophan dimethylallyltransferase	
2.5.1.1	Dimethylallyl transferase Dimethylallyltranstransferase	
2.5.1.35	Dimethylallyltransferase, aspulvinone Aspulvinone dimethylallyltransferase	
2.5.1.39	Dimethylallyltransferase, *p*-hydroxybenzoate poly- 4-Hydroxybenzoate nonaprenyltransferase	
2.5.1.34	Dimethylallyltransferase, tryptophan Tryptophan dimethylallyltransferase	
2.5.1.1	Dimethylallyltranstransferase Dimethylallyltranstransferase	
2.5.1.34	Dimethylallyltryptophan synthetase Tryptophan dimethylallyltransferase	
2.5.1.36	Dimethylallytransferase, trihydroxypterocarpan Trihydroxypterocarpan dimethylallyltransferase	
1.5.99.10	Dimethylamine:(acceptor) oxidoreductase Dimethylamine dehydrogenase	
1.5.99.10	Dimethylamine dehydrogenase Dimethylamine dehydrogenase	
1.6.6.12	4-(Dimethylamino)phenylazobenzene reductase 4-(Dimethylamino)phenylazobenzene reductase	

EC-No.	Synonym Recommended Name	Vol., Page
1.6.6.7	*p*-Dimethylaminoazobenzene azoreductase Azobenzene reductase	
1.6.6.12	Dimethylaminoazobenzene N-oxide reductase 4-(Dimethylamino)phenylazobenzene reductase	
1.6.6.7	Dimethylaminobenzene reductase Azobenzene reductase	
3.5.3.18	Dimethylaminohydrolase, NG,N^g-dimethylarginine Dimethylargininase	
4.1.2.24	Dimethylaniline-N-oxide aldolase Dimethylaniline-N-oxide aldolase	
1.14.13.8	Dimethylaniline monooxygenase (N-oxide-forming) Dimethylaniline monooxygenase (N-oxide-forming)	
1.14.13.8	Dimethylaniline N-oxidase Dimethylaniline monooxygenase (N-oxide-forming)	
1.14.13.8	Dimethylaniline oxidase Dimethylaniline monooxygenase (N-oxide-forming)	
3.5.3.18	Dimethylargininase Dimethylargininase	
3.5.3.18	Dimethylarginine dimethylaminohydrolase Dimethylargininase	
1.5.99.2	Dimethylglycine dehydrogenase Dimethylglycine dehydrogenase	
1.5.3.10	Dimethylglycine oxidase Dimethylglycine oxidase	
2.1.1.44	Dimethylhistidine methyltransferase Dimethylhistidine N-methyltransferase	
2.1.1.44	Dimethylhistidine N-methyltransferase Dimethylhistidine N-methyltransferase	
1.1.1.84	Dimethylmalate dehydrogenase Dimethylmalate dehydrogenase	
1.1.1.84	β,β-Dimethylmalate dehydrogenase Dimethylmalate dehydrogenase	
4.1.3.32	2,3-Dimethylmalate lyase 2,3-Dimethylmalate lyase	
4.1.3.32	2,3-Dimethylmalate pyruvate-lyase 2,3-Dimethylmalate lyase	

EC-No.	Synonym Recommended Name	Vol., Page
4.2.1.85	Dimethylmaleate hydratase Dimethylmaleate hydratase	
1.14.99.28	3,7-Dimethylocta-1,6-dien-3-ol,hydrogen-donor:oxygen oxidoreductase (8-hydroxylating) Linalool 8-monooxygenase	
4.4.1.3	dimethylpropiothetin dethiomethylase dimethylpropiothetin dethiomethylase	
2.1.1.83	3,7-Dimethylquercitin 4'-O-methyltransferase 3,7-Dimethylquercitin 4'-O-methyltransferase	
2.1.1.3	Dimethylsulfonioacetate:L-homocysteine S-methyltransferase Thetin-homocysteine S-methyltransferase	
4.4.1.3	dimethylsulfoniopropionate lyase dimethylpropiothetin dethiomethylase	
2.1.1.3	Dimethylthetin-homocysteine methyltransferase Thetin-homocysteine S-methyltransferase	
3.2.2.24	Dinitrogenase reductase-activating glycohydrolase ADP-ribosyl-[dinitrogen reductase] hydrolase	
2.4.2.37	Dinitrogenase reductase ADP-ribosyltransferase NAD^+-dinitrogen-reductase ADP-D-ribosyltransferase	
2.6.1.26	3,5-Dinitrotyrosine transaminase Thyroid-hormone transaminase	
3.4.22.20	dinorphin-converting enzyme dinorphin-converting enzyme	
2.7.7.53	Dinucleoside oligophosphate α,β-phosphorylase ATP adenylyltransferase	
3.6.1.17	Dinucleoside tetraphosphatase Bis(5'-nucleosyl)-tetraphosphatase (asymmetrical)	
3.6.1.17	Dinucleosidetetraphosphatase (asymmetrical) Bis(5'-nucleosyl)-tetraphosphatase (asymmetrical)	
3.6.1.41	Dinucleosidetetraphosphatase (symmetrical) Bis(5'-nucleosyl)-tetraphosphatase (symmetrical)	
3.6.1.29	Dinucleoside triphosphatase Bis(5'-adenosyl)-triphosphatase	
3.6.1.9	Dinucleotide nucleotidohydrolase Nucleotide diphosphatase	

EC-No.	Synonym Recommended Name	Vol., Page
4.2.1.28	diol dehydrase propanediol dehydratase	
4.2.1.28	dioldehydrase propanediol dehydratase	
4.2.1.28	diol dehydratase propanediol dehydratase	
4.2.1.28	dioldehydratase propanediol dehydratase	
3.7.1.8	2,6-Dioxo-6-phenylhexa-3-enoate benzoylhydrolase 2,6-Dioxo-6-phenylhexa-3-enoate hydrolase	
3.7.1.8	2,6-Dioxo-6-phenylhexa-3-enoate hydrolase 2,6-Dioxo-6-phenylhexa-3-enoate hydrolase	
1.3.1.40	2,6-Dioxo-6-phenylhexanoate:NADP$^+$ Δ^2-oxidoreductase 2-Hydroxy-6-oxo-6-phenylhexa-2,4-dienoate reductase	
1.2.1.26	2,5-Dioxopentanoate:NADP$^+$ 5-oxidoreductase 2,5-Dioxovalerate dehydrogenase	
3.7.1.6	2,4-Dioxopentanoate acetylhydrolase Acetylpyruvate hydrolase	
3.5.2.13	2,5-Dioxopiperazine amidohydrolase 2,5-Dioxopiperazine hydrolase	
3.5.2.13	2,5-Dioxopiperazine hydrolase 2,5-Dioxopiperazine hydrolase	
4.6.1.10	6-(1,2-Dioxopropyl)tetrahydropterin synthetase 6-Pyruvoyltetrahydtropterin synthase	
2.4.2.20	2,4-Dioxotetrahydropyrimidine-nucleotide:pyrophosphate phospho-α-D-ribosyltransferase Dioxotetrahydropyrimidine phosphoribosyltransferase	
2.4.2.20	Dioxotetrahydropyrimidine-ribonucleotide pyrophosphorylase Dioxotetrahydropyrimidine phosphoribosyltransferase	
2.4.2.20	Dioxotetrahydropyrimidine phosphoribosyl transferase Dioxotetrahydropyrimidine phosphoribosyltransferase	
2.4.2.20	Dioxotetrahydropyrimidine phosphoribosyltransferase Dioxotetrahydropyrimidine phosphoribosyltransferase	
2.4.2.20	Dioxotetrahydropyrimidine ribonucleotide pyrophosphorylase Dioxotetrahydropyrimidine phosphoribosyltransferase	

EC-No.	Synonym Recommended Name	Vol., Page
2.6.1.43	4,5-Dioxovalerate aminotransferase Aminolevulinate transaminase	
2.6.1.43	γ,δ-Dioxovalerate aminotransferase Aminolevulinate transaminase	
1.2.1.26	2,5-Dioxovalerate dehydrogenase 2,5-Dioxovalerate dehydrogenase	
2.6.1.43	Dioxovalerate transaminase Aminolevulinate transaminase	
2.6.1.43	4,5-Dioxovaleric acid aminotransferase Aminolevulinate transaminase	
2.6.1.43	4,5-Dioxovaleric acid transaminase Aminolevulinate transaminase	
2.6.1.43	γ,δ-Dioxovaleric acid transaminase Aminolevulinate transaminase	
2.6.1.43	4,5-Dioxovaleric transaminase Aminolevulinate transaminase	
3.4.13.19	Dipeptidase Membrane dipeptidase	
3.4.13.11	dipeptidase dipeptidase	
3.4.19.5	Dipeptidase, β-aspartyl β-Aspartyl-peptidase	
3.4.17.21	Dipeptidase, acetylaspartylglutamate Glutamate carboxypeptidase II	
3.4.13.3	Dipeptidase, aminoacylhistidine Aminoacylhistidine dipeptidase	
3.4.13.5	dipeptidase, aminoacylmethylhistidine X-methyl-His dipeptidase	
3.4.13.18	Dipeptidase, glycylglycine Cytosol nonspecific dipeptidase	
3.4.15.4	Dipeptidase, peptidyl-, B Peptidyl-dipeptidase B	
3.4.13.9	dipeptidase, proline X-Pro dipeptidase	
3.4.13.12	dipeptidase M Met-X-dipeptidase	

EC-No.	Synonym Recommended Name	Vol., Page
3.4.14.1	dipeptide arylamidase I dipeptidyl-peptidase I	
3.4.13.19	Dipeptide hydrolase Membrane dipeptidase	
3.4.14.2	dipeptidyl-aminopeptidase II dipeptidyl-peptidase II	
3.4.14.5	dipeptidyl-aminopeptidase IV dipeptidyl-peptidase IV	
3.4.14.6	dipeptidyl-dipeptidase dipeptidyl-dipeptidase	
3.4.14.1	dipeptidyl-peptidase I dipeptidyl-peptidase I	
3.4.14.2	dipeptidyl-peptidase II dipeptidyl-peptidase II	
3.4.14.4	dipeptidyl-peptidase III dipeptidyl-peptidase III	
3.4.14.5	dipeptidyl-peptidase IV dipeptidyl-peptidase IV	
3.4.14.5	dipeptidyl-peptidase IV (CD26) dipeptidyl-peptidase IV	
3.4.14.1	dipeptidyl-peptide hydrolase dipeptidyl-peptidase I	
3.4.14.2	dipeptidyl-peptide hydrolase dipeptidyl-peptidase II	
3.4.14.4	dipeptidyl-peptide hydrolase dipeptidyl-peptidase III	
3.4.14.5	dipeptidyl-peptide hydrolase dipeptidyl-peptidase IV	
3.4.14.1	dipeptidyl aminopeptidase I dipeptidyl-peptidase I	
3.4.14.2	dipeptidyl aminopeptidase II dipeptidyl-peptidase II	
3.4.14.4	dipeptidyl aminopeptidase III dipeptidyl-peptidase III	
3.4.14.5	dipeptidyl aminopeptidase IV dipeptidyl-peptidase IV	

EC-No.	Synonym Recommended Name	Vol., Page
3.4.14.2	dipeptidyl arylamidase II dipeptidyl-peptidase II	
3.4.14.4	dipeptidyl arylamidase III dipeptidyl-peptidase III	
3.4.15.4	Dipeptidyl carboxyhydrolase Peptidyl-dipeptidase B	
3.4.15.5	Dipeptidyl carboxypeptidase Peptidyl-dipeptidase Dcp	
3.4.15.3	dipeptidyl carboxypeptidase dipeptidyl carboxypeptidase	
3.4.15.1	dipeptidyl carboxypeptidase peptidyl-dipeptidase A	
3.4.15.1	dipeptidyl carboxypeptidase I peptidyl-dipeptidase A	
3.4.13.19	Dipeptidyl hydrolase Membrane dipeptidase	
3.4.14.6	dipeptidyl ligase dipeptidyl-dipeptidase	
3.4.14.2	dipeptidyl peptidase II dipeptidyl-peptidase II	
3.4.14.2	dipeptidylpeptidase II dipeptidyl-peptidase II	
3.4.14.4	dipeptidylpeptidase III dipeptidyl-peptidase III	
3.4.14.6	dipeptidyl tetrapeptide hydrolase dipeptidyl-dipeptidase	
3.4.14.1	dipeptidyl transferase dipeptidyl-peptidase I	
1.14.18.1	o-Diphenol:O_2 oxidoreductase Monophenol monooxygenase	
1.10.3.1	o-Diphenol:oxygen oxidoreductase Catechol oxidase	
1.14.18.1	o-Diphenol:oxygen oxidoreductase Monophenol monooxygenase	
1.10.3.1	o-Diphenolase Catechol oxidase	

EC-No.	Synonym Recommended Name	Vol., Page
1.14.18.1	*o*-Diphenolase Monophenol monooxygenase	
1.10.3.1	Diphenol oxidase Catechol oxidase	
1.14.18.1	Diphenol oxidase Monophenol monooxygenase	
1.14.18.1	*o*-Diphenol oxidase Monophenol monooxygenase	
1.10.3.2	*p*-Diphenol oxidase Laccase	
1.10.3.1	*o*-Diphenol oxidoreductase Catechol oxidase	
1.14.18.1	*o*-Diphenol oxidoreductase Monophenol monooxygenase	
2.7.1.143	Diphosphate-purine nucleoside kinase Diphosphate-purine nucleoside kinase	
4.1.2.13	1,6-Diphosphofructose aldolase Fructose-bisphosphate aldolase	
4.1.2.13	Diphosphofructose aldolase Fructose-bisphosphate aldolase	
2.7.4.17	1,3-Diphosphoglycerate-polyphosphate phosphotransferase 3-Phosphoglyceroyl-phosphate-polyphosphate phosphotransferase	
2.7.4.17	1,3-Diphosphoglycerate:polyphosphate-phosphotransferase 3-Phosphoglyceroyl-phosphate-polyphosphate phosphotransferase	
5.4.2.4	2,3-Diphosphoglycerate mutase Bisphosphoglycerate mutase	Vol. 1, Page 520
5.4.2.4	Diphosphoglycerate mutase Bisphosphoglycerate mutase	Vol. 1, Page 520
3.6.1.7	1,3-Diphosphoglycerate phosphatase Acylphosphatase	
3.1.3.13	2,3-diphosphoglycerate phosphatase bisphosphoglycerate phosphatase	
3.1.3.13	diphosphoglycerate phosphatase bisphosphoglycerate phosphatase	
5.4.2.4	2,3-Diphosphoglycerate synthase Bisphosphoglycerate mutase	Vol. 1, Page 520

EC-No.	Synonym Recommended Name	Vol., Page
3.1.3.13	2,3-diphosphoglyceric acid phosphatase bisphosphoglycerate phosphatase	
5.4.2.4	Diphosphoglyceric mutase Bisphosphoglycerate mutase	Vol. 1, Page 520
5.4.2.4	2,3-Diphosphoglyceromutase Bisphosphoglycerate mutase	Vol. 1, Page 520
5.4.2.4	Diphosphoglyceromutase Bisphosphoglycerate mutase	Vol. 1, Page 520
2.7.1.68	Diphosphoinositide kinase 1-Phosphatidylinositol-4-phosphate 5-kinase	
3.1.3.36	Diphosphoinositide phosphatase Phosphatidylinositol-bisphosphatase	
4.1.1.33	Diphosphomevalonate decarboxylase Diphosphomevalonate decarboxylase	
1.6.99.3	Diphosphopyridine diaphorase NADH dehydrogenase	
3.2.2.5	Diphosphopyridine nucleosidase NAD^+ nucleosidase	
3.2.2.5	Diphosphopyridine nucleotidase NAD^+ nucleosidase	
1.11.1.1	Diphosphopyridine nucleotide peroxidase NADH peroxidase	
2.7.7.1	Diphosphopyridine nucleotide pyrophosphorylase Nicotinamide-nucleotide adenylyltransferase	
6.3.1.5	Diphosphopyridine nucleotide synthetase NAD^+ synthase	Vol. 2, Page 377
4.1.1.39	Diphosphoribulose carboxylase Ribulose-bisphosphate carboxylase	
6.3.2.22	Diphthamide synthase Diphthine-ammonia ligase	Vol. 2, Page 516
6.3.2.22	Diphthine-ammonia ligase Diphthine-ammonia ligase	Vol. 2, Page 516
6.3.2.22	Diphthine:ammonia ligase (ADP-forming) Diphthine-ammonia ligase	Vol. 2, Page 516
2.1.1.98	Diphthine methyltransferase Diphthine synthase	

EC-No.	Synonym Recommended Name	Vol., Page
2.1.1.98	Diphthine synthase Diphthine synthase	
2.5.1.1	Diprenyltransferase Dimethylallyltranstransferase	
4.1.99.3	dipyrimidine photolyase (photosensitive) deoxyribodipyrimidine photolyase	
2.4.1.7	Disaccharide glucosyltransferase Sucrose phosphorylase	
2.5.1.24	Discadenine synthase Discadenine synthase	
2.5.1.24	Discadenine synthetase Discadenine synthase	
1.2.99.4	Dismutase, formaldehyde Formaldehyde dismutase	
1.15.1.1	Dismutase, superoxide Superoxide dismutase	
2.4.1.25	Disproportionating enzyme 4-α-Glucanotransferase	
1.8.99.3	Dissimilatory-type sulfite reductase Hydrogensulfite reductase	
1.8.99.3	Dissimilatory sulfite reductase Hydrogensulfite reductase	
5.3.4.1	Disulfide interchange enzyme Protein disulfide-isomerase	Vol. 1, Page 436
1.6.4.4	Disulfide reductase Protein-disulfide reductase (NAD(P)H)	
3.1.6.11	6,N-Disulfoglucosamine 6-O-sulfohydrolase N-Sulfoglucosamine-6-sulfatase	
3.1.6.11	Disulfoglucosamine 6-sulfatase N-Sulfoglucosamine-6-sulfatase	
2.7.1.92	DKH kinase 5-Dehydro-2-deoxygluconokinase	
4.1.2.29	DKH phosphate aldolase Phospho-5-dehydro-2-deoxygluconate aldolase	
4.2.1.28	DL-1,2-propanediol hydro-lyase propanediol dehydratase	

EC-No.	Synonym Recommended Name	Vol., Page
1.1.1.75	DL-1-Aminopropan-2-ol: NAD$^+$ dehydrogenase (R)-Aminopropanol dehydrogenase	
3.8.1.2	DL-2-Haloacid dehalogenase 2-Haloacid dehalogenase	
4.1.3.16	DL-4-Hydroxy-2-ketoglutarate aldolase 4-Hydroxy-2-oxoglutarate aldolase	
3.1.3.21	DL-glycerol-3-phosphatase glycerol-1-phosphatase	
1.1.99.5	DL-Glycerol 3-phosphate oxidase Glycerol-3-phosphate dehydrogenase	
5.1.99.1	DL-Methylmalonyl-CoA racemase Methylmalonyl-CoA epimerase	Vol. 1, Page 179
5.1.99.1	DL-Methylmalonyl-coenzyme A racemase Methylmalonyl-CoA epimerase	Vol. 1, Page 179
3.2.1.11	DL 2 Dextranase	
3.1.1.45	DLH carboxymethylenebutenolidase	
3.1.1.45	DLHase carboxymethylenebutenolidase	
1.5.99.10	DMADH Dimethylamine dehydrogenase	
1.14.13.8	DMA oxidase Dimethylaniline monooxygenase (N-oxide-forming)	
2.5.1.1	DMAPP:IPP-dimethylallyltransferase Dimethylallyltranstransferase	
2.5.1.34	DMAT synthetase Tryptophan dimethylallyltransferase	
3.5.1.56	DMFase N,N-Dimethylformamidase	
4.4.1.3	DMSP lyase dimethylpropiothetin dethiomethylase	
3.1.3.31	DNA nucleotidase	
2.1.1.37	DNA (cytosine-5-)-methyltransferase DNA (cytosine-5-)-methyltransferase	

EC-No.	Synonym Recommended Name	Vol., Page
3.2.2.15	DNA(hypoxanthine)glycohydrolase DNA-deoxyinosine glycosidase	
4.2.99.18	DNA-(apurinic or apyrimidinic site) 5'-phosphomonoester-lyase DNA-(apurinic or apyrimidinic site) lyase	
4.2.99.18	DNA-(apurinic or apyrimidinic site) lyase DNA-(apurinic or apyrimidinic site) lyase	
3.2.2.20	DNA-3-methyladenine glycosidase I DNA-3-methyladenine glycosidase I	
3.2.2.21	DNA-3-methyladenine glycosidase II DNA-3-methyladenine glycosidase II	
4.3.2.4	DNA-4,6-diamino-5-formamidopyrimidine C^8-N^9-lyase Purine imidazole-ring cyclase	
2.1.1.63	DNA-6-O-methylguanine:protein-L-cysteine S-methyltransferase Methylated-DNA-protein-cysteine S-methyltransferase	
3.1.21.6	DNA-adenine-transferase CC-preferring endodeoxyribonuclease	
3.1.21.2	DNA-adenine-transferase Deoxyribonuclease IV (Phage T_4-induced)	
2.1.1.37	DNA-cytosine 5-methylase DNA (cytosine-5-)-methyltransferase	
2.1.1.37	DNA-cytosine methyltransferase DNA (cytosine-5-)-methyltransferase	
3.2.2.15	DNA-deoxyinosine deoxyribohydrolase DNA-deoxyinosine glycosidase	
3.2.2.15	DNA-deoxyinosine glycosidase DNA-deoxyinosine glycosidase	
2.7.7.7	DNA-dependent DNA polymerase DNA-directed DNA polymerase	
2.7.7.6	DNA-dependent ribonucleate nucleotidyltransferase DNA-directed RNA polymerase	
2.7.7.6	DNA-dependent RNA nucleotidyltransferase DNA-directed RNA polymerase	
2.7.7.6	DNA-dependent RNA polymerase DNA-directed RNA polymerase	
2.7.7.7	DNA-directed DNA polymerase DNA-directed DNA polymerase	

EC-No.	Synonym Recommended Name	Vol., Page
2.7.7.6	DNA-directed RNA polymerase DNA-directed RNA polymerase	
2.4.1.26	DNA α-glucosyltransferase DNA α-glucosyltransferase	
2.4.1.27	DNA β-glucosyltransferase DNA β-glucosyltransferase	
6.5.1.1	DNA-joining enzyme DNA ligase (ATP)	Vol. 2, Page 755
6.5.1.2	DNA-joining enzyme DNA ligase (NAD^+)	Vol. 2, Page 773
4.1.99.3	DNA-photoreactivating enzyme deoxyribodipyrimidine photolyase	
3.1.3.32	DNA 3'-phosphatase polynucleotide 3'-phosphatase	
3.1.4.31	DNA 5'-dinucleotidohydrolase DNA 5'-dinucleotidohydrolase	
2.7.1.78	DNA 5'-hydroxyl kinase Polynucleotide 5'-hydroxy-kinase	
2.1.1.37	DNA 5-cytosine methylase DNA (cytosine-5-)-methyltransferase	
3.1.11.4	DNA 5-dinucleotidohydrolase exodeoxyribonuclease (phage SP3-induced)	
2.1.1.113	DNA[cytosine-N4]methyltransferase Site-specific DNA-methyltransferase (cytosine-N^4-specific)	
3.1.21.1	DNAase Deoxyribonuclease I	
4.1.99.3	DNA cyclobutane dipyrimidine photolyase deoxyribodipyrimidine photolyase	
2.1.1.37	DNA cytosine c^5 methylase DNA (cytosine-5-)-methyltransferase	
2.1.1.37	DNA cytosine methylase DNA (cytosine-5-)-methyltransferase	
3.1.21.1	DNA depolymerase Deoxyribonuclease I	
2.7.7.7	DNA duplicase DNA-directed DNA polymerase	

EC-No.	Synonym Recommended Name	Vol., Page
3.1.21.1	DNA endonuclease Deoxyribonuclease I	
3.2.2.23	DNA glycohydrolase (releasing 2,6-diamino-4-hydroxy-5-(N-methyl)-formamidopyrimidine) Formamidopyrimidine-DNA glycosidase	
5.99.1.3	DNA gyrase DNA topoisomerase (ATP-hydrolysing)	Vol. 1, Page 737
6.5.1.1	DNA joinase DNA ligase (ATP)	Vol. 2, Page 755
6.5.1.2	DNA joinase DNA ligase (NAD^+)	Vol. 2, Page 773
2.7.1.78	DNA kinase Polynucleotide 5'-hydroxy-kinase	
6.5.1.1	DNA ligase DNA ligase (ATP)	Vol. 2, Page 755
6.5.1.2	DNA ligase DNA ligase (NAD^+)	Vol. 2, Page 773
6.5.1.1	DNA ligase (ATP) DNA ligase (ATP)	Vol. 2, Page 755
6.5.1.2	DNA ligase (NAD) DNA ligase (NAD^+)	Vol. 2, Page 773
6.5.1.2	DNA ligase (NAD^+) DNA ligase (NAD^+)	Vol. 2, Page 773
6.5.1.1	DNA ligase I DNA ligase (ATP)	Vol. 2, Page 755
6.5.1.1	DNA ligase II DNA ligase (ATP)	Vol. 2, Page 755
6.5.1.1	DNA ligase III DNA ligase (ATP)	Vol. 2, Page 755
2.1.1.37	DNA methylase DNA (cytosine-5-)-methyltransferase	
2.1.1.37	DNA methyltransferase DNA (cytosine-5-)-methyltransferase	
3.1.21.1	DNA nuclease Deoxyribonuclease I	

EC-No.	Synonym Recommended Name	Vol., Page
2.7.7.31	DNA nucleotidylexotransferase DNA nucleotidylexotransferase	
2.7.7.7	DNA nucleotidyltransferase DNA-directed DNA polymerase	
2.7.7.7	DNA nucleotidyltransferase (DNA-directed) DNA-directed DNA polymerase	
2.7.7.49	DNA nucleotidyltransferase (RNA-directed) RNA-directed DNA polymerase	
4.1.99.3	DNA photolyase deoxyribodipyrimidine photolyase	
2.7.7.7	DNA polmerase β DNA-directed DNA polymerase	
2.7.7.7	DNA polymerase DNA-directed DNA polymerase	
2.7.7.7	DNA polymerase α DNA-directed DNA polymerase	
2.7.7.7	DNA polymerase γ DNA-directed DNA polymerase	
2.7.7.7	DNA polymerase I DNA-directed DNA polymerase	
2.7.7.7	DNA polymerase II DNA-directed DNA polymerase	
2.7.7.7	DNA polymerase III DNA-directed DNA polymerase	
6.5.1.1	DNA repair enzyme DNA ligase (ATP)	Vol. 2, Page 755
6.5.1.2	DNA repair enzyme DNA ligase (NAD^+)	Vol. 2, Page 773
2.7.7.7	DNA replicase DNA-directed DNA polymerase	
3.1.21.1	DNase Deoxyribonuclease I	
3.1.11.4	DNase exodeoxyribonuclease (phage SP3-induced)	
3.1.21.3	Dnase (ATP-dependent) Type I site-specific deoxyribonuclease	

EC-No.	Synonym Recommended Name	Vol., Page
3.1.21.1	DNase I Deoxyribonuclease I	
3.1.22.1	DNase II Deoxyribonuclease II	
3.1.22.3	DNase V Deoxyribonuclease V	
5.99.1.2	DNA topoisomerase DNA topoisomerase	Vol. 1, Page 721
5.99.1.2	DNA topoisomerase DNA topoisomerase	Vol. 1, Page 721
5.99.1.3	DNA topoisomerase (ATP-hydrolysing) DNA topoisomerase (ATP-hydrolysing)	Vol. 1, Page 737
5.99.1.3	DNA topoisomerase (ATP-hydrolysing) DNA topoisomerase (ATP-hydrolysing)	Vol. 1, Page 737
5.99.1.3	DNA topoisomerase II DNA topoisomerase (ATP-hydrolysing)	Vol. 1, Page 737
5.99.1.3	DNA topoisomerase type II DNA topoisomerase (ATP-hydrolysing)	Vol. 1, Page 737
2.1.1.37	DNA transmethylase DNA (cytosine-5-)-methyltransferase	
2.3.1.74	DOCS Naringenin-chalcone synthase	
4.1.1.17	dODC Ornithine decarboxylase	
3.1.2.21	Dodecanoyl-[acyl-carrier-protein] hydrolase Dodecanoyl-[acyl-carrier protein] hydrolase	
3.1.2.21	Dodecanoyl-[acyl-carrier protein] hydrolase Dodecanoyl-[acyl-carrier protein] hydrolase	
3.1.2.21	Dodecanoyl-acyl-carrier-protein hydrolase Dodecanoyl-[acyl-carrier protein] hydrolase	
5.3.3.8	Dodecenoyl-CoA δ-isomerase Dodecenoyl-CoA δ-isomerase	Vol. 1, Page 413
5.3.3.8	Dodecenoyl-CoA Δ^3-cis-Δ^2-trans isomerase Dodecenoyl-CoA δ-isomerase	Vol. 1, Page 413
3.1.2.21	Dodecyl-acyl-carrier protein hydrolase Dodecanoyl-[acyl-carrier protein] hydrolase	

EC-No.	Synonym Recommended Name	Vol., Page
3.1.3.68	2-DOG-6P phosphatase 2-Deoxyglucose-6-phosphatase	
3.1.4.48	Dol-P-Glc phosphodiesterase Dolichyl-phosphate-glucose phosphodiesterase	
3.6.1.43	Dol-P-P phosphohydrolase Dolichyldiphosphatase	
3.1.3.51	Dol-P phosphatase Dolichyl-phosphatase	
2.7.1.108	Dolichol kinase Dolichol kinase	
3.1.3.51	Dolichol monophosphatase Dolichyl-phosphatase	
2.3.1.123	Dolichol O-acyltransferase Dolichol O-acyltransferase	
3.1.3.51	Dolichol phosphatase Dolichyl-phosphatase	
2.4.1.83	Dolichol phosphate mannose synthase Dolichyl-phosphate β-D-mannosyltransferase	
2.7.8.15	Dolichol phosphate N-acetylglucosamine-1-phosphotransferase UDP-N-acetylglucosamine-dolichyl-phosphate N-acetylglucosaminepho-sphotransferase	
3.1.3.51	Dolichol phosphate phosphatase Dolichyl-phosphatase	
2.7.1.108	Dolichol phosphokinase Dolichol kinase	
3.1.4.48	Dolichyl-β-D-glucosyl-phosphate dolichylphosphohydrolase Dolichyl-phosphate-glucose phosphodiesterase	
3.1.4.49	Dolichyl-β-D-mannosyl-phosphate dolichylphosphohydrolase Dolichyl-phosphate-mannose phosphodiesterase	
2.4.2.33	Dolichyl-D-xylosyl-phosphate:protein D-xylosyltransferase Dolichyl-xylosyl-phosphate-protein xylosyltransferase	
2.7.4.20	Dolichyl-diphosphate-polyphosphate phosphotransferase Dolichyl-diphosphate-polyphosphate phosphotransferase	
2.7.4.20	Dolichyl-diphosphate:polyphosphate phosphotransferase Dolichyl-diphosphate-polyphosphate phosphotransferase	

EC-No.	Synonym Recommended Name	Vol., Page
3.6.1.43	Dolichyl-diphosphate phosphatase Dolichyldiphosphatase	
3.6.1.43	Dolichyl-diphosphate phosphohydrolase Dolichyldiphosphatase	
2.4.1.119	Dolichyl-diphosphooligosaccharide-protein glycotransferase Dolichyl-diphosphooligosaccharide-protein glycotransferase	
2.4.1.119	Dolichyl-diphosphooligosaccharide:protein-L-asparagine oligopolysaccharidotransferase Dolichyl-diphosphooligosaccharide-protein glycotransferase	
3.1.3.51	Dolichyl-phosphatase Dolichyl-phosphatase	
2.4.1.130	Dolichyl-phosphate-D-mannose:glycolipid α-D-mannosyltransferase Dolichyl-phosphate-mannose-glycolipid α-mannosyltransferase	
2.4.1.109	Dolichyl-phosphate-D-mannose:protein O-D-mannosyltransferase Dolichyl-phosphate-mannose-protein mannosyltransferase	
2.4.1.83	Dolichyl-phosphate β-D-mannosyltransferase Dolichyl-phosphate β-D-mannosyltransferase	
3.1.4.48	Dolichyl-phosphate-glucose phosphodiesterase Dolichyl-phosphate-glucose phosphodiesterase	
2.4.1.117	Dolichyl-phosphate β-glucosyltransferase Dolichyl-phosphate β-glucosyltransferase	
2.4.1.130	Dolichyl-phosphate-mannose-glycolipid α-mannosyltransferase Dolichyl-phosphate-mannose-glycolipid α-mannosyltransferase	
2.4.1.109	Dolichyl-phosphate-mannose-protein mannosyltransferase Dolichyl-phosphate-mannose-protein mannosyltransferase	
3.1.4.49	Dolichyl-phosphate-mannose phosphodiesterase Dolichyl-phosphate-mannose phosphodiesterase	
2.4.1.153	Dolichyl-phosphate α-N-acetylglucosaminyltransferase Dolichyl-phosphate α-N-acetylglucosaminyltransferase	
2.4.2.32	Dolichyl-phosphate D-xylosyltransferase Dolichyl-phosphate D-xylosyltransferase	
2.4.1.83	Dolichyl-phosphate mannose synthase Dolichyl-phosphate β-D-mannosyltransferase	
3.1.3.51	Dolichyl-phosphate phosphohydrolase Dolichyl-phosphatase	

EC-No.	Synonym Recommended Name Vol., Page
2.4.1.83	Dolichyl-phospho-mannose synthase Dolichyl-phosphate β-D-mannosyltransferase
2.4.2.33	Dolichyl-xylosyl-phosphate-protein xylosyltransferase Dolichyl-xylosyl-phosphate-protein xylosyltransferase
3.6.1.43	Dolichyldiphosphatase Dolichyldiphosphatase
3.6.1.43	Dolichyl diphosphate phosphohydrolase Dolichyldiphosphatase
2.4.1.119	Dolichyldiphosphooligosaccharide-protein oligosaccharyltransferase Dolichyl-diphosphooligosaccharide-protein glycotransferase
2.4.1.83	Dolichyl mannosyl phosphate synthase Dolichyl-phosphate β-D-mannosyltransferase
3.1.3.51	Dolichyl monophosphate phosphatase Dolichyl-phosphatase
2.4.1.153	Dolichyl phosphate acetylglucosaminyltransferase Dolichyl-phosphate α-N-acetylglucosaminyltransferase
2.4.1.83	Dolichyl phosphate mannosyltransferase Dolichyl-phosphate β-D-mannosyltransferase
2.4.1.153	Dolichyl phosphate N-acetylglucosaminyltransferase Dolichyl-phosphate α-N-acetylglucosaminyltransferase
3.1.3.51	Dolichyl phosphate phosphatase Dolichyl-phosphatase
3.6.1.43	Dolichyl pyrophosphatase Dolichyldiphosphatase
2.7.4.20	Dolichylpyrophosphate:polyphosphate phosphotransferase Dolichyl-diphosphate-polyphosphate phosphotransferase
3.6.1.43	Dolichyl pyrophosphate phosphatase Dolichyldiphosphatase
1.97.1.3	(Donor):sulfur oxidoreductase Sulfur reductase
1.11.1.7	Donor:hydrogen-peroxide oxidoreductase Peroxidase
4.1.1.28	DOPA-5-hydroxytryptophan decarboxylase Aromatic-L-amino-acid decarboxylase
1.14.17.1	Dopa β-hydroxylase Dopamine β-monooxygenase

EC-No.	Synonym Recommended Name	Vol., Page
2.6.1.49	Dopa aminotransferase Dihydroxyphenylalanine transaminase	
4.3.1.11	DOPA ammonia-lyase Dihydroxyphenylalanine ammonia-lyase	
5.3.3.12	Dopachrome Δ-isomerase Dopachrome Δ-isomerase	Vol. 1, Page 432
5.3.3.12	Dopachrome-rearranging enzyme Dopachrome Δ-isomerase	Vol. 1, Page 432
5.3.3.12	Dopachrome Δ^7-Δ^2-isomerase Dopachrome Δ-isomerase	Vol. 1, Page 432
5.3.3.12	Dopachrome conversion factor Dopachrome Δ-isomerase	Vol. 1, Page 432
5.3.3.12	Dopachrome isomerase Dopachrome Δ-isomerase	Vol. 1, Page 432
5.3.3.12	Dopachrome oxidoreductase Dopachrome Δ-isomerase	Vol. 1, Page 432
5.3.3.12	Dopachrome tautomerase Dopachrome Δ-isomerase	Vol. 1, Page 432
4.1.1.28	DOPA DC Aromatic-L-amino-acid decarboxylase	
4.1.1.28	DOPA decarboxylase Aromatic-L-amino-acid decarboxylase	
4.1.1.26	DOPA decarboxylase DOPA decarboxylase	
4.1.1.28	L-DOPA decarboxylase Aromatic-L-amino-acid decarboxylase	
1.14.17.1	Dopamine-B-hydroxylase Dopamine β-monooxygenase	
1.14.17.1	Dopamine β-hydroxylase Dopamine β-monooxygenase	
1.14.17.1	Dopamine β-monooxygenase Dopamine β-monooxygenase	
1.14.17.1	Dopamine β-oxidase Dopamine β-monooxygenase	
1.14.17.1	Dopamine hydroxylase Dopamine β-monooxygenase	

EC-No.	Synonym Recommended Name	Vol., Page
2.8.2.1	Dopamine sulfotransferase Aryl sulfotransferase	
1.10.3.1	Dopa oxidase Catechol oxidase	
1.14.18.1	Dopa oxidase Monophenol monooxygenase	
2.6.1.49	Dopa transaminase Dihydroxyphenylalanine transaminase	
2.6.1.49	L-Dopa transaminase Dihydroxyphenylalanine transaminase	
3.1.21.1	Dornase Deoxyribonuclease I	
3.1.21.1	Dornava Deoxyribonuclease I	
3.1.21.1	Dornavac Deoxyribonuclease I	
3.4.21.73	Double-chain urokinase-type plasminogen activator u-Plasminogen activator	
2.6.1.43	DOVA transaminase Aminolevulinate transaminase	
1.3.1.2	DPD Dihydropyrimidine dehydrogenase ($NADP^+$)	
5.4.2.4	DPGM Bisphosphoglycerate mutase	Vol. 1, Page 520
3.2.2.5	DPNase NAD^+ nucleosidase	
1.6.5.3	DPNH-Coenzyme Q reductase NADH dehydrogenase (ubiquinone)	
1.6.99.5	DPNH-menadione reductase NADH dehydrogenase (quinone)	
1.18.1.1	DPNH-rubredoxin reductase Rubredoxin-NAD^+ reductase	
1.6.5.3	DPNH-ubiquinone reductase NADH dehydrogenase (ubiquinone)	
1.6.99.3	DPNH dehydrogenase NADH dehydrogenase	

EC-No.	Synonym Recommended Name	Vol., Page
1.6.99.3	DPNH diaphorase NADH dehydrogenase	
2.7.1.86	DPNH kinase NADH kinase	
1.11.1.1	DPNH peroxidase NADH peroxidase	
3.2.2.5	DPN hydrolase NAD^+ nucleosidase	
2.7.1.23	DPN kinase NAD^+ kinase	
3.1.3.7	DPNPase 3'(2'),5'-bisphosphate nucleotidase	
2.7.7.48	3D polymerase RNA-directed RNA polymerase	
3.4.14.1	DPPI dipeptidyl-peptidase I	
3.4.14.2	DPP II dipeptidyl-peptidase II	
3.4.14.5	DPP IV/CD26 dipeptidyl-peptidase IV	
4.1.2.4	DR aldolase Deoxyribose-phosphate aldolase	
4.1.2.15	DS-Co 2-Dehydro-3-deoxyphosphoheptanoate aldolase	
4.1.2.15	DS-Mn 2-Dehydro-3-deoxyphosphoheptanoate aldolase	
5.3.4.1	DsbA Protein disulfide-isomerase	Vol. 1, Page 436
5.3.4.1	DsbC Protein disulfide-isomerase	Vol. 1, Page 436
5.3.4.1	DsbD Protein disulfide-isomerase	Vol. 1, Page 436
4.2.1.14	DSD D-Serine dehydratase	
4.2.1.14	Dsdase D-Serine dehydratase	

EC-No.	Synonym Recommended Name	Vol., Page
1.1.1.102	DSR 3-Dehydrosphinganine reductase	
1.6.99.2	DT-Diaphorase NAD(P)H dehydrogenase (quinone)	
1.6.5.5	DT-diaphorase NADPH:quinone reductase	
6.3.3.3	DTBS Dethiobiotin synthase	Vol. 2, Page 542
6.3.3.3	DTB synthetase Dethiobiotin synthase	Vol. 2, Page 542
2.6.1.59	dTDP-4,6-dideoxy-D-galactose:2-oxoglutarate aminotransferase dTDP-4-amino-4,6-dideoxygalactose transaminase	
2.6.1.33	dTDP-4-amino-4,6-dideoxy-D-glucose:2-oxoglutarate aminotransferase dTDP-4-amino-4,6-dideoxy-D-glucose transaminase	
2.6.1.33	dTDP-4-amino-4,6-dideoxy-D-glucose transaminase dTDP-4-amino-4,6-dideoxy-D-glucose transaminase	
2.6.1.59	dTDP-4-amino-4,6-dideoxygalactose transaminase dTDP-4-amino-4,6-dideoxygalactose transaminase	
5.1.3.13	dTDP-4-dehydro-6-deoxy-D-glucose 3,5-epimerase dTDP-4-dehydrorhamnose 3,5-epimerase	Vol. 1, Page 152
5.1.3.13	dTDP-4-dehydrorhamnose 3,5-epimerase dTDP-4-dehydrorhamnose 3,5-epimerase	Vol. 1, Page 152
1.1.1.133	dTDP-4-dehydrorhamnose reductase dTDP-4-dehydrorhamnose reductase	
5.1.3.13	dTDP-4-keto-L-rhamnose 3,5-epimerase dTDP-4-dehydrorhamnose 3,5-epimerase	Vol. 1, Page 152
1.1.1.133	dTDP-4-keto-L-rhamnose reductase dTDP-4-dehydrorhamnose reductase	
1.1.1.133	dTDP-4-ketorhamnose reductase dTDP-4-dehydrorhamnose reductase	
1.1.1.133	dTDP-6-deoxy-L-mannose:$NADP^+$ 4-oxidoreductase dTDP-4-dehydrorhamnose reductase	
1.1.1.134	dTDP-6-deoxy-L-talose 4-dehydrogenase dTDP-6-deoxy-L-talose 4-dehydrogenase	
1.1.1.134	dTDP-6-deoxy-L-talose:$NADP^+$ 4-oxidoreductase dTDP-6-deoxy-L-talose 4-dehydrogenase	

EC-No.	Synonym Recommended Name	Vol., Page
1.1.1.134	dTDP-6-deoxy-L-talose dehydrogenase (4-reductase) dTDP-6-deoxy-L-talose 4-dehydrogenase	
1.1.1.186	dTDP-D-galactose:NAD 6-oxidoreductase dTDPgalactose 6-dehydrogenase	
4.2.1.46	dTDP-D-glucose 4,6-dehydratase dTDPglucose 4,6-dehydratase	
4.2.1.46	dTDP-D-glucose oxidoreductase dTDPglucose 4,6-dehydratase	
2.7.7.24	dTDP-glucose-pyrophosphorylase Glucose-1-phosphate thymidylyltransferase	
2.7.7.24	dTDP-glucose synthase Glucose-1-phosphate thymidylyltransferase	
2.4.2.27	dTDP-L-dihydrostreptose:streptidine-6-phosphate dihydrostreptosyltransferase dTDPdihydrostreptose-streptidine-6-phosphate dihydrostreptosyltransferase	
5.1.3.13	dTDP-L-rhamnose synthetase dTDP-4-dehydrorhamnose 3,5-epimerase	Vol. 1, Page 152
2.4.2.27	dTDPdihydrostreptose-streptidine-6-phosphate dihydrostreptosyltransferase dTDPdihydrostreptose-streptidine-6-phosphate dihydrostreptosyltransferase	
1.1.1.186	dTDPgalactose 6-dehydrogenase dTDPgalactose 6-dehydrogenase	
2.7.7.32	dTDPgalactose pyrophosphorylase Galactose-1-phosphate thymidylyltransferase	
4.2.1.46	dTDPglucose 4,6-dehydratase dTDPglucose 4,6-dehydratase	
4.2.1.46	dTDPglucose 4,6-hydro-lyase dTDPglucose 4,6-dehydratase	
2.7.7.24	dTDPglucose pyrophosphorylase Glucose-1-phosphate thymidylyltransferase	
3.1.3.35	dTMPase thymidylate 5'-phosphatase	
2.7.4.9	dTMP kinase dTMP kinase	

EC-No.	Synonym Recommended Name	Vol., Page
2.1.1.45	dTMP synthase Thymidylate synthase	
2.7.7.32	dTTP:α-D-galactose-1-phosphate thymidylyltransferase Galactose-1-phosphate thymidylyltransferase	
2.7.7.24	dTTP:α-D-glucose-1-phosphate thymidylyltransferase Glucose-1-phosphate thymidylyltransferase	
3.6.1.39	dTTPase Thymidine triphosphatase	
3.6.1.39	dTTP nucleotidohydrolase Thymidine triphosphatase	
1.1.1.16	Dulcitol dehydrogenase Galactitol 2-dehydrogenase	
2.7.7.7	Duplicase DNA-directed DNA polymerase	
3.6.1.23	dUTPase dUTP diphosphatase	
3.6.1.23	dUTP diphosphatase dUTP diphosphatase	
3.6.1.23	dUTP nucleotidohydrolase dUTP diphosphatase	
3.6.1.23	dUTP pyrophosphatase dUTP diphosphatase	
3.6.1.33	Dynein adenosine 5'-triphosphatase Dynein ATPase	
3.6.1.33	Dynein ATPase Dynein ATPase	
3.6.1.33	Dynein ATP phosphohydrolase (tubulin translocating) Dynein ATPase	
1.1.1.197	(13E)-(15S)-11α,15-Dihydroxy-9-oxoprost-13-enoate:NADP$^+$ 15-oxidoreductase 15-Hydroxyprostaglandin dehydrogenase (NADP$^+$)	
5.3.3.9	(13E)-(15S)-15-Hydroxy-9-oxoprosta-10,13-dienoate Δ^{10}-Δ^{11}-isomerase Prostaglandin-A1 Δ-isomerase	Vol. 1, Page 423
2.3.1.93	(E)-2-Methylcrotonoyl-CoA:13-hydroxylupanine O-2-methylcrotonyltransferase 13-Hydroxylupanine O-tigloyltransferase	

EC-No.	Synonym Recommended Name Vol., Page
4.1.2.34	(3E)-4-(2-Carboxyphenyl)-2-oxobut-3-enoate 2-carboxybenzaldehyde-lyase 4-(2-Carboxyphenyl)-2-oxobut-3-enoate aldolase
2.8.3.12	(E)-Glutaconate CoA-transferase Glutaconate CoA-transferase
2.5.1.11	(E)-Octaprenyl-diphosphate:isopentenyl-diphosphate octaprenyltranstransferase trans-Octaprenyltranstransferase
2.5.1.1	(2E,6E)-Farnesyl diphosphate synthetase Dimethylallyltranstransferase
1.14.13.34	(7E,9E,11Z,14Z)-(5S,6R)-6-(Cystein-S-yl)-5-hydroxyicosa-7,9,11,14-tetraenoate,NADPH:oxygen oxidoreductase (20-hydroxylating) Leukotriene-E4 20-monooxygenase
2.5.1.37	(7E,9E,11Z,14Z)-(5S,6S)-5,6-Epoxyicosa-7,9,11,14-tetraenoate:glutathione leukotriene-transferase (epoxide-ring-opening) Leukotriene-C_4 synthase
3.3.2.6	(7E,9E,11Z,14Z)-(5S,6S)-5,6-Epoxyicosa-7,9,11,14-tetraenoate hydrolase Leukotriene-A4 hydrolase
2.5.1.40	(E,E)-Farnesyl-diphosphate farnesyltransferase (bicyclizing) Aristolochene synthase
2.3.1.145	(E,E)-Piperoyl-CoA:piperidine N-piperoyltransferase Piperidine N-piperoyltransferase
4.1.1.28	E. C. 4.1.1.26 Aromatic-L-amino-acid decarboxylase
3.4.21.67	E. coli cytoplasmic proteinase Endopeptidase So
4.2.99.18	E. coli endonuclease III DNA-(apurinic or apyrimidinic site) lyase
3.1.21.2	E. coli endonuclease IV Deoxyribonuclease IV (Phage T_4-induced)
3.1.11.1	E. coli exonuclease I Exodeoxyribonuclease I
3.1.11.2	E. coli exonuclease III Exodeoxyribonuclease III
3.1.11.3	E. coli exonuclease IV Exodeoxyribonuclease (Lambda induced)
3.1.11.5	E. coli exonuclease V Exodeoxyribonuclease V

EC-No.	Synonym Recommended Name	Vol., Page
3.1.11.6	E. coli exonuclease VII Exodeoxyribonuclease VII	
3.4.21.87	E. coli protease VII Omptin	
3.1.26.1	E. coli RNase Physarum polycephalum ribonuclease	
6.3.2.19	E1 Ubiquitin-protein ligase	Vol. 2, Page 506
4.3.1.7	EAL ethanolamine ammonia-lyase	
2.3.1.137	Easily solubilized mitochondrial carnitine palmitoyltransferase Carnitine O-octanoyltransferase	
3.4.11.4	EC 3.4.1.3 tripeptide aminopeptidase	
3.4.22.10	EC 3.4.4.18 streptopain	
1.14.99.22	Ecdysone,hydrogen-donor:oxygen oxidoreductase (20-hydroxylating) Ecdysone 20-monooxygenase	
1.14.99.22	Ecdysone 20-hydroxylase Ecdysone 20-monooxygenase	
1.14.99.22	Ecdysone 20-monooxygenase Ecdysone 20-monooxygenase	
1.1.3.16	Ecdysone:oxygen 3-oxidoreductase Ecdysone oxidase	
1.14.99.22	α-Ecdysone C-20 hydroxylase Ecdysone 20-monooxygenase	
2.3.1.139	Ecdysone O-acyltransferase Ecdysone O-acyltransferase	
1.1.3.16	Ecdysone oxidase Ecdysone oxidase	
1.1.3.16	β-Ecdysone oxidase Ecdysone oxidase	
3.4.24.71	ECE-1 Endothelin-converting enzyme 1	
4.2.1.17	ECH enoyl-CoA hydratase	

EC-No.	Synonym Recommended Name	Vol., Page
3.4.99.27	Echis carinatus prothrombin-activating proteinase Echis carinatus prothrombin-activating proteinase	
5.3.3.8	ECI Dodecenoyl-CoA δ-isomerase	Vol. 1, Page 413
5.3.3.8	ECI Dodecenoyl-CoA δ-isomerase	Vol. 1, Page 413
3.1.21.3	EciDXXI Type I site-specific deoxyribonuclease	
3.1.21.3	EcoAI Type I site-specific deoxyribonuclease	
3.1.21.3	EcoB Type I site-specific deoxyribonuclease	
3.1.21.3	EcoBI Type I site-specific deoxyribonuclease	
3.1.21.3	EcoDI Type I site-specific deoxyribonuclease	
3.1.21.3	EcoEI Type I site-specific deoxyribonuclease	
3.1.21.3	EcoK Type I site-specific deoxyribonuclease	
3.1.21.3	EcoKI Type I site-specific deoxyribonuclease	
3.1.21.3	EcoR124/3I Type I site-specific deoxyribonuclease	
3.1.21.3	EcoR124I Type I site-specific deoxyribonuclease	
2.1.1.37	EcoRI methylase DNA (cytosine-5-)-methyltransferase	
1.3.1.52	ECR 2-Methyl-branched-chain-enoyl-CoA reductase	
3.1.3.5	ecto-5'-nucleotidase 5'-nucleotidase	
3.1.3.41	ecto-*p*-nitrophenyl phosphatase 4-nitrophenylphosphatase	
2.7.1.112	Ectoprotein kinase Protein-tyrosine kinase	

EC-No.	Synonym Recommended Name	Vol., Page
3.4.23.40	Edestinase Phytepsin	
3.2.1.123	EGCase Endoglycosylceramidase	
1.14.99.5	Eicosatrienoyl-CoA desaturase Stearoyl-CoA desaturase	
4.1.1.48	eIGPS Indole-3-glycerol phosphate synthase	
3.4.21.37	Elastase Leucocyte elastase	
3.4.21.11	elastase elastase	
3.4.21.36	elastase pancreatic elastase	
3.4.21.2	elastase-associated acidic endopeptidase chymotrypsin C	
3.4.21.71	Elastase 2 Pancreatic elastase II	
6.3.4.5	Elastin-binding protein Argininosuccinate synthase	Vol. 2, Page 595
3.4.21.37	Elaszym Leucocyte elastase	
3.4.21.36	elaszym pancreatic elastase	
3.2.1.21	Elaterase β-glucosidase	
1.5.5.1	Electron-transferring-flavoprotein:ubiquinone oxidoreductase Electron-transferring-flavoprotein dehydrogenase	
1.5.5.1	Electron-transferring-flavoprotein dehydrogenase Electron-transferring-flavoprotein dehydrogenase	
1.6.5.3	Electron transfer complex I NADH dehydrogenase (ubiquinone)	
1.5.5.1	Electron transfer flavoprotein-ubiquinone oxidoreductase Electron-transferring-flavoprotein dehydrogenase	
1.5.5.1	Electron transfer flavoprotein dehydrogenase Electron-transferring-flavoprotein dehydrogenase	

EC-No.	Synonym Recommended Name	Vol., Page
1.5.5.1	Electron transfer flavoprotein Q oxidoreductase Electron-transferring-flavoprotein dehydrogenase	
3.1.1.24	ELH I 3-oxoadipate enol-lactose	
3.1.1.24	ELH II 3-oxoadipate enol-lactose	
3.4.23.1	elixir lactate of pepsin pepsin A	
2.4.1.146	Elongation 3β-GalNAc-transferase β-1,3-Galactosyl-O-glycosyl-glycoprotein β-1,3-N-acetylglucosaminyl-transferase	
3.5.1.1	Elspar Asparaginase	
3.4.24.30	EM 19000 Coccolysin	
3.4.23.34	EMAP Cathepsin E	
3.2.1.21	Emulsin β-glucosidase	
1.3.99.4	4-En-3-oxosteroid:(acceptor)-1-en-oxido-reductase 3-Oxosteroid Δ^1-dehydrogenase	
3.2.1.39	Endo-(1,3)-β-D-glucanase Glucan endo-1,3-β-glucosidase	
3.2.1.71	Endo-(1$\rightarrow$ 2)-β-D-glucanase Glucan endo-1,2-β-glucosidase	
3.2.1.39	Endo-(1$\rightarrow$ 3)-β-D-glucanase Glucan endo-1,3-β-glucosidase	
3.2.1.59	Endo-(1$\rightarrow$3)-α-glucanase Glucan endo-1,3-α-glucosidase	
3.2.1.8	Endo-(1$\rightarrow$ 4)-β-xylanase(1$\rightarrow$ 4)-β-xylan 4-xylanohydrolase Endo-1,4-β-xylanase	
3.2.1.75	Endo-(1$\rightarrow$ 6)-β-D-glucanase Glucan endo-1,6-β-glucosidase	
3.2.1.6	Endo-β-(1$\rightarrow$ 3)-D-glucanase Endo-1,3(4)-β-glucanase	

EC-No.	Synonym Recommended Name	Vol., Page
3.2.1.6	Endo-β-(1-3)-D-glucanase Endo-1,3(4)-β-glucanase	
3.2.1.6	Endo-β-1,3-glucanase IV Endo-1,3(4)-β-glucanase	
3.2.1.32	Endo-β-1,3-xylanase Xylan endo-1,3-β-xylosidase	
3.2.1.89	Endo-β-1,4-galactanase Arabinogalactan endo-1,4-β-galactosidase	
4.2.2.2	endo-α-1,4-polygalacturonic acid lyase pectate lyase	
3.2.1.8	Endo-β-1,4-Xylanase Endo-1,4-β-xylanase	
3.2.1.99	Endo-α-1,5-arabanase Arabinan endo-1,5-α-L-arabinosidase	
3.2.1.75	Endo-β-1,6-glucanase Glucan endo-1,6-β-glucosidase	
3.2.1.101	Endo-α-1→ 6-D-mannanase Mannan endo-1,6-β-mannosidase	
3.2.1.97	Endo-α-acetylgalactosaminidase Glycopeptide α-N-acetylgalactosaminidase	
3.2.1.130	Endo-α-D-mannosidase Glycoprotein endo-α-1,2-mannosidase	
3.2.1.102	Endo-β-galactosidase Blood group-substance endo-1,4-β-galactosidase	
3.2.1.103	Endo-β-galactosidase Keratan-sulfate endo-1,4-β-galactosidase	
3.2.1.78	Endo-β-mannanase Mannan endo-1,4-β-mannosidase	
3.2.1.130	Endo-α-mannosidase Glycoprotein endo-α-1,2-mannosidase	
3.2.1.110	Endo-α-N-acetyl-D-galactosaminidase Mucinaminylserine mucinaminidase	
3.2.1.110	Endo-α-N-acetyl galactosaminidase Mucinaminylserine mucinaminidase	
3.2.1.97	Endo-α-N-acetylgalactosaminidase Glycopeptide α-N-acetylgalactosaminidase	

EC-No.	Synonym Recommended Name	Vol., Page
3.2.1.110	Endo-α-N-acetylgalactosaminidase Mucinaminylserine mucinaminidase	
3.2.1.96	Endo-β-N-acetylglucosaminidase Mannosyl-glycoprotein endo-β-N-acetyl glucosaminidase	
3.2.1.96	Endo-β-N-acetylglucosaminidase D Mannosyl-glycoprotein endo-β-N-acetyl glucosaminidase	
3.2.1.96	endo-β-N-acetylglucosaminidase F Mannosyl-glycoprotein endo-β-N-acetyl glucosaminidase	
3.2.1.96	Endo-β-N-acetylglucosaminidase H Mannosyl-glycoprotein endo-β-N-acetyl glucosaminidase	
3.2.1.96	Endo-β-N-acetylglucosaminidase L Mannosyl-glycoprotein endo-β-N-acetyl glucosaminidase	
3.2.1.129	Endo-α-sialidase Endo-α-sialidase	
3.2.1.6	Endo-.β.,1,3(4)-glucanase Endo-1,3(4)-β-glucanase	
3.2.1.96	Endo-.β.-(1→4)-N-acetylglucosaminidase Mannosyl-glycoprotein endo-β-N-acetyl glucosaminidase	
3.2.1.6	Endo-. β.-1,3-1,4-glucanase Endo-1,3(4)-β-glucanase	
3.2.1.78	Endo-.β.-1,4-mannase Mannan endo-1,4-β-mannosidase	
3.2.1.96	Endo-.β.-acetylglucosaminidase Mannosyl-glycoprotein endo-β-N-acetyl glucosaminidase	
3.2.1.96	Endo-.β.-(1→4)N-acetylglucosaminidase Mannosyl-glycoprotein endo-β-N-acetyl glucosaminidase	
3.2.1.71	Endo-1,2-β-glucanase Glucan endo-1,2-β-glucosidase	
3.2.1.71	Endo-1,2-.β.-glucanase Glucan endo-1,2-β-glucosidase	
3.2.1.6	Endo-1,3(4)-β-glucanase Endo-1,3(4)-β-glucanase	
3.2.1.90	Endo-1,3-β-D-galactanase Arabinogalactan endo-1,3-β-galactosidase	
3.2.1.59	Endo-1,3-α-D-glucanase Glucan endo-1,3-α-glucosidase	

EC-No.	Synonym Recommended Name	Vol., Page
3.2.1.6	Endo-1,3-β-D-glucanase Endo-1,3(4)-β-glucanase	
3.2.1.39	Endo-1,3-β-D-glucanase Glucan endo-1,3-β-glucosidase	
3.2.1.90	Endo-1,3-β-galactanase Arabinogalactan endo-1,3-β-galactosidase	
3.2.1.59	Endo-1,3-α-glucanase Glucan endo-1,3-α-glucosidase	
3.2.1.6	Endo-1,3-β-glucanase Endo-1,3(4)-β-glucanase	
3.2.1.39	Endo-1,3-β-glucanase Glucan endo-1,3-β-glucosidase	
3.2.1.39	Endo-1,3-β-glucosidase Glucan endo-1,3-β-glucosidase	
3.2.1.32	Endo-1,3-β-xylanase Xylan endo-1,3-β-xylosidase	
3.2.1.59	Endo-1,3-.α.-glucanase Glucan endo-1,3-α-glucosidase	
3.2.1.39	Endo-1,3-.β.-glucanase Glucan endo-1,3-β-glucosidase	
3.2.1.32	Endo-1,3-.β.-xylanase Xylan endo-1,3-β-xylosidase	
3.2.1.6	Endo-1,3-1,4-.β.-D-glucanase Endo-1,3(4)-β-glucanase	
3.2.1.32	Endo-1,3-xylanase Xylan endo-1,3-β-xylosidase	
3.2.1.89	Endo-1,4-β-D-galactanase Arabinogalactan endo-1,4-β-galactosidase	
3.2.1.8	Endo-1,4-β-D-xylanase Endo-1,4-β-xylanase	
3.2.1.89	Endo-1,4-β-galactanase Arabinogalactan endo-1,4-β-galactosidase	
3.2.1.4	Endo-1,4-β-glucanase Cellulase	
3.2.1.78	Endo-1,4-β-mannanase Mannan endo-1,4-β-mannosidase	

EC-No.	Synonym Recommended Name	Vol., Page
3.2.1.8	Endo-1,4-β-xylanase Endo-1,4-β-xylanase	
3.2.1.8	Endo-1,4-β-xylanase Endo-1,4-β-xylanase	
3.2.1.4	Endo-1,4-.β.-D-Glucanase Cellulase	
3.2.1.8	Endo-1,4-xylanase Endo-1,4-β-xylanase	
3.2.1.99	Endo-1,5-α-L-arabinanase Arabinan endo-1,5-α-L-arabinosidase	
3.2.1.75	Endo-1,6-β-glucanase Glucan endo-1,6-β-glucosidase	
3.2.1.101	Endo-1,6-β-mannanase Mannan endo-1,6-β-mannosidase	
3.2.1.75	Endo-1,6-.β.-glucanase Glucan endo-1,6-β-glucosidase	
3.2.1.75	Endo-1,6-.β. glucanase Glucan endo-1,6-β-glucosidase	
3.2.1.99	Endo-arabanase Arabinan endo-1,5-α-L-arabinosidase	
3.2.1.15	Endo-D-galacturonanase Polygalacturonase	
3.2.1.15	Endo-D-galacturonase Polygalacturonase	
3.2.1.11	Endo-dextranase Dextranase	
4.6.1.8	(-)-endo-Fenchol cyclase (-)-endo-Fenchol synthase	
4.2.3.10	(-)-endo-fenchol cyclase (-)-endo-fenchol synthase	
4.6.1.8	(-)-endo-Fenchol synthase (-)-endo-Fenchol synthase	
4.2.3.10	(-)-endo-fenchol synthase (-)-endo-fenchol synthase	
1.1.1.241	6-endo-Hydroxycineole:NAD$^+$ 6-oxidoreductase 6-endo-Hydroxycineole dehydrogenase	

EC-No.	Synonym Recommended Name	Vol., Page
1.1.1.241	6-endo-Hydroxycineole dehydrogenase 6-endo-Hydroxycineole dehydrogenase	
3.2.1.7	Endo-inulinase Inulinase	
3.4.21.50	endo-Lys-C protease lysyl endopeptidase	
3.2.1.129	Endo-N Endo-α-sialidase	
3.2.1.96	Endo-N-acetyl-.β.-D-glucosaminidase Mannosyl-glycoprotein endo-β-N-acetyl glucosaminidase	
3.2.1.96	Endo-N-acetyl-.β.-glucosaminidase Mannosyl-glycoprotein endo-β-N-acetyl glucosaminidase	
3.2.1.129	Endo-N-acetylneuraminidase Endo-α-sialidase	
3.4.22.19	endo-oligopeptidase endo-oligopeptidase	
3.4.24.15	endo-oligopeptidase A thimet oligopeptidase	
4.2.2.10	endo-pectin lyase pectin lyase	
3.2.1.15	Endo-polygalacturonase Polygalacturonase	
2.4.1.207	Endo-xyloglucan transferase Xyloglucan:xyloglucosyl transferase	
3.2.1.1	Endoamylase α-amylase	
3.2.1.136	Endoarabinoxylanase Glucuronoarabinoxylan endo-1,4-β-xylanase	
3.2.1.14	Endochitinase Chitinase	
3.1.21.6	Endodeoxyribonuclease CC-preferring endodeoxyribonuclease	
4.2.99.18	Endodeoxyribonuclease DNA-(apurinic or apyrimidinic site) lyase	
3.1.21.2	Endodeoxyribonuclease Deoxyribonuclease IV (Phage T_4-induced)	

EC-No.	Synonym Recommended Name	Vol., Page
3.1.4.30	endodeoxyribonuclease endodeoxyribonuclease	
3.1.21.3	Endodeoxyribonuclease (adenosine triphosphate-hydrolyzing) Type I site-specific deoxyribonuclease	
3.1.25.2	endodeoxyribonuclease (apurinic or apyrimidinic) endodeoxyribonuclease (apurinic or apyrimidinic)	
3.1.4.32	endodeoxyribonuclease (ATP- and S-adenosylmethionine-dependent) endodeoxyribonuclease (ATP- and S-adenosylmethionine-dependent)	
3.1.4.33	endodeoxyribonuclease (ATP-hydrolysing) endodeoxyribonuclease (ATP-hydrolysing)	
3.1.25.1	Endodeoxyribonuclease (pyrimidine dimer) Deoxyribonuclease (pyrimidine dimer)	
3.1.21.1	Endodeoxyribonuclease I Deoxyribonuclease I	
4.2.99.18	Endodeoxyribonuclease III DNA-(apurinic or apyrimidinic site) lyase	
3.1.21.2	Endodeoxyribonuclease IV Deoxyribonuclease IV (Phage T_4-induced)	
3.1.22.3	Endodeoxyribonuclease V Deoxyribonuclease V	
3.1.11.6	Endodeoxyribonuclease VII Exodeoxyribonuclease VII	
3.2.1.11	Endodextranase Dextranase	
3.2.1.109	Endogalactosaminidase Endogalactosaminidase	
3.2.1.15	Endogalacturonase Polygalacturonase	
4.2.2.2	endogalacturonate transeliminase pectate lyase	
3.2.1.4	Endoglucanase D Cellulase	
3.2.1.4	.β.-1,4-Endoglucan hydrolase Cellulase	
3.2.1.123	Endoglycoceramidase Endoglycosylceramidase	

EC-No.	Synonym Recommended Name	Vol., Page
3.2.1.96	Endoglycosidase S Mannosyl-glycoprotein endo-β-N-acetyl glucosaminidase	
3.2.1.123	Endoglycosylceramidase Endoglycosylceramidase	
3.2.1.7	Endoinulinase Inulinase	
3.2.1.130	Endomannosidase Glycoprotein endo-α-1,2-mannosidase	
3.2.1.129	Endoneuraminidase Endo-α-sialidase	
3.1.30.2	Endonuclease (Serratia marcescens) Serratia marcescens nuclease	
3.1.21.6	Endonuclease II CC-preferring endodeoxyribonuclease	
3.1.21.2	Endonuclease II Deoxyribonuclease IV (Phage T4-induced)	
4.2.99.18	Endonuclease III DNA-(apurinic or apyrimidinic site) lyase	
3.1.21.6	Endonuclease NX CC-preferring endodeoxyribonuclease	
3.1.30.1	Endonuclease S_1 Aspergillus nuclease S_1	
3.1.21.6	Endonuclease T CC-preferring endodeoxyribonuclease	
3.2.2.17	Endonuclease V Deoxyribodipyrimidine endonucleosidase	
3.2.1.15	Endopectinase Polygalacturonase	
4.2.2.2	endopectin methyltranseliminase pectate lyase	
3.4.24.18	Endopeptidase-2 Meprin A	
3.4.24.11	endopeptidase-2 neprilysin	
3.4.24.11	endopeptidase-24.11 neprilysin	

EC-No.	Synonym Recommended Name	Vol., Page
3.4.24.15	endopeptidase 24.15 thimet oligopeptidase	
3.4.24.16	endopeptidase 24.16 neurolysin	
3.4.24.16	endopeptidase 24.16B neurolysin	
3.4.21.92	Endopeptidase Clp Endopeptidase Clp	
3.4.19.11	Endopeptidase I γ-D-Glutamyl-meso-diaminopimelate peptidase I	
3.4.21.64	Endopeptidase K Endopeptidase K	
3.4.21.53	Endopeptidase La Endopeptidase La	
3.4.21.50	endopeptidase Lys-C lysyl endopeptidase	
3.4.21.67	Endopeptidase So Endopeptidase So	
3.4.21.92	Endopeptidase Ti Endopeptidase Clp	
5.3.99.3	Endoperoxide isomerase Prostaglandin-E synthetase	Vol. 1, Page 459
3.2.1.15	Endopolygalacturonase Polygalacturonase	
3.2.1.15	Endopolygalacturonate lyase Polygalacturonase	
4.2.2.2	α-1,4-D-endopolygalacturonic acid lyase pectate lyase	
3.6.1.10	Endopolyphosphatase Endopolyphosphatase	
3.4.21.26	endoprolylpeptidase prolyl oligopeptidase	
3.4.24.33	Endoproteinase Asp-N Peptidyl-Asp metalloendopeptidase	
3.4.21.19	endoproteinase Glu-C glutamyl endopeptidase	

EC-No.	Synonym Recommended Name	Vol., Page
3.4.21.61	Endoproteinase Kex2p Kexin	
3.4.21.50	endoproteinase Lys-C lysyl endopeptidase	
3.1.26.4	Endoribonuclease H Calf thymus ribonuclease H	
3.1.4.24	endoribonuclease III endoribonuclease III	
3.1.26.6	Endoribonuclease IV Ribonuclease IV	
2.3.1.88	β-Endorphin acetyltransferase Peptide α-N-acetyltransferase	
3.2.1.129	Endosialidase Endo-α-sialidase	
3.4.15.1	endothelial cell peptidyl dipeptidase peptidyl-dipeptidase A	
3.4.24.71	Endothelin-converting enzyme Endothelin-converting enzyme 1	
3.4.24.71	Endothelin-converting enzyme 1 Endothelin-converting enzyme 1	
1.14.13.39	Endothelium-derived relaxation factor-forming enzyme Nitric-oxide synthase	
1.14.13.39	Endothelium-derived relaxing factor synthase Nitric-oxide synthase	
3.4.23.22	Endothia acid proteinase Endothiapepsin	
3.4.23.10	endothia acid proteinase endothia acid proteinase	
3.4.23.22	Endothia aspartic proteinase Endothiapepsin	
3.4.23.22	Endothia parasitica acid proteinase Endothiapepsin	
3.4.23.22	Endothia parasitica aspartic proteinase Endothiapepsin	
3.4.23.22	Endothiapepsin Endothiapepsin	

EC-No.	Synonym Recommended Name	Vol., Page
2.4.1.207	Endoxyloglucan transferase Xyloglucan:xyloglucosyl transferase	
1.1.1.145	5-Ene-3-β-hydroxysteroid dehydrogenase 3β-Hydroxy-Δ^5-steroid dehydrogenase	
1.3.99.5	4-Ene-3-ketosteroid-5α-oxidoreductase 3-Oxo-5α-steroid Δ^4-dehydrogenase	
1.3.1.22	4-Ene-3-oxosteroid 5α-reductase Cholestenone 5α-reductase	
5.3.3.1	5-Ene-4-ene isomerase Steroid Δ-isomerase	Vol. 1, Page 376
1.3.1.22	4-Ene-5α-reductase Cholestenone 5α-reductase	
1.3.99.4	1-Ene-dehydrogenase 3-Oxosteroid Δ^1-dehydrogenase	
1.6.1.2	Energy-linked transhydrogenase NAD(P)$^+$ transhydrogenase (AB-specific)	
3.4.17.10	enkephalin-precursor endopeptidase carboxypeptidase H	
3.4.24.11	enkephalinase neprilysin	
3.4.14.4	enkephalinase B dipeptidyl-peptidase III	
3.4.17.10	enkephalin convertase carboxypeptidase H	
3.4.17.10	enkephalin precursor carboxypeptidase carboxypeptidase H	
1.3.1.31	2-Enoate reductase 2-Enoate reductase	
1.3.1.31	Enoate reductase 2-Enoate reductase	
4.2.1.17	enol-CoA hydratase enoyl-CoA hydratase	
4.2.1.11	α,α-enolase phosphoenolpyruvate hydratase	
4.2.1.11	β,β-enolase phosphoenolpyruvate hydratase	

EC-No.	Synonym Recommended Name	Vol., Page
4.2.1.11	enolase phosphoenolpyruvate hydratase	
4.2.1.11	γ,γ-enolase phosphoenolpyruvate hydratase	
4.2.1.11	γ-enolase phosphoenolpyruvate hydratase	
4.1.1.32	P-enolpyruvate carboxykinase Phosphoenolpyruvate carboxykinase (GTP)	
2.5.1.19	5-Enolpyruvyl-3-phosphoshikimate synthase 3-Phosphoshikimate 1-carboxyvinyltransferase	
4.6.1.4	5-enolpyruvylshikimate-3-phosphate phospholyase chorismate synthase	
2.5.1.19	5'-Enolpyruvylshikimate-3-phosphate synthase 3-Phosphoshikimate 1-carboxyvinyltransferase	
2.5.1.19	5-Enolpyruvylshikimate-3-phosphate synthase 3-Phosphoshikimate 1-carboxyvinyltransferase	
2.5.1.19	5-Enolpyruvylshikimate-3-phosphate synthetase 3-Phosphoshikimate 1-carboxyvinyltransferase	
2.5.1.19	5-Enolpyruvylshikimate-3-phosphoric acid synthase 3-Phosphoshikimate 1-carboxyvinyltransferase	
2.5.1.19	3-Enolpyruvylshikimate 5-phosphate synthase 3-Phosphoshikimate 1-carboxyvinyltransferase	
2.5.1.19	Enolpyruvylshikimate phosphate synthase 3-Phosphoshikimate 1-carboxyvinyltransferase	
2.5.1.19	5-Enolpyruvylshikimic acid-3-phosphate synthase 3-Phosphoshikimate 1-carboxyvinyltransferase	
2.5.1.19	3-Enolpyruvylshikimic acid-5-phosphate synthetase 3-Phosphoshikimate 1-carboxyvinyltransferase	
1.3.1.9	Enoyl-[acyl-carrier-protein] reductase (NADH) Enoyl-[acyl-carrier-protein] reductase (NADH)	
1.3.1.39	Enoyl-[acyl-carrier-protein] reductase (NADPH, A-specific) Enoyl-[acyl-carrier-protein] reductase (NADPH, A-specific)	
1.3.1.10	Enoyl-[acyl-carrier-protein] reductase (NADPH, B-specific) Enoyl-[acyl-carrier-protein] reductase (NADPH, B-specific)	
1.3.1.9	Enoyl-ACP reductase Enoyl-[acyl-carrier-protein] reductase (NADH)	

EC-No.	Synonym Recommended Name	Vol., Page
1.3.1.10	Enoyl-ACP reductase Enoyl-[acyl-carrier-protein] reductase (NADPH, B-specific)	
1.3.1.39	Enoyl-ACp reductase Enoyl-[acyl-carrier-protein] reductase (NADPH, A-specific)	
4.2.1.17	2-enoyl-CoA hydratase enoyl-CoA hydratase	
4.2.1.17	enoyl-CoA hydratase enoyl-CoA hydratase	
4.2.1.74	2-enoyl-CoA hydratase 2 long-chain-enoyl-CoA hydratase	
5.3.3.8	2,3-Enoyl-CoA isomerase Dodecenoyl-CoA δ-isomerase	Vol. 1, Page 413
5.3.3.8	Δ^3,Δ^2-enoyl-CoA isomerase Dodecenoyl-CoA δ-isomerase	Vol. 1, Page 413
1.3.1.8	2-Enoyl-CoA reductase Acyl-CoA dehydrogenase ($NADP^+$)	
1.3.1.34	4-Enoyl-CoA reductase 2,4-Dienoyl-CoA reductase (NADPH)	
1.3.99.2	Enoyl-coenzyme A reductase Butyryl-CoA dehydrogenase	
1.3.1.10	Enoyl acyl-carrier-protein reductase Enoyl-[acyl-carrier-protein] reductase (NADPH, B-specific)	
4.2.1.58	Enoyl acyl carrier protein hydrase Crotonoyl-[acyl-carrier-protein] hydratase	
4.2.1.55	Enoyl coenzyme A hydrase (D) 3-Hydroxybutyryl-CoA dehydratase	
4.2.1.17	enoyl coenzyme A hydrase (D) enoyl-CoA hydratase	
4.2.1.17	enoyl coenzyme A hydrase (L) enoyl-CoA hydratase	
4.2.1.17	enoyl coenzyme A hydratase enoyl-CoA hydratase	
1.3.1.8	Enoyl coenzyme A reductase Acyl-CoA dehydrogenase ($NADP^+$)	
4.2.1.17	enoyl hydrase enoyl-CoA hydratase	

EC-No.	Synonym Recommended Name	Vol., Page
2.5.1.7	Enoylpyruvate transferase UDP-N-acetylglucosamine 1-carboxyvinyltransferase	
2.5.1.7	Enoylpyruvatetransferase UDP-N-acetylglucosamine 1-carboxyvinyltransferase	
3.4.22.35	Entamoeba histolytica cysteine protease Histolysain	
3.4.22.35	Entamoeba histolytica cysteine proteinase Histolysain	
3.4.22.35	Entamoeba histolytica neutral thiol proteinase Histolysain	
3.1.27.6	Enterobacter ribonuclease Enterobacter ribonuclease	
3.4.21.9	enterokinase enteropeptidase	
3.4.21.9	enteropeptidase enteropeptidase	
3.4.4.8	enteropeptidase enteropeptidase	
1.1.1.49	Entner-Doudoroff Enzyme Glucose-6-phosphate 1-dehydrogenase	
3.4.21.33	Entomophthora collagenolytic proteinase Entomophthora collagenolytic proteinase	
3.4.24.12	envelysin envelysin	
3.4.22.2	enzeco papain papain	
3.4.21.1	Enzeon Chymotrypsin	
2.4.1.25	D-enzyme 4-α-Glucanotransferase	
2.4.1.18	Q-enzyme 1,4-α-Glucan branching enzyme	
3.2.1.41	R-enzyme α-dextrin endo-1,6-α-glucosidase	
3.2.1.142	R-enzyme limit dextrinase	

EC-No.	Synonym Recommended Name	Vol., Page
2.4.1.24	T-Enzyme 1,4-α-Glucan 6-α-glucosyltransferase	
1.8.4.7	Enzyme-thiol transhydrogenase (oxidized-glutathione) Enzyme-thiol transhydrogenase (oxidized-glutathione)	
2.7.3.9	Enzyme I Phosphoenolpyruvate-protein phosphotransferase	
2.7.1.69	Enzyme IIl4ac Protein-Npi-phosphohistidine-sugar phosphotransferase	
2.7.1.69	Enzyme II of the phosphotransferase system Protein-Npi-phosphohistidine-sugar phosphotransferase	
2.7.3.9	Enzyme I of the phosphotransferase system Phosphoenolpyruvate-protein phosphotransferase	
2.4.1.18	Enzyme Q 1,4-α-Glucan branching enzyme	
1.14.15.6	Enzymes, cholesterol side-chain-cleaving Cholesterol monooxygenase (side-chain-cleaving)	
3.1.11.5	Enzymes, gene recBCD Exodeoxyribonuclease V	
5.5.1.1	Enzymes, muconate-lactonizing Muconate cycloisomerase	Vol. 1, Page 660
1.8.2.2	Enzymes, thiosulfate-oxidizing Thiosulfate dehydrogenase	
3.1.1.5	eosinophil Charcot-Leyden crystal protein lysophospholipase	
1.11.1.7	Eosinophil peroxidase Peroxidase	
3.4.24.15	EP24.15 thimet oligopeptidase	
3.4.24.16	EP24.16c neurolysin	
3.4.24.16	EP24.16m neurolysin	
1.5.1.18	(-)-Ephedrine:NAD^+ 2-oxidoreductase Ephedrine dehydrogenase	
1.5.1.18	Ephedrine dehydrogenase Ephedrine dehydrogenase	

EC-No.	Synonym Recommended Name	Vol., Page
5.1.3.2	4-Epimerase UDPglucose 4-epimerase	Vol. 1, Page 97
5.1.99.2	Epimerase, 16-hydroxy steroid 16-Hydroxysteroid epimerase	Vol. 1, Page 183
5.1.2.3	Epimerase, 3-hydroxybutyryl coenzyme A 3-Hydroxybutyryl-CoA epimerase	Vol. 1, Page 80
5.1.3.8	Epimerase, acylglucosamine 2- N-Acylglucosamine 2-epimerase	Vol. 1, Page 140
5.1.3.9	Epimerase, acylglucosamine phosphate 2- N-Acylglucosamine-6-phosphate 2-epimerase	Vol. 1, Page 144
5.1.3.20	Epimerase, adenosine diphosphoglyceromannoheptose 6- ADPglyceromanno-heptose 6-epimerase	Vol. 1, Page 175
5.1.3.20	Epimerase, adenosine diphosphoglyceromannoheptose 6- ADPglyceromanno-heptose 6-epimerase	Vol. 1, Page 175
5.1.3.20	Epimerase, adenosine diphosphoglyceromannoheptose 6- (Escherichia coli clone pCG50 reduced) ADPglyceromanno-heptose 6-epimerase	Vol. 1, Page 175
5.1.3.20	Epimerase, adenosine diphosphoglyceromannoheptose 6- (Neisseria gonorrhoeae gene lis-6) ADPglyceromanno-heptose 6-epimerase	Vol. 1, Page 175
5.1.3.3	Epimerase, aldose-1 Aldose 1-epimerase	Vol. 1, Page 113
5.1.3.11	Epimerase, cellobiose Cellobiose epimerase	Vol. 1, Page 148
5.1.3.19	Epimerase, chondroitin glucuronate 5- Chondroitin-glucuronate 5-epimerase	Vol. 1, Page 172
5.1.3.10	Epimerase, cytidine diphosphoabequose CDPabequose 2-epimerase	Vol. 1, Page 146
5.1.1.7	Epimerase, diaminopimelate Diaminopimelate epimerase	Vol. 1, Page 27
5.1.3.15	Epimerase, glucose 6-phosphate 1- Glucose-6-phosphate 1-epimerase	Vol. 1, Page 162
5.1.3.18	Epimerase, guanosine diphosphomannose GDPmannose 3,5-epimerase	Vol. 1, Page 170
5.1.1.8	Epimerase, hydroxyproline 4-Hydroxyproline epimerase	Vol. 1, Page 33

EC-No.	Synonym Recommended Name	Vol., Page
5.1.2.6	Epimerase, isocitrate Isocitrate epimerase	Vol. 1, Page 89
5.1.1.14	Epimerase, isonocardicin A Nocardicin-A epimerase	Vol. 1, Page 59
5.1.3.4	Epimerase, L-ribulose phosphate 4- L-Ribulose-phosphate 4-epimerase	Vol. 1, Page 123
5.1.3.17	Epimerase, polyglucuronate Heparosan-N-sulfate-glucuronate 5-epimerase	Vol. 1, Page 167
5.1.3.1	Epimerase, ribulose phosphate 3- Ribulose-phosphate 3-epimerase	Vol. 1, Page 91
5.1.2.5	Epimerase, tartrate Tartrate epimerase	Vol. 1, Page 87
5.1.3.13	Epimerase, thymidine diphospho-4-ketorhamnose 3,5- dTDP-4-dehydrorhamnose 3,5-epimerase	Vol. 1, Page 152
5.1.3.7	Epimerase, uridine diphosphoacetylglucosamine UDP-N-acetylglucosamine 4-epimerase	Vol. 1, Page 135
5.1.3.14	Epimerase, uridine diphosphoacetylglucosamine 2- UDP-N-acetylglucosamine 2-epimerase	Vol. 1, Page 154
5.1.3.5	Epimerase, uridine diphosphoarabinose UDParabinose 4-epimerase	Vol. 1, Page 129
5.1.3.2	Epimerase, uridine diphosphoglucose UDPglucose 4-epimerase	Vol. 1, Page 97
5.1.3.6	Epimerase, uridine diphosphoglucuronate UDPglucuronate 4-epimerase	Vol. 1, Page 132
5.1.3.12	Epimerase, uridine diphosphoglucuronate 5' UDPglucuronate 5'-epimerase	Vol. 1, Page 150
1.4.3.4	Epinephrine oxidase Amine oxidase (flavin-containing)	
1.14.11.14	Epoxidase, 6β-hydroxyhyoscyamine 6β-Hydroxyhyoscyamine epoxidase	
1.14.99.20	Epoxidase, phylloquinone Phylloquinone monooxygenase (2,3-epoxidizing)	
3.3.2.3	Epoxide hydratase Epoxide hydrolase	
4.2.1.63	epoxide hydratase epoxide hydratase	

EC-No.	Synonym Recommended Name	Vol., Page
3.3.2.3	Epoxide hydrolase Epoxide hydrolase	
3.3.2.3	Epoxide hydrolase Epoxide hydrolase	
5.4.99.8	2,3-Epoxysqualene-cycloartenol cyclase Cycloartenol synthase	Vol. 1, Page 631
5.4.99.7	2,3-Epoxysqualene-lanosterol cyclase Lanosterol synthase	Vol. 1, Page 624
5.4.99.8	2,3-Epoxysqualene cycloartenol-cyclase Cycloartenol synthase	Vol. 1, Page 631
5.4.99.7	2,3-Epoxysqualene lanosterol-cyclase Lanosterol synthase	Vol. 1, Page 624
2.5.1.19	EPSP synthase 3-Phosphoshikimate 1-carboxyvinyltransferase	
2.7.8.1	EPT Ethanolaminephosphotransferase	
1.11.1.6	Equilase Catalase	
4.2.1.62	5α-Ergosta-7,22-diene-3β,5-diol 5,6-hydro-lyase 5-α-Hydroxysteroid dehydratase	
2.1.1.48	ErmC 23S rRNA methyltransferase rRNA (adenine-N^6-)-methyltransferase	
1.1.1.162	Erythritol:$NADP^+$ oxidoreductase Erythrulose reductase	
1.1.1.9	Erythritol dehydrogenase D-Xylulose reductase	
2.7.1.27	Erythritol kinase Erythritol kinase	
4.1.3.14	erythro-β-Hydroxyaspartate aldolase 3-Hydroxyaspartate aldolase	
4.2.1.38	Erythro-β-hydroxyaspartate dehydratase Erythro-3-hydroxyaspartate dehydratase	
4.1.3.14	erythro-β-Hydroxyaspartate glycine-lyase 3-Hydroxyaspartate aldolase	
4.2.1.19	D-erythro-1-(imidazol-4-yl)glycerol 3 phosphate hydro-lyase imidazoleglycerol-phosphate dehydratase	

EC-No.	Synonym Recommended Name	Vol., Page
1.4.1.11	L-erythro-3,5-Diaminohexanoate:NAD$^+$ oxidoreductase (deaminating) L-erythro-3,5-Diaminohexanoate dehydrogenase	
1.4.1.11	L-erythro-3,5-Diaminohexanoate dehydrogenase L-erythro-3,5-Diaminohexanoate dehydrogenase	
4.1.3.14	erythro-3-Hydroxy-L-aspartate glyoxylate-lyase 3-Hydroxyaspartate aldolase	
4.2.1.38	Erythro-3-hydroxy -LS-aspartate hydro-lyase (deaminating) Erythro-3-hydroxyaspartate dehydratase	
4.2.1.38	Erythro-3-hydroxyaspartate dehydratase Erythro-3-hydroxyaspartate dehydratase	
4.2.1.38	Erythro-3-hydroxyaspartate hydro-lyase (deaminating) Erythro-3-hydroxyaspartate dehydratase	
1.1.1.102	D-erythro-Dihydrosphingosine:NADP$^+$ 3-oxidoreductase 3-Dehydrosphinganine reductase	
1.15.1.1	Erythrocuprein Superoxide dismutase	
3.4.23.34	Erythrocyte membrane aspartic proteinase Cathepsin E	
2.3.1.94	Erythronolide condensing enzyme Erythronolide synthase	
2.3.1.94	Erythronolide synthase Erythronolide synthase	
4.1.2.2	Erythrose-1-phosphate synthase Ketotetrose-phosphate aldolase	
5.3.1.2	erythrose isomerase erythrose isomerase	Vol. 1, Page 248
4.1.2.2	Erythrulose-1-phosphate formaldehyde-lyase Ketotetrose-phosphate aldolase	
4.1.2.2	Erythrulose-1-phosphate synthase Ketotetrose-phosphate aldolase	
1.1.1.162	D-Erythrulose reductase Erythrulose reductase	
1.1.1.162	Erythrulose reductase Erythrulose reductase	
3.1.22.3	Escherichia coli endodeoxyribonuclease V Deoxyribonuclease V	

EC-No.	Synonym Recommended Name	Vol., Page
3.1.22.5	Escherichia coli endodeoxyribonuclease X Deoxyribonuclease X	
3.1.21.1	Escherichia coli endonuclease I Deoxyribonuclease I	
3.1.21.6	Escherichia coli endonuclease II CC-preferring endodeoxyribonuclease	
3.1.21.2	Escherichia coli endonuclease II Deoxyribonuclease IV (Phage T4-induced)	
4.2.99.18	Escherichia coli endonuclease III DNA-(apurinic or apyrimidinic site) lyase	
3.1.13.1	Escherichia coli exo-RNase II Exoribonuclease II	
3.1.11.1	Escherichia coli exonuclease I Exodeoxyribonuclease I	
3.1.11.2	Escherichia coli exonuclease III Exodeoxyribonuclease III	
3.1.11.3	Escherichia coli exonuclease IV Exodeoxyribonuclease (Lambda induced)	
3.1.11.5	Escherichia coli exonuclease V Exodeoxyribonuclease V	
3.1.11.6	Escherichia coli exonuclease VII Exodeoxyribonuclease VII	
3.4.21.89	Escherichia coli leader peptidase Signal peptidase I	
3.4.24.55	Escherichia coli protease III Pitrilysin	
3.4.21.53	Escherichia coli proteinase La Endopeptidase La	
3.1.27.1	Escherichia coli ribonuclease I Ribonuclease T_2	
3.4.21.53	Escherichia coli serine proteinase La Endopeptidase La	
3.4.21.67	Escherichia coli serine proteinase So Endopeptidase So	
3.1.25.1	Escherichia coli UV endonuclease Deoxyribonuclease (pyrimidine dimer)	

EC-No.	Synonym Recommended Name	Vol., Page
3.4.24.40	Escherichia freundii proteinase Serralysin	
3.4.21.62	Esperase Subtilisin	
3.1.8.1	A-esterase Aryldialkylphosphatase	
3.1.1.2	A-esterase arylesterase	
3.1.1.1	B-esterase Carboxylesterase	
3.1.1.72	C-esterase Acetylxylan esterase	
3.1.1.6	C-esterase acetylesterase	
3.1.1.43	esterase, α-amino acid α-amino-acid esterase	
3.1.1.63	Esterase, 11-cis-retinol palmitate 11-cis-Retinyl-palmitate hydrolase	
3.1.1.54	Esterase, acetoxybutynylbithiophene Acetoxybutynylbithiophene deacetylase	
3.1.1.72	Esterase, acetyl Acetylxylan esterase	
3.1.1.7	esterase, acetyl choline acetylcholinesterase	
3.1.1.64	Esterase, all-trans-retinol palmitate all-trans-Retinyl-palmitate hydrolase	
3.1.1.55	Esterase, aspirin Acetylsalicylate deacetylase	
3.1.1.60	Esterase, bis(2-ethylhexyl) phthalate Bis(2-ethylhexyl)phthalate esterase	
3.1.1.8	esterase, butyrylcholine cholinesterase	
3.1.1.72	Esterase, C- Acetylxylan esterase	
3.4.21.42	esterase, C1 complement subcomponent C1sbar	

EC-No.	Synonym Recommended Name	Vol., Page
3.1.1.41	Esterase, cephalosporin C acetyl Cephalosporin-C deacetylase	
3.1.1.70	Esterase, cetraxate benzyl- Cetraxate benzylesterase	
3.1.1.8	esterase, choline cholinesterase	
3.1.1.66	Esterase, diacetoxybutynylbithiophene acetate 5-(3,4-Diacetoxybut-1-ynyl)-2,2'-bithiophene deacetylase	
3.1.1.48	esterase, fusarinine C ornithine fusarinine-C ornithinesterase	
3.1.2.7	esterase, glutathione thiol- glutathione thiolesterase	
3.1.1.50	esterase, jojoba wax wax-ester hydrolase	
3.1.1.59	Esterase, juvenile hormone Juvenile-hormone esterase	
3.1.1.61	Esterase, methyl-accepting chemotaxis protein methyl- Protein-glutamate methylesterase	
3.1.1.44	esterase, methyl oxalacatate 4-methyloxaloacetate esterase	
3.1.8.1	Esterase, organophosphate Aryldialkylphosphatase	
3.1.8.1	Esterase, paraoxon Aryldialkylphosphatase	
3.1.8.1	Esterase, pirimiphos-methyloxon Aryldialkylphosphatase	
3.4.21.46	esterase, properdin factor D complement factor D	
3.1.1.61	Esterase, protein methyl- Protein-glutamate methylesterase	
3.1.1.49	esterase, sinapine sinapine esterase	
3.1.1.10	Esterase, tropine Tropinesterase	
3.1.2.15	Esterase, ubiquitin thiol- Ubiquitin thiolesterase	

EC-No.	Synonym Recommended Name	Vol., Page
3.4.19.12	esterase, ubiquitin thiol- ubiquitinyl hydrolase I	
3.1.1.1	Esterase A Carboxylesterase	
3.1.1.1	Esterase B Carboxylesterase	
3.1.8.1	Esterase B1 Aryldialkylphosphatase	
3.1.1.1	Esterase D Carboxylesterase	
3.1.1.56	Esterase D Methylumbelliferyl-acetate deacetylase	
3.1.8.1	Esterase E4 Aryldialkylphosphatase	
1.14.99.11	Estradiol-17β,hydrogen-donor:oxygen oxidoreductase (6β-hydroxylating) Estradiol 6β-monooxygenase	
1.1.1.62	Estradiol-17β:NAD(P)$^+$ 17-oxidoreductase Estradiol 17β-dehydrogenase	
1.1.1.148	Estradiol 17α-dehydrogenase Estradiol 17α-dehydrogenase	
1.1.1.62	Estradiol 17β-dehydrogenase Estradiol 17β-dehydrogenase	
1.1.1.148	Estradiol 17α-oxidoreductase Estradiol 17α-dehydrogenase	
1.14.99.11	Estradiol 6β-hydroxylase Estradiol 6β-monooxygenase	
1.14.99.11	Estradiol 6β-monooxygenase Estradiol 6β-monooxygenase	
1.1.1.148	17α-Estradiol dehydrogenase Estradiol 17α-dehydrogenase	
1.1.1.62	17β-Estradiol dehydrogenase Estradiol 17β-dehydrogenase	
1.1.1.62	Estradiol dehydrogenase Estradiol 17β-dehydrogenase	
2.4.1.17	Estriol UDPglucuronosyltransferase Glucuronosyltransferase	

EC-No.	Synonym Recommended Name	Vol., Page
1.1.1.62	Estrogen 17-oxidoreductase Estradiol 17β-dehydrogenase	
3.1.6.1	Estrogen sulfatase Arylsulfatase	
2.8.2.2	Estrogen sulfokinase Alcohol sulfotransferase	
2.8.2.2	Estrogen sulfotransferase Alcohol sulfotransferase	
2.8.2.4	Estrogen sulfotransferase Estrone sulfotransferase	
2.8.2.4	Estrogen sulphotransferase Estrone sulfotransferase	
2.8.2.4	Estrone sulfotransferase Estrone sulfotransferase	
2.4.1.17	Estrone UDPglucuronosyltransferase Glucuronosyltransferase	
2.7.7.14	ET Ethanolamine-phosphate cytidylyltransferase	
1.5.5.1	ETF-QO Electron-transferring-flavoprotein dehydrogenase	
1.5.5.1	ETF:ubiquinone oxidoreductase Electron-transferring-flavoprotein dehydrogenase	
2.7.7.14	Ethanolamine-phosphate cytidylyltransferase Ethanolamine-phosphate cytidylyltransferase	
4.2.3.2	ethanolamine-phosphate phospho-lyase ethanolamine-phosphate phospho-lyase	
4.2.99.7	ethanolamine-phosphate phospho-lyase ethanolamine-phosphate phospho-lyase	
4.2.3.2	ethanolamine-phosphate phospho-lyase (deaminating) ethanolamine-phosphate phospho-lyase	
4.2.99.7	ethanolamine-phosphate phospholyase (deaminating) ethanolamine-phosphate phospho-lyase	
1.4.3.8	Ethanolamine:oxygen oxidoreductase (deaminating) Ethanolamine oxidase	
4.3.1.7	ethanolamine ammonia-lyase ethanolamine ammonia-lyase	

EC-No.	Synonym Recommended Name	Vol., Page
4.3.1.7	ethanolamine ammonia-lyase ethanolamine ammonia-lyase	
4.3.1.7	ethanolamine deaminase ethanolamine ammonia-lyase	
2.7.1.82	Ethanolamine kinase Ethanolamine kinase	
1.4.3.8	Ethanolamine oxidase Ethanolamine oxidase	
2.7.7.14	Ethanolamine phosphate cytidylyltransferase Ethanolamine-phosphate cytidylyltransferase	
2.7.1.82	Ethanolamine phosphokinase Ethanolamine kinase	
2.7.8.1	Ethanolamine phosphotransferase Ethanolaminephosphotransferase	
2.7.8.1	Ethanolaminephosphotransferase Ethanolaminephosphotransferase	
2.7.8.1	Ethanolaminephosphotransferase, diacylglycerol Ethanolaminephosphotransferase	
2.7.8.4	Ethanolaminephosphotransferase, serine Serine-phosphoethanolamine synthase	
1.1.1.1	Ethanol dehydrogenase Alcohol dehydrogenase	
1.1.3.13	Ethanol oxidase Alcohol oxidase	
1.2.1.55	Ethyl-(R)-3-hydroxyhexanoate:NADP$^+$ 3-oxidoreductase (R)-3-Hydroxyacid ester dehydrogenase	
1.2.1.56	Ethyl-(S)-3-hydroxyhexanoate:NADP$^+$ 3-oxidoreductase (S)-3-Hydroxyacid ester dehydrogenase	
3.5.99.3	4-(ethylamino)-2-hydroxy-6-(isopropylamino)-1,3,5-triazine ethylaminohydrolase hydroxydechloroatrazine ethylaminohydrolase	
1.3.99.2	Ethylene reductase Butyryl-CoA dehydrogenase	
4.1.3.33	2-Ethylmalate-3-hydroxybutanedioate synthase 2-Ethylmalate synthase	

EC-No.	Synonym Recommended Name	Vol., Page
4.1.3.10	3-ethylmalate glyoxylate-lyase (CoA-butanoylating) 3-ethylmalate synthase	
4.1.3.33	2-Ethylmalate synthase 2-Ethylmalate synthase	
4.1.3.10	3-ethylmalate synthase 3-ethylmalate synthase	
4.1.3.10	β-ethylmalate synthase 3-ethylmalate synthase	
1.1.1.152	Etiocholanolone 3α-dehydrogenase 3α-Hydroxy-5β-androstan-17-one 3α-dehydrogenase	
3.4.21.89	Eukaryotic signal peptidase Signal peptidase I	
3.4.21.89	Eukaryotic signal proteinase Signal peptidase I	
3.4.21.56	euphorbain euphorbain	
3.4.99.7	euphorbain euphorbain	
3.1.25.1	Excision endonuclease UvrABC Deoxyribonuclease (pyrimidine dimer)	
3.1.25.1	Excision nuclease UvrABC Deoxyribonuclease (pyrimidine dimer)	
3.2.1.58	Exo (1→ 3)-β-glucanase Glucan 1,3-β-glucosidase	
4.2.2.13	exo-(1,4)-α-D-Glucan lyase exo-(1,4)-α-D-Glucan lyase	
3.2.1.58	Exo-β-(1-3)-D-glucanase Glucan 1,3-β-glucosidase	
3.2.1.58	Exo-β-(1→ 3)-glucanohydrolase Glucan 1,3-β-glucosidase	
3.2.1.113	Exo-α-1,2-mannanase Mannosyl-oligosaccharide 1,2-α-mannosidase	
3.2.1.72	Exo-β-1,3'-xylanase Xylan 1,3-β-xylosidase	
3.2.1.58	Exo-β-1,3-D-glucanase Glucan 1,3-β-glucosidase	

EC-No.	Synonym Recommended Name	Vol., Page
3.2.1.58	Exo-β-1,3-glucanase Glucan 1,3-β-glucosidase	
3.2.1.74	Exo-β-1,4-glucanase Glucan 1,4-β-glucosidase	
4.2.2.13	Exo-α-1,4-glucan lyase exo-(1,4)-α-D-Glucan lyase	
3.2.1.80	Exo-β-D-fructosidase Fructan β-fructosidase	
3.2.1.31	Exo-β-D-glucuronidase β-glucuronidase	
3.2.1.25	Exo-β-D-mannanase β-mannosidase	
3.2.1.100	Exo-β-Mannanase Mannan 1,4-β-mannobiosidase	
3.2.1.24	Exo-α-mannosidase α-mannosidase	
3.2.1.92	Exo-β-N-acetyl muramidase Peptidoglycan β-N-acetylmuramidase	
3.2.1.92	Exo-β-N-acetylmuramidase Peptidoglycan β-N-acetylmuramidase	
3.2.1.58	Exo-.β-1,3-glucanase Glucan 1,3-β-glucosidase	
3.2.1.58	Exo-.β.-(1→3)-D-glucanase Glucan 1,3-β-glucosidase	
3.2.1.58	Exo-. β.-13-glucanase Glucan 1,3-β-glucosidase	
3.2.1.74	Exo-.β.-1,4-glucanase Glucan 1,4-β-glucosidase	
3.2.1.91	Exo-.β.-1,4-glucan cellobiohydrolase Cellulose 1,4-β-cellobiosidase	
3.2.1.74	Exo-.β.-1,4-glucosidase Glucan 1,4-β-glucosidase	
3.2.1.80	Exo-.β.-fructosidase Fructan β-fructosidase	
3.2.1.100	Exo-.β.-mannanase Mannan 1,4-β-mannobiosidase	

EC-No.	Synonym Recommended Name	Vol., Page
3.2.1.77	Exo-1,2-1,3-α-mannosidase Mannan 1,2-(1,3)-α-mannosidase	
3.2.1.84	Exo-1,3-α-glucanase Glucan 1,3-α-glucosidase	
3.2.1.58	Exo-1,3-β-glucosidase Glucan 1,3-β-glucosidase	
3.2.1.72	Exo-1,3-β-xylosidase Xylan 1,3-β-xylosidase	
3.2.1.58	Exo-1,3-.β.-D-glucanase Glucan 1,3-β-glucosidase	
3.2.1.58	Exo-1,3-.β.-glucanase Glucan 1,3-β-glucosidase	
3.2.1.58	Exo-1,3-.β.-glucosidase Glucan 1,3-β-glucosidase	
3.2.1.72	Exo-1,3-.β.-xylosidase Xylan 1,3-β-xylosidase	
3.2.1.37	Exo-1,4-β-D-xylosidase Xylan 1,4-β-xylosidase	
3.2.1.74	Exo-1,4-β-glucanase Glucan 1,4-β-glucosidase	
3.2.1.3	Exo-1,4-α-glucosidase Glucan 1,4-α-glucosidase	
3.2.1.74	Exo-1,4-β-glucosidase Glucan 1,4-β-glucosidase	
3.2.1.100	Exo-1,4-β-mannobiohydrolase Mannan 1,4-β-mannobiosidase	
3.2.1.37	Exo-1,4-β-xylosidase Xylan 1,4-β-xylosidase	
3.2.1.74	Exo-1,4-.β.-D-glucosidase Glucan 1,4-β-glucosidase	
3.2.1.74	Exo-1,4-.β.-glucosidase Glucan 1,4-β-glucosidase	
3.2.1.37	Exo-1,4-xylosidase Xylan 1,4-β-xylosidase	
3.2.1.70	Exo-1,6-α-glucosidase Glucan 1,6-α-glucosidase	

EC-No.	Synonym Recommended Name	Vol., Page
3.2.1.91	Exo-cellobiohydrolase Cellulose 1,4-β-cellobiosidase	
3.2.1.67	Exo-D-galacturonanase Galacturan 1,4-α-galacturonidase	
3.2.1.67	Exo-D-galacturonase Galacturan 1,4-α-galacturonidase	
3.2.1.94	Exo-isomalto-hydrolase Glucan 1,6-α-isomaltosidase	
3.2.1.95	Exo-isomaltotriohydrolase Dextran 1,6-α-isomaltotriosidase	
3.2.1.98	Exo-maltohexao-hydrolase Glucan 1,4-α-maltohexaosidase	
3.2.1.98	Exo-maltohexaose hydrolase Glucan 1,4-α-maltohexaosidase	
3.2.1.60	Exo-malto tetraohydrolase Glucan 1,4-α-maltotetraohydrolase	
3.2.1.60	Exo-maltotetraohydrolase Glucan 1,4-α-maltotetraohydrolase	
3.2.1.116	Exo-maltotriohydrolase Glucan 1,4-α-maltotriohydrolase	
3.2.1.10	Exo-oligo-1,6-Glucosidase Oligo-1,6-glucosidase	
4.2.2.9	exo-PATE exopolygalacturonate lyase	
4.2.2.9	exo-PGL exopolygalacturonate lyase	
3.2.1.82	Exo-poly-α-galacturonosidase Exo-poly-α-galacturonosidase	
3.2.1.67	Exo-Polygalacturonase Galacturan 1,4-α-galacturonidase	
3.2.1.91	Exocellobiohydrolase Cellulose 1,4-β-cellobiosidase	
3.2.1.74	Exocellulase Glucan 1,4-β-glucosidase	
3.1.21.3	Exodeoxyribonuclease (adenosine triphosphate-hydrolyzing) Type I site-specific deoxyribonuclease	

EC-No.	Synonym Recommended Name	Vol., Page
3.1.11.3	Exodeoxyribonuclease (Lambda induced) Exodeoxyribonuclease (Lambda induced)	
3.1.11.4	exodeoxyribonuclease (phage SP3-induced) exodeoxyribonuclease (phage SP3-induced)	
3.1.11.1	Exodeoxyribonuclease I Exodeoxyribonuclease I	
3.1.4.25	exodeoxyribonuclease I exodeoxyribonuclease I	
3.1.4.26	exodeoxyribonuclease II exodeoxyribonuclease II	
3.1.4.27	exodeoxyribonuclease II exodeoxyribonuclease II	
3.1.11.2	Exodeoxyribonuclease III Exodeoxyribonuclease III	
3.1.11.3	Exodeoxyribonuclease IV Exodeoxyribonuclease (Lambda induced)	
3.1.4.28	exodeoxyribonuclease IV exodeoxyribonuclease IV	
3.1.11.5	Exodeoxyribonuclease V Exodeoxyribonuclease V	
3.1.11.6	Exodeoxyribonuclease VII Exodeoxyribonuclease VII	
3.2.1.70	Exodextranase Glucan 1,6-α-glucosidase	
2.4.2.30	Exoenzyme C3 NAD^+ ADP-ribosyltransferase	
2.4.2.30	Exoenzyme S NAD^+ ADP-ribosyltransferase	
3.2.1.91	Exoglucanase Cellulose 1,4-β-cellobiosidase	
3.2.1.7	Exoinulinase Inulinase	
3.2.1.95	Exoisomaltotriohydrolase Dextran 1,6-α-isomaltotriosidase	
3.2.1.98	Exomaltohexaohydrolase Glucan 1,4-α-maltohexaosidase	

EC-No.	Synonym Recommended Name	Vol., Page
3.2.1.60	Exomaltotetraohydrolase Glucan 1,4-α-maltotetraohydrolase	
3.1.16.1	3'-Exonuclease Spleen exonuclease	
3.1.4.1	5'-Exonuclease Phosphodiesterase I	
3.1.11.1	Exonuclease I Exodeoxyribonuclease I	
3.1.11.3	Exonuclease IV Exodeoxyribonuclease (Lambda induced)	
3.1.11.5	Exonuclease V Exodeoxyribonuclease V	
3.1.11.6	Exonuclease VII Exodeoxyribonuclease VII	
4.2.2.9	exopectate lyase exopolygalacturonate lyase	
4.2.2.9	exopectic acid transeliminase exopolygalacturonate lyase	
3.5.2.6	Exopenicillinase β-lactamase	
3.2.1.67	Exopoly-D-galacturonase Galacturan 1,4-α-galacturonidase	
3.2.1.82	Exopolygalacturanosidase Exo-poly-α-galacturonosidase	
3.2.1.67	Exopolygalacturonase Galacturan 1,4-α-galacturonidase	
4.2.2.9	exopolygalacturonate lyase exopolygalacturonate lyase	
3.2.1.82	Exopolygalacturonosidase Exo-poly-α-galacturonosidase	
3.6.1.11	Exopolyphosphatase Exopolyphosphatase	
3.1.13.4	2',3'-Exoribonuclease Poly(A)-specific ribonuclease	
3.1.13.1	5'-Exoribonuclease Exoribonuclease II	

EC-No.	Synonym Recommended Name	Vol., Page
3.1.13.1	Exoribonuclease Exoribonuclease II	
3.1.4.20	exoribonuclease exoribonuclease	
3.1.13.2	Exoribonuclease H Exoribonuclease H	
3.1.13.1	Exoribonuclease II Exoribonuclease II	
1.11.1.7	Extensin peroxidase Peroxidase	
3.4.21.96	Extracellular lactococcal proteinase Lactocepin	
3.4.22.33	Extranase Fruit bromelain	
1.12.99.1	F420-reducing hydrogenase Coenzyme F420 hydrogenase	
4.1.2.22	F6PPK Fructose-6-phosphate phosphoketolase	
6.2.1.3	FAA1 Long-chain-fatty-acid-CoA ligase	Vol. 2, Page 206
3.4.21.84	Factor C Limulus clotting factor C	
3.4.21.46	factor D complement factor D	
3.4.21.46	factor D (complement) complement factor D	
3.4.21.45	factor I complement factor I	
3.4.21.5	factor IIa thrombin	
3.4.21.22	factor IXa coagulation factor IXa	
3.4.21.22	factor IXaα coagulation factor IXa	
3.4.21.22	factor IXaβ coagulation factor IXa	

EC-No.	Synonym Recommended Name	Vol., Page
3.4.21.22	factor IXaβ' coagulation factor IXa	
3.4.21.22	factor IXaAL coagulation factor IXa	
3.4.21.22	factor IXaCH coagulation factor IXa	
3.4.21.22	factor IXaN coagulation factor IXa	
3.4.21.95	Factor V-activating enzyme Snake venom factor V activator	
3.4.21.95	Factor V activator Snake venom factor V activator	
3.4.21.6	factor Xa coagulation factor Xa	
2.3.2.13	Factor XIIIa Protein-glutamine γ-glutamyltransferase	
1.14.13.8	FAD-containing monooxygenase Dimethylaniline monooxygenase (N-oxide-forming)	
1.6.2.4	FAD-cytochrome c reductase NADPH-ferrihemoprotein reductase	
1.1.99.5	FAD-dependent glycerol-3-phosphate dehydrogenase Glycerol-3-phosphate dehydrogenase	
1.1.99.16	FAD-dependent malate-vitamin K reductase Malate dehydrogenase (acceptor)	
1.1.3.3	FAD-dependent malate oxidase Malate oxidase	
1.1.99.5	FAD-dependent sn-glycerol-3-phosphate dehydrogenase Glycerol-3-phosphate dehydrogenase	
1.1.99.5	FAD-linked glycerol 3-phosphate dehydrogenase Glycerol-3-phosphate dehydrogenase	
1.1.99.5	FAD-linked L-glycerol-3-phosphate dehydrogenase Glycerol-3-phosphate dehydrogenase	
3.6.1.18	FAD diphosphatase FAD diphosphatase	
3.6.1.18	FAD nucleotidohydrolase FAD diphosphatase	

EC-No.	Synonym Recommended Name	Vol., Page
3.6.1.18	FAD pyrophosphatase FAD diphosphatase	
2.7.7.2	FAD pyrophosphorylase FMN adenylyltransferase	
2.7.7.2	FAD synthetase FMN adenylyltransferase	
3.1.1.73	FAE-I feruloyl esterase	
3.1.1.73	FAE-II feruloyl esterase	
3.1.1.73	FAE-III, cinnamoyl ester hydrolase feruloyl esterase	
3.1.1.73	FAEA feruloyl esterase	
3.1.1.67	FAEES Fatty-acyl-ethyl-ester synthase	
3.2.2.23	Fapy-DNA glycosylase Formamidopyrimidine-DNA glycosidase	
5.2.1.9	Farnesol 2-isomerase Farnesol 2-isomerase	Vol. 1, Page 229
1.1.1.216	Farnesol dehydrogenase Farnesol dehydrogenase	
2.5.1.21	Farnesyl-diphosphate:farnesyl-diphosphate farnesyltransferase Farnesyl-diphosphate farnesyltransferase	
2.5.1.21	Farnesyl-diphosphate farnesyltransferase Farnesyl-diphosphate farnesyltransferase	
2.7.4.18	Farnesyl-diphosphate kinase Farnesyl-diphosphate kinase	
2.5.1.10	Farnesyl-diphosphate synthase Geranyltranstransferase	
2.1.1.100	Farnesyl-protein carboxymethyltransferase Protein-S-isoprenylcysteine O-methyltransferase	
2.1.1.100	Farnesylated protein C-terminal O-methyltransferase Protein-S-isoprenylcysteine O-methyltransferase	
2.1.1.100	Farnesyl cysteine C-terminal methyltransferase Protein-S-isoprenylcysteine O-methyltransferase	

EC-No.	Synonym Recommended Name	Vol., Page
2.1.1.100	S-Farnesylcysteine methyltransferase Protein-S-isoprenylcysteine O-methyltransferase	
4.6.1.11	Farnesyl diphosphate-δ-cadinene cyclase (+)-δ-Cadinene synthase	
2.5.1.21	Farnesyldiphosphate:farnesyldiphosphate farnesyltransferase Farnesyl-diphosphate farnesyltransferase	
2.5.1.40	Farnesylpyrophosphate cyclase Aristolochene synthase	
2.7.4.18	Farnesyl pyrophosphate kinase Farnesyl-diphosphate kinase	
2.5.1.10	Farnesyl pyrophosphate synthetase Geranyltranstransferase	
2.5.1.10	Farnesylpyrophosphate synthetase Geranyltranstransferase	
2.5.1.21	Farnesyltransferase Farnesyl-diphosphate farnesyltransferase	
2.5.1.29	Farnesyltransferase Farnesyltranstransferase	
2.5.1.29	Farnesyltranstransferase Farnesyltranstransferase	
1.13.11.12	Fat oxidase Lipoxygenase	
2.1.1.15	Fatty-acid O-methyltransferase Fatty-acid O-methyltransferase	
1.11.1.3	Fatty-acid peroxidase Fatty-acid peroxidase	
2.3.1.85	Fatty-acid synthase Fatty-acid synthase	
2.3.1.142	Fatty-acyl-CoA:mucus-glycoprotein fatty-acyltransferase Glycoprotein O-fatty-acyltransferase	
1.3.99.3	Fatty-acyl-CoA dehydrogenase Acyl-CoA dehydrogenase	
6.2.1.3	Fatty-acyl-CoA ligase Long-chain-fatty-acid-CoA ligase	Vol. 2, Page 206
2.3.1.86	Fatty-acyl-CoA synthase Fatty-acyl-CoA synthase	

EC-No.	Synonym Recommended Name	Vol., Page
3.1.1.67	Fatty-acyl-ethyl-ester synthase Fatty-acyl-ethyl-ester synthase	
1.14.15.3	Fatty acid ω-hydroxylase Alkane 1-monooxygenase	
1.14.99.25	Fatty acid 6-desaturase Linoleoyl-CoA desaturase	
1.14.99.25	Fatty acid Δ^6-desaturase Linoleoyl-CoA desaturase	
1.14.19.3	fatty acid 6-desaturase linoleoyl-CoA desaturase	
1.14.19.3	fatty acid Δ^6-desaturase linoleoyl-CoA desaturase	
1.14.99.5	Fatty acid Δ^9-desaturase Stearoyl-CoA desaturase	
6.2.1.3	Fatty acid CoA ligase Long-chain-fatty-acid-CoA ligase	Vol. 2, Page 206
2.3.1.41	Fatty acid condensing enzyme 3-Oxoacyl-[acyl-carrier-protein] synthase	
1.14.99.1	Fatty acid cyclooxygenase Prostaglandin-endoperoxide synthase	
1.14.99.33	Δ^{12}-fatty acid dehydrogenase Δ^{12}-fatty acid dehydrogenase	
1.14.99.5	Fatty acid desaturase Stearoyl-CoA desaturase	
1.14.19.1	fatty acid desaturase stearoyl-CoA 9-desaturase	
6.2.1.2	Fatty acid elongase Butyrate-CoA ligase	Vol. 2, Page 199
6.2.1.3	Fatty acid elongase Long-chain-fatty-acid-CoA ligase	Vol. 2, Page 206
3.1.1.67	Fatty acid ethyl ester synthase Fatty-acyl-ethyl-ester synthase	
2.1.1.15	Fatty acid methyltransferase Fatty-acid O-methyltransferase	
2.1.1.15	Fatty acid O-methyltransferase Fatty-acid O-methyltransferase	

EC-No.	Synonym Recommended Name	Vol., Page
1.11.1.3	Fatty acid peroxidase Fatty-acid peroxidase	
6.2.1.3	Fatty acid thiokinase (long chain) Long-chain-fatty-acid-CoA ligase	Vol. 2, Page 206
6.2.1.2	Fatty acid thiokinase (medium chain) Butyrate-CoA ligase	Vol. 2, Page 199
2.3.1.139	Fatty acyl-CoA:ecdysone acyltransferase Ecdysone O-acyltransferase	
1.14.99.25	Δ^6-fatty acyl-CoA desaturase Linoleoyl-CoA desaturase	
1.14.19.3	Δ^6-fatty acyl-CoA desaturase linoleoyl-CoA desaturase	
1.14.99.5	Fatty acyl-CoA desaturase Stearoyl-CoA desaturase	
1.3.3.6	Fatty acyl-CoA oxidase Acyl-CoA oxidase	
1.3.3.6	Fatty acyl-coenzyme A oxidase Acyl-CoA oxidase	
6.2.1.3	Fatty acyl-coenzyme A synthetase Long-chain-fatty-acid-CoA ligase	Vol. 2, Page 206
1.14.99.5	Fatty acyl Δ^9-desaturase Stearoyl-CoA desaturase	
3.5.1.4	Fatty acylamidase Amidase	
1.14.99.5	Fatty acyl CoA Δ^9-desaturase Stearoyl-CoA desaturase	
1.3.99.3	Fatty acyl coenzyme A dehydrogenase Acyl-CoA dehydrogenase	
3.1.1.23	fatty acyl monoester lipase acylglycerol lipase	
3.1.2.2	fatty acyl thioesterase I palmitoyl-CoA hydrolase	
1.1.3.20	Fatty alcohol:oxygen oxidoreductase Long-chain-alcohol oxidase	
1.1.3.20	Fatty alcohol oxidase Long-chain-alcohol oxidase	

EC-No.	Synonym Recommended Name	Vol., Page
1.1.1.192	Fatty alcohol oxidoreductase Long-chain-alcohol dehydrogenase	
1.2.1.48	Fatty aldehyde:NAD$^+$ oxidoreductase Long-chain-aldehyde dehydrogenase	
1.5.1.6	FBP-CI proteins Formyltetrahydrofolate dehydrogenase	
3.1.3.11	FBPase Fructose-bisphosphatase	
1.2.1.2	N-FDH Formate dehydrogenase	
1.2.1.2	FDH I Formate dehydrogenase	
1.2.1.2	FDH II Formate dehydrogenase	
1.16.3.1	Fe(II):oxygen oxidoreductase Ferroxidase	
1.15.1.1	Fe-SOD Superoxide dismutase	
1.1.1.182	fenchol dehydronase fenchol dehydronase	
3.2.1.136	Feraxanase Glucuronoarabinoxylan endo-1,4-β-xylanase	
3.2.1.136	Feraxan endoxylanase Glucuronoarabinoxylan endo-1,4-β-xylanase	
1.4.7.1	Ferredoxin-dependent glutamate synthase Glutamate synthase (ferredoxin)	
1.4.7.1	Ferredoxin-glutamate synthase Glutamate synthase (ferredoxin)	
1.18.1.3	Ferredoxin-linked NAD reductase Ferredoxin-NAD$^+$ reductase	
1.18.1.3	Ferredoxin-NAD$^+$ reductase Ferredoxin-NAD$^+$ reductase	
1.18.1.2	Ferredoxin-NADP$^+$ reductase Ferredoxin-NADP$^+$ reductase	
1.6.7.1	ferredoxin-NADP$^+$ reductase ferredoxin-NADP$^+$ reductase	

EC-No.	Synonym Recommended Name	Vol., Page
1.18.1.2	Ferredoxin-NADP-oxidoreductase Ferredoxin-NADP$^+$ reductase	
1.18.1.2	Ferredoxin-NADP oxidoreductase Ferredoxin-NADP$^+$ reductase	
1.18.1.2	Ferredoxin-NADP reductase Ferredoxin-NADP$^+$ reductase	
1.18.1.3	Ferredoxin-NAD reductase Ferredoxin-NAD$^+$ reductase	
1.18.1.2	Ferredoxin-nicotinamide-adenine dinucleotide phosphate (oxidized) reductase Ferredoxin-NADP$^+$ reductase	
1.7.7.2	Ferredoxin-nitrate reductase Ferredoxin-nitrate reductase	
1.7.7.1	Ferredoxin-nitrite reductase Ferredoxin-nitrite reductase	
1.8.7.1	Ferredoxin-sulfite reductase Sulfite reductase (ferredoxin)	
1.18.1.2	Ferredoxin-TPN reductase Ferredoxin-NADP$^+$ reductase	
1.18.99.1	Ferredoxin:H$^+$ oxidoreductase Hydrogenase	
1.18.1.3	Ferredoxin:NAD$^+$ oxidoreductase Ferredoxin-NAD$^+$ reductase	
1.18.1.2	Ferredoxin:NADP$^+$ oxidoreductase Ferredoxin-NADP$^+$ reductase	
1.18.1.2	Ferredoxin:NADP$^+$ oxidoreductase Ferredoxin-NADP$^+$ reductase	
1.18.99.1	Ferredoxin hydrogenase Hydrogenase	
1.12.7.1	ferredoxin hydrogenase ferredoxin hydrogenase	
1.6.99.13	Ferric-chelate reductase Ferric-chelate reductase	
1.6.99.13	Ferric chelate reductase Ferric-chelate reductase	

EC-No.	Synonym Recommended Name	Vol., Page
1.6.2.6	Ferric leghemoglobin reductase Leghemoglobin reductase	
1.6.2.4	Ferrihemprotein P450 reductase NADPH-ferrihemoprotein reductase	
1.15.1.1	Ferrisuperoxide dismutase Superoxide dismutase	
1.16.3.1	Ferro:O_2 oxidoreductase Ferroxidase	
4.99.1.1	ferrochelatase ferrochelatase	
1.9.99.1	Ferrocytochrome-c:Fe^{3+} oxidoreductase Iron-cytochrome-c reductase	
1.11.1.5	Ferrocytochrome-c:hydrogen-peroxide oxidoreductase Cytochrome-c peroxidase	
1.9.3.1	Ferrocytochrome-c:oxygen oxidoreductase Cytochrome-c oxidase	
1.9.3.2	Ferrocytochrome-c:oxygen oxidoreductase Pseudomonas cytochrome oxidase	
1.9.6.1	Ferrocytochrome:nitrate oxidoreductase Nitrate reductase (cytochrome)	
1.9.3.1	Ferrocytochrome c oxidase Cytochrome-c oxidase	
1.16.3.1	Ferroxidase Ferroxidase	
1.16.3.1	Ferroxidase, iron II:oxygen oxidoreductase Ferroxidase	
1.16.3.1	Ferroxidase I Ferroxidase	
3.1.1.73	ferulic acid esterase feruloyl esterase	
2.3.1.130	Feruloyl-CoA:galactarate O-(hydroxycinnamoyl)transferase Galactarate O-hydroxycinnamoyltransferase	
2.3.1.99	Feruloyl-CoA:quinate O-(hydroxycinnamoyl)transferase Quinate O-hydroxycinnamoyltransferase	
2.3.1.110	Feruloyl-CoA:tyramine N-(hydroxycinnamoyl)transferase Tyramine N-feruloyltransferase	

EC-No.	Synonym Recommended Name	Vol., Page
1.2.1.44	Feruloyl-CoA reductase Cinnamoyl-CoA reductase	
2.3.1.110	Feruloyl-CoA tyramine N-feruloyl-CoA transferase Tyramine N-feruloyltransferase	
6.2.1.12	Feruloyl CoA ligase 4-Coumarate-CoA ligase	Vol. 2, Page 256
1.2.1.44	Feruloyl coenzyme A reductase Cinnamoyl-CoA reductase	
6.2.1.12	Feruloyl coenzyme A synthetase 4-Coumarate-CoA ligase	Vol. 2, Page 256
3.1.1.73	feruloyl esterase feruloyl esterase	
3.5.1.71	N-Feruloylglycine amidohydrolase N-Feruloylglycine deacylase	
3.5.1.71	N-Feruloylglycine deacylase N-Feruloylglycine deacylase	
1.2.1.44	Ferulyl-CoA reductase Cinnamoyl-CoA reductase	
2.4.1.100	FFT 1,2-β-Fructan 1F-fructosyltransferase	
6.3.5.3	FGAM synthetase Phosphoribosylformylglycinamidine synthase	Vol. 2, Page 666
6.3.5.3	FGAR amidotransferase Phosphoribosylformylglycinamidine synthase	Vol. 2, Page 666
3.1.2.12	FGH S-formylglutathione hydrolase	
6.3.2.12	FHFS Dihydrofolate synthase	Vol. 2, Page 466
6.3.2.12	FHFS/FPGS Dihydrofolate synthase	Vol. 2, Page 466
3.4.21.7	fibrinase plasmin	
3.4.21.74	α-Fibrinogenase Venombin A	
3.4.21.5	fibrinogenase thrombin	

EC-No.	Synonym Recommended Name	Vol., Page
2.3.2.13	Fibrinoligase Protein-glutamine γ-glutamyltransferase	
3.4.21.7	fibrinolysin plasmin	
3.4.24.72	Fibrinolytic proteinase Fibrolase	
2.3.2.13	Fibrin stabilizing factor Protein-glutamine γ-glutamyltransferase	
3.4.24.72	Fibrolase Fibrolase	
3.4.22.3	ficain ficain	
3.4.22.3	ficin ficain	
3.4.4.12	ficin ficin	
3.4.22.3	ficin E ficain	
3.4.22.3	ficin S ficain	
2.1.2.4	FIG formiminotransferase Glycine formiminotransferase	
1.13.12.7	Firefly luciferase Photinus-luciferin 4-monooxygenase (ATP-hydrolysing)	
1.13.12.7	Firefly luciferin luciferase Photinus-luciferin 4-monooxygenase (ATP-hydrolysing)	
3.4.21.22	FIXa coagulation factor IXa	
3.4.21.22	fIXa coagulation factor IXa	
2.7.1.26	FK Riboflavin kinase	
5.2.1.8	FKBP65 Peptidylprolyl isomerase	Vol. 1, Page 218
1.14.11.9	Flavanone 3-hydroxylase Naringenin 3-dioxygenase	

EC-No.	Synonym / Recommended Name	Vol., Page
1.14.11.9	Flavanone 3β-hydroxylase Naringenin 3-dioxygenase	
1.1.1.234	Flavanone 4-reductase Flavanone 4-reductase	
2.4.1.185	Flavanone 7-O-β-glucosyltransferase Flavanone 7-O-β-glucosyltransferase	
5.5.1.6	Flavanone lyase (decyclizing) Chalcone isomerase	Vol. 1, Page 691
2.3.1.74	Flavanone synthase Naringenin-chalcone synthase	
1.14.11.9	Flavanone synthase I Naringenin 3-dioxygenase	
2.3.1.74	Flavanone synthetase Naringenin-chalcone synthase	
3.4.24.76	flavastacin flavastacin	
1.14.13.8	Flavin-containing monooxygenase Dimethylaniline monooxygenase (N-oxide-forming)	
1.1.99.5	Flavin-linked glycerol-3-phosphate dehydrogenase Glycerol-3-phosphate dehydrogenase	
3.6.1.18	Flavin adenine dinucleotide pyrophosphatase FAD diphosphatase	
3.6.1.18	Flavine adenine dinucleotide pyrophosphatase FAD diphosphatase	
1.6.8.1	Flavine mononucleotide reductase NAD(P)H dehydrogenase (FMN)	
1.6.8.2	Flavine mononucleotide reductase NADPH dehydrogenase (flavin)	
1.6.8.1	Flavin mononucleotide reductase NAD(P)H dehydrogenase (FMN)	
1.6.8.2	Flavin mononucleotide reductase NADPH dehydrogenase (flavin)	
1.14.13.8	Flavin monooxygenase Dimethylaniline monooxygenase (N-oxide-forming)	
1.6.8.2	Flavin reductase NADPH dehydrogenase (flavin)	

EC-No.	Synonym Recommended Name	Vol., Page
3.4.21.91	Flavivirin Flavivirin	
1.1.2.3	Flavocytochrome b2 L-Lactate dehydrogenase (cytochrome)	
2.7.1.26	Flavokinase Riboflavin kinase	
2.3.1.115	Flavone/flavonol 7-O-β-D-glucoside malonyltransferase Isoflavone-7-O-β-glucoside 6-O-malonyltransferase	
2.4.1.81	Flavone 7-O-β-glucosyltransferase Flavone 7-O-β-glucosyltransferase	
2.4.2.25	Flavone apiosyltransferase Flavone apiosyltransferase	
1.14.13.21	Flavonoid,NADPH:oxygen oxidoreductase (3'-hydroxylating) Flavonoid 3'-monooxygenase	
1.14.13.21	Flavonoid 3'-hydroxylase Flavonoid 3'-monooxygenase	
1.14.13.21	Flavonoid 3'-monooxygenase Flavonoid 3'-monooxygenase	
1.14.13.21	Flavonoid 3-hydroxylase Flavonoid 3'-monooxygenase	
2.1.1.75	Flavonoid O-methyltransferase Apigenin 4'-O-methyltransferase	
2.3.1.116	Flavonol-3-O-β-glucoside O-malonyltransferase Flavonol-3-O-β-glucoside O-malonyltransferase	
2.4.1.159	Flavonol-3-O-glucoside L-rhamnosyltransferase Flavonol-3-O-glucoside L-rhamnosyltransferase	
2.4.2.35	Flavonol-3-O-glycoside xylosyltransferase Flavonol-3-O-glycoside xylosyltransferase	
1.13.11.24	Flavonol 2,4-oxygenase Quercetin 2,3-dioxygenase	
2.8.2.26	Flavonol 3'-sulfotransferase Quercetin-3-sulfate 3'-sulfotransferase	
2.4.1.91	Flavonol 3-O-glucosyltransferase Flavonol 3-O-glucosyltransferase	
2.1.1.76	Flavonol 3-O-methyltransferase Quercetin 3-O-methyltransferase	

EC-No.	Synonym Recommended Name	Vol., Page
2.8.2.25	Flavonol 3-sulfotransferase Flavonol 3-sulfotransferase	
2.1.1.83	Flavonol 4'-O-methyltransferase 3,7-Dimethylquercitin 4'-O-methyltransferase	
2.8.2.27	Flavonol 4'-sulfotransferase Quercetin-3-sulfate 4'-sulfotransferase	
2.1.1.84	Flavonol 6-O-methyltransferase Methylquercetagetin 6-O-methyltransferase	
2.1.1.82	Flavonol 7-O-methyltransferase 3-Methylquercitin 7-O-methyltransferase	
2.8.2.28	Flavonol 7-sulfotransferase Quercetin-3,3'-bissulfate 7-sulfotransferase	
2.1.1.88	Flavonol 8-O-methyltransferase 8-Hydroxyquercitin 8-O-methyltransferase	
1.1.99.5	Flavoprotein-linked L-glycerol 3-phosphate dehydrogenase Glycerol-3-phosphate dehydrogenase	
1.14.14.1	Flavoprotein-linked monooxygenase Unspecific monooxygenase	
1.14.14.1	Flavoprotein monooxygenase Unspecific monooxygenase	
1.6.99.2	Flavoprotein NAD(P)H-quinone reductase NAD(P)H dehydrogenase (quinone)	
1.6.5.5	Flavoprotein NAD(P)H:quinone reductase NADPH:quinone reductase	
2.4.1.137	Floridoside-phosphate synthase Sn-Glycerol-3-phosphate 2-α-galactosyltransferase	
2.4.1.137	Floridoside phosphate synthase Sn-Glycerol-3-phosphate 2-α-galactosyltransferase	
3.1.3.48	FLP1 protein-tyrosine-phosphatase	
1.1.1.256	fluoren-9-ol:NAD(P) oxidoreductase fluoren-9-ol dehydrogenase	
1.1.1.256	fluoren-9-ol dehydrogenase fluoren-9-ol dehydrogenase	
2.7.1.40	Fluorokinase Pyruvate kinase	

EC-No.	Synonym Recommended Name	Vol., Page
1.2.99.5	Fmd Formylmethanofuran dehydrogenase	
3.4.19.7	(fMet)-releasing enzyme N-Formylmethionyl-peptidase	
2.7.7.2	FMN adenylyltransferase FMN adenylyltransferase	
1.6.8.1	FMN reductase NAD(P)H dehydrogenase (FMN)	
1.6.8.2	FMN reductase (NADPH) NADPH dehydrogenase (flavin)	
1.14.13.8	FMO Dimethylaniline monooxygenase (N-oxide-forming)	
1.14.13.8	FMO-I Dimethylaniline monooxygenase (N-oxide-forming)	
1.14.13.8	FMO-II Dimethylaniline monooxygenase (N-oxide-forming)	
1.5.1.6	Folate-binding proteins, cytosol I Formyltetrahydrofolate dehydrogenase	
2.1.1.74	Folate-dependent ribothymidyl synthase Methylenetetrahydrofolate-tRNA-(uracil-5-)-methyltransferase ($FADH_2$-oxidizing)	
3.4.19.9	Folate conjugase γ-Glu-X carboxypeptidase	
6.3.2.17	Folate polyglutamate synthetase Tetrahydrofolylpolyglutamate synthase	Vol. 2, Page 488
3.4.19.9	Folic acid conjugase γ-Glu-X carboxypeptidase	
1.5.1.3	Folic acid reductase Dihydrofolate reductase	
1.5.1.3	Folic reductase Dihydrofolate reductase	
6.3.2.17	Folylpoly(.γ.-glutamate) synthase Tetrahydrofolylpolyglutamate synthase	Vol. 2, Page 488
6.3.2.12	Folylpoly-(γ-glutamate) synthetase-dihydrofolate synthase Dihydrofolate synthase	Vol. 2, Page 466

EC-No.	Synonym Recommended Name	Vol., Page
6.3.2.12	Folylpoly-γ-glutamate synthetase-dihydrofolate synthetase Dihydrofolate synthase	Vol. 2, Page 466
6.3.2.17	Folylpoly-γ-glutamate synthetase-dihydrofolate synthetase Tetrahydrofolylpolyglutamate synthase	Vol. 2, Page 488
6.3.2.17	Folylpoly-.γ.-glutamate synthase Tetrahydrofolylpolyglutamate synthase	Vol. 2, Page 488
6.3.2.17	Folylpolyglutamate synthetase Tetrahydrofolylpolyglutamate synthase	Vol. 2, Page 488
6.3.2.17	Folylpolyglutamyl synthetase Tetrahydrofolylpolyglutamate synthase	Vol. 2, Page 488
3.4.22.28	Foot-and-mouth protease 3C Picornain 3C	
1.2.99.4	Formaldehyde:formaldehyde oxidoreductase Formaldehyde dismutase	
1.2.1.46	Formaldehyde:NAD^+ oxidoreductase Formaldehyde dehydrogenase	
1.2.1.1	Formaldehyde:NAD^+ oxidoreductase (glutathione-formylating) Formaldehyde dehydrogenase (glutathione)	
1.2.1.66	formaldehyde:NAD oxidoreductase (mycothiol-formylating) mycothiol-dependent formaldehyde dehydrogenase	
1.2.1.46	Formaldehyde dehydrogenase Formaldehyde dehydrogenase	
1.2.1.1	Formaldehyde dehydrogenase Formaldehyde dehydrogenase (glutathione)	
1.2.1.1	Formaldehyde dehydrogenase (glutathione) Formaldehyde dehydrogenase (glutathione)	
1.2.99.4	Formaldehyde dismutase Formaldehyde dismutase	
2.2.1.3	Formaldehyde transketolase Formaldehyde transketolase	
3.5.1.49	Formamidase Formamidase	
3.5.1.9	Formamidase I Arylformamidase	
3.5.1.9	Formamidase II Arylformamidase	

EC-No.	Synonym Recommended Name	Vol., Page
3.5.1.49	Formamide amidohydrolase Formamidase	
4.2.1.66	formamide dehydratase cyanide hydratase	
4.2.1.66	formamide hydro-lyase cyanide hydratase	
6.3.3.1	2-(Formamido)-N^1-(5-phosphoribosyl)acetamidine cyclo-ligase (ADP-forming) Phosphoribosylformylglycinamidine cyclo-ligase	Vol. 2, Page 530
6.3.5.3	2-Formamido-N-ribosylacetamide 5'-phosphate:L-glutamine amido-ligase (adenosine diphosphate) Phosphoribosylformylglycinamidine synthase	Vol. 2, Page 666
3.2.2.23	Formamidopyrimidine-DNA glycosidase Formamidopyrimidine-DNA glycosidase	
3.2.2.23	Formamidopyrimidine-DNA glycosylase Formamidopyrimidine-DNA glycosidase	
6.3.4.17	Formate-dihydrofolate ligase Formate-dihydrofolate ligase	Vol. 2, Page 649
2.3.1.54	Formate-lyase, pyruvate Formate C-acetyltransferase	
1.97.1.4	Formate-lyase-activating enzyme Formate acetyltransferase activating enzyme	
1.2.1.2	Formate-NAD oxidoreductase Formate dehydrogenase	
6.3.4.3	Formate-tetrahydrofolate ligase Formate-tetrahydrofolate ligase	Vol. 2, Page 567
1.2.2.1	Formate:cytochrome b1 oxidoreductase Formate dehydrogenase (cytochrome)	
6.3.4.17	Formate:dihydrofolate ligase (ADP-forming) Formate-dihydrofolate ligase	Vol. 2, Page 649
1.2.2.1	Formate: ferricytochrome-b1 oxidoreductase Formate dehydrogenase (cytochrome)	
1.2.2.3	Formate:ferricytochrome-c-553 oxidoreductase Formate dehydrogenase (cytochrome-c-553)	
1.2.1.2	Formate:NAD^+ oxidoreductase Formate dehydrogenase	

EC-No.	Synonym Recommended Name	Vol., Page
1.2.1.2	Formate:NAD oxidoreductase Formate dehydrogenase	
1.2.1.43	Formate:NADP$^+$ oxidoreductase Formate dehydrogenase (NADP$^+$)	
6.3.4.3	Formate:tetrahydrofolate ligase (ADP-forming) Formate-tetrahydrofolate ligase	Vol. 2, Page 567
1.97.1.4	Formate acetyltransferase-glycine dihydroflavodoxin:S-adenosyl-L-methionine oxidoreductase (S-adenosyl-L-methionine cleaving) Formate acetyltransferase activating enzyme	
1.97.1.4	Formate acetyltransferase activase Formate acetyltransferase activating enzyme	
1.97.1.4	Formate acetyltransferase activating enzyme Formate acetyltransferase activating enzyme	
1.2.1.2	Formate benzyl-viologen oxidoreductase Formate dehydrogenase	
2.3.1.54	Formate C-acetyltransferase Formate C-acetyltransferase	
1.2.1.2	Formate dehydrogenase Formate dehydrogenase	
1.2.2.1	Formate dehydrogenase Formate dehydrogenase (cytochrome)	
1.2.2.1	Formate dehydrogenase (cytochrome) Formate dehydrogenase (cytochrome)	
1.2.2.3	Formate dehydrogenase (cytochrome-c-553) Formate dehydrogenase (cytochrome-c-553)	
1.2.1.2	Formate dehydrogenase (NAD) Formate dehydrogenase	
1.2.1.43	Formate dehydrogenase (NADP) Formate dehydrogenase (NADP$^+$)	
1.2.1.43	Formate dehydrogenase (NADP$^+$) Formate dehydrogenase (NADP$^+$)	
1.2.1.2	Formate hydrogenlyase Formate dehydrogenase	
2.7.2.6	Formate kinase Formate kinase	

EC-No.	Synonym Recommended Name	Vol., Page
1.2.1.2	Formic acid dehydrogenase Formate dehydrogenase	
1.2.1.1	Formic dehydrogenase Formaldehyde dehydrogenase (glutathione)	
1.2.1.2	Formic hydrogen-lyase Formate dehydrogenase	
3.5.3.5	N-Formimino-L-aspartate iminohydrolase Formiminoaspartate deiminase	
3.5.3.8	N-Formimino-L-glutamate formiminohydrolase Formiminoglutamase	
3.5.3.13	N-Formimino-L-glutamate iminohydrolase Formiminoglutamate deiminase	
4.3.1.4	formimino-THF cyclodeaminase formiminotetrahydrofolate cyclodeaminase	
3.5.3.5	Formiminoaspartate deiminase Formiminoaspartate deiminase	
3.5.3.8	Formiminoglutamase Formiminoglutamase	
3.5.3.13	Formiminoglutamate deiminase Formiminoglutamate deiminase	
3.5.3.13	Formiminoglutamate deiminase Formiminoglutamate deiminase	
3.5.3.8	N-Formiminoglutamate hydrolase Formiminoglutamase	
2.1.2.5	Formiminoglutamic acid transferase Glutamate formiminotransferase	
2.1.2.5	Formiminoglutamic formiminotransferase Glutamate formiminotransferase	
3.5.3.13	Formiminoglutamic iminohydrolase Formiminoglutamate deiminase	
2.1.2.4	Formiminoglycine formiminotransferase Glycine formiminotransferase	
2.1.2.4	5-Formiminotetrahydrofolate:glycine N-formiminotransferase Glycine formiminotransferase	
2.1.2.5	5-Formiminotetrahydrofolate:L-glutamate N-formiminotransferase Glutamate formiminotransferase	

EC-No.	Synonym Recommended Name	Vol., Page
4.3.1.4	5-formiminotetrahydrofolate ammonia-lyase (cyclizing) formiminotetrahydrofolate cyclodeaminase	
4.3.1.4	formiminotetrahydrofolate cyclodeaminase formiminotetrahydrofolate cyclodeaminase	
2.1.2.5	Formiminotransferase, glutamate Glutamate formiminotransferase	
2.1.2.4	Formiminotransferase, glycine Glycine formiminotransferase	
4.3.1.4	formiminotransferase cyclodeaminase formiminotetrahydrofolate cyclodeaminase	
1.14.13.53	Formononetin,NADPH:oxygen oxidoreductase (2'-hydroxylating) Isoflavone 2'-hydroxylase	
1.14.13.52	Formononetin,NADPH:oxygen oxidoreductase (3'-hydroxylating) Isoflavone 3'-hydroxylase	
3.1.2.10	formyl-CoA hydrolase formyl-CoA hydrolase	
3.1.2.10	formyl-CoA hydrolase formyl-CoA hydrolase	
1.5.1.6	10-Formyl-H_2PtGlu:NADP oxidoreductase Formyltetrahydrofolate dehydrogenase	
1.5.1.6	10-Formyl-H_4folate dehydrogenase Formyltetrahydrofolate dehydrogenase	
3.5.1.8	N-Formyl-L-aspartate amidohydrolase Formylaspartate deformylase	
3.5.1.68	N-Formyl-L-glutamate amidohydrolase N-Formylglutamate deformylase	
3.5.1.31	N-Formyl-L-methionine amidohydrolase Formylmethionine deformylase	
3.5.1.27	N-Formyl-L-methionylaminoacyl-tRNA amidohydrolase N-Formylmethionylaminoacyl-tRNA deformylase	
3.5.4.9	Formyl-methenyl-methylenetetrahydrofolate synthetase (combined) Methenyltetrahydrofolate cyclohydrolase	
6.3.4.3	Formyl-THF synthetase Formate-tetrahydrofolate ligase	Vol. 2, Page 567
3.5.1.9	Formylase Arylformamidase	

EC-No.	Synonym Recommended Name	Vol., Page
3.5.1.8	Formylaspartate deformylase Formylaspartate deformylase	
3.5.1.8	Formylaspartic formylase (Formylase I, Formylase II) Formylaspartate deformylase	
1.2.1.62	4-formylbenzenesulfonate:NAD oxidoreductase 4-formylbenzenesulfonate dehydrogenase	
1.2.1.62	4-formylbenzenesulfonate dehydrogenase 4-formylbenzenesulfonate dehydrogenase	
3.1.2.10	formyl coenzyme A hydrolase formyl-CoA hydrolase	
6.3.4.17	Formyl dihydrofolate synthase Formate-dihydrofolate ligase	Vol. 2, Page 649
3.5.1.68	N-Formylglutamate deformylase N-Formylglutamate deformylase	
3.1.2.12	S-formylglutathione hydrolase S-formylglutathione hydrolase	
3.1.2.12	S-formylglutathione hydrolase S-formylglutathione hydrolase	
6.3.5.3	Formylglycinamide ribonucloetide amidotransferase Phosphoribosylformylglycinamidine synthase	Vol. 2, Page 666
3.5.1.9	Formylkynureninase Arylformamidase	
3.5.1.9	Formylkynurenine formamidase Arylformamidase	
2.3.1.101	N-Formylmethanofuran(CHO-MFR):tetrahydromethanopterin(H4MPT) formyltransferase Formylmethanofuran-tetrahydromethanopterin N-formyltransferase	
2.3.1.101	Formylmethanofuran-tetrahydromethanopterin formyltransferase Formylmethanofuran-tetrahydromethanopterin N-formyltransferase	
2.3.1.101	Formylmethanofuran-tetrahydromethanopterin N-formyltransferase Formylmethanofuran-tetrahydromethanopterin N-formyltransferase	
1.2.99.5	Formylmethanofuran:(acceptor) oxidoreductase Formylmethanofuran dehydrogenase	
2.3.1.101	Formylmethanofuran:5,6,7,8-tetrahydromethanopterin N^5-formyltransferase Formylmethanofuran-tetrahydromethanopterin N-formyltransferase	

EC-No.	Synonym Recommended Name	Vol., Page
2.3.1.101	Formylmethanofuran:tetrahydromethanopterin formyltransferase Formylmethanofuran-tetrahydromethanopterin N-formyltransferase	
1.2.99.5	Formylmethanofuran dehydrogenase Formylmethanofuran dehydrogenase	
3.4.19.1	N-formylmethionine (fMet) aminopeptidase acylaminoacyl-peptidase	
3.4.19.7	Formylmethionine aminopeptidase N-Formylmethionyl-peptidase	
3.5.1.31	Formylmethionine deformylase Formylmethionine deformylase	
3.4.19.7	N-Formylmethionyl-peptidase N-Formylmethionyl-peptidase	
2.1.2.9	Formylmethionyl-transfer ribonucleic synthetase Methionyl-tRNA formyltransferase	
3.5.1.27	N-Formylmethionylaminoacyl-tRNA deformylase N-Formylmethionylaminoacyl-tRNA deformylase	
2.1.2.3	10-Formyltetrahydrofolate:5'-phosphoribosyl-5-amino-4-imidazolecarboxamide formyltransferase Phosphoribosylaminoimidazolecarboxamide formyltransferase	
2.1.2.3	10-Formyltetrahydrofolate:5'-phosphoribosyl-5-amino-4-imidazolecarboxamide N-formyltransferase Phosphoribosylaminoimidazolecarboxamide formyltransferase	
2.1.2.2	10-Formyltetrahydrofolate:5'-phosphoribosylglycinamide N-formyltransferase Phosphoribosylglycinamide formyltransferase	
2.1.2.9	10-Formyltetrahydrofolate:L-methionyl-tRNA N-formyltransferase Methionyl-tRNA formyltransferase	
1.5.1.6	10-Formyltetrahydrofolate:$NADP^+$ oxidoreductase Formyltetrahydrofolate dehydrogenase	
1.5.1.6	10-Formyl tetrahydrofolate:NADP oxidoreductase Formyltetrahydrofolate dehydrogenase	
3.5.1.10	10-Formyltetrahydrofolate amidohydrolase Formyltetrahydrofolate deformylase	
6.3.3.2	5-Formyltetrahydrofolate cyclo-ligase 5-Formyltetrahydrofolate cyclo-ligase	Vol. 2, Page 535
6.3.3.2	5-Formyltetrahydrofolate cyclo-ligase (ADP-forming) 5-Formyltetrahydrofolate cyclo-ligase	Vol. 2, Page 535

EC-No.	Synonym Recommended Name	Vol., Page
6.3.3.2	5-Formyltetrahydrofolate cyclodehydrase 5-Formyltetrahydrofolate cyclo-ligase	Vol. 2, Page 535
3.5.1.10	Formyltetrahydrofolate deformylase Formyltetrahydrofolate deformylase	
1.5.1.6	10-Formyltetrahydrofolate dehydrogenase Formyltetrahydrofolate dehydrogenase	
1.5.1.6	Formyltetrahydrofolate dehydrogenase Formyltetrahydrofolate dehydrogenase	
6.3.4.3	10-Formyltetrahydrofolate synthetase Formate-tetrahydrofolate ligase	Vol. 2, Page 567
6.3.4.3	Formyltetrahydrofolate synthetase Formate-tetrahydrofolate ligase	Vol. 2, Page 567
6.3.3.2	Formyltetrahydrofolic cyclodehydrase 5-Formyltetrahydrofolate cyclo-ligase	Vol. 2, Page 535
6.3.2.17	Formyltetrahydropteroyldiglutamate synthetase Tetrahydrofolylpolyglutamate synthase	Vol. 2, Page 488
6.3.4.17	Formyltransferase, dihydrofolate Formate-dihydrofolate ligase	Vol. 2, Page 649
2.3.1.101	Formyltransferase, formylmethanofuran-tetrahydromethanopterin Formylmethanofuran-tetrahydromethanopterin N-formyltransferase	
2.1.2.3	Formyltransferase, phosphoribosylaminoimidazolecarboxamide Phosphoribosylaminoimidazolecarboxamide formyltransferase	
2.1.2.2	Formyltransferase, phosphoribosylglycinamide Phosphoribosylglycinamide formyltransferase	
2.4.1.88	Forssman synthase Globoside α-N-acetylgalactosaminyltransferase	
3.2.2.23	Fpg protein Formamidopyrimidine-DNA glycosidase	
6.3.2.17	FPGS Tetrahydrofolylpolyglutamate synthase	Vol. 2, Page 488
2.5.1.40	FPP-carbocyclase Aristolochene synthase	
2.4.1.137	FPS Sn-Glycerol-3-phosphate 2-α-galactosyltransferase	
2.7.1.112	v-fps Protein-tyrosine kinase Protein-tyrosine kinase	

EC-No.	Synonym Recommended Name	Vol., Page
3.4.24.74	Fragilysin Fragilysin	
3.4.23.23	Fromase 100 Mucorpepsin	
3.4.23.23	Fromase 46TL Mucorpepsin	
6.1.1.20	FRS Phenylalanine-tRNA ligase	Vol. 2, Page 156
4.1.2.13	Fru-P2A Fructose-bisphosphate aldolase	
3.2.1.64	β2,6-fructan-6-levanbiohydrolase 2,6-β-fructan 6-levanbiohydrolase	
3.2.1.80	Fructan β-fructosidase Fructan β-fructosidase	
2.4.1.100	1,2-β-Fructan 1F-fructosyltransferase 1,2-β-Fructan 1F-fructosyltransferase	
3.2.1.64	2,6-β-D-fructan 6-β-D-fructofuranosyl fructohydrolase 2,6-β-fructan 6-levanbiohydrolase	
3.2.1.64	2,6-.β.-Fructan 6-levanbiohydrolase 2,6-β-fructan 6-levanbiohydrolase	
3.2.1.64	2,6-β-fructan 6-levanbiohydrolase 2,6-β-fructan 6-levanbiohydrolase	
2.4.1.100	1,2-β-D-Fructan:1,2-β-D-fructan 1F-β-D-fructosyltransferase 1,2-β-Fructan 1F-fructosyltransferase	
2.4.1.10	β-2,6-Fructan:D-glucose 1-fructosyltransferase Levansucrase	
2.4.1.100	Fructan:fructan fructosyl transferase 1,2-β-Fructan 1F-fructosyltransferase	
3.2.1.80	Fructan exohydrolase Fructan β-fructosidase	
3.2.1.7	2,1-β-D-fructan fructanohydrolase Inulinase	
3.2.1.65	2,6-β-D-fructan fructanohydrolase Levanase	
3.2.1.80	β-D-fructan fructohydrolase Fructan β-fructosidase	

EC-No.	Synonym Recommended Name	Vol., Page
4.1.2.13	Fructoaldolase Fructose-bisphosphate aldolase	
3.2.1.26	.β.-Fructofuranosidase β-fructofuranosidase	
3.2.1.26	β-fructofuranosidase β-fructofuranosidase	
3.2.1.26	Fructofuranosidase,.β.- β-fructofuranosidase	
3.2.1.80	Fructofuranosidase, polysaccharide.β. Fructan β-fructosidase	
3.2.1.26	β-D-fructofuranoside fructohydrolase β-fructofuranosidase	
2.7.1.4	D-Fructokinase Fructokinase	
2.7.1.4	Fructokinase Fructokinase	
2.7.1.7	D-Fructose (D-Mannose) kinase Mannokinase	
2.7.1.4	D-Fructose(D-mannose)kinase Fructokinase	
3.1.3.11	D-Fructose-1,6-bisphosphatase Fructose-bisphosphatase	
3.1.3.11	D-Fructose-1,6-bisphosphate 1-phosphohydrolase Fructose-bisphosphatase	
4.1.2.13	D-Fructose-1,6-bisphosphate D-glyceraldehyde-3-phosphate-lyase Fructose-bisphosphate aldolase	
3.1.3.11	D-Fructose-1,6-bisphosphate phosphatase Fructose-bisphosphatase	
4.1.2.13	Fructose-1,6-bisphosphate triosephosphate-lyase Fructose-bisphosphate aldolase	
2.7.1.56	D-Fructose-1-phosphate kinase 1-Phosphofructokinase	
3.1.3.46	D-fructose-2,6-bisphosphatase fructose-2,6-bisphosphate 2-phosphatase	
3.1.3.46	fructose-2,6-bisphosphatase fructose-2,6-bisphosphate 2-phosphatase	

EC-No.	Synonym Recommended Name	Vol., Page
3.1.3.46	fructose-2,6-bisphosphate 2-phosphatase fructose-2,6-bisphosphate 2-phosphatase	
3.1.3.46	D-fructose-2,6-bisphosphate 2-phosphohydrolase fructose-2,6-bisphosphate 2-phosphatase	
3.1.3.54	Fructose-2,6-bisphosphate 6-phosphatase Fructose-2,6-bisphosphate 6-phosphatase	
3.1.3.54	D-Fructose-2,6-bisphosphate 6-phosphohydrolase Fructose-2,6-bisphosphate 6-phosphatase	
3.1.3.54	Fructose-2,6-bisphosphate 6-phosphohydrolase Fructose-2,6-bisphosphate 6-phosphatase	
3.1.3.46	fructose-2,6-diphosphatase fructose-2,6-bisphosphate 2-phosphatase	
2.7.1.11	D-Fructose-6-phosphate 1-phosphotransferase 6-Phosphofructokinase	
4.1.2.22	D-Fructose-6-phosphate D-erythrose-4-phosphate-lyase (phosphate-acetylating) Fructose-6-phosphate phosphoketolase	
4.1.2.22	Fructose-6-phosphate phosphoketolase Fructose-6-phosphate phosphoketolase	
4.1.2.22	Fructose-6-phosphate phosphoketolase Fructose-6-phosphate phosphoketolase	
3.1.3.11	Fructose-bisphosphatase Fructose-bisphosphatase	
4.1.2.13	Fructose-bisphosphate aldolase Fructose-bisphosphate aldolase	
3.1.3.11	Fructose 1,6-bisphosphatase Fructose-bisphosphatase	
3.1.3.11	Fructose 1,6-bisphosphate 1-phosphatase Fructose-bisphosphatase	
4.1.2.13	Fructose 1,6-bisphosphate aldolase Fructose-bisphosphate aldolase	
3.1.3.11	Fructose 1,6-bisphosphate phosphatase Fructose-bisphosphatase	
3.1.3.11	D-Fructose 1,6-diphosphatase Fructose-bisphosphatase	

EC-No.	Synonym Recommended Name	Vol., Page
3.1.3.11	Fructose 1,6-diphosphatase Fructose-bisphosphatase	
4.1.2.13	Fructose 1,6-diphosphate aldolase Fructose-bisphosphate aldolase	
3.1.3.11	Fructose 1,6-diphosphate phosphatase Fructose-bisphosphatase	
4.1.2.13	Fructose 1-monophosphate aldolase Fructose-bisphosphate aldolase	
4.1.2.13	Fructose 1-phosphate aldolase Fructose-bisphosphate aldolase	
2.7.1.56	Fructose 1-phosphate kinase 1-Phosphofructokinase	
3.1.3.54	Fructose 2,6-bisphosphate-6-phosphohydrolase Fructose-2,6-bisphosphate 6-phosphatase	
1.1.99.11	Fructose 5-dehydrogenase Fructose 5-dehydrogenase	
1.1.1.124	Fructose 5-dehydrogenase ($NADP^+$) Fructose 5-dehydrogenase ($NADP^+$)	
2.7.1.105	Fructose 6-phosphate 2-kinase 6-Phosphofructo-2-kinase	
2.7.1.11	Fructose 6-phosphate kinase 6-Phosphofructokinase	
1.1.1.17	Fructose 6-phosphate reductase Mannitol-1-phosphate 5-dehydrogenase	
2.7.1.11	Fructose 6-phosphokinase 6-Phosphofructokinase	
1.1.99.11	D-Fructose:(acceptor) 5-oxidoreductase Fructose 5-dehydrogenase	
1.1.1.124	D-(-)Fructose:($NADP^+$) 5-oxidoreductase Fructose 5-dehydrogenase ($NADP^+$)	
1.1.1.124	D-Fructose:$NADP^+$ 5-oxidoreductase Fructose 5-dehydrogenase ($NADP^+$)	
4.1.2.13	Fructose bisphosphate aldolase Fructose-bisphosphate aldolase	
3.1.3.11	Fructose bisphosphate phosphatase Fructose-bisphosphatase	

EC-No.	Synonym Recommended Name	Vol., Page
1.1.99.11	D-Fructose dehydrogenase Fructose 5-dehydrogenase	
3.1.3.11	Fructose diphosphatase Fructose-bisphosphatase	
4.1.2.13	Fructose diphosphate aldolase Fructose-bisphosphate aldolase	
3.1.3.11	Fructose diphosphate phosphatase Fructose-bisphosphatase	
3.2.1.26	.β-h-Fructosidase β-fructofuranosidase	
3.2.1.26	.β.-Fructosidase β-fructofuranosidase	
3.2.1.26	Fructosylinvertase β-fructofuranosidase	
2.4.1.10	β-2,6-Fructosyltransferase Levansucrase	
2.4.1.100	Fructosyltransferase, 1,2-β-D-fructan 1F- 1,2-β-Fructan 1F-fructosyltransferase	
2.4.1.99	Fructosyltransferase, sucrose 1F- Sucrose 1F¯fructosyltransferase	
2.4.1.10	Fructosyltransferase, sucrose 6- Levansucrase	
2.4.1.93	Fructotransferase, inulin (depolymerizing) Inulin fructotransferase (depolymerizing, difructofuranose-1,2':2,3'-dianhydride-forming)	
1.1.1.57	Fructuronate reductase Fructuronate reductase	
3.4.22.33	Fruit bromelain Fruit bromelain	
3.4.22.33	Fruit bromelain Fruit bromelain	
3.4.22.33	Fruit bromelain FA2 Fruit bromelain	
3.4.11.1	FTBL proteins leucyl aminopeptidase	

EC-No.	Synonym Recommended Name	Vol., Page
4.3.1.4	FTCD formiminotetrahydrofolate cyclodeaminase	
6.3.4.3	FTHFS Formate-tetrahydrofolate ligase	Vol. 2, Page 567
2.3.1.101	FTR Formylmethanofuran-tetrahydromethanopterin N-formyltransferase	
4.1.2.17	Fuc-1PA L-Fuculose-phosphate aldolase	
3.2.1.44	Fucoidanase Fucoidanase	
2.7.1.52	Fucokinase Fucokinase	
4.2.1.67	D-Fuconate dehydratase D-Fuconate dehydratase	
4.2.1.68	L-fuconate dehydratase L-fuconate hydratase	
4.2.1.68	L-fuconate hydratase L-fuconate hydratase	
4.2.1.67	D-Fuconate hydro-lyase D-Fuconate dehydratase	
4.2.1.68	L-fuconate hydro-lyase L-fuconate hydratase	
3.2.1.63	2-α-L-fucopyranosyl-β-D-galactoside fucohydrolase 1,2-α-L-fucosidase	
1.1.1.122	L-Fucose (D-arabinose) dehydrogenase D-threo-Aldose 1-dehydrogenase	
2.7.7.30	Fucose-1-phosphate guanylyltransferase Fucose-1-phosphate guanylyltransferase	
5.3.1.25	Fucose isomerase L-Fucose isomerase	Vol. 1, Page 359
5.3.1.3	L-Fucose isomerase Arabinose isomerase	Vol. 1, Page 249
5.3.1.25	L-Fucose isomerase L-Fucose isomerase	Vol. 1, Page 359
5.3.1.25	L-Fucose ketol-isomerase L-Fucose isomerase	Vol. 1, Page 359

EC-No.	Synonym Recommended Name	Vol., Page
2.7.1.52	Fucose kinase Fucokinase	
3.2.1.63	1,2-.α.-L-Fucosidase 1,2-α-L-fucosidase	
3.2.1.63	1,2-α-L-fucosidase 1,2-α-L-fucosidase	
3.2.1.111	1,3-α-L-Fucosidase 1,3-α-L-Fucosidase	
3.2.1.127	1,6-α-L-Fucosidase 1,6-α-L-Fucosidase	
3.2.1.63	α-(1→ 2)-L-fucosidase 1,2-α-L-fucosidase	
3.2.1.127	α-L-Fucosidase 1,6-α-L-Fucosidase	
3.2.1.44	α-L-fucosidase Fucoidanase	
3.2.1.51	α-L-fucosidase α-L-fucosidase	
3.2.1.51	α-fucosidase α-L-fucosidase	
3.2.1.38	β-D-fucosidase β-D-fucosidase	
3.2.1.38	β-fucosidase β-D-fucosidase	
3.2.1.63	Fucosidase, 1,2-.α.-L- 1,2-α-L-fucosidase	
3.2.1.111	Fucosidase, 1,3-α-L- 1,3-α-L-Fucosidase	
3.2.1.127	Fucosidase, 1,6-α-L- 1,6-α-L-Fucosidase	
3.2.1.51	α-L-fucoside fucohydrolase α-L-fucosidase	
3.2.1.38	β-D-fucoside fucohydrolase β-D-fucosidase	
4.1.2.33	Fucosterol-epoxide lyase Fucosterol-epoxide lyase	

EC-No.	Synonym Recommended Name	Vol., Page
3.2.1.127	1,6-L-Fucosyl-N-acetyl-D-glucosaminylglycopeptide fucohydrolase 1,6-α-L-Fucosidase	
3.2.1.111	3-α-L-Fucosyl-N-acetylglucosaminyl-glycoprotein fucohydrolase 1,3-α-L-Fucosidase	
2.4.1.40	Fucosylgalactose α-N-acetylgalactosaminyltransferase Fucosylgalactose α-N-acetylgalactosaminyltransferase	
2.4.1.40	Fucosylgalactose acetylgalactosaminyltransferase Fucosylgalactose α-N-acetylgalactosaminyltransferase	
2.4.1.37	Fucosylglycoprotein 3-α-galactosyltransferase Fucosylglycoprotein 3-α-galactosyltransferase	
2.4.1.69	α-2-Fucosyltransferase Galactoside 2-α-L-fucosyltransferase	
2.4.1.69	α-2-L-Fucosyltransferase Galactoside 2-α-L-fucosyltransferase	
2.4.1.65	α-4-L-Fucosyltransferase Galactoside 3(4)-L-fucosyltransferase	
2.4.1.65	Fucosyltransferase, guanosine diphosphofucose-β-acetylglucosaminylsaccharide 4-α-L- Galactoside 3(4)-L-fucosyltransferase	
2.4.1.69	Fucosyltransferase, guanosine diphosphofucose-galactoside 2-L- Galactoside 2-α-L-fucosyltransferase	
2.4.1.152	Fucosyltransferase, guanosine diphosphofucose-glucoside α1→3- Galactoside 3-fucosyltransferase	
2.4.1.68	Fucosyltransferase, guanosine diphosphofucose-glycoprotein Glycoprotein 6-α-L-fucosyltransferase	
2.4.1.69	Fucosyltransferase, guanosine diphosphofucose-glycoprotein 2-α- Galactoside 2-α-L-fucosyltransferase	
2.4.1.65	Fucosyltransferase, guanosine diphosphofucose-glycoprotein 4-α- Galactoside 3(4)-L-fucosyltransferase	
2.4.1.69	Fucosyltransferase, guanosine diphosphofucose-lactose Galactoside 2-α-L-fucosyltransferase	
2.4.1.65	α(1,3/1,4) Fucosyltransferase III Galactoside 3(4)-L-fucosyltransferase	
2.4.1.65	FucT-II Galactoside 3(4)-L-fucosyltransferase	

EC-No.	Synonym Recommended Name	Vol., Page
2.7.1.51	L-Fuculokinase L-Fuculokinase	
4.1.2.17	L-Fuculose-1-P aldolase L-Fuculose-phosphate aldolase	
4.1.2.17	L-Fuculose-1-phosphate lactaldehyde-lyase L-Fuculose-phosphate aldolase	
4.1.2.17	L-Fuculose-phosphate aldolase L-Fuculose-phosphate aldolase	
4.1.2.17	L-Fuculose 1-phosphate aldolase L-Fuculose-phosphate aldolase	
2.7.1.51	L-Fuculose kinase L-Fuculokinase	
4.1.2.17	L-Fuculose phosphate aldolase L-Fuculose-phosphate aldolase	
4.2.1.2	fumarase fumarate hydratase	
4.2.1.2	fumarate hydratase fumarate hydratase	
1.3.99.1	Fumarate reductase Succinate dehydrogenase	
1.3.1.6	Fumarate reductase (NADH) Fumarate reductase (NADH)	
1.3.5.1	Fumarate reductase complex Succinate dehydrogenase (ubiquinone)	
4.3.1.1	fumaric aminase aspartate ammonia-lyase	
1.3.99.1	Fumaric hydrogenase Succinate dehydrogenase	
3.7.1.2	Fumarylacetoacetase Fumarylacetoacetase	
3.7.1.2	4-Fumarylacetoacetate fumarylhydrolase Fumarylacetoacetase	
3.7.1.2	Fumarylacetoacetate hydrolase Fumarylacetoacetase	
3.4.23.1	fundus-pepsin pepsin A	

EC-No.	Synonym Recommended Name	Vol., Page
2.1.1.69	Furanocoumarin 5-O-methyltransferase 5-Hydroxyfuranocoumarin 5-O-methyltransferase	
2.1.1.70	Furanocoumarin 8-methyltransferase 8-Hydroxyfuranocoumarin 8-O-methyltransferase	
2.1.1.70	Furanocoumarin 8-O-methyl-transferase 8-Hydroxyfuranocoumarin 8-O-methyltransferase	
3.4.21.75	Furin Furin	
6.2.1.31	2-Furoate-CoA ligase 2-Furoate-CoA ligase	Vol. 2, Page 334
6.2.1.31	2-Furoate:CoA ligase (AMP-forming) 2-Furoate-CoA ligase	Vol. 2, Page 334
1.3.99.8	2-Furoyl-CoA:(acceptor) oxidoreductase (hydroxylating) 2-Furoyl-CoA dehydrogenase	
1.3.99.8	2-Furoyl-CoA dehydrogenase 2-Furoyl-CoA dehydrogenase	
1.3.99.8	Furoyl-CoA hydroxylase 2-Furoyl-CoA dehydrogenase	
1.3.99.8	2-Furoyl coenzyme A dehydrogenase 2-Furoyl-CoA dehydrogenase	
1.3.99.8	2-Furoyl coenzyme A hydroxylase 2-Furoyl-CoA dehydrogenase	
5.2.1.6	2-(2-Furyl)-3-(5-nitro-2-furyl)-acrylamide cis-trans-isomerase Furylfuramide isomerase	Vol. 1, Page 213
5.2.1.6	Furylfuramide isomerase Furylfuramide isomerase	Vol. 1, Page 213
3.1.1.48	fusarinine-C ornithinesterase fusarinine-C ornithinesterase	
2.4.1.69	α-(1→2)-L-Fucosyltransferase Galactoside 2-α-L-fucosyltransferase	
2.4.1.65	α-(1→2)-L-Fucosyltransferase Galactoside 3(4)-L-fucosyltransferase	
3.4.21.38	FXII coagulation factor XIIa	
3.2.1.94	G2-dextranase Glucan 1,6-α-isomaltosidase	

EC-No.	Synonym Recommended Name	Vol., Page
5.1.3.15	G6P-1-epimerase Glucose-6-phosphate 1-epimerase	Vol. 1, Page 162
3.1.3.5	GA-AMPase 5'-nucleotidase	
3.1.3.5	GA-AMPase 5'-nucleotidase	
3.1.1.3	GA 56 (enzyme) triacylglycerol lipase	
2.6.1.19	GABA-α-ketoglutarate aminotransferase 4-Aminobutyrate transaminase	
2.6.1.19	GABA-α-ketoglutarate transaminase 4-Aminobutyrate transaminase	
2.6.1.19	GABA-α-ketoglutaric acid transaminase 4-Aminobutyrate transaminase	
2.6.1.19	GABA-α-oxoglutarate aminotransferase 4-Aminobutyrate transaminase	
2.6.1.19	GABA-2-oxoglutarate aminotransferase 4-Aminobutyrate transaminase	
2.6.1.19	GABA-2-oxoglutarate transaminase 4-Aminobutyrate transaminase	
2.6.1.19	GABA-oxoglutarate aminotransferase 4-Aminobutyrate transaminase	
2.6.1.19	GABA-oxoglutarate transaminase 4-Aminobutyrate transaminase	
2.6.1.19	GABA aminotransferase 4-Aminobutyrate transaminase	
2.6.1.19	GABA transaminase 4-Aminobutyrate transaminase	
2.6.1.19	GABA transferase 4-Aminobutyrate transaminase	
3.4.21.55	Gabonase Venombin AB	
4.1.1.15	GADCase Glutamate decarboxylase	
1.1.99.3	GADH Gluconate 2-dehydrogenase	

EC-No.	Synonym Recommended Name	Vol., Page
3.4.23.16	Gag Protease Retropepsin	
3.5.4.3	GAH Guanine deaminase	
2.4.1.90	Gal-T N-Acetyllactosamine synthase	
2.4.1.149	Galβ1→4GlcNAc-R β1→3 N-acetylglucosaminyltransferase N-Acetyllactosaminide β-1,3-N-acetylglucosaminyltransferase	
2.4.1.150	Galβ1→4GlcNAc-R β1→6 N-acetylglucosaminyltransferase N-Acetyllactosaminide β-1,6-N-acetylglucosaminyltransferase	
3.4.22.40	Gal6p bleomycin hydrolase	
3.2.1.23	.β.-D-Galactanase β-galactosidase	
3.2.1.90	Galactanase Arabinogalactan endo-1,3-β-galactosidase	
3.2.1.90	Galactanase, endo-1,3-.β. Arabinogalactan endo-1,3-β-galactosidase	
3.2.1.89	Galactanase, endo-1,4-.β.- Arabinogalactan endo-1,4-β-galactosidase	
4.2.1.42	Galactarate dehydrase Galactarate dehydratase	
4.2.1.42	Galactarate dehydratase Galactarate dehydratase	
4.2.1.42	D-Galactarate hydro-lyase Galactarate dehydratase	
2.3.1.130	Galactarate O-hydroxycinnamoyltransferase Galactarate O-hydroxycinnamoyltransferase	
2.4.1.67	Galactinol-raffinose galactosyltransferase Galactinol-raffinose galactosyltransferase	
2.4.1.82	Galactinol-sucrose galactosyltransferase Galactinol-sucrose galactosyltransferase	
2.4.1.123	Galactinol synthase Inositol 1-α-galactosyltransferase	
1.1.1.251	Galactitol-1-phosphate 5-dehydrogenase Galactitol-1-phosphate 5-dehydrogenase	

EC-No.	Synonym Recommended Name	Vol., Page
1.1.1.251	Galactitol-1-phosphate:NAD oxidoreductase Galactitol-1-phosphate 5-dehydrogenase	
1.1.1.251	Galactitol-1-phosphate dehydrogenase (Escherichia coli strain EC3132 gene gatD) Galactitol-1-phosphate 5-dehydrogenase	
1.1.1.16	Galactitol 2-dehydrogenase Galactitol 2-dehydrogenase	
1.1.1.16	Galactitol:NAD$^+$ 2-oxidoreductase Galactitol 2-dehydrogenase	
3.2.1.46	Galactoceramidase Galactoceramidase	
3.2.1.46	.β.-Galactocerebrosidase Galactoceramidase	
3.2.1.46	Galactocerebrosidase Galactoceramidase	
3.2.1.46	Galactocerebroside-.β.-D-galactosidase Galactoceramidase	
3.2.1.46	Galactocerebroside.β.-galactosidase Galactoceramidase	
3.2.1.46	Galactocerebroside galactosidase Galactoceramidase	
2.8.2.11	Galactocerebroside sulfotransferase Galactosylceramide sulfotransferase	
2.4.1.205	Galactogen 6β-galactosyltransferase Galactogen 6β-galactosyltransferase	
3.2.1.87	Galactohydrolase, capsular polysaccharide Capsular-polysaccharide endo-1,3-α-galactosidase	
2.7.1.6	Galactokinase Galactokinase	
3.1.1.26	galactolipase galactolipase	
2.3.1.134	Galactolipid:galactolipid acyltransferase Galactolipid O-acyltransferase	
2.4.1.184	Galactolipid:galactolipid galactosyltransferase Galactolipid galactosyltransferase	

EC-No.	Synonym Recommended Name	Vol., Page
2.4.1.184	Galactolipid galactosyltransferase Galactolipid galactosyltransferase	
2.3.1.134	Galactolipid O-acyltransferase Galactolipid O-acyltransferase	
2.8.2.11	Galactolipid sulfotransferase Galactosylceramide sulfotransferase	
4.2.1.6	D-galactonate dehydrase galactonate dehydratase	
4.2.1.6	D-galactonate dehydratase galactonate dehydratase	
4.2.1.6	galactonate dehydratase galactonate dehydratase	
4.2.1.6	D-galactonate hydro-lyase galactonate dehydratase	
1.3.2.3	L-Galactono-γ-lactone dehydrogenase Galactonolactone dehydrogenase	
1.3.2.3	L-Galactono-1,4-lactone:ferricytochrome-c oxidoreductase Galactonolactone dehydrogenase	
1.1.3.24	L-Galactono-1,4-lactone:oxygen 3-oxidoreductase L-Galactonolactone oxidase	
1.1.3.24	L-Galactono-1,4-lactone oxidase L-Galactonolactone oxidase	
1.3.2.3	Galactonolactone dehydrogenase Galactonolactone dehydrogenase	
1.1.3.24	L-Galactonolactone oxidase L-Galactonolactone oxidase	
2.7.7.32	Galactose-1-phosphate thymidylyltransferase Galactose-1-phosphate thymidylyltransferase	
2.7.7.10	Galactose-1-phosphate uridylyltransferase UTP-hexose-1-phosphate uridylyltransferase	
2.7.8.18	Galactose-1-phosphotransferase UDPgalactose-UDP-N-acetylglucosamine galactosephosphotransferase	
2.7.8.18	Galactose-1-phosphotransferase, uridine diphosphogalactose-uridine diphosphoacetylglucosamine UDPgalactose-UDP-N-acetylglucosamine galactosephosphotransferase	

EC-No.	Synonym Recommended Name	Vol., Page
5.3.1.26	Galactose-6-phosphate isomerase Galactose-6-phosphate isomerase	Vol. 1, Page 364
5.3.1.26	D-Galactose-6-phosphate ketol-isomerase Galactose-6-phosphate isomerase	Vol. 1, Page 364
2.5.1.5	Galactose-6-sulfatase Galactose-6-sulfurylase	
2.5.1.5	D-Galactose-6-sulfate:alkyltransferase (cyclizing) Galactose-6-sulfurylase	
3.1.6.4	Galactose-6-sulfate sulfatase N-Acetylgalactosamine-6-sulfatase	
2.5.1.5	Galactose-6-sulfurylase Galactose-6-sulfurylase	
1.1.1.48	Galactose 1-dehydrogenase Galactose 1-dehydrogenase	
1.1.1.120	Galactose 1-dehydrogenase ($NADP^+$) Galactose 1-dehydrogenase ($NADP^+$)	
2.7.7.32	Galactose 1-phosphate thymidylyl transferase Galactose-1-phosphate thymidylyltransferase	
2.7.7.10	Galactose 1-phosphate uridyltransferase UTP-hexose-1-phosphate uridylyltransferase	
2.7.7.10	Galactose 1-phosphate uridylyltransferase UTP-hexose-1-phosphate uridylyltransferase	
2.7.7.10	α-D-Galactose 1-phosphate uridylyltransferase UTP-hexose-1-phosphate uridylyltransferase	
5.3.1.26	D-galactose 6-phosphate isomerase Galactose-6-phosphate isomerase	Vol. 1, Page 364
2.5.1.5	Galactose 6-sulfatase Galactose-6-sulfurylase	
2.5.1.5	Galactose 6-sulfurylase Galactose-6-sulfurylase	
1.1.1.48	D-Galactose:NAD^+ 1-oxidoreductase Galactose 1-dehydrogenase	
1.1.1.120	D-Galactose:$NADP^+$ 1-oxidoreductase Galactose 1-dehydrogenase ($NADP^+$)	
1.1.3.9	D-Galactose:oxygen 6-oxidoreductase Galactose oxidase	

EC-No.	Synonym Recommended Name	Vol., Page
1.1.1.48	D-Galactose dehydrogenase Galactose 1-dehydrogenase	
1.1.1.48	β-Galactose dehydrogenase Galactose 1-dehydrogenase	
1.1.1.120	D-Galactose dehydrogenase ($NADP^+$) Galactose 1-dehydrogenase ($NADP^+$)	
1.1.3.9	D-Galactose oxidase Galactose oxidase	
1.1.3.9	Galactose oxidase Galactose oxidase	
1.1.3.9	β-Galactose oxidase Galactose oxidase	
2.7.8.6	Galactosephosphotransferase, poly(isoprenol) phosphate Undecaprenyl-phosphate galactosephosphotransferase	
3.2.1.23	β-galactosidase β-galactosidase	
3.2.1.22	α-D-Galactosidase α-Galactosidase	
3.2.1.22	α-Galactosidase α-Galactosidase	
3.2.1.22	α-Galactosidase α-Galactosidase	
3.2.1.23	β-Galactosidase β-galactosidase	
3.2.1.22	Galactosidase,.α. α-Galactosidase	
3.2.1.23	Galactosidase,.β. β-galactosidase	
3.2.1.103	Galactosidase, keratosulfate endo-. β.- Keratan-sulfate endo-1,4-β-galactosidase	
3.2.1.22	α-Galactosidase A α-Galactosidase	
2.4.99.4	β-Galactoside α-2,3-sialyltransferase β-Galactoside α-2,3-sialyltransferase	
2.4.99.1	β-Galactoside α-2,6-sialyltransferase β-Galactoside α-2,6-sialyltransferase	

EC-No.	Synonym Recommended Name	Vol., Page
2.4.99.1	β-Galactoside α-2,6-sialyltransferase β-Galactoside α-2,6-sialyltransferase	
2.4.1.69	β-Galactoside α1→2fucosyltransferase Galactoside 2-α-L-fucosyltransferase	
2.4.1.69	Galactoside 2-α-L-fucosyltransferase Galactoside 2-α-L-fucosyltransferase	
2.4.1.65	Galactoside 3(4)-L-fucosyltransferase Galactoside 3(4)-L-fucosyltransferase	
2.4.1.152	Galactoside 3-fucosyltransferase Galactoside 3-fucosyltransferase	
2.3.1.18	Galactoside acetyltransferase Galactoside O-acetyltransferase	
3.2.1.23	.β.-D-Galactoside galactohydrolase β-galactosidase	
3.2.1.23	β-D-galactoside galactohydrolase β-galactosidase	
3.2.1.22	α-D-Galactoside galactohydrolase α-Galactosidase	
3.2.1.22	α-D-Galactoside galactohydrolase α-Galactosidase	
3.2.1.22	α-Galactoside galactohydrolase α-Galactosidase	
2.3.1.18	Galactoside O-acetyltransferase Galactoside O-acetyltransferase	
3.2.1.23	Galactosyl β-galactosidase	
3.2.1.110	D-Galactosyl-3-(N-acetyl-β-D-galactosaminyl)-L-serine mucinaminohydrolase Mucinaminylserine mucinaminidase	
3.2.1.47	D-Galactosyl-D-galactosyl-D-glucosyl-N-acylsphingosine galactohydrolase Galactosylgalactosylglucosylceramidase	
2.4.1.67	1-α-D-Galactosyl-myo-inositol:raffinose galactosyltransferase Galactinol-raffinose galactosyltransferase	
2.4.1.82	1-α-D-Galactosyl-myo-inositol:sucrose 6-α-D-galactosyltransferase Galactinol-sucrose galactosyltransferase	

EC-No.	Synonym Recommended Name Vol., Page
3.2.1.97	D-Galactosyl-N-acetyl-α-D-galactosamine D-galactosyl-N-acetylgalactosaminohydrolase Glycopeptide α-N-acetylgalactosaminidase
2.4.1.163	β-Galactosyl-N-acetylglucosaminylgalactosylglucosylceramide β-1,3-acetylglucosaminyltransferase β-Galactosyl-N-acetylglucosaminylgalactosylglucosylceramide β-1,3-acetylglucosaminyltransferase
2.4.1.164	Galactosyl-N-acetylglucosaminylgalactosylglucosylceramide β-1,6-N-acetylglucosaminyltransferase Galactosyl-N-acetylglucosaminylgalactosylglucosylceramide β-1,6-N-acetylglucosaminyltransferase
2.4.1.87	β-D-Galactosyl-N-acetylglucosaminylglycopeptide α-1,3-galactosyltransferase β-D-Galactosyl-N-acetylglucosaminylglycopeptide α-1,3-galactosyltransferase
3.2.1.46	D-Galactosyl-N-acylsphingosine galactohydrolase Galactoceramidase
2.4.1.146	β-1,3-Galactosyl-O-glycosyl-glycoprotein β-1,3-N-acetylglucosaminyltransferase β-1,3-Galactosyl-O-glycosyl-glycoprotein β-1,3-N-acetylglucosaminyltransferase
2.4.1.102	β-1,3-Galactosyl-O-glycosyl-glycoprotein β-1,6-N-acetylglucosaminyltransferase β-1,3-Galactosyl-O-glycosyl-glycoprotein β-1,6-N-acetylglucosaminyltransferase
2.4.1.23	Galactosyl-sphingosine transferase Sphingosine β-galactosyltransferase
2.3.1.141	Galactosylacylglycerol O-acyltransferase Galactosylacylglycerol O-acyltransferase
3.2.1.109	Galactosylaminoglycan glycanohydrolase Endogalactosaminidase
3.2.1.46	.β.-Galactosylceramidase Galactoceramidase
3.2.1.46	Galactosylceramidase I Galactoceramidase
3.2.1.46	Galactosylceramide.β.-galactosidase Galactoceramidase
2.8.2.11	Galactosylceramide sulfotransferase Galactosylceramide sulfotransferase

EC-No.	Synonym Recommended Name	Vol., Page
3.2.1.46	Galactosylcerebrosidase Galactoceramidase	
2.4.99.5	Galactosyldiacylglycerol α-2,3-sialyltransferase Galactosyldiacylglycerol α-2,3-sialyltransferase	
3.2.1.47	Galactosylgalactosylglucosylceramidase Galactosylgalactosylglucosylceramidase	
2.4.1.79	Galactosylgalactosylglucosylceramide β-D-acetylgalactosaminyltransferase Galactosylgalactosylglucosylceramide β-D-acetylgalactosaminyltransferase	
2.4.1.135	Galactosylgalactosylxylosylprotein 3-β-glucuronosyltransferase Galactosylgalactosylxylosylprotein 3-β-glucuronosyltransferase	
2.4.1.89	galactosylglucosaminylgalactosylglucosylceramide α-L-fucosyltransferase galactosylglucosaminylgalactosylglucosylceramide α-L-fucosyltransferase	
2.4.1.66	Galactosylhydroxylysyl glucosyltransferase Procollagen glucosyltransferase	
2.7.8.18	Galactosyl phosphotransferase UDPgalactose-UDP-N-acetylglucosamine galactosephosphotransferase	
2.4.1.133	Galactosyltranferase I Xylosylprotein 4-β-galactosyltransferase	
2.4.1.205	1,6-D-Galactosyltransferase Galactogen 6β-galactosyltransferase	
2.4.1.151	Galactosyltransferase N-Acetyllactosaminide α-1,3-galactosyltransferase	
2.4.1.87	α-Galactosyltransferase β-D-Galactosyl-N-acetylglucosaminylglycopeptide α-1,3-galactosyltransferase	
2.4.1.205	β-(1-6)-D-Galactosyltransferase Galactogen 6β-galactosyltransferase	
2.4.1.90	β-1,4-Galactosyltransferase N-Acetyllactosamine synthase	
2.4.1.90	β1-4-Galactosyltransferase N-Acetyllactosamine synthase	
2.4.1.38	β1-4-Galactosyltransferase β-N-Acetylglucosaminylglycopeptide β-1,4-galactosyltransferase	
2.4.1.37	Galactosyltransferase, [blood-group substance] α- Fucosylglycoprotein 3-α-galactosyltransferase	

EC-No.	Synonym Recommended Name	Vol., Page
2.4.1.67	Galactosyltransferase, galactinol-raffinose Galactinol-raffinose galactosyltransferase	
2.4.1.82	Galactosyltransferase, galactinol-sucrose Galactinol-sucrose galactosyltransferase	
2.4.1.184	Galactosyltransferase, galactolipid-galactolipid Galactolipid galactosyltransferase	
2.4.1.96	Galactosyltransferase, glycerol 3-phosphate 1α- Sn-Glycerol-3-phosphate 1-galactosyltransferase	
2.4.1.44	Galactosyltransferase, lipopolysaccharide α, 3- Lipopolysaccharide 3-α-galactosyltransferase	
2.4.1.166	Galactosyltransferase, raffinose (raffinose donor) Raffinose-raffinose α-galactotransferase	
2.4.1.38	Galactosyltransferase, thyroid glycoprotein β- β-N-Acetylglucosaminylglycopeptide β-1,4-galactosyltransferase	
2.4.1.46	Galactosyltransferase, uridine diphosphogalactose-1,2-diacylglycerol 1,2-Diacylglycerol 3-β-galactosyltransferase	
2.4.1.45	Galactosyltransferase, uridine diphosphogalactose-2-hydroxyacylsphingosine 2-Hydroxyacylsphingosine 1-β-galactosyltransferase	
2.4.1.86	Galactosyltransferase, uridine diphosphogalactose-acetyl-glucosaminylgalactosylglucosylceramide Glucosaminylgalactosylglucosylceramide β-galactosyltransferase	
2.4.1.90	Galactosyltransferase, uridine diphosphogalactose-acetylglucosamine N-Acetyllactosamine synthase	
2.4.1.124	Galactosyltransferase, uridine diphosphogalactose-acetyllactosamine N-Acetyllactosamine 3-α-galactosyltransferase	
2.4.1.151	Galactosyltransferase, uridine diphosphogalactose-acetyllactosamine N-Acetyllactosaminide α-1,3-galactosyltransferase	
2.4.1.151	Galactosyltransferase, uridine diphosphogalactose-acetyllactosamine $\alpha 1 \rightarrow 3$- N-Acetyllactosaminide α-1,3-galactosyltransferase	
2.4.1.47	Galactosyltransferase, uridine diphosphogalactose-acylsphingosine N-Acylsphingosine galactosyltransferase	
2.4.1.62	Galactosyltransferase, uridine diphosphogalactose-ceramide Ganglioside galactosyltransferase	
2.4.1.50	Galactosyltransferase, uridine diphosphogalactose-collagen Procollagen galactosyltransferase	

EC-No.	Synonym Recommended Name	Vol., Page
2.4.1.205	Galactosyltransferase, uridine diphosphogalactose-galactogen Galactogen 6β-galactosyltransferase	
2.4.1.87	Galactosyltransferase, uridine diphosphogalactose-galactosylacetylglucosaminylgalactosylglucosylceramide β-D-Galactosyl-N-acetylglucosaminylglycopeptide α-1,3-galactosyltransferase	
2.4.1.134	Galactosyltransferase, uridine diphosphogalactose-galactosylxylose Galactosylxylosylprotein 3-β-galactosyltransferase	
2.4.1.22	Galactosyltransferase, uridine diphosphogalactose-glucose Lactose synthase	
2.4.1.38	Galactosyltransferase, uridine diphosphogalactose-glycoprotein β-N-Acetylglucosaminylglycopeptide β-1,4-galactosyltransferase	
2.4.1.156	Galactosyltransferase, uridine diphosphogalactose-indolylacetylinositol Indolylacetyl-myo-inositol galactosyltransferase	
2.4.1.123	Galactosyltransferase, uridine diphosphogalactose-inositol Inositol 1-α-galactosyltransferase	
2.4.1.179	Galactosyltransferase, uridine diphosphogalactose-lactosylceramide β1→3- Lactosylceramide β-1,3-galactosyltransferase	
2.4.1.44	Galactosyltransferase, uridine diphosphogalactose-lipopolysaccharide α,3- Lipopolysaccharide 3-α-galactosyltransferase	
2.4.1.122	Galactosyltransferase, uridine diphosphogalactose-mucin β-(1→3)- Glycoprotein-N-acetylgalactosamine 3-β-galactosyltransferase	
2.4.1.74	Galactosyltransferase, uridine diphosphogalactose-mucopolysaccharide Glycosaminoglycan galactosyltransferase	
2.4.1.23	Galactosyltransferase, uridine diphosphogalactose-sphingosine β- Sphingosine β-galactosyltransferase	
2.4.1.167	Galactosyltransferase, uridine diphosphogalactose-sucrose 6F-α- Sucrose 6F-α-galactotransferase	
2.4.1.133	Galactosyltransferase, uridine diphosphogalactose-xylose Xylosylprotein 4-β-galactosyltransferase	
2.4.1.134	Galactosyltransferase II Galactosylxylosylprotein 3-β-galactosyltransferase	
2.4.1.96	Galactosyltransferse, uridine diphosphogalactose-glycerol phosphate Sn-Glycerol-3-phosphate 1-galactosyltransferase	
2.4.1.134	Galactosylxylosylprotein 3-β-galactosyltransferase Galactosylxylosylprotein 3-β-galactosyltransferase	

EC-No.	Synonym Recommended Name	Vol., Page
5.1.3.2	Galactowaldenase UDPglucose 4-epimerase	Vol. 1, Page 97
3.2.1.67	Galacturan 1,4-α-galacturonidase Galacturan 1,4-α-galacturonidase	
3.2.1.15	D-Galacturonase Polygalacturonase	
2.7.1.44	D-Galacturonic acid kinase Galacturonokinase	
2.7.1.44	Galacturonokinase Galacturonokinase	
2.4.1.75	Galacturonosyltransferase, uridine diphospho- UDPgalacturonosyltransferase	
2.4.1.43	Galacturonosyltransferase, uridine diphosphogalacturonate-polygalacturonate α- Polygalacturonate 4-α-galacturonosyltransferase	
3.2.1.23	β-Galase β-galactosidase	
2.4.1.136	Gallate 1-β-glucosyltransferase Gallate 1-β-glucosyltransferase	
4.1.1.59	Gallate carboxy-lyase Gallate decarboxylase	
4.1.1.59	Gallate decarboxylase Gallate decarboxylase	
4.1.1.59	Gallic acid decarboxylase Gallate decarboxylase	
2.3.1.143	1-O-Galloyl-β-D-glucose:1,2,3,6-tetrakis-O-galloyl-β-D-glucose 4-O-galloyltransferase β-Glucogallin-tetrakisgalloylglucose O-galloyltransferase	
2.3.1.90	1-O-Galloyl-β-D-glucose:1-O-galloyl-β-D-glucose O-galloyltransferase β-Glucogallin O-galloyltransferase	
2.3.1.90	Galloyltransferase, β-glucogallin (β-glucogallin donor) β-Glucogallin O-galloyltransferase	
2.3.1.143	Galloyltransferase, β-glucogallin-tetragalloylglucose 4- β-Glucogallin-tetrakisgalloylglucose O-galloyltransferase	
2.4.1.154	GalNAc transferase Globotriosylceramide β-1,6-N-acetylgalactosaminyltransferase	

EC-No.	Synonym Recommended Name	Vol., Page
2.4.1.90	β1-4GalT N-Acetyllactosamine synthase	
2.4.1.86	GalT-4 Glucosaminylgalactosylglucosylceramide β-galactosyltransferase	
3.4.24.38	Gamete autolysin Autolysin	
3.4.24.38	Gamete lytic enzyme Autolysin	
2.1.1.2	GA methylpherase Guanidinoacetate N-methyltransferase	
2.4.1.62	Ganglioside galactosyltransferase Ganglioside galactosyltransferase	
2.4.99.8	Ganglioside GD3 synthase α-N-Acetylneuraminate α-2,8-sialyltransferase	
2.4.99.8	Ganglioside GD3 synthetase α-N-Acetylneuraminate α-2,8-sialyltransferase	
2.4.1.92	Ganglioside GM2 synthase (N-Acetylneuraminyl)-galactosylglucosylceramide N-acetylgalactosaminyl-transferase	
2.4.1.92	Ganglioside GM3 acetylgalactosaminyltransferase (N-Acetylneuraminyl)-galactosylglucosylceramide N-acetylgalactosaminyl-transferase	
2.4.99.9	Ganglioside GM3 synthetase Lactosylceramide α-2,3-sialyltransferase	
3.4.23.32	Ganoderma lucidum carboxyl proteinase Scytalidopepsin B	
1.2.1.13	GAPDH Glyceraldehyde-3-phosphate dehydrogenase ($NADP^+$) (phosphorylating)	
1.2.1.59	GAPDH2 Glyceraldehyde-3-phosphate dehydrogenase (NAD(P)) (phosphorylating)	
2.1.2.2	GAR formyltransferase Phosphoribosylglycinamide formyltransferase	
6.3.4.13	GARSase Phosphoribosylamine-glycine ligase	Vol. 2, Page 626
6.3.4.13	GAR synthetase Phosphoribosylamine-glycine ligase	Vol. 2, Page 626

EC-No.	Synonym Recommended Name	Vol., Page
2.1.2.2	GAR TFase Phosphoribosylglycinamide formyltransferase	
2.1.2.2	GAR transformylase Phosphoribosylglycinamide formyltransferase	
3.4.23.3	gastricsin gastricsin	
1.1.1.251	Gat1P-specific NAD-dependent dehydrogenase Galactitol-1-phosphate 5-dehydrogenase	
1.2.1.54	GBAL dehydrogenase γ-Guanidinobutyraldehyde dehydrogenase	
3.5.3.7	GBH Guanidinobutyrase	
6.4.1.5	Gcase Geranoyl-CoA carboxylase	Vol. 2, Page 752
2.4.99.2	GD1a-synthase Monosialoganglioside sialyltransferase	
2.4.99.8	GD3 synthase α-N-Acetylneuraminate α-2,8-sialyltransferase	
4.1.1.15	GDCase Glutamate decarboxylase	
1.8.5.1	GDOR Glutathione dehydrogenase (ascorbate)	
1.1.1.187	GDP-4-dehydro-D-rhamnose reductase GDP-4-dehydro-D-rhamnose reductase	
1.1.1.187	GDP-4-keto-6-deoxy-D-mannose reductase GDP-4-dehydro-D-rhamnose reductase	
1.1.1.187	GDP-4-keto-D-rhamnose reductase GDP-4-dehydro-D-rhamnose reductase	
1.1.1.187	GDP-6-deoxy-D-mannose:NAD(P)$^+$ 4-oxidoreductase GDP-4-dehydro-D-rhamnose reductase	
1.1.1.135	GDP-6-deoxy-D-talose 4-dehydrogenase GDP-6-deoxy-D-talose 4-dehydrogenase	
1.1.1.135	GDP-6-deoxy-D-talose:NAD(P)$^+$ 4-oxidoreductase GDP-6-deoxy-D-talose 4-dehydrogenase	
5.1.3.18	GDP-D-mannose 3,5-epimerase GDPmannose 3,5-epimerase	Vol. 1, Page 170

EC-No.	Synonym Recommended Name	Vol., Page
4.2.1.47	GDP-D-mannose 4,6-dehydratase GDPmannose 4,6-dehydratase	
4.2.1.47	GDP-D-mannose 4-oxido-6-reductase GDPmannose 4,6-dehydratase	
5.1.3.18	GDP-D-mannose:GDP-L-galactose epimerase GDPmannose 3,5-epimerase	Vol. 1, Page 170
2.4.1.54	GDP-D-mannose:lipid phosphate transmannosylase Undecaprenyl-phosphate mannosyltransferase	
1.1.1.132	GDP-D-mannose:NAD$^+$ 6-oxidoreductase GDPmannose 6-dehydrogenase	
4.2.1.47	GDP-D-mannose dehydratase GDPmannose 4,6-dehydratase	
2.4.1.33	GDP-D-mannuronate:alginate D-mannuronyltransferase Alginate synthase	
2.4.1.68	GDP-fucose-glycoprotein fucosyltransferase Glycoprotein 6-α-L-fucosyltransferase	
2.4.1.68	GDP-fucose glycoprotein fucosyltransferase Glycoprotein 6-α-L-fucosyltransferase	
2.7.7.30	GDP-fucose pyrophosphorylase Fucose-1-phosphate guanylyltransferase	
2.4.1.68	GDP-L-Fuc:N-acetyl-β-D-glucosaminide α1→6fucosyltransferase Glycoprotein 6-α-L-fucosyltransferase	
2.4.1.68	GDP-L-fucose-glycoprotein fucosyltransferase Glycoprotein 6-α-L-fucosyltransferase	
2.4.1.65	GDP-L-fucose:β-D-galactosyl-N-acetyl-D-glucosaminyl-R-4-L-fucosyltransferase Galactoside 3(4)-L-fucosyltransferase	
2.4.1.69	GDP-L-fucose:β-D-galactosyl-R 2-α-L-fucosyltransferase Galactoside 2-α-L-fucosyltransferase	
2.4.1.152	GDP-L-fucose:1,4-β-D-galactosyl-N-acetyl-D-glucosaminyl-R 3-L-fucosyltransferase Galactoside 3-fucosyltransferase	
2.4.1.68	GDP-L-fucose:glycoprotein (L-fucose to asparagine-linked N-acetylglucosamine of N-acetyl-β-D-glucosaminyl-1,2-α-D-mannosyl-1,3-(R1-α-1,6)-β-D-mannosyl-1,4-β-N-acetyl-D-glucosaminyl-1,4-N-acetyl-D-glucosaminyl)asparagine 6-α-L-fucosyltransferase Glycoprotein 6-α-L-fucosyltransferase	

EC-No.	Synonym Recommended Name	Vol., Page
2.4.1.69	GDP-L-fucose:lactose fucosyltransferase Galactoside 2-α-L-fucosyltransferase	
2.7.7.30	GDP-L-fucose pyrophosphorylase Fucose-1-phosphate guanylyltransferase	
2.4.1.32	GDP-Man-β-mannan manosyltransferase Glucomannan 4-β-mannosyltransferase	
2.4.1.142	GDP-mannose-dolichol diphosphochitobiose mannosyltransferase Chitobiosyldiphosphodolichol α-mannosyltransferase	
2.4.1.83	GDP-mannose-dolichol phosphate mannosyltransferase Dolichyl-phosphate β-D-mannosyltransferase	
2.4.1.131	GDP-mannose-oligosaccharide-lipid mannosyltransferase Glycolipid 2-α-mannosyltransferase	
2.4.1.132	GDP-mannose-oligosaccharide-lipid mannosyltransferase II Glycolipid 3-α-mannosyltransferase	
2.7.7.22	GDP-mannose 1-phosphate guanylyltransferase Mannose-1-phosphate guanylyltransferase (GDP)	
4.2.1.47	GDP-mannose 4,6-dehydratase GDPmannose 4,6-dehydratase	
4.2.1.47	GDP-mannose dehydratase GDPmannose 4,6-dehydratase	
2.7.7.13	GDP-mannose pyrophosphorylase Mannose-1-phosphate guanylyltransferase	
2.7.7.22	GDP:D-mannose-1-phosphate guanylyltransferase Mannose-1-phosphate guanylyltransferase (GDP)	
3.6.1.42	GDPase Guanosine-diphosphatase	
3.6.1.6	GDPase Nucleoside-diphosphatase	
3.1.4.2	GDPase glycerophosphocholine phosphodiesterase	
2.4.1.69	GDP fucose-lactose fucosyltransferase Galactoside 2-α-L-fucosyltransferase	
2.4.1.68	GDPfucose glycoprotein fucosyltransferase Glycoprotein 6-α-L-fucosyltransferase	
2.7.7.30	GDPfucose pyrophosphorylase Fucose-1-phosphate guanylyltransferase	

EC-No.	Synonym Recommended Name	Vol., Page
2.4.1.36	GDPglucose-glucose-phosphate glucosytransferase α,α-Trehalose-phosphate synthase (GDP-forming)	
2.4.1.36	GDP glucose-glucosephosphate glucosyltransferase α,α-Trehalose-phosphate synthase (GDP-forming)	
2.4.1.29	GDPglucose:1,4-β-D-glucan 4-β-D-glucosyltransferase Cellulose synthase (GDP-forming)	
2.4.1.36	GDPglucose:D-glucose-6-phosphate 1-α-D-glucosyltransferase α,α-Trehalose-phosphate synthase (GDP-forming)	
3.2.1.42	GDPglucose glucohydrolase GDPglucosidase	
2.7.7.34	GDPglucose pyrophosphorylase Glucose-1-phosphate guanylyltransferase	
3.2.1.42	GDPglucosidase GDPglucosidase	
2.7.7.29	GDP hexose pyrophosphorylase Hexose-1-phosphate guanylyltransferase	
2.7.7.29	GDPhexose pyrophosphorylase Hexose-1-phosphate guanylyltransferase	
2.4.1.83	GDPMan:DolP mannosyltransferase Dolichyl-phosphate β-D-mannosyltransferase	
2.4.1.83	GDPmannose-dolichylmonophosphate mannosyltransferase Dolichyl-phosphate β-D-mannosyltransferase	
2.4.1.48	GDP mannose α-mannosyltransferase Heteroglycan α-mannosyltransferase	
2.4.1.57	GDP mannose-phosphatidyl-myo-inositol α-mannosyltransferase Phosphatidyl-myo-inositol α-mannosyltransferase	
2.4.1.54	GDP mannose-undecaprenyl phosphate mannosyltransferase Undecaprenyl-phosphate mannosyltransferase	
5.1.3.18	GDPmannose 3,5-epimerase GDPmannose 3,5-epimerase	Vol. 1, Page 170
4.2.1.47	GDPmannose 4,6-dehydratase GDPmannose 4,6-dehydratase	
4.2.1.47	GDPmannose 4,6-hydro-lyase GDPmannose 4,6-dehydratase	
1.1.1.132	GDPmannose 6-dehydrogenase GDPmannose 6-dehydrogenase	

EC-No.	Synonym Recommended Name	Vol., Page
2.4.1.57	GDPmannose:1-phosphatidyl-myo-inositol α-D-mannosyltransferase Phosphatidyl-myo-inositol α-mannosyltransferase	
2.4.1.142	GDPmannose:chitobiosyldiphosphodolichol α-D-mannosyltransferase Chitobiosyldiphosphodolichol α-mannosyltransferase	
2.4.1.83	GDPmannose:dolichyl-phosphate β-D-mannosyltransferase Dolichyl-phosphate β-D-mannosyltransferase	
2.4.1.83	GDPmannose:dolichyl-phosphate mannosyltransferase Dolichyl-phosphate β-D-mannosyltransferase	
2.4.1.32	GDPmannose:glucomannan 1,4-β-D-mannosyltransferase Glucomannan 4-β-mannosyltransferase	
2.4.1.131	GDPmannose:glycolipid 1,2-α-D-mannosyltransferase Glycolipid 2-α-mannosyltransferase	
2.4.1.132	GDPmannose:glycolipid 1,3-α-D-mannosyltransferase Glycolipid 3-α-mannosyltransferase	
2.4.1.48	GDPmannose:heteroglycan 2-(or 3-)-α-D-mannosyltransferase Heteroglycan α-mannosyltransferase	
2.7.8.9	GDPmannose:phosphomannan mannosephosphotransferase Phosphomannan mannosephosphotransferase	
2.4.1.110	GDPmannose:tRNA^{Asp}-queuosine O-5-β-D-mannosyltransferase tRNA-queuosine β-mannosyltransferase	
2.4.1.54	GDPmannose:undecaprenyl-phosphate D-mannosyltransferase Undecaprenyl-phosphate mannosyltransferase	
1.1.1.132	GDP mannose dehydrogenase GDPmannose 6-dehydrogenase	
2.7.7.22	GDP mannose phosphorylase Mannose-1-phosphate guanylyltransferase (GDP)	
2.7.7.22	GDPmannose phosphorylase Mannose-1-phosphate guanylyltransferase (GDP)	
3.6.1.42	GDP phosphohydrolase Guanosine-diphosphatase	
3.1.1.3	GEH triacylglycerol lipase	
1.3.1.36	Geissoschizine:NADP$^+$ 4,21-oxidoreductase Geissoschizine dehydrogenase	
1.3.1.36	Geissoschizine dehydrogenase Geissoschizine dehydrogenase	

EC-No.	Synonym Recommended Name	Vol., Page
3.4.23.2	gelatinase pepsin B	
3.4.24.24	Gelatinase A Gelatinase A	
3.4.24.35	Gelatinase B Gelatinase B	
3.4.24.35	Gelatinase MMP 9 Gelatinase B	
3.2.2.22	Gelonin rRNA N-glycosidase	
3.1.3.68	GenBank AE000318-derived protein GI 1788630 2-Deoxyglucose-6-phosphatase	
6.3.1.8	Gen Bank AE000381-derived protein GI 1789361 Glutathionylspermidine synthase	Vol. 2, Page 386
5.1.3.20	Genbank AE000440-derived protein GI 1790049 ADPglyceromanno-heptose 6-epimerase	Vol. 1, Page 175
4.2.1.96	GenBank AE000671-derived protein GI 2982796 4a-Hydroxytetrahydrobiopterin dehydratase	
5.1.3.20	GenBank AE000684-derived protein GI 2983012 ADPglyceromanno-heptose 6-epimerase	Vol. 1, Page 175
2.8.1.6	GenBank AE000716-derived protein GI 2983482 Biotin synthase	
2.7.9.3	GenBank AE000719-derived protein GI 2983519 selenide, water dikinase	
5.4.1.2	GenBank AE000809-derived protein GI 2621284 Precorrin-8X methylmutase	Vol. 1, Page 490
2.1.1.130	GenBank AE00098-derived protein GI 2622455 Precorrin-2 C20-methyltransferase	
2.8.1.6	GenBank AE001204-derived protein GI 3322497 Biotin synthase	
2.8.1.6	GenBank AE001343-derived protein GI 3329182 Biotin synthase	
6.3.1.9	GenBank AF006615-derived protein GI 3004644 Trypanothione synthase	Vol. 2, Page 391
2.8.1.6	GenBank AF008220-derived protein GI 2293187 Biotin synthase	

EC-No.	Synonym Recommended Name	Vol., Page
1.3.1.55	GenBank AF009224-derived protein GI 2996624 cis-1,2-Dihydroxycyclohexa-3,5-diene-1-carboxylate dehydrogenase	
3.1.2.22	GenBank AF020543-derived protein GI 2501961 Palmitoyl-(protein) hydrolase	
3.1.3.68	GenBank D90861-derived protein GI 1799667 2-Deoxyglucose-6-phosphatase	
1.14.99.30	genBank D90914-derived protein GI 1653487 Carotene 7,8-desaturase	
2.1.1.132	GenBank L12005-derived protein Precorrin-6Y $C^{5,15}$-methyltransferase (decarboxylating)	
2.1.1.132	Genbank L12006-derived protein Precorrin-6Y $C^{5,15}$-methyltransferase (decarboxylating)	
2.1.1.130	GenBank L12006-derived protein, methyltransferase Precorrin-2 C^{20}-methyltransferase	
3.1.2.22	GenBank L42809-derived protein GI 1160967 Palmitoyl-(protein) hydrolase	
2.4.1.207	GenBank L46792-derived protein GI 950299 Xyloglucan:xyloglucosyl transferase	
4.1.1.74	GenBank L80006-derived protein GI 1507711 Indolepyruvate decarboxylase	
2.8.1.6	GenBank U24147-derived protein Biotin synthase	
2.8.1.6	GenBank U31806-derived protein GI 1403662 Biotin synthase	
2.8.1.6	GenBank U51869-derived protein GI 1277029 Biotin synthase	
6.3.1.8	GenBank U66520-derived protein GI 1813514 Glutathionylspermidine synthase	Vol. 2, Page 386
2.1.1.130	GenBank U67522-derived protein GI 1499591 Precorrin-2 C^{20}-methyltransferase	
5.4.1.2	GenBank U67537-derived protein GI 1499765 Precorrin-8X methylmutase	Vol. 1, Page 490
2.1.1.132	GenBank U67594-derived protein GI 1500413 Precorrin-6Y $C^{5,15}$-methyltransferase (decarboxylating)	
5.1.99.4	GenBank U89905-derived protein GI2145184 α-Methylacyl-CoA racemase	Vol. 1, Page 188

EC-No.	Synonym Recommended Name	Vol., Page
5.1.99.4	GenBank U89906-derived protein GI 2145186 α-Methylacyl-CoA racemase	Vol. 1, Page 188
1.1.1.251	Genbank X79837-derived protein Galactitol-1-phosphate 5-dehydrogenase	
2.4.1.207	GenBank X93173-derived protein GI 1890573 Xyloglucan:xyloglucosyl transferase	
2.4.1.207	GenBank X93174-derived protein GI 1890575 Xyloglucan:xyloglucosyl transferase	
2.4.1.207	Genbank X93175-derived protein GI 1890577 Xyloglucan:xyloglucosyl transferase	
1.14.13.61	GenBank Y12654-derived protein GI 2072729 2-Hydroxyquinoline 8-monooxygenase	
1.14.13.61	GenBank Y12655-derived protein GI 2072732 2-Hydroxyquinoline 8-monooxygenase	
2.9.1.1	GenBank Y14814-derived protein GI 2440135 L-Seryl-tRNASec selenium transferase	
2.4.1.207	GenBank Z97335-derived protein GI Xyloglucan:xyloglucosyl transferase	
4.1.2.39	GenBank Z97337-derived protein GI 2244867 Hydroxynitrilase	
2.8.1.6	GenBank Z99119-derived protein GI 2635504 Biotin synthase	
2.4.1.69	H-Gene-encoded β-galactoside α1→2fucosyltransferase Galactoside 2-α-L-fucosyltransferase	
2.7.1.69	Gene bglC RNA formation factors Protein-N^{π}-phosphohistidine-sugar phosphotransferase	
5.4.1.2	Gene cbiC precorrin-8x isomerase Precorrin-8X methylmutase	Vol. 1, Page 490
2.1.1.132	Gene cbiE precorrin 6y methylase Precorrin-6Y $C^{5,15}$-methyltransferase (decarboxylating)	
2.1.1.132	Gene cbiT precorrin methylase Precorrin-6Y $C^{5,15}$-methyltransferase (decarboxylating)	
5.4.1.2	Gene cobH precorrin-8x isomerase Precorrin-8X methylmutase	Vol. 1, Page 490
2.1.1.131	Gene cobJ enzyme Precorrin-3B C^{17}-methyltransferase	

EC-No.	Synonym Recommended Name	Vol., Page
3.1.2.24	gene dszB-encoded hydrolase 2-(2-hydroxyphenyl)benzenesulfinate hydrolase	
2.7.1.69	Gene glC proteins Protein-Npi-phosphohistidine-sugar phosphotransferase	
3.1.4.46	Gene hpd protein Glycerophosphodiester phosphodiesterase	
3.4.16.6	Gene KEX1 serine carboxypeptidase Carboxypeptidase D	
3.4.21.61	Gene KEX2 dibasic proteinase Kexin	
2.1.1.48	Gene ksgA methyltransferase rRNA (adenine-N^6-)-methyltransferase	
2.7.1.112	Gene lck protein kinase Protein-tyrosine kinase	
2.7.1.112	Gene lck tyrosine kinase Protein-tyrosine kinase	
3.4.21.53	Gene lon protease Endopeptidase La	
3.4.21.53	Gene lon proteins Endopeptidase La	
3.4.21.62	Genenase I Subtilisin	
3.4.21.87	Gene ompT proteins Omptin	
1.3.99.3	General acyl CoA dehydrogenase Acyl-CoA dehydrogenase	
3.4.24.64	General mitochondrial processing peptidase Mitochondrial processing peptidase	
3.1.11.5	Gene recBC DNase Exodeoxyribonuclease V	
3.1.11.5	Gene RecBC endoenzyme Exodeoxyribonuclease V	
2.7.9.3	Gene selD proteins selenide, water dikinase	
3.1.27.5	Gene S glycoproteins Pancreatic ribonuclease	

EC-No.	Synonym Recommended Name	Vol., Page
3.1.27.5	Gene S locus-specific glycoproteins Pancreatic ribonuclease	
2.7.7.46	Gentamicin 2-nucleotidyltransferase Gentamicin 2-nucleotidyltransferase	
2.3.1.59	Gentamicin 2'-N-acetyltransferase Gentamicin 2'-N-acetyltransferase	
2.3.1.60	Gentamicin 3'-N-acetyltransferase Gentamicin 3'-N-acetyltransferase	
2.3.1.60	Gentamicin acetyltransferase I Gentamicin 3'-N-acetyltransferase	
2.3.1.59	Gentamicin acetyltransferase II Gentamicin 2'-N-acetyltransferase	
2.3.1.153	Gentian 5AT Anthocyanin 5-aromatic acyltransferase	
3.2.1.21	Gentiobiase β-glucosidase	
1.13.11.4	Gentisate 1,2-dioxygenase Gentisate 1,2-dioxygenase	
1.13.11.4	Gentisate:oxygen 1,2-oxidoreductase (decyclizing) Gentisate 1,2-dioxygenase	
4.1.1.62	Gentisate carboxy-lyase Gentisate decarboxylase	
4.1.1.62	Gentisate decarboxylase Gentisate decarboxylase	
1.13.11.4	Gentisate dioxygenase Gentisate 1,2-dioxygenase	
1.13.11.4	Gentisate oxygenase Gentisate 1,2-dioxygenase	
1.13.11.4	Gentisic acid oxidase Gentisate 1,2-dioxygenase	
3.4.17.14	G enzyme Zinc D-Ala-D-Ala carboxypeptidase	
1.1.1.183	Geraniol:NADP^{+} oxidoreductase Geraniol dehydrogenase	
1.1.1.183	Geraniol dehydrogenase Geraniol dehydrogenase	

EC-No.	Synonym Recommended Name	Vol., Page
4.2.3.14	β-geraniolene synthase pinene synthase	
6.4.1.5	Geranoyl-CoA:carbon-dioxide ligase (ADP-forming) Geranoyl-CoA carboxylase	Vol. 2, Page 752
6.4.1.5	Geranoyl-CoA carboxylase Geranoyl-CoA carboxylase	Vol. 2, Page 752
6.4.1.5	Geranoyl coenzyme A carboxylase Geranoyl-CoA carboxylase	Vol. 2, Page 752
6.4.1.5	Geranyl-CoA carboxylase Geranoyl-CoA carboxylase	Vol. 2, Page 752
2.5.1.10	Geranyl-diphosphate:isopentenyl-diphosphate geranyltranstransferase Geranyltranstransferase	
5.5.1.8	Geranyl-diphosphate cyclase Geranyl-diphosphate cyclase	Vol. 1, Page 705
4.2.3.11	geranyl-diphosphate diphosphate-lyase (cyclizing, sabinene-hydrate-forming) sabinene-hydrate synthase	
4.2.3.10	geranyl-diphosphate diphosphate-lyase [cyclizing, (-)-endo-fenchol-forming] (-)-endo-fenchol synthase	
4.6.1.8	Geranyl-diphosphate pyrophosphate-lyase (cyclizing,(-)-endo-fenchol-forming) (-)-endo-Fenchol synthase	
4.6.1.9	Geranyl-diphosphate pyrophosphate-lyase (cyclizing, sabinene-hydrate-forming) Sabinene-hydrate synthase	
2.5.1.1	Geranyl-diphosphate synthase Dimethylallyltranstransferase	
4.2.3.16	geranyldiphosphate diphosphate lyase (limonene forming) (-)-(4S)-limonene synthase	
4.2.3.15	geranyldiphosphate diphosphate lyase (myrcene forming) myrcene synthase	
4.2.3.14	geranyldiphosphate diphosphate lyase (pinene forming) pinene synthase	
2.5.1.1	Geranyl diphosphate synthase Dimethylallyltranstransferase	

EC-No.	Synonym Recommended Name	Vol., Page
2.5.1.32	Geranylgeranyl-diphosphate:geranylgeranyl-diphosphate geranylgeranyl-transferase Geranylgeranyl-diphosphate geranylgeranyltransferase	
4.2.3.8	geranylgeranyl-diphosphate diphosphate-lyase (cyclizing) casbene synthase	
2.5.1.32	Geranylgeranyl-diphosphate geranylgeranyltransferase Geranylgeranyl-diphosphate geranylgeranyltransferase	
4.6.1.7	Geranylgeranyl-diphosphate pyrophosphate-lyase (cyclizing) Casbene synthase	
2.5.1.29	Geranylgeranyl-PP synthetase Farnesyltranstransferase	
2.5.1.41	Geranylgeranyl-transferase I Phosphoglycerol geranylgeranyltransferase	
2.5.1.42	Geranylgeranyl diphosphate:sn-3-O-(geranylgeranyl)glycerol 1-phosphate geranylgeranyltransferase Geranylgeranylglycerol-phosphate geranylgeranyltransferase	
2.5.1.41	Geranylgeranyl diphosphate:sn-glyceryl phosphate geranylgeranyltransfer-ase Phosphoglycerol geranylgeranyltransferase	
2.5.1.29	Geranylgeranyl diphosphate synthase Farnesyltranstransferase	
2.5.1.42	Geranylgeranylglycerol-phosphate geranylgeranyltransferase Geranylgeranylglycerol-phosphate geranylgeranyltransferase	
2.5.1.29	Geranylgeranyl pyrophosphate synthase Farnesyltranstransferase	
2.5.1.29	Geranylgeranyl pyrophosphate synthetase Farnesyltranstransferase	
2.5.1.42	Geranylgeranyltransferase, geranylgeranyloxyglycerol phosphate Geranylgeranylglycerol-phosphate geranylgeranyltransferase	
2.5.1.41	Geranylgeranyltransferase, glycerol phosphate Phosphoglycerol geranylgeranyltransferase	
2.5.1.42	Geranylgeranyltransferase II Geranylgeranylglycerol-phosphate geranylgeranyltransferase	
5.5.1.8	Geranyl pyrophosphate:(+)-bornyl pyrophosphate cyclase Geranyl-diphosphate cyclase	Vol. 1, Page 705
5.5.1.8	Geranyl pyrophosphate:(-)-bornyl pyrophosphate cyclase Geranyl-diphosphate cyclase	Vol. 1, Page 705

EC-No.	Synonym Recommended Name	Vol., Page
4.6.1.8	Geranyl pyrophosphate:(-)-endo-fenchol cyclase (-)-endo-Fenchol synthase	
4.2.3.10	geranylpyrophosphate:(-)-endo-fenchol cyclase (-)-endo-fenchol synthase	
2.5.1.1	Geranyl pyrophosphate synthase Dimethylallyltranstransferase	
2.5.1.1	Geranyl pyrophosphate synthetase Dimethylallyltranstransferase	
2.5.1.10	Geranyl transferase I Geranyltranstransferase	
2.5.1.10	Geranyltranstransferase Geranyltranstransferase	
2.6.1.16	GFAT Glutamine-fructose-6-phosphate transaminase (isomerizing)	
1.1.99.28	GFOR Glucose-fructose oxidoreductase	
2.4.1.184	GGGT Galactolipid galactosyltransferase	
6.3.2.18	γ-GHA synthetase γ-Glutamylhistamine synthase	Vol. 2, Page 503
1.14.11.13	(Gibberellin-1),2-oxoglutarate:oxygen oxidoreductase (2 β-hydroxylating) Gibberellin 2β-dioxygenase	
1.14.11.15	(Gibberellin-20),2-oxoglutarate:oxygen oxidoreductase (3β-hydroxylating) Gibberellin 3β-dioxygenase	
1.14.11.12	(Gibberellin-44),2-oxoglutarate:oxygen oxidoreductase Gibberellin-44-dioxygenase	
1.14.11.12	Gibberellin-44-dioxygenase Gibberellin-44-dioxygenase	
2.4.1.176	Gibberellin β-D-glucosyltransferase Gibberellin β-D-glucosyltransferase	
1.14.11.13	Gibberellin 2β-dioxygenase Gibberellin 2β-dioxygenase	
1.14.11.13	Gibberellin 2β-hydroxylase Gibberellin 2β-dioxygenase	
1.14.11.15	Gibberellin 3β-dioxygenase Gibberellin 3β-dioxygenase	

EC-No.	Synonym Recommended Name Vol., Page
1.14.11.15	Gibberellin 3β-hydroxylase Gibberellin 3β-dioxygenase
1.14.11.12	Gibberellin A44 oxidase Gibberellin-44-dioxygenase
1.14.11.13	(Giberellin-1),2-oxoglutarate:oxygen oxidoreductase (2 β-hydroxylating) Gibberellin 2β-dioxygenase
1.14.11.12	(Giberellin-44),2-oxoglutarate:oxygen oxidoreductase Gibberellin-44-dioxygenase
1.14.11.12	Giberellin-44-dioxygenase Gibberellin-44-dioxygenase
2.4.1.176	Giberellin β-D-glucosyltransferase Gibberellin β-D-glucosyltransferase
1.14.11.13	Giberellin 2β-dioxygenase Gibberellin 2β-dioxygenase
1.14.11.13	Giberellin 2β-hydroxylase Gibberellin 2β-dioxygenase
1.14.11.15	Giberellin 3β dioxygenase Gibberellin 3β-dioxygenase
1.14.11.15	(Giberrellin-20),2-oxoglutarate: oxygen oxidoreductase (3β-hydroxylating) Gibberellin 3β-dioxygenase
3.4.22.37	Gingipain-1 Gingipain R
3.4.22.37	Gingipain R Gingipain R
3.4.22.37	Gingivain, arginine-specific Gingipain R
2.7.1.31	GK Glycerate kinase
2.7.1.30	GK Glycerol kinase
3.1.1.17	GL gluconolactonase
3.4.21.22	Gla-domainless factor IXaβ' coagulation factor IXa
3.4.21.35	glandular kallikrein tissue kallikrein

EC-No.	Synonym Recommended Name	Vol., Page
2.7.8.19	Glc-phosphotransferase UDPglucose-glycoprotein glucosephosphotransferase	
3.2.1.106	$Glc_3Man_9NAc_2$oligosaccharide glucosidase Mannosyl-oligosaccharide glucosidase	
3.2.1.45	GlcCer-β-glucosidase Glucosylceramidase	
5.3.1.10	GlcN-6-P isomerase Glucosamine-6-phosphate isomerase	Vol. 1, Page 312
3.1.1.69	GlcNAc-PI deacetylase N-Acetylglucosaminylphosphatidylinositol deacetylase	
2.4.1.94	O-GlcNAc transferase Protein N-acetylglucosaminyltransferase	
3.2.1.114	GlcNAc transferase I-dependent α1,3[α1,6]mannosidase Mannosyl-oligosaccharide 1,3-1,6-α-mannosidase	
4.2.1.26	GlcNA dehydratase aminodeoxygluconate dehydratase	
2.7.8.19	GlcPTase UDPglucose-glycoprotein glucosephosphotransferase	
3.4.99.8	Gliocladium proteinase Gliocladium proteinase	
2.4.2.4	Gliostatins Thymidine phosphorylase	
2.7.7.23	GlmU uridylyltransferase UDP-N-acetylglucosamine pyrophosphorylase	
6.1.1.18	GlnRS Glutamine-tRNA ligase	Vol. 2, Page 139
2.4.1.88	Globoside α-N-acetylgalactosaminyltransferase Globoside α-N-acetylgalactosaminyltransferase	
2.4.1.88	Globoside acetylgalactosaminyltransferase Globoside α-N-acetylgalactosaminyltransferase	
2.4.1.154	Globoside N-acetylgalactosaminyltransferase Globotriosylceramide β-1,6-N-acetylgalactosaminyltransferase	
2.4.1.79	Globoside synthetase Galactosylgalactosylglucosylceramide β-D-acetylgalactosaminyltransferase	
2.4.1.154	Globotriosylceramide β-1,6-N-acetylgalactosaminyltransferase Globotriosylceramide β-1,6-N-acetylgalactosaminyltransferase	

EC-No.	Synonym Recommended Name	Vol., Page
3.2.1.17	Globulin G Lysozyme	
3.2.1.17	Globulin G1 Lysozyme	
3.4.13.7	Glu-Glu dipeptidase Glu-Glu dipeptidase	
3.4.19.9	γ-Glu-X carboxypeptidase γ-Glu-X carboxypeptidase	
2.4.1.34	(1,3)-β-Glucan (callose) synthase 1,3-β-Glucan synthase	
3.2.1.11	1,6-α-D-glucan-6-glucanohydrolase Dextranase	
3.2.1.11	α-D-1,6-glucan-6-glucanohydrolase Dextranase	
2.4.1.18	α-Glucan-branching glycosyltransferase 1,4-α-Glucan branching enzyme	
3.2.1.133	1,4-α-D-Glucan α-maltohydrolase Glucan 1,4-α-maltohydrolase	
2.4.1.113	α-1,4-Glucan-protein synthase (ADP-forming) α-1,4-Glucan-protein synthase (ADP-forming)	
2.4.1.112	α-1,4-Glucan-protein synthase (UDP-forming) α-1,4-Glucan-protein synthase (UDP-forming)	
2.4.1.34	1,3-β-D-Glucan-UDP glucosyltransferase 1,3-β-Glucan synthase	
2.4.1.34	1,3-β-Glucan-uridine diphosphoglucosyltransferase 1,3-β-Glucan synthase	
3.2.1.84	Glucan 1,3-α-glucosidase Glucan 1,3-α-glucosidase	
3.2.1.58	Glucan 1,3-β-glucosidase Glucan 1,3-β-glucosidase	
3.2.1.3	Glucan 1,4-α-glucosidase Glucan 1,4-α-glucosidase	
3.2.1.74	Glucan 1,4-β-glucosidase Glucan 1,4-β-glucosidase	
3.2.1.98	Glucan 1,4-α-maltohexaosidase Glucan 1,4-α-maltohexaosidase	

EC-No.	Synonym Recommended Name	Vol., Page
3.2.1.133	Glucan 1,4-α-maltohydrolase Glucan 1,4-α-maltohydrolase	
3.2.1.60	Glucan 1,4-α-maltotetraohydrolase Glucan 1,4-α-maltotetraohydrolase	
3.2.1.116	Glucan 1,4-α-maltotriohydrolase Glucan 1,4-α-maltotriohydrolase	
4.2.2.13	α-(1,4)-Glucan 1,5-anhydro-D-fructose eliminase exo-(1,4)-α-D-Glucan lyase	
3.2.1.70	Glucan 1,6-α-glucosidase Glucan 1,6-α-glucosidase	
3.2.1.94	Glucan 1,6-α-isomaltosidase Glucan 1,6-α-isomaltosidase	
5.4.99.15	(1→4)-α-Glucan 1-α-glucosylmutase (1→4)-α-Glucan 1-α-glucosylmutase	Vol. 1, Page 652
5.4.99.15	(1→4)-α-Glucan 1-α-glucosylmutase (1→4)-α-Glucan 1-α-glucosylmutase	Vol. 1, Page 652
3.2.1.6	1,3-(1,3;1,4)-β-D-glucan 3(4)glucanohydrolase Endo-1,3(4)-β-glucanase	
3.2.1.39	(1→3)-β-glucan 3-glucanohydrolase Glucan endo-1,3-β-glucosidase	
3.2.1.59	1,3(1,3;1,4)-α-D-glucan 3-glucanohydrolase Glucan endo-1,3-α-glucosidase	
3.2.1.39	1,3-β-D-glucan 3-glucanohydrolase Glucan endo-1,3-β-glucosidase	
3.2.1.84	1,3-α-D-glucan 3-glucohydrolase Glucan 1,3-α-glucosidase	
2.4.1.19	1,4-α-D-Glucan 4-α-D-(1,4-α-D-glucano)-transferase (cyclizing) Cyclomaltodextrin glucanotransferase	
3.2.1.61	1,3-1,4-α-D-glucan 4-glucanohydrolase Mycodextranase	
3.2.1.73	1,3-1,4-β-D-glucan 4-glucanohydrolase Lichenase	
3.2.1.73	1,3-1,4-β-glucan 4-glucanohydrolase Lichenase	
3.2.1.4	1,4-(1,3;1,4)-β-D-Glucan 4-glucanohydrolase Cellulase	

EC-No.	Synonym Recommended Name	Vol., Page
3.2.1.73	β-(1→ 3), (1→ 4)-D-glucan 4-glucanohydrolase Lichenase	
2.4.1.19	α-1,4-Glucan 4-glycosyltransferase, cyclizing Cyclomaltodextrin glucanotransferase	
2.4.1.24	1,4-α-D-Glucan 6-α-D-glucosyltransferase 1,4-α-Glucan 6-α-glucosyltransferase	
2.4.1.24	1,4-α-Glucan 6-α-glucosyltransferase 1,4-α-Glucan 6-α-glucosyltransferase	
3.2.1.75	β-1,6-glucan 6-glucanohydrolase Glucan endo-1,6-β-glucosidase	
2.4.1.18	α-1,4-Glucan:α-1,4-glucan-6-glycosyltransferase 1,4-α-Glucan branching enzyme	
2.4.1.18	α-1,4-Glucan:α-1,4-glucan 6-glycosyltransferase 1,4-α-Glucan branching enzyme	
2.4.1.24	1,4-α-D-Glucan:1,4-α-D-glucan(D-glucose) 6-α-D-glucosyltransferase 1,4-α-Glucan 6-α-glucosyltransferase	
2.4.1.161	1,4-α-D-Glucan:1,4-α-D-glucan 4-α-D-glucosyltransferase Oligosaccharide 4-α-D-glucosyltransferase	
2.4.1.25	1,4-α-D-Glucan:1,4-α-D-glucan 4-α-D-glycosyltransferase 4-α-Glucanotransferase	
2.4.1.18	1,4-α-D-Glucan:1,4-α-D-glucan 6-α-D-(1,4-α-D-glucano)-transferase 1,4-α-Glucan branching enzyme	
2.4.1.161	1,4-α-Glucan:1,4-α-glucan 4-α-glucosyltransferase Oligosaccharide 4-α-D-glucosyltransferase	
2.4.1.2	1,4-α-D-Glucan:1,6-α-D-glucan 6-α-D-glucosyltransferase Dextrin dextranase	
2.4.1.97	1,3-β-D-Glucan:orthophosphate α-D-glucosyltransferase 1,3-β-D-Glucan phosphorylase	
2.4.1.1	1,4-α-D-Glucan:orthophosphate α-D-glucosyltransferase Phosphorylase	
2.4.1.97	1,3-β-D-Glucan:orthophosphate glucosyltransferase 1,3-β-D-Glucan phosphorylase	
3.2.1.74	.β.-1,4-.β.-Glucanase Glucan 1,4-β-glucosidase	
3.2.1.4	.β.-1,4-Glucanase Cellulase	

EC-No.	Synonym Recommended Name	Vol., Page
3.2.1.71	.β.-D-1,2-Glucanase Glucan endo-1,2-β-glucosidase	
3.2.1.6	β-1,3-glucanase Endo-1,3(4)-β-glucanase	
3.2.1.39	β-1,3-glucanase Glucan endo-1,3-β-glucosidase	
3.2.1.6	β-1,3-1,4-Glucanase Endo-1,3(4)-β-glucanase	
3.2.1.54	Glucanase, cyclohepta- Cyclomaltodextrinase	
3.2.1.54	Glucanase, cyclohexa- Cyclomaltodextrinase	
3.2.1.71	Glucanase, endo-1,2-.β. Glucan endo-1,2-β-glucosidase	
3.2.1.6	Glucanase, endo-1,3(4)-.β.- Endo-1,3(4)-β-glucanase	
3.2.1.59	Glucanase, endo-1,3-.α.- Glucan endo-1,3-α-glucosidase	
3.2.1.39	Glucanase, endo-1,3-.β.- Glucan endo-1,3-β-glucosidase	
3.2.1.75	Glucanase, endo-1,6-.β.- Glucan endo-1,6-β-glucosidase	
3.2.1.75	β-1,6-glucanase-pustulanase Glucan endo-1,6-β-glucosidase	
2.4.1.18	1,4-α-Glucan branching enzyme 1,4-α-Glucan branching enzyme	
3.2.1.91	.β.-1,4-Glucan cellobiohydrolase Cellulose 1,4-β-cellobiosidase	
3.2.1.91	1,4-.β.-D-Glucan cellobiohydrolase Cellulose 1,4-β-cellobiosidase	
3.2.1.91	1,4-.β.-Glucan cellobiohydrolase Cellulose 1,4-β-cellobiosidase	
3.2.1.91	1,4-D-Glucan cellobiohydrolase Cellulose 1,4-β-cellobiosidase	
3.2.1.91	β-1,4-glucan cellobiohydrolase Cellulose 1,4-β-cellobiosidase	

EC-No.	Synonym Recommended Name	Vol., Page
3.2.1.91	1,4-.β.-Glucan cellobiosidase Cellulose 1,4-β-cellobiosidase	
3.2.1.91	.β.-1,4-Glucan cellobiosylhydrolase Cellulose 1,4-β-cellobiosidase	
3.2.1.71	Glucan endo-1,2-β-glucosidase Glucan endo-1,2-β-glucosidase	
3.2.1.59	Glucan endo-1,3-α-glucosidase Glucan endo-1,3-α-glucosidase	
3.2.1.39	Glucan endo-1,3-β-glucosidase Glucan endo-1,3-β-glucosidase	
3.2.1.75	Glucan endo-1,6-β-glucosidase Glucan endo-1,6-β-glucosidase	
3.2.1.39	(1→ 3)-β-glucan endohydrolase Glucan endo-1,3-β-glucosidase	
3.2.1.73	1,3-1,4-β-glucan endohydrolase Lichenase	
4.2.2.13	(1,4)-α-D-Glucan exo-4-lyase (1,5-anhydro-D-fructose-forming) exo-(1,4)-α-D-Glucan lyase	
3.2.1.58	β-1,3-glucan exo-hydrolase Glucan 1,3-β-glucosidase	
4.2.2.13	α-1,4-Glucan exo-lyase exo-(1,4)-α-D-Glucan lyase	
3.2.1.74	(1→ 4)-β-D-glucan glucanohydrolase Glucan 1,4-β-glucosidase	
3.2.1.71	1,2-β-D-glucan glucanohydrolase Glucan endo-1,2-β-glucosidase	
3.2.1.73	1,3-1,4-β-D-glucan glucanohydrolase Lichenase	
3.2.1.39	1,3-β-D-glucan glucanohydrolase Glucan endo-1,3-β-glucosidase	
3.2.1.1	1,4-α-D-glucan glucanohydrolase α-amylase	
3.2.1.75	1,6-β-D-glucan glucanohydrolase Glucan endo-1,6-β-glucosidase	
3.2.1.58	1,3-β-D-glucan glucohydrolase Glucan 1,3-β-glucosidase	

EC-No.	Synonym Recommended Name	Vol., Page
3.2.1.3	1,4-α-D-glucan glucohydrolase Glucan 1,4-α-glucosidase	
3.2.1.74	1,4-β-D-glucan glucohydrolase Glucan 1,4-β-glucosidase	
3.2.1.70	1,6-α-D-glucan glucohydrolase Glucan 1,6-α-glucosidase	
3.2.1.3	α-1,4-glucan glucohydrolase Glucan 1,4-α-glucosidase	
3.2.1.75	β-1,6-Glucan hydrolase Glucan endo-1,6-β-glucosidase	
3.2.1.94	1,6-α-D-glucan isomaltohydrolase Glucan 1,6-α-isomaltosidase	
3.2.1.95	1,6-α-D-glucan isomaltotriohydrolase Dextran 1,6-α-isomaltotriosidase	
4.2.2.13	α-1,4-Glucan lyase exo-(1,4)-α-D-Glucan lyase	
3.2.1.98	1,4-α-D-glucan maltohexaohydrolase Glucan 1,4-α-maltohexaosidase	
3.2.1.2	1,4-α-D-glucan maltohydrolase β-amylase	
3.2.1.60	1,4-α-D-glucan maltotetraohydrolase Glucan 1,4-α-maltotetraohydrolase	
3.2.1.116	1,4-α-D-Glucan maltotriohydrolase Glucan 1,4-α-maltotriohydrolase	
2.4.1.25	4-α-Glucanotransferase 4-α-Glucanotransferase	
2.4.1.25	4-α-Glucanotransferase 4-α-Glucanotransferase	
2.4.1.97	1,3-β-D-Glucan phosphorylase 1,3-β-D-Glucan phosphorylase	
2.4.1.1	1,4-α-Glucan phosphorylase Phosphorylase	
2.4.1.1	Glucan phosphorylase Phosphorylase	
2.4.1.1	α-Glucan phosphorylase Phosphorylase	

EC-No.	Synonym Recommended Name	Vol., Page
2.4.1.183	1,3-α-D-Glucan synthase α-1,3-Glucan synthase	
2.4.1.34	1,3-β-D-Glucan synthase 1,3-β-Glucan synthase	
2.4.1.34	1,3-β-Glucan synthase 1,3-β-Glucan synthase	
2.4.1.12	1,4-β-D-Glucan synthase Cellulose synthase (UDP-forming)	
2.4.1.12	1,4-β-Glucan synthase Cellulose synthase (UDP-forming)	
2.4.1.12	Glucan synthase Cellulose synthase (UDP-forming)	
2.4.1.183	α-1,3-Glucan synthase α-1,3-Glucan synthase	
2.4.1.34	β-1,3-Glucan synthase 1,3-β-Glucan synthase	
2.4.1.12	β-1,4-Glucan synthase Cellulose synthase (UDP-forming)	
2.4.1.12	β-Glucan synthase Cellulose synthase (UDP-forming)	
2.4.1.34	1,3-β-D-Glucan synthetase 1,3-β-Glucan synthase	
2.4.1.34	β-1,3-Glucan synthetase 1,3-β-Glucan synthase	
2.4.1.12	β-1,4-Glucan synthetase Cellulose synthase (UDP-forming)	
4.2.1.40	D-Glucarate dehydratase glucarate dehydratase	
4.2.1.40	glucarate dehydratase glucarate dehydratase	
4.2.1.40	D-glucarate hydro-lyase glucarate dehydratase	
2.3.1.131	Glucarate O-hydroxycinnamoyltransferase Glucarate O-hydroxycinnamoyltransferase	
2.3.1.132	Glucarolactone O-hydroxycinnamoyltransferase Glucarolactone O-hydroxycinnamoyltransferase	

EC-No.	Synonym Recommended Name	Vol., Page
4.2.1.40	GlucD glucarate dehydratase	
1.1.1.140	D-Glucitol-6-phosphate dehydrogenase Sorbitol-6-phosphate 2-dehydrogenase	
1.1.1.14	Glucitol dehydrogenase L-Iditol 2-dehydrogenase	
3.2.1.3	Glucoamylase Glucan 1,4-α-glucosidase	
3.2.1.45	Glucocerebrosidase Glucosylceramidase	
3.2.1.45	β-D-Glucocerebrosidase Glucosylceramidase	
3.2.1.45	β-Glucocerebrosidase Glucosylceramidase	
3.2.1.45	Glucocerebroside.β.-glucosidase Glucosylceramidase	
2.8.2.18	Glucocorticoid sulfotransferase Cortisol sulfotransferase	
2.8.2.18	Glucocorticosteroid sulfotransferase Cortisol sulfotransferase	
5.5.1.4	Glucocycloaldolase myo-Inositol-1-phosphate synthase	Vol. 1, Page 674
3.2.1.70	Glucodextranase Glucan 1,6-α-glucosidase	
2.3.1.143	β-Glucogallin-tetrakisgalloylglucose O-galloyltransferase β-Glucogallin-tetrakisgalloylglucose O-galloyltransferase	
2.3.1.143	β-Glucogallin:1,2,3,6-tetra-O-galloyl-β-D-glucose 4-O-galloyltransferase β-Glucogallin-tetrakisgalloylglucose O-galloyltransferase	
2.3.1.143	β-Glucogallin:1,2,3,6-tetra-O-galloylglucose 4-O-galloyltransferase β-Glucogallin-tetrakisgalloylglucose O-galloyltransferase	
2.3.1.90	β-Glucogallin O-galloyltransferase β-Glucogallin O-galloyltransferase	
3.2.1.20	Glucoinvertase α-glucosidase	
2.7.1.2	Glucokinase Glucokinase	

EC-No.	Synonym Recommended Name	Vol., Page
2.7.1.1	Glucokinase Hexokinase	
2.4.1.32	Glucomannan-synthase Glucomannan 4-β-mannosyltransferase	
2.4.1.32	Glucomannan 4-β-mannosyltransferase Glucomannan 4-β-mannosyltransferase	
4.2.1.12	gluconate-6-phosphate dehydratase phosphogluconate dehydratase	
1.1.1.215	Gluconate 2-dehydrogenase Gluconate 2-dehydrogenase	
1.1.99.3	Gluconate 2-dehydrogenase Gluconate 2-dehydrogenase	
1.1.1.69	Gluconate 5-dehydrogenase Gluconate 5-dehydrogenase	
4.2.1.12	gluconate 6-phosphate dehydratase phosphogluconate dehydratase	
1.1.1.43	Gluconate 6-phosphate dehydrogenase Phosphogluconate 2-dehydrogenase	
1.1.99.3	D-Gluconate:(acceptor) 2-oxidoreductase Gluconate 2-dehydrogenase	
1.1.1.69	D-Gluconate:NAD(P)$^+$ 5-oxidoreductase Gluconate 5-dehydrogenase	
1.1.1.215	D-Gluconate:NADP$^+$ oxidoreductase Gluconate 2-dehydrogenase	
4.2.1.39	D-gluconate dehydratase Gluconate dehydratase	
4.2.1.39	Gluconate dehydratase Gluconate dehydratase	
1.1.99.3	D-Gluconate dehydrogenase Gluconate 2-dehydrogenase	
1.1.99.3	Gluconate dehydrogenase Gluconate 2-dehydrogenase	
1.1.99.3	D-Gluconate dehydrogenase, 2-keto-D-gluconate yielding Gluconate 2-dehydrogenase	
4.2.1.39	D-gluconate hydro-lyase Gluconate dehydratase	

EC-No.	Synonym Recommended Name	Vol., Page
2.7.1.12	Gluconate kinase Gluconokinase	
1.1.99.3	Gluconate oxidase Gluconate 2-dehydrogenase	
1.1.99.3	Gluconic acid dehydrogenase Gluconate 2-dehydrogenase	
1.1.99.3	Gluconic dehydrogenase Gluconate 2-dehydrogenase	
3.1.1.17	D-glucono-δ-lacton-hydrolase gluconolactonase	
3.1.1.17	glucono-δ-lactonase gluconolactonase	
3.1.1.17	D-glucono-1,5-lactone lactonohydrolase gluconolactonase	
2.7.1.12	Gluconokinase Gluconokinase	
3.1.1.17	gluconolactonase gluconolactonase	
3.2.1.20	α-glucopyranosidase α-glucosidase	
4.3.1.9	Glucosaminate ammonia-lyase Glucosaminate ammonia-lyase	
4.3.1.9	D-Glucosaminate ammonia-lyase (isomerizing) Glucosaminate ammonia-lyase	
4.2.1.26	D-glucosaminate dehydratase aminodeoxygluconate dehydratase	
2.3.1.4	Glucosamine-6-phosphate acetylase Glucosamine-phosphate N-acetyltransferase	
3.5.99.6	glucosamine-6-phosphate deaminase glucosamine-6-phosphate deaminase	
5.3.1.10	Glucosamine-6-phosphate isomerase Glucosamine-6-phosphate isomerase	Vol. 1, Page 312
3.5.99.6	glucosamine-6-phosphate isomerase glucosamine-6-phosphate deaminase	
2.6.1.16	Glucosamine-6-phosphate isomerase (glutamine-forming) Glutamine-fructose-6-phosphate transaminase (isomerizing)	

EC-No.	Synonym Recommended Name	Vol., Page
2.6.1.16	Glucosamine-6-phosphate synthetase Glutamine-fructose-6-phosphate transaminase (isomerizing)	
2.3.1.4	D-Glucosamine-6-P N-acetyltransferase Glucosamine-phosphate N-acetyltransferase	
2.3.1.4	Glucosamine-phosphate N-acetyltransferase Glucosamine-phosphate N-acetyltransferase	
5.3.1.10	Glucosamine-P isomerase Glucosamine-6-phosphate isomerase	Vol. 1, Page 312
2.3.1.4	Glucosamine 6-phosphate acetylase Glucosamine-phosphate N-acetyltransferase	
5.3.1.10	Glucosamine 6-phosphate deaminase Glucosamine-6-phosphate isomerase	Vol. 1, Page 312
5.3.1.10	Glucosamine 6-phosphate isomerase Glucosamine-6-phosphate isomerase	Vol. 1, Page 312
5.3.1.10	D-Glucosamine 6-phosphate ketol-isomerase (deaminating) Glucosamine-6-phosphate isomerase	Vol. 1, Page 312
2.3.1.4	Glucosamine 6-phosphate N-acetyltransferase Glucosamine-phosphate N-acetyltransferase	
2.6.1.16	Glucosamine 6-phosphate synthase Glutamine-fructose-6-phosphate transaminase (isomerizing)	
2.6.1.16	Glucosamine 6-phosphate synthetase Glutamine-fructose-6-phosphate transaminase (isomerizing)	
2.3.1.3	Glucosamine acetylase Glucosamine N-acetyltransferase	
2.7.1.8	Glucosamine kinase Glucosamine kinase	
2.3.1.3	Glucosamine N-acetyltransferase Glucosamine N-acetyltransferase	
2.3.1.4	Glucosamine phosphate acetyltransferase Glucosamine-phosphate N-acetyltransferase	
5.3.1.10	Glucosamine phosphate deaminase Glucosamine-6-phosphate isomerase	Vol. 1, Page 312
3.5.99.6	glucosamine phosphate deaminase glucosamine-6-phosphate deaminase	
5.3.1.10	Glucosaminephosphate isomerase Glucosamine-6-phosphate isomerase	Vol. 1, Page 312

EC-No.	Synonym Recommended Name	Vol., Page
3.5.99.6	glucosaminephosphate isomerase glucosamine-6-phosphate deaminase	
5.3.1.19	glucosaminephosphate isomerase glucosaminephosphate isomerase	
2.6.1.16	Glucosamine phosphate isomerase (glutamine-forming) Glutamine-fructose-6-phosphate transaminase (isomerizing)	
2.6.1.16	Glucosamine synthase Glutamine-fructose-6-phosphate transaminase (isomerizing)	
4.3.1.9	D-Glucosaminic acid dehydrase Glucosaminate ammonia-lyase	
4.2.1.26	glucosaminic dehydrase aminodeoxygluconate dehydratase	
3.2.1.76	Glucosaminoglycan α-L-iduronohydrolase L-Iduronidase	
2.8.2.23	Glucosaminyl 3-O-sulfotransferase Heparin-glucosamine 3-O-sulfotransferase	
2.4.1.86	Glucosaminylgalactosylglucosylceramide β-galactosyltransferase Glucosaminylgalactosylglucosylceramide β-galactosyltransferase	
2.4.1.1	Glucosan phosphorylase Phosphorylase	
2.4.1.18	Glucosan transglycosylase 1,4-α-Glucan branching enzyme	
5.4.2.7	α-D-Glucose-1,6-bisphosphate:deoxy-D-ribose-1-phosphate phosphotransferase Phosphopentomutase	Vol. 1, Page 535
2.7.1.106	Glucose-1,6-bisphosphate synthase Glucose-1,6-bisphosphate synthase	
2.7.1.106	Glucose-1,6-bisphosphate synthetase Glucose-1,6-bisphosphate synthase	
1.1.3.4	D-Glucose-1-oxidase Glucose oxidase	
3.1.3.10	glucose-1-phosphatase glucose-1-phosphatase	
2.4.1.139	α-D-Glucose-1-phosphate:α-D-glucose-1-phosphate 4-α-D-glucosyltransferase (dephosphorylating) Maltose synthase	

EC-No.	Synonym Recommended Name	Vol., Page
2.7.1.41	D-Glucose-1-phosphate:D-glucose-1-phosphate 6-phosphotransferase Glucose-1-phosphate phosphodismutase	
5.4.2.5	D-Glucose-1-phosphate:D-glucose-6-phosphotransferase Phosphoglucomutase (glucose-cofactor)	Vol. 1, Page 527
2.7.1.42	D-Glucose-1-phosphate:riboflavin 5'-phosphotransferase Riboflavin phosphotransferase	
2.7.7.27	Glucose-1-phosphate adenylyltransferase Glucose-1-phosphate adenylyltransferase	
2.7.7.33	Glucose-1-phosphate cytidylyltransferase Glucose-1-phosphate cytidylyltransferase	
2.7.7.34	Glucose-1-phosphate guanylyltransferase Glucose-1-phosphate guanylyltransferase	
3.1.3.10	glucose-1-phosphate phosphatase glucose-1-phosphatase	
2.7.1.41	Glucose-1-phosphate phosphodismutase Glucose-1-phosphate phosphodismutase	
5.4.2.5	Glucose-1-phosphate phosphotransferase Phosphoglucomutase (glucose-cofactor)	Vol. 1, Page 527
2.7.1.42	Glucose-1-phosphate phosphotransferase Riboflavin phosphotransferase	
2.7.7.24	Glucose-1-phosphate thymidylyltransferase Glucose-1-phosphate thymidylyltransferase	
2.7.7.9	Glucose-1-phosphate uridylyltransferase UTP-glucose-1-phosphate uridylyltransferase	
3.1.4.51	6-(D-Glucose-1-phospho)-D-mannosylglycoprotein glucose-1-phosphohydrolase Glucose-1-phospho-D-mannosylglycoprotein phosphodiesterase	
3.1.4.51	Glucose-1-phospho-D-mannosylglycoprotein phosphodiesterase Glucose-1-phospho-D-mannosylglycoprotein phosphodiesterase	
2.7.8.19	Glucose-1-phosphotransferase, uridine diphosphoglucose-glycoprotein UDPglucose-glycoprotein glucosephosphotransferase	
3.1.3.9	Glucose-6-phosphatase Glucose-6-phosphatase	
5.5.1.4	D-Glucose-6-phosphate,L-myo-inositol-1-phosphate cycloaldolase myo-Inositol-1-phosphate synthase	Vol. 1, Page 674

EC-No.	Synonym Recommended Name	Vol., Page
1.1.1.49	Glucose-6-phosphate 1-dehydrogenase Glucose-6-phosphate 1-dehydrogenase	
5.1.3.15	D-Glucose-6-phosphate 1-epimerase Glucose-6-phosphate 1-epimerase	Vol. 1, Page 162
5.1.3.15	Glucose-6-phosphate 1-epimerase Glucose-6-phosphate 1-epimerase	Vol. 1, Page 162
1.1.1.49	D-Glucose-6-phosphate:NADP$^+$ 1-oxidoreductase Glucose-6-phosphate 1-dehydrogenase	
5.5.1.4	Glucose-6-phosphate inositol monophosphate cycloaldolase myo-Inositol-1-phosphate synthase	Vol. 1, Page 674
5.3.1.9	D-Glucose-6-phosphate isomerase Glucose-6-phosphate isomerase	Vol. 1, Page 298
5.3.1.9	Glucose-6-phosphate isomerase Glucose-6-phosphate isomerase	Vol. 1, Page 298
5.3.1.9	D-Glucose-6-phosphate ketol-isomerase Glucose-6-phosphate isomerase	Vol. 1, Page 298
3.1.3.9	D-Glucose-6-phosphate phosphohydrolase Glucose-6-phosphatase	
1.1.99.28	Glucose-fructose oxidoreductase Glucose-fructose oxidoreductase	
1.1.99.28	Glucose-fructose transhydrogenase Glucose-fructose oxidoreductase	
2.7.1.10	Glucose-phosphate kinase Phosphoglucokinase	
2.7.1.106	Glucose 1,6-diphosphate synthase Glucose-1,6-bisphosphate synthase	
5.4.2.2	α-D-glucose 1,6-phosphomutase Phosphoglucomutase	Vol. 1, Page 506
5.4.2.6	β-D-Glucose 1,6-phosphomutase β-Phosphoglucomutase	Vol. 1, Page 530
5.4.2.5	α-D-glucose 1,6-phosphomutase (glucose-cofactor) Phosphoglucomutase (glucose-cofactor)	Vol. 1, Page 527
1.1.1.47	Glucose 1-dehydrogenase Glucose 1-dehydrogenase	
1.1.1.118	Glucose 1-dehydrogenase (NAD$^+$) Glucose 1-dehydrogenase (NAD$^+$)	

EC-No.	Synonym / Recommended Name	Vol., Page
1.1.1.119	Glucose 1-dehydrogenase ($NADP^+$) / Glucose 1-dehydrogenase ($NADP^+$)	
2.7.7.34	Glucose 1-phosphate guanylyltransferase / Glucose-1-phosphate guanylyltransferase	
3.1.3.10	D-glucose 1-phosphate phosphohydrolase / glucose-1-phosphatase	
2.7.7.24	Glucose 1-phosphate thymidylyltransferase / Glucose-1-phosphate thymidylyltransferase	
2.7.1.41	Glucose 1-phosphate transphosphorylase / Glucose-1-phosphate phosphodismutase	
2.7.7.9	Glucose 1-phosphate uridylyltransferase / UTP-glucose-1-phosphate uridylyltransferase	
1.1.3.10	Glucose 2-oxidase / Pyranose oxidase	
5.5.1.4	D-Glucose 6-phosphate-1L-myoinositol 1-phosphate cyclase / myo-Inositol-1-phosphate synthase	Vol. 1, Page 674
5.5.1.4	D-Glucose 6-phosphate-1L-myoinositol 1-phosphate cycloaldolase / myo-Inositol-1-phosphate synthase	Vol. 1, Page 674
5.5.1.4	D-Glucose 6-phosphate-L-myo-inositol 1-phosphate cyclase / myo-Inositol-1-phosphate synthase	Vol. 1, Page 674
5.1.3.15	Glucose 6-phosphate 1-epimerase / Glucose-6-phosphate 1-epimerase	Vol. 1, Page 162
5.5.1.4	Glucose 6-phosphate cyclase / myo-Inositol-1-phosphate synthase	Vol. 1, Page 674
5.5.1.4	D-Glucose 6-phosphate cycloaldolase / myo-Inositol-1-phosphate synthase	Vol. 1, Page 674
1.1.1.49	D-Glucose 6-phosphate dehydrogenase / Glucose-6-phosphate 1-dehydrogenase	
1.1.1.49	Glucose 6-phosphate dehydrogenase (NADP) / Glucose-6-phosphate 1-dehydrogenase	
5.3.1.9	Glucose 6-phosphate isomerase / Glucose-6-phosphate isomerase	Vol. 1, Page 298
3.1.3.9	Glucose 6-phosphate phosphatase / Glucose-6-phosphatase	
1.1.99.10	D-Glucose:(acceptor) 1-oxidoreductase / Glucose dehydrogenase (acceptor)	

EC-No.	Synonym Recommended Name	Vol., Page
1.1.99.17	D-Glucose:(pyrroloquinoline-quinone) 1-oxidoreductase Glucose dehydrogenase (pyrroloquinoline-quinone)	
1.1.99.28	D-Glucose:D-fructose oxidoreductase Glucose-fructose oxidoreductase	
1.1.1.47	β-D-Glucose:NAD(P)$^+$ 1-oxidoreductase Glucose 1-dehydrogenase	
1.1.1.118	D-Glucose:NAD$^+$ 1-oxidoreductase Glucose 1-dehydrogenase (NAD$^+$)	
1.1.1.118	D-Glucose:NAD oxidoreductase Glucose 1-dehydrogenase (NAD$^+$)	
1.1.1.119	D-Glucose:NADP$^+$ 1-oxidoreductase Glucose 1-dehydrogenase (NADP$^+$)	
1.1.3.4	β-D-Glucose:oxygen 1-oxidoreductase Glucose oxidase	
1.1.3.4	β-D-Glucose:quinone oxidoreductase Glucose oxidase	
1.1.3.4	Glucose aerodehydrogenase Glucose oxidase	
3.2.1.3	Glucose amylase Glucan 1,4-α-glucosidase	
2.7.1.1	Glucose ATP phosphotransferase Hexokinase	
3.2.1.45	Glucose cerebrosidase Glucosylceramidase	
1.1.99.10	Glucose dehydrogenase (acceptor) Glucose dehydrogenase (acceptor)	
1.1.99.10	Glucose dehydrogenase (Aspergillus) Glucose dehydrogenase (acceptor)	
1.1.99.10	Glucose dehydrogenase (decarboxylating) Glucose dehydrogenase (acceptor)	
1.1.1.47	D-Glucose dehydrogenase (NAD(P)) Glucose 1-dehydrogenase	
1.1.99.17	Glucose dehydrogenase (PQQ dependent) Glucose dehydrogenase (pyrroloquinoline-quinone)	
1.1.99.17	Glucose dehydrogenase (pyrroloquinoline-quinone) Glucose dehydrogenase (pyrroloquinoline-quinone)	

EC-No.	Synonym Recommended Name	Vol., Page
5.3.1.18	glucose isomerase glucose isomerase	Vol. 1, Page 340
1.1.3.4	D-Glucose oxidase Glucose oxidase	
1.1.3.4	Glucose oxidase Glucose oxidase	
1.1.3.4	β-D-Glucose oxidase Glucose oxidase	
1.1.3.4	Glucose oxyhydrase Glucose oxidase	
2.7.1.69	Glucose permease Protein-N^{π}-phosphohistidine-sugar phosphotransferase	
5.3.1.9	Glucose phosphate isomerase Glucose-6-phosphate isomerase	Vol. 1, Page 298
5.3.1.9	Glucosephosphate isomerase 2 Glucose-6-phosphate isomerase	Vol. 1, Page 298
5.3.1.9	Glucose phosphoisomerase Glucose-6-phosphate isomerase	Vol. 1, Page 298
5.4.2.2	Glucose phosphomutase Phosphoglucomutase	Vol. 1, Page 506
5.4.2.5	Glucose phosphomutase Phosphoglucomutase (glucose-cofactor)	Vol. 1, Page 527
3.2.1.74	.β.-1,4-Glucosidase Glucan 1,4-β-glucosidase	
3.2.1.20	α-1,4-glucosidase α-glucosidase	
3.2.1.70	α-1,6-glucosidase Glucan 1,6-α-glucosidase	
3.2.1.20	α-D-glucosidase α-glucosidase	
3.2.1.21	β-1,6-glucosidase β-glucosidase	
3.2.1.21	β-D-glucosidase β-glucosidase	
3.2.1.74	β-glucosidase Glucan 1,4-β-glucosidase	

EC-No.	Synonym Recommended Name	Vol., Page
3.2.1.27	α-1,3-glucosidase α-1,3-glucosidase	
3.2.1.20	α-glucosidase α-glucosidase	
3.2.1.21	β-glucosidase β-glucosidase	
3.2.1.112	Glucosidase, 2-deoxy-α 2-Deoxyglucosidase	
3.2.1.117	Glucosidase, amygdalin Amygdalin β-glucosidase	
3.2.1.33	Glucosidase, amylo-1,6- Amylo-1,6-glucosidase	
3.2.1.126	Glucosidase, coniferin β- Coniferin β-glucosidase	
3.2.1.115	Glucosidase, dextran 1,2-α Branched-dextran exo-1,2-α-glucosidase	
3.2.1.58	Glucosidase, exo-1,3-.β.- Glucan 1,3-β-glucosidase	
3.2.1.74	Glucosidase, exo-1,4-.β.- Glucan 1,4-β-glucosidase	
3.2.1.42	Glucosidase, guanosine diphospho- GDPglucosidase	
3.2.1.106	Glucosidase, mannosyloligosaccharide Mannosyl-oligosaccharide glucosidase	
3.2.1.10	Glucosidase, oligo-1,6- Oligo-1,6-glucosidase	
3.2.1.62	Glucosidase, phloretin Glycosylceramidase	
3.2.1.125	Glucosidase, raucaffricine β- Raucaffricine β-glucosidase	
3.2.1.48	Glucosidase, sucrose.α.- Sucrose α-glucosidase	
3.2.3.1	Glucosidase, thio- Thioglucosidase	
3.2.1.84	Glucosidase II Glucan 1,3-α-glucosidase	

EC-No.	Synonym Recommended Name	Vol., Page
1.1.99.13	D-Glucoside 3-dehydrogenase Glucoside 3-dehydrogenase	
1.1.99.13	Glucoside 3-dehydrogenase Glucoside 3-dehydrogenase	
3.2.1.20	α-D-glucoside glucohydrolase α-glucosidase	
3.2.1.21	β-glucoside glucohydrolase β-glucosidase	
3.2.1.21	β-D-glucoside glucohydrolase β-glucosidase	
3.2.1.20	α-glucoside hydrolase α-glucosidase	
2.7.1.85	β-Glucoside kinase β-Glucoside kinase	
3.2.1.20	Glucosidoinvertase α-glucosidase	
3.2.1.20	Glucosidosucrase α-glucosidase	
3.2.1.139	α-Glucosiduronase α-Glucosiduronase	
3.2.1.139	α-D-Glucosiduronate glucuronohydrolase α-Glucosiduronase	
3.2.3.1	Glucosinolase Thioglucosidase	
3.2.1.26	Glucosucrase β-fructofuranosidase	
3.1.6.3	Glucosulfatase Glycosulfatase	
3.2.1.115	1,2-α-D-Glucosyl-branched-dextran 2-glucohydrolase Branched-dextran exo-1,2-α-glucosidase	
2.4.1.31	3-β-D-Glucosyl-D-glucose:orthophosphate α-D-glucosyltransferase Laminaribiose phosphorylase	
2.4.1.28	Glucosyl-DNA β-glucosyltransferase Glucosyl-DNA β-glucosyltransferase	
3.2.1.45	D-Glucosyl-N-acylsphingosine glucohydrolase Glucosylceramidase	

EC-No.	Synonym Recommended Name	Vol., Page
3.5.1.26	Glucosylamidase N^4-(β-N-acetylglucosaminyl)-L-asparaginase	
3.2.1.45	.β.-Glucosylceramidase Glucosylceramidase	
3.2.1.45	Glucosylceramidase Glucosylceramidase	
3.2.1.45	Glucosylceramide.β.-glucosidase Glucosylceramidase	
2.4.1.80	Glucosylceramide synthase Ceramide glucosyltransferase	
3.2.1.45	Glucosylcerebrosidase Glucosylceramidase	
3.2.1.130	Glucosyl mannosidase Glycoprotein endo-α-1,2-mannosidase	
3.2.1.130	Glucosylmannosidase Glycoprotein endo-α-1,2-mannosidase	
2.4.1.158	13-Glucosyloxydocosanoate 2'-β-glucosyltransferase 13-Hydroxydocosanoate 13-β-glucosyltransferase	
3.2.1.45	Glucosylsphingosine.β.-D-glucosidase Glucosylceramidase	
3.2.1.45	Glucosylsphingosine.β.-glucosidase Glucosylceramidase	
2.4.1.24	D-Glucosyltransferase 1,4-α-Glucan 6-α-glucosyltransferase	
5.4.99.11	α-Glucosyltransferase Isomaltulose synthase	Vol. 1, Page 638
2.4.1.12	β-1,4-Glucosyltransferase Cellulose synthase (UDP-forming)	
2.4.1.24	Glucosyltransferase, 1,4-α-glucan 6-α- 1,4-α-Glucan 6-α-glucosyltransferase	
2.4.1.113	Glucosyltransferase, adenosine diphosphoglucose-protein α-1,4-Glucan-protein synthase (ADP-forming)	
2.4.1.21	Glucosyltransferase, adenosine diphosphoglucose-starch Starch synthase	
2.4.1.195	Glucosyltransferase, desulfoglucosinolate-uridine diphosphate Thiohydroximate β-D-glucosyltransferase	

EC-No.	Synonym Recommended Name	Vol., Page
2.4.1.2	Glucosyltransferase, dextrin 6- Dextrin dextranase	
2.4.1.29	Glucosyltransferase, guanosine diphosphoglucose-1,4-β-glucan Cellulose synthase (GDP-forming)	
2.4.1.36	Glucosyltransferase, guanosine diphosphoglucose-glucose phosphate α,α-Trehalose-phosphate synthase (GDP-forming)	
2.4.1.58	Glucosyltransferase, lipopolysaccharide Lipopolysaccharide glucosyltransferase I	
2.4.1.25	Glucosyltransferase, maltose 4- 4-α-Glucanotransferase	
2.4.1.4	Glucosyltransferase, sucrose-1,4-α-glucan Amylosucrase	
2.4.1.140	Glucosyltransferase, sucrose-1,6(3)-α-glucan 6(3)-α- Alternansucrase	
2.4.1.5	Glucosyltransferase, sucrose-1,6-α-glucan Dextransucrase	
2.4.1.125	Glucosyltransferase, sucrose-1,6-α-glucan 3(6)-α- Sucrose-1,6-α-glucan 3(6)-α-glucosyltransferase	
2.4.1.35	Glucosyltransferase, uridine diphospho- Phenol β-glucosyltransferase	
2.4.1.183	Glucosyltransferase, uridine diphosphoglucose-1,3-α-glucan α-1,3-Glucan synthase	
2.4.1.34	Glucosyltransferase, uridine diphosphoglucose-1,3-β-glucan 1,3-β-Glucan synthase	
2.4.1.12	Glucosyltransferase, uridine diphosphoglucose-1,4-β-glucan Cellulose synthase (UDP-forming)	
2.4.1.202	Glucosyltransferase, uridine diphosphoglucose-2,4-dihydroxy-7-methoxy-2H-1,4-benzoxazin-3(4H)-one 2- 2,4-Dihydroxy-7-methoxy-2H-1,4-benzoxazin-3(4H)-one 2-D-glucosyltransferase	
2.4.1.194	Glucosyltransferase, uridine diphosphoglucose-4-hydroxybenzoate 4-Hydroxybenzoate 4-O-β-D-glucosyltransferase	
2.4.1.188	Glucosyltransferase, uridine diphosphoglucose-acetylglucosaminylpyrophosphorylundecaprenol N-Acetylglucosaminyldiphosphoundecaprenol glucosyltransferase	
2.4.1.103	Glucosyltransferase, uridine diphosphoglucose-alizarin Alizarin 2-β-glucosyltransferase	

EC-No.	Synonym Recommended Name	Vol., Page
2.4.1.115	Glucosyltransferase, uridine diphosphoglucose-anthocyanidin 3-O- Anthocyanidin 3-O-glucosyltransferase	
2.4.1.181	Glucosyltransferase, uridine diphosphoglucose-anthraquinone Hydroxyanthraquinone glucosyltransferase	
2.4.1.81	Glucosyltransferase, uridine diphosphoglucose-apigenin 7-O- Flavone 7-O-β-glucosyltransferase	
2.4.1.71	Glucosyltransferase, uridine diphosphoglucose-arylamine Arylamine glucosyltransferase	
2.4.1.80	Glucosyltransferase, uridine diphosphoglucose-ceramide Ceramide glucosyltransferase	
2.4.1.177	Glucosyltransferase, uridine diphosphoglucose-cinnamate Cinnamate β-D-glucosyltransferase	
2.4.1.66	Glucosyltransferase, uridine diphosphoglucose-collagen Procollagen glucosyltransferase	
2.4.1.111	Glucosyltransferase, uridine diphosphoglucose-coniferyl alcohol Coniferyl-alcohol glucosyltransferase	
2.4.1.116	Glucosyltransferase, uridine diphosphoglucose-cyanidin 3-rhamnosylglucoside 5-O- Cyanidin-3-rhamnosylglucoside 5-O-glucosyltransferase	
2.4.1.178	Glucosyltransferase, uridine diphosphoglucose-cyanohydrin Hydroxymandelonitrile glucosyltransferase	
2.4.1.26	Glucosyltransferase, uridine diphosphoglucose-deoxyribonucleate α- DNA α-glucosyltransferase	
2.4.1.27	Glucosyltransferase, uridine diphosphoglucose-deoxyribonucleate β- DNA β-glucosyltransferase	
2.4.1.157	Glucosyltransferase, uridine diphosphoglucose-diacylglycerol 1,2-Diacylglycerol 3-glucosyltransferase	
2.4.1.117	Glucosyltransferase, uridine diphosphoglucose-dolichol Dolichyl-phosphate β-glucosyltransferase	
2.4.1.185	Glucosyltransferase, uridine diphosphoglucose-flavanone 7-O- Flavanone 7-O-β-glucosyltransferase	
2.4.1.91	Glucosyltransferase, uridine diphosphoglucose-flavonol 3-O- Flavonol 3-O-glucosyltransferase	
2.4.1.13	Glucosyltransferase, uridine diphosphoglucose-fructose Sucrose synthase	

EC-No.	Synonym Recommended Name	Vol., Page
2.4.1.14	Glucosyltransferase, uridine diphosphoglucose-fructose phosphate Sucrose-phosphate synthase	
2.4.1.73	Glucosyltransferase, uridine diphosphoglucose-galactosylpolysaccharide Lipopolysaccharide glucosyltransferase II	
2.4.1.176	Glucosyltransferase, uridine diphosphoglucose-gibberellate 3-O- Gibberellin β-D-glucosyltransferase	
2.4.1.176	Glucosyltransferase, uridine diphosphoglucose-gibberellate 7- Gibberellin β-D-glucosyltransferase	
2.4.1.28	Glucosyltransferase, uridine diphosphoglucose-glucosyldeoxyribonucleate β- Glucosyl-DNA β-glucosyltransferase	
2.4.1.11	Glucosyltransferase, uridine diphosphoglucose-glycogen Glycogen (starch) synthase	
2.4.1.126	Glucosyltransferase, uridine diphosphoglucose-hydroxycinnamate Hydroxycinnamate 4-β-glucosyltransferase	
2.4.1.158	Glucosyltransferase, uridine diphosphoglucose-hydroxydocosanoate 13-Hydroxydocosanoate 13-β-glucosyltransferase	
2.4.1.121	Glucosyltransferase, uridine diphosphoglucose-indoleacetate Indole-3-acetate β-glucosyltransferase	
2.4.1.170	Glucosyltransferase, uridine diphosphoglucose-isoflavone 7-O- Isoflavone 7-O-glucosyltransferase	
2.4.1.106	Glucosyltransferase, uridine diphosphoglucose-isovitexin 2- Isovitexin β-glucosyltransferase	
2.4.1.63	Glucosyltransferase, uridine diphosphoglucose-ketone Linamarin synthase	
2.4.1.58	Glucosyltransferase, uridine diphosphoglucose-lipopolysaccharide Lipopolysaccharide glucosyltransferase I	
2.4.1.81	Glucosyltransferase, uridine diphosphoglucose-luteolin Flavone 7-O-β-glucosyltransferase	
2.4.1.171	Glucosyltransferase, uridine diphosphoglucose-methylazoxymethanol Methyl-ONN-azoxymethanol β-D-glucosyltransferase	
2.4.1.208	Glucosyltransferase, uridine diphosphoglucose-monoglucosyldiacylglycerol Diglucosyl diacylglycerol synthase	
2.4.1.127	Glucosyltransferase, uridine diphosphoglucose-monoterpenol Monoterpenol β-glucosyltransferase	

EC-No.	Synonym Recommended Name	Vol., Page
2.4.1.196	Glucosyltransferase, uridine diphosphoglucose-nicotinate N- Nicotinate glucosyltransferase	
2.4.1.192	Glucosyltransferase, uridine diphosphoglucose-nuatigenin Nuatigenin 3*β*-glucosyltransferase	
2.4.1.114	Glucosyltransferase, uridine diphosphoglucose-*o*-coumarate 2-Coumarate O-*β*-glucosyltransferase	
2.4.1.104	Glucosyltransferase, uridine diphosphoglucose-*o*-dihydroxycoumarin 7-O- *o*-Dihydroxycoumarin 7-O-glucosyltransferase	
2.4.1.85	Glucosyltransferase, uridine diphosphoglucose-*p*-hydroxymandelonitrile Cyanohydrin *β*-glucosyltransferase	
2.4.1.52	Glucosyltransferase, uridine diphosphoglucose-poly(glycerol-phosphate) *α*- Poly(glycerol-phosphate) *α*-glucosyltransferase	
2.4.1.53	Glucosyltransferase, uridine diphosphoglucose-poly(ribitol-phosphate) *β*- Poly(ribitol-phosphate) *β*-glucosyltransferase	
2.4.1.78	Glucosyltransferase, uridine diphosphoglucose-polyprenol monophosphate Phosphopolyprenol glucosyltransferase	
2.4.1.173	Glucosyltransferase, uridine diphosphoglucose-poriferasterol Sterol 3*β*-glucosyltransferase	
2.4.1.112	Glucosyltransferase, uridine diphosphoglucose-protein *α*-1,4-Glucan-protein synthase (UDP-forming)	
2.4.1.112	Glucosyltransferase, uridine diphosphoglucose-protein 4-*α*- *α*-1,4-Glucan-protein synthase (UDP-forming)	
2.4.1.160	Glucosyltransferase, uridine diphosphoglucose-pyridoxine 5'-*β*- Pyridoxine 5'-O-*β*-D-glucosyltransferase	
2.4.1.172	Glucosyltransferase, uridine diphosphoglucose-salicyl alcohol 2- Salicyl-alcohol *β*-D-glucosyltransferase	
2.4.1.193	Glucosyltransferase, uridine diphosphoglucose-sarsapogenin Sarsapogenin 3*β*-glucosyltransferase	
2.4.1.128	Glucosyltransferase, uridine diphosphoglucose-scopoletin UDPglucose:scopoletin O-*β*-D-glucosyltransferase	
2.3.1.152	Glucosyltransferase, uridine diphosphoglucose-sinapate Alcohol O-cinnamoyltransferase	
2.4.1.120	Glucosyltransferase, uridine diphosphoglucose-sinapate Sinapate 1-glucosyltransferase	
2.4.1.173	Glucosyltransferase, uridine diphosphoglucose-sterol Sterol 3*β*-glucosyltransferase	

EC-No.	Synonym Recommended Name	Vol., Page
2.4.1.195	Glucosyltransferase, uridine diphosphoglucose-thiohydroximate Thiohydroximate β-D-glucosyltransferase	
2.4.1.136	Glucosyltransferase, uridine diphosphoglucose-vanillate 1- Gallate 1-β-glucosyltransferase	
2.4.1.105	Glucosyltransferase, uridine diphosphoglucose-vitexin 2- Vitexin β-glucosyltransferase	
2.4.1.168	Glucosyltransferase, uridine diphosphoglucose-xyloglucan 4β- Xyloglucan 4-glucosyltransferase	
2.4.1.118	Glucosyltransferase, uridine diphosphoglucose-zeatin 7- Cytokinin 7-β-glucosyltransferase	
2.4.1.203	Glucosyltransferase, uridine diphosphoglucose-zeatin O- Zeatin O-β-D-glucosyltransferase	
2.4.1.15	Glucosyltransferase, uridine diphosphoglucose phosphate α,α-Trehalose-phosphate synthase (UDP-forming)	
1.1.1.70	D-glucronolactone dehydrogenase D-glucronolactone dehydrogenase	
4.2.2.14	(1,4)-β-D-glucuronan lyase glucuronan lyase	
4.2.2.14	glucuronan lyase glucuronan lyase	
2.7.7.44	Glucuronate-1-phosphate uridylyltransferase Glucuronate-1-phosphate uridylyltransferase	
3.1.6.18	Glucuronate-2-sulfatase Glucuronate-2-sulfatase	
3.1.6.18	Glucuronate 2-sulfatase Glucuronate-2-sulfatase	
1.1.1.19	D-Glucuronate dehydrogenase Glucuronate reductase	
5.3.1.12	D-Glucuronate isomerase Glucuronate isomerase	Vol. 1, Page 322
5.3.1.12	Glucuronate isomerase Glucuronate isomerase	Vol. 1, Page 322
5.3.1.12	D-Glucuronate ketol-isomerase Glucuronate isomerase	Vol. 1, Page 322
1.1.1.19	D-Glucuronate reductase Glucuronate reductase	

EC-No.	Synonym Recommended Name	Vol., Page
1.1.1.19	Glucuronate reductase Glucuronate reductase	
3.2.1.131	α-(1→2)-Glucuronidase Xylan α-1,2-glucuronosidase	
3.2.1.139	α-D-Glucuronidase α-Glucosiduronase	
3.2.1.139	α-Glucuronidase α-Glucosiduronase	
3.2.1.31	β-glucuronidase β-glucuronidase	
3.2.1.139	Glucuronidase, α- α-Glucosiduronase	
3.2.1.31	Glucuronidase, β-glucuronide glucuronohydrolase β-glucuronidase	
3.2.1.131	Glucuronidase, 1,2-α- Xylan α-1,2-glucuronosidase	
3.2.1.56	Glucuronidase, glucuronosyldisulfoglucosamine Glucuronosyl-disulfoglucosamine glucuronidase	
3.1.6.18	Glucurono-2-sulfatase Glucuronate-2-sulfatase	
3.1.1.19	D-glucurono-6,2-lactone lactonohydrolase uronolactonase	
3.2.1.136	Glucuronoarabinoxylan 1,4-β-D-xylanohydrolase Glucuronoarabinoxylan endo-1,4-β-xylanase	
3.2.1.136	Glucuronoarabinoxylan endo-1,4-β-xylanase Glucuronoarabinoxylan endo-1,4-β-xylanase	
3.2.1.36	Glucuronoglucosaminoglycan hyaluronate lyase Hyaluronoglucuronidase	
4.2.2.1	glucuronoglycosaminoglycan lyase hyaluronate lyase	
2.7.1.43	Glucuronokinase Glucuronokinase	
2.7.1.43	Glucuronokinase, glucurono- Glucuronokinase	
3.1.1.19	glucuronolactonase uronolactonase	

EC-No.	Synonym Recommended Name	Vol., Page
1.1.1.20	Glucuronolactone reductase Glucuronolactone reductase	
3.2.1.31	β-D-glucuronoside glucuronosohydrolase β-glucuronidase	
3.2.1.56	Glucuronosyl-disulfoglucosamine glucuronidase Glucuronosyl-disulfoglucosamine glucuronidase	
3.2.1.56	3-D-Glucuronosyl-N^2, 6-disulfo-β-D-glucosamine glucuronohydrolase Glucuronosyl-disulfoglucosamine glucuronidase	
2.4.1.17	Glucuronosyltransferase Glucuronosyltransferase	
2.4.1.95	Glucuronosyltransferase, bilirubin glucuronoside Bilirubin-glucuronoside glucuronosyltransferase	
2.4.1.17	Glucuronosyltransferase, uridine diphospho- Glucuronosyltransferase	
2.4.1.17	Glucuronosyltransferase, uridine diphosphoglucuronate-1,2-diacylglycerol Glucuronosyltransferase	
2.4.1.17	Glucuronosyltransferase, uridine diphosphoglucuronate-4-hydroxybiphenyl Glucuronosyltransferase	
2.4.1.17	Glucuronosyltransferase, uridine diphosphoglucuronate-bilirubin Glucuronosyltransferase	
2.4.1.17	Glucuronosyltransferase, uridine diphosphoglucuronate-bilirubin glucuronoside Glucuronosyltransferase	
2.4.1.17	Glucuronosyltransferase, uridine diphosphoglucuronate-estradiol Glucuronosyltransferase	
2.4.1.17	Glucuronosyltransferase, uridine diphosphoglucuronate-estriol Glucuronosyltransferase	
2.4.1.17	Glucuronosyltransferase, uridine diphosphoglucuronate-estriol 16α- Glucuronosyltransferase	
2.4.1.189	Glucuronosyltransferase, uridine diphosphoglucuronate-luteolin 7-O- Luteolin 7-O-glucuronosyltransferase	
2.4.1.191	Glucuronosyltransferase, uridine diphosphoglucuronate-luteolin 7-O-diglucuronide Luteolin-7-O-diglucuronide 4'-O-glucuronosyltransferase	
2.4.1.190	Glucuronosyltransferase, uridine diphosphoglucuronate-luteolin 7-O-glucuronide Luteolin-7-O-glucuronide 7-O-glucuronosyltransferase	

EC-No.	Synonym Recommended Name	Vol., Page
2.4.1.135	Glucuronosyltransferase I Galactosylgalactosylxylosylprotein 3-β-glucuronosyltransferase	
2.1.1.112	Glucuronoxylan 4-O-methyltransferase Glucuronoxylan 4-O-methyltransferase	
3.2.1.136	Glucuronoxylanase Glucuronoarabinoxylan endo-1,4-β-xylanase	
3.2.1.136	Glucuronoxylan xylanohydrolase Glucuronoarabinoxylan endo-1,4-β-xylanase	
3.2.1.136	Glucuronoxylan xylohydrolase Glucuronoarabinoxylan endo-1,4-β-xylanase	
2.4.1.175	Glucuronyl-N-acetylgalactosaminylproteoglycan β-1,4-N-acetylgalactosaminyltransferase Glucuronyl-N-acetylgalactosaminylproteoglycan β-1,4-N-acetylgalactosaminyltransferase	
5.1.3.17	D-Glucuronyl C-5 epimerase Heparosan-N-sulfate-glucuronate 5-epimerase	Vol. 1, Page 167
2.4.1.174	Glucuronylgalactosylproteoglycan β-1,4-N-acetylgalactosaminyltransferase Glucuronylgalactosylproteoglycan β-1,4-N-acetylgalactosaminyltransferase	
2.4.1.17	Glucuronyltransferase, uridine diphospho- Glucuronosyltransferase	
3.4.21.34	glumorin plasma kallikrein	
3.4.21.35	glumorin tissue kallikrein	
6.1.1.17	GluRS Glutamate-tRNA ligase	Vol. 2, Page 128
3.4.21.82	GluSGP Glutamyl endopeptidase II	
2.8.3.12	Glutaconate CoA-transferase Glutaconate CoA-transferase	
4.1.1.70	glutaconyl-CoA decarboxylase glutaconyl-CoA decarboxylase	
1.4.3.15	D-Glutamate(D-aspartate):oxygen oxidoreductase (deaminating) D-Glutamate(D-aspartate) oxidase	
1.4.3.15	D-Glutamate(D-aspartate) oxidase D-Glutamate(D-aspartate) oxidase	

EC-No.	Synonym Recommended Name	Vol., Page
2.6.1.39	Glutamate-α-ketoadipate transaminase 2-Aminoadipate transaminase	
1.2.1.41	Glutamate-.γ.-semialdehyde dehydrogenase Glutamate-5-semialdehyde dehydrogenase	
5.4.3.8	Glutamate-1-semialdehyde 2,1-aminomutase Glutamate-1-semialdehyde 2,1-aminomutase	Vol. 1, Page 575
5.4.3.8	Glutamate-1-semialdehyde aminotransferase Glutamate-1-semialdehyde 2,1-aminomutase	Vol. 1, Page 575
1.2.1.41	L-Glutamate-5-semialdehyde:NADP$^+$ 5-oxidoreductase (phosphorylating) Glutamate-5-semialdehyde dehydrogenase	
1.2.1.41	Glutamate-5-semialdehyde dehydrogenase Glutamate-5-semialdehyde dehydrogenase	
6.3.2.9	D-Glutamate-adding enzyme UDP-N-acetylmuramoylalanine-D-glutamate ligase	Vol. 2, Page 452
2.7.7.42	[Glutamate-ammonia-ligase] adenylyltransferase [Glutamate-ammonia-ligase] adenylyltransferase	
6.3.1.2	Glutamate-ammonia ligase Glutamate-ammonia ligase	Vol. 2, Page 347
2.6.1.42	Glutamate-branched-chain amino acid transaminase Branched-chain-amino-acid transaminase	
6.3.2.2	Glutamate-cysteine ligase Glutamate-cysteine ligase	Vol. 2, Page 399
4.1.1.15	L-Glutamate α-decarboxylase Glutamate decarboxylase	
2.6.1.49	Glutamate-DOPP transaminase (GDT) Dihydroxyphenylalanine transaminase	
6.3.1.6	Glutamate-ethylamine ligase Glutamate-ethylamine ligase	Vol. 2, Page 381
2.1.2.5	Glutamate-formiminotransferase Glutamate formiminotransferase	
6.3.4.12	Glutamate-methylamine ligase Glutamate-methylamine ligase	Vol. 2, Page 624
2.6.1.1	Glutamate-oxalacetate aminotransferase Aspartate transaminase	
2.6.1.1	Glutamate-oxalate transaminase Aspartate transaminase	

EC-No.	Synonym Recommended Name	Vol., Page
3.4.21.19	glutamate-specific proteinase glutamyl endopeptidase	
2.6.1.19	Glutamate-succinic semialdehyde transaminase 4-Aminobutyrate transaminase	
6.1.1.17	Glutamate-tRNA ligase Glutamate-tRNA ligase	Vol. 2, Page 128
6.1.1.17	Glutamate-tRNA synthetase Glutamate-tRNA ligase	Vol. 2, Page 128
4.1.1.15	L-Glutamate 1-carboxy-lyase Glutamate decarboxylase	
2.7.2.13	Glutamate 1-kinase Glutamate 1-kinase	
5.4.3.8	Glutamate 1-semialdehyde aminotransferase Glutamate-1-semialdehyde 2,1-aminomutase	Vol. 1, Page 575
2.7.2.11	Glutamate 5-kinase Glutamate 5-kinase	
6.3.1.2	L-Glutamate:ammonia ligase (ADP-forming) Glutamate-ammonia ligase	Vol. 2, Page 347
6.3.1.6	L-Glutamate:ethylamine ligase (ADP-forming) Glutamate-ethylamine ligase	Vol. 2, Page 381
1.4.7.1	L-Glutamate:ferredoxin oxidoreductase (transaminating) Glutamate synthase (ferredoxin)	
2.6.1.4	L-Glutamate:glyoxylate aminotransferase Glycine transaminase	
6.3.2.18	L-Glutamate:histamine ligase γ-Glutamylhistamine synthase	Vol. 2, Page 503
6.3.2.2	L-Glutamate:L-cysteine γ-ligase (ADP-forming) Glutamate-cysteine ligase	Vol. 2, Page 399
6.3.4.12	L-Glutamate:methylamine ligase (ADP-forming) Glutamate-methylamine ligase	Vol. 2, Page 624
1.4.1.3	L-Glutamate:NAD(P)$^+$ oxidoreductase (deaminating) Glutamate dehydrogenase (NAD(P)$^+$)	
1.4.1.2	L-Glutamate:NAD$^+$ oxidoredcutase (deaminating) Glutamate dehydrogenase	
1.4.1.14	L-Glutamate:NAD$^+$ oxidoreductase (transaminating) Glutamate synthase (NADH)	

EC-No.	Synonym Recommended Name	Vol., Page
1.4.1.4	L-Glutamate:NADP$^+$ oxidoreductase (deaminating) Glutamate dehydrogenase (NADP$^+$)	
1.4.1.13	L-Glutamate:NADP$^+$ oxidoreductase (transaminating) Glutamate synthase (NADPH)	
1.4.3.7	D-Glutamate:oxygen oxidoreductase (deaminating) D-Glutamate oxidase	
1.4.3.11	L-Glutamate:oxygen oxidoreductase (deaminating) L-Glutamate oxidase	
6.1.1.17	L-Glutamate:tRNA^{Glu} ligase (AMP-forming) Glutamate-tRNA ligase	Vol. 2, Page 128
2.3.1.35	Glutamate acetyltransferase Glutamate N-acetyltransferase	
3.4.17.11	Glutamate carboxypeptidase Glutamate carboxypeptidase	
3.4.17.21	Glutamate carboxypeptidase II Glutamate carboxypeptidase II	
4.2.1.48	D-glutamate cyclase D-glutamate cyclase	
4.1.1.15	Glutamate decarboxylase Glutamate decarboxylase	
4.1.1.15	L-Glutamate decarboxylase Glutamate decarboxylase	
4.1.1.15	γ-Glutamate decarboxylase Glutamate decarboxylase	
1.4.1.2	Glutamate dehydrogenase Glutamate dehydrogenase	
1.4.1.2	L-Glutamate dehydrogenase Glutamate dehydrogenase	
1.4.1.3	L-Glutamate dehydrogenase Glutamate dehydrogenase (NAD(P)$^+$)	
1.4.1.4	L-Glutamate dehydrogenase Glutamate dehydrogenase (NADP$^+$)	
1.4.1.3	Glutamate dehydrogenase (NAD(P)$^+$) Glutamate dehydrogenase (NAD(P)$^+$)	
1.4.1.2	Glutamate dehydrogenase (NAD) Glutamate dehydrogenase	

EC-No.	Synonym Recommended Name	Vol., Page
1.4.1.4	Glutamate dehydrogenase ($NADP^+$) Glutamate dehydrogenase ($NADP^+$)	
2.1.2.5	Glutamate formiminotransferase Glutamate formiminotransferase	
2.1.2.5	Glutamate formyltransferase Glutamate formiminotransferase	
2.1.2.6	glutamate formyltransferase glutamate formyltransferase	
4.2.1.48	D-glutamate hydro-lyase (cyclizing) D-glutamate cyclase	
5.4.99.1	Glutamate isomerase Methylaspartate mutase	Vol. 1, Page 582
2.7.2.11	Glutamate kinase Glutamate 5-kinase	
2.7.2.11	γ-Glutamate kinase Glutamate 5-kinase	
5.4.99.1	Glutamate mutase Methylaspartate mutase	Vol. 1, Page 582
2.3.1.35	Glutamate N-acetyltransferase Glutamate N-acetyltransferase	
2.6.1.1	Glutamate oxaloacetate transaminase Aspartate transaminase	
1.4.3.7	D-Glutamate oxidase D-Glutamate oxidase	
1.4.3.11	Glutamate oxidase L-Glutamate oxidase	
1.4.3.11	L-Glutamate oxidase L-Glutamate oxidase	
1.4.1.2	Glutamate oxidoreductase Glutamate dehydrogenase	
5.1.1.3	Glutamate racemase Glutamate racemase	Vol. 1, Page 11
5.1.1.3	Glutamate racemase Glutamate racemase	Vol. 1, Page 11
1.2.1.41	Glutamate semialdehyde dehydrogenase Glutamate-5-semialdehyde dehydrogenase	

EC-No.	Synonym Recommended Name	Vol., Page
3.4.21.19	glutamate specific endopeptidase glutamyl endopeptidase	
1.4.1.13	L-Glutamate synthase Glutamate synthase (NADPH)	
2.6.1.53	glutamate synthase glutamate synthase	
1.4.7.1	Glutamate synthase (ferredoxin) Glutamate synthase (ferredoxin)	
1.4.7.1	Glutamate synthase (ferredoxin-dependent) Glutamate synthase (ferredoxin)	
1.4.1.14	Glutamate synthase (NADH) Glutamate synthase (NADH)	
1.4.1.14	L-Glutamate synthase (NADH) Glutamate synthase (NADH)	
1.4.1.13	Glutamate synthase (NADPH) Glutamate synthase (NADPH)	
1.4.1.14	L-Glutamate synthetase Glutamate synthase (NADH)	
1.4.1.13	L-Glutamate synthetase Glutamate synthase (NADPH)	
1.4.1.13	Glutamate synthetase (NADP) Glutamate synthase (NADPH)	
3.4.21.19	glutamic-acid-specific endopeptidase glutamyl endopeptidase	
2.6.1.2	Glutamic-alanine transaminase Alanine transaminase	
2.6.1.1	Glutamic-aspartic aminotransferase Aspartate transaminase	
1.4.3.15	D-Glutamic-aspartic oxidase D-Glutamate(D-aspartate) oxidase	
2.6.1.1	Glutamic-aspartic transaminase Aspartate transaminase	
2.6.1.4	Glutamic-glyoxylic transaminase Glycine transaminase	
2.6.1.5	Glutamic-hydroxyphenylpyruvic transaminase Tyrosine transaminase	

EC-No.	Synonym Recommended Name	Vol., Page
2.6.1.9	Glutamic-imidazoleacetol phosphate transaminase Histidinol-phosphate transaminase	
2.6.1.39	Glutamic-ketoadipic transaminase 2-Aminoadipate transaminase	
2.6.1.1	Glutamic-oxalacetic transaminase Aspartate transaminase	
2.6.1.1	Glutamic-oxaloacetic transaminase Aspartate transaminase	
2.6.1.2	Glutamic-pyruvic aminotransferase Alanine transaminase	
2.6.1.2	Glutamic-pyruvic transaminase Alanine transaminase	
6.3.2.9	D-Glutamic acid-adding enzyme UDP-N-acetylmuramoylalanine-D-glutamate ligase	Vol. 2, Page 452
2.6.1.2	Glutamic acid-pyruvic acid transaminase Alanine transaminase	
3.4.21.19	glutamic acid-specific proteinase glutamyl endopeptidase	
4.1.1.15	Glutamic acid decarboxylase Glutamate decarboxylase	
4.1.1.15	L-Glutamic acid decarboxylase Glutamate decarboxylase	
1.4.1.2	Glutamic acid dehydrogenase Glutamate dehydrogenase	
1.4.1.3	Glutamic acid dehydrogenase Glutamate dehydrogenase ($NAD(P)^+$)	
1.4.1.4	Glutamic acid dehydrogenase Glutamate dehydrogenase ($NADP^+$)	
1.4.1.3	L-Glutamic acid dehydrogenase Glutamate dehydrogenase ($NAD(P)^+$)	
1.4.1.4	L-Glutamic acid dehydrogenase Glutamate dehydrogenase ($NADP^+$)	
5.4.99.1	Glutamic acid isomerase Methylaspartate mutase	Vol. 1, Page 582
5.4.99.1	Glutamic acid mutase Methylaspartate mutase	Vol. 1, Page 582

EC-No.	Synonym Recommended Name	Vol., Page
1.4.3.7	D-Glutamic acid oxidase D-Glutamate oxidase	
1.4.3.11	Glutamic acid oxidase L-Glutamate oxidase	
1.4.3.11	L-Glutamic acid oxidase L-Glutamate oxidase	
6.1.1.17	Glutamic acid translase Glutamate-tRNA ligase	Vol. 2, Page 128
6.1.1.17	Glutamic acid tRNA ligase Glutamate-tRNA ligase	Vol. 2, Page 128
2.3.2.4	L-Glutamic cyclase γ-Glutamylcyclotransferase	
4.1.1.15	Glutamic decarboxylase Glutamate decarboxylase	
4.1.1.15	L-Glutamic decarboxylase Glutamate decarboxylase	
1.4.1.2	Glutamic dehydrogenase Glutamate dehydrogenase	
1.4.1.3	Glutamic dehydrogenase Glutamate dehydrogenase (NAD(P)$^+$)	
1.4.1.4	Glutamic dehydrogenase Glutamate dehydrogenase (NADP$^+$)	
1.4.3.11	Glutamic dehydrogenase (acceptor) L-Glutamate oxidase	
5.4.99.1	Glutamic isomerase Methylaspartate mutase	Vol. 1, Page 582
5.4.99.1	Glutamic mutase Methylaspartate mutase	Vol. 1, Page 582
2.6.1.1	Glutamic oxalic transaminase Aspartate transaminase	
1.4.3.7	D-Glutamic oxidase D-Glutamate oxidase	
2.6.1.5	Glutamic phenylpyruvic aminotransferase Tyrosine transaminase	
3.5.1.38	Glutamin-(asparagin-)ase Glutamin-(asparagin-)ase	

EC-No.	Synonym Recommended Name	Vol., Page
3.5.1.35	D-Glutaminase D-Glutaminase	
3.5.1.2	Glutaminase Glutaminase	
3.5.1.2	L-Glutaminase Glutaminase	
3.5.1.2	Glutaminase I Glutaminase	
2.6.1.15	Glutaminase II Glutamine-pyruvate transaminase	
3.5.1.38	L-Glutamine(L-asparagine)amidohydrolase Glutamin-(asparagin-)ase	
2.6.1.15	Glutamine-α-keto acid transamidase Glutamine-pyruvate transaminase	
2.6.1.15	Glutamine-α-keto acid transaminase Glutamine-pyruvate transaminase	
6.3.5.5	Glutamine-dependent carbamyl phosphate synthetase Carbamoyl-phosphate synthase (glutamine-hydrolysing)	Vol. 2, Page 689
2.6.1.16	Glutamine-fructose-6-phosphate transaminase (isomerizing) Glutamine-fructose-6-phosphate transaminase (isomerizing)	
2.6.1.16	Glutamine-fructose 6-phosphate amidotransferase Glutamine-fructose-6-phosphate transaminase (isomerizing)	
2.6.1.16	Glutamine-fructose 6-phosphate aminotransferase Glutamine-fructose-6-phosphate transaminase (isomerizing)	
2.6.1.50	L-Glutamine-keto-scyllo-inositol aminotransferase Glutamine-scyllo-inosose transaminase	
2.6.1.15	Glutamine-keto acid aminotransferase Glutamine-pyruvate transaminase	
1.4.1.13	Glutamine-ketoglutaric aminotransferase Glutamate synthase (NADPH)	
2.6.1.15	Glutamine-oxo-acid transaminase Glutamine-pyruvate transaminase	
2.6.1.15	Glutamine-oxo acid aminotransferase Glutamine-pyruvate transaminase	
2.6.1.64	Glutamine-phenylpyruvate transaminase Glutamine-phenylpyruvate transaminase	

EC-No.	Synonym Recommended Name	Vol., Page
2.6.1.15	Glutamine-pyruvate transaminase Glutamine-pyruvate transaminase	
2.6.1.50	Glutamine-scyllo-inosose transaminase Glutamine-scyllo-inosose transaminase	
2.6.1.50	L-Glutamine-scyllo-inosose transaminase Glutamine-scyllo-inosose transaminase	
2.7.7.42	Glutamine-synthetase adenylyltransferase [Glutamate-ammonia-ligase] adenylyltransferase	
6.1.1.18	Glutamine-tRNA ligase Glutamine-tRNA ligase	Vol. 2, Page 139
6.1.1.18	Glutamine-tRNA synthetase Glutamine-tRNA ligase	Vol. 2, Page 139
2.4.2.14	Glutamine 5-phosphoribosylpyrophosphate amidotransferase Amidophosphoribosyltransferase	
2.6.1.50	L-Glutamine:2,4,6/3,5-pentahydroxycyclohexanone aminotransferase Glutamine-scyllo-inosose transaminase	
1.4.1.13	L-Glutamine:2-oxoglutarate aminotransferase, NADPH oxidizing Glutamate synthase (NADPH)	
2.3.2.14	L-Glutamine:D-alanine γ-glutamyltransferase D-Alanine γ-glutamyltransferase	
2.6.1.16	L-Glutamine:D-fructose-6-phosphate aminotransferase (hexose-isomerizing) Glutamine-fructose-6-phosphate transaminase (isomerizing)	
2.3.2.1	Glutamine:D-glutamyl-peptide 5-glutamyltransferase D-Glutamyltransferase	
2.6.1.16	Glutamine:fructose-6-phosphate aminotransferase Glutamine-fructose-6-phosphate transaminase (isomerizing)	
2.6.1.64	L-Glutamine:phenylpyruvate aminotransferase Glutamine-phenylpyruvate transaminase	
2.6.1.15	L-Glutamine:pyruvate aminotransferase Glutamine-pyruvate transaminase	
6.1.1.18	L-Glutamine:tRNA^{Gln} ligase (AMP-forming) Glutamine-tRNA ligase	Vol. 2, Page 139
1.4.1.13	Glutamine amide-2-oxoglutarate aminotransferase (oxidoreductase, NADP) Glutamate synthase (NADPH)	

EC-No.	Synonym Recommended Name	Vol., Page
3.5.1.35	D-Glutamine amidohydrolase D-Glutaminase	
3.5.1.2	L-Glutamine amidohydrolase Glutaminase	
3.5.1.2	Glutamine aminohydrolase Glutaminase	
2.6.1.16	L-Glutamine fructose 6-phosphate transamidase Glutamine-fructose-6-phosphate transaminase (isomerizing)	
2.3.1.68	Glutamine N-acyltransferase Glutamine N-acyltransferase	
2.3.1.14	Glutamine N-phenylacetyltransferase Glutamine N-phenylacetyltransferase	
2.3.1.14	Glutamine phenylacetyltransferase Glutamine N-phenylacetyltransferase	
2.4.2.14	Glutamine phosphoribosylpyrophosphate amidotransferase Amidophosphoribosyltransferase	
2.4.2.14	Glutamine ribosylpyrophosphate 5-phosphate amidotransferase Amidophosphoribosyltransferase	
2.6.1.50	Glutamine scyllo-inosose aminotransferase Glutamine-scyllo-inosose transaminase	
6.3.1.2	Glutamine synthetase Glutamate-ammonia ligase	Vol. 2, Page 347
6.3.1.2	L-Glutamine synthetase Glutamate-ammonia ligase	Vol. 2, Page 347
2.7.7.42	Glutamine synthetase adenylyltransferase [Glutamate-ammonia-ligase] adenylyltransferase	
2.6.1.15	Glutamine transaminase Glutamine-pyruvate transaminase	
2.6.1.64	Glutamine transaminase K Glutamine-phenylpyruvate transaminase	
4.4.1.13	glutamine transaminase K/cysteine conjugate β-lyase cysteine-S-conjugate β-lyase	
2.6.1.15	Glutamine transaminase L Glutamine-pyruvate transaminase	
6.1.1.18	Glutamine translase Glutamine-tRNA ligase	Vol. 2, Page 139

EC-No.	Synonym Recommended Name	Vol., Page
2.3.2.5	L-Glutaminyl-peptide γ-glutamyltransferase (cyclizing) Glutaminyl-peptide cyclotransferase	
2.3.2.13	R-Glutaminyl-peptide:amine γ-glutamyl transferase Protein-glutamine γ-glutamyltransferase	
2.3.2.5	Glutaminyl-peptide cyclotransferase Glutaminyl-peptide cyclotransferase	
3.5.1.44	Glutaminyl-peptide glutaminase Protein-glutamine glutaminase	
6.1.1.18	Glutaminyl-transfer ribonucleate synthetase Glutamine-tRNA ligase	Vol. 2, Page 139
6.1.1.18	Glutaminyl-transfer RNA synthetase Glutamine-tRNA ligase	Vol. 2, Page 139
2.3.2.5	Glutaminyl-tRNA cyclotransferase Glutaminyl-peptide cyclotransferase	
6.1.1.18	Glutaminyl-tRNA synthetase Glutamine-tRNA ligase	Vol. 2, Page 139
2.3.2.5	Glutaminyl cyclase Glutaminyl-peptide cyclotransferase	
2.3.2.13	Glutaminylpeptide γ-glutamyltransferase Protein-glutamine γ-glutamyltransferase	
2.6.1.15	γ-Glutaminyltransferase Glutamine-pyruvate transaminase	
2.3.2.4	(5-L-Glutamyl)-L-amino-acid 5-glutamyltransferase (cyclizing) γ-Glutamylcyclotransferase	
2.3.2.9	(γ-L-Glutamyl)-N^1-(4-hydroxymethylphenyl)hydrazine:(acceptor) γ-glutamyltransferase Agaritine γ-glutamyltransferase	
2.3.2.2	(5-L-Glutamyl)-peptide:amino-acid 5-glutamyltransferase γ-Glutamyltransferase	
2.3.2.4	γ-Glutamyl-amino acid cyclotransferase γ-Glutamylcyclotransferase	
3.4.13.7	α-glutamyl-glutamate dipeptidase Glu-Glu dipeptidase	
6.3.2.23	γ-L-Glutamyl-L-cysteine:β-alanine ligase (ADP-forming) Homoglutathione synthase	Vol. 2, Page 518

EC-No.	Synonym Recommended Name	Vol., Page
6.3.2.3	γ-L-Glutamyl-L-cysteine:glycine ligase (ADP-forming) Glutathione synthase	Vol. 2, Page 410
6.3.2.2	γ-Glutamyl-L-cysteine synthetase Glutamate-cysteine ligase	Vol. 2, Page 399
3.5.1.78	γ-L-Glutamyl-L-cysteinyl-glycine:spermidine amidase Glutathionylspermidine amidase	
6.3.1.8	γ-L-Glutamyl-L-cysteinyl-glycine:spermidine ligase (ADP-forming) Glutathionylspermidine synthase	Vol. 2, Page 386
3.4.19.11	γ-D-Glutamyl-L-meso-diaminopimelate peptidoglycan hydrolase γ-D-Glutamyl-meso-diaminopimelate peptidase I	
3.4.19.11	γ-Glutamyl-L-meso-diaminopimelyl endopeptidase γ-D-Glutamyl-meso-diaminopimelate peptidase I	
3.4.19.11	γ-D-Glutamyl-meso-D-aminopimelic endopeptidase γ-D-Glutamyl-meso-diaminopimelate peptidase I	
3.4.19.11	γ-D-Glutamyl-meso-diaminopimelate endopeptidase γ-D-Glutamyl-meso-diaminopimelate peptidase I	
3.4.19.11	γ-D-Glutamyl-meso-diaminopimelate peptidase I γ-D-Glutamyl-meso-diaminopimelate peptidase I	
3.4.19.11	γ-D-Glutamyl-meso-diaminopimelic endopeptidase γ-D-Glutamyl-meso-diaminopimelate peptidase I	
3.4.19.11	γ-D-Glutamyl-meso-diaminopimelic peptidoglycan hydrolase γ-D-Glutamyl-meso-diaminopimelate peptidase I	
6.1.1.17	Glutamyl-transfer ribonucleate synthetase Glutamate-tRNA ligase	Vol. 2, Page 128
6.1.1.17	Glutamyl-transfer ribonucleic acid synthetase Glutamate-tRNA ligase	Vol. 2, Page 128
6.1.1.17	Glutamyl-transfer RNA synthetase Glutamate-tRNA ligase	Vol. 2, Page 128
6.1.1.17	Glutamyl-tRNA synthetase Glutamate-tRNA ligase	Vol. 2, Page 128
3.4.11.7	glutamyl aminopeptidase glutamyl aminopeptidase	
3.4.11.7	glutamyl aminopeptidase glutamyl aminopeptidase	
3.4.17.11	Glutamyl carboxypeptidase Glutamate carboxypeptidase	

EC-No.	Synonym Recommended Name	Vol., Page
2.3.2.4	γ-Glutamylcyclotransferase γ-Glutamylcyclotransferase	
2.3.2.4	γ-L-Glutamylcyclotransferase γ-Glutamylcyclotransferase	
2.3.2.15	γ-Glutamylcysteine dipeptidyl transpeptidase Glutathione γ-glutamylcysteinyltransferase	
6.3.2.2	γ-Glutamylcysteine synthetase Glutamate-cysteine ligase	Vol. 2, Page 399
6.3.2.2	γ-Glutamylcysteine synthetase Glutamate-cysteine ligase	Vol. 2, Page 399
6.3.2.2	γ-Glutamylcysteinyl-synthetase Glutamate-cysteine ligase	Vol. 2, Page 399
3.4.13.14	γ-glutamyldipeptidase γ-glutamyldipeptidase	
3.4.21.19	glutamyl endopeptidase glutamyl endopeptidase	
3.4.21.82	Glutamyl endopeptidase II Glutamyl endopeptidase II	
6.3.2.18	γ-Glutamylhistamine synthase γ-Glutamylhistamine synthase	Vol. 2, Page 503
6.3.2.18	γ-Glutamylhistamine synthetase γ-Glutamylhistamine synthase	Vol. 2, Page 503
3.4.19.9	γ-Glutamyl hydrolase γ-Glu-X carboxypeptidase	
3.4.22.12	γ-glutamyl hydrolase γ-glutamyl hydrolase	
6.3.1.2	Glutamylhydroxamic synthetase Glutamate-ammonia ligase	Vol. 2, Page 347
2.7.2.11	γ-Glutamyl kinase Glutamate 5-kinase	
6.3.4.12	γ-Glutamylmethylamide synthetase Glutamate-methylamine ligase	Vol. 2, Page 624
3.4.11.7	glutamyl peptidase glutamyl aminopeptidase	
2.3.2.2	γ-Glutamyl peptidyltransferase γ-Glutamyltransferase	

EC-No.	Synonym Recommended Name	Vol., Page
1.2.1.41	.γ.-Glutamyl phosphate reductase Glutamate-5-semialdehyde dehydrogenase	
1.2.1.41	β-glutamylphosphate reductase Glutamate-5-semialdehyde dehydrogenase	
2.3.2.1	D-Glutamyltransferase D-Glutamyltransferase	
2.3.2.2	L-Glutamyltransferase γ-Glutamyltransferase	
2.3.2.2	L-γ-Glutamyltransferase γ-Glutamyltransferase	
2.3.2.2	γ-Glutamyltransferase γ-Glutamyltransferase	
2.3.2.2	Glutamyltransferase, γ- γ-Glutamyltransferase	
2.3.2.9	Glutamyl transferase, agaritine γ- Agaritine γ-glutamyltransferase	
2.3.2.1	Glutamyltransferase, D- D-Glutamyltransferase	
2.3.2.13	Glutamyltransferase, glutaminylpeptide γ- Protein-glutamine γ-glutamyltransferase	
2.3.2.1	D-Glutamyl transpeptidase D-Glutamyltransferase	
2.3.2.1	D-γ-Glutamyl transpeptidase D-Glutamyltransferase	
2.3.2.2	Glutamyl transpeptidase γ-Glutamyltransferase	
2.3.2.2	L-γ-Glutamyl transpeptidase γ-Glutamyltransferase	
2.3.2.2	α-Glutamyl transpeptidase γ-Glutamyltransferase	
2.3.2.2	γ-Glutamyl transpeptidase γ-Glutamyltransferase	
6.1.1.17	Glutamyl tRNA synthetase Glutamate-tRNA ligase	Vol. 2, Page 128
6.2.1.6	Glutarate-CoA ligase Glutarate-CoA ligase	Vol. 2, Page 234

EC-No.	Synonym Recommended Name	Vol., Page
1.2.1.20	Glutarate-semialdehyde:NAD$^+$ oxidoreductase Glutarate-semialdehyde dehydrogenase	
1.2.1.20	Glutarate-semialdehyde dehydrogenase Glutarate-semialdehyde dehydrogenase	
6.2.1.6	Glutarate:CoA ligase (ADP-forming) Glutarate-CoA ligase	Vol. 2, Page 234
1.2.1.20	Glutarate semialdehyde dehydrogenase Glutarate-semialdehyde dehydrogenase	
1.3.99.7	Glutaryl-CoA:(acceptor) 2,3-oxidoreductase (decarboxylating) Glutaryl-CoA dehydrogenase	
1.3.99.7	Glutaryl-CoA dehydrogenase Glutaryl-CoA dehydrogenase	
1.3.99.7	Glutaryl coenzyme A dehydrogenase Glutaryl-CoA dehydrogenase	
6.2.1.6	Glutaryl coenzyme A synthetase Glutarate-CoA ligase	Vol. 2, Page 234
1.8.4.3	Glutathione-CoA-glutathione transhydrogenase Glutathione-CoA-glutathione transhydrogenase	
1.8.4.3	Glutathione-coenzyme A glutathione disulfide transhydrogenase Glutathione-CoA-glutathione transhydrogenase	
1.8.4.4	Glutathione-cystine transhydrogenase Glutathione-cystine transhydrogenase	
1.8.4.7	Glutathione-dependent thiol:disulfide oxidoreductase Enzyme-thiol transhydrogenase (oxidized-glutathione)	
2.8.1.3	Glutathione-dependent thiosulfate reductase Thiosulfate-thiol sulfurtransferase	
2.3.2.15	Glutathione γ-glutamylcysteinyltransferase Glutathione γ-glutamylcysteinyltransferase	
1.8.4.1	Glutathione-homocystine transhydrogenase Glutathione-homocystine transhydrogenase	
1.8.4.2	Glutathione-insulin transhydrogenase Protein-disulfide reductase (glutathione)	
1.8.4.2	Glutathione-protein disulfide oxidoreductase Protein-disulfide reductase (glutathione)	
1.8.4.3	Glutathione:coenzyme A-glutathione transhydrogenase Glutathione-CoA-glutathione transhydrogenase	

EC-No.	Synonym Recommended Name	Vol., Page
1.8.4.4	Glutathione:cystine oxidoreductase Glutathione-cystine transhydrogenase	
1.8.5.1	Glutathione:dehydroascorbate oxidoreductase Glutathione dehydrogenase (ascorbate)	
1.8.5.1	Glutathione:dehydroascorbic acid oxidoreductase Glutathione dehydrogenase (ascorbate)	
1.8.4.1	Glutathione:homocystine oxidoreductase Glutathione-homocystine transhydrogenase	
1.11.1.9	Glutathione:hydrogen-peroxide oxidoreductase Glutathione peroxidase	
1.11.1.12	Glutathione:lipid-hydroperoxide oxidoreductase Phospholipid-hydroperoxide glutathione peroxidase	
1.8.3.3	Glutathione:oxygen oxidoreductase Glutathione oxidase	
2.3.2.15	Glutathione:poly(4-glutamyl-cysteinyl)glycine 4-glutamylcysteinyltransfer-ase Glutathione γ-glutamylcysteinyltransferase	
1.8.4.2	Glutathione:protein-disulfide oxidoreductase Protein-disulfide reductase (glutathione)	
6.3.1.8	Glutathione:spermidine ligase (ADP-forming) Glutathionylspermidine synthase	Vol. 2, Page 386
1.8.4.3	Glutathione coenzyme A-glutathione transhydrogenase Glutathione-CoA-glutathione transhydrogenase	
1.8.5.1	Glutathione dehydroascorbate reductase Glutathione dehydrogenase (ascorbate)	
1.8.5.1	Glutathione dehydrogenase (ascorbate) Glutathione dehydrogenase (ascorbate)	
1.8.3.3	Glutathione oxidase Glutathione oxidase	
1.11.1.9	Glutathione peroxidase Glutathione peroxidase	
1.6.4.2	Glutathione reductase (NADPH) Glutathione reductase (NADPH)	
2.5.1.18	Glutathione S-alkyl transferase Glutathione transferase	

EC-No.	Synonym Recommended Name	Vol., Page
2.5.1.12	glutathione S-alkyltransferase glutathione S-alkyltransferase	
2.5.1.18	Glutathione S-aralkyltransferase Glutathione transferase	
2.5.1.14	glutathione S-aralkyltransferase glutathione S-aralkyltransferase	
2.5.1.18	Glutathione S-aryltransferase Glutathione transferase	
2.5.1.13	glutathione S-aryltransferase glutathione S-aryltransferase	
1.6.4.2	Glutathione S-reductase Glutathione reductase (NADPH)	
2.5.1.18	Glutathione S-transferase Glutathione transferase	
6.3.2.3	Glutathione synthase Glutathione synthase	Vol. 2, Page 410
6.3.2.3	Glutathione synthetase Glutathione synthase	Vol. 2, Page 410
6.3.2.3	Glutathione synthetase (tripeptide) Glutathione synthase	Vol. 2, Page 410
3.1.2.7	glutathione thioesterase glutathione thiolesterase	
3.1.2.7	glutathione thiolesterase glutathione thiolesterase	
2.5.1.18	Glutathione transferase Glutathione transferase	
6.3.1.9	Glutathionylspermidine:glutathione ligase (ADP-forming) Trypanothione synthase	Vol. 2, Page 391
3.5.1.78	Glutathionylspermidine amidase Glutathionylspermidine amidase	
3.5.1.78	Glutathionylspermidine amidohydrolase (spermidine-forming) Glutathionylspermidine amidase	
6.3.1.8	Glutathionylspermidine synthase Glutathionylspermidine synthase	Vol. 2, Page 386
6.3.1.8	Glutathionylspermidine synthetase (Crithidia fasciculata) Glutathionylspermidine synthase	Vol. 2, Page 386

EC-No.	Synonym / Recommended Name	Vol., Page
6.3.1.8	Glutathionylspermidine synthetase (Crithidia fasciculata strain HS6 gene Cf-GSS)	
	Glutathionylspermidine synthase	Vol. 2, Page 386
3.5.1.78	Glutathionylspermidine synthetase/amidase	
	Glutathionylspermidine amidase	
6.3.1.8	Glutathionylspermidine synthetase/amidase	
	Glutathionylspermidine synthase	Vol. 2, Page 386
1.97.1.5	gluthathione:arsenate oxidoreductase	
	arsenate reductase (glutaredoxin)	
1.97.1.7	gluthathione:methylarsonate oxidoreductase	
	methylarsonate reductase	
4.4.1.5	GLXI	
	lactoylglutathione lyase	
3.4.13.18	Gly-Leu hydrolase	
	Cytosol nonspecific dipeptidase	
3.4.14.5	Gly-Pro-naphthylamidase	
	dipeptidyl-peptidase IV	
3.4.17.4	Gly-X carboxypeptidase	
	Gly-X carboxypeptidase	
2.5.1.36	Glyceollin synthase	
	Trihydroxypterocarpan dimethylallyltransferase	
1.2.1.12	Glyceraldehyde-3-P-dehydrogenase	
	Glyceraldehyde-3-phosphate dehydrogenase	
1.2.1.12	D-Glyceraldehyde-3-phosphate: NAD^+ oxidoreductase (phosphorylating)	
	Glyceraldehyde-3-phosphate dehydrogenase	
1.2.1.9	D-Glyceraldehyde-3-phosphate: $NADP^+$ oxidoreductase	
	Glyceraldehyde-3-phosphate dehydrogenase ($NADP^+$)	
1.2.1.13	D-Glyceraldehyde-3-phosphate: $NADP^+$ oxidoreductase (phosphorylating)	
	Glyceraldehyde-3-phosphate dehydrogenase ($NADP^+$) (phosphorylating)	
1.2.1.9	Glyceraldehyde-3-phosphate:NADP reductase	
	Glyceraldehyde-3-phosphate dehydrogenase ($NADP^+$)	
1.2.1.12	Glyceraldehyde-3-phosphate dehydrogenase	
	Glyceraldehyde-3-phosphate dehydrogenase	
1.2.1.59	Glyceraldehyde-3-phosphate dehydrogenase (NAD(P))	
	Glyceraldehyde-3-phosphate dehydrogenase (NAD(P)) (phosphorylating)	

EC-No.	Synonym Recommended Name Vol., Page
1.2.1.59	Glyceraldehyde-3-phosphate dehydrogenase (NAD(P)) (phosphorylating) Glyceraldehyde-3-phosphate dehydrogenase (NAD(P)) (phosphorylating)
1.2.1.12	Glyceraldehyde-3-phosphate dehydrogenase (NAD) Glyceraldehyde-3-phosphate dehydrogenase
1.2.1.9	Glyceraldehyde-3-phosphate dehydrogenase ($NADP^+$) Glyceraldehyde-3-phosphate dehydrogenase ($NADP^+$)
1.2.1.13	Glyceraldehyde-3-phosphate dehydrogenase ($NADP^+$) (phosphorylating) Glyceraldehyde-3-phosphate dehydrogenase ($NADP^+$) (phosphorylating)
5.3.1.1	D-Glyceraldehyde-3-phosphate ketol-isomerase Triose-phosphate isomerase Vol. 1, Page 235
1.2.1.59	D-Glyceraldehyde 3-phosphate:NAD(P) oxidoreductase (phosphorylating) Glyceraldehyde-3-phosphate dehydrogenase (NAD(P)) (phosphorylating)
1.2.1.9	Glyceraldehyde 3-phosphate dehydrogenase (NADP) Glyceraldehyde-3-phosphate dehydrogenase ($NADP^+$)
1.2.1.13	Glyceraldehyde 3-phosphate dehydrogenase (NADP) Glyceraldehyde-3-phosphate dehydrogenase ($NADP^+$) (phosphorylating)
1.2.1.12	Glyceraldehyde phosphate dehydrogenase (NAD) Glyceraldehyde-3-phosphate dehydrogenase
1.2.1.9	Glyceraldehyde phosphate dehydrogenase (NADP) Glyceraldehyde-3-phosphate dehydrogenase ($NADP^+$)
1.2.1.13	Glyceraldehyde phosphate dehydrogenase (nicotinamide adenine dinucleotide phosphate phosphorylating) Glyceraldehyde-3-phosphate dehydrogenase ($NADP^+$) (phosphorylating)
1.1.1.95	Glycerate-1,3-phosphate dehydrogenase Phosphoglycerate dehydrogenase
3.1.3.13	glycerate-2,3-diphosphate phosphatase bisphosphoglycerate phosphatase
3.1.3.20	D-glycerate-2-phosphate phosphohydrolase phosphoglycerate phosphatase
2.7.1.31	Glycerate-3-kinase Glycerate kinase
3.1.3.38	D-glycerate-3-phosphate phosphohydrolase 3-phosphoglycerate phosphatase
2.7.1.31	D-Glycerate 3-kinase Glycerate kinase

EC-No.	Synonym Recommended Name	Vol., Page
1.1.1.95	Glycerate 3-phosphate dehydrogenase Phosphoglycerate dehydrogenase	
2.7.2.3	Glycerate 3-phosphate kinase Phosphoglycerate kinase	
1.1.1.92	D-Glycerate:NAD(P)$^+$ oxidoreductase (carboxylating) Oxaloglycolate reductase (decarboxylating)	
1.1.1.81	D-Glycerate:NADP$^+$ 2-oxidoreductase Hydroxypyruvate reductase	
1.1.1.29	D-Glycerate dehydrogenase Glycerate dehydrogenase	
1.1.1.81	D-Glycerate dehydrogenase Hydroxypyruvate reductase	
1.1.1.29	Glycerate dehydrogenase Glycerate dehydrogenase	
2.7.1.31	D-Glycerate kinase Glycerate kinase	
2.7.1.31	Glycerate kinase Glycerate kinase	
5.4.2.4	Glycerate phosphomutase Bisphosphoglycerate mutase	Vol. 1, Page 520
5.4.2.1	Glycerate phosphomutase (diphosphoglycerate cofactor) Phosphoglycerate mutase	Vol. 1, Page 493
2.7.1.31	D-Glyceric acid kinase Glycerate kinase	
2.7.1.30	Glyceric kinase Glycerol kinase	
1.1.1.177	Glycerin-3-phosphate dehydrogenase Glycerol-3-phosphate 1-dehydrogenase (NADP$^+$)	
1.1.1.6	Glycerin dehydrogenase Glycerol dehydrogenase	
2.7.1.30	Glycerokinase Glycerol kinase	
3.1.4.42	glycerol-1,2-cyclic-phosphate 2-phosphodiesterase glycerol-1,2-cyclic-phosphate 2-phosphodiesterase	
3.1.3.21	glycerol-1-phosphatase glycerol-1-phosphatase	

EC-No.	Synonym Recommended Name	Vol., Page
3.1.3.21	glycerol-1-phosphatase glycerol-1-phosphatase	
1.1.1.261	glycerol-1-phosphate dehydrogenase [NAD(P)] glycerol-1-phosphate dehydrogenase [NAD(P)]	
1.1.3.21	Glycerol-1-phosphate oxidase Glycerol-3-phosphate oxidase	
3.1.3.21	glycerol-1-phosphate phosphohydrolase glycerol-1-phosphatase	
3.1.3.19	glycerol-2-phosphatase glycerol-2-phosphatase	
3.1.3.19	glycerol-2-phosphate phosphohydrolase glycerol-2-phosphatase	
3.1.3.21	glycerol-3-P-phosphatase glycerol-1-phosphatase	
3.1.3.21	glycerol-3-phosphatase glycerol-1-phosphatase	
2.7.1.142	Glycerol-3-phosphate-glucose phosphotransferase Glycerol-3-phosphate-glucose phosphotransferase	
1.1.1.177	Glycerol-3-phosphate 1-dehydrogenase ($NADP^+$) Glycerol-3-phosphate 1-dehydrogenase ($NADP^+$)	
1.1.1.94	L-Glycerol-3-phosphate:NAD(P) oxidoreductase Glycerol-3-phosphate dehydrogenase ($NAD(P)^+$)	
2.7.7.39	Glycerol-3-phosphate cytidylyltransferase Glycerol-3-phosphate cytidylyltransferase	
1.1.99.5	Glycerol-3-phosphate dehydrogenase Glycerol-3-phosphate dehydrogenase	
1.1.99.5	L-Glycerol-3-phosphate dehydrogenase Glycerol-3-phosphate dehydrogenase	
1.1.99.5	Glycerol-3-phosphate dehydrogenase (flavin-linked) Glycerol-3-phosphate dehydrogenase	
1.1.1.94	Glycerol-3-phosphate dehydrogenase ($NAD(P)^+$) Glycerol-3-phosphate dehydrogenase ($NAD(P)^+$)	
1.1.1.8	Glycerol-3-phosphate dehydrogenase (NAD^+) Glycerol-3-phosphate dehydrogenase (NAD^+)	
2.3.1.15	Glycerol-3-phosphate O-acyltransferase Glycerol-3-phosphate O-acyltransferase	

EC-No.	Synonym Recommended Name	Vol., Page
1.1.3.21	Glycerol-3-phosphate oxidase Glycerol-3-phosphate oxidase	
1.1.3.21	L-α-Glycerol-3-phosphate oxidase Glycerol-3-phosphate oxidase	
3.1.3.21	glycerol-3-phosphate phosphohydrolase glycerol-1-phosphatase	
3.1.1.23	glycerol-ester acylhydrolase acylglycerol lipase	
3.1.1.3	glycerol-ester hydrolase triacylglycerol lipase	
3.1.1.23	glycerol-monoester acylhydrolase acylglycerol lipase	
1.1.1.8	Glycerol 1-phosphate dehydrogenase Glycerol-3-phosphate dehydrogenase (NAD^+)	
1.1.1.156	Glycerol 2-dehydrogenase ($NADP^+$) Glycerol 2-dehydrogenase ($NADP^+$)	
2.7.1.142	Glycerol 3-phosphate:glucose transphosphorylase Glycerol-3-phosphate-glucose phosphotransferase	
1.1.1.177	L-Glycerol 3-phosphate:NADP oxidoreductase Glycerol-3-phosphate 1-dehydrogenase ($NADP^+$)	
2.3.1.15	Glycerol 3-phosphate acyltransferase Glycerol-3-phosphate O-acyltransferase	
1.1.99.5	Glycerol 3-phosphate cytochrome c reductase Glycerol-3-phosphate dehydrogenase	
1.1.99.5	Glycerol 3-phosphate dehydrogenase Glycerol-3-phosphate dehydrogenase	
1.1.99.5	L-Glycerol 3-phosphate dehydrogenase Glycerol-3-phosphate dehydrogenase	
1.1.1.94	Glycerol 3-phosphate dehydrogenase (NADP) Glycerol-3-phosphate dehydrogenase ($NAD(P)^+$)	
2.7.8.5	Glycerol 3-phosphate phosphatidyltransferase CDPdiacylglycerol-glycerol-3-phosphate 3-phosphatidyltransferase	
1.1.99.22	Glycerol:(acceptor) 1-oxidoreductase Glycerol dehydrogenase (acceptor)	
1.1.1.6	Glycerol:NAD^+ 2-oxidoreductase Glycerol dehydrogenase	

EC-No.	Synonym Recommended Name	Vol., Page
1.1.1.156	Glycerol:NADP$^+$ 2-oxidoreductase (glycerone-forming) Glycerol 2-dehydrogenase (NADP$^+$)	
1.1.1.72	Glycerol:NADP$^+$ oxidoreductase Glycerol dehydrogenase (NADP$^+$)	
4.2.1.30	glycerol dehydrase glycerol dehydratase	
4.2.1.30	glycerol dehydratase glycerol dehydratase	
1.1.1.6	Glycerol dehydrogenase Glycerol dehydrogenase	
1.1.99.22	Glycerol dehydrogenase (acceptor) Glycerol dehydrogenase (acceptor)	
1.1.1.72	Glycerol dehydrogenase (NADP$^+$) Glycerol dehydrogenase (NADP$^+$)	
3.1.1.3	glycerol ester hydrolase triacylglycerol lipase	
4.2.1.30	glycerol hydro-lyase glycerol dehydratase	
4.2.1.30	glycerol hydrolyase glycerol dehydratase	
2.7.1.30	Glycerol kinase Glycerol kinase	
3.1.3.21	α-glycerol phosphatase glycerol-1-phosphatase	
3.1.3.19	β-glycerolphosphatase glycerol-2-phosphatase	
2.3.1.15	Glycerol phosphate acyltransferase Glycerol-3-phosphate O-acyltransferase	
1.1.99.5	Glycerol phosphate dehydrogenase Glycerol-3-phosphate dehydrogenase	
1.1.1.8	L-Glycerol phosphate dehydrogenase Glycerol-3-phosphate dehydrogenase (NAD$^+$)	
1.1.1.8	L-α-Glycerol phosphate dehydrogenase Glycerol-3-phosphate dehydrogenase (NAD$^+$)	
1.1.2.1	glycerolphosphate dehydrogenase glycerolphosphate dehydrogenase	

EC-No.	Synonym Recommended Name	Vol., Page
1.1.99.5	Glycerol phosphate dehydrogenase (acceptor) Glycerol-3-phosphate dehydrogenase	
1.1.99.5	Glycerol phosphate dehydrogenase (FAD) Glycerol-3-phosphate dehydrogenase	
1.1.1.8	Glycerol phosphate dehydrogenase (NAD) Glycerol-3-phosphate dehydrogenase (NAD^+)	
1.1.1.8	α-Glycerol phosphate dehydrogenase (NAD) Glycerol-3-phosphate dehydrogenase (NAD^+)	
1.1.1.94	Glycerol phosphate dehydrogenase (nicotinamide adenine dinucleotide (phosphate)) Glycerol-3-phosphate dehydrogenase ($NAD(P)^+$)	
1.1.3.21	Glycerol phosphate oxidase Glycerol-3-phosphate oxidase	
3.1.3.21	glycerol phosphate phosphatase glycerol-1-phosphatase	
2.3.1.15	Glycerol phosphate transacylase Glycerol-3-phosphate O-acyltransferase	
3.1.4.2	glycerolphosphorylcholine phosphodiesterase glycerophosphocholine phosphodiesterase	
2.3.1.42	Glycerone-phosphate O-acyltransferase Glycerone-phosphate O-acyltransferase	
4.2.3.3	glycerone-phosphate phospho-lyase methylglyoxal synthase	
4.2.99.11	glycerone-phosphate phospho-lyase methylglyoxal synthase	
2.7.1.29	Glycerone kinase Glycerone kinase	
3.1.3.19	2-glycerophosphatase glycerol-2-phosphatase	
3.1.3.2	Glycerophosphatase Acid phosphatase	
3.1.3.21	α-glycerophosphatase glycerol-1-phosphatase	
3.1.3.1	glycerophosphatase alkaline phosphatase	

EC-No.	Synonym Recommended Name	Vol., Page
3.1.3.19	glycerophosphatase glycerol-2-phosphatase	
1.1.99.5	L-3-Glycerophosphate-ubiquinone oxidoreductase Glycerol-3-phosphate dehydrogenase	
2.3.1.15	3-Glycerophosphate acyltransferase Glycerol-3-phosphate O-acyltransferase	
2.3.1.15	Glycerophosphate acyltransferase Glycerol-3-phosphate O-acyltransferase	
2.3.1.15	α-Glycerophosphate acyltransferase Glycerol-3-phosphate O-acyltransferase	
1.1.99.5	Glycerophosphate dehydrogenase Glycerol-3-phosphate dehydrogenase	
1.1.99.5	L-Glycerophosphate dehydrogenase Glycerol-3-phosphate dehydrogenase	
1.1.1.8	L-Glycerophosphate dehydrogenase Glycerol-3-phosphate dehydrogenase (NAD^+)	
1.1.1.8	L-α-Glycerophosphate dehydrogenase Glycerol-3-phosphate dehydrogenase (NAD^+)	
1.1.99.5	α-Glycerophosphate dehydrogenase Glycerol-3-phosphate dehydrogenase	
1.1.99.5	α-Glycerophosphate dehydrogenase (acceptor) Glycerol-3-phosphate dehydrogenase	
1.1.1.8	Glycerophosphate dehydrogenase (NAD) Glycerol-3-phosphate dehydrogenase (NAD^+)	
1.1.1.8	α-Glycerophosphate dehydrogenase (NAD) Glycerol-3-phosphate dehydrogenase (NAD^+)	
2.7.2.3	Glycerophosphate kinase Phosphoglycerate kinase	
1.1.3.21	L-α-Glycerophosphate oxidase Glycerol-3-phosphate oxidase	
1.1.99.5	α-Glycerophosphate oxidase Glycerol-3-phosphate dehydrogenase	
1.1.3.21	α-Glycerophosphate oxidase Glycerol-3-phosphate oxidase	
3.1.3.19	β-glycerophosphate phosphatase glycerol-2-phosphatase	

EC-No.	Synonym Recommended Name	Vol., Page
3.1.3.20	glycerophosphate phosphatase phosphoglycerate phosphatase	
2.7.8.5	Glycerophosphate phosphatidyltransferase CDPdiacylglycerol-glycerol-3-phosphate 3-phosphatidyltransferase	
2.7.8.12	Glycerophosphate synthetase CDPglycerol glycerophosphotransferase	
2.3.1.15	Glycerophosphate transacylase Glycerol-3-phosphate O-acyltransferase	
3.1.4.2	glycerophosphinicocholine diesterase glycerophosphocholine phosphodiesterase	
3.1.4.38	glycerophosphocholine cholinephosphodiesterase glycerophosphocholine cholinephosphodiesterase	
3.1.4.2	glycerophosphocholinehydrolase glycerophosphocholine phosphodiesterase	
3.1.4.2	glycerophosphocholine phosphodiesterase glycerophosphocholine phosphodiesterase	
3.1.4.2	glycerophosphodiesterase glycerophosphocholine phosphodiesterase	
3.1.4.46	Glycerophosphodiester glycerophosphohydrolase Glycerophosphodiester phosphodiesterase	
3.1.4.46	Glycerophosphodiester phosphodiesterase Glycerophosphodiester phosphodiesterase	
3.1.4.2	glycerophosphodiester phosphodiesterase glycerophosphocholine phosphodiesterase	
3.1.4.44	glycerophosphoinositol glycerophosphodiesterase glycerophosphoinositol glycerophosphodiesterase	
3.1.4.44	glycerophosphoinositol glycerophosphohydrolase glycerophosphoinositol glycerophosphodiesterase	
3.1.4.43	glycerophosphoinositol inositolphosphodiesterase glycerophosphoinositol inositolphosphodiesterase	
2.3.1.148	Glycerophospholipid acyltransferase (CoA-dependent) Glycerophospholipid acyltransferase (CoA-dependent)	
2.3.1.147	Glycerophospholipid arachidonoyl-transferase (CoA-independent) Glycerophospholipid arachidonoyl-transferase (CoA-independent)	
3.1.4.46	Glycerophosphoryl diester phosphodiesterase Glycerophosphodiester phosphodiesterase	

EC-No.	Synonym Recommended Name	Vol., Page
2.7.8.12	Glycerophosphotransferase, cytidine diphosphoglycerol CDPglycerol glycerophosphotransferase	
2.7.8.20	Glycerophosphotransferase, oligosaccharide Phosphatidylglycerol-membrane-oligosaccharide glycerophosphotransfer-ase	
1.14.16.5	Glyceryl-ether cleaving enzyme Glyceryl-ether monooxygenase	
1.14.16.5	Glyceryl-ether monooxygenase Glyceryl-ether monooxygenase	
1.14.99.17	glyceryl-ether monooxygenase glyceryl-ether monooxygenase	
1.14.16.5	Glyceryl ether-cleaving enzyme Glyceryl-ether monooxygenase	
1.14.16.5	Glyceryl etherase Glyceryl-ether monooxygenase	
1.14.16.5	Glyceryl ether monooxygenase Glyceryl-ether monooxygenase	
1.14.16.5	Glyceryl ether oxygenase Glyceryl-ether monooxygenase	
3.1.4.38	L-3-glycerylphosphinicocholine cholinephosphohydrolase glycerophosphocholine cholinephosphodiesterase	
3.1.4.2	glycerylphosphorylcholinediesterase glycerophosphocholine phosphodiesterase	
3.4.19.2	glycinamidase, peptidyl peptidyl-glycinamidase	
6.3.4.13	Glycinamide ribonucleotide synthetase Phosphoribosylamine-glycine ligase	Vol. 2, Page 626
2.1.2.2	Glycinamide ribonucleotide transformylase Phosphoribosylglycinamide formyltransferase	
1.4.4.2	Glycine-cleavage complex Glycine dehydrogenase (decarboxylating)	
1.4.2.1	Glycine-cytochrome c reductase Glycine dehydrogenase (cytochrome)	
2.3.1.13	Glycine-N-acylase Glycine N-acyltransferase	

EC-No.	Synonym Recommended Name	Vol., Page
2.6.1.35	Glycine-oxalacetate aminotransferase Glycine-oxaloacetate transaminase	
2.6.1.35	Glycine-oxaloacetate transaminase Glycine-oxaloacetate transaminase	
2.3.1.65	Glycine-taurine N-acyltransferase Glycine N-choloyltransferase	
6.1.1.14	Glycine-tRNA ligase Glycine-tRNA ligase	Vol. 2, Page 101
2.6.1.4	Glycine:2-oxoglutarate aminotransferase Glycine transaminase	
1.4.2.1	Glycine:ferricytochrome-c oxidoreductase (deaminating) Glycine dehydrogenase (cytochrome)	
1.4.4.2	Glycine:lipoylprotein oxidoreductase (decarboxylating and acceptor-aminomethylating) Glycine dehydrogenase (decarboxylating)	
1.4.1.10	Glycine:NAD^+ oxidoreductase (deaminating) Glycine dehydrogenase	
2.6.1.35	Glycine:oxaloacetate aminotransferase Glycine-oxaloacetate transaminase	
6.1.1.14	Glycine:$tRNA^{Gly}$ ligase (AMP-forming) Glycine-tRNA ligase	Vol. 2, Page 101
2.3.1.29	Glycine acetyltransferase Glycine C-acetyltransferase	
2.3.1.13	Glycine acyltransferase Glycine N-acyltransferase	
6.3.4.13	Glycineamide ribonucleotide synthetase Phosphoribosylamine-glycine ligase	Vol. 2, Page 626
2.1.4.1	Glycine amidinotransferase Glycine amidinotransferase	
2.6.1.4	Glycine aminotransferase Glycine transaminase	
2.3.1.29	Glycine C-acetyltransferase Glycine C-acetyltransferase	
3.4.17.4	glycine carboxypeptidase Gly-X carboxypeptidase	

EC-No.	Synonym Recommended Name	Vol., Page
1.4.4.2	Glycine decarboxylase Glycine dehydrogenase (decarboxylating)	
1.4.4.2	Glycine decarboxylase P-protein Glycine dehydrogenase (decarboxylating)	
1.4.1.10	Glycine dehydrogenase Glycine dehydrogenase	
1.4.1.10	Glycine dehydrogenase Glycine dehydrogenase	
1.4.2.1	Glycine dehydrogenase (cytochrome) Glycine dehydrogenase (cytochrome)	
1.4.4.2	Glycine dehydrogenase (decarboxylating) Glycine dehydrogenase (decarboxylating)	
2.1.2.4	Glycine formiminotransferase Glycine formiminotransferase	
2.1.2.1	Glycine hydroxymethyltransferase Glycine hydroxymethyltransferase	
2.1.1.20	Glycine methyltransferase Glycine N-methyltransferase	
2.3.1.13	Glycine N-acyltransferase Glycine N-acyltransferase	
2.3.1.71	Glycine N-benzoyltransferase Glycine N-benzoyltransferase	
2.3.1.65	Glycine N-choloyltransferase Glycine N-choloyltransferase	
2.1.1.20	Glycine N-methyltransferase Glycine N-methyltransferase	
2.1.4.1	Glycine transamidinase Glycine amidinotransferase	
2.6.1.4	Glycine transaminase Glycine transaminase	
3.5.1.24	Glycocholase Cholylglycine hydrolase	
3.5.3.2	Glycocyaminase Glycocyaminase	
2.7.3.1	Glycocyamine kinase Guanidinoacetate kinase	

EC-No.	Synonym Recommended Name	Vol., Page
2.4.1.11	Glycogen (starch) synthase Glycogen (starch) synthase	
3.1.3.42	[glycogen-synthase-D] phosphatase [glycogen-synthase-D] phosphatase	
3.2.1.68	Glycogen 6-glucanohydrolase Isoamylase	
3.2.1.1	Glycogenase α-amylase	
3.2.1.2	Glycogenase β-amylase	
2.4.1.18	Glycogen branching enzyme 1,4-α-Glucan branching enzyme	
3.1.3.42	glycogen glucosyltransferase phosphatase [glycogen-synthase-D] phosphatase	
2.4.1.186	Glycogenin glucosyltransferase Glycogenin glucosyltransferase	
2.4.1.112	Glycogen initiator synthase α-1,4-Glucan-protein synthase (UDP-forming)	
2.4.1.1	Glycogen phosphorylase Phosphorylase	
2.7.1.38	Glycogen phosphorylase kinase Phosphorylase kinase	
3.1.3.17	glycogen phosphorylase phosphatase [phosphorylase] phosphatase	
3.1.3.42	glycogen synthase D phosphatase [glycogen-synthase-D] phosphatase	
2.7.1.37	Glycogen synthase kinase Protein kinase	
3.1.3.42	glycogen synthase phosphatase [glycogen-synthase-D] phosphatase	
2.4.1.11	Glycogen synthetase (starch) Glycogen (starch) synthase	
3.1.3.42	glycogen synthetase phosphatase [glycogen-synthase-D] phosphatase	
1.1.1.185	L-Glycol:NAD(P)$^+$ oxidoreductase L-Glycol dehydrogenase	

EC-No.	Synonym Recommended Name	Vol., Page
1.1.1.185	L-Glycol:NAD(P) dehydrogenase L-Glycol dehydrogenase	
1.1.1.185	L-(+)-Glycol:NAD(P) oxidoreductase L-Glycol dehydrogenase	
1.2.1.21	Glycolaldehyde: NAD^+ oxidoreductase Glycolaldehyde dehydrogenase	
1.2.1.21	Glycolaldehyde dehydrogenase Glycolaldehyde dehydrogenase	
2.2.1.1	Glycolaldehydetransferase Transketolase	
1.1.99.14	Glycolate:(acceptor) 2-oxidoreductase Glycolate dehydrogenase	
1.1.1.26	Glycolate:NAD^+ oxidoreductase Glyoxylate reductase	
1.1.1.79	Glycolate: $NADP^+$ oxidoreductase Glyoxylate reductase ($NADP^+$)	
1.1.99.14	Glycolate dehydrogenase Glycolate dehydrogenase	
1.1.3.15	Glycolate oxidase (S)-2-Hydroxy-acid oxidase	
1.1.3.1	glycolate oxidase glycolate oxidase	
1.1.99.14	Glycolate oxidoreductase Glycolate dehydrogenase	
3.1.3.18	*p*-Glycolate phosphatase Phosphoglycolate, phosphatase	
1.1.1.185	L-Glycol dehydrogenase L-Glycol dehydrogenase	
1.1.99.14	Glycolic acid dehydrogenase Glycolate dehydrogenase	
2.4.1.131	Glycolipid 2-α-mannosyltransferase Glycolipid 2-α-mannosyltransferase	
2.4.1.132	Glycolipid 3-α-mannosyltransferase Glycolipid 3-α-mannosyltransferase	
2.8.2.11	Glycolipid sulfotransferase Galactosylceramide sulfotransferase	

EC-No.	Synonym Recommended Name	Vol., Page
3.5.1.52	Glycopeptidase Peptide-N^4-(N-acetyl-β-glucosaminyl) asparagine amidase	
3.2.1.96	Glycopeptide-D-mannosyl-N^4-(N-acetyl-D-glucosaminyl)2-asparagine 1,4-N-acetyl-β-glucosaminohydrolase Mannosyl-glycoprotein endo-β-N-acetyl glucosaminidase	
3.2.1.97	Glycopeptide α-N-acetylgalactosaminidase Glycopeptide α-N-acetylgalactosaminidase	
3.5.1.52	Glycopeptide N-glycosidase Peptide-N^4-(N-acetyl-β-glucosaminyl) asparagine amidase	
3.2.2.18	glycopeptide N-glycosidase glycopeptide N-glycosidase	
3.4.24.57	Glycophorin A proteinase O-Sialoglycoprotein endopeptidase	
3.4.24.57	Glycoprotease O-Sialoglycoprotein endopeptidase	
3.4.24.11	glycoprotein, CALLA neprilysin	
2.4.1.38	Glycoprotein β-galactosyltransferase β-N-Acetylglucosaminylglycopeptide β-1,4-galactosyltransferase	
3.1.4.45	Glycoprotein-N-acetyl-D-glucosaminyl-phospho-D-mannose N-acetylglucosaminylphosphohydrolase N-Acetylglucosamine-1-phosphodiester N-acetylglucosaminidase	
2.4.1.122	Glycoprotein-N-acetylgalactosamine 3-β-galactosyltransferase Glycoprotein-N-acetylgalactosamine 3-β-galactosyltransferase	
3.1.4.50	Glycoprotein-phosphatidylinositol phosphatidohydrolase Glycoprotein phospholipase D	
2.4.1.38	Glycoprotein 4-β-galactosyltransferase β-N-Acetylglucosaminylglycopeptide β-1,4-galactosyltransferase	
2.4.1.68	Glycoprotein 6-α-L-fucosyltransferase Glycoprotein 6-α-L-fucosyltransferase	
2.3.1.45	Glycoprotein 7(9)-O-acetyltransferase N-Acetylneuraminate 7-O(or 9-O-)-acetyltransferase	
2.4.1.41	Glycoprotein acetylgalactosaminyltransferase Polypeptide N-acetylgalactosaminyltransferase	
3.4.24.57	Glycoproteinase O-Sialoglycoprotein endopeptidase	

EC-No.	Synonym Recommended Name	Vol., Page
3.2.1.130	Glycoprotein endo-α-1,2-mannosidase Glycoprotein endo-α-1,2-mannosidase	
2.4.1.68	Glycoprotein fucosyltransferase Glycoprotein 6-α-L-fucosyltransferase	
3.2.1.130	Glycoprotein glucosylmannohydrolase Glycoprotein endo-α-1,2-mannosidase	
3.4.14.5	glycoprotein GP110 dipeptidyl-peptidase IV	
3.4.24.36	Glycoprotein gp63 Leishmanolysin	
2.3.1.96	Glycoprotein N-palmitoyltransferase Glycoprotein N-palmitoyltransferase	
2.3.1.142	Glycoprotein O-fatty-acyltransferase Glycoprotein O-fatty-acyltransferase	
3.1.4.50	Glycoprotein phospholipase D Glycoprotein phospholipase D	
3.2.1.113	Glycoprotein processing mannosidase I Mannosyl-oligosaccharide 1,2-α-mannosidase	
3.1.27.5	Glycoproteins, gene S locus-specific Pancreatic ribonuclease	
3.1.27.5	Glycoproteins, S-genotype-asssocd. Pancreatic ribonuclease	
3.1.27.5	Glycoproteins, SLSG Pancreatic ribonuclease	
3.1.27.5	Glycoproteins, specific or class, gene S Pancreatic ribonuclease	
3.1.27.5	Glycoproteins, specific or class, SLSG (gene S locus-specific glycoprotein) Pancreatic ribonuclease	
2.4.1.74	Glycosaminoglycan galactosyltransferase Glycosaminoglycan galactosyltransferase	
3.2.1.21	β-D-glycosidase β-glucosidase	
3.2.2.24	Glycosidase, azoferredoxin ADP-ribosyl-[dinitrogen reductase] hydrolase	
3.2.2.23	Glycosidase, deoxyribonucleate Formamidopyrimidine-DNA glycosidase	

EC-No.	Synonym Recommended Name	Vol., Page
3.2.2.20	Glycosidase, deoxyribonucleate 3-methyladenine, I DNA-3-methyladenine glycosidase I	
3.2.2.21	Glycosidase, deoxyribonucleate 3-methyladenine, II DNA-3-methyladenine glycosidase II	
3.2.2.17	Glycosidase, deoxyribonucleate pyrimidine dimer Deoxyribodipyrimidine endonucleosidase	
3.2.2.22	Glycosidase, ribosomal ribonucleate N- rRNA N-glycosidase	
2.4.1.154	Glycosphingolipid β-N-acetylgalactosaminyltransferase Globotriosylceramide β-1,6-N-acetylgalactosaminyltransferase	
3.5.1.69	Glycosphingolipid amidohydrolase Glycosphingolipid deacylase	
3.5.1.23	Glycosphingolipid ceramide deacylase Acylsphingosine deacylase	
3.5.1.69	Glycosphingolipid ceramide deacylase Glycosphingolipid deacylase	
3.5.1.69	Glycosphingolipid deacylase Glycosphingolipid deacylase	
2.8.2.11	Glycosphingolipid sulfotransferase Galactosylceramide sulfotransferase	
3.1.6.3	Glycosulfatase Glycosulfatase	
3.1.4.47	(Glycosyl)phosphatidylinositol-specific phospholipase C Variant-surface-glycoprotein phospholipase C	
3.2.1.123	Glycosyl-N-acetyl-sphingosine 1,1-β-D-glucanohydrolase Endoglycosylceramidase	
3.2.1.62	Glycosyl-N-acylsphingosine glycohydrolase Glycosylceramidase	
2.4.1.101	N-Glycosyl-oligosaccharide-glycoprotein N-acetylglucosaminyltransferase I α-1,3-Mannosyl-glycoprotein β-1,2-N-acetylglucosaminyltransferase	
2.4.1.102	O-Glycosyl-oligosaccharide-glycoprotein N-acetylglucosaminyltransferase I β-1,3-Galactosyl-O-glycosyl-glycoprotein β-1,6-N-acetylglucosaminyl-transferase	
2.4.1.143	N-Glycosyl-oligosaccharide-glycoprotein N-acetylglucosaminyltransferase II α-1,6-Mannosyl-glycoprotein β-1,2-N-acetylglucosaminyltransferase	

EC-No.	Synonym Recommended Name	Vol., Page
2.4.1.146	O-Glycosyl-oligosaccharide-glycoprotein N-acetylglucosaminyltransferase II β-1,3-Galactosyl-O-glycosyl-glycoprotein β-1,3-N-acetylglucosaminyl-transferase	
2.4.1.144	N-Glycosyl-oligosaccharide-glycoprotein N-acetylglucosaminyltransferase III β-1,4-Mannosyl-glycoprotein β-1,4-N-acetylglucosaminyltransferase	
2.4.1.147	O-Glycosyl-oligosaccharide-glycoprotein N-acetylglucosaminyltransferase III Acetylgalactosaminyl-O-glycosyl-glycoprotein β-1,3-N-acetylglucosaminyl-transferase	
2.4.1.145	N-Glycosyl-oligosaccharide-glycoprotein N-acetylglucosaminyltransferase IV α-1,3-Mannosylglycoprotein β-1,4-N-acetylglucosaminyltransferase	
2.4.1.148	O-Glycosyl-oligosaccharide-glycoprotein N-acetylglucosaminyltransferase IV Acetylgalactosaminyl-O-glycosyl-glycoprotein β-1,6-N-acetylglucosaminyl-transferase	
2.4.1.201	N-Glycosyl-oligosaccharide-glycoprotein N-acetylglucosaminyltransferase VI Mannosyl-glycoprotein β-1,4-N-acetylglucosaminyltransferase	
3.4.13.19	Glycosyl-phosphatidylinositol-anchored renal dipeptidase Membrane dipeptidase	
3.2.1.62	Glycosylceramidase Glycosylceramidase	
3.2.1.62	Glycosylceramidase Glycosylceramidase	
3.2.1.62	Glycosyl ceramide glycosylhydrolase Glycosylceramidase	
3.1.4.47	Glycosyl inositol phospholipid anchor-hydrolyzing enzyme Variant-surface-glycoprotein phospholipase C	
3.1.4.47	Glycosylphosphatidylinositol-phospholipase C Variant-surface-glycoprotein phospholipase C	
3.1.4.47	Glycosylphosphatidylinositol-specific phospholipase C Variant-surface-glycoprotein phospholipase C	
2.4.1.18	Glycosyltransferase, α-glucan-branching 1,4-α-Glucan branching enzyme	
2.4.1.19	Glycosyltransferase, cyclodextrin Cyclomaltodextrin glucanotransferase	

EC-No.	Synonym Recommended Name	Vol., Page
2.4.1.25	Glycosyltransferase, dextrin 4-α-Glucanotransferase	
2.4.1.119	Glycosyltransferase, dolichyldiphosphooligosaccharide-protein Dolichyl-diphosphooligosaccharide-protein glycotransferase	
2.4.1.119	Glycosyltransferase, dolichylpyrophosphodiacetylchitobiose-protein Dolichyl-diphosphooligosaccharide-protein glycotransferase	
2.4.1.129	Glycosyltransferase, peptidoglycan Peptidoglycan glycosyltransferase	
3.2.1.56	Glycuronidase Glucuronosyl-disulfoglucosamine glucuronidase	
3.4.13.18	Glycyl-glycine dipeptidase Cytosol nonspecific dipeptidase	
3.4.13.1	glycyl-glycine dipeptidase glycyl-glycine dipeptidase	
3.4.3.1	glycyl-glycine dipeptidase glycyl-glycine dipeptidase	
3.4.24.75	Glycyl-glycine endopeptidase Lysostaphin	
3.4.13.18	Glycyl-L-leucine dipeptidase Cytosol nonspecific dipeptidase	
3.4.13.18	Glycyl-L-leucine hydrolase Cytosol nonspecific dipeptidase	
3.4.13.18	Glycyl-L-leucine peptidase Cytosol nonspecific dipeptidase	
3.4.13.18	Glycyl-leucine dipeptidase Cytosol nonspecific dipeptidase	
3.4.13.2	glycyl-leucine dipeptidase glycyl-leucine dipeptidase	
3.4.3.2	glycyl-leucine dipeptidase glycyl-leucine dipeptidase	
6.1.1.14	Glycyl-transfer ribonucleate synthetase Glycine-tRNA ligase	Vol. 2, Page 101
6.1.1.14	Glycyl-transfer ribonucleic acid synthetase Glycine-tRNA ligase	Vol. 2, Page 101
6.1.1.14	Glycyl-transfer RNA synthetase Glycine-tRNA ligase	Vol. 2, Page 101

EC-No.	Synonym Recommended Name	Vol., Page
6.1.1.14	Glycyl-tRNA synthetase Glycine-tRNA ligase	Vol. 2, Page 101
3.4.22.25	Glycyl endopeptidase Glycyl endopeptidase	
3.4.24.75	Glycylglycine endopeptidase Lysostaphin	
3.4.13.18	Glycylleucine dipeptidase Cytosol nonspecific dipeptidase	
3.4.13.18	Glycylleucine dipeptide hydrolase Cytosol nonspecific dipeptidase	
3.4.13.18	Glycylleucine hydrolase Cytosol nonspecific dipeptidase	
3.4.13.18	Glycylleucine peptidase Cytosol nonspecific dipeptidase	
2.3.1.97	Glycylpeptide N-tetradecanoyltransferase Glycylpeptide N-tetradecanoyltransferase	
3.4.14.5	glycylproline-dipeptidyl-aminopeptidase dipeptidyl-peptidase IV	
3.4.14.5	glycylproline aminopeptidase dipeptidyl-peptidase IV	
3.4.14.5	glycylprolyl aminopeptidase dipeptidyl-peptidase IV	
3.4.14.5	glycylprolyl dipeptidylaminopeptidase dipeptidyl-peptidase IV	
6.1.1.14	Glycyl translase Glycine-tRNA ligase	Vol. 2, Page 101
3.2.1.128	Glycyrrhizinate β-glucuronidase Glycyrrhizinate β-glucuronidase	
3.2.1.128	Glycyrrhizinate glucuronosylhydrolase Glycyrrhizinate β-glucuronidase	
3.2.1.128	Glycyrrhizin hydrolase Glycyrrhizinate β-glucuronidase	
3.2.1.128	Glycyrrhizinic acid hydrolase Glycyrrhizinate β-glucuronidase	
2.4.1.186	Glyogenin Glycogenin glucosyltransferase	

EC-No.	Synonym Recommended Name	Vol., Page
4.4.1.5	glyoxalase I lactoylglutathione lyase	
3.1.2.6	Glyoxalase II Hydroxyacylglutathione hydrolase	
4.1.1.47	Glyoxalate carboligase Tartronate-semialdehyde synthase	
4.4.1.5	glyoxylase I lactoylglutathione lyase	
4.1.3.15	glyoxylate-2-oxoglutarate carboligase 2-hydroxy-3-oxoadipate synthase	
2.6.1.4	Glyoxylate-glutamate aminotransferase Glycine transaminase	
2.6.1.4	Glyoxylate-glutamic transaminase Glycine transaminase	
1.2.1.17	Glyoxylate:NADP$^+$ oxidoreductase (CoA-oxalylating) Glyoxylate dehydrogenase (acylating)	
1.2.3.5	Glyoxylate:oxygen oxidoreductase Glyoxylate oxidase	
4.1.1.47	Glyoxylate carbo-ligase Tartronate-semialdehyde synthase	
4.1.1.47	Glyoxylate carboligase Tartronate-semialdehyde synthase	
4.1.1.47	Glyoxylate carboxy-lyase Tartronate-semialdehyde synthase	
4.1.1.47	Glyoxylate carboxy-lyase (dimerizing) Tartronate-semialdehyde synthase	
4.1.1.47	Glyoxylate carboxy-lyase (dimerizing and reducing) Tartronate-semialdehyde synthase	
1.2.1.17	Glyoxylate dehydrogenase (acylating) Glyoxylate dehydrogenase (acylating)	
1.2.3.5	Glyoxylate oxidase Glyoxylate oxidase	
1.1.1.26	Glyoxylate reductase Glyoxylate reductase	
1.1.1.79	Glyoxylate reductase (NADP$^+$) Glyoxylate reductase (NADP$^+$)	

EC-No.	Synonym Recommended Name	Vol., Page
4.1.3.2	Glyoxylate transacetase Malate synthase	
1.1.1.26	Glyoxylic acid reductase Glyoxylate reductase	
4.1.1.47	Glyoxylic carbo-ligase Tartronate-semialdehyde synthase	
4.1.3.2	Glyoxylic transacetase Malate synthase	
6.1.1.14	GlyRS Glycine-tRNA ligase	Vol. 2, Page 101
2.4.1.62	GM1-synthase Ganglioside galactosyltransferase	
2.4.1.92	GM2/GD2-synthase (N-Acetylneuraminyl)-galactosylglucosylceramide N-acetylgalactosaminyl-transferase	
2.4.1.92	GM2 synthase (N-Acetylneuraminyl)-galactosylglucosylceramide N-acetylgalactosaminyl-transferase	
2.4.99.9	GM3 synthase Lactosylceramide α-2,3-sialyltransferase	
2.4.99.9	GM3 synthetase Lactosylceramide α-2,3-sialyltransferase	
2.7.4.8	5'-GMP kinase Guanylate kinase	
2.7.4.8	GMP kinase Guanylate kinase	
2.4.2.8	GMP pyrophosphorylase Hypoxanthine phosphoribosyltransferase	
1.6.6.8	GMP reductase GMP reductase	
6.3.4.1	GMPS GMP synthase	Vol. 2, Page 548
6.3.5.2	GMPS GMP synthase (glutamine hydrolysing)	Vol. 2, Page 655
6.3.4.1	GMP synthase GMP synthase	Vol. 2, Page 548

EC-No.	Synonym Recommended Name	Vol., Page
6.3.5.2	GMP synthase GMP synthase (glutamine hydrolysing)	Vol. 2, Page 655
6.3.4.1	GMP synthase (glutamine hydrolysing) GMP synthase	Vol. 2, Page 548
6.3.5.2	GMP synthase (glutamine hydrolysing) GMP synthase (glutamine hydrolysing)	Vol. 2, Page 655
6.3.4.1	GMP synthetase GMP synthase	Vol. 2, Page 548
6.3.5.2	GMP synthetase GMP synthase (glutamine hydrolysing)	Vol. 2, Page 655
6.3.4.1	GMP synthetase (glutamine hydrolysing) GMP synthase	Vol. 2, Page 548
6.3.5.2	GMP synthetase (glutamine hydrolysing) GMP synthase (glutamine hydrolysing)	Vol. 2, Page 655
2.1.1.20	GNMT Glycine N-methyltransferase	
1.1.3.4	GOD Glucose oxidase	
3.2.1.114	Golgi α-mannosidase II Mannosyl-oligosaccharide 1,3-1,6-α-mannosidase	
2.6.1.1	GOT (enzyme) Aspartate transaminase	
3.6.1.7	GP 1-3 Acylphosphatase	
3.4.17.22	gp180 Metallocarboxypeptidase D	
2.7.7.45	Gp4G synthetase Guanosine-triphosphate guanylyltransferase	
3.4.14.2	GPA dipeptidyl-peptidase II	
3.5.3.17	GPase Guanidinopropionase	
3.1.4.2	GPC-PD glycerophosphocholine phosphodiesterase	
3.1.4.2	GPC-PDE glycerophosphocholine phosphodiesterase	

EC-No.	Synonym Recommended Name	Vol., Page
3.1.4.2	GPC-phosphodiesterase glycerophosphocholine phosphodiesterase	
3.1.4.2	GPC:choline phosphodiesterase glycerophosphocholine phosphodiesterase	
3.1.4.2	GPC diesterase glycerophosphocholine phosphodiesterase	
3.5.3.17	GPH Guanidinopropionase	
3.1.3.21	Gpp1p glycerol-1-phosphatase	
3.1.3.21	Gpp2p glycerol-1-phosphatase	
3.4.22.29	Y-G proteinase 2A Picornain 2A	
2.4.2.8	GPRT Hypoxanthine phosphoribosyltransferase	
2.7.6.5	GPSI GTP pyrophosphokinase	
2.7.6.5	GPSII GTP pyrophosphokinase	
2.6.1.2	GPT Alanine transaminase	
2.3.2.2	γ-GPT γ-Glutamyltransferase	
3.6.1.11	Gra-Pase Exopolyphosphatase	
5.1.1.11	Gramicidin S synthetase I Phenylalanine racemase (ATP-hydrolysing)	Vol. 1, Page 48
3.4.21.37	Granulocyte elastase Leucocyte elastase	
2.4.1.1	Granulose phosphorylase Phosphorylase	
3.4.21.78	Granzyme A Granzyme A	
3.4.21.79	Granzyme B Granzyme B	

EC-No.	Synonym Recommended Name	Vol., Page
3.4.21.79	Granzyme G Granzyme B	
3.4.21.79	Granzyme H Granzyme B	
1.1.1.20	GRase Glucuronolactone reductase	
5.4.2.1	GriP mutase Phosphoglycerate mutase	Vol. 1, Page 493
2.7.7.39	Gro-PCT Glycerol-3-phosphate cytidylyltransferase	
3.1.4.2	GroPChoPDE glycerophosphocholine phosphodiesterase	
6.3.1.2	GS Glutamate-ammonia ligase	Vol. 2, Page 347
2.4.1.12	GS-I Cellulose synthase (UDP-forming)	
2.4.1.34	GS-II 1,3-β-Glucan synthase	
5.4.3.8	GSA-AT Glutamate-1-semialdehyde 2,1-aminomutase	Vol. 1, Page 575
5.4.3.8	GSA aminotransferase Glutamate-1-semialdehyde 2,1-aminomutase	Vol. 1, Page 575
3.4.21.69	GSAPC Protein C (activated)	
2.8.2.11	GSase Galactosylceramide sulfotransferase	
3.4.21.19	GSE glutamyl endopeptidase	
1.8.4.4	GSH-cystine transhydrogenase Glutathione-cystine transhydrogenase	
1.8.4.8	GSH-cystine transhydrogenase phosphoadenylyl-sulfate reductase (thioredoxin)	
1.8.4.2	GSH-insulin transhydrogenase Protein-disulfide reductase (glutathione)	
1.11.1.9	GSH peroxidase Glutathione peroxidase	

EC-No.	Synonym Recommended Name	Vol., Page
1.6.4.2	GSH reductase Glutathione reductase (NADPH)	
6.3.2.3	GSH synthetase Glutathione synthase	Vol. 2, Page 410
5.1.1.11	GS I Phenylalanine racemase (ATP-hydrolysing)	Vol. 1, Page 48
6.3.1.8	Gsp synthetase Glutathionylspermidine synthase	Vol. 2, Page 386
3.5.1.78	Gsp synthetase/amidase Glutathionylspermidine amidase	
6.3.2.3	GSS Glutathione synthase	Vol. 2, Page 410
1.6.4.2	GSSG reductase Glutathione reductase (NADPH)	
4.1.3.18	GST-mALS acetolactate synthase	
4.1.3.18	GST-wALS acetolactate synthase	
2.4.1.17	GT Glucuronosyltransferase	
2.3.2.2	γ-GT γ-Glutamyltransferase	
2.4.1.125	GTF⁻S Sucrose-1,6-α-glucan 3(6)-α-glucosyltransferase	
2.4.1.91	GTI Flavonol 3-O-glucosyltransferase	
2.3.2.2	γ-GTP γ-Glutamyltransferase	
3.5.4.25	GTP-8-formylhydrolase GTP cyclohydrolase II	
6.2.1.10	GTP-dependent acyl-CoA synthetase Acid-CoA ligase (GDP-forming)	Vol. 2, Page 249
2.7.7.45	GTP-GTP guanylyltransferase Guanosine-triphosphate guanylyltransferase	
2.7.7.13	GTP-mannose-1-phosphate guanylyltransferase Mannose-1-phosphate guanylyltransferase	

EC-No.	Synonym Recommended Name	Vol., Page
2.7.7.13	GTP-mannose 1-phosphate guanylyltransferase Mannose-1-phosphate guanylyltransferase	
3.5.4.16	GTP 7,8-8,9-dihydrolase GTP cyclohydrolase I	
3.5.4.25	GTP 7,8-8,9-dihydrolase (pyrophosphate-forming) GTP cyclohydrolase II	
3.5.4.16	GTP 8-formylhydrolase GTP cyclohydrolase I	
2.7.7.34	GTP:α-D-glucose-1-phosphate guanylyltransferase Glucose-1-phosphate guanylyltransferase	
2.7.7.29	GTP:α-D-hexose-1-phosphate guanylyltransferase Hexose-1-phosphate guanylyltransferase	
2.7.7.13	GTP:α-D-mannose-1-phosphate guanylyltransferase Mannose-1-phosphate guanylyltransferase	
2.7.2.10	GTP:3-phospho-D-glycerate 1-phosphotransferase Phosphoglycerate kinase (GTP)	
2.7.1.81	GTP:5-hydroxy-L-lysine O-phosphotransferase Hydroxylysine kinase	
2.7.4.10	GTP:AMP phosphotransferase Nucleoside-triphosphate-adenylate kinase	
2.7.7.45	GTP:GTP guanylyltransferase Guanosine-triphosphate guanylyltransferase	
2.7.7.30	GTP:L-fucose-1-phosphate guanylyltransferase Fucose-1-phosphate guanylyltransferase	
2.7.7.50	GTP:mRNA guanylyltransferase mRNA guanylyltransferase	
4.1.1.32	GTP:oxaloacetate carboxy-lyase (transphosphorylating) Phosphoenolpyruvate carboxykinase (GTP)	
3.5.4.16	GTP cyclohydrolase GTP cyclohydrolase I	
3.5.4.16	GTP cyclohydrolase I GTP cyclohydrolase I	
3.5.4.25	GTP cyclohydrolase II GTP cyclohydrolase II	
4.6.1.2	GTP pyrophosphate-lyase (cyclizing) guanylate cyclase	

EC-No.	Synonym Recommended Name	Vol., Page
2.7.6.5	GTP pyrophosphokinase GTP pyrophosphokinase	
1.11.1.7	Guaiacol peroxidase Peroxidase	
3.5.4.3	Guanase Guanine deaminase	
3.1.3.40	1-guanidino-1-deoxy-scyllo-inositol-4-phosphate 4-phosphohydrolase guanidinodeoxy-scyllo-inositol-4-phosphatase	
3.1.3.40	1-guanidino-1-deoxy-scyllo-inositol-4-P phosphohydrolase guanidinodeoxy-scyllo-inositol-4-phosphatase	
4.1.1.75	5-Guanidino-2-oxopentanoate decarboxylase 5-Guanidino-2-oxopentanoate decarboxylase	
2.6.1.56	1D-1-Guanidino-3-amino-1,3-dideoxy-scyllo-inositol:pyruvate aminotransferase 1D-1-Guanidino-3-amino-1,3-dideoxy-scyllo-inositol transaminase	
2.6.1.56	1D-1-Guanidino-3-amino-1,3-dideoxy-scyllo-inositol transaminase 1D-1-Guanidino-3-amino-1,3-dideoxy-scyllo-inositol transaminase	
3.5.3.2	Guanidinoacetate amidinohydrolase Glycocyaminase	
2.7.3.1	Guanidinoacetate kinase Guanidinoacetate kinase	
2.1.1.2	Guanidinoacetate methyltransferase Guanidinoacetate N-methyltransferase	
2.1.1.2	Guanidinoacetate N-methyltransferase Guanidinoacetate N-methyltransferase	
2.1.1.2	Guanidinoacetate transmethylase Guanidinoacetate N-methyltransferase	
1.2.1.54	4-Guanidinobutanal:NAD^+ 1-oxidoreductase γ-Guanidinobutyraldehyde dehydrogenase	
3.5.3.7	4-Guanidinobutanoate amidinohydrolase Guanidinobutyrase	
1.2.1.54	4-Guanidinobutyraldehyde dehydrogenase γ-Guanidinobutyraldehyde dehydrogenase	
1.2.1.19	γ-guanidinobutyraldehyde dehydrogenase Aminobutyraldehyde dehydrogenase	

EC-No.	Synonym Recommended Name	Vol., Page
1.2.1.54	α-Guanidinobutyraldehyde dehydrogenase γ-Guanidinobutyraldehyde dehydrogenase	
1.2.1.54	γ-Guanidinobutyraldehyde dehydrogenase γ-Guanidinobutyraldehyde dehydrogenase	
1.2.1.54	γ-Guanidinobutyraldehyde dehydrogenase γ-Guanidinobutyraldehyde dehydrogenase	
3.5.3.7	Guanidinobutyrase Guanidinobutyrase	
3.5.3.7	4-Guanidinobutyrate amidinobutyrase Guanidinobutyrase	
3.5.3.7	γ-guanidinobutyrate amidinohydrolase Guanidinobutyrase	
3.5.3.7	Guanidinobutyrate ureahydrolase Guanidinobutyrase	
3.1.3.40	guanidinodeoxy-scyllo-inositol-4-phosphatase guanidinodeoxy-scyllo-inositol-4-phosphatase	
3.5.3.17	3-Guanidinopropanoate amidinopropionase Guanidinopropionase	
3.5.3.17	Guanidinopropionase Guanidinopropionase	
2.1.1.2	Guanidoacetate methyltransferase Guanidinoacetate N-methyltransferase	
3.5.3.7	γ-guanidobutyrase Guanidinobutyrase	
2.4.2.29	Guanine, queuine-tRNA transglycosylase Queuine tRNA-ribosyltransferase	
2.1.1.56	Guanine-7-methyltransferase mRNA (guanine-N^7-)-methyltransferase	
2.4.2.8	Guanine-hypoxanthine phosphoribosyltransferase Hypoxanthine phosphoribosyltransferase	
2.1.1.32	Guanine-N^2-methylase tRNA (guanine-N^2-)-methyltransferase	
3.5.4.3	Guanine aminase Guanine deaminase	
3.5.4.3	Guanine aminohydrolase Guanine deaminase	

EC-No.	Synonym Recommended Name	Vol., Page
3.5.4.3	Guanine deaminase Guanine deaminase	
3.5.4.3	Guanine deaminase Guanine deaminase	
2.4.2.29	Guanine insertion enzyme Queuine tRNA-ribosyltransferase	
2.4.2.8	Guanine phosphoribosyltransferase Hypoxanthine phosphoribosyltransferase	
3.1.7.2	Guanosine-3',5'-bis(diphosphate) 3'-pyrophosphatase Guanosine-3',5'-bis(diphosphate) 3'-pyrophosphatase	
3.1.7.2	Guanosine-3',5'-bis(diphosphate) 3'-pyrophosphohydrolase Guanosine-3',5'-bis(diphosphate) 3'-pyrophosphatase	
3.6.1.40	Guanosine-5'-triphosphate,3'-diphosphate 5'-phosphohydrolase Guanosine-5'-triphosphate,3'-diphosphate diphosphatase	
3.6.1.40	Guanosine-5'-triphosphate,3'-diphosphate diphosphatase Guanosine-5'-triphosphate,3'-diphosphate diphosphatase	
3.6.1.40	Guanosine-5'-triphosphate,3'-diphosphate pyrophosphatase Guanosine-5'-triphosphate,3'-diphosphate diphosphatase	
3.6.1.42	Guanosine-diphosphatase Guanosine-diphosphatase	
2.7.7.45	Guanosine-triphosphate guanylyltransferase Guanosine-triphosphate guanylyltransferase	
2.7.6.5	Guanosine 3',5'-polyphosphate synthase GTP pyrophosphokinase	
2.7.6.5	Guanosine 3',5'-polyphosphate synthetase GTP pyrophosphokinase	
2.7.6.5	Guanosine 5',3'-polyphosphate synthetase GTP pyrophosphokinase	
3.6.1.6	Guanosine 5'-diphosphatase Nucleoside-diphosphatase	
4.2.1.47	Guanosine 5'-diphosphate-D-mannose oxidoreductase GDPmannose 4,6-dehydratase	
2.7.7.13	Guanosine 5'-diphospho-D-mannose pyrophosphorylase Mannose-1-phosphate guanylyltransferase	
1.6.6.8	Guanosine 5'-monophosphate reductase GMP reductase	

EC-No.	Synonym Recommended Name	Vol., Page
6.3.4.1	Guanosine 5'-monophosphate synthetase GMP synthase	Vol. 2, Page 548
6.3.5.2	Guanosine 5'-monophosphate synthetase GMP synthase (glutamine hydrolysing)	Vol. 2, Page 655
2.4.2.8	Guanosine 5'-phosphate pyrophosphorylase Hypoxanthine phosphoribosyltransferase	
3.6.1.40	Guanosine 5'-triphosphate-3'-diphosphate 5'-phosphohydrolase Guanosine-5'-triphosphate,3'-diphosphate diphosphatase	
6.3.4.1	Guanosine 5-monophosphate synthetase GMP synthase	Vol. 2, Page 548
6.3.5.2	Guanosine 5-monophosphate synthetase GMP synthase (glutamine hydrolysing)	Vol. 2, Page 655
2.4.2.15	Guanosine:orthophosphate D-ribosyltransferase Guanosine phosphorylase	
3.5.4.15	Guanosine aminase Guanosine deaminase	
3.5.4.15	Guanosine aminohydrolase Guanosine deaminase	
3.1.4.35	Guanosine cyclic 3',5'-phosphate phosphodiesterase 3',5'-Cyclic-GMP phosphodiesterase	
3.5.4.15	Guanosine deaminase Guanosine deaminase	
3.6.1.6	Guanosine diphosphatase Nucleoside-diphosphatase	
1.1.1.187	Guanosine diphosphate-4-keto-D-rhamnose reductase GDP-4-dehydro-D-rhamnose reductase	
2.7.7.22	Guanosine diphosphate-mannose 1-phosphate guanylyltransferase Mannose-1-phosphate guanylyltransferase (GDP)	
3.2.1.42	Guanosine diphosphate D-glucose glucohydrolase GDPglucosidase	
2.7.7.30	Guanosine diphosphate L-fucose pyrophosphorylase Fucose-1-phosphate guanylyltransferase	
1.1.1.135	Guanosine diphospho-6-deoxy-D-talose dehydrogenase GDP-6-deoxy-D-talose 4-dehydrogenase	
1.1.1.132	Guanosine diphospho-D-mannose dehydrogenase GDPmannose 6-dehydrogenase	

EC-No.	Synonym Recommended Name	Vol., Page
2.4.1.69	Guanosine diphospho-L-fucose-lactose fucosyltransferase Galactoside 2-α-L-fucosyltransferase	
2.4.1.69	Guanosine diphosphofucose-β-D-galactosyl-α-2-L-fucosyltransferase Galactoside 2-α-L-fucosyltransferase	
2.4.1.69	Guanosine diphosphofucose-galactosylacetylglucosaminylgalactosylglucosylceramide α-L-fucosyltransferase Galactoside 2-α-L-fucosyltransferase	
2.4.1.69	Guanosine diphosphofucose-glycoprotein 2-α-L-fucosyltransferase Galactoside 2-α-L-fucosyltransferase	
2.4.1.65	Guanosine diphosphofucose-glycoprotein 4-α-L-fucosyltransferase Galactoside 3(4)-L-fucosyltransferase	
2.4.1.68	Guanosine diphosphofucose-glycoprotein fucosyltransferase Glycoprotein 6-α-L-fucosyltransferase	
2.4.1.29	Guanosine diphosphoglucose-1,4-β-glucan glucosyltransferase Cellulose synthase (GDP-forming)	
2.4.1.29	Guanosine diphosphoglucose-cellulose glucosyltransferase Cellulose synthase (GDP-forming)	
2.4.1.36	Guanosine diphosphoglucose-glucose phosphate glucosytransferase α,α-Trehalose-phosphate synthase (GDP-forming)	
2.7.7.34	Guanosine diphosphoglucose pyrophosphorylase Glucose-1-phosphate guanylyltransferase	
3.2.1.42	Guanosine diphosphoglucosidase GDPglucosidase	
2.7.7.29	Guanosine diphosphohexose pyrophosphorylase Hexose-1-phosphate guanylyltransferase	
2.4.1.48	Guanosine diphosphomannose-heteroglycan α-mannosyltransferase Heteroglycan α-mannosyltransferase	
2.4.1.57	Guanosine diphosphomannose-phosphatidyl-inositol α-mannosyltransferase Phosphatidyl-myo-inositol α-mannosyltransferase	
4.2.1.47	Guanosine diphosphomannose 4,6-dehydratase GDPmannose 4,6-dehydratase	
1.1.1.132	Guanosine diphosphomannose dehydrogenase GDPmannose 6-dehydrogenase	
4.2.1.47	Guanosine diphosphomannose oxidoreductase GDPmannose 4,6-dehydratase	

EC-No.	Synonym Recommended Name	Vol., Page
2.7.7.22	Guanosine diphosphomannose phosphorylase Mannose-1-phosphate guanylyltransferase (GDP)	
2.7.7.13	Guanosine diphosphomannose pyrophosphorylase Mannose-1-phosphate guanylyltransferase	
2.7.4.8	Guanosine monophosphate kinase Guanylate kinase	
1.6.6.8	Guanosine monophosphate reductase GMP reductase	
6.3.4.1	Guanosine monophosphate synthetase (glutamine-hydrolyzing) GMP synthase	Vol. 2, Page 548
6.3.5.2	Guanosine monophosphate synthetase (glutamine-hydrolyzing) GMP synthase (glutamine hydrolysing)	Vol. 2, Page 655
3.6.1.40	Guanosine pentaphosphatase Guanosine-5'-triphosphate,3'-diphosphate diphosphatase	
3.6.1.40	Guanosine pentaphosphate phosphatase Guanosine-5'-triphosphate,3'-diphosphate diphosphatase	
3.6.1.40	Guanosine pentaphosphate phosphohydrolase Guanosine-5'-triphosphate,3'-diphosphate diphosphatase	
2.7.6.5	Guanosine pentaphosphate synthetase GTP pyrophosphokinase	
2.4.2.8	Guanosine phosphoribosyltransferase Hypoxanthine phosphoribosyltransferase	
2.4.2.15	Guanosine phosphorylase Guanosine phosphorylase	
2.7.4.10	Guanosine triphosphate-adenylate kinase Nucleoside-triphosphate-adenylate kinase	
2.7.7.45	Guanosine triphosphate-guanose triphosphate guanylyltransferase Guanosine-triphosphate guanylyltransferase	
2.7.7.13	Guanosine triphosphate-mannose 1-phosphate guanylyltransferase Mannose-1-phosphate guanylyltransferase	
3.5.4.16	Guanosine triphosphate 8-deformylase GTP cyclohydrolase I	
2.7.1.81	Guanosine triphosphate:5-hydroxy-L-lysine O-phosphotransferase Hydroxylysine kinase	
3.5.4.16	Guanosine triphosphate cyclohydrolase GTP cyclohydrolase I	

EC-No.	Synonym Recommended Name	Vol., Page
3.5.4.25	Guanosine triphosphate cyclohydrolase II GTP cyclohydrolase II	
3.1.27.3	Guanyl-specific RNase Ribonuclease T1	
4.6.1.2	guanylate cyclase guanylate cyclase	
2.7.4.8	Guanylate kinase Guanylate kinase	
2.4.2.8	Guanylate pyrophosphorylase Hypoxanthine phosphoribosyltransferase	
1.6.6.8	Guanylate reductase GMP reductase	
6.3.4.1	Guanylate synthetase GMP synthase	Vol. 2, Page 548
6.3.5.2	Guanylate synthetase GMP synthase (glutamine hydrolysing)	Vol. 2, Page 655
6.3.4.1	Guanylate synthetase (glutamine-hydrolyzing) GMP synthase	Vol. 2, Page 548
6.3.5.2	Guanylate synthetase (glutamine-hydrolyzing) GMP synthase (glutamine hydrolysing)	Vol. 2, Page 655
4.6.1.2	guanyl cyclase guanylate cyclase	
2.4.2.8	Guanylic pyrophosphorylase Hypoxanthine phosphoribosyltransferase	
3.1.27.3	Guanyloribonuclease Ribonuclease T_1	
2.7.7.26	guanyloribonuclease guanyloribonuclease	
4.6.1.2	guanylyl cyclase guanylate cyclase	
2.7.7.34	Guanylyltransferase, glucose 1-phosphate Glucose-1-phosphate guanylyltransferase	
2.7.7.45	Guanylyltransferase, guanosine triphosphate Guanosine-triphosphate guanylyltransferase	
2.7.7.29	Guanylyltransferase, hexose 1-phosphate Hexose-1-phosphate guanylyltransferase	

EC-No.	Synonym Recommended Name	Vol., Page
2.7.7.13	Guanylyltransferase, mannose 1-phosphate Mannose-1-phosphate guanylyltransferase	
2.7.7.22	Guanylyltransferase, mannose 1-phosphate (guanosine diphosphate) Mannose-1-phosphate guanylyltransferase (GDP)	
1.1.1.45	L-Gulonate 3-dehydrogenase L-Gulonate 3-dehydrogenase	
1.1.1.45	L-Gulonate:NAD$^+$ 3-oxidoreductase L-Gulonate 3-dehydrogenase	
1.1.1.19	L-Gulonate:NADP$^+$ 6-oxidoreductase Glucuronate reductase	
1.1.1.45	L-Gulonic acid dehydrogenase L-Gulonate 3-dehydrogenase	
1.1.3.8	L-Gulono-γ-lactone:O_2 oxidoreductase L-Gulonolactone oxidase	
1.1.3.8	L-Gulono-γ-lactone:oxidoreductase L-Gulonolactone oxidase	
1.1.3.8	L-Gulono-γ-lactone oxidase L-Gulonolactone oxidase	
1.1.1.20	L-Gulono-1,4-lactone:NADP$^+$ 1-oxidoreductase Glucuronolactone reductase	
1.1.3.8	L-Gulono-1,4-lactone:oxygen 2-oxidoreductase L-Gulonolactone oxidase	
3.1.1.17	gulonolactonase gluconolactonase	
1.1.1.20	Gulonolactone dehydrogenase Glucuronolactone reductase	
1.1.3.8	L-Gulonolactone oxidase L-Gulonolactone oxidase	
4.2.2.11	L-Guluronan lyase Poly(α-L-guluronate) lyase	
4.2.2.11	Guluronate lyase Poly(α-L-guluronate) lyase	
4.2.2.11	L-Guluronate lyase Poly(α-L-guluronate) lyase	
1.13.99.5	(1H)-3-Hydroxy-4-oxoquinoline 2,4-dioxygenase 3,4-Dihydroxyquinoline 2,4-dioxygenase	

EC-No.	Synonym Recommended Name	Vol., Page
3.6.1.36	H^+,K^+-ATPase H^+/K^+-transporting ATPase	
3.6.1.35	H^+-ATPase H^+-transporting ATPase	
1.6.1.2	H^+-Thase $NAD(P)^+$ transhydrogenase (AB-specific)	
1.6.1.1	H^+-Thase $NAD(P)^+$ transhydrogenase (B-specific)	
3.6.1.34	H^+-transporting ATPase H^+-transporting ATP synthase	
3.6.1.35	H^+-transporting ATPase H^+-transporting ATPase	
3.6.1.34	H^+-transporting ATP synthase H^+-transporting ATP synthase	
3.6.1.36	H^+/K^+-transporting ATPase H^+/K^+-transporting ATPase	
1.12.99.4	H_2-forming N^5N^{10}-Methenyltetrahydromethanopterin dehydrogenase N^5N^{10}-Methenyltetrahydromethanopterin hydrogenase	
3.4.24.53	H_2-Proteinase Trimerelysin II	
2.7.6.3	H_2-Pteridine-CH_2OH pyrophosphokinase 2-Amino-4-hydroxy-6-hydroxymethyldihydropteridine pyrophosphokinase	
1.12.2.1	H_2:ferricytochrome c_3 oxidoreductase Cytochrome-c3 hydrogenase	
1.12.1.2	H_2:NAD^+ oxidoreductase Hydrogen dehydrogenase	
1.18.99.1	H_2 oxidizing hydrogenase Hydrogenase	
1.18.99.1	H_2 producing hydrogenase Hydrogenase	
1.8.1.2	H_2S-NADP oxidoreductase Sulfite reductase (NADPH)	
3.4.21.74	Habutobin Venombin A	

EC-No.	Synonym Recommended Name	Vol., Page
3.1.1.28	HACH, high activity acylcarnitine hydrolase acylcarnitine hydrolase	
1.14.99.3	Haem oxygenase Heme oxygenase (decyclizing)	
3.4.21.38	Hageman factor coagulation factor XIIa	
3.4.21.38	Hageman factor β-fragment coagulation factor XIIa	
3.4.21.38	Hageman factor fragment HFf coagulation factor XIIa	
3.4.21.38	Hagemann factor coagulation factor XIIa	
4.3.1.3	HAL histidine ammonia-lyase	
4.3.1.3	L-HAL histidine ammonia-lyase	
3.8.1.3	Haloacetate dehalogenase Haloacetate dehalogenase	
3.8.1.3	Haloacetate halidohydrolase Haloacetate dehalogenase	
3.8.1.3	Haloacetate halidohydrolase Haloacetate dehalogenase	
3.8.1.2	2-Haloacid dehalogenase 2-Haloacid dehalogenase	
3.8.1.2	L-2-Haloacid dehalogenase: 2-Haloacid dehalogenase	
3.8.1.2	2-Haloacid halidohydrolase 2-Haloacid dehalogenase	
3.8.1.5	1-Haloalkane dehalogenase Haloalkane dehalogenase	
3.8.1.1	Haloalkane dehalogenase Alkylhalidase	
3.8.1.5	Haloalkane dehalogenase Haloalkane dehalogenase	
3.8.1.5	1-Haloalkane halidohydrolase Haloalkane dehalogenase	

EC-No.	Synonym Recommended Name	Vol., Page
3.8.1.1	Haloalkane halidohydrolase Alkylhalidase	
3.8.1.2	2-Haloalkanoic acid dehalogenase 2-Haloacid dehalogenase	
3.8.1.2	2-Haloalkanoid acid halidohydrolase 2-Haloacid dehalogenase	
6.2.1.33	4-Halobenzoate-coenzyme A ligase 4-Chlorobenzoate-CoA ligase	Vol. 2, Page 339
1.14.12.13	2-Halobenzoate 1,2-dioxygenase 2-Chlorobenzoate 1,2-dioxygenase	
3.8.1.1	Halogenase Alkylhalidase	
2.1.1.136	halogenated phenol O-methyltransferase chlorophenol O-methyltransferase	
2.6.1.24	Halogenated tyrosine aminotransferase Diiodotyrosine transaminase	
2.6.1.24	Halogenated tyrosine transaminase Diiodotyrosine transaminase	
2.7.1.102	Hamamelose kinase Hamamelose kinase	
2.7.1.102	Hamamelosekinase Hamamelose kinase	
1.13.11.6	3HAO 3-Hydroxyanthranilate 3,4-dioxygenase	
1.7.3.4	HAO Hydroxylamine oxidase	
3.4.24.12	hatching enzyme envelysin	
3.1.1.22	3HB-oligomer hydrolase hydroxybutyrate-dimer hydrolase	
1.14.99.23	3-HBA-2-hydroxylase 3-Hydroxybenzoate 2-monooxygenase	
2.4.1.194	HBA glucosyltransferase 4-Hydroxybenzoate 4-O-β-D-glucosyltransferase	
4.1.2.39	HbHNL Hydroxynitrilase	

EC-No.	Synonym Recommended Name	Vol., Page
2.3.1.152	HCA-GT Alcohol O-cinnamoyltransferase	
6.3.2.11	HCarn-Carn synthetase Carnosine synthase	Vol. 2, Page 460
3.4.24.12	HCE envelysin	
3.4.24.12	HCE envelysin	
3.6.1.3	HCO_3^--ATPase Adenosinetriphosphatase	
3.1.3.48	HCP protein-tyrosine-phosphatase	
6.3.4.15	HCS Biotin-[acetyl-CoA-carboxylase] ligase	Vol. 2, Page 638
6.3.4.10	HCS Biotin-[propionyl-CoA-carboxylase (ATP-hydrolysing)] ligase	Vol. 2, Page 617
3.4.21.98	HCV NS3 protease hepacivirin	
4.1.1.22	HDC Histidine decarboxylase	
1.3.1.51	HDR 2'-Hydroxydaidzein reductase	
3.4.24.12	HE envelysin	
3.4.21.47	heat-labile factor alternative-complement-pathway C3/C5 convertase	
3.1.4.3	Heat-labile hemolysin Phospholipase C	
2.5.1.9	Heavy riboflavin synthase Riboflavin synthase	
3.1.26.1	HeLa cell RNase Physarum polycephalum ribonuclease	
1.14.99.3	Heme,hydrogen-donor:oxygen oxidoreductase (α-methene-oxidizing, hydroxylating) Heme oxygenase (decyclizing)	

EC-No.	Synonym Recommended Name	Vol., Page
1.14.99.3	Heme oxygenase (decyclizing) Heme oxygenase (decyclizing)	
1.11.1.7	Heme peroxidase Peroxidase	
4.99.1.1	heme synthase ferrochelatase	
4.99.1.1	heme synthetase ferrochelatase	
3.1.1.73	hemicellulase accessory enzymes feruloyl esterase	
1.15.1.1	Hemocuprein Superoxide dismutase	
3.4.24.52	Hemorrhagic metalloproteinase HR1A Trimerelysin I	
3.4.24.52	Hemorrhagic proteinase HR1A Trimerelysin I	
3.4.24.47	Hemorrhagic proteinase IV Horrilysin	
3.4.24.1	hemorrhagic toxin a atrolysin A	
3.4.24.41	Hemorrhagic toxin b Atrolysin B	
3.4.24.42	Hemorrhagic toxin c and d Atrolysin C	
3.4.24.44	Hemorrhagic toxin e Atrolysin E	
3.4.24.45	Hemorrhagic toxin f Atrolysin F	
3.4.24.48	Hemorrhagic toxin II Ruberlysin	
3.4.21.98	hepacivirin hepacivirin	
2.3.1.78	Heparan-α-glucosaminide N-acetyltransferase Heparan-α-glucosaminide N-acetyltransferase	
2.8.2.12	Heparan sulfate 2-N-sulfotransferase Heparitin sulfotransferase	

EC-No.	Synonym Recommended Name	Vol., Page
4.2.2.8	heparan sulfate lyase heparin-sulfate lyase	
2.8.2.12	Heparan sulfate N-deacetylase/N-sulfotransferase Heparitin sulfotransferase	
2.8.2.12	Heparan sulfate N-sulfotransferase Heparitin sulfotransferase	
2.8.2.12	Heparan sulfate sulfotransferase Heparitin sulfotransferase	
2.8.2.12	N-Heparan sulfate sulfotransferase Heparitin sulfotransferase	
2.8.2.23	Heparin-glucosamine 3-O-sulfotransferase Heparin-glucosamine 3-O-sulfotransferase	
4.2.2.8	heparin-sulfate eliminase heparin-sulfate lyase	
4.2.2.8	heparin-sulfate lyase heparin-sulfate lyase	
4.2.2.7	Heparinase Heparin lyase	
3.2.1.19	heparinase heparinase	
4.2.2.7	Heparin eliminase Heparin lyase	
4.2.2.7	Heparin lyase Heparin lyase	
4.2.2.7	Heparin lyase Heparin lyase	
4.2.2.8	heparin lyase III heparin-sulfate lyase	
2.8.2.8	Heparin N-sulfotransferase Desulfoheparin sulfotransferase	
3.1.1.3	heparin releasable hepatic lipase triacylglycerol lipase	
3.10.1.1	Heparin sulfamidase N-Sulfoglucosamine sulfohydrolase	
4.2.2.8	heparin sulfate eliminase heparin-sulfate lyase	

EC-No.	Synonym Recommended Name	Vol., Page
4.2.2.8	heparitin-sulfate lyase heparin-sulfate lyase	
4.2.2.8	heparitinase heparin-sulfate lyase	
4.2.2.8	heparitinase I heparin-sulfate lyase	
4.2.2.8	heparitinase II heparin-sulfate lyase	
4.2.2.8	heparitin sulfate lyase heparin-sulfate lyase	
2.8.2.12	Heparitin sulfotransferase Heparitin sulfotransferase	
5.1.3.17	Heparosan-N-sulfate-D-glucuronosyl-5-epimerase Heparosan-N-sulfate-glucuronate 5-epimerase	Vol. 1, Page 167
5.1.3.17	Heparosan-N-sulfate-glucuronate 5-epimerase Heparosan-N-sulfate-glucuronate 5-epimerase	Vol. 1, Page 167
5.1.3.17	Heparosan N-sulfate D-glucuronosyl 5-epimerase Heparosan-N-sulfate-glucuronate 5-epimerase	Vol. 1, Page 167
3.1.1.3	hepatic lipase triacylglycerol lipase	
3.1.1.3	hepatic monoacylglycerol acyltransferase triacylglycerol lipase	
3.4.22.28	Hepatitis A virus 3C proteinase Picornain 3C	
3.4.21.98	hepatitis C virus NS3 serine protease hepacivirin	
1.15.1.1	Hepatocuprein Superoxide dismutase	
3.3.2.7	Hepoxilin-epoxide hydrolase Hepoxilin-epoxide hydrolase	
3.3.2.7	Hepoxilin epoxide hydrolase Hepoxilin-epoxide hydrolase	
2.5.1.30	Heptaprenyl diphosphate synthase trans-Hexaprenyltranstransferase	
2.5.1.30	Heptaprenyl pyrophosphate synthetase trans-Hexaprenyltranstransferase	

EC-No.	Synonym Recommended Name	Vol., Page
2.7.1.14	Heptulokinase Sedoheptulokinase	
3.1.1.72	Heroin esterase Acetylxylan esterase	
3.4.21.97	Herpes simplex virus 1 proteinase Pra assemblin	
3.4.21.97	Herpes simplex virus endopeptidase assemblin	
2.4.1.185	Hesperetin 7-O-glucosyl-transferase Flavanone 7-O-β-glucosyltransferase	
1.12.99.2	Heterodisulfide reductase Coenzyme-M-7-mercaptoheptanoylthreonine-phosphate-heterodisulfide hydrogenase	
2.4.1.48	Heteroglycan α-mannosyltransferase Heteroglycan α-mannosyltransferase	
1.2.1.42	Hexadecanal:NAD^+ oxidoreductase (CoA-acylating) Hexadecanal dehydrogenase (acylating)	
1.3.1.27	Hexadecanal: $NADP^+$ Δ^2-oxidoreductase 2-Hexadecenal reductase	
1.3.1.27	Hexadecanal: $NADP^+$ oxidoreductase 2-Hexadecenal reductase	
1.2.1.42	Hexadecanal dehydrogenase (acylating) Hexadecanal dehydrogenase (acylating)	
1.11.1.3	Hexadecanoate:hydrogen-peroxide oxidoreductase Fatty-acid peroxidase	
1.1.1.164	Hexadecanol:NAD^+ oxidoreductase Hexadecanol dehydrogenase	
1.1.1.164	Hexadecanol dehydrogenase Hexadecanol dehydrogenase	
1.3.1.27	2-Hexadecenal reductase 2-Hexadecenal reductase	
2.3.1.125	1-Hexadecyl-2-acetylglycerol acyltransferase 1-Alkyl-2-acetylglycerol O-acyltransferase	
2.1.1.114	Hexaprenyldihydroxybenzoate methyltransferase Hexaprenyldihydroxybenzoate methyltransferase	

EC-No.	Synonym Recommended Name	Vol., Page
2.5.1.33	Hexaprenyl diphosphate synthase trans-Pentaprenyltranstransferase	
2.5.1.33	Hexaprenyl pyrophosphate synthetase trans-Pentaprenyltranstransferase	
2.5.1.33	Hexaprenylpyrophosphate synthetase trans-Pentaprenyltranstransferase	
2.7.1.1	Hexokinase Hexokinase	
2.7.1.1	Hexokinase D Hexokinase	
2.7.1.1	Hexokinase type IV Hexokinase	
1.1.1.19	L-Hexonate:NADP dehydrogenase Glucuronate reductase	
1.1.99.13	Hexopyranoside-cytochrome c oxidoreductase Glucoside 3-dehydrogenase	
3.2.1.52	β-D-hexosaminidase β-N-acetylhexosaminidase	
3.2.1.52	β-hexosaminidase β-N-acetylhexosaminidase	
3.2.1.52	Hexosaminidase β-N-acetylhexosaminidase	
3.2.1.52	Hexosaminidase A β-N-acetylhexosaminidase	
3.1.3.10	hexose-1-phosphatase glucose-1-phosphatase	
2.7.7.29	Hexose-1-phosphate guanylyltransferase Hexose-1-phosphate guanylyltransferase	
2.7.7.12	Hexose-1-phosphate uridylyltransferase UDPglucose-hexose-1-phosphate uridylyltransferase	
2.7.7.29	Hexose 1-phosphate guanylyltransferase Hexose-1-phosphate guanylyltransferase	
2.7.7.28	Hexose 1-phosphate nucleotidyltransferase Nucleoside-triphosphate-hexose-1-phosphate nucleotidyltransferase	
2.7.7.12	Hexose 1-phosphate uridyltransferase UDPglucose-hexose-1-phosphate uridylyltransferase	

EC-No.	Synonym Recommended Name	Vol., Page
2.7.7.12	Hexose 1-phosphate uridylyltransferase UDPglucose-hexose-1-phosphate uridylyltransferase	
5.3.1.9	Hexose 6-phosphate isomerase Glucose-6-phosphate isomerase	Vol. 1, Page 298
1.1.3.5	D-Hexose:oxygen 1-oxidoreductase Hexose oxidase	
3.1.3.11	Hexose bisphosphatase Fructose-bisphosphatase	
3.1.3.11	Hexose diphosphatase Fructose-bisphosphatase	
3.1.3.11	Hexosediphosphatase Fructose-bisphosphatase	
5.3.1.9	Hexose isomerase Glucose-6-phosphate isomerase	Vol. 1, Page 298
5.3.1.9	Hexose monophosphate isomerase Glucose-6-phosphate isomerase	Vol. 1, Page 298
2.7.7.28	Hexose nucleotidylating enzyme Nucleoside-triphosphate-hexose-1-phosphate nucleotidyltransferase	
1.1.3.5	Hexose oxidase Hexose oxidase	
2.7.1.61	Hexose phosphate:hexose phosphotransferase Acyl-phosphate-hexose phosphotransferase	
2.6.1.16	Hexosephosphate aminotransferase Glutamine-fructose-6-phosphate transaminase (isomerizing)	
1.1.1.47	Hexose phosphate dehydrogenase Glucose 1-dehydrogenase	
5.3.1.9	Hexose phosphate isomerase Glucose-6-phosphate isomerase	Vol. 1, Page 298
5.3.1.9	Hexosephosphate isomerase Glucose-6-phosphate isomerase	Vol. 1, Page 298
1.1.1.17	Hexose reductase Mannitol-1-phosphate 5-dehydrogenase	
3.4.24.12	HEz envelysin	
3.4.24.73	HF2-proteinase Jararhagin	

EC-No.	Synonym Recommended Name	Vol., Page
1.16.1.1	Hg:NADP$^+$ oxidoreductase Mercury(II) reductase	
2.6.1.23	HGA 4-Hydroxyglutamate transaminase	
4.3.2.5	HGAD Peptidylamidoglycolate lyase	
2.4.2.8	HGPRTase Hypoxanthine phosphoribosyltransferase	
4.1.1.68	HHDD isomerase/OPET decarboxylase 5-Oxopent-3-ene-1,2,5-tricarboxylate decarboxylase	
3.4.21.97	HHV-6 proteinase assemblin	
3.1.2.4	HIB CoA hydrolase 3-hydroxyisobutyryl-CoA hydrolase	
3.1.2.4	HIBYL-CoA hydrolase 3-hydroxyisobutyryl-CoA hydrolase	
1.1.1.2	High-K_m aldehyde reductase Alcohol dehydrogenase (NADP$^+$)	
2.4.1.197	High-mannose-oligosaccharide β-1,4-N-acetyl-glucosaminyltransferase High-mannose-oligosaccharide β-1,4-N-acetyl-glucosaminyltransferase	
5.3.2.2	High-molecular mass oxalacetate tautomerase Oxalacetate tautomerase	Vol. 1, Page 371
3.4.24.67	High choriolytic enzyme Choriolysin H	
3.4.24.12	high choriolytic enzyme envelysin	
3.4.22.3	higueroxyl Delabarre ficain	
2.1.1.4	HIOMT Acetylserotonin N-methyltransferase	
3.5.1.14	Hippurase Aminoacylase	
3.5.1.32	Hippurate hydrolase Hippurate hydrolase	
3.5.1.14	Hippuricase Aminoacylase	

EC-No.	Synonym Recommended Name	Vol., Page
3.4.17.3	hippuryllysine hydrolase lysine carboxypeptidase	
4.1.1.22	HisDCase Histidine decarboxylase	
6.1.1.21	HisRS Histidine-tRNA ligase	Vol. 2, Page 168
1.4.3.6	Histaminase Amine oxidase (copper-containing)	
2.1.1.8	Histamine-methylating enzyme Histamine N-methyltransferase	
2.1.1.8	Histamine 1-methyltransferase Histamine N-methyltransferase	
1.4.3.6	Histamine deaminase Amine oxidase (copper-containing)	
2.1.1.8	Histamine methyltransferase Histamine N-methyltransferase	
2.1.1.8	Histamine N-methyltransferase Histamine N-methyltransferase	
1.4.3.6	Histamine oxidase Amine oxidase (copper-containing)	
4.3.1.3	L-histidase histidine ammonia-lyase	
4.3.1.3	histidase histidine ammonia-lyase	
4.3.1.3	histidase, Hut histidine ammonia-lyase	
4.3.1.3	histidinase histidine ammonia-lyase	
2.1.1.44	Histidine-α-N-methyltransferase Dimethylhistidine N-methyltransferase	
2.6.1.38	Histidine-2-oxoglutarate aminotransferase Histidine transaminase	
4.3.1.3	histidine α-deaminase histidine ammonia-lyase	
6.1.1.21	Histidine-tRNA ligase Histidine-tRNA ligase	Vol. 2, Page 168

EC-No.	Synonym Recommended Name	Vol., Page
6.3.2.11	L-Histidine:β-alanine ligase (AMP-forming) Carnosine synthase	Vol. 2, Page 460
2.6.1.38	L-Histidine:2-oxoglutarate aminotransferase Histidine transaminase	
2.6.1.9	Histidine:imidazoleacetol phosphate transaminase Histidinol-phosphate transaminase	
2.6.1.58	Histidine:pyruvate aminotransferase Phenylalanine(histidine) transaminase	
6.1.1.21	L-Histidine:$tRNA^{His}$ ligase (AMP-forming) Histidine-tRNA ligase	Vol. 2, Page 168
2.3.1.33	Histidine acetyltransferase Histidine N-acetyltransferase	
2.6.1.38	Histidine aminotransferase Histidine transaminase	
4.3.1.3	L-histidine ammonia-lyase histidine ammonia-lyase	
4.3.1.3	histidine ammonia-lyase histidine ammonia-lyase	
4.1.1.22	L-Histidine carboxy-lyase Histidine decarboxylase	
4.3.1.3	histidine deaminase histidine ammonia-lyase	
4.1.1.22	Histidine decarboxylase Histidine decarboxylase	
4.1.1.22	L-Histidine decarboxylase Histidine decarboxylase	
2.7.3.11	Histidine kinase Protein-histidine pros-kinase	
2.7.3.12	Histidine kinase Protein-histidine tele-kinase	
2.3.1.33	Histidine N-acetyltransferase Histidine N-acetyltransferase	
4.3.1.3	L-histidine NH_3-lyase histidine ammonia-lyase	
2.7.3.11	Histidine protein kinase Protein-histidine pros-kinase	

EC-No.	Synonym Recommended Name	Vol., Page
2.7.3.12	Histidine protein kinase Protein-histidine tele-kinase	
2.6.1.38	Histidine transaminase Histidine transaminase	
6.1.1.21	Histidine translase Histidine-tRNA ligase	Vol. 2, Page 168
3.1.3.15	Histidinol-phosphatase Histidinol-phosphatase	
2.6.1.9	L-Histidinol-phosphate:2-oxoglutarate aminotransferase Histidinol-phosphate transaminase	
2.6.1.9	Histidinol-phosphate aminotransferase Histidinol-phosphate transaminase	
3.1.3.15	L-Histidinol-phosphate phosphohydrolase Histidinol-phosphatase	
2.6.1.9	Histidinol-phosphate transaminase Histidinol-phosphate transaminase	
1.1.1.23	L-Histidinol:NAD^+ oxidoreductase Histidinol dehydrogenase	
1.1.1.23	Histidinol dehydrogenase Histidinol dehydrogenase	
1.1.1.23	L-Histidinol dehydrogenase Histidinol dehydrogenase	
3.1.3.15	Histidinol phosphatase Histidinol-phosphatase	
3.1.3.15	Histidinolphosphatase Histidinol-phosphatase	
2.6.1.9	Histidinol phosphate aminotransferase Histidinol-phosphate transaminase	
2.6.1.9	L-Histidinol phosphate aminotransferase Histidinol-phosphate transaminase	
3.1.3.15	Histidinol phosphate phosphatase Histidinol-phosphatase	
3.1.3.15	Histidinolphosphate phosphatase Histidinol-phosphatase	
3.1.3.15	L-Histidinol phosphate phosphatase Histidinol-phosphatase	

EC-No.	Synonym Recommended Name	Vol., Page
6.1.1.21	Histidyl-transfer ribonucleate synthetase Histidine-tRNA ligase	Vol. 2, Page 168
6.1.1.21	Histidyl-tRNA synthetase Histidine-tRNA ligase	Vol. 2, Page 168
2.4.1.40	Histo-blood group A acetylgalactosaminyltransferase Fucosylgalactose α-N-acetylgalactosaminyltransferase	
2.4.1.40	Histo-blood group A glycosyltransferase (Fucα1→2Galα1→3-N-acetylgalactosaminyltransferase) Fucosylgalactose α-N-acetylgalactosaminyltransferase	
2.4.1.40	Histo-blood group A transferase Fucosylgalactose α-N-acetylgalactosaminyltransferase	
2.4.1.37	Histo-blood group B transferase Fucosylglycoprotein 3-α-galactosyltransferase	
2.4.1.37	Histo-blood substance B-dependent galactosyltransferase Fucosylglycoprotein 3-α-galactosyltransferase	
3.4.22.35	Histolysain Histolysain	
3.4.22.35	Histolysin Histolysain	
2.1.1.125	(Histone)-arginine N-methyltransferase (Histone)-arginine N-methyltransferase	
2.1.1.43	Histone-lysine N-methyltransferase Histone-lysine N-methyltransferase	
2.3.1.48	Histone acetokinase Histone acetyltransferase	
2.3.1.48	Histone acetylase Histone acetyltransferase	
2.3.1.48	Histone acetyltransferase Histone acetyltransferase	
2.1.1.43	Histone H1-specific S-adenosylmethionine:protein-lysine N-methyltransferase Histone-lysine N-methyltransferase	
2.7.1.70	Histone kinase Protamine kinase	
2.1.1.125	Histone protein methylase I (Histone)-arginine N-methyltransferase	

EC-No.	Synonym Recommended Name	Vol., Page
2.3.1.48	Histone transacetylase Histone acetyltransferase	
3.5.1.14	Histozyme Aminoacylase	
3.4.23.16	HIV-1 protease Retropepsin	
3.4.23.16	HIV-2 protease Retropepsin	
3.4.23.16	HIV aspartyl protease Retropepsin	
3.4.23.16	HIV proteinase Retropepsin	
3.4.21.35	hK2 tissue kallikrein	
4.3.1.8	(HMB)-synthase hydroxymethylbilane synthase	
4.3.1.8	HMB-S hydroxymethylbilane synthase	
4.3.1.8	HMBS hydroxymethylbilane synthase	
3.4.21.97	hMCV assemblin	
1.12.99.4	Hmd N^5,N^{10}-Methenyltetrahydromethanopterin hydrogenase	
4.2.1.18	HMG-CoA hydrolyase methylglutaconyl-CoA hydratase	
3.1.3.16	HMG-CoA reductase phosphatase Phosphoprotein phosphatase	
4.1.3.5	HMG-CoA synthase Hydroxymethylglutaryl-CoA synthase	
4.1.3.4	HMG CoA cleavage enzyme Hydroxymethylglutaryl-CoA lyase	
1.1.1.34	HMGCoA reductase-mevalonate:NADP-oxidoreductase (acetylating-CoA) Hydroxymethylglutaryl-CoA reductase (NADPH)	
3.7.1.9	HMSH 2-Hydroxymuconate-semialdehyde hydrolase	

EC-No.	Synonym Recommended Name	Vol., Page
1.3.7.1	HNA reductase 6-Hydroxynicotinate reductase	
4.1.2.11	HNL Hydroxymandelonitrile lyase	
4.1.2.39	HNL Hydroxynitrilase	
4.1.2.39	Hnl Hydroxynitrilase	
1.1.1.252	3HNR Tetrahydroxynaphthalene reductase	
3.6.1.7	Ho 1-3 Acylphosphatase	
3.7.1.9	HOD hydrolase 2-Hydroxymuconate-semialdehyde hydrolase	
3.7.1.8	HOHPDA hydrolase 2,6-Dioxo-6-phenylhexa-3-enoate hydrolase	
2.7.8.7	Holo-[acyl-carrier-protein] synthase Holo-[acyl-carrier-protein] synthase	
2.7.8.7	Holo-ACP synthase Holo-[acyl-carrier-protein] synthase	
2.7.8.7	Holo-ACP synthetase Holo-[acyl-carrier-protein] synthase	
6.3.4.10	Holocarboxylase synthetase Biotin-[propionyl-CoA-carboxylase (ATP-hydrolysing)] ligase	Vol. 2, Page 617
4.4.1.17	Holocytochrome-c apocytochrome-c-lyase Holocytochrome-c synthase	
4.4.1.17	Holocytochrome-c synthase Holocytochrome-c synthase	
4.4.1.17	Holocytochrome c synthetase Holocytochrome-c synthase	
4.2.1.36	Homoaconitase Homoaconitate hydratase	
4.2.1.36	Homoaconitate hydratase Homoaconitate hydratase	

EC-No.	Synonym Recommended Name	Vol., Page
3.4.13.13	homocarnosinase homocarnosinase	
6.3.2.11	Homocarnosine-carnosine synthetase Carnosine synthase	Vol. 2, Page 460
4.1.3.21	Homocitrate-condensing enzyme Homocitrate synthase	
4.1.3.21	Homocitrate synthase Homocitrate synthase	
4.1.3.21	Homocitrate synthetase Homocitrate synthase	
4.1.3.21	Homocondensing enzyme Homocitrate synthase	
4.4.1.2	homocysteine desulfhydrase homocysteine desulfhydrase	
4.4.1.2	homocysteine desulfurase homocysteine desulfhydrase	
4.4.1.2	L-homocysteine hydrogen-sulfide-lyase (deaminating) homocysteine desulfhydrase	
2.1.1.14	Homocysteine methylase 5-Methyltetrahydropteroyltriglutamate-homocysteine S-methyltransferase	
2.1.1.10	Homocysteine methylase Homocysteine S-methyltransferase	
2.1.1.10	Homocysteine methyltransferase Homocysteine S-methyltransferase	
2.1.1.10	Homocysteine S-methyltransferase Homocysteine S-methyltransferase	
2.1.1.10	L-Homocysteine S-methyltransferase Homocysteine S-methyltransferase	
2.1.1.10	Homocysteine transmethylase Homocysteine S-methyltransferase	
1.13.11.5	Homogentisate 1,2-dioxygenase Homogentisate 1,2-dioxygenase	
1.13.11.5	Homogentisate:oxygen 1,2-oxidoreductase (decyclizing) Homogentisate 1,2-dioxygenase	
1.13.11.5	Homogentisate dioxygenase Homogentisate 1,2-dioxygenase	

EC-No.	Synonym Recommended Name	Vol., Page
1.13.11.5	Homogentisate oxidase Homogentisate 1,2-dioxygenase	
1.13.11.5	Homogentisate oxygenase Homogentisate 1,2-dioxygenase	
1.13.11.5	Homogentisic acid oxidase Homogentisate 1,2-dioxygenase	
1.13.11.5	Homogentisic acid oxygenase Homogentisate 1,2-dioxygenase	
1.13.11.5	Homogentisicase Homogentisate 1,2-dioxygenase	
1.13.11.5	Homogentisic oxygenase Homogentisate 1,2-dioxygenase	
6.3.2.23	Homoglutathione synthase Homoglutathione synthase	Vol. 2, Page 518
6.3.2.23	Homoglutathione synthetase Homoglutathione synthase	Vol. 2, Page 518
1.1.1.155	Homoisocitrate dehydrogenase Homoisocitrate dehydrogenase	
1.1.1.155	Homoisocitric dehydrogenase Homoisocitrate dehydrogenase	
1.13.11.15	Homoprotocatechuate 2,3-dioxygenase 3,4-Dihydroxyphenylacetate 2,3-dioxygenase	
2.3.1.31	Homoserine-O-transacetylase Homoserine O-acetyltransferase	
1.1.1.3	L-Homoserine:NAD(P)$^+$ oxidoreductase Homoserine dehydrogenase	
2.3.1.31	Homoserine acetyltransferase Homoserine O-acetyltransferase	
4.4.1.1	homoserine deaminase cystathionine γ-lyase	
4.4.1.1	homoserine deaminase-cystathionase cystathionine γ-lyase	
4.4.1.1	homoserine dehydratase cystathionine γ-lyase	
4.2.1.15	homoserine dehydratase homoserine dehydratase	

EC-No.	Synonym Recommended Name	Vol., Page
1.1.1.3	Homoserine dehydrogenase Homoserine dehydrogenase	
2.7.1.39	Homoserine kinase Homoserine kinase	
2.3.1.31	Homoserine O-acetyltransferase Homoserine O-acetyltransferase	
2.3.1.31	L-Homoserine O-acetyltransferase Homoserine O-acetyltransferase	
2.3.1.46	Homoserine O-succinyltransferase Homoserine O-succinyltransferase	
2.3.1.46	Homoserine O-transsuccinylase Homoserine O-succinyltransferase	
4.2.99.9	homoserine O-transsuccinylase O-succinylhomoserine (thiol)-lyase	
2.3.1.46	Homoserine succinyltransferase Homoserine O-succinyltransferase	
2.3.1.31	Homoserine transacetylase Homoserine O-acetyltransferase	
4.2.99.9	homoserine transsuccinylase O-succinylhomoserine (thiol)-lyase	
2.5.1.44	Homospermidine synthase Homospermidine synthase	
3.4.24.47	Horrilysin Horrilysin	
1.11.1.7	Horseradish peroxidase Peroxidase	
3.4.21.89	HOSP Signal peptidase I	
1.14.13.18	4-HPA 1-hydroxylase 4-Hydroxyphenylacetate 1-monooxygenase	
1.14.13.3	4 HPA 3-hydroxylase 4-Hydroxyphenylacetate 3-monooxygenase	
2.1.1.37	HpaII' methylase DNA (cytosine-5-)-methyltransferase	
2.1.1.37	HpaII methylase DNA (cytosine-5-)-methyltransferase	

EC-No.	Synonym Recommended Name	Vol., Page
1.2.1.53	4-HPAL dehydrogenase 4-Hydroxyphenylacetaldehyde dehydrogenase	
3.5.4.22	HPC deaminase 1-Pyrroline-4-hydroxy-2-carboxylate deaminase	
1.13.11.15	HPC dioxygenase 3,4-Dihydroxyphenylacetate 2,3-dioxygenase	
4.2.1.92	HPI Hydroperoxide dehydratase	
3.1.3.15	HPpase Histidinol-phosphatase	
2.7.6.3	HPPK 2-Amino-4-hydroxy-6-hydroxymethyldihydropteridine pyrophosphokinase	
2.7.1.37	Hpr kinase Protein kinase	
1.1.1.237	HPRP Hydroxyphenylpyruvate reductase	
2.4.2.8	HPRT Hypoxanthine phosphoribosyltransferase	
3.1.3.48	HPTPβ protein-tyrosine-phosphatase	
1.1.1.62	17β-HSD Estradiol 17β-dehydrogenase	
1.1.1.63	17β-HSD Testosterone 17β-dehydrogenase	
1.1.1.149	20α-HSD 20α-Hydroxysteroid dehydrogenase	
1.1.1.53	20β-HSD 3α(or 20β)-Hydroxysteroid dehydrogenase	
1.1.1.239	3α(17β)-HSD 3α(17β)-Hydroxysteroid dehydrogenase (NAD^+)	
1.1.1.3	HSD Homoserine dehydrogenase	
1.1.1.149	20α-HSDH 20α-Hydroxysteroid dehydrogenase	

EC-No.	Synonym Recommended Name Vol., Page
1.1.1.145	3β-HSDH 3β-Hydroxy-Δ^5-steroid dehydrogenase
1.1.1.159	7α-HSDH 7α-Hydroxysteroid dehydrogenase
1.1.1.3	HSDH Homoserine dehydrogenase
2.7.1.39	HSK Homoserine kinase
2.5.1.44	HSPD synthase Homospermidine synthase
2.5.1.44	HSS Homospermidine synthase
2.8.2.12	N-HSST Heparitin sulfotransferase
2.8.2.2	HST Alcohol sulfotransferase
3.4.21.97	HSV-1 assemblin
3.4.21.97	HSV-2 assemblin
3.4.21.97	HSV 1 protease assemblin
3.4.24.48	HT-2 Ruberlysin
3.4.21.97	human cytomegalovirus maturational proteinase assemblin
3.4.21.97	human cytomegalovirus protease assemblin
3.4.21.97	human cytomegalovirus proteinase assemblin
3.4.13.18	Human cytosolic non-specific dipeptidase Cytosol nonspecific dipeptidase
3.4.21.77	Human glandular kallikrein Semenogelase
3.4.24.65	Human macrophage metalloelastase Macrophage elastase

EC-No.	Synonym Recommended Name	Vol., Page
3.4.22.38	Human osteoclast cathepsin K Cathepsin K	
3.4.99.9	hurain hurain	
3.4.21.78	HuTPS Granzyme A	
3.4.23.40	HvAP Phytepsin	
3.2.1.36	Hyaluronate 3-glycanohydrolase Hyaluronoglucuronidase	
3.2.1.35	Hyaluronate 4-glucanohydrolase Hyaluronoglucosaminidase	
4.2.2.1	hyaluronate lyase hyaluronate lyase	
4.2.2.1	hyaluronate lyase hyaluronate lyase	
4.2.99.1	hyaluronate lyase hyaluronate lyase	
3.2.1.35	Hyaluronidase Hyaluronoglucosaminidase	
3.2.1.36	Hyaluronidase Hyaluronoglucuronidase	
4.2.2.1	hyaluronidase hyaluronate lyase	
3.2.1.35	Hyaluronoglucosaminidase Hyaluronoglucosaminidase	
3.2.1.35	Hyaluronoglucosidase Hyaluronoglucosaminidase	
3.2.1.36	Hyaluronoglucuronidase Hyaluronoglucuronidase	
3.2.1.36	Hyaluronoglucuronidase Hyaluronoglucuronidase	
3.1.26.4	Hybridase Calf thymus ribonuclease H	
3.1.26.4	Hybridase (ribonuclease H) Calf thymus ribonuclease H	

EC-No.	Synonym Recommended Name	Vol., Page
3.1.26.4	Hybrid nuclease Calf thymus ribonuclease H	
3.1.4.34	hybrid nuclease hybrid nuclease	
3.1.26.4	Hybrid ribonuclease Calf thymus ribonuclease H	
3.5.2.2	D-Hydantoinase Dihydropyrimidinase	
3.5.2.2	Hydantoinase Dihydropyrimidinase	
3.5.2.4	Hydantoin hydrolase Carboxymethylhydantoinase	
3.5.2.2	Hydantoin peptidase Dihydropyrimidinase	
4.2.1.65	Hydratase, β-cyanoalanine 3-Cyanoalanine hydratase	
4.2.1.72	Hydratase, acetylenedicarboxylate Acetylenedicarboxylate hydratase	
4.2.1.71	Hydratase, acetylenemonocarboxylate Acetylenecarboxylate hydratase	
4.2.1.3	hydratase, aconitate aconitate hydratase	
3.3.2.3	Hydratase, arene oxide Epoxide hydrolase	
4.2.1.35	Hydratase, citraconate (R)-2-Methylmalate dehydratase	
4.2.1.58	Hydratase, crotonoyl-[acyl carrier protein] Crotonoyl-[acyl-carrier-protein] hydratase	
4.2.1.69	hydratase, cyanamide cyanamide hydratase	
4.2.1.66	hydratase, cyanide cyanide hydratase	
4.2.1.85	Hydratase, dimethylmaleate Dimethylmaleate hydratase	
4.2.1.17	hydratase, enoyl coenzyme A enoyl-CoA hydratase	

EC-No.	Synonym Recommended Name	Vol., Page
3.3.2.3	Hydratase, epoxide Epoxide hydrolase	
4.2.1.2	hydratase, fumarate fumarate hydratase	
4.2.1.36	Hydratase, homoaconitate Homoaconitate hydratase	
4.2.1.57	Hydratase, isohexenylglutaconyl coenzyme A Isohexenylglutaconyl-CoA hydratase	
4.2.1.56	Hydratase, itaconyl coenzyme A Itaconyl-CoA hydratase	
4.2.1.95	Hydratase, kievitone Kievitone hydratase	
4.2.1.74	hydratase, long-chain enoyl coenzyme A long-chain-enoyl-CoA hydratase	
4.2.1.34	Hydratase, mesaconate (S)-2-Methylmalate dehydratase	
4.2.1.18	hydratase, methylglutaconyl coenzyme A methylglutaconyl-CoA hydratase	
4.2.1.84	Hydratase, nitrile nitrile hydratase	
4.2.1.53	Hydratase, oleate Oleate hydratase	
4.2.1.83	Hydratase, oxalmesaconate 4-Oxalmesaconate hydratase	
4.2.1.97	Hydratase, phaseollidin Phaseollidin hydratase	
4.2.1.11	hydratase, phosphoenolpyruvate phosphoenolpyruvate hydratase	
3.3.2.4	Hydratase, trans-epoxysuccinate trans-epoxysuccinate hydrolase	
4.2.1.49	Hydratase, urocanate Urocanate hydratase	
1.3.1.55	2-Hydro-1,2-dihydroxybenzoate dehydrogenase cis-1,2-Dihydroxycyclohexa-3,5-diene-1-carboxylate dehydrogenase	
1.18.99.1	Hydrogen-lyase Hydrogenase	

EC-No.	Synonym Recommended Name	Vol., Page
1.11.1.6	Hydrogen-peroxide:hydrogen-peroxide oxidoreductase Catalase	
1.8.7.1	Hydrogen-sulfide:ferredoxin oxidoreductase Sulfite reductase (ferredoxin)	
1.8.1.2	Hydrogen-sulfide:NADP$^+$ oxidoreductase Sulfite reductase (NADPH)	
2.3.1.10	Hydrogen-sulfide S-acetyltransferase Hydrogen-sulfide S-acetyltransferase	
1.8.99.1	Hydrogen-sulfite:(acceptor) oxidoreductase Sulfite reductase	
1.12.99.1	Hydrogen:(acceptor) oxidoreductase Coenzyme F420 hydrogenase	
1.12.99.2	Hydrogen:coenzyme-M-7-mercaptoheptanoylthreonine-phosphate-hetero-disulfide oxidoreductase Coenzyme-M-7-mercaptoheptanoylthreonine-phosphate-heterodisulfide hydrogenase	
1.18.99.1	Hydrogen:ferredoxin oxidoreductase Hydrogenase	
1.12.2.1	Hydrogen:ferricytochrome-c_3 oxidoreductase Cytochrome-c_3 hydrogenase	
1.12.99.3	Hydrogen:menaquinone oxidoreductase Hydrogen:quinone oxidoreductase	
1.12.99.4	Hydrogen:N^5,N^{10}-methenyltetrahydromethanopterin oxidoreductase N^5,N^{10}-Methenyltetrahydromethanopterin hydrogenase	
1.12.1.2	Hydrogen:NAD$^+$ oxidoreductase Hydrogen dehydrogenase	
1.12.99.3	Hydrogen:quinone oxidoreductase Hydrogen:quinone oxidoreductase	
1.12.99.3	Hydrogen:quinone oxidoreductase Hydrogen:quinone oxidoreductase	
1.3.1.3	Δ^4-Hydrogenase Cortisone β-reductase	
1.12.2.1	Hydrogenase Cytochrome-c_3 hydrogenase	
1.12.1.2	Hydrogenase Hydrogen dehydrogenase	

EC-No.	Synonym Recommended Name	Vol., Page
1.18.99.1	Hydrogenase Hydrogenase	
1.18.3.1	hydrogenase hydrogenase	
1.18.99.1	Hydrogenase (ferredoxin) Hydrogenase	
1.12.99.3	hydrogenase (Wolinella succinogenes clone pHyd1 gene hydA subunit precursor reduced) Hydrogen:quinone oxidoreductase	
1.12.99.3	Hydrogenase (Wolinella succinogenes clone pHyd1 gene hydB subunit precursor reduced) Hydrogen:quinone oxidoreductase	
1.12.99.3	Hydrogenase (Wolinella succinogenes clone pHyd1 gene hydC subunit) Hydrogen:quinone oxidoreductase	
1.12.99.1	Hydrogenase, coenzyme F420 Coenzyme F420 hydrogenase	
1.12.2.1	Hydrogenase, cytochrome Cytochrome-c_3 hydrogenase	
1.1.1.222	D-Hydrogenase, D-aryllactate (R)-4-Hydroxyphenyllactate dehydrogenase	
1.18.99.1	Hydrogenase, ferredoxin Hydrogenase	
1.18.99.1	Hydrogenase I Hydrogenase	
1.18.99.1	Hydrogenase II Hydrogenase	
1.12.1.2	Hydrogen dehydrogenase Hydrogen dehydrogenase	
1.2.1.2	Hydrogenlyase Formate dehydrogenase	
1.18.99.1	Hydrogenlyase Hydrogenase	
5.4.1.2	Hydrogenobyrinic acid-binding protein Precorrin-8X methylmutase	Vol. 1, Page 490
2.3.1.10	Hydrogen sulfide acetyltransferase Hydrogen-sulfide S-acetyltransferase	

EC-No.	Synonym Recommended Name	Vol., Page
1.8.99.3	Hydrogensulfite reductase Hydrogensulfite reductase	
1.1.1.8	Hydroglycerophosphate dehydrogenase Glycerol-3-phosphate dehydrogenase (NAD^+)	
3.2.1.23	Hydrolact β-galactosidase	
4.2.1.65	Hydrolase, β-cyanoalanine 3-Cyanoalanine hydratase	
3.4.19.11	Hydrolase, γ-D-glutamyl-(L)-meso-diaminopimelate peptidoglycan γ-D-Glutamyl-meso-diaminopimelate peptidase I	
3.7.1.7	Hydrolase, β-diketone β-Diketone hydrolase	
3.4.19.9	Hydrolase, γ-glutamyl γ-Glu-X carboxypeptidase	
3.7.1.10	Hydrolase, 1,3-cyclohexanedione Cyclohexane-1,3-dione hydrolase	
3.7.1.8	Hydrolase, 2,6-dioxo-6-phenylhexa-3-enoate 2,6-Dioxo-6-phenylhexa-3-enoate hydrolase	
3.7.1.9	Hydrolase, 2-hydroxymuconate semialdehyde 2-Hydroxymuconate-semialdehyde hydrolase	
3.2.1.124	Hydrolase, 2-keto-3-deoxyoctonate 3-Deoxy-2-octulosonidase	
3.1.2.4	hydrolase, 3-hydroxyisobutyryl coenzyme A 3-hydroxyisobutyryl-CoA hydrolase	
3.1.2.23	Hydrolase, 4-hydroxybenzoyl coenzyme A 4-Hydroxybenzoyl-CoA thioesterase	
3.6.1.20	Hydrolase, 5-phosphoadenosine 5'-Acylphosphoadenosine hydrolase	
3.1.2.11	hydrolase, acetoacetyl coenzyme A acetoacetyl-CoA hydrolase	
3.1.2.6	Hydrolase, acetoacetylglutathione Hydroxyacylglutathione hydrolase	
3.1.2.1	hydrolase, acetyl coenzyme A acetyl-CoA hydrolase	
3.7.1.6	Hydrolase, acetylpyruvate Acetylpyruvate hydrolase	

EC-No.	Synonym Recommended Name	Vol., Page
3.1.1.28	hydrolase, acylcarnitine acylcarnitine hydrolase	
3.1.2.20	Hydrolase, acyl coenzyme A Acyl-CoA hydrolase	
3.1.2.22	Hydrolase, acyl protein thioester Palmitoyl-(protein) hydrolase	
3.1.2.22	Hydrolase, acyl protein thioester (cattle clone pBovPPT-18 precursor reduced) Palmitoyl-(protein) hydrolase	
3.1.2.22	Hydrolase, acyl protein thioester (cattle pBovPPT-25 precursor reduced) Palmitoyl-(protein) hydrolase	
3.1.2.22	Hydrolase, acyl protein thioester (human clone B lysosome-associated isoenzyme 2) Palmitoyl-(protein) hydrolase	
3.1.2.22	Hydrolase, acyl protein thioester (human clone pHuPPT-5') Palmitoyl-(protein) hydrolase	
3.1.2.22	Hydrolase, acyl protein thioester (ox clone pBovPPT-18 precursor reduced) Palmitoyl-(protein) hydrolase	
3.1.2.22	Hydrolase, acyl protein thioester (ox pBovPPT-25 precursor reduced) Palmitoyl-(protein) hydrolase	
3.1.2.22	Hydrolase, acyl protein thioester (rat clone pRatPPT-44 precursor reduced) Palmitoyl-(protein) hydrolase	
3.7.1.5	Hydrolase, acylpyruvate Acylpyruvate hydrolase	
3.5.1.54	Hydrolase, allophanate Allophanate hydrolase	
3.1.1.29	hydrolase, aminoacyl-transfer ribonucleate aminoacyl-tRNA hydrolase	
3.5.1.58	Hydrolase, benzyloxycarbonylglycine N-Benzyloxycarbonylglycine hydrolase	
3.5.1.64	Hydrolase, benzyloxycarbonylleucine N^{α}-Benzyloxycarbonylleucine hydrolase	
3.5.1.53	Hydrolase, carbamoylputrescine N-Carbamoylputrescine amidase	
3.5.1.74	Hydrolase, chenodeoxychoyltaurine Chenodeoxycholoyltaurine hydrolase	

EC-No.	Synonym Recommended Name	Vol., Page
3.1.1.42	hydrolase, chlorogenate chlorogenate hydrolase	
4.3.99.1	Hydrolase, cyanate Cyanate lyase	
3.5.2.13	Hydrolase, cyclo(glycylglycine) 2,5-Dioxopiperazine hydrolase	
3.1.4.40	Hydrolase, cytidine monophosphosialate CMP-N-acylneuraminate phosphodiesterase	
3.1.2.17	Hydrolase, D-methylmalonyl coenzyme A (S)-Methylmalonyl-CoA hydrolase	
3.4.14.6	hydrolase, dipeptidyl tetrapeptide dipeptidyl-dipeptidase	
3.1.2.10	hydrolase, formyl coenzyme A formyl-CoA hydrolase	
3.2.1.128	Hydrolase, glycyrrhizin β- Glycyrrhizinate β-glucuronidase	
3.5.4.16	Hydrolase, guanosine triphosphate cyclo- GTP cyclohydrolase I	
3.3.2.7	Hydrolase, hepoxilin epoxide Hepoxilin-epoxide hydrolase	
3.5.1.32	Hydrolase, hippurate Hippurate hydrolase	
3.1.2.6	Hydrolase, hydroxyacylglutathione Hydroxyacylglutathione hydrolase	
3.1.1.22	hydrolase, hydroxybutyrate dimer hydroxybutyrate-dimer hydrolase	
3.1.2.5	hydrolase, hydroxymethylglutaryl coenzyme A hydroxymethylglutaryl-CoA hydrolase	
3.2.1.134	Hydrolase, inulobiose Difructose-anhydride synthase	
3.1.2.21	Hydrolase, lauroyl-[acyl carrier protein] Dodecanoyl-[acyl-carrier protein] hydrolase	
3.3.2.6	Hydrolase, leukotriene A_4 Leukotriene-A_4hydrolase	
3.1.2.19	Hydrolase, medium-chain acyl coenzyme A ADP-dependent medium-chain-acyl-CoA hydrolase	

EC-No.	Synonym Recommended Name	Vol., Page
3.5.4.27	Hydrolase, methenyltetrahydromethanopterin cyclo- Methenyltetrahydromethanopterin cyclohydrolase	
3.5.1.71	Hydrolase, N-feruloylglycine N-Feruloylglycine deacylase	
3.5.1.68	Hydrolase, N-formylglutamate N-Formylglutamate deformylase	
3.6.1.24	Hydrolase, nucleoside phosphoacyl- Nucleoside phosphoacylhydrolase	
3.1.2.14	hydrolase, oleoyl-[acyl carrier protein] oleoyl-[acyl-carrier-protein] hydrolase	
3.2.1.120	Hydrolase, oligoxyloglucan Oligoxyloglucan β-glycosidase	
3.1.1.40	hydrolase, orsellinate depside orsellinate-depside hydrolase	
3.1.2.2	hydrolase, palmitoyl coenzyme A palmitoyl-CoA hydrolase	
3.7.1.4	Hydrolase, phloretin Phloretin hydrolase	
3.11.1.2	Hydrolase, phosphonoacetate Phosphonoacetate hydrolase	
3.11.1.1	Hydrolase, phosphonoacetylaldehyde Phosphonoacetylaldehyde hydrolase	
3.1.2.18	Hydrolase, propionyl coenzyme A ADP-dependent short-chain-acyl-CoA hydrolase	
3.2.1.118	Hydrolase, prunasin Prunasin β-glucosidase	
3.1.2.12	hydrolase, S-formylglutathione S-formylglutathione hydrolase	
3.1.2.13	hydrolase, S-succinylglutathione S-succinylglutathione hydrolase	
3.1.2.18	Hydrolase, short-chain acyl coenzyme A ADP-dependent short-chain-acyl-CoA hydrolase	
3.1.2.3	hydrolase, succinyl coenzyme A succinyl-CoA hydrolase	
3.5.1.65	Hydrolase, theanine Theanine hydrolase	

EC-No.	Synonym Recommended Name	Vol., Page
3.12.1.1	Hydrolase, trithionate Trithionate hydrolase	
3.1.2.15	Hydrolase, ubiquitin carboxy-terminal Ubiquitin thiolesterase	
3.4.19.12	hydrolase, ubiquitin carboxyl-terminal ubiquitinyl hydrolase I	
3.4.19.12	hydrolase, ubiquitin carboxyl-terminal (Aplysia californica gene Ap-uch) ubiquitinyl hydrolase I	
3.5.1.75	Hydrolase, urethane Urethanase	
3.6.1.45	Hydrolase, uridine diphosphoglucose (Escherichia coli precursor reduced) UDP-sugar diphosphatase	
3.6.1.45	Hydrolase, uridine diphosphoglucose (Escherichia coli reduced) UDP-sugar diphosphatase	
3.6.1.45	Hydrolase, uridine diphosphoglucose (Salmonella typhimurium clone pAGS5 gene ushB isoenzyme reduced) UDP-sugar diphosphatase	
3.6.1.45	Hydrolase, uridine diphosphoglucose (Salmonella typhimurium gene ushA0 isoenzyme precursor reduced) UDP-sugar diphosphatase	
3.6.1.45	Hydrolase, uridine diphosphoglucose (Salmonella typhimurium gene ushA0 isoenzyme reduced) UDP-sugar diphosphatase	
3.2.1.119	Hydrolase, vicianin Vicianin β-glucosidase	
3.1.4.15	hydrolase adenylyl(glutamine synthetase) adenylyl-[glutamate-ammonia ligase] hydrolase	
3.4.22.40	hydrolase H bleomycin hydrolase	
4.2.1.92	Hydroperoxide dehydratase Hydroperoxide dehydratase	
1.11.1.12	Hydroperoxide glutathione peroxidase Phospholipid-hydroperoxide glutathione peroxidase	
4.2.1.92	Hydroperoxide isomerase Hydroperoxide dehydratase	
5.3.99.1	hydroperoxide isomerase hydroperoxide isomerase	Vol. 1, Page 450

EC-No.	Synonym Recommended Name	Vol., Page
1.3.1.2	Hydropyrimidine dehydrogenase Dihydropyrimidine dehydrogenase ($NADP^+$)	
3.5.2.2	Hydropyrimidine hydrase Dihydropyrimidinase	
1.2.1.45	α-Hydroxy-γ-carboxymuconic ε-semialdehyde dehydrogenase 2-Hydroxy-4-carboxymuconate-6-semialdehyde dehydrogenase	
4.1.3.26	β-Hydroxy-β-isohexenylglutaryl CoA-lyase 3-Hydroxy-3-isohexenylglutaryl-CoA lyase	
2.7.1.109	β-Hydroxy-β-methylglutaryl-CoA reductase kinase [Hydroxymethylglutaryl-CoA reductase (NADPH)] kinase	
1.1.1.88	β-Hydroxy-β-methylglutaryl CoA-reductase Hydroxymethylglutaryl-CoA reductase	
3.1.2.5	β-hydroxy-β-methylglutaryl coenzyme A deacylase hydroxymethylglutaryl-CoA hydrolase	
3.1.2.5	β-hydroxy-β-methylglutaryl coenzyme A hydrolase hydroxymethylglutaryl-CoA hydrolase	
1.1.1.88	β-Hydroxy-β-methylglutaryl coenzyme A reductase Hydroxymethylglutaryl-CoA reductase	
1.1.1.34	β-Hydroxy-β-methylglutaryl coenzyme A reductase Hydroxymethylglutaryl-CoA reductase (NADPH)	
4.1.3.5	β-Hydroxy-.β.-methylglutaryl-CoA synthase Hydroxymethylglutaryl-CoA synthase	
1.1.1.155	(-)-1-Hydroxy-1,2,4-butanetricarboxylate:NAD^+ oxidoreductase (decarboxylating) Homoisocitrate dehydrogenase	
4.1.2.38	2-Hydroxy-1,2-diphenylethanone benzaldehyde-lyase Benzoin aldolase	
1.6.5.7	2-hydroxy-1,4-benzoquinone reductase 2-hydroxy-1,4-benzoquinone reductase	
1.14.99.27	5-Hydroxy-1,4-naphthoquinone,hydrogen-donor:oxygen oxidoreduc-tase (3-hydroxylating) Juglone 3-monooxygenase	
1.6.6.13	N-Hydroxy-2-acetamidofluorene reductase N-Hydroxy-2-acetamidofluorene reductase	
2.3.1.56	N-Hydroxy-2-acetylaminofluorene N-O acyltransferase Aromatic-hydroxylamine O-acetyltransferase	

EC-No.	Synonym Recommended Name	Vol., Page
1.6.6.13	N-Hydroxy-2-acetylaminofluorene reductase N-Hydroxy-2-acetamidofluorene reductase	
2.3.1.118	N-Hydroxy-2-aminofluorene-O-acetyltransferase N-Hydroxyarylamine O-acetyltransferase	
4.1.3.16	4-Hydroxy-2-ketoglutarate aldolase 4-Hydroxy-2-oxoglutarate aldolase	
4.1.3.16	4-Hydroxy-2-ketoglutaric aldolase 4-Hydroxy-2-oxoglutarate aldolase	
1.1.3.38	4-Hydroxy-2-methoxybenzyl alcohol oxidase Vanillyl-alcohol oxidase	
1.1.4.2	3-Hydroxy-2-methyl-3-phytyl-2,3-dihydronaphthoquinone:oxidized-dithiothreitol oxidoreductase Vitamin-K-epoxide reductase (warfarin-insensitive)	
1.1.1.178	3-Hydroxy-2-methylbutyryl-CoA dehydrogenase 3-Hydroxy-2-methylbutyryl-CoA dehydrogenase	
1.1.1.31	3-Hydroxy-2-methylpropanoate:NAD$^+$ oxidoreductase 3-Hydroxyisobutyrate dehydrogenase	
3.1.2.4	3-hydroxy-2-methylpropanoyl-CoA hydrolase 3-hydroxyisobutyryl-CoA hydrolase	
4.1.1.51	3-Hydroxy-2-methylpyridine-4,5-dicarboxylate 4-carboxy-lyase 3-Hydroxy-2-methylpyridine-4,5-dicarboxylate 4-decarboxylase	
4.1.1.51	3-Hydroxy-2-methylpyridine-4,5-dicarboxylate 4-decarboxylase 3-Hydroxy-2-methylpyridine-4,5-dicarboxylate 4-decarboxylase	
1.14.12.4	3-Hydroxy-2-methylpyridine-5-carboxylate,NAD(P)H:oxygen oxidoreductase (decyclizing) 3-Hydroxy-2-methylpyridine carboxylate dioxygenase	
1.14.12.4	3-Hydroxy-2-methylpyridine carboxylate dioxygenase 3-Hydroxy-2-methylpyridine carboxylate dioxygenase	
1.13.11.38	1-Hydroxy-2-naphthoate-degrading enzyme 1-Hydroxy-2-naphthoate 1,2-dioxygenase	
1.13.11.38	1-Hydroxy-2-naphthoate 1,2-dioxygenase 1-Hydroxy-2-naphthoate 1,2-dioxygenase	
1.13.11.38	1-Hydroxy-2-naphthoate:oxygen 1,2-oxidoreductase (decyclizing) 1-Hydroxy-2-naphthoate 1,2-dioxygenase	
1.13.11.38	1-Hydroxy-2-naphthoic acid dioxygenase 1-Hydroxy-2-naphthoate 1,2-dioxygenase	

EC-No.	Synonym	Recommended Name	Vol., Page
4.1.3.16	4-Hydroxy-2-oxoglutarate aldolase	4-Hydroxy-2-oxoglutarate aldolase	
4.1.3.16	4-Hydroxy-2-oxoglutarate glyoxylate-lyase	4-Hydroxy-2-oxoglutarate aldolase	
4.2.1.80	4-hydroxy-2-oxopentanoate hydro-lyase	2-oxopent-4-enoate hydratase	
1.2.1.33	D-2-Hydroxy-3,3-dimethyl-3-formylpropionate:diphosphopyridine nucleotide(DPN+) oxidoreductase	(R)-Dehydropantoate dehydrogenase	
2.3.1.91	1-O-(4-Hydroxy-3,5-dimethoxycinnamoyl)-β-D-glucose:choline 1-O-(4-hydroxy-3,5-dimethoxycinnamoyl)transferase	Sinapoylglucose-choline O-sinapoyltransferase	
2.3.1.103	1-O-(4-Hydroxy-3,5-dimethoxycinnamoyl)-β-D-glucoside:1-O-(4-hydroxy-3,5-dimethoxycinnamoyl)-β-D-glucoside 1-O-sinapoyltransferase	Sinapoylglucose-sinapoylglucose O-sinapoyltransferase	
4.1.2.41	3-hydroxy-3-(3-hydroxy-4-methoxyphenyl)propionyl-CoA:vanillin lyase (acetyl-CoA-forming)	vanillin synthase	
4.1.3.26	3-Hydroxy-3-(4-methylpent-3-en-1-yl)-glutaryl-CoA isopentenyl-acetoacetyl-CoA-lyase	3-Hydroxy-3-isohexenylglutaryl-CoA lyase	
4.2.1.57	3-Hydroxy-3-(4-methylpent-3-en-1-yl)glutaryl-CoA hydro-lyase	Isohexenylglutaconyl-CoA hydratase	
1.1.1.87	2-Hydroxy-3-carboxyadipate dehydrogenase	3-Carboxy-2-hydroxyadipate dehydrogenase	
4.2.1.36	2-Hydroxy-3-carboxyadipate hydro-lyase	Homoaconitate hydratase	
4.1.2.35	4-Hydroxy-3-hexanone aldolase	Propioin synthase	
4.1.2.35	4-Hydroxy-3-hexanone propanal-lyase	Propioin synthase	
4.2.1.57	3-Hydroxy-3-isohexenylglutaryl-CoA-hydrolase	Isohexenylglutaconyl-CoA hydratase	
4.1.3.26	3-Hydroxy-3-isohexenylglutaryl-CoA lyase	3-Hydroxy-3-isohexenylglutaryl-CoA lyase	
4.1.3.26	3-Hydroxy-3-isohexenylglutaryl coenzyme A lyase	3-Hydroxy-3-isohexenylglutaryl-CoA lyase	

EC-No.	Synonym Recommended Name	Vol., Page
1.1.1.86	2-Hydroxy-3-keto acid reductoisomerase Ketol-acid reductoisomerase	
1.2.3.12	4-hydroxy-3-methoxybenzoate demethylase vanillate demethylase	
3.1.1.73	4-hydroxy-3-methoxycinnamoyl-sugar hydrolase feruloyl esterase	
3.1.2.5	3-hydroxy-3-methylglutaryl-CoA hydrolase hydroxymethylglutaryl-CoA hydrolase	
4.1.3.4	3-Hydroxy-3-methylglutaryl-CoA lyase Hydroxymethylglutaryl-CoA lyase	
1.1.1.88	3-Hydroxy-3-methylglutaryl-CoA reductase Hydroxymethylglutaryl-CoA reductase	
1.1.1.34	3-Hydroxy-3-methylglutaryl-CoA reductase Hydroxymethylglutaryl-CoA reductase (NADPH)	
1.1.1.34	S-3-Hydroxy-3-methylglutaryl-CoA reductase Hydroxymethylglutaryl-CoA reductase (NADPH)	
2.7.1.109	3-Hydroxy-3-methylglutaryl-CoA reductase kinase [Hydroxymethylglutaryl-CoA reductase (NADPH)] kinase	
4.1.3.5	3-Hydroxy-3-methylglutaryl-CoA synthase Hydroxymethylglutaryl-CoA synthase	
4.1.3.5	3-Hydroxy-3-methylglutaryl-coenzyme A synthase Hydroxymethylglutaryl-CoA synthase	
4.1.3.4	3-Hydroxy-3-methylglutaryl CoA cleaving enzyme Hydroxymethylglutaryl-CoA lyase	
1.1.1.34	3-Hydroxy-3-methylglutaryl CoA reductase (NADPH) Hydroxymethylglutaryl-CoA reductase (NADPH)	
4.1.3.5	3-Hydroxy-3-methylglutaryl CoA synthetase Hydroxymethylglutaryl-CoA synthase	
4.1.3.4	3-Hydroxy-3-methylglutaryl coenzyme A lyase Hydroxymethylglutaryl-CoA lyase	
1.1.1.88	3-Hydroxy-3-methylglutaryl coenzyme A reductase (EC 1.1.1.88) Hydroxymethylglutaryl-CoA reductase	
2.7.1.109	3-Hydroxy-3-methylglutaryl coenzyme A reductase kinase [Hydroxymethylglutaryl-CoA reductase (NADPH)] kinase	
4.1.3.5	3-Hydroxy-3-methylglutaryl coenzyme A synthase Hydroxymethylglutaryl-CoA synthase	

EC-No.	Synonym Recommended Name	Vol., Page
4.1.3.5	3-Hydroxy-3-methylglutaryl coenzyme A synthetase Hydroxymethylglutaryl-CoA synthase	
4.1.3.15	2-hydroxy-3-oxoadipate glyoxylate-lyase (carboxylating) 2-hydroxy-3-oxoadipate synthase	
4.1.3.15	2-hydroxy-3-oxoadipate synthase 2-hydroxy-3-oxoadipate synthase	
1.1.1.60	2-Hydroxy-3-oxopropionate reductase 2-Hydroxy-3-oxopropionate reductase	
1.13.99.5	3-Hydroxy-4(1H)-one, 2,4-dioxygenase 3,4-Dihydroxyquinoline 2,4-dioxygenase	
2.3.1.56	N-Hydroxy-4-acetylaminobiphenyl:N-hydroxy-4-aminobiphenyl O-acetyltransferase Aromatic-hydroxylamine O-acetyltransferase	
1.2.1.45	2-Hydroxy-4-carboxy-cis,cis-muconate-6-semialdehyde:NADP$^+$ 6-oxidoreductase 2-Hydroxy-4-carboxymuconate-6-semialdehyde dehydrogenase	
1.2.1.45	2-Hydroxy-4-carboxymuconate-6-semialdehyde dehydrogenase 2-Hydroxy-4-carboxymuconate-6-semialdehyde dehydrogenase	
1.2.1.45	2-Hydroxy-4-carboxymuconate 6-semialdehyde dehydrogenase 2-Hydroxy-4-carboxymuconate-6-semialdehyde dehydrogenase	
4.1.3.17	4-hydroxy-4-methyl-2-ketoglutarate aldolase 4-hydroxy-4-methyl-2-oxoglutarate aldolase	
4.1.3.17	4-hydroxy-4-methyl-2-oxoglutarate aldolase 4-hydroxy-4-methyl-2-oxoglutarate aldolase	
4.1.3.17	4-hydroxy-4-methyl-2-oxoglutarate pyruvate-lyase 4-hydroxy-4-methyl-2-oxoglutarate aldolase	
4.1.3.12	3-hydroxy-4-methyl-3-carboxy-pentanoate 2-oxo-3-methyl-butanoate lyase (CoA-acetylating) 2-isopropylmalate synthase	
1.1.1.170	3β-Hydroxy-4β-methyl-5α-cholest-7-ene-4α-carboxylate:NAD$^+$ 3-oxidoreductase (decarboxylating) 3β-Hydroxy-4β-methylcholestenecarboxylate 3-dehydrogenase (decarboxylating)	
1.1.1.170	3β-Hydroxy-4β-methylcholestenecarboxylate 3-dehydrogenase (decarboxylating) 3β-Hydroxy-4β-methylcholestenecarboxylate 3-dehydrogenase (decarboxylating)	

EC-No.	Synonym Recommended Name Vol., Page
1.1.1.170	3β-Hydroxy-4β-methylcholestenoate dehydrogenase 3β-Hydroxy-4β-methylcholestenecarboxylate 3-dehydrogenase (decarboxylating)
1.13.99.5	3-Hydroxy-4-oxo-1,4-dihydroquinoline 2,4-dioxygenase 3,4-Dihydroxyquinoline 2,4-dioxygenase
1.13.99.5	1H-3-Hydroxy-4-oxo-quinoline oxygenase 3,4-Dihydroxyquinoline 2,4-dioxygenase
4.2.1.83	2-Hydroxy-4-oxobutane-1,2,4-tricarboxylate 2,3-hydro-lyase 4-Oxalmesaconate hydratase
1.13.99.5	1H-3-Hydroxy-4-oxoquinaldine 2,4-dioxygenase 3,4-Dihydroxyquinoline 2,4-dioxygenase
1.13.99.5	1H-3-Hydroxy-4-oxoquinoline 2,4-dioxygenase 3,4-Dihydroxyquinoline 2,4-dioxygenase
1.13.99.5	1H-3-Hydroxy-4-oxoquinoline oxygenase 3,4-Dihydroxyquinoline 2,4-dioxygenase
1.1.1.152	3α-Hydroxy-5β-androstan-17-one 3α-dehydrogenase 3α-Hydroxy-5β-androstan-17-one 3α-dehydrogenase
1.1.1.145	3β-Hydroxy-Δ^5-C_{27}-steroid dehydrogenase/isomerase 3β-Hydroxy-Δ^5-steroid dehydrogenase
1.1.1.145	3β-Hydroxy-Δ^5-C_{27}-steroid oxidoreductase 3β-Hydroxy-Δ^5-steroid dehydrogenase
1.1.1.181	3β-Hydroxy-Δ^5-C_{27}-steroid oxidoreductase Cholest-5-ene-3β,7α-diol 3β-dehydrogenase
1.1.1.52	3α-Hydroxy-5β-cholanate:NAD^+ oxidoreductase 3α-Hydroxycholanate dehydrogenase
1.12.99.1	8-Hydroxy-5-deazaflavin-reducing hydrogenase Coenzyme F420 hydrogenase
1.1.1.145	3β-Hydroxy-5-ene-steroid dehydrogenase 3β-Hydroxy-Δ^5-steroid dehydrogenase
1.1.1.145	3β-Hydroxy-5-ene-steroid oxidoreductase 3β-Hydroxy-Δ^5-steroid dehydrogenase
1.1.1.145	3β-Hydroxy-5-ene steroid dehydrogenase 3β-Hydroxy-Δ^5-steroid dehydrogenase
1.1.1.152	3α-Hydroxy-5β-steroid:NAD^+ 3-oxidoreductase 3α-Hydroxy-5β-androstan-17-one 3α-dehydrogenase

EC-No.	Synonym Recommended Name	Vol., Page
1.1.1.145	3β-Hydroxy-Δ^5-steroid:NAD$^+$ 3-oxidoreductase 3β-Hydroxy-Δ^5-steroid dehydrogenase	
1.1.1.152	3α-Hydroxy-5β-steroid dehydrogenase 3α-Hydroxy-5β-androstan-17-one 3α-dehydrogenase	
1.1.1.145	3β-Hydroxy-Δ^5-steroid dehydrogenase 3β-Hydroxy-Δ^5-steroid dehydrogenase	
1.3.1.40	2-Hydroxy-6-oxo-6-phenylhexa-2,4-dienoate reductase 2-Hydroxy-6-oxo-6-phenylhexa-2,4-dienoate reductase	
3.7.1.9	2-Hydroxy-6-oxohepta-2,4-dienoate hydrolase 2-Hydroxymuconate-semialdehyde hydrolase	
1.1.1.45	L-β-Hydroxy-acid-NAD-oxidoreductase L-Gulonate 3-dehydrogenase	
1.1.99.6	D-2-Hydroxy-acid dehydrogenase D-2-Hydroxy-acid dehydrogenase	
1.1.3.15	Hydroxy-acid oxidase A (S)-2-Hydroxy-acid oxidase	
1.1.3.15	Hydroxy-acid oxidase B (S)-2-Hydroxy-acid oxidase	
1.1.1.52	α-Hydroxy-cholanate dehydrogenase 3α-Hydroxycholanate dehydrogenase	
6.2.1.12	Hydroxy-cinnamate:CoA ligase 4-Coumarate-CoA ligase	Vol. 2, Page 256
1.5.3.6	6-Hydroxy-D-nicotine oxidase (FAD) (R)-6-Hydroxynicotine oxidase	
3.1.2.4	3-hydroxy-isobutyryl CoA hydrolase 3-hydroxyisobutyryl-CoA hydrolase	
4.1.1.16	3-Hydroxy-L-glutamate 1-carboxy-lyase Hydroxyglutamate decarboxylase	
2.6.1.23	4-Hydroxy-L-glutamate:2-oxoglutarate aminotransferase 4-Hydroxyglutamate transaminase	
1.5.3.5	6-Hydroxy-L-nicotine:oxygen oxidoreductase (S)-6-Hydroxynicotine oxidase	
1.5.3.5	6-Hydroxy-L-nicotine oxidase (S)-6-Hydroxynicotine oxidase	
1.1.1.104	4-Hydroxy-L-proline:NAD$^+$ oxidoreductase 4-Oxoproline reductase	

EC-No.	Synonym Recommended Name	Vol., Page
1.1.1.104	Hydroxy-L-proline oxidase 4-Oxoproline reductase	
4.1.1.28	5-Hydroxy-L-tryptophan decarboxylase Aromatic-L-amino-acid decarboxylase	
1.14.11.6	5-Hydroxy-methyluracil dioxygenase Thymine dioxygenase	
2.1.1.116	3'-Hydroxy-N-methyl-(S)-coclaurine 4'-O-methyltransferase 3'-Hydroxy-N-methyl-(S)-coclaurine 4'-O-methyltransferase	
3.5.1.36	5-Hydroxy-N-methyl-pyroglutamate synthase N-Methyl-2-oxoglutaramate hydrolase	
4.2.1.96	4 α-Hydroxy-tetrahydropterin dehydratase 4a-Hydroxytetrahydrobiopterin dehydratase	
3.1.3.16	3-Hydroxy 3-methylglutaryl CoenzymeA reductase phosphatase Phosphoprotein phosphatase	
1.1.1.35	3-Hydroxyacetyl-coenzyme A dehydrogenase 3-Hydroxyacyl-CoA dehydrogenase	
1.1.99.24	Hydroxyacid-oxoacid transhydrogenase Hydroxyacid-oxoacid transhydrogenase	
4.2.1.17	β-hydroxyacid dehydrase enoyl-CoA hydratase	
1.1.99.6	D-2-Hydroxy acid dehydrogenase D-2-Hydroxy-acid dehydrogenase	
1.1.1.45	L-3-Hydroxyacid dehydrogenase L-Gulonate 3-dehydrogenase	
1.1.1.45	L-β-Hydroxyacid dehydrogenase L-Gulonate 3-dehydrogenase	
1.1.1.35	β-Hydroxy acid dehydrogenase 3-Hydroxyacyl-CoA dehydrogenase	
1.1.3.15	L-2-Hydroxy acid oxidase (S)-2-Hydroxy-acid oxidase	
1.1.3.15	L-α-Hydroxy acid oxidase (S)-2-Hydroxy-acid oxidase	
1.1.3.15	Hydroxyacid oxidase A (S)-2-Hydroxy-acid oxidase	
5.1.2.1	Hydroxyacid racemase Lactate racemase	Vol. 1, Page 68

EC-No.	Synonym Recommended Name	Vol., Page
3.1.2.6	S-(2-Hydroxyacyl)glutathione hydrolase Hydroxyacylglutathione hydrolase	
4.2.1.60	β-Hydroxyacyl-acyl carrier protein dehydratase 3-Hydroxydecanoyl-[acyl-carrier-protein] dehydratase	
4.2.1.17	β-hydroxyacyl-CoA dehydrase enoyl-CoA hydratase	
4.2.1.17	D-3-hydroxyacyl-CoA dehydratase enoyl-CoA hydratase	
4.2.1.74	D-3-hydroxyacyl-CoA dehydratase long-chain-enoyl-CoA hydratase	
1.1.1.35	3-Hydroxyacyl-CoA dehydrogenase 3-Hydroxyacyl-CoA dehydrogenase	
1.1.1.211	3-Hydroxyacyl-CoA dehydrogenase Long-chain-3-hydroxyacyl-CoA dehydrogenase	
1.1.1.35	3-L-Hydroxyacyl-CoA dehydrogenase 3-Hydroxyacyl-CoA dehydrogenase	
1.1.1.211	β-Hydroxyacyl-CoA dehydrogenase Long-chain-3-hydroxyacyl-CoA dehydrogenase	
1.1.1.36	D-3-Hydroxyacyl-CoA reductase Acetoacetyl-CoA reductase	
5.1.2.3	3-Hydroxyacyl-coenzyme A epimerase 3-Hydroxybutyryl-CoA epimerase	Vol. 1, Page 80
1.1.1.35	β-Hydroxyacyl-coenzyme A synthetase 3-Hydroxyacyl-CoA dehydrogenase	
1.1.1.35	L-3-Hydroxyacyl CoA dehydrogenase 3-Hydroxyacyl-CoA dehydrogenase	
1.1.1.35	β-Hydroxyacyl CoA dehydrogenase 3-Hydroxyacyl-CoA dehydrogenase	
1.3.99.2	3-Hydroxyacyl CoA reductase Butyryl-CoA dehydrogenase	
1.1.1.36	Hydroxyacyl coenzyme-A dehydrogenase Acetoacetyl-CoA reductase	
1.1.1.35	3-Hydroxyacyl coenzyme A dehydrogenase 3-Hydroxyacyl-CoA dehydrogenase	
1.1.1.35	3β-Hydroxyacyl coenzyme A dehydrogenase 3-Hydroxyacyl-CoA dehydrogenase	

EC-No.	Synonym Recommended Name	Vol., Page
1.1.1.35	L-3-Hydroxyacyl coenzyme A dehydrogenase 3-Hydroxyacyl-CoA dehydrogenase	
1.1.1.35	β-Hydroxyacylcoenzyme A dehydrogenase 3-Hydroxyacyl-CoA dehydrogenase	
1.1.1.35	β-Hydroxyacyl dehydrogenase 3-Hydroxyacyl-CoA dehydrogenase	
3.1.2.6	Hydroxyacylglutathione hydrolase Hydroxyacylglutathione hydrolase	
2.4.1.45	2-Hydroxyacylsphingosine 1-β-galactosyltransferase 2-Hydroxyacylsphingosine 1-β-galactosyltransferase	
1.1.1.172	2-Hydroxyadipate:NAD^+ 2-oxidoreductase 2-Oxoadipate reductase	
2.5.1.18	S-(Hydroxyalkyl)glutathione lyase Glutathione transferase	
4.4.1.7	S-(Hydroxyalkyl)glutathione lyase S-(Hydroxyalkyl)glutathione lyase	
2.7.1.37	Hydroxyalkyl-protein kinase Protein kinase	
1.6.6.11	N-Hydroxy amine reductase Hydroxylamine reductase (NADH)	
4.2.1.14	D-Hydroxy amino acid dehydratase D-Serine dehydratase	
4.2.1.13	L-Hydroxy amino acid dehydratase L-Serine dehydratase	
1.13.11.6	3-Hydroxyanthranilate 3,4-di- 3-Hydroxyanthranilate 3,4-dioxygenase	
1.13.11.6	3-Hydroxyanthranilate 3,4-dioxygenase 3-Hydroxyanthranilate 3,4-dioxygenase	
2.1.1.97	3-Hydroxyanthranilate 4-C-methyltransferase 3-Hydroxyanthranilate 4-C-methyltransferase	
2.1.1.97	3-Hydroxyanthranilate 4-methyltransferase 3-Hydroxyanthranilate 4-C-methyltransferase	
1.13.11.6	3-Hydroxyanthranilate:oxygen 3,4-oxidoreductase (decyclizing) 3-Hydroxyanthranilate 3,4-dioxygenase	
1.10.3.5	3-Hydroxyanthranilate:oxygen oxidoreductase 3-Hydroxyanthranilate oxidase	

EC-No.	Synonym Recommended Name	Vol., Page
1.10.3.5	3-Hydroxyanthranilate oxidase 3-Hydroxyanthranilate oxidase	
1.13.11.6	3-Hydroxyanthranilate oxygenase 3-Hydroxyanthranilate 3,4-dioxygenase	
1.13.11.6	3-Hydroxyanthranilic acid oxidase 3-Hydroxyanthranilate 3,4-dioxygenase	
1.10.3.5	3-Hydroxyanthranilic acid oxidase 3-Hydroxyanthranilate oxidase	
1.13.11.6	3-Hydroxyanthranilic acid oxygenase 3-Hydroxyanthranilate 3,4-dioxygenase	
1.13.11.6	3-Hydroxyanthranilic oxygenase 3-Hydroxyanthranilate 3,4-dioxygenase	
2.4.1.181	Hydroxyanthraquinone glucosyltransferase Hydroxyanthraquinone glucosyltransferase	
2.7.1.112	Hydroxyaryl-protein kinase Protein-tyrosine kinase	
2.3.1.118	N-Hydroxyarylamine O-acetyltransferase N-Hydroxyarylamine O-acetyltransferase	
4.1.3.14	3-Hydroxyaspartate aldolase 3-Hydroxyaspartate aldolase	
4.2.1.38	3-Hydroxyaspartate dehydratase Erythro-3-hydroxyaspartate dehydratase	
4.2.1.38	Hydroxyaspartate dehydratase Erythro-3-hydroxyaspartate dehydratase	
3.5.99.3	hydroxyatrazine ethylaminohydrolase hydroxydechloroatrazine ethylaminohydrolase	
1.2.1.64	3-hydroxybenzaldehyde:NAD oxidoreductase 4-hydroxybenzaldehyde dehydrogenase	
1.2.1.64	4-hydroxybenzaldehyde dehydrogenase 4-hydroxybenzaldehyde dehydrogenase	
1.14.13.68	4-hydroxybenzeneacetaldehyde oxime monooxygenase 4-hydroxyphenylacetaldehyde oxime monooxygenase	
1.14.99.23	3-Hydroxybenzoate,hydrogen-donor:oxygen oxidoreductase (2-hydroxylating) 3-Hydroxybenzoate 2-monooxygenase	

EC-No.	Synonym Recommended Name	Vol., Page
1.14.13.64	4-Hydroxybenzoate, NAD(P)H_2:oxygen oxidoreductase (1-hydroxylating, decarboxylating) 4-Hydroxybenzoate 1-hydroxylase	
1.14.13.33	4-Hydroxybenzoate,NAD(P)H:oxygen oxidoreductase (3-hydroxylating) 4-Hydroxybenzoate 3-monooxygenase (NAD(P)H)	
1.14.13.24	3-Hydroxybenzoate,NADH:oxygen oxidoreductase (6-hydroxylating) 3-Hydroxybenzoate 6-monooxygenase	
1.14.13.2	4-Hydroxybenzoate,NADPH:oxygen oxidoreductase (3-hydroxylating) 4-Hydroxybenzoate 3-monooxygenase	
1.14.13.23	3-Hydroxybenzoate,NADPH:oxygen oxidoreductase (4-hydroxylating) 3-Hydroxybenzoate 4-monooxygenase	
1.14.13.33	4-Hydroxybenzoate-3-hydroxylase 4-Hydroxybenzoate 3-monooxygenase (NAD(P)H)	
1.14.13.2	*p*-Hydroxybenzoate-3-hydroxylase 4-Hydroxybenzoate 3-monooxygenase	
6.2.1.27	4-Hydroxybenzoate-CoA ligase 4-Hydroxybenzoate-CoA ligase	Vol. 2, Page 323
6.2.1.27	4-Hydroxybenzoate-CoA synthetase 4-Hydroxybenzoate-CoA ligase	Vol. 2, Page 323
6.2.1.27	4-Hydroxybenzoate-coenzyme A ligase (AMP-forming) 4-Hydroxybenzoate-CoA ligase	Vol. 2, Page 323
1.14.13.64	4-Hydroxybenzoate 1-hydroxylase 4-Hydroxybenzoate 1-hydroxylase	
1.14.13.64	4-Hydroxybenzoate 1-hydroxylase (decarboxylating) 4-Hydroxybenzoate 1-hydroxylase	
1.14.13.64	4-Hydroxybenzoate 1-monooxygenase 4-Hydroxybenzoate 1-hydroxylase	
1.14.99.23	3-Hydroxybenzoate 2-hydroxylase 3-Hydroxybenzoate 2-monooxygenase	
1.14.99.23	3-Hydroxybenzoate 2-monooxygenase 3-Hydroxybenzoate 2-monooxygenase	
1.14.13.2	4-Hydroxybenzoate 3-hydroxylase 4-Hydroxybenzoate 3-monooxygenase	
1.14.13.33	4-Hydroxybenzoate 3-hydroxylase 4-Hydroxybenzoate 3-monooxygenase (NAD(P)H)	

EC-No.	Synonym Recommended Name	Vol., Page
1.14.13.2	4-Hydroxybenzoate 3-monooxygenase 4-Hydroxybenzoate 3-monooxygenase	
1.14.13.33	4-Hydroxybenzoate 3-monooxygenase (NAD(P)H) 4-Hydroxybenzoate 3-monooxygenase (NAD(P)H)	
1.14.13.23	3-Hydroxybenzoate 4-hydroxylase 3-Hydroxybenzoate 4-monooxygenase	
1.14.13.23	3-Hydroxybenzoate 4-monooxygenase 3-Hydroxybenzoate 4-monooxygenase	
1.14.99.13	3-hydroxybenzoate 4-monooxygenase 3-hydroxybenzoate 4-monooxygenase	
2.4.1.194	4-Hydroxybenzoate 4-O-β-D-glucosyltransferase 4-Hydroxybenzoate 4-O-β-D-glucosyltransferase	
1.14.13.24	3-Hydroxybenzoate 6-hydroxylase 3-Hydroxybenzoate 6-monooxygenase	
1.14.13.24	*m*-Hydroxybenzoate 6-hydroxylase 3-Hydroxybenzoate 6-monooxygenase	
1.14.13.24	3-Hydroxybenzoate 6-monooxygenase 3-Hydroxybenzoate 6-monooxygenase	
6.2.1.27	4-Hydroxybenzoate:CoA ligase (AMP-forming) 4-Hydroxybenzoate-CoA ligase	Vol. 2, Page 323
4.1.1.61	4-Hydroxybenzoate carboxy-lyase 4-Hydroxybenzoate decarboxylase	
4.1.1.61	4-Hydroxybenzoate decarboxylase 4-Hydroxybenzoate decarboxylase	
4.1.1.61	*p*-Hydroxybenzoate decarboxylase 4-Hydroxybenzoate decarboxylase	
2.5.1.39	*p*-Hydroxybenzoate dimethylallyltransferase 4-Hydroxybenzoate nonaprenyltransferase	
2.4.1.194	*p*-Hydroxybenzoate glucosyltransferase 4-Hydroxybenzoate 4-O-β-D-glucosyltransferase	
1.14.13.2	*p*-Hydroxybenzoate hydroxylase 4-Hydroxybenzoate 3-monooxygenase	
1.14.13.12	*p*-Hydroxybenzoate hydroxylase Benzoate 4-monooxygenase	
1.14.13.2	4-Hydroxybenzoate monooxygenase 4-Hydroxybenzoate 3-monooxygenase	

EC-No.	Synonym Recommended Name	Vol., Page
2.5.1.39	4-Hydroxybenzoate nonaprenyltransferase 4-Hydroxybenzoate nonaprenyltransferase	
2.5.1.39	*p*-Hydroxybenzoate polyprenyltransferase 4-Hydroxybenzoate nonaprenyltransferase	
2.5.1.39	4-Hydroxybenzoate transferase 4-Hydroxybenzoate nonaprenyltransferase	
2.5.1.39	*p*-Hydroxybenzoic-polyprenyl transferase 4-Hydroxybenzoate nonaprenyltransferase	
1.14.13.24	3-Hydroxybenzoic acid-6-hydroxylase 3-Hydroxybenzoate 6-monooxygenase	
2.5.1.39	*p*-Hydroxybenzoic acid-polyprenyl transferase 4-Hydroxybenzoate nonaprenyltransferase	
1.14.13.2	*p*-Hydroxybenzoic acid hydrolase 4-Hydroxybenzoate 3-monooxygenase	
1.14.13.2	*p*-Hydroxybenzoic acid hydroxylase 4-Hydroxybenzoate 3-monooxygenase	
1.14.13.2	4-Hydroxybenzoic hydroxylase 4-Hydroxybenzoate 3-monooxygenase	
1.14.13.2	*p*-Hydroxybenzoic hydroxylase 4-Hydroxybenzoate 3-monooxygenase	
1.6.5.7	hydroxybenzoquinone reductase 2-hydroxy-1,4-benzoquinone reductase	
1.13.11.41	(4-Hydroxybenzoyl)methanol oxygenase 2,4'-Dihydroxyacetophenone dioxygenase	
1.3.99.20	4-hydroxybenzoyl-CoA:(acceptor) oxidoreductase 4-hydroxybenzoyl-CoA reductase	
3.1.2.23	4-Hydroxybenzoyl-CoA hydrolase 4-Hydroxybenzoyl-CoA thioesterase	
6.2.1.27	4-Hydroxybenzoyl-CoA ligase 4-Hydroxybenzoate-CoA ligase	Vol. 2, Page 323
1.3.99.20	4-hydroxybenzoyl-CoA reductase 4-hydroxybenzoyl-CoA reductase	
3.1.2.23	4-Hydroxybenzoyl-CoA thioesterase 4-Hydroxybenzoyl-CoA thioesterase	
6.2.1.27	4-Hydroxybenzoyl coenzyme A synthetase 4-Hydroxybenzoate-CoA ligase	Vol. 2, Page 323

EC-No.	Synonym Recommended Name	Vol., Page
1.1.1.97	3-Hydroxybenzyl-alcohol:NADP$^+$ oxidoreductase 3-Hydroxybenzyl-alcohol dehydrogenase	
1.1.1.97	3-Hydroxybenzyl-alcohol dehydrogenase 3-Hydroxybenzyl-alcohol dehydrogenase	
1.1.1.97	*m*-Hydroxybenzyl alcohol dehydrogenase 3-Hydroxybenzyl-alcohol dehydrogenase	
1.1.1.97	*m*-Hydroxybenzylalcohol dehydrogenase 3-Hydroxybenzyl-alcohol dehydrogenase	
1.1.1.90	*p*-Hydroxybenzyl alcohol dehydrogenase Aryl-alcohol dehydrogenase	
1.1.1.97	*m*-Hydroxybenzyl alcohol dehydrogenase (NADP) 3-Hydroxybenzyl-alcohol dehydrogenase	
1.14.13.44	2-Hydroxybiphenyl,NADH:oxygen oxidoreductase (3-hydroxylating) 2-Hydroxybiphenyl 3-monooxygenase	
1.14.13.44	2-Hydroxybiphenyl 3-monooxygenase 2-Hydroxybiphenyl 3-monooxygenase	
2.4.1.17	4-Hydroxybiphenyl UDP-glucuronosyltransferase Glucuronosyltransferase	
2.4.1.17	*p*-Hydroxybiphenyl UDP glucuronyltransferase Glucuronosyltransferase	
2.2.1.4	3-Hydroxybutan-2-one:D-ribose-5-phosphate aldehydetransferase Acetoin-ribose-5-phosphate transaldolase	
4.2.1.79	2-Hydroxybutane-1,2,3-tricarboxylate hydro-lyase 2-Methylcitrate dehydratase	
4.1.3.21	2-Hydroxybutane-1,2,4-tricarboxylate-2-oxoglutarate-lyase (CoA-acetylating) Homocitrate synthase	
1.1.1.61	4-Hydroxybutanoate:NAD$^+$ oxidoreductase 4-Hydroxybutyrate dehydrogenase	
5.1.2.3	3-Hydroxybutanoyl-CoA 3-epimerase 3-Hydroxybutyryl-CoA epimerase	Vol. 1, Page 80
3.1.1.22	D-(-)-3-hydroxybutyrate-dimer hydrolase hydroxybutyrate-dimer hydrolase	
3.1.1.22	hydroxybutyrate-dimer hydrolase hydroxybutyrate-dimer hydrolase	

EC-No.	Synonym Recommended Name	Vol., Page
1.1.1.30	3-D-Hydroxybutyrate dehydrogenase 3-Hydroxybutyrate dehydrogenase	
1.1.1.30	3-Hydroxybutyrate dehydrogenase 3-Hydroxybutyrate dehydrogenase	
1.1.1.61	4-Hydroxybutyrate dehydrogenase 4-Hydroxybutyrate dehydrogenase	
1.1.1.30	D-(-)-3-Hydroxybutyrate dehydrogenase 3-Hydroxybutyrate dehydrogenase	
1.1.1.30	D-3-Hydroxybutyrate dehydrogenase 3-Hydroxybutyrate dehydrogenase	
1.1.1.30	D-β-Hydroxybutyrate dehydrogenase 3-Hydroxybutyrate dehydrogenase	
1.1.1.30	β-Hydroxybutyrate dehydrogenase 3-Hydroxybutyrate dehydrogenase	
1.1.1.61	γ-Hydroxybutyrate dehydrogenase 4-Hydroxybutyrate dehydrogenase	
1.1.1.30	Hydroxybutyrate oxidoreductase 3-Hydroxybutyrate dehydrogenase	
1.1.1.30	β-Hydroxybutyric acid dehydrogenase 3-Hydroxybutyrate dehydrogenase	
1.1.1.30	β-Hydroxybutyric dehydrogenase 3-Hydroxybutyrate dehydrogenase	
4.2.1.55	3-Hydroxybutyryl-CoA dehydratase 3-Hydroxybutyryl-CoA dehydratase	
4.2.1.55	D-3-Hydroxybutyryl-CoA dehydratase 3-Hydroxybutyryl-CoA dehydratase	
1.1.1.35	3-Hydroxybutyryl-CoA dehydrogenase 3-Hydroxyacyl-CoA dehydrogenase	
1.1.1.157	3-Hydroxybutyryl-CoA dehydrogenase 3-Hydroxybutyryl-CoA dehydrogenase	
1.1.1.157	L-(+)-3-Hydroxybutyryl-CoA dehydrogenase 3-Hydroxybutyryl-CoA dehydrogenase	
1.1.1.157	β-Hydroxybutyryl-CoA dehydrogenase 3-Hydroxybutyryl-CoA dehydrogenase	
5.1.2.3	3-Hydroxybutyryl-CoA epimerase 3-Hydroxybutyryl-CoA epimerase	Vol. 1, Page 80

EC-No.	Synonym Recommended Name	Vol., Page
4.2.1.58	β-Hydroxybutyryl acyl carrier protein dehydrase Crotonoyl-[acyl-carrier-protein] hydratase	
4.2.1.58	β-Hydroxybutyryl acyl carrier protein dehydrase Crotonoyl-[acyl-carrier-protein] hydratase	
4.2.1.58	β-Hydroxybutyryl acyl carrier protein dehydrase Crotonoyl-[acyl-carrier-protein] hydratase	
4.2.1.58	3-Hydroxybutyryl acyl carrier protein dehydratase Crotonoyl-[acyl-carrier-protein] hydratase	
6.2.1.2	L-(+)-3-Hydroxybutyryl CoA ligase Butyrate-CoA ligase	Vol. 2, Page 199
4.2.1.55	D-3-Hydroxybutyryl coenzyme A dehydratase 3-Hydroxybutyryl-CoA dehydratase	
1.1.1.157	β-Hydroxybutyryl coenzyme A dehydrogenase 3-Hydroxybutyryl-CoA dehydrogenase	
1.1.1.35	β-Hydroxybutyrylcoenzyme A dehydrogenase 3-Hydroxyacyl-CoA dehydrogenase	
5.1.2.3	3-Hydroxybutyryl coenzyme A epimerase 3-Hydroxybutyryl-CoA epimerase	Vol. 1, Page 80
1.1.1.52	3α-Hydroxycholanate dehydrogenase 3α-Hydroxycholanate dehydrogenase	
1.14.13.13	25-Hydroxycholecalciferol-1-hydroxylase Calcidiol 1-monooxygenase	
1.14.13.13	25-Hydroxycholecalciferol 1-hydroxylase Calcidiol 1-monooxygenase	
1.14.13.13	25-Hydroxycholecalciferol 1α-hydroxylase Calcidiol 1-monooxygenase	
1.14.13.13	25-Hydroxycholecalciferol 1-monooxygenase Calcidiol 1-monooxygenase	
1.14.13.60	27-Hydroxycholesterol, $NADPH_2$:oxygen oxidoreductase (7α-hydroxylating) 27-Hydroxycholesterol 7α-monooxygenase	
1.14.13.60	27-Hydroxycholesterol-7α-hydroxylase 27-Hydroxycholesterol 7α-monooxygenase	
1.14.13.60	27-Hydroxycholesterol 7α-hydroxylase 27-Hydroxycholesterol 7α-monooxygenase	

EC-No.	Synonym Recommended Name	Vol., Page
1.14.13.60	Hydroxycholesterol 7α-hydroxylase 27-Hydroxycholesterol 7α-monooxygenase	
1.14.13.60	27-Hydroxycholesterol 7α-monooxygenase 27-Hydroxycholesterol 7α-monooxygenase	
2.4.1.126	Hydroxycinnamate 4-β-glucosyltransferase Hydroxycinnamate 4-β-glucosyltransferase	
6.2.1.12	Hydroxycinnamate:CoA ligase 4-Coumarate-CoA ligase	Vol. 2, Page 256
6.2.1.12	*p*-Hydroxycinnamic acid:CoA ligase 4-Coumarate-CoA ligase	Vol. 2, Page 256
2.3.1.103	1-(Hydroxycinnamoyl)-glucose:1-(hydroxycinnamoyl)-glucose hydroxycinnamoyltransferase Sinapoylglucose-sinapoylglucose O-sinapoyltransferase	
2.3.1.153	Hydroxycinnamoyl-CoA-anthocyanidin 3,5-diglucoside 5-O-glucoside-6'''-O-hydroxycinnamoyltransferase Anthocyanin 5-aromatic acyltransferase	
2.3.1.153	Hydroxycinnamoyl-CoA:anthocyanidin 3,5-diglucoside 5-O-glucoside-6'''-O-hydroxycinnamoyltransferase Anthocyanin 5-aromatic acyltransferase	
2.3.1.138	Hydroxycinnamoyl-CoA:putrescine hydroxycinnamoyltransferase Putrescine N-hydroxycinnamoyltransferase	
2.3.1.106	Hydroxycinnamoyl-Coenzyme-A:tartronate hydroxycinnamoyltransferase Tartronate O-hydroxycinnamoyltransferase	
6.2.1.12	Hydroxycinnamoyl CoA synthetase 4-Coumarate-CoA ligase	Vol. 2, Page 256
2.3.1.99	Hydroxycinnamoyl coenzyme A-quinate transferase Quinate O-hydroxycinnamoyltransferase	
1.2.1.44	*p*-Hydroxycinnamoyl coenzyme A reductase Cinnamoyl-CoA reductase	
6.2.1.12	*p*-Hydroxycinnamoyl coenzyme A synthetase 4-Coumarate-CoA ligase	Vol. 2, Page 256
3.1.1.73	hydroxycinnamoyl esterase feruloyl esterase	
2.4.1.126	Hydroxycinnamoyl glucosyltransferase Hydroxycinnamate 4-β-glucosyltransferase	

EC-No.	Synonym Recommended Name	Vol., Page
2.3.1.153	Hydroxycinnamoyltransferase, anthocyanidin 3,5-diglucoside 5-O-glucoside-6-O- Anthocyanin 5-aromatic acyltransferase	
2.3.1.153	Hydroxycinnamoyltransferase, anthocyanin Anthocyanin 5-aromatic acyltransferase	
2.3.1.130	Hydroxycinnamoyltransferase, galacturate Galactarate O-hydroxycinnamoyltransferase	
2.3.1.131	Hydroxycinnamoyltransferase, glucarate Glucarate O-hydroxycinnamoyltransferase	
2.3.1.103	Hydroxycinnamoyltransferase, hydroxycinnamoylglucose-hydroxycinnamoylglucose Sinapoylglucose-sinapoylglucose O-sinapoyltransferase	
2.3.1.126	Hydroxycinnamoyltransferase, isocitrate Isocitrate O-dihydroxycinnamoyltransferase	
2.3.1.138	Hydroxycinnamoyltransferase, putrescine Putrescine N-hydroxycinnamoyltransferase	
2.3.1.99	Hydroxycinnamoyltransferase, quinate Quinate O-hydroxycinnamoyltransferase	
2.3.1.133	Hydroxycinnamoyltransferase, shikimate Shikimate O-hydroxycinnamoyltransferase	
2.3.1.106	Hydroxycinnamoyltransferase, tartronate Tartronate O-hydroxycinnamoyltransferase	
1.14.13.66	2-Hydroxycyclohexan-1-one,NADPH$_2$:oxygen 2-oxidoreductase (1,2-lactonizing) 2-Hydroxycyclohexanone 2-monooxygenase	
1.14.13.66	2-Hydroxycyclohexan-1-one monooxygenase 2-Hydroxycyclohexanone 2-monooxygenase	
4.1.3.35	(1-Hydroxycyclohexan-1-yl)acetyl-CoA cyclohexanone-lyase (1-Hydroxycyclohexan-1-yl)acetyl-CoA lyase	
4.1.3.35	(1-Hydroxycyclohexan-1-yl)acetyl-CoA lyase (1-Hydroxycyclohexan-1-yl)acetyl-CoA lyase	
1.1.1.226	4-Hydroxycyclohexanecarboxylate dehydrogenase 4-Hydroxycyclohexanecarboxylate dehydrogenase	
1.1.1.166	Hydroxycyclohexanecarboxylate dehydrogenase Hydroxycyclohexanecarboxylate dehydrogenase	
1.14.13.66	2-Hydroxycyclohexanone 2-monooxygenase 2-Hydroxycyclohexanone 2-monooxygenase	

EC-No.	Synonym Recommended Name	Vol., Page
1.14.12.6	2-hydroxycyclohexanone 2-monooxygenase 2-hydroxycyclohexanone 2-monooxygenase	
1.1.99.26	3-Hydroxycyclohexanone:acceptor 1-oxidoreductase 3-Hydroxycyclohexanone dehydrogenase	
1.1.99.26	3-Hydroxycyclohexanone dehydrogenase 3-Hydroxycyclohexanone dehydrogenase	
1.14.13.13	25-Hydroxy D3-1α-hydroxylase Calcidiol 1-monooxygenase	
1.3.1.51	2'-Hydroxydaidzein:NADP$^+$ 2'-oxidoreductase 2'-Hydroxydaidzein reductase	
1.3.1.51	2'-Hydroxydaidzein reductase 2'-Hydroxydaidzein reductase	
1.1.1.66	10-Hydroxydecanoate:NAD$^+$ 10-oxidoreductase ω-Hydroxydecanoate dehydrogenase	
4.2.1.60	β-Hydroxydecanoate dehydrase 3-Hydroxydecanoyl-[acyl-carrier-protein] dehydratase	
1.1.1.66	ω-Hydroxydecanoate dehydrogenase ω-Hydroxydecanoate dehydrogenase	
4.2.1.60	3-Hydroxydecanoyl-[acyl-carrier-protein] dehydratase 3-Hydroxydecanoyl-[acyl-carrier-protein] dehydratase	
4.2.1.60	β-Hydroxydecanoyl-[acyl carrier protein] dehydrase 3-Hydroxydecanoyl-[acyl-carrier-protein] dehydratase	
4.2.1.60	3-Hydroxydecanoyl-acyl carrier protein dehydrase 3-Hydroxydecanoyl-[acyl-carrier-protein] dehydratase	
4.2.1.60	3-Hydroxydecanoyl-acyl carrier protein dehydratase 3-Hydroxydecanoyl-[acyl-carrier-protein] dehydratase	
4.2.1.60	β-Hydroxydecanoyl coenzyme A dehydrase 3-Hydroxydecanoyl-[acyl-carrier-protein] dehydratase	
4.2.1.60	β-Hydroxydecanoyl thioester dehydrase 3-Hydroxydecanoyl-[acyl-carrier-protein] dehydratase	
3.5.99.3	hydroxydechloroatrazine ethylaminohydrolase hydroxydechloroatrazine ethylaminohydrolase	
2.1.1.120	12-Hydroxydihydrochelirubine 12-O-methyltransferase 12-Hydroxydihydrochelirubine 12-O-methyltransferase	
2.1.1.120	12-Hydroxydihydrochelirubine 12-O-methyltransferase 12-Hydroxydihydrochelirubine 12-O-methyltransferase	

EC-No.	Synonym Recommended Name	Vol., Page
2.1.1.119	10-Hydroxydihydrosanguinarine-10-O-methyltransferase 10-Hydroxydihydrosanguinarine 10-O-methyltransferase	
2.1.1.119	10-Hydroxydihydrosanguinarine 10-O-methyltransferase 10-Hydroxydihydrosanguinarine 10-O-methyltransferase	
2.4.1.158	13-Hydroxydocosanoate 13-β-glucosyltransferase 13-Hydroxydocosanoate 13-β-glucosyltransferase	
1.1.1.232	15-Hydroxyeicosatetraenoate dehydrogenase 15-Hydroxyeicosatetraenoate dehydrogenase	
4.2.1.72	2-Hydroxyethylenedicarboxylate hydro-lyase Acetylenedicarboxylate hydratase	
2.7.1.50	Hydroxyethylthiazole kinase Hydroxyethylthiazole kinase	
1.1.1.98	D-2-Hydroxy fatty acid dehydrogenase (R)-2-Hydroxy-fatty-acid dehydrogenase	
1.1.1.99	L-2-Hydroxy fatty acid dehydrogenase (S)-2-Hydroxy-fatty-acid dehydrogenase	
1.1.1.98	2-Hydroxy fatty acid oxidase (R)-2-Hydroxy-fatty-acid dehydrogenase	
1.1.1.99	2-Hydroxy fatty acid oxidase (S)-2-Hydroxy-fatty-acid dehydrogenase	
2.1.1.69	5-Hydroxyfuranocoumarin 5-O-methyltransferase 5-Hydroxyfuranocoumarin 5-O-methyltransferase	
2.1.1.70	8-Hydroxyfuranocoumarin 8-O-methyltransferase 8-Hydroxyfuranocoumarin 8-O-methyltransferase	
2.6.1.23	4-Hydroxyglutamate aminotransferase 4-Hydroxyglutamate transaminase	
4.1.1.16	Hydroxyglutamate decarboxylase Hydroxyglutamate decarboxylase	
2.6.1.23	4-Hydroxyglutamate transaminase 4-Hydroxyglutamate transaminase	
1.1.99.2	L-α-Hydroxyglutarate:NAD$^+$ 2-oxidoreductase 2-Hydroxyglutarate dehydrogenase	
1.1.99.2	2-Hydroxyglutarate dehydrogenase 2-Hydroxyglutarate dehydrogenase	
1.1.99.2	L-α-Hydroxyglutarate dehydrogenase 2-Hydroxyglutarate dehydrogenase	

EC-No.	Synonym Recommended Name	Vol., Page
1.1.99.2	α-Hydroxyglutarate dehydrogenase 2-Hydroxyglutarate dehydrogenase	
1.1.99.2	α-Hydroxyglutarate dehydrogenase (NAD^+ specific) 2-Hydroxyglutarate dehydrogenase	
4.1.3.9	2-Hydroxyglutarate glyoxylate-lyase (CoA-propanoylating) 2-Hydroxyglutarate synthase	
1.1.99.2	α-Hydroxyglutarate oxidoreductase 2-Hydroxyglutarate dehydrogenase	
4.1.3.9	2-Hydroxyglutarate synthase 2-Hydroxyglutarate synthase	
4.1.3.9	Hydroxyglutarate synthase 2-Hydroxyglutarate synthase	
4.1.3.9	α-Hydroxyglutarate synthase 2-Hydroxyglutarate synthase	
4.1.3.9	2-Hydroxyglutaratic synthetase 2-Hydroxyglutarate synthase	
1.1.99.2	Hydroxyglutaric dehydrogenase 2-Hydroxyglutarate dehydrogenase	
4.1.3.9	2-Hydroxyglutaric synthetase 2-Hydroxyglutarate synthase	
4.3.2.5	α-Hydroxyglycine amidating dealkylase Peptidylamidoglycolate lyase	
1.1.1.230	3α-Hydroxyglycyrrhetinate:$NADP^+$ 3-oxidoreductase 3α-Hydroxyglycyrrhetinate dehydrogenase	
1.1.1.230	3α-Hydroxyglycyrrhetinate dehydrogenase 3α-Hydroxyglycyrrhetinate dehydrogenase	
4.1.1.68	2-Hydroxyhepta-2,4-diene-1,7-dioate isomerase/5-oxopent-3-ene-1,2,5-tricarboxylate decarboxylase 5-Oxopent-3-ene-1,2,5-tricarboxylate decarboxylase	
1.1.1.258	6-hydroxyhexanoate:NAD oxidoreductase 6-hydroxyhexanoate dehydrogenase	
1.1.1.258	6-hydroxyhexanoate dehydrogenase 6-hydroxyhexanoate dehydrogenase	
1.14.11.14	6β-Hydroxyhyoscyamine epoxidase 6β-Hydroxyhyoscyamine epoxidase	

EC-No.	Synonym Recommended Name	Vol., Page
2.1.1.4	Hydroxyindole-O-methyl transferase Acetylserotonin N-methyltransferase	
2.1.1.4	Hydroxyindole methyltransferase Acetylserotonin N-methyltransferase	
2.1.1.4	Hydroxyindole O-methyltransferase Acetylserotonin N-methyltransferase	
1.1.1.31	3-Hydroxyisobutyrate dehydrogenase 3-Hydroxyisobutyrate dehydrogenase	
1.1.1.31	β-Hydroxyisobutyrate dehydrogenase 3-Hydroxyisobutyrate dehydrogenase	
4.1.2.39	2-Hydroxyisobutyronitrile acetone-lyase Hydroxynitrilase	
1.1.1.35	3-Hydroxyisobutyryl-CoA dehydrogenase 3-Hydroxyacyl-CoA dehydrogenase	
3.1.2.4	3-hydroxyisobutyryl-CoA hydrolase 3-hydroxyisobutyryl-CoA hydrolase	
3.1.2.4	β-hydroxyisobutyryl-coenzyme A hydrolase 3-hydroxyisobutyryl-CoA hydrolase	
1.3.1.45	2'-Hydroxyisoflavone reductase 2'-Hydroxyisoflavone reductase	
4.1.3.26	Hydroxyisohexenylglutaryl-CoA:acetatelyase 3-Hydroxy-3-isohexenylglutaryl-CoA lyase	
4.1.3.26	Hydroxyisohexenylglutaryl CoA:acetate lyase 3-Hydroxy-3-isohexenylglutaryl-CoA lyase	
4.1.3.16	Hydroxyketoglutarate aldolase 4-Hydroxy-2-oxoglutarate aldolase	
4.1.3.16	Hydroxyketoglutaric aldolase 4-Hydroxy-2-oxoglutarate aldolase	
3.1.2.6	S-2-Hydroxylacylglutathione hydrolase Hydroxyacylglutathione hydrolase	
1.7.3.4	Hydroxylamine:oxygen oxidoreductase Hydroxylamine oxidase	
1.7.3.4	Hydroxylamine oxidase Hydroxylamine oxidase	
1.7.3.4	Hydroxylamine oxidoreductase Hydroxylamine oxidase	

EC-No.	Synonym Recommended Name	Vol., Page
1.7.99.1	Hydroxylamine reductase Hydroxylamine reductase	
1.6.6.11	Hydroxylamine reductase (NADH) Hydroxylamine reductase (NADH)	
1.14.15.3	1-Hydroxylase Alkane 1-monooxygenase	
1.14.99.10	21-Hydroxylase Steroid 21-monooxygenase	
1.14.15.3	ω-Hydroxylase Alkane 1-monooxygenase	
1.13.12.10	ω-N-hydroxylase Lysine 6-monooxygenase	
1.14.13.13	Hydroxylase, 25-hydroxcholecalciferol 1- Calcidiol 1-monooxygenase	
1.14.13.15	Hydroxylase, 5β-cholestane-3α,7α,12α-triol Cholestanetriol 26-monooxygenase	
1.14.13.12	Hydroxylase, benzoate 4- Benzoate 4-monooxygenase	
1.14.13.14	Hydroxylase, cinnamate 2- trans-Cinnamate 2-monooxygenase	
1.14.13.11	Hydroxylase, cinnamate 4- trans-Cinnamate 4-monooxygenase	
1.14.11.2	Hydroxylase, collagen proline Procollagen-proline dioxygenase	
1.14.99.2	Hydroxylase, kynurenate Kynurenine 7,8-hydroxylase	
1.14.99.2	Hydroxylase, kynurenate 7,8- Kynurenine 7,8-hydroxylase	
1.14.99.7	Hydroxylase, squalene Squalene monooxygenase	
1.14.13.54	Hydroxylase, steroid Ketosteroid monooxygenase	
1.14.13.13	1-Hydroxylase-25-hydroxyvitamin D3 Calcidiol 1-monooxygenase	
1.14.99.9	17α-Hydroxylase-C17,20 lyase Steroid 17α-monooxygenase	

EC-No.	Synonym Recommended Name	Vol., Page
2.7.1.78	5'-Hydroxyl polynucleotide kinase Polynucleotide 5'-hydroxy-kinase	
2.7.1.78	5'-Hydroxyl polyribonucleotide kinase Polynucleotide 5'-hydroxy-kinase	
2.7.1.78	5'-Hydroxyl RNA kinase Polynucleotide 5'-hydroxy-kinase	
1.97.1.2	Hydroxyltransferase, 1,2,3,5-tetrahydroxybenzene Pyrogallol hydroxyltransferase	
2.3.1.93	13-Hydroxylupanine acyltransferase 13-Hydroxylupanine O-tigloyltransferase	
2.3.1.93	13-Hydroxylupanine O-tigloyltransferase 13-Hydroxylupanine O-tigloyltransferase	
2.3.1.102	N^{ε}-Hydroxylysine:acetyl CoA N^{ε}-transacetylase N^{ε}-Hydroxylysine acetylase N^{6}-Hydroxylysine O-acetyltransferase	
2.4.1.50	Hydroxylysine galactosyltransferase Procollagen galactosyltransferase	
2.7.1.81	Hydroxylysine kinase Hydroxylysine kinase	
1.1.1.167	Hydroxymalonate:NAD^{+} oxidoreductase Hydroxymalonate dehydrogenase	
1.1.1.167	Hydroxymalonate dehydrogenase Hydroxymalonate dehydrogenase	
4.1.1.47	Hydroxymalonic semialdehyde carboxylase Tartronate-semialdehyde synthase	
1.1.3.19	4-Hydroxymandelate oxidase 4-Hydroxymandelate oxidase	
1.1.3.19	L-4-Hydroxymandelate oxidase (decarboxylating) 4-Hydroxymandelate oxidase	
2.4.1.178	Hydroxymandelonitrile glucosyltransferase Hydroxymandelonitrile glucosyltransferase	
4.1.2.11	Hydroxymandelonitrile lyase Hydroxymandelonitrile lyase	
2.1.1.108	6-Hydroxymellein O-methyltransferase 6-Hydroxymellein O-methyltransferase	

EC-No.	Synonym Recommended Name	Vol., Page
3.5.1.66	2-(Hydroxymethyl)-3-(acetamidomethylene)succinate amidohydrolase (deaminating, decarboxylating) 2-(Hydroxymethyl)-3-(acetamidomethylene)succinate hydrolase	
3.5.1.66	2-(Hydroxymethyl)-3-(acetamidomethylene)succinate hydrolase 2-(Hydroxymethyl)-3-(acetamidomethylene)succinate hydrolase	
1.1.1.257	4-(hydroxymethyl)benzenesulfonate:NAD oxidoreductase 4-(hydroxymethyl)benzenesulfonate dehydrogenase	
1.1.1.257	4-(hydroxymethyl)benzenesulfonate dehydrogenase 4-(hydroxymethyl)benzenesulfonate dehydrogenase	
3.5.1.66	α-Hydroxymethyl-α'-(N-acetylaminomethylene)succinic acid hydrolase 2-(Hydroxymethyl)-3-(acetamidomethylene)succinate hydrolase	
4.2.1.75	Hydroxymethylbilane hydro-lyase (cyclizing) Uroporphyrinogen-III synthase	
4.3.1.8	hydroxymethylbilane synthase hydroxymethylbilane synthase	
2.1.3.7	3-Hydroxymethylcephem carbamoyltransferase 3-Hydroxymethylcephem carbamoyltransferase	
2.7.6.3	Hydroxymethyldihydropteridine pyrophosphokinase 2-Amino-4-hydroxy-6-hydroxymethyldihydropteridine pyrophosphokinase	
3.1.2.5	hydroxymethylglutaryl-CoA hydrolase hydroxymethylglutaryl-CoA hydrolase	
4.1.3.4	Hydroxymethylglutaryl-CoA lyase Hydroxymethylglutaryl-CoA lyase	
1.1.1.88	Hydroxymethylglutaryl-CoA reductase Hydroxymethylglutaryl-CoA reductase	
1.1.1.34	Hydroxymethylglutaryl-CoA reductase (NADPH) Hydroxymethylglutaryl-CoA reductase (NADPH)	
3.1.3.47	[hydroxymethylglutaryl-CoA reductase (NADPH)]-phosphatase [hydroxymethylglutaryl-CoA reductase (NADPH)]-phosphatase	
3.1.3.47	[hydroxymethylglutaryl-CoA reductase (NADPH)]-phosphate phosphohydrolase [hydroxymethylglutaryl-CoA reductase (NADPH)]-phosphatase	
2.7.1.109	[Hydroxymethylglutaryl-CoA reductase (NADPH)] kinase [Hydroxymethylglutaryl-CoA reductase (NADPH)] kinase	
2.7.1.109	Hydroxymethylglutaryl-CoA reductase kinase [Hydroxymethylglutaryl-CoA reductase (NADPH)] kinase	

EC-No.	Synonym Recommended Name	Vol., Page
4.1.3.5	Hydroxymethylglutaryl-CoA synthase Hydroxymethylglutaryl-CoA synthase	
4.1.3.5	Hydroxymethylglutaryl-coenzyme A synthase Hydroxymethylglutaryl-CoA synthase	
1.1.1.34	Hydroxymethylglutaryl CoA reductase (NADPH) Hydroxymethylglutaryl-CoA reductase (NADPH)	
4.1.3.5	Hydroxymethylglutaryl CoA synthetase Hydroxymethylglutaryl-CoA synthase	
4.1.3.4	Hydroxymethylglutaryl coenzyme A-cleaving enzyme Hydroxymethylglutaryl-CoA lyase	
4.1.3.5	Hydroxymethylglutaryl coenzyme A-condensing enzyme Hydroxymethylglutaryl-CoA synthase	
3.1.2.5	hydroxymethylglutaryl coenzyme A deacylase hydroxymethylglutaryl-CoA hydrolase	
4.1.3.4	Hydroxymethylglutaryl coenzyme A lyase Hydroxymethylglutaryl-CoA lyase	
1.1.1.88	Hydroxymethylglutaryl coenzyme A reductase Hydroxymethylglutaryl-CoA reductase	
1.1.1.34	Hydroxymethylglutaryl coenzyme A reductase (reduced nicotinamide adenine dinucleotide phosphate) Hydroxymethylglutaryl-CoA reductase (NADPH)	
2.7.1.109	Hydroxymethylglutaryl coenzyme A reductase kinase [Hydroxymethylglutaryl-CoA reductase (NADPH)] kinase	
2.7.1.110	Hydroxymethylglutaryl coenzyme A reductase kinase kinase Dephospho-[reductase kinase] kinase	
4.1.3.5	Hydroxymethylglutaryl coenzyme A synthase Hydroxymethylglutaryl-CoA synthase	
2.7.1.49	Hydroxymethylpyrimidine kinase Hydroxymethylpyrimidine kinase	
2.7.4.7	Hydroxymethylpyrimidine phosphokinase Phosphomethylpyrimidine kinase	
2.1.2.7	Hydroxymethyltransferase, 2-methylserine D-Alanine 2-hydroxymethyltransferase	
2.1.2.8	Hydroxymethyltransferase, deoxycytidylate Deoxycytidylate 5-hydroxymethyltransferase	

EC-No.	Synonym Recommended Name	Vol., Page
2.1.2.11	Hydroxymethyltransferase, ketopantoate 3-Methyl-2-oxobutanoate hydroxymethyltransferase	
2.1.2.1	Hydroxymethyltransferase, serine Glycine hydroxymethyltransferase	
1.14.11.5	5-hydroxymethyluracil,2-oxoglutarate dioxygenase 5-hydroxymethyluracil,2-oxoglutarate dioxygenase	
1.14.11.6	5-Hydroxymethyluracil oxygenase Thymine dioxygenase	
3.7.1.9	2-Hydroxymuconate-semialdehyde formylhydrolase 2-Hydroxymuconate-semialdehyde hydrolase	
3.7.1.9	2-Hydroxymuconate-semialdehyde hydrolase 2-Hydroxymuconate-semialdehyde hydrolase	
1.2.1.32	2-Hydroxymuconate semialdehyde dehydrogenase Aminomuconate-semialdehyde dehydrogenase	
3.7.1.9	2-Hydroxymuconate semialdehyde hydrolase 2-Hydroxymuconate-semialdehyde hydrolase	
1.2.1.61	4-hydroxymuconic-semialdehyde:NAD oxidoreductase 4-hydroxymuconic-semialdehyde dehydrogenase	
1.2.1.61	4-hydroxymuconic-semialdehyde dehydrogenase 4-hydroxymuconic-semialdehyde dehydrogenase	
1.2.1.32	α-Hydroxymuconic ε-semialdehyde dehydrogenase Aminomuconate-semialdehyde dehydrogenase	
1.2.1.32	2-Hydroxymuconic acid semialdehyde dehydrogenase Aminomuconate-semialdehyde dehydrogenase	
1.2.1.32	2-Hydroxymuconic semialdehyde dehydrogenase Aminomuconate-semialdehyde dehydrogenase	
3.7.1.9	2-Hydroxymuconic semialdehyde hydrolase 2-Hydroxymuconate-semialdehyde hydrolase	
1.1.1.252	Hydroxynaphthalene reductase Tetrahydroxynaphthalene reductase	
1.3.7.1	6-Hydroxynicotinate reductase 6-Hydroxynicotinate reductase	
1.3.7.1	6-Hydroxynicotinate reductase 6-Hydroxynicotinate reductase	
1.5.3.6	D-6-Hydroxynicotine oxidase (R)-6-Hydroxynicotine oxidase	

EC-No.	Synonym Recommended Name	Vol., Page
1.5.3.5	L-6-Hydroxynicotine oxidase (S)-6-Hydroxynicotine oxidase	
1.3.7.1	6-Hydroxynicotinic reductase 6-Hydroxynicotinate reductase	
4.1.2.39	Hydroxynitrilase Hydroxynitrilase	
4.1.2.10	D-Hydroxynitrile lyase Mandelonitrile lyase	
4.1.2.10	D-α-hydroxynitrile lyase Mandelonitrile lyase	
4.1.2.37	Hydroxynitrile lyase Acetone-cyanhydrin lyase	
4.1.2.11	Hydroxynitrile lyase Hydroxymandelonitrile lyase	
4.1.2.39	Hydroxynitrile lyase Hydroxynitrilase	
4.1.2.10	Hydroxynitrile lyase Mandelonitrile lyase	
4.1.2.37	α-Hydroxynitrile lyase Acetone-cyanhydrin lyase	
4.1.2.39	α-Hydroxynitrile lyase Hydroxynitrilase	
4.2.1.59	3-Hydroxyoctanoyl-[acyl-carrier-protein] dehydratase 3-Hydroxyoctanoyl-[acyl-carrier-protein] dehydratase	
4.2.1.59	β-hydroxyoctanoyl-ACP-dehydrase 3-Hydroxyoctanoyl-[acyl-carrier-protein] dehydratase	
4.2.1.59	β-Hydroxyoctanoyl-acyl carrier protein dehydrase 3-Hydroxyoctanoyl-[acyl-carrier-protein] dehydratase	
4.2.1.59	D-3-Hydroxyoctanoyl-acyl carrier protein dehydratase 3-Hydroxyoctanoyl-[acyl-carrier-protein] dehydratase	
4.1.2.1	hydroxyoxobutyrate aldolase hydroxyoxobutyrate aldolase	
4.2.1.61	3-Hydroxypalmitoyl-[acyl-carrier-protein] dehydratase 3-Hydroxypalmitoyl-[acyl-carrier-protein] dehydratase	
4.2.1.61	D-3-Hydroxypalmitoyl-[acyl carrier protein] dehydratase 3-Hydroxypalmitoyl-[acyl-carrier-protein] dehydratase	

EC-No.	Synonym Recommended Name	Vol., Page
4.2.1.61	β-Hydroxypalmitoyl-acyl carrier protein dehydrase 3-Hydroxypalmitoyl-[acyl-carrier-protein] dehydratase	
4.2.1.61	β-Hydroxypalmityl-ACP dehydrase 3-Hydroxypalmitoyl-[acyl-carrier-protein] dehydratase	
4.2.1.80	2-hydroxypent-2,4-dienoate hydratase 2-oxopent-4-enoate hydratase	
2.8.3.14	5-Hydroxypentanoate CoA-transferase 5-Hydroxypentanoate CoA-transferase	
1.1.1.196	15-Hydroxy PGD_2 dehydrogenase 15-Hydroxyprostaglandin-D dehydrogenase ($NADP^+$)	
4.2.1.88	1-(4-Hydroxyphenyl)-2-(methylamino)ethanol hydro-lyase (methylamine-forming) Synephrine dehydratase	
3.1.2.24	2-(2-hydroxyphenyl) benzenesulfinate:H_2O hydrolase 2-(2-hydroxyphenyl)benzenesulfinate hydrolase	
3.1.2.24	2-(2-hydroxyphenyl)benzenesulfinate hydrolase 2-(2-hydroxyphenyl)benzenesulfinate hydrolase	
1.14.13.4	3-(2-Hydroxyphenyl)propanoate,NADH:oxygen oxidoreductase (3-hydroxylating) Melilotate 3-monooxygenase	
1.3.1.11	3-(2-Hydroxyphenyl)propanoate:NAD^+ oxidoreductase Coumarate reductase	
5.2.1.11	4-Hydroxyphenylacetaldehyde-oxime (E)-(Z)-isomerase 4-Hydroxyphenylacetaldehyde-oxime isomerase	Vol. 1, Page 233
5.2.1.11	4-Hydroxyphenylacetaldehyde-oxime isomerase 4-Hydroxyphenylacetaldehyde-oxime isomerase	Vol. 1, Page 233
1.2.1.53	4-Hydroxyphenylacetaldehyde:NAD^+ oxidoreductase 4-Hydroxyphenylacetaldehyde dehydrogenase	
1.2.1.53	4-Hydroxyphenylacetaldehyde dehydrogenase 4-Hydroxyphenylacetaldehyde dehydrogenase	
4.2.1.78	4-Hydroxyphenylacetaldehyde hydro-lyase (adding dopamine) (S)-Norcoclaurine synthase	
1.14.13.68	4-hydroxyphenylacetaldehyde oxime, $NADPH_2$:oxygen oxidoreductase 4-hydroxyphenylacetaldehyde oxime monooxygenase	
1.14.13.68	4-hydroxyphenylacetaldehyde oxime monooxygenase 4-hydroxyphenylacetaldehyde oxime monooxygenase	

EC-No.	Synonym Recommended Name	Vol., Page
1.14.13.63	3-Hydroxyphenylacetate, NAD(P)H_2:oxygen oxidoreductase (6-hydroxylating) 3-Hydroxyphenylacetate 6-hydroxylase	
1.14.13.18	4-Hydroxyphenylacetate,NAD(P)H:oxygen oxidoreductase (1-hydroxylating) 4-Hydroxyphenylacetate 1-monooxygenase	
1.14.13.3	4-Hydroxyphenylacetate,NADH:oxygen oxidoreductase (3-hydroxylating) 4-Hydroxyphenylacetate 3-monooxygenase	
1.14.13.18	4-Hydroxyphenylacetate 1-hydroxylase 4-Hydroxyphenylacetate 1-monooxygenase	
1.14.13.18	4-Hydroxyphenylacetate 1-monooxygenase 4-Hydroxyphenylacetate 1-monooxygenase	
1.14.13.3	4-Hydroxyphenylacetate 3-hydroxylase 4-Hydroxyphenylacetate 3-monooxygenase	
1.14.13.3	*p*-Hydroxyphenylacetate 3-hydroxylase 4-Hydroxyphenylacetate 3-monooxygenase	
1.14.13.3	4-Hydroxyphenylacetate 3-monooxygenase 4-Hydroxyphenylacetate 3-monooxygenase	
1.14.13.63	3-Hydroxyphenylacetate 6-hydroxylase 3-Hydroxyphenylacetate 6-hydroxylase	
1.14.13.63	*m*-Hydroxyphenylacetate 6-hydroxylase 3-Hydroxyphenylacetate 6-hydroxylase	
1.14.13.63	3-Hydroxyphenylacetate 6-monooxygenase 3-Hydroxyphenylacetate 6-hydroxylase	
1.14.13.3	*p*-Hydroxyphenylacetate hydroxylase 4-Hydroxyphenylacetate 3-monooxygenase	
1.14.13.18	4-Hydroxyphenylacetic 1-hydroxylase 4-Hydroxyphenylacetate 1-monooxygenase	
1.14.13.3	4-Hydroxyphenylacetic acid-3-hydroxylase 4-Hydroxyphenylacetate 3-monooxygenase	
1.14.13.42	4-Hydroxyphenylacetonitrile,NADPH:oxygen oxidoreductase (2-hydroxylating) Hydroxyphenylacetonitrile 2-monooxygenase	
1.14.13.42	Hydroxyphenylacetonitrile 2-monooxygenase Hydroxyphenylacetonitrile 2-monooxygenase	
1.14.13.42	4-Hydroxyphenylacetonitrile hydroxylase Hydroxyphenylacetonitrile 2-monooxygenase	

EC-No.	Synonym Recommended Name Vol., Page
2.6.1.72	D-4-Hydroxyphenylglycine:2-oxoglutarate aminotransferase D-4-Hydroxyphenylglycine transaminase
2.6.1.72	D-4-Hydroxyphenylglycine transaminase D-4-Hydroxyphenylglycine transaminase
1.1.1.237	4-Hydroxyphenyllactate:NAD^+ oxidoreductase Hydroxyphenylpyruvate reductase
1.14.13.4	2-Hydroxyphenylpropionate hydroxylase Melilotate 3-monooxygenase
1.14.13.4	2-Hydroxyphenylpropionic hydroxylase Melilotate 3-monooxygenase
1.2.3.13	4-hydroxyphenylpyruvate:oxygen oxidoreductase (decarboxylating) 4-hydroxyphenylpyruvate oxidase
1.13.11.27	4-Hydroxyphenylpyruvate:oxygen oxidoreductase (hydroxylating, decarboxylating) 4-Hydroxyphenylpyruvate dioxygenase
1.13.11.27	4-Hydroxyphenylpyruvate dioxygenase 4-Hydroxyphenylpyruvate dioxygenase
1.13.11.27	*p*-Hydroxyphenylpyruvate dioxygenase 4-Hydroxyphenylpyruvate dioxygenase
1.13.11.27	*p*-Hydroxyphenylpyruvate hydroxylase 4-Hydroxyphenylpyruvate dioxygenase
1.2.3.13	4-hydroxyphenylpyruvate oxidase 4-hydroxyphenylpyruvate oxidase
1.13.11.27	*p*-Hydroxyphenylpyruvate oxidase 4-Hydroxyphenylpyruvate dioxygenase
1.1.1.237	Hydroxyphenylpyruvate reductase Hydroxyphenylpyruvate reductase
1.3.1.12	Hydroxyphenylpyruvate synthase Prephenate dehydrogenase
1.13.11.27	4-Hydroxyphenylpyruvic acid dioxygenase 4-Hydroxyphenylpyruvate dioxygenase
1.13.11.27	*p*-Hydroxyphenylpyruvic acid hydroxylase 4-Hydroxyphenylpyruvate dioxygenase
1.13.11.27	*p*-Hydroxyphenylpyruvic hydroxylase 4-Hydroxyphenylpyruvate dioxygenase

EC-No.	Synonym Recommended Name	Vol., Page
1.13.11.27	*p*-Hydroxyphenylpyruvic oxidase 4-Hydroxyphenylpyruvate dioxygenase	
1.1.3.27	L-2-Hydroxyphytanate:oxygen 2-oxidoreductase Hydroxyphytanate oxidase	
1.1.3.27	Hydroxyphytanate oxidase Hydroxyphytanate oxidase	
1.1.3.27	L-2-Hydroxyphytanic acid oxidase Hydroxyphytanate oxidase	
1.1.1.259	3-hydroxypimeloyl-CoA:NAD oxidoreductase 3-hydroxypimeloyl-CoA dehydrogenase	
1.1.1.259	3-hydroxypimeloyl-CoA dehydrogenase 3-hydroxypimeloyl-CoA dehydrogenase	
1.14.13.54	17α-Hydroxyprogesterone, $NADPH_2$:oxygen oxidoreductase (20-hydroxylating, side-chain cleaving) Ketosteroid monooxygenase	
4.1.2.30	17α-Hydroxyprogesterone acetaldehyde-lyase 17α-Hydroxyprogesterone aldolase	
4.1.2.30	17α-Hydroxyprogesterone aldolase 17α-Hydroxyprogesterone aldolase	
4.2.1.86	16α-Hydroxyprogesterone dehydratase 16-Dehydroprogesterone hydratase	
4.2.1.98	16α-Hydroxyprogesterone dehydratase 16α-Hydroxyprogesterone dehydratase	
4.2.1.98	16α-Hydroxyprogesterone dehydroxylase 16α-Hydroxyprogesterone dehydratase	
4.2.1.98	Hydroxyprogesterone dehydroxylase 16α-Hydroxyprogesterone dehydratase	
4.2.1.86	16α-Hydroxyprogesterone hydro-lyase 16-Dehydroprogesterone hydratase	
4.2.1.98	16α-Hydroxyprogesterone hydro-lyase 16α-Hydroxyprogesterone dehydratase	
5.1.1.8	4-Hydroxyproline 2-epimerase 4-Hydroxyproline epimerase	Vol. 1, Page 33
5.1.1.8	Hydroxyproline 2-epimerase 4-Hydroxyproline epimerase	Vol. 1, Page 33

EC-No.	Synonym Recommended Name	Vol., Page
5.1.1.8	4-Hydroxyproline epimerase 4-Hydroxyproline epimerase	Vol. 1, Page 33
5.1.1.8	Hydroxyproline epimerase 4-Hydroxyproline epimerase	Vol. 1, Page 33
5.1.1.8	L-Hydroxyproline epimerase 4-Hydroxyproline epimerase	Vol. 1, Page 33
1.1.1.59	3-Hydroxypropanoate:NAD+ oxidoreductase 3-Hydroxypropionate dehydrogenase	
4.2.1.71	3-Hydroxypropenoate hydro-lyase Acetylenecarboxylate hydratase	
1.1.1.202	3-Hydroxypropionaldehyde reductase 1,3-Propanediol 1-dehydrogenase	
1.1.1.59	3-Hydroxypropionate dehydrogenase 3-Hydroxypropionate dehydrogenase	
1.1.1.196	15-Hydroxyprostaglandin-D dehydrogenase (NADP+) 15-Hydroxyprostaglandin-D dehydrogenase (NADP+)	
1.1.1.231	15-Hydroxyprostaglandin-I dehydrogenase (NADP+) 15-Hydroxyprostaglandin-I dehydrogenase (NADP+)	
1.1.1.141	15-Hydroxyprostaglandin dehydrogenase (NAD+) 15-Hydroxyprostaglandin dehydrogenase (NAD+)	
1.1.1.196	15-Hydroxyprostaglandin dehydrogenase (NADP) 15-Hydroxyprostaglandin-D dehydrogenase (NADP+)	
1.1.1.197	15-Hydroxyprostaglandin dehydrogenase (NADP+) 15-Hydroxyprostaglandin dehydrogenase (NADP+)	
1.1.1.141	15-Hydroxyprostanoic dehydrogenase 15-Hydroxyprostaglandin dehydrogenase (NAD+)	
2.4.2.8	6-Hydroxypurine phosphoribosyltransferase Hypoxanthine phosphoribosyltransferase	
1.14.99.26	2-Hydroxypyridine,hydrogen-donor:oxygen oxidoreductase (5-hydroxylating) 2-Hydroxypyridine 5-monooxygenase	
1.14.99.26	2-Hydroxypyridine 5-monooxygenase 2-Hydroxypyridine 5-monooxygenase	
1.14.99.26	2-Hydroxypyridine oxygenase 2-Hydroxypyridine 5-monooxygenase	

EC-No.	Synonym Recommended Name	Vol., Page
2.6.1.51	Hydroxypyruvate:L-alanine transaminase Serine-pyruvate transaminase	
4.1.1.40	Hydroxypyruvate carboxy-lyase Hydroxypyruvate decarboxylase	
4.1.1.40	Hydroxypyruvate decarboxylase Hydroxypyruvate decarboxylase	
5.3.1.22	Hydroxypyruvate isomerase Hydroxypyruvate isomerase	Vol. 1, Page 349
5.3.1.22	Hydroxypyruvate ketol-isomerase Hydroxypyruvate isomerase	Vol. 1, Page 349
1.1.1.29	Hydroxypyruvate reductase Glycerate dehydrogenase	
1.1.1.81	Hydroxypyruvate reductase Hydroxypyruvate reductase	
1.1.1.81	β-Hydroxypyruvate reductase Hydroxypyruvate reductase	
2.6.1.52	Hydroxypyruvic phosphate-glutamic transaminase Phosphoserine transaminase	
2.1.1.88	8-Hydroxyquercitin 8-O-methyltransferase 8-Hydroxyquercitin 8-O-methyltransferase	
1.13.11.37	Hydroxyquinol 1,2-dioxygenase Hydroxyquinol 1,2-dioxygenase	
1.14.13.62	4-Hydroxyquinoline 3-monooxygenase 4-Hydroxyquinoline 3-monooxygenase	
1.14.12.16	2-Hydroxyquinoline 5,6-dioxygenase 2-Hydroxyquinoline 5,6-dioxygenase	
1.14.12.16	2-Hydroxyquinoline 5,6-dioxygenase 2-Hydroxyquinoline 5,6-dioxygenase	
1.14.13.61	2-Hydroxyquinoline 8-monooxygenase 2-Hydroxyquinoline 8-monooxygenase	
1.14.13.65	2-Hydroxyquinoline 8-monooxygenase 2-Hydroxyquinoline 8-monooxygenase	
5.1.99.2	16-Hydroxysteroid 16-epimerase 16-Hydroxysteroid epimerase	Vol. 1, Page 183
1.1.1.147	16α-Hydroxysteroid:NAD(P)$^+$ 16-oxidoreductase 16α-Hydroxysteroid dehydrogenase	

EC-No.	Synonym Recommended Name	Vol., Page
1.1.1.148	17α-Hydroxysteroid:NAD(P)$^+$ 17-oxidoreductase Estradiol 17α-dehydrogenase	
1.1.1.149	20α-Hydroxysteroid:NAD(P)$^+$ 20-oxidoreductase 20α-Hydroxysteroid dehydrogenase	
1.1.1.213	3α-Hydroxysteroid:NAD(P)$^+$ oxidoreductase (A-specific) 3α-Hydroxysteroid dehydrogenase (A-specific)	
1.1.1.50	3α-Hydroxysteroid:NAD(P)$^+$ oxidoreductase (B-specific) 3α-Hydroxysteroid dehydrogenase (B-specific)	
1.1.1.53	3α,20β-Hydroxysteroid:NAD$^+$-oxidoreductase 3α(or 20β)-Hydroxysteroid dehydrogenase	
1.1.1.63	17β-Hydroxysteroid:NAD$^+$ 17-oxidoreductase Testosterone 17β-dehydrogenase	
1.1.1.150	21-Hydroxysteroid:NAD$^+$ 21-oxidoreductase 21-Hydroxysteroid dehydrogenase (NAD$^+$)	
1.1.1.159	7α-Hydroxysteroid:NAD$^+$ 7-oxidoreductase 7α-Hydroxysteroid dehydrogenase	
1.1.1.146	11β-Hydroxysteroid:NADP$^+$ 11-oxidoreductase 11β-Hydroxysteroid dehydrogenase	
1.1.1.176	12α-Hydroxysteroid:NADP$^+$ 12-oxidoreductase 12α-Hydroxysteroid dehydrogenase	
1.1.1.238	12β-Hydroxysteroid:NADP$^+$ 12-oxidoreductase 12β-Hydroxysteroid dehydrogenase	
1.1.1.64	17β-Hydroxysteroid:NADP$^+$ 17-oxidoreductase Testosterone 17β-dehydrogenase (NADP$^+$)	
1.1.1.151	21-Hydroxysteroid:NADP$^+$ 21-oxidoreductase 21-Hydroxysteroid dehydrogenase (NADP$^+$)	
1.1.1.201	7β-Hydroxysteroid:NADP$^+$ 7-oxidoreductase 7β-Hydroxysteroid dehydrogenase (NADP$^+$)	
1.1.3.6	3β-Hydroxysteroid:oxygen oxidoreductase Cholesterol oxidase	
4.2.1.62	5-α-Hydroxysteroid dehydratase 5-α-Hydroxysteroid dehydratase	
1.1.1.146	11β-Hydroxy steroid dehydrogenase 11β-Hydroxysteroid dehydrogenase	
1.1.1.146	11β-Hydroxysteroid dehydrogenase 11β-Hydroxysteroid dehydrogenase	

EC-No.	Synonym Recommended Name	Vol., Page
1.1.1.176	12α-Hydroxy steroid dehydrogenase 12α-Hydroxysteroid dehydrogenase	
1.1.1.176	12α-Hydroxysteroid dehydrogenase 12α-Hydroxysteroid dehydrogenase	
1.1.1.238	12β-Hydroxysteroid dehydrogenase 12β-Hydroxysteroid dehydrogenase	
1.1.1.147	16α-Hydroxysteroid dehydrogenase 16α-Hydroxysteroid dehydrogenase	
1.1.1.148	17α-Hydroxy steroid dehydrogenase Estradiol 17α-dehydrogenase	
1.1.1.62	17β,20α-Hydroxysteroid dehydrogenase Estradiol 17β-dehydrogenase	
1.1.1.51	17β-Hydroxy steroid dehydrogenase 3(or 17)β-Hydroxysteroid dehydrogenase	
1.1.1.149	20α-Hydroxy steroid dehydrogenase 20α-Hydroxysteroid dehydrogenase	
1.1.1.149	20α-Hydroxysteroid dehydrogenase 20α-Hydroxysteroid dehydrogenase	
1.1.1.62	20α-Hydroxysteroid dehydrogenase Estradiol 17β-dehydrogenase	
1.1.1.53	20β-Hydroxysteroid dehydrogenase 3α(or 20β)-Hydroxysteroid dehydrogenase	
1.1.1.209	3(17)α-Hydroxysteroid dehydrogenase 3(or 17)α-Hydroxysteroid dehydrogenase	
1.1.1.51	3β-Hydroxysteroid dehydrogenase 3(or 17)β-Hydroxysteroid dehydrogenase	
1.1.1.159	7α-Hydroxy steroid dehydrogenase 7α-Hydroxysteroid dehydrogenase	
1.1.1.159	7α-Hydroxysteroid dehydrogenase 7α-Hydroxysteroid dehydrogenase	
1.1.1.145	Δ^5-3β-Hydroxysteroid dehydrogenase 3β-Hydroxy-Δ^5-steroid dehydrogenase	
1.1.1.146	β-Hydroxysteroid dehydrogenase 11β-Hydroxysteroid dehydrogenase	
1.1.1.213	3α-Hydroxysteroid dehydrogenase (A-specific) 3α-Hydroxysteroid dehydrogenase (A-specific)	

EC-No.	Synonym Recommended Name	Vol., Page
1.1.1.50	3α-Hydroxysteroid dehydrogenase (B-specific) 3α-Hydroxysteroid dehydrogenase (B-specific)	
1.1.1.150	21-Hydroxysteroid dehydrogenase (NAD^+) 21-Hydroxysteroid dehydrogenase (NAD^+)	
1.1.1.239	3α(17β)-Hydroxysteroid dehydrogenase (NAD^+) 3α(17β)-Hydroxysteroid dehydrogenase (NAD^+)	
1.1.1.151	21-Hydroxysteroid dehydrogenase ($NADP^+$) 21-Hydroxysteroid dehydrogenase ($NADP^+$)	
1.1.1.201	7β-Hydroxysteroid dehydrogenase ($NADP^+$) 7β-Hydroxysteroid dehydrogenase ($NADP^+$)	
1.1.1.151	21-Hydroxy steroid dehydrogenase (nicotinamide adenine dinucleotide phosphate) 21-Hydroxysteroid dehydrogenase ($NADP^+$)	
1.1.1.145	3β-Hydroxy steroid dehydrogenase/isomerase 3β-Hydroxy-Δ^5-steroid dehydrogenase	
5.1.99.2	16-Hydroxy steroid epimerase 16-Hydroxysteroid epimerase	Vol. 1, Page 183
5.1.99.2	16-Hydroxysteroid epimerase 16-Hydroxysteroid epimerase	Vol. 1, Page 183
5.3.3.1	Hydroxysteroid isomerase Steroid Δ-isomerase	Vol. 1, Page 376
1.1.1.148	17α-Hydroxy steroid oxidoreductase Estradiol 17α-dehydrogenase	
1.1.1.148	17α-Hydroxysteroid oxidoreductase Estradiol 17α-dehydrogenase	
1.1.1.50	3α-Hydroxysteroid oxidoreductase 3α-Hydroxysteroid dehydrogenase (B-specific)	
1.1.1.210	3β,20α-Hydroxysteroid oxidoreductase 3β(or 20α)-Hydroxysteroid dehydrogenase	
1.1.3.6	3β-Hydroxy steroid oxidoreductase Cholesterol oxidase	
3.1.6.2	3-β-hydroxysteroid sulfate sulfatase Steryl-sulfatase	
2.8.2.2	Δ^5-3β-Hydroxysteroid sulfokinase Alcohol sulfotransferase	

EC-No.	Synonym Recommended Name	Vol., Page
2.8.2.2	3-Hydroxysteroid sulfotransferase Alcohol sulfotransferase	
2.8.2.2	3β-Hydroxy steroid sulfotransferase Alcohol sulfotransferase	
2.8.2.2	3β-Hydroxysteroid sulfotransferase Alcohol sulfotransferase	
2.8.2.2	Hydroxysteroid sulfotransferase Alcohol sulfotransferase	
2.4.1.17	17β-Hydroxysteroid UDP-glucuronosyltransferase Glucuronosyltransferase	
2.4.1.17	3α-Hydroxysteroid UDP-glucuronosyltransferase Glucuronosyltransferase	
4.2.1.62	5α-Hydroxysterol dehydase 5-α-Hydroxysteroid dehydratase	
4.1.3.29	3-Hydroxytetradecane-1,3,4-tricarboxylate 2-oxoglutarate-lyase (CoA-acylating) Decylhomocitrate synthase	
4.2.1.96	4a-Hydroxytetrahydrobiopterin dehydratase 4a-Hydroxytetrahydrobiopterin dehydratase	
4.2.1.96	4a-Hydroxytetrahydrobiopterin dehydratase 4a-Hydroxytetrahydrobiopterin dehydratase	
4.2.1.96	4a-Hydroxytetrahydrobiopterin hydro-lyase 4a-Hydroxytetrahydrobiopterin dehydratase	
4.2.1.96	4a-Hydroxytetrahydropterin dehydratase 4a-Hydroxytetrahydrobiopterin dehydratase	
1.1.1.262	4-hydroxythreonine-4-phosphate dehydrogenase 4-hydroxythreonine-4-phosphate dehydrogenase	
2.6.1.27	5-Hydroxytryptophan-ketoglutaric transaminase Tryptophan transaminase	
2.6.1.27	Hydroxytryptophan aminotransferase Tryptophan transaminase	
4.1.1.28	5-Hydroxytryptophan decarboxylase Aromatic-L-amino-acid decarboxylase	
4.1.1.28	Hydroxytryptophan decarboxylase Aromatic-L-amino-acid decarboxylase	

EC-No.	Synonym Recommended Name	Vol., Page
4.1.1.28	L-5-Hydroxytryptophan decarboxylase Aromatic-L-amino-acid decarboxylase	
2.8.3.14	5-Hydroxyvalerate CoA-transferase 5-Hydroxypentanoate CoA-transferase	
2.8.3.14	5-Hydroxyvalerate coenzyme A transferase 5-Hydroxypentanoate CoA-transferase	
1.14.13.13	25-Hydroxyvitamin D_3 1α-hydroxylase Calcidiol 1-monooxygenase	
2.7.1.119	Hygromycin-B kinase Hygromycin-B kinase	
2.7.1.119	Hygromycin B phosphotransferase Hygromycin-B kinase	
1.14.11.11	Hyoscyamine (6S)-dioxygenase Hyoscyamine (6S)-dioxygenase	
1.14.11.11	L-Hyoscyamine,2-oxoglutarate:oxygen oxidoreductase ((6S)-hydroxylating) Hyoscyamine (6S)-dioxygenase	
1.14.11.11	Hyoscyamine 6-hydroxylase Hyoscyamine (6S)-dioxygenase	
1.14.11.11	Hyoscyamine 6β-hydroxylase Hyoscyamine (6S)-dioxygenase	
3.4.21.49	hypoderma collagenase hypodermin C	
3.4.21.49	hypodermin C hypodermin C	
1.6.6.6	Hyponitrite reductase Hyponitrite reductase	
1.8.1.3	Hypotaurine:NAD^+ oxidoreductase Hypotaurine dehydrogenase	
1.8.1.3	Hypotaurine dehydrogenase Hypotaurine dehydrogenase	
2.7.3.6	Hypotaurocyamine kinase Hypotaurocyamine kinase	
3.2.2.15	Hypoxanthine-DNA glycosylase DNA-deoxyinosine glycosidase	
2.4.2.8	Hypoxanthine-guanine phosphoribosyltransferase Hypoxanthine phosphoribosyltransferase	

EC-No.	Synonym Recommended Name	Vol., Page
1.1.3.22	Hypoxanthine-xanthine oxidase Xanthine oxidase	
1.1.3.22	Hypoxanthine:oxygen oxidoreductase Xanthine oxidase	
1.1.3.22	Hypoxanthine oxidase Xanthine oxidase	
2.4.2.8	Hypoxanthine phosphoribosyltransferase Hypoxanthine phosphoribosyltransferase	
2.4.1.121	IAA-glucose synthase Indole-3-acetate β-glucosyltransferase	
6.3.2.20	IAA-lysine synthetase Indoleacetate-lysine synthetase	Vol. 2, Page 511
2.4.1.121	IAGlu synthase Indole-3-acetate β-glucosyltransferase	
2.6.1.9	IAP transaminase Histidinol-phosphate transaminase	
3.1.3.48	IAR protein-tyrosine-phosphatase	
3.4.22.36	ICE Interleukin 1β converting enzyme	
2.3.1.119	Icosanoyl-CoA synthase Icosanoyl-CoA synthase	
3.4.24.56	IDE Insulysin	
1.1.1.15	D-Iditol 2-dehydrogenase D-Iditol 2-dehydrogenase	
1.1.1.14	L-Iditol 2-dehydrogenase L-Iditol 2-dehydrogenase	
1.1.1.15	D-Iditol:NAD$^+$ 2-oxidoreductase D-Iditol 2-dehydrogenase	
1.1.1.14	L-Iditol:NAD$^+$ 2-oxidoreductase L-Iditol 2-dehydrogenase	
1.1.1.14	L-Iditol:NAD$^+$ 5-oxidoreductase L-Iditol 2-dehydrogenase	
1.1.1.14	L-Iditol:NAD oxidoreductase L-Iditol 2-dehydrogenase	

EC-No.	Synonym Recommended Name	Vol., Page
1.1.1.14	L-Iditol dehydrogenase (sorbitol) L-Iditol 2-dehydrogenase	
1.1.1.128	L-Idonate 2-dehydrogenase L-Idonate 2-dehydrogenase	
1.1.1.264	L-idonate 5-dehydrogenase L-idonate 5-dehydrogenase	
1.1.1.264	L-idonate:NAD(P) oxidoreductase L-idonate 5-dehydrogenase	
1.1.1.128	L-Idonate:NADP$^+$ 2-oxidoreductase L-Idonate 2-dehydrogenase	
3.6.1.6	IDPase Nucleoside-diphosphatase	
3.1.6.13	Iduronate-2-sulfatase Iduronate 2-sulfatase	
3.1.6.13	L-Iduronate-2-sulfate 2-sulfohydrolase Iduronate 2-sulfatase	
3.1.6.13	Iduronate-2-sulfate sulfatase Iduronate 2-sulfatase	
3.1.6.13	Iduronate 2-sulfatase Iduronate 2-sulfatase	
3.1.6.13	L-Iduronate 2-sulfate sulfatase Iduronate 2-sulfatase	
3.1.6.13	Iduronate sulfatase Iduronate 2-sulfatase	
3.1.6.13	Iduronate sulfate sulfatase Iduronate 2-sulfatase	
3.2.1.76	.α.-L-Iduronidase L-Iduronidase	
3.2.1.76	α-L-iduronidase L-Iduronidase	
3.2.1.76	L-Iduronidase L-Iduronidase	
3.2.1.76	L-Iduronidase L-Iduronidase	
3.2.1.76	Iduronidase,.α.-L- L-Iduronidase	

EC-No.	Synonym Recommended Name	Vol., Page
3.1.6.13	Iduronide-2-sulfate sulfatase Iduronate 2-sulfatase	
3.1.6.13	Idurono-2-sulfatase Iduronate 2-sulfatase	
3.1.6.13	L-Idurono sulfate sulfatase Iduronate 2-sulfatase	
3.4.24.13	IgA-specific metalloendopeptidase IgA-specific metalloendopeptidase	
3.4.21.72	IgA-specific proteinase IgA-specific serine endopeptidase	
3.4.21.72	IgA-specific serine endopeptidase IgA-specific serine endopeptidase	
3.4.24.13	IgA1-specific proteinase IgA-specific metalloendopeptidase	
3.4.24.13	IgA1 protease IgA-specific metalloendopeptidase	
3.4.24.13	IgA1 proteinase IgA-specific metalloendopeptidase	
3.4.21.72	IgA protease IgA-specific serine endopeptidase	
3.4.21.72	IgA proteinase IgA-specific serine endopeptidase	
3.1.4.46	IgD-binding protein D Glycerophosphodiester phosphodiesterase	
4.2.1.19	IGP dehydratase imidazoleglycerol-phosphate dehydratase	
4.1.1.48	IGPS Indole-3-glycerol phosphate synthase	
5.3.1.24	IGPS:PRAI Phosphoribosylanthranilate isomerase	Vol. 1, Page 353
2.7.1.37	βIIPKC Protein kinase	
6.1.1.5	IleRS Isoleucine-tRNA ligase	Vol. 2, Page 33
1.1.1.111	3-(Imidazol-5-yl)lactate dehydrogenase 3-(Imidazol-5-yl)lactate dehydrogenase	

EC-No.	Synonym Recommended Name	Vol., Page
1.1.1.111	Imidazol-5-yl lactate dehydrogenase 3-(Imidazol-5-yl)lactate dehydrogenase	
1.14.13.5	4-Imidazoleacetate,NADH:oxygen oxidoreductase (5-hydroxylating) Imidazoleacetate 4-monooxygenase	
6.3.4.8	Imidazoleacetate-phosphoribosyldiphosphate ligase Imidazoleacetate-phosphoribosyldiphosphate ligase	Vol. 2, Page 611
1.14.13.5	Imidazoleacetate 4-monooxygenase Imidazoleacetate 4-monooxygenase	
6.3.4.8	Imidazoleacetate:5-phosphoribosyl-diphosphate ligase (ADP-and pyrophosphate-forming) Imidazoleacetate-phosphoribosyldiphosphate ligase	Vol. 2, Page 611
1.14.13.5	Imidazoleacetate hydroxylase Imidazoleacetate 4-monooxygenase	
1.14.13.5	Imidazoleacetic hydroxylase Imidazoleacetate 4-monooxygenase	
1.14.13.5	Imidazoleacetic monooxygenase Imidazoleacetate 4-monooxygenase	
2.6.1.9	Imidazoleacetol phosphate transaminase Histidinol-phosphate transaminase	
2.3.1.2	Imidazole acetylase Imidazole N-acetyltransferase	
2.3.1.2	Imidazole acetyltransferase Imidazole N-acetyltransferase	
4.2.1.19	imidazoleglycerol-phosphate dehydratase imidazoleglycerol-phosphate dehydratase	
4.2.1.19	imidazoleglycerolphoshate dehydratase imidazoleglycerol-phosphate dehydratase	
4.2.1.19	imidazoleglycerol phosphate dehydrazase imidazoleglycerol-phosphate dehydratase	
2.1.1.8	Imidazolemethyltransferase Histamine N-methyltransferase	
2.3.1.2	Imidazole N-acetyltransferase Imidazole N-acetyltransferase	
4.2.1.49	4'-Imidazolone-5'-propionate hydro-lyase Urocanate hydratase	

EC-No.	Synonym Recommended Name	Vol., Page
4.2.1.49	4-Imidazolone-5'-propionate hydrolyase Urocanate hydratase	
3.5.2.7	4-Imidazolone-5-propanoate amidohydrolase Imidazolonepropionase	
3.5.2.7	Imidazolonepropionase Imidazolonepropionase	
4.2.1.49	Imidazolonepropionate hydrolase Urocanate hydratase	
2.6.1.9	Imidazolylacetolphosphate aminotransferase Histidinol-phosphate transaminase	
2.6.1.9	Imidazolylacetolphosphate transaminase Histidinol-phosphate transaminase	
3.4.13.9	imidodipeptidase X-Pro dipeptidase	
3.4.3.7	imidodipeptidase, prolidase imidodipeptidase, prolidase	
3.4.11.4	imidoendopeptidase tripeptide aminopeptidase	
3.4.3.6	iminodipeptidase iminodipeptidase	
1.5.1.17	2,2'-Iminodipropanoate:NAD$^+$ oxidoreductase (L-alanine-forming) Alanopine dehydrogenase	
3.4.11.5	iminopeptidas prolyl aminopeptidase	
3.4.24.13	immunoglobulin A1 protease IgA-specific metalloendopeptidase	
3.4.21.72	Immunoglobulin A protease IgA-specific serine endopeptidase	
3.4.21.72	Immunoglobulin A proteinase IgA-specific serine endopeptidase	
6.3.4.4	IMP-aspartate ligase Adenylosuccinate synthase	Vol. 2, Page 579
2.4.2.8	IMP-GMP pyrophosphorylase Hypoxanthine phosphoribosyltransferase	
3.5.4.10	IMP 1,2-hydrolase (decyclizing) IMP cyclohydrolase	

EC-No.	Synonym Recommended Name	Vol., Page
3.1.3.5	IMP 5'-nucleotidase 5'-nucleotidase	
6.3.4.4	IMP:L-aspartate ligase (GDP-forming) Adenylosuccinate synthase	Vol. 2, Page 579
1.1.1.205	IMP:NAD$^+$ oxidoreductase IMP dehydrogenase	
2.4.2.8	IMP:pyrophosphate phospho-D-ribosyltransferase Hypoxanthine phosphoribosyltransferase	
3.5.4.10	IMP cyclohydrolase IMP cyclohydrolase	
1.1.1.205	IMP dehydrogenase IMP dehydrogenase	
1.2.1.14	IMP dehydrogenase IMP dehydrogenase	
1.1.1.205	IMP oxidoreductase IMP dehydrogenase	
2.4.2.8	IMP pyrophosphorylase Hypoxanthine phosphoribosyltransferase	
2.1.1.129	IMT1 protein myo-Inositol 4-O-methyltransferase	
1.1.1.112	Indan-1-ol:NAD(P)$^+$ 1-oxidoreductase Indanol dehydrogenase	
1.1.1.112	Indanol dehydrogenase Indanol dehydrogenase	
4.1.1.74	Indol-3-yl-pyruvate carboxy-lyase Indolepyruvate decarboxylase	
2.4.1.156	Indol-3-ylacetyl-myo-inositol galactoside synthase Indolylacetyl-myo-inositol galactosyltransferase	
2.4.1.121	Indol-3-ylacetylglucose synthase Indole-3-acetate β-glucosyltransferase	
1.13.11.11	Indolamine 2,3-dioxygenase Tryptophan 2,3-dioxygenase	
1.2.3.7	Indole-3-acetaldehyde:oxygen oxidoreductase Indole-3-acetaldehyde oxidase	
1.2.3.7	Indole-3-acetaldehyde oxidase Indole-3-acetaldehyde oxidase	

EC-No.	Synonym Recommended Name	Vol., Page
1.1.1.190	Indole-3-acetaldehyde reductase (NADH) Indole-3-acetaldehyde reductase (NADH)	
1.1.1.191	Indole-3-acetaldehyde reductase (NADPH) Indole-3-acetaldehyde reductase (NADPH)	
2.4.1.121	Indole-3-acetate β-glucosyltransferase Indole-3-acetate β-glucosyltransferase	
2.3.1.72	Indole-3-acetyl-β-1-D-glucoside:myo-inositol indoleacetyltransferase Indoleacetylglucose-inositol O-acyltransferase	
1.13.99.3	Indole-3-alkane α-hydroxylase Tryptophan 2'-dioxygenase	
1.1.1.190	Indole-3-ethanol:NAD^+ oxidoreductase Indole-3-acetaldehyde reductase (NADH)	
1.1.1.191	Indole-3-ethanol:$NADP^+$ oxidoreductase Indole-3-acetaldehyde reductase (NADPH)	
4.1.1.48	Indole-3-glycerol-phosphate synthase Indole-3-glycerol phosphate synthase	
4.1.1.48	Indole-3-glycerol phosphate synthase Indole-3-glycerol phosphate synthase	
4.1.1.48	Indole-3-glycerol phosphate synthetase Indole-3-glycerol phosphate synthase	
4.1.1.48	Indole-3-glycerophosphate synthase Indole-3-glycerol phosphate synthase	
4.1.1.74	Indole-3-pyruvate decarboxylase Indolepyruvate decarboxylase	
1.13.11.17	Indole 2,3-dioxygenase Indole 2,3-dioxygenase	
1.13.11.17	Indole:O2 oxidoreductase Indole 2,3-dioxygenase	
1.13.11.17	Indole:oxygen 2,3-oxidoreductase (decyclizing) Indole 2,3-dioxygenase	
1.2.3.7	Indoleacetaldehyde oxidase Indole-3-acetaldehyde oxidase	
1.1.1.190	Indoleacetaldehyde reductase Indole-3-acetaldehyde reductase (NADH)	
4.2.1.29	Indoleacetaldoxime dehydratase Indoleacetaldoxime dehydratase	

EC-No.	Synonym Recommended Name	Vol., Page
4.2.1.29	3-Indoleacetaldoxime hydro-lyase Indoleacetaldoxime dehydratase	
4.2.1.29	Indoleacetaldoxime hydro-lyase Indoleacetaldoxime dehydratase	
6.3.2.20	Indoleacetate-lysine synthetase Indoleacetate-lysine synthetase	Vol. 2, Page 511
6.3.2.20	Indoleacetate:L-lysine ligase (ADP-forming) Indoleacetate-lysine synthetase	Vol. 2, Page 511
1.14.16.4	Indoleacetic acid-5-hydroxylase Tryptophan 5-monooxygenase	
6.3.2.20	Indoleacetic acid-lysine synthetase Indoleacetate-lysine synthetase	Vol. 2, Page 511
2.3.1.72	Indoleacetylglucose-inositol O-acyltransferase Indoleacetylglucose-inositol O-acyltransferase	
2.3.1.72	Indoleacetyltransferase, indoleacetylglucose-myo-inositol Indoleacetylglucose-inositol O-acyltransferase	
1.13.11.42	Indoleamine-pyrrole 2,3-dioxygenase Indoleamine-pyrrole 2,3-dioxygenase	
1.13.11.17	Indoleamine 2,3-dioxygenase Indole 2,3-dioxygenase	
1.13.11.42	Indoleamine 2,3-dioxygenase Indoleamine-pyrrole 2,3-dioxygenase	
1.13.11.11	Indoleamine 2,3-dioxygenase Tryptophan 2,3-dioxygenase	
1.13.11.42	Indoleamine:oxygen 2,3-dioxygenase (indole-decyclizing) Indoleamine-pyrrole 2,3-dioxygenase	
2.3.1.5	Indoleamine N-acetyltransferase Arylamine N-acetyltransferase	
4.2.1.20	indoleglycerol phosphate aldolase tryptophan synthase	
4.1.2.8	indoleglycerolphosphate aldolase indoleglycerolphosphate aldolase	
4.1.1.48	Indoleglycerol phosphate synthase Indole-3-glycerol phosphate synthase	
4.1.1.48	Indoleglycerol phosphate synthetase Indole-3-glycerol phosphate synthase	

EC-No.	Synonym Recommended Name	Vol., Page
4.1.1.48	Indoleglycerolphosphate synthetase Indole-3-glycerol phosphate synthase	
1.1.1.110	Indolelactate:NAD$^+$ oxidoreductase Indolelactate dehydrogenase	
1.1.1.110	Indolelactate dehydrogenase Indolelactate dehydrogenase	
1.13.11.17	Indole oxidase Indole 2,3-dioxygenase	
2.1.1.47	Indolepyruvate 3-methyltransferase Indolepyruvate C-methyltransferase	
2.1.1.47	Indolepyruvate C-methyltransferase Indolepyruvate C-methyltransferase	
4.1.1.74	Indolepyruvate decarboxylase Indolepyruvate decarboxylase	
4.1.1.74	Indolepyruvate decarboxylase (Enterobacter herbocola strain 299R clone pMB2 gene ipdC reduced) Indolepyruvate decarboxylase	
4.1.1.74	Indolepyruvic acid decarboxylase Indolepyruvate decarboxylase	
2.1.1.47	Indolepyruvic acid methyltransferase Indolepyruvate C-methyltransferase	
1.13.99.3	Indolyl-3-alkan α-hydroxylase Tryptophan 2'-dioxygenase	
2.4.1.156	Indolylacetyl-myo-inositol galactosyltransferase Indolylacetyl-myo-inositol galactosyltransferase	
2.4.2.34	Indolylacetylinositol arabinosyltransferase Indolylacetylinositol arabinosyltransferase	
1.9.3.1	Indophenolase Cytochrome-c oxidase	
1.9.3.1	Indophenol oxidase Cytochrome-c oxidase	
1.2.1.8	betaine-aldehyde:NAD$^+$ oxidoreductase betaine-aldehyde dehydrogenase	
1.2.1.8	betaine-aldehyde dehydrogenase betaine-aldehyde dehydrogenase	

EC-No.	Synonym Recommended Name	Vol., Page
2.1.1.5	betaine-homocysteine methyltransferase betaine-homocysteine S-methyltransferase	
2.1.1.5	betaine-homocysteine S-methyltransferase betaine-homocysteine S-methyltransferase	
2.1.1.5	betaine-homocysteine transmethylase betaine-homocysteine S-methyltransferase	
1.2.1.8	betaine aldehyde dehydrogenase betaine-aldehyde dehydrogenase	
1.2.1.8	betaine aldehyde oxidase betaine-aldehyde dehydrogenase	
3.4.25.1	ingensin proteasome endopeptidase complex	
4.1.1.48	InGPS Indole-3-glycerol phosphate synthase	
4.1.1.48	InGP synthase Indole-3-glycerol phosphate synthase	
4.1.1.48	InGP synthetase Indole-3-glycerol phosphate synthase	
3.6.1.1	Inorganic diphosphatase Inorganic diphosphatase	
3.6.1.1	Inorganic pyrophosphatase Inorganic diphosphatase	
2.7.1.90	Inorganic pyrophosphate-dependent phosphofructokinase diphosphate-fructose-6-phosphate 1-phosphotransferase	
2.7.1.90	Inorganic pyrophosphate-phosphofructokinase diphosphate-fructose-6-phosphate 1-phosphotransferase	
3.6.1.2	Inorganic trimetaphosphatase Trimetaphosphatase	
3.6.1.25	Inorganic triphosphatase Triphosphatase	
2.1.4.2	Inosamine-P amidinotransferase Inosamine-phosphate amidinotransferase	
2.1.4.2	Inosamine-phosphate amidinotransferase Inosamine-phosphate amidinotransferase	
2.1.4.2	Inosamine phosphate amidinotransferase Inosamine-phosphate amidinotransferase	

EC-No.	Synonym Recommended Name	Vol., Page
3.2.2.2	Inosinase Inosine nucleosidase	
3.5.4.10	Inosinate cyclohydrolase IMP cyclohydrolase	
1.1.1.205	Inosinate dehydrogenase IMP dehydrogenase	
3.2.2.12	Inosinate nucleosidase Inosinate nucleosidase	
3.2.2.12	5'-Inosinate phosphoribohydrolase Inosinate nucleosidase	
2.4.2.8	Inosinate pyrophosphorylase Hypoxanthine phosphoribosyltransferase	
1.1.1.205	Inosine-5'-phosphate dehydrogenase IMP dehydrogenase	
2.7.1.73	Inosine-guanosine kinase Inosine kinase	
3.2.2.2	Inosine-guanosine nucleosidase Inosine nucleosidase	
2.4.2.1	Inosine-guanosine phosphorylase Purine-nucleoside phosphorylase	
3.6.1.6	Inosine 5'-diphosphatase Nucleoside-diphosphatase	
1.1.1.205	Inosine 5'-monophosphate dehydrogenase IMP dehydrogenase	
2.4.2.8	Inosine 5'-phosphate pyrophosphorylase Hypoxanthine phosphoribosyltransferase	
3.6.1.6	Inosine diphosphatase Nucleoside-diphosphatase	
2.7.1.73	Inosine kinase Inosine kinase	
1.1.1.205	Inosine monophosphate dehydrogenase IMP dehydrogenase	
1.1.1.205	Inosine monophosphate oxidoreductase IMP dehydrogenase	
3.2.2.2	Inosine nucleosidase Inosine nucleosidase	

EC-No.	Synonym Recommended Name	Vol., Page
2.4.2.1	Inosine phosphorylase Purine-nucleoside phosphorylase	
3.2.2.2	Inosine ribohydrolase Inosine nucleosidase	
3.1.3.56	Inosine triphosphatase Inositol-1,4,5-trisphosphate 5-phosphatase	
1.1.1.205	Inosinic acid dehydrogenase IMP dehydrogenase	
2.4.2.8	Inosinic acid pyrophosphorylase Hypoxanthine phosphoribosyltransferase	
3.5.4.10	Inosinicase IMP cyclohydrolase	
2.4.2.8	Inosinic pyrophosphorylase Hypoxanthine phosphoribosyltransferase	
3.1.3.62	Inositol (1,3,4,5)-tetrakisphosphate 3-phosphatase Inositol-1,3,4,5-tetrakisphosphate 3-phosphatase	
3.1.3.25	inositol-1 (or 4)-monophosphatase inositol-1 (or 4)-monophosphatase	
3.1.3.62	Inositol-1,3,4,5-tetrakisphosphate 3-phosphatase Inositol-1,3,4,5-tetrakisphosphate 3-phosphatase	
3.1.3.65	Inositol-1,3-bisphosphate 3-phosphatase Inositol-1,3-bisphosphate 3-phosphatase	
2.7.1.127	Inositol-1,4,5-trisphosphate-3-kinase 1D-myo-Inositol-trisphosphate 3-kinase	
3.1.3.56	Inositol-1,4,5-trisphosphate/1,3,4,5-tetrakisphosphate 5-phosphatase Inositol-1,4,5-trisphosphate 5-phosphatase	
3.1.3.61	Inositol-1,4,5-trisphosphate 1-phosphatase Inositol-1,4,5-trisphosphate 1-phosphatase	
3.1.3.56	Inositol-1,4,5-trisphosphate 5-phosphatase Inositol-1,4,5-trisphosphate 5-phosphatase	
3.1.3.57	Inositol-1,4-bisphosphate 1-phosphatase Inositol-1,4-bisphosphate 1-phosphatase	
3.1.3.25	inositol-1-phosphatase inositol-1 (or 4)-monophosphatase	
3.1.3.66	Inositol-3,4-bisphosphate 4-phosphatase Inositol-3,4-bisphosphate 4-phosphatase	

EC-No.	Synonym Recommended Name	Vol., Page
3.1.3.57	Inositol-polyphosphate 1-phosphatase Inositol-1,4-bisphosphate 1-phosphatase	
3.1.3.65	Inositol-polyphosphate 3-phosphatase Inositol-1,3-bisphosphate 3-phosphatase	
3.1.4.36	D-Inositol 1,2-cyclic phosphate 2-phosphohydrolase 1,2-Cyclic-inositol-phosphate phosphodiesterase	
3.1.3.62	Inositol 1,3,4,5-tetrakisphosphate-5-phosphomonoesterase Inositol-1,3,4,5-tetrakisphosphate 3-phosphatase	
3.1.3.62	Inositol 1,3,4,5-tetrakisphosphate 3-phosphomonoesterase Inositol-1,3,4,5-tetrakisphosphate 3-phosphatase	
2.7.1.139	Inositol 1,3,4-trisphosphate 5/6-kinase 1D-myo-Inositol-trisphosphate 5-kinase	
2.7.1.133	Inositol 1,3,4-trisphosphate 5/6-kinase 1D-myo-Inositol-trisphosphate 6-kinase	
2.7.1.134	Inositol 1,4,5,6-tetrakisphosphate 3-kinase 1D-myo-Inositol-tetrakisphosphate 1-kinase	
3.1.3.56	Inositol 1,4,5-triphosphate 5-phosphatase Inositol-1,4,5-trisphosphate 5-phosphatase	
3.1.3.36	Inositol 1,4,5-triphosphate 5-phosphomonoesterase Phosphatidylinositol-bisphosphatase	
3.1.3.61	Inositol 1,4,5-trisphosphate 1-phosphatase Inositol-1,4,5-trisphosphate 1-phosphatase	
2.7.1.127	Inositol 1,4,5-trisphosphate 3-kinase 1D-myo-Inositol-trisphosphate 3-kinase	
3.1.3.56	Inositol 1,4,5-trisphosphate 5-monophosphatase Inositol-1,4,5-trisphosphate 5-phosphatase	
3.1.3.56	Inositol 1,4,5-trisphosphate 5-phosphatase Inositol-1,4,5-trisphosphate 5-phosphatase	
2.7.1.127	Inositol 1,4,5-trisphosphate kinase 1D-myo-Inositol-trisphosphate 3-kinase	
3.1.3.56	Inositol 1,4,5-trisphosphate phosphatase Inositol-1,4,5-trisphosphate 5-phosphatase	
2.4.1.123	Inositol 1-α-galactosyltransferase Inositol 1-α-galactosyltransferase	
2.7.1.64	Inositol 1-kinase myo-Inositol 1-kinase	

EC-No.	Synonym Recommended Name	Vol., Page
3.1.3.25	inositol 1-phosphate phosphatase inositol-1 (or 4)-monophosphatase	
5.5.1.4	Inositol 1-phosphate synthase myo-Inositol-1-phosphate synthase	Vol. 1, Page 674
5.5.1.4	Inositol 1-phosphate synthetase myo-Inositol-1-phosphate synthase	Vol. 1, Page 674
1.1.1.18	Inositol 2-dehydrogenase myo-Inositol 2-dehydrogenase	
2.1.1.134	Inositol 6-O-methyltransferase myo-Inositol 6-O-methyltransferase	
2.1.1.40	Inositol D-1-methyltransferase myo-Inositol 3-O-methyltransferase	
1.1.1.18	Inositol dehydrogenase myo-Inositol 2-dehydrogenase	
2.1.1.39	Inositol L-1-methyltransferase myo-Inositol 1-O-methyltransferase	
3.1.3.25	inositol monophosphatase inositol-1 (or 4)-monophosphatase	
3.1.3.25	inositol monophosphate phosphatase inositol-1 (or 4)-monophosphatase	
1.13.99.1	Inositol oxygenase myo-Inositol oxygenase	
3.1.3.25	inositol phosphatase inositol-1 (or 4)-monophosphatase	
3.1.3.56	Inositol phosphate 5-phosphomonoesterase Inositol-1,4,5-trisphosphate 5-phosphatase	
3.1.3.25	inositol phosphohydrolase inositol-1 (or 4)-monophosphatase	
3.1.3.56	Inositol polyphosphate-5-phosphatase Inositol-1,4,5-trisphosphate 5-phosphatase	
2.7.1.133	Inositol polyphosphate-6-kinase 1D-myo-Inositol-trisphosphate 6-kinase	
3.1.3.62	Inositol tetrakisphosphate phosphomonoesterase Inositol-1,3,4,5-tetrakisphosphate 3-phosphatase	
3.1.3.36	Inositol triphosphate 5-phosphomonoesterase Phosphatidylinositol-bisphosphatase	

EC-No.	Synonym Recommended Name	Vol., Page
3.1.3.56	Inositol trisphosphate phosphomonoesterase Inositol-1,4,5-trisphosphate 5-phosphatase	
2.7.7.37	Inosityltransferase, glucose 1-phosphate Aldose-1-phosphate nucleotidyltransferase	
4.2.1.44	Inosose 2,3-dehydratase Myo-inosose-2 dehydratase	
5.5.1.4	INPS1 myo-Inositol-1-phosphate synthase	Vol. 1, Page 674
3.1.3.56	Ins(1,4,5)P3/Ins(1,3,4,5)P4 5-phosphatase Inositol-1,4,5-trisphosphate 5-phosphatase	
3.1.3.56	Ins(1,4,5)P3 5-phosphatase Inositol-1,4,5-trisphosphate 5-phosphatase	
2.4.2.29	Q-insertase Queuine tRNA-ribosyltransferase	
3.4.24.56	Insulin-degrading enzyme Insulysin	
3.4.24.56	Insulin-degrading neutral proteinase Insulysin	
3.4.24.56	Insulin-glucagon protease Insulysin	
1.6.4.4	Insulin-glutathione transhydrogenase Protein-disulfide reductase (NAD(P)H)	
3.4.24.56	Insulin-specific protease Insulysin	
3.4.24.56	Insulinase Insulysin	
3.4.22.11	insulinase insulinase	
3.4.99.10	insulinase insulinase	
3.4.99.45	insulinase insulinase	
3.4.17.10	insulin granule-associated carboxypeptidase carboxypeptidase H	
3.4.24.56	Insulin protease Insulysin	

EC-No.	Synonym Recommended Name	Vol., Page
3.4.24.56	Insulin proteinase Insulysin	
2.7.1.112	Insulin receptor protein-tyrosine kinase Protein-tyrosine kinase	
1.8.4.2	Insulin reductase Protein-disulfide reductase (glutathione)	
3.4.24.56	Insulysin Insulysin	
3.4.22.36	Interleukin-1β convertase Interleukin 1β converting enzyme	
3.4.22.36	Interleukin-1β converting enzyme Interleukin 1β converting enzyme	
3.4.22.36	Interleukin-1β precursor proteinase Interleukin 1β converting enzyme	
3.4.22.36	Interleukin 1β-converting endopeptidase Interleukin 1β converting enzyme	
3.4.22.36	Interleukin 1 converting enzyme Interleukin 1β converting enzyme	
3.4.22.36	Interleukin 1β converting enzyme Interleukin 1β converting enzyme	
2.4.1.184	Interlipid galactosyltransferase Galactolipid galactosyltransferase	
3.4.21.32	interstitial collagenase brachyurin	
3.4.24.7	interstitial collagenase interstitial collagenase	
3.4.24.3	interstitial collagenase microbial collagenase	
3.2.1.48	Intestinal sucrase Sucrose α-glucosidase	
3.1.1.53	intralumenal sialate: 9-O-acetylesterase sialate O-acetylesterase	
3.2.1.7	Inulase Inulinase	

EC-No.	Synonym Recommended Name Vol., Page
2.4.1.93	Inulase II Inulin fructotransferase (depolymerizing, difructofuranose-1,2':2,3'-dianhydride-forming)
3.2.1.7	Inulinase Inulinase
2.4.1.93	Inulinase II Inulin fructotransferase (depolymerizing, difructofuranose-1,2':2,3'-dianhydride-forming)
2.4.1.200	Inulin D-fructosyl-D-fructosyltransferase (1,2':2',1-dianhydride-forming) Inulin fructotransferase (depolymerizing, difructofuranose-1,2':2',1-dianhydride-forming)
2.4.1.93	Inulin D-fructosyl-D-fructosyltransferase (1,2':2,3'-dianhydride-forming) Inulin fructotransferase (depolymerizing, difructofuranose-1,2':2,3'-dianhydride-forming)
2.4.1.93	Inulin fructotransferase (depolymerizing) Inulin fructotransferase (depolymerizing, difructofuranose-1,2':2,3'-dianhydride-forming)
2.4.1.200	Inulin fructotransferase (depolymerizing, difructofuranose-1,2':2',1-dianhydride-forming) Inulin fructotransferase (depolymerizing, difructofuranose-1,2':2',1-dianhydride-forming)
2.4.1.93	Inulin fructotransferase (depolymerizing, difructofuranose-1,2':2,3'-dianhydride-forming) Inulin fructotransferase (depolymerizing, difructofuranose-1,2':2,3'-dianhydride-forming)
2.4.1.200	Inulin fructotransferase (DFA-I-producing) Inulin fructotransferase (depolymerizing, difructofuranose-1,2':2',1-dianhydride-forming)
2.4.1.9	Inulosucrase Inulosucrase
3.2.1.26	Invertase β-fructofuranosidase
3.2.1.26	Invertin β-fructofuranosidase
1.11.1.8	Iodide:hydrogen-peroxide oxidoreductase Iodide peroxidase
1.11.1.8	Iodide peroxidase Iodide peroxidase

EC-No.	Synonym Recommended Name	Vol., Page
1.11.1.8	Iodide peroxidase-tyrosine iodinase Iodide peroxidase	
1.11.1.8	Iodinase Iodide peroxidase	
1.11.1.8	Iodoperoxidase (heme type) Iodide peroxidase	
2.1.1.26	Iodophenol methyltransferase Iodophenol O-methyltransferase	
2.1.1.26	Iodophenol O-methyltransferase Iodophenol O-methyltransferase	
3.8.1.4	Iodothyronine 5'-deiodinase Thyroxine deiodinase	
3.8.1.4	Iodothyronine outer ring monodeiodinase Thyroxine deiodinase	
1.11.1.8	Iodotyrosine deiodase Iodide peroxidase	
1.11.1.8	Iodotyrosine deiodinase Iodide peroxidase	
3.1.3.36	IP3 phosphatase Phosphatidylinositol-bisphosphatase	
4.1.1.74	IpdC Indolepyruvate decarboxylase	
4.1.3.12	IPM-synthase 2-isopropylmalate synthase	
4.2.1.33	IPM dehydratase 3-isopropylmalate dehydratase	
4.1.3.12	IPMS 2-isopropylmalate synthase	
4.1.3.12	IPM synthase 2-isopropylmalate synthase	
4.1.3.12	α-IPM synthase 2-isopropylmalate synthase	
4.1.3.12	α-IPM synthetase 2-isopropylmalate synthase	
5.3.3.2	IPP-isomerase Isopentenyl-diphosphate Δ-isomerase	Vol. 1, Page 386

EC-No.	Synonym Recommended Name	Vol., Page
5.3.3.2	IPP isomerase Isopentenyl-diphosphate Δ-isomerase	Vol. 1, Page 386
4.2.1.3	IRE-BP aconitate hydratase	
2.7.1.112	IRK Protein-tyrosine kinase	
1.16.3.1	Iron(II): oxygen oxidoreductase Ferroxidase	
1.9.99.1	Iron-cytochrome-c reductase Iron-cytochrome-c reductase	
1.9.99.1	Iron-cytochrome c reductase Iron-cytochrome-c reductase	
4.2.1.3	iron-responsive element binding protein aconitate hydratase	
4.99.1.1	iron chelatase ferrochelatase	
1.6.99.13	Iron chelate reductase Ferric-chelate reductase	
3.4.23.29	Irpex lacteus aspartic proteinase Polyporopepsin	
3.4.23.29	Irpex lacteus carboxyl proteinase B Polyporopepsin	
3.1.3.48	islet cell antigen-related PTP protein-tyrosine-phosphatase	
3.2.1.68	Isoamylase Isoamylase	
2.1.1.77	L-Isoaspartyl/D-aspartyl protein carboxyl methyltransferase Protein-L-isoaspartate(D-aspartate) O-methyltransferase	
2.1.1.91	Isobutyraldoxime O-methyltransferase Isobutyraldoxime O-methyltransferase	
2.7.2.14	Isobutyrate kinase Branched-chain-fatty-acid kinase	
5.4.99.13	Isobutyryl-CoA mutase Isobutyryl-CoA mutase	Vol. 1, Page 646
3.3.2.1	Isochorismatase Isochorismatase	

EC-No.	Synonym Recommended Name	Vol., Page
5.4.99.6	Isochorismate hydroxymutase Isochorismate synthase	Vol. 1, Page 618
3.3.2.1	Isochorismate pyruvate-hydrolase Isochorismatase	
5.4.99.6	Isochorismate synthase Isochorismate synthase	Vol. 1, Page 618
5.4.99.6	Isochorismic synthase Isochorismate synthase	Vol. 1, Page 618
4.1.3.1	isocitrase isocitrate lyase	
4.1.3.1	isocitratase isocitrate lyase	
1.1.1.41	Isocitrate:NAD^+ oxidoreductase (decarboxylating) Isocitrate dehydrogenase (NAD^+)	
1.1.1.42	Isocitrate:$NADP^+$ oxidoreductase (decarboxylating) Isocitrate dehydrogenase ($NADP^+$)	
1.1.1.41	Isocitrate dehydrogenase (NAD^+) Isocitrate dehydrogenase (NAD^+)	
1.1.1.42	Isocitrate dehydrogenase (NADP) Isocitrate dehydrogenase ($NADP^+$)	
1.1.1.42	Isocitrate dehydrogenase (NADP) Isocitrate dehydrogenase ($NADP^+$)	
1.1.1.42	Isocitrate dehydrogenase ($NADP^+$) Isocitrate dehydrogenase ($NADP^+$)	
2.7.1.116	[Isocitrate dehydrogenase ($NADP^+$)] kinase [Isocitrate dehydrogenase ($NADP^+$)] kinase	
1.1.1.42	Isocitrate dehydrogenase (NADP-dependent) Isocitrate dehydrogenase ($NADP^+$)	
1.1.1.42	Isocitrate dehydrogenase (nicotinamide adenine dinucleotide phosphate) Isocitrate dehydrogenase ($NADP^+$)	
5.1.2.6	Isocitrate epimerase Isocitrate epimerase	Vol. 1, Page 89
4.1.3.1	isocitrate glyoxylate-lyase isocitrate lyase	
4.1.3.1	isocitrate lyase isocitrate lyase	

EC-No.	Synonym Recommended Name	Vol., Page
2.3.1.126	Isocitrate O-dihydroxycinnamoyltransferase Isocitrate O-dihydroxycinnamoyltransferase	
1.1.1.41	Isocitric acid dehydrogenase Isocitrate dehydrogenase (NAD^+)	
1.1.1.41	Isocitric dehydrogenase Isocitrate dehydrogenase (NAD^+)	
4.1.3.1	isocitric lyase isocitrate lyase	
4.1.3.1	isocitritase isocitrate lyase	
3.5.4.1	Isocytosine deaminase Cytosine deaminase	
2.3.1.115	Isoflavone-7-O-β-glucoside 6-O-malonyltransferase Isoflavone-7-O-β-glucoside 6-O-malonyltransferase	
1.14.13.53	Isoflavone 2'-hydroxylase Isoflavone 2'-hydroxylase	
1.14.13.52	Isoflavone 3'-hydroxylase Isoflavone 3'-hydroxylase	
2.1.1.46	Isoflavone 4'-O-methyltransferase Isoflavone 4'-O-methyltransferase	
2.4.1.170	Isoflavone 7-O-glucosyltransferase Isoflavone 7-O-glucosyltransferase	
2.1.1.46	Isoflavone methyltransferase Isoflavone 4'-O-methyltransferase	
2.1.1.46	Isoflavone O-methyltransferase Isoflavone 4'-O-methyltransferase	
1.3.1.45	Isoflavone reductase 2'-Hydroxyisoflavone reductase	
2.4.1.96	Isofloridoside-phosphate synthase Sn-Glycerol-3-phosphate 1-galactosyltransferase	
4.2.1.57	Isohexenyl-glutaconyl-CoA-hydratase Isohexenylglutaconyl-CoA hydratase	
4.2.1.57	β-isohexenylglutaconyl-CoA-hydratase Isohexenylglutaconyl-CoA hydratase	
4.2.1.57	Isohexenylglutaconyl-CoA hydratase Isohexenylglutaconyl-CoA hydratase	

EC-No.	Synonym Recommended Name	Vol., Page
4.2.1.57	Isohexenylglutaconyl CoA hydratase Isohexenylglutaconyl-CoA hydratase	
4.2.1.57	Isohexenylglutaconyl coenzyme A hydratase Isohexenylglutaconyl-CoA hydratase	
6.1.1.5	Isoleucine-transfer RNA ligase Isoleucine-tRNA ligase	Vol. 2, Page 33
6.1.1.5	Isoleucine-tRNA ligase Isoleucine-tRNA ligase	Vol. 2, Page 33
6.1.1.5	Isoleucine-tRNA synthetase Isoleucine-tRNA ligase	Vol. 2, Page 33
6.1.1.5	L-Isoleucine:tRNA^{Ile} ligase (AMP-forming) Isoleucine-tRNA ligase	Vol. 2, Page 33
6.1.1.5	Isoleucine translase Isoleucine-tRNA ligase	Vol. 2, Page 33
6.1.1.5	Isoleucyl-transfer ribonucleate synthetase Isoleucine-tRNA ligase	Vol. 2, Page 33
6.1.1.5	Isoleucyl-transfer RNA synthetase Isoleucine-tRNA ligase	Vol. 2, Page 33
6.1.1.5	Isoleucyl-tRNA synthetase Isoleucine-tRNA ligase	Vol. 2, Page 33
3.2.1.10	Isomaltase Oligo-1,6-glucosidase	
3.2.1.94	Isomalto-dextranase Glucan 1,6-α-isomaltosidase	
3.2.1.94	Isomaltodextranase Glucan 1,6-α-isomaltosidase	
3.2.1.94	Isomaltohydrolase, exo Glucan 1,6-α-isomaltosidase	
3.2.1.95	Isomaltotriohydrolase, exo- Dextran 1,6-α-isomaltotriosidase	
5.4.99.11	Isomaltulose synthase Isomaltulose synthase	Vol. 1, Page 638
5.4.99.11	Isomaltulose synthetase Isomaltulose synthase	Vol. 1, Page 638
5.4.99.10	isomaltulose synthetase isomaltulose synthetase	Vol. 1, Page 638

EC-No.	Synonym Recommended Name	Vol., Page
5.2.1.10	Isomerase, 2-chlorocarboxymethylenebutenolide 2-Chloro-4-carboxymethylenebut-2-en-1,4-olide isomerase	Vol. 1, Page 231
5.5.1.2	Isomerase, 3-carboxy-cis, cis-muconate cyclo- 3-Carboxy-cis,cis-muconate cycloisomerase	Vol. 1, Page 668
5.3.1.17	Isomerase, 4-deoxy-L-threo-5-hexulose uronate 4-Deoxy-L-threo-5-hexosulose-uronate ketol-isomerase	Vol. 1, Page 338
5.2.1.11	Isomerase, 4-hydroxyphenylacetaldehyde oxime 4-Hydroxyphenylacetaldehyde-oxime isomerase	Vol. 1, Page 233
5.4.99.14	Isomerase, 4-methyl-2-enelactone 4-Carboxymethyl-4-methylbutenolide mutase	Vol. 1, Page 648
5.3.3.10	Isomerase, 5-carboxymethyl-2-hydroxymuconate 5-Carboxymethyl-2-hydroxymuconate Δ-isomerase	Vol. 1, Page 427
5.3.3.7	Isomerase, aconitate Δ- Aconitate Δ-isomerase	Vol. 1, Page 409
5.3.1.3	Isomerase, arabinose Arabinose isomerase	Vol. 1, Page 249
5.3.1.13	Isomerase, arabinose phosphate Arabinose-5-phosphate isomerase	Vol. 1, Page 325
5.5.1.6	Isomerase, chalcone Chalcone isomerase	Vol. 1, Page 691
5.5.1.7	Isomerase, chloromuconate cyclo- Chloromuconate cycloisomerase	Vol. 1, Page 699
5.3.3.5	Isomerase, cholestenol Δ- Cholestenol Δ-isomerase	Vol. 1, Page 404
5.3.1.21	Isomerase, corticosteroid sidechain Corticosteroid side-chain-isomerase	Vol. 1, Page 345
5.5.1.9	Isomerase, cycloeucalenol-obtusifoliol Cycloeucalenol cycloisomerase	Vol. 1, Page 710
5.3.1.15	Isomerase, D-lyxose D-Lyxose ketol-isomerase	Vol. 1, Page 333
5.99.1.2	Isomerase, deoxiribonucleate topo- DNA topoisomerase	Vol. 1, Page 721
5.99.1.3	Isomerase, deoxyribonucleate topo-, II DNA topoisomerase (ATP-hydrolysing)	Vol. 1, Page 737

EC-No.	Synonym Recommended Name	Vol., Page
5.5.1.11	Isomerase, dichloromuconate cyclo- Dichloromuconate cycloisomerase	Vol. 1, Page 716
5.3.3.8	Isomerase, dodecenoyl coenzyme A δ- Dodecenoyl-CoA δ-isomerase	Vol. 1, Page 413
5.3.3.12	Isomerase, dopachrome Δ- Dopachrome Δ-isomerase	Vol. 1, Page 432
5.2.1.9	Isomerase, farnesol Farnesol 2-isomerase	Vol. 1, Page 229
5.2.1.6	Isomerase, furylfuramide Furylfuramide isomerase	Vol. 1, Page 213
5.3.1.26	Isomerase, galactose 6-phosphate Galactose-6-phosphate isomerase	Vol. 1, Page 364
5.3.1.10	Isomerase, glucosamine phosphate Glucosamine-6-phosphate isomerase	Vol. 1, Page 312
2.6.1.16	Isomerase, glucosamine phosphate (glutamine-forming) Glutamine-fructose-6-phosphate transaminase (isomerizing)	
5.3.1.9	Isomerase, glucose phosphate Glucose-6-phosphate isomerase	Vol. 1, Page 298
5.3.1.12	Isomerase, glucuronate Glucuronate isomerase	Vol. 1, Page 322
4.2.1.92	Isomerase, hydroperoxide Hydroperoxide dehydratase	
5.3.3.2	Isomerase, isopentenylpyrophosphate Δ- Isopentenyl-diphosphate Δ-isomerase	Vol. 1, Page 386
5.3.3.11	Isomerase, isopiperitenone Isopiperitenone Δ-isomerase	Vol. 1, Page 430
1.1.1.86	Isomerase, ketol acid reducto- Ketol-acid reductoisomerase	
5.3.1.4	Isomerase, L-arabinose L-Arabinose isomerase	Vol. 1, Page 254
5.3.1.25	Isomerase, L-fucose L-Fucose isomerase	Vol. 1, Page 359
5.3.1.14	Isomerase, L-rhamnose L-Rhamnose isomerase	Vol. 1, Page 328
5.2.1.5	Isomerase, linoleate Linoleate isomerase	Vol. 1, Page 210

EC-No.	Synonym Recommended Name	Vol., Page
5.2.1.1	Isomerase, maleate Maleate isomerase	Vol. 1, Page 192
5.2.1.2	Isomerase, maleylacetoacetate Maleylacetoacetate isomerase	Vol. 1, Page 197
5.2.1.4	Isomerase, maleylpyruvate Maleylpyruvate isomerase	Vol. 1, Page 206
5.3.1.7	Isomerase, mannose Mannose isomerase	Vol. 1, Page 285
5.3.1.8	Isomerase, mannose phosphate Mannose-6-phosphate isomerase	Vol. 1, Page 289
5.3.3.6	Isomerase, methylitaconate Methylitaconate Δ-isomerase	Vol. 1, Page 406
5.3.3.6	Isomerase, methylitaconate Δ- Methylitaconate Δ-isomerase	Vol. 1, Page 406
5.3.1.23	Isomerase, methylthioribose 1-phosphate 5-Methylthioribose-1-phosphate isomerase	Vol. 1, Page 351
5.5.1.1	Isomerase, muconate cyclo- Muconate cycloisomerase	Vol. 1, Page 660
5.3.3.4	Isomerase, muconolactoneΔ- Muconolactone Δ-isomerase	Vol. 1, Page 399
5.3.1.16	Isomerase, N-(phosphoribosylformimino) aminophosphoribosylimidazole-carboxamide N-(5'-Phospho-D-ribosylformimino)-5-amino-1-(5-phosphoribosyl)-4-imidazole carboxamide isomerase	Vol. 1, Page 335
5.1.1.16	Isomerase, peptide serine Protein-serine epimerase	Vol. 1, Page 66
5.2.1.8	Isomerase, peptidylprolyl cis-trans Peptidylprolyl isomerase	Vol. 1, Page 218
5.3.1.24	Isomerase, phosphoribosylanthranilate Phosphoribosylanthranilate isomerase	Vol. 1, Page 353
5.3.1.16	Isomerase, phosphoribosylformiminoaminophosphoribosylimidazolecarb-oxamide N-(5'-Phospho-D-ribosylformimino)-5-amino-1-(5-phosphoribosyl)-4-imidazole carboxamide isomerase	Vol. 1, Page 335
5.4.1.2	Isomerase, precorrin (Methanobacterium thermoautothrophicum strain ΔH gene MTH227) Precorrin-8X methylmutase	Vol. 1, Page 490

EC-No.	Synonym Recommended Name	Vol., Page
5.4.1.2	Isomerase, precorrin (Methanococcus jannaschii gene MJ0930) Precorrin-8X methylmutase	Vol. 1, Page 490
5.3.3.9	Isomerase, prostaglandin $A_1\Delta$- Prostaglandin-A_1 Δ-isomerase	Vol. 1, Page 423
5.3.99.3	Isomerase, prostaglandin R_2 E- Prostaglandin-E synthetase	Vol. 1, Page 459
5.3.99.2	Isomerase, prostaglanin R_2D- Prostaglandin-D synthase	Vol. 1, Page 451
5.2.1.3	Isomerase, retinene Retinal isomerase	Vol. 1, Page 202
5.2.1.7	Isomerase, retinol Retinol isomerase	Vol. 1, Page 215
5.3.1.20	Isomerase, ribose Ribose isomerase	Vol. 1, Page 342
5.3.1.6	Isomerase, ribose phosphate Ribose-5-phosphate isomerase	Vol. 1, Page 277
5.3.3.1	Isomerase, steroid Δ Steroid Δ-isomerase	Vol. 1, Page 376
5.3.99.7	Isomerase, styrene oxide Styrene-oxide isomerase	Vol. 1, Page 486
5.5.1.3	Isomerase, tetrahydroxypteridine cyclo- Tetrahydroxypteridine cycloisomerase	Vol. 1, Page 672
5.99.1.1	Isomerase, thiocyanate Thiocyanate isomerase	Vol. 1, Page 719
5.3.1.1	Isomerase, triose phosphate Triose-phosphate isomerase	Vol. 1, Page 235
4.2.1.75	Isomerase, uroporphyrinogen Uroporphyrinogen-III synthase	
5.3.3.3	Isomerase, vinylacetyl coenzyme A.Δ.- Vinylacetyl-CoA Δ-isomerase	Vol. 1, Page 395
5.3.1.5	Isomerase, xylose Xylose isomerase	Vol. 1, Page 259
1.1.1.86	Isomeroreductase Ketol-acid reductoisomerase	
2.5.1.38	Isonocardicin synthase Isonocardicin synthase	

EC-No.	Synonym Recommended Name	Vol., Page
2.1.1.78	Isoorientin 3'-O-methyltransferase Isoorientin 3'-O-methyltransferase	
5.3.3.2	Isopententenyl diphosphate:dimethylallyl diphosphate isomerase Isopentenyl-diphosphate Δ-isomerase	Vol. 1, Page 386
5.3.3.2	Isopentenyl-diphosphate Δ-isomerase Isopentenyl-diphosphate Δ-isomerase	Vol. 1, Page 386
5.3.3.2	Isopentenyl-diphosphate Δ^3-Δ^2-isomerase Isopentenyl-diphosphate Δ-isomerase	Vol. 1, Page 386
2.5.1.27	2-Isopentenyl-diphosphate:AMP Δ^2-isopentenyltransferase Adenylate isopentenyltransferase	
2.5.1.8	Isopentenyl-diphosphate:tRNA isopentenyltransferase tRNA isopentenyltransferase	
5.3.3.2	Isopentenyldiphosphate Δ-isomerase Isopentenyl-diphosphate Δ-isomerase	Vol. 1, Page 386
2.5.1.8	Δ^2-Isopentenyl pyrophosphate:transfer ribonucleic acid Δ^2-isopentenyl-transferase tRNA isopentenyltransferase	
2.5.1.8	Δ^2-Isopentenyl pyrophosphate:tRNA-Δ^2-isopentenyl transferase tRNA isopentenyltransferase	
2.5.1.20	Isopentenyl pyrophosphate cis-1,4-polyisoprenyl transferase Rubber cis-polyprenylcistransferase	
5.3.3.2	Isopentenyl pyrophosphate isomerase Isopentenyl-diphosphate Δ-isomerase	Vol. 1, Page 386
5.3.3.2	Isopentenylpyrophosphate isomerase Isopentenyl-diphosphate Δ-isomerase	Vol. 1, Page 386
5.3.3.2	Isopentenyl pyrophosphate isomerase:dimethylallyl pyrophosphate isomerase Isopentenyl-diphosphate Δ-isomerase	Vol. 1, Page 386
2.5.1.27	Isopentenyltransferase Adenylate isopentenyltransferase	
2.5.1.27	Isopentenyltransferase, adenylate Adenylate isopentenyltransferase	
2.5.1.8	Isopentenyltransferase, transfer ribonucleate tRNA isopentenyltransferase	
3.1.2.15	Isopeptidase Ubiquitin thiolesterase	

EC-No.	Synonym Recommended Name	Vol., Page
3.4.19.12	isopeptidase ubiquitinyl hydrolase I	
3.1.2.15	Isopeptidase T Ubiquitin thiolesterase	
3.4.19.12	isopeptidase T ubiquitinyl hydrolase I	
1.10.3.4	Isophenoxazine synthase *o*-Aminophenol oxidase	
1.1.1.223	Isopiperitenol dehydrogenase Isopiperitenol dehydrogenase	
5.3.3.11	Isopiperitenone Δ-isomerase Isopiperitenone Δ-isomerase	Vol. 1, Page 430
5.3.3.11	Isopiperitenone Δ^8-Δ^4-isomerase Isopiperitenone Δ-isomerase	Vol. 1, Page 430
5.3.3.11	Isopiperitenone isomerase Isopiperitenone Δ-isomerase	Vol. 1, Page 430
2.7.1.66	Isoprenoid-alcohol kinase Undecaprenol kinase	
2.7.1.66	Isoprenoid alcohol kinase Undecaprenol kinase	
2.7.1.66	Isoprenoid alcohol phosphokinase Undecaprenol kinase	
2.1.1.100	Isoprenylated protein methyltransferase Protein-S-isoprenylcysteine O-methyltransferase	
3.6.1.27	Isoprenyl pyrophosphatase Undecaprenyl-diphosphatase	
3.2.1.120	Isoprimeverose-producing oligoxyloglucan hydrolase Oligoxyloglucan *β*-glycosidase	
1.1.1.80	Isopropanol dehydrogenase ($NADP^+$) Isopropanol dehydrogenase ($NADP^+$)	
3.5.99.4	N-isopropylammelide isopropylaminohydrolase N-isopropylammelide isopropylaminohydrolase	
3.5.99.4	N-isopropylammelide isopropylaminohydrolase N-isopropylammelide isopropylaminohydrolase	
4.2.1.33	3-isopropylmalate dehydratase 3-isopropylmalate dehydratase	

EC-No.	Synonym Recommended Name	Vol., Page
4.2.1.33	β-Isopropylmalate dehydratase 3-isopropylmalate dehydratase	
1.1.1.85	3-Isopropylmalate dehydrogenase 3-Isopropylmalate dehydrogenase	
1.1.1.85	β-Isopropylmalate dehydrogenase 3-Isopropylmalate dehydrogenase	
4.2.1.33	3-isopropylmalate hydro-lyase 3-isopropylmalate dehydratase	
4.2.1.33	α-isopropylmalate isomerase 3-isopropylmalate dehydratase	
4.2.1.33	isopropylmalate isomerase 3-isopropylmalate dehydratase	
4.1.3.12	2-isopropylmalate synthase 2-isopropylmalate synthase	
4.1.3.12	α-isopropylmalate synthase 2-isopropylmalate synthase	
4.1.3.12	isopropylmalate synthase 2-isopropylmalate synthase	
4.1.3.12	α-isopropylmalate synthase I 2-isopropylmalate synthase	
4.1.3.12	α-isopropylmalate synthase II 2-isopropylmalate synthase	
4.1.3.12	α-isopropylmalate synthetase 2-isopropylmalate synthase	
4.1.3.12	isopropylmalate synthetase 2-isopropylmalate synthase	
1.1.1.85	β-Isopropylmalic enzyme 3-Isopropylmalate dehydrogenase	
4.1.3.12	α-isopropylmalic synthetase 2-isopropylmalate synthase	
3.1.8.2	Isopropylphosphorofluoridase Diisopropyl-fluorophosphatase	
3.2.1.57	Isopullulanase Isopullulanase	
1.3.99.16	Isoquinoline 1-oxidoreductase Isoquinoline 1-oxidoreductase	

EC-No.	Synonym Recommended Name	Vol., Page
1.3.99.16	Isoquinoline:acceptor 1-oxidoreductase (hydroxylating) Isoquinoline 1-oxidoreductase	
1.3.99.16	Isoquinoline hydroxylase Isoquinoline 1-oxidoreductase	
1.3.99.16	Isoquinoline oxidase Isoquinoline 1-oxidoreductase	
5.99.1.1	Isothiocyanic isomerase Thiocyanate isomerase	Vol. 1, Page 719
1.3.99.10	Isovaleroyl-coenzyme A dehydrogenase Isovaleryl-CoA dehydrogenase	
2.3.1.156	isovaleryl-CoA:malonyl-CoA acyltransferase phloroisovalerophenone synthase	
1.3.99.10	Isovaleryl-CoA dehydrogenase Isovaleryl-CoA dehydrogenase	
1.3.99.10	Isovaleryl-coenzyme A dehydrogenase Isovaleryl-CoA dehydrogenase	
2.4.1.106	Isovitexin β-glucosyltransferase Isovitexin β-glucosyltransferase	
4.2.99.16	Isowillardiine synthase Uracilylalanine synthase	
2.7.4.10	Isozyme 3 of adenylate kinase Nucleoside-triphosphate-adenylate kinase	
2.8.3.7	Itaconate CoA-transferase Succinate-citramalate CoA-transferase	
4.2.1.56	Itaconyl-CoA hydratase Itaconyl-CoA hydratase	
4.2.1.56	Itaconyl-CoA hydratase Itaconyl-CoA hydratase	
4.2.1.56	Itaconyl coenzyme A hydratase Itaconyl-CoA hydratase	
1.14.13.59	IucD L-Lysine 6-monooxygenase ($NADPH_2$)	
3.5.1.52	Jack-bean glycopeptidase Peptide-N^4-(N-acetyl-β-glucosaminyl) asparagine amidase	
2.7.1.112	JAK protein tyrosine kinase Protein-tyrosine kinase	

EC-No.	Synonym Recommended Name	Vol., Page
1.11.1.7	Japanese radish peroxidase Peroxidase	
3.4.24.73	Jararhagin Jararhagin	
3.4.24.73	JF1 Jararhagin	
3.1.1.59	JH-esterase Juvenile-hormone esterase	
6.1.1.21	Jo-1 Histidine-tRNA ligase	Vol. 2, Page 168
6.1.1.21	Jo-1 antigen Histidine-tRNA ligase	Vol. 2, Page 168
3.4.24.50	J Protease Bothrolysin	
1.14.99.27	Juglone 3-monooxygenase Juglone 3-monooxygenase	
1.14.99.27	Juglone hydroxylase Juglone 3-monooxygenase	
3.4.22.33	Juice bromelain Fruit bromelain	
3.1.1.59	Juvenile-hormone esterase Juvenile-hormone esterase	
3.1.1.59	Juvenile hormone analog esterase Juvenile-hormone esterase	
3.1.1.59	Juvenile hormone carboxyesterase Juvenile-hormone esterase	
3.1.1.59	Juvenile hormone esterase Juvenile-hormone esterase	
3.6.1.36	$(K^+ + H^+)$-ATPase H^+/K^+-transporting ATPase	
4.1.1.75	KAD 5-Guanidino-2-oxopentanoate decarboxylase	
3.4.21.34	kallidinogenase plasma kallikrein	
3.4.21.35	kallidinogenase tissue kallikrein	

EC-No.	Synonym Recommended Name	Vol., Page
3.4.21.8	kallikrein kallikrein	
3.4.4.21	kallikrein kallikrein	
3.4.21.34	kallikrein plasma kallikrein	
3.4.21.35	kallikrein tissue kallikrein	
3.4.21.35	kallikrein 2 tissue kallikrein	
3.4.21.34	kallikrein I plasma kallikrein	
3.4.21.34	kallikrein II plasma kallikrein	
3.4.21.38	kallikreinogen, activator coagulation factor XIIa	
2.3.1.55	kanamycin 6'-N-acetyltransferase kanamycin 6'-N-acetyltransferase	
2.7.1.95	Kanamycin kinase Kanamycin kinase	
3.2.1.83	κ-carrageenan 4-β-D-glycanohydrolase κ-carrageenase	
3.2.1.83	κ-carrageenase κ-carrageenase	
3.2.1.83	κ-Carrageenanase κ-carrageenase	
1.1.1.96	KAR Diiodophenylpyruvate reductase	
2.3.1.16	KAT Acetyl-CoA C-acyltransferase	
3.4.21.62	Kazusase Subtilisin	
4.2.1.43	KDA dehydratase 2-dehydro-3-deoxy-L-arabinonate dehydratase	
3.4.24.24	72 kDa Gelatinase Gelatinase A	

EC-No.	Synonym Recommended Name	Vol., Page
3.4.24.24	72-kDa Gelatinase Gelatinase A	
3.4.24.35	92-kDa Gelatinase Gelatinase B	
3.4.24.24	72 kDa Gelatinase type A Gelatinase A	
5.4.99.16	57-KDa trehalose synthase (Saccharomyces cerevisiae) Maltose α-D-glucosyltransferase	Vol. 1, Page 656
3.4.24.35	92-kDa Type IV collagenase Gelatinase B	
3.4.24.35	95 kDa type IV collagenase/gelatinase Gelatinase B	
4.1.2.20	KDGLucA 2-Dehydro-3-deoxyglucarate aldolase	
4.1.2.20	5-KDGluc aldolase 2-Dehydro-3-deoxyglucarate aldolase	
4.1.2.16	KDO-8-P synthase 2-Dehydro-3-deoxyphosphooctonate aldolase	
4.1.2.16	KDO-8-P synthetase 2-Dehydro-3-deoxyphosphooctonate aldolase	
3.2.1.144	α-Kdo-ase 3-deoxyoctulosonase	
4.1.2.16	Kdo8P synthase 2-Dehydro-3-deoxyphosphooctonate aldolase	
4.1.2.23	KDO aldolase 3-deoxy-D-manno-octulosonate aldolase	
4.1.2.21	KDPGal aldolase 2-Dehydro-3-deoxyphosphogalactonate aldolase	
4.1.2.14	KDPG aldolase 2-Dehydro-3-deoxyphosphogluconate aldolase	
4.1.2.14	KDPG aldolase 2-Dehydro-3-deoxyphosphogluconate aldolase	
4.1.2.15	KDPH synthase 2-Dehydro-3-deoxyphosphoheptanoate aldolase	
4.1.2.15	KDPH synthetase 2-Dehydro-3-deoxyphosphoheptanoate aldolase	

EC-No.	Synonym Recommended Name	Vol., Page
4.1.2.16	KDPO synthetase 2-Dehydro-3-deoxyphosphooctonate aldolase	
3.2.1.103	Keratan-sulfate 1,4-β-D-galactanohydrolase Keratan-sulfate endo-1,4-β-galactosidase	
3.2.1.103	Keratan-sulfate endo-1,4-β-galactosidase Keratan-sulfate endo-1,4-β-galactosidase	
3.2.1.103	Keratanase Keratan-sulfate endo-1,4-β-galactosidase	
3.2.1.103	Keratan sulfate endogalactosidase Keratan-sulfate endo-1,4-β-galactosidase	
2.8.2.21	Keratan sulfate sulfotransferase Keratan sulfotransferase	
2.8.2.21	Keratan sulfotransferase Keratan sulfotransferase	
3.4.4.25	keratinase keratinase	
1.5.1.25	Ketimine-reducing enzyme Thiomorpholine-carboxylate dehydrogenase	
1.5.1.25	Ketimine reductase Thiomorpholine-carboxylate dehydrogenase	
1.1.1.86	α-Keto-β-hydroxylacyl reductoisomerase Ketol-acid reductoisomerase	
1.2.4.4	α-Keto-β-methylvalerate dehydrogenase 3-Methyl-2-oxobutanoate dehydrogenase (lipoamide)	
2.7.1.92	5-Keto-2-deoxygluconokinase 5-Dehydro-2-deoxygluconokinase	
4.1.2.21	2-Keto-3-deoxy-6-phosphogalactonic aldolase 2-Dehydro-3-deoxyphosphogalactonate aldolase	
4.1.2.14	2-Keto-3-deoxy-6-phosphogluconate aldolase 2-Dehydro-3-deoxyphosphogluconate aldolase	
4.1.2.14	2-Keto-3-deoxy-6-phosphogluconic aldolase 2-Dehydro-3-deoxyphosphogluconate aldolase	
4.1.2.16	2-Keto-3-deoxy-8-phosphooctonic acid synthetase 2-Dehydro-3-deoxyphosphooctonate aldolase	
4.1.2.16	2-Keto-3-deoxy-8-phosphooctonic synthetase 2-Dehydro-3-deoxyphosphooctonate aldolase	

EC-No.	Synonym Recommended Name	Vol., Page
4.1.2.15	2-Keto-3-deoxy-D-arabino-heptonic acid 7-phosphate synthetase 2-Dehydro-3-deoxyphosphoheptanoate aldolase	
1.1.1.127	2-Keto-3-deoxy-D-gluconate (3-deoxy-D-glycero-2,5-hexodiulosonic acid) dehydrogenase 2-Dehydro-3-deoxy-D-gluconate 5-dehydrogenase	
1.1.1.127	2-Keto-3-deoxy-D-gluconate dehydrogenase 2-Dehydro-3-deoxy-D-gluconate 5-dehydrogenase	
1.1.1.126	2-Keto-3-deoxy-D-gluconate oxidoreductase 2-Dehydro-3-deoxy-D-gluconate 6-dehydrogenase	
2.7.1.45	2-Keto-3-deoxy-D-gluconic acid kinase 2-Dehydro-3-deoxygluconokinase	
4.1.2.28	2-Keto-3-deoxy-D-pentonate aldolase 2-Dehydro-3-deoxy-D-pentonate aldolase	
4.1.2.18	2-Keto-3-deoxy-D-xylonate aldolase 2-Dehydro-3-deoxy-L-pentonate aldolase	
2.7.1.58	2-Keto-3-deoxy-galactonate kinase 2-Dehydro-3-deoxygalactonokinase	
4.1.2.20	2-Keto-3-deoxy-glucarate aldolase 2-Dehydro-3-deoxyglucarate aldolase	
4.2.1.43	2-keto-3-deoxy-L-arabinonate dehydratase 2-dehydro-3-deoxy-L-arabinonate dehydratase	
4.1.2.18	2-Keto-3-deoxy-L-arabonate aldolase 2-Dehydro-3-deoxy-L-pentonate aldolase	
4.1.2.18	2-Keto-3-deoxy-L-pentonate aldolase 2-Dehydro-3-deoxy-L-pentonate aldolase	
4.2.1.43	L-2-keto-3-deoxyarabonate dehydratase 2-dehydro-3-deoxy-L-arabinonate dehydratase	
4.1.2.21	2-Keto-3-deoxygalactonate-6-phosphate aldolase 2-Dehydro-3-deoxyphosphogalactonate aldolase	
2.7.1.58	2-Keto-3-deoxygalactonate kinase 2-Dehydro-3-deoxygalactonokinase	
2.7.1.58	2-Keto-3-deoxygalactonokinase 2-Dehydro-3-deoxygalactonokinase	
1.1.1.127	2-Keto-3-deoxygluconate-5-dehydrogenase 2-Dehydro-3-deoxy-D-gluconate 5-dehydrogenase	

EC-No.	Synonym / Recommended Name	Vol., Page
4.1.2.14	2-Keto-3-deoxygluconate-6-P-aldolase 2-Dehydro-3-deoxyphosphogluconate aldolase	
4.1.2.14	2-Keto-3-deoxygluconate-6-phosphate aldolase 2-Dehydro-3-deoxyphosphogluconate aldolase	
2.7.1.45	2-Keto-3-deoxygluconate kinase 2-Dehydro-3-deoxygluconokinase	
2.7.1.45	2-Keto-3-deoxygluconokinase 2-Dehydro-3-deoxygluconokinase	
2.7.7.38	2-Keto-3-deoxyoctonate cytidylyltransferase 3-Deoxy-manno-octulosonate cytidylyltransferase	
4.1.2.23	2-keto-3-deoxyoctonic aldolase 3-deoxy-D-manno-octulosonate aldolase	
4.1.2.20	5-Keto-4-deoxy-(D)-glucarate aldolase 2-Dehydro-3-deoxyglucarate aldolase	
4.2.1.41	5-Keto-4-deoxy-glucarate dehydratase 5-Dehydro-4-deoxyglucarate dehydratase	
4.1.3.16	2-Keto-4-hydroxybutyrate aldolase 4-Hydroxy-2-oxoglutarate aldolase	
4.1.3.16	2-Keto-4-hydroxyglutarate aldolase 4-Hydroxy-2-oxoglutarate aldolase	
4.1.3.16	2-Keto-4-hydroxyglutaric aldolase 4-Hydroxy-2-oxoglutarate aldolase	
4.2.1.80	2-keto-4-pentenoate(vinylpyruvate)hydratase 2-oxopent-4-enoate hydratase	
4.2.1.80	2-keto-4-pentenoate hydratase 2-oxopent-4-enoate hydratase	
4.2.1.80	2-keto-4-pentenoic acid hydratase 2-oxopent-4-enoate hydratase	
1.3.99.6	3-Keto-Δ^4-steroid-5β-reductase 3-Oxo-5β-steroid Δ^4-dehydrogenase	
1.3.99.5	3-Keto-δ^4-steroid-5 α-reductase 3-Oxo-5α-steroid Δ^4-dehydrogenase	
5.3.3.1	3-Keto-Δ^5-steroid isomerase Steroid Δ-isomerase	Vol. 1, Page 376
1.1.1.43	2-Keto-6-phosphogluconate reductase Phosphogluconate 2-dehydrogenase	

EC-No.	Synonym Recommended Name	Vol., Page
2.6.1.62	7-Keto-8-aminopelargonic acid-7,8-diaminopelargonic acid aminotransferase Adenosylmethionine-8-amino-7-oxononanoate transaminase	
2.6.1.62	7-Keto-8-aminopelargonic acid aminotransferase Adenosylmethionine-8-amino-7-oxononanoate transaminase	
2.3.1.47	7-Keto-8-aminopelargonic acid synthetase 8-Amino-7-oxononanoate synthase	
2.3.1.47	7-Keto-8-aminopelargonic synthetase 8-Amino-7-oxononanoate synthase	
1.1.1.123	5-Keto-D-fructose reductase Sorbose 5-dehydrogenase ($NADP^+$)	
1.1.1.124	5-Keto-D-fructose reductase ($NADP^+$) Fructose 5-dehydrogenase ($NADP^+$)	
1.1.1.128	5-Keto-D-gluconate 2-reductase L-Idonate 2-dehydrogenase	
1.1.1.69	5-Keto-D-gluconate 5-reductase Gluconate 5-dehydrogenase	
1.1.99.4	2-Keto-D-gluconate dehydrogenase Dehydrogluconate dehydrogenase	
1.1.1.215	2-Keto-D-gluconate reductase Gluconate 2-dehydrogenase	
1.1.1.69	5-Keto-D-gluconate reductase Gluconate 5-dehydrogenase	
4.1.1.34	3-Keto-L-gulonate decarboxylase Dehydro-L-gulonate decarboxylase	
4.1.1.34	Keto-L-gulonate decarboxylase Dehydro-L-gulonate decarboxylase	
1.1.1.130	3-Keto-L-gulonate dehydrogenase 3-Dehydro-L-gulonate 2-dehydrogenase	
1.1.1.189	9-Keto-prostaglandin E_2 reductase Prostaglandin-E_2 9-reductase	
1.1.1.35	β-Keto-reductase 3-Hydroxyacyl-CoA dehydrogenase	
3.5.1.3	α-keto acid-ω-amidase ω-amidase	

EC-No.	Synonym Recommended Name	Vol., Page
4.1.1.1	α-Keto acid carboxylase Pyruvate decarboxylase	
2.8.3.5	3-Ketoacid CoA-transferase 3-Oxoacid CoA-transferase	
2.8.3.5	3-Ketoacid coenzyme A-transferase 3-Oxoacid CoA-transferase	
2.8.3.5	3-Ketoacid coenzyme A transferase 3-Oxoacid CoA-transferase	
1.1.1.100	β-Ketoacyl-[acyl-carrier protein](ACP) reductase 3-Oxoacyl-[acyl-carrier-protein] reductase	
2.3.1.41	β-Ketoacyl-[acyl carrier protein] synthase 3-Oxoacyl-[acyl-carrier-protein] synthase	
1.1.1.100	β-Ketoacyl-ACP reductase 3-Oxoacyl-[acyl-carrier-protein] reductase	
2.3.1.41	β-Ketoacyl-ACP synthetase 3-Oxoacyl-[acyl-carrier-protein] synthase	
1.1.1.100	β-Ketoacyl-acyl carrier protein reductase 3-Oxoacyl-[acyl-carrier-protein] reductase	
2.3.1.41	3-Ketoacyl-acyl carrier protein synthase 3-Oxoacyl-[acyl-carrier-protein] synthase	
2.3.1.41	β-Ketoacyl-acyl carrier protein synthetase 3-Oxoacyl-[acyl-carrier-protein] synthase	
2.3.1.16	Ketoacyl-CoA acyltransferase Acetyl-CoA C-acyltransferase	
1.1.1.35	β-Ketoacyl-CoA reductase 3-Hydroxyacyl-CoA dehydrogenase	
1.1.1.36	β-Ketoacyl-CoA reductase Acetoacetyl-CoA reductase	
2.3.1.16	3-Ketoacyl-CoA thiolase Acetyl-CoA C-acyltransferase	
2.3.1.16	β-Ketoacyl-CoA thiolase Acetyl-CoA C-acyltransferase	
2.3.1.16	Ketoacyl-coenzyme A thiolase Acetyl-CoA C-acyltransferase	
1.1.1.100	β-Ketoacyl acyl carrier protein (ACP) reductase 3-Oxoacyl-[acyl-carrier-protein] reductase	

EC-No.	Synonym Recommended Name	Vol., Page
1.1.1.100	3-Ketoacyl acyl carrier protein reductase 3-Oxoacyl-[acyl-carrier-protein] reductase	
2.3.1.41	β-Ketoacyl acyl carrier protein synthase 3-Oxoacyl-[acyl-carrier-protein] synthase	
2.3.1.16	3-Ketoacyl CoA thiolase Acetyl-CoA C-acyltransferase	
2.3.1.16	3-Ketoacyl coenzyme A thiolase Acetyl-CoA C-acyltransferase	
2.3.1.16	β-Ketoacyl coenzyme A thiolase Acetyl-CoA C-acyltransferase	
4.1.1.56	β-Ketoacyl decarboxylase 3-Oxolaurate decarboxylase	
1.1.1.100	β-Ketoacyl reductase 3-Oxoacyl-[acyl-carrier-protein] reductase	
2.3.1.41	β-Ketoacylsynthase 3-Oxoacyl-[acyl-carrier-protein] synthase	
2.3.1.41	β-Ketoacyl synthetase 3-Oxoacyl-[acyl-carrier-protein] synthase	
1.1.1.100	β-Ketoacyl thioester reductase 3-Oxoacyl-[acyl-carrier-protein] reductase	
2.3.1.16	3-Ketoacyl thiolase Acetyl-CoA C-acyltransferase	
3.1.1.24	β-ketoadipate enol-lactone hydrolase 3-oxoadipate enol-lactose	
1.1.1.172	2-Ketoadipate reductase 2-Oxoadipate reductase	
1.1.1.172	α-Ketoadipate reductase 2-Oxoadipate reductase	
2.3.1.16	β-Ketoadipyl-CoA thiolase Acetyl-CoA C-acyltransferase	
2.3.1.16	β-Ketoadipyl coenzyme A thiolase Acetyl-CoA C-acyltransferase	
1.2.1.23	2-Ketoaldehyde dehydrogenase 2-Oxoaldehyde dehydrogenase (NAD^+)	
1.2.1.49	2-Ketoaldehyde dehydrogenase 2-Oxoaldehyde dehydrogenase ($NADP^+$)	

EC-No.	Synonym Recommended Name	Vol., Page
1.2.1.23	α-Ketoaldehyde dehydrogenase 2-Oxoaldehyde dehydrogenase (NAD^+)	
1.2.1.49	α-Ketoaldehyde dehydrogenase 2-Oxoaldehyde dehydrogenase ($NADP^+$)	
4.1.1.75	α-Ketoarginine decarboxylase 5-Guanidino-2-oxopentanoate decarboxylase	
1.2.7.2	α-Ketobutyrate-ferredoxin oxidoreductase 2-Oxobutyrate synthase	
1.2.7.2	2-Ketobutyrate synthase 2-Oxobutyrate synthase	
1.2.7.2	α-Ketobutyrate synthase 2-Oxobutyrate synthase	
3.2.1.31	Ketodase β-glucuronidase	
2.7.1.45	Ketodeoxygluconokinase 2-Dehydro-3-deoxygluconokinase	
1.1.1.123	5-Ketofructose reductase Sorbose 5-dehydrogenase ($NADP^+$)	
1.1.1.128	5-Ketogluconate 2-reductase L-Idonate 2-dehydrogenase	
1.1.1.69	5-Ketogluconate 5-reductase Gluconate 5-dehydrogenase	
1.1.99.4	Ketogluconate dehydrogenase Dehydrogluconate dehydrogenase	
1.1.99.4	α-Ketogluconate dehydrogenase Dehydrogluconate dehydrogenase	
2.7.1.13	2-Ketogluconate kinase Dehydrogluconokinase	
1.1.1.215	2-Ketogluconate reductase Gluconate 2-dehydrogenase	
1.1.99.3	2-Ketogluconate reductase Gluconate 2-dehydrogenase	
1.1.1.69	5-Ketogluconate reductase Gluconate 5-dehydrogenase	
1.1.1.128	5-Ketoglucono-idono-reductase L-Idonate 2-dehydrogenase	

EC-No.	Synonym Recommended Name	Vol., Page
2.7.1.13	2-Ketogluconokinase Dehydrogluconokinase	
2.7.1.13	Ketogluconokinase Dehydrogluconokinase	
1.14.11.17	α-ketoglutarate-dependent taurine dioxygenase taurine dioxygenase	
1.2.7.3	α-Ketoglutarate-ferredoxin oxidoreductase 2-Oxoglutarate synthase	
4.1.3.15	α-ketoglutarate:glyoxylate carboligase 2-hydroxy-3-oxoadipate synthase	
4.1.1.71	α-Ketoglutarate decarboxylase 2-Oxoglutarate decarboxylase	
1.2.4.2	2-Ketoglutarate dehydrogenase Oxoglutarate dehydrogenase (lipoamide)	
1.2.4.2	α-Ketoglutarate dehydrogenase Oxoglutarate dehydrogenase (lipoamide)	
1.1.99.2	α-Ketoglutarate reductase 2-Hydroxyglutarate dehydrogenase	
1.2.7.3	α-Ketoglutarate synthase 2-Oxoglutarate synthase	
4.1.3.15	α-ketoglutaric-glyoxylic carboligase 2-hydroxy-3-oxoadipate synthase	
1.2.4.2	α-Ketoglutaric acid dehydrogenase Oxoglutarate dehydrogenase (lipoamide)	
4.1.1.71	α-Ketoglutaric decarboxylase 2-Oxoglutarate decarboxylase	
1.2.4.2	Ketoglutaric dehydrogenase Oxoglutarate dehydrogenase (lipoamide)	
1.2.4.2	α-Ketoglutaric dehydrogenase Oxoglutarate dehydrogenase (lipoamide)	
1.1.1.41	β-Ketoglutaricisocitric carboxylase Isocitrate dehydrogenase (NAD^+)	
1.2.1.26	α-Ketoglutaric semialdehyde dehydrogenase 2,5-Dioxovalerate dehydrogenase	
1.1.1.130	3-Ketogulonate dehydrogenase 3-Dehydro-L-gulonate 2-dehydrogenase	

EC-No.	Synonym Recommended Name Vol., Page
2.7.1.3	Ketohexokinase Ketohexokinase
4.2.1.44	Ketoinositol dehydratase Myo-inosose-2 dehydratase
1.2.4.4	α-Ketoisocaproate dehydrogenase 3-Methyl-2-oxobutanoate dehydrogenase (lipoamide)
1.2.4.4	α-Ketoisocaproic-α-keto-β-methylvaleric dehydrogenase 3-Methyl-2-oxobutanoate dehydrogenase (lipoamide)
1.2.4.4	α-Ketoisocaproic dehydrogenase 3-Methyl-2-oxobutanoate dehydrogenase (lipoamide)
1.2.1.25	α-Ketoisovalerate dehydrogenase 2-Oxoisovalerate dehydrogenase (acylating)
1.2.4.4	α-Ketoisovalerate dehydrogenase 3-Methyl-2-oxobutanoate dehydrogenase (lipoamide)
2.1.2.11	α-Ketoisovalerate hydroxymethyltransferase 3-Methyl-2-oxobutanoate hydroxymethyltransferase
1.1.1.86	Ketol-acid reductoisomerase Ketol-acid reductoisomerase
1.14.15.2	Ketolactonase I Camphor 1,2-monooxygenase
4.1.1.56	β-Ketolaurate-decarboxylase 3-Oxolaurate decarboxylase
4.4.1.5	ketone-aldehyde mutase lactoylglutathione lyase
4.1.2.12	Ketopantoaldolase Ketopantoaldolase
2.1.2.11	Ketopantoate hydroxymethyltransferase 3-Methyl-2-oxobutanoate hydroxymethyltransferase
1.1.1.169	2-Ketopantoate reductase 2-Dehydropantoate 2-reductase
1.1.1.169	Ketopantoate reductase 2-Dehydropantoate 2-reductase
1.1.1.169	2-Ketopantoic acid reductase 2-Dehydropantoate 2-reductase
1.1.1.169	Ketopantoic acid reductase 2-Dehydropantoate 2-reductase

EC-No.	Synonym Recommended Name	Vol., Page
1.1.1.168	2-Ketopantoyl lactone reductase 2-Dehydropantoyl-lactone reductase (A-specific)	
1.1.1.214	2-Ketopantoyl lactone reductase 2-Dehydropantoyl-lactone reductase (B-specific)	
1.1.1.168	Ketopantoyl lactone reductase 2-Dehydropantoyl-lactone reductase (A-specific)	
1.1.1.214	Ketopantoyl lactone reductase 2-Dehydropantoyl-lactone reductase (B-specific)	
1.3.1.48	15-Ketoprostaglandin.$\delta.^{13}$-reductase 15-Oxoprostaglandin Δ^{13}-reductase	
1.3.1.48	15-Ketoprostaglandin.$\delta.^{13}$-reductase 15-Oxoprostaglandin Δ^{13}-reductase	
1.3.1.48	15-Ketoprostaglandin Δ^{13}-reductase 15-Oxoprostaglandin Δ^{13}-reductase	
1.3.1.48	.δ.13-15-Ketoprostaglandin reductase 15-Oxoprostaglandin Δ^{13}-reductase	
1.1.1.189	9-Ketoprostaglandin reductase Prostaglandin-E_2 9-reductase	
1.1.1.51	17-Ketoreductase 3(or 17)β-Hydroxysteroid dehydrogenase	
1.1.1.63	17-Ketoreductase Testosterone 17β-dehydrogenase	
1.1.1.64	17-Ketoreductase Testosterone 17β-dehydrogenase ($NADP^+$)	
1.1.1.35	3-Keto reductase 3-Hydroxyacyl-CoA dehydrogenase	
4.1.2.7	ketose-1-phosphate aldolase ketose-1-phosphate aldolase	
4.1.2.13	Ketose 1-phosphate aldolase Fructose-bisphosphate aldolase	
1.1.1.140	Ketosephosphate reductase Sorbitol-6-phosphate 2-dehydrogenase	
1.14.13.54	Ketosteroid, $NADPH_2$:oxygen oxidoreductase (20-hydroxylating, ester-producing/20-hydroxylating, side-chain cleaving/17-hydroxylating, lactonizing) Ketosteroid monooxygenase	

EC-No.	Synonym / Recommended Name	Vol., Page
1.3.99.4	3-Ketosteroid-Δ^1-dehydrogenase 3-Oxosteroid Δ^1-dehydrogenase	
1.3.99.4	3-Ketosteroid-1-en-dehydrogenase 3-Oxosteroid Δ^1-dehydrogenase	
1.3.99.6	Δ-3-ketosteroid-5β-reductase 3-Oxo-5β-steroid Δ^4-dehydrogenase	
1.3.99.6	Δ^4-3-ketosteroid-5β-reductase 3-Oxo-5β-steroid Δ^4-dehydrogenase	
5.3.3.1	3-Ketosteroid $\Delta^5 \rightarrow \Delta^4$-isomerase Steroid Δ-isomerase	Vol. 1, Page 376
1.3.1.22	Δ^4-3-Ketosteroid 5α-oxidoreductase Cholestenone 5α-reductase	
1.3.99.5	δ^4-3-Ketosteroid 5α-oxidoreductase 3-Oxo-5α-steroid Δ^4-dehydrogenase	
1.3.1.23	Δ^4-3-ketosteroid 5β-reductase Cholestenone 5β-reductase	
1.3.1.3	Δ^4-3-ketosteroid 5β-reductase Cortisone β-reductase	
1.3.99.5	δ^4-3-Keto steroid 5α-reductase 3-Oxo-5α-steroid Δ^4-dehydrogenase	
1.1.1.53	Δ^4-3-Ketosteroid hydrogenase 3α(or 20β)-Hydroxysteroid dehydrogenase	
5.3.3.1	Δ-3-ketosteroid isomerase Steroid Δ-isomerase	Vol. 1, Page 376
5.3.3.1	Δ^5-3-keto steroid isomerase Steroid Δ-isomerase	Vol. 1, Page 376
5.3.3.1	Δ^5-3-ketosteroid isomerase Steroid Δ-isomerase	Vol. 1, Page 376
5.3.3.1	Δ^5-ketosteroid isomerase Steroid Δ-isomerase	Vol. 1, Page 376
5.3.3.1	δ 5-3-ketosteroid isomerase Steroid Δ-isomerase	Vol. 1, Page 376
1.14.13.54	Ketosteroid monooxygenase Ketosteroid monooxygenase	
1.3.1.4	Δ^4-3-Ketosteroid reductase (5α) Cortisone α-reductase	

EC-No.	Synonym Recommended Name	Vol., Page
4.1.2.2	Ketotetrose-phosphate aldolase Ketotetrose-phosphate aldolase	
2.3.1.16	3-Ketothiolase Acetyl-CoA C-acyltransferase	
2.3.1.16	β-Ketothiolase Acetyl-CoA C-acyltransferase	
4.3.3.1	3-Ketovalidoxylamine A C-N lyase 3-Ketovalidoxylamine C-N-lyase	
4.3.3.1	3-Ketovalidoxylamine C-N-lyase 3-Ketovalidoxylamine C-N-lyase	
3.4.16.6	KEX1 carboxypeptidase Carboxypeptidase D	
3.4.16.6	KEX1Δp Carboxypeptidase D	
3.4.16.6	KEX1 proteinase Carboxypeptidase D	
3.4.21.61	Kex2-like endoproteinase Kexin	
3.4.21.61	Kex2-like precursor protein processing endoprotease Kexin	
3.4.21.61	Kex2 endopeptidase Kexin	
3.4.21.61	Kex2 endoprotease Kexin	
3.4.21.61	Kex2 endoproteinase Kexin	
3.4.21.61	Kex 2p proteinase Kexin	
3.4.21.61	Kex2 protease Kexin	
3.4.21.61	Kex2 proteinase Kexin	
3.4.21.61	Kexin Kexin	
1.1.1.128	5KGR L-Idonate 2-dehydrogenase	

EC-No.	Synonym Recommended Name	Vol., Page
4.2.1.95	KHase Kievitone hydratase	
2.1.2.11	KHMT 3-Methyl-2-oxobutanoate hydroxymethyltransferase	
3.4.24.11	kidney-brush-border neutral endopeptidase neprilysin	
3.4.24.11	kidney-brush-border neutral peptidase neprilysin	
3.4.24.11	kidney-brush-border neutral proteinase neprilysin	
4.2.1.95	Kievitone-hydrate hydro-lyase Kievitone hydratase	
4.2.1.95	Kievitone hydratase Kievitone hydratase	
2.7.1.37	A-kinase Protein kinase	
2.7.1.126	Kinase, β-adrenergic-receptor (phosphorylating) β-Adrenergic-receptor kinase	
2.7.1.85	Kinase, β-D-glucoside (phosphorylating) β-Glucoside kinase	
2.7.1.107	Kinase, 1,2-diacylglycerol (phosphorylating) Diacylglycerol kinase	
2.7.1.93	Kinase, 1-alkylglycerol (phosphorylating) Alkylglycerol kinase	
2.7.1.56	Kinase, 1-phosphofructo- (phosphorylating) 1-Phosphofructokinase	
2.7.1.58	Kinase, 2-keto-3-deoxygalactonate (phosphorylating) 2-Dehydro-3-deoxygalactonokinase	
2.7.1.45	Kinase, 2-keto-3-deoxyglucono- (phosphorylating) 2-Dehydro-3-deoxygluconokinase	
2.7.1.92	Kinase, 5-keto-2-deoxyglucono- (phosphorylating) 5-Dehydro-2-deoxygluconokinase	
2.7.4.19	Kinase, 5-methyldeoxycytidylate (phosphorylating) 5-Methyldeoxycytidine-5'-phosphate kinase	
2.7.1.100	Kinase, 5-methylthioribose (phosphorylating) 5-Methylthioribose kinase	

EC-No.	Synonym Recommended Name	Vol., Page
2.7.1.105	Kinase, 6-phosphofructo-2- (phosphorylating) 6-Phosphofructo-2-kinase	
2.7.2.1	Kinase, acetate (phosphorylating) Acetate kinase	
2.7.1.29	Kinase, acetol (phosphorylating) Glycerone kinase	
2.7.1.128	Kinase, acetyl coenzyme A carboxylase (phosphorylating) [Acetyl-CoA carboxylase] kinase	
2.7.1.59	Kinase, acetylglucosamine (phosphorylating) N-Acetylglucosamine kinase	
2.7.2.8	Kinase, acetylglutamate (phosphorylating) Acetylglutamate kinase	
2.7.1.60	Kinase, acylmannosamine (phosphorylating) N-Acylmannosamine kinase	
2.7.1.138	Kinase, acylsphingosine (phosphorylating) Ceramide kinase	
2.7.1.20	Kinase, adenosine (phosphorylating) Adenosine kinase	
2.7.4.3	Kinase, adenylate (phosphorylating) Adenylate kinase	
2.7.1.25	Kinase, adenylylsulfate (phosphorylating) Adenylylsulfate kinase	
2.7.3.10	Kinase, agmatine (phosphorylating) Agmatine kinase	
2.7.1.84	Kinase, alkyldihydroxyacetone (phosphorylating) Alkylglycerone kinase	
2.7.1.55	Kinase, allo- (phosphorylating) Allose kinase	
2.7.3.8	Kinase, ammonia (phosphorylating) Ammonia kinase	
2.7.3.3	Kinase, arginine (phosphorylating) Arginine kinase	
2.7.2.4	Kinase, aspartate (phosphorylating) Aspartate kinase	
2.7.2.14	Kinase, branched-chain fatty acid (phosphorylating) Branched-chain-fatty-acid kinase	

EC-No.	Synonym Recommended Name	Vol., Page
2.7.1.115	Kinase, branched-chain oxo acid dehydrogenase (phosphorylating) [3-Methyl-2-oxobutanoate dehydrogenase (lipoamide)] kinase	
2.7.2.7	Kinase, butyrate (phosphorylating) Butyrate kinase	
2.7.1.120	Kinase, caldesmon (phosphorylating) Caldesmon kinase	
2.7.1.37	Kinase, casein (phosphorylating) Protein kinase	
2.7.1.32	Kinase, choline (phosphorylating) Choline kinase	
2.7.3.2	Kinase, creatine (phosphorylating) Creatine kinase	
2.7.4.14	Kinase, cytidylate (phosphorylating) Cytidylate kinase	
2.7.1.54	Kinase, D-arabino- (phosphorylating) D-Arabinokinase	
2.7.1.47	Kinase, D-ribulo- (phosphorylating) D-Ribulokinase	
2.7.1.76	Kinase, deoxyadenosine (phosphorylating) Deoxyadenosine kinase	
2.7.4.11	Kinase, deoxyadenylate (phosphorylating) (Deoxy)adenylate kinase	
2.7.1.74	Kinase, deoxycytidine (phosphorylating) Deoxycytidine kinase	
2.7.1.113	Kinase, deoxyguanosine (phosphorylating) Deoxyguanosine kinase	
2.7.4.13	Kinase, deoxynucleoside monophosphate (phosphorylating) (Deoxy)nucleoside-phosphate kinase	
2.7.4.12	Kinase, deoxynucleotide (phosphorylating, T2-induced) T2-induced deoxynucleotide kinase	
2.7.1.21	Kinase, deoxythymidine (phosphorylating) Thymidine kinase	
2.7.1.24	Kinase, dephosphocoenzyme A (phosphorylating) Dephospho-CoA kinase	
2.7.1.91	Kinase, dihydrosphingosine (phosphorylating) Sphinganine kinase	

EC-No.	Synonym Recommended Name	Vol., Page
2.7.1.88	Kinase, dihydrostreptomycin 6-phosphate (phosphorylating) Dihydrostreptomycin-6-phosphate 3'α-kinase	
2.7.1.108	Kinase, dolichol (phosphorylating) Dolichol kinase	
2.7.1.27	Kinase, erythritol (phosphorylating) Erythritol kinase	
2.7.1.82	Kinase, ethanolamine (phosphorylating) Ethanolamine kinase	
2.7.4.18	Kinase, farnesyl diphosphate (phosphorylating) Farnesyl-diphosphate kinase	
2.7.1.40	Kinase, fluoro- (phosphorylating) Pyruvate kinase	
2.7.2.6	Kinase, formate (phosphorylating) Formate kinase	
2.7.1.4	Kinase, fructo- (phosphorylating) Fructokinase	
2.7.1.52	Kinase, fuco- (phosphorylating) Fucokinase	
2.7.1.6	Kinase, galacto- (phosphorylating) Galactokinase	
2.7.1.44	Kinase, galacturono- (phosphorylating) Galacturonokinase	
2.7.1.2	Kinase, gluco- (phosphorylating) Glucokinase	
2.7.1.12	Kinase, glucono- (phosphorylating) Gluconokinase	
2.7.1.8	Kinase, glucosamine (phosphorylating) Glucosamine kinase	
2.7.1.43	Kinase, glucurono- (phosphorylating) Glucuronokinase	
2.7.2.11	Kinase, glutamate (phosphorylating) Glutamate 5-kinase	
2.7.2.13	Kinase, glutamate 1- (phosphorylating) Glutamate 1-kinase	
2.7.1.31	Kinase, glycerate (phosphorylating) Glycerate kinase	

EC-No.	Synonym Recommended Name	Vol., Page
2.7.1.30	Kinase, glycerol (phosphorylating) Glycerol kinase	
2.7.3.1	Kinase, guanidinoacetate (phosphorylating) Guanidinoacetate kinase	
2.7.4.8	Kinase, guanylate (phosphorylating) Guanylate kinase	
2.7.1.102	Kinase, hamamelose (phosphorylating) Hamamelose kinase	
2.7.1.1	Kinase, hexo- (phosphorylating) Hexokinase	
2.7.1.39	Kinase, homoserine (phosphorylating) Homoserine kinase	
2.7.1.50	Kinase, hydroxyethylthiazole (phosphorylating) Hydroxyethylthiazole kinase	
2.7.1.81	Kinase, hydroxylysine (phosphorylating) Hydroxylysine kinase	
2.7.1.109	Kinase, hydroxymethylglutaryl coenzyme A reductase (phosphorylating) [Hydroxymethylglutaryl-CoA reductase (NADPH)] kinase	
2.7.1.110	Kinase, hydroxymethylglutaryl coenzyme A reductase kinase (phosphorylating) Dephospho-[reductase kinase] kinase	
2.7.1.49	Kinase, hydroxymethylpyrimidine (phosphorylating) Hydroxymethylpyrimidine kinase	
2.7.3.6	Kinase, hypotaurocyamine (phosphorylating) Hypotaurocyamine kinase	
2.7.1.73	Kinase, inosine (phosphorylating) Inosine kinase	
2.7.1.140	Kinase, inositol 1,3,4,6-tetrakisphosphate 5- (phosphorylating) 1D-myo-Inositol-tetrakisphosphate 5-kinase	
2.7.1.139	Kinase, inositol 1,3,4-trisphosphate 5- (phosphorylating) 1D-myo-Inositol-trisphosphate 5-kinase	
2.7.1.133	Kinase, inositol 1,3,4-trisphosphate 6- (phosphorylating) 1D-myo-Inositol-trisphosphate 6-kinase	
2.7.1.134	Kinase, inositol 1,4,5,6-tetrakisphosphate 3- (phosphorylating) 1D-myo-Inositol-tetrakisphosphate 1-kinase	

EC-No.	Synonym Recommended Name	Vol., Page
2.7.1.127	Kinase, inositol 1,4,5-trisphosphate 3- (phosphorylating) 1D-myo-Inositol-trisphosphate 3-kinase	
2.7.1.64	Kinase, inositol 1- (phosphorylating) myo-Inositol 1-kinase	
2.7.1.116	Kinase, isocitrate dehydrogenase (phosphorylating) [Isocitrate dehydrogenase ($NADP^+$)] kinase	
2.7.1.66	Kinase, isoprenoid alcohol (phosphorylating) Undecaprenol kinase	
2.7.1.95	Kinase, kanamycin (phosphorylating) Kanamycin kinase	
2.7.1.13	Kinase, ketoglucono- (phosphorylating) Dehydrogluconokinase	
2.7.1.3	Kinase, ketohexo- (phosphorylating) Ketohexokinase	
2.7.1.46	Kinase, L-arabino- (phosphorylating) L-Arabinokinase	
2.7.1.51	Kinase, L-fuculo- (phosphorylating) L-Fuculokinase	
2.7.1.53	Kinase, L-xylulo- (phosphorylating) L-Xylulokinase	
2.7.1.130	Kinase, lipid A 4'- (phosphorylating) Tetraacyldisaccharide 4'-kinase	
2.7.3.5	Kinase, lombricine (phosphorylating) Lombricine kinase	
2.7.1.131	Kinase, low-density-lipoprotein receptor (phosphorylating) Low-density-lipoprotein kinase	
2.7.1.7	Kinase, manno- (phosphorylating) Mannokinase	
2.7.1.36	Kinase, mevalonate (phosphorylating) Mevalonate kinase	
2.7.1.123	Kinase, microtubule-associated protein 2 (phosphorylating) Ca^{2+}/calmodulin-dependent protein kinase	
2.7.1.94	Kinase, monoacylglycerol (phosphorylating) Acylglycerol kinase	
2.7.4.3	Kinase, myo- (phosphorylating) Adenylate kinase	

EC-No.	Synonym Recommended Name	Vol., Page
2.7.1.129	Kinase, myosin heavy chain (phosphorylating) Myosin-heavy-chain kinase	
2.7.1.117	Kinase, myosin light-chain (phosphorylating) Myosin-light-chain kinase	
2.7.1.23	Kinase, nicotinamide adenine dinucleotide (phosphorylating) NAD^+ kinase	
2.7.4.6	Kinase, nucleoside diphosphate (phosphorylating) Nucleoside-diphosphate kinase	
2.7.4.4	Kinase, nucleoside monophosphate (phosphorylating) Nucleoside-phosphate kinase	
2.7.4.10	Kinase, nucleoside triphosphate-adenylate (phosphorylating) Nucleoside-triphosphate-adenylate kinase	
2.7.3.7	Kinase, opheline (phosphorylating) Opheline kinase	
2.7.1.125	Kinase, opsin (phosphorylating) Rhodopsin kinase	
2.7.1.34	Kinase, pantetheine (phosphorylating) Pantetheine kinase	
2.7.1.33	Kinase, pantothenate (phosphorylating) Pantothenate kinase	
2.7.3.10	Kinase, phosphagen (phosphorylating) Agmatine kinase	
2.7.1.67	Kinase, phosphatidylinositol (phosphorylating) 1-Phosphatidylinositol 4-kinase	
2.7.1.137	Kinase, phosphatidylinositol 3- (phosphorylating) 1-Phosphatidylinositol 3-kinase	
2.7.1.11	Kinase, phosphofructo- (phosphorylating) 6-Phosphofructokinase	
2.7.1.10	Kinase, phosphogluco- (phosphorylating) Phosphoglucokinase	
2.7.2.3	Kinase, phosphoglycerate (phosphorylating) Phosphoglycerate kinase	
2.7.2.10	Kinase, phosphoglycerate (phosphorylating guanosine triphosphate) Phosphoglycerate kinase (GTP)	
2.7.4.7	Kinase, phosphomethylpyrimidine (phosphorylating) Phosphomethylpyrimidine kinase	

EC-No.	Synonym Recommended Name	Vol., Page
2.7.4.2	Kinase, phosphomevalonate (phosphorylating) Phosphomevalonate kinase	
2.7.1.18	Kinase, phosphoribo- (phosphorylating) Phosphoribokinase	
2.7.1.19	Kinase, phosphoribulo- (phosphorylating) Phosphoribulokinase	
2.7.1.38	Kinase, phosphorylase (phosphorylating) Phosphorylase kinase	
2.7.1.78	Kinase, polynucleotide 5'-hydroxyl (phosphorylating) Polynucleotide 5'-hydroxy-kinase	
2.7.4.1	Kinase, polyphosphate (phosphorylating) Polyphosphate kinase	
2.7.1.70	Kinase, protamine (phosphorylating) Protamine kinase	
2.7.1.37	Kinase, protein (phosphorylating) Protein kinase	
2.7.3.11	Kinase, protein (phosphorylating histidine) Protein-histidine pros-kinase	
2.7.3.12	Kinase, protein (phosphorylating histidine) Protein-histidine tele-kinase	
2.7.1.112	Kinase, protein (phosphorylating tyrosine) Protein-tyrosine kinase	
2.7.1.37	Kinase, protein, A (phosphorylating) Protein kinase	
2.7.1.37	Kinase, protein, C (phosphorylating) Protein kinase	
2.7.1.112	Kinase, protein p56lck (phosphorylating) Protein-tyrosine kinase	
2.7.1.83	Kinase, pseudouridine (phosphorylating) Pseudouridine kinase	
2.7.1.143	Kinase, purine nucleoside (phosphorylating) Diphosphate-purine nucleoside kinase	
2.7.1.35	Kinase, pyridoxal (phosphorylating) Pyridoxal kinase	
2.7.1.40	Kinase, pyruvate (phosphorylating) Pyruvate kinase	

EC-No.	Synonym Recommended Name	Vol., Page
2.7.9.1	Kinase, pyruvate-phosphate di- (phosphorylating) phosphate dikinase pyruvate	
2.7.9.2	Kinase, pyruvate-water di- (phosphorylating) water dikinase pyruvate	
2.7.1.99	Kinase, pyruvate dehydrogenase (phosphorylating) [Pyruvate dehydrogenase (lipoamide)] kinase	
2.7.1.86	Kinase, reduced nicotinamide adenine dinucleotide (phosphorylating) NADH kinase	
2.7.1.5	Kinase, rhamnulo- (phosphorylating) Rhamnulokinase	
2.7.1.125	Kinase, rhodopsin (phosphorylating) Rhodopsin kinase	
2.7.1.15	Kinase, ribo- (phosphorylating) Ribokinase	
2.7.1.26	Kinase, riboflavin Riboflavin kinase	
2.7.1.141	Kinase, ribonucleate nucleotidyltransferase II C-terminal domain (phosphorylating) [RNA-polymerase]-subunit kinase	
2.7.1.141	Kinase, ribonucleate nucleotidyltransferase isozyme II IIa subunit (phosphorylating) [RNA-polymerase]-subunit kinase	
2.7.1.22	Kinase, ribosylnicotinamide (phosphorylating) Ribosylnicotinamide kinase	
2.7.1.16	Kinase, ribulo- (phosphorylating) Ribulokinase	
2.7.1.65	Kinase, scyllo-inosamine (phosphorylating) scyllo-Inosamine 4-kinase	
2.7.1.14	Kinase, sedoheptulo- (phosphorylating) Sedoheptulokinase	
2.7.1.71	Kinase, shikimate (phosphorylating) Shikimate kinase	
2.7.1.91	Kinase, sphingosine (phosphorylating) Sphinganine kinase	
2.7.1.72	Kinase, streptidine (phosphorylating) Streptomycin 6-kinase	

EC-No.	Synonym Recommended Name	Vol., Page
2.7.1.87	Kinase, streptomycin 3- (phosphorylating) Streptomycin 3-kinase	
2.7.1.72	Kinase, streptomycin 6- (phosphorylating) Streptomycin 6-kinase	
2.7.1.101	Kinase, tagatose 6-phosphate (phosphorylating) Tagatose kinase	
2.7.1.144	Kinase, tagatose 6-phosphate (phosphorylating) Tagatose-6-phosphate kinase	
2.7.1.135	Kinase, tau factor protein (phosphorylating) τ-protein kinase	
2.7.3.4	Kinase, taurocyamine (phosphorylating) Taurocyamine kinase	
2.7.1.89	Kinase, thiamin (phosphorylating) Thiamine kinase	
2.7.4.15	Kinase, thiamin diphosphate (phosphorylating) Thiamine-diphosphate kinase	
2.7.4.16	Kinase, thiamin monophosphate (phosphorylating) Thiamine-phosphate kinase	
2.7.1.21	Kinase thymidine (phosphorylating) Thymidine kinase	
2.7.4.9	Kinase, thymidine monophosphate (phosphorylating) dTMP kinase	
2.7.4.9	Kinase, thymidylate (phosphorylating) dTMP kinase	
2.7.1.28	Kinase, trio (phosphorylating) Triokinase	
2.7.1.132	Kinase, tropomyosin (phosphorylating) Tropomyosin kinase	
2.7.1.124	Kinase, tyrosine 3-monooxygenase (phosphorylating) [Tyrosine 3-monooxygenase] kinase	
2.7.1.48	Kinase, uridine (phosphorylating) Uridine kinase	
2.7.1.17	Kinase, xylulo- (phosphorylating) Xylulokinase	
3.4.22.17	kinase-activating factor calpain	

EC-No.	Synonym Recommended Name	Vol., Page
2.7.2.2	Kinase carbamate (phosphorylating) Carbamate kinase	
3.4.17.3	kininase I lysine carboxypeptidase	
3.4.17.3	kininase Ia lysine carboxypeptidase	
3.4.15.1	kininase II peptidyl-dipeptidase A	
3.4.21.34	kininogenase plasma kallikrein	
3.4.21.35	kininogenase tissue kallikrein	
3.4.21.34	kininogenin plasma kallikrein	
3.4.21.35	kininogenin tissue kallikrein	
3.2.1.39	Kitalase Glucan endo-1,3-β-glucosidase	
2.7.7.7	Klenow fragment DNA-directed DNA polymerase	
3.4.21.32	kollaza brachyurin	
3.4.24.7	kollaza interstitial collagenase	
3.4.24.3	kollaza microbial collagenase	
2.4.1.19	Konchizaimu Cyclomaltodextrin glucanotransferase	
1.14.99.2	Kynurenate,hydrogen-donor:oxygen oxidoreductase (hydroxylating) Kynurenine 7,8-hydroxylase	
1.3.1.18	Kynurenate-7,8-dihydrodiol dehydrogenase Kynurenate-7,8-dihydrodiol dehydrogenase	
1.3.1.18	Kynurenate 7,8-dihydrodiol dehydrogenase Kynurenate-7,8-dihydrodiol dehydrogenase	
1.14.99.2	Kynurenate 7,8-hydroxylase Kynurenine 7,8-hydroxylase	

EC-No.	Synonym Recommended Name	Vol., Page
1.14.99.2	Kynurenic acid hydroxylase Kynurenine 7,8-hydroxylase	
1.14.99.2	Kynurenic hydroxylase Kynurenine 7,8-hydroxylase	
3.7.1.3	Kynureninase Kynureninase	
1.14.13.9	L-Kynurenine,NADPH:oxygen oxidoreductase (3-hydroxylating) Kynurenine 3-monooxygenase	
1.14.13.9	L-Kynurenine-3-hydroxylase Kynurenine 3-monooxygenase	
2.6.1.63	Kynurenine-glyoxylate transaminase Kynurenine-glyoxylate transaminase	
2.6.1.7	Kynurenine-oxoglutarate transaminase Kynurenine-oxoglutarate transaminase	
2.6.1.7	Kynurenine 2-oxoglutarate transaminase Kynurenine-oxoglutarate transaminase	
1.14.13.9	Kynurenine 3-hydroxylase Kynurenine 3-monooxygenase	
1.14.13.9	Kynurenine 3-monooxygenase Kynurenine 3-monooxygenase	
1.14.99.2	Kynurenine 7,8-hydroxylase Kynurenine 7,8-hydroxylase	
2.6.1.7	L-Kynurenine:2-oxoglutarate aminotransferase Kynurenine-oxoglutarate transaminase	
2.6.1.63	L-Kynurenine:glyoxylate aminotransferase (cyclizing) Kynurenine-glyoxylate transaminase	
2.6.1.7	Kynurenine aminotransferase Kynurenine-oxoglutarate transaminase	
2.6.1.7	L-Kynurenine aminotransferase Kynurenine-oxoglutarate transaminase	
3.5.1.9	Kynurenine formamidase Arylformamidase	
3.7.1.3	L-Kynurenine hydrolase Kynureninase	
1.14.13.9	Kynurenine hydroxylase Kynurenine 3-monooxygenase	

EC-No.	Synonym Recommended Name	Vol., Page
2.6.1.7	Kynurenine transaminase (cyclizing) Kynurenine-oxoglutarate transaminase	
3.4.21.63	Kyorinase Oryzin	
6.3.2.24	Kyotorphin-synthesizing enzyme Tyrosine-arginine ligase	Vol. 2, Page 521
6.3.2.24	Kyotorphin synthase Tyrosine-arginine ligase	Vol. 2, Page 521
1.1.1.75	L(+)-1-Aminopropan-2-ol-NAD/NADP oxidoreductase (R)-Aminopropanol dehydrogenase	
1.1.1.75	L(+)-1-Aminopropan-2-ol:NAD^+ oxidoreductase (R)-Aminopropanol dehydrogenase	
1.1.1.76	L(+)-2,3-Butanediol dehydrogenase (L-acetoin forming) (S,S)-Butanediol dehydrogenase	
1.1.1.157	L(+)-3-Hydroxybutyryl-CoA dehydrogenase 3-Hydroxybutyryl-CoA dehydrogenase	
1.1.2.3	L(+)-Lactate:cytochrome c oxidoreductase L-Lactate dehydrogenase (cytochrome)	
1.1.1.27	L(+)-nLDH L-Lactate dehydrogenase	
4.2.1.32	L(+)-Tartrate dehydrogenase L(+)-Tartrate dehydrogenase	
1.5.1.10	ε-N-(L-Glutaryl-2)-L-lysine:NAD^+(P) oxidoreductase (L-2-aminoadipate-semialdehyde forming) Saccharopine dehydrogenase ($NADP^+$, L-glutamate forming)	
1.5.1.7	ε-N-(L-glutaryl-2)-L-lysine:NAD oxidoreductase (L-lysine forming) Saccharopine dehydrogenase (NAD^+, L-lysine forming)	
5.2.1.8	L.p.Cyp18 Peptidylprolyl isomerase	Vol. 1, Page 218
3.4.22.39	L3 adenain	
3.4.22.39	L3/p23 adenain	
2.4.1.206	LA2 synthase Lactosylceramide 1,3-N-acetyl-β-D-glucosaminyltransferase	

EC-No.	Synonym	Recommended Name	Vol., Page
1.10.3.2	Laccase	Laccase	
1.2.1.22	L-Lactaldehyde:NAD oxidoreductase	Lactaldehyde dehydrogenase	
1.1.1.77	L-Lactaldehyde:propanediol oxidoreductase	Lactaldehyde reductase	
1.1.1.78	D-Lactaldehyde dehydrogenase	D-Lactaldehyde dehydrogenase	
1.2.1.22	Lactaldehyde dehydrogenase	Lactaldehyde dehydrogenase	
1.1.1.77	Lactaldehyde reductase	Lactaldehyde reductase	
1.1.1.55	Lactaldehyde reductase (NADPH)	Lactaldehyde reductase (NADPH)	
3.5.2.6	β-lactamase	β-lactamase	
3.5.2.6	β-lactamase A, B, C	β-lactamase	
3.5.2.6	β-lactamase AME I	β-lactamase	
3.5.2.6	β-lactamase I¯III	β-lactamase	
3.5.2.6	β-lactamhydrolase	β-lactamase	
3.2.1.23	Lactase	β-galactosidase	
3.2.1.108	Lactase	Lactase	
3.2.1.62	Lactase-phlorizin hydrolase	Glycosylceramidase	
3.2.1.108	Lactase-phlorizin hydrolase	Lactase	
3.7.1.4	Lactase-phlorizin hydrolase	Phloretin hydrolase	
3.2.1.62	Lactase/phlorizin hydrolase	Glycosylceramidase	

EC-No.	Synonym Recommended Name	Vol., Page
3.2.1.108	Lactase/phlorizin hydrolase Lactase	
1.13.12.4	L-Lactate-2-monooxygenase Lactate 2-monooxygenase	
3.1.6.17	D-Lactate-2-sulfatase D-Lactate-2-sulfatase	
1.1.2.4	D-Lactate-cytochrome c reductase D-Lactate dehydrogenase (cytochrome)	
1.1.99.7	Lactate-malate transhydrogenase Lactate-malate transhydrogenase	
1.13.12.4	Lactate 2-monooxygenase Lactate 2-monooxygenase	
4.1.2.36	Lactate aldolase Lactate aldolase	
1.1.2.3	L-Lactate cytochrome c oxidoreductase L-Lactate dehydrogenase (cytochrome)	
1.1.2.3	L-Lactate cytochrome c reductase L-Lactate dehydrogenase (cytochrome)	
1.1.1.28	D-Lactate dehydrogenase D-Lactate dehydrogenase	
1.1.1.27	L-(+)-Lactate dehydrogenase L-Lactate dehydrogenase	
1.1.1.27	L-Lactate dehydrogenase L-Lactate dehydrogenase	
1.1.1.27	Lactate dehydrogenase L-Lactate dehydrogenase	
1.1.2.4	D-Lactate dehydrogenase (cytochrome) D-Lactate dehydrogenase (cytochrome)	
1.1.2.3	L-Lactate dehydrogenase (cytochrome) L-Lactate dehydrogenase (cytochrome)	
1.1.2.3	Lactate dehydrogenase (cytochrome) L-Lactate dehydrogenase (cytochrome)	
1.1.2.5	D-Lactate dehydrogenase (cytochrome c-553) D-Lactate dehydrogenase (cytochrome c-553)	
1.1.1.28	D-(-)-Lactate dehydrogenase (NAD) D-Lactate dehydrogenase	

EC-No.	Synonym Recommended Name	Vol., Page
1.1.1.27	Lactate dehydrogenase NAD-dependent L-Lactate dehydrogenase	
3.4.23.1	lactated pepsin pepsin A	
3.4.23.1	lactated pepsin elixir pepsin A	
1.13.12.4	L-Lactate monooxygenase Lactate 2-monooxygenase	
1.13.12.4	Lactate monooxygenase Lactate 2-monooxygenase	
1.13.12.4	Lactate oxidase Lactate 2-monooxygenase	
1.1.3.2	lactate oxidase lactate oxidase	
1.13.12.4	Lactate oxidative decarboxylase Lactate 2-monooxygenase	
1.13.12.4	Lactate oxygenase Lactate 2-monooxygenase	
5.1.2.1	Lactate racemase Lactate racemase	Vol. 1, Page 68
5.1.2.1	Lactate racemase Lactate racemase	Vol. 1, Page 68
1.1.1.28	D-Lactic acid dehydrogenase D-Lactate dehydrogenase	
1.1.1.27	L-Lactic acid dehydrogenase L-Lactate dehydrogenase	
1.1.1.28	Lactic acid dehydrogenase D-Lactate dehydrogenase	
1.1.2.4	Lactic acid dehydrogenase D-Lactate dehydrogenase (cytochrome)	
1.1.1.27	Lactic acid dehydrogenase L-Lactate dehydrogenase	
1.1.2.3	Lactic acid dehydrogenase L-Lactate dehydrogenase (cytochrome)	
5.1.2.1	Lactic acid racemase Lactate racemase	Vol. 1, Page 68

EC-No.	Synonym Recommended Name	Vol., Page
1.1.2.3	Lactic cytochrome c reductase L-Lactate dehydrogenase (cytochrome)	
1.1.1.28	D-Lactic dehydrogenase D-Lactate dehydrogenase	
1.1.1.27	L-Lactic dehydrogenase L-Lactate dehydrogenase	
1.1.1.27	Lactic dehydrogenase L-Lactate dehydrogenase	
5.1.2.1	Lacticoracemase Lactate racemase	Vol. 1, Page 68
1.13.12.4	Lactic oxidase Lactate 2-monooxygenase	
1.13.12.4	Lactic oxygenase Lactate 2-monooxygenase	
3.2.1.140	Lacto-N-biosidase Lacto-N-biosidase	
3.2.1.140	Lacto-N-biosidase Lacto-N-biosidase	
3.2.1.140	Lacto-N-biosidase (Streptomyces strain 142) Lacto-N-biosidase	
3.4.21.96	Lactocepin Lactocepin	
3.4.21.96	Lactocepin I Lactocepin	
3.4.21.96	Lactococcal cell envelope-associated proteinase Lactocepin	
3.4.21.96	Lactococcal cell wall-associated proteinase Lactocepin	
3.4.21.96	Lactococcal PI-type proteinase Lactocepin	
3.4.21.96	Lactococcal PIII-type poteinase Lactocepin	
3.4.21.96	Lactococcal proteinase Lactocepin	
3.4.21.96	Lactococcal proteinase PI Lactocepin	

EC-No.	Synonym Recommended Name	Vol., Page
3.1.1.25	1,4-lactonase 1,4-lactonase	
3.1.1.25	γ-lactonase 1,4-lactonase	
3.1.1.17	lactonase gluconolactonase	
3.1.1.25	lactonase, γ 1,4-lactonase	
3.1.1.57	Lactonase, 2-pyrone-4,6-dicarboxylate 2-Pyrone-4,6-dicarboxylate lactonase	
3.1.1.27	lactonase, 4-pyridoxo 4-pyridoxolactonase	
3.1.1.39	lactonase, actinomycin actinomycin lactonase	
3.1.1.15	lactonase, arabinono L-arabinonolactonase	
3.1.1.30	lactonase, D-arabinono- D-arabinonolactonase	
3.1.1.65	Lactonase, L-rhamno-γ- L-Rhamnono-1,4-lactonase	
3.1.1.36	lactonase, limonin-D-ring limonin-D-ring-lactonase	
3.1.1.31	lactonase, phosphoglucono- 6-phosphogluconolactonase	
3.1.1.37	lactonase, steroid steroid-lactonase	
3.1.1.38	lactonase, triacetate triacetate-lactonase	
3.1.1.19	lactonase, urono uronolactonase	
3.1.1.68	Lactonase, xylono-γ Xylono-1,4-lactonase	
3.1.1.25	1,4-lactone hydroxyacylhydrolase 1,4-lactonase	
1.11.1.7	Lactoperoxidase Peroxidase	

EC-No.	Synonym Recommended Name	Vol., Page
2.4.1.90	Lactosamine synthase N-Acetyllactosamine synthase	
2.4.1.90	Lactosamine synthetase N-Acetyllactosamine synthase	
3.2.1.108	Lactose galactohydrolase Lactase	
2.4.1.22	Lactose synthase Lactose synthase	
2.4.1.22	Lactose synthetase Lactose synthase	
2.4.1.90	Lactose synthetase A protein N-Acetyllactosamine synthase	
3.2.1.23	.β.-D-Lactosidase β-galactosidase	
3.2.1.23	β-Lactosidase β-galactosidase	
3.2.1.46	Lactosylceramidase Galactoceramidase	
3.2.1.46	Lactosylceramidase I Galactoceramidase	
3.2.1.23	Lactosylceramidase II β-galactosidase	
2.4.1.179	Lactosylceramide β-1,3-galactosyltransferase Lactosylceramide β-1,3-galactosyltransferase	
2.4.99.9	Lactosylceramide α-2,3-sialyltransferase Lactosylceramide α-2,3-sialyltransferase	
2.4.99.11	Lactosylceramide α-2,6-N-sialyltransferase Lactosylceramide α-2,6-N-sialyltransferase	
2.4.1.206	Lactosylceramide β-acetylglucosaminyltransferase Lactosylceramide 1,3-N-acetyl-β-D-glucosaminyltransferase	
2.4.1.206	Lactosylceramide 1,3-N-acetyl-β-D-glucosaminyltransferase Lactosylceramide 1,3-N-acetyl-β-D-glucosaminyltransferase	
1.1.1.220	6-Lactoyl-5,6,7,8-tetrahydropterin:NADP$^+$ 2'-oxidoreductase 6-Pyruvoyltetrahydropterin 2'-reductase	
4.2.1.54	lactoyl-CoA dehydratase lactoyl-CoA dehydratase	

EC-No.	Synonym Recommended Name	Vol., Page
4.2.1.54	lactoyl-CoA hydro-lyase lactoyl-CoA dehydratase	
4.2.1.54	lactoyl coenzyme A dehydratase lactoyl-CoA dehydratase	
4.4.1.5	lactoylglutathione lyase lactoylglutathione lyase	
3.2.1.23	Lactozym β-galactosidase	
4.2.1.54	lactyl-CoA dehydrase lactoyl-CoA dehydratase	
4.2.1.54	lactyl-CoA dehydratase lactoyl-CoA dehydratase	
4.2.1.54	lactyl-coenzyme A dehydrase lactoyl-CoA dehydratase	
4.2.1.54	lactyl CoA dehydratase lactoyl-CoA dehydratase	
3.1.1.26	LAH galactolipase	
5.4.3.7	LAM Leucine 2,3-aminomutase	Vol. 1, Page 571
3.1.11.3	Lambda exonuclease Exodeoxyribonuclease (Lambda induced)	
3.2.1.6	Laminaranase Endo-1,3(4)-β-glucanase	
3.2.1.39	Laminaranase Glucan endo-1,3-β-glucosidase	
2.4.1.31	Laminaribiose phosphorylase Laminaribiose phosphorylase	
3.2.1.6	Laminarinase Endo-1,3(4)-β-glucanase	
3.2.1.39	Laminarinase Glucan endo-1,3-β-glucosidase	
2.4.1.97	Laminarin phosphorylase 1,3-β-D-Glucan phosphorylase	
2.4.1.97	Laminarin phosphoryltransferase 1,3-β-D-Glucan phosphorylase	

EC-No.	Synonym Recommended Name	Vol., Page
5.4.99.7	Lanosterol 2,3-oxidosqualene cyclase Lanosterol synthase	Vol. 1, Page 624
5.4.99.7	Lanosterol synthase Lanosterol synthase	Vol. 1, Page 624
3.4.25.1	large multicatalytic protease proteasome endopeptidase complex	
2.3.1.43	LAT Phosphatidylcholine-sterol O-acyltransferase	
1.3.3.2	Lathosterol oxidase Lathosterol oxidase	
1.3.3.2	Lathosterol oxidase Lathosterol oxidase	
1.14.99.21	Latia-luciferin,hydrogen-donor:oxygen oxidoreductase (demethylating) Latia-luciferin monooxygenase (demethylating)	
1.14.99.21	Latia-luciferin monooxygenase (demethylating) Latia-luciferin monooxygenase (demethylating)	
1.14.99.21	Latia luciferin monooxygenase (demethylating) Latia-luciferin monooxygenase (demethylating)	
3.1.2.21	Lauroyl-acyl carrier protein thioesterase Dodecanoyl-[acyl-carrier protein] hydrolase	
3.1.2.21	Lauryl-acyl-carrier-protein hydrolase Dodecanoyl-[acyl-carrier protein] hydrolase	
3.1.3.48	LCA protein-tyrosine-phosphatase	
2.3.1.43	LCAT Phosphatidylcholine-sterol O-acyltransferase	
3.4.24.12	LCE envelysin	
6.2.1.3	LCFA synthetase Long-chain-fatty-acid-CoA ligase	Vol. 2, Page 206
1.1.1.211	LCHAD Long-chain-3-hydroxyacyl-CoA dehydrogenase	
2.7.1.112	Lck Tyrosine kinase Protein-tyrosine kinase	
3.4.17.13	LD-Carboxypeptidase Muramoyltetrapeptide carboxypeptidase	

EC-No.	Synonym Recommended Name	Vol., Page
4.1.1.18	LDC Lysine decarboxylase	
3.1.1.47	LDL-associated phospholipase A_2 1-alkyl-2-acetylglycerophosphocholine esterase	
3.1.1.47	LDL-PLA(2) 1-alkyl-2-acetylglycerophosphocholine esterase	
2.7.1.131	LDL receptor kinase Low-density-lipoprotein kinase	
1.8.1.4	LDP-Glc Dihydrolipoamide dehydrogenase	
1.8.1.4	LDP-Val Dihydrolipoamide dehydrogenase	
3.1.1.36	LDRLase limonin-D-ring-lactonase	
2.4.1.191	LDT Luteolin-7-O-diglucuronide 4'-O-glucuronosyltransferase	
2.4.1.65	(Lea)-dependent α-3/4-fucosyltransferase Galactoside 3(4)-L-fucosyltransferase	
3.4.21.89	Leader peptidase Signal peptidase I	
3.4.99.36	leader peptidase leader peptidase	
3.4.21.89	Leader peptidase I Signal peptidase I	
3.4.23.36	Leader peptidase II Signal peptidase II	
3.4.21.89	Leader peptide hydrolase Signal peptidase I	
3.4.21.89	Leader proteinase Signal peptidase I	
3.1.1.40	lecanorate hydrolase orsellinate-depside hydrolase	
2.3.1.43	Lecithin-cholesterol acyltransferase Phosphatidylcholine-sterol O-acyltransferase	
2.3.1.135	Lecithin-retinol acyltransferase Phosphatidylcholine-retinol O-acyltransferase	

EC-No.	Synonym Recommended Name	Vol., Page
2.3.1.43	Lecithin:cholesterol acyltransferase Phosphatidylcholine-sterol O-acyltransferase	
2.3.1.135	Lecithin:retinol acyltransferase Phosphatidylcholine-retinol O-acyltransferase	
3.1.1.4	Lecithinase A Phospholipase A_2	
3.1.1.5	lecithinase B lysophospholipase	
3.1.4.3	Lecithinase C Phospholipase C	
3.1.4.4	Lecithinase D Phospholipase D	
2.3.1.135	Lecithin retinol acyl transferase Phosphatidylcholine-retinol O-acyltransferase	
3.1.1.5	lecitholipase lysophospholipase	
1.6.2.6	Leghemoglobin reductase Leghemoglobin reductase	
3.4.22.34	Legumain Legumain	
3.4.24.36	Leishmania metalloproteinase Leishmanolysin	
3.4.24.36	Leishmanolysin Leishmanolysin	
3.4.24.5	lens neutral proteinase lens neutral proteinase	
3.4.21.57	Leu-proteinase Leucyl endopeptidase	
3.4.11.22	Leu.AP Aminopeptidase I	
3.4.11.1	leucinamide aminopeptidase leucyl aminopeptidase	
3.4.11.1	leucinaminopeptidase leucyl aminopeptidase	
3.4.11.22	Leucin aminopeptidase V Aminopeptidase I	

EC-No.	Synonym Recommended Name	Vol., Page
2.6.1.6	Leucine-α-ketoglutarate transaminase Leucine transaminase	
2.6.1.12	Leucine-alanine transaminase Alanine-oxo-acid transaminase	
3.4.21.57	Leucine-specific serine proteinase Leucyl endopeptidase	
6.1.1.4	Leucine-tRNA ligase Leucine-tRNA ligase	Vol. 2, Page 23
5.4.3.7	Leucine 2,3-aminomutase Leucine 2,3-aminomutase	Vol. 1, Page 571
2.6.1.6	Leucine 2-oxoglutarate transaminase Leucine transaminase	
2.6.1.6	L-Leucine:2-oxoglutarate aminotransferase Leucine transaminase	
1.4.1.9	L-Leucine:NAD^+ oxidoreductase (deaminating) Leucine dehydrogenase	
1.4.1.9	L-Leucine:NAD^+ oxidoreductase, deaminating Leucine dehydrogenase	
6.1.1.4	L-Leucine:$tRNA^{Leu}$ ligase (AMP-forming) Leucine-tRNA ligase	Vol. 2, Page 23
2.3.1.66	Leucine acyltransferase Leucine N-acetyltransferase	
3.4.11.1	L-leucine aminopeptidase leucyl aminopeptidase	
3.4.1.1	leucine aminopeptidase leucine aminopeptidase	
3.4.11.1	leucine aminopeptidase leucyl aminopeptidase	
3.4.11.22	Leucineaminopeptidase I Aminopeptidase I	
3.4.11.22	Leucine aminopeptidase IV Aminopeptidase I	
2.6.1.6	L-Leucine aminotransferase Leucine transaminase	
2.6.1.6	Leucine aminotransferase Leucine transaminase	

EC-No.	Synonym Recommended Name	Vol., Page
4.1.1.14	Leucine decarboxylase Valine decarboxylase	
1.4.1.9	L-Leucine dehydrogenase Leucine dehydrogenase	
1.4.1.9	Leucine dehydrogenase Leucine dehydrogenase	
3.4.21.57	Leucine endopeptidase Leucyl endopeptidase	
2.3.1.66	Leucine N-acetyltransferase Leucine N-acetyltransferase	
2.6.1.6	Leucine transaminase Leucine transaminase	
6.1.1.4	Leucine translase Leucine-tRNA ligase	Vol. 2, Page 23
3.4.21.37	Leucocyte elastase Leucocyte elastase	
3.4.24.6	leucolysin leucolysin	
3.4.24.6	Leucostoma neutral proteinase leucolysin	
3.4.24.6	Leucostoma peptidase A leucolysin	
2.3.2.6	Leucyl, phenylalanine-tRNA-protein transferase Leucyltransferase	
2.3.2.6	Leucyl-phenylalanine-transfer ribonucleate-protein aminoacyltransferase Leucyltransferase	
2.3.2.6	Leucyl-phenylalanine-transfer ribonucleate-protein transferase Leucyltransferase	
6.1.1.4	Leucyl-transfer ribonucleate synthetase Leucine-tRNA ligase	Vol. 2, Page 23
6.1.1.4	Leucyl-transfer ribonucleic acid synthetase Leucine-tRNA ligase	Vol. 2, Page 23
6.1.1.4	Leucyl-transfer RNA synthetase Leucine-tRNA ligase	Vol. 2, Page 23
2.3.2.6	L-Leucyl-tRNA:protein leucyltransferase Leucyltransferase	

EC-No.	Synonym Recommended Name	Vol., Page
6.1.1.4	Leucyl-tRNA synthetase Leucine-tRNA ligase	Vol. 2, Page 23
3.4.11.1	leucyl aminopeptidase leucyl aminopeptidase	
3.4.21.57	Leucyl endopeptidase Leucyl endopeptidase	
3.4.99.42	leucyllysine endopeptidase leucyllysine endopeptidase	
3.4.11.1	leucyl peptidase leucyl aminopeptidase	
2.3.2.6	Leucyltransferase Leucyltransferase	
1.4.1.9	LeuDH Leucine dehydrogenase	
3.4.21.51	leukocyte-membrane neutral endopeptidase leukocyte-membrane neutral endopeptidase	
3.4.14.5	leukocyte antigen CD26 dipeptidyl-peptidase IV	
3.4.21.76	Leukocyte proteinase 3 Myeloblastin	
3.4.21.76	Leukocyte proteinase 4 Myeloblastin	
3.3.2.6	Leukotriene-A_4 hydrolase Leukotriene-A_4 hydrolase	
1.13.11.34	Leukotriene-A_4 synthase Arachidonate 5-lipoxygenase	
1.14.13.30	Leukotriene-B_4 ω-hydroxylase Leukotriene-B_4 20-monooxygenase	
1.14.13.30	Leukotriene-B_4 20-monooxygenase Leukotriene-B_4 20-monooxygenase	
2.5.1.37	Leukotriene-C_4 synthase Leukotriene-C_4 synthase	
1.14.13.34	Leukotriene-E_4 ω-hydroxylase Leukotriene-E_4 20-monooxygenase	
1.14.13.34	Leukotriene-E_4 20-monooxygenase Leukotriene-E_4 20-monooxygenase	

EC-No.	Synonym Recommended Name	Vol., Page
2.5.1.37	Leukotriene A_4 :glutathione S-leukotrienyltransferase Leukotriene-C_4 synthase	
3.3.2.6	Leukotriene A_4 hydrolase Leukotriene-A_4 hydrolase	
1.13.11.34	Leukotriene A_4 synthase Arachidonate 5-lipoxygenase	
1.13.11.31	Leukotriene A_4 synthase M Arachidonate 12-lipoxygenase	
1.14.13.30	Leukotriene B_4 20-hydroxylase Leukotriene-B_4 20-monooxygenase	
2.5.1.37	Leukotriene C_4 synthase Leukotriene-C_4 synthase	
2.5.1.37	Leukotriene C_4 synthetase Leukotriene-C_4 synthase	
3.5.1.1	Leunase Asparaginase	
3.2.1.65	Levanase Levanase	
3.2.1.64	Levanbiohydrolase, 2,6-.β.-fructan 6- 2,6-β-fructan 6-levanbiohydrolase	
3.2.1.65	Levan hydrolase Levanase	
2.4.1.10	Levansucrase Levansucrase	
4.2.1.24	5-levulinic acid dehydratase porphobilinogen synthase	
2.4.1.65	Lewis(Le) blood group gene-dependent α-3/4-L-fucosyltransferase Galactoside 3(4)-L-fucosyltransferase	
2.4.1.65	Lewis α-3/4-fucosyltransferase Galactoside 3(4)-L-fucosyltransferase	
2.4.1.152	Lewis-negative α-3-fucosyltransferase Galactoside 3-fucosyltransferase	
2.4.1.65	Lewis $\alpha 1 \rightarrow 3/4$ fucosyltransferase Galactoside 3(4)-L-fucosyltransferase	
2.4.1.65	Lewis blood group $\alpha_1 1 \rightarrow 3/4$ fucosyltransferase Galactoside 3(4)-L-fucosyltransferase	

EC-No.	Synonym Recommended Name	Vol., Page
3.4.21.88	LexA repressor Repressor LexA	
2.4.1.189	LGT Luteolin 7-O-glucuronosyltransferase	
3.2.1.73	Lichenase Lichenase	
3.2.1.5	licheninase licheninase	
2.1.1.65	Licodione 2'-O-methyltransferase Licodione 2'-O-methyltransferase	
6.5.1.2	Ligase, polynucleotide (nicotinamide adenine dinucleotide) DNA ligase (NAD^+)	Vol. 2, Page 773
2.5.1.9	Light riboflavin synthase Riboflavin synthase	
1.11.1.14	Ligninase I Diarylpropane peroxidase	
1.13.11.43	Lignostilbene-α,β-dioxygenase Lignostilbene $\alpha\beta$-dioxygenase	
1.13.11.43	Lignostilbene $\alpha\beta$-dioxygenase Lignostilbene $\alpha\beta$-dioxygenase	
3.2.1.21	Limarase β-glucosidase	
3.2.1.10	Limit-dextrinase Oligo-1,6-glucosidase	
3.2.1.10	α-limit dextrinase Oligo-1,6-glucosidase	
3.2.1.10	Limit dextrinase Oligo-1,6-glucosidase	
3.2.1.41	Limit dextrinase α-dextrin endo-1,6-α-glucosidase	
3.2.1.142	limit dextrinase limit dextrinase	
1.14.13.47	(-)-Limonene,NADPH:oxygen oxidoreductase (3-hydroxylating) (-)-Limonene 3-monooxygenase	
1.14.13.48	(-)-Limonene,NADPH:oxygen oxidoreductase (6-hydroxylating) (-)-Limonene 6-monooxygenase	

EC-No.	Synonym Recommended Name	Vol., Page
1.14.13.49	(-)-Limonene,NADPH:oxygen oxidoreductase (7-hydroxylating) (-)-Limonene 7-monooxygenase	
1.14.13.47	(-)-Limonene 3-hydroxylase (-)-Limonene 3-monooxygenase	
1.14.13.47	(-)-Limonene 3-monooxygenase (-)-Limonene 3-monooxygenase	
1.14.13.48	(-)-Limonene 6-hydroxylase (-)-Limonene 6-monooxygenase	
1.14.13.48	(-)-Limonene 6-monooxygenase (-)-Limonene 6-monooxygenase	
1.14.13.49	(-)-Limonene 7-monooxygenase (-)-Limonene 7-monooxygenase	
1.14.13.49	(-)-Limonene hydroxylase (-)-Limonene 7-monooxygenase	
3.1.1.36	limonin-D-ring-lactonase limonin-D-ring-lactonase	
3.1.1.36	limonin-D-ring-lactone hydrolase limonin-D-ring-lactonase	
3.1.1.36	limonin lactone hydrolase limonin-D-ring-lactonase	
3.1.1.36	limonoate-D-ring-lactone lactonohydrolase limonin-D-ring-lactonase	
2.4.1.210	limonoid glucosyltransferase limonoid glucosyltransferase	
3.4.21.86	Limulus clotting enzyme Limulus clotting enzyme	
3.4.21.85	Limulus clotting factor B Limulus clotting factor B	
3.4.21.84	Limulus clotting factor C Limulus clotting factor C	
3.4.21.84	Limulus factor C Limulus clotting factor C	
1.14.99.28	Linalool-8-monooxygenase Linalool 8-monooxygenase	
1.14.99.28	Linalool 8-monooxygenase Linalool 8-monooxygenase	

EC-No.	Synonym Recommended Name	Vol., Page
2.4.1.63	Linamarin synthase Linamarin synthase	
3.5.1.52	N-Linked-glycopeptide-N^4-(N-acetyl-β-D-glucosaminyl)-L-asparagine amidohydrolase Peptide-N^4-(N-acetyl-β-glucosaminyl) asparagine amidase	
1.13.11.44	linoleate (8R)-dioxygenase linoleate diol synthase	
1.14.99.33	linoleate, hydrogen-donor:oxygen oxidoreductase (Δ^{12}-unsaturating) Δ^{12}-fatty acid dehydrogenase	
1.13.11.45	linoleate 11-lipoxygenase linoleate 11-lipoxygenase	
5.2.1.5	Linoleate Δ^{12}-cis,Δ^{11}-trans isomerase Linoleate isomerase	Vol. 1, Page 210
5.2.1.5	Linoleate Δ^{12}-cis-Δ^{11}-trans-isomerase Linoleate isomerase	Vol. 1, Page 210
1.13.11.45	linoleate:oxygen 11S-oxidoreductase linoleate 11-lipoxygenase	
1.13.11.12	Linoleate:oxygen 13-oxidoreductase Lipoxygenase	
1.13.11.44	linoleate:oxygen 7S,8S-oxidoreductase linoleate diol synthase	
1.14.99.33	linoleate D12-fatty acid acetylenase Δ^{12}-fatty acid dehydrogenase	
1.14.99.25	Linoleate desaturase Linoleoyl-CoA desaturase	
1.14.19.3	linoleate desaturase linoleoyl-CoA desaturase	
1.13.11.44	linoleate diol synthase linoleate diol synthase	
1.13.11.45	linoleate dioxygenase linoleate 11-lipoxygenase	
4.2.1.92	Linoleate hydroperoxide isomerase Hydroperoxide dehydratase	
5.2.1.5	Linoleate isomerase Linoleate isomerase	Vol. 1, Page 210

EC-No.	Synonym Recommended Name	Vol., Page
1.3.1.35	Linoleate synthase Phosphatidylcholine desaturase	
1.13.11.33	Linoleic acid ω-6-lipoxygenase Arachidonate 15-lipoxygenase	
1.14.99.25	Linoleic acid desaturase Linoleoyl-CoA desaturase	
1.14.19.3	linoleic acid desaturase linoleoyl-CoA desaturase	
4.2.1.92	Linoleic acid hydroperoxide isomerase Hydroperoxide dehydratase	
5.2.1.5	Linoleic acid isomerase Linoleate isomerase	Vol. 1, Page 210
1.14.99.25	Linoleic desaturase Linoleoyl-CoA desaturase	
1.14.19.3	linoleic desaturase linoleoyl-CoA desaturase	
1.14.99.25	Linoleoyl-CoA,hydrogen-donor:oxygen oxidoreductase Linoleoyl-CoA desaturase	
1.14.19.3	linoleoyl-CoA,hydrogen-donor:oxygen oxidoreductase linoleoyl-CoA desaturase	
1.14.99.25	Linoleoyl-CoA desaturase Linoleoyl-CoA desaturase	
1.14.19.3	linoleoyl-CoA desaturase linoleoyl-CoA desaturase	
1.14.99.25	Linoleoyl-coenzyme A desaturase Linoleoyl-CoA desaturase	
1.14.19.3	linoleoyl-coenzyme A desaturase linoleoyl-CoA desaturase	
1.14.99.25	Linoleoyl CoA desaturase Linoleoyl-CoA desaturase	
1.14.19.3	linoleoyl CoA desaturase linoleoyl-CoA desaturase	
1.13.11.12	Lionoleate:O_2 oxidoreductase Lipoxygenase	
3.1.1.3	lipase triacylglycerol lipase	

EC-No.	Synonym Recommended Name	Vol., Page
3.1.1.3	lipase, triacylglycerol triacylglycerol lipase	
3.1.1.3	lipazin triacylglycerol lipase	
3.1.1.34	lipemia-clearing factor lipoprotein lipase	
2.4.1.182	Lipid-A-disaccharide synthase Lipid-A-disaccharide synthase	
2.7.1.130	Lipid A 4'-kinase Tetraacyldisaccharide 4'-kinase	
3.1.1.26	lipid acyl-hydrolase galactolipase	
2.4.1.182	Lipid A disaccharide synthase Lipid-A-disaccharide synthase	
2.1.1.17	Lipid methyl transferase Phosphatidylethanolamine N-methyltransferase	
1.8.1.4	Lipoamide dehydrogenase (NADH) Dihydrolipoamide dehydrogenase	
1.8.1.4	Lipoamide oxidoreductase (NADH Dihydrolipoamide dehydrogenase	
1.8.1.4	Lipoamide reductase Dihydrolipoamide dehydrogenase	
2.3.1.12	Lipoate acetyltransferase Dihydrolipoamide S-acetyltransferase	
1.8.1.4	Lipoate dehydrogenase Dihydrolipoamide dehydrogenase	
2.3.1.61	Lipoate succinyltransferase Dihydrolipoamide S-succinyltransferase	
2.3.1.12	Lipoate transacetylase Dihydrolipoamide S-acetyltransferase	
2.3.1.12	Lipoic acetyltransferase Dihydrolipoamide S-acetyltransferase	
2.3.1.12	Lipoic acid acetyltransferase Dihydrolipoamide S-acetyltransferase	
1.8.1.4	Lipoic acid dehydrogenase Dihydrolipoamide dehydrogenase	

EC-No.	Synonym Recommended Name	Vol., Page
2.3.1.12	Lipoic transacetylase Dihydrolipoamide S-acetyltransferase	
2.3.1.61	Lipoic transsuccinylase Dihydrolipoamide S-succinyltransferase	
1.13.11.12	Lipoperoxidase Lipoxygenase	
3.1.4.3	Lipophosphodiesterase C Phospholipase C	
3.1.4.3	Lipophosphodiesterase I Phospholipase C	
3.1.4.4	Lipophosphodiesterase II Phospholipase D	
2.4.1.44	Lipopolysaccharide 3-α-galactosyltransferase Lipopolysaccharide 3-α-galactosyltransferase	
2.4.1.58	Lipopolysaccharide glucosyltransferase I Lipopolysaccharide glucosyltransferase I	
2.4.1.73	Lipopolysaccharide glucosyltransferase II Lipopolysaccharide glucosyltransferase II	
2.4.1.56	Lipopolysaccharide N-acetylglucosaminyltransferase Lipopolysaccharide N-acetylglucosaminyltransferase	
2.4.1.180	Lipopolysaccharide N-acetylmannosaminouronosyltransferase Lipopolysaccharide N-acetylmannosaminouronosyltransferase	
3.1.1.34	lipoprotein lipase lipoprotein lipase	
1.13.11.12	Lipoxidase Lipoxygenase	
1.13.11.12	Lipoxydase Lipoxygenase	
1.13.11.31	12-Lipoxygenase Arachidonate 12-lipoxygenase	
1.13.11.31	$^{12}\Delta$-Lipoxygenase Arachidonate 12-lipoxygenase	
1.13.11.31	12S-Lipoxygenase Arachidonate 12-lipoxygenase	
1.13.11.33	15-Lipoxygenase Arachidonate 15-lipoxygenase	

EC-No.	Synonym Recommended Name	Vol., Page
1.13.11.34	$^{5}\Delta$-Lipoxygenase Arachidonate 5-lipoxygenase	
1.13.11.40	8-Lipoxygenase Arachidonate 8-lipoxygenase	
1.13.11.31	C-12 lipoxygenase Arachidonate 12-lipoxygenase	
1.13.11.34	C-5-lipoxygenase Arachidonate 5-lipoxygenase	
1.13.11.31	Δ 12-lipoxygenase Arachidonate 12-lipoxygenase	
1.13.11.34	Δ^{5}-Lipoxygenase Arachidonate 5-lipoxygenase	
1.13.11.12	Lipoxygenase Lipoxygenase	
1.13.11.33	ω-6 Lipoxygenase Arachidonate 15-lipoxygenase	
2.3.1.12	Lipoylacetyltransferase Dihydrolipoamide S-acetyltransferase	
1.8.1.4	Lipoyl dehydrogenase Dihydrolipoamide dehydrogenase	
2.3.1.61	Lipoyl transsuccinylase Dihydrolipoamide S-succinyltransferase	
3.2.1.15	Liquifying polygalacturonase Polygalacturonase	
3.1.1.3	liver lipase triacylglycerol lipase	
5.1.1.7	LL-2,6-Diaminoheptanedioate 2-epimerase Diaminopimelate epimerase	Vol. 1, Page 27
5.1.1.7	LL-Diaminopimelate epimerase Diaminopimelate epimerase	Vol. 1, Page 27
2.4.1.190	LMT Luteolin-7-O-glucuronide 7-O-glucuronosyltransferase	
2.1.1.17	LMTase Phosphatidylethanolamine N-methyltransferase	
3.1.3.48	LMW-PTP protein-tyrosine-phosphatase	

EC-No.	Synonym Recommended Name	Vol., Page
2.1.1.50	Loganate O-methyltransferase Loganate O-methyltransferase	
2.7.3.5	Lombricine kinase Lombricine kinase	
4.2.1.74	long-chain-(3S)-3-hydroxyacyl-CoA hydro-lyase long-chain-enoyl-CoA hydratase	
1.1.1.211	Long-chain-(S)-3-hydroxyacyl-CoA:NAD$^+$ oxidoreductase Long-chain-3-hydroxyacyl-CoA dehydrogenase	
1.1.1.211	Long-chain-3-hydroxyacyl-CoA dehydrogenase Long-chain-3-hydroxyacyl-CoA dehydrogenase	
3.5.1.60	N-(Long-chain-acyl)ethanolamine amidohydrolase N-(Long-chain-acyl)ethanolamine deacylase	
3.5.1.60	N-(Long-chain-acyl)ethanolamine deacylase N-(Long-chain-acyl)ethanolamine deacylase	
1.3.99.13	Long-chain-acyl-CoA:(acceptor) 2,3-oxidoreductase Long-chain-acyl-CoA dehydrogenase	
1.3.99.13	Long-chain-acyl-CoA dehydrogenase Long-chain-acyl-CoA dehydrogenase	
1.1.1.192	Long-chain-alcohol:NAD$^+$ oxidoreductase Long-chain-alcohol dehydrogenase	
1.1.3.20	Long-chain-alcohol:oxygen oxidoreductase Long-chain-alcohol oxidase	
1.1.1.192	Long-chain-alcohol dehydrogenase Long-chain-alcohol dehydrogenase	
2.3.1.75	Long-chain-alcohol O-fatty-acyltransferse Long-chain-alcohol O-fatty-acyltransferse	
1.1.3.20	Long-chain-alcohol oxidase Long-chain-alcohol oxidase	
1.2.1.48	Long-chain-aldehyde:NAD$^+$ oxidoreductase Long-chain-aldehyde dehydrogenase	
1.2.1.50	Long-chain-aldehyde:NADP$^+$ oxidoreductase (acyl-CoA-forming) Long-chain-fatty-acyl-CoA reductase	
1.2.1.48	Long-chain-aldehyde dehydrogenase Long-chain-aldehyde dehydrogenase	
4.2.1.74	long-chain-enoyl-CoA hydratase long-chain-enoyl-CoA hydratase	

EC-No.	Synonym Recommended Name	Vol., Page
6.2.1.20	Long-chain-fatty-acid-[acyl-carrier-protein] ligase Long-chain-fatty-acid-[acyl-carrier-protein] ligase	Vol. 2, Page 296
6.2.1.3	Long-chain-fatty-acid-CoA ligase Long-chain-fatty-acid-CoA ligase	Vol. 2, Page 206
6.2.1.19	Long-chain-fatty-acid-luciferin-component ligase Long-chain-fatty-acid-luciferin-component ligase	Vol. 2, Page 293
6.2.1.20	Long-chain-fatty-acid:[acyl-carrier-protein] ligase (AMP-forming) Long-chain-fatty-acid-[acyl-carrier-protein] ligase	Vol. 2, Page 296
6.2.1.19	Long-chain-fatty-acid:protein ligase (AMP-forming) Long-chain-fatty-acid-luciferin-component ligase	Vol. 2, Page 293
1.2.1.50	Long-chain-fatty-acyl-CoA reductase Long-chain-fatty-acyl-CoA reductase	
3.1.1.67	Long-chain-fatty-acyl-ethyl-ester acylhydrolase Fatty-acyl-ethyl-ester synthase	
3.5.1.55	Long-chain-fatty-acyl-glutamate deacylase Long-chain-fatty-acyl-glutamate deacylase	
3.5.1.55	N-Long-chain-fatty-acyl-L-glutamate amidohydrolase Long-chain-fatty-acyl-glutamate deacylase	
5.3.3.8	Long-chain Δ^3,Δ^2-enoyl-CoA isomerase Dodecenoyl-CoA δ-isomerase	Vol. 1, Page 413
2.3.1.16	Long-chain 3-oxoacyl-CoA thiolase Acetyl-CoA C-acyltransferase	
1.3.99.3	Long-chain acyl-CoA dehydrogenase Acyl-CoA dehydrogenase	
6.2.1.3	Long-chain acyl-CoA synthetase I Long-chain-fatty-acid-CoA ligase	Vol. 2, Page 206
6.2.1.3	Long-chain acyl-CoA synthetase II Long-chain-fatty-acid-CoA ligase	Vol. 2, Page 206
1.3.99.13	Long-chain acyl-coenzyme A dehydrogenase Long-chain-acyl-CoA dehydrogenase	
6.2.1.3	Long-chain acyl-coenzyme A synthetase Long-chain-fatty-acid-CoA ligase	Vol. 2, Page 206
3.1.1.28	long-chain acyl-L-carnitine hydrolase acylcarnitine hydrolase	
6.2.1.3	Long-chain acyl CoA synthetase Long-chain-fatty-acid-CoA ligase	Vol. 2, Page 206

EC-No.	Synonym Recommended Name	Vol., Page
1.3.99.3	Long-chain acyl coenzyme A dehydrogenase Acyl-CoA dehydrogenase	
1.1.1.192	Long-chain alcohol dehydrogenase Long-chain-alcohol dehydrogenase	
1.2.1.48	Long-chain aldehyde dehydrogenase Long-chain-aldehyde dehydrogenase	
3.5.1.55	Long-chain aminoacylase Long-chain-fatty-acyl-glutamate deacylase	
1.14.99.25	Long-chain fatty acid Δ^6-desaturase Linoleoyl-CoA desaturase	
1.14.19.3	long-chain fatty acid Δ^6-desaturase linoleoyl-CoA desaturase	
1.14.99.5	Long-chain fatty acid Δ9-desaturase Stearoyl-CoA desaturase	
1.1.3.20	Long-chain fatty acid oxidase Long-chain-alcohol oxidase	
6.2.1.3	Long-chain fatty acyl coenzyme A synthetase Long-chain-fatty-acid-CoA ligase	Vol. 2, Page 206
1.2.1.48	Long-chain fatty aldehyde dehydrogenase Long-chain-aldehyde dehydrogenase	
3.5.1.14	Long acyl amidoacylase Aminoacylase	
3.1.2.2	long chain acyl-CoA hydrolase palmitoyl-CoA hydrolase	
3.1.2.2	long chain acyl-CoA thioesterase palmitoyl-CoA hydrolase	
3.5.1.82	Long chain acyl aminoacylase N-Acyl-D-glutamate deacylase	
3.1.2.2	long chain fatty-acyl-CoA hydrolase palmitoyl-CoA hydrolase	
3.1.2.2	long chain fatty-acyl-CoA thioesterase palmitoyl-CoA hydrolase	
1.11.1.3	Long chain fatty acid peroxidase Fatty-acid peroxidase	
6.2.1.3	Long chain fatty acyl-CoA synthetase Long-chain-fatty-acid-CoA ligase	Vol. 2, Page 206

EC-No.	Synonym Recommended Name	Vol., Page
6.2.1.3	Long chain fatty acyl CoA ligase Long-chain-fatty-acid-CoA ligase	Vol. 2, Page 206
3.4.21.53	Lon proteinase Endopeptidase La	
3.4.23.13	Lotus aspartic proteinase Lotus aspartic proteinase	
2.7.1.131	Low-density-lipoprotein kinase Low-density-lipoprotein kinase	
2.7.1.131	Low-density lipoprotein receptor kinase Low-density-lipoprotein kinase	
1.1.1.2	Low-K_m aldehyde reductase Alcohol dehydrogenase ($NADP^+$)	
4.1.2.5	Low-specificity L-threonine aldolase Threonine aldolase	
3.4.24.66	Low choriolytic enzyme Choriolysin L	
3.4.24.12	low choriolytic enzyme envelysin	
3.1.1.32	LPC phospholipase A_1	
3.1.1.34	LPL lipoprotein lipase	
3.1.1.5	LPL lysophospholipase	
3.1.1.5	h-LPTA lysophospholipase	
2.3.1.135	LRAT Phosphatidylcholine-retinol O-acyltransferase	
3.1.3.48	LRP protein-tyrosine-phosphatase	
3.1.1.53	LSE sialate O-acetylesterase	
3.1.1.53	Lse sialate O-acetylesterase	
3.3.2.6	LTA_4H Leukotriene-A_4 hydrolase	

EC-No.	Synonym Recommended Name	Vol., Page
3.3.2.6	LTA_4 hydrolase Leukotriene-A_4 hydrolase	
1.13.11.31	LTA_4 synthase Arachidonate 12-lipoxygenase	
1.13.11.34	LTA synthase Arachidonate 5-lipoxygenase	
1.14.13.30	LTB4 ω-hydroxylase Leukotriene-B_4 20-monooxygenase	
1.14.13.30	LTB4 20-hydroxylase Leukotriene-B_4 20-monooxygenase	
2.5.1.37	LTC_4 synthase Leukotriene-C_4 synthase	
2.5.1.37	LTC_4 synthetase Leukotriene-C_4 synthase	
1.14.14.3	Luciferase Alkanal monooxygenase (FMN-linked)	
1.13.12.6	Luciferase (Cypridina luciferin) Cypridina-luciferin 2-monooxygenase	
1.13.12.7	Luciferase (firefly luciferin) Photinus-luciferin 4-monooxygenase (ATP-hydrolysing)	
1.14.99.21	Luciferase (Latia luciferin) Latia-luciferin monooxygenase (demethylating)	
1.13.12.5	Luciferase (Renilla luciferin) Renilla-luciferin 2-monooxygenase	
2.8.2.10	Luciferin sulfokinase Renilla-luciferin sulfotransferase	
2.8.2.10	Luciferin sulfokinase Renilla-luciferin sulfotransferase	
2.8.2.10	Luciferin sulfotransferase Renilla-luciferin sulfotransferase	
4.1.2.37	LuHNL Acetone-cyanhydrin lyase	
3.4.21.59	Lung tryptase Tryptase	
4.2.99.13	lupinic acid synthase β-(9-cytokinin)-alanine synthase	

EC-No.	Synonym Recommended Name	Vol., Page
4.2.99.13	lupinic acid synthetase β-(9-cytokinin)-alanine synthase	
2.4.1.191	Luteolin-7-O-diglucuronide 4'-O-glucuronosyltransferase Luteolin-7-O-diglucuronide 4'-O-glucuronosyltransferase	
2.4.1.190	Luteolin-7-O-glucuronide 7-O-glucuronosyltransferase Luteolin-7-O-glucuronide 7-O-glucuronosyltransferase	
2.4.1.189	Luteolin 7-O-glucuronosyltransferase Luteolin 7-O-glucuronosyltransferase	
2.1.1.42	Luteolin O-methyltransferase Luteolin O-methyltransferase	
4.1.2.30	17,20 lyase 17α-Hydroxyprogesterone aldolase	
4.1.2.30	C-17,20 lyase 17α-Hydroxyprogesterone aldolase	
5.5.1.10	Lyase, α-pinene oxide α-Pinene-oxide decyclase	Vol. 1, Page 713
4.1.3.32	Lyase, 2,3-dimethylmalate 2,3-Dimethylmalate lyase	
4.1.3.26	Lyase, 3-hydroxy-3-isohexenylglutaryl coenzyme A 3-Hydroxy-3-isohexenylglutaryl-CoA lyase	
4.3.3.1	Lyase, 3-ketovalidoxylamine A C-N- 3-Ketovalidoxylamine C-N-lyase	
4.1.2.37	Lyase, acetone-cyanohydrin Acetone-cyanhydrin lyase	
4.1.2.39	Lyase, acetone-cyanohydrin Hydroxynitrilase	
4.1.3.3	Lyase, acetylneuraminate N-acetylneuraminate lyase	
4.3.2.2	lyase, adenylosuccinate adenylosuccinate lyase	
4.2.2.3	lyase, alginate poly(β-D-mannuronate) lyase	
4.99.1.2	lyase, alkylmercury alkylmercury lyase	
4.4.1.4	lyase, alliin alliin lyase	

EC-No.	Synonym Recommended Name	Vol., Page
4.3.2.1	lyase, argininosuccinate argininosuccinate lyase	
4.1.2.38	Lyase, benzaldehyde Benzoin aldolase	
4.2.99.12	lyase, carboxymethyloxysuccinate carboxymethyloxysuccinate lyase	
4.2.2.4	lyase, chondroitin ABC chondroitin ABC lyase	
4.2.2.5	lyase, chondroitin AC chondroitin AC lyase	
4.1.3.22	lyase, citramalate citramalate lyase	
4.1.3.25	lyase, citramalyl coenzyme A citramalyl-CoA lyase	
4.1.3.6	lyase, citrate citrate lyase	
4.4.1.8	lyase, cystathionine β- cystathionine β-lyase	
4.4.1.1	lyase, cystathionine γ- cystathionine γ-lyase	
4.4.1.10	lyase, cysteine (sulfite) cysteine lyase	
4.4.1.13	lyase, cysteine conjugate.β. cysteine-S-conjugate β-lyase	
4.4.1.15	lyase, D-cysteine D-cysteine desulfhydrase	
4.1.99.3	lyase, deoxyribonucleate pyrimidine dimer deoxyribodipyrimidine photolyase	
4.2.2.9	lyase, exopolygalacturonate exopolygalacturonate lyase	
4.1.2.33	Lyase, fucosterol epoxide Fucosterol-epoxide lyase	
4.2.2.1	lyase, glucuronoglycosaminoglycan hyaluronate lyase	
4.2.2.7	Lyase, heparin Heparin lyase	

EC-No.	Synonym Recommended Name	Vol., Page
4.2.2.8	lyase, heparin sulfate heparin-sulfate lyase	
4.2.2.1	lyase, hyaluronate hyaluronate lyase	
4.1.2.11	Lyase, hydroxymandelonitrile Hydroxymandelonitrile lyase	
4.1.3.1	lyase, isocitrate isocitrate lyase	
4.4.1.5	lyase, lactoylglutathione lactoylglutathione lyase	
4.1.3.24	lyase, malyl coenzyme A malyl-CoA lyase	
4.1.2.10	Lyase, mandelonitrile Mandelonitrile lyase	
4.4.1.11	lyase, methionine methionine γ-lyase	
4.1.3.30	Lyase, methylisocitrate Methylisocitrate lyase	
4.2.2.6	lyase, oligogalacturonide oligogalacturonide lyase	
4.1.3.13	Lyase, oxalomalate Oxalomalate lyase	
4.2.2.2	lyase, pectate pectate lyase	
4.2.2.10	lyase, pectin pectin lyase	
4.3.2.5	Lyase, peptidyl-α hydroxyglycine Peptidylamidoglycolate lyase	
4.2.2.11	Lyase, polyguluronate Poly(α-L-guluronate) lyase	
4.4.1.6	lyase, S-alkylcysteine S-alkylcysteine lyase	
4.4.1.16	Lyase, selenocysteine Selenocysteine lyase	
4.4.1.12	lyase, sulfoacetaldehyde sulfoacetaldehyde lyase	

EC-No.	Synonym Recommended Name	Vol., Page
4.3.2.3	lyase, ureidoglycolate ureidoglycolate lyase	
4.2.2.12	Lyase, xanthan Xanthan lyase	
3.4.14.5	lymphocyte, antigen CD26 dipeptidyl-peptidase IV	
3.4.11.4	lymphopeptidase tripeptide aminopeptidase	
2.6.1.71	Lys-AT Lysine-pyruvate 6-transaminase	
3.4.24.38	Lysin Autolysin	
3.5.2.11	L-Lysinamidase L-Lysine-lactamase	
1.14.11.4	Lysine, 2-oxoglutarate 5-dioxygenase Procollagen-lysine 5-dioxygenase	
1.14.13.59	L-Lysine, NADPH2:oxygen oxidoreductase (6-hydroxylating) L-Lysine 6-monooxygenase (NADPH2)	
2.6.1.36	L-Lysine-α-ketoglutarate 6-aminotransferase L-Lysine 6-transaminase	
2.6.1.36	L-Lysine-α-ketoglutarate aminotransferase L-Lysine 6-transaminase	
1.5.1.8	L-Lysine-α-ketoglutarate reductase Saccharopine dehydrogenase ($NADP^+$, L-lysine forming)	
3.5.2.11	L-Lysine-1,6-lactam lactamhydrolase L-Lysine-lactamase	
5.4.3.2	L-Lysine-2,3-aminomutase Lysine 2,3-aminomutase	Vol. 1, Page 553
1.13.12.2	L-Lysine-2-monooxygenase Lysine 2-monooxygenase	
1.14.11.4	Lysine-2-oxoglutarate dioxygenase Procollagen-lysine 5-dioxygenase	
1.5.1.7	Lysine-2-oxoglutarate reductase Saccharopine dehydrogenase (NAD^+, L-lysine forming)	
1.5.1.8	Lysine-2-oxoglutarate reductase Saccharopine dehydrogenase ($NADP^+$, L-lysine forming)	

EC-No.	Synonym Recommended Name	Vol., Page
2.6.1.36	Lysine ε-aminotransferase L-Lysine 6-transaminase	
1.4.1.18	L-Lysine ε-dehydrogenase Lysine 6-dehydrogenase	
1.5.1.8	Lysine-ketoglutarate reductase Saccharopine dehydrogenase ($NADP^+$, L-lysine forming)	
3.5.2.11	L-Lysine-lactamase L-Lysine-lactamase	
1.13.12.10	Lysine-N^6-hydroxylase Lysine 6-monooxygenase	
1.4.3.14	L-Lysine α-oxidase L-Lysine oxidase	
2.6.1.71	Lysine-pyruvate 6-transaminase Lysine-pyruvate 6-transaminase	
3.4.21.50	lysine-specific protease lysyl endopeptidase	
2.6.1.36	Lysine ε-transaminase L-Lysine 6-transaminase	
6.1.1.6	L-Lysine-transfer RNA ligase Lysine-tRNA ligase	Vol. 2, Page 42
6.1.1.6	Lysine-tRNA ligase Lysine-tRNA ligase	Vol. 2, Page 42
6.1.1.6	Lysine-tRNA synthetase Lysine-tRNA ligase	Vol. 2, Page 42
2.6.1.68	Lysine/ornithine:2-oxoglutarate aminotransferase Ornithine(lysine) transaminase	
5.4.3.2	L-Lysine 2,3-aminomutase Lysine 2,3-aminomutase	Vol. 1, Page 553
5.4.3.2	Lysine 2,3-aminomutase Lysine 2,3-aminomutase	Vol. 1, Page 553
1.13.12.2	Lysine 2-monooxygenase Lysine 2-monooxygenase	
5.4.3.4	D-Lysine 5,6-aminomutase D-Lysine 5,6-aminomutase	Vol. 1, Page 562
5.4.3.3	L-β-Lysine 5,6-aminomutase β-Lysine 5,6-aminomutase	Vol. 1, Page 558

EC-No.	Synonym Recommended Name	Vol., Page
5.4.3.3	β-Lysine 5,6-aminomutase β-Lysine 5,6-aminomutase	Vol. 1, Page 558
2.6.1.36	Lysine 6-aminotransferase L-Lysine 6-transaminase	
1.4.1.18	Lysine 6-dehydrogenase Lysine 6-dehydrogenase	
1.13.12.10	Lysine 6-monooxygenase Lysine 6-monooxygenase	
1.14.13.59	L-Lysine 6-monooxygenase ($NADPH_2$) L-Lysine 6-monooxygenase ($NADPH_2$)	
2.6.1.36	L-Lysine 6-transaminase L-Lysine 6-transaminase	
1.5.1.8	Lysine:α-ketoglutarate:TPNH oxidoreductase (ε-N-[gultaryl-2]-L-lysine forming) Saccharopine dehydrogenase ($NADP^+$, L-lysine forming)	
2.6.1.36	Lysine:2-ketoglutarate 6-aminotransferase L-Lysine 6-transaminase	
2.6.1.36	L-Lysine:2-oxoglutarate 6-aminotransferase L-Lysine 6-transaminase	
1.14.13.59	Lysine:N^6-hydroxylase L-Lysine 6-monooxygenase ($NADPH_2$)	
1.4.1.18	L-Lysine:NAD^+ 6-oxidoreductase (deaminating) Lysine 6-dehydrogenase	
1.4.1.15	L-Lysine:NAD^+ oxidoreductase (deaminating, cyclizing) Lysine dehydrogenase	
1.13.12.2	L-Lysine:oxygen 2-oxidoreductase (decarboxylating) Lysine 2-monooxygenase	
1.13.12.10	L-Lysine:oxygen monooxygenase (6-hydroxylating) Lysine 6-monooxygenase	
1.4.3.14	L-Lysine:oxygen oxidoreductase (deaminating) L-Lysine oxidase	
2.6.1.71	L-Lysine:pyruvate aminotransferase Lysine-pyruvate 6-transaminase	
6.1.1.6	L-Lysine:tRNA^{Lys} ligase (AMP-forming) Lysine-tRNA ligase	Vol. 2, Page 42

EC-No.	Synonym Recommended Name	Vol., Page
2.3.1.32	Lysine acetyltransferase Lysine N-acetyltransferase	
3.5.1.17	ε-lysine acylase Acyl-lysine deacylase	
2.1.3.8	Lysine carbamoyltransferase Lysine carbamoyltransferase	
4.1.1.18	L-Lysine carboxy-lyase Lysine decarboxylase	
3.4.17.3	lysine carboxypeptidase lysine carboxypeptidase	
4.1.1.18	Lysine decarboxylase Lysine decarboxylase	
1.4.1.15	Lysine dehydrogenase Lysine dehydrogenase	
1.14.11.4	Lysine hydroxylase Procollagen-lysine 5-dioxygenase	
1.13.12.2	Lysine monooxygenase Lysine 2-monooxygenase	
5.4.3.4	D-α-Lysine mutase D-Lysine 5,6-aminomutase	Vol. 1, Page 562
5.4.3.3	L-β-Lysine mutase β-Lysine 5,6-aminomutase	Vol. 1, Page 558
5.4.3.4	α-Lysine mutase D-Lysine 5,6-aminomutase	Vol. 1, Page 562
5.4.3.3	β-Lysine mutase β-Lysine 5,6-aminomutase	Vol. 1, Page 558
2.3.1.32	Lysine N-acetyltransferase Lysine N-acetyltransferase	
1.14.13.59	Lysine N^6-hydroxylase L-Lysine 6-monooxygenase (NADPH2)	
1.13.12.10	Lysine N^6-hydroxylase Lysine 6-monooxygenase	
1.4.3.14	L-Lysine oxidase L-Lysine oxidase	
1.13.12.2	Lysine oxygenase Lysine 2-monooxygenase	

EC-No.	Synonym Recommended Name	Vol., Page
5.1.1.5	Lysine racemase Lysine racemase	Vol. 1, Page 23
5.1.1.5	Lysine racemase Lysine racemase	Vol. 1, Page 23
3.4.21.50	lysine specific proteinase lysyl endopeptidase	
2.1.3.8	Lysine transcarbamoylase Lysine carbamoyltransferase	
6.1.1.6	Lysine translase Lysine-tRNA ligase	Vol. 2, Page 42
2.3.1.67	Lyso-GPC:acetyl CoA acetyltransferase 1-Alkylglycerophosphocholine O-acetyltransferase	
2.3.1.67	Lyso-platelet-activating factor:acetyl-CoA acetyltransferase 1-Alkylglycerophosphocholine O-acetyltransferase	
2.3.1.67	Lyso-platelet activating factor:acetyl-CoA acetyltransferase 1-Alkylglycerophosphocholine O-acetyltransferase	
5.4.1.1	Lysolecithin 2,3-acylmutase Lysolecithin acylmutase	Vol. 1, Page 488
5.4.1.1	Lysolecithin acylmutase Lysolecithin acylmutase	Vol. 1, Page 488
2.3.1.23	Lysolecithin acyltransferase 1-Acylglycerophosphocholine O-acyltransferase	
2.3.1.43	Lysolecithin acyltransferase Phosphatidylcholine-sterol O-acyltransferase	
3.1.1.5	lysolecithinase lysophospholipase	
5.4.1.1	Lysolecithin migratase Lysolecithin acylmutase	Vol. 1, Page 488
2.3.1.67	LysoPAF:acetyl CoA acetyltransferase 1-Alkylglycerophosphocholine O-acetyltransferase	
3.1.1.5	lysophopholipase L_2 lysophospholipase	
3.1.1.5	lysophosphatidase lysophospholipase	
2.3.1.23	Lysophosphatide acyltransferase 1-Acylglycerophosphocholine O-acyltransferase	

EC-No.	Synonym Recommended Name	Vol., Page
2.3.1.51	Lysophosphatidic acid-acyltransferase 1-Acylglycerol-3-phosphate O-acyltransferase	
3.1.1.5	2-lysophosphatidylcholine acylhydrolase lysophospholipase	
2.3.1.23	Lysophosphatidylcholine acyltransferase 1-Acylglycerophosphocholine O-acyltransferase	
3.1.1.5	lysophosphatidylcholine hydrolase lysophospholipase	
3.1.1.5	lysophospholipase lysophospholipase	
3.1.1.5	lysophospholipase-transacylase lysophospholipase	
3.1.1.5	lysophospholipase-transacylase lysophospholipase	
3.1.1.5	lysophospholipase A_1 lysophospholipase	
3.1.4.39	Lysophospholipase D Alkylglycerophosphoethanolamine phosphodiesterase	
3.1.1.5	lysophospholipase L_2 lysophospholipase	
1.5.1.16	D-Lysopine dehydrogenase D-Lysopine dehydrogenase	
1.5.1.16	Lysopine dehydrogenase D-Lysopine dehydrogenase	
1.5.1.16	D-Lysopine synthase D-Lysopine dehydrogenase	
3.3.2.2	Lysoplasmalogenase Alkenylglycerophosphocholine hydrolase	
2.3.1.25	Lysoplasmenylcholine acyltransferase Plasmalogen synthase	
3.2.1.3	Lysosomal α-glucosidase Glucan 1,4-α-glucosidase	
3.4.19.9	Lysosomal γ-glutamyl carboxypeptidase γ-Glu-X carboxypeptidase	
3.1.4.45	Lysosomal α-N-acetyl-glucosaminidase N-Acetylglucosamine-1-phosphodiester N-acetylglucosaminidase	

EC-No.	Synonym Recommended Name	Vol., Page
3.4.16.2	lysosomal carboxypeptidase lysosomal Pro-X carboxypeptidase	
3.4.16.5	Lysosomal carboxypeptidase A Carboxypeptidase C	
3.4.18.1	lysosomal carboxypeptidase B cysteine-type carboxypeptidase	
3.4.16.2	lysosomal carboxypeptidase C lysosomal Pro-X carboxypeptidase	
3.4.21.37	Lysosomal elastase Leucocyte elastase	
2.7.8.17	Lysosomal enzyme precursor acetylglucosamine-1-phosphotransferase UDP-N-acetylglucosamine-lysosomal-enzyme N-acetylglucosaminephos-photransferase	
3.4.16.2	lysosomal Pro-X carboxypeptidase lysosomal Pro-X carboxypeptidase	
3.4.16.5	Lysosomal protective protein Carboxypeptidase C	
3.1.1.53	lysosomal sialic acid 9-O-acetylesterase sialate O-acetylesterase	
3.1.1.53	lysosomal sialic acid O-acetylesterase sialate O-acetylesterase	
3.4.24.75	Lysostaphin Lysostaphin	
3.4.24.75	Lysostaphin endopeptidase Lysostaphin	
3.2.1.17	Lysozyme Lysozyme	
3.2.1.17	Lysozyme g Lysozyme	
1.4.3.14	L-Lysyl-α-oxidase L-Lysine oxidase	
3.4.17.13	L-Lysyl-D-alanine carboxypeptidase Muramoyltetrapeptide carboxypeptidase	
6.1.1.6	Lysyl-transfer ribonucleate synthetase Lysine-tRNA ligase	Vol. 2, Page 42

EC-No.	Synonym Recommended Name	Vol., Page
6.1.1.6	Lysyl-transfer RNA synthetase Lysine-tRNA ligase	Vol. 2, Page 42
2.3.2.3	L-Lysyl-tRNA:phosphatidylglycerol 3-O-lysyltransferase Lysyltransferase	
6.1.1.6	Lysyl-tRNA synthetase Lysine-tRNA ligase	Vol. 2, Page 42
3.4.21.50	lysyl bond specific proteinase lysyl endopeptidase	
3.4.21.50	lysyl endopeptidase lysyl endopeptidase	
1.14.11.4	Lysyl hydroxylase Procollagen-lysine 5-dioxygenase	
1.14.11.4	Lysylprotocollagen dioxygenase Procollagen-lysine 5-dioxygenase	
2.3.2.3	Lysyltransferase Lysyltransferase	
3.4.21.12	α-lytic endopeptidase α-lytic endopeptidase	
3.4.24.32	β-Lytic metalloendopeptidase β-Lytic metalloendopeptidase	
3.4.21.12	α-lytic protease α-lytic endopeptidase	
3.4.24.32	β-Lytic protease β-Lytic metalloendopeptidase	
3.4.21.12	α-lytic proteinase α-lytic endopeptidase	
3.4.99.13	β-lytic proteinase (Mycobacterium sorangium) β-lytic proteinase (Mycobacterium sorangium)	
5.3.1.15	D-Lyxose isomerase D-Lyxose ketol-isomerase	Vol. 1, Page 333
5.3.1.15	D-Lyxose ketol-isomerase D-Lyxose ketol-isomerase	Vol. 1, Page 333
5.3.1.15	D-Lyxose ketol-isomerase D-Lyxose ketol-isomerase	Vol. 1, Page 333
2.1.1.37	M.BsuRIa DNA (cytosine-5-)-methyltransferase	

EC-No.	Synonym Recommended Name	Vol., Page
2.1.1.37	M.BsuRIb DNA (cytosine-5-)-methyltransferase	
3.1.3.22	M1Pase mannitol 1-phosphatase	
2.1.1.113	m4C-forming MTase Site-specific DNA-methyltransferase (cytosine-N^4-specific)	
2.1.1.134	m6OMT myo-Inositol 6-O-methyltransferase	
3.6.1.30	$m^7G(5)pppN$ diphosphatase $m^7G(5)pppN$ diphosphatase	
3.6.1.30	$m^7G(5)pppN$ pyrophosphatase $m^7G(5)pppN$ diphosphatase	
5.2.1.2	MAAI Maleylacetoacetate isomerase	Vol. 1, Page 197
5.2.1.2	MAc-tI Maleylacetoacetate isomerase	Vol. 1, Page 197
2.1.1.101	Macrocin O-methyltransferase Macrocin O-methyltransferase	
2.7.1.136	Macrolide 2'-kinase Macrolide 2'-kinase	
2.7.1.136	Macrolide 2'-phosphotransferase Macrolide 2'-kinase	
3.4.25.1	macropain proteasome endopeptidase complex	
3.4.24.65	Macrophage elastase Macrophage elastase	
3.4.24.35	Macrophage gelatinase Gelatinase B	
5.3.2.1	Macrophage migration inhibitory factor Phenylpyruvate tautomerase	Vol. 1, Page 367
2.1.1.11	Magnesium protoporphyrin IX methyltransferase Magnesium protoporphyrin O-methyltransferase	
2.1.1.11	Magnesium protoporphyrin O-methyltransferase Magnesium protoporphyrin O-methyltransferase	
3.4.24.62	Magnolysin Magnolysin	

EC-No.	Synonym Recommended Name	Vol., Page
4.3.1.2	MAL methylaspartate ammonia-lyase	
3.4.23.38	Malaria aspartic hemoglobinase Plasmepsin I	
6.2.1.9	Malate-CoA ligase Malate-CoA ligase	Vol. 2, Page 245
1.1.99.7	Malate-lactate transhydrogenase Lactate-malate transhydrogenase	
1.1.1.37	L-Malate-NAD^+ oxidoreductase Malate dehydrogenase	
1.1.99.16	Malate-vitamin K reductase Malate dehydrogenase (acceptor)	
6.2.1.9	Malate:CoA ligase (ADP-forming) Malate-CoA ligase	Vol. 2, Page 245
1.1.1.40	L-Malate:NADP oxidoreductase Malate dehydrogenase (oxaloacetate-decarboxylating, $NADP^+$)	
1.1.1.83	D-Malate dehydrogenase D-Malate dehydrogenase (decarboxylating)	
1.1.1.37	L-Malate dehydrogenase Malate dehydrogenase	
1.1.1.37	Malate dehydrogenase Malate dehydrogenase	
1.1.99.16	Malate dehydrogenase (acceptor) Malate dehydrogenase (acceptor)	
1.1.1.83	D-Malate dehydrogenase (decarboxylating) D-Malate dehydrogenase (decarboxylating)	
1.1.1.39	Malate dehydrogenase (decarboxylating) Malate dehydrogenase (decarboxylating)	
1.1.1.40	Malate dehydrogenase (decarboxylating, NADP) Malate dehydrogenase (oxaloacetate-decarboxylating, $NADP^+$)	
1.1.1.37	Malate dehydrogenase (NAD) Malate dehydrogenase	
1.1.1.82	Malate dehydrogenase ($NADP^+$) Malate dehydrogenase ($NADP^+$)	
1.1.1.40	Malate dehydrogenase (NADP, decarboxylating) Malate dehydrogenase (oxaloacetate-decarboxylating, $NADP^+$)	

EC-No.	Synonym Recommended Name	Vol., Page
1.1.1.38	Malate dehydrogenase (oxaloacetate-decarboxylating) Malate dehydrogenase (oxaloacetate-decarboxylating)	
1.1.1.40	Malate dehydrogenase (oxaloacetate-decarboxylating, $NADP^+$) Malate dehydrogenase (oxaloacetate-decarboxylating, $NADP^+$)	
4.1.3.2	L-Malate glyoxylate-lyase (CoA-acetylating) Malate synthase	
4.2.1.31	D-malate hydro-lyase maleate hydratase	
4.2.1.2	L-malate hydro-lyase fumarate hydratase	
1.1.1.82	Malate NADP dehydrogenase Malate dehydrogenase ($NADP^+$)	
1.1.3.3	Malate oxidase Malate oxidase	
4.1.3.2	Malate synthase Malate synthase	
4.1.3.2	Malate synthetase Malate synthase	
6.2.1.9	Malate thiokinase Malate-CoA ligase	Vol. 2, Page 245
4.2.1.31	malease maleate hydratase	
5.2.1.1	Maleate cis-trans-isomerase Maleate isomerase	Vol. 1, Page 192
4.2.1.31	maleate hydratase maleate hydratase	
5.2.1.1	Maleate isomerase Maleate isomerase	Vol. 1, Page 192
4.2.1.31	maleic-acid hydratase maleate hydratase	
1.3.1.32	Maleolylacetate reductase Maleylacetate reductase	
3.1.1.45	maleylacetate enol-lactonase carboxymethylenebutenolidase	
1.3.1.32	Maleylacetate reductase Maleylacetate reductase	

EC-No.	Synonym Recommended Name	Vol., Page
5.2.1.2	4-Maleylacetoacetate cis-trans-isomerase Maleylacetoacetate isomerase	Vol. 1, Page 197
5.2.1.2	Maleylacetoacetate isomerase Maleylacetoacetate isomerase	Vol. 1, Page 197
5.2.1.2	Maleylacetoacetic isomerase Maleylacetoacetate isomerase	Vol. 1, Page 197
5.2.1.2	Maleylacetone cis-trans-isomerase Maleylacetoacetate isomerase	Vol. 1, Page 197
5.2.1.2	Maleylacetone isomerase Maleylacetoacetate isomerase	Vol. 1, Page 197
5.2.1.4	3-Maleylpyruvate cis-trans-isomerase Maleylpyruvate isomerase	Vol. 1, Page 206
5.2.1.4	Maleylpyruvate isomerase Maleylpyruvate isomerase	Vol. 1, Page 206
1.1.1.40	'Malic'enzyme Malate dehydrogenase (oxaloacetate-decarboxylating, $NADP^+$)	
4.1.3.2	Malic-condensing enzyme Malate synthase	
1.1.1.37	Malic acid dehydrogenase Malate dehydrogenase	
1.1.1.37	Malic dehydrogenase Malate dehydrogenase	
1.1.1.82	Malic dehydrogenase (nicotinamide adenine dinucleotide phosphate) Malate dehydrogenase ($NADP^+$)	
1.1.3.3	Malic dehydrogenase II Malate oxidase	
1.1.1.83	D-Malic enzyme D-Malate dehydrogenase (decarboxylating)	
1.1.1.39	Malic enzyme Malate dehydrogenase (decarboxylating)	
1.1.1.38	Malic enzyme Malate dehydrogenase (oxaloacetate-decarboxylating)	
1.1.3.3	Malic oxidase Malate oxidase	
4.1.3.2	Malic synthetase Malate synthase	

EC-No.	Synonym Recommended Name	Vol., Page
4.2.1.27	malonate-semialdehyde dehydratase malonate-semialdehyde dehydratase	
1.2.1.15	Malonate-semialdehyde dehydrogenase Malonate-semialdehyde dehydrogenase	
1.2.1.18	Malonate-semialdehyde dehydrogenase (acetylating) Malonate-semialdehyde dehydrogenase (acetylating)	
2.8.3.3	Malonate CoA-transferase Malonate CoA-transferase	
2.8.3.3	Malonate coenzyme A-transferase Malonate CoA-transferase	
4.2.1.27	malonate semialdehyde dehydratase malonate-semialdehyde dehydratase	
4.2.1.27	malonate semialdehyde dehydratase, malonate semialdehyde malonate-semialdehyde dehydratase	
1.2.1.18	Malonic semialdehyde oxidative decarboxylase Malonate-semialdehyde dehydrogenase (acetylating)	
2.3.1.39	Malonyl-CoA-acyl carrier protein transacylase [Acyl-carrier-protein] S-malonyltransferase	
2.3.1.114	Malonyl-CoA:3,4-dichloroaniline N-malonyltransferase 3,4-Dichloroaniline N-malonyltransferase	
2.3.1.151	Malonyl-CoA:3-hydroxybenzoyl-CoA malonyltransferase Benzophenone synthase	
2.3.1.74	Malonyl-CoA:4-coumaroyl-CoA malonyltransferase (cyclizing) Naringenin-chalcone synthase	
2.3.1.95	Malonyl-CoA:4-coumaroyl-CoA malonyltransferase (cyclizing) Trihydroxystilbene synthase	
2.3.1.39	Malonyl-CoA:[acyl-carrier-protein] S-malonyltransferase [Acyl-carrier-protein] S-malonyltransferase	
2.3.1.113	Malonyl-CoA:anthranilate N-malonyltransferase Anthranilate N-malonyltransferase	
2.3.1.146	Malonyl-CoA:cinnamoyl-CoA malonyltransferase (cyclizing) Pinosylvin synthase	
2.3.1.112	Malonyl-CoA:D-tryptophan N-malonyltransferase D-Tryptophan N-malonyltransferase	
2.3.1.115	Malonyl-CoA:flavone/flavonol 7-O-glucoside malonyltransferase Isoflavone-7-O-β-glucoside 6-O-malonyltransferase	

EC-No.	Synonym Recommended Name	Vol., Page
2.3.1.116	Malonyl-CoA:flavonol-3-O-β-D-glucoside malonyltransferase Flavonol-3-O-β-glucoside O-malonyltransferase	
2.3.1.115	Malonyl-CoA:isoflavone-7-O-β-D-glucoside 6-O-malonyltransferase Isoflavone-7-O-β-glucoside 6-O-malonyltransferase	
2.3.1.94	Malonyl-CoA:propionyl-CoA malonyltransferase (cyclizing) Erythronolide synthase	
4.1.1.9	Malonyl-CoA carboxy-lyase Malonyl-CoA decarboxylase	
2.1.3.4	malonyl-CoA carboxyltransferase malonyl-CoA carboxyltransferase	
4.1.1.9	Malonyl-CoA decarboxylase Malonyl-CoA decarboxylase	
4.1.1.9	Malonyl-CoA decarboxylase Malonyl-CoA decarboxylase	
2.3.1.115	Malonyl-coenzyme A:flavone/flavonol-7-O-glycoside malonyltransferase Isoflavone-7-O-β-glucoside 6-O-malonyltransferase	
2.3.1.116	Malonyl-coenzyme A:flavonol-3-O-glucoside malonyltransferase Flavonol-3-O-β-glucoside O-malonyltransferase	
2.3.1.115	Malonyl-coenzyme A:isoflavone 7-O-glucoside-6-malonyltransferase Isoflavone-7-O-β-glucoside 6-O-malonyltransferase	
2.3.1.39	Malonyl coenzyme A-acyl carrier protein transacylase [Acyl-carrier-protein] S-malonyltransferase	
4.1.1.9	Malonyl coenzyme A decarboxylase Malonyl-CoA decarboxylase	
2.3.1.39	Malonyl transacylase [Acyl-carrier-protein] S-malonyltransferase	
2.3.1.39	Malonyl transferase [Acyl-carrier-protein] S-malonyltransferase	
2.3.1.39	Malonyltransferase, [acyl-carrier-protein] [Acyl-carrier-protein] S-malonyltransferase	
2.3.1.113	Malonyltransferase, anthranilate Anthranilate N-malonyltransferase	
2.3.1.114	Malonyltransferase, chloroaniline 3,4-Dichloroaniline N-malonyltransferase	
2.3.1.112	Malonyltransferase, D-tryptophan D-Tryptophan N-malonyltransferase	

EC-No.	Synonym Recommended Name	Vol., Page
2.3.1.115	Malonyltransferase, flavone (flavonol) 7-O-glycoside Isoflavone-7-O-β-glucoside 6-O-malonyltransferase	
2.3.1.116	Malonyltransferase, flavonol 3-O-glucoside Flavonol-3-O-β-glucoside O-malonyltransferase	
2.3.1.115	Malonyltransferase, isoflavone 7-O-glucoside 6-O- Isoflavone-7-O-β-glucoside 6-O-malonyltransferase	
3.2.1.20	Maltase α-glucosidase	
3.2.1.20	Maltase-glucoamylase α-glucosidase	
5.4.99.15	Malto-oligosyltrehalose synthase (1->4)-α-Glucan 1-α-glucosylmutase	Vol. 1, Page 652
3.2.1.141	Malto-oligosyltrehalose trehalohydrolase 4-α-D-{(1→4)-α-D-Glucano}trehalose trehalohydrolase	
5.4.99.15	Maltodextrin α-D-glucosyltransferase (1->4)-α-Glucan 1-α-glucosylmutase	Vol. 1, Page 652
2.4.1.25	Maltodextrin glycosyltransferase 4-α-Glucanotransferase	
2.4.1.1	Maltodextrin phosphorylase Phosphorylase	
3.2.1.98	Maltohexaohydrolase, exo- Glucan 1,4-α-maltohexaosidase	
3.2.1.133	Maltohydrolase, glucan 1,4-α- Glucan 1,4-α-maltohydrolase	
5.4.99.15	Maltooligosyl trehalose synthase (Rhizobium strain M-11 clone pBMTU1 gene treY reduced) (1→4)-α-Glucan 1-α-glucosylmutase	Vol. 1, Page 652
3.2.1.141	Maltooligosyltrehalose trehalohydrolase 4-α-D-{(1→4)-α-D-Glucano}trehalose trehalohydrolase	
3.2.1.141	Maltooligosyl trehalose trehalohydrolase (Rhizobium strain M-11 clone pMBTU1 gene treZ reduced) 4-α-D-{(1→4)-α-D-Glucano}trehalose trehalohydrolase	
3.2.1.141	Maltooligosyltrehalose trehalosidase 4-α-D-{(1→4)-α-D-Glucano}trehalose trehalohydrolase	
3.2.1.122	Maltose-6'-phosphate 6-phosphoglucohydrolase Maltose-6'-phosphate glucosidase	

EC-No.	Synonym Recommended Name	Vol., Page
3.2.1.122	Maltose-6'-phosphate glucosidase Maltose-6'-phosphate glucosidase	
5.4.99.16	Maltose α-D-glucosylmutase Maltose α-D-glucosyltransferase	Vol. 1, Page 656
5.4.99.16	Maltose α-D-glucosylmutase Maltose α-D-glucosyltransferase	Vol. 1, Page 656
5.4.99.16	Maltose α-D-glucosyltransferase Maltose α-D-glucosyltransferase	Vol. 1, Page 656
2.4.1.6	maltose 3-glycosyltransferase maltose 3-glycosyltransferase	
2.4.1.25	Maltose 4-glucosyltransferase 4-α-Glucanotransferase	
2.4.1.8	Maltose:orthophosphate 1-β-D-glucosyltransferase Maltose phosphorylase	
5.4.99.16	Maltose glucosylmutase Maltose α-D-glucosyltransferase	Vol. 1, Page 656
2.3.1.79	Maltose O-acetyltransferase Maltose O-acetyltransferase	
2.4.1.8	Maltose phosphorylase Maltose phosphorylase	
2.4.1.139	Maltose synthase Maltose synthase	
2.3.1.79	Maltose transacetylase Maltose O-acetyltransferase	
3.2.1.60	Maltotetraohydrolase, exo- Glucan 1,4-α-maltotetraohydrolase	
3.2.1.116	Maltotriohydrolase Glucan 1,4-α-maltotriohydrolase	
3.2.1.116	Maltotriohydrolase, exo Glucan 1,4-α-maltotriohydrolase	
4.1.3.24	malyl-CoA lyase malyl-CoA lyase	
6.2.1.9	Malyl-CoA synthetase Malate-CoA ligase	Vol. 2, Page 245
4.1.3.24	malyl coenzyme A lyase malyl-CoA lyase	

EC-No.	Synonym Recommended Name	Vol., Page
6.2.1.9	Malyl coenzyme A synthetase Malate-CoA ligase	Vol. 2, Page 245
3.2.1.113	Man9-mannosidase Mannosyl-oligosaccharide 1,2-α-mannosidase	
3.5.1.86	mandelamide amidase mandelamide amidase	
3.5.1.86	mandelamide hydrolase mandelamide amidase	
1.14.16.6	L-Mandelate-4-hydroxylase Mandelate 4-monooxygenase	
1.14.16.6	Mandelate 4-monooxygenase Mandelate 4-monooxygenase	
5.1.2.2	Mandelate racemase Mandelate racemase	Vol. 1, Page 72
5.1.2.2	Mandelate racemase Mandelate racemase	Vol. 1, Page 72
1.14.16.6	Mandelic acid 4-hydroxylase Mandelate 4-monooxygenase	
5.1.2.2	Mandelic acid racemase Mandelate racemase	Vol. 1, Page 72
4.1.2.10	Mandelonitrile benzaldehyde-lyase Mandelonitrile lyase	
4.1.2.10	Mandelonitrile lyase Mandelonitrile lyase	
1.13.11.45	manganese lipoxygenase linoleate 11-lipoxygenase	
1.11.1.13	Manganese peroxidase Manganese peroxidase	
2.4.1.180	ManNAcA transferase Lipopolysaccharide N-acetylmannosaminouronosyltransferase	
3.2.1.101	1,6-β-D-mannan-mannanohydrolase Mannan endo-1,6-β-mannosidase	
3.2.1.77	Mannan 1,2-(1,3)-α-mannosidase Mannan 1,2-(1,3)-α-mannosidase	
3.2.1.100	Mannan 1,4-β-mannobiosidase Mannan 1,4-β-mannobiosidase	

EC-No.	Synonym Recommended Name	Vol., Page
3.2.1.78	β-1, 4-mannan 4-mannanohydrolase Mannan endo-1,4-β-mannosidase	
3.2.1.78	β-D-mannanase Mannan endo-1,4-β-mannosidase	
3.2.1.25	Mannanase β-mannosidase	
3.2.1.78	Mannanase, endo-1,4-.β.- Mannan endo-1,4-β-mannosidase	
3.2.1.78	β-mannanase B Mannan endo-1,4-β-mannosidase	
3.2.1.137	1,2-1,6-α-D-Mannan D-mannohydrolase Mannan exo-1,2-1,6-α-mannosidase	
3.2.1.78	Mannan endo-1,4-β-mannosidase Mannan endo-1,4-β-mannosidase	
3.2.1.101	Mannan endo-1,6-β-mannosidase Mannan endo-1,6-β-mannosidase	
3.2.1.137	Mannan exo-1,2-1,6-α-mannosidase Mannan exo-1,2-1,6-α-mannosidase	
3.2.1.78	1,4-β-D-mannan mannanohydrolase Mannan endo-1,4-β-mannosidase	
3.2.1.100	1,4-β-D-mannan mannobiohydrolase Mannan 1,4-β-mannobiosidase	
3.2.1.77	1,2-1,3-α-D-mannan mannohydrolase Mannan 1,2-(1,3)-α-mannosidase	
3.2.1.25	Mannase β-mannosidase	
1.1.1.17	Mannitol-1-phosphate 5-dehydrogenase Mannitol-1-phosphate 5-dehydrogenase	
1.1.1.17	D-Mannitol-1-phosphate:NAD$^+$ 5-oxidoreductase Mannitol-1-phosphate 5-dehydrogenase	
1.1.1.224	D-Mannitol-1-phosphate:NADP$^+$ 6-oxidoreductase Mannose-6-phosphate 6-reductase	
1.1.1.17	D-Mannitol-1-phosphate dehydrogenase Mannitol-1-phosphate 5-dehydrogenase	
3.1.3.22	D-mannitol 1-phosphatase mannitol 1-phosphatase	

EC-No.	Synonym Recommended Name	Vol., Page
3.1.3.22	mannitol 1-phosphatase mannitol 1-phosphatase	
3.1.3.22	mannitol 1-phosphate phosphatase mannitol 1-phosphatase	
3.1.3.22	D-mannitol 1-phosphate phosphohydrolase mannitol 1-phosphatase	
1.1.1.67	Mannitol 2-dehydrogenase Mannitol 2-dehydrogenase	
1.1.1.138	Mannitol 2-dehydrogenase ($NADP^+$) Mannitol 2-dehydrogenase ($NADP^+$)	
1.1.2.2	D-Mannitol:ferricytochrome-c 2-oxidoreductase Mannitol dehydrogenase (cytochrome)	
1.1.1.255	mannitol:mannose 1-oxidoreductase mannitol dehydrogenase	
1.1.1.67	D-Mannitol:NAD^+ 2-oxidoreductase Mannitol 2-dehydrogenase	
1.1.1.255	mannitol:NAD 1-oxidoreductase mannitol dehydrogenase	
1.1.1.138	D-Mannitol:$NADP^+$ 2-oxidoreductase Mannitol 2-dehydrogenase ($NADP^+$)	
1.1.1.67	D-Mannitol dehydrogenase Mannitol 2-dehydrogenase	
1.1.1.67	Mannitol dehydrogenase Mannitol 2-dehydrogenase	
1.1.1.255	mannitol dehydrogenase mannitol dehydrogenase	
1.1.2.2	Mannitol dehydrogenase (cytochrome) Mannitol dehydrogenase (cytochrome)	
2.7.1.57	mannitol kinase mannitol kinase	
3.2.1.100	Mannobiohydrolase, exo-1,4-.β.- Mannan 1,4-β-mannobiosidase	
2.7.1.7	Mannokinase Mannokinase	
1.1.1.131	D-Mannonate:NAD(P)$^+$ 6-oxidoreductase Mannuronate reductase	

EC-No.	Synonym Recommended Name	Vol., Page
1.1.1.57	D-Mannonate:NAD^+ 5-oxidoreductase Fructuronate reductase	
1.1.1.57	D-Mannonate:NAD oxidoreductase Fructuronate reductase	
1.1.1.131	D-Mannonate:nicotinamide adenine dinucleotide (phosphate oxidoreductase (D-mannuronate-forming)) Mannuronate reductase	
4.2.1.8	Mannonate dehydratase Mannonate dehydratase	
1.1.1.57	D-Mannonate dehydrogenase Fructuronate reductase	
1.1.1.131	Mannonate dehydrogenase Mannuronate reductase	
1.1.1.131	Mannonate dehydrogenase ($NAD(P)^+$) Mannuronate reductase	
1.1.1.180	mannonate dehydrogenase ($NAD(P)^+$) mannonate dehydrogenase ($NAD(P)^+$)	
1.2.1.34	D-mannonate dehydrognase ($NAD(P)^+$) D-mannonate dehydrognase ($NAD(P)^+$)	
4.2.1.8	D-Mannonate hydro-lyase Mannonate dehydratase	
4.2.1.8	D-Mannonate hydrolyase Mannonate dehydratase	
4.2.1.8	Mannonate hydrolyase Mannonate dehydratase	
1.1.1.57	Mannonate oxidoreductase Fructuronate reductase	
1.1.1.57	Mannonic dehydrogenase Fructuronate reductase	
4.2.1.8	Mannonic hydrolase Mannonate dehydratase	
3.2.1.24	α-D-Mannopyranosidase α-mannosidase	
2.7.7.13	Mannose-1-phosphate guanylyltransferase Mannose-1-phosphate guanylyltransferase	

EC-No.	Synonym Recommended Name	Vol., Page
2.7.7.22	Mannose-1-phosphate guanylyltransferase (GDP) Mannose-1-phosphate guanylyltransferase (GDP)	
1.1.1.224	Mannose-6-phosphate 6-reductase Mannose-6-phosphate 6-reductase	
5.3.1.8	Mannose-6-phosphate isomerase Mannose-6-phosphate isomerase	Vol. 1, Page 289
5.3.1.8	D-Mannose-6-phosphate ketol-isomerase Mannose-6-phosphate isomerase	Vol. 1, Page 289
1.1.1.224	Mannose-6-phosphate reductase Mannose-6-phosphate 6-reductase	
3.2.1.113	Mannose-9 processing α-mannosidase Mannosyl-oligosaccharide 1,2-α-mannosidase	
5.4.2.8	D-Mannose 1,6-phosphomutase Phosphomannomutase	Vol. 1, Page 540
2.7.7.22	Mannose 1-phosphate guanylyltransferase Mannose-1-phosphate guanylyltransferase (GDP)	
2.7.7.13	Mannose 1-phosphate guanylyltransferase (guanosine triphosphate) Mannose-1-phosphate guanylyltransferase	
5.3.1.7	D-Mannose isomerase Mannose isomerase	Vol. 1, Page 285
5.3.1.7	Mannose isomerase Mannose isomerase	Vol. 1, Page 285
5.3.1.7	D-Mannose ketol-isomerase Mannose isomerase	Vol. 1, Page 285
5.3.1.8	Mannose phosphate isomerase Mannose-6-phosphate isomerase	Vol. 1, Page 289
2.7.8.9	Mannosephosphotransferase, phosphomannan Phosphomannan mannosephosphotransferase	
3.2.1.24	1,2-α-D-Mannosidase α-mannosidase	
3.2.1.113	1,2-α-Mannosidase Mannosyl-oligosaccharide 1,2-α-mannosidase	
3.2.1.24	1,2-α-Mannosidase α-mannosidase	
3.2.1.25	β-mannosidase β-mannosidase	

EC-No.	Synonym Recommended Name	Vol., Page
3.2.1.113	α-1,2-Mannosidase Mannosyl-oligosaccharide 1,2-α-mannosidase	
3.2.1.24	α-D-Mannosidase α-mannosidase	
3.2.1.24	α-Mannosidase α-mannosidase	
3.2.1.24	α-mannosidase α-mannosidase	
3.2.1.114	α_1-3,6-Mannosidase Mannosyl-oligosaccharide 1,3-1,6-α-mannosidase	
3.2.1.25	β-D-Mannosidase β-mannosidase	
3.2.1.25	β-Mannosidase β-mannosidase	
3.2.1.24	Mannosidase,α- α-mannosidase	
3.2.1.25	Mannosidase,β- β-mannosidase	
3.2.1.130	Mannosidase, endo-α- Glycoprotein endo-α-1,2-mannosidase	
3.2.1.113	Mannosidase, exo-1,2-α- Mannosyl-oligosaccharide 1,2-α-mannosidase	
3.2.1.137	Mannosidase, exo-1,2-1,6-α Mannan exo-1,2-1,6-α-mannosidase	
3.2.1.114	Mannosidase, exo-1,3-1,6-α- Mannosyl-oligosaccharide 1,3-1,6-α-mannosidase	
3.2.1.130	Mannosidase, glucosyl- Glycoprotein endo-α-1,2-mannosidase	
3.2.1.113	(α1,2)-Mannosidase-I Mannosyl-oligosaccharide 1,2-α-mannosidase	
3.2.1.113	Mannosidase 1A Mannosyl-oligosaccharide 1,2-α-mannosidase	
3.2.1.113	Mannosidase 1B Mannosyl-oligosaccharide 1,2-α-mannosidase	
3.2.1.113	Mannosidase I Mannosyl-oligosaccharide 1,2-α-mannosidase	

EC-No.	Synonym Recommended Name	Vol., Page
3.2.1.114	Mannosidase II Mannosyl-oligosaccharide 1,3-1,6-α-mannosidase	
3.2.1.114	α-D-Mannosidase II Mannosyl-oligosaccharide 1,3-1,6-α-mannosidase	
3.2.1.114	α-Mannosidase II Mannosyl-oligosaccharide 1,3-1,6-α-mannosidase	
2.4.1.155	α-Mannoside β-1,6-N-acetylglucosaminyltransferase α-1,3(6)-Mannosylglycoprotein β-1,6-N-acetylglucosaminyltransferase	
3.2.1.25	β-D-mannoside mannohydrolase β-mannosidase	
3.2.1.24	α-D-mannoside mannohydrolase α-mannosidase	
3.2.1.25	β-Mannoside mannohydrolase β-mannosidase	
2.4.1.101	α-1,3-Mannosyl-glycoprotein β-1,2-N-acetylglucosaminyltransferase α-1,3-Mannosyl-glycoprotein β-1,2-N-acetylglucosaminyltransferase	
2.4.1.143	α-1,6-Mannosyl-glycoprotein β-1,2-N-acetylglucosaminyltransferase α-1,6-Mannosyl-glycoprotein β-1,2-N-acetylglucosaminyltransferase	
2.4.1.201	Mannosyl-glycoprotein β-1,4-N-acetylglucosaminyltransferase Mannosyl-glycoprotein β-1,4-N-acetylglucosaminyltransferase	
2.4.1.144	β-1,4-Mannosyl-glycoprotein β-1,4-N-acetylglucosaminyltransferase β-1,4-Mannosyl-glycoprotein β-1,4-N-acetylglucosaminyltransferase	
3.2.1.96	Mannosyl-glycoprotein 1,4-N-acetamidodeoxy-β-D-glycohydrolase Mannosyl-glycoprotein endo-β-N-acetyl glucosaminidase	
3.2.1.96	Mannosyl-glycoprotein endo-β-N-acetyl glucosaminidase Mannosyl-glycoprotein endo-β-N-acetyl glucosaminidase	
3.2.1.113	1,2-α-Mannosyl-oligosaccharide α-D-mannohydrolase Mannosyl-oligosaccharide 1,2-α-mannosidase	
3.2.1.114	1,3-(1,6-)Mannosyl-oligosaccharide α-D-mannohydrolase Mannosyl-oligosaccharide 1,3-1,6-α-mannosidase	
3.2.1.113	Mannosyl-oligosaccharide 1,2-α-mannosidase Mannosyl-oligosaccharide 1,2-α-mannosidase	
3.2.1.114	Mannosyl-oligosaccharide 1,3-1,6-α-mannosidase Mannosyl-oligosaccharide 1,3-1,6-α-mannosidase	
3.2.1.106	Mannosyl-oligosaccharide glucohydrolase Mannosyl-oligosaccharide glucosidase	

EC-No.	Synonym Recommended Name	Vol., Page
3.2.1.106	Mannosyl-oligosaccharide glucosidase Mannosyl-oligosaccharide glucosidase	
2.4.1.145	α-1,3-Mannosylglycoprotein β-1,4-N-acetylglucosaminyltransferase α-1,3-Mannosylglycoprotein β-1,4-N-acetylglucosaminyltransferase	
2.4.1.155	α-1,3(6)-Mannosylglycoprotein β-1,6-N-acetylglucosaminyltransferase α-1,3(6)-Mannosylglycoprotein β-1,6-N-acetylglucosaminyltransferase	
2.4.1.199	β-Mannosylphosphodecaprenol-mannooligosaccharide 6-mannosyltransferase β-Mannosylphosphodecaprenol-mannooligosaccharide 6-mannosyltransferase	
2.4.1.199	β-D-Mannosylphosphodecaprenol:1,6-α-D-mannosyloligosaccharide 1,6-α-D-mannosyltransferase β-Mannosylphosphodecaprenol-mannooligosaccharide 6-mannosyltransferase	
3.1.4.49	Mannosylphosphodolichol phosphodiesterase Dolichyl-phosphate-mannose phosphodiesterase	
2.4.1.83	Mannosylphosphodolichol synthase Dolichyl-phosphate β-D-mannosyltransferase	
2.4.1.83	Mannosylphosphoryldolichol synthase Dolichyl-phosphate β-D-mannosyltransferase	
2.4.1.130	Mannosyltransferase, dolichol phosphomannose-oligosaccharide-lipid Dolichyl-phosphate-mannose-glycolipid α-mannosyltransferase	
2.4.1.109	Mannosyltransferase, dolichol phosphomannose-protein Dolichyl-phosphate-mannose-protein mannosyltransferase	
2.4.1.32	Mannosyltransferase, glucomannan 4-β- Glucomannan 4-β-mannosyltransferase	
2.4.1.142	Mannosyltransferase, guanosine diphosphomannose-dolichol diphosphochitobiose Chitobiosyldiphosphodolichol α-mannosyltransferase	
2.4.1.83	Mannosyltransferase, guanosine diphosphomannose-dolichol phosphate Dolichyl-phosphate β-D-mannosyltransferase	
2.4.1.48	Mannosyltransferase, guanosine diphosphomannose-heteroglycan α- Heteroglycan α-mannosyltransferase	
2.4.1.131	Mannosyltransferase, guanosine diphosphomannose-oligosaccharide-lipid Glycolipid 2-α-mannosyltransferase	
2.4.1.132	Mannosyltransferase, guanosine diphosphomannose-oligosaccharide-lipid II Glycolipid 3-α-mannosyltransferase	

EC-No.	Synonym Recommended Name	Vol., Page
2.4.1.57	Mannosyltransferase, guanosine diphosphomannose-phosphatidyl-inositol α- Phosphatidyl-myo-inositol α-mannosyltransferase	
2.4.1.54	Mannosyltransferase, guanosine diphosphomannose-undecaprenyl phosphate Undecaprenyl-phosphate mannosyltransferase	
2.4.1.199	Mannosyltransferase, mannosylphospholipid-methylmannoside α-1,6- β-Mannosylphosphodecaprenol-mannooligosaccharide 6-mannosyltransferase	
2.4.1.132	Mannosyltransferase II Glycolipid 3-α-mannosyltransferase	
2.4.1.138	Mannotetraose 2-α-N-acetylglucosaminyltransferase Mannotetraose 2-α-N-acetylglucosaminyltransferase	
4.2.2.3	mannuronate alginate lyase poly(β-D-mannuronate) lyase	
1.1.1.131	Mannuronate reductase Mannuronate reductase	
2.4.1.33	Mannuronosyl transferase Alginate synthase	
1.4.3.4	MAO Amine oxidase (flavin-containing)	
3.4.11.18	MAP Methionyl aminopeptidase	
2.7.1.123	MAP-2 kinase Ca^{2+}/calmodulin-dependent protein kinase	
2.7.1.123	MAP-2 protein serine kinase Ca^{2+}/calmodulin-dependent protein kinase	
2.7.1.37	MAPK Protein kinase	
2.7.1.123	MAP kinase Ca^{2+}/calmodulin-dependent protein kinase	
3.4.21.59	Mast cell neutral proteinase Tryptase	
3.4.21.39	mast cell protease I chymase	
3.4.21.59	Mast cell protease II Tryptase	

EC-No.	Synonym Recommended Name	Vol., Page
3.4.21.59	Mast cell proteinase II Tryptase	
3.4.21.59	Mast cell tryptase Tryptase	
2.3.1.116	MAT-3 Flavonol-3-O-β-glucoside O-malonyltransferase	
2.3.1.115	MAT-7 Isoflavone-7-O-β-glucoside 6-O-malonyltransferase	
3.4.24.3	matirx metalloproteinase-18 microbial collagenase	
3.4.24.23	Matrilysin Matrilysin	
3.4.24.23	Matrin Matrilysin	
3.4.21.32	matrix metalloproteinase-1 brachyurin	
3.4.24.7	matrix metalloproteinase-1 interstitial collagenase	
3.4.24.3	matrix metalloproteinase-1 microbial collagenase	
3.4.21.32	matrix metalloproteinase-18 brachyurin	
3.4.24.7	matrix metalloproteinase-18 interstitial collagenase	
3.4.24.23	Matrix metalloproteinase-7 Matrilysin	
3.4.24.34	Matrix metalloproteinase-8 Neutrophil collagenase	
3.4.21.32	matrix metalloproteinase-8 brachyurin	
3.4.24.7	matrix metalloproteinase-8 interstitial collagenase	
3.4.24.3	matrix metalloproteinase-8 microbial collagenase	
3.4.24.7	matrix metalloproteinase 1 interstitial collagenase	

EC-No.	Synonym Recommended Name	Vol., Page
3.4.24.22	Matrix metalloproteinase 10 Stromelysin 2	
3.4.24.24	Matrix metalloproteinase 2 Gelatinase A	
3.4.24.17	Matrix metalloproteinase 3 Stromelysin 1	
3.4.24.23	Matrix metalloproteinase 7 Matrilysin	
3.4.24.34	Matrix metalloproteinase 8 Neutrophil collagenase	
3.4.24.35	Matrix metalloproteinase 9 Gelatinase B	
3.4.24.23	Matrix metalloproteinase pump 1 Matrilysin	
3.4.24.64	Matrix peptidase Mitochondrial processing peptidase	
3.4.24.64	Matrix processing peptidase Mitochondrial processing peptidase	
3.4.24.64	Matrix processing proteinase Mitochondrial processing peptidase	
3.4.21.62	Maxatase Subtilisin	
5.3.1.5	Maxazyme Xylose isomerase	Vol. 1, Page 259
3.2.1.23	Maxilact β-galactosidase	
3.2.1.26	Maxinvert L 1000 β-fructofuranosidase	
2.7.3.2	MB-CK Creatine kinase	
2.1.1.6	MB-COMT Catechol O-methyltransferase	
3.1.1.8	mBuChE I cholinesterase	
3.1.1.8	mBuChE II cholinesterase	

EC-No.	Synonym Recommended Name	Vol., Page
5.5.1.1	MC Muconate cycloisomerase	Vol. 1, Page 660
4.1.3.31	MCA condensing enzyme 2-methylcitrate synthase	
6.4.1.4	Mcase Methylcrotonoyl-CoA carboxylase	Vol. 2, Page 744
4.1.1.9	MCD Malonyl-CoA decarboxylase	
5.4.99.2	MCM Methylmalonyl-CoA mutase	Vol. 1, Page 589
2.1.1.80	MCP methyltransferase I Protein-glutamate O-methyltransferase	
2.1.1.80	MCP methyltransferase II Protein-glutamate O-methyltransferase	
4.1.3.31	MCS 2-methylcitrate synthase	
6.3.2.13	mDAP ligase UDP-N-acetylmuramoylalanyl-D-glutamate-2,6-diaminopimelate ligase	Vol. 2, Page 473
1.6.5.4	MDAsA reductase (NADPH) Monodehydroascorbate reductase (NADH)	
1.14.17.1	MDBH Dopamine β-monooxygenase	
1.1.99.8	MDH Alcohol dehydrogenase (acceptor)	
1.1.1.37	MDH Malate dehydrogenase	
1.6.5.4	MDHA Monodehydroascorbate reductase (NADH)	
4.1.2.10	MDL Mandelonitrile lyase	
3.4.13.19	MDP Membrane dipeptidase	
1.1.1.105	MDR Retinol dehydrogenase	

EC-No.	Synonym Recommended Name	Vol., Page
3.5.3.21	MDUase Methylenediurea deaminase	
1.1.1.246	Medicarpin:$NADP^+$ 2'-oxidoreductase Pterocarpin synthase	
2.3.1.137	Medium-chain/long-chain carnitine acyltransferase Carnitine O-octanoyltransferase	
1.3.99.3	Medium-chain acyl-CoA dehydrogenase Acyl-CoA dehydrogenase	
3.1.2.19	Medium-chain acyl-CoA hydrolase ADP-dependent medium-chain-acyl-CoA hydrolase	
1.3.99.3	Medium-chain acyl-coenzyme A dehydrogenase Acyl-CoA dehydrogenase	
3.1.2.19	Medium-chain acyl-thioester hydrolase ADP-dependent medium-chain-acyl-CoA hydrolase	
3.1.2.19	Medium-chain hydrolase ADP-dependent medium-chain-acyl-CoA hydrolase	
6.2.1.2	Medium chain acyl-coenzyme A synthetase Butyrate-CoA ligase	Vol. 2, Page 199
3.4.24.28	Megateriopeptidase Bacillolysin	
4.1.2.39	MeHNL Hydroxynitrilase	
5.3.3.6	MeIT isomerase Methylitaconate Δ-isomerase	Vol. 1, Page 406
3.1.1.3	meito MY 30 triacylglycerol lipase	
3.1.1.3	meito Sangyo OF lipase triacylglycerol lipase	
3.2.1.22	Melibiase α-Galactosidase	
1.14.13.4	Melilotate 3-monooxygenase Melilotate 3-monooxygenase	
1.3.1.11	Melilotate dehydrogenase Coumarate reductase	
1.14.13.4	Melilotate hydroxylase Melilotate 3-monooxygenase	

EC-No.	Synonym Recommended Name Vol., Page
1.14.13.4	Melilotic hydroxylase Melilotate 3-monooxygenase
3.4.17.10	membrane-bound carboxypeptidase carboxypeptidase H
2.7.8.21	Membrane-derived-oligosaccharide-6-(glycerophospho)-D-glucose:membrane-derived-oligosaccharide-D-glucose glycerophosphotransferase Membrane-oligosaccharide glycerophosphotransferase
2.7.8.21	Membrane-oligosaccharide glycerophosphotransferase Membrane-oligosaccharide glycerophosphotransferase
3.4.11.2	membrane alanyl aminopeptidase membrane alanyl aminopeptidase
3.4.11.7	membrane aminopeptidase II glutamyl aminopeptidase
3.4.13.19	Membrane dipeptidase Membrane dipeptidase
3.4.17.21	Membrane glutamate carboxypeptidase Glutamate carboxypeptidase II
3.4.24.11	membrane metalloendopeptidase neprilysin
3.4.17.16	Membrane Pro-X carboxypeptidase Membrane Pro-X carboxypeptidase
1.6.99.2	Menadione oxidoreductase NAD(P)H dehydrogenase (quinone)
1.6.5.5	Menadione oxidoreductase NADPH:quinone reductase
1.6.99.2	Menadione reductase NAD(P)H dehydrogenase (quinone)
1.6.5.5	Menadione reductase NADPH:quinone reductase
1.6.5.2	menadione reductase menadione reductase
1.3.5.1	Menaquinol: fumarate oxidoreductase Succinate dehydrogenase (ubiquinone)
1.14.13.46	(-)-Menthol,NADPH:oxygen oxidoreductase (8-hydroxylating) (-)-Menthol monooxygenase

EC-No.	Synonym Recommended Name	Vol., Page
1.1.1.207	(-)-Menthol:NADP$^+$ oxidoreductase (-)-Menthol dehydrogenase	
1.1.1.207	(-)-Menthol dehydrogenase (-)-Menthol dehydrogenase	
1.14.13.46	(-)-Menthol monooxygenase (-)-Menthol monooxygenase	
2.3.1.69	Menthol transacetylase Monoterpenol O-acetyltransferase	
3.4.24.18	Meprin Meprin A	
3.4.24.18	Meprin-a Meprin A	
3.4.24.18	Meprin A Meprin A	
3.4.24.63	Meprin B Meprin B	
3.4.24.63	Meprin b Meprin B	
1.13.99.5	MeQDO 3,4-Dihydroxyquinoline 2,4-dioxygenase	
1.16.1.1	Mer A Mercury(II) reductase	
3.4.22.39	mercapto proteinase adenain	
3.4.22.14	mercaptoproteinase A2 actinidain	
2.4.2.8	6-Mercaptopurine phosphoribosyltransferase Hypoxanthine phosphoribosyltransferase	
2.8.1.2	3-Mercaptopyruvate:cyanide sulfurtransferase 3-Mercaptopyruvate sulfurtransferase	
2.8.1.2	3-Mercaptopyruvate sulfurtransferase 3-Mercaptopyruvate sulfurtransferase	
2.8.1.2	β-Mercaptopyruvate sulfurtransferase 3-Mercaptopyruvate sulfurtransferase	
1.16.1.1	Mercurate(II) reductase Mercury(II) reductase	

EC-No.	Synonym Recommended Name	Vol., Page
1.16.1.1	Mercuric ion reductase Mercury(II) reductase	
1.16.1.1	Mercuric reductase Mercury(II) reductase	
1.16.1.1	Mercury(II) reductase Mercury(II) reductase	
1.16.1.1	Mercury reductase Mercury(II) reductase	
4.2.1.34	Mesaconase (S)-2-Methylmalate dehydratase	
4.2.1.34	Mesaconase, mesaconate (S)-2-Methylmalate dehydratase	
4.2.1.34	Mesaconate hydratase (S)-2-Methylmalate dehydratase	
2.4.2.28	MeSAdo/Ado phosphorylase 5'-Methylthioadenosine phosphorylase	
3.2.2.16	MeSAdo nucleosidase Methylthioadenosine nucleosidase	
2.4.2.28	MeSAdo phosphorylase 5'-Methylthioadenosine phosphorylase	
1.4.1.16	meso-α,ε-Diaminopimelate dehydrogenase Diaminopimelate dehydrogenase	
4.2.1.28	meso-2,3-butanediol dehydrase propanediol dehydratase	
1.4.1.16	meso-2,6-Diaminoheptanedioate:$NADP^+$ oxidoreductase (deaminating) Diaminopimelate dehydrogenase	
4.1.1.20	meso-2,6-diaminoheptanedioate carboxy-lyase Diaminopimelate decarboxylase	
4.1.1.20	meso-diaminopimelate decarboxylase Diaminopimelate decarboxylase	
1.4.1.16	meso-Diaminopimelate dehydrogenase Diaminopimelate dehydrogenase	
1.13.99.1	meso-Inositol oxygenase myo-Inositol oxygenase	
1.5.1.17	meso-N-(1-Carboxyethyl)-alanine dehydrogenase Alanopine dehydrogenase	

EC-No.	Synonym Recommended Name	Vol., Page
1.3.1.7	meso-Tartrate:NAD$^+$ oxidoreductase meso-Tartrate dehydrogenase	
1.3.1.7	meso-Tartrate dehydrogenase meso-Tartrate dehydrogenase	
1.11.1.5	Mesocytochrome c peroxidase azide Cytochrome-c peroxidase	
1.11.1.5	Mesocytochrome c peroxidase cyanate Cytochrome-c peroxidase	
1.11.1.5	Mesocytochrome c peroxidase cyanide Cytochrome-c peroxidase	
1.1.1.93	Mesotartrate dehydrogenase Tartrate dehydrogenase	
2.1.1.57	Messenger RNA (nucleoside-2'-)-methyltransferase mRNA (nucleoside-2'-O-)-methyltransferase	
2.1.1.56	Messenger RNA guanine 7-methyltransferase mRNA (guanine-N^7-)-methyltransferase	
2.7.7.50	Messenger RNA guanylyltransferase mRNA guanylyltransferase	
1.8.4.6	Met(O)-peptide reductase Protein-methionine-S-oxide reductase	
3.4.13.12	Met-X-dipeptidase Met-X-dipeptidase	
3.4.11.22	Metallo aminopeptidase Aminopeptidase I	
3.4.17.22	Metallocarboxypeptidase D Metallocarboxypeptidase D	
3.4.21.32	metallocollagenase brachyurin	
3.4.24.7	metallocollagenase interstitial collagenase	
3.4.24.3	metallocollagenase microbial collagenase	
3.4.24.65	Metalloelastase Macrophage elastase	
3.4.24.15	metalloendopeptidase 24.15 thimet oligopeptidase	

EC-No.	Synonym Recommended Name	Vol., Page
3.4.24.56	Metalloinsulinase Insulysin	
3.4.24.15	metallopeptidase thimet oligopeptidase	
3.4.21.32	metalloproteinase-1 brachyurin	
3.4.24.7	metalloproteinase-1 interstitial collagenase	
3.4.24.3	metalloproteinase-1 microbial collagenase	
3.4.24.52	Metalloproteinase HR1A Trimerelysin I	
3.4.24.58	Metalloproteinase RVV-x Russellysin	
3.6.1.10	Metaphosphatase Endopolyphosphatase	
3.6.1.11	Metaphosphatase Exopolyphosphatase	
1.13.11.2	Metapyrocatechase Catechol 2,3-dioxygenase	
1.14.13.25	Methane,NAD(P)H:oxygen oxidoreductase (hydroxylating) Methane monooxygenase	
1.14.13.25	Methane hydroxylase Methane monooxygenase	
1.14.13.25	Methane mono-oxygenase Methane monooxygenase	
1.14.13.25	Methane monooxygenase Methane monooxygenase	
1.8.3.4	Methanethiol:oxygen oxidoreductase Methanethiol oxidase	
1.8.3.4	Methanethiol oxidase Methanethiol oxidase	
2.1.1.90	Methanol-5-hydroxybenzimidazolylcobamide Co-methyltransferase Methanol-5-hydroxybenzimidazolylcobamide Co-methyltransferase	
2.1.1.90	Methanol:5-hydroxybenzimidazolylcobamide Co-methyltransferase Methanol-5-hydroxybenzimidazolylcobamide Co-methyltransferase	

EC-No.	Synonym Recommended Name	Vol., Page
2.1.1.90	Methanol:5-hydroxybenzimidazolylcobamide methyltransferase Methanol-5-hydroxybenzimidazolylcobamide Co-methyltransferase	
1.1.1.244	Methanol:NAD$^+$ oxidoreductase Methanol dehydrogenase	
1.1.3.31	Methanol:oxygen oxidoreductase Methanol oxidase	
2.1.1.90	Methanol cobalamin methyltransferase Methanol-5-hydroxybenzimidazolylcobamide Co-methyltransferase	
1.1.1.244	Methanol dehydrogenase Methanol dehydrogenase	
1.1.3.31	Methanol oxidase Methanol oxidase	
3.5.4.27	Methenyl-H4MPT cyclohydrolase Methenyltetrahydromethanopterin cyclohydrolase	
3.5.4.9	5,10-Methenyltetrahydrofolate 5-hydrolase (decyclizing) Methenyltetrahydrofolate cyclohydrolase	
2.1.2.2	5,10-Methenyltetrahydrofolate:2-amino-N-ribosylacetamide ribonucleotide transformylase Phosphoribosylglycinamide formyltransferase	
3.5.4.9	Methenyltetrahydrofolate cyclohydrolase Methenyltetrahydrofolate cyclohydrolase	
6.3.3.2	5,10-Methenyltetrahydrofolate synthetase 5-Formyltetrahydrofolate cyclo-ligase	Vol. 2, Page 535
6.3.3.2	Methenyltetrahydrofolate synthetase 5-Formyltetrahydrofolate cyclo-ligase	Vol. 2, Page 535
3.5.4.27	5,10-Methenyltetrahydromethanopterin 10-hydrolase (decyclizing) Methenyltetrahydromethanopterin cyclohydrolase	
3.5.4.27	5,10-Methenyltetrahydromethanopterin cyclohydrolase Methenyltetrahydromethanopterin cyclohydrolase	
3.5.4.27	Methenyltetrahydromethanopterin cyclohydrolase Methenyltetrahydromethanopterin cyclohydrolase	
6.3.3.2	5,10-Methenyltetrahydropteroylglutamate synthetase 5-Formyltetrahydrofolate cyclo-ligase	Vol. 2, Page 535
6.3.3.2	Methenyl THFS 5-Formyltetrahydrofolate cyclo-ligase	Vol. 2, Page 535

EC-No.	Synonym Recommended Name	Vol., Page
4.4.1.11	L-methioninase methionine γ-lyase	
4.4.1.11	methioninase methionine γ-lyase	
4.4.1.11	L-methionine-α-deamino-γ-mercaptomethane-lyase methionine γ-lyase	
2.5.1.6	Methionine-activating enzyme Methionine adenosyltransferase	
2.6.1.73	Methionine-glyoxylate aminotransferase Methionine-glyoxylate transaminase	
2.6.1.73	Methionine-glyoxylate transaminase Methionine-glyoxylate transaminase	
2.1.1.2	Methionine-guanidinoacetic transmethylase Guanidinoacetate N-methyltransferase	
4.4.1.11	L-methionine γ-lyase methionine γ-lyase	
4.4.1.11	methionine γ-lyase methionine γ-lyase	
4.4.1.11	methionine γ-lyase methionine γ-lyase	
2.6.1.41	D-Methionine-pyruvate transaminase D-Methionine-pyruvate transaminase	
1.8.4.5	Methionine-S-oxide reductase Methionine-S-oxide reductase	
6.1.1.10	Methionine-tRNA ligase Methionine-tRNA ligase	Vol. 2, Page 68
2.6.1.73	L-Methionine:glyoxylate aminotransferase Methionine-glyoxylate transaminase	
1.8.4.5	L-Methionine:oxidized-thioredoxin S-oxidoreductase Methionine-S-oxide reductase	
2.6.1.41	D-Methionine:pyruvate aminotransferase D-Methionine-pyruvate transaminase	
6.1.1.10	L-Methionine:tRNAMet ligase (AMP-forming) Methionine-tRNA ligase	Vol. 2, Page 68
2.5.1.6	Methionine adenosyltransferase Methionine adenosyltransferase	

EC-No.	Synonym Recommended Name	Vol., Page
3.4.11.18	L-Methionine aminopeptidase Methionyl aminopeptidase	
3.4.11.18	Methionine aminopeptidase Methionyl aminopeptidase	
2.6.1.41	D-Methionine aminotransferase D-Methionine-pyruvate transaminase	
4.1.1.57	L-Methionine carboxy-lyase Methionine decarboxylase	
4.1.1.57	L-Methionine decarboxylase Methionine decarboxylase	
4.1.1.57	Methionine decarboxylase Methionine decarboxylase	
4.4.1.11	methionine dethiomethylase methionine γ-lyase	
4.4.1.11	methionine lyase methionine γ-lyase	
4.4.1.11	L-methionine methanethiol-lyase (deaminating) methionine γ-lyase	
2.1.1.12	Methionine methyltransferase Methionine S-methyltransferase	
5.1.1.2	Methionine racemase Methionine racemase	Vol. 1, Page 10
5.1.1.2	Methionine racemase Methionine racemase	Vol. 1, Page 10
2.5.1.6	Methionine S-adenosyltransferase Methionine adenosyltransferase	
2.1.1.12	Methionine S-methyltransferase Methionine S-methyltransferase	
1.8.4.5	Methionine S-oxide reductase Methionine-S-oxide reductase	
1.8.4.6	Methionine sulfoxide peptide reductase Protein-methionine-S-oxide reductase	
1.8.4.5	Methionine sulfoxide reductase Methionine-S-oxide reductase	
2.1.1.13	Methionine synthase 5-Methyltetrahydrofolate-homocysteine S-methyltransferase	

EC-No.	Synonym Recommended Name	Vol., Page
4.2.99.10	methionine synthase O-acetylhomoserine (thiol)-lyase	
2.1.1.135	[Methionine synthase]-cobalamin methyltransferase (cob(II)alamin reducing) [Methionine synthase]-cobalamin methyltransferase (cob(II)alamin reducing)	
2.1.1.135	Methionine synthase cob(II)alamin reductase (methylating) [Methionine synthase]-cobalamin methyltransferase (cob(II)alamin reducing)	
2.1.1.135	Methionine synthase reductase [Methionine synthase]-cobalamin methyltransferase (cob(II)alamin reducing)	
2.1.1.13	Methionine synthetase 5-Methyltetrahydrofolate-homocysteine S-methyltransferase	
2.6.1.41	D-Methionine transaminase D-Methionine-pyruvate transaminase	
6.1.1.10	Methionine translase Methionine-tRNA ligase	Vol. 2, Page 68
6.1.1.10	Methionyl-transfer ribonucleate synthetase Methionine-tRNA ligase	Vol. 2, Page 68
6.1.1.10	Methionyl-transfer ribonucleic acid synthetase Methionine-tRNA ligase	Vol. 2, Page 68
2.1.2.9	Methionyl-transfer ribonucleic transformylase Methionyl-tRNA formyltransferase	
6.1.1.10	Methionyl-transfer RNA synthetase Methionine-tRNA ligase	Vol. 2, Page 68
2.1.2.9	Methionyl-transfer RNA transformylase Methionyl-tRNA formyltransferase	
2.1.2.9	Methionyl-tRNA formyltransferase Methionyl-tRNA formyltransferase	
2.1.2.9	Methionyl-tRNA Met formyltransferase Methionyl-tRNA formyltransferase	
2.1.2.9	Methionyl-tRNA transformylase Methionyl-tRNA formyltransferase	
3.4.11.18	Methionyl aminopeptidase Methionyl aminopeptidase	

EC-No.	Synonym Recommended Name	Vol., Page
3.4.13.12	methionyl dipeptidase Met-X-dipeptidase	
2.1.2.9	Methionyl ribonucleic formyltransferase Methionyl-tRNA formyltransferase	
6.1.1.10	Methionyl tRNA synthetase Methionine-tRNA ligase	Vol. 2, Page 68
2.1.1.99	16-Methoxy-2,3-dihydro-3-hydroxytabersonine N-methyltransferase 16-Methoxy-2,3-dihydro-3-hydroxytabersonine N-methyltransferase	
1.14.99.15	4-Methoxybenzoate,hydrogen-donor:oxygen oxidoreductase (O-demethylating) 4-Methoxybenzoate monooxygenase (O-demethylating)	
1.14.99.15	4-Methoxybenzoate 4-monooxygenase (O-demethylating) 4-Methoxybenzoate monooxygenase (O-demethylating)	
1.14.99.15	4-Methoxybenzoate monooxygenase (O-demethylating) 4-Methoxybenzoate monooxygenase (O-demethylating)	
1.14.99.15	4-Methoxybenzoate O-demethylase 4-Methoxybenzoate monooxygenase (O-demethylating)	
3.2.1.139	(4-O-Methyl)-α-glucuronidase α-Glucosiduronase	
3.2.1.139	α-(4-O-Methyl)-D-glucuronidase α-Glucosiduronase	
3.1.1.59	Methyl-(2E,6E)-(10R,11S)-10,11-epoxy-3,7,11-trimethyltrideca-2,6-dienoate acylhydrolase Juvenile-hormone esterase	
4.1.3.17	γ-methyl-γ-hydroxy-α-ketoglutaric aldolase 4-hydroxy-4-methyl-2-oxoglutarate aldolase	
2.3.1.156	3-methyl-1-(trihydroxyphenyl)butan-1-one synthase phloroisovalerophenone synthase	
1.2.4.4	3-Methyl-2-oxobutanoate:lipoamide oxidoreductase (decarboxylating and acceptor-2-methyl-propanoylating) 3-Methyl-2-oxobutanoate dehydrogenase (lipoamide)	
1.2.1.25	3-Methyl-2-oxobutanoate:NAD^+ oxidoreductase (CoA-methylpropanoylating) 2-Oxoisovalerate dehydrogenase (acylating)	
1.2.4.4	3-Methyl-2-oxobutanoate dehydrogenase (lipoamide) 3-Methyl-2-oxobutanoate dehydrogenase (lipoamide)	

EC-No.	Synonym Recommended Name	Vol., Page
3.1.3.52	[3-Methyl-2-oxobutanoate dehydrogenase (lipoamide)]-phosphatase [3-Methyl-2-oxobutanoate dehydrogenase (lipoamide)]-phosphatase	
3.1.3.52	[3-Methyl-2-oxobutanoate dehydrogenase (lipoamide)]-phosphate phosphohydrolase [3-Methyl-2-oxobutanoate dehydrogenase (lipoamide)]-phosphatase	
2.7.1.115	[3-Methyl-2-oxobutanoate dehydrogenase (lipoamide)] kinase [3-Methyl-2-oxobutanoate dehydrogenase (lipoamide)] kinase	
2.1.2.11	3-Methyl-2-oxobutanoate hydroxymethyltransferase 3-Methyl-2-oxobutanoate hydroxymethyltransferase	
3.5.1.36	N-Methyl-2-oxoglutaramate hydrolase N-Methyl-2-oxoglutaramate hydrolase	
3.5.1.36	N-Methyl-2-oxoglutaramate methylamidohydrolase N-Methyl-2-oxoglutaramate hydrolase	
5.4.99.14	4-Methyl-3-enelactone methyl isomerase 4-Carboxymethyl-4-methylbutenolide mutase	Vol. 1, Page 648
1.1.1.178	2-Methyl-3-hydroxy-butyryl CoA dehydrogenase 3-Hydroxy-2-methylbutyryl-CoA dehydrogenase	
1.1.1.178	2-Methyl-3-hydroxybutyryl coenzyme A dehydrogenase 3-Hydroxy-2-methylbutyryl-CoA dehydrogenase	
1.14.12.4	2-Methyl-3-hydroxypyridine 5-carboxylic acid dioxygenase 3-Hydroxy-2-methylpyridine carboxylate dioxygenase	
1.2.1.27	2-Methyl-3-oxopropanoate:NAD$^+$ oxidoreductase (CoA-propanoylating) Methylmalonate-semialdehyde dehydrogenase (acylating)	
5.1.99.1	2-Methyl-3-oxopropanoyl-CoA 2-epimerase Methylmalonyl-CoA epimerase	Vol. 1, Page 179
3.1.3.14	S-methyl-3-phospho-1-thio-D-glycerate phosphohydrolase methylphosphothioglycerate phosphatase	
1.1.4.1	2-Methyl-3-phytyl-1,4-naphthoquinone:oxidized-dithiothreitol oxidoreductase Vitamin-K-epoxide reductase (warfarin sensitive)	
2.5.1.3	2-Methyl-4-amino-5-hydroxymethylpyrimidine-diphosphate:4-methyl-5-(2-phosphoethyl)-thiazole 2-methyl-4-aminopyrimidine-5-methenyltransferase Thiamine-phosphate pyrophosphorylase	
1.13.12.11	1-Methyl-4-phenyl-1,2,3,6-tetrahydropyridine:oxygen N-oxidoreductase Methylphenyltetrahydropyridine N-monooxygenase	

EC-No.	Synonym Recommended Name	Vol., Page
2.1.1.86	5-Methyl-5,6,7,8-tetrahydromethanopterin:2-mercaptoethanesulfonate 2-methyltransferase Tetrahydromethanopterin S-methyltransferase	
2.7.1.50	4-Methyl-5-(β-hydroxyethyl)thiazole kinase Hydroxyethylthiazole kinase	
2.1.1.80	Methyl-accepting chemotaxis protein methyltransferase II Protein-glutamate O-methyltransferase	
2.1.1.80	Methyl-accepting chemotaxis protein O-methyltransferase Protein-glutamate O-methyltransferase	
1.3.1.52	2-Methyl-branched-chain-acyl-CoA:NAD$^+$ 2-oxidoreductase 2-Methyl-branched-chain-enoyl-CoA reductase	
1.3.1.52	2-Methyl-branched-chain-enoyl-CoA reductase 2-Methyl-branched-chain-enoyl-CoA reductase	
3.5.4.13	5-Methyl-dCTP deaminase dCTP deaminase	
3.4.13.5	X-methyl-His dipeptidase X-methyl-His dipeptidase	
3.4.13.5	X-methyl-His dipeptidase X-methyl-His dipeptidase	
1.4.1.17	N-Methyl-L-alanine:NADP$^+$ oxidoreductase (demethylating, deaminating) N-Methylalanine dehydrogenase	
1.5.3.2	N-Methyl-L-amino-acid:oxygen oxidoreductase (demethylating) N-Methyl-L-amino-acid oxidase	
1.5.3.2	N-Methyl-L-amino-acid oxidase N-Methyl-L-amino-acid oxidase	
4.3.1.2	3-methyl-L-aspartic acid ammonia-lyase methylaspartate ammonia-lyase	
1.5.99.5	N-Methyl-L-glutamate:(acceptor) oxidoreductase (demethylating) Methylglutamate dehydrogenase	
1.1.1.143	5-O-Methyl-myo-inositol:NAD$^+$ oxidoreductase Sequoyitol dehydrogenase	
2.4.1.171	Methyl-ONN-azoxymethanol β-D-glucosyltransferase Methyl-ONN-azoxymethanol β-D-glucosyltransferase	
2.3.1.9	2-Methylacetoacetyl-CoA thiolase Acetyl-CoA C-acetyltransferase	

EC-No.	Synonym Recommended Name	Vol., Page
5.1.99.4	2-Methylacyl-CoA 2-epimerase α-Methylacyl-CoA racemase	Vol. 1, Page 188
1.3.99.12	2-Methylacyl-CoA dehydrogenase 2-Methylacyl-CoA dehydrogenase	
5.1.99.4	α-Methylacyl-CoA racemase α-Methylacyl-CoA racemase	Vol. 1, Page 188
5.1.99.4	α-Methylacyl CoA racemase α-Methylacyl-CoA racemase	Vol. 1, Page 188
3.2.2.20	3-Methyladenine DNA glycosylase I DNA-3-methyladenine glycosidase I	
3.2.2.21	3-Methyladenine DNA glycosylase II DNA-3-methyladenine glycosidase II	
2.1.1.36	1-Methyladenine transfer RNA methyltransferase tRNA (adenine-N^1-)-methyltransferase	
3.2.2.13	1-Methyladenosine hydrolase 1-Methyladenosine nucleosidase	
3.2.2.13	1-Methyladenosine nucleosidase 1-Methyladenosine nucleosidase	
3.2.2.9	5'-Methyladenosine nucleosidase Adenosylhomocysteine nucleosidase	
3.2.2.13	1-Methyladenosine ribohydrolase 1-Methyladenosine nucleosidase	
1.4.1.17	N-Methylalanine dehydrogenase N-Methylalanine dehydrogenase	
2.1.1.21	Methylamine-glutamate N-methyltransferase Methylamine-glutamate N-methyltransferase	
2.1.1.21	Methylamine:L-glutamate N-methyltransferase Methylamine-glutamate N-methyltransferase	
1.4.99.3	Methylamine dehydrogenase Amine dehydrogenase	
1.5.3.2	N-Methylamino acid oxidase N-Methyl-L-amino-acid oxidase	
1.97.1.7	methylarsonate reductase methylarsonate reductase	
2.1.1.138	methylarsonite methyltransferase methylarsonite methyltransferase	

EC-No.	Synonym Recommended Name	Vol., Page
4.3.1.2	3-methylaspartase methylaspartate ammonia-lyase	
4.3.1.2	β-methylaspartase methylaspartate ammonia-lyase	
4.3.1.2	3-methylaspartate ammonia-lyase methylaspartate ammonia-lyase	
4.3.1.2	methylaspartate ammonia-lyase methylaspartate ammonia-lyase	
5.4.99.1	Methylaspartate mutase Methylaspartate mutase	Vol. 1, Page 582
5.4.99.1	Methylaspartic acid mutase Methylaspartate mutase	Vol. 1, Page 582
2.1.1.63	Methylated-DNA-protein-cysteine S-methyltransferase Methylated-DNA-protein-cysteine S-methyltransferase	
1.2.1.3	*m*-Methylbenzaldehyde dehydrogenase Aldehyde dehydrogenase (NAD^+)	
1.3.1.52	2-Methyl branched-chain acyl-CoA dehydrogenase 2-Methyl-branched-chain-enoyl-CoA reductase	
1.3.1.52	2-Methyl branched-chain enoyl-CoA reductase 2-Methyl-branched-chain-enoyl-CoA reductase	
1.3.1.52	Methyl branched-chain enoyl-CoA reductase 2-Methyl-branched-chain-enoyl-CoA reductase	
1.3.1.52	2-Methyl branched-chain enoyl CoA reductase 2-Methyl-branched-chain-enoyl-CoA reductase	
1.3.99.12	2-Methyl branched chain acyl-CoA dehydrogenase 2-Methylacyl-CoA dehydrogenase	
1.1.1.265	3-methylbutanal reductase 3-methylbutanal reductase	
1.1.1.265	3-methylbutanol:NAD(P) oxidoreductase 3-methylbutanal reductase	
1.3.99.10	3-Methylbutanoyl-CoA:(acceptor) oxidoreductase Isovaleryl-CoA dehydrogenase	
1.3.99.12	2-Methylbutanoyl-CoA:acceptor oxidoreductase 2-Methylacyl-CoA dehydrogenase	
5.3.3.2	Methylbutenylpyrophosphate isomerase Isopentenyl-diphosphate Δ-isomerase	Vol. 1, Page 386

EC-No.	Synonym Recommended Name	Vol., Page
3.1.1.1	Methylbutyrase Carboxylesterase	
3.1.1.1	Methylbutyrate esterase Carboxylesterase	
4.2.1.79	2-Methylcitrate dehydratase 2-Methylcitrate dehydratase	
4.2.1.79	2-Methylcitrate hydro-lyase 2-Methylcitrate dehydratase	
4.1.3.31	2-methylcitrate oxaloacetate-lyase 2-methylcitrate synthase	
4.1.3.31	2-methylcitrate synthase 2-methylcitrate synthase	
4.1.3.31	methylcitrate synthase 2-methylcitrate synthase	
4.1.3.31	methylcitrate synthetase 2-methylcitrate synthase	
6.4.1.4	3-Methylcrotonoyl-CoA:carbon-dioxide ligase (ADP-forming) Methylcrotonoyl-CoA carboxylase	Vol. 2, Page 744
6.4.1.4	Methylcrotonoyl-CoA carboxylase Methylcrotonoyl-CoA carboxylase	Vol. 2, Page 744
6.4.1.4	β-methylcrotonyl CoA carboxylase Methylcrotonoyl-CoA carboxylase	Vol. 2, Page 744
6.4.1.4	Methylcrotonyl coenzyme A carboxylase Methylcrotonoyl-CoA carboxylase	Vol. 2, Page 744
6.4.1.4	β-Methylcrotonyl coenzyme A carboxylase Methylcrotonoyl-CoA carboxylase	Vol. 2, Page 744
6.3.4.11	β-methylcrotonyl coenzyme A holocarboxylase synthetase Biotin-[methylcrotonyl-CoA-carboxylase] ligase	Vol. 2, Page 622
4.2.1.22	Methylcysteine synthase Cystathionine β-synthase	
4.2.1.23	methylcysteine synthase methylcysteine synthase	
2.7.4.19	5-Methyldeoxycytidine-5'-phosphate kinase 5-Methyldeoxycytidine-5'-phosphate kinase	
2.1.1.16	Methylene-fatty-acyl-phospholipid synthase Methylene-fatty-acyl-phospholipid synthase	

EC-No.	Synonym Recommended Name	Vol., Page
1.5.99.9	Methylene-H4MPT:coenzyme F420 oxidoreductase Methylenetetrahydromethanopterin dehydrogenase	
1.12.99.4	5,10-Methylene-H4MPT dehydrogenase N^5,N^{10}-Methenyltetrahydromethanopterin hydrogenase	
1.5.99.11	methylene-H4MPT reductase coenzyme F420-dependent N^5,N^{10}-methenyltetrahydromethanopterin reductase	
6.3.1.7	4-Methylene-L-glutamate:ammonia ligase (AMP-forming) 4-Methyleneglutamate-ammonia ligase	Vol. 2, Page 383
3.5.1.67	4-Methylene-L-glutamine amidohydrolase 4-Methyleneglutaminase	
3.5.3.21	Methylenediurea deaminase Methylenediurea deaminase	
3.5.3.21	Methylenediurea deiminase Methylenediurea deaminase	
3.5.3.21	Methylenediurease Methylenediurea deaminase	
6.3.1.7	4-Methyleneglutamate-ammonia ligase 4-Methyleneglutamate-ammonia ligase	Vol. 2, Page 383
3.5.1.67	4-Methyleneglutaminase 4-Methyleneglutaminase	
3.5.1.67	4-Methyleneglutamine amidohydrolase 4-Methyleneglutaminase	
3.5.1.67	4-Methyleneglutamine deamidase 4-Methyleneglutaminase	
6.3.1.7	4-Methyleneglutamine synthetase 4-Methyleneglutamate-ammonia ligase	Vol. 2, Page 383
5.4.99.4	2-Methyleneglutarate carboxymethylene-methylmutase 2-Methyleneglutarate mutase	Vol. 1, Page 599
5.4.99.4	2-Methyleneglutarate mutase 2-Methyleneglutarate mutase	Vol. 1, Page 599
5.4.99.4	α-Methyleneglutarate mutase 2-Methyleneglutarate mutase	Vol. 1, Page 599
1.14.15.1	Methylene hydroxylase Camphor 5-monooxygenase	

EC-No.	Synonym Recommended Name	Vol., Page
1.14.15.1	Methylene monooxygenase Camphor 5-monooxygenase	
3.5.3.21	Methylenendiurea aminohydrolase Methylenediurea deaminase	
6.3.1.7	4-Methylenenglutamate:ammonia ligase (AMP-forming) 4-Methyleneglutamate-ammonia ligase	Vol. 2, Page 383
1.3.1.17	3-Methyleneoxindole reductase 3-Methyleneoxindole reductase	
1.3.1.17	3-Methyleneoxindole reductase, 3-methyleneoxindole 3-Methyleneoxindole reductase	
2.1.1.74	Methylenetetrahydrofolate-tRNA-(uracil-5-)-methyltransferase ($FADH_2$-oxidizing) Methylenetetrahydrofolate-tRNA-(uracil-5-)-methyltransferase ($FADH_2$-oxidizing)	
2.1.2.11	5,10-Methylene tetrahydrofolate:α-ketoisovalerate hydroxymethyltransferase 3-Methyl-2-oxobutanoate hydroxymethyltransferase	
2.1.2.11	5,10-Methylenetetrahydrofolate:3-methyl-2-oxobutanoate hydroxymethyltransferase 3-Methyl-2-oxobutanoate hydroxymethyltransferase	
2.1.2.7	5,10-Methylenetetrahydrofolate:D-alanine 2-hydroxymethyltransferase D-Alanine 2-hydroxymethyltransferase	
2.1.1.54	5,10-Methylenetetrahydrofolate:dCMP C-methyltransferase Deoxycytidylate C-methyltransferase	
2.1.2.8	5,10-Methylenetetrahydrofolate:deoxycytidylate 5-hydroxymethyltransferase Deoxycytidylate 5-hydroxymethyltransferase	
2.1.1.45	5,10-Methylenetetrahydrofolate:dUMP C-methyltransferase Thymidylate synthase	
2.1.1.45	Methylenetetrahydrofolate:dUMP C-methyltransferase Thymidylate synthase	
2.1.2.1	5,10-Methylenetetrahydrofolate:glycine hydroxymethyltransferase Glycine hydroxymethyltransferase	
1.5.1.15	5,10-Methylenetetrahydrofolate:NAD^+ oxidoreductase Methylenetetrahydrofolate dehydrogenase (NAD^+)	
1.5.1.5	5,10-Methylenetetrahydrofolate:$NADP^+$ oxidoreductase Methylenetetrahydrofolate dehydrogenase ($NADP^+$)	

EC-No.	Synonym Recommended Name	Vol., Page
1.5.1.5	5,10-Methylenetetrahydrofolate:NADP oxidoreductase Methylenetetrahydrofolate dehydrogenase ($NADP^+$)	
2.1.1.74	5,10-Methylenetetrahydrofolate:tRNA-UpsiC (uracil-5-)-methyltransferase Methylenetetrahydrofolate-tRNA-(uracil-5-)-methyltransferase ($FADH_2$-oxidizing)	
1.5.1.5	5,10-Methylenetetrahydrofolate dehydrogenase Methylenetetrahydrofolate dehydrogenase ($NADP^+$)	
1.5.1.5	Methylenetetrahydrofolate dehydrogenase Methylenetetrahydrofolate dehydrogenase ($NADP^+$)	
1.5.1.15	Methylenetetrahydrofolate dehydrogenase (NAD^+) Methylenetetrahydrofolate dehydrogenase (NAD^+)	
1.5.1.5	Methylenetetrahydrofolate dehydrogenase ($NADP^+$) Methylenetetrahydrofolate dehydrogenase ($NADP^+$)	
1.7.99.5	5,10-Methylenetetrahydrofolate reductase 5,10-Methylenetetrahydrofolate reductase ($FADH_2$)	
1.1.1.68	5,10-methylenetetrahydrofolate reductase 5,10-methylenetetrahydrofolate reductase	
1.7.99.5	Methylenetetrahydrofolate reductase 5,10-Methylenetetrahydrofolate reductase ($FADH_2$)	
1.7.99.5	5,10-Methylenetetrahydrofolate reductase ($FADH_2$) 5,10-Methylenetetrahydrofolate reductase ($FADH_2$)	
1.1.99.15	5,10-methylenetetrahydrofolate reductase ($FADH_2$) 5,10-methylenetetrahydrofolate reductase ($FADH_2$)	
1.5.1.20	5,10-Methylenetetrahydrofolate reductase (NADPH) Methylenetetrahydrofolate reductase (NADPH)	
1.5.1.20	Methylenetetrahydrofolate reductase (NADPH) Methylenetetrahydrofolate reductase (NADPH)	
1.1.1.171	methylenetetrahydrofolate reductase ($NADPH_2$) methylenetetrahydrofolate reductase ($NADPH_2$)	
1.5.1.20	5,10-Methylenetetrahydrofolic acid reductase Methylenetetrahydrofolate reductase (NADPH)	
1.7.99.5	Methylenetetrahydrofolic acid reductase 5,10-Methylenetetrahydrofolate reductase ($FADH_2$)	
1.5.99.9	5,10-Methylenetetrahydromethanopterin:coenzyme-F420 oxidoreductase Methylenetetrahydromethanopterin dehydrogenase	

EC-No.	Synonym Recommended Name	Vol., Page
1.5.99.9	Methylene tetrahydromethanopterin:coenzyme F420 oxidoreductase Methylenetetrahydromethanopterin dehydrogenase	
1.5.99.9	Methylenetetrahydromethanopterin dehydrogenase Methylenetetrahydromethanopterin dehydrogenase	
1.12.99.4	Methylenetetrahydromethanopterin dehydrogenase N^5,N^{10}-Methenyltetrahydromethanopterin hydrogenase	
1.7.99.5	5,10-Methylenetetrahydropteroylglutamate reductase 5,10-Methylenetetrahydrofolate reductase ($FADH_2$)	
3.1.1.61	Methylesterase CheB Protein-glutamate methylesterase	
3.2.1.139	4-O-Methylglucuronidase α-Glucosiduronase	
4.2.1.18	methylglutaconase methylglutaconyl-CoA hydratase	
4.2.1.18	methylglutaconyl-CoA hydratase methylglutaconyl-CoA hydratase	
4.2.1.18	3-methylglutaconyl CoA hydratase methylglutaconyl-CoA hydratase	
4.2.1.18	methylglutaconyl coenzyme A hydratase methylglutaconyl-CoA hydratase	
1.5.99.5	Methylglutamate dehydrogenase Methylglutamate dehydrogenase	
1.5.99.5	N-Methylglutamate dehydrogenase Methylglutamate dehydrogenase	
2.1.1.21	N-Methylglutamate synthase Methylamine-glutamate N-methyltransferase	
4.4.1.5	methylglyoxalase lactoylglutathione lyase	
1.2.1.23	Methylglyoxal dehydrogenase 2-Oxoaldehyde dehydrogenase (NAD^+)	
1.2.1.49	Methylglyoxal dehydrogenase 2-Oxoaldehyde dehydrogenase ($NADP^+$)	
1.1.1.78	Methylglyoxal reductase D-Lactaldehyde dehydrogenase	
4.2.3.3	methylglyoxal synthase methylglyoxal synthase	

EC-No.	Synonym Recommended Name	Vol., Page
4.2.99.11	methylglyoxal synthase methylglyoxal synthase	
4.2.3.3	methylglyoxal synthetase methylglyoxal synthase	
4.2.99.11	methylglyoxal synthetase methylglyoxal synthase	
4.4.1.5	methylglyoxylase lactoylglutathione lyase	
3.5.3.16	Methylguanidinase Methylguanidinase	
3.5.3.16	Methylguanidine amidinohydrolase Methylguanidinase	
3.5.3.16	Methylguanidine hydrolase Methylguanidinase	
2.1.1.33	7-Methylguanine transfer ribonucleate methylase tRNA (guanine-N^7-)-methyltransferase	
3.6.1.30	7-Methylguanosine-5'-triphospho-5'-polynucleotide 7-methylguanosine-5'-phosphohydrolase $m^7G(5)pppN$ diphosphatase	
3.5.1.39	N-Methylhexanamide amidohydrolase Alkylamidase	
3.5.2.14	N-Methylhydantoin amidohydrolase N-Methylhydantoinase (ATP-hydrolyzing)	
3.5.2.14	N-Methylhydantoinase N-Methylhydantoinase (ATP-hydrolyzing)	
3.5.2.14	N-Methylhydantoinase (ATP-hydrolyzing) N-Methylhydantoinase (ATP-hydrolyzing)	
3.5.2.14	N-Methylhydantoin hydrolase N-Methylhydantoinase (ATP-hydrolyzing)	
1.14.12.4	Methylhydroxypyridine carboxylate dioxygenase 3-Hydroxy-2-methylpyridine carboxylate dioxygenase	
1.14.12.4	Methylhydroxypyridinecarboxylate oxidase 3-Hydroxy-2-methylpyridine carboxylate dioxygenase	
3.5.2.14	N-Methylimidazolidine-2,4-dione amidohydrolase (ATP-hydrolysing) N-Methylhydantoinase (ATP-hydrolyzing)	

EC-No.	Synonym Recommended Name	Vol., Page
4.2.1.99	2-Methylisocitrate dehydratase 2-Methylisocitrate dehydratase	
4.1.3.30	2-Methylisocitrate lyase Methylisocitrate lyase	
4.1.3.30	Methylisocitrate lyase Methylisocitrate lyase	
5.3.3.6	Methylitaconate Δ-isomerase Methylitaconate Δ-isomerase	Vol. 1, Page 406
5.3.3.6	Methylitaconate Δ^2-Δ^3-isomerase Methylitaconate Δ-isomerase	Vol. 1, Page 406
5.3.3.6	Methylitaconate isomerase Methylitaconate Δ-isomerase	Vol. 1, Page 406
1.5.3.4	ε-N-Methyllysine demethylase N^6-Methyl-lysine oxidase	
1.2.1.27	Methylmalonate-semialdehyde dehydrogenase (acylating) Methylmalonate-semialdehyde dehydrogenase (acylating)	
5.4.99.2	Methylmalonyl-CoA-carbonyl mutase Methylmalonyl-CoA mutase	Vol. 1, Page 589
2.1.3.1	Methylmalonyl-CoA carboxyltransferase Methylmalonyl-CoA carboxyltransferase	
4.1.1.41	Methylmalonyl-CoA decarboxylase Methylmalonyl-CoA decarboxylase	
4.1.1.41	Methylmalonyl-CoA decarboxylase Methylmalonyl-CoA decarboxylase	
5.1.99.1	Methylmalonyl-CoA epimerase Methylmalonyl-CoA epimerase	Vol. 1, Page 179
5.4.99.2	Methylmalonyl-CoA mutase Methylmalonyl-CoA mutase	Vol. 1, Page 589
5.1.99.1	Methylmalonyl-CoA racemase Methylmalonyl-CoA epimerase	Vol. 1, Page 179
2.1.3.1	Methylmalonyl-CoA transcarboxylase Methylmalonyl-CoA carboxyltransferase	
4.1.1.41	Methylmalonyl-coenzyme A decarboxylase Methylmalonyl-CoA decarboxylase	
3.1.2.17	D-Methylmalonyl-coenzyme A hydrolase (S)-Methylmalonyl-CoA hydrolase	

EC-No.	Synonym Recommended Name	Vol., Page
2.1.3.1	Methylmalonyl CoA carboxyltransferase Methylmalonyl-CoA carboxyltransferase	
5.4.99.2	Methylmalonyl CoA mutase Methylmalonyl-CoA mutase	Vol. 1, Page 589
5.4.99.2	Methylmalonyl coenzyme A carbonylmutase Methylmalonyl-CoA mutase	Vol. 1, Page 589
2.1.3.1	Methylmalonyl coenzyme A carboxyltransferase Methylmalonyl-CoA carboxyltransferase	
6.3.4.9	Methylmalonyl coenzyme A holotranscarboxylase synthetase Biotin-[methylmalonyl-CoA-carboxyltransferase] ligase	Vol. 2, Page 613
5.4.99.2	Methylmalonyl coenzyme A mutase Methylmalonyl-CoA mutase	Vol. 1, Page 589
5.1.99.1	Methylmalonyl coenzyme A racemase Methylmalonyl-CoA epimerase	Vol. 1, Page 179
1.8.3.4	Methyl mercaptan oxidase Methanethiol oxidase	
1.8.3.4	Methylmercaptan oxidase Methanethiol oxidase	
3.3.1.2	Methylmethionine-sulfonium-salt hydrolase Adenosylmethionine hydrolase	
2.1.1.10	Methylmethionine:homocysteine methyltransferase Homocysteine S-methyltransferase	
2.1.1.10	S-Methylmethionine homocysteine transmethylase Homocysteine S-methyltransferase	
5.3.3.4	Methylmuconolactone isomerase Muconolactone Δ-isomerase	Vol. 1, Page 399
5.4.99.14	4-Methylmuconolactone methylisomerase 4-Carboxymethyl-4-methylbutenolide mutase	Vol. 1, Page 648
2.1.1.121	(6-O-Methylnorlaudanosoline)-5'-O-methyltransferase 6-O-Methylnorlaudanosoline 5'-O-methyltransferase	
2.1.1.121	6-O-Methylnorlaudanosoline 5'-O-methyltransferase 6-O-Methylnorlaudanosoline 5'-O-methyltransferase	
3.1.1.44	4-methyloxaloacetate esterase 4-methyloxaloacetate esterase	
1.3.1.17	3-Methyloxindole: $NADP^+$ oxidoreductase 3-Methyleneoxindole reductase	

EC-No.	Synonym Recommended Name	Vol., Page
1.13.12.11	Methylphenyltetrahydropyridine N-monooxygenase Methylphenyltetrahydropyridine N-monooxygenase	
2.7.7.57	N-Methylphosphoethanolamine cytidylyltransferase N-Methylphosphoethanolamine cytidylyltransferase	
3.1.3.14	methylphosphothioglycerate phosphatase methylphosphothioglycerate phosphatase	
2.1.1.37	Methylphosphotriester-DNA methyltransferase DNA (cytosine-5-)-methyltransferase	
5.4.99.13	2-Methylpropanoyl-CoA CoA-carbonylmutase Isobutyryl-CoA mutase	Vol. 1, Page 646
2.1.1.84	Methylquercetagetin 6-O-methyltransferase Methylquercetagetin 6-O-methyltransferase	
2.1.1.82	3-Methylquercitin 7-O-methyltransferase 3-Methylquercitin 7-O-methyltransferase	
1.6.6.7	Methyl red azoreductase Azobenzene reductase	
3.1.26.2	2'-O-Methyl RNase Ribonuclease α	
4.1.1.52	6-Methylsalicylate carboxy-lyase 6-Methylsalicylate decarboxylase	
4.1.1.52	6-Methylsalicylate decarboxylase 6-Methylsalicylate decarboxylase	
4.1.1.52	6-Methylsalicylic acid (2,6-cresotic acid) decarboxylase 6-Methylsalicylate decarboxylase	
4.1.1.52	6-Methylsalicylic acid decarboxylase 6-Methylsalicylate decarboxylase	
2.1.2.7	2-Methylserine hydroxymethyltransferase D-Alanine 2-hydroxymethyltransferase	
1.14.99.16	Methylsterol hydroxylase Methylsterol monooxygenase	
1.14.99.16	Methylsterol monooxygenase Methylsterol monooxygenase	
2.1.1.13	5-Methyltetrahydrofolate-homocysteine S-methyltransferase 5-Methyltetrahydrofolate-homocysteine S-methyltransferase	
2.1.1.13	5-Methyltetrahydrofolate-homocysteine transmethylase 5-Methyltetrahydrofolate-homocysteine S-methyltransferase	

EC-No.	Synonym Recommended Name	Vol., Page
2.1.1.13	Methyltetrahydrofolate-homocysteine vitamin B_{12} methyltransferase 5-Methyltetrahydrofolate-homocysteine S-methyltransferase	
1.7.99.5	5-Methyltetrahydrofolate:(acceptor) oxidoreductase 5,10-Methylenetetrahydrofolate reductase ($FADH_2$)	
2.1.1.13	N-Methyltetrahydrofolate:L-homocysteine methyltransferase 5-Methyltetrahydrofolate-homocysteine S-methyltransferase	
2.1.1.13	5-Methyltetrahydrofolate:L-homocysteine S-methyltransferase 5-Methyltetrahydrofolate-homocysteine S-methyltransferase	
1.7.99.5	5-Methyltetrahydrofolate:NAD^+ oxidoreductase 5,10-Methylenetetrahydrofolate reductase ($FADH_2$)	
1.7.99.5	5-Methyltetrahydrofolate:NAD oxidoreductase 5,10-Methylenetetrahydrofolate reductase ($FADH_2$)	
1.5.1.20	5-Methyltetrahydrofolate:$NADP^+$ oxidoreductase Methylenetetrahydrofolate reductase (NADPH)	
1.14.13.37	Methyltetrahydroprotoberberine 14-hydroxylase Methyltetrahydroprotoberberine 14-monooxygenase	
1.14.13.37	Methyltetrahydroprotoberberine 14-monooxygenase Methyltetrahydroprotoberberine 14-monooxygenase	
2.1.1.14	Methyltetrahydropteroylpolyglutamate:homocysteine methyltransferase 5-Methyltetrahydropteroyltriglutamate-homocysteine S-methyltransferase	
2.1.1.14	5-Methyltetrahydropteroyltri-L-glutamate:L-homocysteine S-methyltrans-ferase 5-Methyltetrahydropteroyltriglutamate-homocysteine S-methyltransferase	
2.1.1.14	5-Methyltetrahydropteroyltriglutamate-homocysteine S-methyltransferase 5-Methyltetrahydropteroyltriglutamate-homocysteine S-methyltransferase	
5.3.1.23	5-Methylthio-5-deoxy-D-ribose-1-phosphate ketol-isomerase 5-Methylthioribose-1-phosphate isomerase	Vol. 1, Page 351
2.4.2.28	5'-Methylthioadenosine:orthophosphate methylthio-D-ribosyltransferase 5'-Methylthioadenosine phosphorylase	
3.2.2.16	Methylthioadenosine methylthioribohydrolase Methylthioadenosine nucleosidase	
2.4.2.28	5'-Methylthioadenosine nucleosidase 5'-Methylthioadenosine phosphorylase	
3.2.2.16	5'-Methylthioadenosine nucleosidase Methylthioadenosine nucleosidase	

EC-No.	Synonym Recommended Name	Vol., Page
3.2.2.16	Methylthioadenosine nucleosidase Methylthioadenosine nucleosidase	
3.2.2.16	Methylthioadenosine nucleosidase Methylthioadenosine nucleosidase	
2.4.2.28	Methylthioadenosine nucleoside phosphorylase 5'-Methylthioadenosine phosphorylase	
2.4.2.28	5'-Methylthioadenosine phosphorylase 5'-Methylthioadenosine phosphorylase	
3.1.3.14	methylthiophosphoglycerate phosphatase methylphosphothioglycerate phosphatase	
5.3.1.23	5-Methylthioribose-1-phosphate isomerase 5-Methylthioribose-1-phosphate isomerase	Vol. 1, Page 351
2.7.1.100	5-Methylthioribose kinase 5-Methylthioribose kinase	
2.7.1.100	Methylthioribose kinase 5-Methylthioribose kinase	
2.1.1.41	Δ^{24}-Methyltransferase 24-Sterol C-methyltransferase	
2.1.1.9	S-Methyltransferase Thiol S-methyltransferase	
2.1.1.133	Methyltransferase (Pseudomonas denitrificans clone pXL151 gene cobF reduced) Precorrin-4 C^{11}-methyltransferase	
2.1.1.116	Methyltransferase, (S)-3'-hydroxy-N-methylcoclaurine 4'- 3'-Hydroxy-N-methyl-(S)-coclaurine 4'-O-methyltransferase	
2.1.1.128	Methyltransferase, (S)-coclaurine N- (RS)-Norcoclaurine 6-O-methyltransferase	
2.1.1.117	Methyltransferase, (S)-scoularine 9- (S)-Scoulerine 9-O-methyltransferase	
2.1.1.117	Methyltransferase, (S)-scoularine 9- (Coptis japonica clone pCJSMT) (S)-Scoulerine 9-O-methyltransferase	
2.1.1.95	Methyltransferase, γ-tocopherol Tocopherol O-methyltransferase	
2.1.1.119	Methyltransferase, 10-dihydroxydihydrosanguinarine O- 10-Hydroxydihydrosanguinarine 10-O-methyltransferase	

EC-No.	Synonym Recommended Name	Vol., Page
2.1.1.94	Methyltransferase, 11-demethyl-17-deacetylvindoline 11- 11-O-Demethyl-17-O-deacetylvindoline O-methyltransferase	
2.1.1.120	Methyltransferase, 12-hydroxydihydrochelirubine-12-O 12-Hydroxydihydrochelirubine 12-O-methyltransferase	
2.1.1.99	Methyltransferase, 16-methoxy-2,3-dihydro-3-hydroxytabersonine 16-Methoxy-2,3-dihydro-3-hydroxytabersonine N-methyltransferase	
2.1.1.41	Methyltransferase, Δ^{24}-sterol 24-Sterol C-methyltransferase	
2.1.1.114	Methyltransferase, 3,4-dihydroxy-5-hexaprenylbenzoate (Saccharomyces cerevisiae chromosome XV ORF HRF316) Hexaprenyldihydroxybenzoate methyltransferase	
2.1.1.97	Methyltransferase, 3-hydroxyanthranilate 4- 3-Hydroxyanthranilate 4-C-methyltransferase	
2.1.1.46	Methyltransferase, 4'-hydroxyisoflavone Isoflavone 4'-O-methyltransferase	
2.1.1.64	Methyltransferase, 5-demethylubiquinone-9 3-Demethylubiquinone-9 3-O-methyltransferase	
2.1.1.108	Methyltransferase, 6-hydroxymellein 6-Hydroxymellein O-methyltransferase	
2.1.1.121	Methyltransferase, 6-methylnorlaudanosoline 5'- 6-O-Methylnorlaudanosoline 5'-O-methyltransferase	
2.1.1.4	Methyltransferase, acetylserotonin Acetylserotonin N-methyltransferase	
2.1.1.18	Methyltransferase, acylpolysacharide 6-O Polysaccharide O-methyltransferase	
2.1.1.91	Methyltransferase, aldoxime O- Isobutyraldoxime O-methyltransferase	
2.1.1.111	Methyltransferase, anthranilate N- Anthranilate N-methyltransferase	
2.1.1.92	Methyltransferase, bergaptol Bergaptol O-methyltransferase	
2.1.1.68	Methyltransferase, caffeate Caffeate O-methyltransferase	
2.1.1.104	Methyltransferase, caffeoyl coenzyme A Caffeoyl-CoA O-methyltransferase	

EC-No.	Synonym Recommended Name	Vol., Page
2.1.1.60	Methyltransferase, calmodulin (lysine) Calmodulin-lysine N-methyltransferase	
2.1.1.22	Methyltransferase, carnosine N- Carnosine N-methyltransferase	
2.1.1.6	Methyltransferase, catechol Catechol O-methyltransferase	
2.1.1.118	Methyltransferase, columbamine Columbamine O-methyltransferase	
2.1.1.59	Methyltransferase, cytochrome c (lysine) Cytochrome-c-lysine N-methyltransferase	
2.1.1.123	Methyltransferase, cytochrome c (methionine) [Cytochrome-c]-methionine S-methyltransferase	
2.1.1.102	Methyltransferase, demethylmacrocin Demethylmacrocin O-methyltransferase	
2.1.1.109	Methyltransferase, demethylsterigmatocystin Demethylsterigmatocystin 6-O-methyltransferase	
2.1.1.54	Methyltransferase, deoxycytidylate Deoxycytidylate C-methyltransferase	
2.1.1.37	Methyltransferase, deoxyribonucleate DNA (cytosine-5-)-methyltransferase	
2.1.1.114	Methyltransferase, dihydroxyhexaprenylbenzoate (Saccharomyces cerevisiae clone pRS12A-2.5SB precursor reduced) Hexaprenyldihydroxybenzoate methyltransferase	
2.1.1.3	Methyltransferase, dimethylthetin-homocysteine Thetin-homocysteine S-methyltransferase	
2.1.1.98	Methyltransferase, diphthine Diphthine synthase	
2.1.1.15	Methyltransferase, fatty acid Fatty-acid O-methyltransferase	
2.1.1.75	Methyltransferase, flavonoid Apigenin 4'-O-methyltransferase	
2.1.1.76	Methyltransferase, flavonoid 3- Quercetin 3-O-methyltransferase	
2.1.1.83	Methyltransferase, flavonol 4'- 3,7-Dimethylquercitin 4'-O-methyltransferase	

EC-No.	Synonym Recommended Name	Vol., Page
2.1.1.84	Methyltransferase, flavonol 6- Methylquercetagetin 6-O-methyltransferase	
2.1.1.82	Methyltransferase, flavonol 7- 3-Methylquercitin 7-O-methyltransferase	
2.1.1.88	Methyltransferase, flavonol 8- 8-Hydroxyquercitin 8-O-methyltransferase	
2.1.1.69	Methyltransferase, furanocoumarin 5- 5-Hydroxyfuranocoumarin 5-O-methyltransferase	
2.1.1.70	Methyltransferase, furanocoumarin 8- 8-Hydroxyfuranocoumarin 8-O-methyltransferase	
2.1.1.20	Methyltransferase, glycine Glycine N-methyltransferase	
2.1.1.2	Methyltransferase, guanidinoacetate Guanidinoacetate N-methyltransferase	
2.1.1.8	Methyltransferase, histamine Histamine N-methyltransferase	
2.1.1.44	Methyltransferase, histidine N^{α}- Dimethylhistidine N-methyltransferase	
2.1.1.10	Methyltransferase, homocysteine Homocysteine S-methyltransferase	
2.1.1.47	Methyltransferase, indolepyruvate Indolepyruvate C-methyltransferase	
2.1.1.5	Methyltransferase, betaine-homocysteine *β*ine-homocysteine S-methyltransferase	
2.1.1.40	Methyltransferase, inositol, D-1- myo-Inositol 3-O-methyltransferase	
2.1.1.134	Methyltransferase, inositol 6-O myo-Inositol 6-O-methyltransferase	
2.1.1.39	Methyltransferase, inositol L-1- myo-Inositol 1-O-methyltransferase	
2.1.1.129	Methyltransferase, inositol L-1- (Mesembryanthemum crystallinum clone Imt1 reduced) myo-Inositol 4-O-methyltransferase	
2.1.1.26	Methyltransferase, iodophenol Iodophenol O-methyltransferase	

EC-No.	Synonym Recommended Name	Vol., Page
2.1.1.78	Methyltransferase, isoorientin 3'- Isoorientin 3'-O-methyltransferase	
2.1.1.65	Methyltransferase, licodione 2'-O- Licodione 2'-O-methyltransferase	
2.1.1.50	Methyltransferase, loganate Loganate O-methyltransferase	
2.1.1.42	Methyltransferase, luteolin Luteolin O-methyltransferase	
2.1.1.42	Methyltransferase, luteolin 3'-O- Luteolin O-methyltransferase	
2.1.1.101	Methyltransferase, macrocin Macrocin O-methyltransferase	
2.1.1.11	Methyltransferase, magnesium protoporphyrin Magnesium protoporphyrin O-methyltransferase	
2.1.1.67	Methyltransferase, mercaptopurine Thiopurine S-methyltransferase	
2.1.1.62	Methyltransferase, messenger ribonucleate 2'-O-methyladenosine N^{g}- mRNA (2'-O-methyladenosine-N^{6}-)-methyltransferase	
2.1.1.56	Methyltransferase, messenger ribonucleate guanine 7- mRNA (guanine-N^{7}-)-methyltransferase	
2.1.1.57	Methyltransferase, messenger ribonucleate nucleoside 2'- mRNA (nucleoside-2'-O-)-methyltransferase	
2.1.1.90	Methyltransferase, methanol-cobalamin Methanol-5-hydroxybenzimidazolylcobamide Co-methyltransferase	
2.1.1.12	Methyltransferase, methionine S- Methionine S-methyltransferase	
2.1.2.9	Methyltransferase, methionyl-transfer ribonucleate Methionyl-tRNA formyltransferase	
2.1.1.21	Methyltransferase, methylamine-glutamate Methylamine-glutamate N-methyltransferase	
2.1.1.74	Methyltransferase, methylenetetrahydrofolate-transfer ribonucleate uracil 5- Methylenetetrahydrofolate-tRNA-(uracil-5-)-methyltransferase ($FADH_2$-oxidizing)	
2.1.1.13	Methyltransferase, methyltetrahydrofolate-homocysteine 5-Methyltetrahydrofolate-homocysteine S-methyltransferase	

EC-No.	Synonym Recommended Name	Vol., Page
2.1.1.105	Methyltransferase, N-benzoyl-4-hydroxyanthranilate 4- N-Benzoyl-4-hydroxyanthranilate 4-O-methyltransferase	
2.1.1.1	Methyltransferase, nicotinamide Nicotinamide N-methyltransferase	
2.1.1.7	Methyltransferase, nicotinate Nicotinate N-methyltransferase	
2.1.1.28	Methyltransferase, noradrenaline N- Phenylethanolamine N-methyltransferase	
2.1.1.128	Methyltransferase, norlaudanosoline (RS)-Norcoclaurine 6-O-methyltransferase	
2.1.1.38	Methyltransferase, O-demethylpuromycin O-Demethylpuromycin O-methyltransferase	
2.1.1.42	Methyltransferase, *o*-diphenol *m*-O- Luteolin O-methyltransferase	
2.1.1.25	Methyltransferase, phenol Phenol O-methyltransferase	
2.1.1.17	Methyltransferase, phosphatidylethanolamine Phosphatidylethanolamine N-methyltransferase	
2.1.1.71	Methyltransferase, phosphatidylmonomethylethanolamine Phosphatidyl-N-methylethanolamine N-methyltransferase	
2.1.1.103	Methyltransferase, phosphodimethylethanolamine Phosphoethanolamine N-methyltransferase	
2.1.1.18	Methyltransferase, polysaccharide Polysaccharide O-methyltransferase	
2.1.1.130	Methyltransferase, precorrin 2 Precorrin-2 C^{20}-methyltransferase	
2.1.1.130	Methyltransferase, precorrin 2 (Methanococcus jannaschii gene MJ0771) Precorrin-2 C^{20}-methyltransferase	
2.1.1.130	Methyltransferase, precorrin 2- (Methanobacterium thermoautotrophicum strain ΔH gene MTH1348) Precorrin-2 C^{20}-methyltransferase	
2.1.1.130	Methyltransferase, precorrin 2- (Pseudomonas denitrificans clone pXL151) Precorrin-2 C^{20}-methyltransferase	
2.1.1.133	Methyltransferase, precorrin 2- (Pseudomonas denitrificans clone pXL151) Precorrin-4 C^{11}-methyltransferase	

EC-No.	Synonym Recommended Name	Vol., Page
2.1.1.132	methyltransferase, precorrin 6y Precorrin-6Y $C^{5,15}$-methyltransferase (decarboxylating)	
2.1.1.132	Methyltransferase, precorrin 6y (Methanococcus jannaschii) Precorrin-6Y $C^{5,15}$-methyltransferase (decarboxylating)	
2.1.1.132	Methyltransferase, precorrin 6y (Salmonella typhimurium strain LT2 gene cbiT) Precorrin-6Y $C^{5,15}$-methyltransferase (decarboxylating)	
2.1.1.125	Methyltransferase, protein (arginine) (Histone)-arginine N-methyltransferase	
2.1.1.124	Methyltransferase, protein (arginine) [Cytochrome-c]-arginine N-methyltransferase	
2.1.1.126	Methyltransferase, protein(arginine) [Myelin basic protein]-arginine N-methyltransferase	
2.1.1.77	Methyltransferase, protein (D-aspartate) Protein-L-isoaspartate(D-aspartate) O-methyltransferase	
2.1.1.85	Methyltransferase, protein (histidine) Protein-histidine N-methyltransferase	
2.1.1.43	Methyltransferase, protein (lysine) Histone-lysine N-methyltransferase	
2.1.1.100	Methyltransferase, protein C-terminal farnesylcysteine O- Protein-S-isoprenylcysteine O-methyltransferase	
2.1.1.80	Methyltransferase, protein O- Protein-glutamate O-methyltransferase	
2.1.1.53	Methyltransferase, putrescine Putrescine N-methyltransferase	
2.1.1.87	Methyltransferase, pyridine Pyridine N-methyltransferase	
2.1.1.48	Methyltransferase, ribosomal ribonucleate adenine 6- rRNA (adenine-N^6-)-methyltransferase	
2.1.1.66	Methyltransferase, ribosomal ribonucleate adenosine 2'- rRNA (adenosine-2'-O-)-methyltransferase	
2.1.1.51	Methyltransferase, ribosomal ribonucleate guanine 1- rRNA (guanine-N^1-)-methyltransferase	
2.1.1.52	Methyltransferase, ribosomal ribonucleate guanine 2- rRNA (guanine-N^2-)-methyltransferase	

EC-No.	Synonym Recommended Name	Vol., Page
2.1.1.127	Methyltransferase, ribulose diphosphate carboxylase large subunit(lysine) [Ribulose-bisphosphate-carboxylase]-lysine N-methyltransferase	
2.1.1.110	Methyltransferase, sterigmatocystin Sterigmatocystin 7-O-methyltransferase	
2.1.1.115	Methyltransferase, tetrahydrobenzylisoquinoline (RS)-1-Benzyl-1,2,3,4-tetrahydroisoquinoline N-methyltransferase	
2.1.1.122	Methyltransferase, tetrahydroberberine N- (S)-Tetrahydroprotoberberine N-methyltransferase	
2.1.1.89	Methyltransferase, tetrahydrocolumbamine Tetrahydrocolumbamine 2-O-methyltransferase	
2.1.1.86	Methyltransferase, tetrahydromethanopterin Tetrahydromethanopterin S-methyltransferase	
2.1.1.14	Methyltransferase, tetrahydropteroylglutamate-homocysteine transmethylase 5-Methyltetrahydropteroyltriglutamate-homocysteine S-methyltransferase	
2.1.1.96	Methyltransferase, thioether S- Thioether S-methyltransferase	
2.1.1.9	Methyltransferase, thiol Thiol S-methyltransferase	
2.1.1.61	Methyltransferase, transfer ribonucleate 5-methylaminomethyl-2-thiouridylate 5- tRNA(5-methylaminomethyl-2-thiouridylate)-methyltransferase	
2.1.1.36	Methyltransferase, transfer ribonucleate adenine 1- tRNA (adenine-N^1-)-methyltransferase	
2.1.1.29	Methyltransferase, transfer ribonucleate cytosine 5- tRNA (cytosine-5-)-methyltransferase	
2.1.1.31	Methyltransferase, transfer ribonucleate guanine 1- tRNA (guanine-N^1-)-methyltransferase	
2.1.1.32	Methyltransferase, transfer ribonucleate guanine 2- tRNA (guanine-N^2-)-methyltransferase	
2.1.1.33	Methyltransferase, transfer ribonucleate guanine 7- tRNA (guanine-N^7-)-methyltransferase	
2.1.1.34	Methyltransferase, transfer ribonucleate guanosine 2'- tRNA (guanosine-2'-O-)-methyltransferase	
2.1.1.35	Methyltransferase, transfer ribonucleate uracil 5- tRNA (uracil-5-)-methyltransferase	

EC-No.	Synonym Recommended Name	Vol., Page
2.1.1.19	Methyltransferase, trimethylsulfonium-tetrahydrofolate Trimethylsulfonium-tetrahydrofolate N-methyltransferase	
2.1.1.49	Methyltransferase, tryptamine Amine N-methyltransferase	
2.1.1.106	Methyltransferase, tryptophan 2- Tryptophan 2-C-methyltransferase	
2.1.1.27	Methyltransferase, tyramine N- Tyramine N-methyltransferase	
2.1.1.16	Methyltransferase, unsaturated phospholipid Methylene-fatty-acyl-phospholipid synthase	
2.1.1.107	Methyltransferase, uroporphyrinogen Uroporphyrin-III C-methyltransferase	
2.1.1.93	Methyltransferase, xanthotoxol Xanthotoxol O-methyltransferase	
2.1.1.109	O-Methyltransferase I Demethylsterigmatocystin 6-O-methyltransferase	
2.1.1.71	Methyltransferase II Phosphatidyl-N-methylethanolamine N-methyltransferase	
2.1.1.110	O-Methyltransferase II Sterigmatocystin 7-O-methyltransferase	
2.1.1.132	Methyltransferase precorrin 6y (Salmonella typhimurium strain LT2 gene cbiE) Precorrin-6Y $C^{5,15}$-methyltransferase (decarboxylating)	
2.4.1.17	4-Methylumbelliferone UDP-glucuronosyltransferase Glucuronosyltransferase	
3.1.1.56	4-Methylumbelliferyl-acetate acylhydrolase Methylumbelliferyl-acetate deacetylase	
3.1.1.56	Methylumbelliferyl-acetate deacetylase Methylumbelliferyl-acetate deacetylase	
3.1.6.1	4-Methylumbelliferyl sulfatase Arylsulfatase	
1.7.99.3	Methyl viologen-nitrite reductase Nitrite reductase	
2.1.1.112	Methytransferase, glucuronoxylan Glucuronoxylan 4-O-methyltransferase	

EC-No.	Synonym Recommended Name	Vol., Page
3.4.21.3	metridin metridin	
3.4.21.3	metridium proteinase A metridin	
6.1.1.10	MetRS Methionine-tRNA ligase	Vol. 2, Page 68
1.1.1.32	Mevaldate reductase Mevaldate reductase	
1.1.1.33	Mevaldate reductase (NADPH) Mevaldate reductase (NADPH)	
2.7.4.2	Mevalonate-5-phosphate kinase Phosphomevalonate kinase	
4.1.1.33	Mevalonate-5-pyrophosphate decarboxylase Diphosphomevalonate decarboxylase	
4.1.1.33	Mevalonate 5-diphosphate decarboxylase Diphosphomevalonate decarboxylase	
2.7.1.36	Mevalonate 5-phosphotransferase Mevalonate kinase	
2.7.1.36	Mevalonate kinase Mevalonate kinase	
2.7.4.2	Mevalonate phosphate kinase Phosphomevalonate kinase	
2.7.1.36	Mevalonate phosphokinase Mevalonate kinase	
2.7.1.36	Mevalonic acid kinase Mevalonate kinase	
2.7.4.2	Mevalonic acid phosphate kinase Phosphomevalonate kinase	
1.1.1.32	Mevalonic dehydrogenase Mevaldate reductase	
2.7.1.36	Mevalonic kinase Mevalonate kinase	
3.4.99.14	mexicanain mexicanain	
4.2.1.18	MG-CoA hydratase methylglutaconyl-CoA hydratase	

EC-No.	Synonym Recommended Name	Vol., Page
4.1.1.15	MGAD Glutamate decarboxylase	
2.6.1.73	MGAT Methionine-glyoxylate transaminase	
3.4.17.21	mGCP Glutamate carboxypeptidase II	
2.4.1.46	MGDG synthase 1,2-Diacylglycerol 3-β-galactosyltransferase	
2.4.1.208	MgGlcDAG (1->2) glucosyltransferase Diglucosyl diacylglycerol synthase	
2.7.1.94	MGK Acylglycerol kinase	
3.1.1.23	MGL acylglycerol lipase	
5.4.99.4	α-MG mutase 2-Methyleneglutarate mutase	Vol. 1, Page 599
4.1.3.17	MHK aldolase 4-hydroxy-4-methyl-2-oxoglutarate aldolase	
4.1.3.17	MHKG aldolase 4-hydroxy-4-methyl-2-oxoglutarate aldolase	
5.3.3.4	MI Muconolactone Δ-isomerase	Vol. 1, Page 399
5.5.1.4	MI-1-P synthase myo-Inositol-1-phosphate synthase	Vol. 1, Page 674
2.7.3.2	Mi-CK Creatine kinase	
4.1.3.30	MICL Methylisocitrate lyase	
3.4.24.3	microbial collagenase microbial collagenase	
3.4.24.39	Microbial neutral proteinase II Deuterolysin	
1.1.3.4	Microcid Glucose oxidase	
3.1.31.1	Micrococcal endonuclease Micrococcal nuclease	

EC-No.	Synonym Recommended Name	Vol., Page
3.1.31.1	Micrococcal nuclease Micrococcal nuclease	
3.1.4.7	micrococcal nuclease micrococcal nuclease	
4.2.99.18	Micrococcus luteus UV endonuclease DNA-(apurinic or apyrimidinic site) lyase	
3.4.17.16	Microsomal carboxypeptidase Membrane Pro-X carboxypeptidase	
1.14.14.1	Microsomal monooxygenase Unspecific monooxygenase	
4.1.2.24	Microsomal N-oxide dealkylase Dimethylaniline-N-oxide aldolase	
4.1.2.24	Microsomal oxidase II Dimethylaniline-N-oxide aldolase	
1.14.14.1	Microsomal P-450 Unspecific monooxygenase	
1.1.1.105	Microsomal retinol dehydrogenase Retinol dehydrogenase	
1.3.1.4	Microsomal steroid reductase (5α) Cortisone α-reductase	
2.7.1.123	Microtubule-associated protein 2 kinase Ca^{2+}/calmodulin-dependent protein kinase	
2.7.1.123	Microtubule associated protein kinase Ca^{2+}/calmodulin-dependent protein kinase	
5.3.2.1	MIF Phenylpyruvate tautomerase	Vol. 1, Page 367
3.4.22.17	mikro-calpain calpain	
2.7.3.2	MiMi-CK Creatine kinase	
3.5.1.61	Mimosinase Mimosinase	
3.5.1.61	Mimosine amidohydrolase Mimosinase	
4.2.99.15	L-Mimosine synthase L-Mimosine synthase	

EC-No.	Synonym Recommended Name	Vol., Page
3.4.24.59	MIP Mitochondrial intermediate peptidase	
5.5.1.4	MIPS myo-Inositol-1-phosphate synthase	Vol. 1, Page 674
5.5.1.4	MIP synthase myo-Inositol-1-phosphate synthase	Vol. 1, Page 674
3.2.2.22	Mirabilis antiviral protein rRNA N-glycosidase	
3.1.2.2	mitochondrial acyl-CoA thioesterase palmitoyl-CoA hydrolase	
3.6.1.34	Mitochondrial ATPase H^+-transporting ATP synthase	
3.4.24.64	Mitochondrial chelator-sensitive protease Mitochondrial processing peptidase	
1.6.5.3	Mitochondrial electron transport complex 1 NADH dehydrogenase (ubiquinone)	
1.6.5.3	Mitochondrial electron transport complex I NADH dehydrogenase (ubiquinone)	
1.10.2.2	Mitochondrial electron transport complex III Ubiquinol-cytochrome-c reductase	
3.4.24.59	Mitochondrial intermediate peptidase Mitochondrial intermediate peptidase	
3.4.24.64	Mitochondrial processing peptidase Mitochondrial processing peptidase	
3.4.99.41	mitochondrial processing peptidase mitochondrial processing peptidase	
3.4.24.64	Mitochondrial protein precursor-processing proteinase Mitochondrial processing peptidase	
2.7.1.37	Mitogen-activated protein kinase Protein kinase	
2.7.1.37	Mitogen-activated S6 kinase Protein kinase	
1.14.13.8	Mixed-function amine oxidase Dimethylaniline monooxygenase (N-oxide-forming)	
5.5.1.1	MLE Muconate cycloisomerase	Vol. 1, Page 660

EC-No.	Synonym Recommended Name	Vol., Page
5.5.1.1	MLEI Muconate cycloisomerase	Vol. 1, Page 660
5.5.1.7	MLE II Chloromuconate cycloisomerase	Vol. 1, Page 699
1.8.3.4	(MM)-oxidase Methanethiol oxidase	
2.7.3.2	MM-CK Creatine kinase	
3.4.21.39	MMCP-1 chymase	
3.4.24.23	MMP Matrilysin	
3.4.21.32	MMP-1 brachyurin	
3.4.24.7	MMP-1 interstitial collagenase	
3.4.24.3	MMP-1 microbial collagenase	
3.4.24.17	MMP-3 Stromelysin 1	
3.4.24.34	MMP-8 Neutrophil collagenase	
3.4.21.32	MMP-8 brachyurin	
3.4.24.7	MMP-8 interstitial collagenase	
3.4.24.3	MMP-8 microbial collagenase	
3.4.24.24	MMP 2 Gelatinase A	
3.4.24.23	MMP 7 Matrilysin	
3.4.24.35	MMP 9 Gelatinase B	
1.11.1.13	Mn(II):hydrogen-peroxide oxidoreductase Manganese peroxidase	

EC-No.	Synonym Recommended Name	Vol., Page
1.11.1.13	Mn-dependent peroxidase (NADH-oxidizing) Manganese peroxidase	
1.15.1.1	Mn-SOD Superoxide dismutase	
2.1.1.72	Modification methylase Site-specific DNA-methyltransferase (adenine-specific)	
2.1.1.113	Modification methylase Site-specific DNA-methyltransferase (cytosine-N^4-specific)	
2.1.1.73	Modification methylase Site-specific DNA-methyltransferase (cytosine-specific)	
3.2.2.22	Momorcochin-S rRNA N-glycosidase	
2.4.2.31	Mono(ADPribosyl)transferase $NAD(P)^+$-arginine ADP-ribosyltransferase	
2.4.2.36	Mono(ADPribosyl)transferase NAD^+-diphthamide ADP-ribosyltransferase	
2.4.1.184	Mono-β-D-galactosyldiacylglycerol:mono-β-D-galactosyldiacylglycerol β-D-galactosyltransferase Galactolipid galactosyltransferase	
2.3.1.134	Mono-β-D-galactosyldiacylglycerol:mono-β-D-galactosyldiacylglycerol acyltransferase Galactolipid O-acyltransferase	
2.3.1.22	Monoacylglycerol acyltransferase 2-Acylglycerol O-acyltransferase	
3.1.1.23	monoacylglycerol hydrolase acylglycerol lipase	
2.7.1.94	Monoacylglycerol kinase Acylglycerol kinase	
3.1.1.23	monoacylglycerol lipase acylglycerol lipase	
1.4.3.4	Monoamine:O_2 oxidoreductase (deaminating) Amine oxidase (flavin-containing)	
1.4.3.6	Monoamine oxidase Amine oxidase (copper-containing)	
1.4.3.4	Monoamine oxidase Amine oxidase (flavin-containing)	

EC-No.	Synonym Recommended Name	Vol., Page
1.4.3.15	D-Monoaminodicarboxylic acid oxidase D-Glutamate(D-aspartate) oxidase	
3.1.1.1	Monobutyrase Carboxylesterase	
1.6.5.4	Monodehydroascorbate reductase (NADH) Monodehydroascorbate reductase (NADH)	
4.2.1.51	monofunctional prephenate dehydratase prephenate dehydratase	
2.4.1.208	Monoglucosyl diacylglycerol (1->2) glucosyltransferase Diglucosyl diacylglycerol synthase	
2.4.1.208	Monoglucosyldiacylglycerol glucosyltransferase Diglucosyl diacylglycerol synthase	
3.1.1.23	monoglyceridase acylglycerol lipase	
2.3.1.22	Monoglyceride acyltransferase 2-Acylglycerol O-acyltransferase	
3.1.1.23	monoglyceride hydrolase acylglycerol lipase	
2.7.1.94	Monoglyceride kinase Acylglycerol kinase	
3.1.1.23	monoglyceride lipase acylglycerol lipase	
2.7.1.94	Monoglyceride phosphokinase Acylglycerol kinase	
3.1.1.23	monoglyceridyllipase acylglycerol lipase	
3.8.1.3	Monohaloacetate dehalogenase Haloacetate dehalogenase	
1.11.1.8	Monoiodotyrosine deiodinase Iodide peroxidase	
3.1.1.8	monomeric butyrylcholinesterase I cholinesterase	
3.1.1.8	monomeric butyrylcholinesterase II cholinesterase	
3.1.6.16	Monomethyl-sulfatase Monomethyl-sulfatase	

EC-No.	Synonym Recommended Name	Vol., Page
3.1.6.16	Monomethyl-sulfate sulfohydrolase Monomethyl-sulfatase	
1.5.99.1	Monomethylglycine dehydrogenase Sarcosine dehydrogenase	
3.1.3.6	3'-mononucleotidase 3'-nucleotidase	
3.1.3.5	5'-mononucleotidase 5'-nucleotidase	
1.14.18.1	Monophenol, dihydroxy-L-phenylalanine oxygen oxidoreductase Monophenol monooxygenase	
1.14.18.1	Monophenol,L-dopa:oxygen oxidoreductase Monophenol monooxygenase	
1.14.18.1	Monophenolase Monophenol monooxygenase	
1.14.18.1	Monophenol dihydroxyphenylalanine:oxygen oxidoreductase Monophenol monooxygenase	
1.14.18.1	Monophenol monooxidase Monophenol monooxygenase	
1.14.18.1	Monophenol monooxygenase Monophenol monooxygenase	
1.14.18.1	Monophenol oxidase Monophenol monooxygenase	
3.1.4.17	3': 5'-Monophosphate phosphodiesterase (cyclic CMP) 3',5'-Cyclic-nucleotide phosphodiesterase	
3.1.4.10	Monophosphatidylinositol phosphodiesterase 1-Phosphatidylinositol phosphodiesterase	
5.4.2.1	Monophosphoglycerate mutase Phosphoglycerate mutase	Vol. 1, Page 493
5.4.2.1	Monophosphoglyceromutase Phosphoglycerate mutase	Vol. 1, Page 493
1.14.99.34	monoprenyl isoflavone epoxidase monoprenyl isoflavone epoxidase	
1.14.99.34	monoprenyl isoflavone monooxygenase monoprenyl isoflavone epoxidase	
2.4.99.2	Monosialoganglioside sialyltransferase Monosialoganglioside sialyltransferase	

EC-No.	Synonym Recommended Name	Vol., Page
1.1.1.208	Monoterpenoid dehydrogenase (+)-Neomenthol dehydrogenase	
1.1.1.207	Monoterpenoid dehydrogenase (-)-Menthol dehydrogenase	
2.4.1.127	Monoterpenol β-glucosyltransferase Monoterpenol β-glucosyltransferase	
2.3.1.69	Monoterpenol O-acetyltransferase Monoterpenol O-acetyltransferase	
3.1.7.3	monoterpenyl-diphosphatase monoterpenyl-diphosphatase	
3.1.7.3	monoterpenyl-diphosphate diphosphohydrolase monoterpenyl-diphosphatase	
3.1.7.3	monoterpenyl-pyrophosphatase monoterpenyl-diphosphatase	
5.3.1.1	monoTIM Triose-phosphate isomerase	Vol. 1, Page 235
1.13.99.1	MOO myo-Inositol oxygenase	
3.4.24.16	MOP neurolysin	
1.1.1.218	Morphine 6-dehydrogenase Morphine 6-dehydrogenase	
1.1.1.218	Morphine:NAD(P)$^+$ 6-oxidoreductase Morphine 6-dehydrogenase	
2.4.1.17	Morphine glucuronyltransferase Glucuronosyltransferase	
5.4.2.1	MPGM Phosphoglycerate mutase	Vol. 1, Page 493
2.7.1.136	MPH(2') Macrolide 2'-kinase	
2.7.1.37	M phase-specific cdc2 kinase Protein kinase	
5.2.1.4	MPI Maleylpyruvate isomerase	Vol. 1, Page 206
1.11.1.7	MPO Peroxidase	

EC-No.	Synonym Recommended Name	Vol., Page
3.4.24.64	MPP Mitochondrial processing peptidase	
5.1.2.2	MR Mandelate racemase	Vol. 1, Page 72
2.1.1.62	mRNA (2'-O-methyladenosine-N^6-)-methyltransferase mRNA (2'-O-methyladenosine-N^6-)-methyltransferase	
2.1.1.58	mRNA (adenosine-2'-O-)-methyltransferase mRNA (adenosine-2'-O-)-methyltransferase	
2.1.1.56	mRNA (guanine-N^7-)-methyltransferase mRNA (guanine-N^7-)-methyltransferase	
2.1.1.57	mRNA (nucleoside-2'-O-)-methyltransferase mRNA (nucleoside-2'-O-)-methyltransferase	
2.7.7.50	mRNA capping enzyme mRNA guanylyltransferase	
2.7.7.50	mRNA guanylyltransferase mRNA guanylyltransferase	
6.3.3.2	MS 5-Formyltetrahydrofolate cyclo-ligase	Vol. 2, Page 535
4.1.3.2	MS Malate synthase	
4.1.1.52	6-MSA decarboxylase 6-Methylsalicylate decarboxylase	
3.4.24.70	68000-M Signalpeptide hydrolase Oligopeptidase A	
1.8.3.4	MT-oxidase Methanethiol oxidase	
2.1.1.90	MT 1 Methanol-5-hydroxybenzimidazolylcobamide Co-methyltransferase	
3.2.2.16	MTA nucleosidase Methylthioadenosine nucleosidase	
2.4.2.28	MTA phosphorylase 5'-Methylthioadenosine phosphorylase	
1.5.99.9	MTD Methylenetetrahydromethanopterin dehydrogenase	
1.1.1.255	MTD mannitol dehydrogenase	

EC-No.	Synonym Recommended Name	Vol., Page
3.2.1.141	MTHase 4-α-D-{(1→4)-α-D-Glucano}trehalose trehalohydrolase	
6.3.3.2	MTHFS 5-Formyltetrahydrofolate cyclo-ligase	Vol. 2, Page 535
1.2.4.1	MtPDC Pyruvate dehydrogenase (lipoamide)	
5.4.99.15	MTSase (1->4)-α-Glucan 1-α-glucosylmutase	Vol. 1, Page 652
3.2.1.110	Mucinaminylserine mucinaminidase Mucinaminylserine mucinaminidase	
2.4.1.147	Mucin core 3 β3-GlcNAc-transferase Acetylgalactosaminyl-O-glycosyl-glycoprotein β-1,3-N-acetylglucosaminyl-transferase	
2.4.99.3	Mucin sialyltransferase α-N-Acetylgalactosaminide α-2,6-sialyltransferase	
5.5.1.1	Muconate cycloisomerase Muconate cycloisomerase	Vol. 1, Page 660
5.5.1.1	Muconate cycloisomerase I Muconate cycloisomerase	Vol. 1, Page 660
5.5.1.7	Muconate cycloisomerase II Chloromuconate cycloisomerase	Vol. 1, Page 699
5.5.1.1	Muconate lactonizing enzyme Muconate cycloisomerase	Vol. 1, Page 660
5.3.3.4	Muconolactone Δ-isomerase Muconolactone Δ-isomerase	Vol. 1, Page 399
5.3.3.4	Muconolactone Δ-isomerase Muconolactone Δ-isomerase	Vol. 1, Page 399
5.3.3.4	Muconolactone isomerase Muconolactone Δ-isomerase	Vol. 1, Page 399
3.5.1.28	Mucopeptide amidohydrolase N-Acetylmuramoyl-L-alanine amidase	
3.2.1.17	Mucopeptide glucohydrolase Lysozyme	
3.2.1.17	Mucopeptide N-acetylmuramoylhydrolase Lysozyme	

EC-No.	Synonym Recommended Name	Vol., Page
3.4.23.23	Mucor acid protease Mucorpepsin	
3.4.23.23	Mucor acid proteinase Mucorpepsin	
3.4.23.23	Mucor aspartic proteinase Mucorpepsin	
3.4.23.23	Mucor miehei aspartic protease Mucorpepsin	
3.4.23.23	Mucor miehei aspartic proteinase Mucorpepsin	
3.4.23.23	Mucor miehei rennin Mucorpepsin	
3.4.23.23	Mucorpepsin Mucorpepsin	
3.4.23.23	Mucor pusillus emporase Mucorpepsin	
3.4.23.23	Mucor pusillus rennin Mucorpepsin	
3.4.23.23	Mucor rennin Mucorpepsin	
3.4.24.54	Mucrolysin Mucrolysin	
3.4.24.54	Mucrotoxin A Mucrolysin	
2.3.1.96	Mucus glycoprotein fatty acyltransferase Glycoprotein N-palmitoyltransferase	
3.4.99.46	multicatalytic endopeptidase complex multicatalytic endopeptidase complex	
3.4.25.1	multicatalytic endopeptidase complex proteasome endopeptidase complex	
3.4.25.1	multicatalytic protease proteasome endopeptidase complex	
3.4.25.1	multicatalytic proteinase proteasome endopeptidase complex	
3.2.1.17	Muramidase Lysozyme	

EC-No.	Synonym Recommended Name	Vol., Page
3.4.17.8	muramoylpentapeptide carboxypeptidase muramoylpentapeptide carboxypeptidase	
3.4.17.13	Muramoyltetrapeptide carboxypeptidase Muramoyltetrapeptide carboxypeptidase	
3.5.1.28	Murein hydrolase N-Acetylmuramoyl-L-alanine amidase	
6.3.2.13	MurE synthetase UDP-N-acetylmuramoylalanyl-D-glutamate-2,6-diaminopimelate ligase	Vol. 2, Page 473
2.4.1.1	Muscle phosphorylase Phosphorylase	
2.4.1.1	Muscle phosphorylase a and b Phosphorylase	
3.2.1.59	Mutanase Glucan endo-1,3-α-glucosidase	
5.1.3.3	Mutarotase Aldose 1-epimerase	Vol. 1, Page 113
5.4.99.4	Mutase, 2-methyleneglutarate 2-Methyleneglutarate mutase	Vol. 1, Page 599
5.4.99.3	Mutase, acetolactate 2-Acetolactate mutase	Vol. 1, Page 597
5.4.99.5	Mutase, chorismate Chorismate mutase	Vol. 1, Page 604
5.4.3.4	Mutase, D-α-lysine mutase D-Lysine 5,6-aminomutase	Vol. 1, Page 562
5.4.99.13	Mutase, isobutyryl coenzyme A Isobutyryl-CoA mutase	Vol. 1, Page 646
5.4.3.2	Mutase, lysine 2,3-amino- Lysine 2,3-aminomutase	Vol. 1, Page 553
5.4.1.1	Mutase, lysolecithin acyl- Lysolecithin acylmutase	Vol. 1, Page 488
5.4.99.1	Mutase, methylaspartate Methylaspartate mutase	Vol. 1, Page 582
5.4.99.2	Mutase, methylmalonyl coenzyme A Methylmalonyl-CoA mutase	Vol. 1, Page 589

EC-No.	Synonym Recommended Name	Vol., Page
5.4.1.2	Mutase, precorrin 8x Precorrin-8X methylmutase	Vol. 1, Page 490
5.4.99.9	Mutase, uridine diphosphogalactopyranose UDPgalactopyranose mutase	Vol. 1, Page 635
2.7.1.36	MVA kinase Mevalonate kinase	
3.4.21.12	Mycobacterium sorangium α-lytic proteinase α-lytic endopeptidase	
2.3.1.111	Mycocerosate synthase Mycocerosate synthase	
2.3.1.111	Mycocerosic acid synthase Mycocerosate synthase	
3.2.1.61	Mycodextranase Mycodextranase	
2.3.1.122	Mycolyltransferase, trehalose 6-monomycolate-trehalose Trehalose O-mycolyltransferase	
3.4.24.31	Mycolysin Mycolysin	
1.2.1.66	mycothiol-dependent formaldehyde dehydrogenase mycothiol-dependent formaldehyde dehydrogenase	
2.3.1.100	Myelin-proteolipid O-palmitoyltransferase Myelin-proteolipid O-palmitoyltransferase	
2.1.1.126	[Myelin basic protein]-arginine N-methyltransferase [Myelin basic protein]-arginine N-methyltransferase	
2.1.1.126	Myelin basic protein methylase I [Myelin basic protein]-arginine N-methyltransferase	
2.3.1.100	Myelin PLP acyltransferase Myelin-proteolipid O-palmitoyltransferase	
3.4.21.76	Myeloblastin Myeloblastin	
1.11.1.7	Myeloperoxidase Peroxidase	
3.1.3.56	D-myo-Inositol(1,4,5)/(1,3,4,5)-polyphosphate 5-phosphatase Inositol-1,4,5-trisphosphate 5-phosphatase	
3.1.3.25	myo-inositol-1(or 4)-monophosphate phosphohydrolase inositol-1 (or 4)-monophosphatase	

EC-No.	Synonym Recommended Name	Vol., Page
3.1.4.36	1-D-Myo-inositol-1,2-cyclic phosphate 2-inositolphosphohydrolase 1,2-Cyclic-inositol-phosphate phosphodiesterase	
3.1.3.62	D-myo-Inositol-1,3,4,5-tetrakisphosphate 3-phosphohydrolase Inositol-1,3,4,5-tetrakisphosphate 3-phosphatase	
3.1.3.65	D-myo-Inositol-1,3-bisphosphate 3-phosphohydrolase Inositol-1,3-bisphosphate 3-phosphatase	
3.1.3.61	D-myo-Inositol-1,4,5-trisphosphate 1-phosphohydrolase Inositol-1,4,5-trisphosphate 1-phosphatase	
3.1.3.56	myo-Inositol-1,4,5-trisphosphate 5-phosphatase Inositol-1,4,5-trisphosphate 5-phosphatase	
3.1.3.56	D-myo-Inositol-1,4,5-trisphosphate 5-phosphohydrolase Inositol-1,4,5-trisphosphate 5-phosphatase	
3.1.3.57	D-myo-Inositol-1,4-bisphosphate 1-phosphohydrolase Inositol-1,4-bisphosphate 1-phosphatase	
5.5.1.4	1L-myo-Inositol-1-phosphate lyase (isomerizing) myo-Inositol-1-phosphate synthase	Vol. 1, Page 674
3.1.3.25	L-myo-inositol-1-phosphate phosphatase inositol-1 (or 4)-monophosphatase	
5.5.1.4	1L-myo-Inositol-1-phosphate synthase myo-Inositol-1-phosphate synthase	Vol. 1, Page 674
5.5.1.4	L-myo-Inositol-1-phosphate synthase myo-Inositol-1-phosphate synthase	Vol. 1, Page 674
5.5.1.4	myo-Inositol-1-phosphate synthase myo-Inositol-1-phosphate synthase	Vol. 1, Page 674
5.5.1.4	myo-Inositol-1-P synthase myo-Inositol-1-phosphate synthase	Vol. 1, Page 674
3.1.3.66	D-myo-Inositol-3,4-bisphosphate 4-phosphohydrolase Inositol-3,4-bisphosphate 4-phosphatase	
3.1.3.8	myo-inositol-hexakisphosphate 3-phosphohydrolase 3-phytase	
3.1.3.26	myo-inositol-hexakisphosphate 6-phosphohydrolase 6-phytase	
2.7.1.134	1D-myo-Inositol-tetrakisphosphate 1-kinase 1D-myo-Inositol-tetrakisphosphate 1-kinase	
2.7.1.140	1D-myo-Inositol-tetrakisphosphate 5-kinase 1D-myo-Inositol-tetrakisphosphate 5-kinase	

EC-No.	Synonym Recommended Name	Vol., Page
2.7.1.127	1D-myo-Inositol-trisphosphate 3-kinase 1D-myo-Inositol-trisphosphate 3-kinase	
2.7.1.139	1D-myo-Inositol-trisphosphate 5-kinase 1D-myo-Inositol-trisphosphate 5-kinase	
2.7.1.133	1D-myo-Inositol-trisphosphate 6-kinase 1D-myo-Inositol-trisphosphate 6-kinase	
3.1.3.25	myo-inositol 1 (or 4) -monophosphatase inositol-1 (or 4)-monophosphatase	
3.1.3.56	D-myo-Inositol 1,4,5-triphosphate 5-phosphatase Inositol-1,4,5-trisphosphate 5-phosphatase	
3.1.3.56	L-myo-Inositol 1,4,5-trisphosphate-monoesterase Inositol-1,4,5-trisphosphate 5-phosphatase	
2.7.1.127	D-myo-Inositol 1,4,5-trisphosphate 3-kinase 1D-myo-Inositol-trisphosphate 3-kinase	
3.1.3.56	D-myo-Inositol 1,4,5-trisphosphate 5-phosphatase Inositol-1,4,5-trisphosphate 5-phosphatase	
2.7.1.64	myo-Inositol 1-kinase myo-Inositol 1-kinase	
2.1.1.39	myo-Inositol 1-methyltransferase myo-Inositol 1-O-methyltransferase	
2.1.1.39	myo-Inositol 1-O-methyltransferase myo-Inositol 1-O-methyltransferase	
3.1.3.25	myo-inositol 1-phosphatase inositol-1 (or 4)-monophosphatase	
5.5.1.4	L-myo-Inositol 1-phosphate synthetase myo-Inositol-1-phosphate synthase	Vol. 1, Page 674
3.1.4.36	D-Myo-inositol 1:2-cyclic phosphate 2-phosphohydrolase 1,2-Cyclic-inositol-phosphate phosphodiesterase	
1.1.1.18	myo-Inositol 2-dehydrogenase myo-Inositol 2-dehydrogenase	
2.1.1.40	myo-Inositol 3-O-methyltransferase myo-Inositol 3-O-methyltransferase	
2.1.1.129	myo-Inositol 4-O-methyltransferase myo-Inositol 4-O-methyltransferase	
2.1.1.134	myo-Inositol 6-O-methyltransferase myo-Inositol 6-O-methyltransferase	

EC-No.	Synonym Recommended Name	Vol., Page
2.1.1.134	myo-Inositol 6-O-methyltransferase myo-Inositol 6-O-methyltransferase	
1.1.1.18	myo-Inositol:NAD$^+$ 2-oxidoreductase myo-Inositol 2-dehydrogenase	
1.1.1.18	myo-Inositol:NAD$^+$ oxidoreductase myo-Inositol 2-dehydrogenase	
1.13.99.1	myo-Inositol:oxygen oxidoreductase myo-Inositol oxygenase	
1.1.1.18	myo-Inositol dehydrogenase myo-Inositol 2-dehydrogenase	
3.1.3.25	myo-inositol monophosphatase inositol-1 (or 4)-monophosphatase	
1.13.99.1	myo-Inositol oxygenase myo-Inositol oxygenase	
4.2.1.44	Myo-inosose-2 dehydratase Myo-inosose-2 dehydratase	
3.1.4.36	D-Myoinositol 1,2-cyclic phosphate 2-phosphohydrolase 1,2-Cyclic-inositol-phosphate phosphodiesterase	
2.7.1.64	Myoinositol kinase myo-Inositol 1-kinase	
2.7.4.3	Myokinase Adenylate kinase	
2.4.1.1	Myophosphorylase Phosphorylase	
2.7.1.129	Myosin-heavy-chain kinase Myosin-heavy-chain kinase	
3.1.3.53	Myosin-light-chain-phosphatase Myosin-light-chain-phosphatase	
3.1.3.53	[Myosin-light-chain]-phosphate phosphohydrolase Myosin-light-chain-phosphatase	
2.7.1.117	Myosin-light-chain kinase Myosin-light-chain kinase	
3.6.1.32	Myosin ATPase Myosin ATPase	
3.6.1.32	Myosin ATP phosphohydrolase (actin-translocating) Myosin ATPase	

EC-No.	Synonym Recommended Name	Vol., Page
2.7.1.129	Myosin heavy chain kinase Myosin-heavy-chain kinase	
2.7.1.129	Myosin I heavy-chain kinase Myosin-heavy-chain kinase	
2.7.1.129	Myosin II heavy-chain kinase Myosin-heavy-chain kinase	
2.7.1.117	Myosin kinase Myosin-light-chain kinase	
2.7.1.117	Myosin light-chain kinase Myosin-light-chain kinase	
3.1.3.53	Myosin light chain kinase phosphatase Myosin-light-chain-phosphatase	
2.7.1.117	Myosin light chain protein kinase Myosin-light-chain kinase	
3.1.3.53	Myosin phosphatase Myosin-light-chain-phosphatase	
4.2.3.15	myrcene synthase myrcene synthase	
2.3.1.97	Myristoyl-CoA-protein N-myristoyltransferase Glycylpeptide N-tetradecanoyltransferase	
1.14.99.31	myristoyl-CoA 11-(E) desaturase myristoyl-CoA 11-(E) desaturase	
1.14.99.32	myristoyl-CoA 11-(Z) desaturase myristoyl-CoA 11-(Z) desaturase	
2.3.1.155	myristoyl-CoA:acetylCoA C-myristoyltransferase acetyl-CoA C-myristoyltransferase	
3.1.2.19	Myristoyl-CoA thioesterase ADP-dependent medium-chain-acyl-CoA hydrolase	
2.3.1.97	Myristoyl-coenzyme A:protein N-myristoyl transferase Glycylpeptide N-tetradecanoyltransferase	
2.3.1.97	Myristoylating enzymes Glycylpeptide N-tetradecanoyltransferase	
2.3.1.97	Myristoyltransferase, protein N- Glycylpeptide N-tetradecanoyltransferase	
3.2.3.1	Myrosin Thioglucosidase	

EC-No.	Synonym Recommended Name	Vol., Page
3.2.3.1	Myrosinase Thioglucosidase	
3.4.99.34	mytilidase mytilidase	
3.4.21.12	Myxobacter α-lytic proteinase α-lytic endopeptidase	
3.4.24.32	Myxobacter β-lytic proteinase β-Lytic metalloendopeptidase	
3.4.21.12	Myxobacter 495 α-lytic proteinase α-lytic endopeptidase	
3.4.24.32	Myxobacter495 β-lytic proteinase β-Lytic metalloendopeptidase	
3.4.99.29	Myxobacter AL-1 proteinase I Myxobacter AL-1 proteinase I	
3.4.99.30	Myxobacter AL-1 proteinase II Myxobacter AL-1 proteinase II	
3.4.24.32	Myxobacterium sorangium β-lytic proteinase β-Lytic metalloendopeptidase	
3.1.3.5	e-N 5'-nucleotidase	
1.5.1.22	N(Carboxymethyl)-D-alanine:NAD^+ oxidoreductase (glycine-forming) Strombine dehydrogenase	
2.4.1.141	N,N'-Diacetylchitobiosylpyrophosphoryldolichol synthase N-Acetylglucosaminyldiphosphodolichol N-acetylglucosaminyltransferase	
1.6.6.12	N,N'-Dimethyl-*p*-aminoazobenzene oxide reductase 4-(Dimethylamino)phenylazobenzene reductase	
1.14.13.8	N,N,-Dimethylaniline,NADPH:oxygen oxidoreductase (N-oxide-forming) Dimethylaniline monooxygenase (N-oxide-forming)	
1.6.6.7	N,N-Dimethyl-4-phenylazoaniline azoreductase Azobenzene reductase	
4.1.2.24	N,N-Dimethylaniline-N-oxide formaldehyde-lyase Dimethylaniline-N-oxide aldolase	
1.14.13.8	N,N-Dimethylaniline monooxygenase Dimethylaniline monooxygenase (N-oxide-forming)	
3.5.1.56	N,N-Dimethylformamidase N,N-Dimethylformamidase	

EC-No.	Synonym Recommended Name	Vol., Page
3.5.1.56	N,N-Dimethylformamide amidohydrolase N,N-Dimethylformamidase	
1.5.99.2	N,N-Dimethylglycine:(acceptor) oxidoreductase (demethylating) Dimethylglycine dehydrogenase	
1.5.3.10	N,N-Dimethylglycine:oxygen oxidoreductase (demethylating) Dimethylglycine oxidase	
1.5.99.2	N,N-Dimethylglycine oxidase Dimethylglycine dehydrogenase	
2.3.1.56	N,O-Acetyltransferase Aromatic-hydroxylamine O-acetyltransferase	
3.2.1.17	N,O-Diacetylmuramidase Lysozyme	
3.5.1.77	D-N-α-Carbamoylase N-Carbamoyl-D-amino acid hydrolase	
3.2.1.53	β-D-N-acetylgalactosaminidase β-N-acetylgalactosaminidase	
3.2.1.52	β-D-N-acetylhexosaminidase β-N-acetylhexosaminidase	
3.1.3.5	c-N-I 5'-nucleotidase	
3.1.3.5	c-N-II 5'-nucleotidase	
3.5.4.19	N^1-(5-Phospho-D-ribosyl)-AMP 1,6-hydrolase Phosphoribosyl-AMP cyclohydrolase	
1.5.3.11	N^1-Acetylspermidine:oxygen oxidoreductase (deaminating) Polyamine oxidase	
3.5.1.48	N^1-Acetylspermidine amidohydrolase Acetylspermidine deacetylase	
1.5.1.6	N^{10}-Formyltetrahydrofolate dehydrogenase Formyltetrahydrofolate dehydrogenase	
2.1.2.9	N^{10}-Formyltetrahydrofolic-methionyl-transfer ribonucleic transformylase Methionyl-tRNA formyltransferase	
6.3.2.17	N^{10}-Formyltetrahydropteroyldiglutamate synthetase Tetrahydrofolylpolyglutamate synthase	Vol. 2, Page 488
3.4.13.4	N^2-(4-amino-butyryl)-L-lysine hydrolase X-Arg dipeptidase	

EC-No.	Synonym Recommended Name	Vol., Page
3.2.2.19	N^2-(ADP-D-ribosyl)-L-arginine ADPribosylhydrolase ADPribosylarginine hydrolase	
1.5.1.19	N^2-(D-1,3-Dicarboxypropyl)-L-arginine:NADP$^+$ oxidoreductase (L-arginine-forming) D-Nopaline dehydrogenase	
1.5.1.11	N^2-(D-1-Carboxyethyl)-L-arginine:NAD$^+$ oxidoreductase (L-arginine forming) D-Octopine dehydrogenase	
1.5.1.16	N^2-(D-1-Carboxyethyl)-L-lysine:NADP$^+$ oxidoreductase (L-lysine-forming) D-Lysopine dehydrogenase	
1.5.1.23	N^2-(D-1-Carboxyethyl)taurine:NAD$^+$ oxidoreductase (taurine forming) Tauropine dehydrogenase	
3.4.13.18	N^2-β-Alanylarginine dipeptidase Cytosol nonspecific dipeptidase	
3.4.13.15	N^2-β-alanylarginine dipeptidase N^2-β-alanylarginine dipeptidase	
2.6.1.69	N^2-Acetyl-L-ornithine:2-oxoglutarate 5-aminotransferase N^2-Acetylornithine 5-transaminase	
2.6.1.11	N^2-Acetyl-L-ornithine:2-oxoglutarate aminotransferase Acetylornithine transaminase	
2.3.1.35	N^2-Acetyl-L-ornithine:L-glutamate N-acetyltransferase Glutamate N-acetyltransferase	
3.5.1.16	N^2-Acetyl-L-ornithine amidohydrolase Acetylornithine deacetylase	
2.6.1.69	N^2-Acetylornithine 5-aminotransferase N^2-Acetylornithine 5-transaminase	
2.6.1.69	N^2-Acetylornithine 5-transaminase N^2-Acetylornithine 5-transaminase	
3.1.6.15	N^2-sulfo-3-sulfoglucosamine 3-sulfohydrolase N-sulfoglucosamine-3-sulfatase	
3.1.6.11	N^2-Sulfo-6-sulfo-D-glucosamine 6-sulfohydrolase N-Sulfoglucosamine-6-sulfatase	
1.7.99.6	N_2O reductase Nitrous-oxide reductase	
3.5.1.26	N^4-(β-N-acetyl-D-glucosaminyl)-L-asparagine amidohydrolase N^4-(β-N-acetylglucosaminyl)-L-asparaginase	

EC-No.	Synonym Recommended Name	Vol., Page
3.5.1.26	N^4-(β-N-acetylglucosaminyl)-L-asparaginase N^4-(β-N-acetylglucosaminyl)-L-asparaginase	
1.7.99.5	N^5,N^{10}-Methylenetetrahydrofolate reductase 5,10-Methylenetetrahydrofolate reductase ($FADH_2$)	
3.5.4.27	N^5,N^{10}-Methenyltetrahydromethanopterin cyclohydrolase Methenyltetrahydromethanopterin cyclohydrolase	
1.12.99.4	N^5,N^{10}-Methenyltetrahydromethanopterin hydrogenase N^5,N^{10}-Methenyltetrahydromethanopterin hydrogenase	
1.5.1.5	N^5,N^{10}-Methylenetetrahydrofolate dehydrogenase Methylenetetrahydrofolate dehydrogenase ($NADP^+$)	
1.7.99.5	N^5,N^{10}-Methylenetetrahydrofolate reductase 5,10-Methylenetetrahydrofolate reductase ($FADH_2$)	
1.5.99.11	N^5,N^{10}-methylenetetrahydromethanopterin:coenzyme-F420 oxidoreductase coenzyme F420-dependent N5,N^{10}-methenyltetrahydromethanopterin reductase	
1.5.99.9	N^5,N^{10}-Methylenetetrahydromethanopterin dehydrogenase Methylenetetrahydromethanopterin dehydrogenase	
1.12.99.4	N^5,N^{10}-Methylenetetrahydromethanopterin dehydrogenase N5,N^{10}-Methenyltetrahydromethanopterin hydrogenase	
1.12.99.4	N^5,N^{10}-Methylenetetrahydromethanopterin dehydrogenase (H_2^-forming) N^5,N^{10}-Methenyltetrahydromethanopterin hydrogenase	
1.5.1.24	N^5-(Carboxyethyl)ornithine synthase N^5-(Carboxyethyl)ornithine synthase	
1.5.1.24	N^5-(L-1-Carboxyethyl)-L-ornithine:$NADP^+$ oxidoreductase (L-ornithine-forming) N^5-(Carboxyethyl)ornithine synthase	
3.1.1.48	N^5-acyl-L-ornithine-ester hydrolase fusarinine-C ornithinesterase	
6.3.1.6	N^5-Ethyl-L-glutamine synthetase Glutamate-ethylamine ligase	Vol. 2, Page 381
6.3.3.2	N^5-Formyltetrahydrofolic acid cyclodehydrase 5-Formyltetrahydrofolate cyclo-ligase	Vol. 2, Page 535
2.1.1.13	N^5-Methyltetrahydrofolate-homocysteine cobalamin methyltransferase 5-Methyltetrahydrofolate-homocysteine S-methyltransferase	
2.1.1.13	N^5-Methyltetrahydrofolate methyltransferase 5-Methyltetrahydrofolate-homocysteine S-methyltransferase	

EC-No.	Synonym Recommended Name	Vol., Page
2.1.1.13	N^5-Methyltetrahydrofolic-homocysteine vitamin B_{12} transmethylase 5-Methyltetrahydrofolate-homocysteine S-methyltransferase	
1.14.11.8	N^6,N^6,N^6-Trimethyl-L-lysine,2-oxoglutarate:oxygen oxidoreductase (3-hydroxylating) Trimethyllysine dioxygenase	
1.5.1.7	N^6-(Glutar-2-yl)-L-lysine:NAD oxidoreductase (L-lysine-forming) Saccharopine dehydrogenase (NAD^+, L-lysine forming)	
1.5.1.9	N^6-(L-1,3-Dicarboxypropyl)-L-lysine:NAD^+ oxidoreductase (L-glutamate-forming) Saccharopine dehydrogenase (NAD^+, L-glutamate-forming)	
1.5.1.7	N^6-(L-1,3-Dicarboxypropyl)-L-lysine:NAD^+ oxidoreductase (L-lysine-forming) Saccharopine dehydrogenase (NAD^+, L-lysine forming)	
1.5.1.10	N^6-(L-1,3-Dicarboxypropyl)-L-lysine:$NADP^+$ oxidoreductase (L-glutamate-forming) Saccharopine dehydrogenase ($NADP^+$, L-glutamate forming)	
1.5.1.8	N^6-(L-1,3-Dicarboxypropyl)-L-lysine:$NADP^+$ oxidoreductase (L-lysine forming) Saccharopine dehydrogenase ($NADP^+$, L-lysine forming)	
2.6.1.65	N^6-Acetyl-β-lysine transaminase N^6-Acetyl-β-lysine transaminase	
3.5.1.17	N^6-Acyl-L-lysine amidohydrolase Acyl-lysine deacylase	
2.3.1.102	N^6-Hydroxylysine O-acetyltransferase N^6-Hydroxylysine O-acetyltransferase	
1.5.3.4	N^6-Methyl-L-lysine:oxygen oxidoreductase (demethylating) N^6-Methyl-lysine oxidase	
1.5.3.4	N^6-Methyl-lysine oxidase N^6-Methyl-lysine oxidase	
1.5.3.4	N^6-Methyllysine oxidase N^6-Methyl-lysine oxidase	
2.1.1.33	N^7-Methylguanine methylase tRNA (guanine-N^7-)-methyltransferase	
3.5.1.48	N^8-Acetylspermidine deacetylase Acetylspermidine deacetylase	
3.5.1.48	N^8-Monoacetylspermidine deacetylase Acetylspermidine deacetylase	

EC-No.	Synonym Recommended Name	Vol., Page
3.6.1.37	(Na^+ + K^+)-activated ATPase Na^+/K^+-transporting ATPase	
3.6.1.37	(Na^+ + K^+)-ATPase Na^+/K^+-transporting ATPase	
3.6.1.37	Na^+,K^+-ATPase Na^+/K^+-transporting ATPase	
3.6.1.37	Na^+/K^+-transporting ATPase Na^+/K^+-transporting ATPase	
3.6.1.37	Na,K-activated ATPase Na^+/K^+-transporting ATPase	
3.6.1.37	Na,K-Pump Na^+/K^+-transporting ATPase	
3.4.17.21	NAALADase Glutamate carboxypeptidase II	
3.4.17.21	NAALA dipeptidase, Glutamate carboxypeptidase II	
3.4.13.5	NacHDE X-methyl-His dipeptidase	
1.2.1.59	NAD(NADP)-dependent glyceraldehyde-3-phosphate dehydrogenase Glyceraldehyde-3-phosphate dehydrogenase (NAD(P)) (phosphorylating)	
2.4.2.31	$NAD(P)^+$-arginine ADP-ribosyltransferase $NAD(P)^+$-arginine ADP-ribosyltransferase	
3.2.2.6	$NAD(P)^+$glycohydrolase $NAD(P)^+$nucleosidase	
3.2.2.6	$NAD(P)^+$nucleosidase $NAD(P)^+$nucleosidase	
1.6.1.2	$NAD(P)^+$ transhydrogenase (AB-specific) $NAD(P)^+$ transhydrogenase (AB-specific)	
1.6.1.1	$NAD(P)^+$ transhydrogenase (B-specific) $NAD(P)^+$ transhydrogenase (B-specific)	
1.2.1.59	NAD(P)-dependent G3P dehydrogenase Glyceraldehyde-3-phosphate dehydrogenase (NAD(P)) (phosphorylating)	
1.2.1.59	NAD(P)-dependent glyceraldehyde-3-phosphate dehydrogenase Glyceraldehyde-3-phosphate dehydrogenase (NAD(P)) (phosphorylating)	
1.4.1.3	NAD(P)-glutamate dehydrogenase Glutamate dehydrogenase ($NAD(P)^+$)	

EC-No.	Synonym Recommended Name	Vol., Page
3.2.2.6	NAD(P)-glycohydrolase NAD(P)$^+$nucleosidase	
1.1.99.25	NAD(P)-independent quinate dehydrogenase Quinate dehydrogenase (pyrroloquinoline-quinone)	
1.4.1.19	NAD(P)-L-tryptophan dehydrogenase Tryptophan dehydrogenase	
1.18.1.4	NAD(P)-rubredoxin oxidoreductase rubredoxin-NAD(P) reductase	
3.2.2.6	NAD(P)ase NAD(P)$^+$nucleosidase	
1.4.1.3	NAD(P)H-dependent glutamate dehydrogenase Glutamate dehydrogenase (NAD(P)$^+$)	
1.6.8.1	NAD(P)H-FMN reductase NAD(P)H dehydrogenase (FMN)	
1.6.6.2	NAD(P)H-nitrate reductase Nitrate reductase (NAD(P)H)	
1.6.99.2	NAD(P)H-quinone dehydrogenase NAD(P)H dehydrogenase (quinone)	
1.6.5.5	NAD(P)H-quinone dehydrogenase NADPH:quinone reductase	
1.6.99.2	NAD(P)H-quinone oxidoreductase NAD(P)H dehydrogenase (quinone)	
1.6.5.5	NAD(P)H-quinone oxidoreductase NADPH:quinone reductase	
1.6.5.5	NAD(P)H-quinone reductase NADPH:quinone reductase	
1.6.99.2	NAD(P)H: (quinone-acceptor)oxidoreductase NAD(P)H dehydrogenase (quinone)	
1.6.99.2	NAD(P)H:(quinone-acceptor) oxidoreductase NAD(P)H dehydrogenase (quinone)	
1.6.5.5	NAD(P)H:(quinone-acceptor) oxidoreductase NADPH:quinone reductase	
1.6.6.7	NAD(P)H:1-(4'-sulfophenylazo)-2-naphthol oxidoreductase Azobenzene reductase	
1.6.6.10	NAD(P)H:4-nitroquinoline-N-oxide oxidoreductase Nitroquinoline-N-oxide reductase	

EC-No.	Synonym Recommended Name	Vol., Page
1.6.99.7	NAD(P)H:6,7-dihydropteridine oxidoreductase Dihydropteridine reductase	
1.17.1.1	NAD(P)H:CDP-4-keto-6-deoxy-D-glucose oxidoreductase CDP-4-dehydro-6-deoxyglucose reductase	
1.3.1.5	NAD(P)H: cucurbitacin B Δ^{23}-oxidoreductase Cucurbitacin Δ^{23}-reductase	
1.6.2.6	NAD(P)H:ferrileghemoglobin oxidoreductase Leghemoglobin reductase	
1.6.8.1	NAD(P)H:flavin oxidoreductase NAD(P)H dehydrogenase (FMN)	
1.6.8.1	NAD(P)H:FMN oxidoreductase NAD(P)H dehydrogenase (FMN)	
1.6.8.1	NAD(P)H:FMN oxidoreductase NAD(P)H dehydrogenase (FMN)	
1.6.99.2	NAD(P)H: menadione oxidoreductase NAD(P)H dehydrogenase (quinone)	
1.6.5.5	NAD(P)H:menadione oxidoreductase NADPH:quinone reductase	
1.6.6.13	NAD(P)H:N-hydroxy-2-acetamidofluorene N-oxidoreductase N-Hydroxy-2-acetamidofluorene reductase	
1.6.6.2	NAD(P)H:nitrate oxidoreductase Nitrate reductase (NAD(P)H)	
1.6.6.4	NAD(P)H:nitrite oxidoreductase Nitrite reductase (NAD(P)H)	
1.6.5.5	NAD(P)H:paraquat diaphorase NADPH:quinone reductase	
1.6.4.4	NAD(P)H:protein-disulfide oxidoreductase Protein-disulfide reductase (NAD(P)H)	
1.6.6.2	NAD(P)H bispecific nitrate reductase Nitrate reductase (NAD(P)H)	
1.6.99.2	NAD(P)H dehydrogenase NAD(P)H dehydrogenase (quinone)	
1.6.5.5	NAD(P)H dehydrogenase NADPH:quinone reductase	
1.6.8.1	NAD(P)H dehydrogenase (FMN) NAD(P)H dehydrogenase (FMN)	

EC-No.	Synonym Recommended Name	Vol., Page
1.6.99.2	NAD(P)H dehydrogenase (quinone) NAD(P)H dehydrogenase (quinone)	
1.6.5.5	NAD(P)H dehydrogenase (quinone) NADPH:quinone reductase	
1.6.99.2	NAD(P)H menadione reductase NAD(P)H dehydrogenase (quinone)	
1.6.5.5	NAD(P)H menadione reductase NADPH:quinone reductase	
1.6.5.5	NAD(P)H quinone reductase NADPH:quinone reductase	
3.2.2.6	NAD(P) nucleosidase NAD(P)$^+$nucleosidase	
1.6.1.2	NAD(P) transhydrogenase NAD(P)$^+$ transhydrogenase (AB-specific)	
1.6.1.1	NAD(P) transhydrogenase NAD(P)$^+$ transhydrogenase (B-specific)	
1.1.1.175	(NAD)-linked D-xylose dehydrogenase D-Xylose 1-dehydrogenase	
1.1.1.141	NAD$^+$-dependent 15-hydroxyprostaglandin dehydrogenase (type I) 15-Hydroxyprostaglandin dehydrogenase (NAD$^+$)	
1.5.1.15	NAD$^+$-dependent 5,10-methylenetetrahydrofolate dehydrogenase Methylenetetrahydrofolate dehydrogenase (NAD$^+$)	
1.1.1.14	NAD$^+$-dependent sorbitol dehydrogenase L-Iditol 2-dehydrogenase	
2.4.2.37	NAD$^+$-dinitrogen-reductase ADP-D-ribosyltransferase NAD$^+$-dinitrogen-reductase ADP-D-ribosyltransferase	
2.4.2.36	NAD$^+$-diphthamide ADP-ribosyltransferase NAD$^+$-diphthamide ADP-ribosyltransferase	
2.4.2.37	NAD$^+$:[dinitrogen reductase] (ADP-D-ribosyl)transferase NAD$^+$-dinitrogen-reductase ADP-D-ribosyltransferase	
2.4.2.31	NAD$^+$:L-arginine ADP-D-ribosyltransferase NAD(P)$^+$-arginine ADP-ribosyltransferase	
2.4.2.36	NAD$^+$:peptide-diphthamide N-(ADP-D-ribosyl)transferase NAD$^+$-diphthamide ADP-ribosyltransferase	

EC-No.	Synonym Recommended Name	Vol., Page
2.4.2.30	NAD$^+$:poly(adenine-diphosphate-D-ribosyl)-acceptor ADP-D-ribosyltransferase NAD$^+$ ADP-ribosyltransferase	
2.4.2.30	NAD$^+$ ADP-ribosyltransferase NAD$^+$ ADP-ribosyltransferase	
3.6.1.22	NAD$^+$ diphosphatase NAD$^+$ diphosphatase	
3.2.2.5	NAD$^+$ glycohydrolase NAD$^+$ nucleosidase	
2.7.1.23	NAD$^+$ kinase NAD$^+$ kinase	
3.2.2.5	NAD$^+$ nucleosidase NAD$^+$ nucleosidase	
1.5.1.9	NAD$^+$ oxidoreductase (L-2-aminoadipic-δ-semialdehyde and glutamate forming) Saccharopine dehydrogenase (NAD$^+$, L-glutamate-forming)	
3.6.1.22	NAD$^+$ phosphohydrolase NAD$^+$ diphosphatase	
3.6.1.22	NAD$^+$ pyrophosphatase NAD$^+$ diphosphatase	
2.7.7.1	NAD$^+$ pyrophosphorylase Nicotinamide-nucleotide adenylyltransferase	
6.3.1.5	NAD$^+$ synthase NAD$^+$ synthase	Vol. 2, Page 377
6.3.5.1	NAD$^+$ synthase (glutamine-hydroylysing) NAD$^+$ synthase (glutamine-hydroylysing)	Vol. 2, Page 651
6.3.1.5	NAD$^+$ synthetase NAD$^+$ synthase	Vol. 2, Page 377
6.3.5.1	NAD$^+$ synthetase (glutamine-hydrolysing) NAD$^+$ synthase (glutamine-hydroylysing)	Vol. 2, Page 651
1.1.1.8	NAD-α-glycerophosphate dehydrogenase Glycerol-3-phosphate dehydrogenase (NAD$^+$)	
1.1.1.30	NAD-β-hydroxybutyrate dehydrogenase 3-Hydroxybutyrate dehydrogenase	
1.2.1.3	NAD-aldehyde dehydrogenase Aldehyde dehydrogenase (NAD$^+$)	

EC-No.	Synonym Recommended Name	Vol., Page
2.4.2.31	NAD-arginine ADP-ribosyltransferase NAD(P)$^+$-arginine ADP-ribosyltransferase	
2.4.2.31	NAD-arginine mono-ADP-ribosyltransferase B NAD(P)$^+$-arginine ADP-ribosyltransferase	
2.4.2.37	NAD-azoferredoxin (ADPribose)transferase NAD$^+$-dinitrogen-reductase ADP-D-ribosyltransferase	
1.1.1.175	NAD-D-xylose dehydrogenase D-Xylose 1-dehydrogenase	
1.2.1.23	NAD-dependent α-ketoaldehyde dehydrogenase 2-Oxoaldehyde dehydrogenase (NAD$^+$)	
1.1.1.176	NAD-dependent 12α-hydroxysteroid dehydrogenase 12α-Hydroxysteroid dehydrogenase	
1.2.1.3	NAD-dependent 4-hydroxynonenal dehydrogenase Aldehyde dehydrogenase (NAD$^+$)	
1.5.1.15	NAD-dependent 5,10-methylenetetrahydrofolate dehydrogenase Methylenetetrahydrofolate dehydrogenase (NAD$^+$)	
1.4.1.1	NAD-dependent alanine dehydrogenase Alanine dehydrogenase	
1.1.1.1	NAD-dependent alcohol dehydrogenase Alcohol dehydrogenase	
1.2.1.3	NAD-dependent aldehyde dehydrogenase Aldehyde dehydrogenase (NAD$^+$)	
1.1.1.250	NAD-dependent arabitol dehydrogenase (Candida albicans clone pEM-BLYe23) D-Arabinitol 2-dehydrogenase	
1.1.1.48	NAD-dependent D-galactose dehydrogenase Galactose 1-dehydrogenase	
1.2.1.46	NAD-dependent formaldehyde dehydrogenase Formaldehyde dehydrogenase	
1.2.1.2	NAD-dependent formate dehydrogenase Formate dehydrogenase	
1.1.1.251	NAD-dependent Gat1P-dehydrogenase Galactitol-1-phosphate 5-dehydrogenase	
1.4.1.2	NAD-dependent glutamate dehydrogenase Glutamate dehydrogenase	

EC-No.	Synonym Recommended Name	Vol., Page
1.4.1.2	NAD-dependent glutamic dehydrogenase Glutamate dehydrogenase	
1.2.1.12	NAD-dependent glyceraldehyde phosphate dehydrogenase Glyceraldehyde-3-phosphate dehydrogenase	
1.1.1.8	NAD-dependent glycerol-3-phosphate dehydrogenase Glycerol-3-phosphate dehydrogenase (NAD^+)	
1.1.1.8	NAD-dependent glycerol phosphate dehydrogenase Glycerol-3-phosphate dehydrogenase (NAD^+)	
1.1.1.37	NAD-dependent malate dehydrogenase Malate dehydrogenase	
1.1.1.37	NAD-dependent malic dehydrogenase Malate dehydrogenase	
1.1.1.255	NAD-dependent mannitol dehydrogenase mannitol dehydrogenase	
1.1.1.14	NAD-dependent sorbitol dehydrogenase L-Iditol 2-dehydrogenase	
1.1.1.262	NAD-dependent threonine 4-phosphate dehydrogenase 4-hydroxythreonine-4-phosphate dehydrogenase	
1.1.1.9	NAD-dependent xylitol dehydrogenase D-Xylulose reductase	
2.4.2.36	NAD-diphthamide ADP-ribosyltransferase NAD-elongation factor 2 ADP-ribosyltransferase NAD^+-diphthamide ADP-ribosyltransferase	
1.18.1.3	NAD-ferredoxin reductase Ferredoxin-NAD^+ reductase	
1.2.1.2	NAD-formate dehydrogenase Formate dehydrogenase	
1.4.1.2	NAD-glutamate dehydrogenase Glutamate dehydrogenase	
1.1.99.5	NAD-independent glycerol-phosphate dehydrogenase Glycerol-3-phosphate dehydrogenase	
1.1.99.5	NAD-independent glycerol phosphate dehydrogenase Glycerol-3-phosphate dehydrogenase	
1.1.1.8	NAD-L-glycerol-3-phosphate dehydrogenase Glycerol-3-phosphate dehydrogenase (NAD^+)	

EC-No.	Synonym Recommended Name	Vol., Page
1.1.1.37	NAD-L-malate dehydrogenase Malate dehydrogenase	
1.1.1.27	NAD-lactate dehydrogenase L-Lactate dehydrogenase	
1.2.1.23	NAD-linked α-ketoaldehyde dehydrogenase 2-Oxoaldehyde dehydrogenase (NAD^+)	
1.4.1.1	NAD-linked alanine dehydrogenase Alanine dehydrogenase	
1.2.1.3	NAD-linked aldehyde dehydrogenase Aldehyde dehydrogenase (NAD^+)	
1.2.1.46	NAD-linked formaldehyde dehydrogenase Formaldehyde dehydrogenase	
1.2.1.1	NAD-linked formaldehyde dehydrogenase Formaldehyde dehydrogenase (glutathione)	
1.2.1.2	NAD-linked formate dehydrogenase Formate dehydrogenase	
1.4.1.2	NAD-linked glutamate dehydrogenase Glutamate dehydrogenase	
1.4.1.2	NAD-linked glutamic dehydrogenase Glutamate dehydrogenase	
1.1.1.8	NAD-linked glycerol 3-phosphate dehydrogenase Glycerol-3-phosphate dehydrogenase (NAD^+)	
1.1.1.6	NAD-linked glycerol dehydrogenase Glycerol dehydrogenase	
1.1.1.41	NAD-linked isocitrate dehydrogenase Isocitrate dehydrogenase (NAD^+)	
1.1.1.37	NAD-linked malate dehydrogenase Malate dehydrogenase	
1.1.1.38	NAD-linked malic enzyme Malate dehydrogenase (oxaloacetate-decarboxylating)	
1.1.1.37	NAD-malate dehydrogenase Malate dehydrogenase	
1.1.1.37	NAD-malic dehydrogenase Malate dehydrogenase	
1.1.1.39	NAD-malic enzyme Malate dehydrogenase (decarboxylating)	

EC-No.	Synonym Recommended Name	Vol., Page
1.1.1.38	NAD-malic enzyme Malate dehydrogenase (oxaloacetate-decarboxylating)	
1.1.1.116	NAD-pentose-dehydrogenase D-Arabinose 1-dehydrogenase	
2.4.2.30	NAD-protein ADP-ribosyltransferase NAD^+ ADP-ribosyltransferase	
1.1.1.14	NAD-sorbitol dehydrogenase L-Iditol 2-dehydrogenase	
1.1.1.141	NAD-specific 15-hydroxyprostaglandin dehydrogenase 15-Hydroxyprostaglandin dehydrogenase (NAD^+)	
1.1.1.1	NAD-specific aromatic alcohol dehydrogenase Alcohol dehydrogenase	
1.4.1.2	NAD-specific glutamate dehydrogenase Glutamate dehydrogenase	
1.4.1.2	NAD-specific glutamic dehydrogenase Glutamate dehydrogenase	
1.1.1.41	NAD-specific isocitrate dehydrogenase Isocitrate dehydrogenase (NAD^+)	
1.1.1.37	NAD-specific malate dehydrogenase Malate dehydrogenase	
1.1.1.39	NAD-specific malic enzyme Malate dehydrogenase (decarboxylating)	
1.1.1.38	NAD-specific malic enzyme Malate dehydrogenase (oxaloacetate-decarboxylating)	
1.1.1.204	NAD-xanthine dehydrogenase Xanthine dehydrogenase	
1.2.1.66	NAD/factor-dependent formaldehyde dehydrogenase mycothiol-dependent formaldehyde dehydrogenase	
2.4.2.36	NAD:elongation factor 2-adenosine diphosphate ribose-transferase NAD^+-diphthamide ADP-ribosyltransferase	
1.4.1.2	NAD:glutamate oxidoreductase Glutamate dehydrogenase	
1.5.1.23	NAD: tauropine oxidoreductase Tauropine dehydrogenase	
3.2.2.5	NADase NAD^+ nucleosidase	

EC-No.	Synonym Recommended Name	Vol., Page
1.1.1.41	NAD dependent isocitrate dehydrogenase Isocitrate dehydrogenase (NAD^+)	
3.2.2.5	NAD glycohydrolase NAD^+ nucleosidase	
1.1.1.53	NADH-20β-hydroxysteroid dehydrogenase 3α(or 20β)-Hydroxysteroid dehydrogenase	
1.1.1.1	NADH-alcohol dehydrogenase Alcohol dehydrogenase	
1.1.1.1	NADH-aldehyde dehydrogenase Alcohol dehydrogenase	
1.6.5.3	NADH-coenzyme Q oxidoreductase NADH dehydrogenase (ubiquinone)	
1.6.5.3	NADH-coenzyme Q reductase NADH dehydrogenase (ubiquinone)	
1.6.5.3	NADH-CoQ oxidoreductase NADH dehydrogenase (ubiquinone)	
1.6.5.3	NADH-CoQ reductase NADH dehydrogenase (ubiquinone)	
1.6.2.2	NADH-cytochrome-b_5 reductase Cytochrome-b_5 reductase	
1.6.2.2	NADH-cytochrome b_5 reductase Cytochrome-b_5 reductase	
1.4.1.1	NADH-dependent alanine dehydrogenase Alanine dehydrogenase	
1.6.4.1	NADH-dependent cystine reductase Cystine reductase (NADH)	
1.3.1.6	NADH-dependent fumarate reductase Fumarate reductase (NADH)	
1.4.1.14	NADH-dependent glutamate synthase Glutamate synthase (NADH)	
1.1.1.26	NADH-dependent glyoxylate reductase Glyoxylate reductase	
1.6.6.1	NADH-dependent nitrate reductase Nitrate reductase (NADH)	
1.6.99.7	NADH-dihydropteridine reductase Dihydropteridine reductase	

EC-No.	Synonym Recommended Name	Vol., Page
1.1.1.8	NADH-dihydroxyacetone phosphate reductase Glycerol-3-phosphate dehydrogenase (NAD^+)	
1.3.1.9	NADH-enoyl acyl carrier protein reductase Enoyl-[acyl-carrier-protein] reductase (NADH)	
1.18.1.3	NADH-ferredoxinNAP reductase Ferredoxin-NAD^+ reductase	
1.18.1.3	NADH-ferredoxin oxidoreductase Ferredoxin-NAD^+ reductase	
1.18.1.3	NADH-ferredoxin reductase Ferredoxin-NAD^+ reductase	
1.18.1.3	NADH-ferredoxinTOL reductase Ferredoxin-NAD^+ reductase	
1.6.2.2	NADH-ferricytochrome b_5 oxidoreductase Cytochrome-b_5 reductase	
1.3.1.6	NADH-Fumarate reductase Fumarate reductase (NADH)	
1.4.1.14	NADH-glutamate synthase Glutamate synthase (NADH)	
1.2.1.12	NADH-glyceraldehyde phosphate dehydrogenase Glyceraldehyde-3-phosphate dehydrogenase	
1.1.1.26	NADH-glyoxylate reductase Glyoxylate reductase	
1.6.6.11	NADH-hydroxylamine reductase Hydroxylamine reductase (NADH)	
1.6.99.8	NADH-linked aquacobalamin reductase Aquacobalamin reductase	
1.4.1.2	NADH-linked glutamate dehydrogenase Glutamate dehydrogenase	
1.6.99.3	NADH-menadione oxidoreductase NADH dehydrogenase	
1.6.99.2	NADH-menadione reductase NAD(P)H dehydrogenase (quinone)	
1.6.5.5	NADH-menadione reductase NADPH:quinone reductase	
1.6.5.5	NADH-menaquinone reductase NADPH:quinone reductase	

EC-No.	Synonym Recommended Name	Vol., Page
1.6.1.2	NADH-NADP-transhydrogenase $NAD(P)^+$ transhydrogenase (AB-specific)	
1.6.1.1	NADH-NADP-transhydrogenase $NAD(P)^+$ transhydrogenase (B-specific)	
1.6.6.1	NADH-nitrate reductase Nitrate reductase (NADH)	
1.6.6.4	NADH-nitrite oxidoreductase Nitrite reductase (NAD(P)H)	
1.7.99.3	NADH-nitrite reductase Nitrite reductase	
1.11.1.1	NADH-peroxidase NADH peroxidase	
1.6.5.3	NADH-Q6 oxidoreductase NADH dehydrogenase (ubiquinone)	
1.6.99.5	NADH-quinone oxidoreductase NADH dehydrogenase (quinone)	
1.18.1.1	NADH-rubredoxin oxidoreductase Rubredoxin-NAD^+ reductase	
1.18.1.1	NADH-rubredoxin reductase Rubredoxin-NAD^+ reductase	
1.6.5.4	NADH-semidehydroascorbate oxidoreductase Monodehydroascorbate reductase (NADH)	
1.3.1.9	NADH-specific enoyl-ACP reductase Enoyl-[acyl-carrier-protein] reductase (NADH)	
1.6.5.3	NADH-ubiquinone-1 reductase NADH dehydrogenase (ubiquinone)	
1.6.5.3	NADH-ubiquinone oxidoreductase NADH dehydrogenase (ubiquinone)	
1.6.5.3	NADH-ubiquinone reductase NADH dehydrogenase (ubiquinone)	
1.18.1.3	$NADH_2$-ferredoxin oxidoreductase Ferredoxin-NAD^+ reductase	
1.6.2.1	$NADH_2$ cytochrome c reductase $NADH_2$ cytochrome c reductase	
1.6.2.2	NADH 5α-reductase Cytochrome-b_5 reductase	

EC-No.	Synonym Recommended Name	Vol., Page
1.6.99.3	NADH:(acceptor) oxidoreductase NADH dehydrogenase	
1.6.99.5	NADH:(quinone-acceptor) oxidoreductase NADH dehydrogenase (quinone)	
1.6.5.4	NADH:AFR oxidoreductase Monodehydroascorbate reductase (NADH)	
1.6.5.4	NADH:ascorbate radical oxidoreductase Monodehydroascorbate reductase (NADH)	
1.6.4.7	NADH:asparagusate oxidoreductase Asparagusate reductase (NADH)	
1.6.4.10	NADH:CoA-disulfide oxidoreductase CoA-disulfide reductase (NADH)	
1.6.99.9	NADH:cob(II)alamin oxidoreductase Cob(II)alamin reductase	
1.6.99.8	NADH:cob(III)alamin oxidoreductase Aquacobalamin reductase	
1.6.99.13	NADH:Fe^{3+} oxidoreductase Ferric-chelate reductase	
1.6.2.2	NADH: ferricytochrome-b_5 oxidoreductase Cytochrome-b_5 reductase	
1.4.1.14	NADH: GOGAT Glutamate synthase (NADH)	
1.11.1.1	NADH:hydrogen-peroxide oxidoreductase NADH peroxidase	
1.6.6.11	NADH:hydroxylamine oxidoreductase Hydroxylamine reductase (NADH)	
1.1.1.81	NADH:hydroxypyruvate reductase Hydroxypyruvate reductase	
1.6.6.6	NADH:hyponitrite oxidoreductase Hyponitrite reductase	
1.6.4.1	NADH:L-cystine oxidoreductase Cystine reductase (NADH)	
1.6.5.4	NADH:monodehydroascorbate oxidoreductase Monodehydroascorbate reductase (NADH)	
1.6.6.1	NADH:nitrate oxidoreductase Nitrate reductase (NADH)	

EC-No.	Synonym Recommended Name	Vol., Page
1.6.5.5	NADH:quinone reductase NADPH:quinone reductase	
1.18.1.1	NADH: rubredoxin oxidoreductase Rubredoxin-NAD^+ reductase	
1.6.5.4	NADH:semidehydroascorbic acid oxidoreductase Monodehydroascorbate reductase (NADH)	
1.6.6.9	NADH:trimethylamine-N-oxide oxidoreductase Trimethylamine-N-oxide reductase	
1.6.5.3	NADH:ubiquinone oxidoreductase NADH dehydrogenase (ubiquinone)	
1.6.5.3	NADH:ubiquinone oxidoreductase complex NADH dehydrogenase (ubiquinone)	
1.6.5.3	NADH coenzyme Q1 reductase NADH dehydrogenase (ubiquinone)	
1.9.3.1	NADH cytochrome c oxidase Cytochrome-c oxidase	
1.6.99.3	NADH dehydrogenase NADH dehydrogenase	
1.6.99.3	β-NADH dehydrogenase NADH dehydrogenase	
1.6.99.5	NADH dehydrogenase (quinone) NADH dehydrogenase (quinone)	
1.6.5.3	NADH dehydrogenase (ubiquinone) NADH dehydrogenase (ubiquinone)	
1.6.99.3	NADH diaphorase NADH dehydrogenase	
1.16.1.2	NADH diferric transferrin reductase Diferric-transferrin reductase	
1.18.1.3	NADH flavodoxin oxidoreductase Ferredoxin-NAD^+ reductase	
1.6.99.3	NADH hydrogenase NADH dehydrogenase	
2.7.1.86	NADH kinase NADH kinase	
2.7.1.96	NADH kinase NADH kinase	

EC-No.	Synonym Recommended Name	Vol., Page
1.6.99.3	NADH oxidoreductase NADH dehydrogenase	
1.11.1.1	NADH peroxidase NADH peroxidase	
3.6.1.22	NADH pyrophosphatase NAD^+ diphosphatase	
1.6.1.2	NADH transhydrogenase $NAD(P)^+$ transhydrogenase (AB-specific)	
1.6.1.1	NADH transhydrogenase $NAD(P)^+$ transhydrogenase (B-specific)	
3.2.2.5	NAD hydrolase NAD^+ nucleosidase	
1.1.1.41	NAD isocitrate dehydrogenase Isocitrate dehydrogenase (NAD^+)	
1.1.1.41	NAD isocitric dehydrogenase Isocitrate dehydrogenase (NAD^+)	
2.7.1.23	NADK NAD^+ kinase	
2.7.1.23	NAD kinase NAD^+ kinase	
3.2.2.5	NAD nucleosidase NAD^+ nucleosidase	
1.1.1.231	$NADP^+$-dependent PGI_2-specific 15-hydroxyprostaglandin dehydrogenase 15-Hydroxyprostaglandin-I dehydrogenase ($NADP^+$)	
1.4.1.4	$NADP^+$-glutamate dehydrogenase Glutamate dehydrogenase ($NADP^+$)	
1.1.1.42	$NADP^+$-linked isocitrate dehydrogenase Isocitrate dehydrogenase ($NADP^+$)	
1.1.1.55	NADP-1,2-propanediol dehydrogenase Lactaldehyde reductase (NADPH)	
1.1.1.176	NADP-12α-hydroxysteroid dehydrogenase 12α-Hydroxysteroid dehydrogenase	
1.1.1.151	NADP-21-hydroxysteroid dehydrogenase 21-Hydroxysteroid dehydrogenase ($NADP^+$)	
1.2.1.4	NADP-acetaldehyde dehydrogenase Aldehyde dehydrogenase ($NADP^+$)	

EC-No.	Synonym Recommended Name	Vol., Page
1.1.1.2	NADP-alcohol dehydrogenase Alcohol dehydrogenase ($NADP^+$)	
1.1.1.2	NADP-aldehyde reductase Alcohol dehydrogenase ($NADP^+$)	
1.6.2.4	NADP-cytochrome c reductase NADPH-ferrihemoprotein reductase	
1.6.2.4	NADP-cytochrome reductase NADPH-ferrihemoprotein reductase	
1.2.1.49	NADP-dependent α-ketoaldehyde dehydrogenase 2-Oxoaldehyde dehydrogenase ($NADP^+$)	
1.1.1.197	NADP-dependent 15-hydroxyprostaglandin dehydrogenase 15-Hydroxyprostaglandin dehydrogenase ($NADP^+$)	
1.1.1.196	NADP-dependent 15-hydroxyprostaglandin dehydrogenase 15-Hydroxyprostaglandin-D dehydrogenase ($NADP^+$)	
1.1.1.201	NADP-dependent 7β-hydroxysteroid dehydrogenase 7β-Hydroxysteroid dehydrogenase ($NADP^+$)	
1.2.1.4	NADP-dependent aldehyde dehydrogenase Aldehyde dehydrogenase ($NADP^+$)	
1.1.1.2	NADP-dependent aldehyde reductase Alcohol dehydrogenase ($NADP^+$)	
1.1.1.200	NADP-dependent aldose 6-phosphate reductase Aldose-6-phosphate reductase (NADPH)	
1.2.1.43	NADP-dependent formate dehydrogenase Formate dehydrogenase ($NADP^+$)	
1.1.1.49	NADP-dependent glucose 6-phosphate dehydrogenase Glucose-6-phosphate 1-dehydrogenase	
1.1.1.119	NADP-dependent glucose dehydrogenase Glucose 1-dehydrogenase ($NADP^+$)	
1.4.1.4	NADP-dependent glutamate dehydrogenase Glutamate dehydrogenase ($NADP^+$)	
1.4.1.4	NADP-dependent glutamic dehydrogenase Glutamate dehydrogenase ($NADP^+$)	
1.2.1.13	NADP-dependent glyceraldehyde phosphate dehydrogenase Glyceraldehyde-3-phosphate dehydrogenase ($NADP^+$) (phosphorylating)	
1.1.1.42	NADP-dependent isocitrate dehydrogenase Isocitrate dehydrogenase ($NADP^+$)	

EC-No.	Synonym Recommended Name	Vol., Page
1.1.1.42	NADP-dependent isocitric dehydrogenase Isocitrate dehydrogenase (NADP$^+$)	
1.1.1.224	NADP-dependent mannose-6-P:mannitol-1-P oxidoreductase Mannose-6-phosphate 6-reductase	
1.1.1.64	NADP-dependent testosterone-17β-oxidoreductase Testosterone 17β-dehydrogenase (NADP$^+$)	
1.1.1.216	NADP-Farnesol dehydrogenase Farnesol dehydrogenase	
1.1.1.49	NADP-glucose-6-phosphate dehydrogenase Glucose-6-phosphate 1-dehydrogenase	
1.4.1.4	NADP-glutamate dehydrogenase Glutamate dehydrogenase (NADP$^+$)	
1.2.1.9	NADP-glyceraldehyde-3-phosphate dehydrogenase Glyceraldehyde-3-phosphate dehydrogenase (NADP$^+$)	
1.2.1.13	NADP-glyceraldehyde-3-phosphate dehydrogenase Glyceraldehyde-3-phosphate dehydrogenase (NADP$^+$) (phosphorylating)	
1.2.1.9	NADP-glyceraldehyde phosphate dehydrogenase Glyceraldehyde-3-phosphate dehydrogenase (NADP$^+$)	
1.2.1.13	NADP-glyceraldehyde phosphate dehydrogenase Glyceraldehyde-3-phosphate dehydrogenase (NADP$^+$) (phosphorylating)	
1.1.1.19	NADP-L-gulonate dehydrogenase Glucuronate reductase	
1.2.1.49	NADP-linked α-ketoaldehyde dehydrogenase 2-Oxoaldehyde dehydrogenase (NADP$^+$)	
1.1.1.231	NADP-linked 15-hydroxyprostaglandin (prostacyclin) dehydrogenase 15-Hydroxyprostaglandin-I dehydrogenase (NADP$^+$)	
1.1.1.197	NADP-linked 15-hydroxyprostaglandin dehydrogenase 15-Hydroxyprostaglandin dehydrogenase (NADP$^+$)	
1.1.1.196	NADP-linked 15-hydroxyprostaglandin dehydrogenase 15-Hydroxyprostaglandin-D dehydrogenase (NADP$^+$)	
1.1.1.36	NADP-linked acetoacetyl CoA reductase Acetoacetyl-CoA reductase	
1.1.1.119	NADP-linked aldohexose dehydrogenase Glucose 1-dehydrogenase (NADP$^+$)	
1.2.1.7	NADP-linked benzaldehyde dehydrogenase Benzaldehyde dehydrogenase (NADP$^+$)	

EC-No.	Synonym Recommended Name	Vol., Page
1.1.1.40	NADP-linked decarboxylating malic enzyme Malate dehydrogenase (oxaloacetate-decarboxylating, $NADP^+$)	
1.4.1.4	NADP-linked glutamate dehydrogenase Glutamate dehydrogenase ($NADP^+$)	
1.1.1.42	NADP-linked isocitrate dehydrogenase Isocitrate dehydrogenase ($NADP^+$)	
1.1.1.82	NADP-linked malate dehydrogenase Malate dehydrogenase ($NADP^+$)	
1.1.1.196	NADP-linked prostaglandin D_2 dehydrogenase 15-Hydroxyprostaglandin-D dehydrogenase ($NADP^+$)	
1.1.1.82	NADP-malate dehydrogenase Malate dehydrogenase ($NADP^+$)	
1.1.1.82	NADP-malic enzyme Malate dehydrogenase ($NADP^+$)	
1.1.1.40	NADP-malic enzyme Malate dehydrogenase (oxaloacetate-decarboxylating, $NADP^+$)	
1.1.1.115	NADP-pentose-dehydrogenase Ribose 1-dehydrogenase ($NADP^+$)	
1.1.1.196	NADP-PGD2 dehydrogenase 15-Hydroxyprostaglandin-D dehydrogenase ($NADP^+$)	
1.1.1.197	NADP-specific 15-hydroxyprostaglandin dehydrogenase 15-Hydroxyprostaglandin dehydrogenase ($NADP^+$)	
1.1.1.196	NADP-specific 15-hydroxyprostaglandin dehydrogenase 15-Hydroxyprostaglandin-D dehydrogenase ($NADP^+$)	
1.4.1.4	NADP-specific glutamate dehydrogenase Glutamate dehydrogenase ($NADP^+$)	
1.1.1.42	NADP-specific isocitrate dehydrogenase Isocitrate dehydrogenase ($NADP^+$)	
1.1.1.40	NADP-specific malate dehydrogenase Malate dehydrogenase (oxaloacetate-decarboxylating, $NADP^+$)	
1.1.1.40	NADP-specific malic enzyme Malate dehydrogenase (oxaloacetate-decarboxylating, $NADP^+$)	
1.6.4.5	NADP-thioredoxin reductase Thioredoxin reductase (NADPH)	
1.2.1.13	NADP-triose phosphate dehydrogenase Glyceraldehyde-3-phosphate dehydrogenase ($NADP^+$) (phosphorylating)	

EC-No.	Synonym Recommended Name	Vol., Page
1.18.1.2	NADP:ferredoxin oxidoreductase Ferredoxin-NADP$^+$ reductase	
1.11.1.1	NAD peroxidase NADH peroxidase	
1.4.1.4	NADP glutamic dehydrogenase Glutamate dehydrogenase (NADP$^+$)	
1.1.1.2	NADPH-aldehyde reductase Alcohol dehydrogenase (NADP$^+$)	
1.1.1.21	NADPH-aldopentose reductase Aldehyde reductase	
1.1.1.21	NADPH-aldose reductase Aldehyde reductase	
1.6.2.5	NADPH-cytochrome c_2 reductase NADPH-cytochrome c_2 reductase	
1.6.2.4	NADPH-cytochrome c oxidoreductase NADPH-ferrihemoprotein reductase	
1.6.2.4	NADPH-cytochrome c reductase NADPH-ferrihemoprotein reductase	
1.6.2.4	NADPH-cytochrome *p*-450 reductase NADPH-ferrihemoprotein reductase	
1.14.13.68	NADPH-cytochrome P_{450}reductase 4-hydroxyphenylacetaldehyde oxime monooxygenase	
1.6.99.1	NADPH-dehydrogenase NADPH dehydrogenase	
1.1.1.2	NADPH-dependent aldehyde reductase Alcohol dehydrogenase (NADP$^+$)	
1.1.1.184	NADPH-dependent carbonyl reductase Carbonyl reductase (NADPH)	
1.3.1.37	NADPH-dependent cis-enoyl-CoA reductase cis-2-Enoyl-CoA reductase (NADPH)	
1.6.2.4	NADPH-dependent cytochrome c reductase NADPH-ferrihemoprotein reductase	
1.6.6.12	NADPH-dependent DMAB N-oxide reductase 4-(Dimethylamino)phenylazobenzene reductase	
1.6.8.2	NADPH-dependent FMN reductase NADPH dehydrogenase (flavin)	

EC-No.	Synonym Recommended Name	Vol., Page
1.4.1.4	NADPH-dependent glutamate dehydrogenase Glutamate dehydrogenase ($NADP^+$)	
1.4.1.13	NADPH-dependent glutamate synthase Glutamate synthase (NADPH)	
1.1.1.177	NADPH-dependent glycerin-3-phosphate dehydrogenase Glycerol-3-phosphate 1-dehydrogenase ($NADP^+$)	
1.8.4.4	NADPH-dependent GSH-cystine transhydrogenase Glutathione-cystine transhydrogenase	
1.8.4.8	NADPH-dependent GSH-cystine transhydrogenase phosphoadenylyl-sulfate reductase (thioredoxin)	
1.1.1.224	NADPH-dependent M6P reductase Mannose-6-phosphate 6-reductase	
1.1.1.224	NADPH-dependent mannose 6-phosphate reductase Mannose-6-phosphate 6-reductase	
1.1.1.252	NADPH-dependent naphthol reductase Tetrahydroxynaphthalene reductase	
1.1.1.188	NADPH-dependent prostaglandin D_2 11-keto reductase Prostaglandin-F synthase	
1.8.1.2	NADPH-dependent sulfite reductase Sulfite reductase (NADPH)	
1.3.1.38	NADPH-dependent trans-2-enoyl-CoA reductase trans-2-Enoyl-CoA reductase (NADPH)	
1.6.99.1	NADPH-diaphorase NADPH dehydrogenase	
1.5.1.3	NADPH-dihydrofolate reductase Dihydrofolate reductase	
1.1.1.219	NADPH-dihydromyricetin reductase Dihydrokaempferol 4-reductase	
1.6.99.7	NADPH-dihydropteridine reductase Dihydropteridine reductase	
1.1.1.253	NADPH-dihydropteridine reductase Pteridine reductase	
1.6.2.4	NADPH-ferricytochrome c oxidoreductase NADPH-ferrihemoprotein reductase	
1.6.2.4	NADPH-ferrihemoprotein reductase NADPH-ferrihemoprotein reductase	

EC-No.	Synonym Recommended Name	Vol., Page
1.6.8.2	NADPH-flavin reductase NADPH dehydrogenase (flavin)	
1.6.8.2	NADPH-FMN reductase NADPH dehydrogenase (flavin)	
1.4.1.4	NADPH-glutamate dehydrogenase Glutamate dehydrogenase ($NADP^+$)	
1.4.1.13	NADPH-glutamate synthase Glutamate synthase (NADPH)	
1.6.4.2	NADPH-glutathione reductase Glutathione reductase (NADPH)	
1.1.1.79	NADPH-glyoxylate reductase Glyoxylate reductase ($NADP^+$)	
1.6.4.2	NADPH-GSSG reductase Glutathione reductase (NADPH)	
1.1.1.34	NADPH-hydroxymethylglutaryl-CoA reductase Hydroxymethylglutaryl-CoA reductase (NADPH)	
1.5.1.2	NADPH-L-Δ'-pyrroline carboxylic acid reductase Pyrroline-5-carboxylate reductase	
1.6.99.11	NADPH-linked aquacobalamin reductase Aquacobalamin reductase (NADPH)	
1.1.1.91	NADPH-linked benzaldehyde reductase Aryl-alcohol dehydrogenase ($NADP^+$)	
1.4.1.13	NADPH-linked glutamate synthase Glutamate synthase (NADPH)	
1.1.1.224	NADPH-mannose-6-P reductase Mannose-6-phosphate 6-reductase	
1.6.1.2	NADPH-NAD oxidoreductase $NAD(P)^+$ transhydrogenase (AB-specific)	
1.6.1.1	NADPH-NAD oxidoreductase $NAD(P)^+$ transhydrogenase (B-specific)	
1.6.1.2	NADPH-NAD transhydrogenase $NAD(P)^+$ transhydrogenase (AB-specific)	
1.6.1.1	NADPH-NAD transhydrogenase $NAD(P)^+$ transhydrogenase (B-specific)	
1.6.6.3	NADPH-nitrate reductase Nitrate reductase (NADPH)	

EC-No.	Synonym Recommended Name	Vol., Page
1.6.6.4	NADPH-nitrite reductase Nitrite reductase (NAD(P)H)	
1.3.1.33	NADPH-protochlorophyllide oxidoreductase Protochlorophyllide reductase	
1.3.1.33	NADPH-protochlorophyllide reductase Protochlorophyllide reductase	
1.1.1.100	NADPH-specific 3-oxoacyl-[acylcarrier protein]reductase 3-Oxoacyl-[acyl-carrier-protein] reductase	
1.6.99.7	NADPH-specific dihydropteridine reductase Dihydropteridine reductase	
1.6.8.2	NADPH-specific FMN reductase NADPH dehydrogenase (flavin)	
1.8.1.2	NADPH-sulfite reductase Sulfite reductase (NADPH)	
1.6.4.5	NADPH-thioredoxin reductase Thioredoxin reductase (NADPH)	
1.1.1.113	NADPH-xylose reductase L-Xylose 1-dehydrogenase	
1.5.1.27	NADPH 1,2-dehydroreticuline reductase 1,2-Dehydroreticulinium reductase (NADPH2)	
1.6.99.1	$NADPH_2$-dehydrogenase NADPH dehydrogenase	
1.6.6.7	$NADPH_2^-$dependent azoreductase Azobenzene reductase	
1.3.1.39	NADPH 2-enoyl Co A reductase Enoyl-[acyl-carrier-protein] reductase (NADPH, A-specific)	
1.3.1.10	NADPH 2-enoyl Co A reductase Enoyl-[acyl-carrier-protein] reductase (NADPH, B-specific)	
1.6.5.6	NADPH2:p-benzoquinone oxidoreductase *p*-benzoquinone reductase (NADPH)	
1.6.5.5	NADPH2:quinone oxidoreductase NADPH:quinone reductase	
1.6.99.1	NADPH:(acceptor) oxidoreductase NADPH dehydrogenase	
1.6.99.6	NADPH:(quinone-acceptor)oxidoreductase NADPH dehydrogenase (quinone)	

EC-No.	Synonym Recommended Name	Vol., Page
1.3.1.51	NADPH:2'-hydroxydaidzein oxidoreductase 2'-Hydroxydaidzein reductase	
1.3.1.45	NADPH:2'-hydroxyisoflavone oxidoreductase 2'-Hydroxyisoflavone reductase	
1.6.6.7	NADPH:4(dimethylamino)azobenzene oxidoreductase Azobenzene reductase	
1.6.6.12	NADPH:4-(dimethylamino)phenylazobenzene oxidoreductase 4-(Dimethylamino)phenylazobenzene reductase	
1.3.1.4	NADPH:Δ^4-3-Oxosteroid-5α-oxidoreductase Cortisone α-reductase	
1.1.1.36	NADPH:acetoacetyl-CoA reductase Acetoacetyl-CoA reductase	
1.6.99.11	NADPH:aquacob(III)alamin oxidoreductase Aquacobalamin reductase (NADPH)	
1.6.4.9	NADPH:bis-γ-glutamylcystine oxidoreductase Bis-γ-glutamylcystine reductase (NADPH)	
1.6.4.6	NADPH:CoA-glutathione oxidoreductase CoA-glutathione reductase (NADPH)	
1.6.99.12	NADPH:cyanocob(III)alamin oxidoreductase (cyanide-eliminating) Cyanocobalamin reductase (NADPH, cyanide-eliminating)	
1.18.1.2	NADPH:ferredoxin oxidoreductase Ferredoxin-$NADP^+$ reductase	
1.6.2.5	NADPH:ferricytochrome-c2 oxidoreductase NADPH-cytochrome c2 reductase	
1.6.2.4	NADPH:ferrihemoprotein oxidoreductase NADPH-ferrihemoprotein reductase	
1.6.8.2	NADPH:flavin oxidoreductase NADPH dehydrogenase (flavin)	
1.14.13.21	NADPH:flavonoid-3'-hydroxylase Flavonoid 3'-monooxygenase	
1.6.6.8	NADPH:GMP oxidoreductase (deaminating) GMP reductase	
1.6.6.8	NADPH:guanosine-5'-phosphate oxidoreductase (deaminating) GMP reductase	
1.11.1.2	NADPH:hydrogen-peroxide oxidoreductase NADPH peroxidase	

EC-No.	Synonym Recommended Name	Vol., Page
1.6.1.2	NADPH: NAD^+ oxidoreductase (AB-specific) $NAD(P)^+$ transhydrogenase (AB-specific)	
1.6.1.1	NADPH: NAD^+ oxidoreductase (B-specific) $NAD(P)^+$ transhydrogenase (B-specific)	
1.6.1.2	NADPH:NAD^+ transhydrogenase $NAD(P)^+$ transhydrogenase (AB-specific)	
1.6.1.1	NADPH:NAD^+ transhydrogenase $NAD(P)^+$ transhydrogenase (B-specific)	
1.6.6.3	NADPH:nitrate oxidoreductase Nitrate reductase (NADPH)	
1.6.6.3	NADPH:nitrate reductase Nitrate reductase (NADPH)	
1.6.4.2	NADPH:oxidized-glutathione oxidoreductase Glutathione reductase (NADPH)	
1.6.4.5	NADPH:oxidized-thioredoxin oxidoreductase Thioredoxin reductase (NADPH)	
1.6.5.5	NADPH:quinone oxidoreductase NADPH:quinone reductase	
1.6.5.5	NADPH:quinone reductase NADPH:quinone reductase	
1.6.8.2	NADPH:riboflavin oxidoreductase NADPH dehydrogenase (flavin)	
1.6.4.8	NADPH:trypanothione oxidoreductase Trypanothione reductase	
1.6.99.1	NADPH dehydrogenase NADPH dehydrogenase	
1.6.8.2	NADPH dehydrogenase (flavin) NADPH dehydrogenase (flavin)	
1.6.99.6	NADPH dehydrogenase (quinone) NADPH dehydrogenase (quinone)	
1.6.99.1	NADPH diaphorase NADPH dehydrogenase	
1.6.5.5	NADPH DT-diaphorase NADPH:quinone reductase	
1.6.99.6	NADPH oxidase NADPH dehydrogenase (quinone)	

EC-No.	Synonym Recommended Name	Vol., Page
1.11.1.2	NADPH peroxidase NADPH peroxidase	
1.1.1.42	NADP isocitric dehydrogenase Isocitrate dehydrogenase ($NADP^+$)	
1.4.1.4	NADP L-glutamate dehydrogenase Glutamate dehydrogenase ($NADP^+$)	
1.1.1.82	NADP malate dehydrogenase Malate dehydrogenase ($NADP^+$)	
1.11.1.2	NADP peroxidase NADPH peroxidase	
3.6.1.22	NADP pyrophosphatase NAD^+ diphosphatase	
2.4.2.19	NAD pyrophosphorylase Nicotinate-nucleotide pyrophosphorylase (carboxylating)	
6.3.1.5	NAD synthetase NAD^+ synthase	Vol. 2, Page 377
6.3.5.1	NAD synthetase (glutamine) NAD^+ synthase (glutamine-hydroylysing)	Vol. 2, Page 651
1.6.1.2	NAD transhydrogenase $NAD(P)^+$ transhydrogenase (AB-specific)	
1.6.1.1	NAD transhydrogenase $NAD(P)^+$ transhydrogenase (B-specific)	
3.4.21.62	Nagarse Subtilisin	
4.1.3.3	NALase N-acetylneuraminate lyase	
1.1.1.218	Naloxone reductase Morphine 6-dehydrogenase	
2.4.1.90	NAL synthetase N-Acetyllactosamine synthase	
2.7.7.18	NaMN-ATase Nicotinate-nucleotide adenylyltransferase	
4.1.3.19	NANA condensing enzyme N-acetylneuraminate synthase	
4.1.3.3	NANA lyase N-acetylneuraminate lyase	

EC-No.	Synonym Recommended Name	Vol., Page
3.1.1.72	Naphthal AS-D chloroacetate deacetylase Acetylxylan esterase	
1.14.12.12	Naphthalene,NADH:oxygen oxidoreductase (1,2-hydroxylating) Naphthalene 1,2-dioxygenase	
1.14.12.12	Naphthalene 1,2-dioxygenase Naphthalene 1,2-dioxygenase	
1.3.1.29	Naphthalene dihydrodiol dehydrogenase cis-1,2-Dihydro-1,2-dihydroxynaphthalene dehydrogenase	
1.14.12.12	Naphthalene dioxygenase Naphthalene 1,2-dioxygenase	
1.14.12.12	Naphthalene oxygenase Naphthalene 1,2-dioxygenase	
4.1.3.36	naphthoate synthase naphthoate synthase	
2.4.1.17	1-Naphthol-UDP-glucuronosyltransferase Glucuronosyltransferase	
2.4.1.17	1-Naphthol glucuronyltransferase Glucuronosyltransferase	
2.8.2.1	1-Naphthol phenol sulfotransferase Aryl sulfotransferase	
1.1.1.252	Naphthol reductase Tetrahydroxynaphthalene reductase	
2.8.2.1	2-Naphtholsulfotransferase Aryl sulfotransferase	
1.14.99.27	Naphthoquinone-hydroxylase Juglone 3-monooxygenase	
1.14.99.27	Naphthoquinone hydroxylase Juglone 3-monooxygenase	
1.6.99.2	Naphthoquinone reductase NAD(P)H dehydrogenase (quinone)	
1.6.5.5	Naphthoquinone reductase NADPH:quinone reductase	
2.3.1.5	β-Naphthylamine N-acetyltransferase Arylamine N-acetyltransferase	
3.4.24.61	Nardilysin Nardilysin	

EC-No.	Synonym Recommended Name	Vol., Page
1.14.11.9	Naringenin,2-oxoglutarate:oxygen oxidoreductase (3-hydroxylating) Naringenin 3-dioxygenase	
2.3.1.74	Naringenin-chalcone synthase Naringenin-chalcone synthase	
1.14.11.9	Naringenin 3-dioxygenase Naringenin 3-dioxygenase	
2.4.1.185	Naringenin 7-O-glucosyltransferase Flavanone 7-O-β-glucosyltransferase	
2.5.1.43	NAS Nicotianamine synthase	
3.1.3.5	5'Nase 5'-nucleotidase	
2.5.1.43	NA synthase Nicotianamine synthase	
2.3.1.88	NAT Peptide α-N-acetyltransferase	
1.6.6.7	NC-Reductase Azobenzene reductase	
3.4.22.17	nCL-4 calpain	
3.5.1.53	NCP N-Carbamoylputrescine amidase	
2.7.7.37	NDP:aldose-1-phosphate nucleotidyltransferase Aldose-1-phosphate nucleotidyltransferase	
2.7.7.37	NDPaldose phosphorylase Aldose-1-phosphate nucleotidyltransferase	
3.6.1.6	NDPase Nucleoside-diphosphatase	
2.7.7.28	NDP hexose pyrophosphorylase Nucleoside-triphosphate-hexose-1-phosphate nucleotidyltransferase	
2.7.7.37	NDP sugar phosphorylase Aldose-1-phosphate nucleotidyltransferase	
2.4.99.10	Neolactotetraosylceramide α-2,3-sialyltransferase Neolactotetraosylceramide α-2,3-sialyltransferase	
1.1.1.208	(+)-Neomenthol:NADP$^+$ oxidoreductase (+)-Neomenthol dehydrogenase	

EC-No.	Synonym Recommended Name	Vol., Page
1.1.1.208	(+)-Neomenthol dehydrogenase (+)-Neomenthol dehydrogenase	
2.7.1.95	Neomycin-kanamycin phosphotransferase Kanamycin kinase	
2.7.1.95	Neomycin phosphotransferase Kanamycin kinase	
3.2.1.135	Neopullulanase Neopullulanase	
3.4.24.11	NEP neprilysin	
3.4.23.12	nepenthacin nepenthesin	
3.4.23.12	nepenthasin;aspartyl endopeptidase nepenthesin	
3.4.23.12	nepenthes aspartic protease nepenthesin	
3.4.23.40	Nepenthesin Phytepsin	
3.4.23.12	nepenthesin nepenthesin	
3.4.23.12	nepenthesin nepenthesin	
3.4.24.11	neprilysin neprilysin	
2.5.1.28	Neryl-diphosphate synthase Dimethylallylcistransferase	
2.4.99.4	NeuAc α-2,3-sialyltransferase β-Galactoside α-2,3-sialyltransferase	
3.1.3.29	NeuAc 9-phosphatase N-acylneuraminate-9-phosphatase	
4.1.3.3	NeuAc aldolase N-acetylneuraminate lyase	
4.1.3.19	NeuAc synthase N-acetylneuraminate synthase	
4.1.3.3	Neuraminate aldolase N-acetylneuraminate lyase	

EC-No.	Synonym Recommended Name	Vol., Page
4.1.3.3	Neuraminic acid aldolase N-acetylneuraminate lyase	
4.1.3.3	Neuraminic aldolase N-acetylneuraminate lyase	
3.2.1.18	α-neuraminidase Sialidase	
3.2.1.18	Neuraminidase Sialidase	
3.2.1.138	Neuraminidase, anhydro- Anhydrosialidase	
3.2.1.129	Neuraminidase, endo- Endo-α-sialidase	
3.4.23.21	Neurase Rhizopuspepsin	
3.4.21.93	Neuroendocrine convertase 1 Proprotein convertase 1	
3.4.21.94	Neuroendocrine convertase 2 Proprotein convertase 2	
3.4.24.16	neurolysin neurolysin	
4.2.1.11	neuron-specific enolase phosphoenolpyruvate hydratase	
3.1.30.1	Neurospora crassa single-strand specific endonuclease Aspergillus nuclease S_1	
3.4.24.16	neurotensin endopeptidase neurolysin	
3.4.22.17	neutral, calcium-activated protease calpain	
2.4.1.19	Neutral-cyclodextrin glycosyltransferase Cyclomaltodextrin glucanotransferase	
3.4.24.11	neutral endopeptidase neprilysin	
3.4.24.11	neutral endopeptidase 24.11 neprilysin	
3.4.24.15	neutral endopeptidase 24.15 thimet oligopeptidase	

EC-No.	Synonym Recommended Name	Vol., Page
3.4.24.11	neutral metallendopeptidase neprilysin	
3.4.21.20	neutral proteinase cathepsin G	
3.4.24.39	Neutral proteinase II Deuterolysin	
3.4.24.17	Neutral proteoglycanase Stromelysin 1	
3.5.2.6	Neutrapen β-lactamase	
3.4.24.34	Neutrophil collagenase Neutrophil collagenase	
3.4.21.37	Neutrophil elastase Leucocyte elastase	
1.6.6.7	New Coccine (NC)-reductase Azobenzene reductase	
3.5.3.18	N^g,N^g-Dimethyl-L-arginine dimethylamidohydrolase Dimethylargininase	
3.5.3.18	N^g,N^g-Dimethylarginine dimethylaminohydrolase Dimethylargininase	
6.3.1.1	NH_4^+-dependent asparagine synthetase Aspartate-ammonia ligase	Vol. 2, Page 344
4.2.1.84	H-NHase nitrile hydratase	
4.2.1.84	L-NHase nitrile hydratase	
4.2.1.84	NHase nitrile hydratase	
2.4.2.11	Niacin ribonucleotidase Nicotinate phosphoribosyltransferase	
5.99.1.2	Nicking-closing enzyme DNA topoisomerase	Vol. 1, Page 721
2.5.1.43	Nicotianamine synthase Nicotianamine synthase	
2.5.1.43	Nicotianamine synthase Nicotianamine synthase	

EC-No.	Synonym Recommended Name	Vol., Page
3.5.1.19	Nicotinamidase Nicotinamidase	
2.4.2.12	Nicotinamide-nucleotide:pyrophosphate phospho-α-D-ribosyltransferase Nicotinamide phosphoribosyltransferase	
2.7.7.1	Nicotinamide-nucleotide adenylyltransferase Nicotinamide-nucleotide adenylyltransferase	
3.5.1.42	Nicotinamide-nucleotide amidase Nicotinamide-nucleotide amidase	
3.5.1.42	Nicotinamide-nucleotide amidohydrolase Nicotinamide-nucleotide amidase	
3.2.2.14	Nicotinamide-nucleotide phosphoribohydrolase NMN nucleosidase	
1.2.1.22	Nicotinamide adenine dinucleotide (NAD)-linked dehydrogenase Lactaldehyde dehydrogenase	
1.6.6.7	Nicotinamide adenine dinucleotide (phosphate) azoreductase Azobenzene reductase	
3.2.2.6	Nicotinamide adenine dinucleotide (phosphate) glycohydrolase $NAD(P)^+$nucleosidase	
3.2.2.6	Nicotinamide adenine dinucleotide (phosphate) nucleosidase $NAD(P)^+$nucleosidase	
1.6.1.2	Nicotinamide adenine dinucleotide (phosphate) transhydrogenase $NAD(P)^+$ transhydrogenase (AB-specific)	
1.6.1.1	Nicotinamide adenine dinucleotide (phosphate) transhydrogenase $NAD(P)^+$ transhydrogenase (B-specific)	
3.2.2.5	Nicotinamide adenine dinucleotide glycohydrolase NAD^+ nucleosidase	
2.7.1.23	Nicotinamide adenine dinucleotide kinase NAD^+ kinase	
3.2.2.5	Nicotinamide adenine dinucleotide nucleosidase NAD^+ nucleosidase	
1.11.1.1	Nicotinamide adenine dinucleotide peroxidase NADH peroxidase	
1.1.1.119	Nicotinamide adenine dinucleotide phosphate-linked aldohexose dehydrogenase Glucose 1-dehydrogenase ($NADP^+$)	

EC-No.	Synonym Recommended Name	Vol., Page
1.11.1.2	Nicotinamide adenine dinucleotide phosphate peroxidase NADPH peroxidase	
3.6.1.22	Nicotinamide adenine dinucleotide pyrophosphatase NAD^+ diphosphatase	
2.7.7.1	Nicotinamide adenine dinucleotide pyrophosphorylase Nicotinamide-nucleotide adenylyltransferase	
6.3.1.5	Nicotinamide adenine dinucleotide synthetase NAD^+ synthase	Vol. 2, Page 377
6.3.5.1	Nicotinamide adenine dinucleotide synthetase (glutamine) NAD^+ synthase (glutamine-hydroylysing)	Vol. 2, Page 651
3.5.1.19	Nicotinamide amidase Nicotinamidase	
3.5.1.19	Nicotinamide amidohydrolase Nicotinamidase	
3.5.1.19	Nicotinamide deaminase Nicotinamidase	
3.2.2.14	Nicotinamide mononucleotidase NMN nucleosidase	
2.7.7.1	Nicotinamide mononucleotide adenylyltransferase Nicotinamide-nucleotide adenylyltransferase	
3.5.1.42	Nicotinamide mononucleotide amidohydrolase Nicotinamide-nucleotide amidase	
3.5.1.42	Nicotinamide mononucleotide deamidase Nicotinamide-nucleotide amidase	
2.4.2.12	Nicotinamide mononucleotide pyrophosphorylase Nicotinamide phosphoribosyltransferase	
2.4.2.12	Nicotinamide mononucleotide synthetase Nicotinamide phosphoribosyltransferase	
2.1.1.1	Nicotinamide N-methyltransferase Nicotinamide N-methyltransferase	
1.6.1.2	Nicotinamide nucleotide transhydrogenase $NAD(P)^+$ transhydrogenase (AB-specific)	
1.6.1.1	Nicotinamide nucleotide transhydrogenase $NAD(P)^+$ transhydrogenase (B-specific)	
2.4.2.12	Nicotinamide phosphoribosyltransferase Nicotinamide phosphoribosyltransferase	

EC-No.	Synonym Recommended Name	Vol., Page
2.4.2.21	Nicotinate-nucleotide-dimethylbenzimidazole phosphoribosyltransferase Nicotinate-nucleotide-dimethylbenzimidazole phosphoribosyltransferase	
2.4.2.21	Nicotinate-nucleotide:dimethylbenzimidazole phospho-D-ribosyltransferase Nicotinate-nucleotide-dimethylbenzimidazole phosphoribosyltransferase	
2.4.2.11	Nicotinate-nucleotide:pyrophosphate phospho-α-D-ribosyltransferase Nicotinate phosphoribosyltransferase	
2.4.2.19	Nicotinate-nucleotide:pyrophosphate phospho-α-D-ribosyltransferase (decarboxylating) Nicotinate-nucleotide pyrophosphorylase (carboxylating)	
2.7.7.18	Nicotinate-nucleotide adenylyltransferase Nicotinate-nucleotide adenylyltransferase	
2.4.2.19	Nicotinate-nucleotide pyrophosphorylase (carboxylating) Nicotinate-nucleotide pyrophosphorylase (carboxylating)	
1.5.1.13	Nicotinate:NADP$^+$ 6-oxidoreductase (hydroxylating) Nicotinate dehydrogenase	
1.5.1.13	Nicotinate dehydrogenase Nicotinate dehydrogenase	
1.5.1.13	Nicotinate dehydrogenase Nicotinate dehydrogenase	
2.4.1.196	Nicotinate glucosyltransferase Nicotinate glucosyltransferase	
2.1.1.7	Nicotinate methyltransferase Nicotinate N-methyltransferase	
2.4.2.21	Nicotinate mononucleotide-dimethylbenzimidazole phosphoribosyltransferase Nicotinate-nucleotide-dimethylbenzimidazole phosphoribosyltransferase	
2.4.2.19	Nicotinate mononucleotide pyrophosphorylase (carboxylating) (EC 2.4.2.19) Nicotinate-nucleotide pyrophosphorylase (carboxylating)	
2.1.1.7	Nicotinate N-methyltransferase Nicotinate N-methyltransferase	
2.4.2.11	Nicotinate phosphoribosyltransferase Nicotinate phosphoribosyltransferase	
2.4.2.21	Nicotinate ribonucleotide:benzimidazole (adenine) phosphoribosyltransferase Nicotinate-nucleotide-dimethylbenzimidazole phosphoribosyltransferase	

EC-No.	Synonym Recommended Name	Vol., Page
1.5.99.4	Nicotine:(acceptor) 6-oxidoreductase (hydroxylating) Nicotine dehydrogenase	
1.5.99.4	Nicotine dehydrogenase Nicotine dehydrogenase	
2.1.1.49	Nicotine N-methyltransferase Amine N-methyltransferase	
2.1.1.81	nicotine N-methyltransferase nicotine N-methyltransferase	
1.5.99.4	Nicotine oxidase Nicotine dehydrogenase	
1.5.99.4	d-Nicotine oxidase Nicotine dehydrogenase	
1.5.1.13	Nicotinic acid hydroxylase Nicotinate dehydrogenase	
2.1.1.7	Nicotinic acid methyltransferase Nicotinate N-methyltransferase	
2.7.7.18	Nicotinic acid mononucleotide adenylyltransferase Nicotinate-nucleotide adenylyltransferase	
2.4.2.11	Nicotinic acid mononucleotide glycohydrolase Nicotinate phosphoribosyltransferase	
2.4.2.11	Nicotinic acid mononucleotide pyrophosphorylase Nicotinate phosphoribosyltransferase	
2.4.2.11	Nicotinic acid phosphoribosyltransferase Nicotinate phosphoribosyltransferase	
3.2.2.22	Nigrin b rRNA N-glycosidase	
1.8.6.1	nitrate-ester reductase nitrate-ester reductase	
1.7.99.4	Nitrate reductase Nitrate reductase	
1.7.99.4	Nitrate reductase (acceptor) Nitrate reductase	
1.9.6.1	Nitrate reductase (cytochrome) Nitrate reductase (cytochrome)	
1.6.6.2	Nitrate reductase (NAD(P)H) Nitrate reductase (NAD(P)H)	

EC-No.	Synonym Recommended Name	Vol., Page
1.6.6.1	Nitrate reductase (NADH) Nitrate reductase (NADH)	
1.6.6.3	Nitrate reductase (NADPH) Nitrate reductase (NADPH)	
1.6.6.2	Nitrate reductase (reduced nicotinamide adenine dinucleotide (phosphate)) Nitrate reductase (NAD(P)H)	
1.6.6.2	Nitrate reductase NAD(P)H Nitrate reductase (NAD(P)H)	
1.7.99.3	Nitric-oxide:(acceptor) oxidoreductase Nitrite reductase	
1.7.2.1	Nitric-oxide:ferricytochrome-c oxidoreductase Nitrite reductase (cytochrome)	
1.7.99.7	Nitric-oxide reductase Nitric-oxide reductase	
1.7.99.2	nitric-oxide reductase nitric-oxide reductase	
1.14.13.39	Nitric-oxide synthase Nitric-oxide synthase	
1.14.12.17	nitric oxide, NAD(P)H2:oxygen oxidoreductase nitric oxide dioxygenase	
1.14.12.17	nitric oxide dioxygenase nitric oxide dioxygenase	
1.7.99.7	Nitric oxide reductase Nitric-oxide reductase	
1.14.13.39	Nitric oxide synthase Nitric-oxide synthase	
1.14.13.39	Nitric oxide synthetase Nitric-oxide synthase	
3.5.5.1	Nitrilase Nitrilase	
4.2.1.84	nitrilase nitrile hydratase	
3.5.5.4	Nitrilase, β-cyanoalanine Cyanoalanine nitrilase	
3.5.5.5	Nitrilase, arylaceto- Arylacetonitrilase	

EC-No.	Synonym Recommended Name	Vol., Page
3.5.5.6	Nitrilase, bromoxynil Bromoxynil nitrilase	
3.5.5.1	Nitrile aminohydrolase Nitrilase	
4.2.1.84	nitrile hydratase nitrile hydratase	
4.2.1.84	nitrile hydro-lyase nitrile hydratase	
1.7.99.4	Nitrite:(acceptor) oxidoreductase Nitrate reductase	
1.7.7.2	Nitrite:ferredoxin oxidoreductase Ferredoxin-nitrate reductase	
1.7.99.3	Nitrite reductase Nitrite reductase	
1.6.99.4	nitrite reductase nitrite reductase	
1.7.2.1	Nitrite reductase (cytochrome) Nitrite reductase (cytochrome)	
1.6.6.4	Nitrite reductase (NAD(P)H) Nitrite reductase (NAD(P)H)	
1.3.1.16	β-Nitroacrylate reductase β-Nitroacrylate reductase	
1.13.11.32	Nitroalkane oxidase 2-Nitropropane dioxygenase	
3.1.6.1	Nitrocatechol sulfatase Arylsulfatase	
2.8.2.1	4-Nitrocatechol sulfokinase Aryl sulfotransferase	
1.7.3.1	Nitroethane:oxygen oxidoreductase Nitroethane oxidase	
1.7.3.1	Nitroethane oxidase Nitroethane oxidase	
1.7.99.6	Nitrogen:(acceptor) oxidoreductase (N_2O-forming) Nitrous-oxide reductase	
1.18.6.1	Nitrogenase Nitrogenase	

EC-No.	Synonym Recommended Name	Vol., Page
1.19.6.1	Nitrogenase (flavodoxin) Nitrogenase (flavodoxin)	
1.7.99.7	Nitrogen oxide reductase Nitric-oxide reductase	
1.14.13.29	4-Nitrophenol,NADH:oxygen oxidoreductase (2-hydroxylating) 4-Nitrophenol 2-monooxygenase	
1.14.13.31	2-Nitrophenol,NADPH:oxygen 2-oxidoreductase (2-hydroxylating, nitrite-forming) 2-Nitrophenol 2-monooxygenase	
1.14.13.29	4-Nitrophenol-2-hydroxylase 4-Nitrophenol 2-monooxygenase	
1.14.13.31	2-Nitrophenol 2-monooxygenase 2-Nitrophenol 2-monooxygenase	
1.14.13.29	4-Nitrophenol 2-monooxygenase 4-Nitrophenol 2-monooxygenase	
2.4.1.75	*p*-Nitrophenol conjugating enzyme UDPgalacturonosyltransferase	
1.14.13.29	4-Nitrophenol hydroxylase 4-Nitrophenol 2-monooxygenase	
1.14.13.31	Nitrophenol oxygenase 2-Nitrophenol 2-monooxygenase	
2.8.2.1	*p*-Nitrophenol sulfotransferase Aryl sulfotransferase	
2.4.1.17	*p*-Nitrophenol UDP-glucuronosyltransferase Glucuronosyltransferase	
2.4.1.17	4-Nitrophenol UDP-glucuronyltransferase Glucuronosyltransferase	
2.4.1.17	*p*-Nitrophenol UDP-glucuronyltransferase Glucuronosyltransferase	
2.4.1.17	4-Nitrophenol UDPglucuronosyltransferase Glucuronosyltransferase	
2.4.1.17	4-Nitrophenol UDPGT Glucuronosyltransferase	
3.2.1.24	*p*-Nitrophenyl-.α.-mannosidase α-mannosidase	

EC-No.	Synonym Recommended Name	Vol., Page
4.3.3.1	4-Nitrophenyl-3-ketovalidamine 4-nitroaniline-lyase 3-Ketovalidoxylamine C-N-lyase	
4.3.3.1	*p*-Nitrophenyl-3-ketovalidamine *p*-nitroaniline lyase 3-Ketovalidoxylamine C-N-lyase	
3.2.1.139	*p*-Nitrophenyl α-D-glucuronide-hydrolyzing enzyme α-Glucosiduronase	
3.2.1.21	*p*-Nitrophenyl β-glucosidase β-glucosidase	
3.2.1.23	*p*-Nitrophenyl.β.-galactosidase β-galactosidase	
3.1.1.6	*p*-nitrophenyl acetate esterase acetylesterase	
2.4.1.17	*p*-Nitrophenylglucuronosyltransferase Glucuronosyltransferase	
3.1.3.41	4-nitrophenylphosphatase 4-nitrophenylphosphatase	
3.1.3.41	nitrophenyl phosphatase 4-nitrophenylphosphatase	
3.1.3.41	*p*-nitrophenylphosphatase 4-nitrophenylphosphatase	
3.1.3.41	*p*-nitrophenyl phosphate phosphatase 4-nitrophenylphosphatase	
3.1.3.41	4-nitrophenylphosphate phosphohydrolase 4-nitrophenylphosphatase	
3.1.3.41	*p*-nitrophenylphosphate phosphohydrolase 4-nitrophenylphosphatase	
3.1.4.38	*p*-nitrophenylphosphocholine cholinphosphodiesterase glycerophosphocholine cholinephosphodiesterase	
3.1.6.1	*p*-Nitrophenyl sulfatase Arylsulfatase	
1.13.11.32	2-Nitropropane:oxygen 2-oxidoreductase 2-Nitropropane dioxygenase	
1.13.11.32	2-Nitropropane dioxygenase 2-Nitropropane dioxygenase	
1.3.1.16	3-Nitropropanoate:NADP$^+$ oxidoreductase β-Nitroacrylate reductase	

EC-No.	Synonym Recommended Name	Vol., Page
1.6.6.10	Nitroquinoline-N-oxide reductase Nitroquinoline-N-oxide reductase	
1.6.6.10	4-Nitroquinoline 1-oxide reductase Nitroquinoline-N-oxide reductase	
1.6.6.10	Nitroquinoline N-oxide reductase Nitroquinoline-N-oxide reductase	
1.7.99.7	Nitrous-oxide:(acceptor) oxidoreductase Nitric-oxide reductase	
1.7.99.6	Nitrous-oxide reductase Nitrous-oxide reductase	
1.7.99.6	Nitrous oxide reductase Nitrous-oxide reductase	
2.1.1.128	NLS-OMT (RS)-Norcoclaurine 6-O-methyltransferase	
2.7.7.1	NMN adenylyltransferase Nicotinamide-nucleotide adenylyltransferase	
3.2.2.14	NMNase NMN nucleosidase	
3.5.1.42	NMN deamidase Nicotinamide-nucleotide amidase	
3.2.2.14	NMNGhase NMN nucleosidase	
3.2.2.14	NMN glycohydrolase NMN nucleosidase	
3.2.2.14	NMN nucleosidase NMN nucleosidase	
2.4.2.12	NMN pyrophosphorylase Nicotinamide phosphoribosyltransferase	
2.4.2.12	NMN synthetase Nicotinamide phosphoribosyltransferase	
2.7.4.4	NMP-kinase Nucleoside-phosphate kinase	
2.1.1.115	NMT (RS)-1-Benzyl-1,2,3,4-tetrahydroisoquinoline N-methyltransferase	
2.1.1.99	NMT 16-Methoxy-2,3-dihydro-3-hydroxytabersonine N-methyltransferase	

EC-No.	Synonym Recommended Name	Vol., Page
2.1.1.1	NNMT Nicotinamide N-methyltransferase	
5.1.1.14	Nocardicin-A epimerase Nocardicin-A epimerase	Vol. 1, Page 59
5.1.1.14	Nocardicin-A epimerase Nocardicin-A epimerase	Vol. 1, Page 59
3.1.27.1	Non-base specific ribonuclease Ribonuclease T_2	
1.6.1.1	Non-energy-linked transhydrogenase $NAD(P)^+$ transhydrogenase (B-specific)	
3.4.23.34	Non-pepsin proteinase Cathepsin E	
3.1.3.1	non-specific alkaline phosphatase alkaline phosphatase	
3.1.1.8	non-specific cholinesterase cholinesterase	
3.4.13.18	Non-specific dipeptidase Cytosol nonspecific dipeptidase	
3.4.13.17	non-stereopecific dipeptidase non-stereopecific dipeptidase	
3.7.1.7	Nonane-4,6-dione acylhydrolase β-Diketone hydrolase	
2.5.1.39	Nonaprenyl-4-hydroxybenzoate transferase 4-Hydroxybenzoate nonaprenyltransferase	
2.5.1.11	Nonaprenyl pyrophosphate synthetase trans-Octaprenyltranstransferase	
3.1.27.1	Nonbase-specific RNase Ribonuclease T_2	
1.2.1.9	Nonphosphorylating glyceraldehyde-3-phosphate dehydrogenase Glyceraldehyde-3-phosphate dehydrogenase ($NADP^+$)	
2.7.1.112	Nonreceptor protein tyrosine kinase Protein-tyrosine kinase	
3.1.1.1	Nonspecific carboxylesterase Carboxylesterase	
3.4.13.19	Nonspecific dipeptidase Membrane dipeptidase	

EC-No.	Synonym Recommended Name	Vol., Page
1.1.1.184	Nonspecific NADPH-dependent carbonyl reductase Carbonyl reductase (NADPH)	
2.7.1.77	Nonspecific nucleoside phosphotransferase Nucleoside phosphotransferase	
3.1.27.1	Nonspecific RNase Ribonuclease T_2	
1.1.1.2	Nonspecific succinic semialdehyde reductase (EC 1.1.1.2) Alcohol dehydrogenase ($NADP^+$)	
1.5.1.19	D-Nopaline dehydrogenase D-Nopaline dehydrogenase	
1.5.1.19	Nopaline dehydrogenase D-Nopaline dehydrogenase	
1.5.1.19	D-Nopaline synthase D-Nopaline dehydrogenase	
1.5.1.19	Nopaline synthase D-Nopaline dehydrogenase	
2.1.1.28	Noradrenaline N-methyltransferase Phenylethanolamine N-methyltransferase	
2.1.1.28	Noradrenalin methyltransferase Phenylethanolamine N-methyltransferase	
2.1.1.28	Noradrenalin N-methyltransferase Phenylethanolamine N-methyltransferase	
2.1.1.128	Norcoclaurine 6-O-methyltransferase (RS)-Norcoclaurine 6-O-methyltransferase	
2.1.1.28	Norepinephrine methyltransferase Phenylethanolamine N-methyltransferase	
2.1.1.28	Norepinephrine N-methyltransferase Phenylethanolamine N-methyltransferase	
2.6.1.67	L-Norleucine,leucine: 2-oxoglutarate aminotransferase 2-Aminohexanoate transaminase	
2.6.1.67	Norleucine transaminase 2-Aminohexanoate transaminase	
2.1.1.115	Norreticuline N-methyltransferase (RS)-1-Benzyl-1,2,3,4-tetrahydroisoquinoline N-methyltransferase	
1.5.1.19	NOS D-Nopaline dehydrogenase	

EC-No.	Synonym Recommended Name	Vol., Page
1.14.13.39	NO synthase Nitric-oxide synthase	
1.1.3.4	Notatin Glucose oxidase	
3.5.1.11	Novozym 217 Penicillin amidase	
5.99.1.3	NP170 proteins DNA topoisomerase (ATP-hydrolysing)	Vol. 1, Page 737
4.1.3.3	NPL N-acetylneuraminate lyase	
3.1.3.41	NPPase 4-nitrophenylphosphatase	
3.4.24.61	NRD-convertase Nardilysin	
3.4.21.91	NS2B-3 proteinase Flavivirin	
3.4.21.98	NS2B-NS3 proteinase hepacivirin	
3.4.21.98	NS3-4A serine protease complex hepacivirin	
3.4.21.98	NS3-NS4A protease hepacivirin	
3.4.21.98	NS3 protease hepacivirin	
3.4.21.98	NS3 proteinase hepacivirin	
3.4.21.98	NS3 serine protease hepacivirin	
3.4.21.98	NS3 serine proteinase hepacivirin	
3.4.21.98	NS5A-5B protease hepacivirin	
4.2.1.11	R-NSE phosphoenolpyruvate hydratase	
3.1.3.31	NSP-I nucleotidase	

EC-No.	Synonym Recommended Name	Vol., Page
3.1.3.31	NSP-II nucleotidase	
3.4.21.90	NsP2 proteinase Togavirin	
3.1.3.5	5'-NT 5'-nucleotidase	
3.1.3.6	3'-NT/Nu 3'-nucleotidase	
2.7.1.113	NTP-deoxyguanosine 5'-phosphotransferase Deoxyguanosine kinase	
2.7.1.74	NTP:deoxycytidine 5'-phosphotransferase Deoxycytidine kinase	
2.7.7.46	NTP:gentamicin 2-nucleotidyltransferase Gentamicin 2-nucleotidyltransferase	
2.7.7.28	NTP:hexose-1-phosphate nucleotidyltransferase Nucleoside-triphosphate-hexose-1-phosphate nucleotidyltransferase	
2.7.7.19	NTP polymerase Polynucleotide adenylyltransferase	
2.4.1.192	Nuatigenin 3β-glucosyltransferase Nuatigenin 3β-glucosyltransferase	
3.4.22.44	nuclear-inclusion-a endopeptidase nuclear-inclusion-a endopeptidase	
2.1.1.125	Nuclear protein (histone) N-methyltransferase (Histone)-arginine N-methyltransferase	
5.99.1.3	Nuclear proteins 170,000-mol.wt. DNA topoisomerase (ATP-hydrolysing)	Vol. 1, Page 737
4.2.99.18	Nuclease, apurinic-apyrimidinic endodeoxyribo- DNA-(apurinic or apyrimidinic site) lyase	
4.2.99.18	Nuclease, apurinic endodeoxyribo- DNA-(apurinic or apyrimidinic site) lyase	
3.1.30.1	Nuclease, Aspergillus oryzae S_1 Aspergillus nuclease S_1	
3.1.25.1	Nuclease, bacteriophage T_4 endodeoxyribo-, V Deoxyribonuclease (pyrimidine dimer)	
3.1.22.1	Nuclease, deoxyribo, II Deoxyribonuclease II	

EC-No.	Synonym Recommended Name	Vol., Page
3.1.21.1	Nuclease, deoxyribo- Deoxyribonuclease I	
3.1.21.3	Nuclease, deoxyribo- (adenosine triphosphate-hydrolyzing) Type I site-specific deoxyribonuclease	
3.1.21.3	Nuclease, deoxyribo-, ATP-dependent Type I site-specific deoxyribonuclease	
3.1.21.2	Nuclease, endodeoxyribo Deoxyribonuclease IV (Phage T4-induced)	
3.1.21.6	Nuclease, endodeoxyribo- CC-preferring endodeoxyribonuclease	
4.2.99.18	Nuclease, endodeoxyribo-, III DNA-(apurinic or apyrimidinic site) lyase	
3.1.22.3	Nuclease, endodeoxyribo-, V Deoxyribonuclease V	
3.1.11.2	Nuclease, endoribo-, III Exodeoxyribonuclease III	
3.1.21.1	Nuclease, Escherichia coli endo-, I Deoxyribonuclease I	
3.1.22.5	Nuclease, Escherichia coli endodeoxyribo- Deoxyribonuclease X	
3.1.11.1	Nuclease, Escherichia coli exo, I Exodeoxyribonuclease I	
3.1.11.2	Nuclease, Escherichia coli exo-, III Exodeoxyribonuclease III	
3.1.11.3	Nuclease, Escherichia coli exo-, IV Exodeoxyribonuclease (Lambda induced)	
3.1.27.1	Nuclease, Escherichia coli ribo-, II Ribonuclease T_2	
3.1.25.1	Nuclease, Escherichia coli UV-endodeoxyribo- Deoxyribonuclease (pyrimidine dimer)	
3.1.11.5	Nuclease, exodeoxyribo, V Exodeoxyribonuclease V	
3.1.11.1	Nuclease, exodeoxyribo- Exodeoxyribonuclease I	
3.1.11.3	Nuclease, exodeoxyribo-, IV Exodeoxyribonuclease (Lambda induced)	

EC-No.	Synonym Recommended Name	Vol., Page
3.1.11.6	Nuclease, exodeoxyribo-, VII Exodeoxyribonuclease VII	
3.1.13.1	Nuclease, exoribo- Exoribonuclease II	
3.1.27.3	Nuclease, guanyloribo- Ribonuclease T_1	
3.1.26.4	Nuclease, hybrid ribo- Calf thymus ribonuclease H	
3.1.31.1	Nuclease, micrococcal Micrococcal nuclease	
3.1.30.2	Nuclease, nucleate endo- Serratia marcescens nuclease	
3.1.27.5	Nuclease, ribo- Pancreatic ribonuclease	
3.1.27.1	Nuclease, ribo- (non-base specific) Ribonuclease T_2	
3.1.26.9	Nuclease, ribo- (uracil-specific) Ribonuclease (poly-(U)-specific)	
3.1.27.3	Nuclease, ribo-, Aspergillus oryzae Ribonuclease T_1	
3.1.26.4	Nuclease, ribo-, H Calf thymus ribonuclease H	
3.1.26.5	Nuclease, ribo-, P Ribonuclease P	
2.7.7.56	Nuclease, ribo-, PH tRNA nucleotidyltransferase	
3.1.30.1	Nuclease, single-stranded nucleate endo- Aspergillus nuclease S_1	
3.1.31.1	Nuclease, staphylococcal Micrococcal nuclease	
3.1.27.9	Nuclease, transfer ribonucleate intron endoribo- tRNA-intron endonuclease	
3.1.30.1	Nuclease S_1 Aspergillus nuclease S_1	
3.1.31.1	Nuclease T Micrococcal nuclease	

EC-No.	Synonym Recommended Name	Vol., Page
3.1.31.1	Nuclease T' Micrococcal nuclease	
3.1.30.2	Nucleate endonuclease Serratia marcescens nuclease	
3.1.4.9	nucleate endonuclease nucleate endonuclease	
3.4.21.32	nucleolysin brachyurin	
3.4.24.7	nucleolysin interstitial collagenase	
3.4.24.3	nucleolysin microbial collagenase	
3.2.2.1	Nucleosidase Purine nucleosidase	
3.2.2.13	Nucleosidase, 1-methyladenosine 1-Methyladenosine nucleosidase	
3.2.2.7	Nucleosidase, adenosine Adenosine nucleosidase	
3.2.2.9	Nucleosidase, adenosylhomocysteine Adenosylhomocysteine nucleosidase	
3.2.2.4	Nucleosidase, adenylate AMP nucleosidase	
3.2.2.12	Nucleosidase, inosinate Inosinate nucleosidase	
3.2.2.2	Nucleosidase, inosine Inosine nucleosidase	
3.2.2.16	Nucleosidase, methylthioadenosine Methylthioadenosine nucleosidase	
3.2.2.5	Nucleosidase, nicotinamide adenine dinucleotide NAD^+ nucleosidase	
3.2.2.6	Nucleosidase, nicotinamide adenine dinucleotide (phosphate) $NAD(P)^+$nucleosidase	
3.2.2.14	Nucleosidase, nicotinamide mononucleotide NMN nucleosidase	
3.2.2.8	Nucleosidase, pyrimidine Ribosylpyrimidine nucleosidase	

EC-No.	Synonym Recommended Name	Vol., Page
3.2.2.3	Nucleosidase, uridine Uridine nucleosidase	
3.2.2.1	Nucleosidase g Purine nucleosidase	
3.1.4.37	Nucleoside-2',3'-cyclic-phosohate 2'-nucleotidohydrolase 2',3'-Cyclic-nucleotide 3'-phosphodiesterase	
3.1.4.16	Nucleoside-2',3'-cyclic-phosphate 3'-nucleotidohydrolase 2',3'-Cyclic-nucleotide 2'-phosphodiesterase	
3.1.4.17	Nucleoside-3',5-monophosphate phosphodiesterase 3',5'-Cyclic-nucleotide phosphodiesterase	
3.6.1.24	Nucleoside-5-phosphoacylate acylhydrolase Nucleoside phosphoacylhydrolase	
3.6.1.15	Nucleoside-5-triphosphate phosphohydrolase Nucleoside-triphosphatase	
3.6.1.6	Nucleoside-diphosphatase Nucleoside-diphosphatase	
2.7.4.6	Nucleoside-diphosphate kinase Nucleoside-diphosphate kinase	
3.6.1.6	Nucleoside-diphosphate phosphohydrolase Nucleoside-diphosphatase	
2.7.4.4	Nucleoside-phosphate kinase Nucleoside-phosphate kinase	
3.6.1.15	Nucleoside-triphosphatase Nucleoside-triphosphatase	
2.7.4.10	Nucleoside-triphosphate-adenylate kinase Nucleoside-triphosphate-adenylate kinase	
2.7.7.28	Nucleoside-triphosphate-hexose-1-phosphate nucleotidyltransferase Nucleoside-triphosphate-hexose-1-phosphate nucleotidyltransferase	
2.7.4.10	Nucleoside-triphosphate:AMP phosphotransferase Nucleoside-triphosphate-adenylate kinase	
2.7.7.31	Nucleoside-triphosphate:DNA deoxynucleotidylexotransferase DNA nucleotidylexotransferase	
2.7.7.6	Nucleoside-triphosphate:RNA nucleotidyltransferase (DNA-directed) DNA-directed RNA polymerase	
2.7.7.48	Nucleoside-triphosphate:RNA nucleotidyltransferase (RNA-directed) RNA-directed RNA polymerase	

EC-No.	Synonym Recommended Name	Vol., Page
3.6.1.19	Nucleoside-triphosphate diphosphatase Nucleoside-triphosphate diphosphatase	
3.6.1.19	Nucleoside-triphosphate pyrophosphatase Nucleoside-triphosphate diphosphatase	
3.6.1.19	Nucleoside-triphosphate pyrophosphohydrolase Nucleoside-triphosphate diphosphatase	
3.1.4.17	Nucleoside 3',5'-cyclic phosphate diesterase 3',5'-Cyclic-nucleotide phosphodiesterase	
3.6.1.6	Nucleoside 5'-diphosphatase Nucleoside-diphosphatase	
2.7.4.6	Nucleoside 5'-diphosphate kinase Nucleoside-diphosphate kinase	
3.6.1.15	Nucleoside 5-triphosphatase Nucleoside-triphosphatase	
1.1.3.28	Nucleoside:oxygen 5'-oxidoreductase Nucleoside oxidase	
2.4.2.5	Nucleoside:purine(pyrimidine) D-ribosyltransferase Nucleoside ribosyltransferase	
2.4.2.6	Nucleoside:purine(pyrimidine) deoxy-D-ribosyltransferase Nucleoside deoxyribosyltransferase	
2.4.2.6	Nucleoside deoxyribosyltransferase Nucleoside deoxyribosyltransferase	
2.4.2.6	Nucleoside deoxyribosyltransferase I Nucleoside deoxyribosyltransferase	
2.4.2.6	Nucleoside deoxyribosyltransferase II Nucleoside deoxyribosyltransferase	
2.7.4.6	Nucleoside diphosphate (UDP) kinase Nucleoside-diphosphate kinase	
3.6.1.45	Nucleosidediphosphate-sugar diphosphatase UDP-sugar diphosphatase	
3.6.1.45	Nucleosidediphosphate-sugar pyrophosphatase UDP-sugar diphosphatase	
2.7.7.8	Nucleoside diphosphate:polynucleotidyl transferase Polyribonucleotide nucleotidyltransferase	
3.6.1.6	Nucleoside diphosphate phosphatase Nucleoside-diphosphatase	

EC-No.	Synonym Recommended Name	Vol., Page
3.6.1.6	Nucleoside diphosphate phosphohydrolase Nucleoside-diphosphatase	
1.17.4.1	Nucleoside diphosphate reductase Ribonucleoside-diphosphate reductase	
2.7.7.37	Nucleoside diphosphate sugar:orthophosphate nucleotidyltransferase Aldose-1-phosphate nucleotidyltransferase	
2.7.7.28	Nucleoside diphosphohexose pyrophosphorylase Nucleoside-triphosphate-hexose-1-phosphate nucleotidyltransferase	
2.7.4.6	Nucleoside diphosphokinase Nucleoside-diphosphate kinase	
2.7.7.37	Nucleoside diphosphosugar phosphorylase Aldose-1-phosphate nucleotidyltransferase	
3.2.2.1	Nucleoside hydrolase Purine nucleosidase	
2.4.2.5	Nucleoside N-ribosyltransferase Nucleoside ribosyltransferase	
1.1.3.28	Nucleoside oxidase Nucleoside oxidase	
3.6.1.24	Nucleoside phosphoacylhydrolase Nucleoside phosphoacylhydrolase	
2.7.1.77	Nucleoside phosphotransferase Nucleoside phosphotransferase	
3.2.2.8	Nucleoside ribohydrolase Ribosylpyrimidine nucleosidase	
2.4.2.5	Nucleoside ribosyltransferase Nucleoside ribosyltransferase	
2.4.2.6	Nucleoside trans-N-deoxyribosylase Nucleoside deoxyribosyltransferase	
2.7.4.10	Nucleoside triphosphate-adenosine monophosphate transphosphorylase Nucleoside-triphosphate-adenylate kinase	
2.7.4.10	Nucleoside triphosphate-adenylate kinase Nucleoside-triphosphate-adenylate kinase	
3.6.1.19	Nucleoside triphosphate diphosphatase Nucleoside-triphosphate diphosphatase	
3.6.1.15	Nucleoside triphosphate phosphohydrolase Nucleoside-triphosphatase	

EC-No.	Synonym Recommended Name	Vol., Page
2.3.1.48	Nucleosome-histone acetyltransferase Histone acetyltransferase	
3.1.3.6	3'-nucleotidase 3'-nucleotidase	
3.1.3.31	5'(3')-nucleotidase nucleotidase	
3.1.3.5	5'-nucleotidase 5'-nucleotidase	
3.1.3.31	nucleotidase nucleotidase	
3.1.3.6	nucleotidase, 3'- 3'-nucleotidase	
3.1.3.34	nucleotidase, 3'-deoxyribo deoxynucleotide 3'-phosphatase	
3.1.13.3	Nucleotidase, oligo- Oligonucleotidase	
3.1.3.7	nucleotidase, phosphoadenylate 3'- 3'(2'),5'-bisphosphate nucleotidase	
3.1.3.35	nucleotidase, thymidylate thymidylate 5'-phosphatase	
3.1.3.31	nucleotide-specific phosphatase nucleotidase	
3.6.1.9	Nucleotide-sugar pyrophosphatase Nucleotide diphosphatase	
2.7.6.4	Nucleotide 3'-pyrophosphokinase Nucleotide pyrophosphokinase	
2.7.1.77	Nucleotide:2'-nucleoside 5'-phosphotransferase Nucleoside phosphotransferase	
2.7.1.77	Nucleotide:3'-deoxynucleoside 5'-phosphotransferase Nucleoside phosphotransferase	
3.6.1.9	Nucleotide diphosphatase Nucleotide diphosphatase	
3.1.3.31	nucleotide phosphatase nucleotidase	
2.4.2.1	Nucleotide phosphatase (2.4.2.1) Purine-nucleoside phosphorylase	

EC-No.	Synonym Recommended Name	Vol., Page
2.7.4.6	Nucleotide phosphate kinase Nucleoside-diphosphate kinase	
3.1.4.17	3',5'-Nucleotide phosphodiesterase 3',5'-Cyclic-nucleotide phosphodiesterase	
3.1.16.1	3'-Nucleotide phosphodiesterase Spleen exonuclease	
3.1.4.1	5'-Nucleotide phosphodiesterase Phosphodiesterase I	
3.1.3.31	nucleotide phosphohydrolase nucleotidase	
3.6.1.9	Nucleotide pyrophosphatase Nucleotide diphosphatase	
3.6.1.9	Nucleotide pyrophosphohydrolase Nucleotide diphosphatase	
2.7.6.4	Nucleotide pyrophosphokinase Nucleotide pyrophosphokinase	
2.7.6.4	Nucleotide pyrophosphotransferase Nucleotide pyrophosphokinase	
2.7.1.11	Nucleotide triphosphate-dependent phosphofructokinase 6-Phosphofructokinase	
2.7.7.7	Nucleotidyltransferase, deoxyribonucleate DNA-directed DNA polymerase	
2.7.7.49	Nucleotidyltransferase, deoxyribonucleate, RNA-dependent RNA-directed DNA polymerase	
2.7.7.28	Nucleotidyltransferase, hexose 1-phosphate Nucleoside-triphosphate-hexose-1-phosphate nucleotidyltransferase	
2.7.7.19	Nucleotidyltransferase, polyadenylate Polynucleotide adenylyltransferase	
2.7.7.8	Nucleotidyltransferase, polyribonucleotide Polyribonucleotide nucleotidyltransferase	
2.7.7.6	Nucleotidyltransferase, ribonucleate DNA-directed RNA polymerase	
2.7.7.48	Nucleotidyltransferase, ribonucleate, RNA-dependent RNA-directed RNA polymerase	
2.7.7.37	Nucleotidyltransferase, sugar 1-phosphate Aldose-1-phosphate nucleotidyltransferase	

EC-No.	Synonym Recommended Name	Vol., Page
2.7.7.37	Nucleotidyltransferase, sugar phosphate Aldose-1-phosphate nucleotidyltransferase	
2.7.7.31	Nucleotidyltransferase, terminal deoxyribo- DNA nucleotidylexotransferase	
2.7.7.25	Nucleotidyltransferase, transfer ribonucleate tRNA adenylyltransferase	
2.7.7.21	Nucleotidyltransferase, transfer ribonucleate tRNA cytidylyltransferase	
3.1.3.6	3'-N'ase 3'-nucleotidase	
1.1.1.142	5D-5-O-Methyl-chiro-inositol:NADP$^+$ oxidoreductase D-Pinitol dehydrogenase	
1.14.99.34	7-O -methylluteone:O2 oxidoreductase monoprenyl isoflavone epoxidase	
4.2.99.16	O^3-Acetyl-L-serine acetate-lyase Uracilylalanine synthase	
4.2.99.15	O^3-Acetyl-L-serine acetate-lyase (adding 3,4-dihydroxypyridin-1-yl) L-Mimosine synthase	
4.2.99.14	O3^3-Acetyl-L-serine acetate-lyase (adding pyrazole) β-Pyrazolylalanine synthase (acetylserine)	
4.6.1.4	O^5-(1-carboxyvinyl)-3-phosphoshikimate phosphate-lyase chorismate synthase	
5.3.2.2	OAA tautomerase Oxalacetate tautomerase	Vol. 1, Page 371
4.2.99.10	OAH sulfhydrylase O-acetylhomoserine (thiol)-lyase	
2.6.1.13	OAT Ornithine-oxo-acid transaminase	
5.3.2.2	OAT Oxalacetate tautomerase	Vol. 1, Page 371
5.3.2.2	OAT-1 Oxalacetate tautomerase	Vol. 1, Page 371
5.3.2.2	OAT-2 Oxalacetate tautomerase	Vol. 1, Page 371
4.3.1.12	OCD ornithine cyclodeaminase	

EC-No.	Synonym Recommended Name	Vol., Page
6.2.1.8	OCS Oxalate-CoA ligase	Vol. 2, Page 242
4.1.99.5	Octadecanal alkane-lyase Octadecanal decarbonylase	
4.1.99.5	Octadecanal decarbonylase Octadecanal decarbonylase	
1.1.1.73	Octanol:NAD$^+$ oxidoreductase Octanol dehydrogenase	
1.1.1.73	1-Octanol dehydrogenase Octanol dehydrogenase	
1.1.1.73	Octanol dehydrogenase Octanol dehydrogenase	
2.3.1.137	Octanoyl-CoA:L-carnitine O-octanoyltransferase Carnitine O-octanoyltransferase	
4.2.1.17	2-octenoyl coenzyme A hydrase enoyl-CoA hydratase	
4.2.1.87	Octopamine dehydratase Octopamine dehydratase	
4.2.1.87	Octopamine hydro-lyase (deaminating) Octopamine dehydratase	
4.2.1.87	Octopamine hydrolyase Octopamine dehydratase	
1.5.1.11	Octopine:NAD oxidoreductase D-Octopine dehydrogenase	
1.5.1.11	D-Octopine dehydrogenase D-Octopine dehydrogenase	
1.5.1.11	Octopine dehydrogenase D-Octopine dehydrogenase	
1.5.1.11	D-Octopine synthase D-Octopine dehydrogenase	
3.2.1.124	Octulofuranosylono hydrolase 3-Deoxy-2-octulosonidase	
3.2.1.124	Octulopyranosylonohydrolase 3-Deoxy-2-octulosonidase	
3.2.1.124	Octulosylono hydrolase 3-Deoxy-2-octulosonidase	

EC-No.	Synonym Recommended Name Vol., Page
4.1.1.77	4-OD 4-Oxalocrotonate decarboxylase
2.3.1.58	ODAP synthase 2,3-Diaminopropionate N-oxalyltransferase
4.1.1.17	ODC Ornithine decarboxylase
4.1.1.23	ODCase Orotidine-5'-phosphate decarboxylase
1.5.1.11	ODH D-Octopine dehydrogenase
1.5.1.28	ODH Opine dehydrogenase
3.1.1.40	ODH orsellinate-depside hydrolase
4.1.2.14	ODPG aldolase 2-Dehydro-3-deoxyphosphogluconate aldolase
4.2.1.80	oeh 2-oxopent-4-enoate hydratase
2.8.2.4	Oestrogen sulphotransferase Estrone sulfotransferase
1.2.4.2	OGDC Oxoglutarate dehydrogenase (lipoamide)
4.2.2.6	OGTE oligogalacturonide lyase
1.1.1.141	15-OH-PGDH 15-Hydroxyprostaglandin dehydrogenase (NAD^+)
2.4.1.17	3-OH Androgenic UDPGT Glucuronosyltransferase
2.4.1.17	17-OH Steroid UDPGT Glucuronosyltransferase
3.4.21.55	Okinaxobin II Venombin AB
1.6.99.1	Old yellow enzyme NADPH dehydrogenase
1.14.13.26	Oleate $\delta^1$2-hydroxylase Phosphatidylcholine 12-monooxygenase

EC-No.	Synonym Recommended Name	Vol., Page
1.3.1.35	Oleate desaturase Phosphatidylcholine desaturase	
4.2.1.53	Oleate hydratase Oleate hydratase	
3.1.2.14	oleoyl-[acyl-carrier-protein] hydrolase oleoyl-[acyl-carrier-protein] hydrolase	
3.1.2.14	oleoyl-[acyl-carrier-protein] hydrolase oleoyl-[acyl-carrier-protein] hydrolase	
3.1.2.14	oleoyl-ACP thioesterase oleoyl-[acyl-carrier-protein] hydrolase	
3.1.2.14	oleoyl-acyl carrier protein thioesterase oleoyl-[acyl-carrier-protein] hydrolase	
1.3.1.35	Oleoyl-CoA desaturase Phosphatidylcholine desaturase	
6.2.1.3	Oleoyl-CoA synthetase Long-chain-fatty-acid-CoA ligase	Vol. 2, Page 206
1.3.1.35	Oleoylphosphatidylcholine desaturase Phosphatidylcholine desaturase	
3.2.1.39	Oligo-1,3-glucosidase Glucan endo-1,3-β-glucosidase	
3.2.1.10	Oligo-1,6-glucosidase Oligo-1,6-glucosidase	
3.2.1.10	Oligo-1,6-glucosidase Oligo-1,6-glucosidase	
2.4.1.49	1,4-β-D-Oligo-D-glucan:orthophosphate α-D-glucosyltransferase Cellodextrin phosphorylase	
3.1.4.29	oligodeoxyribonucleate exonuclease oligodeoxyribonucleate exonuclease	
4.2.2.6	oligogalacturonate lyase oligogalacturonide lyase	
4.2.2.6	oligogalacturonide lyase oligogalacturonide lyase	
4.2.2.6	oligogalacturonide lyase oligogalacturonide lyase	
4.2.2.6	oligogalacturonide trans-eliminase oligogalacturonide lyase	

EC-No.	Synonym Recommended Name	Vol., Page
4.2.2.6	oligogalacturonide transeliminase oligogalacturonide lyase	
2.4.1.24	Oligoglucan-branching glycosyltransferase 1,4-α-Glucan 6-α-glucosyltransferase	
2.4.1.30	1,3-β-D-Oligoglucan:orthophosphate α-D-glucosyltransferase 1,3-β-Oligoglucan phosphorylase	
2.4.1.49	β-1,4-Oligoglucan:orthophosphate glucosyltransferase Cellodextrin phosphorylase	
2.4.1.30	β-1,3-Oligoglucan:orthophosphate glucosyltransferase II 1,3-β-Oligoglucan phosphorylase	
2.4.1.30	1,3-β-Oligoglucan phosphorylase 1,3-β-Oligoglucan phosphorylase	
2.4.1.30	β-1,3-Oligoglucan phosphorylase 1,3-β-Oligoglucan phosphorylase	
3.2.1.123	Oligoglycosylglucosylceramide glycohydrolase Endoglycosylceramidase	
2.4.1.130	Oligomannosylsynthase Dolichyl-phosphate-mannose-glycolipid α-mannosyltransferase	
2.4.1.119	Oligomannosyltransferase Dolichyl-diphosphooligosaccharide-protein glycotransferase	
3.1.4.1	Oligonucleate 5'-nucleotidohydrolase Phosphodiesterase I	
3.1.13.3	Oligonucleotidase Oligonucleotidase	
3.1.4.19	oligonucleotidase oligonucleotidase	
3.4.24.70	Oligopeptidase A Oligopeptidase A	
3.4.21.83	Oligopeptidase B Oligopeptidase B	
3.4.24.16	oligopeptidase M neurolysin	
3.1.13.3	Oligoribonuclease Oligonucleotidase	
3.6.1.44	Oligosaccharide-diphosphodolichol diphosphatase Oligosaccharide-diphosphodolichol diphosphatase	

EC-No.	Synonym Recommended Name	Vol., Page
3.6.1.44	Oligosaccharide-diphosphodolichol phosphodolichohydrolase Oligosaccharide-diphosphodolichol diphosphatase	
3.6.1.44	Oligosaccharide-diphosphodolichol pyrophosphatase Oligosaccharide-diphosphodolichol diphosphatase	
2.4.1.131	Oligosaccharide-lipid mannosyltransferase Glycolipid 2-α-mannosyltransferase	
2.4.1.161	Oligosaccharide 4-α-D-glucosyltransferase Oligosaccharide 4-α-D-glucosyltransferase	
3.5.1.52	N-Oligosaccharide glycopeptidase Peptide-N^4-(N-acetyl-β-glucosaminyl) asparagine amidase	
3.2.1.140	Oligosaccharide lacto-N-biosylhydrolase Lacto-N-biosidase	
2.4.1.119	Oligosaccharide transferase Dolichyl-diphosphooligosaccharide-protein glycotransferase	
2.4.1.119	Oligosaccharyltransferase, dolichyldiphosphoryloligosaccharide-protein Dolichyl-diphosphooligosaccharide-protein glycotransferase	
3.2.1.120	Oligoxyloglucan β-glycosidase Oligoxyloglucan β-glycosidase	
3.2.1.120	Oligoxyloglucan hydrolase Oligoxyloglucan β-glycosidase	
3.2.1.120	Oligoxyloglucan xyloglucohydrolase Oligoxyloglucan β-glycosidase	
2.1.1.64	OMHMB-methyltransferase 3-Demethylubiquinone-9 3-O-methyltransferase	
4.1.1.23	OMP-DC Orotidine-5'-phosphate decarboxylase	
4.1.1.23	OMP decarboxylase Orotidine-5'-phosphate decarboxylase	
3.4.21.87	Omptin Omptin	
3.4.21.87	OmpT protease Omptin	
2.1.1.76	3-OMT Quercetin 3-O-methyltransferase	
2.1.1.83	4'-OMT 3,7-Dimethylquercitin 4'-O-methyltransferase	

EC-No.	Synonym Recommended Name	Vol., Page
2.1.1.84	6-OMT Methylquercetagetin 6-O-methyltransferase	
2.1.1.82	7-OMT 3-Methylquercitin 7-O-methyltransferase	
3.4.21.34	onokrein P plasma kallikrein	
3.4.21.35	onokrein P. tissue kallikrein	
3.4.21.63	Onoprose Oryzin	
3.4.21.63	Onoprose SA Oryzin	
3.1.8.1	OPA anhydrase Aryldialkylphosphatase	
3.1.8.2	OPA anhydrase Diisopropyl-fluorophosphatase	
4.1.1.68	OPET decarboxylase 5-Oxopent-3-ene-1,2,5-tricarboxylate decarboxylase	
3.1.8.1	OPH Aryldialkylphosphatase	
2.7.3.7	Opheline kinase Opheline kinase	
1.4.3.2	Ophio-amino-acid oxidase L-Amino-acid oxidase	
3.4.24.51	ophiolysin ophiolysin	
3.4.24.51	Ophiophagus metalloendopeptidase ophiolysin	
1.5.1.28	Opine dehydrogenase Opine dehydrogenase	
2.4.2.10	OPRTase Orotate phosphoribosyltransferase	
2.7.1.125	Opsin kinase Rhodopsin kinase	
2.7.1.97	opsin kinase opsin kinase	

EC-No.	Synonym Recommended Name	Vol., Page
3.4.21.62	Opticlean Subtilisin	
1.11.1.6	Optidase Catalase	
5.3.1.5	Optisweet Xylose isomerase	Vol. 1, Page 259
1.1.1.209	3(or 17)α-Hydroxysteroid:NAD(P)$^+$ oxidoreductase 3(or 17)α-Hydroxysteroid dehydrogenase	
1.1.1.51	3(or 17)β-Hydroxysteroid:NAD(P)$^+$ oxidoreductase 3(or 17)β-Hydroxysteroid dehydrogenase	
1.1.1.239	3α(or 17β)-Hydroxysteroid:NAD$^+$ oxidoreductase 3α(17β)-Hydroxysteroid dehydrogenase (NAD$^+$)	
1.1.1.209	3(or 17)α-Hydroxysteroid dehydrogenase 3(or 17)α-Hydroxysteroid dehydrogenase	
1.1.1.51	3(or 17)β-Hydroxysteroid dehydrogenase 3(or 17)β-Hydroxysteroid dehydrogenase	
1.1.1.53	3α(or 20β)-Hydroxysteroid:NAD$^+$ oxidoreductase 3α(or 20β)-Hydroxysteroid dehydrogenase	
1.1.1.210	3β(or 20α)-Hydroxysteroid:NADP$^+$ oxidoreductase 3β(or 20α)-Hydroxysteroid dehydrogenase	
1.1.1.53	3α(or 20β)-Hydroxysteroid dehydrogenase 3α(or 20β)-Hydroxysteroid dehydrogenase	
1.1.1.210	3β(or 20α)-Hydroxysteroid dehydrogenase 3β(or 20α)-Hydroxysteroid dehydrogenase	
5.3.3.1	Δ^5(or δ^4)-3-keto steroid isomerase Steroid Δ-isomerase	Vol. 1, Page 376
1.6.6.7	Orange I azoreductase Azobenzene reductase	
1.6.6.7	Orange II azoreductase Azobenzene reductase	
1.14.13.6	Orcinol,NADH:oxygen oxidoreductase (2-hydroxylating) Orcinol 2-monooxygenase	
1.14.13.6	Orcinol 2-monooxygenase Orcinol 2-monooxygenase	
1.14.13.6	Orcinol hydroxylase Orcinol 2-monooxygenase	

EC-No.	Synonym Recommended Name	Vol., Page
3.4.25.1	organelle, proteasome proteasome endopeptidase complex	
4.99.1.2	organomercurial lyase alkylmercury lyase	
4.99.1.2	organomercury lyase alkylmercury lyase	
3.1.8.2	Organophosphate acid anhydrase Diisopropyl-fluorophosphatase	
3.1.8.1	Organophosphate hydrolase Aryldialkylphosphatase	
3.1.8.1	Organophosphorus acid anhydrase Aryldialkylphosphatase	
3.1.8.2	Organophosphorus acid anhydrolase Diisopropyl-fluorophosphatase	
3.1.8.1	Organophosphorus hydrolase Aryldialkylphosphatase	
2.3.1.148	1-Organyl-2-acyl-sn-glycero-3-phosphocholine:1-organyl-2-lyso-sn-glycero-3-phosphoethanolamine acyltransferase (CoA-dependent) Glycerophospholipid acyltransferase (CoA-dependent)	
2.3.1.147	1-Organyl-2-arachidonyl-sn-glycero-3-phosphocholine:1-organyl-2-lyso-sn-glycero-3-phosphoethanolamine arachidonoyltransferase (CoA-independent) Glycerophospholipid arachidonoyl-transferase (CoA-independent)	
3.2.1.36	Orgelase Hyaluronoglucuronidase	
3.4.21.62	Orientase 10B Subtilisin	
3.4.13.17	D-(orL-)aminoacyl dipeptidase non-stereopecific dipeptidase	
2.6.1.68	L-Ornithine(L-lysine):2-oxoglutarate ω-aminotransferase Ornithine(lysine) transaminase	
2.6.1.68	Ornithine(lysine) transaminase Ornithine(lysine) transaminase	
2.6.1.13	Ornithine-α-ketoglutarate aminotransferase Ornithine-oxo-acid transaminase	
2.6.1.13	Ornithine-2-oxoacid aminotransferase Ornithine-oxo-acid transaminase	

EC-No.	Synonym Recommended Name	Vol., Page
2.6.1.13	Ornithine-keto acid aminotransferase Ornithine-oxo-acid transaminase	
2.6.1.13	Ornithine-keto acid transaminase Ornithine-oxo-acid transaminase	
2.6.1.13	Ornithine-ketoglutarate aminotransferase Ornithine-oxo-acid transaminase	
2.6.1.13	Ornithine-oxo-acid transaminase Ornithine-oxo-acid transaminase	
2.6.1.13	Ornithine-oxo acid aminotransferase Ornithine-oxo-acid transaminase	
2.6.1.13	Ornithine δ-transaminase Ornithine-oxo-acid transaminase	
5.4.3.5	D-Ornithine 4,5-aminomutase D-Ornithine 4,5-aminomutase	Vol. 1, Page 565
5.4.3.5	D-Ornithine 4,5-aminomutase D-Ornithine 4,5-aminomutase	Vol. 1, Page 565
5.4.3.1	ornithine 4,5-aminomutase ornithine 4,5-aminomutase	Vol. 1, Page 552
2.6.1.13	L-Ornithine 5-aminotransferase Ornithine-oxo-acid transaminase	
2.6.1.13	Ornithine 5-aminotransferase Ornithine-oxo-acid transaminase	
2.6.1.13	L-Ornithine:α-ketoglutarate δ-aminotransferase Ornithine-oxo-acid transaminase	
2.6.1.13	Ornithine:α-oxoglutarate transaminase Ornithine-oxo-acid transaminase	
2.6.1.13	L-Ornithine:2-oxo-acid aminotransferase Ornithine-oxo-acid transaminase	
2.3.1.35	Ornithine acetyltransferase Glutamate N-acetyltransferase	
2.6.1.13	L-Ornithine aminotransferase Ornithine-oxo-acid transaminase	
4.3.1.12	L-ornithine ammonia-lyase (cyclizing) ornithine cyclodeaminase	
2.1.3.3	L-Ornithine carbamoyltransferase Ornithine carbamoyltransferase	

EC-No.	Synonym Recommended Name	Vol., Page
2.1.3.3	Ornithine carbamoyltransferase Ornithine carbamoyltransferase	
2.1.3.3	L-Ornithine carbamyltransferase Ornithine carbamoyltransferase	
2.1.3.3	Ornithine carbamyltransferase Ornithine carbamoyltransferase	
4.1.1.17	L-Ornithine carboxy-lyase Ornithine decarboxylase	
4.3.1.12	ornithine cyclase ornithine cyclodeaminase	
4.3.1.12	ornithine cyclase (deaminating) ornithine cyclodeaminase	
4.3.1.12	ornithine cyclodeaminase ornithine cyclodeaminase	
4.3.1.12	ornithine cyclodeaminase ornithine cyclodeaminase	
4.1.1.17	Ornithine decarboxylase Ornithine decarboxylase	
3.1.1.48	ornithine esterase fusarinine-C ornithinesterase	
2.3.1.127	Ornithine N-acyltransferase Ornithine N-benzoyltransferase	
2.3.1.127	Ornithine N-benzoyltransferase Ornithine N-benzoyltransferase	
5.1.1.12	Ornithine racemase Ornithine racemase	Vol. 1, Page 54
5.1.1.12	Ornithine racemase Ornithine racemase	Vol. 1, Page 54
2.3.1.35	Ornithine transacetylase Glutamate N-acetyltransferase	
2.6.1.13	Ornithine transaminase Ornithine-oxo-acid transaminase	
2.1.3.3	Ornithine transcarbamoylase Ornithine carbamoyltransferase	
2.1.3.3	L-Ornithine transcarbamylase Ornithine carbamoyltransferase	

EC-No.	Synonym Recommended Name	Vol., Page
4.1.1.23	Orotate decarboxylase Orotidine-5'-phosphate decarboxylase	
4.1.1.23	Orotate monophosphate decarboxylase Orotidine-5'-phosphate decarboxylase	
2.4.2.10	Orotate phosphoribosyl pyrophosphate transferase Orotate phosphoribosyltransferase	
2.4.2.10	Orotate phosphoribosyltransferase Orotate phosphoribosyltransferase	
1.3.1.14	Orotate reductase Orotate reductase (NADH)	
1.3.1.14	Orotate reductase (NADH) Orotate reductase (NADH)	
1.3.1.15	Orotate reductase (NADPH) Orotate reductase (NADPH)	
2.4.2.10	Orotic acid phosphoribosyltransferase Orotate phosphoribosyltransferase	
4.1.1.23	Orotic decarboxylase Orotidine-5'-phosphate decarboxylase	
4.1.1.23	Orotidine-5'-monophosphate decarboxylase Orotidine-5'-phosphate decarboxylase	
2.4.2.10	Orotidine-5'-phosphate:pyrophosphate phospho-α-D-ribosyltransferase Orotate phosphoribosyltransferase	
4.1.1.23	Orotidine-5'-phosphate carboxy-lyase Orotidine-5'-phosphate decarboxylase	
4.1.1.23	Orotidine-5'-phosphate decarboxylase Orotidine-5'-phosphate decarboxylase	
2.4.2.10	Orotidine-5'-phosphate pyrophosphorylase Orotate phosphoribosyltransferase	
4.1.1.23	Orotidine 5'-monophosphate decarboxylase Orotidine-5'-phosphate decarboxylase	
2.4.2.10	Orotidine 5'-monophosphate pyrophosphorylase Orotate phosphoribosyltransferase	
4.1.1.23	Orotidine 5'-phosphate decarboxylase Orotidine-5'-phosphate decarboxylase	
2.4.2.10	Orotidine 5'-phosphate pyrophosphorylase Orotate phosphoribosyltransferase	

EC-No.	Synonym Recommended Name	Vol., Page
4.1.1.23	Orotidine monophosphate decarboxylase Orotidine-5'-phosphate decarboxylase	
2.4.2.10	Orotidine monophosphate pyrophosphorylase Orotate phosphoribosyltransferase	
4.1.1.23	Orotidine phosphate decarboxylase Orotidine-5'-phosphate decarboxylase	
2.4.2.10	Orotidine phosphoribosyltransferase Orotate phosphoribosyltransferase	
4.1.1.23	Orotidylate decarboxylase Orotidine-5'-phosphate decarboxylase	
2.4.2.10	Orotidylate phosphoribosyltransferase Orotate phosphoribosyltransferase	
2.4.2.10	Orotidylate pyrophosphorylase Orotate phosphoribosyltransferase	
4.1.1.23	Orotidylic acid decarboxylase Orotidine-5'-phosphate decarboxylase	
2.4.2.10	Orotidylic acid phosphorylase Orotate phosphoribosyltransferase	
2.4.2.10	Orotidylic acid pyrophosphorylase Orotate phosphoribosyltransferase	
4.1.1.23	Orotidylic decarboxylase Orotidine-5'-phosphate decarboxylase	
2.4.2.10	Orotidylic phosphorylase Orotate phosphoribosyltransferase	
2.4.2.10	Orotidylic pyrophosphorylase Orotate phosphoribosyltransferase	
4.1.1.23	Orotodylate decarboxylase Orotidine-5'-phosphate decarboxylase	
1.14.99.3	ORP33 proteins Heme oxygenase (decyclizing)	
3.1.1.40	orsellinate-depside hydrolase orsellinate-depside hydrolase	
3.1.1.40	orsellinate-depside hydrolase orsellinate-depside hydrolase	
4.1.1.58	Orsellinate carboxy-lyase Orsellinate decarboxylase	

EC-No.	Synonym Recommended Name	Vol., Page
4.1.1.58	Orsellinate decarboxylase Orsellinate decarboxylase	
3.1.1.40	orsellinate depside hydrolase orsellinate-depside hydrolase	
4.1.1.31	Orthophosphate: oxaloacetate carboxy-lyase (phosphorylating) Phosphoenolpyruvate carboxylase	
3.1.3.2	Orthophosphoric-monoester phosphohydrolase (acid optimum) Acid phosphatase	
3.1.3.1	orthophosphoric-monoester phosphohydrolase (alkaline optimum) alkaline phosphatase	
3.4.22.39	oryzain adenain	
3.4.23.40	Oryzasin Phytepsin	
3.2.1.23	Oryzatym β-galactosidase	
3.4.21.63	Oryzin Oryzin	
4.4.1.8	osteotoxin cystathionine β-lyase	
2.1.3.3	OTC Ornithine carbamoyltransferase	
3.1.2.14	OTE oleoyl-[acyl-carrier-protein] hydrolase	
2.3.1.137	Overt mitochondrial carnitine palmitoyltransferase Carnitine O-octanoyltransferase	
3.7.1.1	Oxalacetase Oxalacetase	
4.1.1.3	Oxalacetate decarboxylase Oxaloacetate decarboxylase	
5.3.2.2	Oxalacetate tautomerase Oxalacetate tautomerase	Vol. 1, Page 371
4.1.1.3	Oxalacetic β-decarboxylase Oxaloacetate decarboxylase	
4.1.1.3	Oxalacetic acid decarboxylase Oxaloacetate decarboxylase	

EC-No.	Synonym Recommended Name	Vol., Page
4.1.1.3	Oxalacetic carboxylase Oxaloacetate decarboxylase	
3.7.1.1	Oxalacetic hydrolase Oxalacetase	
5.3.2.2	Oxalacetic keto-enol isomerase Oxalacetate tautomerase	Vol. 1, Page 371
4.1.3.7	Oxalacetic transacetase Citrate (si)-synthase	
2.1.3.1	Oxalacetic transcarboxylase Methylmalonyl-CoA carboxyltransferase	
6.2.1.8	Oxalate-CoA ligase Oxalate-CoA ligase	Vol. 2, Page 242
4.1.1.3	Oxalate β-decarboxylase Oxaloacetate decarboxylase	
6.2.1.8	Oxalate:CoA ligase (AMP) Oxalate-CoA ligase	Vol. 2, Page 242
6.2.1.8	Oxalate:CoA ligase (AMP-forming) Oxalate-CoA ligase	Vol. 2, Page 242
1.2.3.4	Oxalate:oxygen oxidoreductase Oxalate oxidase	
4.1.1.2	Oxalate carboxy-lyase Oxalate decarboxylase	
2.8.3.2	Oxalate CoA-transferase Oxalate CoA-transferase	
2.8.3.2	Oxalate coenzyme A-transferase Oxalate CoA-transferase	
4.1.1.2	Oxalate decarboxylase Oxalate decarboxylase	
1.2.3.4	Oxalate oxidase Oxalate oxidase	
1.2.3.4	Oxalic acid oxidase Oxalate oxidase	
4.2.1.83	4-Oxalmesaconate hydratase 4-Oxalmesaconate hydratase	
4.2.1.83	Oxalmesaconate hydratase 4-Oxalmesaconate hydratase	

EC-No.	Synonym Recommended Name	Vol., Page
4.2.1.83	γ-Oxalmesaconate hydratase 4-Oxalmesaconate hydratase	
3.7.1.1	Oxaloacetase Oxalacetase	
3.1.1.44	oxaloacetate-4-methyl-ester oxaloacetohydrolase 4-methyloxaloacetate esterase	
2.6.1.1	Oxaloacetate-aspartate aminotransferase Aspartate transaminase	
3.7.1.1	Oxaloacetate acetylhydrolase Oxalacetase	
4.1.1.3	Oxaloacetate carboxy-lyase Oxaloacetate decarboxylase	
4.1.1.3	Oxaloacetate carboxylyase Oxaloacetate decarboxylase	
4.1.1.3	Oxaloacetate decarboxylase Oxaloacetate decarboxylase	
5.3.2.2	Oxaloacetate keto-enol-isomerase Oxalacetate tautomerase	Vol. 1, Page 371
5.3.2.2	Oxaloacetate keto-enol tautomerase Oxalacetate tautomerase	Vol. 1, Page 371
4.1.1.32	Oxaloacetate kinase (decarboxylating, GDP) Phosphoenolpyruvate carboxykinase (GTP)	
5.3.2.2	Oxaloacetate tautomerase-1 Oxalacetate tautomerase	Vol. 1, Page 371
5.3.2.2	Oxaloacetate tautomerase-2 Oxalacetate tautomerase	Vol. 1, Page 371
4.1.3.7	Oxaloacetate transacetase Citrate (si)-synthase	
2.6.1.1	Oxaloacetate transferase Aspartate transaminase	
4.1.1.77	4-Oxalocrotonate carboxy-lyase 4-Oxalocrotonate decarboxylase	
4.1.1.77	4-Oxalocrotonate decarboxylase 4-Oxalocrotonate decarboxylase	
4.1.1.77	Oxalocrotonate decarboxylase 4-Oxalocrotonate decarboxylase	

EC-No.	Synonym Recommended Name	Vol., Page
1.1.1.92	Oxaloglycolate reductase (decarboxylating) Oxaloglycolate reductase (decarboxylating)	
4.1.3.13	3-Oxalomalate glyoxylate-lyase Oxalomalate lyase	
4.1.3.13	Oxalomalate lyase Oxalomalate lyase	
1.1.1.42	Oxalosuccinate decarboxylase Isocitrate dehydrogenase ($NADP^+$)	
1.1.1.42	Oxalsuccinic decarboxylase Isocitrate dehydrogenase ($NADP^+$)	
2.3.1.58	Oxalyl-CoA:L-α,β-diaminopropionic acid oxalyltransferase 2,3-Diaminopropionate N-oxalyltransferase	
2.3.1.58	Oxalyl-CoA:L-2,3-diaminopropanoate N^3-oxalyltransferase 2,3-Diaminopropionate N-oxalyltransferase	
4.1.1.8	Oxalyl-CoA carboxy-lyase Oxalyl-CoA decarboxylase	
4.1.1.8	Oxalyl-CoA decarboxylase Oxalyl-CoA decarboxylase	
6.2.1.8	Oxalyl-CoA synthetase Oxalate-CoA ligase	Vol. 2, Page 242
6.2.1.8	Oxalyl CoA synthetase Oxalate-CoA ligase	Vol. 2, Page 242
4.1.1.8	Oxalyl coenzyme A decarboxylase Oxalyl-CoA decarboxylase	
6.2.1.8	Oxalyl coenzyme A synthetase Oxalate-CoA ligase	Vol. 2, Page 242
2.3.1.58	Oxalyldiaminopropionate synthase 2,3-Diaminopropionate N-oxalyltransferase	
2.3.1.58	Oxalyldiaminopropionic synthase 2,3-Diaminopropionate N-oxalyltransferase	
2.3.1.58	Oxalyltransferase, 2,3-diaminopropionate 2,3-Diaminopropionate N-oxalyltransferase	
2.1.3.5	Oxamate carbamoyltransferase Oxamate carbamoyltransferase	
2.1.3.5	Oxamic transcarbamylase Oxamate carbamoyltransferase	

EC-No.	Synonym Recommended Name	Vol., Page
4.1.1.2	OXD Oxalate decarboxylase	
1.3.3.8	Oxidase, (S)-tetrahydroberberine Tetrahydroberberine oxidase	
1.10.3.5	Oxidase, 3-hydroxyanthranilate 3-Hydroxyanthranilate oxidase	
1.5.3.6	Oxidase, 6-hydroxy-D-nicotine (R)-6-Hydroxynicotine oxidase	
1.5.3.5	Oxidase, 6-hydroxy-L-nicotine (S)-6-Hydroxynicotine oxidase	
1.7.3.2	Oxidase, acetylindoxyl Acetylindoxyl oxidase	
1.3.3.6	Oxidase, acyl-coenzyme A Acyl-CoA oxidase	
1.14.13.32	Oxidase, albendazole Albendazole monooxygenase	
1.1.3.13	Oxidase, alcohol Alcohol oxidase	
1.2.3.1	Oxidase, aldehyde Aldehyde oxidase	
1.1.3.7	Oxidase, aryl alcohol Aryl-alcohol oxidase	
1.10.3.3	Oxidase, ascorbate L-Ascorbate oxidase	
1.1.3.34	Oxidase, benzyltetrahydroisoquinoline Berbamunine synthase	
1.3.3.5	Oxidase, bilirubin Bilirubin oxidase	
1.2.3.10	Oxidase, carbon monoxide Carbon-monoxide oxidase	
1.1.3.25	Oxidase, cellobiose Cellobiose oxidase	
1.1.3.6	Oxidase, cholesterol Cholesterol oxidase	
1.1.99.1	Oxidase, choline Choline dehydrogenase	

EC-No.	Synonym Recommended Name	Vol., Page
1.1.3.17	Oxidase, choline Choline oxidase	
1.3.3.3	Oxidase, coproporphyrinogen Coproporphyrinogen oxidase	
1.9.3.1	Oxidase, cytochrome Cytochrome-c oxidase	
1.4.3.3	Oxidase, D-amino acid D-Amino-acid oxidase	
1.1.3.37	Oxidase, D-arabinono-γ-lactone D-Arabinono-1,4-lactone oxidase	
1.4.3.1	Oxidase, D-aspartate D-Aspartate oxidase	
1.5.3.12	Oxidase, dihydrobenzophenanthridine Dihydrobenzophenanthridine oxidase	
1.3.3.1	Oxidase, dihydroorotate Dihydroorotate oxidase	
1.3.3.7	Oxidase, dihydrouracil Dihydrouracil oxidase	
1.5.3.10	Oxidase, dimethylglycine Dimethylglycine oxidase	
1.1.3.16	Oxidase, ecdysone Ecdysone oxidase	
1.4.3.8	Oxidase, ethanolamine Ethanolamine oxidase	
1.16.3.1	Oxidase, ferr- Ferroxidase	
1.1.3.9	Oxidase, galactose Galactose oxidase	
1.1.3.4	Oxidase, glucose Glucose oxidase	
1.8.3.3	Oxidase, glutathione Glutathione oxidase	
1.1.3.21	Oxidase, glycerol phosphate Glycerol-3-phosphate oxidase	
1.7.3.4	Oxidase, hydroxylamine Hydroxylamine oxidase	

EC-No.	Synonym Recommended Name	Vol., Page
1.2.3.7	Oxidase, indoleacetaldehyde Indole-3-acetaldehyde oxidase	
1.1.3.15	Oxidase, L-2-hydroxy acid (S)-2-Hydroxy-acid oxidase	
1.4.3.16	Oxidase, L-aspartate L-Aspartate oxidase	
1.1.3.8	Oxidase, L-gulonolactone L-Gulonolactone oxidase	
1.13.12.4	Oxidase, lactate Lactate 2-monooxygenase	
1.3.3.2	Oxidase, lathosterol Lathosterol oxidase	
1.1.3.20	Oxidase, long-chain fatty alcohol Long-chain-alcohol oxidase	
1.1.3.3	Oxidase, malate Malate oxidase	
1.1.3.31	Oxidase, methanol Methanol oxidase	
1.8.3.4	Oxidase, methyl mercaptan Methanethiol oxidase	
1.1.3.29	Oxidase, N-acyl-D-hexosamine N-Acylhexosamine oxidase	
1.5.3.2	Oxidase, N-methylamino acid N-Methyl-L-amino-acid oxidase	
1.5.3.4	Oxidase, N^6-methyllysine N^6-Methyl-lysine oxidase	
1.13.11.32	Oxidase, nitroalkane 2-Nitropropane dioxygenase	
1.7.3.1	Oxidase, nitroethane Nitroethane oxidase	
1.1.3.28	Oxidase, nucleoside Nucleoside oxidase	
1.10.3.4	Oxidase, *o*-aminophenol *o*-Aminophenol oxidase	
1.2.3.4	Oxidase, oxalate Oxalate oxidase	

EC-No.	Synonym Recommended Name	Vol., Page
1.13.12.9	Oxidase, phenylalanine (deaminating, decarboxylating) Phenylalanine 2-monooxygenase	
1.5.3.7	Oxidase, pipecolate L-Pipecolate oxidase	
1.3.3.4	Oxidase, protoporphyrinogen Protoporphyrinogen oxidase	
1.9.3.2	Oxidase, Pseudomonas cytochrome Pseudomonas cytochrome oxidase	
1.17.3.1	Oxidase, pteridine Pteridine oxidase	
1.4.3.10	Oxidase, putrescine Putrescine oxidase	
1.4.3.5	Oxidase, pyridoxamine phosphate Pyridoxamine-phosphate oxidase	
1.1.3.12	Oxidase, pyridoxol 4- Pyridoxine 4-oxidase	
1.2.3.3	Oxidase, pyruvate Pyruvate oxidase	
1.2.3.6	Oxidase, pyruvate (coenzyme A-acetylating) Pyruvate oxidase (CoA-acetylating)	
1.3.99.18	Oxidase, quinaldate 4- Quinaldate 4-oxidoreductase	
1.2.3.11	Oxidase, retinal Retinal oxidase	
1.10.3.6	Oxidase, rifamycin B Rifamycin-B oxidase	
1.5.3.1	Oxidase, sarcosine Sarcosine oxidase	
1.1.3.11	Oxidase, sorbose L-Sorbose oxidase	
1.8.3.1	Oxidase, sulfite Sulfite oxidase	
1.10.3.7	Oxidase, sulochrin Sulochrin oxidase ((+)-bisdechlorogeodin-forming)	
1.10.3.8	Oxidase, sulochrin Sulochrin oxidase ((-)-bisdechlorogeodin-forming)	

EC-No.	Synonym Recommended Name	Vol., Page
1.8.3.2	Oxidase, thiol Thiol oxidase	
1.8.2.2	Oxidase, thiosulfate Thiosulfate dehydrogenase	
1.13.99.3	Oxidase, tryptophan side-chain α,β- Tryptophan 2'-dioxygenase	
1.13.99.3	Oxidase, tryptophan side-chain α,β-, II Tryptophan 2'-dioxygenase	
1.7.3.3	Oxidase, urate Urate oxidase	
1.1.3.38	Oxidase, vanillyl alcohol Vanillyl-alcohol oxidase	
1.1.3.22	Oxidase, xanthine Xanthine oxidase	
3.7.1.7	Oxidized PVA hydrolase β-Diketone hydrolase	
5.4.99.8	2,3-Oxidosqualene-cycloartenol cyclase Cycloartenol synthase	Vol. 1, Page 631
5.4.99.7	2,3-Oxidosqualene-lanosterol cyclase Lanosterol synthase	Vol. 1, Page 624
5.4.99.7	Oxidosqualene-lanosterol cyclase Lanosterol synthase	Vol. 1, Page 624
5.4.99.8	2,3-Oxidosqualene:cycloartenol cyclase Cycloartenol synthase	Vol. 1, Page 631
5.4.99.7	2,3-Oxidosqualene cyclase Lanosterol synthase	Vol. 1, Page 624
5.4.99.7	Oxidosqualene cyclase Lanosterol synthase	Vol. 1, Page 624
5.4.99.8	2,3-Oxidosqualene cycloartenol cyclase Cycloartenol synthase	Vol. 1, Page 631
5.4.99.7	2,3-Oxidosqualene sterol cyclase Lanosterol synthase	Vol. 1, Page 624
2.6.3.1	Oximase Oximinotransferase	
2.6.3.1	Oximinotransferase Oximinotransferase	

EC-No.	Synonym Recommended Name	Vol., Page
1.14.12.16	2-Oxo-1,2-dihydroquinoline 5,6-dioxygenase 2-Hydroxyquinoline 5,6-dioxygenase	
1.14.13.65	2-Oxo-1,2-dihydroquinoline 5,6-dioxygenase 2-Hydroxyquinoline 8-monooxygenase	
1.14.13.61	2-Oxo-1,2-dihydroquinoline 8-monooxygenase 2-Hydroxyquinoline 8-monooxygenase	
1.14.13.61	2-Oxo-1,2-dihydroquinoline 8-monooxygenase (Pseudomonas putida strain 86 gene oxoO subunit) 2-Hydroxyquinoline 8-monooxygenase	
1.14.13.61	2-Oxo-1,2-dihydroquinoline 8-monooxygenase (Pseudomonas putida strain 86 gene oxoR subunit) 2-Hydroxyquinoline 8-monooxygenase	
5.3.3.4	5-Oxo-3,4-dihydrofuran-2-acetate Δ^3-Δ^2-isomerase Muconolactone Δ-isomerase	Vol. 1, Page 399
4.1.2.21	2-Oxo-3-deoxy-6-phosphogalactonate aldolase 2-Dehydro-3-deoxyphosphogalactonate aldolase	
4.1.2.14	2-Oxo-3-deoxy-6-phosphogluconate aldolase 2-Dehydro-3-deoxyphosphogluconate aldolase	
2.7.1.58	2-Oxo-3-deoxygalactonate kinase 2-Dehydro-3-deoxygalactonokinase	
4.1.3.16	2-Oxo-4-hydroxyglutarate aldolase 4-Hydroxy-2-oxoglutarate aldolase	
4.1.2.31	2-oxo-4-hydroxyglutarate aldolase 2-oxo-4-hydroxyglutarate aldolase	
4.1.3.16	2-Oxo-4-hydroxyglutaric aldolase 4-Hydroxy-2-oxoglutarate aldolase	
1.3.99.6	3-Oxo-Δ^4-steroid-5β-reductase 3-Oxo-5β-steroid Δ^4-dehydrogenase	
4.1.1.75	2-Oxo-5-guanidinopentanoate carboxy-lyase 5-Guanidino-2-oxopentanoate decarboxylase	
1.3.99.5	3-Oxo-5α-steroid Δ^4-dehydrogenase 3-Oxo-5α-steroid Δ^4-dehydrogenase	
1.3.99.5	3-Oxo-5α-steroid δ^4-dehydrogenase 3-Oxo-5α-steroid Δ^4-dehydrogenase	
1.3.99.6	3-Oxo-5β-steroid Δ^4-dehydrogenase 3-Oxo-5β-steroid Δ^4-dehydrogenase	

EC-No.	Synonym Recommended Name	Vol., Page
1.3.99.5	3-Oxo-5α-steroid:(acceptor) Δ^4-oxidoreductase 3-Oxo-5α-steroid Δ^4-dehydrogenase	
1.3.99.6	3-Oxo-5β-steroid:(acceptor) Δ^4-oxidoreductase 3-Oxo-5β-steroid Δ^4-dehydrogenase	
1.3.1.22	3-Oxo-5α-steroid:NADP$^+$ Δ-oxidoreductase Cholestenone 5α-reductase	
1.3.1.23	3-Oxo-5β-steroid:NADP$^+$ Δ^4-oxidoreductase Cholestenone 5β-reductase	
5.3.3.1	3-Oxo-δ^5 steroid isomerase Steroid Δ-isomerase	Vol. 1, Page 376
4.1.1.1	2-Oxo-acid carboxy-lyase Pyruvate decarboxylase	
2.8.3.5	3-Oxo-CoA transferase 3-Oxoacid CoA-transferase	
3.5.2.9	5-Oxo-L-prolinase 5-Oxoprolinase (ATP-hydrolysing)	
3.5.2.9	5-Oxo-L-prolinase-Components A (β_5) and B (F8) 5-Oxoprolinase (ATP-hydrolysing)	
3.5.2.9	5-Oxo-L-proline amidohydrolase (ATP-hydrolysing) 5-Oxoprolinase (ATP-hydrolysing)	
1.3.1.42	12-Oxo-phytodienoic acid reductase 12-Oxophytodienoate reductase	
2.8.3.5	3-Oxoacid CoA-transferase 3-Oxoacid CoA-transferase	
2.8.3.5	3-Oxoacid CoA dehydrogenase 3-Oxoacid CoA-transferase	
2.8.3.5	3-Oxoacid coenzyme A-transferase 3-Oxoacid CoA-transferase	
1.1.1.96	2-Oxo acid reductase Diiodophenylpyruvate reductase	
1.1.1.100	3-Oxoacyl-[ACP]reductase 3-Oxoacyl-[acyl-carrier-protein] reductase	
1.1.1.100	3-Oxoacyl-[acyl-carrier-protein] reductase 3-Oxoacyl-[acyl-carrier-protein] reductase	
1.1.1.212	3-Oxoacyl-[acyl-carrier-protein] reductase (NADH) 3-Oxoacyl-[acyl-carrier-protein] reductase (NADH)	

EC-No.	Synonym Recommended Name	Vol., Page
2.3.1.41	3-Oxoacyl-[acyl-carrier-protein] synthase 3-Oxoacyl-[acyl-carrier-protein] synthase	
2.3.1.16	3-Oxoacyl-CoA thiolase Acetyl-CoA C-acyltransferase	
2.3.1.16	6-Oxoacyl-CoA thiolase Acetyl-CoA C-acyltransferase	
2.3.1.16	3-Oxoacyl-coenzyme A thiolase Acetyl-CoA C-acyltransferase	
2.3.1.16	Oxoacyl-coenzyme A thiolase Acetyl-CoA C-acyltransferase	
1.3.1.32	3-Oxoadipate: NAD(P)$^+$ oxidoreductase Maleylacetate reductase	
2.8.3.6	3-Oxoadipate CoA-transferase 3-Oxoadipate CoA-transferase	
2.8.3.6	3-Oxoadipate coenzyme A-transferase 3-Oxoadipate CoA-transferase	
3.1.1.24	3-oxoadipate enol-lactone hydrolase 3-oxoadipate enol-lactose	
3.1.1.24	3-oxoadipate enol-lactose 3-oxoadipate enol-lactose	
1.1.1.172	2-Oxoadipate reductase 2-Oxoadipate reductase	
2.8.3.6	3-Oxoadipate succinyl-CoA transferase 3-Oxoadipate CoA-transferase	
1.2.1.23	2-Oxoaldehyde:NAD$^+$ 2-oxidoreductase 2-Oxoaldehyde dehydrogenase (NAD$^+$)	
1.2.1.49	2-Oxoaldehyde:NADP$^+$ 2-oxidoreductase 2-Oxoaldehyde dehydrogenase (NADP$^+$)	
1.2.1.23	2-Oxoaldehyde dehydrogenase (NAD$^+$) 2-Oxoaldehyde dehydrogenase (NAD$^+$)	
1.2.1.49	2-Oxoaldehyde dehydrogenase (NADP$^+$) 2-Oxoaldehyde dehydrogenase (NADP$^+$)	
1.2.7.2	2-Oxobutanoate: ferredoxin 2-oxidoreductase (CoA-propionylating) 2-Oxobutyrate synthase	
1.2.7.2	2-Oxobutyrate-ferredoxin oxidoreductase 2-Oxobutyrate synthase	

EC-No.	Synonym Recommended Name	Vol., Page
1.2.7.2	2-Oxobutyrate synthase 2-Oxobutyrate synthase	
1.14.13.51	6-Oxocineole,NADPH:oxygen oxidoreductase 6-Oxocineole dehydrogenase	
1.14.13.51	6-Oxocineole dehydrogenase 6-Oxocineole dehydrogenase	
1.1.1.235	8-Oxocoformycin reductase 8-Oxocoformycin reductase	
3.11.1.1	2-Oxoethylphosphonate phosphonohydrolase Phosphonoacetylaldehyde hydrolase	
1.1.99.4	2-Oxogluconate dehydrogenase Dehydrogluconate dehydrogenase	
1.2.7.3	2-Oxoglutarate-ferredoxin oxidoreductase 2-Oxoglutarate synthase	
2.6.1.1	2-Oxoglutarate-glutamate aminotransferase Aspartate transaminase	
1.2.7.3	2-Oxoglutarate:ferredoxin 2-oxidoreductase (CoA-succinylating) 2-Oxoglutarate synthase	
4.1.3.15	2-oxoglutarate: glyoxylate carboligase 2-hydroxy-3-oxoadipate synthase	
4.1.3.15	oxoglutarate: glyoxylate carboligase 2-hydroxy-3-oxoadipate synthase	
1.2.4.2	2-Oxoglutarate:lipoamide 2-oxidoreductase (decarboxylating and acceptor-succinylating) Oxoglutarate dehydrogenase (lipoamide)	
1.2.4.2	2-Oxoglutarate: lipoate oxidoreductase Oxoglutarate dehydrogenase (lipoamide)	
1.2.1.52	2-Oxoglutarate:NADP$^+$ 2-oxidoreductase (CoA-succinylating) Oxoglutarate dehydrogenase (NADP$^+$)	
4.1.1.71	2-Oxoglutarate carboxy-lyase 2-Oxoglutarate decarboxylase	
4.1.1.71	2-Oxoglutarate decarboxylase 2-Oxoglutarate decarboxylase	
4.1.1.71	Oxoglutarate decarboxylase 2-Oxoglutarate decarboxylase	

EC-No.	Synonym Recommended Name	Vol., Page
1.2.4.2	Oxoglutarate decarboxylase Oxoglutarate dehydrogenase (lipoamide)	
1.2.4.2	2-Oxoglutarate dehydrogenase Oxoglutarate dehydrogenase (lipoamide)	
1.2.4.2	Oxoglutarate dehydrogenase Oxoglutarate dehydrogenase (lipoamide)	
1.2.4.2	α-Oxoglutarate dehydrogenase Oxoglutarate dehydrogenase (lipoamide)	
1.2.4.2	Oxoglutarate dehydrogenase (lipoamide) Oxoglutarate dehydrogenase (lipoamide)	
1.2.1.52	Oxoglutarate dehydrogenase ($NADP^+$) Oxoglutarate dehydrogenase ($NADP^+$)	
1.2.1.26	2-Oxoglutarate semialdehyde dehydrogenase 2,5-Dioxovalerate dehydrogenase	
1.2.7.3	2-Oxoglutarate synthase 2-Oxoglutarate synthase	
1.2.7.3	Oxoglutarate synthase 2-Oxoglutarate synthase	
1.2.1.63	6-oxohexanoate:NADP oxidoreductase 6-oxohexanoate dehydrogenase	
1.2.1.63	6-oxohexanoate dehydrogenase 6-oxohexanoate dehydrogenase	
1.2.4.4	2-Oxoisocaproate dehydrogenase 3-Methyl-2-oxobutanoate dehydrogenase (lipoamide)	
1.2.4.3	2-oxoisocaproate dehydrogenase 2-oxoisocaproate dehydrogenase	
1.2.4.4	α-Oxoisocaproate dehydrogenase 3-Methyl-2-oxobutanoate dehydrogenase (lipoamide)	
5.3.1.9	Oxoisomerase Glucose-6-phosphate isomerase	Vol. 1, Page 298
1.2.1.25	2-Oxoisovalerate dehydrogenase 2-Oxoisovalerate dehydrogenase (acylating)	
1.2.1.25	2-Oxoisovalerate dehydrogenase (acylating) 2-Oxoisovalerate dehydrogenase (acylating)	
1.2.4.4	2-Oxoisovalerate dehydrogenase (lipoate) 3-Methyl-2-oxobutanoate dehydrogenase (lipoamide)	

EC-No.	Synonym Recommended Name	Vol., Page
4.1.1.56	3-Oxolaurate carboxy-lyase 3-Oxolaurate decarboxylase	
4.1.1.56	3-Oxolaurate decarboxylase 3-Oxolaurate decarboxylase	
2.3.1.155	3-oxopalmitoyl-CoA-CoA acetyltransferase acetyl-CoA C-myristoyltransferase	
2.3.1.155	3-oxopalmitoyl-CoA hydrolase acetyl-CoA C-myristoyltransferase	
2.1.2.11	Oxopantoate hydroxymethyltransferase 3-Methyl-2-oxobutanoate hydroxymethyltransferase	
4.1.1.68	5-Oxopent-3-ene-1,2,5-tricarboxylate carboxy-lyase 5-Oxopent-3-ene-1,2,5-tricarboxylate decarboxylase	
4.1.1.68	5-Oxopent-3-ene-1,2,5-tricarboxylate decarboxylase 5-Oxopent-3-ene-1,2,5-tricarboxylate decarboxylase	
4.2.1.80	2-oxopent-4-enoate hydratase 2-oxopent-4-enoate hydratase	
4.1.1.68	5-Oxopenta-3-ene-1,2,5-tricarboxylate decarboxylase 5-Oxopent-3-ene-1,2,5-tricarboxylate decarboxylase	
1.3.1.42	12-Oxophytodienoate reductase 12-Oxophytodienoate reductase	
2.3.1.154	3-oxopristanoyl-CoA hydrolase propionyl-CoA C^2-trimethyltridecanoyltransferase	
2.3.1.154	3-oxopristanoyl-CoA thiolase propionyl-CoA C^2-trimethyltridecanoyltransferase	
2.3.1.154	oxopristanoyl-CoA thiolase propionyl-CoA C^2-trimethyltridecanoyltransferase	
3.5.2.9	5-Oxoprolinase 5-Oxoprolinase (ATP-hydrolysing)	
3.5.2.9	Oxoprolinase 5-Oxoprolinase (ATP-hydrolysing)	
3.5.2.9	5-Oxoprolinase (ATP-hydrolysing) 5-Oxoprolinase (ATP-hydrolysing)	
1.1.1.104	4-Oxoproline reductase 4-Oxoproline reductase	
3.4.19.3	5-oxoprolyl-peptidase pyroglutamyl-peptidase I	

EC-No.	Synonym Recommended Name	Vol., Page
4.1.1.45	3-(3-Oxoprop-2-enyl)-2-aminobut-2-endioate carboxy-lyase Aminocarboxymuconate-semialdehyde decarboxylase	
1.2.1.15	3-Oxopropanoate:NAD(P)$^+$ oxidoreductase Malonate-semialdehyde dehydrogenase	
1.2.1.18	3-Oxopropanoate:NAD(P)$^+$ oxidoreductase (decarboxylating, CoA-acetylating) Malonate-semialdehyde dehydrogenase (acetylating)	
4.2.1.27	3-oxopropanoate hydro-lyase malonate-semialdehyde dehydratase	
1.2.3.3	2-Oxopropanoic acid oxidase Pyruvate oxidase	
1.3.1.48	15-Oxoprostaglandin Δ^{13}-reductase 15-Oxoprostaglandin Δ^{13}-reductase	
1.14.13.62	1H-4-Oxoquinoline monooxygenase 4-Hydroxyquinoline 3-monooxygenase	
1.1.1.102	D-3-Oxosphinganine:B-NADPH oxidoreductase 3-Dehydrosphinganine reductase	
1.1.1.102	3-Oxosphinganine:NADPH oxidoreductase 3-Dehydrosphinganine reductase	
1.1.1.102	3-Oxosphinganine reductase 3-Dehydrosphinganine reductase	
1.1.1.102	D-3-Oxosphinganine reductase 3-Dehydrosphinganine reductase	
2.3.1.50	3-Oxosphinganine synthetase Serine C-palmitoyltransferase	
1.3.1.4	Δ^4-3-Oxosteroid-5α-reductase Cortisone α-reductase	
1.3.99.5	δ^4-3-Oxosteroid-5α-reductase 3-Oxo-5α-steroid Δ^4-dehydrogenase	
1.3.99.4	3-Oxosteroid Δ^1-dehydrogenase 3-Oxosteroid Δ^1-dehydrogenase	
1.3.99.4	3-Oxosteroid δ 1-dehydrogenase 3-Oxosteroid Δ^1-dehydrogenase	
5.3.3.1	3-Oxo steroid Δ^4-Δ^5-isomerase Steroid Δ-isomerase	Vol. 1, Page 376

EC-No.	Synonym Recommended Name	Vol., Page
5.3.3.1	3-Oxosteroid Δ^4-Δ^5-isomerase Steroid Δ-isomerase	Vol. 1, Page 376
1.3.99.5	3-Oxosteroid δ^4-dehydrogenase 3-Oxo-5α-steroid Δ^4-dehydrogenase	
1.3.1.22	3-Oxosteroid δ^4-dehydrogenase Cholestenone 5α-reductase	
5.3.3.1	3-Oxosteroid Δ^5-Δ^4-isomerase Steroid Δ-isomerase	Vol. 1, Page 376
5.3.3.1	3-Oxosteroid Δ^5-Δ^4-isomerase Steroid Δ-isomerase	Vol. 1, Page 376
1.3.99.6	Δ^4-3-Oxosteroid 5β-reductase 3-Oxo-5β-steroid Δ^4-dehydrogenase	
1.3.1.22	3-Oxosteroid 5. α-reductase Cholestenone 5α-reductase	
1.3.99.4	3-Oxosteroid:(2,6-dichlorphenolindophenol) Δ1-oxidoreductase 3-Oxosteroid Δ^1-dehydrogenase	
1.3.99.4	3-Oxosteroid:(acceptor) Δ1-oxidoreductase 3-Oxosteroid Δ^1-dehydrogenase	
5.3.3.1	3-Oxosteroid isomerase Steroid Δ-isomerase	Vol. 1, Page 376
5.3.3.1	Δ^5-3-oxosteroid isomerase Steroid Δ-isomerase	Vol. 1, Page 376
1.3.99.5	δ^4-3-Oxo steroid reductase 3-Oxo-5α-steroid Δ^4-dehydrogenase	
1.3.7.1	6-Oxotetrahydronicotinate dehydrogenase 6-Hydroxynicotinate reductase	
2.3.1.9	3-Oxothiolase Acetyl-CoA C-acetyltransferase	
1.14.13.47	Oxygenase, (-)-limonene 3-mono- (-)-Limonene 3-monooxygenase	
1.14.13.48	Oxygenase, (-)-limonene 6-mono- (-)-Limonene 6-monooxygenase	
1.14.13.49	Oxygenase, (-)-limonene mono- (-)-Limonene 7-monooxygenase	
1.14.13.37	Oxygenase, (S)-cis-N-methyltetrahydroberberine 14-mono- Methyltetrahydroprotoberberine 14-monooxygenase	

EC-No.	Synonym Recommended Name	Vol., Page
1.14.11.1	Oxygenase, γ-butyrobetaine di- γ-Butyrobetaine dioxygenase	
1.13.11.21	Oxygenase, β-carotene 15,15'-di- β-Carotene 15,15'-dioxygenase	
1.13.11.38	Oxygenase, 1-hydroxy-2-naphthoate di- 1-Hydroxy-2-naphthoate 1,2-dioxygenase	
1.13.99.5	Oxygenase, 1H-3-hydroxy-4-oxoquinoline 2,4-di 3,4-Dihydroxyquinoline 2,4-dioxygenase	
1.14.13.65	Oxygenase, 2(1H)-quinolinone 5,6-di 2-Hydroxyquinoline 8-monooxygenase	
1.14.12.16	Oxygenase, 2(1H)-quinolinone 5,6-di- 2-Hydroxyquinoline 5,6-dioxygenase	
1.13.11.28	Oxygenase, 2,3-dihydroxybenzoate 2,3-di- 2,3-Dihydroxybenzoate 2,3-dioxygenase	
1.13.11.14	Oxygenase, 2,3-dihydroxybenzoate 3,4-di- 2,3-Dihydroxybenzoate 3,4-dioxygenase	
1.14.13.20	Oxygenase, 2,4-dichlorophenol 6-mono 2,4-Dichlorophenol 6-monooxygenase	
1.14.13.20	Oxygenase, 2,4-dichlorophenol mono- 2,4-Dichlorophenol 6-monooxygenase	
1.13.11.9	Oxygenase, 2,5-dihydroxypyridine 5,6-di- 2,5-Dihydroxypyridine 5,6-dioxygenase	
1.14.12.14	Oxygenase, 2-aminobenzenesulfonate di- 2-Aminobenzenesulfonate 2,3-dioxygenase	
1.14.12.13	Oxygenase, 2-chlorobenzoate 1,2-di- 2-Chlorobenzoate 1,2-dioxygenase	
1.14.13.44	Oxygenase, 2-hydroxybiphenyl 3-mono- 2-Hydroxybiphenyl 3-monooxygenase	
1.14.13.66	Oxygenase, 2-hydroxycyclohexanone 2-mono- 2-Hydroxycyclohexanone 2-monooxygenase	
1.14.99.26	Oxygenase, 2-hydroxypyridine 5-mono- 2-Hydroxypyridine 5-monooxygenase	
1.14.13.31	Oxygenase, 2-nitrophenol 2-Nitrophenol 2-monooxygenase	
1.13.11.32	Oxygenase, 2-nitropropane di- 2-Nitropropane dioxygenase	

EC-No.	Synonym Recommended Name	Vol., Page
1.14.13.61	Oxygenase, 2-oxo-1,2-dihydroquinoline 8-mono- 2-Hydroxyquinoline 8-monooxygenase	
1.14.13.61	Oxygenase, 2-oxo-1,2-dihydroquinoline 8-mono- (Pseudomonas putida strain 86 gene oxoO subunit) 2-Hydroxyquinoline 8-monooxygenase	
1.14.13.61	Oxygenase, 2-oxo-1,2-dihydroquinoline 8-mono- (Pseudomonas putida strain 86 gene oxoR subunit) 2-Hydroxyquinoline 8-monooxygenase	
1.3.99.19	Oxygenase, 2-quinolinecarboxylate 4-mono Quinoline-4-carboxylate 2-oxidoreductase	
1.14.13.13	Oxygenase, 25-hydroxycholecalciferol 1-mono- Calcidiol 1-monooxygenase	
1.14.13.60	Oxygenase, 27-hydroxycholesterol 7α-mono 27-Hydroxycholesterol 7α-monooxygenase	
1.14.13.28	Oxygenase, 3,9-dihydroxypterocarpan 6α-mono- 3,9-Dihydroxypterocarpan 6a-monooxygenase	
1.13.11.6	Oxygenase, 3-hydroxyanthranilate 3,4-di- 3-Hydroxyanthranilate 3,4-dioxygenase	
1.14.99.23	Oxygenase, 3-hydroxybenzoate 2-mono- 3-Hydroxybenzoate 2-monooxygenase	
1.14.13.23	Oxygenase, 3-hydroxybenzoate 4-mono- 3-Hydroxybenzoate 4-monooxygenase	
1.14.13.24	Oxygenase, 3-hydroxybenzoate 6-mono- 3-Hydroxybenzoate 6-monooxygenase	
1.14.13.63	Oxygenase, 3-hydroxyphenylacetate 6-mono 3-Hydroxyphenylacetate 6-hydroxylase	
1.14.13.62	Oxygenase, 4(1H)-oxoquinoline 3-mono- 4-Hydroxyquinoline 3-monooxygenase	
1.14.13.27	Oxygenase, 4-aminobenzoate mono- 4-Aminobenzoate 1-monooxygenase	
1.14.12.9	Oxygenase, 4-chlorophenylacetate 3,4-di- 4-Chlorophenylacetate 3,4-dioxygenase	
1.14.13.64	Oxygenase, 4-hydroxybenzoate 1-mono 4-Hydroxybenzoate 1-hydroxylase	
1.14.13.2	Oxygenase, 4-hydroxybenzoate 3-mono- 4-Hydroxybenzoate 3-monooxygenase	

EC-No.	Synonym Recommended Name	Vol., Page
1.14.13.33	Oxygenase, 4-hydroxybenzoate 3-mono- (reduced nicotinamide adenine dinucleotide (phosphate)) 4-Hydroxybenzoate 3-monooxygenase (NAD(P)H)	
1.14.13.18	Oxygenase, 4-hydroxyphenylacetate 1-mono- 4-Hydroxyphenylacetate 1-monooxygenase	
1.14.13.3	Oxygenase, 4-hydroxyphenylacetate 3-mono- 4-Hydroxyphenylacetate 3-monooxygenase	
1.14.13.42	Oxygenase, 4-hydroxyphenylacetonitrile mono- Hydroxyphenylacetonitrile 2-monooxygenase	
1.13.11.27	Oxygenase, 4-hydroxyphenylpyruvate di- 4-Hydroxyphenylpyruvate dioxygenase	
1.14.99.15	Oxygenase, 4-methoxybenzoate 4-mono- (O-demethylating) 4-Methoxybenzoate monooxygenase (O-demethylating)	
1.14.13.29	Oxygenase, 4-nitrophenol 2-mono- 4-Nitrophenol 2-monooxygenase	
1.14.12.8	Oxygenase, 4-sulfobenzoate di- 4-Sulfobenzoate 3,4-dioxygenase	
1.14.11.6	Oxygenase, 5-hydroxymethyluracil di- Thymine dioxygenase	
1.14.13.51	Oxygenase, 6-oxocineole 6-Oxocineole dehydrogenase	
1.14.15.3	Oxygenase, alkane 1-mono- Alkane 1-monooxygenase	
1.14.99.12	Oxygenase, androstenedione mono- 4-Androstene-3,17-dione monooxygenase	
1.14.13.38	Oxygenase, anhydrotetracycline Anhydrotetracycline monooxygenase	
1.14.12.1	Oxygenase, anthranilate 1,2-di- (deaminating, decarboxylating) Anthranilate 1,2-dioxygenase (deaminating, decarboxylating)	
1.14.13.35	Oxygenase, anthranilate 2,3-di- (deaminating) Anthranilate 3-monooxygenase (deaminating)	
1.14.16.3	Oxygenase, anthranilate 3-mono- Anthranilate 3-monooxygenase	
1.13.11.34	Oxygenase, arachidonate, 5-lip- Arachidonate 5-lipoxygenase	

EC-No.	Synonym Recommended Name	Vol., Page
1.13.11.40	Oxygenase, arachidonate, 8-lip- Arachidonate 8-lipoxygenase	
1.13.11.31	Oxygenase, arachidonate 12-lip- Arachidonate 12-lipoxygenase	
1.13.11.33	Oxygenase, arachidonate 15-lip- Arachidonate 15-lipoxygenase	
1.13.12.1	Oxygenase, arginine 2-mono- Arginine 2-monooxygenase	
1.13.11.13	Oxygenase, ascorbate 2,3-di- Ascorbate 2,3-dioxygenase	
1.14.11.16	Oxygenase, aspartylpeptide β-di- Peptide-aspartate β-dioxygenase	
1.14.12.3	Oxygenase, benzene 1,2-di- Benzene 1,2-dioxygenase	
1.14.12.10	Oxygenase, benzoate 1,2-di- Benzoate 1,2-dioxygenase	
1.14.13.12	Oxygenase, benzoate 4-mono- Benzoate 4-monooxygenase	
1.14.13.58	Oxygenase, benzoyl coenzyme A 3-mono Benzoyl-CoA 3-monooxygenase	
1.13.11.39	Oxygenase, biphenyl-2,3-diol di- Biphenyl-2,3-diol 1,2-dioxygenase	
1.14.15.2	Oxygenase, camphor 1,2-mono Camphor 1,2-monooxygenase	
1.14.15.1	Oxygenase, camphor 5-mono- Camphor 5-monooxygenase	
1.13.11.1	Oxygenase, catechol 1,2-di- Catechol 1,2-dioxygenase	
1.13.11.2	Oxygenase, catechol 2,3-di- Catechol 2,3-dioxygenase	
1.13.11.36	Oxygenase, chloridazoncatechol di- Chloridazon-catechol dioxygenase	
1.14.13.15	Oxygenase, cholestanetriol 26-mono- Cholestanetriol 26-monooxygenase	
1.14.13.17	Oxygenase, cholesterol 7α-mono- Cholesterol 7α-monooxygenase	

EC-No.	Synonym Recommended Name	Vol., Page
1.14.13.14	Oxygenase, cinnamate 2-mono- trans-Cinnamate 2-monooxygenase	
1.14.13.11	Oxygenase, cinnamate 4-mono- trans-Cinnamate 4-monooxygenase	
1.14.15.5	Oxygenase, corticosterone 18-mono- Corticosterone 18-monooxygenase	
1.14.13.36	Oxygenase, coumaroylquinate(coumaroylshikimate) 3'-mono- 5-O-(4-Coumaroyl)-D-quinate 3'-monooxygenase	
1.14.13.22	Oxygenase, cyclohexanone 1, 2-mono- Cyclohexanone monooxygenase	
1.14.13.22	Oxygenase, cyclohexanone mono- Cyclohexanone monooxygenase	
1.13.11.19	Oxygenase, cysteamine Cysteamine dioxygenase	
1.13.11.19	Oxygenase, cysteamine di- Cysteamine dioxygenase	
1.13.11.20	Oxygenase, cysteine di- Cysteine dioxygenase	
1.14.13.45	Oxygenase, cytidine monophosphoacetylneuraminate mono- CMP-N-acetylneuraminate monooxygenase	
1.14.99.29	Oxygenase, deoxyhypusine di- Deoxyhypusine monooxygenase	
1.14.11.10	Oxygenase, deoxyuridine-uridine 1'-di- Pyrimidine-deoxynucleoside 1'-dioxygenase	
1.14.11.3	Oxygenase, deoxyuridine 2'-di- Pyrimidine-deoxynucleoside 2'-dioxygenase	
1.11.1.14	Oxygenase, diarylpropane Diarylpropane peroxidase	
1.14.13.57	Oxygenase, dihydrochelirubine 12-mono Dihydrochelirubine 12-monooxygenase	
1.14.13.56	Oxygenase, dihydrosanguinarine 10-mono Dihydrosanguinarine 10-monooxygenase	
1.14.13.8	Oxygenase, dimethylaniline mono- (N-oxide-forming) Dimethylaniline monooxygenase (N-oxide-forming)	
1.14.17.1	Oxygenase, dopamine β-mono- Dopamine β-monooxygenase	

EC-No.	Synonym Recommended Name	Vol., Page
1.14.99.22	Oxygenase, ecdysone 20-mono- Ecdysone 20-monooxygenase	
1.14.99.11	Oxygenase, estradiol 6β-mono- Estradiol 6β-monooxygenase	
1.14.11.9	Oxygenase, flavanone 3-di- Naringenin 3-dioxygenase	
1.14.13.21	Oxygenase, flavonoid 3'-mono- Flavonoid 3'-monooxygenase	
1.14.13.21	Oxygenase, flavonoid 3-mono- Flavonoid 3'-monooxygenase	
1.14.14.1	Oxygenase, flavoprotein-linked mono- Unspecific monooxygenase	
1.13.11.4	Oxygenase, gentisate 1,2-di- Gentisate 1,2-dioxygenase	
1.14.11.13	Oxygenase, gibberellin 2β-di- Gibberellin 2β-dioxygenase	
1.14.11.15	Oxygenase, gibberellin 3β-di- Gibberellin 3β-dioxygenase	
1.14.11.12	Oxygenase, gibberellin A44 di- Gibberellin-44-dioxygenase	
1.14.16.5	Oxygenase, glyceryl ether mono- Glyceryl-ether monooxygenase	
1.14.99.3	Oxygenase, heme (decyclizing) Heme oxygenase (decyclizing)	
1.13.11.5	Oxygenase, homogentisate 1,2-di- Homogentisate 1,2-dioxygenase	
1.13.11.15	Oxygenase, homoprotocatechuate 2,3-di- 3,4-Dihydroxyphenylacetate 2,3-dioxygenase	
1.13.11.37	Oxygenase, hydroxyquinol 1,2-di- Hydroxyquinol 1,2-dioxygenase	
1.13.11.37	Oxygenase, hydroxyquinol di- Hydroxyquinol 1,2-dioxygenase	
1.14.11.11	Oxygenase, hyoscyamine 6β-di- Hyoscyamine (6S)-dioxygenase	
1.13.11.17	Oxygenase, indole 2,3-di- Indole 2,3-dioxygenase	

EC-No.	Synonym Recommended Name	Vol., Page
1.13.99.1	Oxygenase, inositol myo-Inositol oxygenase	
1.14.13.53	Oxygenase, isoflavone 2'-mono- Isoflavone 2'-hydroxylase	
1.14.13.52	Oxygenase, isoflavone 3'-mono- Isoflavone 3'-hydroxylase	
1.14.99.27	Oxygenase, juglone 3-mono- Juglone 3-monooxygenase	
1.14.13.9	Oxygenase, kynurenine 3-mono- Kynurenine 3-monooxygenase	
1.14.13.46	Oxygenase, l-menthol mono- (-)-Menthol monooxygenase	
1.13.12.4	Oxygenase, lactate 2-mono- Lactate 2-monooxygenase	
1.14.13.30	Oxygenase, leukotriene B_4 20-mono- Leukotriene-B_4 20-monooxygenase	
1.14.13.34	Oxygenase, leukotriene E4 20-mono Leukotriene-E_4 20-monooxygenase	
1.13.11.43	Oxygenase, lignostilbene α,β-di- Lignostilbene $\alpha\beta$-dioxygenase	
1.14.99.28	Oxygenase, linalool 8-mono- Linalool 8-monooxygenase	
1.13.11.12	Oxygenase, lip- Lipoxygenase	
1.13.12.2	Oxygenase, lysine 2-mono Lysine 2-monooxygenase	
1.14.13.59	Oxygenase, lysine N^6-mono- L-Lysine 6-monooxygenase (NADPH2)	
1.13.12.10	Oxygenase, lysine N^6-mono- Lysine 6-monooxygenase	
1.14.16.6	Oxygenase, mandelate 4-mono- Mandelate 4-monooxygenase	
1.14.13.4	Oxygenase, melilotate 3-mono- Melilotate 3-monooxygenase	
1.14.13.25	Oxygenase, methane mono- Methane monooxygenase	

EC-No.	Synonym Recommended Name	Vol., Page
1.13.12.11	Oxygenase, methylphenyltetrahydropyridine N-mono- Methylphenyltetrahydropyridine N-monooxygenase	
1.14.99.16	Oxygenase, methylsterol mono- Methylsterol monooxygenase	
1.14.99.18	Oxygenase, N-acetylneuraminate mono- N-Acetylneuraminate monooxygenase	
1.14.12.12	Oxygenase, naphthalene di- Naphthalene 1,2-dioxygenase	
1.14.13.26	Oxygenase, oleate Δ^{12}-mono Phosphatidylcholine 12-monooxygenase	
1.14.13.6	Oxygenase, orcinol 2-mono Orcinol 2-monooxygenase	
1.14.13.50	Oxygenase, pentachlorophenol 4-mono- Pentachlorophenol monooxygenase	
1.13.11.26	Oxygenase, peptidyltryptophan 2,3-di- Peptide-tryptophan 2,3-dioxygenase	
1.14.13.7	Oxygenase, phenol 2-mono- Phenol 2-monooxygenase	
1.14.16.1	Oxygenase, phenylalanine 4-mono- Phenylalanine 4-monooxygenase	
1.14.12.7	Oxygenase, phthalate 4,5-di- Phthalate 4,5-dioxygenase	
1.14.99.20	Oxygenase, phylloquinone mono-(2,3-epoxidizing) Phylloquinone monooxygenase (2,3-epoxidizing)	
1.14.99.14	Oxygenase, progesterone 11α-mono- Progesterone 11α-monooxygenase	
1.14.99.4	Oxygenase, progesterone mono- Progesterone monooxygenase	
1.13.11.3	Oxygenase, protocatechuate 3,4-di- Protocatechuate 3,4-dioxygenase	
1.13.11.8	Oxygenase, protocatechuate 4,5-di- Protocatechuate 4,5-dioxygenase	
1.14.11.4	Oxygenase, protocollagen lysine, di- Procollagen-lysine 5-dioxygenase	
1.14.11.7	Oxygenase, protocollagen proline 3-di- Procollagen-proline 3-dioxygenase	

EC-No.	Synonym Recommended Name	Vol., Page
1.14.13.55	Oxygenase, protropine 6-mono- Protopine 6-monooxygenase	
1.13.11.35	Oxygenase, pyrogallol 1,2-di- Pyrogallol 1,2-oxygenase	
1.14.13.43	Oxygenase, questin mono- Questin monooxygenase	
1.14.13.1	Oxygenase, salicylate 1-mono- Salicylate 1-monooxygenase	
1.14.99.7	Oxygenase, squalene mono- Squalene monooxygenase	
1.14.15.4	Oxygenase, steroid 11β-mono- Steroid 11β-monooxygenase	
1.14.99.9	Oxygenase, steroid 17α-mono- Steroid 17α-monooxygenase	
1.14.99.10	Oxygenase, steroid 21-mono- Steroid 21-monooxygenase	
1.14.99.24	Oxygenase, steroid 9α-mono- Steroid 9α-monooxygenase	
1.14.13.54	Oxygenase, steroid mono- Ketosteroid monooxygenase	
1.14.12.15	Oxygenase, terephthalate, 1,2-di Terephthalate 1,2-dioxygenase	
1.14.11.3	Oxygenase, thymidine 2'-di- Pyrimidine-deoxynucleoside 2'-dioxygenase	
1.14.11.3	Oxygenase, thymidine di- Pyrimidine-deoxynucleoside 2'-dioxygenase	
1.14.11.6	Oxygenase, thymine di- Thymine dioxygenase	
1.14.12.11	Oxygenase, toluene 2,3-di- Toluene dioxygenase	
1.14.11.8	Oxygenase, trimethyllysine di- Trimethyllysine dioxygenase	
1.13.12.3	Oxygenase, tryptophan 2-mono- Tryptophan 2-monooxygenase	
1.14.16.4	Oxygenase, tryptophan 5-mono- Tryptophan 5-monooxygenase	

EC-No.	Synonym Recommended Name	Vol., Page
1.14.16.2	Oxygenase, tyrosine 3-mono- Tyrosine 3-monooxygenase	
4.1.2.10	D-Oxynitrilase Mandelonitrile lyase	
4.1.2.11	Oxynitrilase Hydroxymandelonitrile lyase	
4.1.2.39	Oxynitrilase Hydroxynitrilase	
4.1.2.10	Oxynitrilase Mandelonitrile lyase	
4.1.2.11	S-Oxynitrilase Hydroxymandelonitrile lyase	
1.11.1.7	Oxyperoxidase Peroxidase	
1.14.13.60	Oxysterol 7α-hydroxylase 27-Hydroxycholesterol 7α-monooxygenase	
1.14.13.60	Oxysterol 7α-hydroxylase 27-Hydroxycholesterol 7α-monooxygenase	
3.4.11.3	oxytocinase cystinyl aminopeptidase	
3.4.11.3	oxytocin peptidase cystinyl aminopeptidase	
3.4.99.28	Oxyuranus scutellatus prothrombin-activating proteinase Oxyuranus scutellatus prothrombin-activating proteinase	
1.6.99.1	OYE NADPH dehydrogenase	
2.7.6.5	(p)ppGpp synthetase I GTP pyrophosphokinase	
2.7.6.5	(p)ppGpp synthetase II GTP pyrophosphokinase	
3.6.1.29	P^1,P^3-Bis(5'-adenosyl)-triphosphate adenylohydrolase Bis(5'-adenosyl)-triphosphatase	
3.6.1.41	P^1,P^4-Bis(5'-nucleosyl)-tetraphosphate nucleosidebisphosphohydrolase Bis(5'-nucleosyl)-tetraphosphatase (symmetrical)	
3.6.1.17	P^1,P^4-Bis(5'-nucleosyl)-tetraphosphate nucleotidohydrolase Bis(5'-nucleosyl)-tetraphosphatase (asymmetrical)	

EC-No.	Synonym Recommended Name	Vol., Page
3.4.22.39	p23 adenain	
1.5.1.21	P2C reductase Δ^1-Piperideine-2-carboxylate reductase	
3.4.24.76	P40 flavastacin	
4.2.1.96	P4aCD 4a-Hydroxytetrahydrobiopterin dehydratase	
3.4.21.63	P 5380 Oryzin	
2.7.1.112	p56lck kinase Protein-tyrosine kinase	
2.7.1.112	p56lck protein kinase Protein-tyrosine kinase	
2.7.1.112	p56lck protein tyrosine kinase Protein-tyrosine kinase	
2.7.1.112	p56lck tyrosine kinase Protein-tyrosine kinase	
5.3.4.1	P58 Protein disulfide-isomerase	Vol. 1, Page 436
2.7.1.37	p82 Kinase Protein kinase	
3.4.21.68	t-PA t-Plasminogen activator	
3.4.21.73	u-PA u-Plasminogen activator	
6.2.1.30	PA-CoA ligase Phenylacetate-CoA ligase	Vol. 2, Page 330
3.4.24.18	PABA-peptide hydrolase Meprin A	
3.4.21.75	PACE Furin	
4.1.1.36	P-PaCysSH decarboxylase Phosphopantothenoylcysteine decarboxylase	
3.4.21.34	padreatin plasma kallikrein	

EC-No.	Synonym Recommended Name	Vol., Page
3.4.21.35	padreatin tissue kallikrein	
3.4.21.34	padutin plasma kallikrein	
3.4.21.35	padutin tissue kallikrein	
3.4.21.35	padutin tissue kallikrein	
3.4.99.15	Paecilomyces proteinase Paecilomyces proteinase	
3.4.23.26	Paecilomyces proteinase Rhodotorulapepsin	
3.1.1.47	PAF¯acetylhydrolase 1-alkyl-2-acetylglycerophosphocholine esterase	
3.1.1.47	PAF-AH 1-alkyl-2-acetylglycerophosphocholine esterase	
3.1.1.47	PAF 2-acylhydrolase 1-alkyl-2-acetylglycerophosphocholine esterase	
2.3.1.149	PAF:acyllyso-GPC transacetylase Platelet-activating factor acetyltransferase	
2.3.1.67	PAF acetyltransferase 1-Alkylglycerophosphocholine O-acetyltransferase	
2.3.1.149	PAF acetyltransferase Platelet-activating factor acetyltransferase	
1.14.16.1	PAH Phenylalanine 4-monooxygenase	
3.4.21.61	Paired-basic endopeptidase Kexin	
3.4.21.75	Paired basic amino acid cleaving enzyme Furin	
3.4.21.75	Paired basic amino acid converting enzyme Furin	
4.3.2.5	PAL Peptidylamidoglycolate lyase	
4.3.1.5	PAL phenylalanine ammonia-lyase	

EC-No.	Synonym Recommended Name	Vol., Page
3.1.2.22	Palmitoyl-(protein) hydrolase Palmitoyl-(protein) hydrolase	
3.1.2.22	Palmitoyl-(protein) hydrolase Palmitoyl-(protein) hydrolase	
2.3.1.20	Palmitoyl-CoA-sn-1,2-diacylglycerol acyltransferase Diacylglycerol O-acyltransferase	
2.3.1.123	Palmitoyl-CoA:dolichol O-palmitoyltransferase Dolichol O-acyltransferase	
2.3.1.139	Palmitoyl-CoA:ecdysone palmitoyltransferase Ecdysone O-acyltransferase	
2.3.1.96	Palmitoyl-CoA:glycoprotein N-palmitoyltransferase Glycoprotein N-palmitoyltransferase	
2.3.1.21	Palmitoyl-CoA:L-carnitine O-palmitoyltransferase Carnitine O-palmitoyltransferase	
2.3.1.50	Palmitoyl-CoA:L-serine C-palmitoyltransferase (decarboxylating) Serine C-palmitoyltransferase	
2.3.1.100	Palmitoyl-CoA:myelin-proteolipid O-palmitoyltransferase Myelin-proteolipid O-palmitoyltransferase	
3.1.2.2	palmitoyl-CoA deacylase palmitoyl-CoA hydrolase	
1.3.99.13	Palmitoyl-CoA dehydrogenase Long-chain-acyl-CoA dehydrogenase	
1.14.99.5	Palmitoyl-CoA desaturase Stearoyl-CoA desaturase	
3.1.2.2	palmitoyl-CoA hydrolase palmitoyl-CoA hydrolase	
3.1.2.2	palmitoyl-CoA hydrolase palmitoyl-CoA hydrolase	
6.2.1.3	Palmitoyl-CoA ligase Long-chain-fatty-acid-CoA ligase	Vol. 2, Page 206
1.3.99.13	Palmitoyl-coenzyme A dehydrogenase Long-chain-acyl-CoA dehydrogenase	
1.1.1.101	Palmitoyl-dihydroxyacetone-phosphate reductase Acylglycerone-phosphate reductase	
3.1.1.28	palmitoyl-L-carnitine hydrolase acylcarnitine hydrolase	

EC-No.	Synonym Recommended Name	Vol., Page
3.1.2.22	Palmitoyl-protein thioesterase Palmitoyl-(protein) hydrolase	
3.1.2.22	Palmitoyl-protein thioesterase (bovine clone pBovPPT-17 precursor) Palmitoyl-(protein) hydrolase	
3.1.2.22	Palmitoyl-protein thioesterase (bovine clone pBovPPT-25 precursor) Palmitoyl-(protein) hydrolase	
3.1.2.22	Palmitoyl-protein thioesterase (human clone pHuPPT-5') Palmitoyl-(protein) hydrolase	
3.1.2.22	Palmitoyl-protein thioesterase (rat clone pRatPPT-44 precursor) Palmitoyl-(protein) hydrolase	
3.1.2.22	Palmitoyl-protein thioesterase PPT2 (human clone B lysosome associated) Palmitoyl-(protein) hydrolase	
3.1.1.28	palmitoyl carnitine hydrolase acylcarnitine hydrolase	
3.1.1.28	palmitoylcarnitine hydrolase acylcarnitine hydrolase	
2.3.1.21	Palmitoylcarnitine transferase Carnitine O-palmitoyltransferase	
1.14.99.5	Palmitoyl CoA desaturase Stearoyl-CoA desaturase	
3.1.2.2	palmitoyl coenzyme A hydrolase palmitoyl-CoA hydrolase	
6.2.1.3	Palmitoyl coenzyme A synthetase Long-chain-fatty-acid-CoA ligase	Vol. 2, Page 206
1.1.1.101	Palmitoyl dihydroxyacetone phosphate reductase Acylglycerone-phosphate reductase	
1.1.1.101	1-Palmitoylglycerol-3-phosphate:NADP$^+$ oxidoreductase Acylglycerone-phosphate reductase	
3.1.2.2	palmitoyl thioesterase palmitoyl-CoA hydrolase	
2.3.1.22	Palmitoyltransferase, acylglycerol 2-Acylglycerol O-acyltransferase	
2.3.1.21	Palmitoyltransferase, carnitine Carnitine O-palmitoyltransferase	
2.3.1.50	Palmitoyltransferase, serine Serine C-palmitoyltransferase	

EC-No.	Synonym Recommended Name	Vol., Page
6.2.1.3	Palmityl-coenzyme A synthetase Long-chain-fatty-acid-CoA ligase	Vol. 2, Page 206
3.1.2.2	palmityl coenzyme A deacylase palmitoyl-CoA hydrolase	
3.1.2.2	palmityl thioesterase palmitoyl-CoA hydrolase	
3.1.2.2	palmityl thioesterase I palmitoyl-CoA hydrolase	
3.1.2.2	palmityl thioesterase II palmitoyl-CoA hydrolase	
1.14.17.3	PAM Peptidylglycine monooxygenase	
1.14.17.3	PAM-A Peptidylglycine monooxygenase	
1.14.17.3	PAM-B Peptidylglycine monooxygenase	
3.2.1.4	Pancellase SS Cellulase	
3.1.21.1	Pancreatic deoxyribonuclease Deoxyribonuclease I	
3.1.21.1	Pancreatic DNase Deoxyribonuclease I	
3.1.22.1	Pancreatic DNase II Deoxyribonuclease II	
3.1.21.1	Pancreatic dornase Deoxyribonuclease I	
3.4.21.36	pancreatic elastase pancreatic elastase	
3.4.21.36	pancreatic elastase I pancreatic elastase	
3.4.21.71	Pancreatic elastase II Pancreatic elastase II	
3.4.21.70	Pancreatic endopeptidase E Pancreatic endopeptidase E	
3.4.21.34	pancreatic kallikrein plasma kallikrein	

EC-No.	Synonym Recommended Name	Vol., Page
3.4.21.35	pancreatic kallikrein tissue kallikrein	
3.4.21.70	Pancreatic protease E Pancreatic endopeptidase E	
3.4.21.70	Pancreatic proteinase E Pancreatic endopeptidase E	
3.1.27.5	Pancreatic ribonuclease Pancreatic ribonuclease	
3.1.27.5	Pancreatic RNase Pancreatic ribonuclease	
3.4.21.36	pancreatopeptidase E pancreatic elastase	
3.4.4.7	pancreatopeptidase E pancreatopeptidase E	
2.7.7.3	Pantetheine-phosphate adenylyltransferase Pantetheine-phosphate adenylyltransferase	
2.7.1.34	Pantetheine kinase Pantetheine kinase	
6.3.2.1	Pantoate-β-alanine ligase Pantoate-β-alanine ligase	Vol. 2, Page 394
1.1.1.106	Pantoate 4-dehydrogenase Pantoate 4-dehydrogenase	
6.3.2.1	D-Pantoate:β-alanine ligase (AMP-forming) Pantoate-β-alanine ligase	Vol. 2, Page 394
1.1.1.106	D-Pantoate:NAD^+ 4-oxidoreductase Pantoate 4-dehydrogenase	
6.3.2.1	Pantoate activating enzyme Pantoate-β-alanine ligase	Vol. 2, Page 394
6.3.2.1	Pantoic-activating enzyme Pantoate-β-alanine ligase	Vol. 2, Page 394
1.1.1.106	Pantothenase Pantoate 4-dehydrogenase	
3.5.1.22	Pantothenase Pantothenase	
6.3.2.5	Pantothenate 4'-phosphate:L-cysteine ligase Phosphopantothenoate-cysteine ligase	Vol. 2, Page 431

EC-No.	Synonym Recommended Name	Vol., Page
3.5.1.22	Pantothenate amidohydrolase Pantothenase	
3.5.1.22	Pantothenate hydrolase Pantothenase	
2.7.1.33	Pantothenate kinase Pantothenate kinase	
6.3.2.1	Pantothenate synthetase Pantoate-β-alanine ligase	Vol. 2, Page 394
2.7.1.33	Pantothenic acid kinase Pantothenate kinase	
4.1.1.30	Pantothenoylcysteine decarboxylase Pantothenoylcysteine decarboxylase	
4.1.1.30	Pantothenylcysteine decarboxylase Pantothenoylcysteine decarboxylase	
3.4.22.2	papain papain	
3.4.4.10	papain papain	
3.4.22.39	papain-like cysteine protease adenain	
3.4.22.2	papaine papain	
3.4.22.30	Papaya peptidase A Caricain	
3.4.22.25	Papaya peptidase B Glycyl endopeptidase	
3.4.22.2	papaya peptidase I papain	
3.4.22.30	Papaya peptidase II Caricain	
3.4.22.30	Papaya proteinase OMEGA Caricain	
3.4.22.30	Papaya proteinase 3 Caricain	
3.4.22.25	Papaya proteinase 4 Glycyl endopeptidase	

EC-No.	Synonym Recommended Name	Vol., Page
3.4.22.30	Papaya proteinase A Caricain	
3.4.22.6	papaya proteinase II chymopapain	
3.4.22.30	Papaya proteinase III Caricain	
3.4.22.25	Papaya proteinase IV Glycyl endopeptidase	
3.4.22.2	papayotin papain	
2.8.2.24	PAPS-desulfoglucosinolate sulfotransferase Desulfoglucosinolate sulfotransferase	
2.8.2.8	PAPS:DSH sulfotransferase Desulfoheparin sulfotransferase	
2.8.2.28	PAPS:flavonol 3,3'/3,4'-disulfate 7-sulfotransferase Quercetin-3,3'-bissulfate 7-sulfotransferase	
2.8.2.26	PAPS:flavonol 3-sulfate 3'-sulfotransferase Quercetin-3-sulfate 3'-sulfotransferase	
2.8.2.27	PAPS:flavonol 3-sulfate 4'-sulfotransferase Quercetin-3-sulfate 4'-sulfotransferase	
2.8.2.8	PAPS:N-desulfoheparin sulfotransferase Desulfoheparin sulfotransferase	
2.8.2.13	PAPS:psychosine sulphotransferase Psychosine sulfotransferase	
1.8.99.4	PAPS reductase Phosphoadenosine-phosphosulfate reductase	
1.8.99.4	PAPS reductase, thioredoxin-dependent Phosphoadenosine-phosphosulfate reductase	
3.6.2.2	PAPS sulfatase Phosphoadenylylsulfatase	
2.8.2.16	PAPS sulfotransferase Thiol sulfotransferase	
1.14.13.2	Para-Hydroxybenzoate hydroxylase 4-Hydroxybenzoate 3-monooxygenase	
3.1.3.41	*para*-nitrophenyl phosphatase 4-nitrophenylphosphatase	

EC-No.	Synonym Recommended Name	Vol., Page
2.4.1.86	Paragloboside synthase Glucosaminylgalactosylglucosylceramide β-galactosyltransferase	
2.4.1.34	Paramylon synthetase 1,3-β-Glucan synthase	
3.1.8.1	Paraoxonase Aryldialkylphosphatase	
3.1.1.2	paraoxonase arylesterase	
3.1.8.1	Paraoxon hydrolase Aryldialkylphosphatase	
3.4.23.2	parapepsin I pepsin B	
3.4.21.4	parenzyme trypsin	
3.4.21.4	parenzymol trypsin	
3.4.11.2	particle-bound aminopeptidase membrane alanyl aminopeptidase	
3.1.3.10	G-1-Pase glucose-1-phosphatase	
2.7.7.48	PB1 proteins RNA-directed RNA polymerase	
2.7.7.48	PB2 proteins RNA-directed RNA polymerase	
4.3.1.8	PBG-D hydroxymethylbilane synthase	
4.3.1.8	PBG-deaminase hydroxymethylbilane synthase	
4.3.1.8	PBGD hydroxymethylbilane synthase	
3.4.21.75	PC1 Furin	
3.4.21.93	PC1 Proprotein convertase 1	
3.4.21.94	PC2 Proprotein convertase 2	

EC-No.	Synonym Recommended Name	Vol., Page
6.4.1.3	Pcase Propionyl-CoA carboxylase	Vol. 2, Page 738
6.4.1.3	PCC Propionyl-CoA carboxylase	Vol. 2, Page 738
4.2.1.96	PCD/DCoH 4a-Hydroxytetrahydrobiopterin dehydratase	
4.2.1.96	PCD/PhhB 4a-Hydroxytetrahydrobiopterin dehydratase	
4.2.1.96	PCDH 4a-Hydroxytetrahydrobiopterin dehydratase	
3.4.23.17	PCE Pro-opiomelanocortin converting enzyme	
3.4.23.33	PCP Xanthomonapepsin	
3.4.16.2	PCP lysosomal Pro-X carboxypeptidase	
1.14.13.50	PCP hydroxylase Pentachlorophenol monooxygenase	
3.2.2.17	PD-DNA glycosylase Deoxyribodipyrimidine endonucleosidase	
1.1.1.202	1,3-PD:NAD$^+$ oxidoreductase 1,3-Propanediol 1-dehydrogenase	
3.4.14.5	X-PDAP dipeptidyl-peptidase IV	
1.5.4.1	PDA synthase Pyrimidodiazepine synthase	
4.1.1.1	PDC Pyruvate decarboxylase	
3.4.17.8	PdcA muramoylpentapeptide carboxypeptidase	
1.3.1.49	PDD DHase cis-3,4-Dihydrophenanthrene-3,4-diol dehydrogenase	
3.1.4.17	PDE 3',5'-Cyclic-nucleotide phosphodiesterase	
3.1.1.51	PDE74H phorbol-diester hydrolase	

EC-No.	Synonym Recommended Name	Vol., Page
1.2.4.1	PDH Pyruvate dehydrogenase (lipoamide)	
3.4.15.1	PDH peptidyl-dipeptidase A	
5.3.4.1	PDI Protein disulfide-isomerase	Vol. 1, Page 436
5.3.4.1	PDIp Protein disulfide-isomerase	Vol. 1, Page 436
1.14.12.7	PDO Phthalate 4,5-dioxygenase	
1.1.1.7	PDP dehydrogenase Propanediol-phosphate dehydrogenase	
4.2.1.51	PDT prephenate dehydratase	
1.1.1.262	PdxA 4-hydroxythreonine-4-phosphate dehydrogenase	
3.1.1.11	PE pectinesterase	
3.1.1.11	pectase pectinesterase	
4.2.2.2	pectate lyase pectate lyase	
4.2.99.3	pectate lyase pectate lyase	
4.2.2.2	pectate transeliminase pectate lyase	
4.2.2.2	pectic acid lyase pectate lyase	
4.2.2.2	pectic acid transeliminase pectate lyase	
3.2.1.15	Pectic depolymerase Polygalacturonase	
3.2.1.15	Pectic hydrolase Polygalacturonase	
4.2.2.2	pectic lyase pectate lyase	

EC-No.	Synonym Recommended Name	Vol., Page
3.2.1.15	Pectinase Polygalacturonase	
3.2.1.15	Pectinase SS Polygalacturonase	
3.1.1.11	pectin demethoxylase pectinesterase	
3.2.1.15	Pectin depolymerase Polygalacturonase	
3.1.1.11	pectinesterase pectinesterase	
3.2.1.15	Pectin hydrolase Polygalacturonase	
4.2.2.10	pectin lyase pectin lyase	
3.1.1.11	pectin methoxylase pectinesterase	
3.1.1.11	pectin methyl esterase pectinesterase	
3.1.1.11	pectin methylesterase pectinesterase	
4.2.2.10	pectin methyltranseliminase pectin lyase	
3.1.1.11	pectinoesterase pectinesterase	
3.1.1.11	pectin pectylhydrolase pectinesterase	
3.1.1.11	pectin pectylhydrolase pectinesterase	
3.2.1.15	Pectin polygalacturonase Polygalacturonase	
4.2.2.10	pectin transeliminase pectin lyase	
3.1.1.11	pectofoetidin pectinesterase	
3.2.1.15	Pectolase Polygalacturonase	

EC-No.	Synonym Recommended Name Vol., Page
4.2.2.10	pectolyase Y23 pectin lyase
3.2.1.15	Pectozyme Polygalacturonase
4.2.2.9	PelL exopolygalacturonate lyase
2.1.1.17	PEMT Phosphatidylethanolamine N-methyltransferase
1.1.3.4	Penatin Glucose oxidase
3.5.2.6	Penicillin β-lactamase β-lactamase
3.5.1.11	Penicillin acylase Penicillin amidase
3.5.1.11	Penicillin amidase Penicillin amidase
3.5.2.6	Penicillin amido-β-lactamhydrolase β-lactamase
3.5.1.11	Penicillin amidohydrolase Penicillin amidase
3.5.2.6	Penicillinase I, II β-lactamase
2.4.1.129	Penicillin binding protein Peptidoglycan glycosyltransferase
3.4.17.8	penicillin binding protein 5 muramoylpentapeptide carboxypeptidase
3.4.16.4	penicillin binding protein 5 serine-type D-Ala-D-Ala carboxypeptidase
3.4.23.20	Penicillium citrinum acid proteinase Penicillopepsin
3.4.23.20	Penicillium cyclopium acid proteinase Penicillopepsin
3.4.23.20	Penicillium expansum acid proteinase Penicillopepsin
3.4.23.7	Penicillium janthinellum acid proteinase Penicillium janthinellum acid proteinase

EC-No.	Synonym Recommended Name	Vol., Page
3.4.23.20	Penicillium janthinellum acid proteinase Penicillopepsin	
3.4.23.20	Penicillium janthinellum aspartic proteinase Penicillopepsin	
3.4.99.16	Penicillium notatum extracellular proteinase Penicillium notatum extracellular proteinase	
3.4.23.20	Penicillium roqueforti acid proteinase Penicillopepsin	
3.4.24.39	Penicillium roqueforti metalloproteinase Deuterolysin	
3.4.24.39	Penicillium roqueforti protease II Deuterolysin	
3.4.23.20	Penicillopepsin Penicillopepsin	
4.1.1.70	pent-2-enoyl-CoA carboxy-lyase glutaconyl-CoA decarboxylase	
1.14.13.50	Pentachlorophenol,NADPH:oxygen oxidoreductase (hydroxylating, dechlorinating) Pentachlorophenol monooxygenase	
1.14.13.50	Pentachlorophenol 4-monooxygenase Pentachlorophenol monooxygenase	
1.14.13.50	Pentachlorophenol dehalogenase Pentachlorophenol monooxygenase	
1.14.13.50	Pentachlorophenol hydroxylase Pentachlorophenol monooxygenase	
1.14.13.50	Pentachlorophenol monooxygenase Pentachlorophenol monooxygenase	
4.2.1.44	2,4,6/3,5-Pentahydroxycyclohexanone hydro-lyase Myo-inosose-2 dehydratase	
4.6.1.5	Pentalenene synthase Pentalenene synthase	
4.2.3.7	pentalenene synthase pentalenene synthase	
4.6.1.5	Pentalenene synthetase Pentalenene synthase	

EC-No.	Synonym Recommended Name	Vol., Page
4.2.3.7	pentalenene synthetase pentalenene synthase	
3.5.1.50	Pentanamidase Pentanamidase	
3.5.1.50	Pentanamide amidohydrolase Pentanamidase	
1.1.1.9	Pentitol-DPN dehydrogenase D-Xylulose reductase	
1.1.1.12	Pentitol-DPN dehydrogenase L-Arabinitol 4-dehydrogenase	
5.1.3.1	Pentose-5-phosphate 3-epimerase Ribulose-phosphate 3-epimerase	Vol. 1, Page 91
4.1.2.3	pentosealdolase pentosealdolase	
4.1.2.9	Pentulose-5-phosphate phosphoketolase Phosphoketolase	
2.7.1.69	PEP-dependent phosphotransferase enzyme II Protein-N^{π}-phosphohistidine-sugar phosphotransferase	
2.7.1.69	PEP-sugar phosphotransferase enzyme II Protein-N^{π}-phosphohistidine-sugar phosphotransferase	
4.1.1.31	PEPC Phosphoenolpyruvate carboxylase	
3.4.22.40	PepC bleomycin hydrolase	
3.4.11.3	PepC cystinyl aminopeptidase	
4.1.1.49	PEP carboxykinase Phosphoenolpyruvate carboxykinase (ATP)	
4.1.1.32	PEP carboxykinase Phosphoenolpyruvate carboxykinase (GTP)	
4.1.1.38	PEP carboxykinase phosphoenolpyruvate carboxykinase (pyrophosphate)	
4.1.1.49	PEP carboxylase Phosphoenolpyruvate carboxykinase (ATP)	
4.1.1.32	PEP carboxylase Phosphoenolpyruvate carboxykinase (GTP)	

EC-No.	Synonym Recommended Name	Vol., Page
4.1.1.31	PEP carboxylase Phosphoenolpyruvate carboxylase	
4.1.1.38	PEP carboxylase phosphoenolpyruvate carboxykinase (pyrophosphate)	
4.1.1.38	PEP carboxyphosphotransferase phosphoenolpyruvate carboxykinase (pyrophosphate)	
4.1.1.49	PEPCK Phosphoenolpyruvate carboxykinase (ATP)	
4.1.1.32	PEPCK Phosphoenolpyruvate carboxykinase (GTP)	
4.1.1.49	PEPCK (ATP) Phosphoenolpyruvate carboxykinase (ATP)	
4.1.1.38	PEPCTrP phosphoenolpyruvate carboxykinase (pyrophosphate)	
3.4.15.2	pepdidyl carboxyamidase pepdidyl carboxyamidase	
4.1.1.49	PEPK Phosphoenolpyruvate carboxykinase (ATP)	
3.1.3.60	PEP phosphatase Phosphoenolpyruvate phosphatase	
5.4.2.9	PEP phosphomutase Phosphoenolpyruvate mutase	Vol. 1, Page 546
5.4.2.9	PEPPM Phosphoenolpyruvate mutase	Vol. 1, Page 546
3.4.4.1	pepsin pepsin	
3.4.23.1	pepsin pepsin A	
3.4.23.18	Pepsin-type aspartic proteinase Aspergillopepsin I	
3.4.23.1	pepsin A pepsin A	
3.4.23.2	pepsin B pepsin B	
3.4.4.2	pepsin B pepsin B	

EC-No.	Synonym Recommended Name	Vol., Page
3.4.23.3	pepsin C gastricsin	
3.4.23.1	pepsin D pepsin A	
3.4.23.1	pepsin fortior pepsin A	
3.4.23.1	pepsin R pepsin A	
3.4.23.37	Pepstatin-insensitive carboxyl proteinase Pseudomonapepsin	
2.7.9.2	PEP synthetase water dikinase pyruvate	
3.4.13.9	γ-peptidase X-Pro dipeptidase	
3.4.19.11	Peptidase, γ-D-glutamyldiaminopimelate endo- γ-D-Glutamyl-meso-diaminopimelate peptidase I	
3.4.21.32	peptidase, clostridio-, A brachyurin	
3.4.24.3	peptidase, clostridio-, A microbial collagenase	
3.4.14.2	peptidase, dipeptidyl, II dipeptidyl-peptidase II	
3.4.14.4	peptidase, dipeptidyl, III dipeptidyl-peptidase III	
3.4.14.5	peptidase, dipeptidyl, IV dipeptidyl-peptidase IV	
3.4.24.11	peptidase, endo- neprilysin	
3.4.21.9	peptidase, entero- enteropeptidase	
3.4.24.11	peptidase, membrane metalloendo- neprilysin	
3.4.24.16	peptidase, neurotensin endo neurolysin	
3.4.24.16	peptidase, neurotensin endo- neurolysin	

EC-No.	Synonym Recommended Name	Vol., Page
3.4.21.36	peptidase, pancreato-, E pancreatic elastase	
3.4.21.26	peptidase, postproline endo- prolyl oligopeptidase	
3.4.24.19	Peptidase, procollagen C-terminal Procollagen C-endopeptidase	
3.4.21.89	Peptidase, signal Signal peptidase I	
3.4.21.62	Peptidase, subtilo-, A Subtilisin	
3.4.24.15	peptidase, thimet oligo- thimet oligopeptidase	
3.4.13.18	Peptidase A Cytosol nonspecific dipeptidase	
3.4.23.20	Peptidase A Penicillopepsin	
3.4.11.4	peptidase B tripeptide aminopeptidase	
3.4.13.9	peptidase D X-Pro dipeptidase	
3.4.11.18	Peptidase M Methionyl aminopeptidase	
3.4.15.1	peptidase P peptidyl-dipeptidase A	
3.4.11.1	peptidase S leucyl aminopeptidase	
3.4.11.4	peptidase T tripeptide aminopeptidase	
1.14.17.3	Peptide α-amidating enzyme Peptidylglycine monooxygenase	
1.14.17.3	Peptide α-amide synthase Peptidylglycine monooxygenase	
1.14.17.3	Peptide α-amide synthetase Peptidylglycine monooxygenase	
1.14.11.16	Peptide-aspartate β-dioxygenase Peptide-aspartate β-dioxygenase	

EC-No.	Synonym Recommended Name	Vol., Page
1.14.11.16	Peptide-L-aspartate,2-oxoglutarate:oxygen oxidoreductase (3-hydroxylating) Peptide-aspartate β-dioxygenase	
2.3.1.88	Peptide α-N-acetyltransferase Peptide α-N-acetyltransferase	
3.5.1.52	Peptide-N^4-(N-acetyl-β-glucosaminyl) asparagine amidase Peptide-N^4-(N-acetyl-β-glucosaminyl) asparagine amidase	
1.13.11.26	Peptide-tryptophan 2,3-dioxygenase Peptide-tryptophan 2,3-dioxygenase	
1.13.11.26	Peptide-tryptophan:oxygen 2,3-oxidoreductase (decyclizing) Peptide-tryptophan 2,3-dioxygenase	
2.3.1.88	Peptide acetyltransferase Peptide α-N-acetyltransferase	
5.2.1.8	Peptide bond isomerase Peptidylprolyl isomerase	Vol. 1, Page 218
5.1.1.16	Peptide D-serine-L-serine isomerase Protein-serine epimerase	Vol. 1, Page 66
3.5.1.43	Peptideglutaminase Peptidyl-glutaminase	
3.4.24.60	Peptide hormone inactivating endopeptidase Dactylysin	
1.8.4.6	Peptide methionine sulfoxide reductase Protein-methionine-S-oxide reductase	
2.3.1.97	Peptide N-myristoyltransferase Glycylpeptide N-tetradecanoyltransferase	
5.1.1.16	Peptide serine isomerase Protein-serine epimerase	Vol. 1, Page 66
3.5.1.43	Peptidoglutaminase Peptidyl-glutaminase	
3.5.1.43	Peptidoglutaminase I Peptidyl-glutaminase	
3.5.1.44	Peptidoglutaminase II Protein-glutamine glutaminase	
3.2.1.92	Peptidoglycan β-N-acetylmuramidase Peptidoglycan β-N-acetylmuramidase	

EC-No.	Synonym Recommended Name	Vol., Page
3.2.1.92	Peptidoglycan β-N-acetylmuramoylexohydrolase Peptidoglycan β-N-acetylmuramidase	
3.4.99.17	peptidoglycan endopeptidase peptidoglycan endopeptidase	
2.4.1.129	Peptidoglycan glycosyltransferase Peptidoglycan glycosyltransferase	
3.2.1.17	Peptidoglycan N-acetylmuramoyl-hydrolase Lysozyme	
2.4.1.129	Peptidoglycan transglycosylase Peptidoglycan glycosyltransferase	
3.4.13.17	peptidy-D-amino acid hydrolase non-stereopecific dipeptidase	
4.3.2.5	Peptidyl-α-hydroxyglycine α-amidating lyase Peptidylamidoglycolate lyase	
1.14.17.3	Peptidyl α-amidating enzyme Peptidylglycine monooxygenase	
3.4.19.2	peptidyl-aminoacylamidase peptidyl-glycinamidase	
3.4.24.33	Peptidyl-Asp metalloendopeptidase Peptidyl-Asp metalloendopeptidase	
3.4.24.33	Peptidyl-Asp metalloproteinase Peptidyl-Asp metalloendopeptidase	
3.4.15.1	peptidyl-dipeptidase A peptidyl-dipeptidase A	
3.4.15.4	Peptidyl-dipeptidase B Peptidyl-dipeptidase B	
3.4.15.5	Peptidyl-dipeptidase Dcp Peptidyl-dipeptidase Dcp	
3.4.15.1	peptidyl-dipeptide hydrolase peptidyl-dipeptidase A	
3.5.1.43	Peptidyl-glutaminase Peptidyl-glutaminase	
3.4.19.2	peptidyl-glycinamidase peptidyl-glycinamidase	
3.4.19.2	peptidyl-glycinamide hydrolase peptidyl-glycinamidase	

EC-No.	Synonym Recommended Name	Vol., Page
3.5.1.43	Peptidyl-L-glutamine amidohydrolase Peptidyl-glutaminase	
3.4.17.10	peptidyl-L-lysine(-L-arginine) hydrolase carboxypeptidase H	
3.4.17.3	peptidyl-L-lysine(-L-arginine) hydrolase lysine carboxypeptidase	
3.4.17.2	peptidyl-L-lysine (L-arginine) hydrolase carboxypeptidase B	
1.14.11.4	Peptidyl-lysine, 2-oxoglutarate: oxygen oxidoreductase Procollagen-lysine 5-dioxygenase	
3.4.24.20	Peptidyl-Lys metalloendopeptidase Peptidyl-Lys metalloendopeptidase	
5.2.1.8	Peptidyl-prolyl cis-trans isomerase Peptidylprolyl isomerase	Vol. 1, Page 218
2.3.2.12	Peptidyl-tRNA:aminoacyl-tRNA N-peptidyltransferase Peptidyltransferase	
3.1.1.29	peptidyl-tRNA hydrolase aminoacyl-tRNA hydrolase	
4.3.2.5	Peptidylamidoglycolate lyase Peptidylamidoglycolate lyase	
4.3.2.5	Peptidylamidoglycolate peptidylamide-lyase Peptidylamidoglycolate lyase	
3.4.19.2	peptidyl amino acid amide hydrolase peptidyl-glycinamidase	
3.5.3.15	Peptidylarginine deiminase Protein-arginine deiminase	
3.4.19.2	peptidyl carboxy-amidase peptidyl-glycinamidase	
3.4.19.2	peptidyl carboxyamidase peptidyl-glycinamidase	
3.4.15.1	peptidyl dipeptidase peptidyl-dipeptidase A	
3.4.15.1	peptidyl dipeptidase-4 peptidyl-dipeptidase A	
3.4.15.1	peptidyl dipeptidase A peptidyl-dipeptidase A	

EC-No.	Synonym Recommended Name	Vol., Page
3.4.15.1	peptidyl dipeptidase I peptidyl-dipeptidase A	
3.4.15.1	peptidyl dipeptide hydrolase peptidyl-dipeptidase A	
3.4.15.1	peptidyldipeptide hydrolase peptidyl-dipeptidase A	
3.5.1.44	Peptidylglutaminase II Protein-glutamine glutaminase	
1.14.17.3	Peptidylglycine,ascorbate:oxygen oxidoreductase (2-hydroxylating) Peptidylglycine monooxygenase	
1.14.17.3	Peptidylglycine α-amidating monooxygenase Peptidylglycine monooxygenase	
1.14.17.3	Peptidylglycine α-hydroxylase Peptidylglycine monooxygenase	
1.14.17.3	Peptidylglycine 2-hydroxylase Peptidylglycine monooxygenase	
1.14.17.3	Peptidylglycine monooxygenase Peptidylglycine monooxygenase	
1.14.11.4	Peptidyllysine, 2-oxoglutarate:oxygen 5-oxidoreductase Procollagen-lysine 5-dioxygenase	
3.4.24.20	Peptidyllysine metalloproteinase Peptidyl-Lys metalloendopeptidase	
5.2.1.8	Peptidylproline cis-trans-isomerase Peptidylprolyl isomerase	Vol. 1, Page 218
1.14.11.2	Peptidyl proline hydroxylase Procollagen-proline dioxygenase	
3.4.16.2	peptidylprolyl-L-amino acid hydrolase lysosomal Pro-X carboxypeptidase	
5.2.1.8	Peptidylprolyl isomerase Peptidylprolyl isomerase	Vol. 1, Page 218
2.3.2.12	Peptidyltransferase Peptidyltransferase	
1.13.11.26	Peptidyltryptophan 2,3-dioxygenase Peptide-tryptophan 2,3-dioxygenase	
3.4.14.5	Pep X dipeptidyl-peptidase IV	

EC-No.	Synonym Recommended Name	Vol., Page
3.4.14.11	PepX X-Pro dipeptidyl-peptidase	
1.1.1.144	Perillyl-alcohol:NAD$^+$ oxidoreductase Perillyl-alcohol dehydrogenase	
1.1.1.144	Perillyl-alcohol dehydrogenase Perillyl-alcohol dehydrogenase	
1.1.1.144	Perillyl alcohol dehydrogenase Perillyl-alcohol dehydrogenase	
3.1.3.10	periplasmic acid glucose-1-phosphatase glucose-1-phosphatase	
2.7.8.21	Periplasmic phosphoglycerotransferase Membrane-oligosaccharide glycerophosphotransferase	
1.11.1.7	Peroxidase Peroxidase	
1.11.1.11	Peroxidase, ascorbate L-Ascorbate peroxidase	
1.11.1.10	Peroxidase, chloride Chloride peroxidase	
1.11.1.5	Peroxidase, cytochrome c Cytochrome-c peroxidase	
1.11.1.3	Peroxidase, fatty acid Fatty-acid peroxidase	
1.11.1.9	Peroxidase, glutathione Glutathione peroxidase	
1.11.1.8	Peroxidase, iodide Iodide peroxidase	
1.11.1.13	Peroxidase, manganese Manganese peroxidase	
1.11.1.1	Peroxidase, nicotinamide adenine dinucleotide NADH peroxidase	
1.11.1.2	Peroxidase, nicotinamide adenine dinucleotide phosphate NADPH peroxidase	
1.11.1.13	Peroxidase-M2 Manganese peroxidase	
1.12.1.1	peroxidase peroxidase peroxidase peroxidase	

EC-No.	Synonym Recommended Name	Vol., Page
1.11.1.12	Peroxidation-inhibiting protein: Peroxidase, glutathione (phospholipid hydroperoxide-reducing) Phospholipid-hydroperoxide glutathione peroxidase	
2.3.1.154	peroxisomal 3-oxoacyl coenzyme A thiolase propionyl-CoA C^2-trimethyltridecanoyltransferase	
2.3.1.154	peroxisome sterol carrier protein thiolase propionyl-CoA C^2-trimethyltridecanoyltransferase	
1.13.11.19	Persulfurase Cysteamine dioxygenase	
3.4.23.39	PFAPD Plasmepsin II	
3.4.23.38	PFAPG Plasmepsin I	
5.2.1.8	PfCyP Peptidylprolyl isomerase	Vol. 1, Page 218
2.7.1.11	PFK 6-Phosphofructokinase	
1.97.1.4	PFL-activating enzyme Formate acetyltransferase activating enzyme	
1.97.1.4	PFL-AE Formate acetyltransferase activating enzyme	
1.97.1.4	PFL-glycine:S-adenosyl-L-methionine H transferase (flavodoxin-oxidizing, S-adenosyl-L-methionine-cleaving) Formate acetyltransferase activating enzyme	
1.97.1.4	PFL activase Formate acetyltransferase activating enzyme	
1.97.1.4	PFL activase Formate acetyltransferase activating enzyme	
5.3.1.1	PfTIM Triose-phosphate isomerase	Vol. 1, Page 235
1.14.99.1	(PG)H synthase Prostaglandin-endoperoxide synthase	
2.4.1.129	PG-II Peptidoglycan glycosyltransferase	
5.3.3.9	PGA isomerase Prostaglandin-A1 Δ-isomerase	Vol. 1, Page 423

EC-No.	Synonym Recommended Name	Vol., Page
4.2.2.2	PGA lyase pectate lyase	
5.4.2.1	PGAM Phosphoglycerate mutase	Vol. 1, Page 493
5.4.2.1	PGAM-d Phosphoglycerate mutase	Vol. 1, Page 493
5.4.2.1	PGAM-i Phosphoglycerate mutase	Vol. 1, Page 493
5.4.2.1	PGA mutase Phosphoglycerate mutase	Vol. 1, Page 493
3.1.3.38	3-PGA phosphatase 3-phosphoglycerate phosphatase	
4.2.1.24	PGBS porphobilinogen synthase	
1.1.1.188	PGD_2 11-ketoreductase Prostaglandin-F synthase	
4.2.1.12	6-PG dehydrase phosphogluconate dehydratase	
1.1.1.141	PGDH 15-Hydroxyprostaglandin dehydrogenase (NAD^+)	
5.3.99.2	PGDS Prostaglandin-D synthase	Vol. 1, Page 451
5.3.99.2	PGD synthase Prostaglandin-D synthase	Vol. 1, Page 451
4.2.1.12	6PGDT phosphogluconate dehydratase	
1.1.1.189	PGE-9-ketoreductase Prostaglandin-E2 9-reductase	
1.1.1.189	PGE_2-9-ketoreductase Prostaglandin-E2 9-reductase	
1.1.1.189	PGE_2-9-OR Prostaglandin-E2 9-reductase	
1.1.1.189	PGE_2 9-oxoreductase Prostaglandin-E2 9-reductase	
5.3.99.3	PGE_2 isomerase Prostaglandin-E synthetase	Vol. 1, Page 459

EC-No.	Synonym Recommended Name	Vol., Page
5.3.99.3	PGE isomerase Prostaglandin-E synthetase	Vol. 1, Page 459
1.1.1.188	$PGF_{2\alpha}$ synthetase Prostaglandin-F synthase	
1.1.1.188	PGF synthetase Prostaglandin-F synthase	
5.3.99.2	PGH-PGD isomerase Prostaglandin-D synthase	Vol. 1, Page 451
5.3.99.3	PGH-PGE isomerase Prostaglandin-E synthetase	Vol. 1, Page 459
5.3.99.3	PGH_2/PGE_2 isomerase Prostaglandin-E synthetase	Vol. 1, Page 459
5.3.1.9	PGI Glucose-6-phosphate isomerase	Vol. 1, Page 298
1.1.1.231	PGI_2 dehydrogenase 15-Hydroxyprostaglandin-I dehydrogenase ($NADP^+$)	
5.3.99.4	PGI_2 synthase Prostacyclin synthase	Vol. 1, Page 465
5.3.99.4	PGI_2 synthetase Prostacyclin synthase	Vol. 1, Page 465
5.3.99.4	PGIS Prostacyclin synthase	Vol. 1, Page 465
2.7.2.3	3-PGK Phosphoglycerate kinase	
2.7.2.3	PGK Phosphoglycerate kinase	
3.1.1.31	6-PGL 6-phosphogluconolactonase	
3.1.1.31	PGL 6-phosphogluconolactonase	
4.3.2.5	PGL Peptidylamidoglycolate lyase	
5.4.2.3	PGlcNAc mutase Phosphoacetylglucosamine mutase	Vol. 1, Page 515
5.4.2.2	PGM Phosphoglucomutase	Vol. 1, Page 506

EC-No.	Synonym Recommended Name	Vol., Page
5.4.2.1	PGM Phosphoglycerate mutase	Vol. 1, Page 493
5.4.2.6	β-PGM β-Phosphoglucomutase	Vol. 1, Page 530
5.4.2.2	PGM1 Phosphoglucomutase	Vol. 1, Page 506
5.4.2.2	PGM2 Phosphoglucomutase	Vol. 1, Page 506
3.1.3.14	PGMTE synthetase methylphosphothioglycerate phosphatase	
3.1.3.27	PGP-A phosphatidylglycerophosphatase	
3.1.3.27	PGP-B phosphatidylglycerophosphatase	
3.1.3.27	PGP-C phosphatidylglycerophosphatase	
3.1.3.27	PGP-tase phosphatidylglycerophosphatase	
3.4.19.12	PGP9.5 ubiquitinyl hydrolase I	
3.1.3.27	PGPA-tase phosphatidylglycerophosphatase	
3.1.3.27	PGPase phosphatidylglycerophosphatase	
3.1.3.27	PGPB-tase phosphatidylglycerophosphatase	
3.1.3.27	PGPC-tase phosphatidylglycerophosphatase	
3.1.3.27	PGP phosphatase phosphatidylglycerophosphatase	
2.7.8.5	PGP synthase CDPdiacylglycerol-glycerol-3-phosphate 3-phosphatidyltransferase	
1.14.99.1	PG synthetase Prostaglandin-endoperoxide synthase	
4.2.99.18	Phage-T_4 UV endonuclease DNA-(apurinic or apyrimidinic site) lyase	

EC-No.	Synonym Recommended Name	Vol., Page
2.7.7.48	Phage f_2 replicase RNA-directed RNA polymerase	
3.1.11.3	Phage lambda-induced exonuclease Exodeoxyribonuclease (Lambda induced)	
3.1.11.4	phage SP_3 exodeoxyribonuclease (phage SP_3-induced)	
3.1.11.4	phage SP_3 DNase exodeoxyribonuclease (phage SP_3-induced)	
4.1.2.10	PhaMDL Mandelonitrile lyase	
3.4.16.5	Phaseolin Carboxypeptidase C	
3.4.22.34	Phaseolin Legumain	
4.2.1.97	Phaseollidin-hydrate hydro-lyase Phaseollidin hydratase	
4.2.1.97	Phaseollidin hydratase Phaseollidin hydratase	
3.4.21.13	phaseolus proteinase phaseolus proteinase	
2.4.1.194	PHB-O-glucosyltransferase 4-Hydroxybenzoate 4-O-β-D-glucosyltransferase	
2.4.1.194	PHB glucosyltransferase 4-Hydroxybenzoate 4-O-β-D-glucosyltransferase	
1.4.1.20	PHD Phenylalanine dehydrogenase	
2.1.1.28	Phenethanolamine methyltransferase Phenylethanolamine N-methyltransferase	
2.1.1.28	Phenethanolamine N-methyltransferase Phenylethanolamine N-methyltransferase	
1.14.13.7	Phenol,NADPH:oxygen oxidoreductase (2-hydroxylating) Phenol 2-monooxygenase	
2.4.1.35	Phenol-β-D-glucosyltransferase Phenol β-glucosyltransferase	
2.4.1.35	Phenol β-glucosyltransferase Phenol β-glucosyltransferase	

EC-No.	Synonym Recommended Name	Vol., Page
4.1.99.2	phenol-lyase, tyrosine tyrosine phenol-lyase	
1.14.13.7	Phenol 2-monooxygenase Phenol 2-monooxygenase	
1.10.3.1	Phenolase Catechol oxidase	
1.14.18.1	Phenolase Monophenol monooxygenase	
4.1.1.61	Phenol carboxylase 4-Hydroxybenzoate decarboxylase	
1.14.13.7	Phenol hydroxylase Phenol 2-monooxygenase	
3.1.6.2	Phenolic steroid sulfatase Steryl-sulfatase	
1.14.13.7	Phenol *o*-hydroxylase Phenol 2-monooxygenase	
2.1.1.25	Phenol O-methyltransferase Phenol O-methyltransferase	
1.14.18.1	Phenol oxidase Monophenol monooxygenase	
3.1.6.1	Phenolsulfatase Arylsulfatase	
2.8.2.1	Phenol sulfokinase Aryl sulfotransferase	
2.8.2.1	Phenol sulfotransferase Aryl sulfotransferase	
2.4.1.17	Phenyl-UDP-glucuronosyltransferase Glucuronosyltransferase	
6.2.1.21	Phenyl/phenoxyacetic acid activating enzyme Phenylacetate-CoA ligase	Vol. 2, Page 301
6.2.1.21	Phenyl/phenoxyacetyl-CoA-ligase Phenylacetate-CoA ligase	Vol. 2, Page 301
1.2.1.39	Phenylacetaldehyde:NAD^+ oxidoreductase Phenylacetaldehyde dehydrogenase	
1.2.1.39	Phenylacetaldehyde dehydrogenase Phenylacetaldehyde dehydrogenase	

EC-No.	Synonym Recommended Name	Vol., Page
1.2.1.39	Phenylacetaldehyde dehydrogenase Phenylacetaldehyde dehydrogenase	
6.2.1.21	Phenylacetate-CoA ligase Phenylacetate-CoA ligase	Vol. 2, Page 301
6.2.1.30	Phenylacetate-CoA ligase Phenylacetate-CoA ligase	Vol. 2, Page 330
6.2.1.21	Phenylacetate:CoA ligase Phenylacetate-CoA ligase	Vol. 2, Page 301
6.2.1.30	Phenylacetate:CoA ligase (AMP-forming) Phenylacetate-CoA ligase	Vol. 2, Page 330
2.3.1.14	Phenylacetyl-CoA:L-glutamine α-N-phenylacetyltransferase Glutamine N-phenylacetyltransferase	
2.3.1.14	Phenylacetyl-CoA:L-glutamine N-acetyltransferase Glutamine N-phenylacetyltransferase	
6.2.1.30	Phenylacetyl-CoA ligase Phenylacetate-CoA ligase	Vol. 2, Page 330
6.2.1.30	Phenylacetyl-CoA ligase (AMP-forming) Phenylacetate-CoA ligase	Vol. 2, Page 330
2.3.1.14	Phenylacetyltransferase, glutamine Glutamine N-phenylacetyltransferase	
1.14.16.1	Phenylalaninase Phenylalanine 4-monooxygenase	
2.6.1.58	Phenylalanine(histidine):pyruvate aminotransferase Phenylalanine(histidine) transaminase	
2.6.1.58	Phenylalanine (histidine) aminotransferase Phenylalanine(histidine) transaminase	
2.6.1.58	Phenylalanine(histidine) transaminase Phenylalanine(histidine) transaminase	
2.6.1.58	L-Phenylalanine(L-histidine):pyruvate aminotransferase Phenylalanine(histidine) transaminase	
1.14.16.1	L-Phenylalanine,tetrahydrobiopterin:oxygen oxidoreductase (4-hydroxylating) Phenylalanine 4-monooxygenase	
2.6.1.5	Phenylalanine-α-ketoglutarate transaminase Tyrosine transaminase	

EC-No.	Synonym Recommended Name	Vol., Page
2.6.1.27	L-Phenylalanine-2-oxoglutarate aminotransferase Tryptophan transaminase	
2.6.1.49	Phenylalanine-DOPP transaminase Dihydroxyphenylalanine transaminase	
6.1.1.20	Phenylalanine-tRNA ligase Phenylalanine-tRNA ligase	Vol. 2, Page 156
6.1.1.20	Phenylalanine-tRNA synthetase Phenylalanine-tRNA ligase	Vol. 2, Page 156
1.13.12.9	Phenylalanine 2-monooxygenase Phenylalanine 2-monooxygenase	
2.6.1.5	L-Phenylalanine 2-oxoglutarate aminotransferase Tyrosine transaminase	
1.14.16.1	Phenylalanine 4-hydroxylase Phenylalanine 4-monooxygenase	
1.14.16.1	Phenylalanine 4-monooxygenase Phenylalanine 4-monooxygenase	
1.4.1.20	L-Phenylalanine:NAD^+ oxidoreductase (deaminating) Phenylalanine dehydrogenase	
1.13.12.9	L-Phenylalanine:oxygen 2-oxidoreductase (decarboxylating) Phenylalanine 2-monooxygenase	
6.1.1.20	L-Phenylalanine:$tRNA^{Phe}$ ligase (AMP-forming) Phenylalanine-tRNA ligase	Vol. 2, Page 156
2.7.7.54	L-Phenylalanine adenylyltransferase Phenylalanine adenylyltransferase	
2.7.7.54	Phenylalanine adenylyltransferase Phenylalanine adenylyltransferase	
2.6.1.5	Phenylalanine aminotransferase Tyrosine transaminase	
4.3.1.5	L-phenylalanine ammonia-lyase phenylalanine ammonia-lyase	
4.3.1.5	phenylalanine ammonia-lyase phenylalanine ammonia-lyase	
4.3.1.5	phenylalanine ammonium-lyase phenylalanine ammonia-lyase	
4.1.1.53	L-Phenylalanine carboxy-lyase Phenylalanine decarboxylase	

EC-No.	Synonym Recommended Name	Vol., Page
4.3.1.5	phenylalanine deaminase phenylalanine ammonia-lyase	
4.1.1.53	L-Phenylalanine decarboxylase Phenylalanine decarboxylase	
4.1.1.53	Phenylalanine decarboxylase Phenylalanine decarboxylase	
1.4.1.20	L-Phenylalanine dehydrogenase Phenylalanine dehydrogenase	
1.4.1.20	Phenylalanine dehydrogenase Phenylalanine dehydrogenase	
1.14.16.1	Phenylalanine hydroxylase Phenylalanine 4-monooxygenase	
4.2.1.96	Phenylalanine hydroxylase-stimulating protein/pterin-4α-carbinolamine dehydratase 4a-Hydroxytetrahydrobiopterin dehydratase	
2.3.1.53	Phenylalanine N-acetyltransferase Phenylalanine N-acetyltransferase	
1.13.12.9	L-Phenylalanine oxidase (deaminating and decarboxylating) Phenylalanine 2-monooxygenase	
5.1.1.11	Phenylalanine racemase Phenylalanine racemase (ATP-hydrolysing)	Vol. 1, Page 48
5.1.1.11	Phenylalanine racemase (adenosine triphosphate-hydrolyzing) Phenylalanine racemase (ATP-hydrolysing)	Vol. 1, Page 48
5.1.1.11	Phenylalanine racemase (ATP-hydrolysing) Phenylalanine racemase (ATP-hydrolysing)	Vol. 1, Page 48
5.1.1.11	Phenylalanine racemase (ATP-hydrolysing) Phenylalanine racemase (ATP-hydrolysing)	Vol. 1, Page 48
5.1.1.11	Phenylalanine racemase (ATP-hydrolyzing) Phenylalanine racemase (ATP-hydrolysing)	Vol. 1, Page 48
2.6.1.5	Phenylalanine transaminase Tyrosine transaminase	
6.1.1.20	Phenylalanine translase Phenylalanine-tRNA ligase	Vol. 2, Page 156
6.1.1.20	Phenylalanyl-transfer ribonucleate synthetase Phenylalanine-tRNA ligase	Vol. 2, Page 156

EC-No.	Synonym Recommended Name	Vol., Page
6.1.1.20	Phenylalanyl-transfer RNA ligase Phenylalanine-tRNA ligase	Vol. 2, Page 156
6.1.1.20	Phenylalanyl-transfer RNA synthetase Phenylalanine-tRNA ligase	Vol. 2, Page 156
6.1.1.20	Phenylalanyl-tRNA ligase Phenylalanine-tRNA ligase	Vol. 2, Page 156
6.1.1.20	L-Phenylalanyl-tRNA synthetase Phenylalanine-tRNA ligase	Vol. 2, Page 156
6.1.1.20	Phenylalanyl-tRNA synthetase Phenylalanine-tRNA ligase	Vol. 2, Page 156
6.1.1.20	Phenylalanyl transfer ribonucleic acid synthetase Phenylalanine-tRNA ligase	Vol. 2, Page 156
1.14.17.1	Phenylamine β-hydroxylase Dopamine β-monooxygenase	
2.1.1.28	Phenylethanolamine N-methyltransferase Phenylethanolamine N-methyltransferase	
1.2.1.58	Phenylglyoxal:NAD$^+$ oxidoreductase (CoA benzoylating) Phenylglyoxylate dehydrogenase (acylating)	
1.2.1.58	Phenylglyoxylate-acceptor oxidoreductase Phenylglyoxylate dehydrogenase (acylating)	
1.2.1.58	Phenylglyoxylate:acceptor oxidoreductase Phenylglyoxylate dehydrogenase (acylating)	
1.2.1.58	Phenylglyoxylate:NAD oxidoreductase Phenylglyoxylate dehydrogenase (acylating)	
4.1.1.7	Phenylglyoxylate decarboxylase Benzoylformate decarboxylase	
1.2.1.58	Phenylglyoxylate dehydrogenase (acylating) Phenylglyoxylate dehydrogenase (acylating)	
1.2.1.58	Phenylglyoxylate dehydrogenase (CoA benzoylating) Phenylglyoxylate dehydrogenase (acylating)	
2.4.1.17	*p*-Phenylphenol glucuronyltransferase Glucuronosyltransferase	
4.1.1.43	Phenylpyruvate carboxy-lyase Phenylpyruvate decarboxylase	
4.1.1.43	Phenylpyruvate decarboxylase Phenylpyruvate decarboxylase	

EC-No.	Synonym Recommended Name	Vol., Page
5.3.2.1	Phenylpyruvate keto-enol-isomerase Phenylpyruvate tautomerase	Vol. 1, Page 367
5.3.2.1	Phenylpyruvate keto-enol tautomerase Phenylpyruvate tautomerase	Vol. 1, Page 367
5.3.2.1	Phenylpyruvate tautomerase Phenylpyruvate tautomerase	Vol. 1, Page 367
2.6.1.5	Phenylpyruvate transaminase Tyrosine transaminase	
2.6.1.5	Phenylpyruvic acid transaminase Tyrosine transaminase	
5.3.2.1	Phenylpyruvic keto-enol isomerase Phenylpyruvate tautomerase	Vol. 1, Page 367
4.1.2.26	Phenylserine aldolase Phenylserine aldolase	
3.1.6.1	Phenylsulfatase Arylsulfatase	
1.11.1.12	PHGPX Phospholipid-hydroperoxide glutathione peroxidase	
5.3.1.9	PHI Glucose-6-phosphate isomerase	Vol. 1, Page 298
3.4.24.60	PHIE Dactylysin	
3.2.1.62	Phloretin-glucosidase Glycosylceramidase	
3.2.1.62	Phloretin glucosidase Glycosylceramidase	
3.7.1.4	Phloretin hydrolase Phloretin hydrolase	
3.2.1.62	Phloridzin.β.-glucosidase Glycosylceramidase	
3.2.1.62	Phloridzin glucosidase Glycosylceramidase	
3.2.1.62	Phlorizin hydrolase Glycosylceramidase	
1.3.1.57	phloroglucinol reductase phloroglucinol reductase	

EC-No.	Synonym Recommended Name	Vol., Page
2.3.1.156	phloroisovalerophenone synthase phloroisovalerophenone synthase	
2.7.9.2	Phoephoenolpyruvate synthetase water dikinase pyruvate	
3.1.1.51	phorbol-12,13-diester 12-ester hydrolase phorbol-diester hydrolase	
3.1.1.51	phorbol-diester hydrolase phorbol-diester hydrolase	
5.3.1.10	Phoshoglucosamine isomerase Glucosamine-6-phosphate isomerase	Vol. 1, Page 312
3.1.3.18	2-Phoshoglycolate phosphohydrolase Phosphoglycolate, phosphatase	
2.7.3.10	Phosphagen phosphokinase Agmatine kinase	
3.9.1.1	Phosphamide hydrolase Phosphoamidase	
3.1.3.6	3'-phosphatase 3'-nucleotidase	
3.1.3.40	phosphatase, 1-guanidino-scyllo-inositol 4- guanidinodeoxy-scyllo-inositol-4-phosphatase	
3.1.3.63	Phosphatase, 2-carboxyarabinitol 1- 2-Carboxy-D-arabinitol-1-phosphatase	
3.1.3.68	Phosphatase, 2-deoxyglucose 6- 2-Deoxyglucose-6-phosphatase	
3.1.3.68	Phosphatase, 2-deoxyglucose 6- (Escherichia coli strain K-12 Kohara clone 405 gene DOG1) 2-Deoxyglucose-6-phosphatase	
3.1.3.68	Phosphatase, 2-deoxyglucose 6- (Saccharomyces cerevisiae clone pGEMT-DOGR1) 2-Deoxyglucose-6-phosphatase	
3.1.3.68	Phosphatase, 2-deoxyglucose 6- (Saccharomyces cerevisiae gene DOGR1) 2-Deoxyglucose-6-phosphatase	
3.1.3.68	Phosphatase, 2-deoxyglucose 6- (Saccharomyces cerevisiae gene DOGR1 isoform) 2-Deoxyglucose-6-phosphatase	

EC-No.	Synonym Recommended Name	Vol., Page
3.1.3.68	Phosphatase, 2-deoxyglucose 6- (Saccharomyces cerevisiae gene DOGR2 isoform) 2-Deoxyglucose-6-phosphatase	
3.1.3.38	phosphatase, 3-phosphoglycerate 3-phosphoglycerate phosphatase	
3.1.3.29	phosphatase, acylneuraminate 9- N-acylneuraminate-9-phosphatase	
3.1.3.28	phosphatase, adenosine diphosphate phosphoglycerate ADPphosphoglycerate phosphatase	
3.6.1.14	Phosphatase, adenosine tetra- Adenosine-tetraphosphatase	
3.6.1.3	Phosphatase, adenosine tri- Adenosinetriphosphatase	
3.1.3.1	phosphatase, alkaline alkaline phosphatase	
3.1.3.59	Phosphatase, alkylacetylglycerophosphate Alkylacetylglycerophosphatase	
3.6.1.41	Phosphatase, bis(5'-nucleosyl) tetra- (symmetrical) Bis(5'-nucleosyl)-tetraphosphatase (symmetrical)	
3.6.1.29	Phosphatase, bis(5-adenosyl) tri- Bis(5'-adenosyl)-triphosphatase	
3.6.1.17	Phosphatase, bis(5-guanosyl)tetra- Bis(5'-nucleosyl)-tetraphosphatase (asymmetrical)	
3.1.3.52	Phosphatase, branched-chain 2-keto acid dehydrogenase [3-Methyl-2-oxobutanoate dehydrogenase (lipoamide)]-phosphatase	
3.1.3.52	Phosphatase, branched-chain oxo acid dehydrogenase [3-Methyl-2-oxobutanoate dehydrogenase (lipoamide)]-phosphatase	
3.6.1.27	Phosphatase, C_{55}-isoprenyl di- Undecaprenyl-diphosphatase	
3.1.3.55	Phosphatase, caldesmon Caldesmon-phosphatase	
3.6.1.12	Phosphatase, deoxycytidine tri- dCTP diphosphatase	
3.1.5.1	Phosphatase, deoxyguanosine tri- dGTPase	

EC-No.	Synonym Recommended Name	Vol., Page
3.6.1.17	Phosphatase, diguanosine tetra- Bis(5'-nucleosyl)-tetraphosphatase (asymmetrical)	
3.6.1.29	Phosphatase, dinucleoside tri- Bis(5'-adenosyl)-triphosphatase	
3.1.3.51	Phosphatase, dolichol Dolichyl-phosphatase	
3.6.1.43	Phosphatase, dolichol di- Dolichyldiphosphatase	
3.6.1.10	Phosphatase, endopoly- Endopolyphosphatase	
3.6.1.11	Phosphatase, exopoly- Exopolyphosphatase	
3.1.3.46	phosphatase, fructose 2,6-di- fructose-2,6-bisphosphate 2-phosphatase	
3.1.3.54	Phosphatase, fructose 2,6-diphosphate 6- Fructose-2,6-bisphosphate 6-phosphatase	
3.1.3.19	phosphatase, glycerol 2- glycerol-2-phosphatase	
3.6.1.40	Phosphatase, guanosine 5'-triphosphate 3'-diphosphate 5'- Guanosine-5'-triphosphate,3'-diphosphate diphosphatase	
3.6.1.42	Phosphatase, guanosine di- Guanosine-diphosphatase	
3.1.3.15	Phosphatase, histidinol- Histidinol-phosphatase	
3.1.3.56	Phosphatase, inosine tri- Inositol-1,4,5-trisphosphate 5-phosphatase	
3.1.3.62	Phosphatase, inositol 1,3,4,5-tetrakisphosphate 3- Inositol-1,3,4,5-tetrakisphosphate 3-phosphatase	
3.1.3.65	Phosphatase, inositol 1,3-bisphosphate Inositol-1,3-bisphosphate 3-phosphatase	
3.1.3.61	Phosphatase, inositol 1,4,5-trisphosphate 1- Inositol-1,4,5-trisphosphate 1-phosphatase	
3.1.3.56	Phosphatase, inositol 1,4,5-trisphosphate 5- Inositol-1,4,5-trisphosphate 5-phosphatase	
3.1.3.57	Phosphatase, inositol 1,4-bisphosphate 1- Inositol-1,4-bisphosphate 1-phosphatase	

EC-No.	Synonym Recommended Name	Vol., Page
3.1.3.25	phosphatase, inositol 1- inositol-1 (or 4)-monophosphatase	
3.1.3.66	Phosphatase, inositol 3,4-bisphosphate 4- Inositol-3,4-bisphosphate 4-phosphatase	
3.1.3.14	phosphatase, methylthiophosphoglycerate methylphosphothioglycerate phosphatase	
3.1.3.53	Phosphatase, myosin Myosin-light-chain-phosphatase	
3.1.3.53	Phosphatase, myosin light-chain kinase Myosin-light-chain-phosphatase	
3.6.1.15	Phosphatase, nucleoside tri- Nucleoside-triphosphatase	
3.1.3.41	phosphatase, *p*-nitrophenyl 4-nitrophenylphosphatase	
3.1.3.64	Phosphatase, phosphatidylinositol 3- Phosphatidylinositol 3-phosphatase	
3.1.3.45	phosphatase, phospho-3-deoxy-2-octulosonate 3-deoxy-manno-octulosonate-8-phosphatase	
3.1.3.60	Phosphatase, phosphoenolpyruvate Phosphoenolpyruvate phosphatase	
3.1.3.20	phosphatase, phosphoglycerate phosphoglycerate phosphatase	
3.1.3.18	Phosphatase, phosphoglycolate Phosphoglycolate, phosphatase	
3.1.3.48	phosphatase, phosphoprotein (phosphotyrosine) protein-tyrosine-phosphatase	
3.1.3.3	phosphatase, phosphoserine phosphoserine phosphatase	
3.1.3.48	phosphatase, phosphotyrosine protein-tyrosine-phosphatase	
3.1.3.32	phosphatase, polynucleotide 3'- polynucleotide 3'-phosphatase	
3.1.3.33	phosphatase, polynucleotide 5- polynucleotide 5'-phosphatase	
3.1.3.43	phosphatase, pyruvate dehydrogenase [pyruvate dehydrogenase (lipoamide)]-phosphatase	

EC-No.	Synonym Recommended Name	Vol., Page
3.1.3.49	Phosphatase, pyruvate kinase [Pyruvate kinase]-phosphatase	
3.1.3.37	Phosphatase, sedoheptulose di- Sedoheptulose-bisphosphatase	
3.1.3.50	phosphatase, sorbitol 6- sorbitol-6-phosphatase	
3.1.3.39	Phosphatase, streptomycin 6- Streptomycin-6-phosphatase	
3.1.3.24	phosphatase, sucrose sucrose-phosphatase	
3.1.3.23	phosphatase, sugar sugar-phosphatase	
3.6.1.28	Phosphatase, thiamine tri- Thiamine-triphosphatase	
3.1.3.36	Phosphatase, triphosphoinositide Phosphatidylinositol-bisphosphatase	
3.1.3.58	Phosphatase, xylitol 5- Sugar-terminal-phosphatase	
3.1.3.32	3'-phosphatase/5'-OH kinase polynucleotide 3'-phosphatase	
3.1.3.16	Phosphatase 2A Phosphoprotein phosphatase	
3.1.3.16	Phosphatase 2B Phosphoprotein phosphatase	
3.1.3.16	Phosphatase C-II Phosphoprotein phosphatase	
3.1.3.16	Phosphatase H-II Phosphoprotein phosphatase	
3.1.3.16	Phosphatase I Phosphoprotein phosphatase	
3.1.3.16	Phosphatase IB Phosphoprotein phosphatase	
3.1.3.16	Phosphatase II Phosphoprotein phosphatase	
3.1.3.16	Phosphatase III Phosphoprotein phosphatase	

EC-No.	Synonym Recommended Name	Vol., Page
3.1.3.16	Phosphatase IV Phosphoprotein phosphatase	
3.1.3.16	Phosphatase SP Phosphoprotein phosphatase	
3.6.1.39	Phosphatase thymidine tri- Thymidine triphosphatase	
2.7.7.56	Phosphate-dependent exonuclease tRNA nucleotidyltransferase	
2.3.1.8	Phosphate acetyltransferase Phosphate acetyltransferase	
2.3.1.19	Phosphate butyryltransferase Phosphate butyryltransferase	
2.7.9.1	phosphate dikinase pyruvate phosphate dikinase pyruvate	
3.1.3.4	Phosphatic acid phosphatase Phosphatidate phosphatase	
3.1.3.4	Phosphatic acid phosphohydrolase Phosphatidate phosphatase	
3.1.1.4	Phosphatidase Phospholipase A_2	
3.1.1.5	phosphatidase B lysophospholipase	
3.1.4.3	Phosphatidase C Phospholipase C	
2.7.7.41	Phosphatidate cytidyltransferase Phosphatidate cytidylyltransferase	
2.7.7.41	Phosphatidate cytidylyltransferase Phosphatidate cytidylyltransferase	
3.1.3.4	Phosphatidate phosphatase Phosphatidate phosphatase	
3.1.1.47	phosphatide 2-acylhydrolase 1-alkyl-2-acetylglycerophosphocholine esterase	
2.7.7.41	Phosphatidic acid cytidylyltransferase Phosphatidate cytidylyltransferase	
3.1.1.4	Phosphatidolipase Phospholipase A_1	

EC-No.	Synonym Recommended Name	Vol., Page
3.1.3.64	Phosphatidyl-3-phosphate 3-phosphohydrolase Phosphatidylinositol 3-phosphatase	
3.1.4.11	1-Phosphatidyl-4,5-bisphosphate phosphodiesterase 1-Phosphatidyl-4,5-bisphosphate phosphodiesterase	
3.1.1.52	1-phosphatidyl-D-myo-inositol 2-acylhydrolase phosphatidylinositol deacylase	
3.1.4.11	1-Phosphatidyl-D-myo-inositol 4,5-bisphosphate inositoltrisphosphohydrolase 1-Phosphatidyl-4,5-bisphosphate phosphodiesterase	
3.1.4.10	1-Phosphatidyl-D-myo-inositol inositolphosphohydrolase (cyclic-phosphate-forming) 1-Phosphatidylinositol phosphodiesterase	
4.1.1.65	Phosphatidyl-L-serine carboxy-lyase Phosphatidylserine decarboxylase	
3.1.3.36	Phosphatidyl-myo-inositol-4,5-bisphosphate 4-phosphohydrolase Phosphatidylinositol-bisphosphatase	
3.1.3.36	Phosphatidyl-myo-inositol-4,5-bisphosphate phosphatase Phosphatidylinositol-bisphosphatase	
2.4.1.57	Phosphatidyl-myo-inositol α-mannosyltransferase Phosphatidyl-myo-inositol α-mannosyltransferase	
2.1.1.71	Phosphatidyl-N-methylethanolamine methyltransferase Phosphatidyl-N-methylethanolamine N-methyltransferase	
2.1.1.71	Phosphatidyl-N-methylethanolamine N-methyltransferase Phosphatidyl-N-methylethanolamine N-methyltransferase	
2.1.1.71	Phosphatidyl-N-monomethylethanolamine methyltransferase Phosphatidyl-N-methylethanolamine N-methyltransferase	
2.7.8.5	3-Phosphatidyl 1'-glycerol-3'-phosphate synthase CDPdiacylglycerol-glycerol-3-phosphate 3-phosphatidyltransferase	
3.1.1.32	phosphatidyl 1-acylhydrolase phospholipase A1	
2.3.1.83	Phosphatidylcholine-dolichol O-acyltransferase Phosphatidylcholine-dolichol O-acyltransferase	
2.3.1.135	Phosphatidylcholine-retinol O-acyltransferase Phosphatidylcholine-retinol O-acyltransferase	
2.3.1.43	Phosphatidylcholine-sterol O-acyltransferase Phosphatidylcholine-sterol O-acyltransferase	

EC-No.	Synonym Recommended Name	Vol., Page
1.14.13.26	Phosphatidylcholine 12-monooxygenase Phosphatidylcholine 12-monooxygenase	
3.1.1.4	Phosphatidylcholine 2-acylhydrolase Phospholipase A_2	
2.3.1.135	Phosphatidylcholine:retinol-(cellular-retinol-binding-protein) O-acyltransferase Phosphatidylcholine-retinol O-acyltransferase	
2.3.1.43	Phosphatidylcholine:sterol O-acyltransferase Phosphatidylcholine-sterol O-acyltransferase	
3.1.4.3	Phosphatidylcholine cholinephosphohydrolase Phospholipase C	
1.3.1.35	Phosphatidylcholine desaturase Phosphatidylcholine desaturase	
3.1.4.4	Phosphatidylcholine phosphatidohydrolase Phospholipase D	
2.1.1.17	Phosphatidylethanolamine-N-methylase Phosphatidylethanolamine N-methyltransferase	
2.1.1.17	Phosphatidylethanolamine-S-adenosylmethionine methyltransferase Phosphatidylethanolamine N-methyltransferase	
2.1.1.17	Phosphatidylethanolamine methyltransferase Phosphatidylethanolamine N-methyltransferase	
2.1.1.71	Phosphatidylethanolamine methyltransferase I Phosphatidyl-N-methylethanolamine N-methyltransferase	
2.1.1.17	Phosphatidylethanolamine N-methyltransferase Phosphatidylethanolamine N-methyltransferase	
2.7.8.20	Phosphatidylglycerol-membrane-oligosaccharide glycerophosphotransferase Phosphatidylglycerol-membrane-oligosaccharide glycerophosphotransferase	
2.7.8.20	Phosphatidylglycerol:membrane-derived-oligosaccharide-D-glucose glycerophosphotransferase Phosphatidylglycerol-membrane-oligosaccharide glycerophosphotransferase	
2.7.8.5	Phosphatidylglycerol phosphate synthase CDPdiacylglycerol-glycerol-3-phosphate 3-phosphatidyltransferase	
2.7.8.5	Phosphatidylglycerolphosphate synthase CDPdiacylglycerol-glycerol-3-phosphate 3-phosphatidyltransferase	

EC-No.	Synonym Recommended Name	Vol., Page
2.7.8.5	Phosphatidylglycerol phosphate synthetase CDPdiacylglycerol-glycerol-3-phosphate 3-phosphatidyltransferase	
3.1.3.27	phosphatidylglycerophosphatase phosphatidylglycerophosphatase	
3.1.3.27	phosphatidylglycero phosphate phosphatase phosphatidylglycerophosphatase	
3.1.3.27	phosphatidylglycerophosphate phosphohydrolase phosphatidylglycerophosphatase	
2.7.8.5	Phosphatidylglycerophosphate synthase CDPdiacylglycerol-glycerol-3-phosphate 3-phosphatidyltransferase	
2.7.8.5	Phosphatidylglycerophosphate synthetase CDPdiacylglycerol-glycerol-3-phosphate 3-phosphatidyltransferase	
3.1.3.67	Phosphatidylinositol-3,4,5-trisphosphate 3-phosphatase Phosphatidylinositol-3,4,5-trisphosphate 3-phosphatase	
3.1.3.67	Phosphatidylinositol-3,4,5-trisphosphate 3-phosphohydrolase Phosphatidylinositol-3,4,5-trisphosphate 3-phosphatase	
3.1.3.64	Phosphatidylinositol-3-phosphatase Phosphatidylinositol 3-phosphatase	
2.7.1.68	1-Phosphatidylinositol-4-phosphate 5-kinase 1-Phosphatidylinositol-4-phosphate 5-kinase	
2.7.1.68	Phosphatidylinositol-4-phosphate 5-kinase 1-Phosphatidylinositol-4-phosphate 5-kinase	
3.1.3.36	Phosphatidylinositol-bisphosphatase Phosphatidylinositol-bisphosphatase	
3.1.4.50	Phosphatidylinositol-specific phospholipase D Glycoprotein phospholipase D	
2.7.1.137	Phosphatidylinositol 3'-kinase 1-Phosphatidylinositol 3-kinase	
2.7.1.137	1-Phosphatidylinositol 3-kinase 1-Phosphatidylinositol 3-kinase	
2.7.1.137	Phosphatidylinositol 3-kinase 1-Phosphatidylinositol 3-kinase	
3.1.3.64	Phosphatidylinositol 3-phosphatase Phosphatidylinositol 3-phosphatase	
3.1.3.36	Phosphatidylinositol 4,5-bisphosphate phosphatase Phosphatidylinositol-bisphosphatase	

EC-No.	Synonym Recommended Name	Vol., Page
2.7.1.67	1-Phosphatidylinositol 4-kinase 1-Phosphatidylinositol 4-kinase	
2.7.1.67	Phosphatidylinositol 4-kinase 1-Phosphatidylinositol 4-kinase	
2.7.1.68	Phosphatidylinositol 4-phosphate kinase 1-Phosphatidylinositol-4-phosphate 5-kinase	
3.1.1.52	phosphatidylinositol deacylase phosphatidylinositol deacylase	
2.7.1.67	Phosphatidylinositol kinase 1-Phosphatidylinositol 4-kinase	
2.4.1.198	Phosphatidylinositol N-acetylglucosaminyltransferase Phosphatidylinositol N-acetylglucosaminyltransferase	
3.1.4.10	1-Phosphatidylinositol phosphodiesterase 1-Phosphatidylinositol phosphodiesterase	
3.1.1.52	phosphatidylinositol phospholipase A_2 phosphatidylinositol deacylase	
3.1.4.10	Phosphatidylinositol phospholipase C 1-Phosphatidylinositol phosphodiesterase	
3.1.4.50	Phosphatidylinositol phospholipase D Glycoprotein phospholipase D	
2.7.8.11	Phosphatidylinositol synthase CDPdiacylglycerol-inositol 3-phosphatidyltransferase	
2.1.1.71	Phosphatidylmonomethylethanolamine methyltransferase Phosphatidyl-N-methylethanolamine N-methyltransferase	
4.1.1.65	Phosphatidylserine decarboxylase Phosphatidylserine decarboxylase	
4.1.1.65	Phosphatidylserine decarboxylase Phosphatidylserine decarboxylase	
2.7.8.8	Phosphatidylserine synthase CDPdiacylglycerol-serine O-phosphatidyltransferase	
2.7.8.8	Phosphatidylserine synthetase CDPdiacylglycerol-serine O-phosphatidyltransferase	
2.7.8.11	Phosphatidyltransferase, cytidine diphosphodiglyceride-inositol CDPdiacylglycerol-inositol 3-phosphatidyltransferase	
2.7.8.8	Phosphatidyltransferase, cytidine diphosphoglyceride-serine O- CDPdiacylglycerol-serine O-phosphatidyltransferase	

EC-No.	Synonym Recommended Name	Vol., Page
2.7.8.5	Phosphatidyltransferase, glycerol phosphate CDPdiacylglycerol-glycerol-3-phosphate 3-phosphatidyltransferase	
3.2.1.85	6-Phospho-β-D-galactosidase 6-Phospho-β-galactosidase	
3.2.1.85	Phospho-β-D-galactosidase 6-Phospho-β-galactosidase	
3.2.1.85	6-Phospho-β-D-galactoside 6-phosphogalactohydrolase 6-Phospho-β-galactosidase	
3.2.1.86	6-Phospho-β-D-glucosyl-(1,4)-D-glucose glucohydrolase 6-Phospho-β-glucosidase	
2.4.2.22	5-Phospho-α-D-ribose-1-diphosphate:xanthine phospho-D-ribosyltransferase Xanthine phosphoribosyltransferase	
5.3.1.24	N-(5-Phospho-β-D-ribosyl)anthranilate ketol-isomerase Phosphoribosylanthranilate isomerase	Vol. 1, Page 353
3.2.1.85	6-Phospho-β-galactosidase 6-Phospho-β-galactosidase	
3.2.1.85	Phospho-β-galactosidase 6-Phospho-β-galactosidase	
3.2.1.86	6-Phospho-β-glucosidase 6-Phospho-β-glucosidase	
3.2.1.122	Phospho-α-glucosidase Maltose-6'-phosphate glucosidase	
3.2.1.86	Phospho-β-glucosidase 6-Phospho-β-glucosidase	
3.2.1.86	Phospho-β-glucosidase A 6-Phospho-β-glucosidase	
2.7.1.11	Phospho-1,6-fructokinase 6-Phosphofructokinase	
4.1.2.15	7-Phospho-2-dehydro-3-deoxy-D-arabino-heptonate D-erythrose-4-phosphate-lyase (pyruvate-phosphorylating) 2-Dehydro-3-deoxyphosphoheptanoate aldolase	
4.1.2.14	6-Phospho-2-dehydro-3-deoxy-D-gluconate D-glyceraldehyde-3-phosphate-lyase 2-Dehydro-3-deoxyphosphogluconate aldolase	

EC-No.	Synonym Recommended Name Vol., Page
4.1.2.16	8-Phospho-2-dehydro-3-deoxy-D-octonate D-arabinose-5-phosphate-lyase (pyruvate-phosporylating) 2-Dehydro-3-deoxyphosphooctonate aldolase
4.1.2.15	Phospho-2-dehydro-3-deoxyheptonate aldolase 2-Dehydro-3-deoxyphosphoheptanoate aldolase
4.1.2.16	Phospho-2-dehydro-3-deoxyoctonate aldolase 2-Dehydro-3-deoxyphosphooctonate aldolase
4.1.2.15	7-Phospho-2-keto-3-deoxy-D-arabino-heptonate D-erythrose-4-phosphate lyase (pyruvate-phosphorylating) 2-Dehydro-3-deoxyphosphoheptanoate aldolase
4.1.2.21	6-Phospho-2-keto-3-deoxygalactonate aldolase 2-Dehydro-3-deoxyphosphogalactonate aldolase
4.1.2.21	Phospho-2-keto-3-deoxygalactonic aldolase 2-Dehydro-3-deoxyphosphogalactonate aldolase
4.1.2.14	6-Phospho-2-keto-3-deoxygluconate aldolase 2-Dehydro-3-deoxyphosphogluconate aldolase
4.1.2.14	Phospho-2-keto-3-deoxygluconate aldolase 2-Dehydro-3-deoxyphosphogluconate aldolase
4.1.2.14	Phospho-2-keto-3-deoxygluconic aldolase 2-Dehydro-3-deoxyphosphogluconate aldolase
4.1.2.15	Phospho-2-keto-3-deoxyheptonate aldolase 2-Dehydro-3-deoxyphosphoheptanoate aldolase
4.1.2.15	Phospho-2-keto-3-deoxyheptonic aldolase 2-Dehydro-3-deoxyphosphoheptanoate aldolase
4.1.2.16	Phospho-2-keto-3-deoxyoctonate aldolase 2-Dehydro-3-deoxyphosphooctonate aldolase
4.1.2.15	Phospho-2-oxo-3-deoxyheptonate aldolase 2-Dehydro-3-deoxyphosphoheptanoate aldolase
4.6.1.3	7-phospho-3-deoxy-D-arabino-heptulosonate phosphate-lyase (cyclizing) 3-dehydroquinate synthase
4.1.2.29	6-Phospho-5-dehydro-2-deoxy-D-gluconate malonate-semialdehyde-lyase Phospho-5-dehydro-2-deoxygluconate aldolase
4.1.2.29	Phospho-5-dehydro-2-deoxygluconate aldolase Phospho-5-dehydro-2-deoxygluconate aldolase
4.1.2.29	Phospho-5-keto-2-deoxygluconate-aldolase Phospho-5-dehydro-2-deoxygluconate aldolase

EC-No.	Synonym Recommended Name	Vol., Page
1.8.99.4	Phospho-adenylylsulfate (PAPS) reductase Phosphoadenosine-phosphosulfate reductase	
1.8.99.4	Phospho-adenylylsulfate reductase Phosphoadenosine-phosphosulfate reductase	
1.1.1.43	6-Phospho-D-gluconate:NAD(P)$^+$ 2-oxidoreductase Phosphogluconate 2-dehydrogenase	
1.1.1.44	6-Phospho-D-gluconate:NADP$^+$ 2-oxidoreductase (decarboxylating) Phosphogluconate dehydrogenase (decarboxylating)	
1.1.1.44	6-Phospho-D-gluconate dehydrogenase Phosphogluconate dehydrogenase (decarboxylating)	
4.2.1.12	6-phospho-D-gluconate hydro-lyase phosphogluconate dehydratase	
3.1.1.31	6-phospho-D-glucono-1,5-lactone lactonohydrolase 6-phosphogluconolactonase	
3.1.1.31	6-phospho-D-glucose-δ-lactone hydrolase 6-phosphogluconolactonase	
5.4.2.4	3-Phospho-D-glycerate 1,2-phosphomutase Bisphosphoglycerate mutase	Vol. 1, Page 520
4.1.1.39	3-Phospho-D-glycerate carboxy-lyase (dimerizing) Ribulose-bisphosphate carboxylase	
4.2.1.11	2-phospho-D-glycerate hydro-lyase phosphoenolpyruvate hydratase	
4.2.1.11	2-phospho-D-glycerate hydro-lyase phosphoenolpyruvate hydratase	
2.7.1.106	3-Phospho-D-glyceroyl-phosphate:D-glucose-1-phosphate 6-phosphotransferase Glucose-1,6-bisphosphate synthase	
2.7.4.17	3-Phospho-D-glyceroyl-phosphate:polyphosphate phosphotransferase 3-Phosphoglyceroyl-phosphate-polyphosphate phosphotransferase	
2.4.2.18	N-(5-Phospho-D-ribosyl)-anthranilate:pyrophosphate phospho-α-D-ribosyltransferase Anthranilate phosphoribosyltransferase	
2.4.2.17	1-(5-Phospho-D-ribosyl)-ATP:pyrophosphate phospho-α-D-ribosyltransferase ATP phosphoribosyltransferase	

EC-No.	Synonym Recommended Name	Vol., Page
5.3.1.16	N-(5'-Phospho-D-ribosylformimino)-5-amino-1-(5-phosphoribosyl)-4-imidazole carboxamide isomerase N-(5'-Phospho-D-ribosylformimino)-5-amino-1-(5-phosphoribosyl)-4-imidazole carboxamide isomerase	Vol. 1, Page 335
5.3.1.16	N-(5'-Phospho-D-ribosylformimino)-5-amino-1-(5-phosphoribosyl)-4-imidazole carboxamide ketol-isomerase N-(5'-Phospho-D-ribosylformimino)-5-amino-1-(5-phosphoribosyl)-4-imidazole carboxamide isomerase	Vol. 1, Page 335
4.2.3.1	O-phospho-L-homoserine phospho-lyase (adding water) threonine synthase	
4.2.99.2	O-phospho-L-homoserine phospho-lyase (adding water) threonine synthase	
2.6.1.52	O-Phospho-L-serine:2-oxoglutarate aminotransferase Phosphoserine transaminase	
4.2.99.7	phospho-lyase, ethanolamine phosphate ethanolamine-phosphate phospho-lyase	
2.7.8.13	Phospho-MurNAc-pentapeptide transferase Phospho-N-acetylmuramoyl-pentapeptide-transferase	
4.1.1.36	4'-Phospho-N-(D-pantothenoyl)-L-cysteine carboxy-lyase Phosphopantothenoylcysteine decarboxylase	
5.4.2.3	Phospho-N-acetylglucosamine mutase Phosphoacetylglucosamine mutase	Vol. 1, Page 515
2.7.8.13	Phospho-N-acetylmuramoyl-pentapeptide-transferase Phospho-N-acetylmuramoyl-pentapeptide-transferase	
2.7.8.13	Phospho-N-acetylmuramoylpentapeptidetransferase Phospho-N-acetylmuramoyl-pentapeptide-transferase	
2.7.8.13	Phospho-N-acetylmuramoyl pentapeptide translocase Phospho-N-acetylmuramoyl-pentapeptide-transferase	
2.7.8.13	Phospho-NAc-muramoyl-pentapeptide translocase (UMP) Phospho-N-acetylmuramoyl-pentapeptide-transferase	
5.4.2.3	Phosphoacetylglucosamine mutase Phosphoacetylglucosamine mutase	Vol. 1, Page 515
2.7.8.13	Phosphoacetylmuramoylpentapeptidetransferase Phospho-N-acetylmuramoyl-pentapeptide-transferase	
2.7.8.13	Phosphoacetylmuramoylpentapeptide translocase Phospho-N-acetylmuramoyl-pentapeptide-transferase	

EC-No.	Synonym Recommended Name	Vol., Page
2.3.1.8	Phosphoacylase Phosphate acetyltransferase	
2.8.2.11	3'-Phosphoadenosine-5'-phosphosulfate-cerebroside sulfotransferase Galactosylceramide sulfotransferase	
2.8.2.24	3'-Phosphoadenosine-5'-phosphosulfate:desulfoglucosinolate sulfotransferase Desulfoglucosinolate sulfotransferase	
1.8.99.4	Phosphoadenosine-phosphosulfate reductase Phosphoadenosine-phosphosulfate reductase	
2.8.2.13	3'-Phosphoadenosine 5'-phosphosulfate-psychosine sulphotransferase Psychosine sulfotransferase	
2.8.2.17	3'-Phosphoadenosine 5'-phosphosulfate:chondroitin sulfate sulfotransferase Chondroitin 6-sulfotransferase	
1.8.99.4	3'-Phosphoadenosine 5'-phosphosulfate reductase Phosphoadenosine-phosphosulfate reductase	
3.6.2.2	3-Phosphoadenosine 5-phosphosulfate sulfatase Phosphoadenylylsulfatase	
2.7.1.25	5'-Phosphoadenosine sulfate kinase Adenylylsulfate kinase	
3.1.3.7	phosphoadenylate 3'-nucleotidase 3'(2'),5'-bisphosphate nucleotidase	
1.8.4.8	phosphoadenylyl-sulfate reductase (thioredoxin) phosphoadenylyl-sulfate reductase (thioredoxin)	
2.8.2.21	3'-Phosphoadenylyl keratan sulfotransferase Keratan sulfotransferase	
3.6.2.2	3-Phosphoadenylyl sulfatase Phosphoadenylylsulfatase	
3.6.2.2	Phosphoadenylylsulfatase Phosphoadenylylsulfatase	
2.8.2.4	3'-Phosphoadenylyl sulfate-estrone 3-sulfotransferase Estrone sulfotransferase	
3.1.3.7	3'-phosphoadenylylsulfate 3'-phosphatase 3'(2'),5'-bisphosphate nucleotidase	
3.1.3.30	3'-phosphoadenylylsulfate 3'-phosphatase 3'-phosphoadenylylsulfate 3'-phosphatase	

EC-No.	Synonym Recommended Name	Vol., Page
2.8.2.2	3'-Phosphoadenylylsulfate:alcohol sulfotransferase Alcohol sulfotransferase	
2.8.2.3	3'-Phosphoadenylylsulfate:amine N-sulfotransferase Amine sulfotransferase	
2.8.2.6	3'-Phosphoadenylylsulfate:choline sulfotransferase Choline sulfotransferase	
2.8.2.5	3'-Phosphoadenylylsulfate:chondroitin 4'-sulfotransferase Chondroitin 4-sulfotransferase	
2.8.2.17	3'-Phosphoadenylylsulfate:chondroitin 6'-sulfotransferase Chondroitin 6-sulfotransferase	
2.8.2.18	3'-Phosphoadenylylsulfate:cortisol 21-sulfotransferase Cortisol sulfotransferase	
2.8.2.24	3'-Phosphoadenylylsulfate:desulfoglucosinolate sulfotransferase Desulfoglucosinolate sulfotransferase	
2.8.2.4	3'-Phosphoadenylylsulfate:estrone 3-sulfotransferase Estrone sulfotransferase	
2.8.2.11	3'-Phosphoadenylylsulfate:galactosylceramide 3'-sulfotransferase Galactosylceramide sulfotransferase	
2.8.2.13	3'-Phosphoadenylylsulfate:galactosylsphingosine sulfotransferase Psychosine sulfotransferase	
2.8.2.23	3'-Phosphoadenylylsulfate:heparin-glucosamine 3-O-sulfotransferase Heparin-glucosamine 3-O-sulfotransferase	
2.8.2.12	3'-Phosphoadenylylsulfate:heparitin N-sulfotransferase Heparitin sulfotransferase	
2.8.2.21	3'-Phosphoadenylylsulfate:keratan 6'-sulfotransferase Keratan sulfotransferase	
2.8.2.21	3'-Phosphoadenylylsulfate:keratan sulfotransferase Keratan sulfotransferase	
2.8.2.9	3'-Phosphoadenylylsulfate:L-tyrosine-methyl-ester sulfotransferase Tyrosine-ester sulfotransferase	
2.8.2.8	3'-Phosphoadenylylsulfate:N-desulfoheparin N-sulfotransferase Desulfoheparin sulfotransferase	
2.8.2.8	3'-Phosphoadenylylsulfate:N-desulfoheparin sulfotransferase Desulfoheparin sulfotransferase	
2.8.2.4	3'-Phosphoadenylylsulfate:oestrone sulfotransferase Estrone sulfotransferase	

EC-No.	Synonym Recommended Name	Vol., Page
2.8.2.15	3'-Phosphoadenylylsulfate:phenolic-steroid sulfotransferase Steroid sulfotransferase	
2.8.2.1	3'-Phosphoadenylylsulfate:phenol sulfotransferase Aryl sulfotransferase	
2.8.2.20	3'-Phosphoadenylylsulfate:protein-tyrosine O-sulfotransferase Protein-tyrosine sulfotransferase	
2.8.2.28	3'-Phosphoadenylylsulfate:quercetin-3,3'-bissulfate 7-sulfotransferase Quercetin-3,3'-bissulfate 7-sulfotransferase	
2.8.2.26	3'-Phosphoadenylylsulfate:quercetin-3-sulfate 3'-sulfotransferase Quercetin-3-sulfate 3'-sulfotransferase	
2.8.2.27	3'-Phosphoadenylylsulfate:quercetin-3-sulfate 4'-sulfotransferase Quercetin-3-sulfate 4'-sulfotransferase	
2.8.2.25	3'-Phosphoadenylylsulfate:quercetin 3-sulfotransferase Flavonol 3-sulfotransferase	
2.8.2.10	3'-Phosphoadenylylsulfate:Renilla luciferin sulfotransferase Renilla-luciferin sulfotransferase	
2.8.2.14	3'-Phosphoadenylylsulfate:taurolithocholate sulfotransferase Bile-salt sulfotransferase	
2.8.2.16	3'-Phosphoadenylylsulfate:thiol S-sulfotransferase Thiol sulfotransferase	
2.8.2.19	3'-Phosphoadenylylsulfate:triglucosyl-1-O-alkyl-2-O-acylglycerol 6-sulfotransferase Triglucosylalkylacylglycerol sulfotransferase	
2.8.2.7	3'-Phosphoadenylylsulfate:UDP-N-acetyl-D-galactosamine-4-sulfate 6-sulfotransferase UDP-N-acetylgalactosamine-4-sulfate sulfotransferase	
1.8.99.4	3'-Phosphoadenylylsulfate reductase Phosphoadenosine-phosphosulfate reductase	
3.6.2.2	3'-Phosphoadenylylsulfate sulfohydrolase Phosphoadenylylsulfatase	
3.6.2.2	Phosphoadenylyl sulphatase Phosphoadenylylsulfatase	
3.9.1.1	Phosphoamidase Phosphoamidase	
5.3.1.13	Phosphoarabinoisomerase Arabinose-5-phosphate isomerase	Vol. 1, Page 325

EC-No.	Synonym Recommended Name	Vol., Page
3.2.1.86	Phosphocellobiase 6-Phospho-β-glucosidase	
2.7.7.15	Phosphocholine cytidylyltransferase Choline-phosphate cytidylyltransferase	
2.7.8.2	Phosphocholine diacylglyceroltransferase Diacylglycerol cholinephosphotransferase	
2.7.3.2	Phosphocreatine kinase Creatine kinase	
4.1.2.4	Phosphodeoxyriboaldolase Deoxyribose-phosphate aldolase	
5.4.2.7	Phosphodeoxyribomutase Phosphopentomutase	Vol. 1, Page 535
3.1.4.1	5'-Phosphodiesterase Phosphodiesterase I	
3.1.4.51	Phosphodiesterase, α-glucose 1-phosphate Glucose-1-phospho-D-mannosylglycoprotein phosphodiesterase	
3.1.4.16	Phosphodiesterase, cyclic 2',3'-nucleotide 2'- 2',3'-Cyclic-nucleotide 2'-phosphodiesterase	
3.1.4.37	Phosphodiesterase, cyclic 2',3'-nucleotide 3'- 2',3'-Cyclic-nucleotide 3'-phosphodiesterase	
3.1.4.17	Phosphodiesterase, cyclic 3',5'-nucleotide 3',5'-Cyclic-nucleotide phosphodiesterase	
3.1.4.48	Phosphodiesterase, dolichol phosphoglucose Dolichyl-phosphate-glucose phosphodiesterase	
3.1.4.42	phosphodiesterase, glycerol cyclic 1,2-phosphate 2- glycerol-1,2-cyclic-phosphate 2-phosphodiesterase	
3.1.4.46	Phosphodiesterase, glycerophosphodiester Glycerophosphodiester phosphodiesterase	
3.1.4.44	phosphodiesterase, glycerophosphoinositol glycerophosphoinositol glycerophosphodiesterase	
3.1.4.43	phosphodiesterase, glycerophosphoinositol glycerophosphoinositol inositolphosphodiesterase	
3.1.4.49	Phosphodiesterase, mannosylphosphodolichol Dolichyl-phosphate-mannose phosphodiesterase	
3.1.4.13	phosphodiesterase, serine-ethanolamine phosphate serine-ethanolaminephosphate phosphodiesterase	

EC-No.	Synonym Recommended Name	Vol., Page
3.1.4.12	Phosphodiesterase, sphingomyelin Sphingomyelin phosphodiesterase	
3.1.4.35	Phosphodiesterase cGMP 3',5'-Cyclic-GMP phosphodiesterase	
3.1.4.1	Phosphodiesterase I Phosphodiesterase I	
3.1.16.1	Phosphodiesterase II Spleen exonuclease	
3.1.4.18	phosphodiesterase II phosphodiesterase II	
3.1.4.45	Phosphodiester glycosidase N-Acetylglucosamine-1-phosphodiester N-acetylglucosaminidase	
2.7.1.41	Phosphodismutase Glucose-1-phosphate phosphodismutase	
2.7.1.41	Phosphodismutase, glucose 1-phosphate Glucose-1-phosphate phosphodismutase	
2.7.1.98	phosphoenolpyruvate-fructose phosphotransferase phosphoenolpyruvate-fructose phosphotransferase	
2.7.1.121	Phosphoenolpyruvate-glycerone phosphotransferase Phosphoenolpyruvate-glycerone phosphotransferase	
2.7.3.9	Phosphoenolpyruvate-protein phosphotransferase Phosphoenolpyruvate-protein phosphotransferase	
2.7.1.69	Phosphoenolpyruvate-sugar phosphotransferase enzyme II Protein-Npi-phosphohistidine-sugar phosphotransferase	
2.5.1.7	Phosphoenolpyruvate-UDP-acetylglucosamine-3-enolpyruvyltransferase UDP-N-acetylglucosamine 1-carboxyvinyltransferase	
5.4.2.9	Phosphoenolpyruvate 2,3-phosphonomutase Phosphoenolpyruvate mutase	Vol. 1, Page 546
2.5.1.19	Phosphoenolpyruvate:3-phosphoshikimate 5-O-(1-carboxyvinyl)transferase 3-Phosphoshikimate 1-carboxyvinyltransferase	
2.7.1.121	Phosphoenolpyruvate:glycerone phosphotransferase Phosphoenolpyruvate-glycerone phosphotransferase	
2.7.3.9	Phosphoenolpyruvate:protein-L-histidine N-pros-phosphotransferase Phosphoenolpyruvate-protein phosphotransferase	

EC-No.	Synonym Recommended Name	Vol., Page
2.5.1.7	Phosphoenolpyruvate:UDP-2-acetamido-2-deoxy-D-glucose 2-enoyl-1-carboxyethyltransferase UDP-N-acetylglucosamine 1-carboxyvinyltransferase	
2.5.1.7	Phosphoenolpyruvate:UDP-N-acetyl-D-glucosamine 1-carboxyvinyltransferase UDP-N-acetylglucosamine 1-carboxyvinyltransferase	
2.5.1.7	Phosphoenolpyruvate:uridine-5'-diphospho-N-acetyl-2-amino-2-deoxyglucose-3-enolpyruvyltransferase UDP-N-acetylglucosamine 1-carboxyvinyltransferase	
2.5.1.7	Phosphoenolpyruvate:uridine diphosphate N-acetylglucosamine enolpyruvyltransferase UDP-N-acetylglucosamine 1-carboxyvinyltransferase	
4.1.1.49	Phosphoenolpyruvate carboxykinase Phosphoenolpyruvate carboxykinase (ATP)	
4.1.1.49	Phosphoenolpyruvate carboxykinase Phosphoenolpyruvate carboxykinase (ATP)	
4.1.1.32	Phosphoenolpyruvate carboxykinase Phosphoenolpyruvate carboxykinase (GTP)	
4.1.1.38	phosphoenolpyruvate carboxykinase phosphoenolpyruvate carboxykinase (pyrophosphate)	
4.1.1.49	Phosphoenolpyruvate carboxykinase (ATP) Phosphoenolpyruvate carboxykinase (ATP)	
4.1.1.32	Phosphoenolpyruvate carboxykinase (GTP) Phosphoenolpyruvate carboxykinase (GTP)	
4.1.1.38	phosphoenolpyruvate carboxykinase (pyrophosphate) phosphoenolpyruvate carboxykinase (pyrophosphate)	
4.1.1.49	Phosphoenolpyruvate carboxylase Phosphoenolpyruvate carboxykinase (ATP)	
4.1.1.32	Phosphoenolpyruvate carboxylase Phosphoenolpyruvate carboxykinase (GTP)	
4.1.1.31	Phosphoenolpyruvate carboxylase Phosphoenolpyruvate carboxylase	
4.1.1.31	Phosphoenolpyruvate carboxylase Phosphoenolpyruvate carboxylase	
4.1.1.38	phosphoenolpyruvate carboxylase phosphoenolpyruvate carboxykinase (pyrophosphate)	

EC-No.	Synonym Recommended Name	Vol., Page
4.1.1.49	Phosphoenolpyruvate carboxylase (ATP) Phosphoenolpyruvate carboxykinase (ATP)	
4.1.1.32	Phosphoenolpyruvate carboxylase (GTP) Phosphoenolpyruvate carboxykinase (GTP)	
4.1.1.38	phosphoenolpyruvate carboxylase (pyrophosphate) phosphoenolpyruvate carboxykinase (pyrophosphate)	
4.1.1.38	phosphoenolpyruvate carboxyphosphotransferase phosphoenolpyruvate carboxykinase (pyrophosphate)	
4.1.1.38	phosphoenolpyruvate carboxytransphosphorylase phosphoenolpyruvate carboxykinase (pyrophosphate)	
4.2.1.11	phosphoenolpyruvate hydratase phosphoenolpyruvate hydratase	
4.2.1.11	phosphoenolpyruvate hydratase phosphoenolpyruvate hydratase	
2.7.1.40	Phosphoenolpyruvate kinase Pyruvate kinase	
5.4.2.9	Phosphoenolpyruvate mutase Phosphoenolpyruvate mutase	Vol. 1, Page 546
3.1.3.60	Phosphoenolpyruvate phosphatase Phosphoenolpyruvate phosphatase	
3.1.3.60	Phosphoenolpyruvate phosphohydrolase Phosphoenolpyruvate phosphatase	
5.4.2.9	Phosphoenolpyruvate phosphomutase Phosphoenolpyruvate mutase	Vol. 1, Page 546
2.7.3.9	Phosphoenolpyruvate sugar phosphotransferase enzyme I Phosphoenolpyruvate-protein phosphotransferase	
2.7.9.2	Phosphoenolpyruvate synthase water dikinase pyruvate	
4.1.1.49	Phosphoenolpyruvic carboxykinase Phosphoenolpyruvate carboxykinase (ATP)	
4.1.1.49	Phosphoenolpyruvic carboxykinase Phosphoenolpyruvate carboxykinase (ATP)	
4.1.1.32	Phosphoenolpyruvic carboxykinase Phosphoenolpyruvate carboxykinase (GTP)	
4.1.1.38	phosphoenolpyruvic carboxykinase phosphoenolpyruvate carboxykinase (pyrophosphate)	

EC-No.	Synonym Recommended Name	Vol., Page
4.1.1.32	Phosphoenolpyruvic carboxykinase (GTP) Phosphoenolpyruvate carboxykinase (GTP)	
4.1.1.32	Phosphoenolpyruvic carboxykinase (inosine triphosphate) Phosphoenolpyruvate carboxykinase (GTP)	
4.1.1.38	phosphoenolpyruvic carboxykinase (pyrophosphate) phosphoenolpyruvate carboxykinase (pyrophosphate)	
4.1.1.49	Phosphoenolpyruvic carboxylase Phosphoenolpyruvate carboxykinase (ATP)	
4.1.1.32	Phosphoenolpyruvic carboxylase Phosphoenolpyruvate carboxykinase (GTP)	
4.1.1.31	Phosphoenolpyruvic carboxylase Phosphoenolpyruvate carboxylase	
4.1.1.38	phosphoenolpyruvic carboxylase phosphoenolpyruvate carboxykinase (pyrophosphate)	
4.1.1.32	Phosphoenolpyruvic carboxylase (GTP) Phosphoenolpyruvate carboxykinase (GTP)	
4.1.1.32	Phosphoenolpyruvic carboxylase (inosine triphosphate) Phosphoenolpyruvate carboxykinase (GTP)	
4.1.1.38	phosphoenolpyruvic carboxylase (pyrophosphate) phosphoenolpyruvate carboxykinase (pyrophosphate)	
4.1.1.38	phosphoenolpyruvic carboxytransphosphorylase phosphoenolpyruvate carboxykinase (pyrophosphate)	
2.7.9.2	Phosphoenolpyruvic synthase water dikinase pyruvate	
2.7.1.40	Phosphoenol transphosphorylase Pyruvate kinase	
4.2.3.2	O-phosphoethanolamine-phospholyase ethanolamine-phosphate phospho-lyase	
4.2.99.7	O-phosphoethanolamine-phospholyase ethanolamine-phosphate phospho-lyase	
2.7.7.14	Phosphoethanolamine cytidylyltransferase Ethanolamine-phosphate cytidylyltransferase	
2.1.1.103	Phosphoethanolamine N-methyltransferase Phosphoethanolamine N-methyltransferase	
2.7.1.105	6-Phosphofructo-2-kinase 6-Phosphofructo-2-kinase	

EC-No.	Synonym Recommended Name	Vol., Page
2.7.1.105	6-Phosphofructo 2-kinase/Fructose 2,6-bisphosphatase 6-Phosphofructo-2-kinase	
4.1.2.13	Phosphofructoaldolase Fructose-bisphosphate aldolase	
2.7.1.56	1-Phosphofructokinase 1-Phosphofructokinase	
2.7.1.11	6-Phosphofructokinase 6-Phosphofructokinase	
2.7.1.11	Phosphofructokinase 6-Phosphofructokinase	
2.7.1.90	6-Phosphofructokinase (pyrophosphate) Pyrophosphate-fructose-6-phosphate 1-phosphotransferase	
2.7.1.56	Phosphofructokinase 1 1-Phosphofructokinase	
2.7.1.11	Phosphofructokinase 1 6-Phosphofructokinase	
2.7.1.105	Phosphofructokinase 2 6-Phosphofructo-2-kinase	
2.7.1.11	6-Phosphofructose-1-kinase 6-Phosphofructokinase	
2.7.1.11	6-Phosphofructose 1-kinase 6-Phosphofructokinase	
2.7.1.105	6-Phosphofructose 2-kinase 6-Phosphofructo-2-kinase	
3.2.1.85	β-D-phosphogalactoside galactohydrolase 6-Phospho-β-galactosidase	
5.3.1.9	Phosphoglucoisomerase Glucose-6-phosphate isomerase	Vol. 1, Page 298
2.7.1.10	Phosphoglucokinase Phosphoglucokinase	
5.4.2.2	Phosphoglucomutase Phosphoglucomutase	Vol. 1, Page 506
5.4.2.6	β-Phosphoglucomutase β-Phosphoglucomutase	Vol. 1, Page 530
2.7.5.1	phosphoglucomutase phosphoglucomutase	

EC-No.	Synonym Recommended Name	Vol., Page
5.4.2.5	Phosphoglucomutase (glucose-cofactor) Phosphoglucomutase (glucose-cofactor)	Vol. 1, Page 527
2.7.5.5	phosphoglucomutase (glucose-cofactor) phosphoglucomutase (glucose-cofactor)	
1.1.1.43	Phosphogluconate 2-dehydrogenase Phosphogluconate 2-dehydrogenase	
4.2.1.12	6-phosphogluconate dehydrase phosphogluconate dehydratase	
4.2.1.12	6-phosphogluconate dehydratase phosphogluconate dehydratase	
4.2.1.12	phosphogluconate dehydratase phosphogluconate dehydratase	
1.1.1.43	Phosphogluconate dehydrogenase Phosphogluconate 2-dehydrogenase	
1.1.1.44	6-Phosphogluconate dehydrogenase (decarboxylating) Phosphogluconate dehydrogenase (decarboxylating)	
1.1.1.44	Phosphogluconate dehydrogenase (decarboxylating) Phosphogluconate dehydrogenase (decarboxylating)	
1.1.1.43	6-Phosphogluconate dehydrogenase (NAD) Phosphogluconate 2-dehydrogenase	
4.2.1.12	6-phosphogluconic dehydrase phosphogluconate dehydratase	
1.1.1.43	6-Phosphogluconic dehydrogenase Phosphogluconate 2-dehydrogenase	
3.1.1.31	6-phosphoglucono-γ-lactonase 6-phosphogluconolactonase	
3.1.1.31	6-phosphogluconolactonase 6-phosphogluconolactonase	
3.1.1.31	phosphogluconolactonase 6-phosphogluconolactonase	
2.3.1.4	Phosphoglucosamine acetylase Glucosamine-phosphate N-acetyltransferase	
3.5.99.6	phosphoglucosamine isomerase glucosamine-6-phosphate deaminase	
2.3.1.4	Phosphoglucosamine N-acetylase Glucosamine-phosphate N-acetyltransferase	

EC-No.	Synonym Recommended Name	Vol., Page
2.3.1.4	Phosphoglucosamine transacetylase Glucosamine-phosphate N-acetyltransferase	
5.3.1.10	Phosphoglucosaminisomerase Glucosamine-6-phosphate isomerase	Vol. 1, Page 312
3.5.99.6	phosphoglucosaminisomerase glucosamine-6-phosphate deaminase	
1.1.1.49	6-Phosphoglucose dehydrogenase Glucose-6-phosphate 1-dehydrogenase	
5.3.1.9	6-Phosphoglucose isomerase Glucose-6-phosphate isomerase	Vol. 1, Page 298
5.3.1.9	Phosphoglucose isomerase Glucose-6-phosphate isomerase	Vol. 1, Page 298
5.4.2.2	Phosphoglucose mutase Phosphoglucomutase	Vol. 1, Page 506
1.2.1.12	3-Phosphoglyceraldehyde dehydrogenase Glyceraldehyde-3-phosphate dehydrogenase	
1.2.1.12	Phosphoglyceraldehyde dehydrogenase Glyceraldehyde-3-phosphate dehydrogenase	
5.4.2.1	D-Phosphoglycerate 2,3-phosphomutase Phosphoglycerate mutase	Vol. 1, Page 493
1.1.1.95	3-Phosphoglycerate:NAD$^+$ 2-oxidoreductase Phosphoglycerate dehydrogenase	
1.1.1.95	D-3-Phosphoglycerate:NAD oxidoreductase Phosphoglycerate dehydrogenase	
4.2.1.11	2-phosphoglycerate dehydratase phosphoenolpyruvate hydratase	
1.1.1.95	3-Phosphoglycerate dehydrogenase Phosphoglycerate dehydrogenase	
1.1.1.95	D-3-Phosphoglycerate dehydrogenase Phosphoglycerate dehydrogenase	
1.1.1.95	Phosphoglycerate dehydrogenase Phosphoglycerate dehydrogenase	
1.1.1.95	α-Phosphoglycerate dehydrogenase Phosphoglycerate dehydrogenase	
4.2.1.11	2-phosphoglycerate enolase phosphoenolpyruvate hydratase	

EC-No.	Synonym Recommended Name	Vol., Page
2.7.2.3	3-Phosphoglycerate kinase Phosphoglycerate kinase	
2.7.2.3	Phosphoglycerate kinase Phosphoglycerate kinase	
2.7.2.10	Phosphoglycerate kinase (GTP) Phosphoglycerate kinase (GTP)	
5.4.2.1	Phosphoglycerate mutase Phosphoglycerate mutase	Vol. 1, Page 493
1.1.1.95	Phosphoglycerate oxidoreductase Phosphoglycerate dehydrogenase	
3.1.3.38	3-phosphoglycerate phosphatase 3-phosphoglycerate phosphatase	
3.1.3.20	D-2-phosphoglycerate phosphatase phosphoglycerate phosphatase	
3.1.3.38	D-3-phosphoglycerate phosphatase 3-phosphoglycerate phosphatase	
3.1.3.20	phosphoglycerate phosphatase phosphoglycerate phosphatase	
2.7.2.3	3-Phosphoglycerate phosphokinase Phosphoglycerate kinase	
5.4.2.1	Phosphoglycerate phosphomutase Phosphoglycerate mutase	Vol. 1, Page 493
5.4.2.1	Phosphoglycerate phosphomutase Phosphoglycerate mutase	Vol. 1, Page 493
1.1.1.95	3-Phosphoglyceric acid dehydrogenase Phosphoglycerate dehydrogenase	
1.1.1.95	Phosphoglyceric acid dehydrogenase Phosphoglycerate dehydrogenase	
2.7.2.3	3-Phosphoglyceric acid kinase Phosphoglycerate kinase	
2.7.2.3	Phosphoglyceric acid kinase Phosphoglycerate kinase	
2.7.2.3	3-Phosphoglyceric acid phosphokinase Phosphoglycerate kinase	
4.2.1.11	2-phosphoglyceric dehydratase phosphoenolpyruvate hydratase	

EC-No.	Synonym Recommended Name	Vol., Page
2.7.2.3	3-Phosphoglyceric kinase Phosphoglycerate kinase	
2.7.2.3	Phosphoglyceric kinase Phosphoglycerate kinase	
2.7.2.3	Phosphoglycerokinase Phosphoglycerate kinase	
2.7.8.21	Phosphoglycerol cyclase Membrane-oligosaccharide glycerophosphotransferase	
2.5.1.41	Phosphoglycerol geranylgeranyltransferase Phosphoglycerol geranylgeranyltransferase	
3.1.3.19	2-phosphoglycerol phosphatase glycerol-2-phosphatase	
2.7.4.17	3-Phosphoglycerol phosphate-polyphosphate phosphotransferase 3-Phosphoglyceroyl-phosphate-polyphosphate phosphotransferase	
2.7.8.20	Phosphoglycerol transferase Phosphatidylglycerol-membrane-oligosaccharide glycerophosphotransferase	
2.7.8.20	Phosphoglycerol transferase I Phosphatidylglycerol-membrane-oligosaccharide glycerophosphotransferase	
5.4.2.1	Phosphoglyceromutase Phosphoglycerate mutase	Vol. 1, Page 493
2.7.5.3	phosphoglyceromutase phosphoglyceromutase	
2.7.4.17	3-Phosphoglyceroyl-phosphate-polyphosphate phosphotransferase 3-Phosphoglyceroyl-phosphate-polyphosphate phosphotransferase	
3.1.3.14	phosphoglyceryl methylthiol ester phosphatase methylphosphothioglycerate phosphatase	
3.1.3.18	Phosphoglycolate, phosphatase Phosphoglycolate, phosphatase	
3.1.3.18	Phosphoglycolate hydrolase Phosphoglycolate, phosphatase	
3.1.3.18	2-Phosphoglycolate phosphatase Phosphoglycolate, phosphatase	
3.1.3.18	Phosphoglycollate phosphatase Phosphoglycolate, phosphatase	

EC-No.	Synonym Recommended Name	Vol., Page
5.3.1.9	Phosphohexoisomerase Glucose-6-phosphate isomerase	Vol. 1, Page 298
5.3.1.8	Phosphohexoisomerase Mannose-6-phosphate isomerase	Vol. 1, Page 289
2.7.1.11	Phosphohexokinase 6-Phosphofructokinase	
5.3.1.9	Phosphohexomutase Glucose-6-phosphate isomerase	Vol. 1, Page 298
5.3.1.9	Phosphohexose isomerase Glucose-6-phosphate isomerase	Vol. 1, Page 298
2.7.1.69	Phosphohistidinoprotein-hexose phosphoribosyltransferase Protein-N^{π}-phosphohistidine-sugar phosphotransferase	
2.7.1.69	Phosphohistidinoprotein-hexose phosphotransferase Protein-N^{π}-phosphohistidine-sugar phosphotransferase	
1.1.1.262	4-(phosphohydroxy)-L-threonine dehydrogenase 4-hydroxythreonine-4-phosphate dehydrogenase	
2.6.1.52	Phosphohydroxypyruvate transaminase Phosphoserine transaminase	
2.6.1.52	Phosphohydroxypyruvic-glutamic transaminase Phosphoserine transaminase	
3.1.4.11	Phosphoinositidase C 1-Phosphatidyl-4,5-bisphosphate phosphodiesterase	
2.7.1.137	Phosphoinositide 3'-kinase 1-Phosphatidylinositol 3-kinase	
2.7.1.137	Phosphoinositide 3-kinase 1-Phosphatidylinositol 3-kinase	
3.1.3.64	Phosphoinositide 3-phosphatase Phosphatidylinositol 3-phosphatase	
4.1.2.9	Phosphoketolase Phosphoketolase	
4.1.2.22	Phosphoketolase, fructose 6-phosphate Fructose-6-phosphate phosphoketolase	
5.1.3.1	Phosphoketopentose 3-epimerase Ribulose-phosphate 3-epimerase	Vol. 1, Page 91
5.1.3.1	Phosphoketopentose epimerase Ribulose-phosphate 3-epimerase	Vol. 1, Page 91

EC-No.	Synonym Recommended Name	Vol., Page
4.1.2.2	Phosphoketotetrose aldolase Ketotetrose-phosphate aldolase	
3.1.1.4	Phospholipase A Phospholipase A_2	
3.1.1.32	phospholipase A_1 phospholipase A_1	
3.1.1.4	Phospholipase A_2 Phospholipase A_2	
3.1.1.52	phospholipase A_2, phosphatidylinositol phosphatidylinositol deacylase	
3.1.1.5	phospholipase B lysophospholipase	
3.1.1.5	phospholipase B/lipase lysophospholipase	
3.1.4.3	Phospholipase C Phospholipase C	
3.1.4.47	Phospholipase C, glycosylphosphatidylinositol Variant-surface-glycoprotein phospholipase C	
3.1.4.4	Phospholipase D Phospholipase D	
3.1.4.50	Phospholipase D, phosphatidylinositol Glycoprotein phospholipase D	
2.3.1.43	Phospholipid-cholesterol acyltransferase Phosphatidylcholine-sterol O-acyltransferase	
1.11.1.12	Phospholipid-hydroperoxide glutathione peroxidase Phospholipid-hydroperoxide glutathione peroxidase	
1.11.1.12	Phospholipid hydroperoxide glutathione peroxidase Phospholipid-hydroperoxide glutathione peroxidase	
2.1.1.71	Phospholipid methyltransferase Phosphatidyl-N-methylethanolamine N-methyltransferase	
2.7.8.9	Phosphomannan mannosephosphotransferase Phosphomannan mannosephosphotransferase	
5.3.1.8	Phosphomannoisomerase Mannose-6-phosphate isomerase	Vol. 1, Page 289
5.4.2.8	Phosphomannomutase Phosphomannomutase	Vol. 1, Page 540

EC-No.	Synonym Recommended Name	Vol., Page
2.7.5.7	phosphomannomutase phosphomannomutase	
5.3.1.8	Phosphomannose isomerase Mannose-6-phosphate isomerase	Vol. 1, Page 289
5.4.2.8	Phosphomannose mutase Phosphomannomutase	Vol. 1, Page 540
2.7.4.7	Phosphomethylpyrimidine kinase Phosphomethylpyrimidine kinase	
2.7.4.2	5-Phosphomevalonate kinase Phosphomevalonate kinase	
2.7.4.2	Phosphomevalonate kinase Phosphomevalonate kinase	
2.7.4.2	Phosphomevalonic kinase Phosphomevalonate kinase	
3.1.3.2	Phosphomonoesterase Acid phosphatase	
3.1.3.1	phosphomonoesterase alkaline phosphatase	
5.4.2.6	Phosphomutase, β-glucose β-Phosphoglucomutase	Vol. 1, Page 530
5.4.2.3	Phosphomutase, acetylglucosamine Phosphoacetylglucosamine mutase	Vol. 1, Page 515
5.4.2.7	Phosphomutase, deoxyribose Phosphopentomutase	Vol. 1, Page 535
5.4.2.2	Phosphomutase, glucose Phosphoglucomutase	Vol. 1, Page 506
5.4.2.5	Phosphomutase, glucose (glucose-cofactor) Phosphoglucomutase (glucose-cofactor)	Vol. 1, Page 527
5.4.2.1	Phosphomutase, glycerate Phosphoglycerate mutase	Vol. 1, Page 493
5.4.2.4	Phosphomutase, glycerate (phosphoglycerate cofactor) Bisphosphoglycerate mutase	Vol. 1, Page 520
5.4.2.8	Phosphomutase, mannose Phosphomannomutase	Vol. 1, Page 540
5.4.2.9	Phosphomutase, phosphoenolpyruvate-phosphonopyruvate Phosphoenolpyruvate mutase	Vol. 1, Page 546

EC-No.	Synonym Recommended Name	Vol., Page
3.11.1.1	Phosphonatase Phosphonoacetylaldehyde hydrolase	
3.11.1.2	Phosphonoacetate hydrolase Phosphonoacetate hydrolase	
3.11.1.2	Phosphonoacetate phosphonohydrolase Phosphonoacetate hydrolase	
3.11.1.1	Phosphonoacetylaldehyde hydrolase Phosphonoacetylaldehyde hydrolase	
3.11.1.1	2 Phosphonoacetylaldehyde phosphonohydrolase Phosphonoacetylaldehyde hydrolase	
2.7.8.23	Phosphonomutase, carboxyphosphonoenolpyruvate Carboxyvinyl-carboxyphosphonate phosphorylmutase	
1.1.1.262	4-(phosphonooxy)threonine:NAD oxidoreductase 4-hydroxythreonine-4-phosphate dehydrogenase	
6.3.2.5	Phosphopantothenoate-cysteine ligase Phosphopantothenoate-cysteine ligase	Vol. 2, Page 431
6.3.2.5	Phosphopantothenoyl-cysteine ligase Phosphopantothenoate-cysteine ligase	Vol. 2, Page 431
4.1.1.36	4'-Phosphopantothenoyl-L-cysteine decarboxylase Phosphopantothenoylcysteine decarboxylase	
4.1.1.36	Phosphopantothenoylcysteine decarboxylase Phosphopantothenoylcysteine decarboxylase	
4.1.1.36	Phosphopantothenoylcysteine decarboxylase Phosphopantothenoylcysteine decarboxylase	
6.3.2.5	Phosphopantothenoylcysteine synthetase Phosphopantothenoate-cysteine ligase	Vol. 2, Page 431
4.1.1.36	4'-Phosphopantotheoylcysteine decarboxylase Phosphopantothenoylcysteine decarboxylase	
5.3.1.6	Phosphopentoisomerase Ribose-5-phosphate isomerase	Vol. 1, Page 277
2.7.1.19	Phosphopentokinase Phosphoribulokinase	
5.4.2.7	Phosphopentomutase Phosphopentomutase	Vol. 1, Page 535
2.7.5.6	phosphopentomutase phosphopentomutase	

EC-No.	Synonym Recommended Name	Vol., Page
5.3.1.6	Phosphopentose isomerase Ribose-5-phosphate isomerase	Vol. 1, Page 277
2.4.1.78	Phosphopolyprenol glucosyltransferase Phosphopolyprenol glucosyltransferase	
2.7.1.69	Phosphoprotein factor-hexose phosophotransferase Protein-Npi-phosphohistidine-sugar phosphotransferase	
3.1.3.16	Phosphoprotein phosphatase Phosphoprotein phosphatase	
3.1.3.48	phosphoprotein phosphatase (phosphotyrosine) protein-tyrosine-phosphatase	
3.1.3.16	Phosphoprotein phosphohydrolase Phosphoprotein phosphatase	
2.7.3.9	Phosphopyruvate-protein factor phosphotransferase Phosphoenolpyruvate-protein phosphotransferase	
2.7.3.9	Phosphopyruvate-protein phosphotransferase Phosphoenolpyruvate-protein phosphotransferase	
2.5.1.7	Phosphopyruvate-uridine diphosphoacetylglucosamine pyruvatetransferase UDP-N-acetylglucosamine 1-carboxyvinyltransferase	
4.1.1.49	Phosphopyruvate carboxykinase Phosphoenolpyruvate carboxykinase (ATP)	
4.1.1.32	Phosphopyruvate carboxykinase Phosphoenolpyruvate carboxykinase (GTP)	
4.1.1.38	phosphopyruvate carboxykinase phosphoenolpyruvate carboxykinase (pyrophosphate)	
4.1.1.49	Phosphopyruvate carboxykinase (adenosine triphosphate) Phosphoenolpyruvate carboxykinase (ATP)	
4.1.1.32	Phosphopyruvate carboxylase Phosphoenolpyruvate carboxykinase (GTP)	
4.1.1.38	phosphopyruvate carboxylase phosphoenolpyruvate carboxykinase (pyrophosphate)	
4.1.1.32	Phosphopyruvate carboxylase (GTP) Phosphoenolpyruvate carboxykinase (GTP)	
4.1.1.38	phosphopyruvate carboxylase (pyrophosphate) phosphoenolpyruvate carboxykinase (pyrophosphate)	
3.1.3.16	Phosphopyruvate dehydrogenase phosphatase Phosphoprotein phosphatase	

EC-No.	Synonym Recommended Name	Vol., Page
3.1.3.43	phosphopyruvate dehydrogenase phosphatase [pyruvate dehydrogenase (lipoamide)]-phosphatase	
2.7.9.2	Phosphopyruvate synthetase, phosphopyruvate water dikinase pyruvate	
2.7.3.8	Phosphoramidate-adenosine diphosphate phosphotransferase Ammonia kinase	
2.7.3.8	Phosphoramidate-ADP-phosphotransferase Ammonia kinase	
2.7.1.62	Phosphoramidate-hexose phosphotransferase Phosphoramidate-hexose phosphotransferase	
2.7.1.62	Phosphoramidate-hexose transphosphorylase Phosphoramidate-hexose phosphotransferase	
2.7.1.62	Phosphoramidate:hexose 1-phosphotransferase Phosphoramidate-hexose phosphotransferase	
2.7.1.62	Phosphoramidic-hexose transphosphorylase Phosphoramidate-hexose phosphotransferase	
5.3.1.6	Phosphoriboisomerase Ribose-5-phosphate isomerase	Vol. 1, Page 277
2.7.1.18	Phosphoribokinase Phosphoribokinase	
5.4.2.7	Phosphoribomutase Phosphopentomutase	Vol. 1, Page 535
5.3.1.6	5-Phosphoribose isomerase Ribose-5-phosphate isomerase	Vol. 1, Page 277
2.4.2.14	Phosphoribose pyrophosphate amidotransferase Amidophosphoribosyltransferase	
2.7.6.1	5-Phosphoribose pyrophosphorylase Ribose-phosphate pyrophosphokinase	
6.3.2.6	1-(5-Phosphoribosyl)-5-amino-4-carboxyimidazole:L-aspartate ligase (ADP-forming) Phosphoribosylaminoimidazolesuccinocarboxamide synthase	Vol. 2, Page 434
4.1.1.21	1-(5-Phosphoribosyl)-5-amino-4-imidazolecarboxylate carboxy-lyase Phosphoribosylaminoimidazole carboxylase	
2.7.6.1	5-Phosphoribosyl-α-1-pyrophosphate synthetase Ribose-phosphate pyrophosphokinase	

EC-No.	Synonym Recommended Name	Vol., Page
2.4.2.14	5-Phosphoribosyl-1-pyrophosphate amidotransferase Amidophosphoribosyltransferase	
2.4.2.14	α-5-Phosphoribosyl-1-pyrophosphate amidotransferase Amidophosphoribosyltransferase	
2.7.6.1	5-Phosphoribosyl-1-pyrophosphate synthetase Ribose-phosphate pyrophosphokinase	
2.1.2.3	5'-Phosphoribosyl-5-amino-4-imidazolecarboxamide formyltransferase Phosphoribosylaminoimidazolecarboxamide formyltransferase	
4.1.1.21	5-Phosphoribosyl-5-aminoimidazole carboxylase Phosphoribosylaminoimidazole carboxylase	
6.3.3.1	5'-Phosphoribosyl-5-aminoimidazole synthetase Phosphoribosylformylglycinamidine cyclo-ligase	Vol. 2, Page 530
3.5.4.19	Phosphoribosyl-AMP cyclohydrolase Phosphoribosyl-AMP cyclohydrolase	
2.4.2.18	Phosphoribosyl-anthranilate pyrophosphorylase Anthranilate phosphoribosyltransferase	
2.4.2.17	Phosphoribosyl-ATP:pyrophosphate-phosphoribosyl phosphotransferase ATP phosphoribosyltransferase	
3.6.1.31	Phosphoribosyl-ATP diphosphatase Phosphoribosyl-ATP diphosphatase	
3.6.1.31	Phosphoribosyl-ATP pyrophosphatase Phosphoribosyl-ATP diphosphatase	
3.6.1.31	Phosphoribosyl-ATP pyrophosphohydrolase Phosphoribosyl-ATP diphosphatase	
2.4.2.17	Phosphoribosyl-ATP pyrophosphorylase ATP phosphoribosyltransferase	
2.7.6.1	Phosphoribosyl-diphosphate synthetase Ribose-phosphate pyrophosphokinase	
6.3.1.3	phosphoribosyl-glycinamide synthetase phosphoribosyl-glycinamide synthetase	Vol. 2, Page 626
2.7.6.1	5-Phosphoribosyl 1-pyrophosphate synthetase Ribose-phosphate pyrophosphokinase	
3.5.4.19	Phosphoribosyladenosine monophosphate cyclohydrolase Phosphoribosyl-AMP cyclohydrolase	
2.4.2.17	Phosphoribosyladenosine triphosphate pyrophosphorylase ATP phosphoribosyltransferase	

EC-No.	Synonym Recommended Name	Vol., Page
2.4.2.17	Phosphoribosyladenosine triphosphate synthetase ATP phosphoribosyltransferase	
6.3.4.13	Phosphoribosylamine-glycine ligase Phosphoribosylamine-glycine ligase	Vol. 2, Page 626
6.3.4.13	5-Phosphoribosylamine:glycine ligase (ADP-forming) Phosphoribosylamine-glycine ligase	Vol. 2, Page 626
2.4.2.14	5-Phosphoribosylamine:pyrophosphate phospho-α-D-ribosyltransferase (glutamate-amidating) Amidophosphoribosyltransferase	
6.3.4.7	5-Phosphoribosylamine synthetase Ribose-5-phosphate-ammonia ligase	Vol. 2, Page 608
2.1.2.3	Phosphoribosylaminoimidazolecarboxamide formyltransferase Phosphoribosylaminoimidazolecarboxamide formyltransferase	
4.1.1.21	Phosphoribosylaminoimidazole carboxylase Phosphoribosylaminoimidazole carboxylase	
6.3.2.6	Phosphoribosylaminoimidazolesuccinocarboxamide synthase Phosphoribosylaminoimidazolesuccinocarboxamide synthase	Vol. 2, Page 434
6.3.2.6	Phosphoribosylaminoimidazolesuccinocarboxamide synthetase Phosphoribosylaminoimidazolesuccinocarboxamide synthase	Vol. 2, Page 434
6.3.3.1	Phosphoribosylaminoimidazole synthetase Phosphoribosylformylglycinamidine cyclo-ligase	Vol. 2, Page 530
5.3.1.24	Phosphoribosylanthranilate isomerase Phosphoribosylanthranilate isomerase	Vol. 1, Page 353
4.1.1.48	Phosphoribosylanthranilate isomerase-indoleglycerol phosphate synthetase Indole-3-glycerol phosphate synthase	
2.4.2.18	Phosphoribosylanthranilate pyrophosphorylase Anthranilate phosphoribosyltransferase	
2.4.2.18	Phosphoribosylanthranilate transferase Anthranilate phosphoribosyltransferase	
2.4.2.17	Phosphoribosyl ATP:pyrophosphate phosphoribosyltransferase ATP phosphoribosyltransferase	
2.4.2.17	Phosphoribosyl ATP synthetase ATP phosphoribosyltransferase	
2.4.2.14	Phosphoribosyldiphosphate 5-amidotransferase Amidophosphoribosyltransferase	

EC-No.	Synonym Recommended Name	Vol., Page
6.3.5.3	5'-Phosphoribosylformylglycinamide:L-glutamine amido-ligase (ADP-forming) Phosphoribosylformylglycinamidine synthase	Vol. 2, Page 666
6.3.3.1	Phosphoribosylformylglycinamidine cyclo-ligase Phosphoribosylformylglycinamidine cyclo-ligase	Vol. 2, Page 530
6.3.5.3	Phosphoribosylformylglycinamidine synthase Phosphoribosylformylglycinamidine synthase	Vol. 2, Page 666
6.3.5.3	Phosphoribosylformylglycinamidine synthetase Phosphoribosylformylglycinamidine synthase	Vol. 2, Page 666
6.3.5.3	Phosphoribosylformylglycinamidine synthetase Phosphoribosylformylglycinamidine synthase	Vol. 2, Page 666
6.3.5.3	Phosphoribosylformylglycineamidine synthetase Phosphoribosylformylglycinamidine synthase	Vol. 2, Page 666
2.1.2.2	Phosphoribosylglycinamide formyltransferase Phosphoribosylglycinamide formyltransferase	
6.3.4.13	5'-Phosphoribosylglycinamide synthetase Phosphoribosylamine-glycine ligase	Vol. 2, Page 626
6.3.4.13	Phosphoribosylglycinamide synthetase Phosphoribosylamine-glycine ligase	Vol. 2, Page 626
6.3.4.13	Phosphoribosylglycineamide synthetase Phosphoribosylamine-glycine ligase	Vol. 2, Page 626
6.3.4.8	5-Phosphoribosylimidazoleacetate synthetase Imidazoleacetate-phosphoribosyldiphosphate ligase	Vol. 2, Page 611
2.4.2.14	5'-Phosphoribosylpyrophosphate amidotransferase Amidophosphoribosyltransferase	
2.4.2.14	5-Phosphoribosylpyrophosphate amidotransferase Amidophosphoribosyltransferase	
2.4.2.14	Phosphoribosyl pyrophosphate amidotransferase Amidophosphoribosyltransferase	
2.4.2.14	Phosphoribosylpyrophosphate glutamyl amidotransferase Amidophosphoribosyltransferase	
2.7.6.1	Phosphoribosylpyrophosphate synthase Ribose-phosphate pyrophosphokinase	
2.7.6.1	Phosphoribosyl pyrophosphate synthetase Ribose-phosphate pyrophosphokinase	

EC-No.	Synonym Recommended Name	Vol., Page
2.7.6.1	Phosphoribosylpyrophosphate synthetase Ribose-phosphate pyrophosphokinase	
2.4.2.8	Phosphoribosyltransferase, 6-mercaptopurine Hypoxanthine phosphoribosyltransferase	
2.4.2.7	Phosphoribosyltransferase, adenine Adenine phosphoribosyltransferase	
2.4.2.17	Phosphoribosyltransferase, adenosine triphosphate ATP phosphoribosyltransferase	
2.4.2.18	Phosphoribosyltransferase, anthranilate Anthranilate phosphoribosyltransferase	
2.4.2.20	Phosphoribosyltransferase, dioxotetrahydropyrimidine Dioxotetrahydropyrimidine phosphoribosyltransferase	
2.4.2.8	Phosphoribosyltransferase, hypoxanthine Hypoxanthine phosphoribosyltransferase	
2.4.2.12	Phosphoribosyltransferase, nicotinamide Nicotinamide phosphoribosyltransferase	
2.4.2.11	Phosphoribosyltransferase, nicotinate Nicotinate phosphoribosyltransferase	
2.4.2.21	Phosphoribosyltransferase, nicotinate mononucleotide-dimethylbenzimidazole Nicotinate-nucleotide-dimethylbenzimidazole phosphoribosyltransferase	
2.4.2.10	Phosphoribosyltransferase, orotate Orotate phosphoribosyltransferase	
2.4.2.9	Phosphoribosyltransferase, uracil Uracil phosphoribosyltransferase	
2.4.2.22	Phosphoribosyltransferase, xanthine Xanthine phosphoribosyltransferase	
2.7.1.19	Phosphoribulokinase Phosphoribulokinase	
5.1.3.1	Phosphoribulose epimerase Ribulose-phosphate 3-epimerase	Vol. 1, Page 91
5.1.3.4	Phosphoribulose isomerase L-Ribulose-phosphate 4-epimerase	Vol. 1, Page 123
2.7.1.19	5-Phosphoribulose kinase Phosphoribulokinase	

EC-No.	Synonym Recommended Name	Vol., Page
2.4.2.14	5-Phosphororibosyl-1-pyrophosphate amidotransferase Amidophosphoribosyltransferase	
2.4.1.1	Phosphorylase Phosphorylase	
2.4.1.1	Phosphorylase, α-glucan Phosphorylase	
2.4.1.97	Phosphorylase, 1,3-β-glucan 1,3-β-D-Glucan phosphorylase	
2.4.1.30	Phosphorylase, 1,3-β-oligoglucan 1,3-β-Oligoglucan phosphorylase	
2.4.1.20	Phosphorylase, cellobiose Cellobiose phosphorylase	
2.4.1.49	Phosphorylase, cellodextrin Cellodextrin phosphorylase	
2.4.2.23	Phosphorylase, deoxyuridine Deoxyuridine phosphorylase	
2.4.2.15	Phosphorylase, guanosine Guanosine phosphorylase	
2.4.1.31	Phosphorylase, laminaribiose Laminaribiose phosphorylase	
2.4.1.8	Phosphorylase, maltose Maltose phosphorylase	
2.4.2.28	Phosphorylase, methylthioadenosine 5'-Methylthioadenosine phosphorylase	
2.4.2.1	Phosphorylase, purine nucleoside Purine-nucleoside phosphorylase	
2.4.1.7	Phosphorylase, sucrose Sucrose phosphorylase	
2.4.2.4	Phosphorylase, thymidine Thymidine phosphorylase	
2.4.1.64	Phosphorylase, trehalose α,α-Trehalose phosphorylase	
2.4.2.16	Phosphorylase, urate ribonucleotide Urate-ribonucleotide phosphorylase	
2.4.2.3	Phosphorylase, uridine Uridine phosphorylase	

EC-No.	Synonym Recommended Name	Vol., Page
3.1.3.17	[phosphorylase] phosphatase [phosphorylase] phosphatase	
3.1.3.17	[phosphorylase a] phosphohydrolase [phosphorylase] phosphatase	
3.1.3.17	phosphorylase a phosphatase [phosphorylase] phosphatase	
2.7.1.38	Phosphorylase b kinase Phosphorylase kinase	
2.7.1.37	Phosphorylase b kinase kinase Protein kinase	
2.7.1.38	Phosphorylase kinase Phosphorylase kinase	
2.7.8.3	Phosphorylcholine-ceramide transferase Ceramide cholinephosphotransferase	
2.7.8.2	Phosphorylcholine-glyceride transferase Diacylglycerol cholinephosphotransferase	
2.7.8.10	Phosphorylcholine-sphingosine transferase Sphingosine cholinephosphotransferase	
2.7.7.15	Phosphorylcholine:CTP cytidylyltransferase Choline-phosphate cytidylyltransferase	
2.7.7.15	Phosphorylcholine cytidylyltransferase Choline-phosphate cytidylyltransferase	
2.7.7.15	Phosphorylcholine transferase Choline-phosphate cytidylyltransferase	
4.2.3.2	O-phosphorylethanol-amine phospho-lyase ethanolamine-phosphate phospho-lyase	
4.2.99.7	O-phosphorylethanol-amine phospho-lyase ethanolamine-phosphate phospho-lyase	
2.7.8.1	Phosphorylethanolamine-glyceride transferase Ethanolaminephosphotransferase	
2.7.7.14	Phosphorylethanolamine transferase Ethanolamine-phosphate cytidylyltransferase	
5.3.1.9	Phosphosaccharomutase Glucose-6-phosphate isomerase	Vol. 1, Page 298
2.6.1.52	3-Phosphoserine aminotransferase Phosphoserine transaminase	

EC-No.	Synonym Recommended Name	Vol., Page
2.6.1.52	L-Phosphoserine aminotransferase Phosphoserine transaminase	
2.6.1.52	Phosphoserine aminotransferase Phosphoserine transaminase	
3.1.3.3	3-phosphoserine phosphatase phosphoserine phosphatase	
3.1.3.3	phosphoserine phosphatase phosphoserine phosphatase	
3.1.3.3	O-phosphoserine phosphohydrolase phosphoserine phosphatase	
2.6.1.52	Phosphoserine transaminase Phosphoserine transaminase	
2.5.1.19	3-Phosphoshikimate 1-carboxyvinyltransferase 3-Phosphoshikimate 1-carboxyvinyltransferase	
3.1.3.16	Phosphospectrin phosphatase Phosphoprotein phosphatase	
2.3.1.8	Phosphotransacetylase Phosphate acetyltransferase	
2.3.1.19	Phosphotransbutyrylase Phosphate butyryltransferase	
2.7.1.61	Phosphotransferase, acyl phosphate-hexose Acyl-phosphate-hexose phosphotransferase	
2.7.1.118	Phosphotransferase, adenosine diphosphate-thymidine ADP-thymidine kinase	
2.7.1.103	Phosphotransferase, capreomycin Viomycin kinase	
2.7.4.17	Phosphotransferase, diphosphoglycerate-polyphosphate 3-Phosphoglyceroyl-phosphate-polyphosphate phosphotransferase	
2.7.4.20	Phosphotransferase, dolichol diphosphate-polyphosphate Dolichyl-diphosphate-polyphosphate phosphotransferase	
2.7.1.142	Phosphotransferase, glycerol phosphate-glucose Glycerol-3-phosphate-glucose phosphotransferase	
2.7.1.119	Phosphotransferase, hygromycin B Hygromycin-B kinase	
2.7.1.136	Phosphotransferase, macrolide 2'- Macrolide 2'-kinase	

EC-No.	Synonym Recommended Name	Vol., Page
2.7.1.77	Phosphotransferase, nucleoside Nucleoside phosphotransferase	
2.7.3.9	Phosphotransferase, phosphoenolpyruvate-protein Phosphoenolpyruvate-protein phosphotransferase	
2.7.1.121	Phosphotransferase, phosphohistidinoprotein-dihydroxyacetone Phosphoenolpyruvate-glycerone phosphotransferase	
2.7.1.69	Phosphotransferase, phosphohistidinoprotein-hexose Protein-N^{π}-phosphohistidine-sugar phosphotransferase	
2.7.1.62	Phosphotransferase, phosphoramidate-hexose Phosphoramidate-hexose phosphotransferase	
2.7.1.63	Phosphotransferase, polyphosphate-glucose Polyphosphate-glucose phosphotransferase	
2.7.2.12	Phosphotransferase, pyrophosphate-acetate Acetate kinase (diphosphate)	
2.7.1.90	Phosphotransferase, pyrophosphate-fructose 6-phosphate 1- Pyrophosphate-fructose-6-phosphate 1-phosphotransferase	
2.7.1.79	Phosphotransferase, pyrophosphate-glycerol Pyrophosphate-glycerol phosphotransferase	
2.7.1.104	Phosphotransferase, pyrophosphate-protein Pyrophosphate-protein phosphotransferase	
2.7.1.80	Phosphotransferase, pyrophosphate-serine Pyrophosphate-serine phosphotransferase	
2.7.1.42	Phosphotransferase, riboflavin Riboflavin phosphotransferase	
2.7.1.122	Phosphotransferase, xylitol Xylitol kinase	
3.2.1.93	Phosphotrehalase α,α-phosphotrehalase	
3.2.1.93	α,α-phosphotrehalase α,α-phosphotrehalase	
2.4.1.15	Phosphotrehalose-uridine diphosphate transglucosylase α,α-Trehalose-phosphate synthase (UDP-forming)	
3.1.8.1	Phosphotriesterase Aryldialkylphosphatase	
5.3.1.1	Phosphotriose isomerase Triose-phosphate isomerase	Vol. 1, Page 235

EC-No.	Synonym Recommended Name	Vol., Page
3.1.3.48	[phosphotyrosine]protein phosphatase protein-tyrosine-phosphatase	
3.1.3.48	phosphotyrosine histone phosphatase protein-tyrosine-phosphatase	
3.1.3.48	phosphotyrosine phosphatase protein-tyrosine-phosphatase	
3.1.3.48	phosphotyrosine phosphatase 1B protein-tyrosine-phosphatase	
3.1.3.48	phosphotyrosine protein phosphatase protein-tyrosine-phosphatase	
2.7.1.112	Phosphotyrosyl-protein kinase Protein-tyrosine kinase	
3.1.3.48	phosphotyrosylprotein phosphatase protein-tyrosine-phosphatase	
5.3.1.8	Phosphphexomutase Mannose-6-phosphate isomerase	Vol. 1, Page 289
1.13.12.7	Photinus-luciferin 4-monooxygenase (ATP-hydrolysing) Photinus-luciferin 4-monooxygenase (ATP-hydrolysing)	
1.13.12.7	Photinus-luciferin:oxygen 4-oxidoreductase (decarboxylating, ATP-hydrolysing) Photinus-luciferin 4-monooxygenase (ATP-hydrolysing)	
1.13.12.7	Photinus luciferin 4-monooxygenase (adenosine triphosphate-hydrolyzing) Photinus-luciferin 4-monooxygenase (ATP-hydrolysing)	
1.13.12.7	Photinus pyralis luciferase Photinus-luciferin 4-monooxygenase (ATP-hydrolysing)	
4.1.99.3	photolyase deoxyribodipyrimidine photolyase	
4.1.99.3	photoreactivating enzyme deoxyribodipyrimidine photolyase	
4.1.99.3	phr A photolyase deoxyribodipyrimidine photolyase	
4.1.99.3	PhrB photolyase deoxyribodipyrimidine photolyase	
4.2.1.96	PHS/PCD 4a-Hydroxytetrahydrobiopterin dehydratase	

EC-No.	Synonym Recommended Name	Vol., Page
2.3.1.138	PHT Putrescine N-hydroxycinnamoyltransferase	
1.14.12.7	Phthalate,NADH:oxygen oxidoreductase (4,5-hydroxylating) Phthalate 4,5-dioxygenase	
1.3.1.64	phthalate 4,5-cis-dihydrodiol dehydrogenase phthalate 4,5-cis-dihydrodiol dehydrogenase	
1.14.12.7	Phthalate 4,5-dioxygenase Phthalate 4,5-dioxygenase	
1.14.12.7	Phthalate dioxygenase Phthalate 4,5-dioxygenase	
3.5.1.79	Phthalyl-amide amidohydrolase Phthalyl amidase	
3.5.1.79	Phthalyl amidase Phthalyl amidase	
3.5.1.79	*o*-Phthalyl amidase Phthalyl amidase	
3.2.1.15	Phylendonase Polygalacturonase	
1.14.99.20	Phylloquinone,hydrogen-donor:oxygen oxidoreductase (2,3-epoxidizing) Phylloquinone monooxygenase (2,3-epoxidizing)	
1.14.99.20	Phylloquinone epoxidase Phylloquinone monooxygenase (2,3-epoxidizing)	
1.1.4.1	Phylloquinone epoxide reductase Vitamin-K-epoxide reductase (warfarin sensitive)	
1.14.99.20	Phylloquinone monooxygenase (2,3-epoxidizing) Phylloquinone monooxygenase (2,3-epoxidizing)	
1.6.99.2	Phylloquinone reductase NAD(P)H dehydrogenase (quinone)	
1.6.5.5	Phylloquinone reductase NADPH:quinone reductase	
3.4.23.27	Physaropepsin Physaropepsin	
3.4.23.27	Physarum aspartic proteinase Physaropepsin	
3.1.26.1	Physarum polycephalum ribonuclease Physarum polycephalum ribonuclease	

EC-No.	Synonym Recommended Name	Vol., Page
6.2.1.24	Phytanate-CoA ligase Phytanate-CoA ligase	Vol. 2, Page 311
6.2.1.24	Phytanate:CoA ligase (AMP-forming) Phytanate-CoA ligase	Vol. 2, Page 311
1.14.11.18	phytanoyl-CoA, 2-oxoglutarate:oxygen oxidoreductase (2-hydroxylating) phytanoyl-CoA dioxygenase	
1.14.11.18	phytanoyl-CoA dioxygenase phytanoyl-CoA dioxygenase	
1.14.11.18	phytanoyl-CoA hydroxylase phytanoyl-CoA dioxygenase	
6.2.1.24	Phytanoyl-CoA ligase Phytanate-CoA ligase	Vol. 2, Page 311
6.2.1.24	Phytanoyl-CoA synthetase Phytanate-CoA ligase	Vol. 2, Page 311
3.1.3.8	1-phytase 3-phytase	
3.1.3.8	3-phytase 3-phytase	
3.1.3.26	6-phytase 6-phytase	
3.1.3.8	phytase 3-phytase	
3.1.3.26	phytase 6-phytase	
3.1.3.8	phytate 1-phosphatase 3-phytase	
3.1.3.8	phytate 6-phosphatase 3-phytase	
3.1.3.26	phytate 6-phosphatase 6-phytase	
3.4.23.40	Phytepsin Phytepsin	
2.3.2.15	Phytochelatin synthase Glutathione γ-glutamylcysteinyltransferase	
2.5.1.32	Phytoene-synthetase Geranylgeranyl-diphosphate geranylgeranyltransferase	

EC-No.	Synonym Recommended Name	Vol., Page
2.5.1.32	Phytoene synthase Geranylgeranyl-diphosphate geranylgeranyltransferase	
2.5.1.32	Phytoene synthetase Geranylgeranyl-diphosphate geranylgeranyltransferase	
3.4.23.1	P I pepsin A	
2.7.1.137	PI3K 1-Phosphatidylinositol 3-kinase	
2.7.1.67	PI 4-kinase 1-Phosphatidylinositol 4-kinase	
3.1.3.1	PiALP alkaline phosphatase	
4.1.1.45	Picolinic acid carboxylase Aminocarboxymuconate-semialdehyde decarboxylase	
4.1.1.45	Picolinic acid decarboxylase Aminocarboxymuconate-semialdehyde decarboxylase	
4.1.1.45	Picolinic decarboxylase Aminocarboxymuconate-semialdehyde decarboxylase	
3.4.22.29	Picornain 2A Picornain 2A	
3.4.22.28	Picornain 3C Picornain 3C	
3.4.22.29	Picornaviral 2A proteinase Picornain 2A	
3.4.22.29	Picornavirus 2A proteinase Picornain 2A	
3.4.22.29	Picornavirus endopeptidase 2A Picornain 2A	
3.4.22.28	Picornavirus endopeptidase 3C Picornain 3C	
3.1.26.1	Pig liver nuclease Physarum polycephalum ribonuclease	
3.4.23.3	pig parapepsin II gastricsin	
3.4.23.1	P II pepsin A	

EC-No.	Synonym Recommended Name	Vol., Page
2.7.7.59	PII uridylyl-transferase [Protein-PII]uridylyltransferase	
2.7.7.59	PII uridylyltransferase [Protein-PII]uridylyltransferase	
2.7.7.59	PII uridylyltransferase (Escherichia coli gene glnD) [Protein-PII]uridylyltransferase	
2.7.1.67	PI kinase 1-Phosphatidylinositol 4-kinase	
3.4.21.89	Pilin leader peptidase Signal peptidase I	
2.7.7.13	PIM-GMP Mannose-1-phosphate guanylyltransferase	
3.4.21.53	PIM1 protease Endopeptidase La	
3.4.21.53	PIM1 proteinase Endopeptidase La	
1.3.1.62	pimeloyl-CoA:NAD oxidoreductase pimeloyl-CoA dehydrogenase	
1.3.1.62	pimeloyl-CoA dehydrogenase pimeloyl-CoA dehydrogenase	
6.2.1.14	Pimeloyl-CoA synthase 6-Carboxyhexanoate-CoA ligase	Vol. 2, Page 273
6.2.1.14	Pimeloyl-CoA synthetase 6-Carboxyhexanoate-CoA ligase	Vol. 2, Page 273
6.2.1.14	Pimelyl-CoA synthetase 6-Carboxyhexanoate-CoA ligase	Vol. 2, Page 273
3.4.22.33	Pinase Fruit bromelain	
3.4.22.33	Pineapple enzyme Fruit bromelain	
3.4.22.32	Pineapple stem bromelain Stem bromelain	
5.5.1.10	α-Pinene-oxide decyclase α-Pinene-oxide decyclase	Vol. 1, Page 713
5.5.1.10	α-Pinene-oxide lyase (decyclizing) α-Pinene-oxide decyclase	Vol. 1, Page 713

EC-No.	Synonym Recommended Name	Vol., Page
5.5.1.10	α-Pinene oxide lyase α-Pinene-oxide decyclase	Vol. 1, Page 713
4.2.3.14	pinene synthase pinene synthase	
2.3.1.146	Pine stilbene synthase Pinosylvin synthase	
3.4.99.18	pinguinain pinguinain	
1.1.1.142	D-Pinitol dehydrogenase D-Pinitol dehydrogenase	
2.3.1.146	Pinosylvin synthase Pinosylvin synthase	
1.5.99.3	L-Pipecolate:(acceptor) 1,6-oxidoreductase L-Pipecolate dehydrogenase	
1.5.1.21	L-Pipecolate:$NADP^+$ 2-oxidoreductase Δ1-Piperideine-2-carboxylate reductase	
1.5.3.7	L-Pipecolate:oxygen 1,6-oxidoreductase L-Pipecolate oxidase	
1.5.99.3	L-Pipecolate dehydrogenase L-Pipecolate dehydrogenase	
1.5.3.7	L-Pipecolate oxidase L-Pipecolate oxidase	
1.5.3.7	Pipecolate oxidase L-Pipecolate oxidase	
1.5.3.7	L-Pipecolic acid oxidase L-Pipecolate oxidase	
1.5.1.21	Δ^1-Piperideine-2-carboxylate reductase Δ^1-Piperideine-2-carboxylate reductase	
2.3.1.145	Piperidine N-piperoyltransferase Piperidine N-piperoyltransferase	
2.3.1.145	Piperidine piperoyltransferase Piperidine N-piperoyltransferase	
1.14.99.15	Piperonylate-4-O-demethylase 4-Methoxybenzoate monooxygenase (O-demethylating)	
2.3.1.145	Piperoyl-CoA:piperidine N-piperoyltransferase Piperidine N-piperoyltransferase	

EC-No.	Synonym Recommended Name	Vol., Page
2.3.1.145	Piperoyltransferase, piperidine N- Piperidine N-piperoyltransferase	
2.7.1.68	PIP kinase 1-Phosphatidylinositol-4-phosphate 5-kinase	
3.4.24.55	Pitrilysin Pitrilysin	
3.4.99.44	pitrilysin pitrilysin	
3.4.21.59	Pituitary tryptase Tryptase	
2.7.1.37	PKA Protein kinase	
2.7.1.37	PKC Protein kinase	
2.7.1.37	εPKC Protein kinase	
2.7.1.19	PKK Phosphoribulokinase	
4.2.2.10	PL pectin lyase	
3.4.21.87	Pla Omptin	
1.8.4.9	plant-type 5'-adenylylsulfate reductase adenylyl-sulfate reductase (glutathione)	
2.4.1.18	Plant branching enzyme 1,4-α-Glucan branching enzyme	
3.4.21.57	Plant Leu-proteinase Leucyl endopeptidase	
3.1.30.2	Plant nuclease I Serratia marcescens nuclease	
2.4.1.152	Plasma α-3-fucosyltransferase Galactoside 3-fucosyltransferase	
3.4.17.20	Plasma carboxypeptidase B Carboxypeptidase U	
3.4.17.3	plasma carboxypeptidase B lysine carboxypeptidase	

EC-No.	Synonym Recommended Name	Vol., Page
3.4.21.34	plasma kallikrein plasma kallikrein	
2.3.1.25	Plasmalogen synthase Plasmalogen synthase	
1.14.99.19	Plasmanylethanolamine desaturase Plasmanylethanolamine desaturase	
3.4.21.6	plasma thromboplastin coagulation factor Xa	
3.4.21.27	plasma thromboplastin antecedent coagulation factor XIa	
3.4.23.38	Plasmepsin I Plasmepsin I	
3.4.23.39	Plasmepsin II Plasmepsin II	
3.4.21.7	plasmin plasmin	
3.4.4.14	plasmin plasmin	
3.4.21.68	t-Plasminogen activator t-Plasminogen activator	
3.4.21.73	u-Plasminogen activator u-Plasminogen activator	
3.4.21.68	Plasminogen activator, tissue-type t-Plasminogen activator	
3.4.21.73	Plasminogen activator, urokinase-type u-Plasminogen activator	
1.10.99.1	Plastoquinol-plastocyanin reductase Plastoquinol-plastocyanin reductase	
1.10.99.1	Plastoquinol/plastocyanin oxidoreductase Plastoquinol-plastocyanin reductase	
1.10.99.1	Plastoquinol:oxidized-plastocyanin oxidoreductase Plastoquinol-plastocyanin reductase	
2.3.1.67	Platelet-activating factor-synthesizing enzyme 1-Alkylglycerophosphocholine O-acetyltransferase	
3.1.1.47	platelet-activating factor acetylhydrolase 1-alkyl-2-acetylglycerophosphocholine esterase	

EC-No.	Synonym Recommended Name	Vol., Page
3.1.1.47	platelet-activating factor acetylhydrolyase 1-alkyl-2-acetylglycerophosphocholine esterase	
2.3.1.149	Platelet-activating factor acetyltransferase Platelet-activating factor acetyltransferase	
2.3.1.67	Platelet-activating factor acylhydrolase 1-Alkylglycerophosphocholine O-acetyltransferase	
3.1.1.5	PLB lysophospholipase	
3.1.1.5	PLB/LIP lysophospholipase	
3.1.27.4	Pleospora RNase Ribonuclease U_2	
2.1.1.71	PLMT Phosphatidyl-N-methylethanolamine N-methyltransferase	
3.1.3.5	PM-AMPase 5'-nucleotidase	
3.1.1.61	PME Protein-glutamate methylesterase	
3.1.1.11	PME pectinesterase	
5.3.1.8	PMI Mannose-6-phosphate isomerase	Vol. 1, Page 289
5.4.2.8	PMM Phosphomannomutase	Vol. 1, Page 540
3.4.24.34	PMNL collagenase Neutrophil collagenase	
3.4.21.76	PMNL proteinase Myeloblastin	
1.4.3.5	PMP oxidase Pyridoxamine-phosphate oxidase	
2.1.1.25	PMT Phenol O-methyltransferase	
5.3.1.23	1-PMTR isomerase 5-Methylthioribose-1-phosphate isomerase	Vol. 1, Page 351
3.5.1.52	PNGase A Peptide-N^4-(N-acetyl-β-glucosaminyl) asparagine amidase	

EC-No.	Synonym Recommended Name	Vol., Page
3.5.1.52	PNGase F Peptide-N^4-(N-acetyl-β-glucosaminyl) asparagine amidase	
2.7.1.78	PNK Polynucleotide 5'-hydroxy-kinase	
3.1.3.32	PNKP polynucleotide 3'-phosphatase	
4.2.2.10	PNL pectin lyase	
3.2.1.139	PNP-GAase α-Glucosiduronase	
2.4.1.17	PNP-UDPGT Glucuronosyltransferase	
2.7.7.8	PNPase Polyribonucleotide nucleotidyltransferase	
2.4.2.1	PNPase Purine-nucleoside phosphorylase	
3.1.3.41	PNPPase 4-nitrophenylphosphatase	
3.1.3.41	k-pNPPase 4-nitrophenylphosphatase	
3.4.22.29	Poliovirus protease 2A Picornain 2A	
3.4.22.29	Poliovirus protease 2Apro Picornain 2A	
3.4.22.28	Poliovirus protease 3C Picornain 3C	
3.4.22.28	Poliovirus proteinase 3C Picornain 3C	
3.2.1.129	Poly(α-2,8-sialoside) α-2,8-sialosylhydrolase Endo-α-sialidase	
3.2.1.129	Poly(α-2,8-sialosyl) endo-N-acetylneuraminidase Endo-α-sialidase	
4.2.2.3	poly(β-D-1,4-mannuronide) lyase poly(β-D-mannuronate) lyase	
4.2.2.3	poly(β-D-mannuronate) lyase poly(β-D-mannuronate) lyase	

EC-No.	Synonym Recommended Name	Vol., Page
3.4.19.9	Poly(γ-glutamic acid) endohydrolase γ-Glu-X carboxypeptidase	
4.2.2.11	Poly(α-L-1,4-guluronide) exo-lyase Poly(α-L-guluronate) lyase	
4.2.2.11	Poly(α-L-guluronate) lyase Poly(α-L-guluronate) lyase	
3.2.1.44	Poly(1,2-α-L-fucoside-4-sulfate) glycanohydrolase Fucoidanase	
3.2.1.14	Poly(1,4-(N-acetyl-β-D-glucosaminide)) glycanohydrolase Chitinase	
3.2.1.82	Poly(1,4-α-D-galactosiduronate) digalacturonohydrolase Exo-poly-α-galacturonosidase	
4.2.2.9	poly(1,4-α-D-galacturonide) exo-lyase exopolygalacturonate lyase	
3.2.1.67	Poly(1,4-α-D-galacturonide)galacturonohydrolase Galacturan 1,4-α-galacturonidase	
3.2.1.15	Poly(1,4-α-D-galacturonide)glycanohydrolase Polygalacturonase	
4.2.2.2	poly(1,4-α-D-galacturonide) lyase pectate lyase	
4.2.2.3	poly(1,4-β-D-mannuronide) lyase poly(β-D-mannuronate) lyase	
3.1.13.4	Poly(A)-specific ribonuclease Poly(A)-specific ribonuclease	
2.7.7.19	Poly(A) hydrolase Polynucleotide adenylyltransferase	
2.7.7.19	Poly(A) polymerase Polynucleotide adenylyltransferase	
2.7.7.19	Poly(A) synthetase Polynucleotide adenylyltransferase	
3.2.1.143	poly(ADP-ribose) glycohydrolase poly(ADP-ribose) glycohydrolase	
2.4.2.30	Poly(ADP-ribose) polymerase NAD^+ ADP-ribosyltransferase	
2.4.2.30	Poly(ADP-ribose) synthase NAD^+ ADP-ribosyltransferase	

EC-No.	Synonym Recommended Name	Vol., Page
2.4.2.30	Poly(ADP-ribose) transferase NAD^+ ADP-ribosyltransferase	
2.4.2.30	Poly(ADP-ribosyl)transferase NAD^+ ADP-ribosyltransferase	
6.5.1.1	Poly(deoxyribonucleotide):poly(deoxyribonucleotide) ligase (AMP-forming) DNA ligase (ATP)	Vol. 2, Page 755
6.5.1.2	Poly(deoxyribonucleotide):poly(deoxyribonucleotide) ligase (AMP-forming, NMN-forming) DNA ligase (NAD^+)	Vol. 2, Page 773
3.2.1.67	Poly(galacturonate)hydrolase Galacturan 1,4-α-galacturonidase	
3.4.19.9	Poly(glutamic acid) hydrolase II γ-Glu-X carboxypeptidase	
2.4.1.52	Poly(glycerol-phosphate) α-glucosyltransferase Poly(glycerol-phosphate) α-glucosyltransferase	
2.7.8.12	Poly(glycerol phosphate) polymerase CDPglycerol glycerophosphotransferase	
2.7.8.6	Poly(isoprenol)-phosphate galactosephosphotransferase Undecaprenyl-phosphate galactosephosphotransferase	
2.7.8.6	Poly(isoprenyl)phosphate galactosephosphatetransferase Undecaprenyl-phosphate galactosephosphotransferase	
4.2.2.3	poly(mana)alginate lyase poly(β-D-mannuronate) lyase	
3.2.1.121	Poly(mannuronide) mannuronohydrolase Polymannuronate hydrolase	
4.2.2.10	poly(methoxygalacturonide) lyase pectin lyase	
2.4.1.53	Poly(ribitol-phosphate) β-glucosyltransferase Poly(ribitol-phosphate) β-glucosyltransferase	
2.4.1.70	Poly(ribitol-phosphate) N-acetylglucosaminyltransferase Poly(ribitol-phosphate) N-acetylglucosaminyltransferase	
2.7.8.14	Poly(ribitol phosphate) synthetase CDPribitol ribitolphosphotransferase	
6.5.1.3	Poly(ribonucleotide):poly(ribonucleotide) ligase (AMP-forming) RNA ligase (ATP)	Vol. 2, Page 787

EC-No.	Synonym Recommended Name	Vol., Page
3.2.1.129	Poly(sialoside) α-2,8-sialosylhydrolase Endo-α-sialidase	
3.1.26.10	Poly(U)- and poly(C)-specific endoribonuclease Ribonuclease IX	
3.2.1.15	Poly-α-1,4-galacturonide glycanohydrolase Polygalacturonase	
3.2.1.14	Poly-β-glucosaminidase Chitinase	
4.2.2.11	Poly-α-L-guluronate lyase Poly(α-L-guluronate) lyase	
2.5.1.20	Poly-cis-Polyprenyl-diphosphate:isopentenyl-diphosphate polyprenylcis-transferase Rubber cis-polyprenylcistransferase	
3.2.1.14	β-1,4-poly-N-acetyl glucosamidinase Chitinase	
3.2.1.14	1,4-β-poly-N-acetylglucosaminidase Chitinase	
2.4.1.149	Poly-N-acetyllactosamine extension enzyme N-Acetyllactosaminide β-1,3-N-acetylglucosaminyltransferase	
2.4.1.163	Poly-N-acetyllactosamine extension enzyme β-Galactosyl-N-acetylglucosaminylgalactosylglucosylceramide β-1,3-acetylglucosaminyltransferase	
5.1.3.17	Poly[(1,4)-β-D-glucuronyl-(1,4)-N-sulfo-α-D-glucosaminyl]glucurono-5-epimerase Heparosan-N-sulfate-glucuronate 5-epimerase	Vol. 1, Page 167
2.7.7.19	Polyadenylate nucleotidyltransferase Polynucleotide adenylyltransferase	
2.7.7.19	Polyadenylate polymerase Polynucleotide adenylyltransferase	
2.7.7.19	Polyadenylate synthetase Polynucleotide adenylyltransferase	
2.7.7.19	Polyadenylic acid polymerase Polynucleotide adenylyltransferase	
2.7.7.19	Polyadenylic polymerase Polynucleotide adenylyltransferase	
1.4.3.4	Polyamine oxidase Amine oxidase (flavin-containing)	

EC-No.	Synonym Recommended Name	Vol., Page
1.5.3.11	Polyamine oxidase Polyamine oxidase	
2.3.2.13	Polyamine transglutaminase Protein-glutamine γ-glutamyltransferase	
1.14.18.1	Polyaromatic oxidase Monophenol monooxygenase	
3.1.3.16	Polycation modulated (PCM-) phosphatase Phosphoprotein phosphatase	
6.5.1.1	Polydeoxyribonucleotide synthase (ATP) DNA ligase (ATP)	Vol. 2, Page 755
6.5.1.2	Polydeoxyribonucleotide synthase (NAD^+) DNA ligase (NAD^+)	Vol. 2, Page 773
1.1.99.20	Polyethylene glycol dehydrogenase Alkan-1-ol dehydrogenase (acceptor)	
3.2.1.15	Polygalacturonase Polygalacturonase	
2.4.1.43	Polygalacturonate 4-α-galacturonosyltransferase Polygalacturonate 4-α-galacturonosyltransferase	
4.2.2.2	polygalacturonate lyase pectate lyase	
4.2.2.2	polygalacturonic acid lyase pectate lyase	
4.2.2.2	polygalacturonic transeliminase pectate lyase	
5.1.3.19	Polyglucuronate 5-epimerase Chondroitin-glucuronate 5-epimerase	Vol. 1, Page 172
4.2.99.5	Polyglucuronide lyase Polyglucuronide lyase	
3.4.19.9	Polyglutamate hydrolase γ-Glu-X carboxypeptidase	
4.2.2.11	Polyguluronate-specific alginate lyase Poly(α-L-guluronate) lyase	
2.7.1.66	Polyisoprenol kinase Undecaprenol kinase	
3.1.3.51	Polyisoprenyl phosphate phosphatase Dolichyl-phosphatase	

EC-No.	Synonym Recommended Name	Vol., Page
3.2.1.121	Polymannuronate hydrolase Polymannuronate hydrolase	
3.2.1.121	Polymannuronic acid polymerase Polymannuronate hydrolase	
2.7.7.48	Polymerase L RNA-directed RNA polymerase	
3.6.1.10	Polymetaphosphatase Endopolyphosphatase	
4.2.2.10	polymethylgalacturonic transeliminase pectin lyase	
4.1.1.4	Polymyxin MI Acetoacetate decarboxylase	
3.1.3.32	2'(3')-polynucleotidase polynucleotide 3'-phosphatase	
3.1.3.33	5'-polynucleotidase polynucleotide 5'-phosphatase	
3.1.3.32	polynucleotide 3'-phosphatase polynucleotide 3'-phosphatase	
3.1.3.32	polynucleotide 3'-phosphohydrolase polynucleotide 3'-phosphatase	
2.7.1.78	Polynucleotide 5'-hydroxy-kinase Polynucleotide 5'-hydroxy-kinase	
2.7.1.78	Polynucleotide 5'-hydroxyl-kinase Polynucleotide 5'-hydroxy-kinase	
3.1.3.33	polynucleotide 5'-phosphatase polynucleotide 5'-phosphatase	
3.1.3.33	polynucleotide 5'-phosphohydrolase polynucleotide 5'-phosphatase	
2.7.7.19	Polynucleotide adenylyltransferase Polynucleotide adenylyltransferase	
2.7.1.78	Polynucleotide kinase Polynucleotide 5'-hydroxy-kinase	
3.1.3.32	polynucleotide kinase 3'-phosphatase polynucleotide 3'-phosphatase	
3.1.3.32	5' polynucleotidekinase 3' phosphatase polynucleotide 3'-phosphatase	

EC-No.	Synonym Recommended Name	Vol., Page
6.5.1.1	Polynucleotide ligase DNA ligase (ATP)	Vol. 2, Page 755
6.5.1.2	Polynucleotide ligase DNA ligase (NAD^+)	Vol. 2, Page 773
2.7.7.8	Polynucleotide phosphorylase Polyribonucleotide nucleotidyltransferase	
6.5.1.2	Polynucleotide synthetase DNA ligase (NAD^+)	Vol. 2, Page 773
6.5.1.3	Polynucleotide synthetase RNA ligase (ATP)	Vol. 2, Page 787
6.5.1.2	Polynucleotide synthetase (nicotinamide adenine dinucleotide) DNA ligase (NAD^+)	Vol. 2, Page 773
1.1.1.14	Polyol dehydrogenase L-Iditol 2-dehydrogenase	
1.1.2.2	Polyol dehydrogenase Mannitol dehydrogenase (cytochrome)	
1.1.1.21	Polyol dehydrogenase ($NADP^+$) Aldehyde reductase	
1.1.1.139	polyol dehydrogenase ($NADP^+$) polyol dehydrogenase ($NADP^+$)	
2.4.1.41	Polypeptide-N-acetylgalactosamine transferase Polypeptide N-acetylgalactosaminyltransferase	
2.4.1.41	Polypeptide N-acetylgalactosaminyltransferase Polypeptide N-acetylgalactosaminyltransferase	
1.14.18.1	Polyphenolase Monophenol monooxygenase	
1.10.3.1	Polyphenol oxidase Catechol oxidase	
1.14.18.1	Polyphenol oxidase Monophenol monooxygenase	
3.6.1.10	Polyphosphatase Endopolyphosphatase	
2.7.1.63	Polyphosphate-D-(+)-glucose-6-phosphotransferase Polyphosphate-glucose phosphotransferase	
2.7.1.63	Polyphosphate-glucose 6-phosphotransferase Polyphosphate-glucose phosphotransferase	

EC-No.	Synonym Recommended Name	Vol., Page
2.7.1.63	Polyphosphate-glucose phosphotransferase Polyphosphate-glucose phosphotransferase	
2.7.1.63	Polyphosphate:D-glucose 6-phosphotransferase Polyphosphate-glucose phosphotransferase	
3.6.1.10	Polyphosphate depolymerase Endopolyphosphatase	
2.7.1.63	Polyphosphate glucokinase Polyphosphate-glucose phosphotransferase	
2.7.4.1	Polyphosphate kinase Polyphosphate kinase	
3.6.1.11	Polyphosphate phosphohydrolase Exopolyphosphatase	
3.6.1.10	Polyphosphate polyphosphohydrolase Endopolyphosphatase	
2.7.4.1	Polyphosphoric acid kinase Polyphosphate kinase	
2.4.1.1	Polyphosphorylase Phosphorylase	
3.4.23.29	Polyporopepsin Polyporopepsin	
3.4.23.29	Polyporus aspartic proteinase Polyporopepsin	
2.4.1.117	Polyprenyl phosphate:UDP-D-glucose glucosyltransferase Dolichyl-phosphate β-glucosyltransferase	
3.1.3.51	Polyprenylphosphate phosphatase Dolichyl-phosphatase	
2.5.1.11	Polyprenylpyrophosphate synthetase trans-Octaprenyltranstransferase	
3.4.21.98	polyprotein-processing proteinase NS3 hepacivirin	
2.7.8.14	Polyribitol phosphate polymerase CDPribitol ribitolphosphotransferase	
2.7.8.14	Polyribitol phosphate synthetase CDPribitol ribitolphosphotransferase	
2.7.7.8	Polyribonucleotide:orthophosphate nucleotidyltransferase Polyribonucleotide nucleotidyltransferase	

EC-No.	Synonym Recommended Name	Vol., Page
6.5.1.3	Polyribonucleotide ligase RNA ligase (ATP)	Vol. 2, Page 787
2.7.7.8	Polyribonucleotide nucleotidyltransferase Polyribonucleotide nucleotidyltransferase	
2.7.7.8	Polyribonucleotide phosphorylase Polyribonucleotide nucleotidyltransferase	
6.5.1.3	Polyribonucleotide synthase (ATP) RNA ligase (ATP)	Vol. 2, Page 787
3.1.6.18	Polysaccharide-2-O-sulfo-D-glucuronate 2-sulfohydrolase Glucuronate-2-sulfatase	
3.2.1.55	Polysaccharide α-L-arabinofuranosidase α-L-arabinofuranosidase	
3.2.1.80	Polysaccharide.β.-fructofuranosidase Fructan β-fructosidase	
3.1.1.58	Polysaccharide deacetylase N-Acetylgalactosaminoglycan deacetylase	
3.2.1.87	Polysaccharide depolymerase Capsular-polysaccharide endo-1,3-α-galactosidase	
2.1.1.18	Polysaccharide O-methyltransferase Polysaccharide O-methyltransferase	
2.3.1.136	Polysialic-acid O-acetyltransferase Polysialic-acid O-acetyltransferase	
1.1.99.23	Polyvinyl-alcohol:(acceptor) oxidoreductase Polyvinyl-alcohol dehydrogenase (acceptor)	
1.1.3.30	Polyvinyl-alcohol:oxygen oxidoreductase Polyvinyl-alcohol oxidase	
1.1.99.23	Polyvinyl-alcohol dehydrogenase (acceptor) Polyvinyl-alcohol dehydrogenase (acceptor)	
1.1.3.30	Polyvinyl-alcohol oxidase Polyvinyl-alcohol oxidase	
1.1.3.18	Polyvinyl alcohol oxidase Secondary-alcohol oxidase	
4.3.1.8	porphobilinogen ammonia-lyase (polymerizing) hydroxymethylbilane synthase	
4.2.1.75	Porphobilinogenase Uroporphyrinogen-III synthase	

EC-No.	Synonym Recommended Name	Vol., Page
4.3.1.8	porphobilinogen deaminase hydroxymethylbilane synthase	
4.2.1.24	porphobilinogen synthase porphobilinogen synthase	
4.2.1.24	porphobilinogen synthetase porphobilinogen synthase	
2.5.1.5	Porphyran sulfatase Galactose-6-sulfurylase	
4.1.1.37	Porphyrinogen carboxy-lyase Uroporphyrinogen decarboxylase	
3.1.1.3	post-heparin plasma protamine-resistant lipase triacylglycerol lipase	
3.4.21.26	post-proline cleaving enzyme prolyl oligopeptidase	
3.4.21.26	post-proline endopeptidase prolyl oligopeptidase	
3.1.1.34	postheparin esterase lipoprotein lipase	
3.1.1.34	postheparin lipase lipoprotein lipase	
3.4.21.26	postproline-cleaving enzyme prolyl oligopeptidase	
3.4.14.5	postproline dipeptidyl aminopeptidase IV dipeptidyl-peptidase IV	
3.4.21.26	postproline endopeptidase prolyl oligopeptidase	
2.4.1.1	Potato phosphorylase Phosphorylase	
3.4.22.44	potyvirus NIa protease nuclear-inclusion-a endopeptidase	
3.4.22.30	Ppω Caricain	
2.7.6.1	PP-ribose P synthetase Ribose-phosphate pyrophosphokinase	
3.1.3.17	ppase-1 [phosphorylase] phosphatase	

EC-No.	Synonym Recommended Name	Vol., Page
4.2.2.2	PPase-N pectate lyase	
4.1.1.36	PPC-decarboxylase Phosphopantothenoylcysteine decarboxylase	
2.7.9.1	PPDK phosphate dikinase pyruvate	
5.1.3.1	PPE Ribulose-phosphate 3-epimerase	Vol. 1, Page 91
3.1.7.2	PpGpp-3'-pyrophosphohydrolase Guanosine-3',5'-bis(diphosphate) 3'-pyrophosphatase	
3.1.7.2	PpGpp phosphohydrolase Guanosine-3',5'-bis(diphosphate) 3'-pyrophosphatase	
3.4.24.18	PPH Meprin A	
1.1.1.220	6PPH4(2'-oxo) reductase 6-Pyruvoyltetrahydropterin 2'-reductase	
2.7.1.42	G-1-P phosphotransferase Riboflavin phosphotransferase	
2.7.1.143	PPi-dependent nucleoside kinase Diphosphate-purine nucleoside kinase	
2.7.1.79	PPi-glycerol phosphotransferase Pyrophosphate-glycerol phosphotransferase	
5.2.1.8	PPIase Peptidylprolyl isomerase	Vol. 1, Page 218
5.2.1.8	PPIase Peptidylprolyl isomerase	Vol. 1, Page 218
3.4.22.25	PPIV Glycyl endopeptidase	
3.4.21.35	PPK tissue kallikrein	
3.1.1.3	PPL triacylglycerol lipase	
3.6.1.40	pppGpp-5'-phosphohydrolase Guanosine-5'-triphosphate,3'-diphosphate diphosphatase	
3.6.1.40	pppGpp 5'-phosphohydrolase Guanosine-5'-triphosphate,3'-diphosphate diphosphatase	

EC-No.	Synonym Recommended Name	Vol., Page
2.7.6.1	PPRibP synthetase Ribose-phosphate pyrophosphokinase	
5.4.99.5	P protein Chorismate mutase	Vol. 1, Page 604
3.1.2.22	PPT Palmitoyl-(protein) hydrolase	
3.1.3.48	PPT-phosphatase protein-tyrosine-phosphatase	
3.1.2.22	PPT1 Palmitoyl-(protein) hydrolase	
4.2.1.50	β-Ppyrazol-1-ylalanine synthetase Pyrazolylalanine synthase	
4.6.1.10	6-Ppyruvoyltetrahydropterin synthase [16-cysteine] (human clone lamda HSY2 gene PCBD subunit 6-Pyruvoyltetrahydtropterin synthase	
3.1.3.17	PR-enzyme [phosphorylase] phosphatase	
3.2.1.17	PR1-Lysozyme Lysozyme	
3.4.23.25	PRA Saccharopepsin	
5.3.1.24	PRAI Phosphoribosylanthranilate isomerase	Vol. 1, Page 353
4.1.1.48	PRAI-InGPS Indole-3-glycerol phosphate synthase	
5.3.1.24	PRAI[ML256-452] Phosphoribosylanthranilate isomerase	Vol. 1, Page 353
5.3.1.24	PRA isomerase Phosphoribosylanthranilate isomerase	Vol. 1, Page 353
3.5.4.19	PRAMP-cyclohydrolase Phosphoribosyl-AMP cyclohydrolase	
4.1.99.3	PRE deoxyribodipyrimidine photolyase	
4.1.1.71	Pre-2-oxoglutarate decarboxylase 2-Oxoglutarate decarboxylase	

EC-No.	Synonym Recommended Name	Vol., Page
4.3.1.8	pre-uroporphyrinogen synthase hydroxymethylbilane synthase	
3.4.21.38	prealbumin activator coagulation factor XIIa	
2.1.1.132	Precorrhin-6y methyltransferase Precorrin-6Y $C^{5,15}$-methyltransferase (decarboxylating)	
2.1.1.130	Precorrin-2-methyltransferase Precorrin-2 C^{20}-methyltransferase	
2.1.1.130	Precorrin-2 C^{20}-methyltransferase Precorrin-2 C^{20}-methyltransferase	
2.1.1.131	Precorrin-3B C^{17}-methyltransferase Precorrin-3B C^{17}-methyltransferase	
2.1.1.131	Precorrin-3B methyltransferase Precorrin-3B C^{17}-methyltransferase	
2.1.1.131	Precorrin-3 $C\text{-}^{17}$ methyltransferase Precorrin-3B C^{17}-methyltransferase	
2.1.1.133	Precorrin-3 methylase Precorrin-4 C^{11}-methyltransferase	
2.1.1.131	Precorrin-3 methyltransferase Precorrin-3B C^{17}-methyltransferase	
2.1.1.133	Precorrin-4 C^{11}-methyltransferase Precorrin-4 C^{11}-methyltransferase	
2.1.1.131	Precorrin-4 synthase Precorrin-3B C17-methyltransferase	
2.1.1.132	Precorrin-6 methyltransferase Precorrin-6Y C5,15-methyltransferase (decarboxylating)	
1.3.1.54	Precorrin-6X reductase Precorrin-6X reductase	
1.3.1.54	Precorrin-6Y:NADP oxidoreductase Precorrin-6X reductase	
2.1.1.132	Precorrin-6Y $C^{5,15}$-methyltransferase (decarboxylating) Precorrin-6Y $C^{5,15}$-methyltransferase (decarboxylating)	
2.1.1.132	Precorrin-6Y methylase Precorrin-6Y $C^{5,15}$-methyltransferase (decarboxylating)	
5.4.1.2	Precorrin-8X 11,12-methylmutase Precorrin-8X methylmutase	Vol. 1, Page 490

EC-No.	Synonym Recommended Name	Vol., Page
5.4.1.2	Precorrin-8X methylmutase Precorrin-8X methylmutase	Vol. 1, Page 490
2.1.1.130	Precorrin 2- (Salmonella typhimurium strain LT2 gene cbiL isoenzyme) Precorrin-2 C^{20}-methyltransferase	
5.4.1.2	Precorrin isomerase Precorrin-8X methylmutase	Vol. 1, Page 490
3.4.22.36	Precursor interleukin-1β converting enzyme Interleukin 1β converting enzyme	
1.3.1.30	5α-Pregnane-3,20-dione:$NADP^+$ oxidoreductase Progesterone 5α-reductase	
5.3.3.1	5-Pregnene-3,20-dione isomerase Steroid Δ-isomerase	Vol. 1, Page 376
3.1.6.2	Pregnenolone sulfatase Steryl-sulfatase	
3.4.21.38	prekallikrein activator coagulation factor XIIa	
3.4.23.36	Premurein-leader peptidase Signal peptidase II	
3.4.99.35	premurein-leader peptidase premurein-leader peptidase	
3.4.23.36	Premurein leader peptidase Signal peptidase II	
3.4.23.36	Premurein leader proteinase Signal peptidase II	
3.1.7.1	prenol-pyrophosphatase prenyl-diphosphatase	
3.1.7.1	prenol pyrophosphate pyrophosphohydrolase prenyl-diphosphatase	
3.1.7.1	prenyl-diphosphatase prenyl-diphosphatase	
3.1.7.1	prenyl-diphosphate diphosphohydrolase prenyl-diphosphatase	
4.4.1.18	prenyl-L-cysteine:L-cysteine-lyase prenylcysteine lyase	
3.1.7.1	prenyl-pyrophosphatase prenyl-diphosphatase	

EC-No.	Synonym Recommended Name	Vol., Page
2.1.1.100	Prenylated protein methyltransferase Protein-S-isoprenylcysteine O-methyltransferase	
4.4.1.18	prenylcysteine lyase prenylcysteine lyase	
3.1.7.1	prenylphosphatase prenyl-diphosphatase	
2.5.1.1	Prenyltransferase Dimethylallyltranstransferase	
2.5.1.10	Prenyltransferase Geranyltranstransferase	
2.5.1.41	Prenyltransferase Phosphoglycerol geranylgeranyltransferase	
1.3.1.12	Prephenate: NAD^+ oxidoreductase (decarboxylating) Prephenate dehydrogenase	
1.3.1.13	Prephenate: $NADP^+$ oxidoreductase (decarboxylating) Prephenate dehydrogenase ($NADP^+$)	
4.2.1.51	prephenate dehydratase prephenate dehydratase	
1.3.1.12	Prephenate dehydrogenase Prephenate dehydrogenase	
1.3.1.13	Prephenate dehydrogenase Prephenate dehydrogenase ($NADP^+$)	
1.3.1.13	Prephenate dehydrogenase ($NADP^+$) Prephenate dehydrogenase ($NADP^+$)	
4.2.1.51	prephenate hydro-lyase (decarboxylating) prephenate dehydratase	
2.5.1.32	Prephytoene-diphosphate synthase Geranylgeranyl-diphosphate geranylgeranyltransferase	
2.5.1.21	Presqualene-diphosphate synthase Farnesyl-diphosphate farnesyltransferase	
2.5.1.21	Presqualene synthase Farnesyl-diphosphate farnesyltransferase	
1.3.1.43	Pretyrosine dehydrogenase Pretyrosine dehydrogenase	
4.3.1.8	preuroporphyrinogen synthetase hydroxymethylbilane synthase	

EC-No.	Synonym Recommended Name	Vol., Page
6.3.4.13	PRG synthetase Phosphoribosylamine-glycine ligase	Vol. 2, Page 626
1.4.99.3	Primary-amine:(acceptor) oxidoreductase (deaminating) Amine dehydrogenase	
1.4.99.3	Primary-amine dehydrogenase Amine dehydrogenase	
1.1.1.1	Primary alcohol dehydrogenase Alcohol dehydrogenase	
1.1.99.8	Primary alcohol dehydrogenase Alcohol dehydrogenase (acceptor)	
3.2.1.21	Primeverosidase β-glucosidase	
2.4.1.186	Priming glucosyltransferase Glycogenin glucosyltransferase	
6.2.1.3	Pristanoyl-CoA synthetase Long-chain-fatty-acid-CoA ligase	Vol. 2, Page 206
2.7.1.19	PRK Phosphoribulokinase	
2.3.1.16	Pro-3-ketoacyl-CoA thiolase Acetyl-CoA C-acyltransferase	
3.4.99.40	pro-gonadoliberin proteinase pro-gonadoliberin proteinase	
3.4.22.36	Pro-interleukin 1β proteinase caspase-1	
3.4.23.17	Pro-opiomelanocortin-converting enzyme Pro-opiomelanocortin converting enzyme	
3.4.23.17	Pro-opiomelanocortin converting enzyme Pro-opiomelanocortin converting enzyme	
3.4.99.38	pro-opiomelanotropin-converting proteinase pro-opiomelanotropin-converting proteinase	
3.4.24.62	Pro-oxytocin/neurophysin convertase Magnolysin	
3.4.24.62	Pro-oxytocin convertase Magnolysin	
3.4.17.20	pro-pCPB, Carboxypeptidase U	

EC-No.	Synonym Recommended Name	Vol., Page
6.1.1.15	Pro-tRNA synthetase Proline-tRNA ligase	Vol. 2, Page 111
3.4.11.5	Pro-X aminopeptidase prolyl aminopeptidase	
3.4.13.8	Pro-X dipeptidase Pro-X dipeptidase	
3.4.11.9	X-Pro aminopeptidase X-Pro aminopeptidase	
3.1.1.1	Procaine esterase Carboxylesterase	
3.4.17.20	Procarboxypeptidase B Carboxypeptidase U	
3.4.24.64	Processing enhancing peptidase Mitochondrial processing peptidase	
1.14.11.4	Procollagen-L-lysine,2-oxoglutarate:oxygen oxidoreductase (5-hydroxylating) Procollagen-lysine 5-dioxygenase	
1.14.11.7	Procollagen-L-proline,2-oxoglutarate:oxygen oxidoreductase (3-hydroxylating) Procollagen-proline 3-dioxygenase	
1.14.11.2	Procollagen-L-proline,2-oxoglutarate:oxygen oxidoreductase (4-hydroxylating) Procollagen-proline dioxygenase	
1.14.11.4	Procollagen-lysine 5-dioxygenase Procollagen-lysine 5-dioxygenase	
1.14.11.7	Procollagen-proline 3-dioxygenase Procollagen-proline 3-dioxygenase	
1.14.11.2	Procollagen-proline dioxygenase Procollagen-proline dioxygenase	
3.4.24.14	procollagen aminopeptidase procollagen N-endopeptidase	
3.4.24.14	procollagen aminoterminal protease procollagen N-endopeptidase	
3.4.24.17	Procollagenase activator Stromelysin 1	
3.4.24.19	Procollagen C-endopeptidase Procollagen C-endopeptidase	

EC-No.	Synonym Recommended Name	Vol., Page
3.4.24.19	Procollagen C-proteinase Procollagen C-endopeptidase	
3.4.24.19	Procollagen C-terminal proteinase Procollagen C-endopeptidase	
3.4.24.19	Procollagen carboxy-terminal proteinase Procollagen C-endopeptidase	
3.4.24.19	Procollagen carboxypeptidase Procollagen C-endopeptidase	
2.4.1.50	Procollagen galactosyltransferase Procollagen galactosyltransferase	
2.4.1.66	Procollagen glucosyltransferase Procollagen glucosyltransferase	
3.4.24.14	procollagen N-endopeptidase procollagen N-endopeptidase	
3.4.24.14	procollagen N-protease procollagen N-endopeptidase	
3.4.24.14	procollagen N-proteinase procollagen N-endopeptidase	
3.4.24.14	procollagen N-terminal peptidase procollagen N-endopeptidase	
3.4.24.14	procollagen N-terminal proteinase procollagen N-endopeptidase	
3.4.24.19	Procollagen peptidase Procollagen C-endopeptidase	
3.4.24.14	procollagen peptidase procollagen N-endopeptidase	
3.4.23.18	Proctase Aspergillopepsin I	
3.4.23.19	Proctase A Aspergillopepsin II	
3.4.23.18	Proctase B Aspergillopepsin I	
3.4.23.18	Proctase P Aspergillopepsin I	
3.4.13.9	X-Pro dipeptidase X-Pro dipeptidase	

EC-No.	Synonym Recommended Name	Vol., Page
3.4.14.11	X-Pro dipeptidyl-peptidase X-Pro dipeptidyl-peptidase	
3.4.21.47	proenzyme factor B alternative-complement-pathway C3/C5 convertase	
1.14.99.14	Progesterone,hydrogen-donor:oxygen oxidoreductase (11α-hydroxylating) Progesterone 11α-monooxygenase	
1.14.99.4	Progesterone,hydrogen-donor:oxygen oxidoreductase (hydroxylating) Progesterone monooxygenase	
1.14.13.54	Progesterone, $NADPH_2$:oxygen oxidoreductase (20-hydroxylating, ester-producing) Ketosteroid monooxygenase	
1.14.99.14	Progesterone 11α-hydroxylase Progesterone 11α-monooxygenase	
1.14.99.14	Progesterone 11α-monooxygenase Progesterone 11α-monooxygenase	
1.3.1.30	Progesterone 5α-reductase Progesterone 5α-reductase	
1.3.1.30	Progesterone 5α-reductase Progesterone 5α-reductase	
1.14.99.4	Progesterone hydroxylase Progesterone monooxygenase	
1.14.99.4	Progesterone monooxygenase Progesterone monooxygenase	
1.1.1.210	Progesterone reductase 3β(or 20α)-Hydroxysteroid dehydrogenase	
1.1.1.145	Progesterone reductase 3β-Hydroxy-Δ^5-steroid dehydrogenase	
2.4.1.112	Proglycogen synthase α-1,4-Glucan-protein synthase (UDP-forming)	
3.4.21.61	Prohormone-processing endoprotease Kexin	
3.4.21.61	Prohormone-processing KEX2 proteinase Kexin	
3.4.21.61	Prohormone-processing proteinase Kexin	

EC-No.	Synonym Recommended Name	Vol., Page
3.4.21.75	Prohormone convertase Furin	
3.4.21.93	Prohormone convertase 3 Proprotein convertase 1	
3.4.23.17	Prohormone converting enzyme Pro-opiomelanocortin converting enzyme	
3.4.21.58	prohormone serine proteinase prohormone serine proteinase	
3.4.22.36	Prointerleukin 1β protease caspase-1	
3.4.21.89	Prokaryotic leader peptidase Signal peptidase I	
3.4.21.89	Prokaryotic signal peptidase Signal peptidase I	
3.4.21.89	Prokaryotic signal proteinase Signal peptidase I	
3.4.13.9	prolidase X-Pro dipeptidase	
1.14.11.7	Proline,2-oxoglutarate 3-dioxygenase Procollagen-proline 3-dioxygenase	
1.14.11.2	Proline,2-oxoglutarate 4-dioxygenase Procollagen-proline dioxygenase	
1.14.11.2	Proline, 2-oxoglutarate dioxygenase Procollagen-proline dioxygenase	
1.5.1.2	L-Proline-NAD(P)$^+$ 5-oxidoreductase Pyrroline-5-carboxylate reductase	
3.4.16.2	proline-specific carboxypeptidase P lysosomal Pro-X carboxypeptidase	
3.4.21.26	proline-specific endopeptidase prolyl oligopeptidase	
6.1.1.15	Proline-tRNA ligase Proline-tRNA ligase	Vol. 2, Page 111
1.5.99.8	L-Proline:(acceptor) oxidoreductase Proline dehydrogenase	
1.5.1.1	L-Proline: NAD(P)$^+$ 2-oxidoreductase Pyrroline-2-carboxylate reductase	

EC-No.	Synonym Recommended Name	Vol., Page
1.5.1.2	L-Proline: NAD(P)$^+$ 5-oxidoreductase Pyrroline-5-carboxylate reductase	
6.1.1.15	L-Proline:tRNA^{Pro} ligase (AMP-forming) Proline-tRNA ligase	Vol. 2, Page 111
3.4.11.5	L-proline aminopeptidase prolyl aminopeptidase	
3.4.11.9	proline aminopeptidase X-Pro aminopeptidase	
3.4.11.5	proline aminopeptidase prolyl aminopeptidase	
3.4.16.2	proline carboxypeptidase lysosomal Pro-X carboxypeptidase	
1.5.99.8	L-Proline dehydrogenase Proline dehydrogenase	
1.5.99.8	Proline dehydrogenase Proline dehydrogenase	
3.4.21.26	proline endopeptidase prolyl oligopeptidase	
1.14.11.2	Proline hydroxylase Procollagen-proline dioxygenase	
3.4.1.4	proline iminopeptidase proline iminopeptidase	
1.5.1.2	L-Proline oxidase Pyrroline-5-carboxylate reductase	
1.5.1.2	Proline oxidase Pyrroline-5-carboxylate reductase	
1.14.11.2	Proline protocollagen hydroxylase Procollagen-proline dioxygenase	
5.1.1.4	Proline racemase Proline racemase	Vol. 1, Page 19
5.1.1.4	Proline racemase Proline racemase	Vol. 1, Page 19
1.4.1.6	D-proline reductase D-proline reductase	
1.4.4.1	D-Proline reductase (dithiol) D-Proline reductase (dithiol)	

EC-No.	Synonym Recommended Name	Vol., Page
6.1.1.15	Proline translase Proline-tRNA ligase	Vol. 2, Page 111
3.4.23.36	Prolipoprotein signal peptidase Signal peptidase II	
1.14.11.7	Prolyl-4-hydroxyprolyl-glycyl-peptide, 2-oxoglutarate: oxygen oxidoreductase, 3-hydroxylating Procollagen-proline 3-dioxygenase	
3.4.14.11	X-Prolyl-dipeptidyl aminopeptidase X-Pro dipeptidyl-peptidase	
1.14.11.2	Prolyl-glycyl-peptide, 2-oxoglutarate:oxygen oxidoreductase, 4-hydroxylating Procollagen-proline dioxygenase	
3.4.11.5	L-prolyl-peptide hydrolase prolyl aminopeptidase	
6.1.1.15	Prolyl-transfer ribonucleate synthetase Proline-tRNA ligase	Vol. 2, Page 111
6.1.1.15	Prolyl-transfer ribonucleic acid synthetase Proline-tRNA ligase	Vol. 2, Page 111
6.1.1.15	Prolyl-transfer RNA synthetase Proline-tRNA ligase	Vol. 2, Page 111
6.1.1.15	Prolyl-tRNA synthetase Proline-tRNA ligase	Vol. 2, Page 111
6.1.1.15	Prolyl-tRNA synthetase Proline-tRNA ligase	Vol. 2, Page 111
1.14.11.7	Prolyl 3-hydroxylase Procollagen-proline 3-dioxygenase	
1.14.11.2	Prolyl 4-hydroxylase Procollagen-proline dioxygenase	
3.4.11.5	prolyl aminopeptidase prolyl aminopeptidase	
3.4.16.2	prolyl carboxypeptidase lysosomal Pro-X carboxypeptidase	
3.4.16.2	prolylcarboxypeptidase lysosomal Pro-X carboxypeptidase	
3.4.14.11	X-Prolyl dipeptidyl aminopeptidase X-Pro dipeptidyl-peptidase	

EC-No.	Synonym Recommended Name	Vol., Page
3.4.14.5	X-prolyl dipeptidyl aminopeptidase dipeptidyl-peptidase IV	
3.4.14.11	X-Prolyl dipeptidyl peptidase X-Pro dipeptidyl-peptidase	
3.4.21.27	prolyl endopeptidase coagulation factor XIa	
3.4.21.26	prolyl endopeptidase prolyl oligopeptidase	
3.4.22.18	prolyl endopeptidase (thiol-dependent) prolyl endopeptidase (thiol-dependent)	
1.14.11.2	Prolyl hydroxylase Procollagen-proline dioxygenase	
3.4.21.26	prolyl oligopeptidase prolyl oligopeptidase	
1.14.11.2	Prolylprotocollagen dioxygenase Procollagen-proline dioxygenase	
1.14.11.2	Prolylprotocollagen hydroxylase Procollagen-proline dioxygenase	
6.1.1.15	Prolyl RNA synthetase Proline-tRNA ligase	Vol. 2, Page 111
3.4.24.36	Promastigote surface endopeptidase Leishmanolysin	
3.4.24.36	Promastigote surface protease Leishmanolysin	
3.4.21.63	Promelase Oryzin	
3.4.21.81	Pronase B Streptogrisin B	
3.4.24.31	Pronase component Mycolysin	
3.4.23.17	Proopiomelanocortin proteinase Pro-opiomelanocortin converting enzyme	
1.1.1.80	Propan-2-ol:NADP$^+$ oxidoreductase Isopropanol dehydrogenase (NADP$^+$)	
1.1.1.7	Propane-1,2-diol-1-phosphate:NAD$^+$ oxidoreductase Propanediol-phosphate dehydrogenase	

EC-No.	Synonym Recommended Name	Vol., Page
1.1.1.55	Propane-1,2-diol:$NADP^+$ oxidoreductase Lactaldehyde reductase (NADPH)	
1.1.1.202	Propane-1,3-diol:NAD^+ oxidoreductase 1,3-Propanediol 1-dehydrogenase	
1.1.1.7	1,2-Propanediol-1-phosphate:NAD^+ oxidoreductase Propanediol-phosphate dehydrogenase	
1.1.1.7	Propanediol-phosphate dehydrogenase Propanediol-phosphate dehydrogenase	
1.1.1.202	1,3-Propanediol 1-dehydrogenase 1,3-Propanediol 1-dehydrogenase	
1.1.1.202	1,3-Propanediol:NAD^+ oxidoreductase 1,3-Propanediol 1-dehydrogenase	
1.1.1.55	1,2-Propanediol:$NADP^+$ oxidoreductase Lactaldehyde reductase (NADPH)	
1.1.1.77	Propanediol:nicotinamide adenine dinucleotide oxidoreductase (NAD) Lactaldehyde reductase	
4.2.1.28	Propanediol dehydrase propanediol dehydratase	
4.2.1.28	1,2-propanediol dehydratase propanediol dehydratase	
4.2.1.28	propanediol dehydratase propanediol dehydratase	
1.1.1.202	1,3-Propanediol dehydrogenase 1,3-Propanediol 1-dehydrogenase	
1.1.1.55	Propanediol dehydrogenase Lactaldehyde reductase (NADPH)	
4.2.1.28	1,2-propanediol hydro-lyase propanediol dehydratase	
1.1.1.7	Propanediol phosphate dehydrogenase Propanediol-phosphate dehydrogenase	
6.2.1.17	Propanoate:CoA ligase (AMP-forming) Propionate-CoA ligase	Vol. 2, Page 286
6.4.1.3	Propanoyl-CoA:carbon-dioxide ligase (ADP-forming) Propionyl-CoA carboxylase	Vol. 2, Page 738
3.4.21.89	Propeptidase Signal peptidase I	

EC-No.	Synonym Recommended Name	Vol., Page
3.4.21.47	properdin factor B alternative-complement-pathway C3/C5 convertase	
4.1.2.35	Propioin synthase Propioin synthase	
1.2.1.3	Propionaldehyde dehydrogenase Aldehyde dehydrogenase (NAD^+)	
1.7.3.5	Propionate-3-nitronate oxidase 3-aci-Nitropropanoate oxidase	
2.8.3.1	Propionate-CoA:lactoyl-CoA transferase Propionate CoA-transferase	
6.2.1.17	Propionate-CoA ligase Propionate-CoA ligase	Vol. 2, Page 286
6.2.1.17	Propionate-CoA synthetase Propionate-CoA ligase	Vol. 2, Page 286
2.8.3.1	Propionate CoA-transferase Propionate CoA-transferase	
2.8.3.1	Propionate coenzyme A-transferase Propionate CoA-transferase	
2.3.1.154	propionyl-CoA C^2-trimethyltridecanoyltransferase propionyl-CoA C^2-trimethyltridecanoyltransferase	
4.1.1.41	Propionyl-CoA carboxylase Methylmalonyl-CoA decarboxylase	
6.4.1.3	Propionyl-CoA carboxylase Propionyl-CoA carboxylase	Vol. 2, Page 738
3.1.2.18	Propionyl-CoA hydrolase ADP-dependent short-chain-acyl-CoA hydrolase	
6.2.1.17	Propionyl-CoA synthase Propionate-CoA ligase	Vol. 2, Page 286
6.2.1.17	Propionyl-CoA synthetase Propionate-CoA ligase	Vol. 2, Page 286
3.1.2.18	Propionyl-CoA thioesterase ADP-dependent short-chain-acyl-CoA hydrolase	
2.8.3.1	Propionyl-CoA transferase Propionate CoA-transferase	
6.2.1.17	Propionyl-coenzyme A synthetase Propionate-CoA ligase	Vol. 2, Page 286

EC-No.	Synonym Recommended Name	Vol., Page
3.1.1.8	propionylcholinesterase cholinesterase	
2.8.3.1	Propionyl CoA:acetate CoA transferase Propionate CoA-transferase	
6.2.1.17	Propionyl CoA synthetase Propionate-CoA ligase	Vol. 2, Page 286
4.1.1.41	Propionyl coenzyme A carboxylase Methylmalonyl-CoA decarboxylase	
6.4.1.3	Propionyl coenzyme A carboxylase Propionyl-CoA carboxylase	Vol. 2, Page 738
6.4.1.3	Propionyl coenzyme A carboxylase (adenosine triphosphate-hydrolyzing) Propionyl-CoA carboxylase	Vol. 2, Page 738
6.4.1.3	Propionyl coenzyme A carboxylase (ATP-hydrolyzing) Propionyl-CoA carboxylase	Vol. 2, Page 738
6.3.4.10	Propionyl coenzyme A holocarboxylase synthetase Biotin-[propionyl-CoA-carboxylase (ATP-hydrolysing)] ligase	Vol. 2, Page 617
3.1.1.1	Propionyl esterase Carboxylesterase	
3.4.21.75	Proprotein convertase Furin	
3.4.21.61	Proprotein convertase Kexin	
3.4.21.93	Proprotein convertase 1 Proprotein convertase 1	
3.4.21.94	Proprotein convertase 2 Proprotein convertase 2	
4.1.3.11	3-(n-propyl)-Malate synthase 3-Propylmalate synthase	
4.1.3.11	3-Propylmalate glyoxylate-lyase (CoA-pentanoylating) 3-Propylmalate synthase	
4.1.3.11	3-Propylmalate synthase 3-Propylmalate synthase	
4.1.3.33	Propylmalate synthase 2-Ethylmalate synthase	

EC-No.	Synonym Recommended Name	Vol., Page
4.1.3.11	β-n-Propylmalate synthase 3-Propylmalate synthase	
4.1.3.11	n-Propylmalate synthase 3-Propylmalate synthase	
4.1.3.33	Propylmalic synthase 2-Ethylmalate synthase	
3.4.21.35	prorenin converting enzyme tissue kallikrein	
3.4.25.1	prosome proteasome endopeptidase complex	
1.1.1.231	Prostacyclin dehydrogenase 15-Hydroxyprostaglandin-I dehydrogenase ($NADP^+$)	
5.3.99.4	Prostacycline synthetase Prostacyclin synthase	Vol. 1, Page 465
5.3.99.4	Prostacyclin synthase Prostacyclin synthase	Vol. 1, Page 465
5.3.99.4	Prostacyclin synthetase Prostacyclin synthase	Vol. 1, Page 465
5.3.3.9	Prostaglandin-$A_1\Delta$-isomerase Prostaglandin-A_1 Δ-isomerase	Vol. 1, Page 423
1.1.1.196	Prostaglandin-D 15-dehydrogenase ($NADP^+$) 15-Hydroxyprostaglandin-D dehydrogenase ($NADP^+$)	
1.1.1.188	Prostaglandin-D_2 11-reductase Prostaglandin-F synthase	
5.3.99.2	Prostaglandin-D synthase Prostaglandin-D synthase	Vol. 1, Page 451
1.1.1.189	Prostaglandin-E_2 9-reductase Prostaglandin-E_2 9-reductase	
1.14.99.1	Prostaglandin-endoperoxide synthase Prostaglandin-endoperoxide synthase	
5.3.99.3	Prostaglandin-E synthetase Prostaglandin-E synthetase	Vol. 1, Page 459
1.1.1.188	Prostaglandin-F synthase Prostaglandin-F synthase	
5.3.99.2	Prostaglandin-H_2 D-isomerase Prostaglandin-D synthase	Vol. 1, Page 451

EC-No.	Synonym Recommended Name	Vol., Page
5.3.99.3	Prostaglandin-H_2 E-isomerase Prostaglandin-E synthetase	Vol. 1, Page 459
5.3.99.2	Prostaglandin-R-prostaglandin D isomerase Prostaglandin-D synthase	Vol. 1, Page 451
1.3.1.48	Prostaglandin.δ.13-reductase 15-Oxoprostaglandin Δ^{13}-reductase	
1.1.1.188	Prostaglandin 11-keto reductase Prostaglandin-F synthase	
1.1.1.188	Prostaglandin 11-ketoreductase Prostaglandin-F synthase	
1.3.1.48	Prostaglandin 13-reductase 15-Oxoprostaglandin Δ^{13}-reductase	
1.1.1.184	Prostaglandin 9-ketoreductase Carbonyl reductase (NADPH)	
1.1.1.189	Prostaglandin 9-ketoreductase Prostaglandin-E_2 9-reductase	
5.3.3.9	Prostaglandin A isomerase Prostaglandin-A_1 Δ-isomerase	Vol. 1, Page 423
1.1.1.188	Prostaglandin D_2-ketoreductase Prostaglandin-F synthase	
1.1.1.188	Prostaglandin D_2 11-ketoreductase Prostaglandin-F synthase	
1.1.1.196	Prostaglandin D_2 dehydrogenase 15-Hydroxyprostaglandin-D dehydrogenase ($NADP^+$)	
1.1.1.141	Prostaglandin dehydrogenase 15-Hydroxyprostaglandin dehydrogenase (NAD^+)	
5.3.99.2	Prostaglandin D synthase Prostaglandin-D synthase	Vol. 1, Page 451
1.1.1.189	Prostaglandin E_2-9-oxoreductase Prostaglandin-E_2 9-reductase	
5.3.99.3	Prostaglandin E_2 isomerase Prostaglandin-E synthetase	Vol. 1, Page 459
1.1.1.189	Prostaglandin E 9-ketoreductase Prostaglandin-E_2 9-reductase	
5.3.99.3	Prostaglandin endoperoxide-E isomerase Prostaglandin-E synthetase	Vol. 1, Page 459

EC-No.	Synonym Recommended Name	Vol., Page
5.3.99.3	Prostaglandin endoperoxide E_2 isomerase Prostaglandin-E synthetase	Vol. 1, Page 459
5.3.99.3	Prostaglandin endoperoxide E isomerase Prostaglandin-E synthetase	Vol. 1, Page 459
1.14.99.1	Prostaglandin endoperoxide synthetase Prostaglandin-endoperoxide synthase	
5.3.99.3	Prostaglandin E synthase Prostaglandin-E synthetase	Vol. 1, Page 459
1.1.1.188	Prostaglandin F synthase Prostaglandin-F synthase	
1.1.1.188	Prostaglandin F synthetase Prostaglandin-F synthase	
1.14.99.1	Prostaglandin G/H synthase Prostaglandin-endoperoxide synthase	
5.3.99.3	Prostaglandin H-E isomerase Prostaglandin-E synthetase	Vol. 1, Page 459
5.3.99.3	Prostaglandin H-prostaglandin E isomerase Prostaglandin-E synthetase	Vol. 1, Page 459
5.3.99.4	Prostaglandin I_2 synthetase Prostacyclin synthase	Vol. 1, Page 465
5.3.99.3	Prostaglandin R-prostaglandin E isomerase Prostaglandin-E synthetase	Vol. 1, Page 459
1.14.99.1	Prostaglandin synthase Prostaglandin-endoperoxide synthase	
1.14.99.1	Prostaglandin synthetase Prostaglandin-endoperoxide synthase	
3.4.21.77	Prostate-specific antigen Semenogelase	
3.4.17.21	Prostrate-specific membrane antigen Glutamate carboxypeptidase II	
3.4.17.2	protaminase carboxypeptidase B	
3.4.17.3	protaminase lysine carboxypeptidase	
2.7.1.70	Protamine kinase Protamine kinase	

EC-No.	Synonym Recommended Name	Vol., Page
3.4.22.29	Protease 2A Picornain 2A	
3.4.22.28	Protease 3C Picornain 3C	
3.4.22.36	Protease A Interleukin 1β converting enzyme	
3.4.21.87	Protease A Omptin	
3.4.21.80	Protease A Streptogrisin A	
3.4.21.50	protease I lysyl endopeptidase	
3.4.21.83	Protease II Oligopeptidase B	
3.4.21.61	Protease KEX2 Kexin	
3.4.21.53	Protease La Endopeptidase La	
3.4.21.63	Protease P Oryzin	
3.4.24.55	Protease Pi Pitrilysin	
3.4.21.62	Protease S Subtilisin	
3.4.21.92	Protease Ti Endopeptidase Clp	
3.4.21.19	protease V8 glutamyl endopeptidase	
3.4.22.36	Protease VII Interleukin 1β converting enzyme	
3.4.21.87	Protease VII Omptin	
3.4.21.62	Protease VIII Subtilisin	
3.4.21.62	Protease XXVII Subtilisin	

EC-No.	Synonym Recommended Name	Vol., Page
3.4.25.1	proteasome proteasome endopeptidase complex	
3.4.25.1	proteasome endopeptidase complex proteasome endopeptidase complex	
4.2.1.11	14-3-2 protein phosphoenolpyruvate hydratase	
1.4.4.2	P-protein Glycine dehydrogenase (decarboxylating)	
4.2.1.51	P-protein prephenate dehydratase	
2.1.2.10	T-protein Aminomethyltransferase	
5.99.1.2	ω-Protein DNA topoisomerase	Vol. 1, Page 721
2.1.1.125	Protein (arginine) methyltransferase (Histone)-arginine N-methyltransferase	
2.1.1.124	Protein (arginine) methyltransferase [Cytochrome-c]-arginine N-methyltransferase	
2.1.1.126	Protein (arginine) methyltransferase [Myelin basic protein]-arginine N-methyltransferase	
2.1.1.125	Protein (arginine) N-methyltransferase (Histone)-arginine N-methyltransferase	
2.1.1.124	Protein (arginine) N-methyltransferase [Cytochrome-c]-arginine N-methyltransferase	
2.1.1.126	Protein (arginine) N-methyltransferase [Myelin basic protein]-arginine N-methyltransferase	
2.1.1.80	Protein(aspartate)methyltransferase Protein-glutamate O-methyltransferase	
2.1.1.80	Protein(carboxyl)methyltransferase Protein-glutamate O-methyltransferase	
3.1.3.68	Protein (Escherichia coli strain K12-MG1655 gene b2293) 2-Deoxyglucose-6-phosphatase	
6.3.1.8	Protein (Escherichia coli strain K12-MG1655 gene gsp) Glutathionylspermidine synthase	Vol. 2, Page 386
2.1.1.77	Protein (L-isoaspartate) O-methyltransferase Protein-L-isoaspartate(D-aspartate) O-methyltransferase	

EC-No.	Synonym Recommended Name	Vol., Page
2.1.1.43	Protein (lysine) methyltransferase Histone-lysine N-methyltransferase	
1.8.4.6	Protein (methionine sulfoxide) reductase Protein-methionine-S-oxide reductase	
2.1.1.133	Protein (Pseudomonas denitrificans clone pXL151 gene cobF reduced) Precorrin-4 C^{11}-methyltransferase	
2.1.1.114	Protein (Saccharomyces cerevisiae chromosome XV ORF HRF316) Hexaprenyldihydroxybenzoate methyltransferase	
5.4.99.16	Protein (Saccharomyces cerevisiae clone pMB14 gene CIF reduced) Maltose α-D-glucosyltransferase	Vol. 1, Page 656
5.4.99.16	Protein (Saccharomyces cerevisiae gene CIF1 reduced) Maltose α-D-glucosyltransferase	Vol. 1, Page 656
2.1.1.130	Protein (Salmonella typhimurium clone pJO26 gene cbiL reduced) Precorrin-2 C^{20}-methyltransferase	
2.1.1.132	Protein (Salmonella typhimurium clone pJO26 gene cbiT reduced) Precorrin-6Y $C^{5,15}$-methyltransferase (decarboxylating)	
2.1.1.132	Protein (Salmonella typhimurium clone pJO gene cbiE reduced) Precorrin-6Y $C^{5,15}$-methyltransferase (decarboxylating)	
1.14.99.30	Protein (Synechocystis strain PCC 6803 clone cs0223/cs0128/ps0014/ cs0681/cs0294 open reading frame slr0940 reduced) Carotene 7,8-desaturase	
2.7.1.69	Protein, specific or class, gene bglC Protein-N^{π}-phosphohistidine-sugar phosphotransferase	
2.1.1.77	Protein-β-aspartate O-methyltransferase Protein-L-isoaspartate(D-aspartate) O-methyltransferase	
3.2.1.107	Protein-α-D-glycosyl-1,2-β-D-galactosyl-L-hydroxylysine glucohydrolase Protein-glucosylgalactosylhydroxylysine glucosidase	
2.1.1.24	protein-γ-glutamate O-methyltransferase protein-γ-glutamate O-methyltransferase	
3.5.3.15	Protein-arginine deiminase Protein-arginine deiminase	
2.1.1.23	protein-arginine N-methyltransferase protein-arginine N-methyltransferase	
2.7.1.37	Protein-aspartyl kinase Protein kinase	

EC-No.	Synonym Recommended Name	Vol., Page
2.7.1.37	Protein-cysteine kinase Protein kinase	
1.8.4.2	Protein-disulfide interchange enzyme Protein-disulfide reductase (glutathione)	
1.8.4.2	Protein-disulfide isomerase/oxidoreductase Protein-disulfide reductase (glutathione)	
1.8.4.2	Protein-disulfide reductase (glutathione) Protein-disulfide reductase (glutathione)	
1.6.4.4	Protein-disulfide reductase (NAD(P)H) Protein-disulfide reductase (NAD(P)H)	
3.2.1.107	Protein-glucosylgalactosylhydroxylysine glucosidase Protein-glucosylgalactosylhydroxylysine glucosidase	
3.1.1.61	Protein-glutamate methylesterase Protein-glutamate methylesterase	
2.1.1.80	Protein-glutamate O-methyltransferase Protein-glutamate O-methyltransferase	
2.3.2.13	Protein-glutamine γ-glutamyltransferase Protein-glutamine γ-glutamyltransferase	
2.3.2.13	Protein-glutamine:amine γ-glutamyltransferase Protein-glutamine γ-glutamyltransferase	
3.5.1.44	Protein-glutamine glutaminase Protein-glutamine glutaminase	
2.1.1.85	Protein-histidine N-methyltransferase Protein-histidine N-methyltransferase	
2.7.3.11	Protein-histidine pros-kinase Protein-histidine pros-kinase	
2.7.3.12	Protein-histidine tele-kinase Protein-histidine tele-kinase	
3.5.3.15	Protein-L-arginine iminohydrolase Protein-arginine deiminase	
3.1.1.61	Protein-L-glutamate-O4-methyl-ester acylhydrolase Protein-glutamate methylesterase	
3.5.1.44	Protein-L-glutamine amidohydrolase Protein-glutamine glutaminase	
2.1.1.77	Protein-L-isoaspartate(D-aspartate) O-methyltransferase Protein-L-isoaspartate(D-aspartate) O-methyltransferase	

EC-No.	Synonym Recommended Name	Vol., Page
2.1.1.77	Protein-L-isoaspartate O-methyltransferase Protein-L-isoaspartate(D-aspartate) O-methyltransferase	
1.4.3.13	Protein-L-lysine:oxygen 6-oxidoreductase (deaminating) Protein-lysine 6-oxidase	
1.8.4.6	Protein-L-methionine:oxidized-thioredoxin S-oxidoreductase Protein-methionine-S-oxide reductase	
1.4.3.13	Protein-lysine 6-oxidase Protein-lysine 6-oxidase	
2.1.1.43	Protein-lysine N-methyltransferase Histone-lysine N-methyltransferase	
1.8.4.6	Protein-methionine-S-oxide reductase Protein-methionine-S-oxide reductase	
2.7.1.69	Protein-N^{π}pi-phosphohistidine-sugar phosphotransferase Protein-N^{π}-phosphohistidine-sugar phosphotransferase	
2.7.1.69	Protein-N^{π}-phosphohistidine:sugar N-pros-phosphotransferase Protein-N^{π}-phosphohistidine-sugar phosphotransferase	
2.7.7.59	[Protein-PII]uridylyltransferase [Protein-PII]uridylyltransferase	
2.1.1.100	Protein-S-isoprenylcysteine O-methyltransferase Protein-S-isoprenylcysteine O-methyltransferase	
2.7.1.37	Protein-serine/threonine kinase Protein kinase	
5.1.1.16	Protein-serine epimerase Protein-serine epimerase	Vol. 1, Page 66
2.7.1.37	Protein-serine kinase Protein kinase	
5.1.1.16	Protein-serine racemase Protein-serine epimerase	Vol. 1, Page 66
3.1.3.48	protein-tyrosine-phosphatase protein-tyrosine-phosphatase	
3.1.3.48	protein-tyrosine-phosphate phosphohydrolase protein-tyrosine-phosphatase	
2.7.1.112	Protein-tyrosine kinase Protein-tyrosine kinase	
2.8.2.20	Protein-tyrosine sulfotransferase Protein-tyrosine sulfotransferase	

EC-No.	Synonym Recommended Name	Vol., Page
2.4.1.41	Protein-UDP acetylgalactosaminyltransferase Polypeptide N-acetylgalactosaminyltransferase	
5.1.1.16	[Protein]-serine epimerase Protein-serine epimerase	Vol. 1, Page 66
3.4.21.87	Protein a Omptin	
3.4.22.30	Proteinase ω Caricain	
3.4.21.50	proteinase, Achromobacter lyticus alkaline I lysyl endopeptidase	
3.4.23.28	Proteinase, Acrocylindricum Acrocylindropepsin	
3.4.21.10	proteinase, acrosomal acrosin	
3.4.24.25	Proteinase, Aeromonas proteolytica neutral Aeromonolysin	
3.4.24.72	Proteinase, Agkistrodon contortrix contortrix venom metallo- Fibrolase	
3.4.23.18	Proteinase, Aspergillus acid Aspergillopepsin I	
3.4.21.63	Proteinase, Aspergillus alkaline Oryzin	
3.4.23.18	Proteinase, Aspergillus awamori acid Aspergillopepsin I	
3.4.21.63	Proteinase, Aspergillus flavus alkaline Oryzin	
3.4.23.18	Proteinase, Aspergillus kawachii aspartic Aspergillopepsin I	
3.4.21.63	Proteinase, Aspergillus melleus alkaline Oryzin	
3.4.21.63	Proteinase, Aspergillus oryzae alkaline Oryzin	
3.4.21.63	Proteinase, Aspergillus parasiticus alkaline Oryzin	
3.4.23.18	Proteinase, Aspergillus saitoi acid Aspergillopepsin I	

EC-No.	Synonym Recommended Name	Vol., Page
3.4.21.63	Proteinase, Aspergillus soya alkaline Oryzin	
3.4.21.63	Proteinase, Aspergillus sulphureus alkaline Oryzin	
3.4.21.63	Proteinase, Aspergillus sydowi alkaline Oryzin	
3.4.21.97	proteinase, assembly protein precursor-processing assemblin	
3.4.21.62	Proteinase, Bacillus subtilis alkaline Subtilisin	
3.4.24.27	Proteinase, Bacillus thermoproteolyticus neutral Thermolysin	
3.4.21.55	Proteinase, Bitis gabonica venom serine Venombin AB	
3.4.24.49	Proteinase, Bothrops jararaca venom metallo- Bothropasin	
3.4.22.17	proteinase, calcium activated neutral calpain	
3.4.23.24	Proteinase, Candida albicans aspartic Candidapepsin	
3.4.23.24	Proteinase, Candida aspartic Candidapepsin	
3.4.23.24	Proteinase, Candida olea aspartic Candidapepsin	
3.4.21.79	Proteinase, CCP1 Granzyme B	
3.4.21.70	Proteinase, cholesterol-binding serine Pancreatic endopeptidase E	
3.4.24.12	proteinase, chorion-digesting envelysin	
3.4.23.26	Proteinase, Cladosporium aspartic Rhodotorulapepsin	
3.4.21.32	proteinase, Clostridium histolyticum, A brachyurin	
3.4.24.3	proteinase, Clostridium histolyticum, A microbial collagenase	

EC-No.	Synonym Recommended Name	Vol., Page
3.4.22.8	proteinase, Clostridium histolyticum, B clostripain	
3.4.24.46	Proteinase, Crotalus adamanteus venom, II Adamalysin	
3.4.24.1	proteinase, Crotalus atrox atrolysin A	
3.4.24.47	Proteinase, Crotalus horridus horridus venom hemorrhagic Horrilysin	
3.4.23.27	Proteinase, Dictyostelium discoideum aspartic Physaropepsin	
3.4.23.27	Proteinase, Dictyostelium discoideum aspartic, E Physaropepsin	
3.4.21.70	Proteinase, E Pancreatic endopeptidase E	
3.4.23.22	Proteinase, Endothia aspartic Endothiapepsin	
3.4.22.35	Proteinase, Entamoeba histolytica cysteine Histolysain	
3.4.21.83	Proteinase, Escherichia coli alkaline, II Oligopeptidase B	
3.4.24.55	Proteinase, Escherichia coli metallo-, Pi Pitrilysin	
3.4.21.53	Proteinase, Escherichia coli serine, La Endopeptidase La	
3.4.21.67	Proteinase, Escherichia coli serine, So Endopeptidase So	
3.4.21.89	Proteinase, eukaryotic signal Signal peptidase I	
3.4.22.28	Proteinase, foot-and-mouth-disease virus, 3C Picornain 3C	
3.4.23.32	Proteinase, Ganoderma lucidum aspartic Scytalidopepsin B	
3.4.21.19	proteinase, glutamate-specific glutamyl endopeptidase	
3.4.22.25	Proteinase, glycine-specific Glycyl endopeptidase	

EC-No.	Synonym Recommended Name	Vol., Page
3.4.24.57	Proteinase, glyco- O-Sialoglycoprotein endopeptidase	
3.4.21.72	Proteinase, immunoglobulin A IgA-specific serine endopeptidase	
3.4.24.13	proteinase, immunoglobulin A1 IgA-specific metalloendopeptidase	
3.4.22.36	Proteinase, interleukin 1β precursor Interleukin 1β converting enzyme	
3.4.23.29	Proteinase, Irpex lacteus aspartic Polyporopepsin	
3.4.21.53	Proteinase, La Endopeptidase La	
3.4.21.57	Proteinase, leucine-specific Leucyl endopeptidase	
3.4.21.50	proteinase, lysine specific lysyl endopeptidase	
3.4.24.32	Proteinase, β lytic metallo- β-Lytic metalloendopeptidase	
3.4.21.65	Proteinase, Malbranchea pulchella sulfurea extracellular Thermomycolin	
3.4.21.59	Proteinase, mast cell neutral Tryptase	
3.4.21.39	proteinase, mast cell serine, chymase chymase	
3.4.21.59	Proteinase, mast cell serine, II Tryptase	
3.4.21.59	Proteinase, mast cell serine, tryptase Tryptase	
3.4.24.23	Proteinase, metallo-, pump-1 Matrilysin	
3.4.21.3	proteinase, Metridium A metridin	
3.4.24.59	Proteinase, mitochondrial intermediate precursor-processing Mitochondrial intermediate peptidase	
3.4.24.64	Proteinase, mitochondrial protein precursor-processing Mitochondrial processing peptidase	

EC-No.	Synonym Recommended Name	Vol., Page
3.4.23.23	Proteinase, Mucor aspartic Mucorpepsin	
3.4.21.12	proteinase, Mycobacterium sorangium α-lytic α-lytic endopeptidase	
3.4.21.12	proteinase, Myxobacter α-lytic α-lytic endopeptidase	
3.4.24.32	Proteinase, Myxobacterium sorangium β-lytic β-Lytic metalloendopeptidase	
3.4.23.12	proteinase, nepenthes acid nepenthesin	
3.4.21.60	Proteinase, Oxyuranus scutellatus prothrombin-activating Scutelarin	
3.4.23.26	Proteinase, Paecilomyces Rhodotorulapepsin	
3.4.21.70	Proteinase, pancreas E Pancreatic endopeptidase E	
3.4.22.30	Proteinase, papaya, III Caricain	
3.4.22.30	Proteinase, papaya A Caricain	
3.4.23.20	Proteinase, Penicillium aspartic Penicillopepsin	
3.4.23.20	Proteinase, Penicillium caseicolum aspartic Penicillopepsin	
3.4.23.20	Proteinase, Penicillium citrinum aspartic Penicillopepsin	
3.4.23.20	Proteinase, Penicillium cyclopium acid Penicillopepsin	
3.4.23.20	Proteinase, Penicillium duponti aspartic Penicillopepsin	
3.4.23.20	Proteinase, Penicillium expansum aspartic Penicillopepsin	
3.4.23.20	Proteinase, Penicillium janthinellum acid Penicillopepsin	
3.4.23.20	Proteinase, Penicillium roqueforti aspartic Penicillopepsin	

EC-No.	Synonym Recommended Name	Vol., Page
3.4.24.39	Proteinase, Penicillium roqueforti metallo- Deuterolysin	
3.4.24.33	Proteinase, peptidyl-Asp metallo- Peptidyl-Asp metalloendopeptidase	
3.4.24.20	Proteinase, peptidyllysine metallo- Peptidyl-Lys metalloendopeptidase	
3.4.23.27	Proteinase, Physarum flavicomum aspartic Physaropepsin	
3.4.23.27	Proteinase, Physarum polycephalum acid Physaropepsin	
3.4.22.28	Proteinase, picornavirus, 3C Picornain 3C	
3.4.22.29	Proteinase, poliovirus, 2A Picornain 2A	
3.4.22.28	Proteinase, poliovirus, 3C Picornain 3C	
3.4.21.98	proteinase, polyprotein-processing, NS3 hepacivirin	
3.4.23.36	Proteinase, premurein leader Signal peptidase II	
3.4.24.19	Proteinase, procollagen C-terminal Procollagen C-endopeptidase	
3.4.21.61	Proteinase, prohormone-processing Kexin	
3.4.23.17	Proteinase, proopiomelanocortin Pro-opiomelanocortin converting enzyme	
3.4.24.62	Proteinase, prooxyphysin Magnolysin	
3.4.21.50	proteinase, Pseudomonas lyticus alkaline , I lysyl endopeptidase	
3.4.24.23	Proteinase, PUMP-1 Matrilysin	
3.4.23.30	Proteinase, Pycnoporus coccineus aspartic Pycnoporopepsin	
3.4.22.28	Proteinase, rhinovirus, 3C Picornain 3C	

EC-No.	Synonym Recommended Name	Vol., Page
3.4.23.21	Proteinase, Rhizopus acid Rhizopuspepsin	
3.4.23.26	Proteinase, Rhodotorula glutinis aspartic Rhodotorulapepsin	
3.4.23.26	Proteinase, Rhodotorula glutinis aspartic, II Rhodotorulapepsin	
3.4.24.37	Proteinase, Saccharomyces cerevisiae, yscD Saccharolysin	
3.4.23.31	Proteinase, Scytalidium lignicolum aspartic Scytalidopepsin A	
3.4.23.31	Proteinase, Scytalidium lignicolum aspartic, A-2 Scytalidopepsin A	
3.4.23.31	Proteinase, Scytalidium lignicolum aspartic, A-I Scytalidopepsin A	
3.4.23.32	Proteinase, Scytalidium lignicolum aspartic, B Scytalidopepsin B	
3.4.23.31	Proteinase, Scytalidium lignicolum aspartic, C- Scytalidopepsin A	
3.4.21.3	proteinase, sea anemone metridin	
3.4.24.40	Proteinase, Serratia marcescens metallo- Serralysin	
3.4.21.89	Proteinase, signal Signal peptidase I	
3.4.21.57	Proteinase, spinach serine (leucine specific) Leucyl endopeptidase	
3.4.21.19	proteinase, staphylococcal serine glutamyl endopeptidase	
3.4.24.29	Proteinase, Staphylococcus aureus neutral Aureolysin	
3.4.22.10	proteinase, streptococcal streptopain	
3.4.21.80	Proteinase, Streptomyces griseus serine, A Streptogrisin A	
3.4.21.81	Proteinase, Streptomyces griseus serine B Streptogrisin B	

EC-No.	Synonym Recommended Name	Vol., Page
3.4.21.66	Proteinase, Thermoactinomyces vulgaris serine Thermitase	
3.4.23.30	Proteinase, Trametes acid Pycnoporopepsin	
3.4.21.64	Proteinase, Tritirachium album serine Endopeptidase K	
3.4.24.58	Proteinase, Vipera russelli Russellysin	
3.4.23.33	Proteinase, Xanthomonas aspartic Xanthomonapepsin	
3.4.23.25	Proteinase, yeast A Saccharopepsin	
3.4.21.48	proteinase, yeast B cerevisin	
3.4.21.76	Proteinase-3 Myeloblastin	
3.4.22.29	Proteinase 2Apro Picornain 2A	
3.4.23.19	Proteinase A Aspergillopepsin II	
3.4.23.25	Proteinase A Saccharopepsin	
3.4.21.80	Proteinase A Streptogrisin A	
3.4.22.14	proteinase A_2 of Actinidia chinensis actinidain	
3.4.23.18	Proteinase B Aspergillopepsin I	
3.4.22.34	Proteinase B Legumain	
3.4.24.53	Proteinase H2 Trimerelysin II	
3.4.23.30	Proteinase Ia Pycnoporopepsin	
3.4.24.46	Proteinase I and II Adamalysin	

EC-No.	Synonym Recommended Name	Vol., Page
3.4.21.64	Proteinase K Endopeptidase K	
3.4.21.61	Proteinase Kex2p Kexin	
3.4.21.53	Proteinase La Endopeptidase La	
3.4.21.98	proteinase NS3 hepacivirin	
3.4.24.55	Proteinase Pi Pitrilysin	
3.4.21.76	Proteinase PR-3 Myeloblastin	
3.4.21.67	Proteinase So Endopeptidase So	
3.4.23.25	Proteinase yscA Saccharopepsin	
3.4.24.37	Proteinase yscD Saccharolysin	
3.4.21.61	Proteinase yscF Kexin	
3.4.11.1	proteinates FTBL leucyl aminopeptidase	
2.7.4.15	Protein bound thiamin diphosphate:ATP phosphoryltransferase Thiamine-diphosphate kinase	
3.4.21.69	Protein C (activated) Protein C (activated)	
3.4.21.69	Protein Ca Protein C (activated)	
2.1.1.80	Protein carboxyl-methylase Protein-glutamate O-methyltransferase	
2.1.1.80	Protein carboxyl-O-methyltransferase Protein-glutamate O-methyltransferase	
3.1.1.61	Protein carboxyl methylesterase Protein-glutamate methylesterase	
2.1.1.80	Protein carboxylmethyltransferase II Protein-glutamate O-methyltransferase	

EC-No.	Synonym Recommended Name	Vol., Page
2.1.1.80	Protein carboxymethylase Protein-glutamate O-methyltransferase	
2.1.1.80	Protein carboxymethyltransferase Protein-glutamate O-methyltransferase	
2.1.1.77	Protein D-aspartate methyltransferase Protein-L-isoaspartate(D-aspartate) O-methyltransferase	
5.3.4.1	Protein disulfide-isomerase Protein disulfide-isomerase	Vol. 1, Page 436
5.3.4.1	Protein disulfide-isomerase Protein disulfide-isomerase	Vol. 1, Page 436
1.6.4.4	Protein disulfide reductase Protein-disulfide reductase (NAD(P)H)	
1.8.4.2	Protein disulfide reductase (glutathione) Protein-disulfide reductase (glutathione)	
1.8.4.2	Protein disulfide transhydrogenase Protein-disulfide reductase (glutathione)	
5.3.4.1	Protein disulphide isomerase Protein disulfide-isomerase	Vol. 1, Page 436
3.1.3.16	Protein D phosphatase Phosphoprotein phosphatase	
1.1.1.249	[protein eIF-5a]-deoxyhypusine synthase [protein eIF-5a]-deoxyhypusine synthase	
2.7.1.37	Protein glutamyl kinase Protein kinase	
2.7.3.11	Protein histidine kinase Protein-histidine pros-kinase	
2.7.3.12	Protein histidine kinase Protein-histidine tele-kinase	
2.7.1.37	Protein kinase Protein kinase	
2.7.3.11	Protein kinase (histidine) Protein-histidine pros-kinase	
2.7.3.12	Protein kinase (histidine) Protein-histidine tele-kinase	
2.7.1.37	Protein kinase (phosphorylating) Protein kinase	

EC-No.	Synonym Recommended Name	Vol., Page
2.7.1.112	Protein kinase (tyrosine-phosphorylating) Protein-tyrosine kinase	
2.7.1.37	Protein kinase-C Protein kinase	
2.7.1.37	Protein kinase A Protein kinase	
2.7.1.37	Protein kinase C Protein kinase	
2.7.1.37	Protein kinase CK2 Protein kinase	
2.7.1.112	Protein kinase Lck Protein-tyrosine kinase	
2.7.1.112	Protein kinase p56-LCK Protein-tyrosine kinase	
2.7.1.112	Protein kinase p56lck Protein-tyrosine kinase	
2.7.1.37	Protein kinase p58 Protein kinase	
2.1.1.77	Protein L-isoaspartate methyltransferase Protein-L-isoaspartate(D-aspartate) O-methyltransferase	
2.1.1.77	Protein L-isoaspartyl methyltransferase Protein-L-isoaspartate((D-aspartate) O-methyltransferase	
2.7.7.50	Protein lambda2 mRNA guanylyltransferase	
2.1.1.125	Protein methylase I (Histone)-arginine N-methyltransferase	
2.1.1.125	Protein methylase I (Histone)-arginine N-methyltransferase	
2.1.1.124	Protein methylase I [Cytochrome-c]-arginine N-methyltransferase	
2.1.1.126	Protein methylase I [Myelin basic protein]-arginine N-methyltransferase	
2.1.1.126	Protein methylase I [Myelin basic protein]-arginine N-methyltransferase	
2.1.1.80	Protein methylase II Protein-glutamate O-methyltransferase	

EC-No.	Synonym Recommended Name	Vol., Page
2.1.1.43	Protein methylase III Histone-lysine N-methyltransferase	
2.1.1.85	Protein methylase IV Protein-histidine N-methyltransferase	
3.1.1.61	Protein methylesterase Protein-glutamate methylesterase	
2.1.1.125	Protein methyltransferase I (Histone)-arginine N-methyltransferase	
2.1.1.124	Protein methyltransferase I [Cytochrome-c]-arginine N-methyltransferase	
2.1.1.126	Protein methyltransferase I [Myelin basic protein]-arginine N-methyltransferase	
2.1.1.43	Protein methyltransferase II Histone-lysine N-methyltransferase	
2.1.1.80	Protein methyltransferase II Protein-glutamate O-methyltransferase	
2.4.1.94	Protein N-acetylglucosaminyltransferase Protein N-acetylglucosaminyltransferase	
2.3.1.97	Protein N-myristoyltransferase Glycylpeptide N-tetradecanoyltransferase	
2.3.1.88	Protein N-terminal acetyltransferase Peptide α-N-acetyltransferase	
2.4.1.109	Protein O-D-mannosyltransferase Dolichyl-phosphate-mannose-protein mannosyltransferase	
2.1.1.80	Protein O-methyltransferase Protein-glutamate O-methyltransferase	
1.4.4.2	Protein P1 Glycine dehydrogenase (decarboxylating)	
2.7.1.112	Protein p56c-lck kinase Protein-tyrosine kinase	
2.7.1.112	Protein p56lck tyrosine kinase Protein-tyrosine kinase	
4.1.1.49	Protein p60 Phosphoenolpyruvate carboxykinase (ATP)	
3.1.3.16	Protein phosphatase Phosphoprotein phosphatase	

EC-No.	Synonym Recommended Name	Vol., Page
3.1.3.53	Protein phosphatase 2A Myosin-light-chain-phosphatase	
3.1.3.17	protein phosphatase C [phosphorylase] phosphatase	
2.7.1.37	Protein phosphokinase Protein kinase	
3.1.3.48	protein phosphotyrosine phosphatase protein-tyrosine-phosphatase	
5.2.1.8	Proteins, cyclophilins Peptidylprolyl isomerase	Vol. 1, Page 218
1.5.1.6	Proteins, folate-binding cytosol I Formyltetrahydrofolate dehydrogenase	
3.4.21.53	Proteins, gene lon Endopeptidase La	
2.7.9.3	Proteins, gene selD selenide, water dikinase	
2.7.9.3	Proteins, gene selD (specific proteins and subclasses) selenide, water dikinase	
5.99.1.3	Proteins, NP170 (specific proteins and subclasses nuclear protein, 170,000-mol.-wt.) DNA topoisomerase (ATP-hydrolysing)	Vol. 1, Page 737
2.7.7.48	Proteins, PB 2 RNA-directed RNA polymerase	
1.1.1.27	Proteins, specific or class, anoxic stress response, p34 L-Lactate dehydrogenase	
5.2.1.8	Proteins, specific or class, cyclophilins Peptidylprolyl isomerase	Vol. 1, Page 218
3.4.11.1	proteins, specific or class, FTBL leucyl aminopeptidase	
3.4.21.53	Proteins, specific or class, gene lon Endopeptidase La	
1.6.99.3	Proteins, specific or class, gene MURF3 NADH dehydrogenase	
3.4.21.87	Proteins, specific or class, gene ompT Omptin	

EC-No.	Synonym Recommended Name	Vol., Page
2.7.7.48	Proteins, specific or class, lambda3, of reovirus RNA-directed RNA polymerase	
1.14.99.3	Proteins, specific or class, ORP33 (oxygen-regulated protein 33,000-mol.-wt.) Heme oxygenase (decyclizing)	
2.7.7.48	Proteins, specific or class, PB 1 RNA-directed RNA polymerase	
2.7.7.48	Proteins, specific or class, PB 2 RNA-directed RNA polymerase	
2.1.1.100	Protein S-farnesylcysteine C-terminal methyltransferase Protein-S-isoprenylcysteine O-methyltransferase	
2.7.1.37	Protein serine-threonine kinase Protein kinase	
2.7.1.37	Protein serine kinase Protein kinase	
2.7.7.48	Proteins PB1 RNA-directed RNA polymerase	
2.7.1.135	Protein tau kinase τ-protein kinase	
2.7.1.112	Protein tyrosine kinase Protein-tyrosine kinase	
2.7.1.112	Protein tyrosine kinase lck Protein-tyrosine kinase	
2.7.1.112	Protein tyrosine kinase p56lck Protein-tyrosine kinase	
2.7.1.112	Protein tyrosine kinase pp56lck Protein-tyrosine kinase	
3.1.3.48	protein tyrosine phosphatase xi protein-tyrosine-phosphatase	
2.4.2.26	Protein xylosyltransferase Protein xylosyltransferase	
3.4.24.17	Proteoglycanase Stromelysin 1	
3.4.24.22	Proteoglycanase 2 Stromelysin 2	

EC-No.	Synonym Recommended Name	Vol., Page
3.1.27.2	Proteus mirabilis RNase Bacillus subtilis ribonuclease	
3.4.21.6	prothrombase coagulation factor Xa	
3.4.21.6	prothrombinase coagulation factor Xa	
3.4.21.62	Protin A 3L Subtilisin	
4.2.1.73	Protoaphin-aglucone dehydratase (cyclizing) Protoaphin-aglucone dehydratase (cyclizing)	
4.2.1.73	Protoaphin-aglucone hydro-lyase (cyclizing) Protoaphin-aglucone dehydratase (cyclizing)	
4.2.1.73	Protoaphin dehydratase Protoaphin-aglucone dehydratase (cyclizing)	
4.2.1.73	Protoaphin dehydratase (cyclizing) Protoaphin-aglucone dehydratase (cyclizing)	
4.2.1.73	Protoaphin dehydratase (cyclizing) Protoaphin-aglucone dehydratase (cyclizing)	
1.13.11.3	Protocatechuate 3,4-dioxygenase Protocatechuate 3,4-dioxygenase	
1.13.11.8	Protocatechuate 4,5-dioxygenase Protocatechuate 4,5-dioxygenase	
1.13.11.8	Protocatechuate 4,5-oxygenase Protocatechuate 4,5-dioxygenase	
1.13.11.3	Protocatechuate:oxygen 3,4-oxidoreductase (decyclizing) Protocatechuate 3,4-dioxygenase	
1.13.11.8	Protocatechuate:oxygen 4,5-oxidoreductase (decyclizing) Protocatechuate 4,5-dioxygenase	
4.1.1.63	Protocatechuate carboxy-lyase Protocatechuate decarboxylase	
4.1.1.63	Protocatechuate decarboxylase Protocatechuate decarboxylase	
1.13.11.3	Protocatechuate oxygenase Protocatechuate 3,4-dioxygenase	
1.13.11.3	Protocatechuic 3,4-dioxygenase Protocatechuate 3,4-dioxygenase	

EC-No.	Synonym Recommended Name	Vol., Page
1.13.11.3	Protocatechuic 3,4-oxygenase Protocatechuate 3,4-dioxygenase	
1.13.11.8	Protocatechuic 4,5-dioxygenase Protocatechuate 4,5-dioxygenase	
1.13.11.8	Protocatechuic 4,5-oxygenase Protocatechuate 4,5-dioxygenase	
1.13.11.3	Protocatechuic acid oxidase Protocatechuate 3,4-dioxygenase	
1.3.1.33	Protochlorophyllide oxidoreductase Protochlorophyllide reductase	
1.3.1.33	Protochlorophyllide photooxidoreductase Protochlorophyllide reductase	
1.3.1.33	Protochlorophyllide reductase Protochlorophyllide reductase	
1.14.11.2	Protocollagen hydroxylase Procollagen-proline dioxygenase	
1.14.11.4	Protocollagen lysine hydroxylase Procollagen-lysine 5-dioxygenase	
1.14.11.4	Protocollagen lysyl hydroxylase Procollagen-lysine 5-dioxygenase	
1.14.11.7	Protocollagen proline 3-hydroxylase Procollagen-proline 3-dioxygenase	
1.14.11.2	Protocollagen proline 4-hydroxylase Procollagen-proline dioxygenase	
1.14.11.2	Protocollagen proline dioxygenase Procollagen-proline dioxygenase	
1.14.11.2	Protocollagen proline hydroxylase Procollagen-proline dioxygenase	
1.14.11.2	Protocollagen prolyl hydroxylase Procollagen-proline dioxygenase	
4.99.1.1	protoheme ferro-lyase ferrochelatase	
1.11.1.7	Protoheme peroxidase Peroxidase	
3.6.1.35	Proton-translocating ATPase H^+-transporting ATPase	

EC-No.	Synonym Recommended Name	Vol., Page
1.14.13.55	Protopine, $NADPH_2$:oxygen oxidoreductase (6-hydroxylating) Protopine 6-monooxygenase	
1.14.13.55	Protopine-6-hydroxylase Protopine 6-monooxygenase	
1.14.13.55	Protopine 6-monooxygenase Protopine 6-monooxygenase	
1.3.3.4	Protoporphyrinogen-IX:oxygen oxidoreductase Protoporphyrinogen oxidase	
1.3.3.4	Protoporphyrinogenase Protoporphyrinogen oxidase	
1.3.3.4	Protoporphyrinogen IX oxidase Protoporphyrinogen oxidase	
1.3.3.4	Protoporphyrinogen oxidase Protoporphyrinogen oxidase	
3.4.21.63	Prozyme Oryzin	
3.4.21.63	Prozyme 10 Oryzin	
2.7.6.1	PRPP synthase Ribose-phosphate pyrophosphokinase	
2.7.6.1	PRPP synthetase Ribose-phosphate pyrophosphokinase	
2.4.2.18	PRT Anthranilate phosphoribosyltransferase	
3.4.21.96	PrtP Lactocepin	
2.7.1.19	PRuK Phosphoribulokinase	
3.2.1.118	Prunasin β-D-glucohydrolase Prunasin β-glucosidase	
3.2.1.118	Prunasin β-glucosidase Prunasin β-glucosidase	
3.2.1.118	Prunasin hydrolase Prunasin β-glucosidase	
3.1.3.43	pryruvate dehydrogenase phosphate phosphatase [pyruvate dehydrogenase (lipoamide)]-phosphatase	

EC-No.	Synonym Recommended Name	Vol., Page
2.6.1.52	PSAT Phosphoserine transaminase	
4.1.1.65	PSD Phosphatidylserine decarboxylase	
4.1.1.65	PS decarboxylase Phosphatidylserine decarboxylase	
3.5.1.13	Pseudocholinesterase Arylacylamidase	
3.1.1.8	pseudocholinesterase cholinesterase	
3.4.24.26	pseudolysin pseudolysin	
3.4.23.37	Pseudomonapepsin Pseudomonapepsin	
3.4.24.40	Pseudomonas aeruginosa alk. protease Serralysin	
3.4.24.40	Pseudomonas aeruginosa alkaline proteinase Serralysin	
3.4.21.50	Pseudomonas aeruginosa lysine-specific protease lysyl endopeptidase	
3.4.24.26	Pseudomonas aeruginosa neutral metalloproteinase pseudolysin	
3.4.23.33	Pseudomonas carboxyl proteinase Xanthomonapepsin	
1.9.3.2	Pseudomonas cytochrome oxidase Pseudomonas cytochrome oxidase	
3.4.24.26	Pseudomonas elastase pseudolysin	
3.5.1.86	Pseudomonas mandelamide hydrolase mandelamide amidase	
3.4.23.37	Pseudomonas sp. pepstatin-insensitive carboxyl proteinase Pseudomonapepsin	
3.4.99.39	pseudomurein endopeptidase pseudomurein endopeptidase	
1.1.1.236	Pseudotropine:NADP$^+$ 3-oxidoreductase Tropinone reductase	

EC-No.	Synonym Recommended Name	Vol., Page
1.1.1.236	Pseudotropine forming tropinone reductase Tropinone reductase	
3.4.21.4	pseudotrypsin trypsin	
2.7.1.83	Pseudouridine kinase Pseudouridine kinase	
4.2.1.70	pseudouridine monophosphate synthetase pseudouridylate synthase	
5.4.99.12	Pseudouridine synthase tRNA-pseudouridine synthase I	Vol. 1, Page 642
4.2.1.70	pseudouridine synthase pseudouridylate synthase	
4.2.1.70	pseudouridine synthase I pseudouridylate synthase	
4.2.1.70	pseudouridylate synthase pseudouridylate synthase	
4.2.1.70	pseudouridylate synthetase pseudouridylate synthase	
4.2.1.70	pseudouridylate synthetase I pseudouridylate synthase	
4.2.1.70	pseudouridylate synthetase II pseudouridylate synthase	
4.2.1.70	pseudouridylic acid synthetase pseudouridylate synthase	
3.4.21.10	ψ-acrosin acrosin	
4.2.1.70	ψMP synthetase pseudouridylate synthase	
4.2.1.70	ψsynthetase pseudouridylate synthase	
4.2.1.70	ψUMP synthetase pseudouridylate synthase	
3.4.17.21	PSMA Glutamate carboxypeptidase II	
3.4.17.21	PSM antigen, Glutamate carboxypeptidase II	

EC-No.	Synonym Recommended Name	Vol., Page
3.1.3.3	PSP phosphoserine phosphatase	
3.1.3.3	PSPase phosphoserine phosphatase	
2.7.8.8	PS synthase CDPdiacylglycerol-serine O-phosphatidyltransferase	
2.8.2.1	PST Aryl sulfotransferase	
4.2.1.70	PSUI pseudouridylate synthase	
2.4.1.23	Psychosine-UDP galactosyltransferase Sphingosine β-galactosyltransferase	
2.4.1.23	Psychosine-uridine diphosphate galactosyltransferase Sphingosine β-galactosyltransferase	
2.8.2.13	Psychosine sulfotransferase Psychosine sulfotransferase	
2.3.1.8	PTA Phosphate acetyltransferase	
3.4.17.20	pTAFI Carboxypeptidase U	
2.1.3.6	PTCase Putrescine carbamoyltransferase	
3.1.3.36	PtdIns(4,5)P2phosphatase Phosphatidylinositol-bisphosphatase	
2.7.1.137	PtdIns 3'-kinase 1-Phosphatidylinositol 3-kinase	
3.5.4.11	Pteridin deaminase Pteridin deaminase	
1.17.3.1	Pteridine oxidase Pteridine oxidase	
1.1.1.253	Pteridine reductase Pteridine reductase	
1.1.1.253	Pteridine reductase 1 Pteridine reductase	
1.5.1.3	Pteridine reductase:dihydrofolate reductase Dihydrofolate reductase	

EC-No.	Synonym Recommended Name Vol., Page
4.2.1.96	Pterin-4-α-carbinolamine dehydratase 4a-Hydroxytetrahydrobiopterin dehydratase
4.2.1.96	Pterin-4 α-carbinolamine dehydratase 4a-Hydroxytetrahydrobiopterin dehydratase
4.2.1.96	Pterin-4α-carbinolamine dehydratase 4a-Hydroxytetrahydrobiopterin dehydratase
4.2.1.96	Pterin-4α-carbinolamine dehydratase (PCD)/dimerization cofactor for the transcription factor HBF-1α 4a-Hydroxytetrahydrobiopterin dehydratase
4.2.1.96	Pterin-4a-carbinolamine dehydratase 4a-Hydroxytetrahydrobiopterin dehydratase
4.2.1.96	Pterin-4a-carbinolamine dehydratase (Aquifex aeolicus gene phhB) 4a-Hydroxytetrahydrobiopterin dehydratase
4.2.1.96	Pterin-4a-carbinolamine dehydratase/dimerization cofactor of HNF1 4a-Hydroxytetrahydrobiopterin dehydratase
1.1.1.246	Pterocarpan synthase Pterocarpin synthase
1.1.1.246	Pterocarpin synthase Pterocarpin synthase
3.4.17.11	N-Pteroyl-L-glutamate hydrolase Glutamate carboxypeptidase
3.4.19.9	Pteroyl-poly-γ-glutamate hydrolase γ-Glu-X carboxypeptidase
3.4.19.9	Pteroylpoly-γ-glutamyl hydrolase γ-Glu-X carboxypeptidase
3.1.1.29	PTH aminoacyl-tRNA hydrolase
3.1.1.29	Pth aminoacyl-tRNA hydrolase
3.1.3.48	PTPγ protein-tyrosine-phosphatase
3.1.3.48	PTP-phosphatase protein-tyrosine-phosphatase
3.1.3.48	PTP-SH2β protein-tyrosine-phosphatase

EC-No.	Synonym Recommended Name	Vol., Page
3.1.3.48	PTP1C protein-tyrosine-phosphatase	
3.1.3.48	PTP1D protein-tyrosine-phosphatase	
3.1.3.48	PTP2C protein-tyrosine-phosphatase	
3.1.3.48	PTP36 protein-tyrosine-phosphatase	
3.1.3.48	PtpA protein-tyrosine-phosphatase	
3.1.3.48	PTPase protein-tyrosine-phosphatase	
3.1.3.48	PTPMEG protein-tyrosine-phosphatase	
3.1.3.48	PTPN6 protein-tyrosine-phosphatase	
3.1.3.48	PTPP protein-tyrosine-phosphatase	
3.1.3.48	PTPPase protein-tyrosine-phosphatase	
4.6.1.10	PTPS 6-Pyruvoyltetrahydtropterin synthase	
4.6.1.10	PTP synthase 6-Pyruvoyltetrahydtropterin synthase	
3.1.3.48	PTPxi/RPTPβ protein-tyrosine-phosphatase	
3.4.24.55	PTR Pitrilysin	
1.1.1.253	PTR1 Pteridine reductase	
2.7.1.69	PTS permease Protein-N$^{\pi}$-phosphohistidine-sugar phosphotransferase	
4.1.2.9	Pu5PPK Phosphoketolase	
3.4.21.89	PuIO prepilin peptidase Signal peptidase I	

EC-No.	Synonym Recommended Name	Vol., Page
3.2.1.135	Pullulan 4-D-glucanohydrolase (panose-forming) Neopullulanase	
3.2.1.57	Pullulan 4-glucanohydrolase Isopullulanase	
3.2.1.41	Pullulanase α-dextrin endo-1,6-α-glucosidase	
3.2.1.57	Pullulanase, iso- Isopullulanase	
3.2.1.135	Pullulanase, neo- Neopullulanase	
3.2.1.135	Pullulanase II Neopullulanase	
3.4.24.23	PUMP Matrilysin	
2.4.2.1	PUNPI Purine-nucleoside phosphorylase	
2.4.2.1	PUNPII Purine-nucleoside phosphorylase	
2.4.2.6	Purine(pyrimidine) nucleoside:purine(pyrimidine) deoxyribosyl transfer-ase Nucleoside deoxyribosyltransferase	
2.4.2.8	Purine-6-thiol phosphoribosyltransferase Hypoxanthine phosphoribosyltransferase	
2.7.1.76	Purine-deoxyribonucleoside kinase Deoxyadenosine kinase	
2.4.2.1	Purine-nucleoside:orthophosphate ribosyltransferase Purine-nucleoside phosphorylase	
2.4.2.1	Purine-nucleoside phosphorylase Purine-nucleoside phosphorylase	
3.1.27.4	Purine-specific ribonuclease Ribonuclease U_2	
3.1.27.4	Purine-specific RNase Ribonuclease U_2	
3.2.2.1	Purine.β.-ribosidase Purine nucleosidase	

EC-No.	Synonym Recommended Name	Vol., Page
2.4.2.1	Purine deoxynucleoside phosphorylase Purine-nucleoside phosphorylase	
2.4.2.1	Purine deoxyribonucleoside phosphorylase Purine-nucleoside phosphorylase	
4.3.2.4	Purine imidazole-ring cyclase Purine imidazole-ring cyclase	
3.2.2.1	Purine nucleosidase Purine nucleosidase	
3.2.2.1	Purine nucleoside hydrolase Purine nucleosidase	
2.7.1.143	Purine nucleoside kinase Diphosphate-purine nucleoside kinase	
2.4.2.1	Purine nucleoside phosphorylase Purine-nucleoside phosphorylase	
3.2.2.1	Purine ribonucleosidase Purine nucleosidase	
2.4.2.1	Purine ribonucleoside phosphorylase Purine-nucleoside phosphorylase	
3.1.27.4	Purine specific endoribonuclease Ribonuclease U_2	
3.4.11.14	puromycin-sensitive aminopeptidase aminopeptidase, human liver	
3.4.24.23	Putative (or punctuated) metalloproteinase-1 Matrilysin	
3.4.24.23	Putative metalloproteinase Matrilysin	
2.3.1.57	Putrescine (diamine)-acetylating enzyme Diamine N-acetyltransferase	
1.4.3.10	Putrescine:oxygen oxidoreductase (deaminating) Putrescine oxidase	
2.5.1.44	Putrescine:putrescine 4-aminobutyltransferase (ammonia-forming) Homospermidine synthase	
2.3.1.57	Putrescine acetylase Diamine N-acetyltransferase	
2.3.1.57	Putrescine acetyltransferase Diamine N-acetyltransferase	

EC-No.	Synonym Recommended Name	Vol., Page
2.5.1.16	Putrescine aminopropyltransferase Spermidine synthase	
2.1.3.6	Putrescine carbamoyltransferase Putrescine carbamoyltransferase	
2.3.1.138	Putrescine hydroxycinnamoyl transferase Putrescine N-hydroxycinnamoyltransferase	
2.3.1.138	Putrescine hydroxycinnamoyltransferase Putrescine N-hydroxycinnamoyltransferase	
2.1.1.53	Putrescine methyltransferase Putrescine N-methyltransferase	
2.3.1.57	Putrescine N-acetyltransferase Diamine N-acetyltransferase	
2.3.1.138	Putrescine N-hydroxycinnamoyltransferase Putrescine N-hydroxycinnamoyltransferase	
2.1.1.53	Putrescine N-methyltransferase Putrescine N-methyltransferase	
1.4.3.10	Putrescine oxidase Putrescine oxidase	
2.7.2.2	Putrescine synthase Carbamate kinase	
2.1.3.6	Putrescine synthase Putrescine carbamoyltransferase	
2.1.3.6	Putrescine transcarbamylase Putrescine carbamoyltransferase	
1.1.99.23	PVA dehydrogenase Polyvinyl-alcohol dehydrogenase (acceptor)	
1.1.3.30	PVA oxidase Polyvinyl-alcohol oxidase	
3.4.23.30	Pycnoporopepsin Pycnoporopepsin	
1.1.3.10	Pyranose-2-oxidase Pyranose oxidase	
1.1.3.10	Pyranose:oxygen 2-oxidoreductase Pyranose oxidase	
1.1.3.10	Pyranose oxidase Pyranose oxidase	

EC-No.	Synonym Recommended Name	Vol., Page
3.4.19.3	PYRase pyroglutamyl-peptidase I	
4.2.1.50	β-(Pyrazol-1-yl)-L-alanine /L-cysteine synthase Pyrazolylalanine synthase	
4.2.1.50	β-(1-Pyrazol-1-yl)-L-alanine synthase Pyrazolylalanine synthase	
4.2.1.50	Pyrazolealanine synthase Pyrazolylalanine synthase	
4.2.99.14	Pyrazolealanine synthase β-Pyrazolylalanine synthase (acetylserine)	
4.2.99.14	β-Pyrazolealanine synthase β-Pyrazolylalanine synthase (acetylserine)	
4.2.1.50	β-(1-Pyrazolyl)alanine synthase Pyrazolylalanine synthase	
4.2.99.14	β-(1-Pyrazolyl)alanine synthase β-Pyrazolylalanine synthase (acetylserine)	
4.2.99.14	Pyrazolylalaninase β-Pyrazolylalanine synthase (acetylserine)	
4.2.1.50	β-Pyrazolylalaninase Pyrazolylalanine synthase	
4.2.1.50	Pyrazolylalanine synthase Pyrazolylalanine synthase	
4.2.99.14	β-Pyrazolylalanine synthase (acetylserine) β-Pyrazolylalanine synthase (acetylserine)	
1.13.11.9	Pyridine-2,5-diol dioxygenase 2,5-Dihydroxypyridine 5,6-dioxygenase	
3.2.2.17	Pyridine dimer-DNA glycosylase Deoxyribodipyrimidine endonucleosidase	
2.1.1.87	Pyridine N-methyltransferase Pyridine N-methyltransferase	
1.1.99.5	Pyridine nucleotide-independent L-glycerol 3-phosphate dehydrogenase Glycerol-3-phosphate dehydrogenase	
1.6.1.2	Pyridine nucleotide transferase $NAD(P)^+$ transhydrogenase (AB-specific)	
1.6.1.1	Pyridine nucleotide transferase $NAD(P)^+$ transhydrogenase (B-specific)	

EC-No.	Synonym Recommended Name	Vol., Page
1.6.1.2	Pyridine nucleotide transhydrogenase $NAD(P)^+$ transhydrogenase (AB-specific)	
1.6.1.1	Pyridine nucleotide transhydrogenase $NAD(P)^+$ transhydrogenase (B-specific)	
1.1.99.9	Pyridoxal-5-dehydrogenase Pyridoxine 5-dehydrogenase	
1.1.1.107	Pyridoxal 4-dehydrogenase Pyridoxal 4-dehydrogenase	
2.7.1.35	Pyridoxal 5-phosphate-kinase Pyridoxal kinase	
1.1.1.107	Pyridoxal:NAD^+ 4-oxidoreductase Pyridoxal 4-dehydrogenase	
1.2.3.8	Pyridoxal:oxygen 4-oxidoreductase Pyridoxal oxidase	
1.1.1.107	Pyridoxal dehydrogenase Pyridoxal 4-dehydrogenase	
2.7.1.35	Pyridoxal kinase Pyridoxal kinase	
1.2.3.8	Pyridoxal oxidase Pyridoxal oxidase	
2.7.1.35	Pyridoxal phosphokinase Pyridoxal kinase	
2.6.1.54	Pyridoxamine-5'-phosphate:2-oxoglutarate aminotransferase (D-glutamate-forming) Pyridoxamine-phosphate transaminase	
1.4.3.5	Pyridoxamine-5'-phosphate:oxygen oxidoreductase (deaminating) Pyridoxamine-phosphate oxidase	
2.6.1.31	Pyridoxamine-oxalacetate aminotransferase Pyridoxamine-oxaloacetate transaminase	
2.6.1.31	Pyridoxamine-oxaloacetate transaminase Pyridoxamine-oxaloacetate transaminase	
1.4.3.5	Pyridoxamine-phosphate oxidase Pyridoxamine-phosphate oxidase	
2.6.1.54	Pyridoxamine-phosphate transaminase Pyridoxamine-phosphate transaminase	

EC-No.	Synonym Recommended Name	Vol., Page
2.6.1.30	Pyridoxamine-pyruvate aminotransferase Pyridoxamine-pyruvate transaminase	
2.6.1.30	Pyridoxamine-pyruvate transaminase Pyridoxamine-pyruvate transaminase	
2.6.1.30	Pyridoxamine-pyruvic transaminase Pyridoxamine-pyruvate transaminase	
2.6.1.54	Pyridoxamine 5'-phosphate-α-ketoglutarate transaminase Pyridoxamine-phosphate transaminase	
1.4.3.5	Pyridoxamine 5'-phosphate oxidase Pyridoxamine-phosphate oxidase	
2.6.1.54	Pyridoxamine 5'-phosphate transaminase Pyridoxamine-phosphate transaminase	
2.6.1.31	Pyridoxamine:oxaloacetate aminotransferase Pyridoxamine-oxaloacetate transaminase	
2.6.1.30	Pyridoxamine:pyruvate aminotransferase Pyridoxamine-pyruvate transaminase	
1.4.3.5	Pyridoxamine phosphate oxidase Pyridoxamine-phosphate oxidase	
1.4.3.5	Pyridoxaminephosphate oxidase (EC 1.4.3.5: deaminating) Pyridoxamine-phosphate oxidase	
1.14.12.5	5-Pyridoxate,NADPH:oxygen oxidoreductase (decyclizing) 5-Pyridoxate dioxygenase	
1.14.12.5	5-Pyridoxate dioxygenase 5-Pyridoxate dioxygenase	
1.14.12.5	5-Pyridoxate oxidase 5-Pyridoxate dioxygenase	
1.1.3.12	Pyridoxin 4-oxidase Pyridoxine 4-oxidase	
1.1.99.9	Pyridoxin 5-dehydrogenase Pyridoxine 5-dehydrogenase	
1.1.1.65	Pyridoxin dehydrogenase Pyridoxine 4-dehydrogenase	
1.4.3.5	Pyridoxine (pyridoxamine) 5'-phosphate oxidase Pyridoxamine-phosphate oxidase	
1.4.3.5	Pyridoxine (pyridoxamine) phosphate oxidase Pyridoxamine-phosphate oxidase	

EC-No.	Synonym Recommended Name	Vol., Page
1.1.1.65	Pyridoxine 4-dehydrogenase Pyridoxine 4-dehydrogenase	
1.1.3.12	Pyridoxine 4-oxidase Pyridoxine 4-oxidase	
1.1.99.9	Pyridoxine 5'-dehydrogenase Pyridoxine 5-dehydrogenase	
2.4.1.160	Pyridoxine 5'-O-β-D-glucosyltransferase Pyridoxine 5'-O-β-D-glucosyltransferase	
1.1.99.9	Pyridoxine 5-dehydrogenase Pyridoxine 5-dehydrogenase	
1.1.99.9	Pyridoxine:(acceptor) 5-oxidoreductase Pyridoxine 5-dehydrogenase	
1.1.1.65	Pyridoxine:NADP$^+$ 4-oxidoreductase Pyridoxine 4-dehydrogenase	
1.1.3.12	Pyridoxine:oxygen 4-oxidoreductase Pyridoxine 4-oxidase	
1.1.1.65	Pyridoxine dehydrogenase Pyridoxine 4-dehydrogenase	
1.1.99.9	Pyridoxine dehydrogenase Pyridoxine 5-dehydrogenase	
2.7.1.35	Pyridoxine kinase Pyridoxal kinase	
1.1.3.12	Pyridoxol 4-oxidase Pyridoxine 4-oxidase	
1.1.99.9	Pyridoxol 5-dehydrogenase Pyridoxine 5-dehydrogenase	
3.1.1.27	4-pyridoxolactonase 4-pyridoxolactonase	
3.1.1.27	4-pyridoxolactone lactonohydrolase 4-pyridoxolactonase	
1.1.1.65	Pyridoxol dehydrogenase Pyridoxine 4-dehydrogenase	
3.2.2.10	Pyrimidine-5'-nucleotide nucleosidase Pyrimidine-5'-nucleotide nucleosidase	
3.2.2.10	Pyrimidine-5'-nucleotide phosphoribo(deoxyribo)hydrolase Pyrimidine-5'-nucleotide nucleosidase	

EC-No.	Synonym Recommended Name	Vol., Page
1.14.11.10	Pyrimidine-deoxynucleoside 1'-dioxygenase Pyrimidine-deoxynucleoside 1'-dioxygenase	
1.14.11.3	Pyrimidine-deoxynucleoside 2'-dioxygenase Pyrimidine-deoxynucleoside 2'-dioxygenase	
2.4.2.2	Pyrimidine-nucleoside:orthophosphate α-D-ribosyltransferase Pyrimidine-nucleoside phosphorylase	
2.4.2.2	Pyrimidine-nucleoside phosphorylase Pyrimidine-nucleoside phosphorylase	
2.4.2.4	Pyrimidine deoxynucleoside phosphorylase Thymidine phosphorylase	
1.14.11.3	Pyrimidine deoxyribonucleoside 2'-hydroxylase Pyrimidine-deoxynucleoside 2'-dioxygenase	
3.2.2.17	Pyrimidine dimer DNA-glycosylase Deoxyribodipyrimidine endonucleosidase	
3.2.2.17	Pyrimidine dimer DNA glycosylase Deoxyribodipyrimidine endonucleosidase	
3.5.2.2	Pyrimidine hydrase Dihydropyrimidinase	
3.2.2.8	Pyrimidine nucleosidase Ribosylpyrimidine nucleosidase	
2.7.4.14	Pyrimidine nucleoside monophosphate kinase Cytidylate kinase	
3.2.2.10	Pyrimidine nucleotide N-ribosidase Pyrimidine-5'-nucleotide nucleosidase	
2.4.2.4	Pyrimidine phosphorylase Thymidine phosphorylase	
2.4.2.3	Pyrimidine phosphorylase Uridine phosphorylase	
1.3.1.1	Pyrimidine reductase Dihydrouracil dehydrogenase (NAD^+)	
2.7.1.48	Pyrimidine ribonucleoside kinase Uridine kinase	
2.5.1.2	Pyrimidine transferase Thiamine pyridinylase	

EC-No.	Synonym Recommended Name	Vol., Page
1.5.4.1	Pyrimidodiazepine:oxidized-glutathione oxidoreductase (ring-opening, cyclizing) Pyrimidodiazepine synthase	
1.5.4.1	Pyrimidodiazepine synthase Pyrimidodiazepine synthase	
3.5.4.20	Pyrithiamin deaminase Pyrithiamin deaminase	
3.5.4.20	Pyrithiamine deaminase Pyrithiamin deaminase	
1.13.11.1	1,2-Pyrocatechase Catechol 1,2-dioxygenase	
1.13.11.2	2,3-Pyrocatechase Catechol 2,3-dioxygenase	
1.13.11.1	Pyrocatechase Catechol 1,2-dioxygenase	
1.13.11.1	Pyrocatechol 1,2-dioxygenase Catechol 1,2-dioxygenase	
1.13.11.2	Pyrocatechol 2,3-dioxygenase Catechol 2,3-dioxygenase	
1.10.3.1	Pyrocatechol oxidase Catechol oxidase	
1.14.18.1	Pyrocatechol oxidase Monophenol monooxygenase	
1.11.1.7	Pyrocatechol peroxidase Peroxidase	
4.1.1.46	*o*-Pyrocatechuate decarboxylase *o*-Pyrocatechuate decarboxylase	
1.13.11.14	*o*-Pyrocatechuate oxygenase 2,3-Dihydroxybenzoate 3,4-dioxygenase	
4.1.1.46	*o*-Pyrocatechuic acid carboxy-lyase *o*-Pyrocatechuate decarboxylase	
1.13.11.35	Pyrogallol 1,2-dioxygenase Pyrogallol 1,2-oxygenase	
1.13.11.35	Pyrogallol 1,2-oxygenase Pyrogallol 1,2-oxygenase	

EC-No.	Synonym Recommended Name	Vol., Page
1.97.1.2	Pyrogallol hydroxyltransferase Pyrogallol hydroxyltransferase	
3.5.2.9	Pyroglutamase 5-Oxoprolinase (ATP-hydrolysing)	
3.5.2.9	Pyroglutamase (ATP-hydrolysing) 5-Oxoprolinase (ATP-hydrolysing)	
3.4.19.3	pyroglutamate aminopeptidase pyroglutamyl-peptidase I	
3.4.19.6	Pyroglutamate aminopeptidase II Pyroglutamyl-peptidase II	
3.5.2.9	L-Pyroglutamate hydrolase 5-Oxoprolinase (ATP-hydrolysing)	
3.5.2.9	Pyroglutamate hydrolase 5-Oxoprolinase (ATP-hydrolysing)	
3.5.2.9	Pyroglutamic hydrolase 5-Oxoprolinase (ATP-hydrolysing)	
3.4.19.3	pyroglutamidase pyroglutamyl-peptidase I	
3.4.19.3	pyroglutamyl-peptidase I pyroglutamyl-peptidase I	
3.4.19.6	Pyroglutamyl-peptidase II Pyroglutamyl-peptidase II	
3.4.11.8	pyroglutamyl aminopeptidase pyroglutamyl aminopeptidase	
3.4.19.3	pyroglutamyl aminopeptidase pyroglutamyl-peptidase I	
3.4.19.3	pyroglutamylaminopeptidase pyroglutamyl-peptidase I	
3.4.19.6	Pyroglutamyl aminopeptidase II Pyroglutamyl-peptidase II	
3.4.19.6	Pyroglutamyl peptidase II Pyroglutamyl-peptidase II	
3.1.1.57	2-Pyrone-4,6-dicarboxylate lactonase 2-Pyrone-4,6-dicarboxylate lactonase	
3.1.1.57	2-Pyrone-4,6-dicarboxylate lactonohydrolase 2-Pyrone-4,6-dicarboxylate lactonase	

EC-No.	Synonym Recommended Name	Vol., Page
3.6.1.13	Pyrophosphatase, adenosine diphosphoribose ADPribose diphosphatase	
3.6.1.8	Pyrophosphatase, adenosine triphosphate ATP diphosphatase	
3.6.1.26	Pyrophosphatase, cytidine diphosphodiacylglycerol CDPdiacylglycerol diphosphatase	
3.6.1.16	Pyrophosphatase, cytidine diphosphoglycerol CDP glycerol diphosphatase	
3.6.1.1	Pyrophosphatase, inorganic Inorganic diphosphatase	
3.6.1.19	Pyrophosphatase, nucleoside triphosphate Nucleoside-triphosphate diphosphatase	
3.6.1.44	Pyrophosphatase, oligosaccharide pyrophosphodolichol Oligosaccharide-diphosphodolichol diphosphatase	
3.6.1.31	Pyrophosphatase, phosphoribosyladenosine triphosphate Phosphoribosyl-ATP diphosphatase	
3.6.1.18	Pyrophosphatase, riboflavin adenine dinucleotide FAD diphosphatase	
2.7.2.12	Pyrophosphate-acetate phosphotransferase Acetate kinase (pyrophosphate)	
2.7.1.90	Pyrophosphate-D-fructose 6-phosphate 1-phosphotransferase Pyrophosphate-fructose-6-phosphate 1-phosphotransferase	
2.7.1.90	Pyrophosphate-dependent phosphofructo-1-kinase Pyrophosphate-fructose-6-phosphate 1-phosphotransferase	
2.7.1.90	Pyrophosphate-dependent phosphofructokinase Pyrophosphate-fructose-6-phosphate 1-phosphotransferase	
2.7.1.90	Pyrophosphate-fructose-6-phosphate 1-phosphotransferase Pyrophosphate-fructose-6-phosphate 1-phosphotransferase	
2.7.1.90	Pyrophosphate-fructose 6-phosphate phosphotransferase Pyrophosphate-fructose-6-phosphate 1-phosphotransferase	
2.7.1.79	Pyrophosphate-glycerol phosphotransferase Pyrophosphate-glycerol phosphotransferase	
2.7.1.80	Pyrophosphate-L-serine phosphotransferase Pyrophosphate-serine phosphotransferase	
2.7.1.104	Pyrophosphate-protein phosphotransferase Pyrophosphate-protein phosphotransferase	

EC-No.	Synonym Recommended Name Vol., Page
2.7.1.143	Pyrophosphate-purine nucleoside phosphotransferase Diphosphate-purine nucleoside kinase
2.7.1.80	Pyrophosphate-serine phosphotransferase Pyrophosphate-serine phosphotransferase
2.7.2.12	Pyrophosphate:acetate phosphotransferase Acetate kinase (pyrophosphate)
2.7.1.90	Pyrophosphate:D-fructose-6-phosphate 1-phosphotransferase Pyrophosphate-fructose-6-phosphate 1-phosphotransferase
2.7.1.79	Pyrophosphate:glycerol 1-phosphotransferase Pyrophosphate-glycerol phosphotransferase
2.7.1.80	Pyrophosphate:L-serine O-phosphotransferase Pyrophosphate-serine phosphotransferase
2.7.1.80	Pyrophosphate:L-serine phosphotransferase Pyrophosphate-serine phosphotransferase
2.7.1.104	Pyrophosphate:microsomal-membrane-protein O-phosphotransferase Pyrophosphate-protein phosphotransferase
4.1.1.38	pyrophosphate:oxaloacetate carboxy-lyase (transphosphorylating) phosphoenolpyruvate carboxykinase (pyrophosphate)
2.7.1.104	Pyrophosphate:protein phosphotransferase Pyrophosphate-protein phosphotransferase
2.7.1.90	Pyrophosphate D-fructose-6-phosphate 1-phosphotransferase Pyrophosphate-fructose-6-phosphate 1-phosphotransferase
3.1.7.3	pyrophosphate hydrolase monoterpenyl-diphosphatase
3.6.1.1	Pyrophosphate phosphohydrolase Inorganic diphosphatase
2.7.6.3	Pyrophosphokinase, 2-amino-4-hydroxy-6-hydroxymethyldihydropteridine 2-Amino-4-hydroxy-6-hydroxymethyldihydropteridine pyrophosphokinase
2.7.6.4	Pyrophosphokinase, nucleotide Nucleotide pyrophosphokinase
2.7.6.1	Pyrophosphokinase, ribose phosphate Ribose-phosphate pyrophosphokinase
2.7.6.2	Pyrophosphokinase, thiamin Thiamine pyrophosphokinase

EC-No.	Synonym Recommended Name	Vol., Page
4.1.1.33	5-Pyrophosphomevalonate decarboxylase Diphosphomevalonate decarboxylase	
4.1.1.33	Pyrophosphomevalonate decarboxylase Diphosphomevalonate decarboxylase	
4.1.1.33	Pyrophosphomevalonic acid decarboxylase Diphosphomevalonate decarboxylase	
2.7.6.1	Pyrophosphoribosylphosphate synthetase Ribose-phosphate pyrophosphokinase	
2.4.2.19	Pyrophosphorylase, nicotinate mononucleotide (carboxylating) Nicotinate-nucleotide pyrophosphorylase (carboxylating)	
2.5.1.3	Pyrophosphorylase, thiamin phosphate Thiamine-phosphate pyrophosphorylase	
3.4.19.3	pyrrolidone-carboxylate peptidase pyroglutamyl-peptidase I	
3.4.19.3	L-pyrrolidonecarboxylate peptidase pyroglutamyl-peptidase I	
3.4.19.3	pyrrolidone carboxyl peptidase pyroglutamyl-peptidase I	
3.4.19.3	pyrrolidonecarboxylyl peptidase pyroglutamyl-peptidase I	
3.4.19.3	pyrrolidonecarboxy peptidase pyroglutamyl-peptidase I	
3.4.19.3	pyrrolidonyl peptidase pyroglutamyl-peptidase I	
1.5.1.1	Δ^1-pyrroline-2-carboxylate reductase Pyrroline-2-carboxylate reductase	
1.5.1.1	Pyrroline-2-carboxylate reductase Pyrroline-2-carboxylate reductase	
3.5.4.22	1-Pyrroline-4-hydroxy-2-carboxylate aminohydrolase (decyclizing) 1-Pyrroline-4-hydroxy-2-carboxylate deaminase	
3.5.4.22	1-Pyrroline-4-hydroxy-2-carboxylate deaminase 1-Pyrroline-4-hydroxy-2-carboxylate deaminase	
1.5.1.12	L-Pyrroline-5-carboxylate-NAD^+ oxidoreductase 1-Pyrroline-5-carboxylate dehydrogenase	
1.5.1.12	1-Pyrroline-5-carboxylate: NAD^+ oxidoreductase 1-Pyrroline-5-carboxylate dehydrogenase	

EC-No.	Synonym Recommended Name Vol., Page
1.5.1.12	1-Pyrroline-5-carboxylate dehydrogenase 1-Pyrroline-5-carboxylate dehydrogenase
1.5.1.12	Δ^1-pyrroline-5-carboxylate dehydrogenase 1-Pyrroline-5-carboxylate dehydrogenase
1.5.1.12	Pyrroline-5-carboxylate dehydrogenase 1-Pyrroline-5-carboxylate dehydrogenase
1.5.1.2	1-Pyrroline-5-carboxylate reductase Pyrroline-5-carboxylate reductase
1.5.1.2	Pyrroline-5-carboxylate reductase Pyrroline-5-carboxylate reductase
1.5.1.12	Pyrroline-5-carboxylic acid dehydrogenase 1-Pyrroline-5-carboxylate dehydrogenase
1.5.1.12	1-Pyrroline dehydrogenase 1-Pyrroline-5-carboxylate dehydrogenase
1.13.11.26	Pyrrolooxygenase Peptide-tryptophan 2,3-dioxygenase
2.7.9.1	Pyruvate, orthophosphate dikinase phosphate dikinase pyruvate
2.7.9.1	Pyruvate, phosphate dikinase phosphate dikinase pyruvate
2.7.9.1	Pyruvate, *Π* dikinase phosphate dikinase pyruvate
2.7.9.2	Pyruvate, water dikinase water dikinase pyruvate
2.6.3.1	Pyruvate-acetone oximinotransferase Oximinotransferase
2.6.1.2	Pyruvate-alanine aminotransferase Alanine transaminase
4.2.1.52	pyruvate-aspartic semialdehyde condensing enzyme dihydrodipicolinate synthase
2.6.1.2	Pyruvate-glutamate transaminase Alanine transaminase
2.7.9.1	Pyruvate-inorganic phosphate dikinase phosphate dikinase pyruvate
2.6.3.1	Pyruvate-oxime:acetone oximinotransferase Oximinotransferase

EC-No.	Synonym Recommended Name	Vol., Page
2.7.9.1	Pyruvate-phosphate dikinase phosphate dikinase pyruvate	
2.7.9.1	Pyruvate-phosphate ligase phosphate dikinase pyruvate	
2.5.1.7	Pyruvate-UDP-acetylglucosamine transferase UDP-N-acetylglucosamine 1-carboxyvinyltransferase	
2.5.1.7	Pyruvate-uridine diphospho-N-acetyl-glucosamine transferase UDP-N-acetylglucosamine 1-carboxyvinyltransferase	
2.5.1.7	Pyruvate-uridine diphospho-N-acetylglucosamine transferase UDP-N-acetylglucosamine 1-carboxyvinyltransferase	
6.4.1.1	Pyruvate:carbon-dioxide ligase (ADP-forming) Pyruvate carboxylase	Vol. 2, Page 708
1.2.7.1	Pyruvate:ferredoxin 2-oxidoreductase (CoA-acetylating) Pyruvate synthase	
1.2.7.1	Pyruvate:ferredoxin oxidoreductase Pyruvate synthase	
1.2.2.2	Pyruvate:ferricytochrome-b1 oxidoreductase Pyruvate dehydrogenase (cytochrome)	
1.2.4.1	Pyruvate:lipoamide 2-oxidoreductase (decarboxylating and acceptor-acetylating) Pyruvate dehydrogenase (lipoamide)	
1.2.1.51	Pyruvate:NADP$^+$ 2-oxidoreductase (CoA-acetylating) Pyruvate dehydrogenase (NADP$^+$)	
1.2.3.6	Pyruvate:oxygen 2-oxidoreductase (CoA-acetylating) Pyruvate oxidase (CoA-acetylating)	
1.2.3.3	Pyruvate:oxygen 2-oxidoreductase (phosphorylating) Pyruvate oxidase	
1.2.2.2	Pyruvate:ubiquinone-8-oxidoreductase Pyruvate dehydrogenase (cytochrome)	
4.1.3.17	pyruvate aldolase 4-hydroxy-4-methyl-2-oxoglutarate aldolase	
6.4.1.1	Pyruvate carboxylase Pyruvate carboxylase	Vol. 2, Page 708
4.1.1.1	Pyruvate decarboxylase Pyruvate decarboxylase	

EC-No.	Synonym Recommended Name	Vol., Page
1.2.4.1	Pyruvate decarboxylase Pyruvate dehydrogenase (lipoamide)	
1.2.2.2	Pyruvate dehydrogenase Pyruvate dehydrogenase (cytochrome)	
1.2.4.1	Pyruvate dehydrogenase Pyruvate dehydrogenase (lipoamide)	
1.2.2.2	Pyruvate dehydrogenase (cytochrome) Pyruvate dehydrogenase (cytochrome)	
1.2.4.1	Pyruvate dehydrogenase (lipoamide) Pyruvate dehydrogenase (lipoamide)	
3.1.3.43	[pyruvate dehydrogenase (lipoamide)]-phosphatase [pyruvate dehydrogenase (lipoamide)]-phosphatase	
3.1.3.43	[pyruvate dehydrogenase (lipoamide)]-phosphate phosphohydrolase [pyruvate dehydrogenase (lipoamide)]-phosphatase	
2.7.1.99	[Pyruvate dehydrogenase (lipoamide)] kinase [Pyruvate dehydrogenase (lipoamide)] kinase	
1.2.1.51	Pyruvate dehydrogenase ($NADP^+$) Pyruvate dehydrogenase ($NADP^+$)	
1.2.4.1	Pyruvate dehydrogenase complex Pyruvate dehydrogenase (lipoamide)	
2.7.1.99	Pyruvate dehydrogenase kinase [Pyruvate dehydrogenase (lipoamide)] kinase	
2.7.1.99	Pyruvate dehydrogenase kinase activator protein [Pyruvate dehydrogenase (lipoamide)] kinase	
3.1.3.43	pyruvate dehydrogenase phosphatase [pyruvate dehydrogenase (lipoamide)]-phosphatase	
2.3.1.54	Pyruvate formate-lyase Formate C-acetyltransferase	
1.97.1.4	Pyruvate formate-lyase activase Formate acetyltransferase activating enzyme	
1.97.1.4	Pyruvate formate-lyase activating enzyme Formate acetyltransferase activating enzyme	
2.7.1.40	Pyruvate kinase Pyruvate kinase	
3.1.3.49	[Pyruvate kinase]-phosphatase [Pyruvate kinase]-phosphatase	

EC-No.	Synonym Recommended Name Vol., Page
3.1.3.49	Pyruvate kinase phosphatase [Pyruvate kinase]-phosphatase
1.2.2.2	Pyruvate oxidase Pyruvate dehydrogenase (cytochrome)
1.2.3.3	Pyruvate oxidase Pyruvate oxidase
1.2.3.6	Pyruvate oxidase (CoA-acetylating) Pyruvate oxidase (CoA-acetylating)
1.2.7.1	Pyruvate oxidoreductase Pyruvate synthase
2.7.1.40	Pyruvate phosphotransferase (EC 2.7.1.40) Pyruvate kinase
1.2.7.1	Pyruvate synthase Pyruvate synthase
1.2.7.1	Pyruvate synthetase Pyruvate synthase
2.6.1.2	Pyruvate transaminase Alanine transaminase
2.5.1.7	Pyruvatetransferase, phosphoenolpyruvate-uridine diphosphoacetylglucosamine UDP-N-acetylglucosamine 1-carboxyvinyltransferase
1.2.7.1	Pyruvic-ferredoxin oxidoreductase Pyruvate synthase
1.1.1.39	Pyruvic-malic carboxylase Malate dehydrogenase (decarboxylating)
1.1.1.38	Pyruvic-malic carboxylase Malate dehydrogenase (oxaloacetate-decarboxylating)
1.1.1.40	Pyruvic-malic carboxylase Malate dehydrogenase (oxaloacetate-decarboxylating, $NADP^+$)
2.7.9.1	Pyruvic-phosphate dikinase phosphate dikinase pyruvate
2.7.9.1	Pyruvic-phosphate ligase phosphate dikinase pyruvate
2.5.1.7	Pyruvic-uridine diphospho-N-acetylglucosaminyltransferase UDP-N-acetylglucosamine 1-carboxyvinyltransferase

EC-No.	Synonym Recommended Name	Vol., Page
1.2.4.1	Pyruvic acid dehydrogenase Pyruvate dehydrogenase (lipoamide)	
1.2.3.3	Pyruvic acid oxidase Pyruvate oxidase	
6.4.1.1	Pyruvic carboxylase Pyruvate carboxylase	Vol. 2, Page 708
4.1.1.1	Pyruvic decarboxylase Pyruvate decarboxylase	
1.2.4.1	Pyruvic dehydrogenase Pyruvate dehydrogenase (lipoamide)	
1.2.2.2	Pyruvic dehydrogenase (cytochrome b1) Pyruvate dehydrogenase (cytochrome)	
2.3.1.54	Pyruvic formate-lyase Formate C-acetyltransferase	
2.7.1.40	Pyruvic kinase Pyruvate kinase	
1.2.3.3	Pyruvic oxidase Pyruvate oxidase	
1.1.1.220	6-Pyruvoyl-tetrahydropterin 2'-reductase 6-Pyruvoyltetrahydropterin 2'-reductase	
1.1.1.220	Pyruvoyl-tetrahydropterin reductase 6-Pyruvoyltetrahydropterin 2'-reductase	
1.1.1.220	6-Pyruvoyl tetrahydropterin (2'-oxo)reductase 6-Pyruvoyltetrahydropterin 2'-reductase	
1.1.1.220	6-Pyruvoyltetrahydropterin 2'-reductase 6-Pyruvoyltetrahydropterin 2'-reductase	
4.6.1.10	6-pyruvoyltetrahydropterin synthase 6-Pyruvoyltetrahydtropterin synthase	
4.2.3.12	6-pyruvoyltetrahydropterin synthase 6-pyruvoyltetrahydropterin synthase	
4.6.1.10	6-Pyruvoyltetrahydropterin synthase [25-glutamine] (human clone lambdaHSY2 gene PCBD subunit) 6-Pyruvoyltetrahydtropterin synthase	
4.6.1.10	6-Pyruvoyltetrahydropterin synthase [de-57-valine] (human clone lambdaHSY2 gene PCBD subunit) 6-Pyruvoyltetrahydtropterin synthase	

EC-No.	Synonym Recommended Name	Vol., Page
4.6.1.10	6-Pyruvoyltetrahydtropterin synthase 6-Pyruvoyltetrahydtropterin synthase	
3.4.24.15	Pz-peptidase thimet oligopeptidase	
2.4.2.19	QAPRTase Nicotinate-nucleotide pyrophosphorylase (carboxylating)	
1.13.99.5	QDO 3,4-Dihydroxyquinoline 2,4-dioxygenase	
1.10.2.2	QH_2:cytochrome c oxidoreductase Ubiquinol-cytochrome-c reductase	
1.3.99.17	Qor Quinoline 2-oxidoreductase	
2.8.2.28	Quercetin-3,3'-bissulfate 7-sulfotransferase Quercetin-3,3'-bissulfate 7-sulfotransferase	
2.8.2.26	Quercetin-3-sulfate 3'-sulfotransferase Quercetin-3-sulfate 3'-sulfotransferase	
2.8.2.27	Quercetin-3-sulfate 4'-sulfotransferase Quercetin-3-sulfate 4'-sulfotransferase	
1.13.11.24	Quercetin 2,3-dioxygenase Quercetin 2,3-dioxygenase	
2.1.1.76	Quercetin 3-O-methyltransferase Quercetin 3-O-methyltransferase	
1.13.11.24	Quercetin:oxygen 2,3-oxidoreductase (decyclizing) Quercetin 2,3-dioxygenase	
1.13.11.24	Quercetinase Quercetin 2,3-dioxygenase	
3.2.1.66	Quercitrin 3-rhamnohydrolase Quercitrinase	
3.2.1.66	Quercitrinase Quercitrinase	
1.14.13.43	Questin,NADPH:oxygen oxidoreductase (hydroxylating, anthraquinone-ring-opening) Questin monooxygenase	
1.14.13.43	Questin monooxygenase Questin monooxygenase	

EC-No.	Synonym Recommended Name	Vol., Page
1.14.13.43	Questin oxygenase Questin monooxygenase	
2.4.2.29	Queuine tRNA-ribosyltransferase Queuine tRNA-ribosyltransferase	
2.4.2.29	Queuine tRNA ribosyltransferase Queuine tRNA-ribosyltransferase	
3.4.13.9	quiescent cell proline dipeptidase X-Pro dipeptidase	
3.4.21.1	Quimar Chymotrypsin	
3.4.21.1	Quimotrase Chymotrypsin	
1.3.99.19	Quinaldate 4-monooxygenase Quinoline-4-carboxylate 2-oxidoreductase	
1.3.99.18	Quinaldate 4-oxidoreductase Quinaldate 4-oxidoreductase	
1.3.99.19	Quinaldic acid 4-monooxygenase Quinoline-4-carboxylate 2-oxidoreductase	
1.3.99.18	Quinaldic acid 4-oxidoreductase Quinaldate 4-oxidoreductase	
1.3.99.19	Quinaldic acid 4-oxidoreductase Quinoline-4-carboxylate 2-oxidoreductase	
1.1.1.24	Quinate 5-dehydrogenase Quinate 5-dehydrogenase	
1.1.1.24	Quinate:NAD$^+$ 5-oxidoreductase Quinate 5-dehydrogenase	
1.1.1.24	Quinate:NAD oxidoreductase Quinate 5-dehydrogenase	
1.1.99.25	Quinate:pyrroloquinoline-quinone 5-oxidoreductase Quinate dehydrogenase (pyrroloquinoline-quinone)	
1.1.99.25	Quinate dehydrogenase (pyrroloquinoline-quinone) Quinate dehydrogenase (pyrroloquinoline-quinone)	
1.1.1.24	Quinate dehydrogenase Dehydrogenase, quinate Quinate 5-dehydrogenase	
2.3.1.99	Quinate O-hydroxycinnamoyltransferase Quinate O-hydroxycinnamoyltransferase	

EC-No.	Synonym Recommended Name	Vol., Page
1.1.1.24	Quinic dehydrogenase Quinate 5-dehydrogenase	
1.14.13.67	quinine, $NADPH_2$:oxygen oxidoreductase quinine 3-monooxygenase	
1.14.13.67	quinine 3-hydroxylase quinine 3-monooxygenase	
1.14.13.67	quinine 3-monooxygenase quinine 3-monooxygenase	
1.1.99.8	Quinohemoprotein alcohol dehydrogenase Alcohol dehydrogenase (acceptor)	
1.14.13.61	Quinolin-2(1H)-one, NADH:oxygen oxidoreductase (8-oxygenating) 2-Hydroxyquinoline 8-monooxygenase	
1.14.12.16	Quinolin-2(1H)-one 5,6-dioxygenase 2-Hydroxyquinoline 5,6-dioxygenase	
1.14.13.65	Quinolin-2-ol, $NADH_2$:oxygen oxidoreductase (8-hydroxylating) 2-Hydroxyquinoline 8-monooxygenase	
1.14.12.16	Quinolin-2-ol, NADH:oxygen oxidoreductase (5,6-hydroxylating) 2-Hydroxyquinoline 5,6-dioxygenase	
1.14.12.16	Quinolin-2-ol 5,6-dioxygenase 2-Hydroxyquinoline 5,6-dioxygenase	
1.14.13.62	Quinolin-4(1H)-one, $NADH_2$:oxygen oxidoreductase (3-oxygenating) 4-Hydroxyquinoline 3-monooxygenase	
1.14.13.62	Quinolin-4(1H)-one 3-monooxygenase 4-Hydroxyquinoline 3-monooxygenase	
2.4.2.19	Quinolinate phosphoribosyltransferase (decarboxylating) Nicotinate-nucleotide pyrophosphorylase (carboxylating)	
1.3.99.18	Quinoline-2-carboxylate:acceptor 4-oxidoreductase (hydroxylating) Quinaldate 4-oxidoreductase	
1.13.99.5	Quinoline-3,4-diol 2,4-dioxygenase 3,4-Dihydroxyquinoline 2,4-dioxygenase	
1.13.99.5	Quinoline-3,4-diol 2,4-dioxygenase (carbon monoxide-forming) 3,4-Dihydroxyquinoline 2,4-dioxygenase	
1.3.99.19	Quinoline-4-carboxylate 2-oxidoreductase Quinoline-4-carboxylate 2-oxidoreductase	
1.3.99.19	Quinoline-4-carboxylate:acceptor 2-oxidoreductase (hydroxylating) Quinoline-4-carboxylate 2-oxidoreductase	

EC-No.	Synonym Recommended Name	Vol., Page
1.3.99.19	Quinoline-4-carboxylic acid oxidoreductase Quinoline-4-carboxylate 2-oxidoreductase	
1.3.99.17	Quinoline 2-oxidoreductase Quinoline 2-oxidoreductase	
1.3.99.17	Quinoline:acceptor 2-oxidoreductase (hydroxylating) Quinoline 2-oxidoreductase	
1.2.3.1	Quinoline oxidase Aldehyde oxidase	
1.3.99.17	Quinoline oxidoreductase Quinoline 2-oxidoreductase	
2.4.2.19	Quinolinic acid phosphoribosyltransferase Nicotinate-nucleotide pyrophosphorylase (carboxylating)	
2.4.2.19	Quinolinic phosphoribosyltransferase Nicotinate-nucleotide pyrophosphorylase (carboxylating)	
1.6.99.2	Quinone reductase NAD(P)H dehydrogenase (quinone)	
1.6.5.5	Quinone reductase NADPH:quinone reductase	
1.6.5.1	quinone reductase quinone reductase	
1.1.99.8	Quinoprotein alcohol dehydrogenase Alcohol dehydrogenase (acceptor)	
1.1.99.17	Quinoprotein D-glucose dehydrogenase Glucose dehydrogenase (pyrroloquinoline-quinone)	
1.1.99.8	Quinoprotein ethanol dehydrogenase Alcohol dehydrogenase (acceptor)	
1.1.99.17	Quinoprotein glucose dehydrogenase Glucose dehydrogenase (pyrroloquinoline-quinone)	
1.3.99.17	QuinOr Quinoline 2-oxidoreductase	
4.1.2.10	(R)-(+)-Mandelonitrile lyase Mandelonitrile lyase	
1.5.1.28	N-[1-(R)-(Carboxy)ethyl]-(S)-norvaline: NAD^+ oxidoreductase (L-norvaline forming) Opine dehydrogenase	

EC-No.	Synonym Recommended Name	Vol., Page
1.1.1.75	(R)-1-Aminopropan-2-ol:NAD^+ oxidoreductase (R)-Aminopropanol dehydrogenase	
4.2.1.53	(R)-10-Hydroxystearate 10-hydro-lyase Oleate hydratase	
1.1.1.4	(R)-2,3-Butanediol dehydrogenase (R,R)-Butanediol dehydrogenase	
1.1.1.86	(R)-2,3-Dihydroxy-3-methylbutanoate:$NADP^+$ oxidoreductase (isomerizing) Ketol-acid reductoisomerase	
4.1.3.33	(R)-2-Ethylmalate 2-oxobutanoyl-lyase (CoA-acetylating) 2-Ethylmalate synthase	
1.1.99.6	(R)-2-Hydroxy-acid:(acceptor) 2-oxidoreductase D-2-Hydroxy-acid dehydrogenase	
1.1.1.98	(R)-2-Hydroxy-fatty-acid dehydrogenase (R)-2-Hydroxy-fatty-acid dehydrogenase	
1.1.1.98	(R)-2-Hydroxystearate:NAD^+ oxidoreductase (R)-2-Hydroxy-fatty-acid dehydrogenase	
5.4.99.2	(R)-2-Methyl-3-oxopropanoyl-CoA CoA-carbonylmutase Methylmalonyl-CoA mutase	Vol. 1, Page 589
4.2.1.35	(R)-2-Methylmalate dehydratase (R)-2-Methylmalate dehydratase	
4.2.1.35	(R)-2-Methylmalate hydro-lyase (R)-2-Methylmalate dehydratase	
1.1.1.53	(R)-20-Hydroxysteroid dehydrogenase 3α(or 20β)-Hydroxysteroid dehydrogenase	
1.1.1.84	(R)-3,3-Dimethylmalate:NAD^+ oxidoreductase (decarboxylating) Dimethylmalate dehydrogenase	
3.1.1.22	(R)-3-((R)-3-hydroxybutanoyloxy)butanoate hydroxybutanoylhydrolase hydroxybutyrate-dimer hydrolase	
1.1.1.222	(R)-3-(4-Hydroxyphenyl)lactate:NAD(P)$^+$ 2-oxidoreductase (R)-4-Hydroxyphenyllactate dehydrogenase	
2.6.1.61	(R)-3-Amino-2-methylpropanoate:2-oxoglutarate aminotransferase (R)-3-Amino-2-methylpropionate transaminase	
2.6.1.40	(R)-3-Amino-2-methylpropanoate:pyruvate aminotransferase (R)-3-Amino-2-methylpropionate-pyruvate transaminase	

EC-No.	Synonym Recommended Name	Vol., Page
2.6.1.40	(R)-3-Amino-2-methylpropionate-pyruvate transaminase (R)-3-Amino-2-methylpropionate-pyruvate transaminase	
2.6.1.61	(R)-3-Amino-2-methylpropionate transaminase (R)-3-Amino-2-methylpropionate transaminase	
1.2.1.55	(R)-3-Hydroxyacid ester dehydrogenase (R)-3-Hydroxyacid ester dehydrogenase	
1.1.1.212	(3R)-3-Hydroxyacyl-[acyl-carrier-protein]:NAD$^+$ oxidoreductase 3-Oxoacyl-[acyl-carrier-protein] reductase (NADH)	
1.1.1.100	(3R)-3-Hydroxyacyl-[acyl-carrier-protein]:NADP$^+$ oxidoreductase 3-Oxoacyl-[acyl-carrier-protein] reductase	
1.1.1.36	(R)-3-Hydroxyacyl-CoA:NADP$^+$ oxidoreductase Acetoacetyl-CoA reductase	
1.1.1.36	(R)-3-Hydroxyacyl-CoA dehydrogenase Acetoacetyl-CoA reductase	
1.1.1.30	(R)-3-Hydroxybutanoate:NAD$^+$ oxidoreductase 3-Hydroxybutyrate dehydrogenase	
4.2.1.58	(3R)-3-Hydroxybutanoyl-[acyl-carrier protein] hydro-lyase Crotonoyl-[acyl-carrier-protein] hydratase	
4.2.1.55	(3R)-3-Hydroxybutanoyl-CoA hydro-lyase 3-Hydroxybutyryl-CoA dehydratase	
4.2.1.60	(3R)-3-Hydroxydecanoyl-[acyl-carrier-protein] hydro-lyase 3-Hydroxydecanoyl-[acyl-carrier-protein] dehydratase	
4.2.1.59	(3R)-3-Hydroxyoctanoyl-[acyl-carrier-protein] hydro-lyase 3-Hydroxyoctanoyl-[acyl-carrier-protein] dehydratase	
4.2.1.61	(3R)-3-Hydroxypalmitoyl-[acyl-carrier-protein] hydro-lyase 3-Hydroxypalmitoyl-[acyl-carrier-protein] dehydratase	
2.3.1.129	(R)-3-Hydroxytetradecanoyl-[acyl-carrier-protein]:UDP-N-acetylglucosamine 3-O-(3-hydroxytetradecanoyl)transferase Acyl-[acyl-carrier-protein]-UDP-N-acetylglucosamine O-acyltransferase	
1.2.1.33	(R)-4-Dehydropantoate:NAD$^+$ 4-oxidoreductase (R)-Dehydropantoate dehydrogenase	
1.1.1.222	(R)-4-Hydroxyphenyllactate dehydrogenase (R)-4-Hydroxyphenyllactate dehydrogenase	
4.1.1.36	N-((R)-4-Phosphopantothenoyl)-L-cysteine carboxy-lyase Phosphopantothenoylcysteine decarboxylase	

EC-No.	Synonym Recommended Name	Vol., Page
6.3.2.5	(R)-4′-Phosphopantothenate:L-cysteine ligase Phosphopantothenoate-cysteine ligase	Vol. 2, Page 431
1.5.3.6	(R)-6-Hydroxynicotine:oxygen oxidoreductase (R)-6-Hydroxynicotine oxidase	
1.5.3.6	(R)-6-Hydroxynicotine oxidase (R)-6-Hydroxynicotine oxidase	
1.1.1.75	(R)-Aminopropanol dehydrogenase (R)-Aminopropanol dehydrogenase	
1.1.1.222	(R)-Aromatic lactate dehydrogenase (R)-4-Hydroxyphenyllactate dehydrogenase	
4.1.3.28	(R)-Citrate, synthase Citrate (re)-synthase	
1.2.1.33	(R)-Dehydropantoate dehydrogenase (R)-Dehydropantoate dehydrogenase	
1.1.1.4	(R)-Diacetyl reductase (R,R)-Butanediol dehydrogenase	
1.1.1.60	(R)-Glycerate:NAD(P)$^+$ oxidoreductase 2-Hydroxy-3-oxopropionate reductase	
1.1.1.29	(R)-Glycerate:NAD$^+$ oxidoreductase Glycerate dehydrogenase	
1.1.1.78	(R)-Lactaldehyde:NAD$^+$ oxidoreductase D-Lactaldehyde dehydrogenase	
1.1.2.5	(R)-Lactate:ferricytochrome-c-553 2-oxidoreductase D-Lactate dehydrogenase (cytochrome c-553)	
1.1.2.4	(R)-Lactate:ferricytochrome-c 2-oxidoreductase D-Lactate dehydrogenase (cytochrome)	
1.1.1.28	(R)-Lactate:NAD$^+$ oxidoreductase D-Lactate dehydrogenase	
4.2.1.54	(R)-lactyl-CoA dehydratase lactoyl-CoA dehydratase	
1.13.11.40	8(R)-Lipoxygenase Arachidonate 8-lipoxygenase	
1.1.1.83	(R)-Malate:NAD$^+$ oxidoreductase (decarboxylating) D-Malate dehydrogenase (decarboxylating)	
4.2.1.31	(R)-malate hydro-lyase maleate hydratase	

EC-No.	Synonym Recommended Name	Vol., Page
4.1.2.10	(R)-Mandelonitrile lyase Mandelonitrile lyase	
1.1.1.32	(R)-Mevalonate:NAD^+ oxidoreductase Mevaldate reductase	
1.1.1.33	(R)-Mevalonate:$NADP^+$ oxidoreductase Mevaldate reductase (NADPH)	
1.1.1.34	(R)-Mevalonate:$NADP^+$ oxidoreductase (CoA-acylating) Hydroxymethylglutaryl-CoA reductase (NADPH)	
4.1.2.37	(R)-Oxynitrilase Acetone-cyanhydrin lyase	
4.1.2.10	(R)-Oxynitrilase Mandelonitrile lyase	
6.3.2.1	(R)-Pantoate:β-alanine ligase (AMP-forming) Pantoate-β-alanine ligase	Vol. 2, Page 394
1.1.1.106	(R)-Pantoate:NAD^+ 4-oxidoreductase Pantoate 4-dehydrogenase	
1.1.1.169	(R)-Pantoate:$NADP^+$ 2-oxidoreductase 2-Dehydropantoate 2-reductase	
1.1.99.27	(R)-Pantolactone dehydrogenase (flavin) (R)-Pantolactone dehydrogenase (flavin)	
1.1.99.27	(R)-Pantonolactone:acceptor oxidoreductase (flavin-containing) (R)-Pantolactone dehydrogenase (flavin)	
4.1.1.30	N-((R)-Pantothenoyl)-L-cysteine carboxy-lyase Pantothenoylcysteine decarboxylase	
1.1.1.168	(R)-Pantoyl-lactone:$NADP^+$ oxidoreductase (A-specific) 2-Dehydropantoyl-lactone reductase (A-specific)	
1.1.1.214	(R)-Pantoyl-lactone:$NADP^+$ oxidoreductase (B-specific) 2-Dehydropantoyl-lactone reductase (B-specific)	
1.1.99.27	(R)-Pantoyllactone dehydrogenase (flavin) (R)-Pantolactone dehydrogenase (flavin)	
1.1.3.35	(R)-Reticuline ,NADPH:oxygen oxidoreductase (C-C phenol-coupling) Salutaridine synthase	
1.5.1.27	(R)-Reticuline:NADP oxidoreductase 1,2-Dehydroreticulinium reductase ($NADPH_2$)	
1.1.3.35	(R)-Reticuline dehydrogenase Salutaridine synthase	

EC-No.	Synonym Recommended Name	Vol., Page
1.1.3.35	(R)-Reticuline oxidase (C-C phenol-coupling) Salutaridine synthase	
4.4.1.5	(R)-S-lactoylglutathione methylglyoxal-lyase (isomerizing) lactoylglutathione lyase	
3.2.1.119	(R)-Vicianin β-D-glucohydrolase Vicianin β-glucosidase	
1.1.1.77	[(R) or (S)]-Propane-1,2-diol:NAD$^+$ oxidoreductase Lactaldehyde reductase	
4.1.2.33	(24R,24'R)-Fucosterol-epoxide acetaldehyde-lyase Fucosterol-epoxide lyase	
1.3.1.42	8-[(1R,2R)-3-Oxo-2-((Z)-pent-2-enyl)cyclopentyl]octanoate: NADP$^+$ 4-oxidoreductase 12-Oxophytodienoate reductase	
1.3.1.53	(1R,2S)-1,2-Dihydroxy-3,5-cyclohexadiene-1,4-dicarboxylate dehydrogenase (3S,4R)3,4-Dihydroxycyclohexa-1,5-diene-1,4-dicarboxylate dehydrogenase	
5.1.2.6	(1R, 2S)-1-Hydroxypropane-1,2,3 tricarboxylate 1-epimerase Isocitrate epimerase	Vol. 1, Page 89
1.3.1.53	(1R,2S)-Dihydroxy-3,5-cyclohexadiene-1,4-dicarboxylate dehydrogenase (3S,4R)3,4-Dihydroxycyclohexa-1,5-diene-1,4-dicarboxylate dehydrogenase	
4.2.1.85	(2R,3S)-2,3-Dimethylmalate hydro-lyase Dimethylmaleate hydratase	
1.1.1.4	(R,R)-Butane-2,3-diol:NAD$^+$ oxidoreductase (R,R)-Butanediol dehydrogenase	
1.1.1.4	(R,R)-Butanediol dehydrogenase (R,R)-Butanediol dehydrogenase	
4.1.1.73	(R,R)-Tartrate carboxy-lyase Tartrate decarboxylase	
4.2.1.32	(R,R)-Tartrate hydro-lyase L(+)-Tartrate dehydrogenase	
2.1.1.115	(R,S)-Tetrahydrobenzylisoquinoline-N-methyltransferase (RS)-1-Benzyl-1,2,3,4-tetrahydroisoquinoline N-methyltransferase	
5.1.3.1	R5P3E Ribulose-phosphate 3-epimerase	Vol. 1, Page 91

EC-No.	Synonym Recommended Name	Vol., Page
3.1.4.42	rac-glycerol-1,2-cyclic-phosphate 2-glycerophosphohydrolase glycerol-1,2-cyclic-phosphate 2-phosphodiesterase	
3.1.4.42	rac-glycerol 1:2-cyclic phosphate 2-phosphodiesterase glycerol-1,2-cyclic-phosphate 2-phosphodiesterase	
5.1.1.15	Racemase, α-amino-ε-caprolactam 2-Aminohexano-6-lactam racemase	Vol. 1, Page 61
5.1.1.15	Racemase, α-amino-ε-caprolactam (Achromobacter obae reduced) 2-Aminohexano-6-lactam racemase	Vol. 1, Page 61
5.1.99.4	Racemase, α-methylacyl coenzyme A α-Methylacyl-CoA racemase	Vol. 1, Page 188
5.1.99.4	Racemase, α-methylacyl coenzyme A (Mus musculus clone 3) α-Methylacyl-CoA racemase	Vol. 1, Page 188
5.1.99.4	Racemase, α-methylacyl coenzyme A (Rattus norvegicus clone 11) α-Methylacyl-CoA racemase	Vol. 1, Page 188
5.1.2.4	Racemase, acetoin Acetoin racemase	Vol. 1, Page 85
5.1.1.1	Racemase, alanine Alanine racemase	Vol. 1, Page 1
5.1.99.3	Racemase, allantoin Allantoin racemase	Vol. 1, Page 185
5.1.1.10	Racemase, amino acid Amino-acid racemase	Vol. 1, Page 41
5.1.1.9	Racemase, arginine Arginine racemase	Vol. 1, Page 37
5.1.1.13	Racemase, aspartate Aspartate racemase	Vol. 1, Page 55
5.1.1.3	Racemase, glutamate Glutamate racemase	Vol. 1, Page 11
5.1.2.1	Racemase, lactate Lactate racemase	Vol. 1, Page 68
5.1.1.5	Racemase, lysine Lysine racemase	Vol. 1, Page 23
5.1.2.2	Racemase, mandelate Mandelate racemase	Vol. 1, Page 72
5.1.1.2	Racemase, methionine Methionine racemase	Vol. 1, Page 10

EC-No.	Synonym Recommended Name	Vol., Page
5.1.99.1	Racemase, methylmalonyl coenzyme A Methylmalonyl-CoA epimerase	Vol. 1, Page 179
5.1.99.1	Racemase, methylmalonyl coenzyme A Methylmalonyl-CoA epimerase	Vol. 1, Page 179
5.1.1.12	Racemase, ornithine Ornithine racemase	Vol. 1, Page 54
5.1.1.11	Racemase, phenylalanine (adenosine triphosphate-hydrolyzing) Phenylalanine racemase (ATP-hydrolysing)	Vol. 1, Page 48
5.1.1.4	Racemase, proline Proline racemase	Vol. 1, Page 19
5.1.1.6	Racemase, threonine Threonine racemase	Vol. 1, Page 25
2.7.1.37	Raf-1 Protein kinase	
2.4.1.166	Raffinose-raffinose α-galactotransferase Raffinose-raffinose α-galactotransferase	
2.4.1.166	Raffinose:raffinose α-D-galactosyltransferase Raffinose-raffinose α-galactotransferase	
2.4.1.166	Raffinose:raffinose α-galactosyltransferase Raffinose-raffinose α-galactotransferase	
2.7.1.37	Raf kinase Protein kinase	
5.2.1.3	RalPI Retinal isomerase	Vol. 1, Page 202
3.4.21.59	Rat mast cell protease II Tryptase	
3.4.17.21	Rat NAAG peptidase, Glutamate carboxypeptidase II	
3.2.1.125	Raucaffricine β-D-glucohydrolase Raucaffricine β-glucosidase	
3.2.1.125	Raucaffricine β-D-glucosidase Raucaffricine β-glucosidase	
3.2.1.125	Raucaffricine β-glucosidase Raucaffricine β-glucosidase	
3.2.1.125	Raucaffricine glucosidase Raucaffricine β-glucosidase	

EC-No.	Synonym Recommended Name	Vol., Page
4.2.99.18	X-ray endonuclease III DNA-(apurinic or apyrimidinic site) lyase	
2.7.7.48	RDRP RNA-directed RNA polymerase	
4.1.3.28	Re-Citrate-synthase Citrate (re)-synthase	
5.3.4.1	Rearrangease Protein disulfide-isomerase	Vol. 1, Page 436
3.4.99.37	RecA peptidase RecA peptidase	
3.1.11.5	RecBC Deoxyribonuclease Exodeoxyribonuclease V	
2.7.1.112	Receptor-activated Janus kinase Protein-tyrosine kinase	
3.4.22.17	receptor-transforming factor calpain	
1.11.1.4	Recommended Name never specified Recommended Name never specified	
1.6.6.5	Recommended Name never specified Recommended Name never specified	
1.8.2.3	Recommended Name never specified Recommended Name never specified	
2.4.2.13	Recommended Name never specified Recommended Name never specified	
2.6.2.1	Recommended Name never specified Recommended Name never specified	
3.4.19.8	Recommended Name never specified Recommended Name never specified	
3.4.4.22	Recommended Name never specified Recommended Name never specified	
3.4.4.23	Recommended Name never specified Recommended Name never specified	
3.4.4.24	Recommended Name never specified Recommended Name never specified	
3.4.14.4	red cell angiotensinase dipeptidyl-peptidase III	

EC-No.	Synonym Recommended Name	Vol., Page
3.4.21.24	red cell neutral endopeptidase red cell neutral endopeptidase	
3.1.21.6	Redoxyendonuclease CC-preferring endodeoxyribonuclease	
3.1.21.2	Redoxyendonuclease Deoxyribonuclease IV (Phage T_4-induced)	
1.1.1.199	Reduced-(S)-usnate:NAD^+ oxidoreductase (ether-bond-forming) (S)-Usnate reductase	
1.10.2.2	Reduced coenzyme Q-cytochrome c reductase Ubiquinol-cytochrome-c reductase	
1.6.99.3	Reduced diphosphopyridine nucleotide diaphorase NADH dehydrogenase	
2.7.1.86	Reduced diphosphopyridine nucleotide kinase NADH kinase	
1.18.6.1	Reduced ferredoxin:dinitrogen oxidoreductase (ATP-hydrolyzing) Nitrogenase	
1.19.6.1	Reduced flavodoxin:dinitrogen oxidoreductase (ATP-hydrolyzing) Nitrogenase (flavodoxin)	
1.11.1.9	Reduced glutathione peroxidase Glutathione peroxidase	
1.6.99.2	Reduced NAD(P)H dehydrogenase NAD(P)H dehydrogenase (quinone)	
1.6.5.5	Reduced NAD(P)H dehydrogenase NADPH:quinone reductase	
1.16.1.1	Reduced NADP:mercuric ion oxidoreductase Mercury(II) reductase	
1.6.99.2	Reduced nicotinamide-adenine dinucleotide (phosphate) dehydrogenase NAD(P)H dehydrogenase (quinone)	
1.6.5.3	Reduced nicotinamide adenine dinucleotide-coenzyme Q reductase NADH dehydrogenase (ubiquinone)	
1.6.2.2	Reduced nicotinamide adeninedinucleotide-cytochrome b_5 reductase Cytochrome-b_5 reductase	
1.18.1.1	Reduced nicotinamide adenine dinucleotide-rubredoxin reductase Rubredoxin-NAD^+ reductase	

EC-No.	Synonym Recommended Name	Vol., Page
1.18.1.2	Reduced nicotinamide adenine dinucleotide phosphate-adrenodoxin reductase Ferredoxin-$NADP^+$ reductase	
1.6.2.4	Reduced nicotinamide adenine dinucleotide phosphate-cytochrome c reductase NADPH-ferrihemoprotein reductase	
1.1.1.123	Reduced nicotinamide adenine dinucleotide phosphate-linked reductase Sorbose 5-dehydrogenase ($NADP^+$)	
1.3.1.22	Reduced nicotinamide adenine dinucleotide phosphate:Δ^4-3-ketosteroid 5α-oxidoreductase Cholestenone 5α-reductase	
1.6.99.1	Reduced nicotinamide adenine dinucleotide phosphate dehydrogenase NADPH dehydrogenase	
5.3.4.1	Reduced ribonuclease reactivating enzyme Protein disulfide-isomerase	Vol. 1, Page 436
1.10.2.2	Reduced ubiquinone-cytochrome c oxidoreductase Ubiquinol-cytochrome-c reductase	
1.10.2.2	Reduced ubiquinone-cytochrome c reductase, Complex III (mitochondrial electron transport) Ubiquinol-cytochrome-c reductase	
1.3.1.23	5.β.-Reductase Cholestenone 5β-reductase	
1.3.99.5	5α-Reductase 3-Oxo-5α-steroid Δ^4-dehydrogenase	
1.3.1.22	5α-Reductase Cholestenone 5α-reductase	
1.3.99.6	5β-Reductase 3-Oxo-5β-steroid Δ^4-dehydrogenase	
1.3.1.23	Δ 4-5β-reductase Cholestenone 5β-reductase	
1.3.1.4	Δ^4-5α-reductase Cortisone α-reductase	
1.3.1.47	Reductase,-α-santonin 1,2- α-Santonin 1,2-reductase	
1.3.1.16	Reductase, β-nitroacrylate β-Nitroacrylate reductase	

EC-No.	Synonym Recommended Name	Vol., Page
1.2.1.41	Reductase,.γ.-glutamyl phosphate Glutamate-5-semialdehyde dehydrogenase	
1.5.1.27	Reductase, 1,2-dehydroreticuline 1,2-Dehydroreticulinium reductase ($NADPH_2$)	
1.5.1.21	Reductase, Δ 1-piperideine-2-carboxylate Δ1-Piperideine-2-carboxylate reductase	
1.1.1.188	Reductase, 15-hydroxy-11-oxoprostaglandin Prostaglandin-F synthase	
1.1.1.189	Reductase, 15-hydroxy-9-oxoprostaglandin Prostaglandin-E2 9-reductase	
1.3.1.48	Reductase, 15-oxo-.δ.13-prostaglandin 15-Oxoprostaglandin Δ^{13}-reductase	
4.2.1.86	Reductase, 16-dehydroprogesterone 16-Dehydroprogesterone hydratase	
1.3.1.45	Reductase, 2',7-dihydroxy-4',5'-methylenedioxyisoflavone 2'-Hydroxyisoflavone reductase	
1.3.1.51	Reductase, 2'-hydroxydaidzein 2'-Hydroxydaidzein reductase	
1.3.1.45	Reductase, 2'-hydroxyisoflavone 2'-Hydroxyisoflavone reductase	
1.3.1.27	Reductase, 2-alkenal 2-Hexadecenal reductase	
1.3.1.31	Reductase, 2-enoate 2-Enoate reductase	
1.3.1.40	Reductase,2-hydroxy-6-oxo-phenylhexa-2,4-dienoate (reduced nicotinamide adenine dinucleotide phosphate) 2-Hydroxy-6-oxo-6-phenylhexa-2,4-dienoate reductase	
1.1.1.172	Reductase, 2-ketoadipate 2-Oxoadipate reductase	
1.1.1.215	Reductase, 2-ketogluconate Gluconate 2-dehydrogenase	
1.1.1.169	Reductase, 2-oxopantoate 2-Dehydropantoate 2-reductase	
1.1.1.168	Reductase, 2-oxopantoyl lactone 2-Dehydropantoyl-lactone reductase (A-specific)	

EC-No.	Synonym Recommended Name	Vol., Page
1.1.1.214	Reductase, 2-oxopantoyl lactone 2-Dehydropantoyl-lactone reductase (B-specific)	
1.8.99.4	Reductase, 3'-phosphoadenosine 5'-phosphosulfate Phosphoadenosine-phosphosulfate reductase	
1.1.1.102	Reductase, 3-dehydrosphinganine 3-Dehydrosphinganine reductase	
1.3.1.17	Reductase, 3-methyleneoxindole 3-Methyleneoxindole reductase	
1.1.1.100	Reductase, 3-oxoacyl-[acyl carrier protein] 3-Oxoacyl-[acyl-carrier-protein] reductase	
1.1.1.212	Reductase, 3-oxoacyl-[acyl carrier protein] (reduced nicotinamide adenine dinucleotide) 3-Oxoacyl-[acyl-carrier-protein] reductase (NADH)	
1.2.1.55	Reductase, 3-oxo ester (R)- (R)-3-Hydroxyacid ester dehydrogenase	
1.2.1.56	Reductase, 3-oxo ester (S)- (S)-3-Hydroxyacid ester dehydrogenase	
1.3.1.34	Reductase, 4-enoyl coenzyme A (reduced nicotinamide adenine dinucleotide phosphate) 2,4-Dienoyl-CoA reductase (NADPH)	
1.1.1.69	Reductase, 5-ketogluconate 5- Gluconate 5-dehydrogenase	
1.1.1.128	Reductase, 5-ketogluconate 5- (L-idonate-forming) L-Idonate 2-dehydrogenase	
1.3.7.1	Reductase, 6-hydroxynicotinate 6-Hydroxynicotinate reductase	
1.1.1.224	Reductase, 6-phosphomannose Mannose-6-phosphate 6-reductase	
1.1.1.220	Reductase, 6-pyruvoyltetrahydropterin 6-Pyruvoyltetrahydropterin 2'-reductase	
1.3.1.21	Reductase, 7-dehydrocholesterol 7-Dehydrocholesterol reductase	
1.1.1.235	Reductase, 8-ketodeoxycoformycin 8-Oxocoformycin reductase	
1.2.1.38	Reductase, acetyl-.γ.-glutamyl phosphate N-Acetyl-γ-glutamyl-phosphate reductase	

EC-No.	Synonym Recommended Name	Vol., Page
1.2.1.50	Reductase, acyl coenzyme A Long-chain-fatty-acyl-CoA reductase	
1.8.99.2	Reductase, adenylylsulfate Adenylylsulfate reductase	
1.1.1.200	Reductase, aldose 6-phosphate Aldose-6-phosphate reductase (NADPH)	
1.14.13.40	Reductase, anthraniloyl coenzyme A Anthraniloyl-CoA monooxygenase	
1.6.99.11	Reductase, aquacobalamin (reduced nicotinamide adenine dinucleotide phosphate) Aquacobalamin reductase (NADPH)	
1.10.2.1	Reductase, ascorbate-cytochrome b_5 L-Ascorbate-cytochrome-b_5 reductase	
1.6.6.7	Reductase, azobenzene Azobenzene reductase	
1.3.99.15	Reductase, benzoyl coenzyme A (dearomatizing) Benzoyl-CoA reductase	
1.3.1.24	Reductase, biliverdin Biliverdin reductase	
1.3.1.46	Reductase, biochanin A Biochanin-A reductase	
1.6.4.9	Reductase, bis-γ-glutamylcystine Bis-γ-glutamylcystine reductase (NADPH)	
1.8.99.3	Reductase, bisulfite Hydrogensulfite reductase	
1.1.1.184	Reductase, carbonyl Carbonyl reductase (NADPH)	
1.2.99.6	Reductase, carboxylate Carboxylate reductase	
1.97.1.1	Reductase, chlorate Chlorate reductase	
1.1.1.225	Reductase, chlordecone Chlordecone reductase	
1.3.1.22	Reductase, cholestenone 5.α.- Cholestenone 5α-reductase	

EC-No.	Synonym Recommended Name	Vol., Page
1.3.1.23	Reductase, cholestenone 5.β.- Cholestenone 5β-reductase	
1.2.1.44	Reductase, cinnamoyl coenzyme A Cinnamoyl-CoA reductase	
1.3.1.37	Reductase, cis-2-enoyl coenzyme A cis-2-Enoyl-CoA reductase (NADPH)	
1.6.99.9	Reductase, cob(II)alamin Cob(II)alamin reductase	
1.1.1.247	Reductase, codeinone (reduced nicotinamide adenine dinucleotide phosphate) Codeinone reductase (NADPH)	
1.12.99.2	Reductase, coenzyme M-7-mercaptoheptanoylthreonine phosphate heterodisulfide Coenzyme-M-7-mercaptoheptanoylthreonine-phosphate-heterodisulfide hydrogenase	
1.3.1.3	Reductase, cortisone Δ^4-5β- Cortisone β-reductase	
1.3.1.4	Reductase, cortisone Δ: 4-5α- Cortisone α-reductase	
1.3.1.5	Reductase, cucurbitacin.δ.23- Cucurbitacin Δ^{23}-reductase	
1.6.99.12	Reductase, cyanocobalamin Cyanocobalamin reductase (NADPH, cyanide-eliminating)	
1.6.4.1	Reductase, cystine Cystine reductase (NADH)	
1.17.1.1	Reductase, cytidine diphospho-4-keto-6-deoxy-D-glucose CDP-4-dehydro-6-deoxyglucose reductase	
1.6.2.2	Reductase, cytochrome b_5 Cytochrome-b_5 reductase	
1.6.2.4	Reductase, cytochrome c (reduced nicotinamide adenine dinucleotide phosphate) NADPH-ferrihemoprotein reductase	
1.6.2.5	Reductase, cytochrome c_2 (reduced nicotinamide adenine dinucleotide phosphate) NADPH-cytochrome c_2 reductase	
1.1.1.102	Reductase, D-3-oxosphinganine 3-Dehydrosphinganine reductase	

EC-No.	Synonym Recommended Name	Vol., Page
1.1.1.114	Reductase, D-apiose Apiose 1-reductase	
1.1.1.162	Reductase, D-erythrulose Erythrulose reductase	
1.4.4.1	Reductase, D-proline (dithiol) D-Proline reductase (dithiol)	
1.1.1.9	Reductase, D-xylulose D-Xylulose reductase	
1.1.1.229	Reductase, diethyl 2-methyl-3-oxosuccinate Diethyl 2-methyl-3-oxosuccinate reductase	
1.3.1.26	Reductase, dihydrodipicolinate Dihydrodipicolinate reductase	
1.5.1.3	Reductase, dihydrofolate Dihydrofolate reductase	
1.1.1.219	Reductase, dihydromyricetin Dihydrokaempferol 4-reductase	
1.6.99.7	Reductase, dihydropteridine (reduced nicotinamide adenine dinucleotide) Dihydropteridine reductase	
1.1.1.253	Reductase, dihydropteridine (reduced nuicotinamide adenine dinucleotide phosphate) Pteridine reductase	
1.1.1.219	Reductase, dihydroquercetin Dihydrokaempferol 4-reductase	
1.1.1.156	Reductase, dihydroxyacetone (reduced nicotinamide adenine dinucleotide phosphate) Glycerol 2-dehydrogenase ($NADP^+$)	
1.1.1.96	Reductase, diiodophenylpyruvate Diiodophenylpyruvate reductase	
1.5.5.1	Reductase, electron transfer flavoprotein Electron-transferring-flavoprotein dehydrogenase	
1.3.1.9	Reductase, enoyl-[acyl carrier protein] Enoyl-[acyl-carrier-protein] reductase (NADH)	
1.3.1.39	Reductase, enoyl-[acyl carrier protein] (reduced nicotinamide adenine dinucleotide phosphate) Enoyl-[acyl-carrier-protein] reductase (NADPH, A-specific)	

EC-No.	Synonym Recommended Name	Vol., Page
1.3.1.10	Reductase, enoyl-[acyl carrier protein] (reduced nicotinamide adenine dinucleotide phosphate) Enoyl-[acyl-carrier-protein] reductase (NADPH, B-specific)	
1.18.1.3	Reductase, ferredoxin Ferredoxin-NAD^+ reductase	
1.18.1.3	Reductase, ferredoxin-nicotinamide adenine dinucleotide Ferredoxin-NAD^+ reductase	
1.18.1.2	Reductase, ferredoxin-nicotinamide adenine dinucleotide phosphate Ferredoxin-$NADP^+$ reductase	
1.7.7.1	Reductase, ferredoxin-nitrite Ferredoxin-nitrite reductase	
1.1.1.234	Reductase, flavanone 4- Flavanone 4-reductase	
1.1.1.57	Reductase, fructuronate Fructuronate reductase	
1.3.1.6	Reductase, fumarate Fumarate reductase (NADH)	
1.1.1.19	Reductase, glucuronate Glucuronate reductase	
1.1.1.20	Reductase, glucuronolactone Glucuronolactone reductase	
1.6.4.2	Reductase, glutathione Glutathione reductase (NADPH)	
1.4.2.1	Reductase, glycine-cytochrome c Glycine dehydrogenase (cytochrome)	
1.1.1.26	Reductase, glyoxylate Glyoxylate reductase	
1.6.6.8	Reductase, guanylate GMP reductase	
1.6.6.11	Reductase, hydroxylamine Hydroxylamine reductase (NADH)	
1.7.99.1	Reductase, hydroxylamine (acceptor) Hydroxylamine reductase	
1.1.1.88	Reductase, hydroxymethylglutaryl coenzyme A Hydroxymethylglutaryl-CoA reductase	

EC-No.	Synonym Recommended Name	Vol., Page
1.1.1.237	Reductase, hydroxyphenylpyruvate Hydroxyphenylpyruvate reductase	
1.1.1.81	Reductase, hydroxypyruvate Hydroxypyruvate reductase	
1.1.1.190	Reductase, indoleacetaldehyde Indole-3-acetaldehyde reductase (NADH)	
1.1.1.191	Reductase, indoleacetaldehyde (reduced nicotinamide adenine dinucleotide phosphate) Indole-3-acetaldehyde reductase (NADPH)	
1.9.99.1	Reductase, iron-cytochrome c Iron-cytochrome-c reductase	
1.6.99.13	Reductase, iron chelate Ferric-chelate reductase	
1.3.99.16	Reductase, isoquinoline Isoquinoline 1-oxidoreductase	
1.5.1.25	Reductase, ketimine Thiomorpholine-carboxylate dehydrogenase	
1.1.1.10	Reductase, L-xylulose L-Xylulose reductase	
1.1.1.77	Reductase, lactaldehyde Lactaldehyde reductase	
1.1.1.55	Reductase, lactaldehyde (reduced nicotinamide adenine dinucleotide phosphate) Lactaldehyde reductase (NADPH)	
1.3.1.32	Reductase, maleylacetate Maleylacetate reductase	
1.1.1.131	Reductase, mannuronate Mannuronate reductase	
1.16.1.1	Reductase, mercurate(II) Mercury(II) reductase	
1.8.4.5	Reductase, methionine sulfoxide Methionine-S-oxide reductase	
1.8.4.6	Reductase, methionine sulfoxide (protein) Protein-methionine-S-oxide reductase	
2.1.1.135	Reductase, methionine synthase [Methionine synthase]-cobalamin methyltransferase (cob(II)alamin reducing)	

EC-No.	Synonym Recommended Name	Vol., Page
1.3.1.52	Reductase, methyl branched-chain enoyl coenzyme A 2-Methyl-branched-chain-enoyl-CoA reductase	
1.5.1.20	Reductase, methylenetetrahydrofolate (reduced nicotinamide adenine dinucleotide phosphate) Methylenetetrahydrofolate reductase (NADPH)	
1.7.99.5	Reductase, methylenetetrahydrofolate (reduced riboflavin adenine dinucleotide) 5,10-Methylenetetrahydrofolate reductase ($FADH_2$)	
1.1.1.33	Reductase, mevaldate (reduced nicotinamide adenine dinucleotide phosphate) Mevaldate reductase (NADPH)	
1.1.1.218	Reductase, naloxone Morphine 6-dehydrogenase	
1.1.1.252	Reductase, naphthol Tetrahydroxynaphthalene reductase	
1.1.1.252	Reductase, naphthol (Magnaporthe grisea clone pAV501 precursor reduced) Tetrahydroxynaphthalene reductase	
1.6.6.1	Reductase, nitrate Nitrate reductase (NADH)	
1.7.99.4	Reductase, nitrate (acceptor) Nitrate reductase	
1.9.6.1	Reductase, nitrate (cytochrome) Nitrate reductase (cytochrome)	
1.7.7.2	Reductase, nitrate (ferredoxin) Ferredoxin-nitrate reductase	
1.6.6.2	Reductase, nitrate (reduced nicotinamide adenine dinucleotide (phosphate)) Nitrate reductase (NAD(P)H)	
1.7.99.7	Reductase, nitric oxide Nitric-oxide reductase	
1.7.99.3	Reductase, nitrite Nitrite reductase	
1.7.2.1	Reductase, nitrite (cytochrome) Nitrite reductase (cytochrome)	
1.6.6.4	Reductase, nitrite (reduced nicotinamide adenine dinucleotide (phosphate)) Nitrite reductase (NAD(P)H)	

EC-No.	Synonym Recommended Name	Vol., Page
1.6.6.10	Reductase, nitroquinoline N-oxide Nitroquinoline-N-oxide reductase	
1.7.99.6	Reductase, nitrous oxide Nitrous-oxide reductase	
1.3.1.14	Reductase, orotate Orotate reductase (NADH)	
1.3.1.15	Reductase, orotate (reduced nicotinamide adenine dinucleotide phosphate) Orotate reductase (NADPH)	
1.1.1.101	Reductase, palmitoyl dihydroxyacetone phosphate Acylglycerone-phosphate reductase	
1.1.4.1	Reductase, phylloquinone epoxide Vitamin-K-epoxide reductase (warfarin sensitive)	
1.10.99.1	Reductase, plastoquinol-plastocyanin Plastoquinol-plastocyanin reductase	
1.3.1.54	Reductase, precorrin 6x (Pseudomonas denitrificans clone pXL367 gene cobK) Precorrin-6X reductase	
1.3.1.30	Reductase, progesterone 5α- Progesterone 5α-reductase	
1.6.4.4	Reductase, protein disulfide Protein-disulfide reductase (NAD(P)H)	
1.8.4.2	Reductase, protein disulfide (glutathione) Protein-disulfide reductase (glutathione)	
1.3.1.33	Reductase, protochlorophyllide Protochlorophyllide reductase	
1.5.1.1	Reductase, pyrroline-2-carboxylate Pyrroline-2-carboxylate reductase	
1.5.1.2	Reductase, pyrroline-5-carboxylate Pyrroline-5-carboxylate reductase	
1.18.1.3	Reductase, reduced nicotinamide adenine dinucleotide-ferredoxin Ferredoxin-NAD$^+$ reductase	
1.6.8.1	Reductase, riboflavin mononucleotide (reduced nicotinamide adenine dinucleotide (phosphate)) NAD(P)H dehydrogenase (FMN)	
1.6.8.2	Reductase, riboflavin mononucleotide (reduced nicotinamide adenine dinucleotide phosphate) NADPH dehydrogenase (flavin)	

EC-No.	Synonym Recommended Name	Vol., Page
1.17.4.1	Reductase, ribonucleoside diphosphate Ribonucleoside-diphosphate reductase	
1.17.4.2	Reductase, ribonucleoside triphosphate Ribonucleoside-triphosphate reductase	
1.18.1.1	Reductase, rubredoxin-nicotinamide adenine dinucleotide Rubredoxin-NAD^+ reductase	
1.1.1.248	Reductase, salutaridine 7- Salutaridine reductase (NADPH)	
1.1.1.153	Reductase, sepiapterin Sepiapterin reductase	
1.8.99.1	Reductase, sulfite Sulfite reductase	
1.8.7.1	Reductase, sulfite (ferredoxin) Sulfite reductase (ferredoxin)	
1.8.1.2	Reductase, sulfite (reduced nicotinamide adenine dinucleotide phosphate) Sulfite reductase (NADPH)	
1.97.1.3	Reductase, sulfur Sulfur reductase	
1.1.1.58	Reductase, tagaturonate Tagaturonate reductase	
1.6.4.5	Reductase, thioredoxin Thioredoxin reductase (NADPH)	
2.8.1.5	Reductase, thiosulfate Thiosulfate-dithiol sulfurtransferase	
1.1.1.133	Reductase, thymidine diphospho-4-ketorhamnose dTDP-4-dehydrorhamnose reductase	
1.3.1.38	Reductase, trans-enoyl coenzyme A trans-2-Enoyl-CoA reductase (NADPH)	
1.6.6.9	Reductase, trimethylamine N-oxide Trimethylamine-N-oxide reductase	
1.1.1.236	Reductase, tropinone (ψ-tropine-forming) Tropinone reductase	
1.10.2.2	Reductase, ubiquinol-cytochrome c Ubiquinol-cytochrome-c reductase	
1.6.5.3	Reductase, ubiquinone NADH dehydrogenase (ubiquinone)	

EC-No.	Synonym Recommended Name	Vol., Page
1.1.1.158	Reductase, uridine diphosphoacetylpyruvoylglucosamine UDP-N-acetylmuramate dehydrogenase	
1.6.99.8	Reductase, vitamin B_{12a} Aquacobalamin reductase	
1.6.99.9	Reductase, vitamin B_{12r} Cob(II)alamin reductase	
1.1.4.2	Reductase, vitamin K epoxide (warfarin-insensitive) Vitamin-K-epoxide reductase (warfarin-insensitive)	
1.3.1.41	Reductase, xanthommatin Xanthommatin reductase	
1.1.1.242	Reductase, zeatin Zeatin reductase	
1.6.99.8	Reductase aquacobalamin Aquacobalamin reductase	
2.7.1.109	Reductase kinase [Hydroxymethylglutaryl-CoA reductase (NADPH)] kinase	
2.7.1.110	Reductase kinase kinase Dephospho-[reductase kinase] kinase	
3.1.3.47	reductase phosphatase [hydroxymethylglutaryl-CoA reductase (NADPH)]-phosphatase	
1.1.1.86	Reductoisomerase Ketol-acid reductoisomerase	
5.99.1.2	Relaxing enzyme DNA topoisomerase	Vol. 1, Page 721
3.2.1.15	Remanase Polygalacturonase	
3.4.13.19	Renal dipeptidase Membrane dipeptidase	
1.13.12.5	Renilla-luciferin 2-monooxygenase Renilla-luciferin 2-monooxygenase	
1.13.12.5	Renilla-luciferin:oxygen 2-oxidoreductase (decarboxylating) Renilla-luciferin 2-monooxygenase	
2.8.2.10	Renilla-luciferin sulfotransferase Renilla-luciferin sulfotransferase	
1.13.12.5	Renilla-type luciferase Renilla-luciferin 2-monooxygenase	

EC-No.	Synonym Recommended Name	Vol., Page
1.13.12.5	Renilla luciferin 2-monooxygenase Renilla-luciferin 2-monooxygenase	
3.4.21.54	γ-Renin γ-Renin	
3.4.23.15	renin renin	
3.4.4.15	renin renin	
3.4.99.19	renin renin	
3.4.21.54	Renin, γ- γ-Renin	
5.1.3.8	Renin-binding protein N-Acylglucosamine 2-epimerase	Vol. 1, Page 140
3.4.23.4	rennin chymosin	
3.4.4.3	rennin rennin	
2.7.7.48	Q-β replicase RNA-directed RNA polymerase	
2.7.7.48	Qβ-replicase RNA-directed RNA polymerase	
2.7.7.48	Replicase, phage f2 RNA-directed RNA polymerase	
2.7.7.48	Replicase, Qβ RNA-directed RNA polymerase	
3.4.21.88	Repressor LexA Repressor LexA	
1.7.99.4	Respiratory nitrate reductase Nitrate reductase	
1.9.6.1	Respiratory nitrate reductase Nitrate reductase (cytochrome)	
2.1.1.72	Restriction-Modification system Site-specific DNA-methyltransferase (adenine-specific)	
2.1.1.113	Restriction-Modification system Site-specific DNA-methyltransferase (cytosine-N^4-specific)	

EC-No.	Synonym Recommended Name	Vol., Page
2.1.1.73	Restriction-Modification system Site-specific DNA-methyltransferase (cytosine-specific)	
3.1.21.3	Restriction-modification system Type I site-specific deoxyribonuclease	
3.1.21.5	Restriction-modification system Type III site-specific deoxyribonuclease	
2.3.1.95	Resveratrol synthase Trihydroxystilbene synthase	
1.5.3.9	Reticuline oxidase Reticuline oxidase	
1.2.1.36	Retinal:NAD^+ oxidoreductase Retinal dehydrogenase	
1.2.3.11	Retinal:oxygen oxidoreductase Retinal oxidase	
1.2.1.36	Retinal dehydrogenase Retinal dehydrogenase	
5.2.1.3	Retinal isomerase Retinal isomerase	Vol. 1, Page 202
1.2.3.11	Retinal oxidase Retinal oxidase	
5.2.1.3	Retinal photoisomerase Retinal isomerase	Vol. 1, Page 202
1.1.1.71	Retinal reductase Alcohol dehydrogenase ($NAD(P)^+$)	
1.1.1.105	Retinal reductase Retinol dehydrogenase	
5.2.1.3	Retinene isomerase Retinal isomerase	Vol. 1, Page 202
1.2.3.11	Retinene oxidase Retinal oxidase	
1.1.1.105	Retinene reductase Retinol dehydrogenase	
5.2.1.3	Retinochrome Retinal isomerase	Vol. 1, Page 202
5.2.1.3	Retinoid isomerase Retinal isomerase	Vol. 1, Page 202

EC-No.	Synonym Recommended Name	Vol., Page
1.1.1.105	Retinol:NAD$^+$ oxidoreductase Retinol dehydrogenase	
2.3.1.76	Retinol acyltransferase Retinol O-fatty-acyltransferase	
1.1.1.105	Retinol dehydrogenase Retinol dehydrogenase	
1.1.1.105	Retinol dehydrogenase (vitamin A_1) Retinol dehydrogenase	
2.3.1.76	Retinol fatty-acyltransferase Retinol O-fatty-acyltransferase	
5.2.1.7	Retinol isomerase Retinol isomerase	Vol. 1, Page 215
2.3.1.76	Retinol O-fatty-acyltransferase Retinol O-fatty-acyltransferase	
3.1.1.21	Retinyl-palmitate esterase Retinyl-palmitate esterase	
3.1.1.21	Retinyl-palmitate palmitohydrolase Retinyl-palmitate esterase	
3.1.1.21	Retinyl ester hydrolase Retinyl-palmitate esterase	
2.3.1.135	Retinyl ester synthase Phosphatidylcholine-retinol O-acyltransferase	
3.1.1.63	Retinyl palmitate hydrolase 11-cis-Retinyl-palmitate hydrolase	
3.1.1.21	Retinyl palmitate hydrolase Retinyl-palmitate esterase	
3.1.1.21	Retinyl palmitate hydrolyase Retinyl-palmitate esterase	
3.4.23.16	Retropepsin Retropepsin	
3.4.23.16	Retroproteinase Retropepsin	
3.1.13.2	Retroviral Reverse Transcriptase RNaseH Exoribonuclease H	
5.99.1.2	Reverse gyrase DNA topoisomerase	Vol. 1, Page 721

EC-No.	Synonym Recommended Name	Vol., Page
2.7.7.49	Reverse transcriptase RNA-directed DNA polymerase	
2.7.7.49	Revertase RNA-directed DNA polymerase	
3.4.22.37	RGP Gingipain R	
3.4.22.37	RGP-1 Gingipain R	
5.3.1.14	RhaI L-Rhamnose isomerase	Vol. 1, Page 328
4.1.2.19	Rhamn-1PA Rhamnulose-1-phosphate aldolase	
1.1.1.173	L-Rhamnofuranose:NAD^+ 1-oxidoreductase L-Rhamnose 1-dehydrogenase	
4.2.1.90	L-Rhamnonate dehydratase L-Rhamnonate dehydratase	
3.1.1.65	L-Rhamnonate dehydratase L-Rhamnono-1,4-lactonase	
4.2.1.90	L-Rhamnonate hydro-lyase L-Rhamnonate dehydratase	
3.1.1.65	L-Rhamnono-γ-lactonase L-Rhamnono-1,4-lactonase	
3.1.1.65	L-Rhamnono-1,4-lactonase L-Rhamnono-1,4-lactonase	
3.1.1.65	L-Rhamnono-1,4-lactone lactonohydrolase L-Rhamnono-1,4-lactonase	
1.1.1.173	L-Rhamnose 1-dehydrogenase L-Rhamnose 1-dehydrogenase	
5.3.1.14	L-Rhamnose isomerase L-Rhamnose isomerase	Vol. 1, Page 328
5.3.1.14	Rhamnose isomerase L-Rhamnose isomerase	Vol. 1, Page 328
5.3.1.14	L-Rhamnose ketol-isomerase L-Rhamnose isomerase	Vol. 1, Page 328
3.2.1.40	.α.-L-Rhamnosidase α-L-rhamnosidase	

EC-No.	Synonym Recommended Name	Vol., Page
3.2.1.43	.β.-L-Rhamnosidase β-L-rhamnosidase	
3.2.1.40	α-L-rhamnosidase α-L-rhamnosidase	
3.2.1.43	β-L-rhamnosidase β-L-rhamnosidase	
3.2.1.40	Rhamnosidase,.α.-L- α-L-rhamnosidase	
3.2.1.43	Rhamnosidase,.β.-L- β-L-rhamnosidase	
3.2.1.40	α-L-rhamnosidase N α-L-rhamnosidase	
3.2.1.40	α-L-rhamnosidase T α-L-rhamnosidase	
3.2.1.40	α-L-rhamnoside rhamnohydrolase α-L-rhamnosidase	
3.2.1.43	β-L-rhamnoside rhamnohydrolase β-L-rhamnosidase	
2.4.1.159	Rhamnosyltransferase, uridine diphosphorhamnose-flavonol 3-O-glucoside Flavonol-3-O-glucoside L-rhamnosyltransferase	
2.7.1.5	L-Rhamnulokinase Rhamnulokinase	
2.7.1.5	Rhamnulokinase Rhamnulokinase	
4.1.2.19	L-Rhamnulose-1-phosphate aldolase Rhamnulose-1-phosphate aldolase	
4.1.2.19	Rhamnulose-1-phosphate aldolase Rhamnulose-1-phosphate aldolase	
4.1.2.19	L-Rhamnulose-1-phosphate L-lactaldehyde-lyase Rhamnulose-1-phosphate aldolase	
4.1.2.19	L-Rhamnulose-1-phosphate lactaldehyde-lyase Rhamnulose-1-phosphate aldolase	
2.7.1.5	L-Rhamnulose kinase Rhamnulokinase	
2.7.1.5	Rhamnulose kinase Rhamnulokinase	

EC-No.	Synonym Recommended Name	Vol., Page
4.1.2.19	Rhamnulose phosphate aldolase Rhamnulose-1-phosphate aldolase	
3.4.22.38	rhCK Cathepsin K	
3.4.22.29	Rhinovirus protease 2A Picornain 2A	
3.4.22.28	Rhinovirus protease 3C Picornain 3C	
3.4.23.21	Rhizopus acid protease Rhizopuspepsin	
3.4.23.21	Rhizopus acid proteinase Rhizopuspepsin	
3.4.23.9	rhizopus acid proteinase rhizopus acid proteinase	
3.4.23.21	Rhizopus aspartic proteinase Rhizopuspepsin	
3.4.23.21	Rhizopuspepsin Rhizopuspepsin	
2.8.1.1	Rhodanase Thiosulfate sulfurtransferase	
2.8.1.1	Rhodanese Thiosulfate sulfurtransferase	
2.7.1.125	Rhodopsin kinase Rhodopsin kinase	
3.4.23.26	Rhodotorula acid proteinase Rhodotorulapepsin	
3.4.23.26	Rhodotorula aspartic proteinase Rhodotorulapepsin	
3.4.23.26	Rhodotorula glutinis acid proteinase Rhodotorulapepsin	
3.4.23.26	Rhodotorulapepsin Rhodotorulapepsin	
2.7.1.5	RhuK Rhamnulokinase	
5.3.1.14	L-RI L-Rhamnose isomerase	Vol. 1, Page 328

EC-No.	Synonym Recommended Name	Vol., Page
1.1.1.137	Ribitol-5-phosphate 2-dehydrogenase Ribitol-5-phosphate 2-dehydrogenase	
1.1.1.137	D-Ribitol-5-phosphate:NAD(P)$^+$ 2-oxidoreductase Ribitol-5-phosphate 2-dehydrogenase	
2.7.7.40	D-Ribitol-5-phosphate cytidylyltransferase D-Ribitol-5-phosphate cytidylyltransferase	
1.1.1.56	Ribitol 2-dehydrogenase Ribitol 2-dehydrogenase	
2.7.7.40	Ribitol 5-phosphate cytidylyltransferase D-Ribitol-5-phosphate cytidylyltransferase	
1.1.1.56	Ribitol:NAD$^+$ 2-oxidoreductase Ribitol 2-dehydrogenase	
1.1.1.56	Ribitol dehydrogenase A Ribitol 2-dehydrogenase	
1.1.1.56	Ribitol dehydrogenase B Ribitol 2-dehydrogenase	
1.1.1.56	Ribitol dehydrogenase D Ribitol 2-dehydrogenase	
3.6.1.18	Riboflavin adenine dinucleotide pyrophosphatase FAD diphosphatase	
2.7.7.2	Riboflavin adenine dinucleotide pyrophosphorylase FMN adenylyltransferase	
3.5.99.1	Riboflavinase Riboflavinase	
2.7.7.2	Riboflavine adenine dinucleotide adenylyltransferase FMN adenylyltransferase	
3.6.1.18	Riboflavine adenine dinucleotide pyrophosphatase FAD diphosphatase	
3.5.99.1	Riboflavine hydrolase Riboflavinase	
2.7.1.26	Riboflavine kinase Riboflavin kinase	
1.6.8.1	Riboflavine mononucleotide reductase NAD(P)H dehydrogenase (FMN)	
1.6.8.2	Riboflavine mononucleotide reductase NADPH dehydrogenase (flavin)	

EC-No.	Synonym / Recommended Name	Vol., Page
2.7.1.42	Riboflavine phosphotransferase Riboflavin phosphotransferase	
2.5.1.9	Riboflavine synthase Riboflavin synthase	
2.5.1.9	Riboflavine synthetase Riboflavin synthase	
3.5.99.1	Riboflavin hydrolase Riboflavinase	
2.7.1.26	Riboflavin kinase Riboflavin kinase	
1.6.8.1	Riboflavin mononucleotide reductase NAD(P)H dehydrogenase (FMN)	
1.6.8.2	Riboflavin mononucleotide reductase NADPH dehydrogenase (flavin)	
2.7.1.42	Riboflavin phosphotransferase Riboflavin phosphotransferase	
2.5.1.9	Riboflavin synthase Riboflavin synthase	
2.5.1.9	Riboflavin synthetase Riboflavin synthase	
2.7.1.15	D-Ribokinase Ribokinase	
2.7.1.15	Ribokinase Ribokinase	
3.1.27.1	Ribonnuclease (non-base specific) Ribonuclease T_2	
3.1.27.5	Ribonuclease Pancreatic ribonuclease	
3.1.26.2	Ribonuclease α Ribonuclease α	
2.7.7.16	ribonuclease ribonuclease	
2.7.7.17	ribonuclease ribonuclease	
3.1.26.9	Ribonuclease (poly-(U)-specific) Ribonuclease (poly-(U)-specific)	

EC-No.	Synonym Recommended Name	Vol., Page
3.1.27.4	Ribonuclease (purine) Ribonuclease U_2	
3.1.26.9	Ribonuclease (uracil-specific) Ribonuclease (poly-(U)-specific)	
3.1.27.3	Ribonuclease C_2 Ribonuclease T_1	
3.1.27.3	Ribonuclease Ch Ribonuclease T_1	
3.1.26.3	Ribonuclease D Ribonuclease III	
3.1.27.7	Ribonuclease F Ribonuclease F	
3.1.27.7	Ribonuclease F Ribonuclease F	
3.1.27.3	Ribonuclease F_1 Ribonuclease T_1	
3.1.27.3	Ribonuclease guaninenucleotido-2'-transferase (cyclizing) Ribonuclease T_1	
3.1.26.4	Ribonuclease H Calf thymus ribonuclease H	
3.1.27.5	Ribonuclease I Pancreatic ribonuclease	
3.1.4.22	ribonuclease I ribonuclease I	
3.1.13.1	Ribonuclease II Exoribonuclease II	
3.1.27.1	Ribonuclease II Ribonuclease T_2	
3.1.4.23	ribonuclease II ribonuclease II	
3.1.26.3	Ribonuclease III Ribonuclease III	
3.1.26.6	Ribonuclease IV Ribonuclease IV	
3.1.26.10	Ribonuclease IX Ribonuclease IX	

EC-No.	Synonym Recommended Name	Vol., Page
3.1.27.1	Ribonuclease M Ribonuclease T_2	
3.1.26.8	Ribonuclease M_5 Ribonuclease M_5	
3.1.27.3	Ribonuclease N_1 Ribonuclease T_1	
3.1.27.1	Ribonuclease N_2 Ribonuclease T_2	
3.1.27.3	Ribonuclease N_3 Ribonuclease T_1	
3.1.26.5	Ribonuclease P Ribonuclease P	
3.1.26.5	Ribonuclease P Ribonuclease P	
3.1.26.7	Ribonuclease P_4 Ribonuclease P_4	
2.7.7.56	Ribonuclease PH tRNA nucleotidyltransferase	
3.1.27.3	Ribonuclease PP_1 Ribonuclease T_1	
3.1.27.1	Ribonuclease PP_2 Ribonuclease T_2	
3.1.27.1	Ribonuclease PP_3 Ribonuclease T_2	
3.1.27.3	Ribonuclease SA Ribonuclease T_1	
3.1.27.3	Ribonuclease T_1 Ribonuclease T_1	
3.1.27.1	Ribonuclease T_2 Ribonuclease T_2	
3.1.27.3	Ribonuclease U_1 Ribonuclease T_1	
3.1.27.4	Ribonuclease U_2 Ribonuclease U_2	
3.1.27.4	Ribonuclease U_3 Ribonuclease U_2	

EC-No.	Synonym Recommended Name	Vol., Page
3.1.27.1	Ribonuclease U_4 Ribonuclease T_2	
3.1.27.8	Ribonuclease V Ribonuclease V	
3.1.31.1	Ribonucleate (deoxynucleate) 3'-nucleotidohydrolase Micrococcal nuclease	
3.1.27.1	Ribonucleate 3'-oligonucleotide hydrolase Ribonuclease T_2	
3.1.27.5	Ribonucleate 3'-pyrimidino-oligonucleotidohydrolase Pancreatic ribonuclease	
3.1.27.2	Ribonucleate nucleotido-2'-transferase (cyclizing) Bacillus subtilis ribonuclease	
3.1.27.1	Ribonucleate nucleotido-2'-transferase (cyclizing) Ribonuclease T_2	
2.7.7.6	Ribonucleate nucleotidyltransferase DNA-directed RNA polymerase	
2.7.7.6	Ribonucleate polymerase DNA-directed RNA polymerase	
2.1.1.48	Ribonucleic acid-adenine (N^6) methylase rRNA (adenine-N^6-)-methyltransferase	
2.7.7.48	Ribonucleic acid-dependent ribonucleate nucleotidyltransferase RNA-directed RNA polymerase	
2.7.7.48	Ribonucleic acid-dependent ribonucleic acid polymerase RNA-directed RNA polymerase	
2.7.1.69	Ribonucleic acid formation factor, gene glC Protein-Npi-phosphohistidine-sugar phosphotransferase	
2.7.7.6	Ribonucleic acid formation factors, C DNA-directed RNA polymerase	
2.7.7.6	Ribonucleic acid nucleotidyltransferase DNA-directed RNA polymerase	
2.7.7.6	Ribonucleic acid polymerase DNA-directed RNA polymerase	
2.7.7.48	Ribonucleic acid replicase RNA-directed RNA polymerase	
2.7.7.6	Ribonucleic acid transcriptase DNA-directed RNA polymerase	

EC-No.	Synonym Recommended Name	Vol., Page
2.7.7.25	Ribonucleic cytidylic cytidylic adenylic pyrophosphorylase tRNA adenylyltransferase	
2.7.7.21	Ribonucleic cytidylic cytidylic adenylic pyrophosphorylase tRNA cytidylyltransferase	
2.7.7.25	Ribonucleic cytidylyltransferase tRNA adenylyltransferase	
2.7.7.21	Ribonucleic cytidylyltransferase tRNA cytidylyltransferase	
6.5.1.3	Ribonucleic ligase RNA ligase (ATP)	Vol. 2, Page 787
3.1.27.5	Ribonucleic phosphatase Pancreatic ribonuclease	
2.7.7.6	Ribonucleic polymerase DNA-directed RNA polymerase	
2.7.7.48	Ribonucleic replicase RNA-directed RNA polymerase	
2.7.7.48	Ribonucleic synthetase RNA-directed RNA polymerase	
2.7.7.6	Ribonucleic transcriptase DNA-directed RNA polymerase	
1.17.4.1	Ribonucleoside-diphosphate reductase Ribonucleoside-diphosphate reductase	
1.17.4.2	Ribonucleoside-triphosphate reductase Ribonucleoside-triphosphate reductase	
3.1.4.16	Ribonucleoside 2',3'-cyclic phosphate diesterase 2',3'-Cyclic-nucleotide 2'-phosphodiesterase	
1.17.4.1	Ribonucleoside 5'-diphosphate reductase Ribonucleoside-diphosphate reductase	
1.17.4.1	Ribonucleoside diphosphate reductase Ribonucleoside-diphosphate reductase	
3.2.2.1	Ribonucleoside hydrolase Purine nucleosidase	
1.17.4.2	Ribonucleoside triphosphate reductase Ribonucleoside-triphosphate reductase	
3.1.3.6	3'-ribonucleotidase 3'-nucleotidase	

EC-No.	Synonym Recommended Name	Vol., Page
1.17.4.1	Ribonucleotide diphosphate reductase Ribonucleoside-diphosphate reductase	
3.1.3.6	3'-ribonucleotide phosphohydrolase 3'-nucleotidase	
3.1.3.5	5'-ribonucleotide phosphohydrolase 5'-nucleotidase	
1.17.4.1	Ribonucleotide reductase Ribonucleoside-diphosphate reductase	
1.17.4.2	Ribonucleotide reductase Ribonucleoside-triphosphate reductase	
2.7.6.1	Ribophosphate pyrophosphokinase Ribose-phosphate pyrophosphokinase	
6.3.4.7	Ribose-5-phosphate-ammonia ligase Ribose-5-phosphate-ammonia ligase	Vol. 2, Page 608
6.3.4.7	Ribose-5-phosphate:ammonia ligase (ADP-forming) Ribose-5-phosphate-ammonia ligase	Vol. 2, Page 608
2.7.7.35	Ribose-5-phosphate adenylyltransferase Ribose-5-phosphate adenylyltransferase	
5.3.1.6	Ribose-5-phosphate isomerase Ribose-5-phosphate isomerase	Vol. 1, Page 277
5.3.1.6	D-Ribose-5-phosphate ketol-isomerase Ribose-5-phosphate isomerase	Vol. 1, Page 277
2.7.6.1	Ribose-5-phosphate pyrophosphokinase Ribose-phosphate pyrophosphokinase	
5.3.1.6	Ribose-5-P isomerase Ribose-5-phosphate isomerase	Vol. 1, Page 277
2.7.6.1	Ribose-phosphate pyrophosphokinase Ribose-phosphate pyrophosphokinase	
5.4.2.7	D-Ribose 1,5-phosphomutase Phosphopentomutase	Vol. 1, Page 535
1.1.1.115	Ribose 1-dehydrogenase ($NADP^+$) Ribose 1-dehydrogenase ($NADP^+$)	
6.3.4.7	Ribose 5-phosphate aminotransferase Ribose-5-phosphate-ammonia ligase	Vol. 2, Page 608
5.3.1.6	D-Ribose 5-phosphate isomerase Ribose-5-phosphate isomerase	Vol. 1, Page 277

EC-No.	Synonym Recommended Name	Vol., Page
1.1.1.115	D-Ribose:NADP$^+$ 1-oxidoreductase Ribose 1-dehydrogenase (NADP$^+$)	
1.1.1.115	D-Ribose dehydrogenase (NADP$^+$) Ribose 1-dehydrogenase (NADP$^+$)	
5.3.1.20	D-Ribose isomerase Ribose isomerase	Vol. 1, Page 342
5.3.1.20	Ribose isomerase Ribose isomerase	Vol. 1, Page 342
5.3.1.20	D-Ribose ketol-isomerase Ribose isomerase	Vol. 1, Page 342
5.3.1.6	Ribose phosphate isomerase Ribose-5-phosphate isomerase	Vol. 1, Page 277
5.3.1.6	Ribosephosphate isomerase A Ribose-5-phosphate isomerase	Vol. 1, Page 277
5.3.1.6	Ribosephosphate isomerase B Ribose-5-phosphate isomerase	Vol. 1, Page 277
3.4.11.10	ribosomal-bound aminopeptidase bacterial leucyl aminopeptidase	
2.3.1.128	Ribosomal-protein-alanine N-acetyltransferase Ribosomal-protein-alanine N-acetyltransferase	
2.3.2.12	Ribosomal peptidyltransferase Peptidyltransferase	
2.7.1.37	Ribosomal protein S6 kinase II Protein kinase	
2.1.1.52	Ribosomal ribonucleate guanine-2-methyltransferase rRNA (guanine-N^2-)-methyltransferase	
2.1.1.51	Ribosomal ribonucleate guanine 1-methyltransferase rRNA (guanine-N^1-)-methyltransferase	
2.7.1.37	Ribosomal S6 protein kinase Protein kinase	
3.3.1.3	S-Ribosyl-L-homocysteine ribohydrolase Ribosylhomocysteinase	
3.2.2.7	N-Ribosyladenine ribohydrolase Adenosine nucleosidase	
3.3.1.3	Ribosylhomocysteinase Ribosylhomocysteinase	

EC-No.	Synonym Recommended Name	Vol., Page
3.3.1.3	S-Ribosylhomocysteine hydrolase Ribosylhomocysteinase	
2.7.1.22	Ribosylnicotinamide kinase Ribosylnicotinamide kinase	
3.2.2.1	N-Ribosyl purine ribohydrolase Purine nucleosidase	
3.2.2.8	N-Ribosylpyrimidine nucleosidase Ribosylpyrimidine nucleosidase	
3.2.2.8	Ribosylpyrimidine nucleosidase Ribosylpyrimidine nucleosidase	
3.2.2.8	N-Ribosylpyrimidine ribohydrolase Ribosylpyrimidine nucleosidase	
2.4.2.5	Ribosyltransferase, nucleoside Nucleoside ribosyltransferase	
2.4.2.29	Ribosyltransferase, queuine transfer ribonucleate Queuine tRNA-ribosyltransferase	
4.2.1.70	5-ribosyluracil 5'-phosphate synthetase pseudouridylate synthase	
2.1.1.35	Ribothymidyl synthase tRNA (uracil-5-)-methyltransferase	
2.7.1.47	D-Ribulokinase D-Ribulokinase	
2.7.1.16	L-Ribulokinase Ribulokinase	
2.7.1.16	Ribulokinase Ribulokinase	
4.1.1.39	D-Ribulose-1,5-bisphosphate carboxylase Ribulose-bisphosphate carboxylase	
2.1.1.127	Ribulose-1,5-bisphosphate carboxylase/oxygenase large subunit $^{\varepsilon}$N-methyltransferase [Ribulose-bisphosphate-carboxylase]-lysine N-methyltransferase	
2.1.1.127	Ribulose-1,5-bisphosphate carboxylase/oxygenase large subunit $^{\varepsilon}$N-methyltransferase [Ribulose-bisphosphate-carboxylase]-lysine N-methyltransferase	
5.1.3.1	D-Ribulose-5-P 3-epimerase Ribulose-phosphate 3-epimerase	Vol. 1, Page 91

EC-No.	Synonym Recommended Name	Vol., Page
5.1.3.1	D-Ribulose-5-phosphate 3-epimerase Ribulose-phosphate 3-epimerase	Vol. 1, Page 91
5.1.3.4	L-Ribulose-5-phosphate 4-epimerase L-Ribulose-phosphate 4-epimerase	Vol. 1, Page 123
5.1.3.4	L-Ribulose-5-phosphate 4-epimerase L-Ribulose-phosphate 4-epimerase	Vol. 1, Page 123
5.1.3.1	D-Ribulose-5-phosphate epimerase Ribulose-phosphate 3-epimerase	Vol. 1, Page 91
2.7.1.19	Ribulose-5-phosphate kinase Phosphoribulokinase	
2.1.1.127	Ribulose-bisphosphate-carboxylase/oxygenase N-methyltransferase [Ribulose-bisphosphate-carboxylase]-lysine N-methyltransferase	
2.1.1.127	[Ribulose-bisphosphate-carboxylase]-lysine N-methyltransferase [Ribulose-bisphosphate-carboxylase]-lysine N-methyltransferase	
4.1.1.39	Ribulose-bisphosphate carboxylase Ribulose-bisphosphate carboxylase	
5.1.3.1	Ribulose-phosphate 3-epimerase Ribulose-phosphate 3-epimerase	Vol. 1, Page 91
5.1.3.4	L-Ribulose-phosphate 4-epimerase L-Ribulose-phosphate 4-epimerase	Vol. 1, Page 123
4.1.1.39	Ribulose 1,5-bisphosphate carboxylase Ribulose-bisphosphate carboxylase	
4.1.1.39	Ribulose 1,5-bisphosphate carboxylase-oxygenase Ribulose-bisphosphate carboxylase	
2.1.1.127	Ribulose 1,5-bisphosphate carboxylase/oxygenase large subunit N^{ε}-methyltransferase [Ribulose-bisphosphate-carboxylase]-lysine N-methyltransferase	
4.1.1.39	D-Ribulose 1,5-diphosphate carboxylase Ribulose-bisphosphate carboxylase	
4.1.1.39	Ribulose 1,5-diphosphate carboxylase Ribulose-bisphosphate carboxylase	
4.1.1.39	Ribulose 1,5-diphosphate carboxylase/oxygenase Ribulose-bisphosphate carboxylase	
5.1.3.1	Ribulose 5-phosphate 3-epimerase Ribulose-phosphate 3-epimerase	Vol. 1, Page 91

EC-No.	Synonym Recommended Name	Vol., Page
2.7.1.19	Ribulose 5-phosphate kinase Phosphoribulokinase	
4.1.1.39	Ribulose bisphosphate carboxylase-oxygenase Ribulose-bisphosphate carboxylase	
2.1.1.127	Ribulose bisphosphate carboxylase large subunit methyltransferase [Ribulose-bisphosphate-carboxylase]-lysine N-methyltransferase	
4.1.1.39	Ribulose diphosphate carboxylase Ribulose-bisphosphate carboxylase	
4.1.1.39	Ribulose diphosphate carboxylase/oxygenase Ribulose-bisphosphate carboxylase	
5.1.3.1	D-Ribulose phosphate-3-epimerase Ribulose-phosphate 3-epimerase	Vol. 1, Page 91
5.1.3.1	Ribulose phosphate 3-epimerase Ribulose-phosphate 3-epimerase	Vol. 1, Page 91
5.1.3.4	L-Ribulose phosphate 4-epimerase L-Ribulose-phosphate 4-epimerase	Vol. 1, Page 123
5.1.3.4	Ribulose phosphate 4-epimerase L-Ribulose-phosphate 4-epimerase	Vol. 1, Page 123
2.7.1.19	Ribulose phosphate kinase Phosphoribulokinase	
3.2.2.22	Ricin rRNA N-glycosidase	
3.5.5.2	Ricinine aminohydrolase Ricinine nitrilase	
3.5.5.2	Ricinine nitrilase Ricinine nitrilase	
1.14.13.26	Ricinoleic acid synthase Phosphatidylcholine 12-monooxygenase	
1.10.3.6	Rifamycin-B:oxygen oxidoreductase Rifamycin-B oxidase	
1.10.3.6	Rifamycin-B oxidase Rifamycin-B oxidase	
1.10.3.6	Rifamycin B oxidase Rifamycin-B oxidase	
2.8.2.1	Ritodrine sulfotransferase Aryl sulfotransferase	

EC-No.	Synonym Recommended Name	Vol., Page
3.4.21.39	RMCP I chymase	
3.1.26.4	RNA*DNA hybrid ribonucleotidohydrolase Calf thymus ribonuclease H	
6.5.1.4	RNA-3'-phosphate:RNA ligase (cyclizing, AMP-forming) RNA-3'-phosphate cyclase	Vol. 2, Page 793
6.5.1.4	RNA-3'-phosphate cyclase RNA-3'-phosphate cyclase	Vol. 2, Page 793
2.7.7.49	RNA-dependent DNA polymerase RNA-directed DNA polymerase	
2.7.7.48	RNA-dependent ribonucleate nucleotidyltransferase RNA-directed RNA polymerase	
2.7.7.48	RNA-dependent RNA polymerase RNA-directed RNA polymerase	
2.7.7.48	RNA-dependent RNA replicase RNA-directed RNA polymerase	
2.7.7.49	RNA-directed DNA polymerase RNA-directed DNA polymerase	
2.7.7.48	RNA-directed RNA polymerase RNA-directed RNA polymerase	
2.7.7.49	RNA-instructed DNA polymerase RNA-directed DNA polymerase	
2.1.1.66	RNA-pentose methylase rRNA (adenosine-2'-O-)-methyltransferase	
2.7.1.141	[RNA-polymerase]-subunit kinase [RNA-polymerase]-subunit kinase	
6.5.1.4	RNA 3'-terminal phosphate cyclase RNA-3'-phosphate cyclase	Vol. 2, Page 793
6.5.1.4	RNA 3'-terminal phosphate cyclase RNA-3'-phosphate cyclase	Vol. 2, Page 793
2.7.7.19	RNA adenylating enzyme Polynucleotide adenylyltransferase	
3.1.27.1	RNAase CL Ribonuclease T_2	
6.5.1.4	RNA cyclase RNA-3'-phosphate cyclase	Vol. 2, Page 793

EC-No.	Synonym Recommended Name	Vol., Page
2.7.7.6	RNA formation factors, C DNA-directed RNA polymerase	
2.7.7.19	RNA formation factors, PF1 Polynucleotide adenylyltransferase	
6.5.1.3	RNA ligase RNA ligase (ATP)	Vol. 2, Page 787
6.5.1.3	RNA ligase (ATP) RNA ligase (ATP)	Vol. 2, Page 787
3.2.2.22	RNA N-glycosidase rRNA N-glycosidase	
2.7.7.6	RNA nucleotidyltransferase DNA-directed RNA polymerase	
2.7.7.6	RNA nucleotidyltransferase (DNA-directed) DNA-directed RNA polymerase	
2.7.7.48	RNA nucleotidyltransferase (RNA-directed) RNA-directed RNA polymerase	
2.7.7.6	RNA polymerase DNA-directed RNA polymerase	
2.7.7.6	RNA polymerase I DNA-directed RNA polymerase	
2.7.7.6	RNA polymerase II DNA-directed RNA polymerase	
2.7.1.141	RNA polymerase II CTD kinase [RNA-polymerase]-subunit kinase	
2.7.7.6	RNA polymerase III DNA-directed RNA polymerase	
2.7.7.48	RNA replicase RNA-directed RNA polymerase	
2.7.7.49	RNA revertase RNA-directed DNA polymerase	
3.1.27.5	RNase Pancreatic ribonuclease	
3.1.27.1	Rnase (non-base specific) Ribonuclease T_2	
3.1.27.5	RNase A Pancreatic ribonuclease	

EC-No.	Synonym Recommended Name	Vol., Page
3.1.26.3	RNase D Ribonuclease III	
3.1.27.3	RNase F1 Ribonuclease T_1	
3.1.27.3	RNase G Ribonuclease T_1	
3.1.26.4	RNase H Calf thymus ribonuclease H	
3.1.27.5	RNase I Pancreatic ribonuclease	
3.1.13.1	RNase II Exoribonuclease II	
3.1.27.1	RNase II Ribonuclease T_2	
3.1.27.1	RNase M Ribonuclease T_2	
3.1.26.8	RNase M5 Ribonuclease M_5	
3.1.27.1	RNase Ms Ribonuclease T_2	
3.1.27.3	RNase N_1 Ribonuclease T_1	
3.1.27.3	RNase N_2 Ribonuclease T_1	
3.1.26.3	RNase O Ribonuclease III	
3.1.26.5	RNase P Ribonuclease P	
2.7.7.56	RNase PH tRNA nucleotidyltransferase	
3.1.27.3	RNase Sa Ribonuclease T_1	
3.1.27.3	RNase T_1 Ribonuclease T_1	
3.1.27.1	RNase T_2 Ribonuclease T_2	

EC-No.	Synonym Recommended Name	Vol., Page
3.1.27.4	RNase U_2 Ribonuclease U_2	
3.1.27.4	RNase U_3 Ribonuclease U_2	
2.8.1.4	RNA sulfurtransferase tRNA sulfurtransferase	
2.7.7.48	RNA synthetase RNA-directed RNA polymerase	
2.7.7.6	RNA transcriptase DNA-directed RNA polymerase	
2.7.7.48	RNA transcriptase RNA-directed RNA polymerase	
2.7.7.52	RNA uridylyltransferase RNA uridylyltransferase	
2.3.1.140	Rosmarinate synthase Rosmarinate synthase	
2.3.1.140	Rosmarinic acid synthase Rosmarinate synthase	
5.2.1.8	Rotamase Peptidylprolyl isomerase	Vol. 1, Page 218
5.1.3.1	RPE Ribulose-phosphate 3-epimerase	Vol. 1, Page 91
3.1.1.63	RPH 11-cis-Retinyl-palmitate hydrolase	
5.3.1.6	RPI Ribose-5-phosphate isomerase	Vol. 1, Page 277
3.1.3.48	RPTPα protein-tyrosine-phosphatase	
2.1.1.48	rRNA (adenine-N^6-)-methyltransferase rRNA (adenine-N^6-)-methyltransferase	
2.1.1.66	rRNA (adenosine-2'-O-)-methyltransferase rRNA (adenosine-2'-O-)-methyltransferase	
2.1.1.51	rRNA (guanine-N^1-)-methyltransferase rRNA (guanine-N^1-)-methyltransferase	
2.1.1.52	rRNA (guanine-N^2-)-methyltransferase rRNA (guanine-N^2-)-methyltransferase	

EC-No.	Synonym Recommended Name	Vol., Page
2.1.1.66	rRNA adenosine 2'-methylase rRNA (adenosine-2'-O-)-methyltransferase	
3.1.27.10	rRNA endonuclease rRNA endonuclease	
3.2.2.22	rRNA N-glycohydrolase rRNA N-glycosidase	
3.2.2.22	rRNA N-glycosidase rRNA N-glycosidase	
2.1.1.115	(RS)-1-Benzyl-1,2,3,4-tetrahydroisoquinoline N-methyltransferase (RS)-1-Benzyl-1,2,3,4-tetrahydroisoquinoline N-methyltransferase	
2.1.1.128	(RS)-Norcoclaurine 6-O-methyltransferase (RS)-Norcoclaurine 6-O-methyltransferase	
2.7.7.49	RT RNA-directed DNA polymerase	
3.4.17.20	rTAFI Carboxypeptidase U	
2.5.1.20	Rubber allyltransferase Rubber cis-polyprenylcistransferase	
2.5.1.20	Rubber cis-polyprenylcistransferase Rubber cis-polyprenylcistransferase	
2.5.1.20	Rubber polymerase Rubber cis-polyprenylcistransferase	
2.5.1.20	Rubber prenyltransferase Rubber cis-polyprenylcistransferase	
2.5.1.20	Rubber transferase Rubber cis-polyprenylcistransferase	
3.4.24.48	Ruberlysin Ruberlysin	
4.1.1.39	Rubisco Ribulose-bisphosphate carboxylase	
2.1.1.127	Rubisco large subunit $^{\varepsilon}$N-methyltransferase [Ribulose-bisphosphate-carboxylase]-lysine N-methyltransferase	
2.1.1.127	Rubisco LSMT [Ribulose-bisphosphate-carboxylase]-lysine N-methyltransferase	
2.1.1.127	Rubisco methyltransferase [Ribulose-bisphosphate-carboxylase]-lysine N-methyltransferase	

EC-No.	Synonym Recommended Name	Vol., Page
4.1.1.39	RuBP carboxylase Ribulose-bisphosphate carboxylase	
1.18.1.4	rubredoxin-NAD(P) reductase rubredoxin-NAD(P) reductase	
1.18.1.1	Rubredoxin-NAD$^+$ reductase Rubredoxin-NAD$^+$ reductase	
1.6.7.2	rubredoxin-NAD$^+$ reductase rubredoxin-NAD$^+$ reductase	
1.18.1.1	Rubredoxin-NAD reductase Rubredoxin-NAD$^+$ reductase	
1.18.1.4	rubredoxin-nicotinamide adenine dinucleotide (phosphate) reductase rubredoxin-NAD(P) reductase	
1.18.1.4	rubredoxin-nicotinamide adenine dinucleotide phosphate reductase rubredoxin-NAD(P) reductase	
1.18.1.4	rubredoxin:NAD(P) oxidoreductase rubredoxin-NAD(P) reductase	
1.18.1.1	Rubredoxin:NAD$^+$ oxidoreductase Rubredoxin-NAD$^+$ reductase	
1.18.1.1	Rubredoxin reductase Rubredoxin-NAD$^+$ reductase	
3.4.21.35	RUK tissue kallikrein	
3.4.24.58	Russell's viper blood coagulation factor X activator Russellysin	
3.4.24.58	Russell's viper venom factor X activator, RVV-X Russellysin	
3.4.24.58	Russellysin Russellysin	
3.4.24.58	RVV-V Russellysin	
3.4.21.95	RVV-V Snake venom factor V activator	
2.5.1.18	RX:glutathione R-transferase Glutathione transferase	

EC-No.	Synonym Recommended Name	Vol., Page
4.2.1.93	(6S)-β-6-Hydroxy-1,4,5,6-tetrahydronicotinamide-adenine dinucleotide hydro-lyase (ATP-hydrolysing) ATP-dependent H4NAD(P)OH dehydratase	
1.1.1.232	(15S)-15-Hydroxy-5,8,11-cis-13-trans-eicosatetraenoate:NAD(P)$^+$ 15-oxidoreductase 15-Hydroxyeicosatetraenoate dehydrogenase	
5.4.99.8	(S)-2,3-Epoxysqualene mutase (cyclizing, cycloartenol-forming) Cycloartenol synthase	Vol. 1, Page 631
5.4.99.7	(S)-2,3-Epoxysqualene mutase (cyclizing, lanosterol-forming) Lanosterol synthase	Vol. 1, Page 624
1.1.3.19	(S)-2-Hydroxy-2-(4-hydroxyphenyl)acetate:oxygen 1-oxidoreductase 4-Hydroxymandelate oxidase	
4.1.1.5	(S)-2-Hydroxy-2-methyl-3-oxobutanoate carboxy-lyase Acetolactate decarboxylase	
1.14.16.6	(S)-2-Hydroxy-2-phenylacetate,tetrahydrobiopterin:oxygen oxidoreductase (4-hydroxylating) Mandelate 4-monooxygenase	
1.1.3.15	(S)-2-Hydroxy-acid:oxygen 2-oxidoreductase (S)-2-Hydroxy-acid oxidase	
1.1.3.15	(S)-2-Hydroxy-acid oxidase (S)-2-Hydroxy-acid oxidase	
1.1.1.99	(S)-2-Hydroxy-fatty-acid dehydrogenase (S)-2-Hydroxy-fatty-acid dehydrogenase	
1.1.99.2	(S)-2-Hydroxyglutarate:(acceptor) 2-oxidoreductase 2-Hydroxyglutarate dehydrogenase	
1.1.1.99	(S)-2-Hydroxystearate:NAD$^+$ oxidoreductase (S)-2-Hydroxy-fatty-acid dehydrogenase	
2.1.3.1	(S)-2-Methyl-3-oxopropanoyl-CoA:pyruvate carboxyltransferase Methylmalonyl-CoA carboxyltransferase	
4.1.1.41	(S)-2-Methyl-3-oxopropanoyl-CoA carboxy-lyase Methylmalonyl-CoA decarboxylase	
4.2.1.34	(S)-2-Methylmalate dehydratase (S)-2-Methylmalate dehydratase	
4.2.1.34	(S)-2-Methylmalate hydro-lyase (S)-2-Methylmalate dehydratase	
3.1.6.17	(S)-2-O-Sulfolactate 2-sulfohydrolase D-Lactate-2-sulfatase	

EC-No.	Synonym Recommended Name	Vol., Page
1.5.1.28	(2S)-2-{[1-(R)-carboxyethyl]amino}pentanoate dehydrogenase (NAD, L-aminopentanoate-forming) Opine dehydrogenase	
5.4.3.3	(3S)-3,6-Diaminohexanoate 5,6-aminomutase β-Lysine 5,6-aminomutase	Vol. 1, Page 558
1.1.1.111	(S)-3-(Imidazol-5-yl)lactate:NAD(P)$^+$ oxidoreductase 3-(Imidazol-5-yl)lactate dehydrogenase	
2.6.1.22	(S)-3-Amino-2-methylpropanoate:2-oxoglutarate aminotransferase (S)-3-Amino-2-methylpropionate transaminase	
2.6.1.22	(S)-3-Amino-2-methylpropionate transaminase (S)-3-Amino-2-methylpropionate transaminase	
4.1.3.24	(3S)-3-carboxy-3-hydroxypropanoyl-CoA glyoxylate-lyase malyl-CoA lyase	
4.1.3.4	(S)-3-Hydroxy-3-methylglutaryl-CoA acetoacetate-lyase Hydroxymethylglutaryl-CoA lyase	
4.1.3.5	(S)-3-Hydroxy-3-methylglutaryl-CoA acetoacetyl-CoA-lyase (CoA-acetylating) Hydroxymethylglutaryl-CoA synthase	
4.2.1.18	(S)-3-hydroxy-3-methylglutaryl-CoA hydro-lyase methylglutaconyl-CoA hydratase	
3.1.2.5	(S)-3-hydroxy-3-methylglutaryl-CoA hydrolase hydroxymethylglutaryl-CoA hydrolase	
1.2.1.56	(S)-3-Hydroxyacid ester dehydrogenase (S)-3-Hydroxyacid ester dehydrogenase	
1.1.1.35	(S)-3-Hydroxyacyl-CoA:NAD$^+$ oxidoreductase 3-Hydroxyacyl-CoA dehydrogenase	
4.2.1.17	(3S)-3-hydroxyacyl-CoA hydro-lyase enoyl-CoA hydratase	
1.1.99.24	(S)-3-Hydroxybutanoate:2-oxoglutarate oxidoreductase Hydroxyacid-oxoacid transhydrogenase	
1.1.1.157	(S)-3-Hydroxybutanoyl-CoA:NADP$^+$ oxidoreductase 3-Hydroxybutyryl-CoA dehydrogenase	
4.1.1.72	(3S)-3-Methyl-2-oxopentanoate carboxy-lyase Branched-chain-2-oxoacid decarboxylase	
5.4.3.8	(S)-4-Amino-5-oxopentanoate 4,5-aminomutase Glutamate-1-semialdehyde 2,1-aminomutase	Vol. 1, Page 575

EC-No.	Synonym Recommended Name	Vol., Page
4.1.2.11	(S)-4-Hydroxymandelonitrile hydroxybenzaldehyde-lyase Hydroxymandelonitrile lyase	
1.14.11.14	(6S)-6-Hydroxyhyoscyamine,2-oxoglutarate oxidoreductase (epoxide-forming) 6β-Hydroxyhyoscyamine epoxidase	
1.5.3.5	(S)-6-Hydroxynicotine:oxygen oxidoreductase (S)-6-Hydroxynicotine oxidase	
1.5.3.5	(S)-6-Hydroxynicotine oxidase (S)-6-Hydroxynicotine oxidase	
5.3.99.2	(5,13)-(15S)-9α,11α-Epidioxy-15-hydroxyprosta-5,13-dienoate D-isomerase Prostaglandin-D synthase	Vol. 1, Page 451
2.1.1.41	(S)-Adenosyl-4-methionine:sterol $\Delta^2$4-methyltransferase 24-Sterol C-methyltransferase	
2.1.1.117	(S)-Adenosyl-L-methionine:(S)-scoularine 9-O-methyltransferase (S)-Scoulerine 9-O-methyltransferase	
1.1.3.36	(S)-Canadine synthase (S)-Canadine synthase	
1.1.1.254	(S)-Carnitine 3-dehydrogenase (S)-Carnitine 3-dehydrogenase	
1.1.1.254	(S)-Carnitine:NAD oxidoreductase (S)-Carnitine 3-dehydrogenase	
1.1.3.32	(S)-Cheilanthifoline, $NADPH_2$:oxygen oxidoreductase (methylenedioxy-bridge-forming) (S)-Stylopine synthase	
1.1.3.32	(S)-Cheilanthifoline oxidase (methylenedioxy-bridge-forming) (S)-Stylopine synthase	
1.14.13.37	(S)-cis-N-Methyltetrahydroprotoberberine-14-hydroxylase Methyltetrahydroprotoberberine 14-monooxygenase	
4.1.3.22	(S)-citramalate lyase citramalate lyase	
4.1.3.22	(3S)-citramalate pyruvate-lyase citramalate lyase	
4.1.3.25	(3S)-citramalyl-CoA pyruvate-lyase citramalyl-CoA lyase	
4.1.3.34	(3S)-citryl-CoA oxaloacetate-lyase citryl-CoA lyase	

EC-No.	Synonym Recommended Name	Vol., Page
2.1.1.128	(S)-Coclaurine N-methyltransferase (RS)-Norcoclaurine 6-O-methyltransferase	
1.3.99.11	(S)-Dihydroorotate:(acceptor) oxidoreductase Dihydroorotate dehydrogenase	
1.3.1.14	(S)-Dihydroorotate:NAD^+ oxidoreductase Orotate reductase (NADH)	
1.3.1.15	(S)-Dihydroorotate:$NADP^+$ oxidoreductase Orotate reductase (NADPH)	
1.3.3.1	(S)-Dihydroorotate:oxygen oxidoreductase Dihydroorotate oxidase	
3.5.2.3	(S)-Dihydroorotate amidohydrolase Dihydroorotase	
1.1.1.234	(2S)-Flavan-4-ol:$NADP^+$ 4-oxidoreductase Flavanone 4-reductase	
1.14.11.9	(2S)-Flavanone 3-hydroxylase Naringenin 3-dioxygenase	
4.1.2.11	(S)-HNL Hydroxymandelonitrile lyase	
4.1.2.39	(S)-HNL Hydroxynitrilase	
4.1.2.11	(S)-Hydroxynitrile lyase Hydroxymandelonitrile lyase	
1.2.1.22	(S)-Lactaldehyde:NAD^+ oxidoreductase Lactaldehyde dehydrogenase	
1.1.2.3	(S)-Lactate:ferricytochrome-c 2-oxidoreductase L-Lactate dehydrogenase (cytochrome)	
1.1.1.27	(S)-Lactate:NAD^+ oxidoreductase L-Lactate dehydrogenase	
1.1.99.7	(S)-Lactate:oxaloacetate oxidoreductase Lactate-malate transhydrogenase	
1.13.12.4	(S)-Lactate:oxygen 2-oxidoreductase (decarboxylating) Lactate 2-monooxygenase	
4.1.2.36	(S)-Lactate acetaldehyde-lyase Lactate aldolase	
5.4.3.7	(2S)α-Leucine 2,3-aminomutase Leucine 2,3-aminomutase	Vol. 1, Page 571

EC-No.	Synonym Recommended Name	Vol., Page
4.2.3.16	(-)-(4S)-limonene synthase (-)-(4S)-limonene synthase	
1.1.99.16	(S)-Malate:(acceptor) oxidoreductase Malate dehydrogenase (acceptor)	
1.1.1.37	(S)-Malate:NAD^+ oxidoreductase Malate dehydrogenase	
1.1.1.39	(S)-Malate:NAD^+ oxidoreductase (decarboxylating) Malate dehydrogenase (decarboxylating)	
1.1.1.38	(S)-Malate:NAD^+ oxidoreductase (oxaloacetate-decarboxylating) Malate dehydrogenase (oxaloacetate-decarboxylating)	
1.1.1.82	(S)-Malate:$NADP^+$ oxidoreductase Malate dehydrogenase ($NADP^+$)	
1.1.1.40	(S)-Malate:$NADP^+$ oxidoreductase (oxaloacetate-decarboxylating) Malate dehydrogenase (oxaloacetate-decarboxylating, $NADP^+$)	
1.1.3.3	(S)-Malate:oxygen oxidoreductase Malate oxidase	
4.2.1.2	(S)-malate hydro-lyase fumarate hydratase	
3.1.2.17	(S)-Methylmalonyl-CoA hydrolase (S)-Methylmalonyl-CoA hydrolase	
3.1.2.17	(S)-Methylmalonyl-CoA hydrolase (S)-Methylmalonyl-CoA hydrolase	
5.4.99.2	(S)-Methylmalonyl-CoA mutase Methylmalonyl-CoA mutase	Vol. 1, Page 589
1.1.1.88	(S)-Mevalonate:NAD^+ oxidoreductase (CoA-acylating) Hydroxymethylglutaryl-CoA reductase	
3.5.1.85	(S)-N-acetyl-1-phenylethylamine amidohydrolase (S)-N-acetyl-1-phenylethylamine hydrolase	
3.5.1.85	(S)-N-acetyl-1-phenylethylamine hydrolase (S)-N-acetyl-1-phenylethylamine hydrolase	
3.5.1.85	(S)-N-acetylphenylethylamine:H_2O hydrolase (S)-N-acetyl-1-phenylethylamine hydrolase	
1.14.13.37	(S)-N-Methylcanadine,NADPH:oxygen oxidoreductase (14-hydroxylating) Methyltetrahydroprotoberberine 14-monooxygenase	

EC-No.	Synonym Recommended Name	Vol., Page
1.1.3.34	(S)-N-Methylcoclaurine, NADPH$_2$:oxygen oxidoreductase (C-O phenol-coupling) Berbamunine synthase	
1.1.3.34	(S)-N-Methylcoclaurine oxidase (C-O phenol-couling) Berbamunine synthase	
4.2.1.78	(S)-Norcoclaurine synthase (S)-Norcoclaurine synthase	
4.2.1.78	(S)-Norlaudanosoline synthase (S)-Norcoclaurine synthase	
4.1.2.39	(S)-Oxynitrilase Hydroxynitrilase	
4.1.2.11	(S)-*p*-Hydroxy-mandelonitrile lyase Hydroxymandelonitrile lyase	
1.5.3.9	(S)-Reticuline:oxygen oxidoreductase (methylene-bridge forming) Reticuline oxidase	
1.1.3.33	(S)-Scoulerine, NADPH$_2$:oxygen oxidoreductase (methylenedioxy-bridge-forming) S-Cheilanthifoline synthase	
2.1.1.117	(S)-Scoulerine 9-O-methyltransferase (S)-Scoulerine 9-O-methyltransferase	
1.1.3.33	(S)-Scoulerine oxidase (methylenedioxy-bridge-forming) S-Cheilanthifoline synthase	
4.3.3.2	3-α(S)-Strictosidine tryptamine-lyase Strictosidine synthase	
1.1.3.32	(S)-Stylopine synthase (S)-Stylopine synthase	
1.3.3.8	(S)-Tetrahydroberberine:oxygen oxidoreductase Tetrahydroberberine oxidase	
1.3.3.8	(S)-Tetrahydroberberine oxidase Tetrahydroberberine oxidase	
1.1.3.36	(S)-Tetrahydroberberine synthase (S)-Canadine synthase	
1.1.3.36	(S)-Tetrahydrocolumbamine, NADPH$_2$:oxygen oxidoreductase (methylene-dioxy-bridge-forming) (S)-Canadine synthase	
1.1.3.36	(S)-Tetrahydrocolumbamine oxidase (methylenedioxy-bridge-forming) (S)-Canadine synthase	

EC-No.	Synonym Recommended Name	Vol., Page
2.1.1.122	(S)-Tetrahydroprotoberberine-cis-N-methyltransferase (S)-Tetrahydroprotoberberine N-methyltransferase	
2.1.1.122	(S)-Tetrahydroprotoberberine N-methyltransferase (S)-Tetrahydroprotoberberine N-methyltransferase	
1.5.3.8	(S)-tetrahydroprotoberberine oxidase (S)-tetrahydroprotoberberine oxidase	
1.3.3.8	(S)-THB oxidase Tetrahydroberberine oxidase	
1.1.1.154	(S)-Ureidoglycolate:NAD(P)$^+$ oxidoreductase Ureidoglycolate dehydrogenase	
1.1.1.199	(S)-Usnate reductase (S)-Usnate reductase	
4.1.3.23	(2S,3R)-2-Hydroxytridecane-1,2,3-tricarboxylate oxaloacetate-lyase (CoA-acylating) Decylcitrate synthase	
4.2.1.99	(2S,3R)-3-Hydroxybutane-1,2,3-tricarboxylate hydro-lyase 2-Methylisocitrate dehydratase	
4.2.1.99	(2S,3R)-3-Hydroxybutane-1,2,3-tricarboxylate hydro-lyase 2-Methylisocitrate dehydratase	
4.1.3.30	(2S,3R)-3-Hydroxybutane-1,2,3-tricarboxylate pyruvate-lyase Methylisocitrate lyase	
1.1.1.122	(2S,3R)-Aldose dehydrogenase D-threo-Aldose 1-dehydrogenase	
1.1.1.166	(1S,3R,4S)-3,4-Dihydroxycyclohexane-1-carboxylate:NAD$^+$ 3-oxidoreductase Hydroxycyclohexanecarboxylate dehydrogenase	
1.1.1.122	2S,3R-Aldose-dehydrogenase D-threo-Aldose 1-dehydrogenase	
1.1.1.178	(2S,3S)-3-Hydroxy-2-methylbutanoyl-CoA:NAD$^+$ oxidoreductase 3-Hydroxy-2-methylbutyryl-CoA dehydrogenase	
1.3.1.53	(3S,4R)3,4-Dihydroxycyclohexa-1,5-diene-1,4-dicarboxylate:NADoxidoreductase (3S,4R)3,4-Dihydroxycyclohexa-1,5-diene-1,4-dicarboxylate dehydrogenase	
1.3.1.53	(3S,4R)3,4-Dihydroxycyclohexa-1,5-diene-1,4-dicarboxylate dehydrogenase (3S,4R)3,4-Dihydroxycyclohexa-1,5-diene-1,4-dicarboxylate dehydrogenase	

EC-No.	Synonym Recommended Name	Vol., Page
4.2.3.14	(-)-(1S,5S)-pinene synthase pinene synthase	
1.1.1.76	(S,S)-Butane-2,3-diol:NAD$^+$ oxidoreductase (S,S)-Butanediol dehydrogenase	
1.1.1.76	(S,S)-Butanediol dehydrogenase (S,S)-Butanediol dehydrogenase	
4.2.1.81	(S,S)-Tartrate hydro-lyase D(-)-Tartrate dehydratase	
4.4.1.3	S,S-dimethyl-β-propiothetin dimethyl-sulfide-lyase dimethylpropiothetin dethiomethylase	
3.1.30.1	S_1 Nuclease Aspergillus nuclease S_1	
3.2.1.23	S 2107 β-galactosidase	
4.6.1.9	Sabinene-hydrate synthase Sabinene-hydrate synthase	
4.2.3.11	sabinene-hydrate synthase sabinene-hydrate synthase	
4.6.1.9	Sabinene hydrate cyclase Sabinene-hydrate synthase	
4.2.3.11	sabinene hydrate cyclase sabinene-hydrate synthase	
1.1.1.228	(+)-Sabinol dehydrogenase (+)-Sabinol dehydrogenase	
3.2.1.26	Saccharase β-fructofuranosidase	
3.2.1.2	Saccharogen amylase β-amylase	
3.2.1.2	Saccharogenamylase β-amylase	
3.4.24.37	Saccharolysin Saccharolysin	
3.4.23.25	Saccharomyces aspartic proteinase Saccharopepsin	
3.4.23.25	Saccharomyces cerevisiae aspartic proteinase A Saccharopepsin	

EC-No.	Synonym Recommended Name	Vol., Page
3.4.16.6	Saccharomyces cerevisiae KEX1 gene product Carboxypeptidase D	
3.4.23.25	Saccharopepsin Saccharopepsin	
1.5.1.9	Saccharopin dehydrogenase Saccharopine dehydrogenase (NAD^+, L-glutamate-forming)	
1.5.1.9	Saccharopine dehydrogenase (NAD^+, L-glutamate-forming) Saccharopine dehydrogenase (NAD^+, L-glutamate-forming)	
1.5.1.7	Saccharopine dehydrogenase (NAD^+, L-lysine forming) Saccharopine dehydrogenase (NAD^+, L-lysine forming)	
1.5.1.10	Saccharopine dehydrogenase ($NADP^+$, L-glutamate forming) Saccharopine dehydrogenase ($NADP^+$, L-glutamate forming)	
1.5.1.8	Saccharopine dehydrogenase ($NADP^+$, L-lysine forming) Saccharopine dehydrogenase ($NADP^+$, L-lysine forming)	
1.5.1.10	Saccharopine reductase Saccharopine dehydrogenase ($NADP^+$, L-glutamate forming)	
3.3.1.1	SAHase Adenosylhomocysteinase	
6.3.2.6	SAICAR synthetase Phosphoribosylaminoimidazolesuccinocarboxamide synthase	Vol. 2, Page 434
3.2.1.21	Salicilinase β-glucosidase	
2.4.1.172	Salicyl-alcohol β-D-glucosyltransferase Salicyl-alcohol β-D-glucosyltransferase	
1.2.1.65	salicylaldehyde:NAD oxidoreductase salicylaldehyde dehydrogenase	
1.2.1.65	salicylaldehyde dehydrogenase salicylaldehyde dehydrogenase	
1.14.13.1	Salicylate,NADH:oxygen oxidoreductase (1-hydroxylating, decarboxylating) Salicylate 1-monooxygenase	
1.14.13.1	Salicylate 1-hydroxylase Salicylate 1-monooxygenase	
1.14.13.1	Salicylate 1-monooxygenase Salicylate 1-monooxygenase	

EC-No.	Synonym Recommended Name	Vol., Page
1.14.13.1	Salicylate hydroxylase (decarboxylating) Salicylate 1-monooxygenase	
1.14.13.1	Salicylate monooxygenase Salicylate 1-monooxygenase	
1.14.13.1	Salicylic hydroxylase Salicylate 1-monooxygenase	
3.4.21.35	salivary kallikrein tissue kallikrein	
3.1.1.3	salt-resistant post-heparin lipase triacylglycerol lipase	
1.1.1.248	Salutaridine 7-reductase Salutaridine reductase (NADPH)	
1.1.1.248	Salutaridine reductase (NADPH) Salutaridine reductase (NADPH)	
1.1.3.35	Salutaridine synthase Salutaridine synthase	
2.3.1.150	Salutaridinol 7-O-acetyltransferase Salutaridinol 7-O-acetyltransferase	
1.1.1.248	Salutaridinol:NADP 7-oxidoreductase Salutaridine reductase (NADPH)	
2.1.1.131	SAM-dependent C-17 methyl transferase Precorrin-3B C^{17}-methyltransferase	
2.1.1.120	SAM:12-hydroxydihydrochelirubine-12-O-methyltransferase 12-Hydroxydihydrochelirubine 12-O-methyltransferase	
2.1.1.116	SAM:3'-hydroxy-N-methyl-(S)-coclaurine-4'-O-methyltransferase 3'-Hydroxy-N-methyl-(S)-coclaurine 4'-O-methyltransferase	
4.1.1.50	SAMDC Adenosylmethionine decarboxylase	
4.1.1.50	SAM decarboxylase Adenosylmethionine decarboxylase	
1.3.1.47	α-Santonin 1,2-reductase α-Santonin 1,2-reductase	
3.2.2.22	Saporins rRNA N-glycosidase	
3.1.27.10	α-Sarcin rRNA endonuclease	

EC-No.	Synonym Recommended Name	Vol., Page
1.5.99.1	Sarcosine:(acceptor) oxidoreductase (demethylating) Sarcosine dehydrogenase	
1.5.3.1	Sarcosine:oxygen oxidoreductase (demethylating) Sarcosine oxidase	
1.5.99.1	Sarcosine dehydrogenase Sarcosine dehydrogenase	
1.5.99.1	Sarcosine N-demethylase Sarcosine dehydrogenase	
1.5.3.1	Sarcosine oxidase Sarcosine oxidase	
2.4.1.193	Sarsapogenin 3β-glucosyltransferase Sarsapogenin 3β-glucosyltransferase	
2.4.99.8	SAT-2 α-N-Acetylneuraminate α-2,8-sialyltransferase	
2.4.99.10	SAT-3 Neolactotetraosylceramide α-2,3-sialyltransferase	
2.4.99.2	SAT-4 Monosialoganglioside sialyltransferase	
2.4.99.9	SAT 1 Lactosylceramide α-2,3-sialyltransferase	
2.3.1.30	SATase Serine O-acetyltransferase	
3.4.21.62	Savinase Subtilisin	
3.4.21.62	Savinase 16.0L Subtilisin	
3.4.21.62	Savinase 32.0 L EX Subtilisin	
3.4.21.62	Savinase 4.0T Subtilisin	
3.4.21.62	Savinase 8.0L Subtilisin	
4.1.2.11	SbHNL Hydroxymandelonitrile lyase	
3.1.3.37	SBPase Sedoheptulose-bisphosphatase	

EC-No.	Synonym Recommended Name	Vol., Page
1.1.3.22	Schardinger enzyme Xanthine oxidase	
2.4.1.128	Scopoletin glucosyltransferase UDPglucose:scopoletin O-β-D-glucosyltransferase	
1.11.1.7	Scopoletin peroxidase Peroxidase	
3.4.99.20	scopulariopsis proteinase scopulariopsis proteinase	
3.4.22.10	SCP streptopain	
2.3.1.154	SCPx propionyl-CoA C^2-trimethyltridecanoyltransferase	
6.2.1.5	A-SCS Succinate-CoA ligase (ADP-forming)	Vol. 2, Page 224
6.2.1.4	SCS Succinate-CoA ligase (GDP-forming)	Vol. 2, Page 219
3.4.21.60	Scutelarin Scutelarin	
2.7.1.65	scyllo-Inosamine 4-kinase scyllo-Inosamine 4-kinase	
2.7.1.65	scyllo-Inosamine kinase scyllo-Inosamine 4-kinase	
3.4.23.31	Scytalidium aspartic proteinase A Scytalidopepsin A	
3.4.23.32	Scytalidium aspartic proteinase B Scytalidopepsin B	
3.4.23.31	Scytalidium lignicolum acid proteinase Scytalidopepsin A	
3.4.23.31	Scytalidium lignicolum aspartic proteinase Scytalidopepsin A	
3.4.23.31	Scytalidium lignicolum carboxyl proteinase Scytalidopepsin A	
3.4.23.31	Scytalidopepsin A Scytalidopepsin A	
3.4.23.32	Scytalidopepsin B Scytalidopepsin B	

EC-No.	Synonym Recommended Name	Vol., Page
4.2.1.94	Scytalone 7,8-hydro-lyase Scytalone dehydratase	
1.1.1.252	Scytalone:NADP Δ^5-oxidoreductase Tetrahydroxynaphthalene reductase	
4.2.1.94	Scytalone dehydratase Scytalone dehydratase	
4.2.1.13	L-SD L-Serine dehydratase	
4.2.1.13	SD L-Serine dehydratase	
1.6.5.4	SDA reductase Monodehydroascorbate reductase (NADH)	
1.14.17.1	SDBH Dopamine β-monooxygenase	
1.5.1.22	SDH Strombine dehydrogenase	
3.4.24.12	sea-urchin-hatching proteinase envelysin	
3.4.21.3	sea anemone protease A metridin	
3.4.21.3	sea anemone proteinase A metridin	
6.5.1.1	Sealase DNA ligase (ATP)	Vol. 2, Page 755
3.4.21.63	Seaprose S Oryzin	
3.4.24.12	sea urchin embryo hatching enzyme envelysin	
1.1.1.184	Secondary-alcohol:NADP$^+$ oxidoreductase Carbonyl reductase (NADPH)	
1.1.3.18	Secondary-alcohol:oxygen oxidoreductase Secondary-alcohol oxidase	
1.1.3.18	Secondary-alcohol oxidase Secondary-alcohol oxidase	
1.1.3.18	Secondary alcohol oxidase Secondary-alcohol oxidase	

EC-No.	Synonym Recommended Name	Vol., Page
2.4.1.69	Secretor-type β-galactoside $\alpha1\rightarrow2$fucosyltransferase Galactoside 2-α-L-fucosyltransferase	
2.7.1.14	Sedoheptulokinase Sedoheptulokinase	
3.1.3.37	Sedoheptulose-1,7-bisphosphate 1-phosphohydrolase Sedoheptulose-bisphosphatase	
2.2.1.2	Sedoheptulose-7-phosphate:D-glyceraldehyde-3-phosphate glyceronetransferase Transaldolase	
2.2.1.1	Sedoheptulose-7-phosphate:D-glyceraldehyde-3-phosphate glycolaldehydetransferase Transketolase	
3.1.3.37	Sedoheptulose-bisphosphatase Sedoheptulose-bisphosphatase	
3.1.3.37	Sedoheptulose 1,7-bisphosphatase Sedoheptulose-bisphosphatase	
3.1.3.37	Sedoheptulose 1,7-diphospate phosphatase Sedoheptulose-bisphosphatase	
3.1.3.37	Sedoheptulose 1,7-diphosphatase Sedoheptulose-bisphosphatase	
3.1.3.37	Sedoheptulose bisphosphatase Sedoheptulose-bisphosphatase	
3.1.3.37	Sedoheptulose diphosphatase Sedoheptulose-bisphosphatase	
2.7.9.3	SELD protein selenide, water dikinase	
2.7.9.3	selenide, water dikinase selenide, water dikinase	
1.11.1.9	Selenium-glutathione peroxidase Glutathione peroxidase	
4.4.1.16	Selenocysteine β-lyase Selenocysteine lyase	
2.9.1.1	L-Selenocysteine-tRNASec synthase selenocysteine synthase L-Seryl-tRNASec selenium transferase	
2.9.1.1	L-Selenocysteine-tRNASel synthase L-Seryl-tRNASec selenium transferase	

EC-No.	Synonym Recommended Name	Vol., Page
2.9.1.1	Selenocysteine-tRNA synthase L-Seryl-tRNASec selenium transferase	
4.4.1.16	Selenocysteine lyase Selenocysteine lyase	
4.4.1.16	Selenocysteine reductase Selenocysteine lyase	
4.4.1.16	L-Selenocysteine selenide-lyase Selenocysteine lyase	
2.9.1.1	Selenocysteine synthase L-Seryl-tRNASec selenium transferase	
2.9.1.1	Selenocysteine synthase (Moorella thermoacetica strain DSM521 clone pCTA 100/pCTAB1 gene selA) L-Seryl-tRNASec selenium transferase	
2.9.1.1	Selenocysteinyl-tRNA synthase L-Seryl-tRNASec selenium transferase	
2.9.1.1	Selenophosphate:L-seryl-tRNAselenium transferase L-Seryl-tRNASec selenium transferase	
2.7.9.3	Selenophosphate synthase selenide, water dikinase	
2.7.9.3	Selenophosphate synthase (Aquifex aeolicus gene selD) selenide, water dikinase	
2.7.9.3	Selenophosphate synthetase selenide, water dikinase	
3.5.1.11	Semacylase Penicillin amidase	
3.4.21.77	Semenogelase Semenogelase	
3.4.21.63	Semi-alkaline protease Oryzin	
1.6.5.4	Semidehydroascorbate reductase Monodehydroascorbate reductase (NADH)	
3.4.21.77	Seminin Semenogelase	
3.4.21.77	γ-Seminoglycoprotein (human protein moiety reduced) Semenogelase	

EC-No.	Synonym Recommended Name	Vol., Page
3.4.21.77	γ-Seminoprotein Semenogelase	
3.1.4.13	SEP diesterase serine-ethanolaminephosphate phosphodiesterase	
3.4.24.2	Sepia proteinase Sepia proteinase	
3.5.4.24	Sepiapterin aminohydrolase Sepiapterin deaminase	
3.5.4.24	Sepiapterin deaminase Sepiapterin deaminase	
1.1.1.153	Sepiapterin reductase Sepiapterin reductase	
2.7.7.7	Sequenase DNA-directed DNA polymerase	
1.1.1.143	Sequoyitol dehydrogenase Sequoyitol dehydrogenase	
2.7.1.37	Serine(threonine) protein kinase Protein kinase	
3.1.4.13	serine-ethanolamine phosphate phosphodiesterase serine-ethanolaminephosphate phosphodiesterase	
3.1.4.13	serine-ethanolaminephosphate phosphodiesterase serine-ethanolaminephosphate phosphodiesterase	
2.6.1.45	Serine-glyoxylate aminotransferase Serine-glyoxylate transaminase	
2.6.1.45	Serine-glyoxylate transaminase Serine-glyoxylate transaminase	
4.3.1.10	L-serine-O-sulfate ammmonia-lyase (pyruvate-forming) serine-sulfate ammonia-lyase	
2.3.1.50	Serine-palmitoyltransferase Serine C-palmitoyltransferase	
2.7.8.4	Serine-phosphinico-ethanolamine synthase Serine-phosphoethanolamine synthase	
3.1.4.13	serine-phosphoethanolamine ethanolaminephosphohydrolase serine-ethanolaminephosphate phosphodiesterase	
2.7.8.4	Serine-phosphoethanolamine synthase Serine-phosphoethanolamine synthase	

EC-No.	Synonym Recommended Name	Vol., Page
2.6.1.51	Serine-pyruvate transaminase Serine-pyruvate transaminase	
2.7.1.37	Serine-specific protein kinase Protein kinase	
4.3.1.10	serine-sulfate ammonia-lyase serine-sulfate ammonia-lyase	
6.1.1.11	Serine-tRNA ligase Serine-tRNA ligase	Vol. 2, Page 77
3.4.16.4	serine-type D-Ala-D-Ala carboxypeptidase serine-type D-Ala-D-Ala carboxypeptidase	
2.7.1.37	Serine/threonine kinase Protein kinase	
2.7.1.37	Serine/threonine protein kinase Protein kinase	
2.6.1.45	L-Serine:glyoxylate aminotransferase Serine-glyoxylate transaminase	
1.4.1.7	L-Serine:NAD^+ oxidoreductase (deaminating) Serine dehydrogenase	
2.6.1.51	L-Serine:pyruvate aminotransferase Serine-pyruvate transaminase	
6.1.1.11	L-Serine:$tRNA^{Ser}$ ligase (AMP-forming) Serine-tRNA ligase	Vol. 2, Page 77
2.3.1.30	L-Serine acetyltransferase Serine O-acetyltransferase	
2.3.1.30	Serine acetyltransferase Serine O-acetyltransferase	
2.1.2.1	Serine aldolase Glycine hydroxymethyltransferase	
2.3.1.50	Serine C-palmitoyltransferase Serine C-palmitoyltransferase	
3.4.16.1	serine carboxypeptidase serine carboxypeptidase	
3.4.16.5	Serine carboxypeptidase I Carboxypeptidase C	
4.2.1.14	D-Serine deaminase D-Serine dehydratase	

EC-No.	Synonym Recommended Name	Vol., Page
4.2.1.13	L-Serine deaminase L-Serine dehydratase	
4.2.1.13	Serine deaminase L-Serine dehydratase	
4.2.1.14	D-Serine dehydrase D-Serine dehydratase	
4.2.1.14	D-Serine dehydratase D-Serine dehydratase	
4.2.1.13	L-Serine dehydratase L-Serine dehydratase	
4.2.1.13	Serine dehydratase L-Serine dehydratase	
1.4.1.7	Serine dehydrogenase Serine dehydrogenase	
3.4.21.37	Serine elastase Leucocyte elastase	
3.4.21.36	serine elastase pancreatic elastase	
3.1.1.1	Serine esterase Carboxylesterase	
2.7.8.4	Serine ethanolamine phosphate synthetase Serine-phosphoethanolamine synthase	
3.1.4.13	serine ethanolamine phosphodiester phosphodiesterase serine-ethanolaminephosphate phosphodiesterase	
2.7.8.4	Serine ethanolamine phosphodiester synthase Serine-phosphoethanolamine synthase	
2.7.8.4	Serine ethanolaminephosphotransferase Serine-phosphoethanolamine synthase	
2.6.1.45	L-Serine glyoxylate aminotransferase Serine-glyoxylate transaminase	
4.2.1.22	L-Serine hydro-lyase (adding homocysteine) Cystathionine β-synthase	
4.2.1.20	L-serine hydro-lyase (adding indoleglycerol-phosphate) tryptophan synthase	
4.2.1.50	L-Serine hydro-lyase (adding pyrazole) Pyrazolylalanine synthase	

EC-No.	Synonym Recommended Name	Vol., Page
4.2.1.14	D-Serine hydro-lyase (deaminating) D-Serine dehydratase	
4.2.1.13	L-Serine hydro-lyase (deaminating) L-Serine dehydratase	
4.2.1.14	D-Serine hydrolase (deaminating) D-Serine dehydratase	
2.1.2.1	Serine hydroxymethylase Glycine hydroxymethyltransferase	
2.1.2.1	L-Serine hydroxymethyltransferase Glycine hydroxymethyltransferase	
2.1.2.1	Serine hydroxymethyltransferase Glycine hydroxymethyltransferase	
5.1.1.16	Serine isomerase Protein-serine epimerase	Vol. 1, Page 66
2.7.1.37	Serine kinase Protein kinase	
2.3.1.30	Serine O-acetyltransferase Serine O-acetyltransferase	
2.3.1.50	Serine palmitoyltransferase Serine C-palmitoyltransferase	
2.7.8.4	Serinephosphoethanolamine synthase Serine-phosphoethanolamine synthase	
3.4.21.53	Serine protease La Endopeptidase La	
3.4.21.81	Serine proteinase B Streptogrisin B	
3.4.21.98	serine proteinase NS3 hepacivirin	
3.4.21.75	Serine proteinase PACE Furin	
2.7.1.37	Serine protein kinase Protein kinase	
4.2.1.22	Serine sulfhydrase Cystathionine β-synthase	
4.2.1.22	Serine sulfhydrylase Cystathionine β-synthase	

EC-No.	Synonym Recommended Name	Vol., Page
4.2.1.22	Serine sulphhydrase Cystathionine β-synthase	
2.3.1.30	Serine transacetylase Serine O-acetyltransferase	
6.1.1.11	Serine transfer RNA synthetase Serine-tRNA ligase	Vol. 2, Page 77
2.1.2.1	Serine transhydroxymethylase Glycine hydroxymethyltransferase	
6.1.1.11	Serine translase Serine-tRNA ligase	Vol. 2, Page 77
2.3.1.87	Serotonin acetylase Aralkylamine N-acetyltransferase	
2.3.1.87	Serotonin acetyltransferase Aralkylamine N-acetyltransferase	
2.3.1.5	Serotonin acetyltransferase Arylamine N-acetyltransferase	
1.4.3.4	Serotonin deaminase Amine oxidase (flavin-containing)	
2.3.1.5	Serotonin N-acetyltransferase Arylamine N-acetyltransferase	
3.4.24.40	Serralysin Serralysin	
3.4.24.40	Serratia marcescens extracellular proteinase Serralysin	
3.4.24.40	Serratia marcescens metalloprotease Serralysin	
3.4.24.40	Serratia marcescens metalloproteinase Serralysin	
3.1.30.2	Serratia marcescens nuclease Serratia marcescens nuclease	
6.1.1.11	SERSEC Serine-tRNA ligase	Vol. 2, Page 77
3.4.13.20	Serum carnosinase β-Ala-His dipeptidase	
3.1.1.8	serum cholinesterase cholinesterase	

EC-No.	Synonym Recommended Name	Vol., Page
3.4.21.34	serum kallikrein plasma kallikrein	
3.4.21.7	serum tryptase plasmin	
6.1.1.11	Seryl-transfer ribonucleate synthetase Serine-tRNA ligase	Vol. 2, Page 77
6.1.1.11	Seryl-transfer ribonucleic acid synthetase Serine-tRNA ligase	Vol. 2, Page 77
6.1.1.11	Seryl-transfer RNA synthetase Serine-tRNA ligase	Vol. 2, Page 77
2.9.1.1	L-Seryl-tRNASec selenium transferase L-Seryl-tRNASec selenium transferase	
6.1.1.11	Seryl-tRNA synthetase Serine-tRNA ligase	Vol. 2, Page 77
2.5.1.40	Sesquiterpene cyclase Aristolochene synthase	
4.2.3.9	sesquiterpene cyclase aristolochene synthase	
4.2.3.6	sesquiterpene cyclase trichodiene synthase	
2.4.1.5	SGE Dextransucrase	
3.4.24.31	SGNPI Mycolysin	
3.4.21.19	SGPE glutamyl endopeptidase	
3.4.21.4	SGT trypsin	
2.4.1.128	SGTase UDPglucose:scopoletin O-β-D-glucosyltransferase	
3.1.3.48	SH-PTP1 protein-tyrosine-phosphatase	
3.1.3.48	SH-PTP2 protein-tyrosine-phosphatase	
3.1.3.48	SH-PTP3 protein-tyrosine-phosphatase	

EC-No.	Synonym Recommended Name	Vol., Page
1.1.1.25	Shikimate 5-dehydrogenase Shikimate 5-dehydrogenase	
1.1.1.25	Shikimate:NADP$^+$ 5-oxidoreductase Shikimate 5-dehydrogenase	
1.1.1.25	Shikimate:NADP$^+$ oxidoreductase Shikimate 5-dehydrogenase	
2.3.1.133	Shikimate hydroxycinnamoyltransferase Shikimate O-hydroxycinnamoyltransferase	
2.7.1.71	Shikimate kinase Shikimate kinase	
2.7.1.71	Shikimate kinase II Shikimate kinase	
2.3.1.133	Shikimate O-hydroxycinnamoyltransferase Shikimate O-hydroxycinnamoyltransferase	
1.1.1.25	Shikimate oxidoreductase Shikimate 5-dehydrogenase	
3.1.2.18	Short-chain acyl-CoA hydrolase ADP-dependent short-chain-acyl-CoA hydrolase	
6.2.1.2	Short-chain acyl-CoA synthetase Butyrate-CoA ligase	Vol. 2, Page 199
3.1.2.18	Short-chain acyl-CoA thioesterase ADP-dependent short-chain-acyl-CoA hydrolase	
1.3.99.2	Short-chain acyl-coenzyme A dehydrogenase Butyryl-CoA dehydrogenase	
6.2.1.1	Short-chain acyl-coenzyme A synthetase Acetate-CoA ligase	Vol. 2, Page 186
1.3.99.2	Short-chain acyl CoA dehydrogenase Butyryl-CoA dehydrogenase	
4.2.1.17	short-chain enoyl-CoA hydratase enoyl-CoA hydratase	
4.2.1.17	short-chain enoyl-CoA hydratase enoyl-CoA hydratase	
3.5.1.14	Short acyl amidoacylase Aminoacylase	
1.1.1.36	Short chain β-ketoacetyl(acetoacetyl)-CoA reductase Acetoacetyl-CoA reductase	

EC-No.	Synonym Recommended Name	Vol., Page
5.3.3.8	Short chain Δ^3,Δ^2-enoyl-CoA isomerase Dodecenoyl-CoA δ-isomerase	Vol. 1, Page 413
4.2.1.17	short chain enoyl coenzyme A hydratase enoyl-CoA hydratase	
6.2.1.1	Short chain fatty acyl-CoA synthetase Acetate-CoA ligase	Vol. 2, Page 186
3.1.3.48	SHP protein-tyrosine-phosphatase	
3.1.1.53	sialate 9-O-acetylesterase sialate O-acetylesterase	
4.1.3.3	Sialate lyase N-acetylneuraminate lyase	
3.1.1.53	sialate O-acetylesterase sialate O-acetylesterase	
2.3.1.44	Sialate O-acetyltransferase N-Acetylneuraminate 4-O-acetyltransferase	
2.3.1.45	Sialate O-acetyltransferase N-Acetylneuraminate 7-O(or 9-O-)-acetyltransferase	
3.2.1.138	Sialglycoconjugate N-acylneuraminylhydrolase (2,7-cyclizing) Anhydrosialidase	
4.1.3.20	Sialic acid 9-phosphate synthase, N-acylneuraminate 9-phosphate N-Acetylneuraminate-9-phosphate synthase	
4.1.3.20	Sialic acid 9-phosphate synthetase N-Acetylneuraminate-9-phosphate synthase	
4.1.3.3	Sialic acid aldolase N-acetylneuraminate lyase	
3.1.1.53	sialic acid O-acetylesterase sialate O-acetylesterase	
4.1.3.3	Sialic aldolase N-acetylneuraminate lyase	
3.2.1.18	Sialidase Sialidase	
3.2.1.138	Sialidase L Anhydrosialidase	
3.4.24.57	Sialoglycoprotease O-Sialoglycoprotein endopeptidase	

EC-No.	Synonym Recommended Name	Vol., Page
3.4.24.57	Sialoglycoproteinase O-Sialoglycoprotein endopeptidase	
3.4.24.57	O-Sialoglycoprotein endopeptidase O-Sialoglycoprotein endopeptidase	
2.4.99.1	Sialotransferase β-Galactoside α-2,6-sialyltransferase	
2.4.99.7	Sialyltransferase (α-N-Acetylneuraminyl-2,3-β-galactosyl-1,3)-N-acetylgalactosaminide α-2,6-sialyltransferase	
2.4.99.6	Sialyltransferase N-Acetyllactosaminide α-2,3-sialyltransferase	
2.4.99.10	Sialyltransferase Neolactotetraosylceramide α-2,3-sialyltransferase	
2.4.99.1	Sialyltransferase β-Galactoside α-2,6-sialyltransferase	
2.4.99.8	α-2,8-Sialyltransferase α-N-Acetylneuraminate α-2,8-sialyltransferase	
2.4.99.6	$\alpha_2\rightarrow 3$ Sialyltransferase N-Acetyllactosaminide α-2,3-sialyltransferase	
2.4.99.4	$\alpha_2\rightarrow 3$ Sialyltransferase β-Galactoside α-2,3-sialyltransferase	
2.4.99.1	α2-6 Sialyltransferase β-Galactoside α-2,6-sialyltransferase	
2.4.99.7	Sialyltransferase, cytidine monophosphoacetylneuraminate-(α-N-acetylneuraminyl-2,3-β-galactosyl-1,3)-N-acetylgalactosaminide-α-2,6- (α-N-Acetylneuraminyl-2,3-β-galactosyl-1,3)-N-acetylgalactosaminide α-2,6-sialyltransferase	
2.4.99.3	Sialyltransferase, cytidine monophosphoacetylneuraminate-α-acetylgalactosaminide $\alpha 2\rightarrow 6$- α-N-Acetylgalactosaminide α-2,6-sialyltransferase	
2.4.99.4	Sialyltransferase, cytidine monophosphoacetylneuraminate-β-galactoside $\alpha 2\rightarrow 3$- β-Galactoside α-2,3-sialyltransferase	
2.4.99.6	Sialyltransferase, cytidine monophosphoacetylneuraminate-β-galactosyl($1\rightarrow 4$)acetylglucosaminide $\alpha 2\rightarrow 3$- N-Acetyllactosaminide α-2,3-sialyltransferase	

EC-No.	Synonym Recommended Name Vol., Page
2.4.99.5	Sialyltransferase, cytidine monophosphoacetylneuraminate-galactosyldiacylglycerol Galactosyldiacylglycerol α-2,3-sialyltransferase
2.4.99.1	Sialyltransferase, cytidine monophosphoacetylneuraminate-galactosylglycoprotein β-Galactoside α-2,6-sialyltransferase
2.4.99.8	Sialyltransferase, cytidine monophosphoacetylneuraminate-ganglioside GM3 α-N-Acetylneuraminate α-2,8-sialyltransferase
2.4.99.11	Sialyltransferase, cytidine monophosphoacetylneuraminate-lactosylceramide Lactosylceramide α-2,6-N-sialyltransferase
2.4.99.9	Sialyltransferase, cytidine monophosphoacetylneuraminate-lactosylceramide α2,3- Lactosylceramide α-2,3-sialyltransferase
2.4.99.2	Sialyltransferase, cytidine monophosphoacetylneuraminate-monosialoganglioside Monosialoganglioside sialyltransferase
2.4.99.10	Sialyltransferase, cytidine monophosphoacetylneuraminate-neolactotetraosylceramide Neolactotetraosylceramide α-2,3-sialyltransferase
2.4.99.10	Sialyltransferase 3 Neolactotetraosylceramide α-2,3-sialyltransferase
3.4.21.89	Signalase Signal peptidase I
3.4.21.89	Signal peptidase Signal peptidase I
3.4.21.89	Signal peptidase I Signal peptidase I
3.4.23.36	Signal peptidase II Signal peptidase II
3.4.21.89	Signal peptide hydrolase Signal peptidase I
3.4.21.89	Signal peptide peptidase Signal peptidase I
4.1.1.48	sIGPS Indole-3-glycerol phosphate synthase

EC-No.	Synonym Recommended Name	Vol., Page
2.4.1.120	Sinapate 1-glucosyltransferase Sinapate 1-glucosyltransferase	
2.3.1.152	Sinapate glucosyltransferase Alcohol O-cinnamoyltransferase	
2.3.1.152	Sinapic acid glucosyltransferase Alcohol O-cinnamoyltransferase	
3.1.1.49	sinapine esterase sinapine esterase	
2.3.1.91	Sinapine synthase Sinapoylglucose-choline O-sinapoyltransferase	
2.3.1.92	1-O-Sinapoyl-β-D-glucose:(S)-malate O-sinapoyltransferase Sinapoylglucose-malate O-sinapoyltransferase	
2.3.1.106	Sinapoyl-CoA:2-hydroxymalonate O-(hydroxycinnamoyl)transferase Tartronate O-hydroxycinnamoyltransferase	
2.3.1.131	Sinapoyl-CoA:glucarate O-(hydroxycinnamoyl)transferase Glucarate O-hydroxycinnamoyltransferase	
2.3.1.132	Sinapoyl-CoA:glucarolactone O-(hydroxycinnamoyl)transferase Glucarolactone O-hydroxycinnamoyltransferase	
3.1.1.49	sinapoylcholine sinapohydrolase sinapine esterase	
6.2.1.12	Sinapoyl coenzyme A snthetase 4-Coumarate-CoA ligase	Vol. 2, Page 256
2.3.1.91	Sinapoylglucose-choline O-sinapoyltransferase Sinapoylglucose-choline O-sinapoyltransferase	
2.3.1.92	1-Sinapoylglucose-L-malate sinapoyltransferase Sinapoylglucose-malate O-sinapoyltransferase	
2.3.1.92	Sinapoylglucose-malate O-sinapoyltransferase Sinapoylglucose-malate O-sinapoyltransferase	
2.3.1.92	Sinapoylglucose-malate sinapoyltransferase Sinapoylglucose-malate O-sinapoyltransferase	
2.3.1.103	Sinapoylglucose-sinapoylglucose O-sinapoyltransferase Sinapoylglucose-sinapoylglucose O-sinapoyltransferase	
2.3.1.92	Sinapoylglucose:malate sinapoyltransferase Sinapoylglucose-malate O-sinapoyltransferase	
2.3.1.91	Sinapoyltransferase, sinapoylglucose-choline Sinapoylglucose-choline O-sinapoyltransferase	

EC-No.	Synonym Recommended Name	Vol., Page
2.3.1.92	Sinapoyltransferase, sinapoylglucose-malate Sinapoylglucose-malate O-sinapoyltransferase	
3.4.21.90	Sindbis virus core protein Togavirin	
3.4.21.90	Sindbis virus protease Togavirin	
3.1.30.1	Single-strand-specific endodeoxyribonuclease Aspergillus nuclease S_1	
3.1.4.21	single-stranded-nucleate endonuclease single-stranded-nucleate endonuclease	
3.1.30.1	Single-stranded DNA specific endonuclease Aspergillus nuclease S_1	
3.1.30.1	Single-stranded nucleate endonuclease Aspergillus nuclease S_1	
3.1.30.1	Single-strand endodeoxyribonuclease Aspergillus nuclease S_1	
3.1.30.1	Single strand-specific DNase Aspergillus nuclease S_1	
3.2.3.1	Sinigrase Thioglucosidase	
3.2.3.1	Sinigrinase Thioglucosidase	
3.1.6.5	sinigrin sulfohydrolase; myrosulfatase sinigrin sulfohydrolase; myrosulfatase	
2.1.1.72	Site-specific DNA-methyltransferase (adenine-specific) Site-specific DNA-methyltransferase (adenine-specific)	
2.1.1.113	Site-specific DNA-methyltransferase (cytosine-N^4-specific) Site-specific DNA-methyltransferase (cytosine-N^4-specific)	
2.1.1.73	Site-specific DNA-methyltransferase (cytosine-specific) Site-specific DNA-methyltransferase (cytosine-specific)	
3.4.21.39	skeletal muscle protease chymase	
3.4.21.39	skin chymotryptic proteinase chymase	
3.4.21.59	Skin tryptase Tryptase	

EC-No.	Synonym Recommended Name	Vol., Page
3.4.23.32	SLB Scytalidopepsin B	
3.4.23.34	Slow-moving proteinase Cathepsin E	
3.1.27.5	SLSG glycoproteins Pancreatic ribonuclease	
3.4.21.77	γ-SM Semenogelase	
2.7.1.72	SM 6-kinase Streptomycin 6-kinase	
4.1.2.13	SMALDO Fructose-bisphosphate aldolase	
5.2.1.8	SmCYP A Peptidylprolyl isomerase	Vol. 1, Page 218
5.2.1.8	SmCYP B Peptidylprolyl isomerase	Vol. 1, Page 218
2.7.1.117	Smooth-muscle-myosin-light-chain kinase Myosin-light-chain kinase	
3.4.23.34	SMP Cathepsin E	
3.1.3.55	SMP-I Caldesmon-phosphatase	
2.1.1.117	SMT (S)-Scoulerine 9-O-methyltransferase	
2.7.8.2	Sn-1,2-Diacylglycerol cholinephosphotransferase Diacylglycerol cholinephosphotransferase	
2.7.1.107	Sn-1,2-Diacylglycerol kinase Diacylglycerol kinase	
2.7.1.94	Sn-2-Monoacylglycerol kinase Acylglycerol kinase	
3.1.4.44	sn-glycero(3)phosphoinositol glycerophosphohydrolase glycerophosphoinositol glycerophosphodiesterase	
3.1.4.44	sn-glycero-3-phospho-1-inositol glycerophosphohydrolase glycerophosphoinositol glycerophosphodiesterase	
3.1.4.43	sn-glycero-3-phospho-1-inositol inositolphosphohydrolase glycerophosphoinositol inositolphosphodiesterase	

EC-No.	Synonym Recommended Name	Vol., Page
3.1.4.38	sn-glycero-3-phosphocholine cholinephosphohydrolase glycerophosphocholine cholinephosphodiesterase	
3.1.4.2	sn-glycero 3-phosphocholine diesterase glycerophosphocholine phosphodiesterase	
3.1.4.2	sn-glycero 3-phosphocholine glycerophosphohydrolase glycerophosphocholine phosphodiesterase	
3.1.4.2	sn-glycero 3-phosphorylcholine diesterase glycerophosphocholine phosphodiesterase	
1.1.1.261	sn-glycerol-1-phosphate:NAD(P) 2-oxidoreductase glycerol-1-phosphate dehydrogenase [NAD(P)]	
2.4.1.96	Sn-Glycerol-3-phosphate 1-galactosyltransferase Sn-Glycerol-3-phosphate 1-galactosyltransferase	
2.4.1.137	Sn-Glycerol-3-phosphate 2-α-galactosyltransferase Sn-Glycerol-3-phosphate 2-α-galactosyltransferase	
1.1.99.5	Sn-Glycerol-3-phosphate:(acceptor) 2-oxidoreductase Glycerol-3-phosphate dehydrogenase	
2.7.1.142	Sn-Glycerol-3-phosphate:D-glucose 6-phosphotransferase Glycerol-3-phosphate-glucose phosphotransferase	
1.1.1.94	Sn-Glycerol-3-phosphate:NAD(P)$^+$ 2-oxidoreductase Glycerol-3-phosphate dehydrogenase (NAD(P)$^+$)	
1.1.1.8	Sn-Glycerol-3-phosphate:NAD$^+$ 2-oxidoreductase Glycerol-3-phosphate dehydrogenase (NAD$^+$)	
1.1.1.177	Sn-Glycerol-3-phosphate:NADP$^+$ 1-oxidoreductase Glycerol-3-phosphate 1-dehydrogenase (NADP$^+$)	
1.1.3.21	Sn-Glycerol-3-phosphate:oxygen 2-oxidoreductase Glycerol-3-phosphate oxidase	
2.3.1.15	Sn-Glycerol-3-phosphate acyltransferase (EC 2.3.1.15) Glycerol-3-phosphate O-acyltransferase	
1.1.99.5	Sn-Glycerol-3-phosphate dehydrogenase Glycerol-3-phosphate dehydrogenase	
1.1.99.5	Sn-Glycerol-3-phosphate oxidase (EC 1.1.99.5) Glycerol-3-phosphate dehydrogenase	
2.7.8.5	Sn-Glycerol-3-phosphate phosphatidyltransferase CDPdiacylglycerol-glycerol-3-phosphate 3-phosphatidyltransferase	
2.3.1.15	Sn-Glycerol 3-phosphate acyltransferase Glycerol-3-phosphate O-acyltransferase	

EC-No.	Synonym Recommended Name	Vol., Page
1.1.99.5	sn-Glycerol 3-phosphate oxidase Glycerol-3-phosphate dehydrogenase	
3.1.3.4	3-sn-Phosphatidate phosphohydrolase Phosphatidate phosphatase	
2.3.1.83	3-sn-Phosphatidylcholine:dolichol O-acyltransferase Phosphatidylcholine-dolichol O-acyltransferase	
3.1.3.5	snake venom 5'-nucleotidase 5'-nucleotidase	
3.4.21.95	Snake venom factor V activator Snake venom factor V activator	
3.1.3.32	SNQI-PNK polynucleotide 3'-phosphatase	
1.15.1.1	SOD Superoxide dismutase	
1.15.1.1	SOD-1 Superoxide dismutase	
1.15.1.1	SOD-2 Superoxide dismutase	
1.15.1.1	SOD-3 Superoxide dismutase	
1.15.1.1	SOD-4 Superoxide dismutase	
1.15.1.1	SODF Superoxide dismutase	
3.6.1.37	Sodium pump Na^+/K^+-transporting ATPase	
4.1.1.41	Sodium pump methylmalonyl-CoA decarboxylase Methylmalonyl-CoA decarboxylase	
1.15.1.1	SODS Superoxide dismutase	
5.3.99.7	SOI Styrene-oxide isomerase	Vol. 1, Page 486
3.4.99.21	solanain solanain	
2.5.1.11	Solanesyl-diphosphate synthase trans-Octaprenyltranstransferase	

EC-No.	Synonym Recommended Name	Vol., Page
2.5.1.39	Solanesyl-pyrophosphate:4-hydroxybenzoate nonaprenyltransferase 4-Hydroxybenzoate nonaprenyltransferase	
2.5.1.11	Solanesyl pyrophosphate synthetase trans-Octaprenyltranstransferase	
3.4.11.14	soluble alanyl aminopeptidase aminopeptidase, human liver	
3.4.17.17	Soluble carboxypeptidase Tubulinyl-Tyr carboxypeptidase	
3.4.24.15	soluble metallo-endopeptidase thimet oligopeptidase	
3.4.24.15	soluble metallopeptidase thimet oligopeptidase	
3.1.8.2	Somanase Diisopropyl-fluorophosphatase	
1.6.5.4	SOR Monodehydroascorbate reductase (NADH)	
1.1.1.140	Sorbitol-6-P-dehydrogenase Sorbitol-6-phosphate 2-dehydrogenase	
3.1.3.50	sorbitol-6-phosphatase sorbitol-6-phosphatase	
1.1.1.140	Sorbitol-6-phosphate 2-dehydrogenase Sorbitol-6-phosphate 2-dehydrogenase	
1.1.1.140	D-Sorbitol-6-phosphate:NAD$^+$ 2-oxidoreductase Sorbitol-6-phosphate 2-dehydrogenase	
1.1.1.140	D-Sorbitol-6-phosphate dehydrogenase Sorbitol-6-phosphate 2-dehydrogenase	
3.1.3.50	sorbitol-6-phosphate phosphatase sorbitol-6-phosphatase	
3.1.3.50	sorbitol-6-phosphate phosphohydrolase sorbitol-6-phosphatase	
1.1.99.21	D-Sorbitol:(acceptor) 1-oxidoreductase D-Sorbitol dehydrogenase (acceptor)	
1.1.1.15	D-Sorbitol dehydrogenase D-Iditol 2-dehydrogenase	
1.1.1.14	Sorbitol dehydrogenase L-Iditol 2-dehydrogenase	

EC-No.	Synonym Recommended Name	Vol., Page
1.1.99.21	D-Sorbitol dehydrogenase (acceptor) D-Sorbitol dehydrogenase (acceptor)	
1.1.1.123	Sorbose 5-dehydrogenase ($NADP^+$) Sorbose 5-dehydrogenase ($NADP^+$)	
1.1.99.12	L-Sorbose:(acceptor) 5-oxidoreductase Sorbose dehydrogenase	
1.1.1.123	L-Sorbose:$NADP^+$ 5-oxidoreductase Sorbose 5-dehydrogenase ($NADP^+$)	
1.1.3.11	L-Sorbose:oxygen 5-oxidoreductase L-Sorbose oxidase	
1.1.99.12	Sorbose dehydrogenase Sorbose dehydrogenase	
1.1.3.11	L-Sorbose oxidase L-Sorbose oxidase	
3.4.23.14	sorghum aspartic proteinase sorghum aspartic proteinase	
4.1.2.11	Sorghum hydroxynitrile lyase Hydroxymandelonitrile lyase	
4.3.1.10	(L-SOS) lyase serine-sulfate ammonia-lyase	
3.4.21.32	soycollagestin brachyurin	
3.4.24.7	soycollagestin interstitial collagenase	
3.4.24.3	soycollagestin microbial collagenase	
4.2.2.3	SP2 poly(β-D-mannuronate) lyase	
3.4.21.62	SP 266 Subtilisin	
2.1.1.130	SP2MT Precorrin-2 C^{20}-methyltransferase	
3.4.21.89	SPC Signal peptidase I	
3.4.21.75	SPC3 Furin	

EC-No.	Synonym Recommended Name	Vol., Page
3.4.22.10	SPE B streptopain	
3.4.22.10	Spe B streptopain	
1.1.1.35	1-Specific DPN-linked β-hydroxybutyric dehydrogenase 3-Hydroxyacyl-CoA dehydrogenase	
1.1.1.28	D-Specific lactic dehydrogenase D-Lactate dehydrogenase	
2.3.1.57	Spermidine/spermine N^1-acetyltransferase Diamine N-acetyltransferase	
1.5.99.6	Spermidine:(acceptor) oxidoreductase Spermidine dehydrogenase	
2.3.1.57	Spermidine acetyltransferase Diamine N-acetyltransferase	
2.5.1.22	Spermidine aminopropyltransferase Spermine synthase	
1.5.99.6	Spermidine dehydrogenase Spermidine dehydrogenase	
2.3.1.57	Spermidine N^1-acetyltransferase Diamine N-acetyltransferase	
1.4.3.4	Spermidine oxidase Amine oxidase (flavin-containing)	
2.5.1.16	Spermidine synthase Spermidine synthase	
2.5.1.16	Spermidine synthetase Spermidine synthase	
1.4.3.4	Spermine oxidase Amine oxidase (flavin-containing)	
1.5.3.3	spermine oxidase spermine oxidase	
2.5.1.22	Spermine synthase Spermine synthase	
2.5.1.22	Spermine synthetase Spermine synthase	
3.4.21.4	sperm receptor hydrolase trypsin	

EC-No.	Synonym Recommended Name	Vol., Page
5.3.1.5	Spezyme Xylose isomerase	Vol. 1, Page 259
4.1.2.27	Sphinganine-1-phosphate-alkanal-lyase Sphinganine-1-phosphate aldolase	
4.1.2.27	Sphinganine-1-phosphate aldolase Sphinganine-1-phosphate aldolase	
4.1.2.27	Sphinganine-1-phosphate aldolase Sphinganine-1-phosphate aldolase	
4.1.2.27	Sphinganine-1-phosphate palmitaldehyde-lyase Sphinganine-1-phosphate aldolase	
4.1.2.27	Sphinganine 1-phosphate lyase Sphinganine-1-phosphate aldolase	
2.7.1.91	Sphinganine kinase Sphinganine kinase	
3.1.4.12	Sphingomyelinase Sphingomyelin phosphodiesterase	
3.1.4.41	Sphingomyelinase D Sphingomyelin phosphodiesterase D	
3.1.4.41	Sphingomyelin ceramide-phosphohydrolase Sphingomyelin phosphodiesterase D	
3.1.4.12	Sphingomyelin cholinephosphohydrolase Sphingomyelin phosphodiesterase	
3.1.4.12	Sphingomyelin phosphodiesterase Sphingomyelin phosphodiesterase	
3.1.4.41	Sphingomyelin phosphodiesterase D Sphingomyelin phosphodiesterase D	
4.1.2.27	Sphingosine-1-phosphate lyase Sphinganine-1-phosphate aldolase	
4.1.2.27	Sphingosine-1-phosphate lyase Sphinganine-1-phosphate aldolase	
2.4.1.23	Sphingosine β-galactosyltransferase Sphingosine β-galactosyltransferase	
4.1.2.27	Sphingosine-phosphate lyase Sphinganine-1-phosphate aldolase	
2.3.1.24	Sphingosine acyltransferase Sphingosine N-acyltransferase	

EC-No.	Synonym Recommended Name	Vol., Page
2.7.8.10	Sphingosine cholinephosphotransferase Sphingosine cholinephosphotransferase	
2.3.1.24	Sphingosine N-acyltransferase Sphingosine N-acyltransferase	
3.4.21.57	Spinach leucine-specific serine proteinase Leucyl endopeptidase	
4.1.2.27	SPL Sphinganine-1-phosphate aldolase	
3.1.31.1	Spleen endonuclease Micrococcal nuclease	
3.1.16.1	Spleen exonuclease Spleen exonuclease	
3.1.31.1	Spleen phosphodiesterase Micrococcal nuclease	
3.1.16.1	Spleen phosphodiesterase Spleen exonuclease	
3.1.27.9	Splicing endonuclease tRNA-intron endonuclease	
3.1.3.24	SPP sucrose-phosphatase	
2.5.1.11	SPP synthase trans-Octaprenyltranstransferase	
3.4.25.1	26S protease proteasome endopeptidase complex	
2.4.1.14	SPS Sucrose-phosphate synthase	
2.3.1.50	SPT Serine C-palmitoyltransferase	
2.6.1.51	SPT Serine-pyruvate transaminase	
1.14.99.7	Squalene,hydrogen-donor:oxygen oxidoreductase (2,3-epoxidizing) Squalene monooxygenase	
1.14.99.7	Squalene-2,3-epoxidase Squalene monooxygenase	
1.14.99.7	Squalene-2,3-epoxide cyclase Squalene monooxygenase	

EC-No.	Synonym Recommended Name	Vol., Page
5.4.99.7	Squalene-2,3-oxide-lanosterol cyclase Lanosterol synthase	Vol. 1, Page 624
5.4.99.7	Squalene 2,3-epoxide:lanosterol cyclase Lanosterol synthase	Vol. 1, Page 624
5.4.99.7	Squalene 2,3-oxide-lanosterol cyclase Lanosterol synthase	Vol. 1, Page 624
1.14.99.7	Squalene 2,3-oxidocyclase Squalene monooxygenase	
1.14.99.7	Squalene epoxidase Squalene monooxygenase	
5.4.99.7	Squalene epoxidase-cyclase Lanosterol synthase	Vol. 1, Page 624
1.14.99.7	Squalene hydroxylase Squalene monooxygenase	
1.14.99.7	Squalene monooxygenase Squalene monooxygenase	
1.14.99.7	Squalene oxydocyclase Squalene monooxygenase	
2.5.1.21	Squalene synthase Farnesyl-diphosphate farnesyltransferase	
2.5.1.21	Squalene synthetase Farnesyl-diphosphate farnesyltransferase	
5.3.4.1	S-S rearrangase Protein disulfide-isomerase	Vol. 1, Page 436
3.1.26.8	5 S Ribosomal maturation nuclease Ribonuclease M5	
3.1.26.8	5S Ribosomal RNA maturation endonuclease Ribonuclease M5	
2.7.7.20	sRNA nucleotidyl transferase sRNA nucleotidyl transferase	
1.2.1.24	SSAH Succinate semialdehyde dehydrogenase	
2.4.1.99	SST Sucrose 1F-fructosyltransferase	
2.4.1.67	Stachyose synthetase Galactinol-raffinose galactosyltransferase	

EC-No.	Synonym Recommended Name	Vol., Page
3.4.22.13	staphylococcal cysteine proteinase staphylococcal cysteine proteinase	
3.1.31.1	Staphylococcal nuclease Micrococcal nuclease	
3.4.21.19	staphylococcal serine proteinase glutamyl endopeptidase	
3.4.24.29	Staphylococcus aureus neutral protease Aureolysin	
3.4.24.29	Staphylococcus aureus neutral proteinase Aureolysin	
3.1.31.1	Staphylococcus aureus nuclease Micrococcal nuclease	
3.1.31.1	Staphylococcus aureus nuclease B Micrococcal nuclease	
3.4.99.22	staphylokinase staphylokinase	
2.4.1.18	Starch branching enzyme 1,4-α-Glucan branching enzyme	
2.4.1.1	Starch phosphorylase Phosphorylase	
2.4.1.21	Starch synthase Starch synthase	
2.4.1.21	Starch synthetase Starch synthase	
2.1.1.139	staurosporine synthase 3'-demethylstaurosporine O-methyltransferase	
1.5.1.22	StDH Strombine dehydrogenase	
3.1.2.14	STE oleoyl-[acyl-carrier-protein] hydrolase	
3.1.1.3	steapsin triacylglycerol lipase	
1.14.99.6	Stearoyl-[acyl carrier protein] desaturase Acyl-[acyl-carrier-protein] desaturase	
1.14.99.6	Stearoyl-ACP desaturase Acyl-[acyl-carrier-protein] desaturase	

EC-No.	Synonym Recommended Name	Vol., Page
6.2.1.20	Stearoyl-ACP synthetase Long-chain-fatty-acid-[acyl-carrier-protein] ligase	Vol. 2, Page 296
1.14.99.5	Stearoyl-CoA,hydrogen-donor:oxygen oxidoreductase Stearoyl-CoA desaturase	
1.14.19.1	stearoyl-CoA, hydrogen-donor:oxygen oxidoreductase stearoyl-CoA 9-desaturase	
1.14.19.1	stearoyl-CoA 9-desaturase stearoyl-CoA 9-desaturase	
2.3.1.119	Stearoyl-CoA:malonyl-CoA C-acyltransferase (decarboxylating, oxoacyl- and enoyl-reducing) Icosanoyl-CoA synthase	
1.14.99.5	Stearoyl-CoA desaturase Stearoyl-CoA desaturase	
2.3.1.119	Stearoyl-CoA elongase Icosanoyl-CoA synthase	
6.2.1.3	Stearoyl-CoA synthetase Long-chain-fatty-acid-CoA ligase	Vol. 2, Page 206
1.14.99.5	Stearoyl coenzyme A desaturase Stearoyl-CoA desaturase	
1.14.99.6	Stearyl-ACP desaturase Acyl-[acyl-carrier-protein] desaturase	
1.14.19.2	stearyl-ACP desaturase acyl-[acyl-carrier-protein] desaturase	
1.14.99.6	Stearyl-acyl carrier protein desaturase Acyl-[acyl-carrier-protein] desaturase	
1.14.99.5	Stearyl-CoA desaturase Stearoyl-CoA desaturase	
1.14.19.2	stearyl acyl carrier protein desaturase acyl-[acyl-carrier-protein] desaturase	
3.4.22.32	Stem bromelain Stem bromelain	
3.4.22.10	Steptococcus proteinase streptopain	
3.4.11.19	D-Stereospecific aminopeptidase D-Stereospecific aminopeptidase	

EC-No.	Synonym Recommended Name	Vol., Page
2.1.1.110	Sterigmatocystin 7-O-methyltransferase Sterigmatocystin 7-O-methyltransferase	
2.1.1.110	Sterigmatocystin methyltransferase Sterigmatocystin 7-O-methyltransferase	
1.1.1.50	Sterognost 3α 3α-Hydroxysteroid dehydrogenase (B-specific)	
1.14.99.9	Steroid,hydrogen-donor:oxygen oxidoreductase (17α-hydroxylating) Steroid 17α-monooxygenase	
1.14.99.10	Steroid,hydrogen-donor:oxygen oxidoreductase (21-hydroxylating) Steroid 21-monooxygenase	
1.14.99.24	Steroid,hydrogen-donor:oxygen oxidoreductase (9-epoxidizing) Steroid 9α-monooxygenase	
1.14.15.4	Steroid,reduced-adrenal-ferredoxin:oxygen oxidoreductase (11β-hydroxylating) Steroid 11β-monooxygenase	
1.1.1.145	Steroid-Δ^5-3β-ol dehydrogenase 3β-Hydroxy-Δ^5-steroid dehydrogenase	
1.3.99.6	Steroid Δ-5β-dehydrogenase 3-Oxo-5β-steroid Δ^4-dehydrogenase	
5.3.3.1	Steroid Δ-isomerase Steroid Δ-isomerase	Vol. 1, Page 376
1.14.13.54	Steroid-ketone monooxygenase Ketosteroid monooxygenase	
3.1.1.37	steroid-lactonase steroid-lactonase	
1.14.15.4	Steroid 11β-hydroxylase Steroid 11β-monooxygenase	
1.14.15.4	Steroid 11β-monooxygenase Steroid 11β-monooxygenase	
1.14.15.4	Steroid 11β/18-hydroxylase Steroid 11β-monooxygenase	
1.14.99.9	Steroid 17α-hydroxylase Steroid 17α-monooxygenase	
1.14.99.9	Steroid 17α-monooxygenase Steroid 17α-monooxygenase	

EC-No.	Synonym Recommended Name	Vol., Page
1.14.15.6	Steroid 20-22-lyase Cholesterol monooxygenase (side-chain-cleaving)	
1.14.15.6	Steroid 20-22 desmolase Cholesterol monooxygenase (side-chain-cleaving)	
1.14.99.10	Steroid 21-hydroxylase Steroid 21-monooxygenase	
1.14.99.10	Steroid 21-monooxygenase Steroid 21-monooxygenase	
3.1.6.2	Steroid 3-sulfatase Steryl-sulfatase	
1.13.11.25	Steroid 4,5-dioxygenase 3,4-Dihydroxy-9,10-secoandrosta-1,3,5(10)-triene-9,17-dione 4,5-dioxygenase	
1.3.99.5	Steroid δ^4-5α-reductase 3-Oxo-5α-steroid Δ^4-dehydrogenase	
1.3.1.30	Steroid 5-α-reductase Progesterone 5α-reductase	
5.3.3.1	Steroid 5$\rightarrow$4-isomerase Steroid Δ-isomerase	Vol. 1, Page 376
1.3.99.5	Steroid 5α-reductase 3-Oxo-5α-steroid Δ^4-dehydrogenase	
1.3.1.22	Steroid 5α-reductase Cholestenone 5α-reductase	
1.3.1.30	Δ^4-Steroid 5α-reductase (progesterone) Progesterone 5α-reductase	
1.3.1.22	Steroid 5.α.-hydrogenase Cholestenone 5α-reductase	
1.3.1.23	Steroid 5.β.-reductase Cholestenone 5β-reductase	
1.3.99.6	Steroid 5β reductase 3-Oxo-5β-steroid Δ^4-dehydrogenase	
1.14.99.24	Steroid 9α-hydroxylase Steroid 9α-monooxygenase	
1.14.99.24	Steroid 9α-monooxygenase Steroid 9α-monooxygenase	

EC-No.	Synonym Recommended Name	Vol., Page
2.4.1.39	Steroid acetylglucosaminyltransferase Steroid N-acetylglucosaminyltransferase	
2.8.2.2	Steroid alcohol sulfotransferase Alcohol sulfotransferase	
2.8.2.15	Steroid alcohol sulfotransferase Steroid sulfotransferase	
4.1.2.30	Steroid C17(20) lyase 17α-Hydroxyprogesterone aldolase	
1.14.13.54	Steroid hormone hydroxylase Ketosteroid monooxygenase	
1.14.13.54	Steroid hydroxylase Ketosteroid monooxygenase	
5.3.3.1	Δ^5-steroid isomerase Steroid Δ-isomerase	Vol. 1, Page 376
5.3.3.1	Steroid isomerase Steroid Δ-isomerase	Vol. 1, Page 376
3.1.1.37	steroid lactonase steroid-lactonase	
1.14.13.54	Steroid monooxygenase Ketosteroid monooxygenase	
2.4.1.39	Steroid N-acetylglucosaminyltransferase Steroid N-acetylglucosaminyltransferase	
1.3.99.4	Δ^1-Steroid reductase 3-Oxosteroid Δ^1-dehydrogenase	
3.1.6.2	Steroid sulfatase Steryl-sulfatase	
3.1.6.2	Steroid sulfate sulfohydrolase Steryl-sulfatase	
2.8.2.2	Steroid sulfokinase Alcohol sulfotransferase	
2.8.2.2	Steroid sulfotransferase Alcohol sulfotransferase	
2.8.2.15	Steroid sulfotransferase Steroid sulfotransferase	
2.4.1.173	Sterol-β-D-glucosyltransferase Sterol 3β-glucosyltransferase	

EC-No.	Synonym Recommended Name	Vol., Page
3.1.1.13	Sterol-ester acylhydrolase Cholesterol esterase	
2.3.1.26	Sterol-ester synthase Sterol O-acyltransferase	
2.3.1.26	Sterol-ester synthetase Sterol O-acyltransferase	
2.4.1.173	Sterol 3β-glucosyltransferase Sterol 3β-glucosyltransferase	
1.1.1.170	Sterol 4α-carboxylic decarboxylase 3β-Hydroxy-4β-methylcholestenecarboxylate 3-dehydrogenase (decarboxylating)	
1.3.3.2	δ^7-Sterol δ^5-dehydrogenase Lathosterol oxidase	
1.3.3.2	δ^7-Sterol 5-desaturase Lathosterol oxidase	
2.4.1.173	Sterol:UDPG glucosyltransferase Sterol 3β-glucosyltransferase	
2.1.1.41	24-Sterol C-methyltransferase 24-Sterol C-methyltransferase	
2.3.1.154	sterol carrier protein propionyl-CoA C^2-trimethyltridecanoyltransferase	
3.1.1.13	Sterol ester hydrolase Cholesterol esterase	
2.4.1.173	Sterol glucosyltransferase Sterol 3β-glucosyltransferase	
2.1.1.41	Δ^{24}-Sterol methyltransferase 24-Sterol C-methyltransferase	
2.3.1.26	Sterol O-acyltransferase Sterol O-acyltransferase	
3.1.6.2	Sterol sulfatase Steryl-sulfatase	
2.8.2.2	Sterol sulfokinase Alcohol sulfotransferase	
2.8.2.2	Sterol sulfotransferase Alcohol sulfotransferase	

EC-No.	Synonym Recommended Name	Vol., Page
3.2.1.104	Steryl-β-glucosidase Steryl-β-glucosidase	
3.1.6.2	Steryl-sulfatase Steryl-sulfatase	
3.1.6.2	Steryl-sulfate sulfohydrolase Steryl-sulfatase	
2.3.1.146	Stilbene synthase Pinosylvin synthase	
2.3.1.95	Stilbene synthase Trihydroxystilbene synthase	
4.1.1.60	Stipitatonate carboxy-lyase (decyclizing) Stipitatonate decarboxylase	
4.1.1.60	Stipitatonate decarboxylase Stipitatonate decarboxylase	
1.13.11.29	Stizolobate synthase Stizolobate synthase	
1.13.11.30	Stizolobinate synthase Stizolobinate synthase	
6.2.1.5	A-STK Succinate-CoA ligase (ADP-forming)	Vol. 2, Page 224
6.2.1.4	G-STK Succinate-CoA ligase (GDP-forming)	Vol. 2, Page 219
6.2.1.5	STK Succinate-CoA ligase (ADP-forming)	Vol. 2, Page 224
2.7.1.72	Streptidine kinase Streptomycin 6-kinase	
3.4.22.10	Streptococcal cysteine proteinase streptopain	
3.4.22.10	streptococcal erythrogenic toxin B streptopain	
3.4.22.10	streptococcal pyrogenic exotoxin B streptopain	
3.4.22.10	Streptococcus peptidase A streptopain	
3.4.4.18	streptococcus peptidase A streptococcus peptidase A	

EC-No.	Synonym Recommended Name Vol., Page
3.4.22.10	Streptococcus protease streptopain
3.4.24.30	Streptococcus thermophilus intracellular proteinase Coccolysin
3.4.21.80	Streptogrisin A Streptogrisin A
3.4.21.81	Streptogrisin B Streptogrisin B
3.4.99.11	Streptomyces alkalophilic keratinase Streptomyces alkalophilic keratinase
3.1.21.6	Streptomyces glaucescens exocytoplasmic dodeoxyribonuclease CC-preferring endodeoxyribonuclease
3.4.24.31	Streptomyces griseus neutral proteinase Mycolysin
3.4.21.80	Streptomyces griseus protease A Streptogrisin A
3.4.21.81	Streptomyces griseus protease B Streptogrisin B
3.4.21.81	Streptomyces griseus proteinase 1 Streptogrisin B
3.4.21.80	Streptomyces griseus proteinase A Streptogrisin A
3.4.21.81	Streptomyces griseus proteinase B Streptogrisin B
3.4.21.80	Streptomyces griseus serine proteinase 3 Streptogrisin A
3.4.21.80	Streptomyces griseus serine proteinase A Streptogrisin A
3.1.3.39	Streptomycin-6-phosphatase Streptomycin-6-phosphatase
3.1.3.39	Streptomycin-6-phosphate phosphohydrolase Streptomycin-6-phosphatase
3.1.3.39	Streptomycin-6-P phosphohydrolase Streptomycin-6-phosphatase
2.7.7.47	Streptomycin-spectinomycin adenylyltransferase Streptomycin 3-adenylyltransferase

EC-No.	Synonym Recommended Name	Vol., Page
2.7.7.47	Streptomycin 3-adenylyltransferase Streptomycin 3-adenylyltransferase	
2.7.1.87	Streptomycin 3-kinase Streptomycin 3-kinase	
2.7.1.87	Streptomycin 3-phosphotransferase Streptomycin 3-kinase	
2.7.1.72	Streptomycin 6-kinase Streptomycin 6-kinase	
2.7.1.72	Streptomycin 6-O-phosphotransferase Streptomycin 6-kinase	
3.1.3.39	Streptomycin 6-phosphate phosphatase Streptomycin-6-phosphatase	
3.1.3.39	Streptomycin 6-phosphate phosphohydrolase Streptomycin-6-phosphatase	
2.7.1.72	Streptomycin 6-phosphotransferase Streptomycin 6-kinase	
2.7.7.47	Streptomycin adenylate synthetase Streptomycin 3-adenylyltransferase	
2.7.7.47	Streptomycin adenyltransferase Streptomycin 3-adenylyltransferase	
2.7.7.47	Streptomycin adenylylase Streptomycin 3-adenylyltransferase	
2.7.7.47	Streptomycin adenylyltransferase Streptomycin 3-adenylyltransferase	
3.4.22.10	streptopain streptopain	
3.2.1.105	Strictosidine β-D-glucohydrolase Strictosidine β-glucosidase	
3.2.1.105	Strictosidine β-glucosidase Strictosidine β-glucosidase	
4.3.3.2	Strictosidine synthase Strictosidine synthase	
4.3.3.2	Strictosidine synthetase Strictosidine synthase	
2.7.6.5	Stringent factor GTP pyrophosphokinase	

EC-No.	Synonym Recommended Name	Vol., Page
1.5.1.22	Strombine dehydrogenase Strombine dehydrogenase	
3.4.24.17	Stromelysin Stromelysin 1	
3.4.24.17	Stromelysin 1 Stromelysin 1	
3.4.24.22	Stromelysin 2 Stromelysin 2	
5.3.99.7	Styrene-oxide isomerase Styrene-oxide isomerase	Vol. 1, Page 486
5.3.99.7	Styrene-oxide isomerase (epoxide-cleaving) Styrene-oxide isomerase	Vol. 1, Page 486
5.3.99.7	Styrene oxide isomerase Styrene-oxide isomerase	Vol. 1, Page 486
3.1.21.3	StySBI Type I site-specific deoxyribonuclease	
3.1.21.3	StySJI Type I site-specific deoxyribonuclease	
3.1.21.3	StySPI Type I site-specific deoxyribonuclease	
3.1.21.3	StySQI Type I site-specific deoxyribonuclease	
3.4.21.40	submandibular proteinase A submandibular proteinase A	
3.1.1.29	N-substituted aminoacyl transfer RNA hydrolase aminoacyl-tRNA hydrolase	
1.14.14.1	Substrate,reduced-flavoprotein:oxygen oxidoreductase (RH-hydroxylating or-epoxidizing) Unspecific monooxygenase	
3.4.21.62	Subtilisin Subtilisin	
3.4.21.62	Subtilisin amylosacchariticus Subtilisin	
3.4.21.62	Subtilisin BL Subtilisin	

EC-No.	Synonym Recommended Name	Vol., Page
3.4.21.62	Subtilisin DY Subtilisin	
3.4.21.62	Subtilisin E Subtilisin	
3.4.21.62	Subtilisin GX Subtilisin	
3.4.21.62	Subtilisin S41 Subtilisin	
3.4.21.62	Subtilisin Sendai Subtilisin	
3.4.21.62	Subtilisn J Subtilisin	
3.4.21.62	Subtilopeptidase Subtilisin	
3.4.4.16	subtilopeptidase A subtilopeptidase A	
3.1.3.24	Suc-phosphate phosphatase sucrose-phosphatase	
2.8.3.7	Succinate-citramalate CoA-transferase Succinate-citramalate CoA-transferase	
6.2.1.5	Succinate-CoA ligase (ADP-forming) Succinate-CoA ligase (ADP-forming)	Vol. 2, Page 224
6.2.1.4	Succinate-CoA ligase (GDP-forming) Succinate-CoA ligase (GDP-forming)	Vol. 2, Page 219
2.8.3.13	Succinate-hydroxymethylglutarate CoA-transferase Succinate-hydroxymethylglutarate CoA-transferase	
1.2.1.16	Succinate-semialdehyde:NAD(P)$^+$ oxidoreductase Succinate-semialdehyde dehydrogenase (NAD(P)$^+$)	
1.2.1.24	Succinate-semialdehyde:NAD$^+$ oxidoreductase Succinate semialdehyde dehydrogenase	
1.2.1.16	Succinate-semialdehyde dehydrogenase (NAD(P)$^+$) Succinate-semialdehyde dehydrogenase (NAD(P)$^+$)	
1.3.99.1	Succinate: (acceptor) oxidoreductase Succinate dehydrogenase	
2.8.3.13	Succinate:(S)-3-hydroxy-3-methylglutarate CoA-transferase Succinate-hydroxymethylglutarate CoA-transferase	

EC-No.	Synonym Recommended Name	Vol., Page
6.2.1.5	Succinate:CoA ligase (ADP-forming) Succinate-CoA ligase (ADP-forming)	Vol. 2, Page 224
6.2.1.4	Succinate:CoA ligase (GDP-forming) Succinate-CoA ligase (GDP-forming)	Vol. 2, Page 219
1.3.1.6	Succinate: NAD+ oxidoreductase Fumarate reductase (NADH)	
1.3.5.1	Succinate: ubiquinone oxidoreductase Succinate dehydrogenase (ubiquinone)	
1.3.99.1	Succinate dehydrogenase Succinate dehydrogenase	
1.3.5.1	Succinate dehydrogenase (ubiquinone) Succinate dehydrogenase (ubiquinone)	
1.3.5.1	Succinate dehydrogenase complex Succinate dehydrogenase (ubiquinone)	
1.3.99.1	Succinate oxidoreductase Succinate dehydrogenase	
1.2.1.24	Succinate semialdehyde:NAD+ oxidoreductase Succinate semialdehyde dehydrogenase	
1.2.1.24	Succinate semialdehyde dehydrogenase Succinate semialdehyde dehydrogenase	
1.2.1.24	Succinate semialdehyde dehydrogenase Succinate semialdehyde dehydrogenase	
1.2.1.16	Succinate semialdehyde dehydrogenase (nicotinamide adenine dinucleotide (phosphate)) Succinate-semialdehyde dehydrogenase (NAD(P)+)	
6.2.1.5	Succinate thiokinase Succinate-CoA ligase (ADP-forming)	Vol. 2, Page 224
6.2.1.4	Succinate thiokinase Succinate-CoA ligase (GDP-forming)	Vol. 2, Page 219
1.3.99.1	Succinic acid dehydrogenase Succinate dehydrogenase	
1.3.99.1	Succinic dehydrogenase Succinate dehydrogenase	
1.3.5.1	Succinic dehydrogenase Succinate dehydrogenase (ubiquinone)	

EC-No.	Synonym Recommended Name	Vol., Page
1.2.1.24	Succinic semialdehyde dehydrogenase Succinate semialdehyde dehydrogenase	
6.2.1.5	Succinic thiokinase Succinate-CoA ligase (ADP-forming)	Vol. 2, Page 224
6.2.1.4	Succinic thiokinase Succinate-CoA ligase (GDP-forming)	Vol. 2, Page 219
6.3.2.6	Succino-AICAR synthetase Phosphoribosylaminoimidazolesuccinocarboxamide synthase	Vol. 2, Page 434
4.3.2.2	succino-AMP lyase adenylosuccinate lyase	
6.3.4.4	Succino-AMP synthetase Adenylosuccinate synthase	Vol. 2, Page 579
6.3.4.4	Succinoadenylic kinosynthetase Adenylosuccinate synthase	Vol. 2, Page 579
4.3.2.2	succino AMP-lyase adenylosuccinate lyase	
1.3.99.1	Succinodehydrogenase Succinate dehydrogenase	
2.8.3.2	Succinyl-β-ketoacyl-CoA transferase Oxalate CoA-transferase	
2.3.1.117	Succinyl-CoA:2,3,4,5-tetrahydropyridine-2-carboxylate N-succinyltransferase 2,3,4,5-Tetrahydropyridine-2-carboxylate N-succinyltransferase	
2.8.3.5	Succinyl-CoA:3-oxo-acid CoA-transferase 3-Oxoacid CoA-transferase	
2.8.3.6	Succinyl-CoA:3-oxoadipate CoA-transferase 3-Oxoadipate CoA-transferase	
2.8.3.8	Succinyl-CoA:acetate CoA transferase Acetate CoA-transferase	
2.8.3.7	Succinyl-CoA:citramalate CoA-transferase Succinate-citramalate CoA-transferase	
2.3.1.61	Succinyl-CoA:dihydrolipoamide S-succinyltransferase Dihydrolipoamide S-succinyltransferase	
2.3.1.61	Succinyl-CoA:dihydrolipoate S-succinyltransferase Dihydrolipoamide S-succinyltransferase	

EC-No.	Synonym Recommended Name	Vol., Page
2.3.1.37	Succinyl-CoA:glycine C-succinyltransferase (decarboxylating) 5-Aminolevulinate synthase	
2.3.1.109	Succinyl-CoA:L-arginine N^2-succinyltransferase Arginine N-succinyltransferase	
2.3.1.46	Succinyl-CoA:L-homoserine O-succinyltransferase Homoserine O-succinyltransferase	
2.8.3.2	Succinyl-CoA:oxalate CoA-transferase Oxalate CoA-transferase	
2.3.1.117	Succinyl-CoA:tetrahydrodipicolinate N-succinyltransferase 2,3,4,5-Tetrahydropyridine-2-carboxylate N-succinyltransferase	
3.1.2.3	succinyl-CoA acylase succinyl-CoA hydrolase	
3.1.2.3	succinyl-CoA hydrolase succinyl-CoA hydrolase	
3.1.2.3	succinyl-CoA hydrolase succinyl-CoA hydrolase	
6.2.1.5	Succinyl-CoA synthetase Succinate-CoA ligase (ADP-forming)	Vol. 2, Page 224
6.2.1.5	Succinyl-CoA synthetase (ADP-forming) Succinate-CoA ligase (ADP-forming)	Vol. 2, Page 224
6.2.1.4	Succinyl-CoA synthetase (GDP-forming) Succinate-CoA ligase (GDP-forming)	Vol. 2, Page 219
2.8.3.5	Succinyl-CoA transferase 3-Oxoacid CoA-transferase	
3.5.1.18	N-Succinyl-L-α,ε-diaminopimelic acid deacylase Succinyldiaminopimelate desuccinylase	
2.6.1.17	N-Succinyl-L-2,6-diaminoheptanedioate:2-oxoglutarate aminotransferase Succinyldiaminopimelate transaminase	
2.6.1.17	N-Succinyl-L-diaminopimelic glutamic transaminase Succinyldiaminopimelate transaminase	
4.2.99.9	O-succinyl-L-homoserine succinate-lyase (adding cysteine) O-succinylhomoserine (thiol)-lyase	
3.5.1.18	N-Succinyl-LL-2,6-diaminoheptanedioate amidohydrolase Succinyldiaminopimelate desuccinylase	
6.2.1.26	O-Succinylbenzoate-CoA ligase O-Succinylbenzoate-CoA ligase	Vol. 2, Page 320

EC-No.	Synonym Recommended Name	Vol., Page
6.2.1.26	O-Succinylbenzoate:CoA ligase (AMP-forming) O-Succinylbenzoate-CoA ligase	Vol. 2, Page 320
4.1.3.36	O-succinylbenzoyl-CoA 1,4-dihydroxy-2-naphthoate-lyase (cyclizing) naphthoate synthase	
6.2.1.26	*o*-Succinylbenzoyl-coenzyme A synthetase O-Succinylbenzoate-CoA ligase	Vol. 2, Page 320
6.2.1.4	Succinyl CoA synthetase Succinate-CoA ligase (GDP-forming)	Vol. 2, Page 219
2.8.3.5	Succinyl coenzyme A-acetoacetyl coenzyme A-transferase 3-Oxoacid CoA-transferase	
2.8.3.7	Succinyl coenzyme A-citramalyl coenzyme A transferase Succinate-citramalate CoA-transferase	
3.1.2.3	succinyl coenzyme A deacylase succinyl-CoA hydrolase	
3.1.2.3	succinyl coenzyme A hydrolase succinyl-CoA hydrolase	
6.2.1.5	Succinyl coenzyme A synthetase Succinate-CoA ligase (ADP-forming)	Vol. 2, Page 224
6.2.1.4	Succinyl coenzyme A synthetase Succinate-CoA ligase (GDP-forming)	Vol. 2, Page 219
6.2.1.5	Succinyl coenzyme A synthetase (adenosine diphosphate-forming) Succinate-CoA ligase (ADP-forming)	Vol. 2, Page 224
6.2.1.4	Succinyl coenzyme A synthetase (GDP-forming) Succinate-CoA ligase (GDP-forming)	Vol. 2, Page 219
6.2.1.4	Succinyl coenzyme A synthetase (guanosine diphosphate-forming) Succinate-CoA ligase (GDP-forming)	Vol. 2, Page 219
1.3.99.1	Succinyl dehydrogenase Succinate dehydrogenase	
2.6.1.17	Succinyldiaminopimelate aminotransferase Succinyldiaminopimelate transaminase	
3.5.1.18	Succinyldiaminopimelate desuccinylase Succinyldiaminopimelate desuccinylase	
2.6.1.17	Succinyldiaminopimelate transaminase Succinyldiaminopimelate transaminase	
3.1.2.13	S-succinylglutathione hydrolase S-succinylglutathione hydrolase	

EC-No.	Synonym Recommended Name	Vol., Page
3.1.2.13	S-succinylglutathione hydrolase S-succinylglutathione hydrolase	
4.2.99.9	O-succinylhomoserine (thiol)-lyase O-succinylhomoserine (thiol)-lyase	
4.2.99.9	O-succinylhomoserine synthase O-succinylhomoserine (thiol)-lyase	
4.2.99.9	O-succinylhomoserine synthetase O-succinylhomoserine (thiol)-lyase	
2.6.1.69	Succinylornithine aminotransferase N^2-Acetylornithine 5-transaminase	
1.2.1.24	Succinyl semialdehyde dehydrogenase Succinate semialdehyde dehydrogenase	
2.3.1.109	Succinyltransferase, arginine Arginine N-succinyltransferase	
2.3.1.46	Succinyltransferase, homoserine Homoserine O-succinyltransferase	
2.3.1.61	Succinyltransferase, lipoate Dihydrolipoamide S-succinyltransferase	
2.3.1.117	Succinyltransferase, tetrahydrodipicolinate 2,3,4,5-Tetrahydropyridine-2-carboxylate N-succinyltransferase	
3.2.1.26	Sucrase β-fructofuranosidase	
3.2.1.48	Sucrase Sucrose α-glucosidase	
3.2.1.48	Sucrase(invertase) Sucrose α-glucosidase	
3.2.1.10	Sucrase-isomaltase Oligo-1,6-glucosidase	
3.2.1.48	Sucrase-isomaltase Sucrose α-glucosidase	
2.4.1.125	Sucrose-1,6-α-glucan 3(6)-α-glucosyltransferase Sucrose-1,6-α-glucan 3(6)-α-glucosyltransferase	
3.1.3.24	sucrose-6-phosphatase sucrose-phosphatase	
3.1.3.24	sucrose-6F¯phosphate phosphohydrolase sucrose-phosphatase	

EC-No.	Synonym Recommended Name	Vol., Page
3.2.1.48	Sucrose α-D-glucohydrolase Sucrose α-glucosidase	
2.4.1.4	Sucrose-glucan glucosyltransferase Amylosucrase	
3.2.1.48	Sucrose α-glucohydrolase Sucrose α-glucosidase	
3.2.1.48	Sucrose α-glucosidase Sucrose α-glucosidase	
5.4.99.11	Sucrose α-glucosyltransferase Isomaltulose synthase	Vol. 1, Page 638
3.1.3.24	sucrose-phosphatase sucrose-phosphatase	
3.1.3.24	sucrose-phosphate hydrolase sucrose-phosphatase	
3.1.3.24	sucrose-phosphate phosphohydrolase sucrose-phosphatase	
2.4.1.14	Sucrose-phosphate synthase Sucrose-phosphate synthase	
2.4.1.99	Sucrose-sucrose 1-fructosyltransferase Sucrose 1F-fructosyltransferase	
2.4.1.13	Sucrose-UDP glucosyltransferase Sucrose synthase	
2.4.1.13	Sucrose-uridine diphosphate glucosyltransferase Sucrose synthase	
3.2.1.48	Sucrose.α.-glucohydrolase Sucrose α-glucosidase	
2.4.1.9	Sucrose 1-fructosyltransferase Inulosucrase	
2.4.1.99	Sucrose 1F-fructosyltransferase Sucrose 1F-fructosyltransferase	
2.4.1.10	Sucrose 6-fructosyltransferase Levansucrase	
5.4.99.11	Sucrose 6-glucosylmutase Isomaltulose synthase	Vol. 1, Page 638
2.4.1.5	Sucrose 6-glucosyltransferase Dextransucrase	

EC-No.	Synonym Recommended Name	Vol., Page
3.1.3.24	sucrose 6-phosphate hydrolase sucrose-phosphatase	
2.4.1.14	Sucrose 6-phosphate synthase Sucrose-phosphate synthase	
2.4.1.167	Sucrose 6F-α-galactotransferase Sucrose 6Fα-galactotransferase	
2.4.1.4	Sucrose:1,4-α-D-glucan 4-α-D-glucosyltransferase Amylosucrase	
2.4.1.140	Sucrose:1,6(1,3)-α-D-glucan 6(3)-α-D-glucosyltransferase Alternansucrase	
2.4.1.140	Sucrose:1,6-, 1,3-α-D-glucan 3-α- and 6-α-D-glucosyltransferase Alternansucrase	
2.4.1.125	Sucrose:1,6-, 1,3-α-D-glucan 3-α- and 6-α-D-glucosyltransferase Sucrose-1,6-α-glucan 3(6)-α-glucosyltransferase	
2.4.1.125	Sucrose:1,6-α-D-glucan 3(6)-α-D-glucosyltransferase Sucrose-1,6-α-glucan 3(6)-α-glucosyltransferase	
2.4.1.125	Sucrose:1,6-α-D-glucan 3-α- and 6-α-glucosyltransferase Sucrose-1,6-α-glucan 3(6)-α-glucosyltransferase	
2.4.1.5	Sucrose:1,6-α-D-glucan 6-α-D-glucosyltransferase Dextransucrase	
2.4.1.9	Sucrose:2,1-β-D-fructan 1-β-D-fructosyltransferase Inulosucrase	
2.4.1.10	Sucrose:2,6-β-D-fructan 6-β-D-fructosyltransferase Levansucrase	
2.4.1.7	Sucrose:orthophosphate α-D-glucosyltransferase Sucrose phosphorylase	
2.4.1.99	Sucrose:sucrose 1-fructosyltransferase Sucrose 1F-fructosyltransferase	
2.4.1.99	Sucrose:sucrose 1F-β-D-fructosyltransferase Sucrose 1F-fructosyltransferase	
2.4.1.99	Sucrose:sucrose fructosyltransferase Sucrose 1F-fructosyltransferase	
5.4.99.11	Sucrose glucosylmutase Isomaltulose synthase	Vol. 1, Page 638
2.4.1.7	Sucrose glucosyltransferase Sucrose phosphorylase	

EC-No.	Synonym Recommended Name	Vol., Page
5.4.99.11	Sucrose mutase Isomaltulose synthase	Vol. 1, Page 638
3.1.3.24	sucrose phosphatase sucrose-phosphatase	
3.1.3.24	sucrose phosphatase sucrose-phosphatase	
2.4.1.14	Sucrosephosphate-UDP glucosyltransferase Sucrose-phosphate synthase	
2.4.1.14	Sucrose phosphate-uridine diphosphate glucosyltransferase Sucrose-phosphate synthase	
2.4.1.14	Sucrose phosphate synthetase Sucrose-phosphate synthase	
2.4.1.7	Sucrose phosphorylase Sucrose phosphorylase	
2.7.1.69	Sucrose phosphotransferase system II Protein-Npi-phosphohistidine-sugar phosphotransferase	
2.4.1.13	Sucrose synthase Sucrose synthase	
2.4.1.13	Sucrose synthetase Sucrose synthase	
3.1.3.58	Sugar-ω-phosphate phosphohydrolase Sugar-terminal-phosphatase	
2.7.7.36	Sugar-1-phosphate adenylyltransferase Aldose-1-phosphate adenylyltransferase	
2.7.7.37	Sugar-1-phosphate nucleotidyltransferase Aldose-1-phosphate nucleotidyltransferase	
2.7.3.9	Sugar-PEP phosphotransferase enzyme I Phosphoenolpyruvate-protein phosphotransferase	
3.1.3.23	sugar-phosphatase sugar-phosphatase	
3.1.3.23	sugar-phosphate phosphohydrolase sugar-phosphatase	
3.1.6.3	Sugar-sulfate sulfohydrolase Glycosulfatase	
3.1.3.58	Sugar-terminal-phosphatase Sugar-terminal-phosphatase	

EC-No.	Synonym Recommended Name	Vol., Page
2.7.7.37	Sugar nucleotide phosphorylase Aldose-1-phosphate nucleotidyltransferase	
3.1.3.23	sugar phosphatase sugar-phosphatase	
3.1.3.23	sugar phosphate phosphohydrolase sugar-phosphatase	
1.1.1.260	sulcatol:NAD oxidoreductase sulcatone reductase	
1.1.1.260	sulcatone reductase sulcatone reductase	
2.8.1.3	Sulfane reductase Thiosulfate-thiol sulfurtransferase	
2.8.1.3	Sulfane sulfurtransferase Thiosulfate-thiol sulfurtransferase	
3.1.6.1	Sulfatase Arylsulfatase	
3.1.6.12	Sulfatase, acetylgalactosamine 4- N-Acetylgalactosamine-4-sulfatase	
3.1.6.4	Sulfatase, acetylgalactosamine 6- N-Acetylgalactosamine-6-sulfatase	
3.1.6.14	Sulfatase, acetylglucosamine 6- N-Acetylglucosamine-6-sulfatase	
3.6.2.1	Sulfatase, adenylyl Adenylylsulfatase	
3.1.6.1	Sulfatase, aryl- Arylsulfatase	
3.1.6.7	sulfatase, cellulose poly- cellulose polysulfatase	
3.1.6.8	Sulfatase, cerebroside Cerebroside-sulfatase	
3.1.6.6	Sulfatase, choline Cholinesulfatase	
3.1.6.9	Sulfatase, chondro-4- Chondro-4-sulfatase	
3.1.6.10	Sulfatase, chondro-6- Chondro-6-sulfatase	

EC-No.	Synonym Recommended Name	Vol., Page
3.1.6.4	Sulfatase, chondroitin N-Acetylgalactosamine-6-sulfatase	
3.1.6.17	Sulfatase, D-lactate 2- D-Lactate-2-sulfatase	
3.1.6.11	Sulfatase, disulfoglucosamine 6- N-Sulfoglucosamine-6-sulfatase	
3.1.6.18	Sulfatase, glucuronate 2- Glucuronate-2-sulfatase	
3.1.6.3	Sulfatase, glyco- Glycosulfatase	
3.1.6.13	Sulfatase, L-idurono- Iduronate 2-sulfatase	
3.6.2.2	Sulfatase, phosphoadenylyl Phosphoadenylylsulfatase	
2.5.1.5	Sulfatase, porphyran Galactose-6-sulfurylase	
3.1.6.2	Sulfatase, sterol Steryl-sulfatase	
2.7.7.4	Sulfate adenylyltransferase Sulfate adenylyltransferase	
2.7.7.5	Sulfate adenylyltransferase (ADP) Sulfate adenylyltransferase (ADP)	
3.4.22.39	sulfhydryl esterase adenain	
1.8.3.2	Sulfhydryl oxidase Thiol oxidase	
3.4.22.39	sulfhydryl protease adenain	
3.4.22.39	sulfhydryl proteinase adenain	
4.1.1.29	3-Sulfino-L-alanine carboxy-lyase Sulfinoalanine decarboxylase	
4.1.1.29	Sulfinoalanine decarboxylase Sulfinoalanine decarboxylase	
1.8.2.1	Sulfite-cytochrome c oxidoreductase Sulfite dehydrogenase	

EC-No.	Synonym Recommended Name	Vol., Page
1.8.2.1	Sulfite:ferricytochrome-c oxidoreductase Sulfite dehydrogenase	
1.8.3.1	Sulfite:oxygen oxidoreductase Sulfite oxidase	
1.8.2.1	Sulfite cytochrome c reductase Sulfite dehydrogenase	
1.8.2.1	Sulfite dehydrogenase Sulfite dehydrogenase	
1.8.2.1	Sulfite oxidase Sulfite dehydrogenase	
1.8.3.1	Sulfite oxidase Sulfite oxidase	
1.8.99.1	Sulfite reductase Sulfite reductase	
1.8.7.1	Sulfite reductase (ferredoxin) Sulfite reductase (ferredoxin)	
1.8.1.2	Sulfite reductase (NADPH) Sulfite reductase (NADPH)	
3.10.1.1	N-Sulfo-D-glucosamine sulfohydrolase N-Sulfoglucosamine sulfohydrolase	
3.1.6.13	2-Sulfo-L-iduronate 2-sulfatase Iduronate 2-sulfatase	
3.1.6.13	Sulfo-L-iduronate sulfatase Iduronate 2-sulfatase	
4.4.1.12	sulfo-lyase, sulfoacetaldehyde sulfoacetaldehyde lyase	
4.4.1.12	sulfoacetaldehyde lyase sulfoacetaldehyde lyase	
4.4.1.12	sulfoacetaldehyde sulfo-lyase sulfoacetaldehyde lyase	
1.14.12.8	4-Sulfobenzoate,NADH:oxygen oxidoreductase (3,4-hydroxylating, sulfite-forming) 4-Sulfobenzoate 3,4-dioxygenase	
1.14.12.8	4-Sulfobenzoate 3,4-dioxygenase 4-Sulfobenzoate 3,4-dioxygenase	

EC-No.	Synonym Recommended Name	Vol., Page
1.14.12.8	4-Sulfobenzoate 3,4-dioxygenase system 4-Sulfobenzoate 3,4-dioxygenase	
4.2.99.8	S-sulfocysteine synthase Cysteine synthase	
3.1.6.15	N-sulfoglucosamine-3-sulfatase N-sulfoglucosamine-3-sulfatase	
3.1.6.11	N-Sulfoglucosamine-6-sulfatase N-Sulfoglucosamine-6-sulfatase	
3.10.1.1	Sulfoglucosamine sulfamidase N-Sulfoglucosamine sulfohydrolase	
3.10.1.1	N-Sulfoglucosamine sulfohydrolase N-Sulfoglucosamine sulfohydrolase	
3.1.6.13	Sulfoiduronate sulfohydrolase Iduronate 2-sulfatase	
2.8.2.1	Sulfokinase Aryl sulfotransferase	
2.8.2.26	3'-Sulfotransferase Quercetin-3-sulfate 3'-sulfotransferase	
2.8.2.28	7-Sulfotransferase Quercetin-3,3'-bissulfate 7-sulfotransferase	
2.8.2.2	Sulfotransferase, 3β-hydroxy steroid Alcohol sulfotransferase	
2.8.2.1	Sulfotransferase, aryl Aryl sulfotransferase	
2.8.2.3	Sulfotransferase, aryl amine Amine sulfotransferase	
2.8.2.22	Sulfotransferase, arylsulfate Arylsulfate sulfotransferase	
2.8.2.14	Sulfotransferase, bile salt Bile-salt sulfotransferase	
2.8.2.6	Sulfotransferase, choline Choline sulfotransferase	
2.8.2.5	Sulfotransferase, chondroitin Chondroitin 4-sulfotransferase	
2.8.2.5	Sulfotransferase, chondroitin 4- Chondroitin 4-sulfotransferase	

EC-No.	Synonym Recommended Name	Vol., Page
2.8.2.17	Sulfotransferase, chondroitin 6- Chondroitin 6-sulfotransferase	
2.8.2.24	Sulfotransferase, desulfoglucosinolate Desulfoglucosinolate sulfotransferase	
2.8.2.8	Sulfotransferase, desulfoheparin Desulfoheparin sulfotransferase	
2.8.2.4	Sulfotransferase, estrone Estrone sulfotransferase	
2.8.2.26	Sulfotransferase, flavonol 3'- Quercetin-3-sulfate 3'-sulfotransferase	
2.8.2.25	Sulfotransferase, flavonol 3- Flavonol 3-sulfotransferase	
2.8.2.27	Sulfotransferase, flavonol 4'- Quercetin-3-sulfate 4'-sulfotransferase	
2.8.2.28	Sulfotransferase, flavonol 7- Quercetin-3,3'-bissulfate 7-sulfotransferase	
2.8.2.11	Sulfotransferase, galactocerebroside Galactosylceramide sulfotransferase	
2.8.2.18	Sulfotransferase, glucocorticoid Cortisol sulfotransferase	
2.8.2.23	Sulfotransferase, glucosaminyl 3-O- Heparin-glucosamine 3-O-sulfotransferase	
2.8.2.12	Sulfotransferase, heparitin Heparitin sulfotransferase	
2.8.2.21	Sulfotransferase, keratan Keratan sulfotransferase	
2.8.2.10	Sulfotransferase, luciferin Renilla-luciferin sulfotransferase	
2.8.2.16	Sulfotransferase, phosphoadenylylsulfate-thiol Thiol sulfotransferase	
2.8.2.20	Sulfotransferase, protein (tyrosine) Protein-tyrosine sulfotransferase	
2.8.2.13	Sulfotransferase, psychosine Psychosine sulfotransferase	
2.8.2.19	Sulfotransferase, triglucosylmonoalkylmonoacyl Triglucosylalkylacylglycerol sulfotransferase	

EC-No.	Synonym Recommended Name	Vol., Page
2.8.2.9	Sulfotransferase, tyrosine ester Tyrosine-ester sulfotransferase	
2.8.2.7	Sulfotransferase, uridine diphosphoacetylgalactosamine 4-sulfate UDP-N-acetylgalactosamine-4-sulfate sulfotransferase	
1.13.11.18	Sulfur:oxygen oxidoreductase Sulfur dioxygenase	
1.13.11.18	Sulfur dioxygenase Sulfur dioxygenase	
1.13.11.18	Sulfur oxygenase Sulfur dioxygenase	
1.97.1.3	Sulfur reductase Sulfur reductase	
2.8.1.2	Sulfurtransferase, 3-mercaptopyruvate 3-Mercaptopyruvate sulfurtransferase	
2.8.1.4	Sulfurtransferase, ribonucleate tRNA sulfurtransferase	
2.8.1.3	Sulfurtransferase, sulfane Thiosulfate-thiol sulfurtransferase	
2.8.1.1	Sulfurtransferase, thiosulfate Thiosulfate sulfurtransferase	
2.8.1.4	Sulfurtransferase, transfer ribonucleate tRNA sulfurtransferase	
2.7.7.4	Sulfurylase Sulfate adenylyltransferase	
1.10.3.7	Sulochrin:oxygen oxidoreductase (cyclizing, (+)-specific) Sulochrin oxidase ((+)-bisdechlorogeodin-forming)	
1.10.3.8	Sulochrin:oxygen oxidoreductase (cyclizing, (-)-specific) Sulochrin oxidase ((-)-bisdechlorogeodin-forming)	
1.10.3.7	Sulochrin oxidase Sulochrin oxidase ((+)-bisdechlorogeodin-forming)	
1.10.3.7	Sulochrin oxidase ((+)-bisdechlorogeodin-forming) Sulochrin oxidase ((+)-bisdechlorogeodin-forming)	
1.10.3.8	Sulochrin oxidase ((-)-bisdechlorogeodin-forming) Sulochrin oxidase ((-)-bisdechlorogeodin-forming)	
1.8.3.1	Sulphite oxidase Sulfite oxidase	

EC-No.	Synonym Recommended Name	Vol., Page
3.2.1.23	Sumiklat β-galactosidase	
3.4.23.18	Sumizyme AP Aspergillopepsin I	
3.4.21.63	Sumizyme MP Oryzin	
3.4.22.2	summetrin papain	
3.4.21.62	Superase Subtilisin	
1.15.1.1	Superoxidase dismutase Superoxide dismutase	
1.15.1.1	Superoxide:superoxide oxidoreductase Superoxide dismutase	
1.15.1.1	Superoxide dismutase Superoxide dismutase	
1.15.1.1	Superoxide dismutase I Superoxide dismutase	
1.15.1.1	Superoxide dismutase II Superoxide dismutase	
3.4.24.36	Surface acid proteinase Leishmanolysin	
3.6.1.3	SV40 T-antigen Adenosinetriphosphatase	
5.3.1.5	Sweetzyme Xylose isomerase	Vol. 1, Page 259
5.3.1.5	Sweetzyme Q Xylose isomerase	Vol. 1, Page 259
5.3.1.5	Swetase Xylose isomerase	Vol. 1, Page 259
5.99.1.2	Swivelase DNA topoisomerase	Vol. 1, Page 721
2.5.1.23	Sym-Norspermidine synthase Sym-Norspermidine synthase	
3.6.1.41	Symmetrical diadenosine tetraphosphate hydrolase Bis(5'-nucleosyl)-tetraphosphatase (symmetrical)	

EC-No.	Synonym Recommended Name	Vol., Page
4.2.1.88	Synephrine dehydratase Synephrine dehydratase	
4.1.3.28	Synthase, (R)-citrate Citrate (re)-synthase	
1.1.3.33	Synthase, (S)-cheilanthifoline S-Cheilanthifoline synthase	
4.2.1.78	Synthase, (S)-norlaudanosoline (S)-Norcoclaurine synthase	
1.1.3.32	Synthase, (S)-stylopine (S)-Stylopine synthase	
4.1.3.12	synthase, α-isopropylmalate 2-isopropylmalate synthase	
4.4.1.9	synthase,.β.-cyanoalanine L-3-cyanoalanine synthase	
2.4.2.24	Synthase, 1,4-β-xylan 1,4-β-D-Xylan synthase	
4.1.3.36	synthase, 1,4-dihydroxy-2-naphthoate naphthoate synthase	
4.4.1.14	synthase, 1-aminocyclopropanecarboxylate 1-aminocyclopropane-1-carboxylate synthase	
2.2.1.4	Synthase, 1-deoxy-D-altro-heptulose 7-phosphate Acetoin-ribose-5-phosphate transaldolase	
4.1.3.33	Synthase, 2-ethylmalate 2-Ethylmalate synthase	
4.1.3.10	synthase, 2-ethylmalate 3-ethylmalate synthase	
4.1.3.9	Synthase, 2-hydroxyglutarate 2-Hydroxyglutarate synthase	
4.1.3.12	synthase, 2-isopropylmalate 2-isopropylmalate synthase	
1.2.7.2	Synthase, 2-oxobutyrate 2-Oxobutyrate synthase	
1.2.7.3	Synthase, 2-oxoglutarate 2-Oxoglutarate synthase	
2.3.1.41	Synthase, 3-oxoacyl-[acyl-carrier-protein] 3-Oxoacyl-[acyl-carrier-protein] synthase	

EC-No.	Synonym Recommended Name	Vol., Page
4.1.3.11	Synthase, 3-propylmalate 3-Propylmalate synthase	
4.6.1.3	synthase, 5-dehydroquinate 3-dehydroquinate synthase	
2.5.1.19	Synthase, 5-enolpyruvoylshikimate 3-phosphate 3-Phosphoshikimate 1-carboxyvinyltransferase	
4.6.1.10	Synthase, 6-pyruvoyltetrahydropterin 6-Pyruvoyltetrahydtropterin synthase	
4.6.1.10	Synthase, 6-pyruvoyltetrahydropterin [16-cysteine] (human clone lambda HSY2 gene PCBD subunit) 6-Pyruvoyltetrahydtropterin synthase	
4.6.1.10	Synthase, 6-pyruvoyltetrahydropterin [25-glutamine] (human clone lambdaHSY2 gene PCBD subunit) 6-Pyruvoyltetrahydtropterin synthase	
4.6.1.10	Synthase, 6-pyruvoyltetrahydropterin [87-leucine] (human clone lambdaHSY2 gene PCBD subunit) 6-Pyruvoyltetrahydtropterin synthase	
4.6.1.10	Synthase, 6-pyruvoyltetrahydropterin [de-57-valine] (human clone lambdaHSY2 gene PCBD subunit) 6-Pyruvoyltetrahydtropterin synthase	
2.3.1.47	Synthase, 7-oxo-8-aminononanoate 8-Amino-7-oxononanoate synthase	
4.1.3.18	synthase, acetolactate acetolactate synthase	
6.1.1.7	Synthase, alanyl-transfer ribonucleate Alanine-tRNA ligase	Vol. 2, Page 51
2.4.1.33	Synthase, alginate Alginate synthase	
2.5.1.26	Synthase, alkylglycerone phosphate Alkylglycerone-phosphate synthase	
2.3.1.37	Synthase, aminolevulinate 5-Aminolevulinate synthase	
4.1.3.27	synthase, anthranilate anthranilate synthase	
2.5.1.40	Synthase, aristolochene Aristolochene synthase	

EC-No.	Synonym Recommended Name	Vol., Page
2.3.1.151	Synthase, benzophenone Benzophenone synthase	
1.1.3.26	Synthase, berberine Columbamine oxidase	
6.3.5.5	Synthase, carbamoylphosphate (glutamine) Carbamoyl-phosphate synthase (glutamine-hydrolysing)	Vol. 2, Page 689
4.6.1.7	Synthase, casbene Casbene synthase	
4.1.3.7	Synthase, citrate Citrate (si)-synthase	
4.2.99.9	synthase, cystathionine. γ.- O-succinylhomoserine (thiol)-lyase	
4.2.99.8	synthase, cysteine Cysteine synthase	
4.1.3.23	Synthase, decylcitrate Decylcitrate synthase	
4.1.3.29	Synthase, decylhomocitrate Decylhomocitrate synthase	
2.6.1.62	Synthase, diaminopelargonate Adenosylmethionine-8-amino-7-oxononanoate transaminase	
4.2.1.52	synthase, dihydrodipicolinate dihydrodipicolinate synthase	
2.5.1.15	Synthase, dihydropteroate Dihydropteroate synthase	
2.2.1.3	Synthase, dihydroxyacetone Formaldehyde transketolase	
2.5.1.24	Synthase, discadenine Discadenine synthase	
2.3.1.94	Synthase, erythronolide Erythronolide synthase	
3.1.1.67	Synthase, fatty acid ethyl ester Fatty-acyl-ethyl-ester synthase	
2.3.1.110	Synthase, feruloyltyramine Tyramine N-feruloyltransferase	
2.3.1.74	Synthase, flavanone Naringenin-chalcone synthase	

EC-No.	Synonym Recommended Name	Vol., Page
2.7.1.106	Synthase, glucose 1,6-diphosphate Glucose-1,6-bisphosphate synthase	
1.4.1.14	Synthase, glutamate (reduced nicotinamide adenine dinucleotide) Glutamate synthase (NADH)	
1.4.1.13	Synthase, glutamate (reduced nicotinamide adenine dinucleotide phosphate) Glutamate synthase (NADPH)	
2.1.2.10	Synthase, glycine Aminomethyltransferase	
2.5.1.30	Synthase, heptaprenyl pyrophosphate trans-Hexaprenyltranstransferase	
2.7.8.7	Synthase, holo- Holo-[acyl-carrier-protein] synthase	
4.1.3.21	Synthase, homocitrate Homocitrate synthase	
2.5.1.44	Synthase, homospermidine Homospermidine synthase	
4.1.3.5	Synthase, hydroxymethylglutaryl coenzyme A Hydroxymethylglutaryl-CoA synthase	
2.3.1.119	Synthase, icosanoyl coenzyme A Icosanoyl-CoA synthase	
4.1.1.48	Synthase, indole-3-glycerol phosphate Indole-3-glycerol phosphate synthase	
5.4.99.6	Synthase, isochorismate Isochorismate synthase	Vol. 1, Page 618
5.4.99.11	Synthase, isomaltulose Isomaltulose synthase	Vol. 1, Page 638
4.2.99.15	Synthase, L-mimosine L-Mimosine synthase	
4.1.2.36	Synthase, lactate Lactate aldolase	
2.5.1.37	Synthase, leukotriene C_4 Leukotriene-C_4 synthase	
2.4.1.182	Synthase, lipid A disaccharide Lipid-A-disaccharide synthase	

EC-No.	Synonym Recommended Name	Vol., Page
5.4.99.15	Synthase, maltooligosyltrehalose (Rhizobium strain M-11 clone pBMTU1 gene treY reduced) (1->4)-α-Glucan 1-α-glucosylmutase	Vol. 1, Page 652
5.4.99.15	Synthase, maltooligosyltrehalose (Sulfolobus acidocaldarius clone pBST1 gene treY) (1->4)-α-Glucan 1-α-glucosylmutase	Vol. 1, Page 652
2.4.1.139	Synthase, maltose Maltose synthase	
4.2.99.10	synthase, methionine O-acetylhomoserine (thiol)-lyase	
4.1.3.31	synthase, methylcitrate 2-methylcitrate synthase	
4.2.99.11	synthase, methylglyoxal methylglyoxal synthase	
2.3.1.111	Synthase, mycocerosate Mycocerosate synthase	
5.5.1.4	Synthase, myo-inositol 1-phosphate myo-Inositol-1-phosphate synthase	Vol. 1, Page 674
4.1.3.19	synthase, N-acetylneuraminate N-acetylneuraminate synthase	
4.1.3.20	Synthase, N-acylneuraminate 9-phosphate N-Acetylneuraminate-9-phosphate synthase	
1.5.1.24	Synthase, N^5-(carboxy ethyl)ornithine N^5-(Carboxyethyl)ornithine synthase	
2.5.1.43	Synthase, nicotianamine Nicotianamine synthase	
2.3.2.11	Synthase, O-alanylphosphatidylglycerol Alanylphosphatidylglycerol synthase	
1.14.17.3	Synthase, peptide α-amide Peptidylglycine monooxygenase	
2.3.2.15	Synthase, phytochelatin Glutathione γ-glutamylcysteinyltransferase	
2.3.1.146	Synthase, pinosylvin Pinosylvin synthase	
2.3.1.25	Synthase, plasmalogen Plasmalogen synthase	

EC-No.	Synonym Recommended Name	Vol., Page
4.2.1.24	synthase, porphobilinogen porphobilinogen synthase	
2.1.1.131	Synthase, precorrin-4 Precorrin-3B C^{17}-methyltransferase	
4.1.2.35	Synthase, propioin Propioin synthase	
1.14.99.1	Synthase, prostaglandin Prostaglandin-endoperoxide synthase	
4.2.1.70	synthase, pseudouridylate pseudouridylate synthase	
1.1.1.246	Synthase, pterocarpan Pterocarpin synthase	
4.2.99.14	Synthase, pyrazolealanine *β*-Pyrazolylalanine synthase (acetylserine)	
1.5.4.1	Synthase, pyrimidodiazepine Pyrimidodiazepine synthase	
1.2.7.1	Synthase, pyruvate Pyruvate synthase	
2.3.1.95	Synthase, resveratrol Trihydroxystilbene synthase	
2.5.1.9	Synthase, riboflavin Riboflavin synthase	
2.3.1.140	Synthase, rosmarinate Rosmarinate synthase	
2.9.1.1	Synthase, selenocysteine (Moorella thermoacetica clone pCTA100/pCTAB1 gene selA) L-Seryl-tRNASec selenium transferase	
2.9.1.1	Synthase, selenocysteinyl-transfer ribonucleate L-Seryl-tRNASec selenium transferase	
2.5.1.16	Synthase, spermidine Spermidine synthase	
2.5.1.22	Synthase, spermine Spermine synthase	
2.5.1.21	Synthase, squalene Farnesyl-diphosphate farnesyltransferase	

EC-No.	Synonym Recommended Name	Vol., Page
4.3.3.2	Synthase, strictosidine Strictosidine synthase	
4.1.1.47	Synthase, tartronate semialdehyde Tartronate-semialdehyde synthase	
4.2.99.2	synthase, threonine threonine synthase	
2.1.1.45	Synthase, thymidylate Thymidylate synthase	
5.4.99.16	Synthase, trehalose Maltose α-D-glucosyltransferase	Vol. 1, Page 656
5.4.99.16	Synthase, trehalose (Pimelobacter strain R48 clone pBRM8 gene treS precursor reduced) Maltose α-D-glucosyltransferase	Vol. 1, Page 656
5.4.99.16	Synthase, trehalose (Saccharomyces cerevisiae gene TPS1 subunit) Maltose α-D-glucosyltransferase	Vol. 1, Page 656
5.4.99.16	Synthase, trehalose (Saccharomyces cerevisiae gene TSL1 subunit) Maltose α-D-glucosyltransferase	Vol. 1, Page 656
5.4.99.16	Synthase, trehalose (Thermus aquaticus strain ATCC33923 clone pBTM5) Maltose α-D-glucosyltransferase	Vol. 1, Page 656
4.2.3.6	synthase, trichodiene trichodiene synthase	
4.2.1.20	synthase, tryptophan tryptophan synthase	
4.2.99.16	Synthase, uracilylalanine Uracilylalanine synthase	
4.3.1.8	synthase, uroporphyrinogen I hydroxymethylbilane synthase	
4.2.1.75	Synthase, uroporphyrinogen III co- Uroporphyrinogen-III synthase	
2.7.7.58	Synthetase, (2,3-dihydroxybenzoyl)adenylate (2,3-Dihydroxybenzoyl)adenylate synthase	
6.3.2.2	Synthetase, γ-glutamylcysteine Glutamate-cysteine ligase	Vol. 2, Page 399
6.3.4.12	Synthetase, γ-glutamylmethylamide Glutamate-methylamine ligase	Vol. 2, Page 624

EC-No.	Synonym Recommended Name	Vol., Page
6.3.2.14	Synthetase, 2,3-dihydroxybenzoylserine 2,3-Dihydroxybenzoate-serine ligase	Vol. 2, Page 478
6.2.1.31	Synthetase, 2-furoyl coenzyme A 2-Furoate-CoA ligase	Vol. 2, Page 334
6.2.1.29	Synthetase, 3α,7α,12α-trihydroxy-5β-cholestanoyl coenzyme A 3α,7α,12α-Trihydroxy-5β-cholestanate-CoA ligase	Vol. 2, Page 328
6.2.1.28	Synthetase, 3α,7α-dihydroxy-5β-cholestanoyl coenzyme A 3α,7α-Dihydroxy-5β-cholestanate-CoA ligase	Vol. 2, Page 326
6.2.1.33	Synthetase, 4-chlorobenzoyl coenzyme A 4-Chlorobenzoate-CoA ligase	Vol. 2, Page 339
6.2.1.27	Synthetase, 4-hydroxybenzoyl coenzyme A 4-Hydroxybenzoate-CoA ligase	Vol. 2, Page 323
2.3.1.47	Synthetase, 7-keto-8-aminopelargonate 8-Amino-7-oxononanoate synthase	
6.2.1.16	Synthetase, acetoacetyl coenzyme A Acetoacetate-CoA ligase	Vol. 2, Page 282
6.2.1.1	Synthetase, acetyl coenzyme A Acetate-CoA ligase	Vol. 2, Page 186
6.2.1.1	Synthetase, acetyl coenzyme A Acetate-CoA ligase	Vol. 2, Page 186
6.2.1.13	Synthetase, acetyl coenzyme A (adenosine diphosphate-forming) Acetate-CoA ligase (ADP-forming)	Vol. 2, Page 267
6.2.1.20	Synthetase, acyl-[acyl carrier protein] Long-chain-fatty-acid-[acyl-carrier-protein] ligase	Vol. 2, Page 296
6.2.1.10	Synthetase, acyl coenzyme A (guanosine diphosphate-forming) Acid-CoA ligase (GDP-forming)	Vol. 2, Page 249
2.3.1.100	Synthetase, acyl protein Myelin-proteolipid O-palmitoyltransferase	
6.2.1.19	synthetase, acyl protein Long-chain-fatty-acid-luciferin-component ligase	Vol. 2, Page 293
2.5.1.26	Synthetase, alkyldihydroxyacetone phosphate Alkylglycerone-phosphate synthase	
2.3.1.37	Synthetase, aminolevulinate 5-Aminolevulinate synthase	
4.1.3.27	synthetase, anthranilate anthranilate synthase	

EC-No.	Synonym Recommended Name	Vol., Page
6.2.1.32	Synthetase, anthraniloyl coenzyme A Anthranilate-CoA ligase	Vol. 2, Page 336
6.3.4.5	Synthetase, argininosuccinate Argininosuccinate synthase	Vol. 2, Page 595
6.1.1.19	Synthetase, arginyl-transfer ribonucleate Arginine-tRNA ligase	Vol. 2, Page 146
6.3.5.4	Synthetase, Asn (glutamine) Asparagine synthase (glutamine-hydrolysing)	Vol. 2, Page 672
6.3.1.1	Synthetase, asparagine Aspartate-ammonia ligase	Vol. 2, Page 344
6.3.1.4	Synthetase, asparagine (adenosine diphosphate-forming) Aspartate-ammonia ligase	Vol. 2, Page 375
6.1.1.22	Synthetase, asparaginyl-transfer ribonucleate Asparagine-tRNA ligase	Vol. 2, Page 178
6.1.1.12	Synthetase, aspartyl-transfer ribonucleate Aspartate-tRNA ligase	Vol. 2, Page 86
6.2.1.25	Synthetase, benzoyl coenzyme A Benzoate-CoA ligase	Vol. 2, Page 314
2.8.1.6	Synthetase, biotin Biotin synthase	
2.8.1.6	Synthetase, biotin (Arabidopsis thaliana clone lambdaBIO2 gene bioB) Biotin synthase	
2.8.1.6	Synthetase, biotin (Arabidopsis thaliana clone pMP101 gene BIO2 reduced) Biotin synthase	
2.8.1.6	Synthetase, biotin (Arabidopsis thaliana clone $pYESCBS_1$ gene bioB) Biotin synthase	
2.8.1.6	Synthetase, biotin (Bacillus subtilis clone pBIO100/pBIO350/pBIO201 gene bioB) Biotin synthase	
2.8.1.6	Synthetase, biotin (Bacillus subtilis gene bioB) Biotin synthase	
2.8.1.6	Synthetase, biotin (Saccharomyces cerevisiae strain 20B-12 clone pUCH2.4 gene BIO2 reduced) Biotin synthase	
6.3.4.11	Synthetase, biotin-β-methylcrotonyl coenzyme A carboxylase Biotin-[methylcrotonyl-CoA-carboxylase] ligase	Vol. 2, Page 622

EC-No.	Synonym Recommended Name	Vol., Page
6.3.4.15	Synthetase, biotin-[acetyl coenzyme A carboxylase] Biotin-[acetyl-CoA-carboxylase] ligase	Vol. 2, Page 638
6.3.4.9	Synthetase, biotin-methylmalonyl coenzyme A carboxyltransferase Biotin-[methylmalonyl-CoA-carboxyltransferase] ligase	Vol. 2, Page 613
6.3.4.10	Synthetase, biotin-propionyl coenzyme A carboxylase Biotin-[propionyl-CoA-carboxylase (ATP-hydrolysing)] ligase	Vol. 2, Page 617
6.2.1.11	Synthetase, biotinyl coenzyme A Biotin-CoA ligase	Vol. 2, Page 252
5.5.1.8	Synthetase, bornyl pyrophosphate Geranyl-diphosphate cyclase	Vol. 1, Page 705
6.2.1.2	Synthetase, butyryl conzyme A Butyrate-CoA ligase	Vol. 2, Page 199
6.3.5.5	Synthetase, carbamoylphosphate (glutamine-hydrolyzing) Carbamoyl-phosphate synthase (glutamine-hydrolysing)	Vol. 2, Page 689
6.3.2.11	Synthetase, carnosine Carnosine synthase	Vol. 2, Page 460
6.2.1.7	Synthetase, choloyl coenzyme A Cholate-CoA ligase	Vol. 2, Page 236
2.1.1.79	Synthetase, cyclopropane fatty acid Cyclopropane-fatty-acyl-phospholipid synthase	
6.1.1.16	Synthetase, cysteinyl-transfer ribonucleate Cysteine-tRNA ligase	Vol. 2, Page 121
6.3.4.2	Synthetase, cytidine triphosphate CTP synthase	Vol. 2, Page 559
6.3.2.4	Synthetase, D-alanylalanine D-Alanine-D-alanine ligase	Vol. 2, Page 423
6.3.2.16	Synthetase, D-alanylalanylpoly(phosphoglycerol) D-Alanine-alanyl-poly(glycerolphosphate) ligase	Vol. 2, Page 485
6.3.3.3	Synthetase, dethiobiotin Dethiobiotin synthase	Vol. 2, Page 542
2.7.7.45	Synthetase, diguanosine tetraphosphate Guanosine-triphosphate guanylyltransferase	
6.3.2.12	Synthetase, dihydrofolate Dihydrofolate synthase	Vol. 2, Page 466

EC-No.	Synonym Recommended Name	Vol., Page
6.3.2.22	Synthetase, diphthamide Diphthine-ammonia ligase	Vol. 2, Page 516
2.4.1.137	Synthetase, floridoside phosphate Sn-Glycerol-3-phosphate 2-α-galactosyltransferase	
6.3.2.17	Synthetase, folylpolyglutamate Tetrahydrofolylpolyglutamate synthase	Vol. 2, Page 488
6.3.4.3	Synthetase, formyl tetrahydrofolate Formate-tetrahydrofolate ligase	Vol. 2, Page 567
2.5.1.29	Synthetase, geranylgeranyl pyrophosphate Farnesyltranstransferase	
6.3.1.2	Synthetase, glutamine Glutamate-ammonia ligase	Vol. 2, Page 347
6.1.1.18	Synthetase, glutaminyl-transfer ribonucleate Glutamine-tRNA ligase	Vol. 2, Page 139
6.2.1.6	Synthetase, glutaryl coenzyme A Glutarate-CoA ligase	Vol. 2, Page 234
6.3.2.3	Synthetase, glutathione Glutathione synthase	Vol. 2, Page 410
6.3.1.8	Synthetase, glutathionylspermidine Glutathionylspermidine synthase	Vol. 2, Page 386
6.3.1.8	Synthetase, glutathionylspermidine (Crithidia fasciculata) Glutathionylspermidine synthase	Vol. 2, Page 386
6.3.1.8	Synthetase, glutathionylspermidine (Crithidia fasciculata fragment) Glutathionylspermidine synthase	Vol. 2, Page 386
6.3.1.8	Synthetase, glutathionylspermidine (Crithidia fasciculata strain HS6 gene Cf-GSS) Glutathionylspermidine synthase	Vol. 2, Page 386
6.3.1.8	Synthetase, glutathionylspermidine (Escherichia coli clone pJBM1 gene gsp) Glutathionylspermidine synthase	Vol. 2, Page 386
6.1.1.14	Synthetase, glycyl-transfer ribonucleate Glycine-tRNA ligase	Vol. 2, Page 101
6.3.4.1	Synthetase, guanylate GMP synthase	Vol. 2, Page 548
6.3.5.2	Synthetase, guanylate GMP synthase (glutamine hydrolysing)	Vol. 2, Page 655

EC-No.	Synonym Recommended Name	Vol., Page
2.5.1.33	Synthetase, hexaprenyl pyrophosphate trans-Pentaprenyltranstransferase	
6.1.1.21	Synthetase, histidyl-transfer ribonucleate Histidine-tRNA ligase	Vol. 2, Page 168
4.4.1.17	Synthetase, holocytochrome c Holocytochrome-c synthase	
6.3.2.23	Synthetase, homoglutathione Homoglutathione synthase	Vol. 2, Page 518
6.1.1.5	Synthetase, isoleucyl-transfer ribonucleate Isoleucine-tRNA ligase	Vol. 2, Page 33
6.3.2.24	Synthetase, kyotorphin Tyrosine-arginine ligase	Vol. 2, Page 521
6.1.1.4	Synthetase, leucyl-transfer ribonucleate Leucine-tRNA ligase	Vol. 2, Page 23
4.2.99.13	synthetase, lupinate β-(9-cytokinin)-alanine synthase	
6.1.1.6	Synthetase, lysyl-transfer ribonucleate Lysine-tRNA ligase	Vol. 2, Page 42
6.2.1.9	Synthetase, malyl coenzyme A Malate-CoA ligase	Vol. 2, Page 245
6.3.3.2	Synthetase, methenyltetrahydrofolate 5-Formyltetrahydrofolate cyclo-ligase	Vol. 2, Page 535
6.1.1.10	Synthetase, methionyl-transfer ribonucleate Methionine-tRNA ligase	Vol. 2, Page 68
6.3.1.7	Synthetase, methyleneglutamine 4-Methyleneglutamate-ammonia ligase	Vol. 2, Page 383
4.1.3.20	Synthetase, N-acylneuraminate 9-phosphate N-Acetylneuraminate-9-phosphate synthase	
6.3.1.6	Synthetase, N^5-ethylglutamine Glutamate-ethylamine ligase	Vol. 2, Page 381
6.3.1.5	Synthetase, nicotinamide adenine dinucleotide NAD^+ synthase	Vol. 2, Page 377
6.3.5.1	Synthetase, nicotinamide adenine dinucleotide (glutamine) NAD^+ synthase (glutamine-hydroylysing)	Vol. 2, Page 651
1.14.13.39	Synthetase, nitric oxide Nitric-oxide synthase	

EC-No.	Synonym Recommended Name	Vol., Page
6.2.1.26	Synthetase, *o*-succinylbenzoyl coenzyme A O-Succinylbenzoate-CoA ligase	Vol. 2, Page 320
6.2.1.8	Synthetase, oxalyl coenzyme A Oxalate-CoA ligase	Vol. 2, Page 242
6.2.1.12	Synthetase, *p*-coumaroyl coenzyme A 4-Coumarate-CoA ligase	Vol. 2, Page 256
6.3.2.1	Synthetase, pantothenate Pantoate-β-alanine ligase	Vol. 2, Page 394
4.6.1.5	Synthetase, pentalenene Pentalenene synthase	
6.2.1.30	Synthetase, phenacyl coenzyme A Phenylacetate-CoA ligase	Vol. 2, Page 330
6.1.1.20	Synthetase, phenylalanyl-transfer ribonucleate Phenylalanine-tRNA ligase	Vol. 2, Page 156
6.3.2.5	Synthetase, phosphopantothenoylcysteine Phosphopantothenoate-cysteine ligase	Vol. 2, Page 431
2.7.9.2	Synthetase, phosphopyruvate water dikinase pyruvate	
6.3.4.7	Synthetase, phosphoribosylamine Ribose-5-phosphate-ammonia ligase	Vol. 2, Page 608
6.3.3.1	Synthetase, phosphoribosylaminoimidazole Phosphoribosylformylglycinamidine cyclo-ligase	Vol. 2, Page 530
6.3.2.6	Synthetase, phosphoribosylaminoimidazolesuccinocarboxamide Phosphoribosylaminoimidazolesuccinocarboxamide synthase	Vol. 2, Page 434
6.3.5.3	Synthetase, phosphoribosylformylglycinamide Phosphoribosylformylglycinamidine synthase	Vol. 2, Page 666
6.3.4.13	Synthetase, phosphoribosylglycinamide Phosphoribosylamine-glycine ligase	Vol. 2, Page 626
6.3.4.8	Synthetase, phosphoribosylimidazoleacetate Imidazoleacetate-phosphoribosyldiphosphate ligase	Vol. 2, Page 611
6.2.1.24	Synthetase, phytanoyl coenzyme A Phytanate-CoA ligase	Vol. 2, Page 311
2.5.1.32	Synthetase, phytoene Geranylgeranyl-diphosphate geranylgeranyltransferase	

EC-No.	Synonym Recommended Name	Vol., Page
6.2.1.14	Synthetase, pimelyl coenzyme A 6-Carboxyhexanoate-CoA ligase	Vol. 2, Page 273
6.5.1.2	Synthetase, polydeoxyribonucleotide (nicotinamide adenine dinucleotide) DNA ligase (NAD^+)	Vol. 2, Page 773
6.5.1.3	Synthetase, polyribonucleotide RNA ligase (ATP)	Vol. 2, Page 787
6.2.1.17	Synthetase, propionyl coenzyme A Propionate-CoA ligase	Vol. 2, Page 286
5.3.99.4	Synthetase, prostacyclin Prostacyclin synthase	Vol. 1, Page 465
1.1.1.188	Synthetase, prostaglandin $F_{2\alpha}$ Prostaglandin-F synthase	
2.7.9.3	Synthetase, selenophosphate selenide, water dikinase	
6.1.1.11	Synthetase, seryl-transfer ribonucleate Serine-tRNA ligase	Vol. 2, Page 77
2.5.1.11	Synthetase, solanesyl pyrophosphate trans-Octaprenyltranstransferase	
2.7.7.47	Synthetase, streptomycin adenylate Streptomycin 3-adenylyltransferase	
6.2.1.5	Synthetase, succinyl coenzyme A (adenosine diphosphate-forming) Succinate-CoA ligase (ADP-forming)	Vol. 2, Page 224
6.2.1.4	Synthetase, succinyl coenzyme A (guanosine diphosphate-forming) Succinate-CoA ligase (GDP-forming)	Vol. 2, Page 219
2.7.8.14	Synthetase, teichoate CDPribitol ribitolphosphotransferase	
6.1.1.3	Synthetase, threonyl-transfer ribonucleate Threonine-tRNA ligase	Vol. 2, Page 17
5.3.99.5	Synthetase, thromboxane Thromboxane-A synthase	Vol. 1, Page 472
5.4.99.12	Synthetase, transfer ribonucleate pseudouridine tRNA-pseudouridine synthase I	Vol. 1, Page 642
6.3.1.9	Synthetase, trypanothione Trypanothione synthase	Vol. 2, Page 391
6.3.1.9	Synthetase, trypanothione (Crithidia fasciculata strain HS6 gene Cf-TS) Trypanothione synthase	Vol. 2, Page 391

EC-No.	Synonym Recommended Name	Vol., Page
6.1.1.2	Synthetase, tryptophanyl-transfer ribonucleate Tryptophan-tRNA ligase	Vol. 2, Page 9
6.3.2.25	Synthetase, tubulin-tyrosine Tubulin-tyrosine ligase	Vol. 2, Page 524
6.3.2.21	Synthetase, ubiquitin-calmodulin Ubiquitin-calmodulin ligase	Vol. 2, Page 513
6.3.2.19	Synthetase, ubiquitin-protein Ubiquitin-protein ligase	Vol. 2, Page 506
2.5.1.31	Synthetase, undecaprenyl pyrophosphate Di-trans,poly-cis-Decaprenylcistransferase	
6.3.2.8	Synthetase, uridine diphospho-N-acetylmuramoylalanine UDP-N-acetylmuramate-alanine ligase	Vol. 2, Page 442
6.3.2.9	Synthetase, uridine diphospho-N-acetylmuramoylalanyl-D-glutamate UDP-N-acetylmuramoylalanine-D-glutamate ligase	Vol. 2, Page 452
6.3.2.13	Synthetase, uridine diphospho-N-acetylmuramoylalanyl-D-glutamyl-meso-2,6-diaminopimelate UDP-N-acetylmuramoylalanyl-D-glutamate-2,6-diaminopimelate ligase	Vol. 2, Page 473
6.3.2.7	Synthetase, uridine diphospho-N-acetylmuramoylalanyl-D-glutamyllysine UDP-N-acetylmuramoyl-L-alanyl-D-glutamate-lysine ligase	Vol. 2, Page 439
6.3.2.15	Synthetase, uridine diphosphoacetylmuramoylalanyl-D-glutamyl-meso-2,6-diaminopimeloyl-D-alanyl-D-alanine UDP-N-acetylmuramoylalanyl-D-glutamyl-2,6-diaminopimelate-D-alanyl-D-alanine ligase	Vol. 2, Page 480
6.3.2.10	Synthetase, uridine diphosphoacetylmuramoylpentapeptide UDP-N-acetylmuramoylalanyl-D-glutamyl-lysine-D-alanyl-D-alanine ligase	Vol. 2, Page 458
6.1.1.9	Synthetase, valyl-transfer ribonucleate Valine-tRNA ligase	Vol. 2, Page 59
4.1.1.19	Synthetic arginine decarboxylase Arginine decarboxylase	
3.1.3.48	Syp protein-tyrosine-phosphatase	
5.4.99.15	Sythase, maltooligosyltrehalose (Arthrobacter clone pBRT4 gene treY) (1->4)-α-Glucan 1-α-glucosylmutase	Vol. 1, Page 652
1.1.1.166	(-)t-3,t-4-Dihydroxycyclohexane-c-1-carboxylate-NAD oxidoreductase Hydroxycyclohexanecarboxylate dehydrogenase	

EC-No.	Synonym Recommended Name	Vol., Page
2.4.1.26	T_2-HMC-α-glucosyl transferase DNA α-glucosyltransferase	
2.7.4.12	T_2-induced deoxynucleotide kinase T_2-induced deoxynucleotide kinase	
2.4.1.27	T_4-β-glucosyl transferase DNA β-glucosyltransferase	
2.4.1.26	T_4-HMC-α-glucosyl transferase DNA α-glucosyltransferase	
2.4.1.27	T_4-HMC-β-glucosyl transferase DNA β-glucosyltransferase	
3.2.2.17	T_4-induced UV endonuclease Deoxyribodipyrimidine endonucleosidase	
3.1.25.1	T_4 Endonuclease V Deoxyribonuclease (pyrimidine dimer)	
2.4.1.27	T_4 phage β-glucosyltransferase DNA β-glucosyltransferase	
2.4.1.28	T_6-β-glucosyl transferase Glucosyl-DNA β-glucosyltransferase	
2.4.1.28	T_6-glucosyl-HMC-β-glucosyl transferase Glucosyl-DNA β-glucosyltransferase	
2.4.1.26	T_6-HMC-α-glucosyl transferase DNA α-glucosyltransferase	
3.4.99.23	tabernamontanain tabernamontanain	
3.1.8.2	Tabunase Diisopropyl-fluorophosphatase	
1.5.1.23	TaDH Tauropine dehydrogenase	
3.4.17.20	TAFI Carboxypeptidase U	
3.4.17.20	TAFIa Carboxypeptidase U	
4.1.2.40	D-Tagatose-1,6-bisphosphate aldolase (class I) Tagatose-bisphosphate aldolase	
4.1.2.40	D-Tagatose-1,6-bisphosphate aldolase (class II) Tagatose-bisphosphate aldolase	

EC-No.	Synonym Recommended Name	Vol., Page
4.1.2.40	D-Tagatose-1,6-bisphosphate triosephosphate lyase Tagatose-bisphosphate aldolase	
4.1.2.40	Tagatose-1,6-diphosphate aldolase Tagatose-bisphosphate aldolase	
2.7.1.101	Tagatose-6-phosphate kinase Tagatose kinase	
2.7.1.144	Tagatose-6-phosphate kinase Tagatose-6-phosphate kinase	
4.1.2.40	Tagatose-bisphosphate aldolase Tagatose-bisphosphate aldolase	
4.1.2.40	Tagatose 1,6-bisphosphate aldolase Tagatose-bisphosphate aldolase	
4.1.2.40	D-Tagatose 1,6-diphosphate aldolase Tagatose-bisphosphate aldolase	
2.7.1.101	D-Tagatose 6-phosphate kinase Tagatose kinase	
2.7.1.144	D-Tagatose 6-phosphate kinase Tagatose-6-phosphate kinase	
4.1.2.40	Tagatose bisphosphate aldolase Tagatose-bisphosphate aldolase	
2.7.1.101	Tagatose kinase Tagatose kinase	
1.1.1.58	D-Tagaturonate reductase Tagaturonate reductase	
1.1.1.58	Tagaturonate reductase Tagaturonate reductase	
1.1.1.58	TagUAR Tagaturonate reductase	
3.4.21.60	Taipan activator Scutelarin	
3.2.1.1	Taka-amylase A α-amylase	
3.1.1.3	takedo 1969-4-9 triacylglycerol lipase	
3.1.1.38	TAL hydrolase triacetate-lactonase	

EC-No.	Synonym Recommended Name	Vol., Page
3.1.1.20	tannase tannase	
3.1.1.20	tannase I tannase	
3.1.1.20	tannase II tannase	
3.1.1.20	tannin acyl-hydrolase tannase	
3.1.1.20	tannin acyl hydrolase tannase	
3.1.1.20	tannin acylhydrolase tannase	
2.7.7.7	Taq DNA polymerase DNA-directed DNA polymerase	
2.7.7.7	Taq Pol I DNA-directed DNA polymerase	
4.2.1.32	L-(+)-Tartaric acid dehydrase L(+)-Tartrate dehydrogenase	
4.2.1.32	Tartaric acid dehydrase L(+)-Tartrate dehydrogenase	
5.1.2.5	Tartaric racemase Tartrate epimerase	Vol. 1, Page 87
1.1.1.93	Tartrate:NAD$^+$ oxidoreductase Tartrate dehydrogenase	
4.1.1.73	Tartrate decarboxylase Tartrate decarboxylase	
4.2.1.81	D-Tartrate dehydratase D(-)-Tartrate dehydratase	
4.2.1.32	L-(+)-Tartrate dehydratase L(+)-Tartrate dehydrogenase	
4.2.1.32	L-Tartrate dehydratase L(+)-Tartrate dehydrogenase	
4.2.1.32	Tartrate dehydratase L(+)-Tartrate dehydrogenase	
1.1.1.93	Tartrate dehydrogenase Tartrate dehydrogenase	

EC-No.	Synonym Recommended Name	Vol., Page
5.1.2.5	Tartrate epimerase Tartrate epimerase	Vol. 1, Page 87
5.1.2.5	Tartrate epimerase Tartrate epimerase	Vol. 1, Page 87
3.3.2.4	Tartrate epoxydase trans-epoxysuccinate hydrolase	
4.1.1.47	Tartronate-semialdehyde synthase Tartronate-semialdehyde synthase	
2.3.1.106	Tartronate O-hydroxycinnamoyltransferase Tartronate O-hydroxycinnamoyltransferase	
4.1.1.47	Tartronate semialdehyde carboxylase Tartronate-semialdehyde synthase	
1.1.1.60	Tartronate semialdehyde reductase 2-Hydroxy-3-oxopropionate reductase	
2.3.1.106	Tartronate sinapoyltransferase Tartronate O-hydroxycinnamoyltransferase	
4.1.1.47	Tartronic semialdehyde carboxylase Tartronate-semialdehyde synthase	
4.1.1.47	Tartronic semialdehyde synthase Tartronate-semialdehyde synthase	
2.3.1.108	TAT Tubulin N-acetyltransferase	
2.7.1.135	τ-protein kinase τ-protein kinase	
2.7.1.135	*T* kinase τ-protein kinase	
2.7.1.135	*T* protein kinase τ-protein kinase	
2.7.1.135	*T* protein kinase I τ-protein kinase	
1.14.11.17	taurine, 2-oxoglutarate:O2 oxidoreductase (sulfite-forming) taurine dioxygenase	
2.6.1.55	Taurine-α-ketoglutarate aminotransferase Taurine transaminase	
2.6.1.55	Taurine-glutamate transaminase Taurine transaminase	

EC-No.	Synonym Recommended Name	Vol., Page
1.4.99.2	Taurine:(acceptor) oxidoreductase (deaminating) Taurine dehydrogenase	
2.6.1.55	Taurine:α-ketoglutarate aminotransferase Taurine transaminase	
2.6.1.55	Taurine:2-oxoglutarate aminotransferase Taurine transaminase	
1.4.99.2	Taurine dehydrogenase Taurine dehydrogenase	
1.14.11.17	taurine dioxygenase taurine dioxygenase	
2.6.1.55	Taurine transaminase Taurine transaminase	
2.7.3.4	Taurocyamine kinase Taurocyamine kinase	
2.7.3.4	Taurocyamine phosphotransferase Taurocyamine kinase	
1.5.1.23	Tauropine: NAD oxidoreductase Tauropine dehydrogenase	
1.5.1.23	Tauropine dehydrogenase Tauropine dehydrogenase	
5.3.2.2	Tautomerase, oxalacetate Oxalacetate tautomerase	Vol. 1, Page 371
5.3.2.1	Tautomerase, phenylpyruvate Phenylpyruvate tautomerase	Vol. 1, Page 367
1.14.13.19	Taxifolin,NAD(P)H:oxygen oxidoreductase (8-hydroxylating) Taxifolin 8-monooxygenase	
1.14.13.19	Taxifolin 8-monooxygenase Taxifolin 8-monooxygenase	
1.14.13.19	Taxifolin hydroxylase Taxifolin 8-monooxygenase	
2.7.7.7	Tca DNA polymerase DNA-directed DNA polymerase	
3.4.14.5	T cell triggering molecule Tp103 dipeptidyl-peptidase IV	
5.5.1.1	TcMLE Muconate cycloisomerase	Vol. 1, Page 660

EC-No.	Synonym Recommended Name	Vol., Page
4.2.1.16	TD threonine dehydratase	
4.1.1.25	TDC Tyrosine decarboxylase	
4.2.1.16	L-TDH threonine dehydratase	
1.4.1.19	TDH Tryptophan dehydrogenase	
2.6.1.33	TDP-4-keto-6-deoxy-D-glucose transaminase dTDP-4-amino-4,6-dideoxy-D-glucose transaminase	
5.1.3.13	TDP-4-keto-L-rhamnose-3,5-epimerase dTDP-4-dehydrorhamnose 3,5-epimerase	Vol. 1, Page 152
1.1.1.133	TDP-4-keto-rhamnose reductase dTDP-4-dehydrorhamnose reductase	
5.1.3.13	TDP-4-ketorhamnose 3,5-epimerase dTDP-4-dehydrorhamnose 3,5-epimerase	Vol. 1, Page 152
1.1.1.134	TDP-6-deoxy-L-talose dehydrogenase dTDP-6-deoxy-L-talose 4-dehydrogenase	
4.2.1.46	TDP-D-glucose 4,6-dehydratase dTDPglucose 4,6-dehydratase	
4.2.1.46	TDP-glucose oxidoreductase dTDPglucose 4,6-dehydratase	
2.7.7.24	TDP-glucose pyrophosphorylase Glucose-1-phosphate thymidylyltransferase	
4.1.2.40	TDP aldolase Tagatose-bisphosphate aldolase	
2.7.4.15	TDP kinase Thiamine-diphosphate kinase	
2.7.7.31	TdT DNA nucleotidylexotransferase	
3.1.1.3	teenesterase triacylglycerol lipase	
1.1.1.161	TEHC-NAD oxidoreductase Cholestanetetraol 26-dehydrogenase	
2.7.8.14	Teichoate synthase CDPribitol ribitolphosphotransferase	

EC-No.	Synonym Recommended Name	Vol., Page
2.7.8.12	Teichoic-acid synthase CDPglycerol glycerophosphotransferase	
2.7.8.14	Teichoic-acid synthase CDPribitol ribitolphosphotransferase	
2.4.1.55	teichoic-acid synthase teichoic-acid synthase	
2.7.8.12	Teichoic acid glycerol transferase CDPglycerol glycerophosphotransferase	
2.7.8.14	Teichoic acid synthase CDPribitol ribitolphosphotransferase	
3.4.24.67	Teleost hatching enzyme (component) Choriolysin H	
3.4.24.66	Teleost hatching enzyme (component) Choriolysin L	
3.4.21.18	Tenebrio α-proteinase Tenebrio α-proteinase	
3.4.99.24	Tenebrio α-proteinase Tenebrio α-proteinase	
3.4.24.68	Tentoxilysin Tentoxilysin	
1.14.12.15	TER dioxygenase system Terephthalate 1,2-dioxygenase	
1.14.12.15	TERDOS Terephthalate 1,2-dioxygenase	
1.3.1.61	terephthalate 1,2-cis-dihydrodiol dehydrogenase terephthalate 1,2-cis-dihydrodiol dehydrogenase	
1.14.12.15	Terephthalate 1,2-dioxygenase Terephthalate 1,2-dioxygenase	
2.8.2.17	Terminal 6-sulfotransferase Chondroitin 6-sulfotransferase	
2.7.7.31	Terminal addition enzyme DNA nucleotidylexotransferase	
2.7.7.31	Terminal deoxynucleotide transferase DNA nucleotidylexotransferase	
2.7.7.31	Terminal deoxynucleotidyltransferase DNA nucleotidylexotransferase	

EC-No.	Synonym Recommended Name	Vol., Page
2.7.7.31	Terminal deoxyribonucleotidyltransferase DNA nucleotidylexotransferase	
2.7.1.141	C-terminal repeat domain kinase [RNA-polymerase]-subunit kinase	
2.7.7.19	Terminal riboadenylate transferase Polynucleotide adenylyltransferase	
2.7.7.52	Terminal uridylyltransferase RNA uridylyltransferase	
2.5.1.11	Terpenoidallyltransferase trans-Octaprenyltranstransferase	
2.5.1.11	Terpenyl pyrophosphate synthetase trans-Octaprenyltranstransferase	
1.1.1.160	(+,-)-5-[(tert-Butylamino)-2'-hydroxypropoxy]-1,2,3,4-tetrahydro-1-naphthol:NADP$^+$ oxidoreductase Dihydrobunolol dehydrogenase	
3.1.1.37	testololactone lactonohydrolase steroid-lactonase	
1.3.1.22	Testosterone.δ. 4-hydrogenase Cholestenone 5α-reductase	
1.1.1.63	Testosterone 17β-dehydrogenase Testosterone 17β-dehydrogenase	
1.1.1.64	Testosterone 17β-dehydrogenase (NADP$^+$) Testosterone 17β-dehydrogenase (NADP$^+$)	
1.3.1.22	Testosterone δ^4-5α-reductase Cholestenone 5α-reductase	
1.3.99.5	Testosterone 5α-reductase 3-Oxo-5α-steroid Δ^4-dehydrogenase	
1.3.1.22	Testosterone 5α-reductase Cholestenone 5α-reductase	
1.3.99.6	Testosterone 5β-reductase 3-Oxo-5β-steroid Δ^4-dehydrogenase	
1.3.1.23	Testosterone 5β-reductase Cholestenone 5β-reductase	
1.3.1.23	Testosterone 5.β.-reductase Cholestenone 5β-reductase	

EC-No.	Synonym Recommended Name	Vol., Page
3.4.24.68	Tetanus neurotoxin Tentoxilysin	
2.7.1.130	Tetraacyldisaccharide 4'-kinase Tetraacyldisaccharide 4'-kinase	
1.14.99.31	n-tetradecanoyl-CoA, $NADPH_2$:O_2 oxidoreductase [11-(E) desaturating] myristoyl-CoA 11-(E) desaturase	
1.14.99.32	n-tetradecanoyl-CoA, $NADPH_2$:O_2 oxidoreductase [11-(Z) desaturating] myristoyl-CoA 11-(Z) desaturase	
2.3.1.97	Tetradecanoyl-CoA:glycylpeptide N-tetradecanoyltransferase Glycylpeptide N-tetradecanoyltransferase	
1.3.7.1	1,4,5,6-Tetrahydro-6-oxonicotinate:ferredoxin oxidoreductase 6-Hydroxynicotinate reductase	
2.1.1.122	Tetrahydroberberine cis-N-methyltransferase (S)-Tetrahydroprotoberberine N-methyltransferase	
1.3.3.8	Tetrahydroberberine oxidase Tetrahydroberberine oxidase	
1.1.1.253	5,6,7,8-Tetrahydrobiopterin:NADP oxidoreductase Pteridine reductase	
2.1.1.89	Tetrahydrocolumbamine 2-O-methyltransferase Tetrahydrocolumbamine 2-O-methyltransferase	
2.3.1.89	Tetrahydrodipicolinate:acetyl-CoA acetyltransferase Tetrahydrodipicolinate N-acetyltransferase	
1.3.1.26	2,3,4,5-Tetrahydrodipicolinate: $NAD(P)^+$ oxidoreductase Dihydrodipicolinate reductase	
2.3.1.89	Tetrahydrodipicolinate acetylase Tetrahydrodipicolinate N-acetyltransferase	
2.3.1.89	Tetrahydrodipicolinate N-acetyltransferase Tetrahydrodipicolinate N-acetyltransferase	
2.3.1.117	Tetrahydrodipicolinate N-succinyltransferase 2,3,4,5-Tetrahydropyridine-2-carboxylate N-succinyltransferase	
2.3.1.117	Tetrahydrodipicolinate succinylase 2,3,4,5-Tetrahydropyridine-2-carboxylate N-succinyltransferase	
2.3.1.117	Tetrahydrodipicolinate succinyltransferase 2,3,4,5-Tetrahydropyridine-2-carboxylate N-succinyltransferase	
6.3.2.17	Tetrahydrofolate:L-glutamate γ-ligase (ADP-forming) Tetrahydrofolylpolyglutamate synthase	Vol. 2, Page 488

EC-No.	Synonym Recommended Name	Vol., Page
1.5.1.3	5,6,7,8-Tetrahydrofolate:$NADP^+$ oxidoreductase Dihydrofolate reductase	
2.1.2.10	Tetrahydrofolate aminomethyltransferase Aminomethyltransferase	
1.5.1.3	Tetrahydrofolate dehydrogenase Dihydrofolate reductase	
6.3.4.3	Tetrahydrofolate formylase Formate-tetrahydrofolate ligase	Vol. 2, Page 567
2.1.1.13	Tetrahydrofolate methyltransferase 5-Methyltetrahydrofolate-homocysteine S-methyltransferase	
6.3.4.3	Tetrahydrofolic formylase Formate-tetrahydrofolate ligase	Vol. 2, Page 567
6.3.2.17	Tetrahydrofolylpolyglutamate synthase Tetrahydrofolylpolyglutamate synthase	Vol. 2, Page 488
2.1.1.86	Tetrahydromethanopterin methyltransferase Tetrahydromethanopterin S-methyltransferase	
2.1.1.86	Tetrahydromethanopterin S-methyltransferase Tetrahydromethanopterin S-methyltransferase	
1.5.3.9	Tetrahydroprotoberberine synthase Reticuline oxidase	
2.1.1.13	Tetrahydropteroylglutamate methyltransferase 5-Methyltetrahydrofolate-homocysteine S-methyltransferase	
2.1.1.13	Tetrahydropteroylglutamic methyltransferase 5-Methyltetrahydrofolate-homocysteine S-methyltransferase	
2.1.1.14	Tetrahydropteroyltriglutamate methyltransferase 5-Methyltetrahydropteroyltriglutamate-homocysteine S-methyltransferase	
2.3.1.117	2,3,4,5-Tetrahydropyridine-2-carboxylate N-succinyltransferase 2,3,4,5-Tetrahydropyridine-2-carboxylate N-succinyltransferase	
1.97.1.2	1,2,3,5-Tetrahydroxybenzene-pyrogallol hydroxyltransferase Pyrogallol hydroxyltransferase	
1.97.1.2	1,2,3,5-Tetrahydroxybenzene:1,2,3-trihydroxybenzene hydroxyltransferase Pyrogallol hydroxyltransferase	
1.97.1.2	1,2,3,5-Tetrahydroxybenzene:pyrogallol transhydroxylase Pyrogallol hydroxyltransferase	
2.3.1.151	2,3',4,6-Tetrahydroxybenzophenone synthase Benzophenone synthase	

EC-No.	Synonym Recommended Name	Vol., Page
3.7.1.4	2,4,4,6-Tetrahydroxydehydrochalcone 1,3,5-trihydroxybenzene hydrolase Phloretin hydrolase	
1.1.1.252	1,3,6,8-Tetrahydroxynaphthalene reductase Tetrahydroxynaphthalene reductase	
1.1.1.252	Tetrahydroxynaphthalene reductase Tetrahydroxynaphthalene reductase	
1.3.1.50	tetrahydroxynaphthalene reductase tetrahydroxynaphthalene reductase	
1.1.1.252	Tetrahydroxynaphthalene reductase (Magnaporthe grisea clone pAV501) Tetrahydroxynaphthalene reductase	
5.5.1.3	Tetrahydroxypteridine cycloisomerase Tetrahydroxypteridine cycloisomerase	Vol. 1, Page 672
5.5.1.3	Tetrahydroxypteridine lyase (isomerizing) Tetrahydroxypteridine cycloisomerase	Vol. 1, Page 672
3.4.14.7	tetralysine endopeptidase tetralysine endopeptidase	
3.4.14.6	tetrapeptide dipeptidase dipeptidyl-dipeptidase	
1.8.2.2	Tetrathionate synthase Thiosulfate dehydrogenase	
2.4.2.29	TGT Queuine tRNA-ribosyltransferase	
1.2.1.40	THAL-NAD oxidoreductase 3α,7α,12α-Trihydroxycholestan-26-al 26-dehydrogenase	
6.2.1.29	THCA-CoA ligase 3α,7α,12α-Trihydroxy-5β-cholestanate-CoA ligase	Vol. 2, Page 328
3.5.1.65	L-Theanine amidohydrolase Theanine hydrolase	
3.5.1.65	Theanine hydrolase Theanine hydrolase	
6.3.1.6	Theanine synthetase Glutamate-ethylamine ligase	Vol. 2, Page 381
3.4.21.66	Thermitase Thermitase	
3.4.21.66	Thermoactinomyces vulgaris serine proteinase Thermitase	

EC-No.	Synonym Recommended Name	Vol., Page
3.4.21.62	Thermoase Subtilisin	
3.4.24.27	Thermoase Thermolysin	
3.4.21.62	Thermoase PC 10 Subtilisin	
3.4.24.27	Thermoase Y10 Thermolysin	
3.4.24.27	Thermolysin Thermolysin	
3.4.21.65	Thermomycolase Thermomycolin	
3.4.21.65	Thermomycolin Thermomycolin	
3.1.31.1	Thermonuclease Micrococcal nuclease	
3.4.11.12	thermophilic aminopeptidase thermophilic aminopeptidase	
3.4.21.66	Thermophilic Streptomyces serine proteinase Thermitase	
3.4.23.42	thermopsin thermopsin	
3.4.99.43	thermopsin thermopsin	
4.2.99.17	thermopsin thermopsin	
2.1.1.3	Thetin-homocysteine methylpherase Thetin-homocysteine S-methyltransferase	
2.1.1.3	Thetin-homocysteine S-methyltransferase Thetin-homocysteine S-methyltransferase	
6.3.4.3	THFS Formate-tetrahydrofolate ligase	Vol. 2, Page 567
2.7.4.15	Thiamin-diphosphate kinase Thiamine-diphosphate kinase	
2.7.4.16	Thiamin-monophosphate kinase Thiamine-phosphate kinase	

EC-No.	Synonym Recommended Name	Vol., Page
2.7.4.16	Thiamin-phosphate kinase Thiamine-phosphate kinase	
2.5.1.3	Thiamin-phosphate pyrophosphorylase Thiamine-phosphate pyrophosphorylase	
2.7.6.2	Thiamin:ATP pyrophosphotransferase Thiamine pyrophosphokinase	
2.5.1.2	Thiamin:base 2-methyl-4-aminopyrimidine-5-methenyltransferase Thiamine pyridinylase	
1.1.3.23	Thiamin:oxygen 5-oxidoreductase Thiamine oxidase	
3.5.99.2	Thiaminase Thiaminase	
2.5.1.2	Thiaminase Thiamine pyridinylase	
2.5.1.2	Thiaminase I Thiamine pyridinylase	
3.5.99.2	Thiaminase II Thiaminase	
2.7.4.15	Thiamin diphosphate kinase Thiamine-diphosphate kinase	
2.7.4.15	Thiamin diphosphate phosphotransferase Thiamine-diphosphate kinase	
2.7.4.15	Thiamine-diphosphate kinase Thiamine-diphosphate kinase	
2.7.4.16	Thiamine-phosphate kinase Thiamine-phosphate kinase	
2.5.1.3	Thiamine-phosphate pyrophosphorylase Thiamine-phosphate pyrophosphorylase	
3.6.1.28	Thiamine-triphosphatase Thiamine-triphosphatase	
3.6.1.28	Thiamine-triphosphate phosphohydrolase Thiamine-triphosphatase	
2.5.1.2	Thiamine:base 2-methyl-4-aminopyrimidine-5-methenyltransferase Thiamine pyridinylase	
1.1.3.23	Thiamine:oxygen 5-oxidoreductase Thiamine oxidase	

EC-No.	Synonym Recommended Name	Vol., Page
1.1.3.23	Thiamine dehydrogenase Thiamine oxidase	
2.7.4.15	Thiamine diphosphate kinase Thiamine-diphosphate kinase	
2.7.1.89	Thiamine kinase Thiamine kinase	
2.7.6.2	Thiamine kinase Thiamine pyrophosphokinase	
2.7.4.16	Thiamine monophosphate kinase Thiamine-phosphate kinase	
2.5.1.3	Thiamine monophosphate pyrophosphorylase Thiamine-phosphate pyrophosphorylase	
2.7.4.16	Thiamine monophosphokinase Thiamine-phosphate kinase	
1.1.3.23	Thiamine oxidase Thiamine oxidase	
2.5.1.3	Thiamine phosphate pyrophosphorylase Thiamine-phosphate pyrophosphorylase	
2.5.1.3	Thiaminephosphate pyrophosphorylase Thiamine-phosphate pyrophosphorylase	
2.5.1.2	Thiamine pyridinolase Thiamine pyridinylase	
2.5.1.2	Thiamine pyridinylase Thiamine pyridinylase	
2.7.6.2	Thiamine pyrophosphokinase Thiamine pyrophosphokinase	
3.6.1.28	Thiamine triphosphatase Thiamine-triphosphatase	
3.5.99.2	Thiamin hydrolase Thiaminase	
2.5.1.2	Thiamin hydrolase Thiamine pyridinylase	
2.7.1.89	Thiamin kinase Thiamine kinase	
2.7.6.2	Thiamin kinase Thiamine pyrophosphokinase	

EC-No.	Synonym Recommended Name	Vol., Page
2.7.4.16	Thiamin monophosphatase Thiamine-phosphate kinase	
2.7.4.16	Thiamin monophosphate kinase Thiamine-phosphate kinase	
2.7.4.16	Thiamin monophosphokinase Thiamine-phosphate kinase	
2.7.6.2	Thiaminokinase Thiamine pyrophosphokinase	
1.1.3.23	Thiamin oxidase Thiamine oxidase	
2.7.4.16	Thiamin phosphate kinase Thiamine-phosphate kinase	
2.5.1.3	Thiamin phosphate pyrophosphorylase Thiamine-phosphate pyrophosphorylase	
2.7.1.89	Thiamin phosphokinase Thiamine kinase	
2.5.1.2	Thiamin pyridinolase Thiamine pyridinylase	
2.5.1.2	Thiamin pyridinylase Thiamine pyridinylase	
3.6.1.6	Thiaminpyrophosphatase Nucleoside-diphosphatase	
2.7.4.15	Thiamin pyrophosphate kinase Thiamine-diphosphate kinase	
2.7.6.2	Thiamin pyrophosphokinase Thiamine pyrophosphokinase	
2.7.6.2	Thiamin pyrophosphotransferase Thiamine pyrophosphokinase	
3.4.24.15	thimet oligopeptidase thimet oligopeptidase	
3.4.24.15	thimet peptidase thimet oligopeptidase	
3.4.24.16	thimet peptidase II neurolysin	
3.5.5.8	thiocyanate aminohydrolase thiocyanate hydrolase	

EC-No.	Synonym Recommended Name	Vol., Page
3.5.5.8	thiocyanate hydrolase thiocyanate hydrolase	
5.99.1.1	Thiocyanate isomerase Thiocyanate isomerase	Vol. 1, Page 719
1.11.1.7	Thiocyanate peroxidase Peroxidase	
3.1.2.20	Thioesterase B Acyl-CoA hydrolase	
3.1.2.20	Thioesterase II Acyl-CoA hydrolase	
2.3.1.11	Thioethanolamine acetyltransferase Thioethanolamine S-acetyltransferase	
2.3.1.11	Thioethanolamine S-acetyltransferase Thioethanolamine S-acetyltransferase	
2.1.1.96	Thioether methyltransferase Thioether S-methyltransferase	
2.1.1.96	Thioether S-methyltransferase Thioether S-methyltransferase	
2.3.1.18	Thiogalactoside acetyltransferase Galactoside O-acetyltransferase	
2.3.1.18	Thiogalactoside transacetylase Galactoside O-acetyltransferase	
3.2.3.1	Thioglucosidase Thioglucosidase	
3.2.3.1	β-Thioglucosidase Thioglucosidase	
3.2.3.1	Thioglucoside glucohydrolase Thioglucosidase	
3.2.3.1	β-Thioglucoside glucohydrolase Thioglucosidase	
2.4.1.195	Thiohydroximate β-D-glucosyltransferase Thiohydroximate β-D-glucosyltransferase	
6.2.1.3	Thiokinase Long-chain-fatty-acid-CoA ligase	Vol. 2, Page 206
3.4.24.15	thiol-dependent metalloendopeptidase thimet oligopeptidase	

EC-No.	Synonym Recommended Name	Vol., Page
1.8.4.2	Thiol-protein disulphide oxidoreductase Protein-disulfide reductase (glutathione)	
1.8.4.7	Thiol:disulphide oxidoreductase Enzyme-thiol transhydrogenase (oxidized-glutathione)	
1.8.3.2	Thiol:oxygen oxidoreductase Thiol oxidase	
1.8.4.2	Thiol:protein-disulfide oxidoreductase Protein-disulfide reductase (glutathione)	
2.3.1.16	Thiolase I Acetyl-CoA C-acyltransferase	
2.3.1.9	Thiolase II Acetyl-CoA C-acetyltransferase	
3.4.22.39	thiol endopeptidase adenain	
3.4.22.39	thiol endoproteinase adenain	
1.8.3.2	Thiol oxidase Thiol oxidase	
3.4.22.39	thiol protease adenain	
3.4.22.39	thiol proteinase adenain	
2.1.1.9	Thiol S-methyltransferase Thiol S-methyltransferase	
2.8.2.16	Thiol sulfotransferase Thiol sulfotransferase	
2.3.1.12	Thioltransacetylase A Dihydrolipoamide S-acetyltransferase	
2.3.1.11	Thioltransacetylase B Thioethanolamine S-acetyltransferase	
1.5.1.25	Thiomorpholine-3-carboxylate:NAD(P)$^+$ 5,6-oxidoreductase Thiomorpholine-carboxylate dehydrogenase	
1.5.1.25	Thiomorpholine-carboxylate dehydrogenase Thiomorpholine-carboxylate dehydrogenase	
4.2.1.22	β-thionase Cystathionine β-synthase	

EC-No.	Synonym Recommended Name	Vol., Page
1.14.99.35	thiophene-2-carbonyl-CoA, hydrogen-donor:oxygen oxidoreductase thiophene-2-carbonyl-CoA monooxygenase	
1.14.99.35	thiophene-2-carbonyl-CoA monooxygenase thiophene-2-carbonyl-CoA monooxygenase	
1.14.99.35	thiophene-2-carboxyl-CoA dehydrogenase thiophene-2-carbonyl-CoA monooxygenase	
1.14.99.35	thiophene-2-carboxyl-CoA hydroxylase thiophene-2-carbonyl-CoA monooxygenase	
1.14.99.35	thiophene-2-carboxyl-CoA monooxygenase thiophene-2-carbonyl-CoA monooxygenase	
3.4.22.39	thioprotease adenain	
2.1.1.67	Thiopurine methyltransferase Thiopurine S-methyltransferase	
2.1.1.67	Thiopurine S-methyltransferase Thiopurine S-methyltransferase	
2.1.1.67	6-Thiopurine transmethylase Thiopurine S-methyltransferase	
1.8.99.4	Thioredoxin: 3'-phospho-adenylylsulfate reductase Phosphoadenosine-phosphosulfate reductase	
1.8.99.4	Thioredoxin:adenosine 3'-phosphate 5'-phosphosulfate reductase Phosphoadenosine-phosphosulfate reductase	
1.6.4.5	Thioredoxin reductase (NADPH) Thioredoxin reductase (NADPH)	
2.1.1.66	Thiostrepton-resistance methylase rRNA (adenosine-2'-O-)-methyltransferase	
1.8.2.2	Thiosulfate-acceptor oxidoreductase Thiosulfate dehydrogenase	
2.8.1.5	Thiosulfate-dithiol sulfurtransferase Thiosulfate-dithiol sulfurtransferase	
1.8.2.2	Thiosulfate-oxidizing enzyme Thiosulfate dehydrogenase	
2.8.1.3	Thiosulfate-thiol sulfurtransferase Thiosulfate-thiol sulfurtransferase	
2.8.1.1	Thiosulfate:cyanide sulfurtransferase Thiosulfate sulfurtransferase	

EC-No.	Synonym Recommended Name	Vol., Page
2.8.1.5	Thiosulfate:dithioerythritol sulfurtransferase Thiosulfate-dithiol sulfurtransferase	
1.8.2.2	Thiosulfate:ferricytochrome-c oxidoreductase Thiosulfate dehydrogenase	
2.8.1.3	Thiosulfate:thiol sulfurtransferase Thiosulfate-thiol sulfurtransferase	
2.8.1.1	Thiosulfate cyanide transsulfurase Thiosulfate sulfurtransferase	
1.8.2.2	Thiosulfate dehydrogenase Thiosulfate dehydrogenase	
1.8.2.2	Thiosulfate oxidase Thiosulfate dehydrogenase	
2.8.1.5	Thiosulfate reductase Thiosulfate-dithiol sulfurtransferase	
2.8.1.1	Thiosulfate sulfurtransferase Thiosulfate sulfurtransferase	
2.8.1.1	Thiosulfate thiotransferase Thiosulfate sulfurtransferase	
1.1.1.252	THNR Tetrahydroxynaphthalene reductase	
1.1.1.162	D-Threitol:NADP$^+$ oxidoreductase Erythrulose reductase	
4.3.1.2	threo-3-methyl-L-aspartate ammonia-lyase methylaspartate ammonia-lyase	
4.3.1.2	threo-3-methylaspartase ammonia-lyase methylaspartate ammonia-lyase	
4.3.1.2	L-threo-3-methylaspartate ammonia-lyase methylaspartate ammonia-lyase	
5.4.99.1	L-threo-3-Methylaspartate carboxy-aminomethylmutase Methylaspartate mutase	Vol. 1, Page 582
4.1.2.26	L-threo-3-Phenylserine benzaldehyde-lyase Phenylserine aldolase	
1.1.1.122	D-threo-Aldose 1-dehydrogenase D-threo-Aldose 1-dehydrogenase	
1.1.1.122	D-threo-Aldose:NAD$^+$ 1-oxidoreductase D-threo-Aldose 1-dehydrogenase	

EC-No.	Synonym Recommended Name	Vol., Page
1.1.1.85	Threo-Ds-3-Isopropylmalate dehydrogenase 3-Isopropylmalate dehydrogenase	
4.1.3.1	threo-DS-Isocitrate glyoxylate-lyase isocitrate lyase	
1.1.1.129	L-Threonate 3-dehydrogenase L-Threonate 3-dehydrogenase	
1.1.1.129	L-Threonate:NAD$^+$ 3-oxidoreductase L-Threonate 3-dehydrogenase	
1.1.1.129	L-Threonic acid dehydrogenase L-Threonate 3-dehydrogenase	
5.1.1.6	Threonine α-epimerase Threonine racemase	Vol. 1, Page 25
2.7.1.37	Threonine-specific protein kinase Protein kinase	
6.1.1.3	Threonine-transfer ribonucleate synthetase Threonine-tRNA ligase	Vol. 2, Page 17
6.1.1.3	Threonine-tRNA ligase Threonine-tRNA ligase	Vol. 2, Page 17
1.1.1.103	L-Threonine 3-dehydrogenase L-Threonine 3-dehydrogenase	
1.1.1.103	Threonine 3-dehydrogenase L-Threonine 3-dehydrogenase	
1.1.1.262	L-threonine 4-phosphate dehydrogenase 4-hydroxythreonine-4-phosphate dehydrogenase	
1.1.1.103	L-Threonine:NAD$^+$ oxidoreductase L-Threonine 3-dehydrogenase	
6.1.1.3	L-Threonine:tRNA^{Thr} ligase (AMP-forming) Threonine-tRNA ligase	Vol. 2, Page 17
4.1.2.5	L-Threonine acetaldehyde-lyase Threonine aldolase	
2.1.2.1	L-Threonine aldolase Glycine hydroxymethyltransferase	
4.1.2.5	L-Threonine aldolase Threonine aldolase	
2.1.2.1	Threonine aldolase Glycine hydroxymethyltransferase	

EC-No.	Synonym Recommended Name	Vol., Page
4.1.2.5	Threonine aldolase Threonine aldolase	
4.2.1.16	L-threonine deaminase threonine dehydratase	
4.2.1.16	threonine dehydrase threonine dehydratase	
4.2.1.16	L-threonine dehydratase threonine dehydratase	
4.2.1.16	threonine dehydratase threonine dehydratase	
1.1.1.103	L-Threonine dehydrogenase L-Threonine 3-dehydrogenase	
1.1.1.103	Threonine dehydrogenase L-Threonine 3-dehydrogenase	
4.2.1.16	L-threonine hydro-lyase (deaminating) threonine dehydratase	
5.1.1.6	Threonine racemase Threonine racemase	Vol. 1, Page 25
5.1.1.6	Threonine racemase Threonine racemase	Vol. 1, Page 25
4.2.3.1	threonine synthase threonine synthase	
4.2.99.2	threonine synthase threonine synthase	
4.2.3.1	threonine synthetase threonine synthase	
4.2.99.2	threonine synthetase threonine synthase	
6.1.1.3	Threonine translase Threonine-tRNA ligase	Vol. 2, Page 17
6.1.1.3	Threonyl-ribonucleic synthetase Threonine-tRNA ligase	Vol. 2, Page 17
6.1.1.3	Threonyl-transfer ribonucleate synthetase Threonine-tRNA ligase	Vol. 2, Page 17
6.1.1.3	Threonyl-transfer ribonucleic acid synthetase Threonine-tRNA ligase	Vol. 2, Page 17

EC-No.	Synonym Recommended Name	Vol., Page
6.1.1.3	Threonyl-transfer RNA synthetase Threonine-tRNA ligase	Vol. 2, Page 17
6.1.1.3	Threonyl-tRNA synthetase Threonine-tRNA ligase	Vol. 2, Page 17
3.4.21.5	thrombase thrombin	
3.4.21.5	thrombin thrombin	
3.4.4.13	thrombin thrombin	
3.4.21.5	thrombin, E thrombin	
3.4.17.20	Thrombin-activable fibrinolysis inhibitor Carboxypeptidase U	
3.4.21.5	thrombin-C thrombin	
3.4.21.5	thrombofort thrombin	
3.4.21.6	thrombokinase coagulation factor Xa	
3.4.21.7	thrombolysin plasmin	
3.4.21.6	thromboplastin coagulation factor Xa	
3.4.21.6	thromboplastin, plasma coagulation factor Xa	
5.3.99.5	Thromboxane-A synthase Thromboxane-A synthase	Vol. 1, Page 472
5.3.99.5	Thromboxane A_2 synthetase Thromboxane-A synthase	Vol. 1, Page 472
5.3.99.5	Thromboxane synthetase Thromboxane-A synthase	Vol. 1, Page 472
5.3.99.5	Thromboxan synthetase Thromboxane-A synthase	Vol. 1, Page 472
6.1.1.3	ThrRS Threonine-tRNA ligase	Vol. 2, Page 17

EC-No.	Synonym Recommended Name	Vol., Page
1.1.1.186	thymidine-diphosphate-galactose dehydrogenase dTDPgalactose 6-dehydrogenase	
2.4.2.4	Thymidine-orthophosphate deoxyribosyltransferase Thymidine phosphorylase	
1.14.11.3	Thymidine 2'-hydroxylase Pyrimidine-deoxynucleoside 2'-dioxygenase	
1.14.11.3	Thymidine 2-oxoglutarate dioxygenase Pyrimidine-deoxynucleoside 2'-dioxygenase	
2.7.4.9	Thymidine 5'-monophosphate kinase dTMP kinase	
2.4.2.4	Thymidine:orthophosphate deoxy-D-ribosyltransferase Thymidine phosphorylase	
2.6.1.59	Thymidine diphosphate 4-keto-6-deoxy-D-glucose transaminase dTDP-4-amino-4,6-dideoxygalactose transaminase	
4.2.1.46	Thymidine diphosphate D-glucose oxidoreductase dTDPglucose 4,6-dehydratase	
2.7.7.24	Thymidine diphosphate glucose pyrophosphorylase Glucose-1-phosphate thymidylyltransferase	
2.6.1.33	Thymidine diphospho-4-amino-6-deoxyglucose aminotransferase dTDP-4-amino-4,6-dideoxy-D-glucose transaminase	
2.6.1.33	Thymidine diphospho-4-keto-6-deoxy-D-glucose-glutamic transaminase dTDP-4-amino-4,6-dideoxy-D-glucose transaminase	
2.6.1.33	Thymidine diphospho-4-keto-6-deoxy-D-glucose transaminase dTDP-4-amino-4,6-dideoxy-D-glucose transaminase	
5.1.3.13	Thymidine diphospho-4-ketorhamnose 3,5-epimerase dTDP-4-dehydrorhamnose 3,5-epimerase	Vol. 1, Page 152
1.1.1.133	Thymidine diphospho-4-ketorhamnose reductase dTDP-4-dehydrorhamnose reductase	
1.1.1.134	Thymidine diphospho-6-deoxy-L-talose dehydrogenase dTDP-6-deoxy-L-talose 4-dehydrogenase	
4.2.1.46	Thymidine diphospho-D-glucose 4,6-dehydratase dTDPglucose 4,6-dehydratase	
2.7.7.32	Thymidine diphosphogalactose pyrophosphorylase Galactose-1-phosphate thymidylyltransferase	
4.2.1.46	Thymidine diphosphoglucose oxidoreductase dTDPglucose 4,6-dehydratase	

EC-No.	Synonym Recommended Name	Vol., Page
2.7.7.24	Thymidine diphosphoglucose pyrophosphorylase Glucose-1-phosphate thymidylyltransferase	
2.7.1.21	Thymidine kinase Thymidine kinase	
2.7.1.75	thymidine kinase thymidine kinase	
2.7.4.9	Thymidine monophosphate kinase dTMP kinase	
3.1.3.5	thymidine monophosphate nucleotidase 5'-nucleotidase	
2.4.2.4	Thymidine phosphorylase Thymidine phosphorylase	
2.7.1.114	Thymidine phosphotransferase AMP-thymidine kinase	
3.6.1.39	Thymidine triphosphatase Thymidine triphosphatase	
2.7.7.32	Thymidine triphosphate:α-D-galactose 1-phosphate thymidylyltransferase Galactose-1-phosphate thymidylyltransferase	
3.6.1.39	Thymidine triphosphate nucleotidohydrolase Thymidine triphosphatase	
3.1.3.35	thymidylate 5'-nucleotidase thymidylate 5'-phosphatase	
3.1.3.35	thymidylate 5'-phosphatase thymidylate 5'-phosphatase	
3.1.3.35	thymidylate 5'-phosphohydrolase thymidylate 5'-phosphatase	
2.7.4.9	Thymidylate kinase dTMP kinase	
2.7.4.9	Thymidylate monophosphate kinase dTMP kinase	
2.1.1.45	Thymidylate synthase Thymidylate synthase	
2.1.1.45	Thymidylate synthetase Thymidylate synthase	
1.5.1.3	Thymidylate synthetase-dihydrofolate reductase Dihydrofolate reductase	

EC-No.	Synonym Recommended Name	Vol., Page
2.7.4.9	Thymidylic acid kinase dTMP kinase	
2.7.4.9	Thymidylic kinase dTMP kinase	
2.7.7.32	Thymidylyltransferase, galactose 1-phosphate Galactose-1-phosphate thymidylyltransferase	
2.7.7.24	Thymidylyltransferase, glucose 1-phosphate Glucose-1-phosphate thymidylyltransferase	
1.14.11.6	Thymine,2-oxoglutarate:oxygen oxidoreductase (7-hydroxylating) Thymine dioxygenase	
1.14.11.6	Thymine 7-hydroxylase Thymine dioxygenase	
1.14.11.6	Thymine dioxygenase Thymine dioxygenase	
1.14.11.6	Thymine dioxygenase Thymine dioxygenase	
1.3.1.1	Thymine reductase Dihydrouracil dehydrogenase (NAD^+)	
3.1.21.1	Thymonuclease Deoxyribonuclease I	
2.6.1.26	Thyroid-hormone transaminase Thyroid-hormone transaminase	
3.4.23.11	thyroid aspartic proteinase thyroid aspartic proteinase	
2.4.1.38	Thyroid galactosyltransferase β-N-Acetylglucosaminylglycopeptide β-1,4-galactosyltransferase	
5.3.4.1	Thyroid hormone-binding protein Protein disulfide-isomerase	Vol. 1, Page 436
1.11.1.8	Thyroid peroxidase Iodide peroxidase	
3.4.19.6	Thyroliberin-hydrolyzing pyroglutamate aminopeptidase Pyroglutamyl-peptidase II	
3.4.19.6	Thyroliberinase Pyroglutamyl-peptidase II	
1.11.1.8	Thyroperoxidase Iodide peroxidase	

EC-No.	Synonym Recommended Name	Vol., Page
3.4.19.6	Thyrotropin-releasing hormone-degrading peptidase Pyroglutamyl-peptidase II	
3.4.19.6	Thyrotropin-releasing hormone-degrading pyroglutamate aminopeptidase Pyroglutamyl-peptidase II	
3.8.1.4	Thyroxine 5-deiodinase Thyroxine deiodinase	
3.8.1.4	Thyroxine deiodinase Thyroxine deiodinase	
3.8.1.4	L-Thyroxine iodohydrolase (reducing) Thyroxine deiodinase	
2.6.1.25	thyroxine transaminase thyroxine transaminase	
3.1.1.3	tiacetinase triacylglycerol lipase	
3.1.1.3	tibutyrin esterase triacylglycerol lipase	
2.3.1.93	Tigloyl-CoA:13-hydroxylupanine O-tigloyltransferase 13-Hydroxylupanine O-tigloyltransferase	
2.3.1.93	Tigloyltransferase, 13-hydroxylupanine 13-Hydroxylupanine O-tigloyltransferase	
5.3.1.1	TIM Triose-phosphate isomerase	Vol. 1, Page 235
3.4.21.68	Tissue-type plasminogen activator t-Plasminogen activator	
3.4.99.31	Tissue endopeptidase degrading collagenase synthetic substrate Tissue endopeptidase degrading collagenase synthetic substrate	
3.4.21.35	tissue kallikrein tissue kallikrein	
3.4.21.68	Tissue plasminogen activator t-Plasminogen activator	
2.3.2.13	Tissue transglutaminase Protein-glutamine γ-glutamyltransferase	
3.4.21.4	TLE trypsin	
3.4.24.27	TLN Thermolysin	

EC-No.	Synonym Recommended Name	Vol., Page
4.1.2.32	TMAO demethylase Trimethylamine-oxide aldoase	
1.6.6.9	TMAO reductase Trimethylamine-N-oxide reductase	
1.14.11.8	TML-α-ketoglutarate dioxygenase Trimethyllysine dioxygenase	
1.14.11.8	TML hydroxylase Trimethyllysine dioxygenase	
2.5.1.3	TMP-PPase Thiamine-phosphate pyrophosphorylase	
2.7.4.9	TMPK dTMP kinase	
2.1.1.45	TMP synthetase Thymidylate synthase	
2.1.1.9	TMT Thiol S-methyltransferase	
2.1.1.95	γ-Tocopherol methyltransferase Tocopherol O-methyltransferase	
2.1.1.95	Tocopherol O-methyltransferase Tocopherol O-methyltransferase	
3.4.21.90	Togavirin Togavirin	
1.14.12.11	Toluene,NADH:oxygen oxidoreductase (1,2-hydroxylating) Toluene dioxygenase	
1.14.12.11	Toluene dioxygenase Toluene dioxygenase	
3.4.22.28	Tomato ringspot nepovirus 3C-related protease Picornain 3C	
3.4.24.15	TOP thimet oligopeptidase	
3.4.21.5	topical thrombin	
5.99.1.2	Topo I DNA topoisomerase	Vol. 1, Page 721
5.99.1.2	Topoisomerase DNA topoisomerase	Vol. 1, Page 721

EC-No.	Synonym Recommended Name	Vol., Page
5.99.1.2	Topoisomerase I DNA topoisomerase	Vol. 1, Page 721
5.99.1.3	Topoisomerase II DNA topoisomerase (ATP-hydrolysing)	Vol. 1, Page 737
5.99.1.3	Topoisomerase type II DNA topoisomerase (ATP-hydrolysing)	Vol. 1, Page 737
1.6.6.9	TOR Trimethylamine-N-oxide reductase	
3.1.4.3	α-toxin Phospholipase C	
3.6.1.45	Toxin A UDP-sugar diphosphatase	
3.6.1.45	Toxin B UDP-sugar diphosphatase	
3.4.21.68	tPA t-Plasminogen activator	
4.1.99.1	Tpase tryptophanase	
5.99.1.2	TpI DNA topoisomerase	Vol. 1, Page 721
4.1.99.2	TPL tyrosine phenol-lyase	
2.1.1.67	TPMT Thiopurine S-methyltransferase	
1.1.1.19	TPN-L-Gulonate dehydrogenase Glucuronate reductase	
1.6.2.4	TPNH-cytochrome c reductase NADPH-ferrihemoprotein reductase	
1.6.99.1	TPNH-diaphorase NADPH dehydrogenase	
1.18.1.2	TPNH-ferredoxin reductase Ferredoxin-$NADP^+$ reductase	
1.6.2.4	$TPNH_2$ cytochrome c reductase NADPH-ferrihemoprotein reductase	
1.6.99.1	TPNH dehydrogenase NADPH dehydrogenase	

EC-No.	Synonym Recommended Name	Vol., Page
1.11.1.2	TPNH peroxidase NADPH peroxidase	
1.11.1.2	TPN peroxidase NADPH peroxidase	
3.4.14.10	TPP Tripeptidyl-peptidase II	
2.1.2.10	T proteins Aminomethyltransferase	
2.7.6.2	TPTase Thiamine pyrophosphokinase	
5.3.99.2	β-Trace Prostaglandin-D synthase	Vol. 1, Page 451
3.4.23.30	Trametes acid proteinase Pycnoporopepsin	
3.4.99.25	trametes acid proteinase trametes acid proteinase	
5.2.1.9	2-trans,6-trans-Farnesol 2-cis-trans-isomerase Farnesol 2-isomerase	Vol. 1, Page 229
1.1.1.216	2-trans,6-trans-Farnesol:NADP$^+$ 1-oxidoreductase Farnesol dehydrogenase	
4.6.1.11	2-trans,6-trans-Farnesyl-diphosphate diphosphate-lyase (cyclizing, (+)-α-cadinene-forming) (+)-δ-Cadinene synthase	
4.2.3.13	2-trans,6-trans-farnesyl-diphosphate diphosphate-lyase (cyclizing, (+)-δ-cadinene-forming) (+)-d-cadinene synthase	
4.2.3.7	2-trans,6-trans-farnesyldiphosphate diphosphate-lyase (cyclizing, pentalenene-forming) pentalenene synthase	
4.6.1.5	2-trans,6-trans-Farnesyldiphosphate pyrophosphate-lyase (cyclizing, pentalenene-forming) Pentalenene synthase	
2.5.1.29	trans,trans-Farnesyl-diphosphate:isopentenyl-diphosphate farnesyltranstransferase Farnesyltranstransferase	
4.2.3.6	trans,trans-farnesyl-diphosphate sesquiterpenoid-lyase trichodiene synthase	

EC-No.	Synonym Recommended Name	Vol., Page
4.2.3.9	trans,trans-farnesyl diphosphate aristolochene-lyase aristolochene synthase	
1.10.1.1	trans-1,2-Acenaphthenediol dehydrogenase trans-Acenaphthene-1,2-diol dehydrogenase	
1.3.1.20	trans-1,2-Dihydrobenzene-1,2-diol:$NADP^+$ oxidoreductase trans-1,2-Dihydrobenzene-1,2-diol dehydrogenase	
1.3.1.20	trans-1,2-Dihydrobenzene-1,2-diol dehydrogenase trans-1,2-Dihydrobenzene-1,2-diol dehydrogenase	
1.3.1.34	trans-2,3-Didehydroacyl-CoA: $NADP^+$ 4-oxidoreductase 2,4-Dienoyl-CoA reductase (NADPH)	
3.3.2.4	trans-2,3-epoxysuccinate hydrolase trans-epoxysuccinate hydrolase	
4.2.1.17	trans-2-enoyl-CoA hydratase enoyl-CoA hydratase	
1.3.1.44	trans-2-Enoyl-CoA reductase (NAD^+) trans-2-Enoyl-CoA reductase (NAD^+)	
1.3.1.38	trans-2-Enoyl-CoA reductase (NADPH) trans-2-Enoyl-CoA reductase (NADPH)	
1.1.1.226	trans-4-Hydroxycyclohexanecarboxylate:NAD^+ 4-oxidoreductase 4-Hydroxycyclohexanecarboxylate dehydrogenase	
1.14.13.36	trans-5-O-(4-Coumaroyl)-D-quinate,NADPH:oxygen oxidoreductase (3'-hydroxylating) 5-O-(4-Coumaroyl)-D-quinate 3'-monooxygenase	
1.10.1.1	(+,-)-trans-Acenaphthene-1,2-diol:$NADP^+$ oxidoreductase trans-Acenaphthene-1,2-diol dehydrogenase	
1.10.1.1	trans-Acenaphthene-1,2-diol dehydrogenase trans-Acenaphthene-1,2-diol dehydrogenase	
2.1.1.104	trans-Caffeoyl-CoA 3-O-methyltransferase Caffeoyl-CoA O-methyltransferase	
1.1.1.243	(-)-trans-Carveol:$NADP^+$ oxidoreductase Carveol dehydrogenase	
1.14.13.14	trans-Cinnamate,NADPH:oxygen oxidoreductase (2-hydroxylating) trans-Cinnamate 2-monooxygenase	
1.14.13.11	trans-Cinnamate,NADPH:oxygen oxidoreductase (4-hydroxylating) trans-Cinnamate 4-monooxygenase	

EC-No.	Synonym Recommended Name	Vol., Page
1.14.13.14	trans-Cinnamate 2-monooxygenase trans-Cinnamate 2-monooxygenase	
1.14.13.11	trans-Cinnamate 4-hydroxylase trans-Cinnamate 4-monooxygenase	
1.14.13.11	trans-Cinnamate 4-monooxygenase trans-Cinnamate 4-monooxygenase	
1.14.13.14	trans-Cinnamic acid 2-hydroxylase trans-Cinnamate 2-monooxygenase	
1.14.13.11	trans-Cinnamic acid 4-hydroxylase trans-Cinnamate 4-monooxygenase	
2.3.1.152	1-O-trans-Cinnamoyl-β-D-glucopyranose:alcohol O-cinnamoyltransferase Alcohol O-cinnamoyltransferase	
1.1.1.174	trans-Cyclohexane-1,2-diol:NAD^+ 1-oxidoreductase Cyclohexane-1,2-diol dehydrogenase	
2.4.2.6	trans-Deoxyribosylase Nucleoside deoxyribosyltransferase	
5.3.3.8	3,2-trans-Enoyl-CoA isomerase Dodecenoyl-CoA δ-isomerase	Vol. 1, Page 413
5.3.3.8	3-2trans-Enoyl-CoA isomerase Dodecenoyl-CoA δ-isomerase	Vol. 1, Page 413
4.2.1.37	trans-epoxysuccinate hydratase trans-epoxysuccinate hydratase	
3.3.2.4	trans-epoxysuccinate hydratase trans-epoxysuccinate hydrolase	
3.3.2.4	trans-epoxysuccinate hydrolase trans-epoxysuccinate hydrolase	
2.5.1.1	trans-Farnesyl pyrophosphate synthetase Dimethylallyltranstransferase	
6.2.1.34	trans-ferulate:CoASH ligase (ATP-hydrolysing) trans-feruloyl-CoA synthase	
4.2.1.101	trans-feruloyl-CoA hydratase trans-feruloyl-CoA hydratase	
4.2.1.101	trans-feruloyl-CoA hydro-lyase trans-feruloyl-CoA hydratase	
6.2.1.34	trans-feruloyl-CoA synthase trans-feruloyl-CoA synthase	

EC-No.	Synonym Recommended Name	Vol., Page
6.2.1.34	trans-feruloyl-CoA synthetase trans-feruloyl-CoA synthase	
2.5.1.30	trans-Hexaprenyltranstransferase trans-Hexaprenyltranstransferase	
1.1.1.223	(-)-trans-Isopiperitenol:NAD$^+$ oxidoreductase Isopiperitenol dehydrogenase	
4.2.1.77	trans-L-3-Hydroxyproline dehydratase trans-L-3-Hydroxyproline dehydratase	
4.2.1.77	trans-L-3-Hydroxyproline hydro-lyase trans-L-3-Hydroxyproline dehydratase	
2.4.1.16	trans-N-Acetylglucosaminosylase Chitin synthase	
2.4.2.6	trans-N-Deoxyribosylase Nucleoside deoxyribosyltransferase	
2.4.2.6	trans-N-Glycosidase Nucleoside deoxyribosyltransferase	
2.5.1.11	trans-Octaprenyltranstransferase trans-Octaprenyltranstransferase	
2.5.1.11	trans-Octaprenyltranstransferase trans-Octaprenyltranstransferase	
2.5.1.33	trans-Pentaprenyltranstransferase trans-Pentaprenyltranstransferase	
2.5.1.11	trans-Prenyltransferase trans-Octaprenyltranstransferase	
3.3.2.3	Trans-stilbene oxide hydrolase Epoxide hydrolase	
2.3.1.12	Transacetylase X Dihydrolipoamide S-acetyltransferase	
2.2.1.2	Transaldolase Transaldolase	
2.6.1.1	Transaminase A Aspartate transaminase	
2.6.1.42	Transaminase B Branched-chain-amino-acid transaminase	
2.6.1.66	Transaminase C Valine-pyruvate transaminase	

EC-No.	Synonym Recommended Name	Vol., Page
2.1.3.1	Transcarboxylase Methylmalonyl-CoA carboxyltransferase	
2.7.7.6	Transcriptase DNA-directed RNA polymerase	
2.7.7.48	Transcriptase RNA-directed RNA polymerase	
2.4.2.6	Transdeoxyribosylase Nucleoside deoxyribosyltransferase	
2.7.7.25	transfer-RNA nucleotidyltransferase tRNA adenylyltransferase	
2.7.7.21	transfer-RNA nucleotidyltransferase tRNA cytidylyltransferase	
2.4.1.40	A-transferase Fucosylgalactose α-N-acetylgalactosaminyltransferase	
2.1.1.36	Transfer ribonucleate adenine 1-methyltransferase tRNA (adenine-N^1-)-methyltransferase	
2.7.7.25	Transfer ribonucleate adenyltransferase tRNA adenylyltransferase	
2.7.7.21	Transfer ribonucleate adenyltransferase tRNA cytidylyltransferase	
2.7.7.25	Transfer ribonucleate cytidylyltransferase tRNA adenylyltransferase	
2.7.7.21	Transfer ribonucleate cytidylyltransferase tRNA cytidylyltransferase	
2.1.1.29	Transfer ribonucleate cytosine 5-methyltransferase tRNA (cytosine-5-)-methyltransferase	
2.4.2.29	Transfer ribonucleate glycosyltransferase Queuine tRNA-ribosyltransferase	
2.1.1.31	Transfer ribonucleate guanine 1-methyltransferase tRNA (guanine-N^1-)-methyltransferase	
2.1.1.32	Transfer ribonucleate guanine 2-methyltransferase tRNA (guanine-N^2-)-methyltransferase	
2.1.1.33	Transfer ribonucleate guanine 7-methyltransferase tRNA (guanine-N^7-)-methyltransferase	
2.1.1.32	Transfer ribonucleate guanine N^2-methyltransferase tRNA (guanine-N^2-)-methyltransferase	

EC-No.	Synonym Recommended Name	Vol., Page
2.1.1.34	Transfer ribonucleate guanosine 2'-methyltransferase tRNA (guanosine-2'-O-)-methyltransferase	
2.5.1.8	Transfer ribonucleate isopentenyltransferase tRNA isopentenyltransferase	
2.7.7.25	Transfer ribonucleate nucleotidyltransferase tRNA adenylyltransferase	
2.7.7.21	Transfer ribonucleate nucleotidyltransferase tRNA cytidylyltransferase	
2.7.7.25	Transfer ribonucleic-terminal trinucleotide nucleotidyltransferase tRNA adenylyltransferase	
2.7.7.21	Transfer ribonucleic-terminal trinucleotide nucleotidyltransferase tRNA cytidylyltransferase	
2.7.7.25	Transfer ribonucleic acid nucleotidyl transferase tRNA adenylyltransferase	
2.7.7.21	Transfer ribonucleic acid nucleotidyl transferase tRNA cytidylyltransferase	
2.7.7.25	Transfer ribonucleic adenylyl (cytidylyl) transferase tRNA adenylyltransferase	
2.7.7.21	Transfer ribonucleic adenylyl (cytidylyl) transferase tRNA cytidylyltransferase	
1.16.1.2	Transferrin[Fe(II)]2:NAD^+ oxidoreductase Diferric-transferrin reductase	
1.16.1.2	Transferrin reductase Diferric-transferrin reductase	
2.1.1.36	Transfer RNA (adenine-1) methyltransferase tRNA (adenine-N^1-)-methyltransferase	
2.7.7.21	Transfer RNA adenylyltransferase tRNA cytidylyltransferase	
2.7.7.25	transfer RNA adenylyltransferase tRNA adenylyltransferase	
2.1.1.29	Transfer RNA cytosine 5-methyltransferase tRNA (cytosine-5-)-methyltransferase	
2.1.1.32	Transfer RNA guanine 2-methyltransferase tRNA (guanine-N^2-)-methyltransferase	
5.4.99.12	Transfer RNA pseudouridine synthetase tRNA-pseudouridine synthase I	Vol. 1, Page 642

EC-No.	Synonym Recommended Name	Vol., Page
2.8.1.4	Transfer RNA sulfurtransferase tRNA sulfurtransferase	
2.8.1.4	TransferRNA thiolase tRNA sulfurtransferase	
2.1.1.35	Transfer RNA uracil 5-methyltransferase tRNA (uracil-5-)-methyltransferase	
2.1.1.35	Transfer RNA uracil methylase tRNA (uracil-5-)-methyltransferase	
3.1.27.9	Transfer splicing endonuclease tRNA-intron endonuclease	
2.4.1.15	Transglucosylase α,α-Trehalose-phosphate synthase (UDP-forming)	
2.3.2.13	Transglutaminase Protein-glutamine γ-glutamyltransferase	
1.6.1.2	Transhydrogenase $NAD(P)^+$ transhydrogenase (AB-specific)	
1.6.1.1	Transhydrogenase $NAD(P)^+$ transhydrogenase (B-specific)	
1.1.99.28	Transhydrogenase, glucose-fructose Glucose-fructose oxidoreductase	
1.8.4.3	Transhydrogenase, glutathione-coenzyme A glutathione disulfide Glutathione-CoA-glutathione transhydrogenase	
1.8.4.4	Transhydrogenase, glutathione-cystine Glutathione-cystine transhydrogenase	
1.8.4.1	Transhydrogenase, glutathione-homocystine Glutathione-homocystine transhydrogenase	
1.1.99.24	Transhydrogenase, hydroxy acid-oxo acid Hydroxyacid-oxoacid transhydrogenase	
1.1.99.7	Transhydrogenase, lactate-malate Lactate-malate transhydrogenase	
1.6.1.2	Transhydrogenase, nicotinamide adenine dinucleotide (phosphate) $NAD(P)^+$ transhydrogenase (AB-specific)	
1.6.1.1	Transhydrogenase, nicotinamide adenine dinucleotide (phosphate) $NAD(P)^+$ transhydrogenase (B-specific)	
3.4.24.17	Transin Stromelysin 1	

EC-No.	Synonym Recommended Name	Vol., Page
3.4.24.22	Transin 2 Stromelysin 2	
3.4.24.22	Transins, 2 Stromelysin 2	
2.2.1.1	Transketolase Transketolase	
2.2.1.3	Transketolase, formaldehyde Formaldehyde transketolase	
2.6.3.1	Transoximase Oximinotransferase	
2.6.3.1	Transoximinase Oximinotransferase	
2.3.2.12	Transpeptidase Peptidyltransferase	
2.4.2.7	Transphosphoribosidase Adenine phosphoribosyltransferase	
2.4.2.8	Transphosphoribosidase Hypoxanthine phosphoribosyltransferase	
3.4.22.33	Traumanase Fruit bromelain	
3.2.1.28	Trehalase α,α-trehalase	
3.2.1.28	α,α'-trehalase α,α-trehalase	
3.2.1.28	α,α-trehalase α,α-trehalase	
2.3.1.122	α,α'-Trehalose-6-mycolate:α,α'-trehalose-6-mycolate 6'-mycolyltransferase Trehalose O-mycolyltransferase	
3.2.1.93	α,α-trehalose-6-phosphate phosphoglucohydrolase α,α-phosphotrehalase	
3.1.3.12	Trehalose-6-phosphate phosphohydrolase Trehalose-phosphatase	
3.1.3.12	Trehalose-phosphatase Trehalose-phosphatase	
2.4.1.36	α,α-Trehalose-phosphate synthase (GDP-forming) α,α-Trehalose-phosphate synthase (GDP-forming)	

EC-No.	Synonym Recommended Name	Vol., Page
2.4.1.15	α,α-Trehalose-phosphate synthase (UDP-forming) α,α-Trehalose-phosphate synthase (UDP-forming)	
2.4.1.15	Trehalose-P synthetase α,α-Trehalose-phosphate synthase (UDP-forming)	
2.3.1.122	α,α'-Trehalose 6-monomycolate:α,α'-trehalose mycolyltransferase Trehalose O-mycolyltransferase	
3.1.3.12	Trehalose 6-phosphatase Trehalose-phosphatase	
3.1.3.12	Trehalose 6-phosphate phosphatase Trehalose-phosphatase	
2.4.1.15	Trehalose 6-phosphate synthase α,α-Trehalose-phosphate synthase (UDP-forming)	
2.4.1.15	Trehalose 6-phosphate synthetase α,α-Trehalose-phosphate synthase (UDP-forming)	
2.4.1.64	α,α-Trehalose:orthophosphate β-D-glucosyltransferase α,α-Trehalose phosphorylase	
3.2.1.28	α,α-trehalose glucohydrolase α,α-trehalase	
2.3.1.122	Trehalose O-mycolyltransferase Trehalose O-mycolyltransferase	
2.4.1.15	Trehalosephosphate-UDP glucosyl transferase α,α-Trehalose-phosphate synthase (UDP-forming)	
2.4.1.15	Trehalose phosphate-uridine diphosphate glucosyltransferase α,α-Trehalose-phosphate synthase (UDP-forming)	
2.4.1.15	Trehalose phosphate synthase α,α-Trehalose-phosphate synthase (UDP-forming)	
2.4.1.36	Trehalose phosphate synthase (GDP-forming) α,α-Trehalose-phosphate synthase (GDP-forming)	
2.4.1.15	α,α-Trehalose phosphate synthase (UDP-forming) α,α-Trehalose-phosphate synthase (UDP-forming)	
2.4.1.15	Trehalose phosphate synthetase α,α-Trehalose-phosphate synthase (UDP-forming)	
2.4.1.64	Trehalose phosphorylase α,α-Trehalose phosphorylase	
2.4.1.64	α,α-Trehalose phosphorylase α,α-Trehalose phosphorylase	

EC-No.	Synonym Recommended Name	Vol., Page
5.4.99.16	Trehalose synthase Maltose α-D-glucosyltransferase	Vol. 1, Page 656
5.4.99.16	Trehalose synthase (Pimelobacter strain R48 clone pBRM8 gene treS precursor reduced) Maltose α-D-glucosyltransferase	Vol. 1, Page 656
5.4.99.16	Trehalose synthetase Maltose α-D-glucosyltransferase	Vol. 1, Page 656
3.2.1.141	Trehalosidase, maltooligosyltrehalose 4-α-D-{(1→4)-α-D-Glucano}trehalose trehalohydrolase	
3.2.1.141	Trehalosidase, maltooligosyltrehalose (Rhizobium strain M-11 clone pBMTU1 gene treZ reduced) 4-α-D-{(1→4)-α-D-Glucano}trehalose trehalohydrolase	
5.4.99.11	Trehalulose synthase Isomaltulose synthase	Vol. 1, Page 638
3.4.19.6	TRH aminopeptidase Pyroglutamyl-peptidase II	
3.1.1.38	triacetate-lactonase triacetate-lactonase	
3.1.1.38	triacetate lactonase triacetate-lactonase	
3.1.1.38	triacetate lactone hydrolase triacetate-lactonase	
3.1.1.38	triacetic acid lactone hydrolase triacetate-lactonase	
3.1.1.38	triacetic lactone hydrolase triacetate-lactonase	
3.1.1.1	Triacetin esterase Carboxylesterase	
3.1.1.38	triacetolactone lactonohydrolase triacetate-lactonase	
3.1.1.34	triacylglycero-protein acylhydrolase lipoprotein lipase	
2.3.1.77	Triacylglycerol-sterol O-acyltransferase Triacylglycerol-sterol O-acyltransferase	
2.3.1.77	Triacylglycerol:3β-hydroxysterol O-acyltransferase Triacylglycerol-sterol O-acyltransferase	

EC-No.	Synonym Recommended Name	Vol., Page
2.3.1.77	Triacylglycerol:sterol acyltransferase Triacylglycerol-sterol O-acyltransferase	
3.1.1.3	triacylglycerol acylhydrolase triacylglycerol lipase	
3.1.1.3	triacylglycerol ester hydrolase triacylglycerol lipase	
3.1.1.3	triacylglycerol lipase triacylglycerol lipase	
3.1.1.3	tributyrase triacylglycerol lipase	
3.1.1.3	tributyrinase triacylglycerol lipase	
4.5.1.1	1,1,1-trichloro-2,2-bis(4-chlorophenyl)ethane hydrogen-chloride-lyase DDT-dehydrochlorinase	
2.1.1.136	trichlorophenol O-methyltransferase chlorophenol O-methyltransferase	
3.1.27.4	Trichoderma koningi RNase III Ribonuclease U_2	
4.1.99.6	Trichodiene synthase Trichodiene synthase	
4.2.3.6	trichodiene synthase trichodiene synthase	
4.2.3.6	trichodiene synthetase trichodiene synthase	
3.4.24.10	Trichophyton mentagrophytes keratinase Trichophyton mentagrophytes keratinase	
3.4.99.12	Trichophyton mentagrophytes keratinase Trichophyton mentagrophytes keratinase	
3.4.24.9	Trichophyton schoenleinii collagenase Trichophyton schoenleinii collagenase	
3.4.25.1	tricorn protease proteasome endopeptidase complex	
3.4.25.1	tricorn proteinase proteasome endopeptidase complex	
3.4.22.39	trigger peptidase adenain	

EC-No.	Synonym Recommended Name	Vol., Page
2.8.2.19	Triglucosylalkylacylglycerol sulfotransferase Triglucosylalkylacylglycerol sulfotransferase	
3.1.1.3	triglyceridase triacylglycerol lipase	
3.1.1.3	triglyceride hydrolase triacylglycerol lipase	
3.1.1.3	triglyceride lipase triacylglycerol lipase	
2.4.1.60	Trihexose diphospholipid abequosyltransferase Abequosyltransferase	
3.2.1.47	Trihexosylceramide α-galactosidase Galactosylgalactosylglucosylceramidase	
3.2.1.47	Trihexosylceramide.α.-galactosidase Galactosylgalactosylglucosylceramidase	
3.2.1.47	Trihexosyl ceramide galactosidase Galactosylgalactosylglucosylceramidase	
3.5.1.24	3α,7α,12α-trihydroxy-5β-cholan-24-oylglycine amidohydrolase Cholylglycine hydrolase	
1.2.1.40	3α,7α,12α-Trihydroxy-5β-cholestan-26-al:NAD$^+$ 26-oxidoreductase 3α,7α,12α-Trihydroxycholestan-26-al 26-dehydrogenase	
1.2.1.40	3α,7α,12α-Trihydroxy-5β-cholestan-26-al dehydrogenase 3α,7α,12α-Trihydroxycholestan-26-al 26-dehydrogenase	
6.2.1.29	3α,7α,12α-Trihydroxy-5β-cholestanate-CoA ligase 3α,7α,12α-Trihydroxy-5β-cholestanate-CoA ligase	Vol. 2, Page 328
6.2.1.29	3α,7α,12α-Trihydroxy-5β-cholestanate:CoA ligase (AMP-forming) 3α,7α,12α-Trihydroxy-5β-cholestanate-CoA ligase	Vol. 2, Page 328
6.2.1.29	3α,7α,12α-Trihydroxy-5β-cholestanoate-CoA ligase 3α,7α,12α-Trihydroxy-5β-cholestanate-CoA ligase	Vol. 2, Page 328
6.2.1.29	3α,7α,12α-Trihydroxy-5β-cholestanoate-CoA synthetase 3α,7α,12α-Trihydroxy-5β-cholestanate-CoA ligase	Vol. 2, Page 328
1.13.11.35	1,2,3-Trihydroxybenzene:oxygen 1,2-oxidoreductase (decyclizing) Pyrogallol 1,2-oxygenase	
1.2.1.40	3α,7α,12α-Trihydroxycholestan-26-al 26-dehydrogenase 3α,7α,12α-Trihydroxycholestan-26-al 26-dehydrogenase	
1.2.1.40	Trihydroxydeoxycoprostanal dehydrogenase 3α,7α,12α-Trihydroxycholestan-26-al 26-dehydrogenase	

EC-No.	Synonym Recommended Name	Vol., Page
1.1.1.252	1,3,8-Trihydroxynaphthalene reductase Tetrahydroxynaphthalene reductase	
1.1.1.252	Trihydroxynaphthalene reductase Tetrahydroxynaphthalene reductase	
2.5.1.36	Trihydroxypterocarpan dimethylallyltransferase Trihydroxypterocarpan dimethylallyltransferase	
2.3.1.95	Trihydroxystilbene synthase Trihydroxystilbene synthase	
2.6.1.26	L-3,5,3'-Triiodothyronine:2-oxoglutarate aminotransferase Thyroid-hormone transaminase	
3.2.1.23	Trilactase β-galactosidase	
3.4.24.52	Trimerelysin I Trimerelysin I	
3.4.24.53	Trimerelysin II Trimerelysin II	
3.4.24.54	Trimeresurus metalloendopeptidase A Mucrolysin	
3.4.24.52	Trimeresurus metalloendopeptidase I Trimerelysin I	
3.4.24.53	Trimeresurus metalloendopeptidase II Trimerelysin II	
3.6.1.2	Trimetaphosphatase Trimetaphosphatase	
3.6.1.2	Trimetaphosphate hydrolase Trimetaphosphatase	
4.1.2.32	Trimethylamine-N-oxide formaldehyde-lyase Trimethylamine-oxide aldoase	
1.6.6.9	Trimethylamine-N-oxide reductase Trimethylamine-N-oxide reductase	
4.1.2.32	Trimethylamine-oxide aldoase Trimethylamine-oxide aldoase	
1.5.99.7	Trimethylamine:(acceptor) oxidoreductase (demethylating) Trimethylamine dehydrogenase	
1.5.99.7	Trimethylamine dehydrogenase Trimethylamine dehydrogenase	

EC-No.	Synonym Recommended Name	Vol., Page
4.1.2.32	Trimethylamine N-oxide aldolase Trimethylamine-oxide aldoase	
4.1.2.32	Trimethylamine N-oxide demethylase Trimethylamine-oxide aldoase	
4.1.2.32	Trimethylamine N-oxide formaldehyde-lyase Trimethylamine-oxide aldoase	
1.6.6.9	Trimethylamine N-oxide reductase Trimethylamine-N-oxide reductase	
1.6.6.9	Trimethylamine oxide reductase Trimethylamine-N-oxide reductase	
1.2.1.47	4-N-Trimethylaminobutyraldehyde dehydrogenase 4-Trimethylammoniobutyraldehyde dehydrogenase	
1.2.1.47	4-Trimethylaminobutyraldehyde dehydrogenase 4-Trimethylammoniobutyraldehyde dehydrogenase	
2.1.1.5	Trimethylammonioacetate:L-homocysteine S-methyltransferase betaine-homocysteine S-methyltransferase	
1.2.1.47	4-Trimethylammoniobutanal:NAD^+ 1-oxidoreductase 4-Trimethylammoniobutyraldehyde dehydrogenase	
1.14.11.1	4-Trimethylammoniobutanoate,2-oxoglutarate:oxygen oxidoreductase (3-hydroxylating) γ-Butyrobetaine dioxygenase	
1.2.1.47	4-Trimethylammoniobutyraldehyde dehydrogenase 4-Trimethylammoniobutyraldehyde dehydrogenase	
1.14.11.8	Trimethyllysine α-ketoglutarate dioxygenase Trimethyllysine dioxygenase	
1.14.11.8	Trimethyllysine dioxygenase Trimethyllysine dioxygenase	
2.1.1.19	Trimethylsulfonium-tetrahydrofolate methyltransferase Trimethylsulfonium-tetrahydrofolate N-methyltransferase	
2.1.1.19	Trimethylsulfonium-tetrahydrofolate N-methyltransferase Trimethylsulfonium-tetrahydrofolate N-methyltransferase	
2.1.1.19	Trimethylsulfonium:tetrahydrofolate N-methyltransferase Trimethylsulfonium-tetrahydrofolate N-methyltransferase	
2.3.1.154	4,8,12-trimethyltridecanoyl-CoA:propanoyl-CoA C^2-4,8,12-trimethyltridecanoyltransferase propionyl-CoA C2-trimethyltridecanoyltransferase	

EC-No.	Synonym Recommended Name	Vol., Page
3.2.1.106	Trimming glucosidase I Mannosyl-oligosaccharide glucosidase	
2.7.1.28	D-Triokinase Triokinase	
2.7.1.28	Triokinase Triokinase	
3.1.1.3	triolein hydrolase triacylglycerol lipase	
5.3.1.1	Triose-phosphate isomerase Triose-phosphate isomerase	Vol. 1, Page 235
2.7.1.28	Triose kinase Triokinase	
1.2.1.12	Triose phosphate dehydrogenase Glyceraldehyde-3-phosphate dehydrogenase	
1.2.1.9	Triose phosphate dehydrogenase Glyceraldehyde-3-phosphate dehydrogenase ($NADP^+$)	
1.2.1.59	Triosephosphate dehydrogenase (NAD(P)) Glyceraldehyde-3-phosphate dehydrogenase (NAD(P)) (phosphorylating)	
1.2.1.13	Triosephosphate dehydrogenase ($NADP^+$) Glyceraldehyde-3-phosphate dehydrogenase ($NADP^+$) (phosphorylating)	
5.3.1.1	Triose phosphate isomerase Triose-phosphate isomerase	Vol. 1, Page 235
5.3.1.1	Triosephosphate isomerase Triose-phosphate isomerase	Vol. 1, Page 235
5.3.1.1	Triose phosphate mutase Triose-phosphate isomerase	Vol. 1, Page 235
5.3.1.1	Triosephosphate mutase Triose-phosphate isomerase	Vol. 1, Page 235
5.3.1.1	Triose phosphoisomerase Triose-phosphate isomerase	Vol. 1, Page 235
3.4.21.4	tripcellim trypsin	
3.4.11.4	tripeptidase tripeptide aminopeptidase	
3.4.11.4	tripeptide aminopeptidase tripeptide aminopeptidase	

EC-No.	Synonym Recommended Name	Vol., Page
3.4.14.9	Tripeptidyl-peptidase I Tripeptidyl-peptidase I	
3.4.14.10	Tripeptidyl-peptidase II Tripeptidyl-peptidase II	
3.4.14.9	Tripeptidyl aminopeptidase Tripeptidyl-peptidase I	
3.4.14.10	Tripeptidyl aminopeptidase Tripeptidyl-peptidase II	
3.4.11.4	tripeptidyl aminopeptidase tripeptide aminopeptidase	
3.4.14.9	Tripeptidyl aminopeptidase I Tripeptidyl-peptidase I	
3.4.14.10	Tripeptidyl aminopeptidase II Tripeptidyl-peptidase II	
3.4.14.9	Tripeptidyl peptidase Tripeptidyl-peptidase I	
3.4.14.10	Tripeptidyl peptidase Tripeptidyl-peptidase II	
3.4.14.8	tripeptidyl peptidase tripeptidyl peptidase	
3.4.14.9	Tripeptidyl peptidase I Tripeptidyl-peptidase I	
3.4.14.10	Tripeptidyl peptidase II Tripeptidyl-peptidase II	
3.6.1.3	Triphosphatase Adenosinetriphosphatase	
3.6.1.25	Triphosphatase Triphosphatase	
3.6.1.25	Triphosphate phosphohydrolase Triphosphatase	
3.1.3.36	Triphosphoinositide phosphatase Phosphatidylinositol-bisphosphatase	
3.1.4.11	Triphosphoinositide phosphodiesterase 1-Phosphatidyl-4,5-bisphosphate phosphodiesterase	
3.1.3.36	Triphosphoinositide phosphomonoesterase Phosphatidylinositol-bisphosphatase	

EC-No.	Synonym Recommended Name	Vol., Page
1.3.1.15	Triphosphopyridine-linked dihydroorotic dehydrogenase Orotate reductase (NADPH)	
1.6.99.1	Triphosphopyridine diaphorase NADPH dehydrogenase	
3.2.2.6	Triphosphopyridine nucleotidase $NAD(P)^+$nucleosidase	
1.6.6.3	Triphosphopyridine nucleotide-nitrate reductase Nitrate reductase (NADPH)	
1.6.99.1	Triphosphopyridine nucleotide diaphorase NADPH dehydrogenase	
1.11.1.2	Triphosphopyridine nucleotide peroxidase NADPH peroxidase	
1.8.99.3	Trithionate:(acceptor) oxidoreductase Hydrogensulfite reductase	
3.12.1.1	Trithionate hydrolase Trithionate hydrolase	
3.12.1.1	Trithionate thiosulfohydrolase Trithionate hydrolase	
3.4.21.64	Tritirachium album proteinase K Endopeptidase K	
3.4.21.64	Tritirachium alkaline proteinase Endopeptidase K	
2.1.1.61	tRNA(5-methylaminomethyl-2-thiouridylate)-methyltransferase tRNA(5-methylaminomethyl-2-thiouridylate)-methyltransferase	
2.1.1.36	tRNA (adenine-N^1-)-methyltransferase tRNA (adenine-N^1-)-methyltransferase	
2.1.1.55	tRNA (adenine-N^6-)-methyltransferase tRNA (adenine-N^6-)-methyltransferase	
2.1.1.29	tRNA (cytosine-5-)-methyltransferase tRNA (cytosine-5-)-methyltransferase	
2.1.1.31	tRNA (guanine-N^1-)-methyltransferase tRNA (guanine-N^1-)-methyltransferase	
2.1.1.32	tRNA (guanine-N^2-)-methyltransferase tRNA (guanine-N^2-)-methyltransferase	
2.1.1.33	tRNA (guanine-N^7-)-methyltransferase tRNA (guanine-N^7-)-methyltransferase	

EC-No.	Synonym Recommended Name	Vol., Page
2.1.1.34	tRNA (guanosine-2'-O-)-methyltransferase tRNA (guanosine-2'-O-)-methyltransferase	
2.1.1.30	tRNA (purine-2- or -6-)-methyltransferase tRNA (purine-2- or -6-)-methyltransferase	
2.1.1.35	tRNA (uracil-5-)-methyltransferase tRNA (uracil-5-)-methyltransferase	
2.4.2.29	tRNA-guanine:queuine tRNA-D-ribosyltransferase Queuine tRNA-ribosyltransferase	
2.4.2.29	tRNA-guanine transglycosylase Queuine tRNA-ribosyltransferase	
3.1.27.9	tRNA-intron endonuclease tRNA-intron endonuclease	
2.7.7.25	tRNA-nucleotidyltransferase tRNA adenylyltransferase	
2.7.7.21	tRNA-nucleotidyltransferase tRNA cytidylyltransferase	
5.4.99.12	tRNA-pseudouridine synthase I tRNA-pseudouridine synthase I	Vol. 1, Page 642
2.4.1.110	tRNA-queuosine β-mannosyltransferase tRNA-queuosine β-mannosyltransferase	
2.5.1.25	tRNA-uridine aminocarboxypropyltransferase tRNA-uridine aminocarboxypropyltransferase	
5.4.99.12	tRNA-uridine isomerase tRNA-pseudouridine synthase I	Vol. 1, Page 642
5.4.99.12	tRNA-uridine uracilmutase tRNA-pseudouridine synthase I	Vol. 1, Page 642
2.1.1.61	tRNA 5-methylaminomethyl-2-thiouridylate 5'-methyltransferase tRNA(5-methylaminomethyl-2-thiouridylate)-methyltransferase	
2.7.7.56	tRNA:orthophosphate nucleotidyltransferase tRNA nucleotidyltransferase	
2.7.7.25	tRNA adenylyl(cytidylyl)transferase tRNA adenylyltransferase	
2.7.7.21	tRNA adenylyl(cytidylyl)transferase tRNA cytidylyltransferase	
2.7.7.25	tRNA adenylyltransferase tRNA adenylyltransferase	

EC-No.	Synonym Recommended Name	Vol., Page
2.7.7.21	tRNA adenylyltransferase tRNA cytidylyltransferase	
2.7.7.25	tRNA CCA-pyrophosphorylase tRNA adenylyltransferase	
2.7.7.21	tRNA CCA-pyrophosphorylase tRNA cytidylyltransferase	
2.7.7.25	tRNA cytidylyltransferase tRNA adenylyltransferase	
2.7.7.21	tRNA cytidylyltransferase tRNA cytidylyltransferase	
2.1.1.29	tRNA cytosine 5-methyltransferase tRNA (cytosine-5-)-methyltransferase	
2.1.1.31	tRNA guanine 1-methyltransferase tRNA (guanine-N^1-)-methyltransferase	
2.1.1.33	tRNA guanine 7-methyltransferase tRNA (guanine-N^7-)-methyltransferase	
2.4.2.29	tRNA guanine transglycosidase Queuine tRNA-ribosyltransferase	
2.4.2.29	tRNA guanine transglycosylase Queuine tRNA-ribosyltransferase	
2.1.1.34	tRNA guanosine 2'-methyltransferase tRNA (guanosine-2'-O-)-methyltransferase	
3.1.27.9	tRNA intron endonuclease tRNA-intron endonuclease	
2.5.1.8	tRNA isopentenyltransferase tRNA isopentenyltransferase	
2.7.7.56	tRNA nucleotidyltransferase tRNA nucleotidyltransferase	
5.4.99.12	tRNA pseudouridine synthase tRNA-pseudouridine synthase I	Vol. 1, Page 642
5.4.99.12	tRNA pseudouridylate synthase I tRNA-pseudouridine synthase I	Vol. 1, Page 642
3.1.27.9	tRNA splicing endonuclease tRNA-intron endonuclease	
2.8.1.4	tRNA sulfurtransferase tRNA sulfurtransferase	

EC-No.	Synonym Recommended Name	Vol., Page
2.4.2.29	tRNA transglycosylase Queuine tRNA-ribosyltransferase	
3.1.27.9	tRNATRPintron endonuclease tRNA-intron endonuclease	
2.1.1.35	tRNA uracil 5-methyltransferase tRNA (uracil-5-)-methyltransferase	
1.1.1.206	Tropine:NADP$^+$ 3α-oxidoreductase Tropine dehydrogenase	
1.1.1.206	Tropine dehydrogenase Tropine dehydrogenase	
3.1.1.10	Tropinesterase Tropinesterase	
1.1.1.236	Tropinone reductase Tropinone reductase	
1.1.1.236	Tropinone reductase II Tropinone reductase	
2.7.1.132	Tropomyosin kinase Tropomyosin kinase	
3.4.21.5	tropostasin thrombin	
1.4.1.19	L-Trp-dehydrogenase Tryptophan dehydrogenase	
5.3.3.12	TRP2 Dopachrome Δ-isomerase	Vol. 1, Page 432
6.1.1.2	TrpRS Tryptophan-tRNA ligase	Vol. 2, Page 9
6.1.1.3	TRS Threonine-tRNA ligase	Vol. 2, Page 17
3.1.1.7	true cholinesterase acetylcholinesterase	
1.6.4.8	Trypanothione reductase Trypanothione reductase	
6.3.1.9	Trypanothione synthase Trypanothione synthase	Vol. 2, Page 391
6.3.1.9	Trypanothione synthetase Trypanothione synthase	Vol. 2, Page 391

EC-No.	Synonym Recommended Name Vol., Page
6.3.1.9	Trypanothione synthetase (Crithidia fasciculata strain HS6 gene Cf-TS) Trypanothione synthase Vol. 2, Page 391
3.4.21.4	α-trypsin trypsin
3.4.21.4	β-trypsin trypsin
3.4.21.4	trypsin trypsin
3.4.4.4	trypsin trypsin
3.4.23.18	Trypsinogen kinase Aspergillopepsin I
1.13.11.11	Tryptamine 2,3-dioxygenase Tryptophan 2,3-dioxygenase
2.1.1.49	Tryptamine methyltransferase Amine N-methyltransferase
2.1.1.49	Tryptamine N-methyltransferase Amine N-methyltransferase
3.4.21.4	tryptar trypsin
3.4.21.59	Tryptase Tryptase
3.4.21.4	tryptase trypsin
3.4.21.59	Tryptase M Tryptase
1.4.3.17	L-tryptophan α,β-dehydrogenase tryptophan α,β-oxidase
1.4.3.17	tryptophan α,β-oxidase tryptophan α,β-oxidase
1.14.16.4	L-Tryptophan,tetrahydrobiopterin:oxygen oxidoreductase (5-hydroxylating) Tryptophan 5-monooxygenase
2.6.1.28	L-Tryptophan-α-ketoisocaproate aminotransferase Tryptophan-phenylpyruvate transaminase
2.6.1.28	Tryptophan-phenylpyruvate transaminase Tryptophan-phenylpyruvate transaminase

EC-No.	Synonym Recommended Name	Vol., Page
6.1.1.2	Tryptophan-tRNA ligase Tryptophan-tRNA ligase	Vol. 2, Page 9
6.1.1.2	L-Tryptophan-tRNA^{Thr} ligase (AMP-forming) Tryptophan-tRNA ligase	Vol. 2, Page 9
1.13.99.3	Tryptophan 2'-dioxygenase Tryptophan 2'-dioxygenase	
1.13.11.11	Tryptophan 2,3-dioxygenase Tryptophan 2,3-dioxygenase	
2.1.1.106	Tryptophan 2-C-methyltransferase Tryptophan 2-C-methyltransferase	
1.13.12.3	Tryptophan 2-monooxygenase Tryptophan 2-monooxygenase	
1.14.16.4	Tryptophan 5-hydroxylase Tryptophan 5-monooxygenase	
1.14.16.4	Tryptophan 5-monooxygenase Tryptophan 5-monooxygenase	
2.6.1.27	L-Tryptophan:2-oxoglutarate aminotransferase Tryptophan transaminase	
1.4.1.19	L-Tryptophan:NAD(P)$^+$ oxidoreductase (deaminating) Tryptophan dehydrogenase	
1.4.3.17	L-tryptophan:oxygen α,β-oxidoreductase tryptophan α,β-oxidase	
1.13.99.3	L-Tryptophan:oxygen 2'-oxidoreductase (side-chain-cleaving) Tryptophan 2'-dioxygenase	
1.13.11.11	L-Tryptophan: oxygen 2,3-oxidoreductase (decyclizing) Tryptophan 2,3-dioxygenase	
1.13.12.3	L-Tryptophan:oxygen 2-oxidoreductase (decarboxylating) Tryptophan 2-monooxygenase	
2.6.1.28	L-Tryptophan:phenylpyruvate aminotransferase Tryptophan-phenylpyruvate transaminase	
2.3.1.34	D-Tryptophan acetyltransferase D-Tryptophan N-acetyltransferase	
3.5.1.57	Tryptophanamidase Tryptophanamidase	
3.5.1.57	L-Tryptophanamide amidohydrolase Tryptophanamidase	

EC-No.	Synonym Recommended Name	Vol., Page
3.5.1.57	L-Tryptophan aminopeptidase Tryptophanamidase	
3.4.11.17	L-tryptophan aminopeptidase tryptophanyl aminopeptidase	
3.5.1.57	Tryptophan aminopeptidase Tryptophanamidase	
3.4.11.17	tryptophan aminopeptidase tryptophanyl aminopeptidase	
2.6.1.27	L-Tryptophan aminotransferase Tryptophan transaminase	
2.6.1.27	Tryptophan aminotransferase Tryptophan transaminase	
4.1.99.1	L-tryptophanase tryptophanase	
1.13.11.11	Tryptophanase Tryptophan 2,3-dioxygenase	
4.1.99.1	tryptophanase tryptophanase	
4.1.1.28	L-Tryptophan decarboxylase Aromatic-L-amino-acid decarboxylase	
4.1.1.28	Tryptophan decarboxylase Aromatic-L-amino-acid decarboxylase	
4.1.1.27	tryptophan decarboxylase tryptophan decarboxylase	
1.4.1.19	L-Tryptophan dehydrogenase Tryptophan dehydrogenase	
1.4.1.19	Tryptophan dehydrogenase Tryptophan dehydrogenase	
4.2.1.20	tryptophan desmolase tryptophan synthase	
2.5.1.34	Tryptophan dimethylallyltransferase Tryptophan dimethylallyltransferase	
1.14.16.4	L-Tryptophan hydroxylase Tryptophan 5-monooxygenase	
1.14.16.4	Tryptophan hydroxylase Tryptophan 5-monooxygenase	

EC-No.	Synonym Recommended Name	Vol., Page
4.1.99.1	tryptophan indole-lyase tryptophanase	
4.1.99.1	L-tryptophan indole-lyase (deaminating) tryptophanase	
2.3.1.34	D-Tryptophan N-acetyltransferase D-Tryptophan N-acetyltransferase	
2.3.1.112	D-Tryptophan N-malonyltransferase D-Tryptophan N-malonyltransferase	
1.13.11.11	Tryptophan oxygenase Tryptophan 2,3-dioxygenase	
1.13.11.11	Tryptophan peroxidase Tryptophan 2,3-dioxygenase	
1.13.11.11	L-Tryptophan pyrrolase Tryptophan 2,3-dioxygenase	
1.13.11.11	Tryptophan pyrrolase Tryptophan 2,3-dioxygenase	
1.13.11.26	Tryptophan pyrrolooxygenase Peptide-tryptophan 2,3-dioxygenase	
1.13.99.3	Tryptophan side-chain α,β-oxidase Tryptophan 2'-dioxygenase	
1.13.99.3	Tryptophan side-chain oxidase Tryptophan 2'-dioxygenase	
1.13.99.3	Tryptophan side chain oxidase Tryptophan 2'-dioxygenase	
1.13.99.3	Tryptophan side chain oxidase II Tryptophan 2'-dioxygenase	
1.13.99.3	Tryptophan side chain oxidase type I Tryptophan 2'-dioxygenase	
4.2.1.20	tryptophan synthase tryptophan synthase	
4.2.1.20	L-tryptophan synthetase tryptophan synthase	
4.2.1.20	tryptophan synthetase tryptophan synthase	
2.6.1.27	L-Tryptophan transaminase Tryptophan transaminase	

EC-No.	Synonym Recommended Name	Vol., Page
2.6.1.27	Tryptophan transaminase Tryptophan transaminase	
6.1.1.2	Tryptophan translase Tryptophan-tRNA ligase	Vol. 2, Page 9
6.1.1.2	Tryptophanyl-transfer ribonucleate synthetase Tryptophan-tRNA ligase	Vol. 2, Page 9
6.1.1.2	Tryptophanyl-transfer ribonucleic acid synthetase Tryptophan-tRNA ligase	Vol. 2, Page 9
6.1.1.2	Tryptophanyl-transfer ribonucleic synthetase Tryptophan-tRNA ligase	Vol. 2, Page 9
6.1.1.2	Tryptophanyl-transfer RNA synthetase Tryptophan-tRNA ligase	Vol. 2, Page 9
6.1.1.2	Tryptophanyl-tRNA synthase Tryptophan-tRNA ligase	Vol. 2, Page 9
6.1.1.2	Tryptophanyl-tRNA synthetase Tryptophan-tRNA ligase	Vol. 2, Page 9
3.4.11.17	tryptophanyl aminopeptidase tryptophanyl aminopeptidase	
6.1.1.2	Tryptophanyl ribonucleic synthetase Tryptophan-tRNA ligase	Vol. 2, Page 9
3.4.21.4	trypure trypsin	
6.3.1.9	TSH synthetase Trypanothione synthase	Vol. 2, Page 391
1.13.99.3	TSO Tryptophan 2'-dioxygenase	
1.13.99.3	TSO I Tryptophan 2'-dioxygenase	
1.13.99.3	TSO II Tryptophan 2'-dioxygenase	
3.4.21.78	TSP-1 Granzyme A	
2.8.1.5	TSR Thiosulfate-dithiol sulfurtransferase	
3.4.17.17	TTCPase Tubulinyl-Tyr carboxypeptidase	

EC-No.	Synonym Recommended Name	Vol., Page
4.2.1.32	L-Ttd L(+)-Tartrate dehydrogenase	
6.3.2.25	TTL Tubulin-tyrosine ligase	Vol. 2, Page 524
6.3.2.25	TTLase Tubulin-tyrosine ligase	Vol. 2, Page 524
6.3.2.25	α-Tubulin-L-tyrosine ligase (ADP-forming) Tubulin-tyrosine ligase	Vol. 2, Page 524
3.4.17.17	Tubulin-tyrosine carboxypeptidase Tubulinyl-Tyr carboxypeptidase	
6.3.2.25	Tubulin-tyrosine ligase Tubulin-tyrosine ligase	Vol. 2, Page 524
6.3.2.25	Tubulin:tyrosine ligase Tubulin-tyrosine ligase	Vol. 2, Page 524
2.3.1.108	α-Tubulin acetylase Tubulin N-acetyltransferase	
2.3.1.108	α-Tubulin acetyltransferase Tubulin N-acetyltransferase	
3.4.17.17	Tubulin carboxypeptidase Tubulinyl-Tyr carboxypeptidase	
2.3.1.108	Tubulin N-acetyltransferase Tubulin N-acetyltransferase	
3.4.17.17	Tubulinyl-Tyr carboxypeptidase Tubulinyl-Tyr carboxypeptidase	
3.4.17.17	Tubulinyltyrosine carboxypeptidase Tubulinyl-Tyr carboxypeptidase	
2.7.7.52	TUT RNA uridylyltransferase	
3.1.1.3	tween-hydrolyzing esterase triacylglycerol lipase	
3.1.1.3	Tweenase triacylglycerol lipase	
3.1.1.3	tween hydrolase triacylglycerol lipase	
2.7.1.37	Twitchin kinase Protein kinase	

EC-No.	Synonym Recommended Name	Vol., Page
3.4.21.73	Two-chain urokinase-type plasminogen activator u-Plasminogen activator	
5.3.99.5	TXAS Thromboxane-A synthase	Vol. 1, Page 472
4.1.1.28	TYDC Aromatic-L-amino-acid decarboxylase	
4.1.1.25	TYDC/DODC Tyrosine decarboxylase	
2.7.1.37	Type-2 casein kinase Protein kinase	
3.1.3.17	type 1 protein phosphatase [phosphorylase] phosphatase	
3.6.1.6	Type B nucleoside diphosphatase Nucleoside-diphosphatase	
3.4.24.14	type I/II procollagen N-proteinase procollagen N-endopeptidase	
3.1.6.1	Type I Arylsulfatase Arylsulfatase	
1.6.99.3	Type I dehydrogenase NADH dehydrogenase	
1.6.5.3	Type I dehydrogenase NADH dehydrogenase (ubiquinone)	
4.2.1.10	Type I dehydroquinase 3-dehydroquinate dehydratase	
5.99.1.2	Type I DNA topoisomerase DNA topoisomerase	Vol. 1, Page 721
5.99.1.3	Type II-DNA-topoisomerase DNA topoisomerase (ATP-hydrolysing)	Vol. 1, Page 737
1.1.1.197	Type II 15-hydroxyprostaglandin dehydrogenase 15-Hydroxyprostaglandin dehydrogenase ($NADP^+$)	
3.1.6.1	Type II Arylsulfatase Arylsulfatase	
4.2.1.10	Type II dehydroquinase 3-dehydroquinate dehydratase	
3.1.21.5	Type III restriction enzyme Type III site-specific deoxyribonuclease	

EC-No.	Synonym Recommended Name	Vol., Page
3.1.21.5	Type III site-specific deoxyribonuclease Type III site-specific deoxyribonuclease	
2.7.1.67	Type II phosphatidylinositol kinase 1-Phosphatidylinositol 4-kinase	
3.1.21.4	Type II restriction enzyme Type II site-specific deoxyribonuclease	
3.1.21.4	Type II site-specific deoxyribonuclease Type II site-specific deoxyribonuclease	
3.4.24.19	Type I procollagen C-proteinase Procollagen C-endopeptidase	
3.1.21.3	Type I restriction enzyme Type I site-specific deoxyribonuclease	
3.1.21.3	Type I site-specific deoxyribonuclease Type I site-specific deoxyribonuclease	
3.4.24.24	Type IV collagenase Gelatinase A	
3.4.24.35	Type IV collagenase Gelatinase B	
3.4.24.24	Type IV collagenase/gelatinase Gelatinase A	
3.4.24.35	Type IV collagenase/gelatinase Gelatinase B	
3.4.24.24	Type IV collagen metalloproteinase Gelatinase A	
3.4.24.35	Type IV collagen metalloproteinase Gelatinase B	
3.6.1.6	Type L nucleoside diphosphatase Nucleoside-diphosphatase	
3.4.24.35	Type V collagenase Gelatinase B	
2.1.1.37	Typ II DNA methylase DNA (cytosine-5-)-methyltransferase	
1.4.3.4	Tyraminase Amine oxidase (flavin-containing)	
1.4.99.4	Tyramine dehydrogenase Aralkylamine dehydrogenase	

EC-No.	Synonym Recommended Name	Vol., Page
2.3.1.110	Tyramine feruloyltransferase Tyramine N-feruloyltransferase	
2.1.1.27	Tyramine methylpherase Tyramine N-methyltransferase	
2.3.1.110	Tyramine N-feruloyl-CoA transferase Tyramine N-feruloyltransferase	
2.3.1.110	Tyramine N-feruloyltransferase Tyramine N-feruloyltransferase	
2.1.1.27	Tyramine N-methyltransferase Tyramine N-methyltransferase	
1.4.3.4	Tyramine oxidase Amine oxidase (flavin-containing)	
1.4.3.9	tyramine oxidase tyramine oxidase	
4.3.1.5	tyrase phenylalanine ammonia-lyase	
1.10.3.1	Tyrosinase Catechol oxidase	
1.14.18.1	Tyrosinase Monophenol monooxygenase	
4.1.99.2	β-tyrosinase tyrosine phenol-lyase	
5.4.3.6	Tyrosine α,β-amino mutase Tyrosine 2,3-aminomutase	Vol. 1, Page 568
5.4.3.6	Tyrosine α,β-mutase Tyrosine 2,3-aminomutase	Vol. 1, Page 568
1.14.13.41	L-Tyrosine,NADPH:oxygen oxidoreductase (N-hydroxylating) Tyrosine N-monooxygenase	
1.14.16.2	L-Tyrosine,tetrahydrobiopterin:oxygen oxidoreductase (3-hydroxylating) Tyrosine 3-monooxygenase	
2.6.1.5	Tyrosine-α-ketoglutarate aminotransferase Tyrosine transaminase	
2.6.1.5	Tyrosine-α-ketoglutarate transaminase Tyrosine transaminase	
2.6.1.5	Tyrosine-2-ketoglutarate aminotransferase Tyrosine transaminase	

EC-No.	Synonym Recommended Name	Vol., Page
2.6.1.5	L-Tyrosine-2-oxoglutarate aminotransferase Tyrosine transaminase	
2.6.1.5	Tyrosine-2-oxoglutarate aminotransferase Tyrosine transaminase	
6.3.2.24	Tyrosine-arginine ligase Tyrosine-arginine ligase	Vol. 2, Page 521
1.14.18.1	Tyrosine-dopa oxidase Monophenol monooxygenase	
2.8.2.9	Tyrosine-ester sulfotransferase Tyrosine-ester sulfotransferase	
2.6.1.20	tyrosine-pyruvate transaminase tyrosine-pyruvate transaminase	
2.7.1.112	Tyrosine-specific protein kinase Protein-tyrosine kinase	
6.1.1.1	Tyrosine-transfer ribonucleate synthetase Tyrosine-tRNA ligase	Vol. 2, Page 1
6.1.1.1	Tyrosine-transfer RNA ligase Tyrosine-tRNA ligase	Vol. 2, Page 1
6.1.1.1	Tyrosine-tRNA ligase Tyrosine-tRNA ligase	Vol. 2, Page 1
4.1.1.25	Tyrosine/Dopa decarboxylase Tyrosine decarboxylase	
4.1.1.28	Tyrosine/dopa decarboxylase Aromatic-L-amino-acid decarboxylase	
5.4.3.6	L-Tyrosine 2,3-aminomutase Tyrosine 2,3-aminomutase	Vol. 1, Page 568
5.4.3.6	Tyrosine 2,3-aminomutase Tyrosine 2,3-aminomutase	Vol. 1, Page 568
1.14.16.2	Tyrosine 3-hydroxylase Tyrosine 3-monooxygenase	
1.14.16.2	Tyrosine 3-monooxygenase Tyrosine 3-monooxygenase	
2.7.1.124	[Tyrosine 3-monooxygenase] kinase [Tyrosine 3-monooxygenase] kinase	
2.6.1.5	L-Tyrosine:2-oxoglutarate aminotransferase Tyrosine transaminase	

EC-No.	Synonym Recommended Name	Vol., Page
6.3.2.24	L-Tyrosine:L-arginine ligase (AMP-forming) Tyrosine-arginine ligase	Vol. 2, Page 521
6.1.1.1	L-Tyrosine:tRNA^{Tyr} ligase (AMP-forming) Tyrosine-tRNA ligase	Vol. 2, Page 1
2.6.1.5	L-Tyrosine aminotransferase Tyrosine transaminase	
2.6.1.5	Tyrosine aminotransferase Tyrosine transaminase	
4.3.1.5	L-tyrosine ammonia-lyase phenylalanine ammonia-lyase	
4.3.1.5	tyrosine ammonia-lyase phenylalanine ammonia-lyase	
4.1.1.25	L-(-)-Tyrosine apodecarboxylase Tyrosine decarboxylase	
4.1.1.25	L-Tyrosine carboxy-lyase Tyrosine decarboxylase	
3.4.16.3	tyrosine carboxypeptidase tyrosine carboxypeptidase	
4.1.1.25	L-Tyrosine decarboxylase Tyrosine decarboxylase	
4.1.1.25	Tyrosine decarboxylase Tyrosine decarboxylase	
2.8.2.9	Tyrosine ester sulfotransferase Tyrosine-ester sulfotransferase	
1.14.16.2	L-Tyrosine hydroxylase Tyrosine 3-monooxygenase	
1.14.16.2	Tyrosine hydroxylase Tyrosine 3-monooxygenase	
1.11.1.8	Tyrosine iodinase Iodide peroxidase	
2.7.1.112	Tyrosine kinase Protein-tyrosine kinase	
2.7.1.112	Tyrosine kinase lck Protein-tyrosine kinase	
2.7.1.112	Tyrosine kinase p56lck Protein-tyrosine kinase	

EC-No.	Synonym Recommended Name	Vol., Page
2.8.2.9	L-Tyrosine methyl ester sulfotransferase Tyrosine-ester sulfotransferase	
1.14.13.41	Tyrosine N-hydroxylase Tyrosine N-monooxygenase	
1.14.13.41	Tyrosine N-monooxygenase Tyrosine N-monooxygenase	
3.1.3.48	tyrosine O-phosphate phosphatase protein-tyrosine-phosphatase	
4.1.99.2	L-tyrosine phenol-lyase tyrosine phenol-lyase	
4.1.99.2	tyrosine phenol-lyase tyrosine phenol-lyase	
4.1.99.2	L-tyrosine phenol-lyase (deaminating) tyrosine phenol-lyase	
2.7.1.112	Tyrosine phosphokinase Protein-tyrosine kinase	
2.7.1.112	Tyrosine protein kinase Protein-tyrosine kinase	
2.7.1.112	Tyrosine protein kinase p56lck Protein-tyrosine kinase	
2.6.1.5	Tyrosine transaminase Tyrosine transaminase	
6.1.1.1	Tyrosine translase Tyrosine-tRNA ligase	Vol. 2, Page 1
6.1.1.1	Tyrosine tRNA synthetase Tyrosine-tRNA ligase	Vol. 2, Page 1
3.4.17.17	Tyrosinotubulin carboxypeptidase Tubulinyl-Tyr carboxypeptidase	
6.3.2.24	Tyrosyl-arginine synthase Tyrosine-arginine ligase	Vol. 2, Page 521
6.1.1.1	Tyrosyl-transfer ribonucleate synthetase Tyrosine-tRNA ligase	Vol. 2, Page 1
6.1.1.1	Tyrosyl-transfer ribonucleic acid synthetase Tyrosine-tRNA ligase	Vol. 2, Page 1
6.1.1.1	Tyrosyl-transfer RNA synthetase Tyrosine-tRNA ligase	Vol. 2, Page 1

EC-No.	Synonym Recommended Name	Vol., Page
6.1.1.1	Tyrosyl-tRNA ligase Tyrosine-tRNA ligase	Vol. 2, Page 1
6.1.1.1	Tyrosyl-tRNA synthetase Tyrosine-tRNA ligase	Vol. 2, Page 1
2.7.1.112	Tyrosylprotein kinase Protein-tyrosine kinase	
3.1.3.48	tyrosylprotein phosphatase protein-tyrosine-phosphatase	
2.8.2.20	Tyrosylprotein sulfotransferase Protein-tyrosine sulfotransferase	
3.4.17.17	Tyrosyltubulin carboxypeptidase Tubulinyl-Tyr carboxypeptidase	
6.3.2.25	Tyrosyltubulin ligase Tubulin-tyrosine ligase	Vol. 2, Page 524
6.3.4.6	UALase Urea carboxylase	Vol. 2, Page 603
2.4.2.16	UAR phosphorylase Urate-ribonucleotide phosphorylase	
1.10.2.2	Ubihydroquinol:cytochrome c oxidoreductase Ubiquinol-cytochrome-c reductase	
1.10.2.2	Ubiquinol-cytochrome-c reductase Ubiquinol-cytochrome-c reductase	
1.10.2.2	Ubiquinol-cytochrome c-2 oxidoreductase Ubiquinol-cytochrome-c reductase	
1.10.2.2	Ubiquinol-cytochrome c_1 oxidoreductase Ubiquinol-cytochrome-c reductase	
1.10.2.2	Ubiquinol-cytochrome c_2 reductase Ubiquinol-cytochrome-c reductase	
1.10.2.2	Ubiquinol-cytochrome c oxidoreductase Ubiquinol-cytochrome-c reductase	
1.10.2.2	Ubiquinol:ferricytochrome-c oxidoreductase Ubiquinol-cytochrome-c reductase	
1.10.2.2	Ubiquinone-cytochrome b-c_1 oxidoreductase Ubiquinol-cytochrome-c reductase	
1.10.2.2	Ubiquinone-cytochrome c oxidoreductase Ubiquinol-cytochrome-c reductase	

EC-No.	Synonym Recommended Name	Vol., Page
1.10.2.2	Ubiquinone-cytochrome c reductase Ubiquinol-cytochrome-c reductase	
1.6.5.3	Ubiquinone reductase NADH dehydrogenase (ubiquinone)	
6.3.2.19	Ubiquitin-activating enzyme Ubiquitin-protein ligase	Vol. 2, Page 506
3.1.2.15	Ubiquitin-C-terminal-thiolester hydrolase Ubiquitin thiolesterase	
6.3.2.21	Ubiquitin-calmodulin ligase Ubiquitin-calmodulin ligase	Vol. 2, Page 513
3.4.19.12	ubiquitin-carboxyl-terminal hydrolase PGP-9.5 ubiquitinyl hydrolase I	
6.3.2.19	Ubiquitin-protein ligase Ubiquitin-protein ligase	Vol. 2, Page 506
6.3.2.19	Ubiquitin-protein synthetase Ubiquitin-protein ligase	Vol. 2, Page 506
6.3.2.19	Ubiquitin:protein-lysine N-ligase (AMP-forming) Ubiquitin-protein ligase	Vol. 2, Page 506
3.1.2.15	Ubiquitin C-terminal hydrolase Ubiquitin thiolesterase	
3.4.19.12	ubiquitin C-terminal hydrolase ubiquitinyl hydrolase I	
3.4.19.12	ubiquitin C-terminal hydrolase (Aplysia californica gene Ap-uch) ubiquitinyl hydrolase I	
3.1.2.15	Ubiquitin carboxy-terminal esterase Ubiquitin thiolesterase	
3.1.2.15	Ubiquitin carboxy-terminal hydrolase Ubiquitin thiolesterase	
3.4.19.12	ubiquitin carboxy-terminal hydrolase ubiquitinyl hydrolase I	
3.4.19.12	ubiquitin carboxyl-terminal hydrolase ubiquitinyl hydrolase I	
3.1.2.15	Ubiquitin thiolesterase Ubiquitin thiolesterase	
3.4.19.12	ubiquitinyl hydrolase I ubiquitinyl hydrolase I	

EC-No.	Synonym Recommended Name	Vol., Page
6.3.2.21	Ubiquityl-calmodulin synthase Ubiquitin-calmodulin ligase	Vol. 2, Page 513
6.3.2.21	Ubiquityl-calmodulin synthetase Ubiquitin-calmodulin ligase	Vol. 2, Page 513
6.3.2.21	uCaM-synthetase Ubiquitin-calmodulin ligase	Vol. 2, Page 513
3.4.21.32	uca pugilator collagenolytic proteinase brachyurin	
3.4.19.12	UCH-1 ubiquitinyl hydrolase I	
3.4.19.12	UCH-L1 ubiquitinyl hydrolase I	
3.4.19.12	UCH-L2 ubiquitinyl hydrolase I	
3.4.19.12	UCH-L3 ubiquitinyl hydrolase I	
2.4.1.90	UDP-β-1,4-galactosyltransferase N-Acetyllactosamine synthase	
1.1.1.22	UDP-α-D-glucose:NAD oxidoreductase UDPglucose 6-dehydrogenase	
2.4.1.182	UDP-2,3-bis(3-hydroxytetradecanoyl)glucosamine:2,3-bis(3-hydroxytetra-decanoyl)-β-D-glucosaminyl-1-phosphate 2,3-bis(3-hydroxytetradecanoyl)-glucosaminyltransferase Lipid-A-disaccharide synthase	
1.1.1.136	UDP-2-acetamido-2-deoxy-D-glucose:NAD oxidoreductase UDP-N-acetylglucosamine 6-dehydrogenase	
2.6.1.34	UDP-2-acetamido-4-amino-2,4,6-trideoxyglucose:2-oxoglutarate amino-transferase UDP-2-acetamido-4-amino-2,4,6-trideoxyglucose transaminase	
2.6.1.34	UDP-2-acetamido-4-amino-2,4,6-trideoxyglucose transaminase UDP-2-acetamido-4-amino-2,4,6-trideoxyglucose transaminase	
2.4.1.41	UDP-acetylgalactosamine-glycoprotein acetylgalactosaminyltransferase Polypeptide N-acetylgalactosaminyltransferase	
2.4.1.41	UDP-acetylgalactosamine:peptide-N-galactosaminyltransferase Polypeptide N-acetylgalactosaminyltransferase	

EC-No.	Synonym Recommended Name	Vol., Page
2.7.8.15	UDP-acetylglucosamine-dolichol phosphate acetylglucosamine-1-phosphotransferase UDP-N-acetylglucosamine-dolichyl-phosphate N-acetylglucosamine-phosphotransferase	
2.7.8.15	UDP-acetylglucosamine-dolichol phosphate acetylglucosamine phosphotransferase UDP-N-acetylglucosamine-dolichyl-phosphate N-acetylglucosamine-phosphotransferase	
2.7.8.17	UDP-acetylglucosamine:lysosomal enzyme N-acetylglucosamine-1-phosphotransferase UDP-N-acetylglucosamine-lysosomal-enzyme N-acetylglucosamine-phosphotransferase	
6.3.2.8	UDP-acetylmuramoyl-L-alanine synthetase UDP-N-acetylmuramate-alanine ligase	Vol. 2, Page 442
5.1.3.5	UDP-arabinose 4-epimerase UDParabinose 4-epimerase	Vol. 1, Page 129
2.4.2.34	UDP-arabinose:indol-3-ylacetyl-myo-inositol arabinosyl transferase Indolylacetylinositol arabinosyltransferase	
5.4.99.9	UDP-D-galactopyranose furanomutase UDPgalactopyranose mutase	Vol. 1, Page 635
5.1.3.2	UDP-D-galactose 4-epimerase UDPglucose 4-epimerase	Vol. 1, Page 97
2.4.1.133	UDP-D-galactose:D-xylose galactosyltransferase Xylosylprotein 4-β-galactosyltransferase	
2.4.1.123	UDP-D-galactose:inositol galactosyltransferase Inositol 1-α-galactosyltransferase	
2.4.1.133	UDP-D-galactose:xylose galactosyltransferase Xylosylprotein 4-β-galactosyltransferase	
2.4.1.43	UDP-D-galacturonate:1,4-α-poly-D-galacturonate 4-α-D-galacturonosyltransferase Polygalacturonate 4-α-galacturonosyltransferase	
4.1.1.67	UDP-D-galacturonate carboxy-lyase UDPgalacturonate decarboxylase	
5.1.3.6	UDP-D-galacturonic acid 4-epimerase UDPglucuronate 4-epimerase	Vol. 1, Page 132
4.2.1.76	UDP-D-glucose-4,6-hydrolyase UDPglucose 4,6-dehydratase	

EC-No.	Synonym Recommended Name	Vol., Page
4.2.1.76	UDP-D-glucose 5,6-dehydratase UDPglucose 4,6-dehydratase	
2.4.1.188	UDP-D-glucose:N-acetylglucosaminyl pyrophosphorylundecaprenol glucosyltransferase N-Acetylglucosaminyldiphosphoundecaprenol glucosyltransferase	
2.4.1.53	UDP-D-glucose:polyribitol phosphate glucosyl transferase Poly(ribitol-phosphate) β-glucosyltransferase	
1.1.1.22	UDP-D-glucose dehydrogenase UDPglucose 6-dehydrogenase	
2.4.1.53	UDP-D-glucose polyribitol phosphate glucosyl transferase Poly(ribitol-phosphate) β-glucosyltransferase	
2.7.7.44	UDP-D-glucuronic acid pyrophosphorylase Glucuronate-1-phosphate uridylyltransferase	
2.7.8.15	UDP-D-N-acetylglucosamine N-acetylglucosamine 1-phosphate transferase UDP-N-acetylglucosamine-dolichyl-phosphate N-acetylglucosamine-phosphotransferase	
5.1.3.5	UDP-D-xylose 4-epimerase UDParabinose 4-epimerase	Vol. 1, Page 129
2.4.2.24	UDP-D-xylose:1,4-β-D-xylan 4-β-D-xylosyltransferase 1,4-β-D-Xylan synthase	
2.4.2.26	UDP-D-xylose:core protein β-D-xylosyltransferase Protein xylosyltransferase	
2.4.2.26	UDP-D-xylose:core protein xylosyltransferase Protein xylosyltransferase	
2.4.2.32	UDP-D-xylose:dolichyl-phosphate D-xylosyltransferase Dolichyl-phosphate D-xylosyltransferase	
2.4.2.35	UDP-D-xylose:flavonol-3-O-glycoside D-xylosyltransferase Flavonol-3-O-glycoside xylosyltransferase	
2.4.2.26	UDP-D-xylose:protein β-D-xylosyltransferase Protein xylosyltransferase	
2.4.2.26	UDP-D-xylose:proteoglycan core protein β-D-xylosyltransferase Protein xylosyltransferase	
2.4.1.169	UDP-D-xylose:xyloglucan 1,6-α-D-xylosyltransferase Xyloglucan 6-xylosyltransferase	
2.4.1.204	UDP-D-xylose:zeatin O-β-D-xylosyltransferase Zeatin O-β-D-xylosyltransferase	

EC-No.	Synonym Recommended Name	Vol., Page
2.4.1.151	UDP-Gal:β-D-Gal(1,4)-D-GlcNAc α(1,3)-galactosyltransferase N-Acetyllactosaminide α-1,3-galactosyltransferase	
2.4.1.151	UDP-Gal:Galβ1→4GlcNAc-R α1→3-galactosyltransferase N-Acetyllactosaminide α-1,3-galactosyltransferase	
2.4.1.90	UDP-Gal:N-acetylglucosamine β1-4-galactosyltransferase N-Acetyllactosamine synthase	
2.4.1.151	UDP-Gal:N-acetyllactosaminide α(1,3)-galactosyltransferase N-Acetyllactosaminide α-1,3-galactosyltransferase	
2.4.1.124	UDP-Gal:N-acetyllactosaminide α-1,3-D-galactosyltransferase N-Acetyllactosamine 3-α-galactosyltransferase	
2.4.1.151	UDP-Gal:N-acetyllactosaminide α-1,3-D-galactosyltransferase N-Acetyllactosaminide α-1,3-galactosyltransferase	
2.4.1.96	UDP-Gal:sn-glycero-3-phosphoric acid 1-α-galactosyl-transferase Sn-Glycerol-3-phosphate 1-galactosyltransferase	
5.4.99.9	UDP-galactopyranose mutase UDPgalactopyranose mutase	Vol. 1, Page 635
2.4.1.137	UDP-galactose, sn-3-glycerol phosphate:1→2' galactosyltransferase Sn-Glycerol-3-phosphate 2-α-galactosyltransferase	
2.4.1.90	UDP-galactose-acetylglucosamine galactosyltransferase N-Acetyllactosamine synthase	
2.4.1.124	UDP-galactose-acetyllactosamine α-D-galactosyltransferase N-Acetyllactosamine 3-α-galactosyltransferase	
2.4.1.151	UDP-galactose-acetyllactosamine α-D-galactosyltransferase N-Acetyllactosaminide α-1,3-galactosyltransferase	
2.4.1.46	UDP-galactose-diacylglyceride galactosyltransferase 1,2-Diacylglycerol 3-β-galactosyltransferase	
2.4.1.22	UDP-galactose-glucose galactosyltransferase Lactose synthase	
2.4.1.38	UDP-galactose-glycoprotein galactosyltransferase β-N-Acetylglucosaminylglycopeptide β-1,4-galactosyltransferase	
2.4.1.62	UDP-galactose-GM2 galactosyltransferase Ganglioside galactosyltransferase	
2.4.1.62	UDP-galactose-GM2 ganglioside galactosyltransferase Ganglioside galactosyltransferase	
2.4.1.90	UDP-galactose-N-acetylglucosamine β-1,4-galactosyltransferase N-Acetyllactosamine synthase	

EC-No.	Synonym Recommended Name	Vol., Page
2.4.1.90	UDP-galactose-N-acetylglucosamine galactosyltransferase N-Acetyllactosamine synthase	
5.1.3.2	UDP-galactose 4-epimerase UDPglucose 4-epimerase	Vol. 1, Page 97
2.4.1.46	UDP-galactose:diacylglycerol galactosyltransferase 1,2-Diacylglycerol 3-β-galactosyltransferase	
2.4.1.44	UDP-galactose:lipopolysaccharide α,3-galactosyltransferase Lipopolysaccharide 3-α-galactosyltransferase	
2.4.1.62	UDP-galactose:N-acetylgalactosaminyl-(N-acetylneuraminyl) galactosyl-glucosyl-ceramide galactosyltransferase Ganglioside galactosyltransferase	
2.4.1.90	UDP-galactose:N-acetylglucosaminide β1-4-galactosyltransferase N-Acetyllactosamine synthase	
2.4.1.44	UDP-galactose:polysaccharide galactosyltransferase Lipopolysaccharide 3-α-galactosyltransferase	
2.4.1.137	UDP-galactose:sn-glycerol-3-phosphate-2-D-galactosyl transferase Sn-Glycerol-3-phosphate 2-α-galactosyltransferase	
2.4.1.90	UDP-galactose N-acetylglucosamine β-4-galactosyltransferase N-Acetyllactosamine synthase	
5.1.3.6	UDP-galacturonate 4-epimerase UDPglucuronate 4-epimerase	Vol. 1, Page 132
4.1.1.67	UDP-galacturonic acid decarboxylase UDPgalacturonate decarboxylase	
2.4.1.40	UDP-GalNAc:Fucα1$\rightarrow$2Galα1$\rightarrow$3-N-acetylgalactosaminyltransferase Fucosylgalactose α-N-acetylgalactosaminyltransferase	
2.4.1.41	UDP-GalNAc:polypeptide N-acetylgalactosaminyl transferase Polypeptide N-acetylgalactosaminyltransferase	
2.4.1.41	UDP-GalNAc:polypeptide N-acetylgalactosaminyltransferase Polypeptide N-acetylgalactosaminyltransferase	
1.1.1.158	UDP-GlcNAc-enoylpyruvate reductase UDP-N-acetylmuramate dehydrogenase	
5.1.3.14	UDP-GlcNAc 2'-epimerase UDP-N-acetylglucosamine 2-epimerase	Vol. 1, Page 154
2.7.8.15	UDP-GlcNAc:dolichyl-phosphate GlcNAc-1-phosphate transferase UDP-N-acetylglucosamine-dolichyl-phosphate N-acetylglucosaminephosphotransferase	

EC-No.	Synonym Recommended Name Vol., Page
2.4.1.141	UDP-GlcNAc:dolichyl-pyrophosphoryl-GlcNAc GlcNAc transferase N-Acetylglucosaminyldiphosphodolichol N-acetylglucosaminyltransferase
2.4.1.150	UDP-GlcNAc:Gal-R, β-D-6-N-acetylglucosaminyltransferase N-Acetyllactosaminide β-1,6-N-acetylglucosaminyltransferase
2.4.1.149	UDP-GlcNAc:Galβ1→4GlcNAcβ-Rβ1→3-N-acetylglucosaminyltransferase N-Acetyllactosaminide β-1,3-N-acetylglucosaminyltransferase
2.4.1.149	UDP-GlcNAc:GalR, β-D-3-N-acetylglucosaminyltransferase N-Acetyllactosaminide β-1,3-N-acetylglucosaminyltransferase
2.7.8.17	UDP-GlcNAc:glycoprotein N-acetylglucosamine-1-phosphotransferase UDP-N-acetylglucosamine-lysosomal-enzyme N-acetylglucosamine-phosphotransferase
2.7.8.17	UDP-GlcNAc:lysosomal enzyme N-acetylglucosamine-1-phosphotransferase UDP-N-acetylglucosamine-lysosomal-enzyme N-acetylglucosamine-phosphotransferase
2.4.1.143	UDP-GlcNAc:mannoside α1-6 acetylglucosaminyltransferase α-1,6-Mannosyl-glycoprotein β-1,2-N-acetylglucosaminyltransferase
2.4.1.197	UDP-GlcNAc:oligosaccharide β-N-acetylglucosaminyltransferase High-mannose-oligosaccharide β-1,4-N-acetyl-glucosaminyltransferase
1.1.1.136	UDP-GLcNAc dehydrogenase UDP-N-acetylglucosamine 6-dehydrogenase
2.7.7.23	UDP-GlcNAc pyrophosphorylase UDP-N-acetylglucosamine pyrophosphorylase
2.4.1.34	UDP-glucose-β-glucan glucosyltransferase 1,3-β-Glucan synthase
2.4.1.34	UDP-glucose-1,3-β-glucan glucosyltransferase 1,3-β-Glucan synthase
2.4.1.12	UDP-glucose-1,4-β-glucan glucosyltransferase Cellulose synthase (UDP-forming)
2.4.1.158	UDP-glucose-13-hydroxydocosanoate glucosyltransferase 13-Hydroxydocosanoate 13-β-glucosyltransferase
2.4.1.66	UDP-glucose-collagen glucosyltransferase Procollagen glucosyltransferase
2.4.1.157	UDP-glucose-diacylglycerol glucosyltransferase 1,2-Diacylglycerol 3-glucosyltransferase
2.4.1.13	UDP-glucose-fructose glucosyltransferase Sucrose synthase

EC-No.	Synonym Recommended Name	Vol., Page
2.4.1.11	UDP-glucose-glycogen glucosyltransferase Glycogen (starch) synthase	
2.4.1.126	UDP-glucose-hydroxycinnamate glucosyltransferase Hydroxycinnamate 4-β-glucosyltransferase	
2.4.1.85	UDP-glucose-*p*-hydroxymandelonitrile glucosyltransferase Cyanohydrin β-glucosyltransferase	
2.4.1.112	UDP-glucose-protein glucosyltransferase α-1,4-Glucan-protein synthase (UDP-forming)	
2.4.1.160	UDP-glucose-pyridoxine glucosyltransferase Pyridoxine 5'-O-β-D-glucosyltransferase	
2.4.1.173	UDP-glucose-sterol β-glucosyltransferase Sterol 3β-glucosyltransferase	
2.4.1.173	UDP-glucose-sterol glucosyltransferase Sterol 3β-glucosyltransferase	
2.4.1.34	UDP-glucose:(1,3)β-glucan synthase 1,3-β-Glucan synthase	
2.4.1.208	UDP-glucose:1,2-diacyl-3-O-(α-D-glucopyranosyl)-sn-glycerol (1->2) glucosyltransferase Diglucosyl diacylglycerol synthase	
2.4.1.157	UDP-glucose:1,2-diacylglycerol glucosyltransferase 1,2-Diacylglycerol 3-glucosyltransferase	
2.4.1.12	UDP-glucose:1,4-β-D-glucan 4-β-D-glucosyltransferase Cellulose synthase (UDP-forming)	
2.4.1.158	UDP-glucose:13-hydroxydocosanoic acid glucosyltransferase 13-Hydroxydocosanoate 13-β-glucosyltransferase	
2.4.1.194	UDP-glucose:4-(β-D-glucopyranosyloxy)benzoic acid glucosyltransferase 4-Hydroxybenzoate 4-O-β-D-glucosyltransferase	
2.4.1.115	UDP-glucose:anthocyanidin/flavonol 3-O-glucosyltransferase Anthocyanidin 3-O-glucosyltransferase	
2.4.1.80	UDP-glucose:ceramide glucosyltransferase Ceramide glucosyltransferase	
2.4.1.209	UDP-glucose:cis-*p*-coumarate β-D-glucosyltransferase cis-*p*-coumarate glucosyltransferase	
2.4.1.115	UDP-glucose:cyanidin-3-O-glucosyltransferase Anthocyanidin 3-O-glucosyltransferase	

EC-No.	Synonym Recommended Name	Vol., Page
2.4.1.117	UDP-glucose:dolichol phosphate glucosyltransferase Dolichyl-phosphate β-glucosyltransferase	
2.4.1.117	UDP-glucose:dolicholphosphoryl glucosyltransferase Dolichyl-phosphate β-glucosyltransferase	
2.4.1.117	UDP-glucose:dolichyl monophosphate glucosyltransferase Dolichyl-phosphate β-glucosyltransferase	
2.4.1.117	UDP-glucose:dolichyl phosphate glucosyltransferase Dolichyl-phosphate β-glucosyltransferase	
2.4.1.117	UDP-glucose:dolichylphosphate glucosyltransferase Dolichyl-phosphate β-glucosyltransferase	
2.4.1.91	UDP-glucose:flavonol 3-O-glucosyltransferase Flavonol 3-O-glucosyltransferase	
2.7.8.19	UDP-glucose:glycoprotein glucose-1-phosphotransferase UDPglucose-glycoprotein glucosephosphotransferase	
2.4.1.121	UDP-glucose:indol-3-ylacetate glucosyl-transferase Indole-3-acetate β-glucosyltransferase	
2.4.1.121	UDP-glucose:indol-3-ylacetate glucosyltransferase Indole-3-acetate β-glucosyltransferase	
2.4.1.196	UDP-glucose:nicotinic acid-N-glucosyltransferase Nicotinate glucosyltransferase	
2.4.1.104	UDP-glucose:o-dihydroxycoumarin glucosyltransferase *o*-Dihydroxycoumarin 7-O-glucosyltransferase	
2.4.1.160	UDP-glucose:pyridoxine 5'-O-β-glucosyltransferase Pyridoxine 5'-O-β-D-glucosyltransferase	
2.3.1.152	UDP-glucose:sinapic acid glucosyltransferase Alcohol O-cinnamoyltransferase	
2.4.1.111	UDP-glucose coniferyl alcohol glucosyltransferase Coniferyl-alcohol glucosyltransferase	
2.4.1.117	UDP-glucose dolichyl-phosphate glucosyltransferase Dolichyl-phosphate β-glucosyltransferase	
5.1.3.2	UDP-glucose epimerase UDPglucose 4-epimerase	Vol. 1, Page 97
2.4.1.35	UDP-glucose glucosyltransferase Phenol β-glucosyltransferase	
2.4.1.112	UDP-glucose protein transglucosylase α-1,4-Glucan-protein synthase (UDP-forming)	

EC-No.	Synonym Recommended Name	Vol., Page
2.7.7.9	UDP-glucose pyrophosphorylase UTP-glucose-1-phosphate uridylyltransferase	
2.4.1.81	UDP-glucosyltransferase Flavone 7-O-β-glucosyltransferase	
2.4.1.17	UDP-glucuronate-4-hydroxybiphenyl glucuronosyltransferase Glucuronosyltransferase	
2.4.1.17	UDP-glucuronate-bilirubin glucuronyltransferase Glucuronosyltransferase	
2.4.1.191	UDP-glucuronate:luteolin 7-O-diglucuronide-glucuronosyltransferase Luteolin-7-O-diglucuronide 4'-O-glucuronosyltransferase	
2.4.1.190	UDP-glucuronate:luteolin 7-O-glucuronide-glucuronosyltransferase Luteolin-7-O-glucuronide 7-O-glucuronosyltransferase	
4.1.1.35	UDP-glucuronate decarboxylase UDPglucuronate decarboxylase	
2.7.7.44	UDP-glucuronate pyrophosphorylase Glucuronate-1-phosphate uridylyltransferase	
5.1.3.12	UDP-glucuronic acid 5'-epimerase UDPglucuronate 5'-epimerase	Vol. 1, Page 150
2.7.7.44	UDP-glucuronic acid pyrophosphorylase Glucuronate-1-phosphate uridylyltransferase	
2.4.1.17	UDP-glucuronosyltransferase Glucuronosyltransferase	
2.4.1.17	UDP-glucuronyltransferase Glucuronosyltransferase	
3.1.3.42	UDP-glycogen glucosyltransferase phosphatase [glycogen-synthase-D] phosphatase	
2.4.1.11	UDP-glycogen synthase Glycogen (starch) synthase	
5.1.3.5	UDP-L-arabinose 4-epimerase UDParabinose 4-epimerase	Vol. 1, Page 129
2.4.2.34	UDP-L-arabinose:indol-3-ylacetyl-myo-inositol L-arabinosyltransferase Indolylacetylinositol arabinosyltransferase	
2.4.1.159	UDP-L-rhamnose:flavonol-3-O-D-glucoside L-rhamnosyltransferase Flavonol-3-O-glucoside L-rhamnosyltransferase	
6.3.2.9	UDP-Mur-Nac-L-Ala:L-Glu ligase UDP-N-acetylmuramoylalanine-D-glutamate ligase	Vol. 2, Page 452

EC-No.	Synonym Recommended Name	Vol., Page
2.7.8.13	UDP-MurNAc-Ala-γDGlu-Lys-DAla-DAla:undecaprenylphosphate transferase Phospho-N-acetylmuramoyl-pentapeptide-transferase	
2.7.8.13	UDP-MurNAc-L-Ala-D-γ-Glu-L-Lys-D-Ala-D-Ala:C_{55}-isoprenoid alcohol transferase Phospho-N-acetylmuramoyl-pentapeptide-transferase	
6.3.2.15	UDP-MurNAc-L-Ala-D-Glu-meso-Dpm:D-Ala-D-Ala ligase UDP-N-acetylmuramoylalanyl-D-glutamyl-2,6-diaminopimelate-D-alanyl-D-alanine ligase	Vol. 2, Page 480
6.3.2.8	UDP-MurNAc:L-alanine ligase UDP-N-acetylmuramate-alanine ligase	Vol. 2, Page 442
2.4.1.41	UDP-N-acetyl-α-D-galactosamine:polypeptide N-acetylgalactosaminyltransferase Polypeptide N-acetylgalactosaminyltransferase	
2.4.1.180	UDP-N-acetyl-β-D-mannosaminouronate:lipopolysaccharide N-acetyl-β-D-mannosaminouronosyltransferase Lipopolysaccharide N-acetylmannosaminouronosyltransferase	
2.4.1.154	UDP-N-acetyl-D-galactosamine-globotriosylceramide β-1,6-N-acetylgalactosaminyltransferase Globotriosylceramide β-1,6-N-acetylgalactosaminyltransferase	
2.4.1.92	UDP-N-acetyl-D-galactosamine:(N-acetylneuraminyl)-D-galactosyl-D-glucosylceramide N-acetyl-D-galactosaminyltransferase (N-Acetylneuraminyl)-galactosylglucosylceramide N-acetylgalactosaminyltransferase	
2.4.1.40	UDP-N-acetyl-D-galactosamine:α-L-fucosyl-(1,2)-D-galactose 3-N-acetyl-D-galactosaminyltransferase Fucosylgalactose α-N-acetylgalactosaminyltransferase	
2.4.1.40	UDP-N-acetyl-D-galactosamine:α-L-fucosyl-1,2-D-galactose 3-N-acetyl-D-galactosaminyltransferase Fucosylgalactose α-N-acetylgalactosaminyltransferase	
2.4.1.79	UDP-N-acetyl-D-galactosamine:D-galactosyl-1,4-D-galactosyl-1,4-D-glucosylceramide β-N-acetyl-D-galactosaminyltransferase Galactosylgalactosylglucosylceramide β-D-acetylgalactosaminyltransferase	
2.4.1.174	UDP-N-acetyl-D-galactosamine:D-glucuronyl-1,3-β-D-galactosylproteoglycan β-1,4-N-acetylgalactosaminyltransferase Glucuronylgalactosylproteoglycan β-1,4-N-acetylgalactosaminyltransferase	
2.4.1.175	UDP-N-acetyl-D-galactosamine:D-glucuronyl-N-acetyl-1,3-β-D-galactosaminylproteoglycan β-1,4-N-acetylgalactosaminyltransferase Glucuronyl-N-acetylgalactosaminylproteoglycan β-1,4-N-acetylgalactosaminyltransferase	

EC-No.	Synonym Recommended Name	Vol., Page
2.4.1.88	UDP-N-acetyl-D-galactosamine:N-acetyl-D-galactosaminyl-1,3-D-galactosyl-1,4-D-galactosyl-1,4-D-glucosylceramide α-N-acetyl-D-galactosaminyltransferase Globoside α-N-acetylgalactosaminyltransferase	
2.4.1.165	UDP-N-acetyl-D-galactosamine:N-acetylneuraminyl-2,3-α-D-galactosyl-1,4-β-D-glucosylceramide β-1,4-N-acetylgalactosaminyltransferase N-Acetylneuraminylgalactosylglucosylceramide β-1,4-N-acetylgalactosaminyltransferase	
2.4.1.41	UDP-N-acetyl-D-galactosamine:polypeptide N-acetylgalactosaminyltransferase Polypeptide N-acetylgalactosaminyltransferase	
5.1.3.14	UDP-N-acetyl-D-glucosamine 2-epimerase UDP-N-acetylglucosamine 2-epimerase	Vol. 1, Page 154
5.1.3.7	UDP-N-acetyl-D-glucosamine 4-epimerase UDP-N-acetylglucosamine 4-epimerase	Vol. 1, Page 135
2.4.1.163	UDP-N-acetyl-D-glucosamine:β-D-galactosyl-1,4-N-acetyl-β-D-glucosaminyl-1,3-β-D-galactosyl-1,4-β-D-glucosylceramide β-1,3-acetylglucosaminyltransferase β-Galactosyl-N-acetylglucosaminylgalactosylglucosylceramide β-1,3-acetylglucosaminyltransferase	
2.4.1.149	UDP-N-acetyl-D-glucosamine:β-D-galactosyl-1,4-N-acetyl-D-glucosamine β-1,3-acetyl-D-glucosaminyltransferase N-Acetyllactosaminide β-1,3-N-acetylglucosaminyltransferase	
2.4.1.150	UDP-N-acetyl-D-glucosamine:β-D-galactosyl-1,4-N-acetyl-D-glucosaminide β-1,6-N-acetyl-D-glucosaminyltransferase N-Acetyllactosaminide β-1,6-N-acetylglucosaminyltransferase	
2.4.1.16	UDP-N-acetyl-D-glucosamine:chitin 4-β-N-acetylglucosaminyltransferase Chitin synthase	
2.4.1.206	UDP-N-acetyl-D-glucosamine:D-galactosyl-1,4-β-D-glucosylceramide β-1,3-acetylglucosaminyltransferase Lactosylceramide 1,3-N-acetyl-β-D-glucosaminyltransferase	
2.4.1.164	UDP-N-acetyl-D-glucosamine:D-galactosyl-1,4-N-acetyl-β-D-glucosaminyl-1,3-β-D-galactosyl-1,4-β-D-glucosylceramide β-1,6-N-acetylglucosaminyltransferase Galactosyl-N-acetylglucosaminylgalactosylglucosylceramide β-1,6-N-acetylglucosaminyltransferase	
2.7.8.15	UDP-N-acetyl-D-glucosamine:dolichol phosphate N-acetyl-D-glucosamine-1-phosphate transferase UDP-N-acetylglucosamine-dolichyl-phosphate N-acetylglucosaminephosphotransferase	

EC-No.	Synonym Recommended Name Vol., Page
2.4.1.153	UDP-N-acetyl-D-glucosamine:dolichyl-phosphate α-N-acetyl-D-glucosaminyltransferase Dolichyl-phosphate α-N-acetylglucosaminyltransferase
2.7.8.15	UDP-N-acetyl-D-glucosamine:dolichyl-phosphate N-acetyl-D-glucosamine-phosphotransferase UDP-N-acetylglucosamine-dolichyl-phosphate N-acetylglucosaminephosphotransferase
2.4.1.39	UDP-N-acetyl-D-glucosamine:estradiol-17α-3-D-glucuronoside 17α-N-acetylglucosaminyltransferase Steroid N-acetylglucosaminyltransferase
2.4.1.155	UDP-N-acetyl-D-glucosamine:glycoprotein (N-acetyl-D-glucosamine to α-D-mannosyl-1,3(6)-N-acetyl-β-D-glucosaminyl-R) β-1,6-N-acetylglucosaminyltransferase α-1,3(6)-Mannosylglycoprotein β-1,6-N-acetylglucosaminyltransferase
2.4.1.101	UDP-N-acetyl-D-glucosamine:glycoprotein (N-acetyl-D-glucosamine to α-D-mannosyl-1,3-(R^1)-β-D-mannosyl-R^2) β-1,2-N-acetyl-D-glucosaminyltransferase α-1,3-Mannosyl-glycoprotein β-1,2-N-acetylglucosaminyltransferase
2.4.1.145	UDP-N-acetyl-D-glucosamine:glycoprotein (N-acetyl-D-glucosamine to α-D-mannosyl-1,3- of R^1-α-D-mannosyl-1,3-(R^2)-β-D-mannosyl-R^3) β-1,4-N-acetyl-D-glucosaminyltransferase α-1,3-Mannosylglycoprotein β-1,4-N-acetylglucosaminyltransferase
2.4.1.143	UDP-N-acetyl-D-glucosamine:glycoprotein (N-acetyl-D-glucosamine to α-D-mannosyl-1,6-(R^1)-β-D-mannosyl-R^2) β-1,2-N-acetyl-D-glucosaminyltransferase α-1,6-Mannosyl-glycoprotein β-1,2-N-acetylglucosaminyltransferase
2.4.1.144	UDP-N-acetyl-D-glucosamine:glycoprotein (N-acetyl-D-glucosamine to β-D-mannosyl of R^1(R^2)-β-D-mannosyl-1,4-N-acetyl-β-D-glucosaminyl-R^3) β-1,4-N-acetyl-D-glucosaminyltransferase β-1,4-Mannosyl-glycoprotein β-1,4-N-acetylglucosaminyltransferase
2.4.1.201	UDP-N-acetyl-D-glucosamine:glycoprotein (N-acetyl-D-glucosamine to D-mannose of N-acetyl-β-D-glucosaminyl-1,6-β-D-(N-acetyl-D-glucosaminyl-1,2-)-β-D-mannosyl-R) β-1,4-N-acetyl-D-glucosaminyltransferase Mannosyl-glycoprotein β-1,4-N-acetylglucosaminyltransferase
2.4.1.197	UDP-N-acetyl-D-glucosamine:high-mannose-oligosaccharide β-1,4-N-acetylglucosaminyltransferase High-mannose-oligosaccharide β-1,4-N-acetyl-glucosaminyltransferase
2.4.1.56	UDP-N-acetyl-D-glucosamine:lipopolysaccharide N-acetyl-D-glucosaminyltransferase Lipopolysaccharide N-acetylglucosaminyltransferase

EC-No.	Synonym Recommended Name Vol., Page
2.7.8.17	UDP-N-acetyl-D-glucosamine:lysosomal-enzyme N-acetylglucosamine-phosphotransferase UDP-N-acetylglucosamine-lysosomal-enzyme N-acetylglucosamine-phosphotransferase
2.4.1.138	UDP-N-acetyl-D-glucosamine:mannotetraose α-N-acetyl-D-glucosaminyl-transferase Mannotetraose 2-α-N-acetylglucosaminyltransferase
2.4.1.141	UDP-N-acetyl-D-glucosamine:N-acetyl-D-glucosaminyl-diphosphodolichol N-acetyl-D-glucosaminyltransferase N-Acetylglucosaminyldiphosphodolichol N-acetylglucosaminyltransferase
1.1.1.136	UDP-N-acetyl-D-glucosamine:NAD^+ 6-oxidoreductase UDP-N-acetylglucosamine 6-dehydrogenase
2.4.1.146	UDP-N-acetyl-D-glucosamine:O-glycosyl-glycoprotein (N-acetyl-D-glucosamine to D-galactose of β-D-galactosyl-1,3-(N-acetyl-D-glucosaminyl-1,6)-N-acetyl-D-galactosaminyl-R) β-1,3-N-acetyl-D-glucosaminyltransferase β-1,3-Galactosyl-O-glycosyl-glycoprotein β-1,3-N-acetylglucosaminyl-transferase
2.4.1.102	UDP-N-acetyl-D-glucosamine:O-glycosyl-glycoprotein (N-acetyl-D-glucosamine to N-acetyl-D-galactosamine of β-D-galactosyl-1,3-N-acetyl-D-galactosaminyl-R) β-1,6-N-acetyl-D-glucosaminyltransferase β-1,3-Galactosyl-O-glycosyl-glycoprotein β-1,6-N-acetylglucosaminyl-transferase
2.4.1.147	UDP-N-acetyl-D-glucosamine:O-glycosyl-glycoprotein (N-acetyl-D-glucosamine to N-acetyl-D-galactosaminyl-R) β-1,3-N-acetyl-D-glucosaminyltransferase Acetylgalactosaminyl-O-glycosyl-glycoprotein β-1,3-N-acetylglucosaminyl-transferase
2.4.1.148	UDP-N-acetyl-D-glucosamine:O-oligosaccharide-glycoprotein (N-acetyl-D-glucosamine to N-acetyl-D-galactosamine of N-acetyl-β-D-glucosaminyl-1,3-N-acetyl-D-galactosaminyl-R) β-1,6-N-acetyl-D-glucosaminyltransferase Acetylgalactosaminyl-O-glycosyl-glycoprotein β-1,6-N-acetylglucosaminyl-transferase
2.4.1.198	UDP-N-acetyl-D-glucosamine:phosphatidylinositol N-acetyl-D-glucosaminyltransferase Phosphatidylinositol N-acetylglucosaminyltransferase
2.4.1.70	UDP-N-acetyl-D-glucosamine:poly(ribitol-phosphate) N-acetyl-D-glucosaminyltransferase Poly(ribitol-phosphate) N-acetylglucosaminyltransferase
2.4.1.94	UDP-N-acetyl-D-glucosamine:protein β-N-acetyl-D-glucosaminyltransferase Protein N-acetylglucosaminyltransferase

EC-No.	Synonym Recommended Name Vol., Page
2.4.1.187	UDP-N-acetyl-D-mannosamine:N-acetyl-β-D-glucosaminyldiphosphoundecaprenol β-1,4-N-acetylmannosaminyltransferase N-Acetylglucosaminyldiphosphoundecaprenol N-acetyl-β-D-mannosaminyltransferase
1.1.1.158	UDP-N-acetylenol-pyruvoylglucosamine reductase UDP-N-acetylmuramate dehydrogenase
1.1.1.158	UDP-N-acetylenolpyruvoylglucosamine reductase UDP-N-acetylmuramate dehydrogenase
2.8.2.7	UDP-N-acetylgalactosamine-4-sulfate sulfotransferase UDP-N-acetylgalactosamine-4-sulfate sulfotransferase
2.4.1.41	UDP-N-acetylgalactosamine-glycoprotein N-acetylgalactosaminyltransferase Polypeptide N-acetylgalactosaminyltransferase
2.4.1.41	UDP-N-acetylgalactosamine-protein N-acetylgalactosaminyltransferase Polypeptide N-acetylgalactosaminyltransferase
2.4.1.79	UDP-N-acetylgalactosamine:globotriaosylceramide β-3-N-acetylgalactosaminyltransferase Galactosylgalactosylglucosylceramide β-D-acetylgalactosaminyltransferase
2.4.1.41	UDP-N-acetylgalactosamine:κ-casein polypeptide N-acetylgalactosaminyltransferase Polypeptide N-acetylgalactosaminyltransferase
2.4.1.41	UDP-N-acetylgalactosamine:polypeptide N-acetylgalactosaminyltransferase Polypeptide N-acetylgalactosaminyltransferase
2.4.1.41	UDP-N-acetylgalactosamine:protein N-acetylgalactosaminyl transferase Polypeptide N-acetylgalactosaminyltransferase
2.4.1.92	UDP-N-acetylgalactosamine GM3 N-acetylgalactosaminyltransferase (N-Acetylneuraminyl)-galactosylglucosylceramide N-acetylgalactosaminyltransferase
2.4.1.153	UDP-N-acetylglucosamine-dolichol phosphate N-acetylglucosaminyltransferase Dolichyl-phosphate α-N-acetylglucosaminyltransferase
2.7.8.15	UDP-N-acetylglucosamine-dolichyl-phosphate N-acetylglucosamine-phosphotransferase UDP-N-acetylglucosamine-dolichyl-phosphate N-acetylglucosamine-phosphotransferase
1.1.1.158	UDP-N-acetylglucosamine-enoylpyruvate reductase UDP-N-acetylmuramate dehydrogenase
2.4.1.51	UDP-N-acetylglucosamine-glycoprotein N-acetylglucosaminyltransferase UDP-N-acetylglucosamine-glycoprotein N-acetylglucosaminyltransferase

EC-No.	Synonym Recommended Name	Vol., Page
2.4.1.56	UDP-N-acetylglucosamine-lipopolysaccharide N-acetylglucosaminyltransferase Lipopolysaccharide N-acetylglucosaminyltransferase	
2.7.8.17	UDP-N-acetylglucosamine-lysosomal-enzyme N-acetylglucosamine-phosphotransferase UDP-N-acetylglucosamine-lysosomal-enzyme N-acetylglucosamine-phosphotransferase	
2.5.1.7	UDP-N-acetylglucosamine 1-carboxyvinyl-transferase UDP-N-acetylglucosamine 1-carboxyvinyltransferase	
2.5.1.7	UDP-N-acetylglucosamine 1-carboxyvinyltransferase UDP-N-acetylglucosamine 1-carboxyvinyltransferase	
5.1.3.14	UDP-N-acetylglucosamine 2'-epimerase UDP-N-acetylglucosamine 2-epimerase	Vol. 1, Page 154
5.1.3.14	UDP-N-acetylglucosamine 2-epimerase UDP-N-acetylglucosamine 2-epimerase	Vol. 1, Page 154
5.1.3.7	UDP-N-acetylglucosamine 4-epimerase UDP-N-acetylglucosamine 4-epimerase	Vol. 1, Page 135
1.1.1.136	UDP-N-acetylglucosamine 6-dehydrogenase UDP-N-acetylglucosamine 6-dehydrogenase	
2.4.1.155	UDP-N-acetylglucosamine:α-mannoside-β-1,6-N-acetylglucosaminyltransferase α-1,3(6)-Mannosylglycoprotein β-1,6-N-acetylglucosaminyltransferase	
2.7.8.17	UDP-N-acetylglucosamine:glycoprotein N-acetylglucosamine-1-phosphotransferase UDP-N-acetylglucosamine-lysosomal-enzyme N-acetylglucosamine-phosphotransferase	
2.7.8.17	UDP-N-acetylglucosamine:glycoprotein N-acetylglucosaminyl-1-phosphotransferase UDP-N-acetylglucosamine-lysosomal-enzyme N-acetylglucosamine-phosphotransferase	
2.7.8.17	UDP-N-acetylglucosamine:lysosomal enzyme N-acetylglucosamine-1-phosphotransferase UDP-N-acetylglucosamine-lysosomal-enzyme N-acetylglucosamine-phosphotransferase	
2.3.1.129	UDP-N-acetylglucosamine acyltransferase Acyl-[acyl-carrier-protein]-UDP-N-acetylglucosamine O-acyltransferase	
2.5.1.7	UDP-N-acetylglucosamine enoylpyruvyltransferase UDP-N-acetylglucosamine 1-carboxyvinyltransferase	

EC-No.	Synonym Recommended Name	Vol., Page
2.7.7.23	UDP-N-acetylglucosamine pyrophosphorylase UDP-N-acetylglucosamine pyrophosphorylase	
2.4.1.101	UDP-N-acetylglucosaminyl: α-1,3-D-mannoside-β-1,2-N-acetylglucosaminyltransferase I α-1,3-Mannosyl-glycoprotein β-1,2-N-acetylglucosaminyltransferase	
2.4.1.101	UDP-N-acetylglucosaminyl:α-3-D-mannoside β-1,2-N-acetylglucosaminyltransferase I α-1,3-Mannosyl-glycoprotein β-1,2-N-acetylglucosaminyltransferase	
2.4.1.187	UDP-N-Acetylmannosamine:N-acetylglucosaminyl diphosphorylundecaprenol N-acetylmannosaminyltransferase N-Acetylglucosaminyldiphosphoundecaprenol N-acetyl-β-D-mannosaminyltransferase	
6.3.2.8	UDP-N-acetylmuramate-alanine ligase UDP-N-acetylmuramate-alanine ligase	Vol. 2, Page 442
6.3.2.8	UDP-N-acetylmuramate:L-alanine ligase (ADP-forming) UDP-N-acetylmuramate-alanine ligase	Vol. 2, Page 442
1.1.1.158	UDP-N-acetylmuramate:$NADP^+$ oxidoreductase UDP-N-acetylmuramate dehydrogenase	
1.1.1.158	UDP-N-acetylmuramate dehydrogenase UDP-N-acetylmuramate dehydrogenase	
6.3.2.9	UDP-N-acetylmuramoyl-L-alanine:D-glutamate ligase UDP-N-acetylmuramoylalanine-D-glutamate ligase	Vol. 2, Page 452
6.3.2.9	UDP-N-acetylmuramoyl-L-alanine:glutamate ligase (ADP-forming) UDP-N-acetylmuramoylalanine-D-glutamate ligase	Vol. 2, Page 452
6.3.2.8	UDP-N-acetylmuramoyl-L-alanine synthetase UDP-N-acetylmuramate-alanine ligase	Vol. 2, Page 442
6.3.2.7	UDP-N-acetylmuramoyl-L-alanyl-D-glutamate-lysine ligase UDP-N-acetylmuramoyl-L-alanyl-D-glutamate-lysine ligase	Vol. 2, Page 439
6.3.2.7	UDP-N-acetylmuramoyl-L-alanyl-D-glutamate:L-lysine ligase (ADP-forming) UDP-N-acetylmuramoyl-L-alanyl-D-glutamate-lysine ligase	Vol. 2, Page 439
6.3.2.13	UDP-N-acetylmuramoyl-L-alanyl-D-glutamate:meso-2,6-diaminoheptanedioate ligase (ADP-forming) UDP-N-acetylmuramoylalanyl-D-glutamate-2,6-diaminopimelate ligase	Vol. 2, Page 473
6.3.2.9	UDP-N-acetylmuramoyl-L-alanyl-D-glutamate synthetase UDP-N-acetylmuramoylalanine-D-glutamate ligase	Vol. 2, Page 452

EC-No.	Synonym Recommended Name	Vol., Page
6.3.2.10	UDP-N-acetylmuramoyl-L-alanyl-D-glutamyl-L-lysine:alanyl-D-alanine ligase (ADP forming) UDP-N-acetylmuramoylalanyl-D-glutamyl-lysine-D-alanyl-D-alanine ligase	Vol. 2, Page 458
6.3.2.7	UDP-N-acetylmuramoyl-L-alanyl-D-glutamyl-L-lysine synthetase UDP-N-acetylmuramoyl-L-alanyl-D-glutamate-lysine ligase	Vol. 2, Page 439
6.3.2.10	UDP-N-acetylmuramoyl-L-alanyl-D-glutamyl-L-lysyl-D-alanyl-D-alanine synthetase UDP-N-acetylmuramoylalanyl-D-glutamyl-lysine-D-alanyl-D-alanine ligase	Vol. 2, Page 458
6.3.2.15	UDP-N-acetylmuramoyl-L-alanyl-D-glutamyl-meso-2,6-diaminoheptanedioate:D-alanyl-D-alanine ligase (ADP-forming) UDP-N-acetylmuramoylalanyl-D-glutamyl-2,6-diaminopimelate-D-alanyl-D-alanine ligase	Vol. 2, Page 480
6.3.2.13	UDP-N-acetylmuramoyl-L-alanyl-D-glutamyl-meso-2,6-diaminopimelate synthetase UDP-N-acetylmuramoylalanyl-D-glutamate-2,6-diaminopimelate ligase	Vol. 2, Page 473
6.3.2.15	UDP-N-acetylmuramoyl-L-alanyl-D-glutamyl-meso-2,6-diaminopimeloyl-D-alanyl-D-alanine synthetase UDP-N-acetylmuramoylalanyl-D-glutamyl-2,6-diaminopimelate-D-alanyl-D-alanine ligase	Vol. 2, Page 480
3.4.17.8	UDP-N-acetylmuramoyl-tetrapeptidyl-D-alanine alanine-hydrolase muramoylpentapeptide carboxypeptidase	
6.3.2.9	UDP-N-acetylmuramoylalanine-D-glutamate ligase UDP-N-acetylmuramoylalanine-D-glutamate ligase	Vol. 2, Page 452
6.3.2.8	UDP-N-acetylmuramoylalanine synthetase UDP-N-acetylmuramate-alanine ligase	Vol. 2, Page 442
6.3.2.13	UDP-N-acetylmuramoylalanyl-D-glutamate-2,6-diaminopimelate ligase UDP-N-acetylmuramoylalanyl-D-glutamate-2,6-diaminopimelate ligase	Vol. 2, Page 473
6.3.2.15	UDP-N-acetylmuramoylalanyl-D-glutamyl-2,6-diaminopimelate-D-alanyl-D-alanine ligase UDP-N-acetylmuramoylalanyl-D-glutamyl-2,6-diaminopimelate-D-alanyl-D-alanine ligase	Vol. 2, Page 480
6.3.2.10	UDP-N-acetylmuramoylalanyl-D-glutamyl-lysine-D-alanyl-D-alanine ligase UDP-N-acetylmuramoylalanyl-D-glutamyl-lysine-D-alanyl-D-alanine ligase	Vol. 2, Page 458

EC-No.	Synonym Recommended Name	Vol., Page
2.3.2.10	UDP-N-acetylmuramoylpentapeptide-lysine N^6-alanyltransferase UDP-N-acetylmuramoylpentapeptide-lysine N^6-alanyltransferase	
2.3.2.10	UDP-N-acetylmuramoylpentapeptide lysine N^6-alanyltransferase UDP-N-acetylmuramoylpentapeptide-lysine N^6-alanyltransferase	
6.3.2.8	UDP-N-acetylmuramyl:L-alanine ligase UDP-N-acetylmuramate-alanine ligase	Vol. 2, Page 442
2.4.1.159	UDP-rhamnose:flavonol 3-O-glucoside rhamnosyltransferase Flavonol-3-O-glucoside L-rhamnosyltransferase	
3.6.1.45	UDP-sugar diphosphatase UDP-sugar diphosphatase	
3.6.1.45	UDP-sugar hydrolase UDP-sugar diphosphatase	
3.6.1.45	UDP-sugar pyrophosphatase UDP-sugar diphosphatase	
3.6.1.45	UDP-sugar sugarphosphohydrolase UDP-sugar diphosphatase	
2.4.2.26	UDP-xylose-core protein β-D-xylosyltransferase Protein xylosyltransferase	
2.4.2.35	UDP-xylose:flavonol 3-glycoside xylosyltransferase Flavonol-3-O-glycoside xylosyltransferase	
2.4.1.92	UDP acetylgalactosamine-(N-acetylneuraminyl)-D-galactosyl-D-glucosylceramide acetylgalactosaminyltransferase (N-Acetylneuraminyl)-galactosylglucosylceramide N-acetylgalactosaminyltransferase	
2.4.1.70	UDP acetylglucosamine-poly(ribitol phosphate) acetylglucosaminyltransferase Poly(ribitol-phosphate) N-acetylglucosaminyltransferase	
1.1.1.136	UDPacetylglucosamine dehydrogenase UDP-N-acetylglucosamine 6-dehydrogenase	
5.1.3.7	UDP acetylglucosamine epimerase UDP-N-acetylglucosamine 4-epimerase	Vol. 1, Page 135
2.7.7.23	UDPacetylglucosamine pyrophosphorylase UDP-N-acetylglucosamine pyrophosphorylase	
6.3.2.10	UDPacetylmuramoylpentapeptide synthetase UDP-N-acetylmuramoylalanyl-D-glutamyl-lysine-D-alanyl-D-alanine ligase	Vol. 2, Page 458

EC-No.	Synonym Recommended Name	Vol., Page
2.4.2.25	UDPapiose:7-O-(β-D-glucosyl)-flavone apiosyltransferase Flavone apiosyltransferase	
2.4.2.25	UDPapiose:7-O-β-D-glucosyl-5,7,4'-trihydroxyflavone β-D-apiofuranosyl-transferase Flavone apiosyltransferase	
5.1.3.5	UDParabinose 4-epimerase UDParabinose 4-epimerase	Vol. 1, Page 129
5.1.3.5	UDP arabinose epimerase UDParabinose 4-epimerase	Vol. 1, Page 129
3.6.1.6	UDPase Nucleoside-diphosphatase	
5.1.3.2	UDPG-4-epimerase UDPglucose 4-epimerase	Vol. 1, Page 97
2.4.1.11	UDPG-glycogen synthetase Glycogen (starch) synthase	
2.4.1.11	UDPG-glycogen transglucosylase Glycogen (starch) synthase	
2.4.1.121	UDPG-indol-3-ylacetyl glucosyl transferase Indole-3-acetate β-glucosyltransferase	
2.4.1.173	UDPG-SGTase Sterol 3β-glucosyltransferase	
2.4.1.91	UDPG:flavonoid-3-O-glucosyltransferase Flavonol 3-O-glucosyltransferase	
1.1.1.22	UDPG:NAD oxidoreductase UDPglucose 6-dehydrogenase	
2.4.1.114	UDPG:o-coumaric acid O-glucosyltransferase 2-Coumarate O-β-glucosyltransferase	
2.4.1.173	UDPG:sterol glucosyltransferase Sterol 3β-glucosyltransferase	
2.4.1.177	UDPG:t-Cinnamate glucosyltransferase Cinnamate β-D-glucosyltransferase	
2.4.1.17	UDPGA-glucuronyltransferase Glucuronosyltransferase	
5.1.3.16	UDPGal-4-epimerase UDPglucosamine 4-epimerase	Vol. 1, Page 165

EC-No.	Synonym Recommended Name	Vol., Page
5.4.99.9	UDPgalactopyranose mutase UDPgalactopyranose mutase	Vol. 1, Page 635
2.4.1.46	UDP galactose-1,2-diacylglycerol galactosyltransferase 1,2-Diacylglycerol 3-β-galactosyltransferase	
2.4.1.45	UDPgalactose-2-hydroxyacylsphingosine galactosyltransferase 2-Hydroxyacylsphingosine 1-β-galactosyltransferase	
2.4.1.62	UDPgalactose-ceramide galactosyltransferase Ganglioside galactosyltransferase	
2.4.1.50	UDP galactose-collagen galactosyltransferase Procollagen galactosyltransferase	
2.4.1.22	UDPgalactose-glucose galactosyltransferase Lactose synthase	
2.4.1.38	UDPgalactose-glycoprotein galactosyltransferase β-N-Acetylglucosaminylglycopeptide β-1,4-galactosyltransferase	
2.4.1.62	UDP galactose-LAC Tet-ceramide α-galactosyltransferase Ganglioside galactosyltransferase	
2.4.1.98	UDPgalactose-N-acetylglucosamine β-D-galactosyl-transferase UDPgalactose-N-acetylglucosamine β-D-galactosyl-transferase	
2.4.1.90	UDPgalactose-N-acetylglucosamine β-D-galactosyltransferase N-Acetyllactosamine synthase	
2.4.1.47	UDP galactose-N-acylsphingosine galactosyltransferase N-Acylsphingosine galactosyltransferase	
2.7.8.18	UDPgalactose-UDP-N-acetylglucosamine galactosephosphotransferase UDPgalactose-UDP-N-acetylglucosamine galactosephosphotransferase	
5.1.3.2	UDPgalactose 4-epimerase UDPglucose 4-epimerase	Vol. 1, Page 97
2.4.1.87	UDPgalactose:β-D-galactosyl-β-1,4-N-acetyl-D-glucosaminyl-glycopeptide α-1,3-D-galactosyltransferase β-D-Galactosyl-N-acetylglucosaminylglycopeptide α-1,3-galactosyltransferase	
2.4.1.151	UDPgalactose:β-D-galactosyl-1,4-acetyl-D-glucosaminide α-1,3-D-galactosyltransferase N-Acetyllactosaminide α-1,3-galactosyltransferase	
2.4.1.124	UDPgalactose:β-D-galactosyl-1,4-N-acetyl-D-glucosamine 3-α-D-galactosyltransferase N-Acetyllactosamine 3-α-galactosyltransferase	

EC-No.	Synonym Recommended Name	Vol., Page
2.4.1.46	UDPgalactose:1,2-diacylglycerol 3-β-D-galactosyltransferase 1,2-Diacylglycerol 3-β-galactosyltransferase	
2.4.1.45	UDPgalactose:2-(2-hydroxyacyl)sphingosine 1-β-D-galactosyltransferase 2-Hydroxyacylsphingosine 1-β-galactosyltransferase	
2.4.1.45	UDPgalactose:2-2-hydroxyacylsphingosine galactosyltransferase 2-Hydroxyacylsphingosine 1-β-galactosyltransferase	
2.4.1.134	UDPgalactose:4-β-D-galactosyl-O-β-D-xylosylprotein 3-β-D-galactosyltransferase Galactosylxylosylprotein 3-β-galactosyltransferase	
2.4.1.50	UDPgalactose:5-hydroxylysine-collagen galactosyltransferase Procollagen galactosyltransferase	
2.4.1.45	UDPgalactose:ceramide galactosyltransferase 2-Hydroxyacylsphingosine 1-β-galactosyltransferase	
2.4.1.179	UDPgalactose:D-galactosyl-1,4-β-D-glucosyl-R β-1,3-galactosyltransferase Lactosylceramide β-1,3-galactosyltransferase	
2.4.1.22	UDPgalactose:D-glucose 4-β-D-galactotransferase Lactose synthase	
2.4.1.205	UDPgalactose:galactogen β-1,6-D-galactosyltransferase Galactogen 6β-galactosyltransferase	
2.4.1.37	UDPgalactose:glycoprotein-α-L-fucosyl-(1,2)-D-galactose 3-α-D-galactosyltransferase Fucosylglycoprotein 3-α-galactosyltransferase	
2.4.1.122	UDPgalactose:glycoprotein-N-acetyl-D-galactosamine 3-β-D-galactosyltransferase Glycoprotein-N-acetylgalactosamine 3-β-galactosyltransferase	
2.4.1.74	UDPgalactose:glycosaminoglycan D-galactosyltransferase Glycosaminoglycan galactosyltransferase	
2.4.1.156	UDPgalactose:indol-3-ylacetyl-myo-inositol 5-O-D-galactosyltransferase Indolylacetyl-myo-inositol galactosyltransferase	
2.4.1.44	UDPgalactose:lipopolysaccharide 3-α-D-galactosyltransferase Lipopolysaccharide 3-α-galactosyltransferase	
2.4.1.123	UDPgalactose:myo-inositol 1-α-D-galactosyltransferase Inositol 1-α-galactosyltransferase	
2.4.1.38	UDPgalactose:N-acetyl-β-D-glucosaminylglycopeptide β-1,4-galactosyltransferase β-N-Acetylglucosaminylglycopeptide β-1,4-galactosyltransferase	

EC-No.	Synonym Recommended Name Vol., Page
2.4.1.62	UDPgalactose:N-acetyl-D-galactosaminyl-(N-acetylneuraminyl)-D-galactosyl-D-glucosyl-N-acylsphingosine β-1,3-D-galactosyltransferase Ganglioside galactosyltransferase
2.4.1.90	UDPgalactose:N-acetyl-D-glucosamine 4-β-D-galactosyltransferase N-Acetyllactosamine synthase
2.4.1.86	UDPgalactose:N-acetyl-D-glucosaminyl-1,3-D-galactosyl-1,4-D-glucosylceramide β-D-galactosyltransferase Glucosaminylgalactosylglucosylceramide β-galactosyltransferase
2.4.1.90	UDPgalactose:N-acetylglucosaminyl(β1-4)galactosyltransferase N-Acetyllactosamine synthase
2.4.1.47	UDPgalactose:N-acylsphingosine D-galactosyltransferase N-Acylsphingosine galactosyltransferase
2.4.1.133	UDPgalactose:O-β-D-xylosylprotein 4-β-D-galactosyltransferase Xylosylprotein 4-β-galactosyltransferase
2.4.1.37	UDPgalactose:O-α-L-fucosyl(1→2)D-galactose α-D-galactosyltransferase Fucosylglycoprotein 3-α-galactosyltransferase
2.4.1.74	UDPgalactose:polysaccharide transferase Glycosaminoglycan galactosyltransferase
2.4.1.50	UDPgalactose:procollagen-5-hydroxy-L-lysine D-galactosyltransferase Procollagen galactosyltransferase
2.4.1.96	UDPgalactose:sn-glycerol-3-phosphate α-D-galactosyltransferase sn-Glycerol-3-phosphate 1-galactosyltransferase
2.4.1.96	UDPgalactose:sn-glycerol-3-phosphate 1-α-D-galactosyltransferase sn-Glycerol-3-phosphate 1-galactosyltransferase
2.4.1.137	UDPgalactose:sn-glycerol-3-phosphate 2-α-D-galactosyltransferase sn-Glycerol-3-phosphate 2-α-galactosyltransferase
2.4.1.23	UDPgalactose:sphingosine 1-β-galactotransferase Sphingosine β-galactosyltransferase
2.4.1.23	UDPgalactose:sphingosine O-galactosyl transferase Sphingosine β-galactosyltransferase
2.4.1.167	UDPgalactose:sucrose 6F-α-D-galactosyltransferase Sucrose 6F^{-} α-galactotransferase
2.4.1.167	UDPgalactose:sucrose 6fru-α-galactosyltransferase Sucrose 6F^{-} α-galactotransferase
2.7.8.18	UDPgalactose:UDP-N-acetyl-D-glucosamine galactosephosphotransferase UDPgalactose-UDP-N-acetylglucosamine galactosephosphotransferase

EC-No.	Synonym Recommended Name	Vol., Page
2.7.8.6	UDPgalactose:undecaprenyl-phosphate galactosephosphotransferase Undecaprenyl-phosphate galactosephosphotransferase	
2.7.7.10	UDPgalactose pyrophosphorylase UTP-hexose-1-phosphate uridylyltransferase	
2.4.1.75	UDPgalacturonate β-D-galacturonosyltransferase (acceptor-unspecific) UDPgalacturonosyltransferase	
2.4.1.43	UDP galacturonate-polygalacturonate α-galacturonosyltransferase Polygalacturonate 4-α-galacturonosyltransferase	
4.1.1.67	UDPgalacturonate decarboxylase UDPgalacturonate decarboxylase	
2.4.1.75	UDPgalacturonosyltransferase UDPgalacturonosyltransferase	
4.1.1.67	UDPGalUA carboxy lyase UDPgalacturonate decarboxylase	
2.4.1.17	UDPGA transferase Glucuronosyltransferase	
1.1.1.22	UDPG dehydrogenase UDPglucose 6-dehydrogenase	
2.4.1.112	UDPGlc:protein transglucosylase α-1,4-Glucan-protein synthase (UDP-forming)	
5.1.3.16	UDPglucosamine 4-epimerase UDPglucosamine 4-epimerase	Vol. 1, Page 165
5.1.3.16	UDPglucosamine 4-epimerase UDPglucosamine 4-epimerase	Vol. 1, Page 165
2.4.1.12	UDPglucose-β-glucan glucosyltransferase Cellulose synthase (UDP-forming)	
2.4.1.34	UDPglucose-1,3-β-D-glucan glucosyltransferase 1,3-β-Glucan synthase	
2.4.1.81	UDPglucose-apigenin β-glucosyltransferase Flavone 7-O-β-glucosyltransferase	
2.4.1.71	UDP glucose-arylamine glucosyltransferase Arylamine glucosyltransferase	
2.4.1.12	UDPglucose-cellulose glucosyltransferase Cellulose synthase (UDP-forming)	
2.4.1.80	UDP glucose-ceramide glucosyltransferase Ceramide glucosyltransferase	

EC-No.	Synonym Recommended Name	Vol., Page
2.4.1.27	UDP glucose-DNA β-glucosyltransferase DNA β-glucosyltransferase	
2.4.1.26	UDPglucose-DNA α-glucosyltransferase DNA α-glucosyltransferase	
2.4.1.170	UDPglucose-favonoid 7-O-glucosyltransferase Isoflavone 7-O-glucosyltransferase	
2.4.1.14	UDPglucose-fructose-phosphate glucosyltransferase Sucrose-phosphate synthase	
2.4.1.15	UDPglucose-glucose-phosphate glucosyltransferase α,α-Trehalose-phosphate synthase (UDP-forming)	
2.4.1.11	UDPglucose-glycogen glucosyltransferase Glycogen (starch) synthase	
3.1.3.42	UDPglucose-glycogen glucosyltransferase phosphatase [glycogen-synthase-D] phosphatase	
2.7.8.19	UDPglucose-glycoprotein glucosephosphotransferase UDPglucose-glycoprotein glucosephosphotransferase	
2.7.7.12	UDPglucose-hexose-1-phosphate uridylyltransferase UDPglucose-hexose-1-phosphate uridylyltransferase	
2.4.1.81	UDPglucose-luteolin β-D-glucosyltransferase Flavone 7-O-β-glucosyltransferase	
2.4.1.171	UDPglucose-methylazoxymethanol glucosyltransferase Methyl-ONN-azoxymethanol β-D-glucosyltransferase	
2.4.1.52	UDP glucose-poly(glycerol-phosphate) α-glucosyltransferase Poly(glycerol-phosphate) α-glucosyltransferase	
2.4.1.53	UDP glucose-poly(ribitol-phosphate) β-glucosyltransferase Poly(ribitol-phosphate) β-glucosyltransferase	
2.4.1.136	UDPglucose-vanillate 1-glucosyltransferase Gallate 1-β-glucosyltransferase	
4.2.1.76	UDPglucose 4,6-dehydratase UDPglucose 4,6-dehydratase	
4.2.1.76	UDPglucose 4,6-hydro-lyase UDPglucose 4,6-dehydratase	
5.1.3.2	UDPglucose 4-epimerase UDPglucose 4-epimerase	Vol. 1, Page 97
5.1.3.2	UDPglucose 4-epimerase UDPglucose 4-epimerase	Vol. 1, Page 97

EC-No.	Synonym Recommended Name	Vol., Page
1.1.1.22	UDPglucose 6-dehydrogenase UDPglucose 6-dehydrogenase	
2.4.1.127	UDPglucose:(-)-menthol O-β-D-glucosyltransferase Monoterpenol β-glucosyltransferase	
2.4.1.192	UDPglucose:(20S,22S,25S)-22,25-epoxyfurost-5-ene-3β,26-diol 3-O-β-D-glucosyltransferase Nuatigenin 3β-glucosyltransferase	
2.4.1.193	UDPglucose:(25S)-5β-spirostan-3β-ol 3-O-β-D-glucosyltransferase Sarsapogenin 3β-glucosyltransferase	
2.4.1.85	UDPglucose:(S)-4-hydroxymandelonitrile β-D-glucosyltransferase Cyanohydrin β-glucosyltransferase	
2.4.1.183	UDPglucose:α-D-(1-3)-glucan 3-α-D-glucosyltransferase α-1,3-Glucan synthase	
2.7.7.12	UDPglucose:α-D-galactose-1-phosphate uridylyltransferase UDPglucose-hexose-1-phosphate uridylyltransferase	
2.4.1.157	UDPglucose:1,2-diacylglycerol 3-D-glucosyltransferase 1,2-Diacylglycerol 3-glucosyltransferase	
2.4.1.103	UDPglucose:1,2-dihydroxy-9,10-anthraquinone 2-O-β-D-glucosyltransferase Alizarin 2-β-glucosyltransferase	
2.4.1.34	UDPglucose:1,3-β-D-glucan 3-β-D-glucosyltransferase 1,3-β-Glucan synthase	
2.4.1.12	UDPglucose:1,4-β-D-glucan 4-β-D-glucosyltransferase Cellulose synthase (UDP-forming)	
2.4.1.158	UDPglucose:13-hydroxydocosanoate 13-β-D-glucosyltransferase 13-Hydroxydocosanoate 13-β-glucosyltransferase	
2.4.1.202	UDPglucose:2,4-dihydroxy-7-methoxy-2H-1,4-benzoxazin-3(4H)-one 2-D-glucosyltransferase 2,4-Dihydroxy-7-methoxy-2H-1,4-benzoxazin-3(4H)-one 2-D-glucosyltransferase	
2.4.1.63	UDPglucose:2-hydroxy-2-methylpropanenitrile β-D-glucosyltransferase Linamarin synthase	
2.4.1.194	UDPglucose:4-hydroxybenzoate 4-O-β-D-glucosyltransferase 4-Hydroxybenzoate 4-O-β-D-glucosyltransferase	
2.4.1.178	UDPglucose:4-hydroxymandelonitrile glucosyltransferase Hydroxymandelonitrile glucosyltransferase	

EC-No.	Synonym Recommended Name	Vol., Page
2.4.1.81	UDPglucose:5,7,3',4'-tetrahydroxyflavone 7-O-β-D-glucosyltransferase Flavone 7-O-β-glucosyltransferase	
2.4.1.66	UDPglucose:5-(D-galactosyloxy)-L-lysine-procollagen D-glucosyltransferase Procollagen glucosyltransferase	
2.4.1.104	UDPglucose:7,8-dihydroxycoumarin 7-O-β-D-glucosyltransferase *o*-Dihydroxycoumarin 7-O-glucosyltransferase	
2.4.1.115	UDPglucose:anthocyanidin 3-O-D-glucosyltransferase Anthocyanidin 3-O-glucosyltransferase	
2.4.1.71	UDPglucose:arylamine N-D-glucosyltransferase Arylamine glucosyltransferase	
2.4.1.111	UDPglucose:coniferyl-alcohol 4'-β-D-glucosyltransferase Coniferyl-alcohol glucosyltransferase	
2.4.1.116	UDPglucose:cyanidin-3-O-D-rhamnosyl-1,6-D-glucoside 5-O-D-glucosyl-transferase Cyanidin-3-rhamnosylglucoside 5-O-glucosyltransferase	
2.4.1.14	UDPglucose:D-fructose-6-phosphate 2-α-D-glucosyltransferase Sucrose-phosphate synthase	
2.4.1.13	UDPglucose:D-fructose 2-α-D-glucosyltransferase Sucrose synthase	
2.4.1.15	UDPglucose:D-glucose-6-phosphate 1-α-D-glucosyltransferase α,α-Trehalose-phosphate synthase (UDP-forming)	
2.4.1.28	UDPglucose:D-glucosyl-DNA β-D-glucosyltransferase Glucosyl-DNA β-glucosyltransferase	
2.4.1.157	UDPglucose:diacylglycerol glucosyltransferase 1,2-Diacylglycerol 3-glucosyltransferase	
2.4.1.26	UDPglucose:DNA α-D-glucosyltransferase DNA α-glucosyltransferase	
2.4.1.27	UDPglucose:DNA β-D-glucosyltransferase DNA β-glucosyltransferase	
2.4.1.117	UDPglucose:dolichyl-phosphate β-D-glucosyltransferase Dolichyl-phosphate β-glucosyltransferase	
2.4.1.185	UDPglucose:flavanone 7-O-β-D-glucosyltransferase Flavanone 7-O-β-glucosyltransferase	
2.4.1.91	UDPglucose:flavonol 3-O-D-glucosyltransferase Flavonol 3-O-glucosyltransferase	

EC-No.	Synonym Recommended Name Vol., Page
2.4.1.73	UDPglucose:galactosyl-lipopolysaccharide α-D-glucosyltransferase Lipopolysaccharide glucosyltransferase II
2.4.1.136	UDPglucose:gallate β-D-glucosyltransferase Gallate 1-β-glucosyltransferase
2.4.1.136	UDPglucose:gallate glucosyltransferase Gallate 1-β-glucosyltransferase
2.4.1.176	UDPglucose:gibberellin 2-O-β-D-glucosyltransferase Gibberellin β-D-glucosyltransferase
2.4.1.11	UDPglucose:glycogen 4-α-D-glucosyltransferase Glycogen (starch) synthase
3.1.3.42	[UDPglucose:glycogen 4-α-D-glucosyltransferase-D] phosphohydrolase [glycogen-synthase-D] phosphatase
2.4.1.186	UDPglucose:glycogenin glucosyltransferase Glycogenin glucosyltransferase
2.7.8.19	UDPglucose:glycoprotein-D-mannose glucosephosphotransferase UDPglucose-glycoprotein glucosephosphotransferase
2.4.1.181	UDPglucose:hydroxyanthraquinone O-glucosyltransferase Hydroxyanthraquinone glucosyltransferase
2.4.1.121	UDPglucose:indole-3-acetate β-D-glucosyltransferase Indole-3-acetate β-glucosyltransferase
2.4.1.170	UDPglucose:isoflavone 7-O-β-D-glucosyltransferase Isoflavone 7-O-glucosyltransferase
2.4.1.170	UDPglucose:isoflavone 7-O-glucosyltransferase Isoflavone 7-O-glucosyltransferase
2.4.1.106	UDPglucose:isovitexin 2-O-β-D-glucosyltransferase Isovitexin β-glucosyltransferase
2.4.1.63	UDPglucose:ketone cyanohydrin β-glucosyltransferase Linamarin synthase
2.4.1.58	UDPglucose:lipopolysaccharide glucosyltransferase Lipopolysaccharide glucosyltransferase I
2.4.1.58	UDPglucose:lipopolysaccharide glucosyltransferase I Lipopolysaccharide glucosyltransferase I
2.4.1.171	UDPglucose:methyl-ONN-azoxymethanol β-D-glucosyltransferase Methyl-ONN-azoxymethanol β-D-glucosyltransferase
2.4.1.127	UDPglucose:monoterpenol glucosyltransferase Monoterpenol β-glucosyltransferase

EC-No.	Synonym Recommended Name	Vol., Page
2.4.1.188	UDPglucose:N-acetyl-D-glucosaminyldiphosphoundecaprenol 4-β-D-glucosyltransferase N-Acetylglucosaminyldiphosphoundecaprenol glucosyltransferase	
2.4.1.80	UDPglucose:N-acylsphingosine D-glucosyltransferase Ceramide glucosyltransferase	
1.1.1.22	UDPglucose:NAD$^+$ 6-oxidoreductase UDPglucose 6-dehydrogenase	
1.1.1.22	UDPglucose:NAD$^+$ oxidoreductase UDPglucose 6-dehydrogenase	
2.4.1.196	UDPglucose:nicotinate N-glucosyltransferase Nicotinate glucosyltransferase	
2.4.1.35	UDPglucose:phenol β-D-glucosyltransferase Phenol β-glucosyltransferase	
2.4.1.78	UDPglucose:phosphopolyprenol D-glucosyltransferase Phosphopolyprenol glucosyltransferase	
2.4.1.52	UDPglucose:poly(glycerol-phosphate) α-D-glucosyltransferase Poly(glycerol-phosphate) α-glucosyltransferase	
2.4.1.53	UDPglucose:poly(ribitol-phosphate) β-D-glucosyltransferase Poly(ribitol-phosphate) β-glucosyltransferase	
2.4.1.78	UDPglucose:polyprenol monophosphate glucosyltransferase Phosphopolyprenol glucosyltransferase	
2.4.1.112	UDPglucose:protein 4-α-glucosyltransferase α-1,4-Glucan-protein synthase (UDP-forming)	
2.4.1.112	UDPglucose:protein glucosyltransferase α-1,4-Glucan-protein synthase (UDP-forming)	
2.4.1.160	UDPglucose:pyridoxine 5'-O-β-D-glucosyltransferase Pyridoxine 5'-O-β-D-glucosyltransferase	
2.4.1.172	UDPglucose:salicyl-alcohol β-D-glucosyltransferase Salicyl-alcohol β-D-glucosyltransferase	
2.4.1.172	UDPglucose:salicyl alcohol phenyl-glucosyltransferase Salicyl-alcohol β-D-glucosyltransferase	
2.4.1.128	UDPglucose:scopoletin glucosyltransferase UDPglucose:scopoletin O-β-D-glucosyltransferase	
2.4.1.128	UDPglucose:scopoletin O-β-D-glucosyltransferase UDPglucose:scopoletin O-β-D-glucosyltransferase	

EC-No.	Synonym Recommended Name	Vol., Page
2.4.1.120	UDPglucose:sinapate D-glucosyltransferase Sinapate 1-glucosyltransferase	
2.4.1.120	UDPglucose:sinapic acid glucosyltransferase Sinapate 1-glucosyltransferase	
2.4.1.173	UDPglucose:sterol 3-O-β-D-glucosyltransferase Sterol 3β-glucosyltransferase	
2.4.1.195	UDPglucose:thiohydroximate S-β-D-glucosyltransferase Thiohydroximate β-D-glucosyltransferase	
2.4.1.114	UDPglucose:trans-2-hydroxycinnamate O-β-D-glucosyltransferase 2-Coumarate O-β-glucosyltransferase	
2.4.1.126	UDPglucose:trans-4-hydroxycinnamate 4-O-β-D-glucosyltransferase Hydroxycinnamate 4-β-glucosyltransferase	
2.4.1.177	UDPglucose:trans-cinnamate β-D-glucosyltransferase Cinnamate β-D-glucosyltransferase	
2.4.1.136	UDPglucose:vanillate 1-O-glucosyltransferase Gallate 1-β-glucosyltransferase	
2.4.1.105	UDPglucose:vitexin 2-O-β-D-glucosyltransferase Vitexin β-glucosyltransferase	
2.4.1.168	UDPglucose:xyloglucan 1,4-β-D-glucosyltransferase Xyloglucan 4-glucosyltransferase	
2.4.1.118	UDPglucose:zeatin 7-glucosyltransferase Cytokinin 7-β-glucosyltransferase	
2.4.1.203	UDPglucose:zeatin O-β-D-glucosyltransferase Zeatin O-β-D-glucosyltransferase	
1.1.1.22	UDPglucose dehydrogenase UDPglucose 6-dehydrogenase	
2.4.1.63	UDP glucose ketone cyanohydrin glucosyltransferase Linamarin synthase	
2.7.7.9	UDP glucose pyrophosphorylase UTP-glucose-1-phosphate uridylyltransferase	
2.7.7.9	UDPglucose pyrophosphorylase UTP-glucose-1-phosphate uridylyltransferase	
2.4.1.35	UDP glucosyltransferase Phenol β-glucosyltransferase	
2.4.1.35	UDPglucosyltransferase Phenol β-glucosyltransferase	

EC-No.	Synonym Recommended Name	Vol., Page
2.4.1.84	UDPglucuronate-1,2-diacylglycerol glucuronosyltransferase UDPglucuronate-1,2-diacylglycerol glucuronosyltransferase	
2.4.1.77	UDPglucuronate-bilirubin-glucuronoside glucuronosyltransferase UDPglucuronate-bilirubin-glucuronoside glucuronosyltransferase	
2.4.1.76	UDPglucuronate-bilirubin glucuronosyltransferase UDPglucuronate-bilirubin glucuronosyltransferase	
2.4.1.17	UDPglucuronate β-D-glucuronosyltransferase (acceptor-unspecific) Glucuronosyltransferase	
2.4.1.17	UDP glucuronate-estradiol-glucuronosyltransferase Glucuronosyltransferase	
2.4.1.59	UDPglucuronate-estradiol glucuronosyltransferase UDPglucuronate-estradiol glucuronosyltransferase	
2.4.1.61	UDPglucuronate-estriol 16α-glucuronosyltransferase UDPglucuronate-estriol 16α-glucuronosyltransferase	
2.4.1.42	UDPglucuronate-estriol 17β-D-glucuronosyltransferase UDPglucuronate-estriol 17β-D-glucuronosyltransferase	
2.4.1.17	UDP glucuronate-estriol glucuronosyltransferase Glucuronosyltransferase	
2.4.1.108	UDPglucuronate-phenol glucuronosyltransferase UDPglucuronate-phenol glucuronosyltransferase	
2.4.1.107	UDPglucuronate-testosterone glucuronosyltransferase UDPglucuronate-testosterone glucuronosyltransferase	
5.1.3.6	UDPglucuronate 4-epimerase UDPglucuronate 4-epimerase	Vol. 1, Page 132
5.1.3.6	UDPglucuronate 4-epimerase UDPglucuronate 4-epimerase	Vol. 1, Page 132
5.1.3.12	UDP glucuronate 5'-epimerase UDPglucuronate 5'-epimerase	Vol. 1, Page 150
5.1.3.12	UDPglucuronate 5'-epimerase UDPglucuronate 5'-epimerase	Vol. 1, Page 150
5.1.3.12	UDPglucuronate 5'-epimerase UDPglucuronate 5'-epimerase	Vol. 1, Page 150
2.4.1.135	UDPglucuronate:3-β-D-galactosyl-4-β-D-galactosyl-O-β-D-xylosylprotein D-glucuronosyltransferase Galactosylgalactosylxylosylprotein 3-β-glucuronosyltransferase	

EC-No.	Synonym Recommended Name	Vol., Page
2.4.1.191	UDPglucuronate:luteolin-7-O-β-D-diglucuronide 4'-O-glucuronosyltransferase Luteolin-7-O-diglucuronide 4'-O-glucuronosyltransferase	
2.4.1.190	UDPglucuronate:luteolin-7-O-β-D-glucuronide 7-O-glucuronosyltransferase Luteolin-7-O-glucuronide 7-O-glucuronosyltransferase	
2.4.1.191	UDPglucuronate:luteolin 7-O-diglucuronide-4'-O-glucuronosyl-transferase Luteolin-7-O-diglucuronide 4'-O-glucuronosyltransferase	
2.4.1.189	UDPglucuronate:luteolin 7-O-glucuronosyltransferase Luteolin 7-O-glucuronosyltransferase	
4.1.1.35	UDPglucuronate carboxy-lyase UDPglucuronate decarboxylase	
4.1.1.35	UDPglucuronate decarboxylase UDPglucuronate decarboxylase	
2.4.1.17	UDP glucuronic acid transferase Glucuronosyltransferase	
5.1.3.6	UDP glucuronic epimerase UDPglucuronate 4-epimerase	Vol. 1, Page 132
2.4.1.17	UDP glucuronosyltransferase Glucuronosyltransferase	
2.4.1.17	UDPglucuronosyltransferase Glucuronosyltransferase	
2.4.1.17	UDP glucuronyltransferase Glucuronosyltransferase	
2.7.7.9	UDPG phosphorylase UTP-glucose-1-phosphate uridylyltransferase	
2.7.7.9	UDPG pyrophosphorylase UTP-glucose-1-phosphate uridylyltransferase	
2.4.1.17	UDPGT Glucuronosyltransferase	
2.7.4.6	UDP kinase Nucleoside-diphosphate kinase	
2.7.8.13	UDPMurAc(oyl-L-Ala-D-γ-Glu-L-Lys-D-Ala-D-Ala):undecaprenyl-phosphate phospho-N-acetylmuramoyl-pentapeptide-transferase Phospho-N-acetylmuramoyl-pentapeptide-transferase	
6.3.2.9	UDPMurNAc-L-alanine:D-glutamate ligase (ADP-forming) UDP-N-acetylmuramoylalanine-D-glutamate ligase	Vol. 2, Page 452

EC-No.	Synonym Recommended Name	Vol., Page
6.3.2.13	UDPMurNAc-L-alanyl-D-glutamate:mDAP ligase (ADP-forming) UDP-N-acetylmuramoylalanyl-D-glutamate-2,6-diaminopimelate ligase	Vol. 2, Page 473
1.17.4.1	UDP reductase Ribonucleoside-diphosphate reductase	
2.7.7.11	UDPxylose pyrophosphorylase UTP-xylose-1-phosphate uridylyltransferase	
4.3.2.3	UGL ureidoglycolate lyase	
4.2.2.10	Ultrazym-100 pectin lyase	
4.2.2.10	Ultrazym-20 pectin lyase	
4.2.2.10	Ultrazym-40 pectin lyase	
2.7.4.14	UMP-CMP kinase Cytidylate kinase	
2.4.2.9	UMP:pyrophosphate phospho-α-D-ribosyltransferase Uracil phosphoribosyltransferase	
2.4.2.9	UMP:pyrophosphate phosphoribosyltransferase Uracil phosphoribosyltransferase	
3.1.3.5	UMPase 5'-nucleotidase	
2.4.2.9	UMP pyrophosphorylase Uracil phosphoribosyltransferase	
2.7.1.66	Undecaprenol kinase Undecaprenol kinase	
3.6.1.27	Undecaprenyl-diphosphatase Undecaprenyl-diphosphatase	
3.6.1.27	Undecaprenyl-diphosphate phosphohydrolase Undecaprenyl-diphosphatase	
2.5.1.31	Undecaprenyl-diphosphate synthase Di-trans,poly-cis-Decaprenylcistransferase	
2.7.8.6	Undecaprenyl-phosphate galactosephosphotransferase Undecaprenyl-phosphate galactosephosphotransferase	

EC-No.	Synonym Recommended Name	Vol., Page
2.4.1.54	Undecaprenyl-phosphate mannosyltransferase Undecaprenyl-phosphate mannosyltransferase	
2.5.1.31	Undecaprenyl diphosphate synthase Di-trans,poly-cis-Decaprenylcistransferase	
2.5.1.31	Undecaprenyl diphosphate synthetase Di-trans,poly-cis-Decaprenylcistransferase	
2.7.8.6	Undecaprenyl phosphate galactosyl-1-phosphate transferase Undecaprenyl-phosphate galactosephosphotransferase	
2.5.1.31	Undecaprenyl pyrophosphate synthetase Di-trans,poly-cis-Decaprenylcistransferase	
2.1.1.79	Unsaturated-phospholipid methyltransferase Cyclopropane-fatty-acyl-phospholipid synthase	
2.1.1.16	Unsaturated-phospholipid methyltransferase Methylene-fatty-acyl-phospholipid synthase	
4.2.1.17	unsaturated acyl-CoA hydratase enoyl-CoA hydratase	
1.3.99.2	Unsaturated acyl-CoA reductase Butyryl-CoA dehydrogenase	
1.3.99.2	Unsaturated acyl coenzyme A reductase Butyryl-CoA dehydrogenase	
4.2.2.6	unsaturated oligogalacturonate transeliminase oligogalacturonide lyase	
2.1.1.16	Unsaturated phospholipid methyltransferase Methylene-fatty-acyl-phospholipid synthase	
3.6.1.15	Unspecific diphosphate phosphohydrolase Nucleoside-triphosphatase	
1.14.14.1	Unspecific monooxygenase Unspecific monooxygenase	
5.99.1.2	Untwisting enzyme DNA topoisomerase	Vol. 1, Page 721
3.4.21.73	uPA u-Plasminogen activator	
2.4.2.3	UPase Uridine phosphorylase	
4.3.1.8	UPGI-S hydroxymethylbilane synthase	

EC-No.	Synonym Recommended Name	Vol., Page
2.4.2.3	UPH Uridine phosphorylase	
2.5.1.31	UPP synthetase Di-trans,poly-cis-Decaprenylcistransferase	
2.4.2.9	UPRTase Uracil phosphoribosyltransferase	
1.18.99.1	Uptake hydrogenase Hydrogenase	
2.4.1.112	UPTG α-1,4-Glucan-protein synthase (UDP-forming)	
4.1.1.66	Uracil-5-carboxylate carboxy-lyase Uracil-5-carboxylate decarboxylase	
4.1.1.66	Uracil-5-carboxylate decarboxylase Uracil-5-carboxylate decarboxylase	
4.1.1.66	Uracil-5-carboxylic acid decarboxylase Uracil-5-carboxylate decarboxylase	
3.1.26.9	Uracil-specific endoribonuclease Ribonuclease (poly-(U)-specific)	
3.1.26.9	Uracil-specific RNase Ribonuclease (poly-(U)-specific)	
1.1.99.19	Uracil:(acceptor) oxidoreductase Uracil dehydrogenase	
1.1.99.19	Uracil dehydrogenase Uracil dehydrogenase	
1.2.99.1	uracil dehydrogenase uracil dehydrogenase	
4.2.1.70	uracil hydro-lyase (adding D-ribose 5-phosphate) pseudouridylate synthase	
1.1.99.19	Uracil oxidase Uracil dehydrogenase	
2.4.2.9	Uracil phosphoribosyltransferase Uracil phosphoribosyltransferase	
1.3.1.1	Uracil reductase Dihydrouracil dehydrogenase (NAD^+)	
4.2.99.16	Uracilylalanine synthase Uracilylalanine synthase	

EC-No.	Synonym Recommended Name	Vol., Page
2.4.2.16	Urate-ribonucleotide:orthophosphate D-ribosyltransferase Urate-ribonucleotide phosphorylase	
2.4.2.16	Urate-ribonucleotide phosphorylase Urate-ribonucleotide phosphorylase	
1.7.3.3	Urate:oxygen oxidoreductase Urate oxidase	
1.7.3.3	Urate oxidase Urate oxidase	
2.4.2.16	Urate ribonucleotide phosphorylase Urate-ribonucleotide phosphorylase	
2.4.2.3	UrdPase Uridine phosphorylase	
3.5.1.54	Urea-1-carboxylate amidohydrolase Allophanate hydrolase	
6.3.4.6	Urea:carbon-dioxide ligase (ADP-forming) Urea carboxylase	Vol. 2, Page 603
6.3.4.6	Urea amido-lyase Urea carboxylase	Vol. 2, Page 603
3.5.1.5	Urea amidohydrolase Urease	
6.3.4.6	Urea carboxylase Urea carboxylase	Vol. 2, Page 603
6.3.4.6	Urea carboxylase (hydrolysing) Urea carboxylase	Vol. 2, Page 603
4.2.1.69	urea hydro-lyase cyanamide hydratase	
3.5.1.5	Urease Urease	
6.3.4.6	Urease (ATP-hydrolysing) Urea carboxylase	Vol. 2, Page 603
3.5.1.45	urease (ATP-hydrolysing) urease (ATP-hydrolysing)	
4.3.2.3	ureidoglycolase ureidoglycolate lyase	
4.3.2.3	ureidoglycolatase ureidoglycolate lyase	

EC-No.	Synonym Recommended Name	Vol., Page
3.5.3.19	Ureidoglycolate amidohydrolase (decarboxylating) Ureidoglycolate hydrolase	
1.1.1.154	Ureidoglycolate dehydrogenase Ureidoglycolate dehydrogenase	
3.5.3.19	Ureidoglycolate hydrolase Ureidoglycolate hydrolase	
4.3.2.3	ureidoglycolate hydrolase ureidoglycolate lyase	
4.3.2.3	ureidoglycolate lyase ureidoglycolate lyase	
4.3.2.3	(-)-ureidoglycolate urea-lyase ureidoglycolate lyase	
3.5.1.6	*β*-ureidopropionase *β*-ureidopropionase	
3.5.1.7	Ureidosuccinase Ureidosuccinase	
3.5.1.75	Urethanase Urethanase	
3.5.1.75	Urethane amidohydrolase (decarboxylating) Urethanase	
3.5.1.75	Urethane hydrolase Urethanase	
1.7.3.3	Uric acid oxidase Urate oxidase	
1.7.3.3	Uricase Urate oxidase	
1.7.3.3	Uricase II Urate oxidase	
1.1.1.158	Uridine-5'-diphospho-N-acetyl-2-amino-2-deoxy-3-O-lactylglucose:NADP-oxidoreductase UDP-N-acetylmuramate dehydrogenase	
2.7.1.48	Uridine-cytidine kinase Uridine kinase	
6.3.2.8	Uridine-diphosphate-N-acetylmuramate:L-alanine ligase UDP-N-acetylmuramate-alanine ligase	Vol. 2, Page 442

EC-No.	Synonym Recommended Name	Vol., Page
6.3.2.13	Uridine-diphosphate-N-acetylmuramyl-L-alanyl-D-glutamate:meso-2,6-diaminopimelate synthetase UDP-N-acetylmuramoylalanyl-D-glutamate-2,6-diaminopimelate ligase	Vol. 2, Page 473
2.7.7.9	Uridine-diphosphate glucose pyrophosphorylase UTP-glucose-1-phosphate uridylyltransferase	
4.1.1.35	Uridine-diphospho glucuronate decarboxylase UDPglucuronate decarboxylase	
3.6.1.6	Uridine 5'-diphosphatase Nucleoside-diphosphatase	
5.1.3.5	Uridine 5'-diphosphate-D-xylose 4-epimerase UDParabinose 4-epimerase	Vol. 1, Page 129
5.1.3.7	Uridine 5'-diphospho-N-acetylglucosamine-4-epimerase UDP-N-acetylglucosamine 4-epimerase	Vol. 1, Page 135
2.3.1.152	Uridine 5'-diphosphoglucose-hydroxycinnamic acid acylglucosyltransferase Alcohol O-cinnamoyltransferase	
2.4.1.120	Uridine 5'-diphosphoglucose-hydroxycinnamic acid acylglucosyltransferase Sinapate 1-glucosyltransferase	
2.3.1.152	Uridine 5'-diphosphoglucose:hydroxycinnamic acid acyl-glucosyltransferase Alcohol O-cinnamoyltransferase	
2.7.7.9	Uridine 5'-diphosphoglucose pyrophosphorylase UTP-glucose-1-phosphate uridylyltransferase	
2.4.1.17	Uridine 5'-diphosphoglucuronyltransferase Glucuronosyltransferase	
3.1.3.5	uridine 5'-nucleotidase 5'-nucleotidase	
2.4.2.9	Uridine 5'-phosphate pyrophosphorylase Uracil phosphoribosyltransferase	
6.3.2.8	Uridine 5′-diphosphate-N-acetylmuramyl-L-alanine synthetase UDP-N-acetylmuramate-alanine ligase	Vol. 2, Page 442
2.4.2.3	Uridine:orthophosphate α-D-ribosyltransferase Uridine phosphorylase	
3.6.1.6	Uridine diphosphatase Nucleoside-diphosphatase	
2.7.7.9	Uridine diphosphate-D-glucose pyrophosphorylase UTP-glucose-1-phosphate uridylyltransferase	

EC-No.	Synonym Recommended Name	Vol., Page
5.1.3.14	Uridine diphosphate-N-acetylglucosamine-2'-epimerase UDP-N-acetylglucosamine 2-epimerase	Vol. 1, Page 154
2.7.7.23	Uridine diphosphate-N-acetylglucosamine pyrophosphorylase UDP-N-acetylglucosamine pyrophosphorylase	
2.4.1.62	Uridine diphosphate D-galactose:glycolipid galactosyltransferase Ganglioside galactosyltransferase	
1.1.1.22	Uridine diphosphate D-glucose dehydrogenase UDPglucose 6-dehydrogenase	
5.1.3.2	Uridine diphosphate galactose 4-epimerase UDPglucose 4-epimerase	Vol. 1, Page 97
2.4.1.44	Uridine diphosphate galactose:lipopolysaccharide α-3-galactosyltransferase Lipopolysaccharide 3-α-galactosyltransferase	
2.7.7.10	Uridine diphosphate galactose pyrophosphorylase UTP-hexose-1-phosphate uridylyltransferase	
2.4.1.112	Uridine diphosphate glucose-protein transglucosylase I α-1,4-Glucan-protein synthase (UDP-forming)	
5.1.3.2	Uridine diphosphate glucose 4-epimerase UDPglucose 4-epimerase	Vol. 1, Page 97
1.1.1.22	Uridine diphosphate glucose dehydrogenase UDPglucose 6-dehydrogenase	
4.1.1.35	Uridine diphosphate glucuronate carboxy-lyase UDPglucuronate decarboxylase	
2.4.1.17	Uridine diphosphate glucuronyltransferase Glucuronosyltransferase	
2.7.4.6	Uridine diphosphate kinase Nucleoside-diphosphate kinase	
5.1.3.7	Uridine diphosphate N-acetylglucosamine-4-epimerase UDP-N-acetylglucosamine 4-epimerase	Vol. 1, Page 135
6.3.2.8	Uridine diphosphate N-acetylmuramate:L-alanine ligase UDP-N-acetylmuramate-alanine ligase	Vol. 2, Page 442
6.3.2.9	Uridine diphosphate N-acetylmuramoyl-L-alanine:D-glutamate ligase UDP-N-acetylmuramoylalanine-D-glutamate ligase	Vol. 2, Page 452
6.3.2.13	Uridine diphosphate N-acetylmuramyl-L-alanyl-D-glutamate:meso-2,6-diaminopimelate ligase UDP-N-acetylmuramoylalanyl-D-glutamate-2,6-diaminopimelate ligase	Vol. 2, Page 473

EC-No.	Synonym Recommended Name	Vol., Page
5.1.3.6	Uridine diphospho-D-galacturonic acid UDPglucuronate 4-epimerase	Vol. 1, Page 132
5.1.3.2	Uridine diphospho-galactose-4-epimerase UDPglucose 4-epimerase	Vol. 1, Page 97
2.8.2.7	Uridine diphospho-N-acetylgalactosamine 4-sulfate sulfotransferase UDP-N-acetylgalactosamine-4-sulfate sulfotransferase	
1.1.1.158	Uridine diphospho-N-acetylglucosamine-enolpyruvate reductase UDP-N-acetylmuramate dehydrogenase	
5.1.3.14	Uridine diphospho-N-acetylglucosamine 2'-epimerase UDP-N-acetylglucosamine 2-epimerase	Vol. 1, Page 154
2.4.1.94	Uridine diphospho-N-acetylglucosamine:polypeptide β-N-acetylglucosaminyltransferase Protein N-acetylglucosaminyltransferase	
6.3.2.8	Uridine diphospho-N-acetylmuramoylalanine synthetase UDP-N-acetylmuramate-alanine ligase	Vol. 2, Page 442
6.3.2.9	Uridine diphospho-N-acetylmuramoylalanyl-D-glutamate synthetase UDP-N-acetylmuramoylalanine-D-glutamate ligase	Vol. 2, Page 452
6.3.2.7	Uridine diphospho-N-acetylmuramoylalanyl-D-glutamyllysine synthetase UDP-N-acetylmuramoyl-L-alanyl-D-glutamate-lysine ligase	Vol. 2, Page 439
2.4.1.92	Uridine diphosphoacetylgalactosamine-acetylneuraminylgalactosylglucosylceramide acetylgalactosaminyltransferase (N-Acetylneuraminyl)-galactosylglucosylceramide N-acetylgalactosaminyltransferase	
2.4.1.92	Uridine diphosphoacetylgalactosamine-hematoside acetylgalactosaminyltransferase (N-Acetylneuraminyl)-galactosylglucosylceramide N-acetylgalactosaminyltransferase	
2.8.2.7	Uridine diphosphoacetylgalactosamine 4-sulfate sulfotransferase UDP-N-acetylgalactosamine-4-sulfate sulfotransferase	
2.4.1.56	Uridine diphosphoacetylglucosamine-lipopolysaccharide acetylglucosaminyltransferase Lipopolysaccharide N-acetylglucosaminyltransferase	
2.4.1.70	Uridine diphosphoacetylglucosamine-poly(ribitol phosphate) acetylglucosaminyltransferase Poly(ribitol-phosphate) N-acetylglucosaminyltransferase	
2.4.1.39	Uridine diphosphoacetylglucosamine-steroid acetylglucosaminyltransferase Steroid N-acetylglucosaminyltransferase	

EC-No.	Synonym Recommended Name	Vol., Page
5.1.3.14	Uridine diphosphoacetylglucosamine 2'-epimerase UDP-N-acetylglucosamine 2-epimerase	Vol. 1, Page 154
5.1.3.7	Uridine diphosphoacetylglucosamine epimerase UDP-N-acetylglucosamine 4-epimerase	Vol. 1, Page 135
2.7.7.23	Uridine diphosphoacetylglucosamine phosphorylase UDP-N-acetylglucosamine pyrophosphorylase	
2.7.7.23	Uridine diphosphoacetylglucosamine pyrophosphorylase UDP-N-acetylglucosamine pyrophosphorylase	
2.3.2.10	Uridine diphosphoacetylmuramoylpentapeptide lysine N^6-alanyltransferase UDP-N-acetylmuramoylpentapeptide-lysine N^6-alanyltransferase	
6.3.2.10	Uridine diphosphoacetylmuramoylpentapeptide synthetase UDP-N-acetylmuramoylalanyl-D-glutamyl-lysine-D-alanyl-D-alanine ligase	Vol. 2, Page 458
5.1.3.5	Uridine diphosphoarabinose epimerase UDParabinose 4-epimerase	Vol. 1, Page 129
2.4.1.46	Uridine diphosphogalactose-1,2-diacylglycerol galactosyltransferase 1,2-Diacylglycerol 3-β-galactosyltransferase	
5.1.3.16	Uridine diphosphogalactose-4-epimerase UDPglucosamine 4-epimerase	Vol. 1, Page 165
2.4.1.90	Uridine diphosphogalactose-acetylglucosamine galactosyltransferase N-Acetyllactosamine synthase	
2.4.1.47	Uridine diphosphogalactose-acylsphingosine galactosyltransferase N-Acylsphingosine galactosyltransferase	
2.4.1.50	Uridine diphosphogalactose-collagen galactosyltransferase Procollagen galactosyltransferase	
2.4.1.38	Uridine diphosphogalactose-glycoprotein galactosyltransferase β-N-Acetylglucosaminylglycopeptide β-1,4-galactosyltransferase	
2.4.1.62	Uridine diphosphogalactose-GM2 galactosyltransferase Ganglioside galactosyltransferase	
2.4.1.23	Uridine diphosphogalactose-sphingosine β-galactosyltransferase Sphingosine β-galactosyltransferase	
2.7.7.10	Uridine diphosphogalactose pyrophosphorylase UTP-hexose-1-phosphate uridylyltransferase	
2.4.1.43	Uridine diphosphogalacturonate-polygalacturonate α-galacturonosyltransferase Polygalacturonate 4-α-galacturonosyltransferase	

EC-No.	Synonym Recommended Name	Vol., Page
2.4.1.34	Uridine diphosphoglucose-1,3-β-glucan glucosyltransferase 1,3-β-Glucan synthase	
2.4.1.81	Uridine diphosphoglucose-apigenin 7-O-glucosyltransferase Flavone 7-O-β-glucosyltransferase	
2.4.1.71	Uridine diphosphoglucose-arylamine glucosyltransferase Arylamine glucosyltransferase	
2.4.1.12	Uridine diphosphoglucose-cellulose glucosyltransferase Cellulose synthase (UDP-forming)	
2.4.1.66	Uridine diphosphoglucose-collagen glucosyltransferase Procollagen glucosyltransferase	
2.4.1.85	Uridine diphosphoglucose-cyanohydrin glucosyltransferase Cyanohydrin β-glucosyltransferase	
2.4.1.26	Uridine diphosphoglucose-deoxyribonucleate α-glucosyltransferase DNA α-glucosyltransferase	
2.4.1.27	Uridine diphosphoglucose-deoxyribonucleate β-glucosyltransferase DNA β-glucosyltransferase	
2.4.1.13	Uridine diphosphoglucose-fructose glucosyltransferase Sucrose synthase	
2.4.1.28	Uridine diphosphoglucose-glucosyldeoxyribonucleate β-glucosyltransferase Glucosyl-DNA β-glucosyltransferase	
2.4.1.11	Uridine diphosphoglucose-glycogen glucosyltransferase Glycogen (starch) synthase	
3.1.3.42	uridine diphosphoglucose-glycogen glucosyltransferase phosphatase [glycogen-synthase-D] phosphatase	
2.4.1.63	Uridine diphosphoglucose-ketone cyanohydrin glucosyltransferase Linamarin synthase	
2.4.1.210	uridine diphosphoglucose-limonoid glucosyltransferase limonoid glucosyltransferase	
2.4.1.52	Uridine diphosphoglucose-poly(glycerol-phosphate) α-glucosyltransferase Poly(glycerol-phosphate) α-glucosyltransferase	
2.4.1.53	Uridine diphosphoglucose-poly(ribitol-phosphate) β-glucosyltransferase Poly(ribitol-phosphate) β-glucosyltransferase	
2.4.1.173	Uridine diphosphoglucose-sterol glucosyltransferase Sterol 3β-glucosyltransferase	
5.1.3.2	Uridine diphosphoglucose 4-epimerase UDPglucose 4-epimerase	Vol. 1, Page 97

EC-No.	Synonym Recommended Name	Vol., Page
2.4.1.85	Uridine diphosphoglucose:aldehyde cyanohydrin β-glucosyltransferase Cyanohydrin β-glucosyltransferase	
1.1.1.22	Uridine diphosphoglucose dehydrogenase UDPglucose 6-dehydrogenase	
5.1.3.2	Uridine diphosphoglucose epimerase UDPglucose 4-epimerase	Vol. 1, Page 97
2.4.1.112	Uridine diphosphoglucose protein transglucosylase I α-1,4-Glucan-protein synthase (UDP-forming)	
2.7.7.9	Uridine diphosphoglucose pyrophosphorylase UTP-glucose-1-phosphate uridylyltransferase	
2.4.1.35	Uridine diphosphoglucosyltransferase Phenol β-glucosyltransferase	
2.4.1.17	Uridine diphosphoglucuronate-bilirubin glucuronosyltransferase Glucuronosyltransferase	
5.1.3.12	Uridine diphosphoglucuronate 5'-epimerase UDPglucuronate 5'-epimerase	Vol. 1, Page 150
5.1.3.6	Uridine diphosphoglucuronate epimerase UDPglucuronate 4-epimerase	Vol. 1, Page 132
5.1.3.6	Uridine diphosphoglucuronic epimerase UDPglucuronate 4-epimerase	Vol. 1, Page 132
2.7.7.44	Uridine diphosphoglucuronic pyrophosphorylase Glucuronate-1-phosphate uridylyltransferase	
2.4.1.17	Uridine diphosphoglucuronosyltransferase Glucuronosyltransferase	
2.4.1.17	Uridine diphosphoglucuronyltransferase Glucuronosyltransferase	
2.4.2.26	Uridine diphosphoxylose-protein xylosyltransferase Protein xylosyltransferase	
2.7.7.11	Uridine diphosphoxylose pyrophosphorylase UTP-xylose-1-phosphate uridylyltransferase	
3.2.2.3	Uridine hydrolase Uridine nucleosidase	
2.7.1.48	Uridine kinase Uridine kinase	
2.4.2.9	Uridine monophosphate pyrophosphorylase Uracil phosphoribosyltransferase	

EC-No.	Synonym Recommended Name	Vol., Page
3.2.2.3	Uridine nucleosidase Uridine nucleosidase	
2.7.1.48	Uridine phosphokinase Uridine kinase	
2.4.2.3	Uridine phosphorylase Uridine phosphorylase	
3.2.2.3	Uridine ribohydrolase Uridine nucleosidase	
6.3.4.2	Uridine triphosphate aminase CTP synthase	Vol. 2, Page 559
2.4.2.9	Uridylate pyrophosphorylase Uracil phosphoribosyltransferase	
2.4.2.9	Uridylic pyrophosphorylase Uracil phosphoribosyltransferase	
2.7.7.59	Uridyl removing enzyme [Protein-PII]uridylyltransferase	
2.7.7.12	Uridyl transferase UDPglucose-hexose-1-phosphate uridylyltransferase	
2.7.7.12	Uridyltransferase UDPglucose-hexose-1-phosphate uridylyltransferase	
2.7.7.59	Uridyltransferase, glutamine synthetase adenylyltransferase-regulating protein [Protein-PII]uridylyltransferase	
2.7.7.59	Uridylyl removing enzyme [Protein-PII]uridylyltransferase	
2.7.7.23	Uridylyltransferase, acetylglucosamine 1-phosphate UDP-N-acetylglucosamine pyrophosphorylase	
2.7.7.10	Uridylyltransferase, galactose 1-phosphate UTP-hexose-1-phosphate uridylyltransferase	
2.7.7.9	Uridylyltransferase, glucose 1-phosphate UTP-glucose-1-phosphate uridylyltransferase	
2.7.7.44	Uridylyltransferase, glucuronate 1-phosphate Glucuronate-1-phosphate uridylyltransferase	
2.7.7.59	Uridylyltransferase, glutamine synthetase adenylyltransferase-regulating protein (Escherichia coli clone 21C8 gene glnD reduced) [Protein-PII]uridylyltransferase	

EC-No.	Synonym Recommended Name	Vol., Page
2.7.7.12	Uridylyltransferase, hexose 1-phosphate UDPglucose-hexose-1-phosphate uridylyltransferase	
2.7.7.52	Uridylyltransferase, terminal RNA uridylyltransferase	
2.7.7.11	Uridylyltransferase, xylose 1-phosphate UTP-xylose-1-phosphate uridylyltransferase	
2.7.7.59	Uridylyltransferase/uridylyl-removing enzyme [Protein-PII]uridylyltransferase	
2.7.7.59	Uridylyltransferase/uridylyl removing enzyme [Protein-PII]uridylyltransferase	
3.4.21.73	Urinary esterase A u-Plasminogen activator	
3.4.21.34	urinary kallikrein plasma kallikrein	
3.4.21.35	urinary kallikrein tissue kallikrein	
3.4.21.73	Urinary plasminogen activator u-Plasminogen activator	
1.10.3.2	Urishiol oxidase Laccase	
4.1.1.37	URO-D Uroporphyrinogen decarboxylase	
4.3.1.8	URO-S hydroxymethylbilane synthase	
4.2.1.75	URO-synthase Uroporphyrinogen-III synthase	
4.2.1.49	Urocanase Urocanate hydratase	
4.2.1.49	Urocanate hydratase Urocanate hydratase	
4.3.1.8	urogenI synthase hydroxymethylbilane synthase	
3.4.21.34	urokallikrein plasma kallikrein	
3.4.21.35	urokallikrein tissue kallikrein	

EC-No.	Synonym Recommended Name	Vol., Page
3.4.21.73	Urokinase u-Plasminogen activator	
3.4.21.31	urokinase urokinase	
3.4.99.26	urokinase urokinase	
3.4.21.73	Urokinase-type plasminogen activator u-Plasminogen activator	
3.4.21.73	Urokinase plasminogen activator u-Plasminogen activator	
1.1.1.203	Uronate:NAD^+ 1-oxidoreductase Uronate dehydrogenase	
1.1.1.203	Uronate: NAD-oxidoreductase Uronate dehydrogenase	
1.1.1.203	Uronate dehydrogenase Uronate dehydrogenase	
1.2.1.35	uronate dehydrogenase uronate dehydrogenase	
5.3.1.12	Uronate isomerase Glucuronate isomerase	Vol. 1, Page 322
1.1.1.203	Uronic acid dehydrogenase Uronate dehydrogenase	
5.3.1.12	Uronic isomerase Glucuronate isomerase	Vol. 1, Page 322
3.1.1.19	uronolactonase uronolactonase	
5.1.3.12	C-5-Uronosylepimerase UDPglucuronate 5'-epimerase	Vol. 1, Page 150
2.1.1.107	Uroporphyrin-III C-methyltransferase Uroporphyrin-III C-methyltransferase	
4.1.1.37	Uroporphyrinogen-III carboxy-lyase Uroporphyrinogen decarboxylase	
4.2.1.75	Uroporphyrinogen-III cosynthase Uroporphyrinogen-III synthase	
4.2.1.75	Uroporphyrinogen-III synthase Uroporphyrinogen-III synthase	

EC-No.	Synonym Recommended Name	Vol., Page
4.1.1.37	Uroporphyrinogen decarboxylase Uroporphyrinogen decarboxylase	
4.2.1.75	Uroporphyrinogen III co-synthase Uroporphyrinogen-III synthase	
4.2.1.75	Uroporphyrinogen III cosynthase Uroporphyrinogen-III synthase	
4.1.1.37	Uroporphyrinogen III decarboxylase Uroporphyrinogen decarboxylase	
4.2.1.75	Uroporphyrinogen III synthase Uroporphyrinogen-III synthase	
4.2.1.75	Uroporphyrinogen isomerase Uroporphyrinogen-III synthase	
4.3.1.8	uroporphyrinogen I synthase hydroxymethylbilane synthase	
4.3.1.8	uroporphyrinogen I synthetase hydroxymethylbilane synthase	
4.3.1.8	uroporphyrinogen synthase hydroxymethylbilane synthase	
4.3.1.8	uroporphyrinogen synthetase hydroxymethylbilane synthase	
1.10.3.2	Urushiol oxidase Laccase	
3.6.1.45	UshB protein UDP-sugar diphosphatase	
1.1.1.199	L-Usnic acid dehydrogenase (S)-Usnate reductase	
2.7.7.59	Utase/UR [Protein-PII]uridylyltransferase	
3.4.24.23	Uterine metalloendopeptidase Matrilysin	
3.1.3.2	Uteroferrin Acid phosphatase	
6.3.4.2	UTP-ammonia ligase CTP synthase	Vol. 2, Page 559
2.7.7.9	UTP-glucose-1-phosphate uridylyltransferase UTP-glucose-1-phosphate uridylyltransferase	

EC-No.	Synonym Recommended Name	Vol., Page
2.7.7.10	UTP-hexose-1-phosphate uridylyltransferase UTP-hexose-1-phosphate uridylyltransferase	
2.7.7.11	UTP-xylose-1-phosphate uridylyltransferase UTP-xylose-1-phosphate uridylyltransferase	
2.7.7.9	UTP:α-D-glucose-1-phosphate uridylyltransferase UTP-glucose-1-phosphate uridylyltransferase	
2.7.7.10	UTP:α-D-hexose-1-phosphate uridylyltransferase UTP-hexose-1-phosphate uridylyltransferase	
2.7.7.11	UTP:α-D-xylose-1-phosphate uridylyltransferase UTP-xylose-1-phosphate uridylyltransferase	
2.7.7.44	UTP:1-phospho-α-D-glucuronate uridylyltransferase Glucuronate-1-phosphate uridylyltransferase	
2.7.7.23	UTP:2-acetamido-2-deoxy-α-D-glucose-1-phosphate uridylyltransferase UDP-N-acetylglucosamine pyrophosphorylase	
2.7.7.59	UTP:[protein-PII] uridylyltransferase [Protein-PII]uridylyltransferase	
6.3.4.2	UTP:ammonia ligase (ADP-forming) CTP synthase	Vol. 2, Page 559
2.7.7.23	UTP:N-acetyl-α-D-glucosamine-1-phosphate uridylyltransferase UDP-N-acetylglucosamine pyrophosphorylase	
2.7.7.59	UTP:PII protein uridylyltransferase [Protein-PII]uridylyltransferase	
2.7.7.52	UTP:RNA uridylyltransferase RNA uridylyltransferase	
3.4.21.19	V8 proteinase glutamyl endopeptidase	
3.5.1.50	Valeramidase Pentanamidase	
2.3.1.156	valerophenone synthase phloroisovalerophenone synthase	
2.6.1.32	Valine-2-keto-methylvalerate aminotransferase Valine-3-methyl-2-oxovalerate transaminase	
2.6.1.32	Valine-3-methyl-2-oxovalerate transaminase Valine-3-methyl-2-oxovalerate transaminase	
2.6.1.32	Valine-isoleucine aminotransferase Valine-3-methyl-2-oxovalerate transaminase	

EC-No.	Synonym Recommended Name	Vol., Page
2.6.1.32	Valine-isoleucine transaminase Valine-3-methyl-2-oxovalerate transaminase	
2.6.1.66	Valine-pyruvate transaminase Valine-pyruvate transaminase	
6.1.1.9	Valine-tRNA ligase Valine-tRNA ligase	Vol. 2, Page 59
2.6.1.32	L-Valine:(S)-3-methyl-2-oxopentanoate aminotransferase Valine-3-methyl-2-oxovalerate transaminase	
1.4.1.8	L-Valine:$NADP^+$ oxidoreductase (deaminating) Valine dehydrogenase ($NADP^+$)	
2.6.1.66	L-Valine:pyruvate aminotransferase Valine-pyruvate transaminase	
6.1.1.9	L-Valine:$tRNA^{Val}$ ligase (AMP-forming) Valine-tRNA ligase	Vol. 2, Page 59
4.1.1.14	L-Valine carboxy-lyase Valine decarboxylase	
4.1.1.14	L-Valine carboxy-lyase Valine decarboxylase	
4.1.1.14	Valine decarboxylase Valine decarboxylase	
1.4.1.8	Valine dehydrogenase ($NADP^+$) Valine dehydrogenase ($NADP^+$)	
1.4.1.8	Valine dehydrogenase (nicotinanide adenine dinucleotide phosphate) Valine dehydrogenase ($NADP^+$)	
6.1.1.9	Valine transfer ribonucleate ligase Valine-tRNA ligase	Vol. 2, Page 59
6.1.1.9	Valine translase Valine-tRNA ligase	Vol. 2, Page 59
6.1.1.9	ValRS Valine-tRNA ligase	Vol. 2, Page 59
6.1.1.9	Valyl-transfer ribonucleate synthetase Valine-tRNA ligase	Vol. 2, Page 59
6.1.1.9	Valyl-transfer RNA synthetase Valine-tRNA ligase	Vol. 2, Page 59
6.1.1.9	Valyl-tRNA ligase Valine-tRNA ligase	Vol. 2, Page 59

EC-No.	Synonym Recommended Name	Vol., Page
6.1.1.9	Valyl-tRNA synthetase Valine-tRNA ligase	Vol. 2, Page 59
6.1.1.9	Valyl transfer ribonucleic acid synthetase Valine-tRNA ligase	Vol. 2, Page 59
6.3.2.4	VanA D-Alanine-D-alanine ligase	Vol. 2, Page 423
6.3.2.4	VanB ligase D-Alanine-D-alanine ligase	Vol. 2, Page 423
1.2.3.12	vanillate:oxygen oxidoreductase (demethylating) vanillate demethylase	
1.2.3.12	vanillate demethylase vanillate demethylase	
1.2.1.67	vanillin:NAD oxidoreductase vanillin dehydrogenase	
1.2.1.67	vanillin dehydrogenase vanillin dehydrogenase	
4.1.2.41	vanillin synthase vanillin synthase	
1.1.3.38	Vanillyl-alcohol oxidase Vanillyl-alcohol oxidase	
1.1.3.38	Vanillyl alcohol:oxygen oxidoreductase Vanillyl-alcohol oxidase	
3.4.17.8	VanX muramoylpentapeptide carboxypeptidase	
3.4.16.4	VanX serine-type D-Ala-D-Ala carboxypeptidase	
3.4.17.8	VanY muramoylpentapeptide carboxypeptidase	
3.4.16.4	VanY serine-type D-Ala-D-Ala carboxypeptidase	
1.1.3.38	VAO Vanillyl-alcohol oxidase	
3.1.4.47	Variant-surface-glycoprotein-1,2-didecanoyl-sn-phosphatidylinositol inositolphosphohydrolase Variant-surface-glycoprotein phospholipase C	

EC-No.	Synonym Recommended Name	Vol., Page
3.1.4.47	Variant-surface-glycoprotein phospholipase C Variant-surface-glycoprotein phospholipase C	
3.4.21.97	Varicella-zoster virus gene 33 proteinase assemblin	
3.4.11.3	vasopresssinase cystinyl aminopeptidase	
1.4.1.8	VDH Valine dehydrogenase ($NADP^+$)	
3.4.22.2	velardon papain	
3.4.21.74	Venombin A Venombin A	
3.4.21.55	Venombin AB Venombin AB	
3.1.15.1	Venom exonuclease Venom exonuclease	
3.1.15.1	Venom phosphodiesterase Venom exonuclease	
1.1.3.7	Veratryl alcohol oxidase Aryl-alcohol oxidase	
1.11.1.7	Verdoperoxidase Peroxidase	
3.1.1.72	Versiconal hemiacetal acetate esterase Acetylxylan esterase	
3.4.24.7	vertebrate collagenase interstitial collagenase	
3.1.2.2	very long chain acyl-CoA thioesterase palmitoyl-CoA hydrolase	
1.3.1.45	Vestitone:$NADP^+$ oxidoreductase 2'-Hydroxyisoflavone reductase	
3.1.1.58	Vi-Polysaccharide deacetylase N-Acetylgalactosaminoglycan deacetylase	
1.14.14.3	Vibrio fischeri luciferase Alkanal monooxygenase (FMN-linked)	
3.2.1.119	Vicianin β-glucosidase Vicianin β-glucosidase	

EC-No.	Synonym Recommended Name	Vol., Page
3.2.1.119	Vicianin hydrolase Vicianin β-glucosidase	
3.2.1.88	Vicianosidase β-L-arabinosidase	
3.4.22.34	Vicilin peptidohydrolase Legumain	
5.3.3.3	Vinylacetyl-CoA Δ-isomerase Vinylacetyl-CoA Δ-isomerase	Vol. 1, Page 395
5.3.3.3	Vinylacetyl-CoA δ^3-δ^2-isomerase Vinylacetyl-CoA Δ-isomerase	Vol. 1, Page 395
5.3.3.3	Vinylacetyl coenzyme A isomerase Vinylacetyl-CoA Δ-isomerase	Vol. 1, Page 395
1.6.99.2	Viologen accepting pyridine nucleotide oxidoreductase NAD(P)H dehydrogenase (quinone)	
1.6.5.5	Viologen accepting pyridine nucleotide oxidoreductase NADPH:quinone reductase	
2.7.1.103	Viomycin kinase Viomycin kinase	
2.7.1.103	Viomycin phosphotransferase Viomycin kinase	
3.4.21.23	Vipera russelli proteinase Vipera russelli proteinase	
3.1.21.6	VirD1/D2 endonuclease CC-preferring endodeoxyribonuclease	
1.1.4.2	Vitamin-K-epoxide reductase (warfarin-insensitive) Vitamin-K-epoxide reductase (warfarin-insensitive)	
1.1.4.1	Vitamin-K-epoxide reductase (warfarin sensitive) Vitamin-K-epoxide reductase (warfarin sensitive)	
1.6.99.2	Vitamin-K reductase NAD(P)H dehydrogenase (quinone)	
3.1.1.1	Vitamin A esterase Carboxylesterase	
3.1.1.12	vitamin A esterase vitamin A esterase	
1.6.99.8	Vitamin B_{12a} reductase Aquacobalamin reductase	

EC-No.	Synonym Recommended Name	Vol., Page
2.1.1.13	Vitamin B_{12} methyltransferase 5-Methyltetrahydrofolate-homocysteine S-methyltransferase	
1.6.99.9	Vitamin B_{12r} reductase Cob(II)alamin reductase	
2.5.1.17	Vitamin B_{12s} adenosyltransferase Cob(I)alamin adenosyltransferase	
1.14.99.20	Vitamin K_1 epoxidase Phylloquinone monooxygenase (2,3-epoxidizing)	
1.1.4.1	Vitamin K_1 epoxide reductase Vitamin-K-epoxide reductase (warfarin sensitive)	
1.14.99.20	Vitamin K 2,3-epoxidase Phylloquinone monooxygenase (2,3-epoxidizing)	
1.1.4.2	Vitamin K 2,3 epoxide reductase Vitamin-K-epoxide reductase (warfarin-insensitive)	
1.14.99.20	Vitamin K epoxidase Phylloquinone monooxygenase (2,3-epoxidizing)	
1.1.4.1	Vitamin K epoxide reductase Vitamin-K-epoxide reductase (warfarin sensitive)	
1.1.4.2	Vitamin K epoxide reductase (warfarin–insensitive) Vitamin-K-epoxide reductase (warfarin-insensitive)	
1.1.4.2	Vitamin KO reductase Vitamin-K-epoxide reductase (warfarin-insensitive)	
1.6.99.2	Vitamin K reductase NAD(P)H dehydrogenase (quinone)	
1.6.5.5	Vitamin K reductase NADPH:quinone reductase	
2.4.1.105	Vitexin β-glucosyltransferase Vitexin β-glucosyltransferase	
1.1.1.221	Vomifoliol 4'-dehydrogenase Vomifoliol 4'-dehydrogenase	
1.1.1.221	Vomifoliol:NAD$^+$ 4'-oxidoreductase Vomifoliol 4'-dehydrogenase	
3.1.4.47	VSG-lipase Variant-surface-glycoprotein phospholipase C	
5.3.1.1	vTIM Triose-phosphate isomerase	Vol. 1, Page 235

EC-No.	Synonym Recommended Name	Vol., Page
1.9.3.1	Warburg's respiratory enzyme Cytochrome-c oxidase	
1.13.12.8	Watasenia-luciferin 2-monooxygenase Watasenia-luciferin 2-monooxygenase	
1.13.12.8	Watasenia-luciferin:oxygen 2-oxidoreductase (decarboxylating) Watasenia-luciferin 2-monooxygenase	
1.13.12.8	Watasenia-type luciferase Watasenia-luciferin 2-monooxygenase	
2.4.1.125	Water-soluble-glucan synthase Sucrose-1,6-α-glucan 3(6)-α-glucosyltransferase	
2.7.9.2	water dikinase pyruvate water dikinase pyruvate	
3.1.1.50	wax-ester acylhydrolase wax-ester hydrolase	
3.1.1.50	wax-ester hydrolase wax-ester hydrolase	
2.3.1.75	Wax-ester synthase Long-chain-alcohol O-fatty-acyltransferse	
3.1.1.50	wax ester hydrolase wax-ester hydrolase	
2.3.1.75	Wax synthase Long-chain-alcohol O-fatty-acyltransferse	
3.4.21.76	Wegener's granulomatosis autoantigen Myeloblastin	
3.1.1.50	WEH wax-ester hydrolase	
4.2.99.16	Willardiine synthase Uracilylalanine synthase	
3.4.13.5	Xaa-methyl-His dipeptidase X-methyl-His dipeptidase	
3.4.14.5	Xaa-Pro-dipeptidyl-aminopeptidase dipeptidyl-peptidase IV	
3.4.11.9	Xaa-Pro aminopeptidase X-Pro aminopeptidase	
2.4.2.22	Xan phosphoribosyltransferase Xanthine phosphoribosyltransferase	

EC-No.	Synonym Recommended Name Vol., Page
4.2.2.12	Xanthan lyase Xanthan lyase
4.2.2.12	Xanthan lyase Xanthan lyase
1.8.4.7	[Xanthine-dehydrogenase]:oxidized-glutathione S-oxidoreductase Enzyme-thiol transhydrogenase (oxidized-glutathione)
1.1.1.204	Xanthine-NAD oxidoreductase Xanthine dehydrogenase
1.1.1.204	Xanthine/NAD^+ oxidoreductase Xanthine dehydrogenase
1.1.1.204	Xanthine:NAD^+ oxidoreductase Xanthine dehydrogenase
1.1.3.22	Xanthine:O_2 oxidoreductase Xanthine oxidase
1.1.3.22	Xanthine:oxygen oxidoreductase Xanthine oxidase
1.1.3.22	Xanthine:xanthine oxidase Xanthine oxidase
1.1.1.204	Xanthine dehydrogenase Xanthine dehydrogenase
1.2.1.37	xanthine dehydrogenase xanthine dehydrogenase
1.1.3.22	Xanthine oxidase Xanthine oxidase
1.2.3.2	xanthine oxidase xanthine oxidase
1.1.1.204	Xanthine oxidoreductase Xanthine dehydrogenase
1.1.3.22	Xanthine oxidoreductase Xanthine oxidase
2.4.2.22	Xanthine phosphoribosyltransferase Xanthine phosphoribosyltransferase
1.3.1.41	Xanthommatin reductase Xanthommatin reductase
3.4.23.33	Xanthomonapepsin Xanthomonapepsin

EC-No.	Synonym Recommended Name	Vol., Page
3.4.23.33	Xanthomonas aspartic proteinase Xanthomonapepsin	
6.3.4.1	Xanthosine-5'-phosphate-ammonia ligase GMP synthase	Vol. 2, Page 548
6.3.5.2	Xanthosine-5'-phosphate-ammonia ligase GMP synthase (glutamine hydrolysing)	Vol. 2, Page 655
6.3.4.1	Xanthosine-5'-phosphate:ammonia ligase (AMP-forming) GMP synthase	Vol. 2, Page 548
6.3.5.2	Xanthosine-5'-phosphate:ammonia ligase (AMP-forming) GMP synthase (glutamine hydrolysing)	Vol. 2, Page 655
6.3.4.1	Xanthosine-5'-phosphate:L-glutamine amido-ligase (AMP-forming) GMP synthase	Vol. 2, Page 548
6.3.5.2	Xanthosine-5'-phosphate:L-glutamine amido-ligase (AMP-forming) GMP synthase (glutamine hydrolysing)	Vol. 2, Page 655
6.3.4.1	Xanthosine 5'-phosphate amidotransferase GMP synthase	Vol. 2, Page 548
6.3.5.2	Xanthosine 5'-phosphate amidotransferase GMP synthase (glutamine hydrolysing)	Vol. 2, Page 655
2.4.2.22	Xanthosine 5'-phosphate pyrophosphorylase Xanthine phosphoribosyltransferase	
6.3.4.1	Xanthosine 5-monophosphate aminase GMP synthase	Vol. 2, Page 548
6.3.5.2	Xanthosine 5-monophosphate aminase GMP synthase (glutamine hydrolysing)	Vol. 2, Page 655
2.1.1.93	Xanthotoxol O-methyltransferase Xanthotoxol O-methyltransferase	
2.4.2.22	Xanthylate pyrophosphorylase Xanthine phosphoribosyltransferase	
2.4.2.22	Xanthylic pyrophosphorylase Xanthine phosphoribosyltransferase	
3.3.2.3	Xenobiotic epoxide hydrolase Epoxide hydrolase	
1.1.1.184	Xenobiotic ketone reductase Carbonyl reductase (NADPH)	
1.14.14.1	Xenobiotic monooxygenase Unspecific monooxygenase	

EC-No.	Synonym Recommended Name	Vol., Page
2.4.1.207	XET Xyloglucan:xyloglucosyl transferase	
5.3.1.5	XI Xylose isomerase	Vol. 1, Page 259
6.3.4.1	XMP aminase GMP synthase	Vol. 2, Page 548
6.3.5.2	XMP aminase GMP synthase (glutamine hydrolysing)	Vol. 2, Page 655
2.4.2.22	XMP pyrophosphorylase Xanthine phosphoribosyltransferase	
5.3.1.24	xPRAI Phosphoribosylanthranilate isomerase	Vol. 1, Page 353
3.2.1.131	Xylan α-1,2-glucuronosidase Xylan α-1,2-glucuronosidase	
3.2.1.131	Xylan α-D-1,2-(4-O-methyl)glucuronohydrolase Xylan α-1,2-glucuronosidase	
3.2.1.8	1,4-β-D-xylan-xylanohydrolase Endo-1,4-β-xylanase	
3.2.1.72	Xylan 1,3-β-xylosidase Xylan 1,3-β-xylosidase	
3.2.1.37	Xylan 1,4-β-xylosidase Xylan 1,4-β-xylosidase	
3.2.1.32	.β.-1,3-Xylanase Xylan endo-1,3-β-xylosidase	
3.2.1.32	1,3-.β.-Xylanase Xylan endo-1,3-β-xylosidase	
3.2.1.32	1,3-Xylanase Xylan endo-1,3-β-xylosidase	
3.2.1.72	β-1,3'-xylanase Xylan 1,3-β-xylosidase	
3.2.1.8	β-1,4-xylanase Endo-1,4-β-xylanase	
3.2.1.8	β-D-xylanase Endo-1,4-β-xylanase	
3.2.1.8	Xylanase Endo-1,4-β-xylanase	

EC-No.	Synonym Recommended Name	Vol., Page
3.2.1.32	Xylanase Xylan endo-1,3-β-xylosidase	
3.2.1.8	β-1,4-D-Xylanase Endo-1,4-β-xylanase	
3.2.1.8	β-Xylanase Endo-1,4-β-xylanase	
3.2.1.32	Xylanase, endo-1,3- Xylan endo-1,3-β-xylosidase	
3.2.1.8	Xylanase, endo-1,4- Endo-1,4-β-xylanase	
3.2.1.136	Xylanase, glucuronoarabinoxylan endo-1,4-β- Glucuronoarabinoxylan endo-1,4-β-xylanase	
3.2.1.32	Xylan endo-1,3-β-xylosidase Xylan endo-1,3-β-xylosidase	
2.4.2.24	1,4-β-D-Xylan synthase 1,4-β-D-Xylan synthase	
2.4.2.24	1,4-β-Xylan synthase 1,4-β-D-Xylan synthase	
2.4.1.72	1,4-β-xylan synthase 1,4-β-xylan synthase	
2.4.2.24	Xylan synthase 1,4-β-D-Xylan synthase	
2.4.2.24	Xylan synthetase 1,4-β-D-Xylan synthase	
3.2.1.32	1,3-β-D-xylan xylanohydrolase Xylan endo-1,3-β-xylosidase	
3.2.1.37	1,4-β-D-xylan xylanohydrolase Xylan 1,4-β-xylosidase	
3.2.1.8	1,4-β-xylan xylanohydrolase Endo-1,4-β-xylanase	
3.2.1.8	β-1,4-Xylan xylanohydrolase Endo-1,4-β-xylanase	
3.2.1.72	1,3-β-D-xylan xylohydrolase Xylan 1,3-β-xylosidase	
1.1.1.9	Xylitol-2-dehydrogenase D-Xylulose reductase	

EC-No.	Synonym Recommended Name	Vol., Page
3.1.3.58	Xylitol-5-phosphatase Sugar-terminal-phosphatase	
1.1.1.9	Xylitol:NAD+ 2-oxidoreductase (D-xylulose-forming) D-Xylulose reductase	
1.1.1.10	Xylitol:NADP+ 4-oxidoreductase (L-xylulose-forming) L-Xylulose reductase	
1.1.1.9	Xylitol dehydrogenase D-Xylulose reductase	
1.1.1.10	Xylitol dehydrogenase L-Xylulose reductase	
2.7.1.122	Xylitol kinase Xylitol kinase	
2.7.1.122	Xylitol phosphotransferase Xylitol kinase	
4.2.1.82	D-xylo-Aldonate dehydratase Xylonate dehydratase	
3.2.1.37	Xylobiase Xylan 1,4-β-xylosidase	
2.4.1.207	Xyloglucan-specific endo-(1->4)-β-D-glucanase Xyloglucan:xyloglucosyl transferase	
2.4.1.168	Xyloglucan 4β-D-glucosyltransferase Xyloglucan 4-glucosyltransferase	
2.4.1.168	Xyloglucan 4-glucosyltransferase Xyloglucan 4-glucosyltransferase	
2.4.1.169	Xyloglucan 6-α-D-xylosyltransferase Xyloglucan 6-xylosyltransferase	
2.4.1.169	Xyloglucan 6-xylosyltransferase Xyloglucan 6-xylosyltransferase	
2.4.1.207	Xyloglucan:xyloglucanotransferase Xyloglucan:xyloglucosyl transferase	
2.4.1.207	Xyloglucan:xyloglucan xyloglucanotransferase Xyloglucan:xyloglucosyl transferase	
2.4.1.207	Xyloglucan:xyloglucosyl transferase Xyloglucan:xyloglucosyl transferase	
2.4.1.207	Xyloglucan endotransglycosylase Xyloglucan:xyloglucosyl transferase	

EC-No.	Synonym Recommended Name	Vol., Page
2.4.1.207	Xyloglucan endotransglycosylase (Actinia deliciosa strain Hayward pericarp clone AdXET-5 gene XET precursor) Xyloglucan:xyloglucosyl transferase	
2.4.1.207	Xyloglucan endotransglycosylase (Arabidopsis thaliana gene TCH4 precursor reduced) Xyloglucan:xyloglucosyl transferase	
2.4.1.207	Xyloglucan endotransglycosylase (barley clone PM5 gene HVPM5) Xyloglucan:xyloglucosyl transferase	
2.4.1.207	Xyloglucan endotransglycosylase (barley clone XEA gene HVXEA) Xyloglucan:xyloglucosyl transferase	
2.4.1.207	Xyloglucan endotransglycosylase (barley clone XEB gene HVXEB) Xyloglucan:xyloglucosyl transferase	
2.4.1.207	Xyloglucan endotransglycosylase-related protein XTR-7 (Arabidopsis thaliana) Xyloglucan:xyloglucosyl transferase	
2.4.1.168	Xyloglucan glucosyltransferase Xyloglucan 4-glucosyltransferase	
2.4.1.207	Xyloglucanotransferase, xyloglucan (xyloglucan donor) Xyloglucan:xyloglucosyl transferase	
2.4.1.207	Xyloglucanotransferase, xyloglucan (xyloglucan donor Actinia chinensis clone AdXET-5 gene XET precursor) Xyloglucan:xyloglucosyl transferase	
2.4.1.207	Xyloglucanotransferase, xyloglucan (xyloglucan donor Arabidopsis thaliana gene TCH4 precursor reduced) Xyloglucan:xyloglucosyl transferase	
2.4.1.207	Xyloglucanotransferase, xyloglucan (xyloglucan donor barley clone PM5 gene HVPM5) Xyloglucan:xyloglucosyl transferase	
2.4.1.207	Xyloglucanotransferase, xyloglucan (xyloglucan donor barley clone XEA gene HVXEA) Xyloglucan:xyloglucosyl transferase	
2.4.1.207	Xyloglucanotransferase, xyloglucan (xyloglucan donor barley clone XEB gene HVXEB) Xyloglucan:xyloglucosyl transferase	
2.4.1.207	Xyloglucan X93174-derived protein GI 1890575 Xyloglucan:xyloglucosyl transferase	
4.2.1.82	Xylonate dehydratase Xylonate dehydratase	

EC-No.	Synonym Recommended Name	Vol., Page
4.2.1.82	D-Xylonate hydro-lyase Xylonate dehydratase	
3.1.1.68	Xylono-1,4-lactonase Xylono-1,4-lactonase	
3.1.1.68	D-Xylono-1,4-lactone lactonohydrolase Xylono-1,4-lactonase	
3.1.1.68	Xylonolactonase Xylono-1,4-lactonase	
3.2.1.37	.β.-D-Xylopyranosidase Xylan 1,4-β-xylosidase	
2.7.7.11	Xylose-1-phosphate uridylyltransferase UTP-xylose-1-phosphate uridylyltransferase	
1.1.1.179	D-Xylose-NADP dehydrogenase D-Xylose 1-dehydrogenase ($NADP^+$)	
1.1.1.175	D-Xylose 1-dehydrogenase D-Xylose 1-dehydrogenase	
1.1.1.113	L-Xylose 1-dehydrogenase L-Xylose 1-dehydrogenase	
1.1.1.179	D-Xylose 1-dehydrogenase ($NADP^+$) D-Xylose 1-dehydrogenase ($NADP^+$)	
2.7.7.11	Xylose 1-phosphate uridylyltransferase UTP-xylose-1-phosphate uridylyltransferase	
1.1.1.175	D-Xylose:NAD^+ 1-oxidoreductase D-Xylose 1-dehydrogenase	
1.1.1.179	D-Xylose:$NADP^+$ 1-oxidoreductase D-Xylose 1-dehydrogenase ($NADP^+$)	
1.1.1.113	L-Xylose:$NADP^+$ 1-oxidoreductase L-Xylose 1-dehydrogenase	
1.1.1.179	D-Xylose:$NADP^+$ oxidoreductase D-Xylose 1-dehydrogenase ($NADP^+$)	
1.1.1.175	D-Xylose dehydrogenase D-Xylose 1-dehydrogenase	
5.3.1.5	D-Xylose isomerase Xylose isomerase	Vol. 1, Page 259
5.3.1.5	Xylose isomerase Xylose isomerase	Vol. 1, Page 259

EC-No.	Synonym Recommended Name	Vol., Page
5.3.1.5	D-Xylose ketoisomerase Xylose isomerase	Vol. 1, Page 259
5.3.1.5	D-Xylose ketol-isomerase Xylose isomerase	Vol. 1, Page 259
3.2.1.37	.β.-Xylosidase Xylan 1,4-β-xylosidase	
3.2.1.37	β-xylosidase Xylan 1,4-β-xylosidase	
3.2.1.37	Xylosidase, exo-1,4-.β.- Xylan 1,4-β-xylosidase	
3.2.1.37	β-xylosidase/β-glucosidase Xylan 1,4-β-xylosidase	
2.4.1.133	Xylosylprotein 4-β-galactosyltransferase Xylosylprotein 4-β-galactosyltransferase	
2.4.2.24	Xylosyltransferase, uridine diphosphoxylose-1,4-β-xylan 1,4-β-D-Xylan synthase	
2.4.2.26	Xylosyltransferase, uridine diphosphoxylose-core protein β- Protein xylosyltransferase	
2.4.2.35	Xylosyltransferase, uridine diphosphoxylose-flavonol 3-glycoside Flavonol-3-O-glycoside xylosyltransferase	
2.4.1.169	Xylosyltransferase, uridine diphosphoxylose-xyloglucan 6α- Xyloglucan 6-xylosyltransferase	
2.4.1.204	Xylosyltransferase, uridine diphosphoxylose-zeatin Zeatin O-β-D-xylosyltransferase	
2.7.1.17	D-Xylulokinase Xylulokinase	
2.7.1.53	L-Xylulokinase L-Xylulokinase	
2.7.1.17	Xylulokinase Xylulokinase	
5.1.3.1	D-Xylulose-5-phosphate 3-epimerase Ribulose-phosphate 3-epimerase	Vol. 1, Page 91
2.2.1.3	D-Xylulose-5-phosphate:formaldehyde glycolaldehydetransferase Formaldehyde transketolase	
4.1.2.9	D-Xylulose-5-phosphate D-glyceraldehyde-3-phosphate-lyase Phosphoketolase	

EC-No.	Synonym Recommended Name	Vol., Page
4.1.2.9	D-Xylulose-5-phosphate D-glyceraldehyde-3-phosphate-lyase (phosphate-acetylating) Phosphoketolase	
4.1.2.9	D-Xylulose-5-phosphate phosphoketolase Phosphoketolase	
5.1.3.1	Xylulose phosphate 3-epimerase Ribulose-phosphate 3-epimerase	Vol. 1, Page 91
1.1.1.9	D-Xylulose reductase D-Xylulose reductase	
1.1.1.10	L-Xylulose reductase L-Xylulose reductase	
3.4.23.41	Yap3 gene product yapsin I	
3.4.23.41	yapsin I yapsin I	
3.4.23.41	yapsin I yapsin I	
1.1.1.1	Yeast alcohol dehydrogenase Alcohol dehydrogenase	
3.4.11.22	Yeast aminopeptidase I Aminopeptidase I	
3.4.23.41	yeast aspartic protease yapsin I	
3.4.23.41	yeast aspartic protease 3 yapsin I	
3.4.2.3	yeast carboxypeptidase yeast carboxypeptidase	
3.4.22.22	yeast cysteine proteinase D yeast cysteine proteinase D	
3.4.22.21	yeast cysteine proteinase E yeast cysteine proteinase E	
3.4.22.23	yeast cysteine proteinase F yeast cysteine proteinase F	
3.4.23.25	Yeast endopeptidase A Saccharopepsin	

EC-No.	Synonym Recommended Name	Vol., Page
2.3.1.86	Yeast fatty acid synthase Fatty-acyl-CoA synthase	
3.4.21.61	Yeast KEX2 protease Kexin	
3.4.23.25	Yeast proteinase A Saccharopepsin	
3.4.23.8	yeast proteinase A yeast proteinase A	
3.4.21.48	yeast proteinase B cerevisin	
3.4.22.9	yeast proteinase B yeast proteinase B	
3.4.21.48	yeast proteinase yscB cerevisin	
3.1.14.1	Yeast ribonuclease Yeast ribonuclease	
3.1.1.20	yeast tannase tannase	
3.4.19.12	yeast ubiquitin hydrolase ubiquitinyl hydrolase I	
3.4.21.91	Yellow fever virus (flavivirus) protease Flavivirin	
3.5.1.19	YNDase Nicotinamidase	
3.4.11.3	yscXVI cystinyl aminopeptidase	
3.4.19.12	YUH1 ubiquitinyl hydrolase I	
5.3.99.6	(9Z)-(13S)-12,13-Epoxyoctadeca-9,11,15-trienoate isomerase (cyclizing) Allene-oxide cyclase	Vol. 1, Page 483
1.3.1.48	(5Z)-(15S)-11α-Hydroxy-9,15-dioxoprostanoate:NAD(P)$^+$ Δ^{13}-oxidoreductase 15-Oxoprostaglandin Δ13-reductase	
4.2.1.92	(9Z,11E,14Z)-(13S)-Hydroperoxyoctadeca-(9,11,14)-trienoate 12,13-hydro-lyase Hydroperoxide dehydratase	

EC-No.	Synonym Recommended Name	Vol., Page
1.1.1.141	(5Z,13E)-(15S)-11α,15-Dihydroxy-9-oxoprost-13-enoate:NAD^+ 15-oxidoreductase 15-Hydroxyprostaglandin dehydrogenase (NAD^+)	
1.1.1.231	(5Z,13E)-(15S)-6,9α-Epoxy-11α,15-dihydroxyprosta-5,13-dienoate:$NADP^+$ 15-oxidoreductase 15-Hydroxyprostaglandin-I dehydrogenase ($NADP^+$)	
1.1.1.188	(5Z,13E)-(15S)-9α,11α,15-Trihydroxyprosta-5,13-dienoate:$NADP^+$ 11-oxidoreductase Prostaglandin-F synthase	
1.1.1.189	(5Z,13E)-(15S)-9α,11α,15-Trihydroxyprosta-5,13-dienoate:$NADP^+$ 9-oxidoreductase Prostaglandin-E_2 9-reductase	
5.3.99.4	(5Z,13E)-(15S)-9α,11α-Epidioxy-15-hydroxyprosta-5,13-dienoate 6-isomerase Prostacyclin synthase	Vol. 1, Page 465
5.3.99.3	(5Z,13E)-(15S)-9α,11α-Epidioxy-15-hydroxyprosta-5,13-dienoate E-isomerase Prostaglandin-E synthetase	Vol. 1, Page 459
5.3.99.5	(5Z,13E)-(15S)-9α,11α-Epidioxy-15-hydroxyprosta-5,13-dienoate thromboxane-A2-isomerase Thromboxane-A synthase	Vol. 1, Page 472
1.1.1.196	(5Z,13E)-(15S)-9α,15-Dihydroxy-11-oxoprosta-5,13-dienoate:$NADP^+$ 15-oxidoreductase 15-Hydroxyprostaglandin-D dehydrogenase ($NADP^+$)	
1.14.13.30	(6Z,8E,10E,14Z)-(5S,12R)-5,12-Dihydroxyicosa-6,8,10,14-tetraenoate,NADPH:oxygen oxidoreductase (20-hydroxylating) Leukotriene-B_4 20-monooxygenase	
1.14.99.1	(5Z,8Z,11Z,14Z)-Icosa-5,8,11,14-tetraenoate,hydrogen-donor:oxygen oxidoreductase Prostaglandin-endoperoxide synthase	
3.3.2.7	(5Z,9E,14Z)-(8xi,11R,12S)-11,12-Epoxy-8-hydroxyicosa-5,9,14-trienoate hydrolase Hepoxilin-epoxide hydrolase	
2.4.1.203	Zeatin O-β-D-glucosyltransferase Zeatin O-β-D-glucosyltransferase	
2.4.1.204	Zeatin O-β-D-xylosyltransferase Zeatin O-β-D-xylosyltransferase	
2.4.1.203	Zeatin O-glucosyltransferase Zeatin O-β-D-glucosyltransferase	

EC-No.	Synonym Recommended Name	Vol., Page
2.4.1.204	Zeatin O-xylosyltransferase Zeatin O-β-D-xylosyltransferase	
1.1.1.242	Zeatin reductase Zeatin reductase	
1.14.99.30	ζ-Carotene desaturase Carotene 7,8-desaturase	
1.14.99.30	ζ-Carotene desaturase (Anabaena strain PCC 7120 clone pZDS1A) Carotene 7,8-desaturase	
1.14.99.30	ζ-Carotene desaturase (Capsicum annuum clone pCapZDS precursor reduced) Carotene 7,8-desaturase	
1.14.99.30	ζ-Carotene desaturase (Synechocystis strain PCC 6803 gene crtQ-2) Carotene 7,8-desaturase	
1.6.5.5	ζ-Crystallin NADPH:quinone reductase	
1.6.5.5	ζ-Crystallin/NADPH:quinone oxidoreductase NADPH:quinone reductase	
1.6.5.5	ζ-Crystallin/quinone reductase NADPH:quinone reductase	
3.4.17.14	Zinc D-Ala-D-Ala carboxypeptidase Zinc D-Ala-D-Ala carboxypeptidase	
3.4.21.74	Zinc metalloproteinase Cbfib1.1 Venombin A	
3.4.21.74	Zinc metalloproteinase Cbfib1.2 Venombin A	
3.4.21.74	Zinc metalloproteinase Cbfib2 Venombin A	
3.1.4.38	Zn^{2+}-glycerophosphocholine cholinphosphodiesterase glycerophosphocholine cholinephosphodiesterase	
3.1.4.38	Zn^{2+}-GPC cholinphosphodiesterase glycerophosphocholine cholinephosphodiesterase	
3.4.17.14	Zn^{2+} G peptidase Zinc D-Ala-D-Ala carboxypeptidase	
1.1.1.49	Zwischenferment Glucose-6-phosphate 1-dehydrogenase	

EC-No.	Synonym Recommended Name	Vol., Page
4.1.2.13	Zymohexase Fructose-bisphosphate aldolase	
2.1.1.41	Zymosterol-24-methyltransferase 24-Sterol C-methyltransferase	

1.14.99.9	EC 1. 99.1.9 Steroid 17α-monooxygenase	
1.3.1.28	EC 1.1.1.109 2,3-Dihydro-2,3-dihydroxybenzoate dehydrogenase	
1.5.1.20	EC 1.1.1.171 Methylenetetrahydrofolate reductase (NADPH)	
1.1.1.131	EC 1.1.1.180 Mannuronate reductase	
1.7.99.5	EC 1.1.1.68 5,10-Methylenetetrahydrofolate reductase ($FADH_2$)	
1.1.1.4	EC 1.1.1.74 (R,R)-Butanediol dehydrogenase	
1.1.99.5	EC 1.1.2.1 Glycerol-3-phosphate dehydrogenase	
1.1.3.15	EC 1.1.3.1 (S)-2-Hydroxy-acid oxidase	
1.13.12.4	EC 1.1.3.2 Lactate 2-monooxygenase	
1.7.99.5	EC 1.1.99.15 5,10-Methylenetetrahydrofolate reductase ($FADH_2$)	
1.14.18.1	EC 1.10.3.1 Monophenol monooxygenase	
1.13.11.11	EC 1.11.1.4 Tryptophan 2,3-dioxygenase	
1.18.99.1	EC 1.12.1.1 Hydrogenase	
1.18.99.1	EC 1.12.7.1 Hydrogenase	
1.13.11.1	EC 1.13.1.1 Catechol 1,2-dioxygenase	

EC-No.	Synonym Recommended Name	Vol., Page
1.13.11.10	EC 1.13.1.10 7,8-Dihydroxykynurenate 8,8a-dioxygenase	
1.13.99.1	EC 1.13.1.11 myo-Inositol oxygenase	
1.13.11.11	EC 1.13.1.12 Tryptophan 2,3-dioxygenase	
1.13.11.12	EC 1.13.1.13 Lipoxygenase	
1.13.11.2	EC 1.13.1.2 Catechol 2,3-dioxygenase	
1.13.11.3	EC 1.13.1.3 Protocatechuate 3,4-dioxygenase	
1.13.11.4	EC 1.13.1.4 Gentisate 1,2-dioxygenase	
1.13.11.5	EC 1.13.1.5 Homogentisate 1,2-dioxygenase	
1.13.11.6	EC 1.13.1.6 3-Hydroxyanthranilate 3,4-dioxygenase	
1.13.11.8	EC 1.13.1.8 Protocatechuate 4,5-dioxygenase	
1.13.11.9	EC 1.13.1.9 2,5-Dihydroxypyridine 5,6-dioxygenase	
1.14.12.10	EC 1.13.99.2 Benzoate 1,2-dioxygenase	
1.14.12.9	EC 1.13.99.4 4-Chlorophenylacetate 3,4-dioxygenase	
1.14.14.1	EC 1.14.1.1 Unspecific monooxygenase	
1.14.99.11	EC 1.14.1.10 Estradiol 6β-monooxygenase	
1.14.13.9	EC 1.14.1.2 Kynurenine 3-monooxygenase	
1.14.99.2	EC 1.14.1.4 Kynurenine 7,8-hydroxylase	
1.14.13.5	EC 1.14.1.5 Imidazoleacetate 4-monooxygenase	

EC-No.	Synonym Recommended Name	Vol., Page
1.14.15.4	EC 1.14.1.6 Steroid 11β-monooxygenase	
1.14.99.9	EC 1.14.1.7 Steroid 17α-monooxygenase	
1.14.99.10	EC 1.14.1.8 Steroid 21-monooxygenase	
1.14.11.6	EC 1.14.11.5 Thymine dioxygenase	
1.14.13.35	EC 1.14.12.2 Anthranilate 3-monooxygenase (deaminating)	
1.14.14.1	EC 1.14.14.2 Unspecific monooxygenase	
1.14.17.1	EC 1.14.2.1 Dopamine β-monooxygenase	
1.13.11.27	EC 1.14.2.2 4-Hydroxyphenylpyruvate dioxygenase	
1.14.16.1	EC 1.14.3.1 Phenylalanine 4-monooxygenase	
1.14.13.23	EC 1.14.99.13 3-Hydroxybenzoate 4-monooxygenase	
1.14.16.5	EC 1.14.99.17 Glyceryl-ether monooxygenase	
1.14.19.1	EC 1.14.99.5 stearoyl-CoA 9-desaturase	
1.14.19.2	EC 1.14.99.6 acyl-[acyl-carrier-protein] desaturase	
1.14.14.1	EC 1.14.99.8 Unspecific monooxygenase	
1.18.6.1	EC 1.18.2.1 Nitrogenase	
1.18.99.1	EC 1.18.3.1 Hydrogenase	
1.19.6.1	EC 1.19.2.1 Nitrogenase (flavodoxin)	
1.1.1.205	EC 1.2.1.14 IMP dehydrogenase	

EC-No.	Synonym Recommended Name	Vol., Page
1.2.1.33	EC 1.2.1.33, D-aldopantoate (R)-Dehydropantoate dehydrogenase	
1.1.1.131	EC 1.2.1.34 Mannuronate reductase	
1.1.1.203	EC 1.2.1.35 Uronate dehydrogenase	
1.2.1.36	EC 1.2.1.36, retinal Retinal dehydrogenase	
1.1.1.204	EC 1.2.1.37 Xanthine dehydrogenase	
1.1.3.22	EC 1.2.3.2. Xanthine oxidase	
1.2.4.4	EC 1.2.4.3 3-Methyl-2-oxobutanoate dehydrogenase (lipoamide)	
1.1.99.19	EC 1.2.99.1 Uracil dehydrogenase	
1.3.1.19	EC 1.3.1.19, cis-1,2-hydrobenzene-1,2-diol cis-1,2-Dihydrobenzene-1,2-diol dehydrogenase	
1.1.1.252	EC 1.3.1.50 Tetrahydroxynaphthalene reductase	
1.3.1.7	EC 1.3.1.7, meso-Tartrate meso-Tartrate dehydrogenase	
1.3.99.2	EC 1.3.2.1 Butyryl-CoA dehydrogenase	
1.3.99.3	EC 1.3.2.2 Acyl-CoA dehydrogenase	
1.4.4.1	EC 1.4.1.6 D-Proline reductase (dithiol)	
1.5.1.21	EC 1.5.1.14 Δ^1-Piperideine-2-carboxylate reductase	
1.5.1.3	EC 1.5.1.4 Dihydrofolate reductase	
1.12.99.4	EC 1.5.99.9 N^5,N^{10}-Methenyltetrahydromethanopterin hydrogenase	
1.6.99.3	EC 1.6.2.1 NADH dehydrogenase	

EC-No.	Synonym Recommended Name	Vol., Page
1.8.1.4	EC 1.6.4.3 Dihydrolipoamide dehydrogenase	
1.6.99.2	EC 1.6.5.2 NAD(P)H dehydrogenase (quinone)	
1.7.99.3	EC 1.6.6.5 Nitrite reductase	
1.18.1.2	EC 1.6.7.1 Ferredoxin-$NADP^+$ reductase	
1.18.1.1	EC 1.6.7.2 Rubredoxin-NAD^+ reductase	
1.6.99.7	EC 1.6.99.10 Dihydropteridine reductase	
1.6.2.4	EC 1.6.99.2 NADPH-ferrihemoprotein reductase	
1.6.5.5	EC 1.6.99.2 NADPH:quinone reductase	
1.18.1.2	EC 1.6.99.4 Ferredoxin-$NADP^+$ reductase	
1.7.99.7	EC 1.7.99.2 Nitric-oxide reductase	
1.18.99.1	EC 1.98.1.1 Hydrogenase	
1.14.14.1	EC 1.99.1.1 Unspecific monooxygenase	
1.14.99.10	EC 1.99.1.11 Steroid 21-monooxygenase	
1.14.16.1	EC 1.99.1.2 Phenylalanine 4-monooxygenase	
1.13.11.27	EC 1.99.1.4 4-Hydroxyphenylpyruvate dioxygenase	
1.14.13.9	EC 1.99.1.5 Kynurenine 3-monooxygenase	
1.14.15.4	EC 1.99.1.7 Steroid 11β-monooxygenase	
1.14.99.11	EC 1.99.1.8 Estradiol 6β-monooxygenase	

EC-No.	Synonym Recommended Name	Vol., Page
1.13.11.12	EC 1.99.2.1 Lipoxygenase	
1.13.11.1	EC 1.99.2.2 Catechol 1,2-dioxygenase	
1.13.11.3	EC 1.99.2.3 Protocatechuate 3,4-dioxygenase	
1.13.11.4	EC 1.99.2.4 Gentisate 1,2-dioxygenase	
1.13.11.5	EC 1.99.2.5 Homogentisate 1,2-dioxygenase	
1.13.99.1	EC 1.99.2.6 myo-Inositol oxygenase	
5.4.2.5	E.C. 2. 7.5.5 Phosphoglucomutase (glucose-cofactor)	Vol. 1, Page 527
2.1.1.125	EC 2.1.1.23 (Histone)-arginine N-methyltransferase	
2.1.1.124	EC 2.1.1.23 [Cytochrome-c]-arginine N-methyltransferase	
2.1.1.126	EC 2.1.1.23 [Myelin basic protein]-arginine N-methyltransferase	
2.1.1.80	E.C. 2.1.1.24 Protein-glutamate O-methyltransferase	
2.1.1.77	EC 2.1.1.24 Protein-L-isoaspartate(D-aspartate) O-methyltransferase	
2.1.1.57	EC 2.1.1.58 mRNA (nucleoside-2'-O-)-methyltransferase	
2.1.2.5	EC 2.1.2.6 Glutamate formiminotransferase	
2.3.1.74	EC 2.3.1.120 Naringenin-chalcone synthase	
2.3.1.20	EC 2.3.1.124 Diacylglycerol O-acyltransferase	
2.4.1.17	EC 2.4.1.107 Glucuronosyltransferase	
2.4.1.17	EC 2.4.1.108 Glucuronosyltransferase	

EC-No.	Synonym Recommended Name	Vol., Page
2.3.1.152	EC 2.4.1.120 Alcohol O-cinnamoyltransferase	
2.4.1.25	EC 2.4.1.3 4-α-Glucanotransferase	
2.4.1.17	EC 2.4.1.42 Glucuronosyltransferase	
2.7.8.14	EC 2.4.1.55 CDPribitol ribitolphosphotransferase	
2.4.1.17	EC 2.4.1.59 Glucuronosyltransferase	
2.4.1.17	EC 2.4.1.61 Glucuronosyltransferase	
2.4.2.24	EC 2.4.1.72 1,4-β-D-Xylan synthase	
2.4.1.17	EC 2.4.1.76 Glucuronosyltransferase	
2.4.1.17	EC 2.4.1.77 Glucuronosyltransferase	
2.4.1.17	EC 2.4.1.84 Glucuronosyltransferase	
2.4.1.69	EC 2.4.1.89 Galactoside 2-α-L-fucosyltransferase	
2.4.1.90	EC 2.4.1.98 N-Acetyllactosamine synthase	
2.5.1.6	EC 2.4.2.13 Methionine adenosyltransferase	
4.2.3.9	EC 2.5.1.40 aristolochene synthase	
1.4.1.13	EC 2.6.1.53 Glutamate synthase (NADPH)	
2.1.4.1	EC 2.6.2.1 Glycine amidinotransferase	
2.7.1.21	EC 2.7.1.75 Thymidine kinase	
6.3.4.16	EC 2.7.2.5 Carbamoyl-phosphate synthase (ammonia)	Vol. 2, Page 641

EC-No.	Synonym Recommended Name	Vol., Page
6.3.5.5	E.C. 2.7.2.9 Carbamoyl-phosphate synthase (glutamine-hydrolysing)	Vol. 2, Page 689
5.4.2.2	E.C. 2.7.5.1 Phosphoglucomutase	Vol. 1, Page 506
5.4.2.3	EC 2.7.5.2. Phosphoacetylglucosamine mutase	Vol. 1, Page 515
5.4.2.4	E.C. 2.7.5.4 Bisphosphoglycerate mutase	Vol. 1, Page 520
5.4.2.7	E.C. 2.7.5.6 Phosphopentomutase	Vol. 1, Page 535
5.4.2.8	E.C. 2.7.5.7 Phosphomannomutase	Vol. 1, Page 540
3.1.27.5	E.C. 2.7.7.16 Pancreatic ribonuclease	
3.1.27.1	E.C. 2.7.7.17 Ribonuclease T_2	
2.7.7.25	EC 2.7.7.21 tRNA adenylyltransferase	
2.7.7.21	EC 2.7.7.25 tRNA cytidylyltransferase	
3.1.27.3	E.C. 2.7.7.26 Ribonuclease T_1	
2.7.8.2	EC 2.7.8.16 Diacylglycerol cholinephosphotransferase	
3.1.21.3	E.C. 3.1.23 Type I site-specific deoxyribonuclease	
3.1.21.4	E.C. 3.1.23 Type II site-specific deoxyribonuclease	
3.1.21.5	E.C. 3.1.23 Type III site-specific deoxyribonuclease	
3.1.21.4	E.C. 3.1.24 Type II site-specific deoxyribonuclease	
3.1.21.5	E.C. 3.1.24 Type III site-specific deoxyribonuclease	
3.1.21.3	E.C. 3.1.24._ Type I site-specific deoxyribonuclease	

EC-No.	Synonym Recommended Name	Vol., Page
4.2.99.18	EC 3.1.25.2 DNA-(apurinic or apyrimidinic site) lyase	
3.1.16.1	EC 3.1.4.18 Spleen exonuclease	
3.1.13.3	E.C. 3.1.4.19 Oligonucleotidase	
3.1.13.1	E.C. 3.1.4.20 Exoribonuclease II	
3.1.30.1	E.C. 3.1.4.21 Aspergillus nuclease S_1	
3.1.27.5	E.C. 3.1.4.22 Pancreatic ribonuclease	
3.1.27.1	E.C. 3.1.4.23 Ribonuclease T_2	
3.1.11.1	E.C. 3.1.4.25 Exodeoxyribonuclease I	
3.1.11.2	E.C. 3.1.4.27 Exodeoxyribonuclease III	
3.1.11.3	E.C. 3.1.4.28 Exodeoxyribonuclease (Lambda induced)	
3.1.21.2	E.C. 3.1.4.30 Deoxyribonuclease IV (Phage T_4-induced)	
3.1.21.6	EC 3.1.4.30 CC-preferring endodeoxyribonuclease	
3.1.21.3	E.C. 3.1.4.33 Type I site-specific deoxyribonuclease	
3.1.26.4	E.C. 3.1.4.34 Calf thymus ribonuclease H	
3.1.21.1	E.C. 3.1.4.5 Deoxyribonuclease I	
3.1.22.1	E.C. 3.1.4.6 Deoxyribonuclease II	
3.1.31.1	E.C. 3.1.4.7 Micrococcal nuclease	
3.1.27.3	E.C. 3.1.4.8 Ribonuclease T_1	

EC-No.	Synonym Recommended Name	Vol., Page
3.1.30.2	E.C. 3.1.4.9 Serratia marcescens nuclease	
3.2.1.54	E.C. 3.2.1.12 Cyclomaltodextrinase	
3.2.1.54	E.C. 3.2.1.13 Cyclomaltodextrinase	
3.2.1.35	E.C. 3.2.1.34 Hyaluronoglucosaminidase	
3.2.1.41	EC 3.2.1.69 α-dextrin endo-1,6-α-glucosidase	
3.2.1.55	EC 3.2.1.79 α-L-arabinofuranosidase	
3.2.1.33	E.C. 3.2.1.9 Amylo-1,6-glucosidase	
3.4.11.2	EC 3.4.1. 2 membrane alanyl aminopeptidase	
3.4.11.1	EC 3.4.1.1 leucyl aminopeptidase	
3.4.11.5	EC 3.4.1.4 prolyl aminopeptidase	
3.4.19.3	EC 3.4.11.8 pyroglutamyl-peptidase I	
3.4.16.5	EC 3.4.12.1 Carboxypeptidase C	
3.4.16.6	EC 3.4.12.1 Carboxypeptidase D	
3.4.19.9	EC 3.4.12.10 γ-Glu-X carboxypeptidase	
3.4.17.6	E.C. 3.4.12.11. alanine carboxypeptidase	
3.4.17.1	E.C. 3.4.12.2 carboxypeptidase A	
3.4.17.2	EC 3.4.12.3 carboxypeptidase B	
3.4.16.2	EC 3.4.12.4 lysosomal Pro-X carboxypeptidase	

EC-No.	Synonym Recommended Name	Vol., Page
3.4.17.8	EC 3.4.12.6 muramoylpentapeptide carboxypeptidase	
3.4.17.3	E.C. 3.4.12.7 lysine carboxypeptidase	
3.4.17.4	EC 3.4.12.8 Gly-X carboxypeptidase	
3.4.13.18	EC 3.4.13.1 Cytosol nonspecific dipeptidase	
3.4.19.5	EC 3.4.13.10 β-Aspartyl-peptidase	
3.4.13.18	EC 3.4.13.11 Cytosol nonspecific dipeptidase	
3.4.13.19	EC 3.4.13.11 Membrane dipeptidase	
3.4.13.18	EC 3.4.13.15 Cytosol nonspecific dipeptidase	
3.4.13.18	EC 3.4.13.2 Cytosol nonspecific dipeptidase	
3.4.13.18	EC 3.4.13.8 Cytosol nonspecific dipeptidase	
3.4.15.5	EC 3.4.15.1 Peptidyl-dipeptidase Dcp	
3.4.15.5	EC 3.4.15.3 Peptidyl-dipeptidase Dcp	
3.4.16.5	EC 3.4.16.1 Carboxypeptidase C	
3.4.16.6	EC 3.4.16.1 Carboxypeptidase D	
3.4.17.4	EC 3.4.17.9 Gly-X carboxypeptidase	
3.4.17.1	E.C. 3.4.2.1 carboxypeptidase A	
3.4.17.2	EC 3.4.2.2 carboxypeptidase B	
3.4.17.4	EC 3.4.2.3 Gly-X carboxypeptidase	

EC-No.	Synonym Recommended Name	Vol., Page
3.4.21.37	E.C. 3.4.21.11 Leucocyte elastase	
3.4.21.36	E.C. 3.4.21.11 pancreatic elastase	
3.4.16.5	EC 3.4.21.13 Carboxypeptidase C	
3.4.16.6	EC 3.4.21.13 Carboxypeptidase D	
3.4.21.64	EC 3.4.21.14 Endopeptidase K	
3.4.21.67	EC 3.4.21.14 Endopeptidase So	
3.4.21.63	EC 3.4.21.14 Oryzin	
3.4.21.62	EC 3.4.21.14 Subtilisin	
3.4.21.66	EC 3.4.21.14 Thermitase	
3.4.21.65	EC 3.4.21.14 Thermomycolin	
3.4.21.63	EC 3.4.21.15 Oryzin	
3.4.21.2	EC 3.4.21.2 chymotrypsin C	
3.4.24.58	EC 3.4.21.23 Russellysin	
3.4.21.74	EC 3.4.21.28 Venombin A	
3.4.21.74	EC 3.4.21.29 Venombin A	
3.4.21.74	EC 3.4.21.30 Venombin A	
3.4.21.68	EC 3.4.21.31 t-Plasminogen activator	
3.4.21.73	EC 3.4.21.31 u-Plasminogen activator	

EC-No.	Synonym Recommended Name	Vol., Page
3.4.21.34	E.C. 3.4.21.8 plasma kallikrein	
3.4.21.35	E.C. 3.4.21.8 tissue kallikrein	
3.4.24.56	EC 3.4.22.11 Insulysin	
3.4.19.9	EC 3.4.22.12 γ-Glu-X carboxypeptidase	
3.4.24.15	EC 3.4.22.19 thimet oligopeptidase	
3.4.24.37	EC 3.4.22.22 Saccharolysin	
3.4.21.61	EC 3.4.22.23 Kexin	
3.4.22.33	EC 3.4.22.4 Fruit bromelain	
3.4.22.32	EC 3.4.22.4 Stem bromelain	
3.4.22.33	EC 3.4.22.5 Fruit bromelain	
3.4.21.48	EC 3.4.22.9 cerevisin	
3.4.23.22	EC 3.4.23.10 Endothiapepsin	
3.4.23.28	EC 3.4.23.6 Acrocylindropepsin	
3.4.23.18	EC 3.4.23.6 Aspergillopepsin I	
3.4.23.19	EC 3.4.23.6 Aspergillopepsin II	
3.4.23.24	EC 3.4.23.6 Candidapepsin	
3.4.23.22	EC 3.4.23.6 Endothiapepsin	
3.4.23.23	EC 3.4.23.6 Mucorpepsin	

EC-No.	Synonym Recommended Name	Vol., Page
3.4.23.20	EC 3.4.23.6 Penicillopepsin	
3.4.23.27	EC 3.4.23.6 Physaropepsin	
3.4.23.39	EC 3.4.23.6 Plasmepsin II	
3.4.23.30	EC 3.4.23.6 Pycnoporopepsin	
3.4.23.21	EC 3.4.23.6 Rhizopuspepsin	
3.4.23.26	EC 3.4.23.6 Rhodotorulapepsin	
3.4.23.25	EC 3.4.23.6 Saccharopepsin	
3.4.23.20	EC 3.4.23.7 Penicillopepsin	
3.4.23.25	EC 3.4.23.8 Saccharopepsin	
3.4.23.21	EC 3.4.23.9 Rhizopuspepsin	
3.4.24.18	EC 3.4.24.11 Meprin A	
3.4.21.32	EC 3.4.24.3 brachyurin	
3.4.21.32	EC 3.4.24.34 brachyurin	
3.4.24.25	EC 3.4.24.4 Aeromonolysin	
3.4.24.29	EC 3.4.24.4 Aureolysin	
3.4.24.28	EC 3.4.24.4 Bacillolysin	
3.4.24.30	EC 3.4.24.4 Coccolysin	
3.4.24.39	EC 3.4.24.4 Deuterolysin	

EC-No.	Synonym Recommended Name	Vol., Page
3.4.24.31	EC 3.4.24.4 Mycolysin	
3.4.24.27	EC 3.4.24.4 Thermolysin	
3.4.24.26	EC 3.4.24.4 pseudolysin	
3.4.24.3	E.C. 3.4.24.7 microbial collagenase	
3.4.24.34	EC 3.4.24.7 Neutrophil collagenase	
3.4.21.32	EC 3.4.24.7 brachyurin	
3.4.13.3	E.C. 3.4.3. 3 Aminoacylhistidine dipeptidase	
3.4.13.18	EC 3.4.3.1 Cytosol nonspecific dipeptidase	
3.4.13.18	EC 3.4.3.2 Cytosol nonspecific dipeptidase	
3.4.13.5	E.C. 3.4.3.4 X-methyl-His dipeptidase	
3.4.13.9	EC 3.4.3.7 X-Pro dipeptidase	
3.4.23.1	E.C. 3.4.4.1 pepsin A	
3.4.22.2	EC 3.4.4.10 papain	
3.4.22.6	EC 3.4.4.11 chymopapain	
3.4.22.3	EC 3.4.4.12 ficain	
3.4.21.5	E.C. 3.4.4.13 thrombin	
3.4.21.7	E.C. 3.4.4.14 plasmin	
3.4.23.15	EC 3.4.4.15 renin	

EC-No.	Synonym Recommended Name	Vol., Page
3.4.21.62	EC 3.4.4.16 Subtilisin	
3.4.24.7	E.C. 3.4.4.19 interstitial collagenase	
3.4.24.3	E.C. 3.4.4.19 microbial collagenase	
3.4.21.32	EC 3.4.4.19 brachyurin	
3.4.23.2	E.C. 3.4.4.2 pepsin B	
3.4.22.8	EC 3.4.4.20 clostripain	
3.4.21.35	E.C. 3.4.4.21 tissue kallikrein	
3.4.21.34	EC 3.4.4.21 plasma kallikrein	
3.4.23.3	E.C. 3.4.4.22 gastricsin	
3.4.23.5	E.C. 3.4.4.23 cathepsin D	
3.4.22.32	EC 3.4.4.24 Stem bromelain	
3.4.23.4	E.C. 3.4.4.3 chymosin	
3.4.21.4	E.C. 3.4.4.4 trypsin	
3.4.21.1	E.C. 3.4.4.5 Chymotrypsin	
3.4.21.1	E.C. 3.4.4.6 Chymotrypsin	
3.4.21.37	E.C. 3.4.4.7 Leucocyte elastase	
3.4.21.36	E.C. 3.4.4.7 pancreatic elastase	
3.4.21.9	E.C. 3.4.4.8 enteropeptidase	

EC-No.	Synonym Recommended Name	Vol., Page
3.4.14.1	E.C. 3.4.4.9 dipeptidyl-peptidase I	
3.4.23.28	EC 3.4.99.1 Acrocylindropepsin	
3.4.24.56	EC 3.4.99.10 Insulysin	
3.4.23.26	EC 3.4.99.15 Rhodotorulapepsin	
3.4.23.15	EC 3.4.99.19 renin	
3.4.24.29	EC 3.4.99.22 Aureolysin	
3.4.23.30	EC 3.4.99.25 Pycnoporopepsin	
3.4.21.68	EC 3.4.99.26 t-Plasminogen activator	
3.4.21.73	EC 3.4.99.26 u-Plasminogen activator	
3.4.21.60	EC 3.4.99.28 Scutelarin	
3.4.24.20	EC 3.4.99.30 Peptidyl-Lys metalloendopeptidase	
3.4.24.15	EC 3.4.99.31 thimet oligopeptidase	
3.4.24.20	EC 3.4.99.32 Peptidyl-Lys metalloendopeptidase	
3.4.23.36	EC 3.4.99.35 Signal peptidase II	
3.4.21.89	EC 3.4.99.36 Signal peptidase I	
3.4.21.88	EC 3.4.99.37 Repressor LexA	
3.4.23.17	EC 3.4.99.38 Pro-opiomelanocortin converting enzyme	
3.4.23.12	EC 3.4.99.4 nepenthesin	

EC-No.	Synonym Recommended Name	Vol., Page
3.4.24.55	EC 3.4.99.44 Pitrilysin	
3.4.24.56	EC 3.4.99.45 Insulysin	
3.4.24.7	E.C. 3.4.99.5 interstitial collagenase	
3.4.24.3	E.C. 3.4.99.5 microbial collagenase	
3.4.21.32	EC 3.4.99.5 brachyurin	
3.4.24.21	EC 3.4.99.6 Astacin	
3.5.1.13	E.C. 3.5.1.13 AAA-2 Arylacylamidase	
3.5.1.13	E.C. 3.5.1.13, AAA-1 Arylacylamidase	
3.5.1.83	EC 3.5.1.15 N-Acyl-D-aspartate deacylase	
3.4.13.5	E.C. 3.5.1.34 X-methyl-His dipeptidase	
3.5.1.82	EC 3.5.1.55 N-Acyl-D-glutamate deacylase	
3.5.2.6	EC 3.5.2.8 β-lactamase	
4.3.99.1	EC 3.5.5.3 Cyanate lyase	
3.6.1.38	E.C. 3.6.1.3 Ca^{2+}-transporting ATPase	
3.6.1.33	E.C. 3.6.1.3 Dynein ATPase	
3.6.1.34	E.C. 3.6.1.3 H^+-transporting ATP synthase	
3.6.1.35	E.C. 3.6.1.3 H^+-transporting ATPase	
3.6.1.36	E.C. 3.6.1.3 H^+/K^+-transporting ATPase	

EC-No.	Synonym Recommended Name	Vol., Page
3.6.1.32	E.C. 3.6.1.3 Myosin ATPase	
3.6.1.37	E.C. 3.6.1.3 Na^+/K^+-transporting ATPase	
3.1.8.2	EC 3.8.2.1 Diisopropyl-fluorophosphatase	
4.1.1.12	E.C. 4.1.1.10 Aspartate 4-decarboxylase	
4.1.1.28	E.C. 4.1.1.27 Aromatic-L-amino-acid decarboxylase	
4.1.3.16	E.C. 4.1.2.31 4-Hydroxy-2-oxoglutarate aldolase	
2.1.2.1	EC 4.1.2.6 Glycine hydroxymethyltransferase	
4.1.2.13	E.C. 4.1.2.7 Fructose-bisphosphate aldolase	
4.2.3.16	EC 4.1.99.10 (-)-(4S)-limonene synthase	
4.2.3.6	EC 4.1.99.6 trichodiene synthase	
4.2.3.9	EC 4.1.99.7 aristolochene synthase	
4.2.3.14	EC 4.1.99.8 pinene synthase	
4.2.3.15	EC 4.1.99.9 myrcene synthase	
4.4.1.1	E.C. 4.2.1.15 cystathionine γ-lyase	
3.3.2.4	E.C. 4.2.1.37 trans-epoxysuccinate hydrolase	
4.2.1.59	EC 4.2.1.59, D-3-hydroxyoctanoyl-[acyl carrier protein] 3-Hydroxyoctanoyl-[acyl-carrier-protein] dehydratase	
3.3.2.3	E.C. 4.2.1.63 Epoxide hydrolase	
3.3.2.3	E.C. 4.2.1.64 Epoxide hydrolase	

EC-No.	Synonym Recommended Name	Vol., Page
4.1.1.78	EC 4.2.1.72 acetylenedicarboxylate decarboxylase	
4.2.2.1	EC 4.2.99.1 hyaluronate lyase	
4.2.3.3	EC 4.2.99.11 methylglyoxal synthase	
4.2.3.1	EC 4.2.99.2 threonine synthase	
4.2.2.2	EC 4.2.99.3 pectate lyase	
4.2.2.3	EC 4.2.99.4 poly(β-D-mannuronate) lyase	
4.2.2.5	EC 4.2.99.6 chondroitin AC lyase	
4.2.3.2	EC 4.2.99.7 ethanolamine-phosphate phospho-lyase	
4.2.3.12	EC 4.6.1.10 6-pyruvoyltetrahydropterin synthase	
4.2.3.13	EC 4.6.1.11 (+)-d-cadinene synthase	
4.2.3.4	EC 4.6.1.3 3-dehydroquinate synthase	
4.2.3.5	EC 4.6.1.4 chorismate synthase	
4.2.3.7	EC 4.6.1.5 pentalenene synthase	
4.2.3.8	EC 4.6.1.7 casbene synthase	
4.2.3.10	EC 4.6.1.8 (-)-endo-fenchol synthase	
4.2.3.11	EC 4.6.1.9 sabinene-hydrate synthase	
5.3.1.3	EC 5.1.3.25 Arabinose isomerase	Vol. 1, Page 249
3.5.99.6	EC 5.3.1.10 glucosamine-6-phosphate deaminase	

EC-No.	Synonym Recommended Name	Vol., Page
2.6.1.16	EC 5.3.1.19 Glutamine-fructose-6-phosphate transaminase (isomerizing)	
5.3.1.25	EC 5.3.1.3 L-Fucose isomerase	Vol. 1, Page 359
5.3.99.6	EC 5.3.99.1 Allene-oxide cyclase	Vol. 1, Page 483
4.2.1.92	EC 5.3.99.1 Hydroperoxide dehydratase	
6.3.4.13	E.C. 6.3.1.3 Phosphoribosylamine-glycine ligase	Vol. 2, Page 626
3.2.1.141	4-α-D-{(1→4)-α-D-Glucano}trehalose glucanohydrolase (trehalose-producing) 4-α-D-{(1→4)-α-D-Glucano}trehalose trehalohydrolase	
3.2.1.141	4-α-D-{(1→4)-α-D-Glucano}trehalose trehalohydrolase 4-α-D-{(1→4)-α-D-Glucano}trehalose trehalohydrolase	
6.3.2.5	4´-Phosphopantothenoylcysteine synthetase Phosphopantothenoate-cysteine ligase	Vol. 2, Page 431

Printing: Mercedes-Druck, Berlin
Binding: Buchbinderei Lüderitz & Bauer, Berlin